T0325882

Sampling Theory

Beyond Bandlimited Systems

Covering the fundamental mathematical underpinnings together with key principles and applications, this book provides a comprehensive guide to the theory and practice of sampling from an engineering perspective. Beginning with traditional ideas such as uniform sampling in shift-invariant spaces and working through to the more recent fields of compressed sensing and sub-Nyquist sampling, the key concepts are addressed in a unified and coherent way. Emphasis is given to applications in signal processing and communications, as well as hardware considerations, throughout.

The book is divided into three main sections: first is a comprehensive review of linear algebra, Fourier analysis, and prominent signal classes figuring in the context of sampling, followed by coverage of sampling with subspace or smoothness priors, including nonlinear sampling and sample rate conversion. Finally, sampling over union of subspaces is discussed, including a detailed introduction to the field of compressed sensing and the theory and applications of sub-Nyquist sampling.

With 200 worked examples and over 250 end-of-chapter problems, this is an ideal course textbook for senior undergraduate and graduate students. It is also an invaluable reference or self-study guide for engineers and students across industry and academia.

Yonina C. Eldar is a Professor in the Department of Electrical Engineering at the Technion – Israel Institute of Technology, and holds the Edwards Chair in Engineering. She is a Research Affiliate with the Research Laboratory of Electronics at the Massachusetts Institute of Technology, and was a Visiting Professor at Stanford University. She has received numerous awards for excellence in research and teaching, including the Wolf Foundation Krill Prize for Excellence in Scientific Research, the Hershel Rich Innovation Award, the Michael Bruno Memorial Award from the Rothschild Foundation, the Weizmann Prize for Exact Sciences, the Muriel and David Jacknow Award for Excellence in Teaching, the IEEE Signal Processing Society Technical Achievement Award, and the IEEE/AESS Fred Nathanson Memorial Radar Award. She is the Editor in Chief of *Foundations and Trends in Signal Processing* and was an Associate Editor for several journals in the areas of signal processing and mathematics and a Signal Processing Distinguished Lecturer. She is an IEEE Fellow, a member of the Young Israel Academy of Science and the Israel Committee for Higher Education.

Sampling Theory

Beyond Bandlimited Systems

YONINA C. ELDAR

Department of Electrical Engineering
Technion - Israel Institute of Technology

CAMBRIDGE
UNIVERSITY PRESS

University Printing House, Cambridge CB2 8BS, United Kingdom

One Liberty Plaza, 20th Floor, New York, NY 10006, USA

477 Williamstown Road, Port Melbourne, VIC 3207, Australia

314-321, 3rd Floor, Plot 3, Splendor Forum, Jasola District Centre, New Delhi-110025, India

79 Anson Road, #06-04/06, Singapore 079906

Cambridge University Press is part of the University of Cambridge.

It furthers the University's mission by disseminating knowledge in the pursuit of
education, learning and research at the highest international levels of excellence.

www.cambridge.org
Information on this title: www.cambridge.org/9781107003392

© Cambridge University Press 2015

First published 2015

A catalogue record for this publication is available from the British Library

Library of Congress Cataloging in Publication data
Eldar, Yonina C.
Sampling theory : beyond bandlimited systems / Yonina C. Eldar.
    pages   cm
ISBN 978-1-107-00339-2 (Hardback)
1. Signal processing–Digital techniques–Study and teaching (Higher)   2. Signal
processing–Digital techniques–Study and teaching (Graduate)   3. Signal processing–
Statistical methods–Study and teaching (Higher)   4. Signal processing–Statistical
methods–Study and teaching (Graduate)   5. Sampling (Statistics)   I. Title.
TK5102.9.E435 2014
621.382′23–dc23   2014014930

ISBN 978-1-107-00339-2 Hardback

Additional resources for this publication at www.cambridge.org/9781107003392

To my parents

To Shalomi, Yonatan, Moriah, Tal, Noa and Roei

The beginning of wisdom is to acquire wisdom;
And with all your means acquire understanding.
Proverbs 4:7

Contents

Preface

Digital signal processing (DSP) is one of the most prominent areas in engineering, including subfields such as speech and image processing, statistical data processing, spectral estimation, biomedical applications, and many others. As the name suggests, the goal is to perform various signal processing tasks (e.g., filtering, amplification, and more) in the digital domain where design, verification, and implementation are considerably simplified compared with analog signal processing. DSP is the basis of many areas of technology, and is one of the most powerful technologies that have shaped science and engineering in the past century.

In order to represent and process analog signals on a computer the signals must be sampled with an analog-to-digital converter (ADC) which converts the signal to a sequence of numbers. After processing, the samples are converted back to the analog domain via a digital-to-analog converter (DAC). Consequently, the theory and practice of sampling are at the heart of DSP. Evidently, any technology advances in ADCs and DACs have a huge impact on a vast array of applications.

The goal of this book is to provide a comprehensive treatment of the theory and practice of sampling from an engineering perspective. Although there are many excellent mathematical textbooks on signal expansions and harmonic analysis, our aim is to present an up-to-date engineering textbook on the topic by combining the fundamental mathematical underpinnings of sampling with practical engineering applications and principles. A large part of the book is also devoted to the more recent fields of compressed sensing and sub-Nyquist sampling which are not covered in standard linear algebra or harmonic analysis books. Throughout, we focus on various applications in signal processing and communications. We assume that the reader is familiar with basic signals processing concepts such as filtering and convolution. The intended audience is a senior undergraduate or first-year graduate level class; however, some background in digital signal processing and Fourier analysis should be enough to follow the material. Required background needed in linear algebra is covered in the text. The book can also be used as a reference for engineers, students working in related areas, and researchers from industry and academia. We also believe that the book is suited for self-study as it is largely self-contained.

Sampling theory is a broad and deep subject, and a vivid area of research, with roots going back over a century. It is therefore impossible to cover all the advances and results in this rich area in a single textbook. The point of this book is not to do justice to the beautiful mathematical theory underlying sampling, but rather to bring forth some of the important engineering concepts in a coherent way. We have chosen to focus primarily on uniform sampling in shift-invariant spaces, and on deterministic signals. The important topics of nonuniform sampling, Gabor and wavelet expansions, errors in sampling due to noise, quantization, implicit sampling, and other approximations are only briefly touched upon. Many of these subjects are covered in other textbooks focused on these specific topics, or in the many references provided at the end of the book.

Organization of the book

The book can be broadly divided into three sections:

- Introductory material including motivation, review of linear algebra and Fourier analysis, and survey of signal classes (Chapters 1–5);
- Sampling with subspace or smoothness priors, including nonlinear sampling and sampling rate conversion (Chapters 6–9);
- Sampling over union of subspaces, including a detailed introduction to the field of compressed sensing and the theory and applications of sub-Nyquist sampling (Chapters 10–15).

We begin in Chapter 1 with a brief introduction to the topic of sampling in general, its importance, and the necessity to move beyond the traditional Shannon–Nyquist theorem. Chapter 2 contains a comprehensive review of the linear algebra background needed in order to develop the mathematical notions underlying sampling theory. In it we have attempted to summarize the main mathematical machinery required for the rest of the book. A fundamental understanding of linear algebra is key to developing sampling theories, and therefore this chapter is quite extensive. Chapter 3 summarizes important notions regarding linear time-invariant systems and Fourier transforms. We review both the continuous-time and discrete-time Fourier transforms, and discuss the relationship between the two in the context of sampled signals. The classes of signals that we will focus on throughout the book are introduced in Chapter 4, along with some of the fundamental mathematical properties associated with such signal sets. In particular, we discuss the celebrated Shannon–Nyquist theorem, and its extension to more general shift-invariant subspaces. We briefly consider Gabor and wavelet expansions, and introduce union of subspaces and smoothness priors. Our primary focus in this book is on signal models involving shift-invariant (SI) spaces. We therefore devote Chapter 5 to studying some of the mathematical properties associated with these spaces. Examples include bandlimited signals, splines, and many classes of digital communication signals.

In Chapter 6 we turn to treat specific sampling theorems. We begin by considering linear sampling with subspace priors. As we show, in many cases perfect recovery of the signal from the given samples is possible, based on the subspace prior, even when the input signal is not bandlimited or the sampling rate is lower than the Nyquist rate. We also treat the case in which constraints are imposed on the recovery process and consider different criteria to recover or approximate the original signal in these settings. In particular, we develop the well-known Papoulis' generalized sampling theorem as a special case of our framework. These ideas are extended in Chapter 7 to include smoothness priors, namely, when all we know about the signal is that it is smooth in some sense. An interesting special case that we treat in this context is super-resolution: obtaining a high-resolution image from several low-resolution images by using ideas of sampling and reconstruction. Nonlinear sampling is considered in Chapter 8, assuming a subspace prior. Surprisingly, we will see that many types of nonlinearities that are encountered in practice can be completely compensated for without having to increase the sampling rate, even though typically nonlinearities lead to an increase in bandwidth. Although sampling theory is focused on recovery of continuous-time signals from their discrete samples, in Chapter 9 we demonstrate that sampling also plays a crucial role in the design of fully discrete-time algorithms in the context of sampling rate conversion. We will discuss several methods for converting between signals or images at different rates. This allows us, in particular, to efficiently vary the size of an image or an audio file.

Chapters 10–15 are devoted to sub-Nyquist sampling and compressed sensing. In Chapter 10 we introduce the union of subspaces (UoS) model, which underlies many sub-Nyquist sampling paradigms. This model allows for nonlinear signal classes which can describe, for example, streams of pulses with unknown delays and amplitudes, multiband signals with unknown carrier frequencies, and more. One of the most well-studied examples of a UoS is that of a vector that is sparse in an appropriate subspace. This model is the basis of the rapidly growing field of compressed sensing, which we review in detail in Chapter 11. This material is based on the chapter "Introduction to compressed sensing," co-authored by M. Davenport, M. Duarte, Y. C. Eldar, and G. Kutyniok, which appears in the book *Compressed Sensing* (Cambridge, 2012). Chapter 12 considers an extension of the basic sparsity model to block sparsity, which can be used to describe more general finite-dimensional unions. This chapter also discusses how to learn the subspaces from subsampled data, when they are not known a priori. Unions of shift-invariant spaces are treated in Chapter 13 along with applications to low-complexity detectors in various settings. The class of multiband signals is considered in Chapter 14. These are signals whose Fourier transform comprises a small number of bands, spread over a wide frequency range. We present a variety of different methods that allow such signals to be sampled at sub-Nyquist rates proportional to the actual band occupancy, even though the carrier frequencies are unknown, and not to the high Nyquist rate associated with the largest frequency. Along with developing the theoretical concepts, we also address practical considerations and demonstrate a hardware realization of a sub-Nyquist sensing board for multiband signals. Chapter 15 is focused on sub-Nyquist sampling of pulse streams which appear in applications such as radar, ultrasound, and multipath channel identification. Example hardware prototypes for problems in radar and ultrasound are also presented.

The appendices cover basic material used in various parts of the book. Specifically, Appendix A summarizes key results related to matrix algebra, and Appendix B reviews basic concepts from probability theory and random processes.

Not all theorems in the book are proven in detail. When proofs are not included, we provide references to where they can be found. Furthermore, in some places, mathematical rigor has been replaced by emphasis on the main ideas.

Matlab implementations and examples

The book contains many worked examples in order to provide deeper understanding and greater intuition for the material, and to illustrate the main points, as well as to explore the behavior of the different methods, and various tradeoffs relevant to the problems at hand. Numerical results are also sometimes used to illustrate points that are not developed rigorously in the text. The numerical experiments have all been programmed in Matlab using standard toolboxes. Numerical examples and computational figures can be reproduced using the m-files available on the author's web page.

At the end of each chapter there is also a list of homework exercises which further expand on and demonstrate the various concepts introduced, and provide an opportunity to practice the material. Some of the exercises are used to derive proofs of theorems that were omitted in the text itself. The order of the exercises follows the presentation of the material in the chapter.

Teaching

This book is intended as a senior year or graduate textbook. It has emerged from teaching "Generalized Sampling Methods" at the Technion – Israel Institute of Technology, and from several tutorials delivered and written on these topics.

Electrical engineering students are often deterred by the vector space formulation of linear algebra used throughout the book. We are accustomed to filtering and convolutions, and manipulation of finite-dimensional matrices. However, much of the beauty of the results in this field comes from the Hilbert space structure. Once these structures are understood, the rest of the results follow naturally and simply. As we will see, proper understanding of these concepts also ultimately leads to simple and efficient hardware. It is therefore very worthwhile to go through the experience of truly comprehending and appreciating linear algebra. Accordingly, the book begins by providing an overview of the essential ingredients in linear algebra needed for the presentation of the material. When teaching this course at the Technion, we dedicate the first few weeks to covering linear algebra basics in depth before delving into sampling theory. In our opinion, beginning with a review of linear algebra is essential. Although all engineering students take basic linear algebra, such courses are typically taught from a matrix-oriented point of view. The more abstract viewpoint advocated here is essential for the chapters to follow, and often new to the students.

The chapter on Fourier analysis can typically be skipped, with only a short reminder of the essential results. In particular, discrete – continuous relations, which are often overlooked but key in the development of sampling results, may be emphasized.

The rest of the book is designed to provide flexibility in how to present the material. The book can be used as a basis for a broad class in sampling theory which covers all the topics in the book – focusing in class on the main results and relying on the book to fill in the details regarding proofs, examples, and applications. On the other hand, one can choose to cover only a subset of the chapters, in greater detail.

As we outlined in the section discussing the book structure, the book is conveniently divided into three sections. The first provides a comprehensive overview of the basic building blocks needed in order to understand and develop subsequent material. These chapters are provided mainly for reference. In a course, most of this material can be skipped, focusing only on the essential concepts which the students in the course may be lacking. As an example, in teaching this course at the Technion, we devote about four classes to linear algebra and shift-invariant spaces; subspace sampling is covered in one class, two classes are dedicated to smoothness priors and interpolation methods, and one class is dedicated to nonlinear sampling. The remaining six weeks of the course focus on compressed sensing and sub-Nyquist sampling, of which about one week is devoted to each of Chapters 10, 11, 13, and 14, and two classes to Chapter 15 and some of its applications.

Alternatively, a semester-length course can focus on the core material in Chapters 5–9, complemented by selected material from Chapters 10–15 as time permits. Most of these chapters can be taught independently of each other.

The book may also be used for a course focused more on the recently growing field of compressed sensing and sub-Nyquist sampling. In this case, the course can begin with a brief introduction to linear algebra and concepts of shift-invariant spaces, and then go through the last unit of the book, i.e. Chapters 10–15, in more detail.

Thanks

Completing this book would not have been possible without the help of many people throughout the multiple stages of the book's evolution. During my years in academia I have been surrounded by good friends and colleagues who have encouraged and supported me. I am very grateful to my colleagues whom I had the pleasure to work with and from whom I learned a great deal about sampling theory and compressed sensing in particular and about research and teaching more generally. I am also indebted to my friends and family, who do not share my passion and interest in engineering and math, and have therefore made sure to provide ample opportunities to be reminded of the many other aspects of life, giving me the energy to continue and the distraction I needed at the many stumbling points during this project.

I would like to thank my students at the Technion for their course participation and feedback on the course notes which evolved into this book. My dedicated PhD student, Tomer Michaeli, was the first teaching assistant for the course on generalized sampling methods and is responsible for the examples and simulations in the first part of the book. He provided many new perspectives and insights on the various parts of the book. I thank him sincerely for his time and dedication to this project. Several of my graduate students and course students helped with examples and simulations in the second part of the book, focused on compressed sensing and sub-Nyquist sampling. In particular I would like to thank Kfir Aberman, Tanya Chernyakova, Deborah Cohen, Tomer Hammam, Etgar Israeli, Ori Kats, Saman Mousazadeh, and Shahar Tsiper, for their work on examples in these chapters. I would also like to thank Douglas Adams, Omer Bar-Ilan, Zvika Ben-Haim, Yuxin Chen, Kfir Cohen, Pier Luigi Dragotti, Tsvi Dvorkind, Nikolaus Hammler, Moshe Mishali, Tomer Peleg, Volker Pohl, Danny Rosenfeld, Igal Rozenberg, Andreas Tillmann, and Lior Weizman for proofreading many of the chapters and providing important feedback, and Kfir Gedalyahu, Moshe Mishali, Ronen Tur, and Noam Wagner for sharing Matlab simulations from their theses. I am grateful to my current and former graduate students for their contributions to this book through their research results, and for the opportunity to learn from each of them during this process. I apologize for any errors and inconsistencies remaining in the book, and for any omitted subjects which deserved better coverage.

I would like to thank several friends and colleagues for their early and ongoing support of my professional activities: Arye Yeredor and Udi Weinstein inspired my original interest in digital signal processing and taught me the value of seeking a simple and intuitive explanation to even the most complicated algorithm. Al Oppenheim incited my interest in sampling theory and inspired the abstract linear algebra viewpoint of sampling theory presented in this book. I thank him for his support over the years and for his creative approach and passion towards research which he instilled in his students. Several colleagues supported my early steps into the world of sampling theory. Special thanks to Michael Unser, P. P. Vaidyanathan, Akram Aldroubi, Ole Christensen, Hans Feichtinger, John Benedetto, Stephane Mallat, Abdul Jerri, and Ahmed Zayed, who welcomed me into the world of sampling and its applications, were always appreciative and encouraging, and helped in completing my mathematical education. The sampling theory research community is a warm and welcoming group, and I feel very fortunate to be a part of it.

In recent years we have been working extensively on applications of sampling theory in a wide variety of areas. I have been very fortunate to have brilliant and dedicated colleagues to collaborate with, who are experts in the respective application domains. They have been a tremendous source of inspiration and support and have made research a fun and rewarding experience. Special thanks to Amir Beck, Emmanuel Candes, Israel Cidon, Oren Cohen, Alex Gershman, Andrea Goldsmith, Alex Haimovich, Arye Nehorai, Guillermo Sapiro, Anna Scaglione, Moti Segev, Shlomo Shamai, and Joshua Zeevi. I am very grateful to my excellent hosts during my Sabbatical at Stanford University – Emmanuel Candes at the Statistics department and Andrea Goldsmith at the Electrical Engineering department. My Sabbatical provided many opportunities for working on the book and was full of fun, stimulating, and interesting discussions. Many of my colleagues mentioned above are now personal friends with whom I share more than just our joint passion for research. I would also like to mention my colleagues at the Technion, Gitti Frey, Idit Keidar, Ayellet Tal, and Lihi Zelnik-Manor, who have provided a safety net that helped keep my sanity while trying to balance family life with a demanding career. I am further grateful to all my collaborators throughout the years from whom I have learned a lot about research in general, and signal processing in particular. The Electrical Engineering Department at the Technion has provided an exciting and stimulating environment for both research and teaching during the past 10 years.

In 2013 we established the SAMPL laboratory – Sampling, Acquisition, Modeling and Processing Lab – at the Electrical Engineering Department in the Technion. The sub-Nyquist prototypes presented in this book, as well as many other sub-Nyquist projects, were all developed in the laboratory. I have been extremely fortunate to have the support and expertise of many truly talented engineers. Special thanks to Yoram Or-Chen, Alon Eilam, Rolf Hilgendorf, Alex Reysenson, Idan Shmuel, and Eli Shoshan. The laboratory would not have been established and operative without the support of Peretz

Lavie – the Technion President, Gadi Schuster – Executive Vice President for Academic Affairs, Moti Segev, Joshua Zeevi, and Gadi Eisenstein. The hardware and experiments in the laboratory have been conducted in collaboration with National Instruments, General Electric, and Agilent. We gratefully acknowledge their support and partnership. Many thanks to my administrative assistant over the past two years, Sasha Azimov, for help with various aspects of the book.

I would like to thank the copy editor Lindsay Nightingale for her care for detail, Vania Cunha for supervising the book production, and Phil Meyler, from Cambridge University Press for supporting and overseeing this project throughout.

Special thanks to my parents who inspired me from an early age to follow my ambitions and who have taught me values through a constant living example: to my mother for instilling in me a passion for life, the energy to pursue my goals and for finding solutions to many of life's problems, and to my father for implanting in me the love of knowledge, drive for perfection and excellence, and for his constant sound advice and inspiration. My in-laws have brought me into their family as one of their own and have taken pride in all my accomplishments. I am sincerely grateful to my parents and in-laws as well as to my extended family for their continuing love and support.

My deepest gratitude and unbounded love goes to my husband Shalomi and children, Yonatan, Moriah, Tal, Noa and Roei, who will probably not read this book, but with whom life is far more exciting and rich than sampling. They have provided many opportunities for welcome breaks from writing and editing and many reasons to smile. Their boundless love and encouragement, emotional support, and pride in me have filled my life with happiness, making it all worthwhile. Shalomi has stood beside me throughout my career, providing infinite support, helpful advice, and encouragement. He is my rock I can lean on, and my constant compass always pointing in the right direction and values. He has been my inspiration and motivation to continue to improve in all aspects and has made sure that our family and home are rich with values and activities beyond our professions. We have been partners on many life journeys, far from the world of engineering and research. Thanks for having the patience with me while taking on the challenge of writing this book! I dedicate this book to them.

Abbreviations

ADC	Analog-to-digital converter
AM	Amplitude modulation
AWG	Arbitrary waveform generator
AWR	Applied wave research
BCS	Blind compressed sensing
BIBO	Bounded-input, bounded-output
BK-SVD	Block K-SVD
BMP	Block matching pursuit
BOMP	Block orthogonal matching pursuit
BP	Basis pursuit
BPDN	Basis pursuit denoising
C-HiLasso	Collaborative HiLasso
CLS	Constrained least squares
CPI	Coherent processing interval
CPM	Continuous-phase modulation
CR	Cognitive radio
CRB	Cramér–Rao bound
CS	Compressed sensing
CTF	Continuous to finite
CTFT	Continuous-time Fourier transform
DAC	Digital-to-analog converter
DC	Direct-current
DCT	Discrete cosine transform
DFT	Discrete Fourier transform
DL	Dictionary learning
DSP	Digital signal processing
DTFT	Discrete-time Fourier transform
ESPRIT	Estimation of signal parameters by rotational invariance
FIR	Finite-impulse response
FM	Frequency-modulation
FRI	Finite rate of innovation
FUS	Finite union of subspaces
GHz	Gigahertz
IHT	Iterative hard thresholding

IIR	Infinite impulse response
IMV	Infinite measurement vector
ISI	Intersymbol interference
LPF	Low-pass filter
LS	Least squares
LTI	Linear time-invariant
MF	Matched filter
MIMO	Multiple-input, multiple-output
MMSE	Minimum mean-squared error
MMV	Multiple measurement vector
MOD	Method of direction
MP	Matching pursuit
MSE	Mean-squared error
MUD	Multiuser detection
MUSIC	Multiple signal classification
MWC	Modulated wideband converter
NI	National instruments
NSP	Null space property
OBD-BCS	Orthonormal block diagonal BCS
OMP	Orthogonal matching pursuit
PAM	Pulse AM
PRI	Pulse repetition interval
PSD	Power spectrum density
PSF	Point-spread function
PSNR	Peak SNR
QAM	Quadratic amplitude modulation
RDD	Reduced-dimension
RDDF	Reduced-dimension decision-feedback
RD-MUD	Reduced-dimension multiuser detector
RF	Radio frequency
RIP	Restricted isometry property
RKHS	Reproducing kernel Hilbert space
SAC	Sparse agglomerative clustering
SI	Shift-invariant
SIC	Successive interference cancelation
SMI	Shift and modulation invariant
SNR	Signal to noise ratio
SOCP	Second-order cone program
SVD	Singular value decomposition
TLS	Total least squares
ULS	Underspread linear system
UoS	Union of subspaces
WSS	Wide-sense stationary

Chapter 1

Introduction

We live in an analog world, but we would like our digital computers to interact with it. Indeed, digital signal processing (DSP) has become pervasive. It is the basis for most modern consumer electronics, medical imaging devices, cell phones, internet protocol telephony, multimedia standards, speech processing, and a myriad of other products. Digital algorithms, implemented with microprocessors, are less pricey, easier to control, more robust, and more flexible than their analog counterparts, so that nowadays analog circuits are often replaced by digital chips. Digital data is also far easier to store, transmit, and manipulate than analog data. Therefore, in modern applications, an increasing number of functions are being pushed forward to sophisticated software algorithms, leaving only delicate finely tuned tasks for the circuit level. Nowadays, it feels natural that a media player shows our favorite movie, or that our surround system synthesizes pure acoustics, as if sitting in the orchestra, and not in the living room. The digital world plays a fundamental role in our everyday routine, to such a point that we almost forget that we cannot "hear" or "watch" these streams of bits, running behind the scenes. The world around us is analog, yet most modern man-made means for exchanging information are digital. "I am an analog girl in a digital world," sings Judy Gorman [One Sky, 1998], capturing the essence of the digital revolution.

Whether recording sounds, capturing images, or processing an electromagnetic wave, many sources of information are of analog or continuous-time nature. Therefore, DSP inherently relies on a sampling mechanism which converts continuous signals to discrete sequences of numbers, while preserving the information present in those signals. This conversion is performed using a device known as an *analog-to-digital converter (ADC)*. ADC devices translate physical information into a stream of numbers, enabling digital processing by sophisticated software algorithms. After processing, the samples are converted back to the analog domain via a *digital-to-analog converter (DAC)*. Consequently, sampling theories lie at the heart of DSP and play a major role in enabling the digital revolution.

The ADC task is inherently intricate: its hardware must hold a snapshot of a fast-varying input signal steady, while acquiring measurements. Since these measurements are spaced in time, the values between consecutive snapshots are lost. In general, therefore, there is no way to recover the analog input unless some prior information on its structure is incorporated.

1.1   Standard sampling

The simplest way to record an analog signal $x(t)$ is to sample its values $x(nT)$ at intervals of length T, as depicted in Fig. 1.1(a). This type of sampling is referred to as *pointwise* sampling. We use the block diagram in Fig. 1.1(b) to illustrate this operation.

Given the samples, an approximation of $x(t)$ can be obtained by using an appropriate interpolating function, which we denote by $w(t)$. Figure 1.2 demonstrates several possible interpolations of the samples in Fig. 1.1 using different functions $w(t)$: zero-order-hold, linear interpolation, cubic spline interpolation (third-order polynomial), and sinc interpolation. In each case recovery is obtained by modulating the chosen $w(t)$ by the sample values:

$$\hat{x}(t) = \sum_{n \in \mathbb{Z}} d[n] w(t - nT), \qquad (1.1)$$

where in our setting $d[n] = x(nT)$ are the given samples and T is the sampling period. More generally, we can choose $d[n]$ to be a function of the samples $x(nT)$,

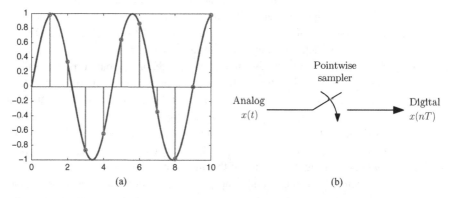

Figure 1.1 Pointwise sampling. (a) A continuous-time signal $x(t) = \sin(4\pi t/9)$ and its pointwise samples with period $T = 1$. (b) Block diagram of a pointwise sampler.

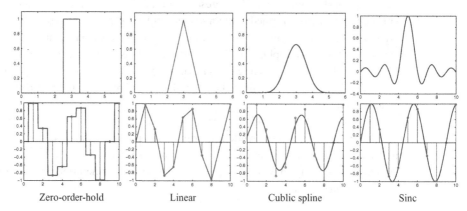

Figure 1.2 Signal reconstruction by various interpolation functions. Top: Interpolating function $w(t)$. Bottom: Recovery $\hat{x}(t)$.

designed to optimize the recovery process. In this case we refer to $d[n]$ as the *corrected samples*.

Clearly, the approximation quality depends on the interpolator chosen and on how well it matches the properties of the original input, and on the samples. Thus, a key component in any sampling theory is the information we have about our signal. Without incorporating prior knowledge, the problem of recovery from samples is ill-posed; there are always many curves that can pass through a set of points, as illustrated in Fig. 1.2. A challenge in practice is to find the "best" curve in some sense consistent with our prior information. Consequently, a large part of sampling theory deals with methods for optimizing $d[n]$ and for selecting $w(t)$ based on the input properties. An additional important design consideration is how small T has to be in order to ensure perfect recovery for certain classes of signals.

While here we have treated pointwise sampling, other more elaborate methods of sampling exist which we will discuss throughout the text. A straightforward generalization of pointwise sampling is depicted in Fig. 1.3, in which $x(t)$ is first filtered with a sampling filter $s(-t)$ that allows us to incorporate imperfections in the ideal sampler. The output is then pointwise sampled on a uniform grid leading to *generalized samples*, denoted by $c[n]$. We consider such sampling mechanisms in detail throughout the book. In particular, we will discuss methods for optimizing the sampling filter $s(t)$ based on the input properties.

Undoubtedly, the most-studied sampling theorem that has had a major influence on signal processing is the well-known Shannon–Nyquist theorem. This theorem was introduced formally into the information theory community by Shannon in [1], but Nyquist had already brought it to the attention of communication engineers in [2]. Kotelnikov is credited with introducing the theorem in the Russian literature [3]. In mathematics, the theorem was developed as part of the study of cardinal series in the works of E. T. Whittaker and his son J. M. Whittaker [4, 5]. The basic idea behind the bandlimited sampling theorem has also been attributed to Cauchy [6], who stated the essential result although without proof. A very illuminating review of the history and mathematics surrounding this theorem can be found in [7].

The Shannon–Nyquist theorem has become a landmark in both the mathematical and engineering literature and has had one of the most profound impacts on industrial development of DSP systems. It provides a method to compute $x(t)$ exactly from its pointwise samples, as long as the signal is sufficiently smooth. More precisely, to allow perfect recovery in (1.1), the sampling frequency, $1/T$, must be at least twice the highest frequency in the signal $x(t)$. This minimal rate is referred to as the *Nyquist rate*. The signal

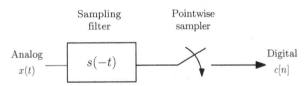

Figure 1.3 Generalized sampling. The signal $x(t)$ is first filtered with a sampling filter $s(-t)$ and then sampled by a pointwise sampler.

is then interpolated from its samples using shifts of the sinc function $w(t) = \text{sinc}(t/T)$, with $\text{sinc}(t) = \sin(\pi t)/(\pi t)$. This theorem assumes that the class of input functions is bandlimited to an appropriate frequency. The interpolating function $w(t)$ in (1.1) is then also a bandlimited function. Capitalizing on this result, much of signal processing has moved from the analog to the digital domain as it allows a continuous-time bandlimited function to be replaced by a discrete set of its samples without any loss of information.

To accommodate high operating rates while retaining low computational cost, efficient ADCs must be developed. While the Shannon–Nyquist theorem is extremely elegant and has had a major impact on DSP, it has several drawbacks. Unfortunately, real-world signals are rarely truly bandlimited. Even signals which are approximately bandlimited may have to be sampled at a fairly high Nyquist rate, requiring expensive sampling hardware and high-throughput digital machinery. Many natural signals, even if they are bandlimited, are often better represented (using fewer coefficients) in bases other than the Fourier basis. Bandlimiting also tends to introduce Gibbs oscillations which can be visually disturbing, for example in images. Finally, many classes of signals possess further structure that can be exploited in order to reduce sampling rates. Classical sampling theory, however, necessitates a high sampling rate whenever a signal has large bandwidth, even if the actual information content in the signal is small. For example, a piecewise linear signal is nondifferentiable; it is therefore not bandlimited, and moreover, its Fourier transform decays at a fairly slow rate. Nonetheless, such a signal is completely described by the location of its knots (transitions between linear segments) and the signal values at those positions, which can be far fewer parameters than the number of samples required by the Shannon–Nyquist theorem. It would therefore be more efficient to have a variety of sampling techniques, tailored to different signal models, such as bandlimited or piecewise linear signals. Such an approach echoes the fundamental quest of the recent field of compressive sampling, which is to capture only the essential information embedded in a signal.

Two other difficulties with the Shannon–Nyquist theorem are the assumptions of ideal pointwise sampling, and of sinc interpolation. Practical ADCs are usually not ideal, that is, they do not produce the exact signal values at the sampling locations. A common situation is that the ADC integrates the signal, usually over small neighborhoods surrounding the sampling points. Moreover, nonlinear distortions are often introduced during the sampling process. These various distortions need to be accounted for in the reconstruction. In addition, implementing the infinite sinc interpolating kernel as part of the DAC required by the Shannon–Nyquist theorem is difficult, since it has slow decay. In practice, much simpler kernels are used, such as linear interpolation. Another major obstacle in the context of modern imaging and communication systems is that signals today can be modulated up to several gigahertz (GHz), while standard ADCs have difficulty in accommodating such large analog bandwidths.

Therefore, to design sampling and interpolation methods that are adapted to practical scenarios, there are several issues that need to be properly addressed:

1. The sampling mechanism should be adequately modeled;
2. Relevant prior knowledge about the class of input signals should be taken into account;

3. Limitations should be imposed on the reconstruction algorithm in order to ensure robust and efficient recovery.

Throughout the book we treat each of these three essential components of the sampling scheme. For each of the elements we focus on several models, which commonly arise in signal processing and communication systems.

1.2 Beyond bandlimited signals

Following the introduction of the bandlimited sampling theorem by Shannon [1], sampling theory became an active area of research, reaching quite a mature state by the 1980s. Several thorough and beautiful tutorials were written on the topic around that time [8, 9]. At that point, research in the area of sampling became quite mathematical with less immediate impact on signal processing and communication applications, or on the actual design of ADCs. In the 1990s, sampling theory benefited from a surge of research due to the intense interest in wavelet theory and the connections made between the two fields. A lot of the theory developed for wavelet analysis is immediately applicable to sampling theorems. This led to many interesting interpretations of existing results along with new methods for sampling and processing signals that move away from the bandlimited paradigm, and instead consider more general signal models and sampling devices. An excellent summary of this perspective can be found in [10] (see also [11]).

In the past few years sampling has again been revived, this time by the vast interest in the area of compressed sensing (CS) [12, 13, 14] which suggests methods for reducing the number of measurements needed to represent sparse signals, or signals with certain types of structure. This framework has focused primarily on sampling of discrete-time signals and reconstruction techniques from a finite number of samples. Works in this area have shown that a high-dimensional vector with only a few nonzero elements is recoverable from a properly chosen underdetermined set of equations. Recovery can be obtained using a variety of different polynomial-time algorithms under appropriate conditions. These results suggest that sparse signals may be sampled at sub-Nyquist rates, which is crucial in modern communications settings.

The CS framework is mostly focused on discrete and finite settings, while sampling inherently deals with continuous-time signals. There are of course examples of analog signals that naturally possess finite representations, such as trigonometric polynomials. However, extending the ideas of CS to acquisition of more general continuous-time signals using practical hardware devices remains a difficult challenge despite the widespread literature in this area. Nonetheless, we will see that by combining analog sampling results with ideas from CS, a variety of efficient sub-Nyquist systems can be developed, leading to low-rate sampling of a broad set of analog signals. In addition, we show that often the acquired samples can be directly processed without having to interpolate them back to the high Nyquist grid, resulting in low-rate processing as well. Beyond developing the fundamental theory, we also discuss practical aspects of reduced-rate sampling and demonstrate prototype sub-Nyquist hardware realizations for a variety of applications. These devices allow practical sampling and processing of many classes of signals at sub-Nyquist rates.

The key to developing low-rate analog sensing methods is relying on structure in the input. Signal processing algorithms have a long history of leveraging structure for various tasks. As an example, MUSIC [15] and ESPRIT [16] are popular techniques for spectrum estimation that exploit signal structure. Model-order selection methods in estimation [17], parametric estimation and parametric feature detection [18] are further examples where structure is heavily used. In our context, we are interested in utilizing signal models in order to reduce sampling rate. Classic approaches to sub-Nyquist sampling include carrier demodulation [19] and bandpass undersampling [20], which assume a linear model corresponding to a bandlimited input with predefined frequency support and fixed carrier frequencies. In the spirit of CS, where unknown nonzero locations result in a nonlinear model, we extend these classical results to analog inputs with unknown frequency support, as well as more broadly to scenarios that involve nonlinear input structures. The approach we take in this book follows the recently proposed Xampling framework [21, 22], which treats a nonlinear model of union of subspaces. In this structure, the input signal belongs to a single subspace out of multiple, possibly even infinitely many, candidate subspaces. The exact subspace to which the signal belongs is unknown a priori. This model encompasses a large variety of structured analog signals and paves the way to the development of practical sub-Nyquist sampling systems.

The importance of sampling theory is likely to continue to grow with the ongoing demand for more sophisticated and efficient DSP systems. The connection with CS offers yet another new perspective on sampling, and ways to better exploit the signal degrees of freedom. This relationship has also brought to the surface the need not only to develop sound mathematical frameworks but also to tie them to concrete hardware implementations so that the benefits predicted by the theory, such as reduced sampling rates, can be met in practice and have an impact on the ADC market.

1.3   Outline and outlook

In this book we consider many extensions of the Shannon–Nyquist theorem, which treat a wide class of input signals as well as nonideal sampling and nonlinear distortions. Our exposition is based on a Hilbert-space interpretation of sampling techniques, where the aim is to develop more traditional sampling theories, along with modern techniques emerging from the field of CS, in a unified framework. The roots of this framework can be found in the fundamentals of linear algebra, and rely on familiar engineering building blocks such as filters and Fourier analysis. This unification has provided new understandings of classical interpolation methods, and has set the stage for new and exciting frontiers.

The framework we consider is based on viewing sampling in a broader sense of projection onto appropriate subspaces, and then choosing the subspaces to yield interesting new possibilities. For example, the results we present can be used to uniformly sample nonbandlimited signals, and to compensate perfectly for nonlinear effects.

Chapter 2 is therefore dedicated to a detailed exposition of the linear algebra basics needed to derive our extended sampling framework, followed by a brief summary of relevant Fourier analysis tools in Chapter 3. The more recent concepts of sub-Nyquist sampling, or extensions of CS to the analog setting, require more background on CS which we will provide in Chapter 11. Here again we retain the subspace approach by viewing these problems within the broader framework of a union of subspaces.

In order to develop ADCs for a particular problem, we must have accurate models for the signals of interest. We devote Chapters 4 and 5 to a detailed exposition of the signal models we will be focusing on throughout the book, along with some of the fundamental mathematical properties associated with such signal sets. Much of classical signal processing is based on the notion that signals can be modeled as vectors living in an appropriate subspace. Chapter 6 is focused on sampling theorems for signals confined to an arbitrary subspace in the presence of possibly nonideal sampling. The methods we develop can also be used to reconstruct a signal using a given interpolation kernel that is easy to implement, with often only a minor loss in signal quality with respect to the optimal kernel matched to the input subspace properties. In Chapter 8 we extend this basic framework to include nonlinear distortions in the sampling process. Surprisingly, many types of nonlinearities that are encountered in practice do not pose any technical difficulty and can be completely compensated for despite their effect of bandwidth increase, without requiring higher sampling rates.

A more general and less restrictive formulation of the sampling problem is considered in Chapter 7 in which our prior knowledge on the signal is that it is smooth in some sense. Unlike subspace priors, a one-to-one correspondence between smooth signals and their sampled version does not exist since smoothness is a far less restrictive constraint than confining the signal to a subspace. Perfect recovery is therefore generally impossible. Instead, we focus on approximating the input as well as possible under several different design objectives. These concepts can also be used to develop effective rate conversion techniques between digital formats, as we discuss in Chapter 9.

Although linear models are very popular in sampling theory, and more generally in DSP, such simple models often fail to capture much of the structure present in many common classes of signals. For example, while it may be reasonable to model signals as vectors, in many cases not all possible vectors in the space represent valid signals. In response to these challenges, there has been a surge of interest in recent years, across many fields, in a variety of *low-dimensional signal models* that quantify the notion that the number of degrees of freedom in high-dimensional signals is often quite small compared with their ambient dimension. One path to developing a framework for sampling and processing of such signals is by using the *union of subspaces model* which is introduced more formally in Chapter 10. Probably the most well-studied example of a union of subspaces is that of a vector $\mathbf{x}$ that is sparse in an appropriate basis. This model underlies the rapidly growing field of CS, which has attracted considerable attention in signal processing, statistics, and computer science, as well as the broader scientific community. A review of the essential CS concepts is provided in Chapter 11. In Chapters 12–15 we study how the fundamentals of CS can be expanded and extended to include richer structures in both analog and discrete-time signals, ultimately leading to sub-Nyquist sampling techniques for a broad class of continuous-time signals. A more detailed out-line of the book chapters can be found in the Preface.

The need and importance of sub-Nyquist techniques stems from the phenomenal success of DSP, thanks in large part to the Shannon–Nyquist theorem. This has spurred the digital revolution that is driving the development and deployment of new kinds of sensing systems with ever-increasing fidelity and resolution. As a result of this success, the amount of data generated by sensing systems has grown substantially. Unfortunately, in many important and emerging applications, the resulting sampling rate is so high that we end up with far too many samples that need to be transmitted, stored, and processed. In addition, in applications involving very wideband inputs it is often very costly, and sometimes even physically impossible, to build devices capable of acquiring samples at the necessary rate. Thus, despite extraordinary advances in sampling theory as well as computational power, the acquisition and processing of signals in application areas such as radar, wideband communications, imaging, video, medical imaging, remote surveillance, spectroscopy, and genomic data analysis continue to pose a tremendous challenge. Today, we are witnessing the outset of an interesting trend. Advances in related fields, such as wideband communication and radio-frequency technology, open a considerable gap with ADC devices. Conversion speeds which are twice the signal's maximal frequency component have become more and more difficult to obtain. Consequently, alternatives to high-rate sampling are drawing considerable attention in both academia and industry.

Over the years, theory and practice in the field of sampling have developed in parallel routes. Contributions by many research groups have suggested a multitude of methods, other than uniform sampling, to acquire analog signals. The math has deepened, leading to abstract signal spaces and innovative sampling techniques, with the ability to treat a large class of input signals far beyond the standard bandlimited model associated with the Shannon–Nyquist theorem. At the same time, the market has adhered to the Nyquist paradigm; the footprints of Shannon–Nyquist are evident whenever conversion to digital takes place in commercial applications.

Throughout the book, wherever possible, we try to put an emphasis on practical aspects of sampling beyond the fundamental theory. Our goal is to bridge theory and practice, and to try to highlight where advances in sampling theory can have and already have had an impact on ADC design and on applications. This is particularly relevant in the second half of the book, which is targeted at solving a practical problem: reducing sampling and processing rates which are too costly in many modern applications. The exposition is aimed at trying to pinpoint the potential of sub-Nyquist strategies to emerge from the math to the hardware. In this spirit, we integrate contemporary theoretical viewpoints, which study signal modeling in a union of subspaces, together with a taste of practical aspects, including basic circuit design features. Our hope is that this combination of theory and practice will serve to further promote both academic and industrial advances in sampling theory.

As a final note before beginning our theoretical journey into linear algebra: We have very much enjoyed gathering and presenting the ideas in a unified way. We hope that the reader will share some of our enthusiasm for the material!

Chapter 2

Introduction to linear algebra

The process of sampling and reconstruction can be viewed as an expansion of a signal onto a set of vectors that span the space. Suppose we have a signal x that is defined on some domain, and has a series representation there of the form

$$x = \sum_n a[n]x_n, \tag{2.1}$$

where $\{a[n]\}$ is a countable set of coefficients, which depend on the input signal x, and $\{x_n\}$ is a fixed set of signals (or vectors). The expression in (2.1) implies that x is completely specified in terms of the coefficients $\{a[n]\}$, which we may think of as samples of x. We can therefore interpret (2.1) as the statement that x can be reconstructed from the samples $\{a[n]\}$ using the known vectors $\{x_n\}$. Series of the form (2.1), their generalizations and extensions, are the subject of this book.

When we consider sampling, or signal expansions, we need to clearly identify the class of possible inputs x, the expansion vectors $\{x_n\}$, and the relationship between the samples $\{a[n]\}$ and the original signal x. In this chapter we describe the mathematical machinery needed to explain series of the form (2.1). In particular we introduce vector spaces and Hilbert spaces which provide the setting for describing the class of input signals x and the domain in which the expansions take place. Some important concepts we will consider in detail are the linear transformation, its adjoint and the subspaces associated with it, projection operators, and the pseudoinverse, which are all essential for computing the representation coefficients. We also define bases in Hilbert spaces and focus on stable expansions which leads to the notion of a Riesz basis. At the end of the chapter we briefly discuss overcomplete representations and frames.

2.1 Signal expansions: some examples

Before delving into the mathematical notions underlying sampling theory, we begin by studying several simple examples that highlight some of the issues that arise when considering signal expansions.

One of the main concepts central to sampling theory is that prior knowledge regarding the signal structure is essential in order to enable recovery from a given set of samples. As we explained in more detail in the Introduction, the process of sampling reduces the continuous-time signal to a countable set of coefficients, so that without any prior knowledge it is impossible to recover the full degrees of freedom describing the signal. To compensate for this dimensionality reduction, we must exploit knowledge regarding the signal structure. The following examples show how such information is incorporated into the recovery process. In all of the examples below we assume that the prior knowledge takes on the form of a subspace prior. The mathematics associated with such settings will be studied in much greater detail in Chapter 6.

Example 2.1 Suppose we are given two values $x(0)$ and $x(1)$ of a linear function $x(t) = at + b$, where a and b are unknown. Our goal is to evaluate $x(t)$ for any time t. Since the structure of $x(t)$ is known, to accomplish this goal all that is needed is to determine a and b. Therefore, our problem becomes that of computing a and b from the given samples.

It is easy to see that $b = x(0)$, and $a = x(1) - x(0)$. Therefore, for all t,

$$x(t) = x(0)(1 - t) + x(1)t = \sum_{n=0}^{1} a[n]x_n(t) \tag{2.2}$$

where $a[n] = x(n)$, and the expansion vectors are $x_0(t) = 1 - t$ and $x_1(t) = t$. Thus $x(t)$ can be represented by its samples $x(0)$ and $x(1)$. This example can be easily extended to allow recovery of a piecewise linear signal over the real line.

Although the previous example is very simple and almost trivial, it highlights some important features shared by many sampling theorems. First, we note that any function $x(t)$ from the given class of signals, in our case polynomials of degree 1, can be represented in the form (2.2) regardless of the values of a and b. For more general expansions, this corresponds to the statement that the expansion vectors $\{x_n\}$ are independent of the input x. Second, the ability to reconstruct the signal depends on our prior knowledge. In the example, we were able to recover $x(t)$ from its samples $x(0)$ and $x(1)$ since we knew that $x(t)$ is a polynomial of degree 1. On the other hand, note that the right-hand side of (2.2) always specifies a polynomial of degree 1; therefore, if $x(t)$ does not have this form, then it cannot be reconstructed using (2.2).

In Example 2.1 there were a finite number of expansion coefficients over each interval. Similar expansions are possible when there are infinitely many coefficients, as we show in the next example. The function we consider in the example below belongs to the well-known class of bandlimited signals, which leads to the Shannon–Nyquist theorem. We will revisit this example with more mathematical rigor in Chapter 4.

Example 2.2 Consider a signal $x(t)$ bandlimited to the frequency π/T. The famous Shannon–Nyquist theorem states that such a signal can be reconstructed from its samples $x(nT)$ using the expansion

$$x(t) = \sum_{n \in \mathbb{Z}} x(nT) \frac{\sin(\pi(t - nT)/T)}{\pi(t - nT)/T}. \tag{2.3}$$

This statement will be formally proven in Chapter 4 (along with a precise definition of bandlimitedness). We can associate (2.3) with an expansion of the form (2.1) where $x_n(t) = \sin(\pi(t - nT)/T)/(\pi(t - nT)/T)$ and $a[n] = x(nT)$. Consequently, any function bandlimited to π/T is represented by its samples $x(nT)$.

In both of the examples so far, the samples $a[n]$ associated with the signal x are equal to its values at certain time instances. We refer to this type of sampling as *pointwise (or ideal) sampling*. However, this does not necessarily have to be the case. In later chapters of the book we treat *generalized sampling*, in which the samples are no longer pointwise signal values. The following example illustrates this point.

Example 2.3 Suppose we know that a signal $x(t)$ has the form

$$x(t) = \sum_{n \in \mathbb{Z}} c[n]\phi(t - n) \tag{2.4}$$

for some function $\phi(t)$ and coefficients $c[n]$. In Chapter 4 we will discuss in detail classes of signals, popular in image processing and communication systems, that can be expressed this way. The representation in (2.4) fits the model (2.1) with $a[n] = c[n]$ and $x_n = \phi(t - n)$. However, it is no longer true in general that $c[n] = x(t = n)$. Therefore, given a signal $x(t)$ obeying (2.4), an interesting question is to determine the relationship between the generalized samples $a[n] = c[n]$ and the input signal $x(t)$.

In Chapter 6 we will see that there are many possible ways to characterize the samples $c[n]$. Below we choose one route, which is relatively straightforward. We will revisit this example in Chapter 6 and provide several complementary derivations.

Substituting $t = n$ in (2.4) we have that

$$x[n] = \sum_{k \in \mathbb{Z}} c[k]\phi(n - k), \tag{2.5}$$

which is a discrete-time convolution between the sequences $c[n]$ and $\phi[n] = \phi(t = n)$. Denoting by $X(e^{j\omega})$, $C(e^{j\omega})$, and $\Phi(e^{j\omega})$ the discrete-time Fourier transforms (DTFTs)[1] of $x[n]$, $c[n]$, and $\phi[n]$, we can write (2.5) in the Fourier domain as

$$X(e^{j\omega}) = C(e^{j\omega})\Phi(e^{j\omega}). \tag{2.6}$$

The notation $X(e^{j\omega})$ is a reminder that the DTFT is 2π-periodic. Assuming that $\Phi(e^{j\omega})$ is nonzero for all $\omega \in [-\pi, \pi]$, we have that $C(e^{j\omega}) = X(e^{j\omega})B(e^{j\omega})$ where $B(e^{j\omega}) = 1/\Phi(e^{j\omega})$. In the time domain, this relation becomes

$$c[n] = \sum_{k \in \mathbb{Z}} x[k]b[n - k], \tag{2.7}$$

---

[1] We will formally define convolution and the DTFT in Chapter 3.

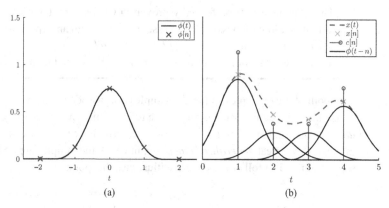

Figure 2.1 (a) A pulse $\phi(t)$ and its samples at the integers $\phi[n]$. (b) A signal $x(t)$ formed by linear combinations of shifts of $\phi(t)$ with coefficients $c[n]$.

where we relied on the convolution property of the DTFT, discussed in Chapter 3. Here $b[n]$ is the inverse DTFT of $B(e^{j\omega})$. Equation (2.7) relates the expansion coefficients $c[n]$ to uniform samples at times $t = n$ of $x(t)$, and the discrete-time series $b[n]$. It is evident from (2.7) that $c[n]$ is not equal to $x[n] = x(t = n)$ unless $b[n] = \delta[n]$, where $\delta[n] = 1$ if $n = 0$ and is 0 otherwise.

Figure 2.1 illustrates these concepts. The function $\phi(t)$ and its samples at the integers $\phi[n]$ are shown in Fig. 2.1(a). In this example, $\phi(t) = \beta^2(t)$ is chosen as a B-spline function of degree 2. B-splines are very popular in signal and image processing, and will be discussed in detail in Chapter 4. Figure 2.1(b) depicts the signal $x(t)$ resulting from linearly combining integer shifts of $\phi(t)$ with a given set of coefficients $c[n]$. As can be seen, the coefficients $c[n]$ of the combination, marked in circles, do not coincide with the samples $x[n]$ of $x(t)$, shown in X-marks. The relationship between these two sequences is given by (2.7), where $b[n]$ is the convolutional inverse of $\phi[n]$.

The examples above illustrate that in order to understand signal expansions we need a framework in which to describe classes of inputs, and the types of generalized samples that are allowed. The framework we consider throughout the book is a signal space viewpoint in which signals are regarded as vectors in an abstract Hilbert space referred to as the *signal space*.

In this chapter we will therefore focus on Hilbert spaces and some of their properties, with the intention of rendering the book as self-contained as possible. Our objective is to lay the groundwork for subsequent chapters focused on sampling theories, and establish the signal space notation used throughout. Background material on Hilbert spaces and linear algebra helpful to understanding the material presented in this chapter can be found in [23, 24, 25, 26]. Many of the results herein are stated without proof; we refer the reader to standard textbooks for the detailed derivations. In places where the proof is insightful and/or simple, we have included it for completeness.

2.2   Vector spaces

A vector space $\mathcal{V}$ over the complex numbers $\mathbb{C}$ or the real numbers $\mathbb{R}$ is a set of elements called vectors, together with vector addition and scalar multiplication by elements of $\mathbb{C}$ or $\mathbb{R}$ such that $\mathcal{V}$ is closed under both operations, namely:

1. $x + y \in \mathcal{V}$ for any $x, y \in \mathcal{V}$.
2. $ax \in \mathcal{V}$ for any $x \in \mathcal{V}$ and $a \in \mathbb{C}$ or $\mathbb{R}$.

For any $x, y \in \mathcal{V}$ and $a, b \in \mathbb{C}$ or $\mathbb{R}$ the addition and multiplication must satisfy:

1. Commutativity: $x + y = y + x$.
2. Associativity: $(x + y) + z = x + (y + z)$, $(ab)x = a(bx)$.
3. Distributivity: $a(x + y) = ax + ay$, $(a + b)x = ax + bx$.
4. Additive identity: there exists $0 \in \mathcal{V}$ such that $x + 0 = x$ for any $x \in \mathcal{V}$.
5. Additive inverse: for all $x \in \mathcal{V}$ there exists $-x \in \mathcal{V}$ such that $x + (-x) = 0$.
6. Multiplicative identity: $1 \cdot x = x$ for any $x \in \mathcal{V}$.

In the sequel we will typically consider the space $\mathcal{V}$ to be over $\mathbb{C}$, unless specifically stated otherwise.

    The elements of $\mathcal{V}$ can be arbitrary. For example, x can be a finite length vector $x = \mathbf{x} = [x_1 \ x_2 \ x_3]^T$ (where $[\ ]^T$ denotes the transpose), an infinite sequence $x = x[n], n \in \mathbb{Z}$, or a function over the real line $x = x(t), t \in \mathbb{R}$. The essential point is that we can consider these varied types of signals as vectors, and develop a unified theory that is applicable to all of these cases without having to study each setting individually.

2.2.1   Subspaces

Typically when we treat sampling theorems we are interested in representing a subset of all signals in the space. A useful concept in this context is that of a subspace.

Definition 2.1. *A set $\mathcal{W}$ of vectors in $\mathcal{V}$ is a subspace of $\mathcal{V}$, denoted $\mathcal{W} \subseteq \mathcal{V}$, if it is closed under addition and scalar multiplication:*

1. $x + y \in \mathcal{W}$ for any $x, y \in \mathcal{W}$.
2. $ax \in \mathcal{W}$ for any $x \in \mathcal{W}$ and $a \in \mathbb{C}$.

It is easy to see that a subspace is itself a vector space.

    An important example of a subspace is the *span* of a set of vectors.

Definition 2.2. *Given a set of vectors $S = \{x_1, \ldots, x_m\}$ where $x_n \in \mathcal{V}$ and m may be infinite, the span of S is the subspace of $\mathcal{V}$ consisting of all finite linear combinations of vectors in S:*

$$\mathrm{span}(S) = \left\{ \sum_{n=1}^{N} a[n]x_n \ \middle| \ a[n] \in \mathbb{C}, x_n \in S \right\}. \tag{2.8}$$

When there is a finite number m of vectors, $N = m$ in (2.8). Otherwise, N varies over all possible integers.

The span will be used frequently throughout the book.

2.2.2 Properties of subspaces

Two useful relations between subspaces are those of disjoint spaces and sums.

We begin with the notion of disjoint spaces. Since a subspace is closed under multiplication, it must contain the zero vector (as we can choose the multiplier to be $a = 0$). Therefore, two subspaces cannot technically be disjoint, since they both contain the zero vector. With slight abuse of terminology we therefore define two subspaces $\mathcal{V}$ and $\mathcal{W}$ as *disjoint* if they intersect only at the zero vector, i.e., $\mathcal{V} \cap \mathcal{W} = \{0\}$.

The *sum* of two subspaces $\mathcal{V}$ and $\mathcal{W}$ of a vector space $\mathcal{H}$, denoted $\mathcal{V} + \mathcal{W}$, is the set of all vectors of the form $x = v + w$ where $v \in \mathcal{V}$ and $w \in \mathcal{W}$. The *direct sum* of $\mathcal{V}$ and $\mathcal{W}$, denoted $\mathcal{V} \oplus \mathcal{W}$, is the sum of two *disjoint* subspaces. If a vector space $\mathcal{H}$ can be written as the direct sum of two subspaces, $\mathcal{H} = \mathcal{V} \oplus \mathcal{W}$, then $\mathcal{V} \cap \mathcal{W} = \{0\}$ and any $x \in \mathcal{H}$ can be decomposed *uniquely* as $x = v + w$, with $v \in \mathcal{V}$ and $w \in \mathcal{W}$. A useful property that can be used to check whether a set of subspaces is disjoint is given by the following proposition.

Proposition 2.1. *If $\mathcal{V}_1 \ldots \mathcal{V}_n$ are subspaces of $\mathcal{H}$, then $\mathcal{H} = \mathcal{V}_1 \oplus \ldots \oplus \mathcal{V}_n$ if and only if $\mathcal{H} = \mathcal{V}_1 + \ldots + \mathcal{V}_n$, and $0 = v_1 + \ldots + v_n$ where each $v_j \in \mathcal{V}_j$ implies $v_j = 0$ for every j. In other words, there is only one way to represent the zero vector.*

Example 2.4 Consider the vector space $\mathcal{H}$ of discrete-time signals $x[n]$ defined for $-N \leq n \leq N$. Let $\mathcal{V}$ be the subspace of $\mathcal{H}$ consisting of strictly causal signals $v[n]$ such that $v[n] = 0, -N \leq n \leq 0$, and let $\mathcal{W}$ be the subspace of $\mathcal{H}$ containing symmetric signals $w[n]$ satisfying $w[n] = w[-n], -N \leq n \leq N$. We now examine whether $\mathcal{H} = \mathcal{V} \oplus \mathcal{W}$.

It is easy to see that $\mathcal{H} = \mathcal{V} + \mathcal{W}$, i.e., any signal in $\mathcal{H}$ can be written as a linear combination of a signal in $\mathcal{V}$ and a signal in $\mathcal{W}$. To see this let $x[n]$ be arbitrary. Choose $w[n] = x[n]$ for $-N \leq n \leq 0$ and $w[n] = w[-n]$ for $1 \leq n \leq N$. Then let $v[n]$ be the strictly causal sequence defined by $v[n] = x[n] - w[n], 1 \leq n \leq N$. It is also immediately obvious that $\mathcal{V} \cap \mathcal{W} = \{0\}$, so that $\mathcal{H} = \mathcal{V} \oplus \mathcal{W}$. This follows from the fact that for any $x[n] \in \mathcal{V}$, we have $x[n] = 0, -N \leq n \leq 0$. If $x[n]$ is also in $\mathcal{W}$, then $x[n] = x[-n] = 0$ for $1 \leq n \leq N$, and $x[n] = 0$. This decomposition is illustrated in Fig. 2.2.

Now, suppose we choose a new space $\mathcal{U}$ of all causal sequences $u[n] = 0, -N \leq n \leq -1$. We then still have that $\mathcal{H} = \mathcal{U} + \mathcal{W}$. However, $\mathcal{U} \cap \mathcal{W} \neq \{0\}$ so that we no longer have a direct sum decomposition. Indeed, any signal of the form $x[n] = a\delta[n]$ where a is a scalar, is both in $\mathcal{U}$ and in $\mathcal{W}$. Equivalently, we can write 0 in many different ways in terms of a signal in $\mathcal{U}$ and a signal in $\mathcal{W}$: $0 = a\delta[n] - a\delta[n]$ where $a\delta[n] \in \mathcal{U}$ and $-a\delta[n] \in \mathcal{W}$ for any a.

Direct sum decompositions are very important for the developments later in the book. We will return to this concept in Section 2.6 when we discuss projection operators. As we will see, projections and direct sum decompositions are two sides of the same coin:

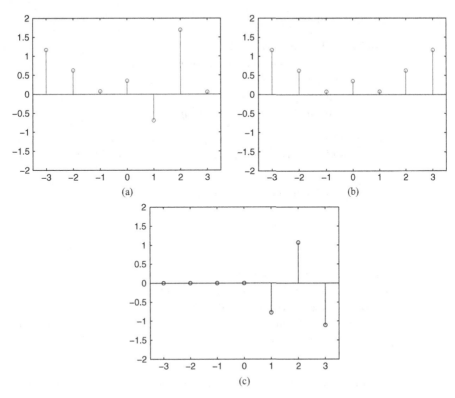

Figure 2.2 Unique decomposition of a sequence into a symmetric part and a strictly causal part. (a) A sequence $x[n]$. (b) The symmetric part $w[n]$ of $x[n]$. (c) The strictly causal part $v[n]$ of $x[n]$.

projection operators can be used to decompose the space into subspaces defined by the range and null spaces of the projectors.

2.3   Inner product spaces

Until now we have focused on the algebra of a set of vectors. We now add geometric structure to a vector space in the form of an *inner product* relation between pairs of vectors, which also induces a distance measure or metric on the space. The inner product is fundamental for defining norms and orthogonality. Equipped with a norm, we can further develop calculus analysis over the underlying space, defining important concepts such as Cauchy sequences and a subspace closure. Combining geometry and calculus will lead to the definition of a Hilbert space, which is instrumental in the development of sampling theorems.

2.3.1   The inner product

Definition 2.3. *An inner product on a vector space* $\mathcal{V}$*, denoted* $\langle x, y \rangle$*, is a mapping from* $\mathcal{V} \times \mathcal{V}$ *to* $\mathbb{C}$ *that satisfies the following properties:*
1. $\langle x, y \rangle = \overline{\langle y, x \rangle}$*;*
2. $\langle x, ay + bz \rangle = a\langle x, y \rangle + b\langle x, z \rangle$*;*
3. $\langle x, x \rangle \geq 0$ *and* $\langle x, x \rangle = 0$ *if and only if* $x = 0$*,*

where $\overline{(\cdot)}$ *denotes the complex conjugate. The* norm *of a vector* x *is defined by* $\|x\| = \sqrt{\langle x, x \rangle}$*, and the* distance *between* x *and* y *is the norm of the difference* $\|x - y\|$*.*

An immediate implication of properties (1)–(2) is that

$$\langle ax + bz, y \rangle = \overline{a}\langle x, y \rangle + \overline{b}\langle z, y \rangle. \tag{2.9}$$

The proof is left as a homework exercise (see Exercise 3). A vector space equipped with an inner product is called an *inner product space*.

Any mapping satisfying (1)–(3) above is a valid inner product. The choice of inner product will depend on the properties of the underlying vector space $\mathcal{V}$, as well as the particular application. We now give some examples of inner products.

Example 2.5 Let $\mathcal{V} = L_2(\mathbb{R})$ be the vector space of all continuous-time finite-energy signals. In this case a vector $x \in \mathcal{V}$ represents a signal $x(t)$ with $\int_{-\infty}^{\infty} |x(t)|^2 dt < \infty$. We can immediately verify that $\langle x, y \rangle = \int_{-\infty}^{\infty} \overline{x(t)}y(t)dt$ is a valid inner product on this space, namely properties (1)–(3) are satisfied.[2] Another legitimate choice is $\langle x, y \rangle = \int_{-\infty}^{\infty} w(t)\overline{x(t)}y(t)dt$ for a real, bounded, positive function $w(t)$. The positivity requirement ensures that the norm of $x(t)$ is not zero unless $x(t) = 0$ almost everywhere (namely, the set over which $x(t) \neq 0$ has zero measure).

Example 2.6 Suppose now that $\mathcal{V} = \ell_2(\mathbb{Z})$ is the space of discrete-time finite-energy sequences so that $x = x[n]$. We can then define an inner product by $\langle x, y \rangle = \sum_{n \in \mathbb{Z}} w[n]\overline{x[n]}y[n]$ for any sequence $w[n] > 0$ that is bounded.

In the previous two examples we introduced two of the subspaces that will be used extensively throughout the book: $L_2(\mathbb{R})$ and $\ell_2(\mathbb{Z})$. When referring to these subspaces, we will often omit the arguments and use the shorthand notation L_2 and ℓ_2.

---

[2] Technically speaking, this inner product does not satisfy property (3) since $\int_{-\infty}^{\infty} |x(t)|^2 dt = 0$ even if $x(t)$ is nonzero over a set of measure zero. However, it is customary to identify all signals in $L_2(\mathbb{R})$ that differ over a set of measure zero; in other words, if $x(t) = y(t)$ almost everywhere then we will say that $x(t) = y(t)$. With this definition, property (3) holds. For a more thorough treatment of this delicate point, see [27, Chapter 2.6].

Inner products obey some useful properties, several of which are listed below:

1. *Cauchy–Schwarz inequality*:

$$|\langle x, y \rangle| \leq \|x\| \|y\| \tag{2.10}$$

with equality if and only if $x = ay$ for some $a \in \mathbb{C}$.

2. *Triangle inequality*:

$$\|x + y\| \leq \|x\| + \|y\| \tag{2.11}$$

with equality if and only if $x = ay$ where $a \geq 0$. This result is called the triangle inequality because of its geometric interpretation that the length of any side of a triangle is less than the sum of the lengths of the other two sides.

3. *Parallelogram law*:

$$\|x\|^2 + \|y\|^2 = \frac{1}{2} \left(\|x + y\|^2 + \|x - y\|^2 \right). \tag{2.12}$$

This equality states that the sum of the squares of the lengths of the four sides of a parallelogram equals the sum of the squares of the lengths of its two diagonals.

4. *Polarization identity*:

$$\Re\{\langle x, y \rangle\} = \frac{1}{4} \left(\|x + y\|^2 - \|x - y\|^2 \right), \tag{2.13}$$

where $\Re\{x\}$ denotes the real part of the complex vector x.

2.3.2 Orthogonality

A very important concept related to inner products is that of orthogonality.

Definition 2.4. *Two vectors x, y are said to be* orthogonal *in $\mathcal{V}$ if $\langle x, y \rangle = 0$. Two subspaces $\mathcal{W}$ and $\mathcal{V}$ are orthogonal if $\langle w, v \rangle = 0$ for any $w \in \mathcal{W}$ and $v \in \mathcal{V}$.*

If two vectors x, y are orthogonal, then

$$\|x + y\|^2 = \|x\|^2 + \|y\|^2. \tag{2.14}$$

This relation is known as *Pythagoras' theorem*. To prove the result, note that

$$\|x + y\|^2 = \langle x + y, x + y \rangle = \|x\|^2 + \langle x, y \rangle + \langle y, x \rangle + \|y\|^2, \tag{2.15}$$

where we used property (2) of inner products. Since $\langle x, y \rangle = 0$ and $\langle y, x \rangle = \overline{\langle x, y \rangle} = 0$, the result follows.

Another important concept of orthogonality related to subspaces is that of the orthogonal complement:

Definition 2.5. *Let $\mathcal{W}$ be a subspace of $\mathcal{V}$. Then the* orthogonal complement *of $\mathcal{W}$ in $\mathcal{V}$, denoted $\mathcal{W}^\perp$, is the subspace of vectors in $\mathcal{V}$ that are orthogonal to all vectors in $\mathcal{W}$:*

$$\mathcal{W}^\perp = \{x \in \mathcal{V} | \langle x, y \rangle = 0 \text{ for all } y \in \mathcal{W}\}. \tag{2.16}$$

It is easy to see that $W \cap W^\perp = \{0\}$, so that the only vector that lies both in W and $W^\perp$ is the zero vector. To prove this statement formally, suppose that x lies in both W and $W^\perp$. Then, because x is in $W^\perp$, $\langle x, y \rangle = 0$ for any $y \in W$. In particular, since $x \in W$, we can choose $y = x$ leading to $\langle x, x \rangle = 0$ which implies that $x = 0$ from property (3) of the inner product.

Example 2.7 Consider the space V of real sequences $x[n]$ defined over $-N \le n \le N$ and let W be the subspace of V comprising all symmetric sequences $w[n]$, namely those that satisfy $w[n] = w[-n]$. We would like to determine the orthogonal complement $W^\perp$ of W under the inner product $\langle a, b \rangle = \sum_{|n| \le N} a[n]b[n]$.

By definition, any $v \in W^\perp$ has to satisfy $\sum_{|n| \le N} v[n]w[n] = 0$ for every $w \in W$. Taking into account that $w[n] = w[-n]$, this condition can be written as

$$w[0]v[0] + \sum_{n=1}^{N} w[n](v[n] + v[-n]) = 0. \tag{2.17}$$

Since (2.17) must be satisfied for all $w[n] \in W$, we can obtain necessary conditions on $v[n]$ by enforcing (2.17) for specific choices of $w[n]$. Letting $w[n] = \delta[n]$, (2.17) reduces to $v[0] = 0$. For $w[n] = \delta[n - n_0] + \delta[n + n_0]$ with n_0 an arbitrary integer in $[1, N]$, (2.17) becomes $v[n_0] + v[-n_0] = 0$. Thus, $v[n]$ must be an antisymmetric sequence satisfying $v[n] = -v[-n]$. It is easy to see that this condition is also sufficient: when $v[n]$ is antisymmetric (2.17) will be satisfied for all $w[n]$. We therefore conclude that $W^\perp$ is the space of all antisymmetric sequences.

A set of vectors $x_1, \ldots, x_n$ is *orthogonal* if $\langle x_i, x_j \rangle = 0$ for all $i \ne j$. If in addition the vectors all have unit norm, then $\langle x_i, x_j \rangle = \delta_{ij}$ and the vectors are said to be *orthonormal*. Here δ_{ij} is the *Kronecker delta* function defined by

$$\delta_{ij} = \begin{cases} 1, & i = j \\ 0, & i \ne j. \end{cases} \tag{2.18}$$

Note that $\delta[n] = \delta_{0n}$.

Gram–Schmidt orthogonalization

Given a nonorthonormal finite number of linearly independent vectors[3] $\{v_n, 1 \le n \le N\}$, it is always possible to construct an orthonormal set $\{w_n, 1 \le n \le N\}$ spanning the same space. One such procedure is known as the *Gram–Schmidt* process [28] (see Exercise 5). The Gram–Schmidt orthogonal vectors $\{w_n\}$ are chosen such that w_1 is in the direction of v_1, w_2 is in the direction of the component of v_2 which is perpendicular to v_1, and so forth. Thus, w_1 is perfectly aligned with v_1, but w_N may be relatively "far" from v_N. It is immediately obvious that the Gram–Schmidt vectors depend on the order in which the vectors $\{v_n\}$ are arranged. Furthermore, this approach is not known

---

[3] A formal definition of linear independence is given below in Definition 2.13.

to be optimal in any sense – in particular, the orthogonal vectors are not guaranteed to be "close" to the original vectors.

Least squares orthogonalization

Instead of choosing an orthogonal set in a greedy fashion, we can design the vectors to be orthogonal, and closest in a least squares sense to the original (nonorthogonal) vector set. More specifically, given a set of vectors $\{v_n, 1 \le n \le N\}$ we may design a set of orthonormal vectors $\{w_n, 1 \le n \le N\}$ such that the squared error

$$E = \sum_{n=1}^{N} \|v_n - w_n\|^2 \qquad (2.19)$$

is minimized. The resulting vectors are referred to as the *least squares vectors* [29]. An important attribute of these vectors is that they do not depend on the order of the vectors $\{v_n\}$. More generally, we can seek an orthogonal set $\{w_n\}$ without requiring the vectors to be normalized, but rather select their norms in an optimal way to minimize the error defined by (2.19) [30].

A closed-form solution to (2.19) is derived in [29, 30, 31]. The expression for the least squares vectors depends on some definitions and concepts that we have not yet discussed in full. For completeness, we provide the solution here together with pointers to where the appropriate definitions may be found. We begin by defining the Gram-matrix $\mathbf{G}$ of inner products $g_{nm} = \langle v_n, v_m \rangle$. The least squares vectors can then be written as $w_n = \sum_{m=1}^{N} a_{nm} v_m$ where the coefficients a_{nm} are the elements of the matrix $\mathbf{A}$ given by $\mathbf{A}^{-1} = \mathbf{G}^{1/2}$. The square-root of $\mathbf{G}$ is the unique Hermitian positive-definite matrix $\mathbf{B}$ such that $\mathbf{B}^2 = \mathbf{G}$; see Appendix A for exact definitions. The matrix inverse will be formally defined in Section 2.4.1 below. In that context, we show that the Gram-matrix associated with a set of linearly independent vectors is always invertible. For more details on the derivation and applications of the least squares vectors we refer the reader to [29, 30, 31].

2.3.3 Calculus in inner product spaces

Although our focus in this book is primarily on the algebraic and geometric properties of vector spaces, we will need some basic notions of calculus in order to define Hilbert spaces. In subsequent chapters we also rely on various concepts from functional analysis in addressing different classes of sampling theorems. Therefore, we now turn to introduce some basic elements of calculus in inner product spaces.

We begin by defining the *closure* $\mathcal{V}^c$ of an inner product space $\mathcal{V}$. A vector x is a point of closure of $\mathcal{V}$ if for every $\varepsilon > 0$ there exists a vector $v \in \mathcal{V}$ such that $\|x - v\| < \varepsilon$. Intuitively, points of closure can be approximated arbitrarily closely by a vector in the space. The *closure* $\mathcal{V}^c$ of a subspace $\mathcal{V}$ is the set of all points of closure of $\mathcal{V}$. A subspace $\mathcal{V}$ is *closed* if $\mathcal{V} = \mathcal{V}^c$. In our development of sampling theorems, we will always work with closed subspaces. This is achieved by using well-behaved expansion vectors, as we discuss in the context of Riesz bases in Section 2.5.3.

Example 2.8 A useful example of a closure of a subspace is the closure of a span of vectors. The span of a set of vectors $\{x_n \in \mathcal{V}\}$ was introduced in Definition 2.2. The closure of the span is the set of vectors that can be approximated arbitrarily closely by a vector in the span. Mathematically, a vector x is in the closure of the span of $\{x_n\}$ if, given an $\varepsilon > 0$, we can find an N and a sequence of constants $a[n]$ such that $\|x - \sum_{n=1}^{N} a[n]x_n\| < \varepsilon$. In the sequel, with a slight abuse of terminology, when we refer to the span of an infinite set of vectors, we explicitly mean the closure of the span, as defined in this example.

The next definition we introduce is that of a *convergent sequence*. A sequence of vectors $\{x_n \in \mathcal{V}\}$ is said to converge to a vector $x \in \mathcal{V}$ if for every $\varepsilon > 0$ there exists an integer N such that $\|x_n - x\| < \varepsilon$ for all $n > N$. We use shorthand notation to denote convergence of a sequence to x by $x_n \to x$. We also write this as $\lim_{n \to \infty} x_n = x$.

The vectors $\{x_n\}$ are called a *Cauchy sequence* if for every $\varepsilon > 0$ there exists an integer N such that for all $m, n > N$ we have that $\|x_m - x_n\| < \varepsilon$. A *complete* vector space $\mathcal{V}$ is one in which every Cauchy sequence converges to a vector in $\mathcal{V}$.

2.3.4   Hilbert spaces

With the notion of a complete vector space at hand, we are now ready to define a Hilbert space.

Definition 2.6. *A Hilbert space is a complete inner product space.*

In most of the book the signals we are concerned with will lie in Hilbert spaces, and the analysis will be performed over such spaces. In fact, many of the commonly used spaces in engineering are Hilbert spaces. Some examples that are used throughout the book are considered below.

Example 2.9 The Hilbert space $\mathcal{H} = \mathbb{C}^m$ denotes the set of all m-dimensional vectors $x = \mathbf{x}$ with components in $\mathbb{C}$. The standard inner product on $\mathbb{C}^m$ is defined by $\langle x, y \rangle = \mathbf{x}^H \mathbf{y} = \sum_{n=1}^{m} \overline{x[n]} y[n]$, where $x[n]$ and $y[n]$ denote the nth components of $\mathbf{x}$ and $\mathbf{y}$, respectively, and $\mathbf{x}^H$ denotes the Hermitian conjugate.

Example 2.10 The set of all sequences $x = \{x[n]\}$ with $x[n] \in \mathbb{C}$ that are absolutely square summable, i.e., $\sum_{n \in \mathbb{Z}} |x[n]|^2 < \infty$, is the Hilbert space $\mathcal{H} = \ell_2(\mathbb{Z})$. The standard inner product on $\ell_2(\mathbb{Z})$ is defined by $\langle x, y \rangle = \sum_{n \in \mathbb{Z}} \overline{x[n]} y[n]$.

We have seen in Example 2.6 that a whole class of inner products can be defined over $\ell_2(\mathbb{Z})$ by choosing a positive, bounded, weighting function $w[n]$. Removing the requirement that $w[n]$ is bounded, we can define a new Hilbert space of weighted $\ell_2(\mathbb{Z})$ functions that is not equal to $\ell_2(\mathbb{Z})$.

Example 2.11 Consider the set of all sequences $x = \{x[n]\}$ for which $\sum_{n \in \mathbb{Z}} w[n]|x[n]|^2 < \infty$, where $w[n] > 0$ is an arbitrary weighting function. We define the inner product on the space as $\langle x, y \rangle = \sum_{n \in \mathbb{Z}} w[n]\overline{x[n]}y[n]$, and denote the resulting Hilbert space by $\ell_2(\mathbb{Z}, w)$. Since $w[n]$ is not bounded it is easy to see that $\ell_2(\mathbb{Z}, w)$ is not equal to $\ell_2(\mathbb{Z})$.

We can repeat Examples 2.10 and 2.11 for the class of continuous-time bounded signals:

Example 2.12 The continuous-time counterpart of Example 2.10 is the Hilbert space $\mathcal{H} = L_2(\mathbb{R})$ consisting of all functions $x = x(t)$ that are absolutely square integrable, i.e. $\int_{-\infty}^{\infty} |x(t)|^2 dt < \infty$. The standard inner product on $L_2(\mathbb{R})$ is $\langle x, y \rangle = \int_{-\infty}^{\infty} \overline{x(t)}y(t)dt$. Similarly to Example 2.11, we can define a weighted $L_2(\mathbb{R})$ space $L_2(\mathbb{R}, w)$ as the space of all functions satisfying $\int_{-\infty}^{\infty} w(t)|x(t)|^2 dt < \infty$ for a weighting function $w(t) > 0$, with the inner product $\langle x, y \rangle = \int_{-\infty}^{\infty} w(t)\overline{x(t)}y(t)dt$. Note that if $w(t)$ is bounded, then $L_2(\mathbb{R}, w)$ and $L_2(\mathbb{R})$ are identical.

A *signal space* is a Hilbert space whose elements are signals; we refer to these elements as vectors or signals interchangeably. Despite the fact that we treat all signal spaces within the same framework, it is still useful at times to use different notation for vectors, continuous-time signals, and sequences. Throughout, we denote vectors in $\mathbb{C}^m$ (m arbitrary) by boldface lower case letters, e.g. $\mathbf{x}$. The nth component of $\mathbf{x}$ is written as $x[n]$ (or x_n where there is no chance of confusion with the nth element in a series of vectors). For sequences in ℓ_2 we use lowercase letters, e.g. a, and denote their elements similarly by $a[n]$.

2.4 Linear transformations

Having defined the Hilbert space structure, we turn to operations performed within the space or between spaces. A large part of sampling theory is concerned with linear operations and linear expansions. We therefore focus our attention now on linear transformations. Of particular importance are the set transformation defined in Section 2.5, and the projection operator discussed in Section 2.6.

Definition 2.7. *Let $\mathcal{H}$ and $\mathcal{S}$ be Hilbert spaces. T is a linear transformation from $\mathcal{H}$ to $\mathcal{S}$, denoted $T: \mathcal{H} \to \mathcal{S}$, if every $x \in \mathcal{H}$ is mapped to one and only one $y \in \mathcal{S}$, and $T(c_1 x_1 + c_2 x_2) = c_1 T(x_1) + c_2 T(x_2)$ for all $x_1, x_2 \in \mathcal{H}$ and $c_1, c_2 \in \mathbb{C}$ (or $\mathbb{R}$).*

Since most transformations we consider in this chapter, and in the book, are linear, we will often omit the word "linear" and simply refer to a linear transform as a transform.

We use uppercase letters to denote general transformations. When $\mathcal{H} = \mathbb{C}^m$ and $\mathcal{S} = \mathbb{C}^n$, T can be written as a matrix-vector multiplication by an $n \times m$ matrix $\mathbf{T}$; the convention we adopt is to write matrices in boldface upper case letters. Two important transformations are the identity and the zero transform. The identity on $\mathcal{H}$ is the operator defined by $I_{\mathcal{H}}x = x$ for all $x \in \mathcal{H}$, or simply I. When $\mathcal{H} = \mathbb{C}^m$ we write the identity as $\mathbf{I}_m$, or $\mathbf{I}$ when the dimension is clear from the context. Similarly, the zero operator, denoted 0, is such that $0x = 0$ for all $x \in \mathcal{H}$. For $\mathcal{H} = \mathbb{C}^m$ we use the boldface matrix notation $\mathbf{0}$.

Two desirable properties of linear transforms are boundedness and continuity. As we will see, these properties are actually the same.

Definition 2.8. *A linear transformation* $T: \mathcal{H} \to \mathcal{S}$ *is bounded if* $\|Tx\| \leq \alpha\|x\|$ *for all* $x \in \mathcal{H}$ *and some* $\alpha > 0$. *We say that* $T: \mathcal{H} \to \mathcal{S}$ *is continuous if* $x_n \to x$ *implies that* $Tx_n \to Tx$ *for every* $x \in \mathcal{H}$.

The following proposition establishes the relation between continuity and boundedness [23, p. 94]:

Proposition 2.2. *A linear transformation* $T: \mathcal{H} \to \mathcal{S}$ *is continuous if and only if it is bounded.*

In most of this book we will focus our attention on bounded operators. Therefore, we assume (unless otherwise stated) that T is bounded (and continuous).

Example 2.13 Let $\mathcal{H}$ be a Hilbert space, and define the transformation $Tx = \langle y, x \rangle$ for a fixed $y \in \mathcal{H}$ with bounded norm. It is easy to see that T is bounded (and therefore continuous). Indeed, using the Cauchy–Schwarz inequality we have that for any $x \in \mathcal{H}$

$$\|Tx\| = |\langle y, x \rangle| \leq \|y\|\|x\| = \alpha\|x\|, \tag{2.20}$$

with $\alpha = \|y\|$.

The previous example shows that the inner product is a continuous function. Therefore, if a sequence of vectors x_n converges to x, then

$$\lim_{n\to\infty} \langle y, x_n \rangle = \langle y, x \rangle. \tag{2.21}$$

2.4.1 Subspaces associated with a linear transformation

A useful way to analyze transformations in general, and sampling operators in particular, is to understand their action on smaller subspaces of the underlying Hilbert space. As we will see, this allows important characteristics of the transformations to be unveiled. A convenient choice of subspaces to associate with a given linear transformation $T: \mathcal{H} \to \mathcal{S}$ is the following four spaces:

- The null space (kernel) $\mathcal{N}(T)$;
- The orthogonal complement $\mathcal{N}(T)^{\perp}$ of $\mathcal{N}(T)$ in $\mathcal{H}$;
- The range space (image) $\mathcal{R}(T)$;
- The orthogonal complement $\mathcal{R}(T)^{\perp}$ of $\mathcal{R}(T)$ in $\mathcal{S}$.

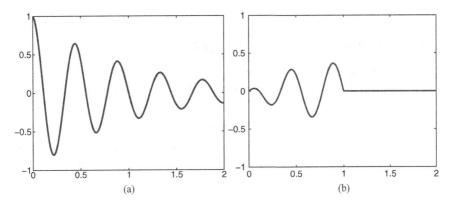

Figure 2.3 (a) The signal $x(t) = e^{-t}\cos(4.5\pi t)$. (b) The signal $y(t) = T\{x\}(t)$ with T given by (2.22).

The first two spaces are defined in $\mathcal{H}$, while the latter two are subspaces of $\mathcal{S}$.

The null space of T is the set of vectors $x \in \mathcal{H}$ satisfying $Tx = 0$. This space can never be empty since it always contains the zero vector. The range of T is the set of vectors $y \in \mathcal{S}$ for which there exists an $x \in \mathcal{H}$ such that $y = Tx$. Choosing $x = 0$ shows that the range also contains the zero vector. From Definition 2.7, it is easy to see that $\mathcal{N}(T)$ and $\mathcal{R}(T)$ are linear subspaces (see Exercise 7). The definitions of $\mathcal{N}(T)^{\perp}$ and $\mathcal{R}(T)^{\perp}$ follow immediately from (2.16).

Example 2.14 Consider the space $\mathcal{H}$ of real signals $x(t)$ defined over $t \in [0, 2]$, whose energy $\int_0^2 x^2(t)dt$ is finite. We denote this space by $\mathcal{H} = L_2([0, 2])$. Let $T : \mathcal{H} \to \mathcal{H}$ be the transformation given by

$$y(t) = T\{x\}(t) = \begin{cases} tx(t), & t \in [0, 1] \\ 0, & t \in (1, 2]. \end{cases} \tag{2.22}$$

An example of the operation of T on a signal $x(t)$ is depicted in Fig. 2.3.

We now determine the null space $\mathcal{N}(T)$ and range space $\mathcal{R}(T)$ of T. By definition, any signal $x \in \mathcal{N}(T)$ must satisfy $tx(t) = 0$ for every $t \in [0, 1]$, and can take on arbitrary values for $t \in (1, 2]$. Thus, $\mathcal{N}(T)$ comprises all signals that vanish almost everywhere on $t \in [0, 1]$.

To determine the range space, we note that every signal $y \in \mathcal{R}(T)$ vanishes on $t \in (1, 2]$ and can be expressed as $tx(t)$ on $t \in [0, 1]$ for some finite energy signal $x(t)$. In other words, for every signal $y \in \mathcal{R}(T)$ the (improper) integral

$$\int_0^1 \left(\frac{y(t)}{t}\right)^2 dt \tag{2.23}$$

must converge (that is, take on a finite value). For this to happen, $y(t)$ must tend to 0 as $t \to 0$ faster than $t^{1/2}$.

An important concept associated with $\mathcal{R}(T)$ is that of the rank of T:

Definition 2.9.   *The* rank *of T is equal to the dimension of $\mathcal{R}(T)$.*

When $\mathcal{R}(T)$ has infinite dimension, the rank is infinite. We will formally define the notion of dimension in Section 2.5.2, after introducing bases. Linear independence of vectors is introduced in Definition 2.13.

If $T = \mathbf{T}$ is an $m \times n$ matrix, then:

1. The rank r is equal to the number of columns of $\mathbf{T}$ that are linearly independent.
2. $\mathcal{R}(\mathbf{T})$ is spanned by the linearly independent columns of $\mathbf{T}$.
3. The dimension of $\mathcal{N}(\mathbf{T})$ is equal to $n - r$.

Classical linear algebra textbooks discuss in detail how the null space and range space of a matrix $\mathbf{T}$ can be determined by appropriate matrix manipulations such as Gauss–Jordan elimination and Gaussian elimination. In finite-dimensional linear algebra, the null space is often referred to as the kernel, while the range is defined as the column space of the matrix. Since our emphasis is on sampling of continuous-time signals, the linear operators we consider are typically defined over infinite-dimensional spaces and cannot necessarily be represented using a finite matrix. Therefore, we focus on the geometric and algebraic meanings of $\mathcal{N}(T)$ and $\mathcal{R}(T)$ rather than concrete algorithms for finding them via matrix reduction.

2.4.2   Invertibility

A fundamental property of a linear transformation T is whether or not it is invertible; if it is not, then we cannot determine an arbitrary input vector x from its output Ty unless we have further prior information about x.

Definition 2.10.   *A transformation $T \colon \mathcal{H} \to \mathcal{S}$ is injective if for any $x \neq y$ we have that $Tx \neq Ty$. It is surjective if $\mathcal{R}(T) = \mathcal{S}$. Finally, T is invertible if and only if it is bijective, namely both injective and surjective. In this case the inverse is denoted by T^{-1}.*

Note that if $T \colon \mathcal{H} \to \mathcal{S}$ is a bounded invertible transformation, then $T^{-1} \colon \mathcal{S} \to \mathcal{H}$ is also bounded.

The first property of injectivity is related to the null space of T, while the property of surjectivity is associated with the range of T. The connection between T being injective and its null space is incorporated in the following proposition:

Proposition 2.3.   *A transformation $T \colon \mathcal{H} \to \mathcal{S}$ is injective if and only if $\mathcal{N}(T) = \{0\}$.*

Proof: To prove the proposition, suppose first that the null space of T contains only the zero vector, and let x and y be two vectors such that $Tx = Ty$. Clearly this implies that $Tv = 0$ where $v = x - y$, so that v is in $\mathcal{N}(T)$. But since $\mathcal{N}(T) = \{0\}$ we immediately conclude that $v = 0$ or $x = y$, proving that T is injective.

Next, assume that T is injective, and let v be an arbitrary vector in $\mathcal{N}(T)$ with $Tv = 0$. Writing v as $v = av + (1 - a)v$ for any $0 < a < 1$, we have that

$$T(av) = T((a - 1)v). \tag{2.24}$$

Denoting $x = av$ and $y = (a - 1)v$, (2.24) can be written as $Tx = Ty$. Since T is injective, we conclude that $x = y$, or in terms of v,

$$av = (a - 1)v, \tag{2.25}$$

which implies that $v = 0$. We have therefore shown that any v satisfying $Tv = 0$ must be the zero vector. $\qquad\square$

Example 2.15 The transformation T of (2.22) is not injective because its null space is nontrivial. Indeed, as we have seen, $\mathcal{N}(T)$ contains all signals that vanish on $t \in [0, 1]$, regardless of their value on $t \in (1, 2]$. Since T zeros out the signal content in $(1, 2]$ this part cannot be recovered, and consequently, T is not invertible.

Let us, then, restrict attention to signals defined over $t \in [0, 1]$. Specifically, let $\widetilde{T} : L_2([0, 1]) \to L_2([0, 1])$ be the transformation defined by

$$y(t) = \widetilde{T}\{x\}(t) = tx(t). \tag{2.26}$$

This transformation is clearly injective because for every $x_1(t) \neq x_2(t)$ almost everywhere, we have that $y_1(t) = tx_1(t) \neq tx_2(t) = y_2(t)$ almost everywhere. However, $\widetilde{T}$ is not surjective because not every signal $y \in L_2([0, 1])$ can be obtained by application of $\widetilde{T}$ on a signal $x(t)$ in $L_2([0, 1])$. For example, any signal $y \in L_2([0, 1])$ that does not tend to zero when $t \to 0$ is not in the range of $\widetilde{T}$. Thus, $\mathcal{R}(\widetilde{T}) \neq L_2([0, 1])$, which implies that $\widetilde{T}$ is not invertible.

To obtain a transformation that is both injective and surjective, we can modify $\widetilde{T}$ to operate on signals defined over $t \in [\varepsilon, 1]$ for some constant $0 < \varepsilon < 1$. This results in a transformation $\widetilde{T} : L_2([\varepsilon, 1]) \to L_2([\varepsilon, 1])$ that is invertible.

2.4.3 Direct-sum decompositions

The four spaces we associated with a transformation $T \colon \mathcal{H} \to \mathcal{S}$ can be used to decompose $\mathcal{H}$ and $\mathcal{S}$ into direct sums of smaller subspaces. We can then study the action of T on each of these subspaces separately, which reveals some of its underlying properties. To arrive at such a decomposition, we rely on the following pair of propositions [23].

Proposition 2.4. *If $\mathcal{V}$ is a closed linear subspace of a Hilbert space $\mathcal{H}$, then $\mathcal{H} = \mathcal{V} \oplus \mathcal{V}^{\perp}$.*

Proposition 2.5. *The null space $\mathcal{N}(T)$ of a linear transformation $T \colon \mathcal{H} \to \mathcal{S}$ is a closed subspace of $\mathcal{H}$. The orthogonal complement $\mathcal{V}^{\perp}$ of an arbitrary subspace $\mathcal{V} \subseteq \mathcal{H}$ is also a closed subspace of $\mathcal{H}$.*

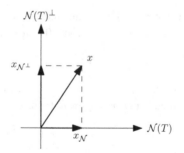

Figure 2.4 Decomposition of a vector into its orthogonal components $x_\mathcal{N}$ and $x_{\mathcal{N}^\perp}$.

Combining Propositions 2.4 and 2.5, we can decompose $\mathcal{H}$ as

$$\mathcal{H} = \mathcal{N}(T) \oplus \mathcal{N}(T)^\perp. \tag{2.27}$$

Using this decomposition, any $x \in \mathcal{H}$ can be written uniquely as a sum of two components $x = x_\mathcal{N} + x_{\mathcal{N}^\perp}$ where $x_\mathcal{N}$ is in $\mathcal{N}(T)$, and $x_{\mathcal{N}^\perp}$ is in $\mathcal{N}(T)^\perp$. From the definition of $\mathcal{N}(T)^\perp$ it follows immediately that these vectors are orthogonal: $\langle x_\mathcal{N}, x_{\mathcal{N}^\perp} \rangle = 0$. Figure 2.4 shows a geometric interpretation of the components $x_\mathcal{N}$ and $x_{\mathcal{N}^\perp}$ of x.

Example 2.16 As we have seen, the null space of the transformation T of (2.22) consists of all signals that vanish on $t \in [0, 1]$. Consequently, under the standard inner product $\langle a, b \rangle = \int_0^2 a(t)b(t)dt$, the space $\mathcal{N}(T)^\perp$ comprises all signals that vanish on $t \in (1, 2]$ (see Example 2.7 for a similar derivation). Now, it is easy to see that every signal $x(t)$ in $L_2([0, 2])$ can be written as $x(t) = x_\mathcal{N}(t) + x_{\mathcal{N}^\perp}(t)$, where

$$x_\mathcal{N}(t) = \begin{cases} 0, & 0 \le t \le 1 \\ x(t), & 1 < t \le 2, \end{cases} \qquad x_{\mathcal{N}^\perp}(t) = \begin{cases} x(t), & 0 \le t \le 1 \\ 0, & 1 < t \le 2. \end{cases} \tag{2.28}$$

Next, we would like to obtain a similar decomposition of $\mathcal{S}$ in terms of $\mathcal{R}(T)$ and $\mathcal{R}(T)^\perp$. However, since $\mathcal{R}(T)$ is not necessarily closed we cannot apply Proposition 2.4 directly. Instead, we substitute $V = \mathcal{R}(T)^c$ in Proposition 2.4, where $\mathcal{R}(T)^c$ denotes the closure of $\mathcal{R}(T)$. This leads to the decomposition

$$\mathcal{S} = \mathcal{R}(T)^c \oplus \mathcal{R}(T)^\perp. \tag{2.29}$$

Any $y \in \mathcal{S}$ can then be expressed uniquely as $y = y_\mathcal{R} + y_{\mathcal{R}^\perp}$ where $y_\mathcal{R} \in \mathcal{R}(T)^c$, $y_{\mathcal{R}^\perp} \in \mathcal{R}(T)^\perp$, and the two vectors are orthogonal: $\langle y_\mathcal{R}, y_{\mathcal{R}^\perp} \rangle = 0$.

Throughout the book, the transformations we consider will typically have closed range so that we will not have to be concerned with the closure. This is a result of the following proposition.

Proposition 2.6. *If T is bounded (i.e., continuous), then the range of T is closed if and only if $\|Tx\| \ge a\|x\|$ for all $x \in \mathcal{N}(T)^\perp$, and some $a > 0$.*

The transformations we will deal with in the context of sampling typically arise from signal expansions that are bounded, and therefore satisfy the conditions of the proposition. We will therefore assume throughout that the range of T is closed. However, the mapping given by (2.22) does not have a closed range, as we show in the next example.

Example 2.17 To show that the range of T given by (2.22) is not closed, let $x_n(t)$, $n = 1, 2, \ldots$, be the sequence of signals in $\mathcal{N}(T)^\perp$ defined by

$$x_n(t) = \begin{cases} \sqrt{n}, & 0 \leq t \leq \frac{1}{n} \\ 0, & \frac{1}{n} < t \leq 2. \end{cases} \tag{2.30}$$

These signals all have unit norm, since $\|x_n\|^2 = \int_0^2 x_n^2(t)dt = 1$. Now, let $y_n = Tx_n$ be the sequence in $\mathcal{R}(T)$ obtained by applying T to each of the signals $\{x_n\}_{n=1}^\infty$. Then

$$\|y_n\|^2 = \int_0^2 y_n^2(t)dt = \int_0^{\frac{1}{n}} \left(t\sqrt{n}\right)^2 dt = \frac{1}{3n^2}. \tag{2.31}$$

Therefore, $\|y_n\|$ becomes arbitrarily small as n increases, implying that there does not exist a positive scalar a such that $\|Tx\| \geq a\|x\|$ for all $x \in \mathcal{N}(T)^\perp$. Since T is bounded, we conclude from Proposition 2.6 that $\mathcal{R}(T)$ is not closed.

We now demonstrate the decomposition (2.29) on the transformation T of (2.22).

Example 2.18 The closure of the range of T defined by (2.22) is the set $\mathcal{A}$ of signals in $L_2([0, 2])$ that vanish on $t \in (1, 2]$. To see why this is true, notice first that $\mathcal{R}(T)$ is contained in $\mathcal{A}$. This is because, by definition of T, every $y \in \mathcal{R}(T)$ vanishes on $t \in (1, 2]$ and satisfies

$$\int_0^2 y^2(t)dt = \int_0^1 (tx(t))^2 dt \leq \int_0^1 x^2(t)dt \leq \int_0^2 x^2(t)dt < \infty, \tag{2.32}$$

implying that y is also in $L_2([0, 2])$.

It remains to establish that every $z \in L_2([0, 2])$ which vanishes on $t \in (1, 2]$ can be approximated arbitrarily well by functions in $\mathcal{R}(T)$. Given a function $z(t)$ in $L_2([0, 2])$ that is equal to 0 on $t \in (1, 2]$, we construct the sequence

$$x_n(t) = \begin{cases} 0, & 0 \leq t \leq \frac{1}{n} \\ \frac{z(t)}{t}, & \frac{1}{n} < t \leq 1 \end{cases} \tag{2.33}$$

for $n = 1, 2, \ldots$ These functions are all in $L_2([0, 2])$ since

$$\int_0^2 x_n^2(t)dt = \int_{\frac{1}{n}}^1 \left(\frac{z(t)}{t}\right)^2 dt \leq n^2 \int_{\frac{1}{n}}^1 z^2(t)dt \leq n^2 \int_0^2 z^2(t)dt < \infty, \tag{2.34}$$

for any finite value of n. Now, let $y_n = Tx_n$ be the sequence in $\mathcal{R}(T)$ obtained by applying T on $\{x_n\}_{n=1}^{\infty}$. Then,

$$y_n(t) = \begin{cases} 0, & 0 \le t \le \frac{1}{n} \\ z(t), & \frac{1}{n} < t \le 1 \\ 0, & 1 < t \le 2, \end{cases} \tag{2.35}$$

and therefore

$$\int_0^2 (y_n(t) - z(t))^2 dt = \int_0^{\frac{1}{n}} z^2(t) dt. \tag{2.36}$$

Consequently $\|y_n - z\|^2 \to 0$ as $n \to \infty$, demonstrating that the sequence $y_n \in \mathcal{R}(T)$ approximates $z \in \mathcal{A}$ arbitrarily well.

Having identified the subspace $\mathcal{R}(T)^c$, it is easy to see that every signal $y(t)$ in $L_2([0, 2])$ can be written as $y(t) = y_{\mathcal{R}}(t) + y_{\mathcal{R}\perp}(t)$, where the signals $y_{\mathcal{R}} \in R(T)^c$ and $y_{\mathcal{R}\perp} \in \mathcal{R}(T)^{\perp}$ are defined by

$$y_{\mathcal{R}}(t) = \begin{cases} y(t), & 0 \le t \le 1 \\ 0, & 1 < t \le 2, \end{cases} \qquad y_{\mathcal{R}\perp}(t) = \begin{cases} 0, & 0 \le t \le 1 \\ y(t), & 1 < t \le 2. \end{cases} \tag{2.37}$$

Decomposing $\mathcal{H}$ and $\mathcal{S}$ as in (2.27) and (2.29) respectively, we may describe the action of T on each of these subspaces. By definition, T maps $\mathcal{N}(T)$ to 0 and $\mathcal{N}(T)^{\perp}$ into $\mathcal{R}(T)$, since for any x in $\mathcal{N}(T)^{\perp}$, we have that $Tx \in \mathcal{R}(T)$. The next proposition shows that to every $y \in \mathcal{R}(T)$ there corresponds a unique $x \in \mathcal{N}(T)^{\perp}$ such that $y = Tx$ so that there is a one-to-one correspondence between $\mathcal{N}(T)^{\perp}$ and $\mathcal{R}(T)$.

Proposition 2.7. *The linear transformation* $T \colon \mathcal{H} \to \mathcal{S}$ *with closed range is invertible when restricted to* $\mathcal{N}(T)^{\perp}$ *and* $\mathcal{R}(T)$. *In other words, for each* $y \in \mathcal{R}(T)$ *there exists a unique* $x \in \mathcal{N}(T)^{\perp}$ *such that* $y = Tx$.

Proof: We first show that if $Tx_1 = Tx_2 = y$ with both x_1 and x_2 in $\mathcal{N}(T)^{\perp}$, then $x_1 = x_2$. Clearly we have that $T(x_1 - x_2) = 0$, so that $v = x_1 - x_2$ lies in $\mathcal{N}(T)$. But since both x_1 and x_2 are in $\mathcal{N}(T)^{\perp}$, so is their difference v. We therefore have shown that v lies in both $\mathcal{N}(T)$ and $\mathcal{N}(T)^{\perp}$. However, the only vector in the intersection of any subspace $\mathcal{W}$ and its orthogonal complement $\mathcal{W}^{\perp}$ is the zero vector, from which we conclude that $v = 0$, and $x_1 = x_2$.

To complete the proof we have to show that for each $y \in \mathcal{R}(T)$ there is some $x \in \mathcal{N}(T)^{\perp}$ such that $y = Tx$. By definition of $\mathcal{R}(T)$, we know that there exists a vector $z \in \mathcal{H}$, such that $y = Tz$. Using the decomposition (2.27) we can write z as $z = x + w$ where x is in $\mathcal{N}(T)^{\perp}$ and w lies in $\mathcal{N}(T)$. But since w is in $\mathcal{N}(T)$, we have that $Tw = 0$. Therefore, $y = Tz = Tx$, with $x \in \mathcal{N}(T)^{\perp}$. $\square$

The action of T is illustrated in Fig. 2.5. Since we have seen that there is an invertible mapping between $\mathcal{N}(T)^{\perp}$ and $\mathcal{R}(T)$ it is natural to define an inverse of T, when restricted to these spaces. This is precisely the intuition behind the pseudoinverse, which we introduce in Section 2.7.

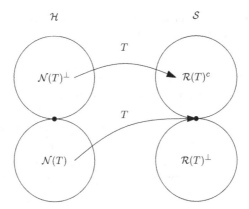

Figure 2.5 The action of T on the subspaces $\mathcal{N}(T)$ and $\mathcal{N}(T)^{\perp}$.

The direct sum decompositions of (2.27) and (2.29) are not unique. Specifically, there are many choices of subspaces $\mathcal{V} \subseteq \mathcal{H}$ with $\mathcal{V} \neq \mathcal{N}(T)^{\perp}$ such that $\mathcal{H} = \mathcal{N}(T) \oplus \mathcal{V}$. Then, any $x \in \mathcal{H}$ can be decomposed uniquely into its components in $\mathcal{N}(T)$ and $\mathcal{V}$; however, these components are not necessarily orthogonal. Similarly, there are many choices of subspaces $\mathcal{W} \subseteq \mathcal{S}$ with $\mathcal{W} \neq \mathcal{R}(T)^{\perp}$ such that $\mathcal{S} = \mathcal{R}(T)^{c} \oplus \mathcal{W}$, so that any $y \in \mathcal{S}$ can be decomposed uniquely into its components in $\mathcal{R}(T)^{c}$ and $\mathcal{W}$, where these components are generally not orthogonal. When we discuss projection operators in Section 2.6 we will see that the decompositions of (2.27) and (2.29) correspond to orthogonal projection operators, while more general decompositions such as $\mathcal{H} = \mathcal{N}(T) \oplus \mathcal{V}$, where $\mathcal{V}$ is not orthogonal to $\mathcal{N}(T)$, are associated with oblique projections.

Figure 2.5 illustrates how T maps the input space $\mathcal{H}$ to the output space $\mathcal{S}$. We would also like to obtain a dual map, i.e. a transformation that maps $\mathcal{R}(T)$ onto $\mathcal{N}(T)^{\perp}$ and $\mathcal{R}(T)^{\perp}$ to 0. One such choice is the pseudoinverse, which we discuss after introducing projection operators. An additional transformation with this property is the *adjoint*, defined next.

2.4.4 The adjoint

The adjoint plays a central role in linear algebra at large and in sampling theory in particular. As we will see, the adjoint of an appropriately defined transformation serves as the sampling operator whose output is the desired sequence of samples and is instrumental in determining basis expansions.

Definition 2.11. *The* adjoint *of a continuous linear transformation* $T \colon \mathcal{H} \to \mathcal{S}$ *is the unique continuous linear transformation* $T^{*} \colon \mathcal{S} \to \mathcal{H}$ *such that* $\langle Tx, y \rangle_{\mathcal{S}} = \langle x, T^{*}y \rangle_{\mathcal{H}}$ *for all* $x \in \mathcal{H}$, $y \in \mathcal{S}$.

Note that in the definition we indicated in which space the inner products are defined: for example, $\langle z, y \rangle_{\mathcal{S}}$ denotes the inner product over $\mathcal{S}$. This distinction is important since the inner products on both spaces can be different. In particular, the adjoint will depend on the choice of inner product on each space.

Example 2.19 A simple setting in which to consider the adjoint is when $\mathcal{H} = \mathbb{C}^m$ and $\mathcal{S} = \mathbb{C}^n$ with the standard inner product on both spaces (namely $\langle \mathbf{x}, \mathbf{y} \rangle = \mathbf{x}^H \mathbf{y}$ for two vectors of the same dimension). In this case, $T = \mathbf{T}$ is an $n \times m$ matrix. It is easy to see that the adjoint is simply the Hermitian conjugate $\mathbf{T}^* = \mathbf{T}^H$. The matrix $\mathbf{M} = \mathbf{T}^H$ is the $m \times n$ matrix which results from transposing $\mathbf{T}$, and then taking the complex conjugate of each entry. Formally, $\mathbf{M}_{ij} = \overline{\mathbf{T}}_{ji}$.

The next example treats a more general scenario in which both $\mathcal{H}$ and $\mathcal{S}$ can be infinite-dimensional. The transformation defined in this example will play an important role in the development of sampling theory, and is referred to as the *set transformation*. We study such transformations in greater detail in Section 2.5.

Example 2.20 Let $T\colon \ell_2 \to \mathcal{H}$ be the transformation defined by $Ta = \sum_n a[n] t_n$ where a is a sequence in ℓ_2 and $\{t_n\}$ is a set of vectors in an arbitrary Hilbert space $\mathcal{H}$. Thus, applying T is equivalent to taking linear combinations of the given set of vectors $\{t_n\}$. We assume that the inner product on ℓ_2 is the standard one, while the inner product on $\mathcal{H}$ is arbitrary. Clearly, our description of T^* will depend on the choice of inner product on $\mathcal{H}$.

To compute the adjoint we need to calculate the inner products $\langle Ta, y \rangle$ on $\mathcal{H}$, and $\langle a, T^* y \rangle$ on ℓ_2. From the properties of the inner product,

$$\langle Ta, y \rangle = \langle \sum_n a[n] t_n, y \rangle = \sum_n \overline{a[n]} \langle t_n, y \rangle, \tag{2.38}$$

for any $y \in \mathcal{H}$ and $a \in \ell_2$. This is true for any choice of inner product on $\mathcal{H}$. We now turn to evaluate $\langle a, T^* y \rangle$. For convenience, let us denote by b the sequence in ℓ_2 given by $b = T^* y$, and let its elements be written as $b[n]$. Then, from the definition of the inner product on ℓ_2:

$$\langle a, T^* y \rangle = \langle a, b \rangle = \sum_n \overline{a[n]} b[n]. \tag{2.39}$$

It is easy to see that we can equate (2.38) and (2.39) by choosing $b[n] = \langle t_n, y \rangle$. We therefore conclude that the adjoint of T is the transform defined such that if $b = T^* y$, then the nth element of b is given by $b[n] = \langle t_n, y \rangle$.

Classes of operators

Now that we have defined the adjoint, it can be used to characterize several important classes of transformations. We begin by defining a *linear operator* as a continuous linear transformation of a Hilbert space onto itself $T\colon \mathcal{H} \to \mathcal{H}$. We then introduce the following classes of operators:

Definition 2.12. *Let* $T\colon \mathcal{H} \to \mathcal{H}$ *be a linear operator on* $\mathcal{H}$. *Then*

1. T *is* unitary *if* $T^*T = TT^* = I_{\mathcal{H}}$;
2. T *is* Hermitian (self-adjoint) *if* $T^* = T$;
3. T *is* positive semidefinite, *denoted* $T \succeq 0$, *if it is Hermitian, and* $\langle Tx, x \rangle \geq 0$ *for any* $x \in \mathcal{H}$.

Note that if T is Hermitian, then the inner product $\langle Tx, x \rangle$ is real for all x (see Exercise 12).

The definitions above are easily adapted to matrices $\mathbf{T}$ over $\mathbb{C}^{m \times m}$ by replacing the adjoint with the Hermitian conjugate $\mathbf{T}^H$.

Example 2.21 In this example we show that the transformation T of (2.22) is Hermitian.

From the definition of Tx, for every $x, y \in L_2([0, 2])$,

$$\langle Tx, y \rangle = \int_0^1 (tx(t))y(t)dt = \int_0^1 x(t)(ty(t))dt. \qquad (2.40)$$

Denoting $z = T^*y$,

$$\langle x, T^*y \rangle = \langle x, z \rangle = \int_0^2 x(t)z(t)dt. \qquad (2.41)$$

Choosing

$$z(t) = T^*\{y\}(t) = \begin{cases} ty(t), & t \in [0, 1] \\ 0, & t \in (1, 2], \end{cases} \qquad (2.42)$$

results in $\langle Tx, y \rangle = \langle x, T^*y \rangle$, proving that $T = T^*$.

Properties of the adjoint

The subspaces associated with T^* are closely related to those defined for T, as given in the following proposition [23]:

Proposition 2.8. *Let* $T\colon \mathcal{H} \to \mathcal{S}$ *be a continuous linear transformation. Then*

1. $\mathcal{N}(T) = \mathcal{R}(T^*)^{\perp}$;
2. $\mathcal{N}(T)^{\perp} = \mathcal{R}(T^*)^c$;
3. $\mathcal{N}(T^*) = \mathcal{R}(T)^{\perp}$;
4. $\mathcal{N}(T^*)^{\perp} = \mathcal{R}(T)^c$.

If T *is Hermitian, then* $\mathcal{H} = \mathcal{S}$ *and* $\mathcal{N}(T) = \mathcal{R}(T)^{\perp}$.

In analogy to Proposition 2.7, we can show that to every vector y in $\mathcal{R}(T^*) = \mathcal{N}(T)^{\perp}$ there corresponds a unique vector x in $\mathcal{N}(T^*)^{\perp} = \mathcal{R}(T)$ such that $y = T^*x$. The

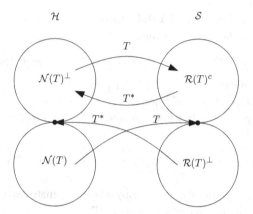

Figure 2.6 The action of T and T^* on the subspaces $\mathcal{N}(T)$, $\mathcal{N}(T)^\perp$, $\mathcal{R}(T)^c$, and $\mathcal{R}(T)^\perp$.

proof follows that of Proposition 2.7, while using the properties of $\mathcal{R}(T^*)$ and $\mathcal{N}(T^*)$ incorporated in Proposition 2.8; see Exercise 13.

The action of T and T^* is illustrated in Fig. 2.6. Note that T^* does not invert T on $\mathcal{R}(T)$. Although it maps any vector in $\mathcal{R}(T)$ to a vector in $\mathcal{N}(T)^\perp$, in general we do not have that $T^*Tx = x$.

The properties of T and T^* illustrated in Fig. 2.6 are essential for the development of signal expansions. As we will see, many of the features of basis expansions can be understood and proven by using the relations incorporated in the figure. In our future derivations we will often refer back to this figure, and therefore strongly recommend understanding its implications.

From the properties of T and T^* it is easy to establish the following useful relations (see Exercise 14):

1. $\mathcal{N}(T^*T) = \mathcal{N}(T)$;
2. $\mathcal{R}(T^*T) = \mathcal{R}(T^*) = \mathcal{N}(T)^\perp$.

Two additional properties of the adjoint that we will use throughout are:

1. $(AB)^* = B^*A^*$,
2. $(A^{-1})^* = (A^*)^{-1}$.

In the first equation we defined $A\colon \mathcal{H} \to \mathcal{S}$ and $B\colon \mathcal{W} \to \mathcal{H}$, while in the second A is assumed to be invertible.

2.5   Basis expansions

Sampling theory deals primarily with linear combinations of vectors of the form $x = \sum_n a[n]x_n$. Often, the vector set $\{x_n\}$ is chosen as a *basis* for the underlying space. A basis is a set of vectors satisfying *linear independence* and *completeness*. The first property ensures that the expansion is unique. The second guarantees that any vector in the space can be expanded in terms of the given vector set. Therefore, to properly define and characterize linear expansions we first consider the notions of linear independence and completeness. Next, we introduce the *set transformation*, which is a useful mathematical tool for describing linear combinations of vectors. Equipped with the set transformation, we define general bases representations, and then turn to study stable expansions which are at the heart of modern sampling methods.

Definition 2.13. *A set of vectors $x_1, \ldots, x_m$ is* linearly independent *if $\sum_n a[n]x_n = 0$ is true only if $a[n] = 0$ for all n. Otherwise, the vectors are* linearly dependent. *If there are infinitely many vectors $x_1, x_2, \ldots$, then they are ω-linearly independent if $\sum_n a[n]x_n = 0$ implies $a[n] = 0$ for all n.*

The sum in the definition should be interpreted in the limit sense, namely, $\|\sum_{n=1}^{N} a[n]x_n\| \to 0$ as $N \to \infty$. Throughout, when we refer to linearly independent vectors we will mean ω-linearly independence.[4]

A useful property of orthogonal vectors is that they are always linearly independent. Indeed, suppose that the vectors $\{x_n\}$ are orthogonal with $\|x_n\| > 0$, and $\sum_n a[n]x_n = 0$. Then, for any j,

$$0 = \langle x_j, \sum_n a[n]x_n \rangle = \sum_n a[n]\langle x_j, x_n \rangle = a[j]\|x_j\|^2, \tag{2.43}$$

which implies that $a[j] = 0$ for all j.

A set of vectors $\{x_n\}$ in $\mathcal{H}$ is *complete* in $\mathcal{H}$ if the closure of the span of $\{x_n\}$ equals $\mathcal{H}$. That is, for each vector x and $\varepsilon > 0$ there is a finite linear combination such that $\|x - \sum_{n=1}^{m} a[n]x_n\| < \varepsilon$ so that x can be approximated arbitrarily closely by a linear combination of vectors x_n. The following characterization is useful for determining completeness of a set of vectors.

Proposition 2.9. *A set of vectors $\{x_n\}$ in $\mathcal{H}$ is complete in $\mathcal{H}$ if and only if $\langle x, x_n \rangle = 0$ for all n implies that $x = 0$.*

2.5.1   Set transformations

One of the nice features of finite-dimensional linear algebra is that, given a finite set of m vectors $\mathbf{x}_n, 1 \leq n \leq m$ in $\mathbb{C}^N$, we can create an $N \times m$ matrix $\mathbf{X} = [\mathbf{x}_1 \ \mathbf{x}_2 \cdots \mathbf{x}_m]$ whose columns are the given vectors. This allows simple manipulation of the vector set, and leads to easy interpretation of its basic properties. For example, it is well known that the vectors $\{\mathbf{x}_n\}$ are linearly independent if and only if $\mathbf{X}$ has full column-rank. If the vectors are linearly dependent, then studying the null space of $\mathbf{X}$ reveals combinations $\sum_n a[n]x_n$ that are equal to 0. The set transformation, defined below (and introduced earlier in Example 2.20), is a generalization of this concept to sets in which each vector

---

[4] One can also define the notion of finite linear independence which means that every finite subset of $\{x_n\}$ is linearly independent. If a set of vectors is ω-linearly independent then it is also finitely linearly independent. The reverse implication is not necessarily true.

can be an arbitrary element in a Hilbert space (for example a signal $x(t)$ in L_2). Furthermore, there may be an infinite (countable) number of such vectors.

Definition 2.14. *Let $\{x_n: n \in \mathcal{I}\}$ be a countable set of vectors in a Hilbert space $\mathcal{H}$. The* set transformation $X: \ell_2 \to \mathcal{H}$ *corresponding to these vectors is defined by* $Xa = \sum_{n \in \mathcal{I}} a[n]x_n$ *for any $a \in \ell_2$.*

The definition above assumes that the sum Xa converges. When the vector sets are chosen as Riesz bases or frames, which are defined in Sections 2.5.3 and 2.8 respectively, convergence is automatically guaranteed. As we will be dealing almost exclusively with such sets, we do not concern ourselves here with the issue of convergence.

In Example 2.20 we introduced the set transformation and computed its adjoint. Specifically, we showed that if $a = X^*y$, then

$$a[n] = \langle x_n, y \rangle_{\mathcal{H}}. \tag{2.44}$$

This property of the adjoint will be used throughout the book when considering sampling operations.

Many properties of vector sets can be described in terms of their corresponding set transformations, as incorporated in the following proposition.

Proposition 2.10. *Let $X: \ell_2 \to \mathcal{H}$ be a bounded set transformation corresponding to a set of vectors $\{x_n\}$ in $\mathcal{H}$. Then*

1. *The vectors x_n are linearly independent if and only if $\mathcal{N}(X) = \{0\}$;*
2. *$X^*X = I_{\ell_2}$ if and only if the vectors x_n are orthonormal;*
3. *If $XX^* = I_{\mathcal{H}}$ then the vectors $\{x_n\}$ are complete in $\mathcal{H}$.*

Proof: To prove the first part, note that the vectors $\{x_n\}$ are linearly independent if and only if $\sum_n a[n]x_n = 0$ implies that $a[n]$ is equal to 0 for all n. Now, any sequence $a \in \ell_2$ in the null space of X satisfies

$$Xa = \sum_n a[n]x_n = 0. \tag{2.45}$$

Therefore, linear independence is equivalent to the requirement that $Xa = 0$ implies $a = 0$ or that the null space of X contains only the zero vector.

To prove the second part, suppose that the vectors $\{x_n\}$ are orthonormal, and let $b = X^*Xa$ with $a \in \ell_2$. Then, from (2.44),

$$b[k] = \langle x_k, Xa \rangle = \langle x_k, \sum_n a[n]x_n \rangle = \sum_n a[n]\langle x_k, x_n \rangle = \sum_n a[n]\delta_{kn} = a[k] \tag{2.46}$$

for all k, where δ_{kn} is defined in (2.18). Therefore, $X^*Xa = a$, for any $a \in \ell_2$, and $X^*X = I_{\ell_2}$. Conversely, if $X^*X = I_{\ell_2}$ then for any $a \in \ell_2$ we have that $X^*Xa = a$. Let $e^k \in \ell_2$ denote the sequence with nth element $e^k[n] = \delta_{kn}$. Then, $Xe^k = x_k$. Combining this with $e^k = X^*Xe^k$ we have that $e^k = X^*x_k$ or $e^k[n] = \langle x_n, x_k \rangle$. Since $e^k[n] = \delta_{kn}$, the vectors $\{x_n\}$ are orthonormal.

Finally, suppose that $XX^* = I_{\mathcal{H}}$, and let y be a vector in $\mathcal{H}$ orthogonal to all the vectors $\{x_n\}$: $\langle x_n, y \rangle = 0$ for all n. Since $y = XX^*y$, we have that

$$y = \sum_n \langle x_n, y \rangle x_n = 0, \tag{2.47}$$

where we used that fact that if $b = X^*y$, then $b[n] = \langle x_n, y \rangle$. From (2.47) we conclude that the only vector orthogonal to all the vectors x_n is the zero vector, which by Proposition 2.9 proves completeness. $\qquad\square$

In the next section we use set transformations to determine the coefficients in a basis expansion of a signal.

2.5.2 Bases

One of the useful features of a signal space $\mathcal{H}$ is that every signal $x \in \mathcal{H}$ can be represented by a unique sequence of scalars using a set of vectors that form a *basis* for $\mathcal{H}$. Equipped with the notions of linear independence, completeness and the set transformation, we are now ready to define a basis:

Definition 2.15. *A set of vectors $\{x_n \in \mathcal{H}, n \in \mathcal{I}\}$ is a* Schauder basis[5] *for $\mathcal{H}$ if to each vector $x \in \mathcal{H}$ there corresponds a unique sequence of scalars $a[n] \in \mathbb{C}$ such that $x = \sum_{n \in \mathcal{I}} a[n]x_n$. The dimension of $\mathcal{H}$ is equal to the cardinality of $\mathcal{I}$.*

In the infinite-dimensional case, the equality in the sum $x = \sum_{n=0}^{\infty} a[n]x_n$ is to be interpreted in the limit sense; namely, $\|x - \sum_{n=0}^{N} a[n]x_n\| \to 0$ as $N \to \infty$. This is why in order to define such a basis we must operate in a normed space.

It is easy to see that there will be a unique set of scalars $a[n]$ such that $x = \sum_n a[n]x_n$ for any $x \in \mathcal{H}$ if and only if the vectors $\{x_n\}$ are linearly independent and complete. Therefore a set of vectors forms a basis if it is linearly independent and complete. Since any orthonormal vector set is linearly independent, it follows that orthonormal sets form a basis for their span. Finally, note that although a Hilbert space can have many bases, they all have the same cardinality so that the dimension is well defined.

Example 2.22 Consider the space $\mathbb{P}^d([a, b])$ of polynomials $p(t)$ on the interval $[a, b]$ whose degree is less than or equal to d. A natural choice of a basis for this space is the set of functions $\{1, t, t^2, \ldots, t^d\}$. To see that this set constitutes a basis, first note that by definition every polynomial $p \in \mathbb{P}^d([a, b])$ can be expressed as a linear combination of these functions: $p(t) = a[0] + a[1]t + a[2]t^2 + \cdots + a[d]t^d$. Therefore, this set is complete. In addition, there is no nontrivial linear combination of these monomials which yields $p(t) = 0$ for every $t \in [a, b]$, implying that they are also linearly independent. The dimension of $\mathbb{P}^d([a, b])$ is equal to $d + 1$, i.e. the number of basis functions.

---

[5] Throughout the book, we will use the term basis to denote a Schauder basis.

Example 2.23 As another example, consider the set $L_2([-\pi, \pi])$ of complex functions $x(t)$ on the interval $[-\pi, \pi]$ satisfying $\int_{-\pi}^{\pi} |x(t)|^2 dt < \infty$. We define the inner product on $L_2([-\pi, \pi])$ by $\langle x, y \rangle = \int_{-\pi}^{\pi} \overline{x(t)} y(t) dt$, so that the induced norm is given by $\|x\|^2 = \int_{-\pi}^{\pi} |x(t)|^2 dt$. With this norm, a popular choice of basis for $L_2([-\pi, \pi])$ is the complex Fourier basis $\{e^{jnt}\}_{n \in \mathbb{Z}}$.

To prove that this set is a basis, we invoke the Fourier theorem [32]. This theorem states that for every function $f \in L_2([-\pi, \pi])$ and scalar $\varepsilon > 0$ there exists an integer N and coefficients $\{a[n]\}_{n=-N}^{N}$ such that $\int_{-\pi}^{\pi} |f(t) - \sum_{n=-N}^{N} a[n] e^{jnt}| dt < \varepsilon$. Furthermore, for any N, the coefficients minimizing the error are unique. In other words, any finite-energy signal $f(t)$ on the interval $[-\pi, \pi]$ can be approximated arbitrarily well by a unique linear combination of the form $\sum_{n \in \mathbb{Z}} a[n] e^{jnt}$, meaning that $\{e^{jnt}\}_{n \in \mathbb{Z}}$ is a basis for $L_2([-\pi, \pi])$.

We are particularly interested in Hilbert spaces that are *separable*. A Hilbert space is separable if and only if it contains a countable orthonormal basis. Throughout the book, the Hilbert spaces we will be dealing with are separable; therefore, we implicitly assume that this property is satisfied.

If the vectors $\{x_n, n \in \mathcal{I}\}$ form a basis for $\mathcal{H}$, then any $x \in \mathcal{H}$ has a unique decomposition of the form $x = \sum_{n \in \mathcal{I}} a[n] x_n$, where $a[n] \in \mathbb{C}$. However, the coefficients $a[n]$ are in general not guaranteed to be in ℓ_2, leading to unstable expansions, with coefficients whose norm may grow without bound. In the context of sampling, stability is an important issue. A sampling method that is not stable can lead to large noise enhancement. Since noise is always present in real-world engineering systems, ensuring stability is of paramount importance. Mathematically, stability is guaranteed by the use of Riesz bases.

2.5.3   Riesz bases

To develop the notion of a Riesz basis, suppose we have a set of basis vectors $\{x_n\}$ and consider the vector $x = \sum_n a[n] x_n$. For a basis expansion to be numerically stable, we expect that when the norm of a is small, the norm of x will be small, and vice versa. If the vectors x_n are orthonormal, then

$$\|x\|^2 = \langle \sum_i a[i] x_i, \sum_j a[j] x_j \rangle = \sum_{ij} \overline{a[i]} a[j] \langle x_i, x_j \rangle = \sum_i |a[i]|^2 = \|a\|^2, \quad (2.48)$$

which of course guarantees this property. However, requiring an expansion to be orthonormal imposes stringent conditions on the basis vectors, leading to limited flexibility in their design.

The notion of a Riesz basis extends the energy-preserving property (2.48) to a larger class of bases. Instead of constraining the norms $\|x\|^2$ and $\|a\|^2$ to be equal, we require that

$$\alpha \|a\|^2 \leq \|x\|^2 \leq \beta \|a\|^2, \quad (2.49)$$

for some $\alpha > 0$ and $\beta < \infty$. In this energy sense, a Riesz basis is the "best" thing after an orthonormal basis: although the energy in the original vector and the expansion coefficients are not equal, they are not too far apart. In fact, it can be shown that a set of vectors $\{x_n\}$ forms a Riesz basis for $\mathcal{H}$ if and only if there is a bounded invertible linear transformation T such that $x_n = Te_n$ where $\{e_n\}$ is an orthonormal basis for $\mathcal{H}$. In some books this property is used to define a Riesz basis; see, for example, [26, Definition 3.6.1], [25, p. 26].

Definition 2.16. *A sequence $\{x_n \in \mathcal{H}, n \in \mathcal{I}\}$ is a Riesz basis for $\mathcal{H}$ if it is complete and there exist constants $\alpha > 0$ and $\beta < \infty$ such that*

$$\alpha \sum_{n \in \mathcal{I}} |a[n]|^2 \leq \left\| \sum_{n \in \mathcal{I}} a[n] x_n \right\|^2 \leq \beta \sum_{n \in \mathcal{I}} |a[n]|^2, \tag{2.50}$$

for all $a \in \ell_2$.

Note that (2.50) is identical to (2.49) with $x = \sum_n a[n] x_n$. Clearly, any orthonormal basis for $\mathcal{H}$ is also a Riesz basis.

The lower bound in (2.50) implies that if $\sum_n a[n] x_n = 0$, then $a[n] = 0$ for all n which means that the vectors $\{x_n\}$ are linearly independent. Together with the assumption of completeness, this justifies the fact that we refer to the set as a basis. Choosing $a[n]$ in (2.50) to be equal to δ_{kn} results in $\alpha \leq \|x_k\|^2 \leq \beta$, for all k. Thus, the norm of each basis vector cannot be arbitrarily large or small.

It is easy to see that any basis for a finite-dimensional space is a Riesz basis. Indeed, suppose that $|\mathcal{I}| = m$, and let $A = \max_n \|x_n\|^2$. Since $|\langle x_n, x_\ell \rangle| \leq \|x_n\| \|x_\ell\| \leq A$, it follows that

$$\left\| \sum_{n=1}^m a[n] x_n \right\|^2 \leq \sum_{n,\ell=1}^m |\overline{a[n]} a[\ell]| |\langle x_n, x_\ell \rangle| \leq A \left| \sum_{n=1}^m |a[n]| \right|^2 \leq mA \sum_{n=1}^m |a[n]|^2. \tag{2.51}$$

The last inequality follows from applying Cauchy–Schwarz to $\sum_{n=1}^m 1 \cdot |a[n]|$. Therefore, the upper bound in (2.50) is always satisfied with $\beta = mA$. The lower bound holds when the vectors are linearly independent since in that case there is no nontrivial sequence $a[n]$ for which $\|\sum_{n=1}^m a[n] x_n\|^2 = 0$.

It is useful to reformulate (2.50) in terms of the set transformation X corresponding to $\{x_n\}$. Using operator notation will allow us to gain further insight into the properties of Riesz bases, and is also convenient when developing an expression for the coefficients in a Riesz basis expansion, which is the subject of the next section. Noting that $\sum_{n \in \mathcal{I}} a[n] x_n = Xa$ and $\sum_{n \in \mathcal{I}} |a[n]|^2 = \|a\|^2$, and using the relation $\langle Xa, Xa \rangle = \langle a, X^* X a \rangle$, which follows from the definition of the adjoint, we may rewrite (2.50) as

$$\alpha \langle a, a \rangle \leq \langle a, X^* X a \rangle \leq \beta \langle a, a \rangle. \tag{2.52}$$

This inequality implies that $X^* X$ is bounded above and below:

$$\alpha I_{\ell_2} \preceq X^* X \preceq \beta I_{\ell_2}, \tag{2.53}$$

where the notation $A \succeq B$ means that $A - B \succeq 0$, namely the difference $A - B$ is positive semidefinite. Indeed, (2.52) is precisely the meaning of the operator inequality (2.53). An important conclusion from (2.53) is that X^*X is invertible, and its inverse satisfies

$$\frac{1}{\beta}I_{\ell_2} \preceq (X^*X)^{-1} \preceq \frac{1}{\alpha}I_{\ell_2}. \tag{2.54}$$

For a formal proof of this result see [33, p. 58]. We can also use (2.53) together with Proposition 2.6 to conclude that the range of X is closed. Therefore, when treating transformations that result from a Riesz basis expansion, the range is automatically closed and we do not need to be concerned with the closure.

An additional implication of the Riesz basis definition is that the sequence of expansion coefficients with respect to such a basis is bounded:

Proposition 2.11. *Let $\{x_n\}$ be a Riesz basis for a Hilbert space $\mathcal{H}$. Then, for any $x \in \mathcal{H}$,*

$$\alpha\|x\|^2 \leq \sum_n |\langle x, x_n \rangle|^2 \leq \beta\|x\|^2. \tag{2.55}$$

Proof: To prove the proposition we begin by expressing (2.55) in operator form. Let $b[n] = \langle x_n, x \rangle$ so that $b = X^*x$. Then, the inner expression in (2.55) is equal to $\|b\|^2 = \langle X^*x, X^*x \rangle$. From the properties of the adjoint, $\langle X^*x, X^*x \rangle = \langle x, XX^*x \rangle$. Therefore, we can write (2.55) as

$$\alpha\langle x, x \rangle \leq \langle x, XX^*x \rangle \leq \beta\langle x, x \rangle, \tag{2.56}$$

for all $x \in \mathcal{H}$.

We now establish that (2.50), or equivalently (2.53), implies (2.56). Let x be an arbitrary vector in $\mathcal{H}$. Since the vectors $\{x_n\}$ are complete, we can express x as $x = \sum_n c[n]x_n = Xc$ for some sequence c. Define $a = (X^*X)^{1/2}c$ (see Appendix A for a definition of the square root); since X^*X is positive definite and bounded, a is well defined. Substituting a into (2.52), and noting that

$$\langle a, a \rangle = \langle (X^*X)^{1/2}c, (X^*X)^{1/2}c \rangle = \langle c, X^*Xc \rangle = \langle Xc, Xc \rangle, \tag{2.57}$$

and $\langle a, X^*Xa \rangle = \langle c, (X^*X)^2c \rangle$, we have

$$\alpha\langle Xc, Xc \rangle \leq \langle c, (X^*X)^2c \rangle \leq \beta\langle Xc, Xc \rangle. \tag{2.58}$$

Finally, recalling that $x = Xc$ the inner expression in (2.58) can be expressed in terms of x as

$$\langle c, (X^*X)^2c \rangle = \langle Xc, XX^*Xc \rangle = \langle x, XX^*x \rangle, \tag{2.59}$$

and $\langle Xc, Xc \rangle = \langle x, x \rangle$, therefore, (2.58) becomes

$$\alpha\langle x, x \rangle \leq \langle x, XX^*x \rangle \leq \beta\langle x, x \rangle. \tag{2.60}$$

Since x was arbitrary, this establishes that (2.56) holds for all x.   □

We now consider some examples of Riesz bases.

Example 2.24 To demonstrate the concept of a Riesz basis, we revisit Example 2.23 and ask whether the basis $\{x_n(t) = e^{jnt}\}_{n\in\mathbb{Z}}$ constitutes a Riesz basis for $L_2([-\pi, \pi])$. Since these vectors are complete, it is enough to check whether the inequalities in (2.50), or equivalently (2.52), are satisfied.

To this end we compute the inner products $\langle x_m, x_n \rangle$ over $L_2([-\pi, \pi])$:

$$\langle x_m, x_n \rangle = \int_{-\pi}^{\pi} e^{-jmt} e^{jnt} dt = 2\pi\delta[m - n]. \tag{2.61}$$

Consequently, $X^*X = 2\pi I$, and $\{e^{jnt}\}_{n\in\mathbb{Z}}$ is a Riesz basis with bounds $\alpha = \beta = 2\pi$, or equivalently, a scaled orthonormal basis.

Example 2.25 As another example, consider the set of functions $\{x_n(t) = e^{jnt}/(1+|n|)\}_{n\in\mathbb{Z}}$. It is easy to see that this set also constitutes a basis for $L_2([-\pi, \pi])$. Indeed, since $\{e^{jnt}\}$ is a basis for $L_2([-\pi, \pi])$, for any $x(t) \in L_2([-\pi, \pi])$ there exist coefficients $a[n]$ such that $x(t) = \sum_n a[n]e^{jnt}$. Defining $\tilde{a}[n] = a[n](1+|n|)$, we can write $x(t) = \sum_n \tilde{a}[n]x_n(t)$. However, these vectors do not form a Riesz basis. The easiest way to see this is by noting that the norm of $x_n(t)$ is not bounded below: $\|x_n(t)\| = \sqrt{2\pi}/(1 + |n|)$. As $|n| \to \infty$, $\|x_n(t)\| \to 0$, so that the lower bound in (2.50) is violated.

Example 2.26 Consider the set of signals $\{x_n(t)\}_{n\in\mathbb{Z}}$ in L_2 defined by

$$x_n(t) = \begin{cases} \alpha^n, & 0 \le t < \alpha^{-2n} \\ 0, & \alpha^{-2n} \le t \end{cases} \tag{2.62}$$

for some scalar $\alpha > 1$. In contrast with the two previous examples, this set of functions is not orthogonal with respect to the standard inner product on L_2. Indeed,

$$\langle x_m, x_n \rangle = \int_0^{\alpha^{-2\max\{m,n\}}} \alpha^m \alpha^n dt = \alpha^{-|m-n|}. \tag{2.63}$$

In order to determine whether or not these functions form a Riesz basis, we consider bounding the operator X^*X. The analysis below relies on the DTFT and its properties, which will be studied in detail in the next chapter. We assume that the reader has a basic familiarity with these topics, so that the derivations below should not be hard to follow. Otherwise, the example can be skipped for now, and returned to after these concepts are better understood.

Let $b = X^*Xa$. Then,

$$b[m] = \langle x_m, Xa \rangle = \langle x_m, \sum_{n\in\mathbb{Z}} a[n]x_n \rangle = \sum_{n\in\mathbb{Z}} a[n]\langle x_m, x_n \rangle = \sum_{n\in\mathbb{Z}} a[n]\alpha^{-|m-n|}. \tag{2.64}$$

Defining the sequence $c[n] = \alpha^{-|n|}$, we can express $b[n]$ as $b[n] = (a * c)[n]$, where $*$ denotes the discrete-time convolution between two sequences. We discuss the convolution operator in detail in Section 3.3. For now, we recall one of the key properties of convolution that we use below: The DTFT $B(e^{j\omega})$ of the convolution is equal to the product of the DTFTs $A(e^{j\omega})$, $C(e^{j\omega})$ of the sequences $a[n]$, $c[n]$. Combining these observations with Parseval's theorem (see (3.66) in Chapter 3) we conclude that

$$\langle a, X^*Xa \rangle = \sum_{n \in \mathbb{Z}} \overline{a[n]}(a*c)[n]$$

$$= \frac{1}{2\pi} \int_{-\pi}^{\pi} \overline{A(e^{j\omega})} A(e^{j\omega}) C(e^{j\omega}) d\omega$$

$$= \frac{1}{2\pi} \int_{-\pi}^{\pi} \left| A(e^{j\omega}) \right|^2 C(e^{j\omega}) d\omega. \tag{2.65}$$

To bound the inner product $\langle a, X^*Xa \rangle$, we use the fact that

$$C(e^{j\omega}) = \frac{\alpha^2 - 1}{\alpha^2 - 2\alpha \cos(\omega) + 1}. \tag{2.66}$$

Since $|\cos(\omega)| \le 1$,

$$\frac{\alpha - 1}{\alpha + 1} \le C(e^{j\omega}) \le \frac{\alpha + 1}{\alpha - 1}. \tag{2.67}$$

Therefore,

$$\langle a, X^*Xa \rangle \le \frac{\alpha+1}{\alpha-1} \frac{1}{2\pi} \int_{-\pi}^{\pi} \left| A(e^{j\omega}) \right|^2 d\omega = \frac{\alpha+1}{\alpha-1} \|a\|^2, \tag{2.68}$$

$$\langle a, X^*Xa \rangle \ge \frac{\alpha-1}{\alpha+1} \frac{1}{2\pi} \int_{-\pi}^{\pi} \left| A(e^{j\omega}) \right|^2 d\omega = \frac{\alpha-1}{\alpha+1} \|a\|^2, \tag{2.69}$$

implying that (2.52) is satisfied. We thus conclude that the set $\{x_n(t)\}_{n \in \mathbb{Z}}$ constitutes a Riesz basis for its closed linear span.

Until now we discussed several basic properties of Riesz bases and have seen some examples. In the next section we address how to expand a given vector into a Riesz basis expansion. As we will see, (2.55) implies that the sequence of expansion coefficients is in ℓ_2.

2.5.4   Riesz basis expansions

Given a Riesz basis $\{x_n, n \in \mathcal{I}\}$ for $\mathcal{H}$, any vector x in $\mathcal{H}$ can be expressed uniquely as

$$x = \sum_{n \in \mathcal{I}} a[n] x_n, \tag{2.70}$$

where $a[n] \in \mathbb{C}$. With X denoting the set transformation corresponding to the vectors $\{x_n\}$ we may write (2.70) as $x = Xa$. The question is how we find a, when we are given only x and X. To answer this question, we start with the simplest example of a Riesz basis: an orthonormal basis.

A sequence $\{x_n \in \mathcal{H}, n \in \mathcal{I}\}$ is an *orthonormal basis* for $\mathcal{H}$ if it is complete and orthonormal, i.e. $\langle x_n, x_k \rangle = \delta_{nk}$. In this case

$$\left\| \sum_{n \in \mathcal{I}} a[n] x_n \right\|^2 = \left\langle \sum_{n \in \mathcal{I}} a[n] x_n, \sum_{k \in \mathcal{I}} a[k] x_k \right\rangle = \sum_{n,k \in \mathcal{I}} \overline{a[n]} a[k] \langle x_n, x_k \rangle = \sum_{n \in \mathcal{I}} |a[n]|^2,$$

$$(2.71)$$

so that $\alpha = \beta = 1$ in (2.50). One of the nice features of an orthonormal basis is that it is particularly easy to determine the basis expansion coefficients. Suppose that $x = \sum_n a[n] x_n$ and $\{x_n\}$ are orthonormal. Then, taking the inner product of x with x_k, we have:

$$\langle x_k, x \rangle = \sum_n a[n] \langle x_k, x_n \rangle = a[k], \qquad (2.72)$$

so that $a[n] = \langle x_n, x \rangle$. The reason that finding the expansion coefficients is straightforward is the fact that we can exploit the orthonormality property $\langle x_n, x_k \rangle = \delta_{nk}$ in order to isolate each of the coefficients.

Determining the coefficients in more general Riesz bases representations is more involved. In order to retain the flavor of the orthonormal basis expansion, we define a new set of vectors $\{\tilde{x}_n\}$ that have the inner product property

$$\langle \tilde{x}_n, x_k \rangle = \delta_{nk}. \qquad (2.73)$$

We will see later on that such a set always exists. Once (2.73) is established, the coefficients can be found exactly as in an orthonormal expansion, where in (2.72) we replace x_k by $\tilde{x}_k$, leading to $a[n] = \langle \tilde{x}_n, x \rangle$. Then, any $x \in \mathcal{H}$ can be expressed as

$$x = \sum_{n \in \mathcal{I}} \langle \tilde{x}_n, x \rangle x_n. \qquad (2.74)$$

The set $\{\tilde{x}_n\}$ is referred to as the dual or *biorthogonal* basis of $\{x_n\}$. It is easy to see that if such a set exists then it is unique and complete, as incorporated in the following proposition.

Proposition 2.12. *Let the vectors $\{x_n, n \in \mathcal{I}\}$ form a Riesz basis for a Hilbert space $\mathcal{H}$ and let $\{\tilde{x}_n, n \in \mathcal{I}\}$ be biorthogonal vectors defined by (2.73). Then the biorthogonal set is complete, and is the unique set of vectors satisfying (2.73).*

Proof: To prove completeness, we rely on Proposition 2.9. Suppose there exists a $y \in \mathcal{H}$ such that $\langle y, \tilde{x}_n \rangle = 0$ for all $n \in \mathcal{I}$. Since the vectors $\{x_n\}$ are complete, we can write y as $y = \sum_n a[n] x_n$ for some coefficient sequence a. We then have that for every n

$$0 = \langle y, \tilde{x}_n \rangle = \sum_k \overline{a[k]} \langle x_k, \tilde{x}_n \rangle = \overline{a[n]}, \qquad (2.75)$$

where we used the biorthogonality property. Thus, $a[n] = 0$ for all n and consequently $y = 0$, proving completeness.

To show uniqueness, suppose that the vectors $\{z_k\}$ also form a biorthogonal set so that $\langle z_k, x_n \rangle = \langle \tilde{x}_k, x_n \rangle = \delta_{kn}$. Then $\langle z_k - \tilde{x}_k, x_n \rangle = 0$ for all k, n. Since the vectors

$\{x_n\}$ are complete, we have by Proposition 2.9 that $z_k - \tilde{x}_k = 0$, or $z_k = \tilde{x}_k$, for each k. $\qquad\square$

Completeness of the vectors $\{\tilde{x}_n\}$ means that any vector x can be expressed as $x = \sum_n b[n]\tilde{x}_n$ for some sequence b. Taking the inner products with x_n leads to $b[n] = \langle x_n, x \rangle$. Combining this observation with (2.74) we conclude that

$$x = \sum_{n\in\mathcal{I}} \langle \tilde{x}_n, x \rangle x_n = \sum_{n\in\mathcal{I}} \langle x_n, x \rangle \tilde{x}_n. \tag{2.76}$$

For an orthonormal basis, the biorthogonal vectors are equal to the vectors themselves and we get the familiar orthonormal decomposition $x = \sum_{n\in\mathcal{I}} \langle x_n, x \rangle x_n$. In summary, we have shown that if a biorthogonal set of vectors satisfying (2.73) exists, then x can be written as a series expansion of the form (2.76).

It remains to derive an explicit expression for the biorthogonal vectors. This can be done conveniently by using set transformation notation. Let X and $\tilde{X}$ denote the set transformations corresponding to the vector sets $\{x_n\}$ and $\{\tilde{x}_n\}$, respectively. We then have the following lemma:

Lemma 2.1. *Condition (2.73) is equivalent to*

$$\tilde{X}^* X = I_{\ell_2}. \tag{2.77}$$

Proof: To prove the lemma, note that (2.77) means that $\tilde{X}^* X a = a$ for any $a \in \ell_2$. Suppose that (2.77) holds. Denoting $y = Xa$, we have that $a[n] = \langle \tilde{x}_n, y \rangle$. Writing out y explicitly leads to

$$a[n] = a[n]\langle \tilde{x}_n, x_n \rangle + \sum_{k\neq n} a[k]\langle \tilde{x}_n, x_k \rangle. \tag{2.78}$$

The only way (2.78) can hold for all choices of $a[n]$ is if $\langle \tilde{x}_n, x_n \rangle = 1$ and $\langle \tilde{x}_n, x_k \rangle = 0$ for $k \neq n$. Therefore, (2.77) implies (2.73).

To show the converse, let $b = \tilde{X}^* X a$ and suppose that (2.73) holds. Following the same steps as before establishes that

$$b[n] = a[n]\langle \tilde{x}_n, x_n \rangle + \sum_{k\neq n} a[k]\langle \tilde{x}_n, x_k \rangle. \tag{2.79}$$

Using the biorthogonality property, (2.79) becomes $b[n] = a[n]$ for all $a[n]$, so that $\tilde{X}^* X = I_{\ell_2}$. $\qquad\square$

To solve (2.77) for $\tilde{X}$ we recall that since $\{x_n\}$ is a Riesz basis, $X^* X$ is invertible (see (2.54)). It is then easy to see that

$$\tilde{X} = X(X^* X)^{-1} \tag{2.80}$$

satisfies (2.77): indeed, $\tilde{X}^* X = (X^* X)^{-1} X^* X = I_{\ell_2}$.
Since $\tilde{X}^* \tilde{X} = (X^* X)^{-1}$ we have from (2.54) that

$$\frac{1}{\beta} I_{\ell_2} \preceq \tilde{X}^* \tilde{X} \preceq \frac{1}{\alpha} I_{\ell_2}. \tag{2.81}$$

Together with the fact that the vectors $\{\tilde{x}_n\}$ are complete, this implies that the set $\{\tilde{x}_n\}$ also forms a Riesz basis for $\mathcal{H}$. In particular, for any $x \in \mathcal{H}$, (2.55) holds, and

$$\|a\|^2 = \sum_n |\langle \tilde{x}_n, x \rangle|^2 \le \frac{1}{\alpha} \|x\|^2, \tag{2.82}$$

where $a[n] = \langle \tilde{x}_n, x \rangle$ are the expansion coefficients in (2.74). Consequently, the expansion coefficient sequence a is in ℓ_2.

Our discussion on Riesz bases is summarized in the following theorem:

Theorem 2.1 (Riesz basis expansion).   *Let the vectors $\{x_n, n \in \mathcal{I}\}$ form a Riesz basis for a Hilbert space $\mathcal{H}$ with bounds α, β in (2.50) and let X be the corresponding set transformation. Then the vectors $\{\tilde{x}_n, n \in \mathcal{I}\}$ corresponding to*

$$\tilde{X} = X(X^*X)^{-1}$$

are the unique set of vectors in $\mathcal{H}$ biorthogonal to $\{x_n\}$, namely, such that $\langle \tilde{x}_k, x_n \rangle = \delta_{kn}$ for all k, n. These vectors also form a Riesz basis for $\mathcal{H}$ with bounds $1/\beta, 1/\alpha$. Any $x \in \mathcal{H}$ can then be expressed uniquely as $x = \sum_{n \in \mathcal{I}} \langle \tilde{x}_n, x \rangle x_n$ where the sequence of coefficients $\langle \tilde{x}_n, x \rangle$ is in ℓ_2. Alternatively, $x = \sum_{n \in \mathcal{I}} \langle x_n, x \rangle \tilde{x}_n$, where the sequence of coefficients $\langle x_n, x \rangle$ is in ℓ_2.

Example 2.27 In this example we compute the biorthogonal basis to the Riesz basis $\{x_n(t)\}_{n \in \mathbb{Z}}$ of Example 2.26.

From Theorem 2.1, the set transformation corresponding to the biorthogonal basis is given by $\tilde{X} = X(X^*X)^{-1}$, which implies that $\tilde{x}_n(t) = \tilde{X}e_n = Xa_n$ where $a_n = (X^*X)^{-1}e_n$. Here, e_n is the sequence with elements $e_n[m] = \delta[m - n]$. Therefore, we can write

$$\tilde{x}_n(t) = \sum_{m \in \mathbb{Z}} a_n[m] x_m(t). \tag{2.83}$$

Since $X^*Xa_n = e_n$, the sequence $a_n[m]$ has to satisfy $\sum_{\ell \in \mathbb{Z}} \langle x_m, x_\ell \rangle a_n[\ell] = e_n[m] = \delta[m - n]$. As shown in (2.64) within Example 2.26, this is equivalent to requiring that $(a_n * c)[m] = \delta[m - n]$, where $c[m] = \alpha^{-|m|}$. Using (2.66), this condition can be written in the frequency domain in terms of the DTFT $A_n(e^{j\omega})$ of $a_n[m]$ as

$$A_n(e^{j\omega})C(e^{j\omega}) = A_n(e^{j\omega}) \frac{\alpha^2 - 1}{\alpha^2 - 2\alpha \cos(\omega) + 1} = e^{-j\omega n}, \tag{2.84}$$

implying that

$$A_n(e^{j\omega}) = e^{-j\omega n} \frac{\alpha^2 - 2\alpha \cos(\omega) + 1}{\alpha^2 - 1}. \tag{2.85}$$

Taking the inverse DTFT,

$$a_n[m] = \frac{1}{\alpha^2 - 1} \begin{cases} \alpha^2 + 1, & m = n \\ -\alpha, & |m - n| = 1 \\ 0, & |m - n| \ge 2. \end{cases} \tag{2.86}$$

Therefore, combining (2.83) and (2.62) we conclude that

$$\tilde{x}_n(t) = \frac{1}{\alpha^2 - 1} \left((\alpha^2 + 1)\, x_n(t) - \alpha x_{n+1}(t) - \alpha x_{n-1}(t) \right)$$

$$= \frac{1}{\alpha^2 - 1} \begin{cases} 0, & 0 \leq t < \alpha^{-2(n+1)} \\ \alpha^{n+2}, & \alpha^{-2(n+1)} \leq t < \alpha^{-2n} \\ -\alpha^n, & \alpha^{-2n} \leq t < \alpha^{-2(n-1)} \\ 0, & t \geq \alpha^{-2(n-1)}. \end{cases} \tag{2.87}$$

2.6   Projection operators

In the previous section we focused on expansions of signals within a given space using bases, or more specifically, Riesz bases. We now turn our attention to decompositions of vectors into components in two different subspaces. We already encountered the topic of direct-sum decompositions when studying the subspaces associated with an arbitrary linear transformation T in Section 2.4.3. Here, we complete these ideas by considering the projection operator, which leads to such decompositions. Projection operators play a very important role in the development of modern sampling methods since they allow us to focus our attention on the parts of the signal that can be recovered from the given samples.

Definition 2.17.   *A linear operator* $T\colon \mathcal{H} \to \mathcal{H}$ *is a* projection *if* $T = T^2$.

We distinguish between two different types of projections:

- Hermitian projections (orthogonal projections) for which $T = T^*$;
- Non-Hermitian projections (oblique projections).

Orthogonal projections have enjoyed widespread use in the signal processing literature; oblique projections have received much less attention. The reason Hermitian projections are called orthogonal is because their symmetry properties result in the fact that $\mathcal{N}(T) = \mathcal{R}(T)^\perp$ (see Proposition 2.8). As we show below, this implies that this class of projections decomposes the space into orthogonal subspaces, while non-Hermitian projections result in decompositions that are not orthogonal.

From the definition of a projection we can already learn a lot about how it acts on arbitrary vectors in $\mathcal{H}$. Our first important observation is the following proposition.

Proposition 2.13.   *Given a projection* T *on a Hilbert space* $\mathcal{H}$, *we have that*

$$\mathcal{H} = \mathcal{R}(T) \oplus \mathcal{N}(T). \tag{2.88}$$

In addition, for any y in $\mathcal{R}(T)$, $Ty = y$.

Note that the decomposition (2.88) is not true for general transformations T. However, the fact that $T^2 = T$ is enough to ensure this property. An immediate consequence of the proposition is that once we define $\mathcal{R}(T)$ and $\mathcal{N}(T)$, the operator T is completely specified: For $x \in \mathcal{N}(T)$, $Tx = 0$, and for $x \in \mathcal{R}(T)$, $Tx = x$. From (2.88), this describes the operation of T on any vector in $\mathcal{H}$.

Proof: To prove (2.88) note that any vector $x \in \mathcal{H}$ can be written as

$$x = Tx + (I_\mathcal{H} - T)x. \tag{2.89}$$

This, of course, is true for all T. What is special about the projection operator is that $y = (I_\mathcal{H} - T)x$ lies in its null space $\mathcal{N}(T)$. Indeed,

$$Ty = T(I_\mathcal{H} - T)x = (T - T^2)x = 0, \tag{2.90}$$

since $T = T^2$. Obviously, Tx is in the range $\mathcal{R}(T)$. Therefore, (2.89) shows that an arbitrary $x \in \mathcal{H}$ can be written as a sum of a component in $\mathcal{N}(T)$ and a component in $\mathcal{R}(T)$ which proves that $\mathcal{H} = \mathcal{N}(T) + \mathcal{R}(T)$.

To establish the direct sum we need to show that these two subspaces are disjoint. Let y be an arbitrary vector in $\mathcal{R}(T)$. By definition, $y = Tx$ for some x, so that

$$Ty = T(Tx) = T^2 x = Tx = y. \tag{2.91}$$

Now, suppose that y is also in the null space of T. Then $Ty = 0$, and from (2.91), $y = 0$. Therefore, $\mathcal{R}(T) \cap \mathcal{N}(T) = \{0\}$. Finally, (2.91) shows that for any $y \in \mathcal{R}(T)$, $Ty = y$.   □

Throughout the book, we denote a projection with range equal to $\mathcal{V}$ and null space equal to $\mathcal{W}$ by $E_{\mathcal{V}\mathcal{W}}$, and refer to this operator as a *projection onto $\mathcal{V}$ along $\mathcal{W}$*. By definition, $E_{\mathcal{V}\mathcal{W}}$ is the unique operator satisfying

1. $E_{\mathcal{V}\mathcal{W}}v = v$ for any $v \in \mathcal{V}$
2. $E_{\mathcal{V}\mathcal{W}}w = 0$ for any $w \in \mathcal{W}$.

When T is Hermitian, $\mathcal{R}(T) = \mathcal{N}(T)^\perp$ so that $\mathcal{V} = \mathcal{W}^\perp$. In this case, we refer to T as an *orthogonal projection* and denote it simply by $P_\mathcal{V}$. Otherwise, the projection is called an *oblique projection* [34, 35, 36].

Given a projection $E_{\mathcal{V}\mathcal{W}}$, it follows from (2.88) that any $x \in \mathcal{H}$ can be written uniquely as

$$x = x_\mathcal{V} + x_\mathcal{W}, \tag{2.92}$$

where $x_\mathcal{V}$ is in $\mathcal{V}$ and $x_\mathcal{W}$ is in $\mathcal{W}$. In addition, from property (1) above, we have that $x_\mathcal{V} = E_{\mathcal{V}\mathcal{W}}x_\mathcal{V}$. Since property (2) implies that $E_{\mathcal{V}\mathcal{W}}x_\mathcal{W} = 0$ we can write

$$x_\mathcal{V} = E_{\mathcal{V}\mathcal{W}}x. \tag{2.93}$$

It then follows that

$$x_\mathcal{W} = x - x_\mathcal{V} = (I_\mathcal{H} - E_{\mathcal{V}\mathcal{W}})x. \tag{2.94}$$

It is easy to see that $T = I_\mathcal{H} - E_{\mathcal{V}\mathcal{W}}$ is also a projection. For any $v \in \mathcal{V}$, $Tv = 0$; for $w \in \mathcal{W}$, $Tw = w$. Therefore, $T = E_{\mathcal{W}\mathcal{V}}$ is a projection onto $\mathcal{W}$ along $\mathcal{V}$, leading to the following proposition.

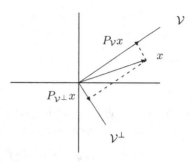

Figure 2.7 Decomposition of x into its orthogonal components in $\mathcal{V}$ and $\mathcal{V}^\perp$ given by $P_{\mathcal{V}}x$ and $P_{\mathcal{V}^\perp}x$, respectively.

Proposition 2.14. *Let $E_{\mathcal{V}\mathcal{W}}$ be a projection on a Hilbert space $\mathcal{H}$. Then, any $x \in \mathcal{H}$ can be written uniquely as*

$$x = x_{\mathcal{V}} + x_{\mathcal{W}}, \tag{2.95}$$

with $x_{\mathcal{V}} = E_{\mathcal{V}\mathcal{W}}x$ and $x_{\mathcal{W}} = (I_{\mathcal{H}} - E_{\mathcal{V}\mathcal{W}})x = E_{\mathcal{W}\mathcal{V}}x$. The vectors $x_{\mathcal{V}}$ and $x_{\mathcal{W}}$ are called the projections onto $\mathcal{V}$ and $\mathcal{W}$ respectively.

2.6.1   Orthogonal projection operators

As we have seen, if $T = T^*$, then $\mathcal{R}(T) = \mathcal{N}(T)^\perp$ so that an orthogonal projection is completely specified by its range space $\mathcal{V}$. Therefore, an orthogonal projection $P_{\mathcal{V}}$ is the unique operator satisfying $P_{\mathcal{V}}v = v$ for any $v \in \mathcal{V}$ and $P_{\mathcal{V}}w = 0$ for any $w \in \mathcal{V}^\perp$.

Orthogonal projections can be used to decompose a signal space into orthogonal subspaces. Specifically, given an orthogonal projection $P_{\mathcal{V}}$ on $\mathcal{H}$, it follows from (2.88) that $\mathcal{H} = \mathcal{V} \oplus \mathcal{V}^\perp$. Proposition 2.14 then implies that any $x \in \mathcal{H}$ can be expressed uniquely as $x = x_{\mathcal{V}} + x_{\mathcal{V}^\perp}$, where $x_{\mathcal{V}} = P_{\mathcal{V}}x \in \mathcal{V}$ and $x_{\mathcal{V}^\perp} = (I_{\mathcal{H}} - P_{\mathcal{V}})x = P_{\mathcal{V}^\perp}x \in \mathcal{V}^\perp$. These projections have the additional property that they are orthogonal, as illustrated in Fig. 2.7. Owing to the orthogonality of the components, $\|x\|^2 = \|x_{\mathcal{V}}\|^2 + \|x_{\mathcal{V}^\perp}\|^2$, from which it follows that the norm of the projection of x is never greater than that of x:

$$\|P_{\mathcal{V}}x\|^2 = \|x_{\mathcal{V}}\|^2 \leq \|x\|^2. \tag{2.96}$$

This property does not necessarily hold for an oblique projection onto $\mathcal{V}$ [37]. The importance of (2.96) is that if x contains noise, then the projection will not amplify it.

The orthogonal projection $x_{\mathcal{V}} = P_{\mathcal{V}}x$ has another well-known characterization; it is the closest vector to x in $\mathcal{V}$.

Proposition 2.15. *Let $\mathcal{V} \subseteq \mathcal{H}$ be a closed subspace of $\mathcal{H}$, let $P_{\mathcal{V}}$ denote the orthogonal projection onto $\mathcal{V}$, and let x be an arbitrary vector in $\mathcal{H}$. Then*

$$x_{\mathcal{V}} = P_{\mathcal{V}}x = \arg\min_{v \in \mathcal{V}} \|x - v\|^2. \tag{2.97}$$

Proof: Express x as $x = x_V + x_{V^\perp}$ where $x_V = P_V x \in V$ and $x_{V^\perp} \in V^\perp$. Noting that $x_V - v \in V$ for any $v \in V$, we have that $\langle x_V - v, x_{V^\perp} \rangle = 0$. Therefore,

$$\|x - v\|^2 = \|x_{V^\perp} + x_V - v\|^2 = \|x_{V^\perp}\|^2 + \|x_V - v\|^2 \geq \|x_{V^\perp}\|^2, \qquad (2.98)$$

with equality if and only if $v = x_V$. □

We conclude our discussion on orthogonal projections by providing an explicit construction of P_V using a Riesz basis for V.

Proposition 2.16. *Let $\{v_n\}$ be a Riesz basis for a Hilbert space V, and let V be the corresponding set transformation. Then, the orthogonal projection onto V can be written as*

$$P_V = V(V^*V)^{-1}V^*. \qquad (2.99)$$

Note that since V corresponds to a Riesz basis, $(V^*V)^{-1}$ is well defined (see (2.54)).

Proof: To prove the proposition it is enough to show that $P_V v = v$ for any $v \in V$ and $P_V x = 0$ for $x \in V^\perp$. The latter follows immediately since for any $x \in V^\perp$, $\langle v_n, x \rangle = 0$, and consequently $V^* x = 0$. Next we note that any $v \in V$ can be written as $v = Va$ for some $a \in \ell_2$. Thus,

$$V(V^*V)^{-1}V^* v = V(V^*V)^{-1}V^*(Va) = Va = v, \qquad (2.100)$$

completing the proof. □

The representation (2.99) can also be interpreted in terms of the biorthogonal basis vectors corresponding to the set transformation

$$\widetilde{V} = V(V^*V)^{-1}. \qquad (2.101)$$

Writing (2.99) as

$$P_V = V\widetilde{V}^* \qquad (2.102)$$

we see that for any x in $\mathcal{H}$,

$$x_V = P_V x = V\widetilde{V}^* x = \sum_n \langle \tilde{v}_n, x \rangle v_n. \qquad (2.103)$$

If the vectors $\{v_n\}$ are orthonormal, then from Proposition 2.10 we have that $V^*V = I$ and $P_V = VV^*$. The orthogonal projection x_V is then given by

$$x_V = \sum_n \langle v_n, x \rangle v_n. \qquad (2.104)$$

Expansion (2.103) is identical to (2.76), which we developed in the context of Riesz bases (with $v_n, \tilde{v}_n$ replacing $x_n, \tilde{x}_n$). The difference is in the left-hand side. When we discussed Riesz bases, which led to (2.76), we focused on signals x lying in the space spanned by the basis vectors x_n. In contrast, here the vector x to be expanded can be an arbitrary vector in a larger Hilbert space $\mathcal{H}$, while the basis vectors span a lower-dimensional subspace V. If x lies in V, then $P_V x = x$ and the two expansions are identical. However, if x is not contained in V, then (2.103) shows that a standard Riesz basis expansion results in the orthogonal projection of x onto V. From Proposition 2.15, this is the closest vector in V to the original signal x.

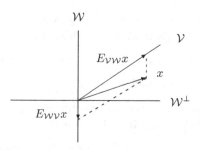

Figure 2.8 Decomposition of x into its components in V and in W given by $E_{VW}x$ and $E_{WV}x$, respectively.

2.6.2   Oblique projection operators

Oblique projection operators can also be used to decompose a signal space into smaller subspaces; however, when using oblique projections, these spaces are no longer constrained to be orthogonal.

Given a projection E_{VW} on $\mathcal{H}$, it follows from (2.88) and (2.89) that $\mathcal{H}$ can be decomposed as $\mathcal{H} = V \oplus W$. Every $x \in \mathcal{H}$ can then be expressed uniquely in the form $x = x_V + x_W$, where $x_V = E_{VW}x \in V$ and $x_W = (I_{\mathcal{H}} - E_{VW})x = E_{WV}x \in W$, as illustrated in Fig. 2.8. Note, however, that x_V and x_W are not necessarily orthogonal. The fact that the projection is not orthogonal means that the norm of each one of the components x_V and x_W can in principle be larger than that of x, as can be seen in Fig. 2.8 (where the vector $x_V = E_{VW}x$ is longer than x). This has important consequences in sampling theory. If the vector x represents a signal contaminated by noise, then applying an oblique projection to it can increase the norm of the noise.

Constructing an oblique projection is more complicated than an orthogonal projection since it involves two subspaces. To this end, we begin with two Riesz bases, one for the space V and one for $W^\perp$, the orthogonal complement of the null space of E_{VW}. We then rely on the following lemma.

Lemma 2.2.   *Let the vectors $\{v_n, n \in \mathcal{I}\}$ form a Riesz basis for a subspace V of a Hilbert space $\mathcal{H}$, and let the vectors $\{w_n, n \in \mathcal{I}\}$ form a Riesz basis for a subspace $W^\perp$ of $\mathcal{H}$ such that $V + W = \mathcal{H}$. Let $V: \ell_2 \to \mathcal{H}$ and $W: \ell_2 \to \mathcal{H}$ denote the set transformations corresponding to the vectors $\{v_n, n \in \mathcal{I}\}$ and $\{w_n, n \in \mathcal{I}\}$, respectively. Then W^*V is invertible if and only if $V \cap W = \{0\}$.*

Proof: We first show that if there is a nonzero vector in the intersection of V and W, then W^*V is not invertible. Suppose that x is a nonzero vector in $V \cap W$. Since the vectors v_n form a basis for V, we can write $x = Va$ for some nonzero $a \in \ell_2$. In addition, using the fact that $\mathcal{N}(W^*) = \mathcal{R}(W)^\perp = W$ (see Proposition 2.8), we have $W^*x = 0$ because x is in W. Substituting $x = Va$, we conclude that $W^*x = W^*Va = 0$ for a nonzero $a \in \ell_2$, and W^*V is not invertible.

Next, we assume that $V \cap W = \{0\}$ and show that W^*V is invertible, or equivalently, that W^*V is both injective and surjective. Suppose to the contrary that W^*V is not injective, namely, there exists a nonzero a satisfying $W^*Va = 0$. Since the vectors $\{v_n\}$ are linearly independent, $Va \neq 0$ (this follows from Proposition 2.10). Therefore, $W^*Va = 0$ implies that $W^*v = 0$ where $v = Va \neq 0$, or equivalently, $v \in \mathcal{N}(W^*)$. Since $\mathcal{N}(W^*) = \mathcal{R}(W)^{\perp} = \mathcal{W}$, $v \neq 0$ must lie in $\mathcal{W}$. But by definition, $v = Va$ lies in $\mathcal{V}$, which contradicts the assumption that $\mathcal{V} \cap \mathcal{W} = \{0\}$.

We now show that W^*V is surjective. This means that for every $a \in \ell_2$ there exists some $b \in \ell_2$ such that $a = W^*Vb$. To show this, let $a \in \ell_2$ be arbitrary, and define the vector $w \in \mathcal{W}^{\perp}$ by $w = \sum_n a[n]\tilde{w}_n$, where $\{\tilde{w}_n\}$ are the biorthogonal vectors to $\{w_n\}$. Since $\langle w_k, \tilde{w}_n \rangle = \delta_{kn}$, we have that $a = W^*w$. Using the fact that $\mathcal{H} = \mathcal{W} + \mathcal{V}$ we can write any $w \in \mathcal{H}$ as $w = w_1 + v$ where $w_1 \in \mathcal{W}$ and $v \in \mathcal{V}$. Since $\mathcal{N}(W^*) = \mathcal{W}$, clearly $W^*w_1 = 0$. Therefore, $a = W^*w = W^*v$. But since $v \in \mathcal{V}$, we can write $v = Vb$ for some $b \in \ell_2$. We conclude that there exists $b \in \ell_2$ such that $a = W^*v = W^*Vb$, proving that W^*V is surjective. $\qquad\square$

Using Lemma 2.2 we can now construct an oblique projection operator:

Theorem 2.2 (Oblique projection). *Let the vectors $\{v_n, n \in \mathcal{I}\}$ form a Riesz basis for a subspace $\mathcal{V}$ of a Hilbert space $\mathcal{H}$, and let the vectors $\{w_n, n \in \mathcal{I}\}$ form a Riesz basis for a subspace $\mathcal{W}^{\perp}$ such that $\mathcal{H} = \mathcal{V} \oplus \mathcal{W}$. Let $V: \ell_2 \to \mathcal{H}$ and $W: \ell_2 \to \mathcal{H}$ denote the set transformations corresponding to the vectors $\{v_n, n \in \mathcal{I}\}$ and $\{w_n, n \in \mathcal{I}\}$, respectively. Then the oblique projection onto $\mathcal{V}$ along $\mathcal{W}$ can be written as*

$$E_{\mathcal{V}\mathcal{W}} = V(W^*V)^{-1}W^*.$$

Proof: Let $T = V(W^*V)^{-1}W^*$. From Lemma 2.2, W^*V is invertible so that T is well defined.

We now need to show that $Tw = 0$ for any $w \in \mathcal{W}$ and $Tv = v$ for any $v \in \mathcal{V}$. The first part is straightforward: Since $w_n \in \mathcal{W}^{\perp}$, $W^*w = 0$ for any $w \in \mathcal{W}$, and $Tw = 0$. Equivalently, as we have seen in the proof of Lemma 2.2, $\mathcal{N}(W^*) = \mathcal{W}$. To prove the second part, note that any $v \in \mathcal{V}$ can be expressed as $v = Va$ for some $a \in \ell_2$. Therefore, $Tv = TVa = Va = v$. $\qquad\square$

When $\mathcal{W} = \mathcal{V}^{\perp}$, we can choose $W = V$ and the expression for $E_{\mathcal{V}\mathcal{W}}$ reduces to that of $P_{\mathcal{V}}$ given by (2.99).

The construction of Theorem 2.2 will be used in the context of sampling, when the sampling and reconstruction spaces are not constrained to be equal.

To summarize our discussion on projections, both oblique and orthogonal projections can be used to decompose a signal into components in disjoint subspaces as illustrated in Figs. 2.7 and 2.8. However, contrary to decompositions using orthogonal projections, when using oblique projections the components are not necessarily orthogonal. Furthermore, while the norm of the elements in an orthogonal projection are no larger than the norm of the original vector, this is not necessarily true when using oblique projections, as can be seen in Fig. 2.8.

Example 2.28 Consider the space V of length-4 vectors whose first two elements are equal and whose last two elements are equal. To compute the orthogonal projection of an arbitrary vector x onto V, we construct a basis for V as

$$\mathbf{v}_1 = [1\ 1\ 0\ 0]^T, \quad \mathbf{v}_2 = [0\ 0\ 1\ 1]^T. \tag{2.105}$$

With this choice, the set transformation V corresponds to the matrix

$$\mathbf{V} = \begin{bmatrix} 1 & 0 \\ 1 & 0 \\ 0 & 1 \\ 0 & 1 \end{bmatrix}. \tag{2.106}$$

We can now use Proposition 2.16 to compute the orthogonal projection P_V, which is given by the matrix

$$\mathbf{V}(\mathbf{V}^T\mathbf{V})^{-1}\mathbf{V}^T = \mathbf{V}(2\mathbf{I})^{-1}\mathbf{V}^T = \frac{1}{2}\begin{bmatrix} 1 & 1 & 0 & 0 \\ 1 & 1 & 0 & 0 \\ 0 & 0 & 1 & 1 \\ 0 & 0 & 1 & 1 \end{bmatrix}. \tag{2.107}$$

For example, if $x = [2\ 4\ 6\ 8]^T$ then $x_V = P_V x = [3\ 3\ 7\ 7]^T$ and $x_{V\perp} = P_{V\perp}x = x - P_V x = [-1\ 1\ -1\ 1]^T$. As we expect, x_V consists of the local averages of x.

Next, we wish to compute the oblique projection onto the space V along the space $W^\perp$, where W consists of length-4 sequences of the form $x[n] = a + bn$, $1 \le n \le 4$ for arbitrary scalars a and b. To do this, we choose the following basis for W:

$$\mathbf{w}_1 = [1\ 1\ 1\ 1]^T, \quad \mathbf{w}_2 = [1\ 2\ 3\ 4]^T. \tag{2.108}$$

Clearly, any $x[n]$ in W can be written as $x = a\mathbf{w}_1 + b\mathbf{w}_2$. The corresponding set transformation is the matrix

$$\mathbf{W} = \begin{bmatrix} 1 & 1 \\ 1 & 2 \\ 1 & 3 \\ 1 & 4 \end{bmatrix}. \tag{2.109}$$

From Theorem 2.2,

$$E_{VW^\perp} = \mathbf{V}(\mathbf{W}^T\mathbf{V})^{-1}\mathbf{W}^T = \frac{1}{8}\begin{bmatrix} 5 & 3 & 1 & -1 \\ 5 & 3 & 1 & -1 \\ -1 & 1 & 3 & 5 \\ -1 & 1 & 3 & 5 \end{bmatrix}. \tag{2.110}$$

Using this operator, the decomposition of $x = [2\ 4\ 6\ 8]^T$ takes the form $x_V = E_{VW^\perp}x = [2.5\ 2.5\ 7.5\ 7.5]^T$ and $x_{W^\perp} = x - E_{VW^\perp}x = [-0.5\ 1.5\ -1.5\ 0.5]^T$.

The orthogonal and oblique decompositions of x are shown in Fig. 2.9. Both projections have the same range V. However, the value of the projection x_V in V changes when the null space of the projection (i.e. $W^\perp$) varies.

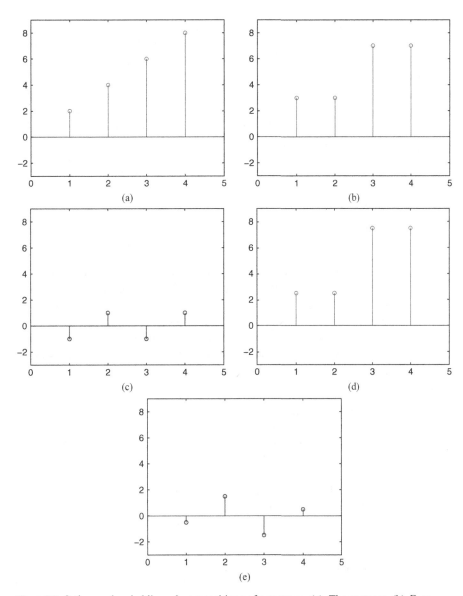

Figure 2.9 Orthogonal and oblique decompositions of a vector $\mathbf{x}$. (a) The vector $\mathbf{x}$. (b) $P_{\mathcal{V}}\mathbf{x}$. (c) $P_{\mathcal{V}^\perp}\mathbf{x}$. (d) $E_{\mathcal{V}\mathcal{W}^\perp}\mathbf{x}$. (e) $E_{\mathcal{W}^\perp\mathcal{V}}\mathbf{x}$.

2.7 Pseudoinverse of a transformation

Equipped with the notion of an orthogonal projection, we can now formally define the *pseudoinverse*, which was mentioned in Section 2.4.3.

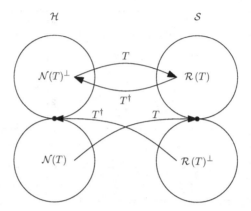

Figure 2.10 The action of T and $T^\dagger$ on the subspaces $\mathcal{N}(T)$, $\mathcal{N}(T)^\perp$, $\mathcal{R}(T)$, and $\mathcal{R}(T)^\perp$.

A linear transformation between spaces of different dimension (e.g. a rectangular matrix), or an operator that is not bijective (e.g. a singular matrix), does not have an inverse in the usual sense. Nevertheless, we have seen in Section 2.4.1 that we can invert an arbitrary transformation T if we restrict our attention to $\mathcal{N}(T)^\perp$ and $\mathcal{R}(T)$. This leads to the notion of the pseudoinverse, which serves as a restricted inverse. As we show below, applying combinations of T and its pseudoinverse results in orthogonal projections onto the original and image spaces.

2.7.1   Definition and properties

Let T be a continuous linear transformation $T\colon \mathcal{H} \to \mathcal{S}$ with closed range. We have seen in Proposition 2.7 that to each $y \in \mathcal{R}(T)$ there corresponds a unique $x \in \mathcal{N}(T)^\perp$ such that $y = Tx$. Thus, if we restrict T to the subspace $\mathcal{N}(T)^\perp$ then T is invertible: its inverse is defined as the pseudoinverse, and denoted by $T^\dagger$. The pseudoinverse operates on vectors in $\mathcal{S}$ such that $T^\dagger y = x$ for $y = Tx$, that is for y in the range of T, and $x \in \mathcal{N}(T)^\perp$. To complete the description of $T^\dagger$ we require that it is zero on $\mathcal{R}(T)^\perp$, as illustrated in Fig. 2.10. Formally, we have the following definition.

Definition 2.18. *Let $T\colon \mathcal{H} \to \mathcal{S}$ be a continuous linear transformation with closed range. The* pseudoinverse *of T, denoted $T^\dagger$, is the unique transformation satisfying*

$$T^\dagger T x = x, \quad \forall x \in \mathcal{N}(T)^\perp; \tag{2.111}$$
$$T^\dagger y = 0, \quad \forall y \in \mathcal{R}(T)^\perp. \tag{2.112}$$

An immediate consequence of Definition 2.18 is that if T is restricted to $\mathcal{N}(T)^\perp \subseteq \mathcal{H}$ and $T^\dagger$ to $\mathcal{R}(T) \subseteq \mathcal{S}$, then T is invertible and its inverse is $T^\dagger$.

We can replace the definition of $T^\dagger$ by an equivalent set of conditions that relies on orthogonal projection operators, and provides further insight into the effect of $T^\dagger$:

Theorem 2.3 (Pseudoinverse). *Let T be a continuous linear transformation $T\colon \mathcal{H} \to \mathcal{S}$ with closed range. Then the pseudoinverse of T, denoted $T^\dagger$, is the unique* *transformation satisfying the following conditions:*

$$TT^\dagger = P_{\mathcal{R}(T)}; \tag{2.113}$$

$$T^\dagger T = P_{\mathcal{N}(T)^\perp}; \tag{2.114}$$

$$\mathcal{R}(T^\dagger) = \mathcal{N}(T)^\perp. \tag{2.115}$$

Conditions (2.113)–(2.115) may be interpreted similarly to Definition 2.18. If we apply T to an $x \in \mathcal{N}(T)^\perp$, then we can invert this mapping by applying $T^\dagger$ to the result: $T^\dagger T x = x$. Similarly, if we apply $T^\dagger$ to a $y \in \mathcal{R}(T)$, then it is invertible by applying T to the result: $TT^\dagger y = y$.

Proof: We first prove that any $T^\dagger$ satisfying (2.113)–(2.115) obeys (2.111)–(2.112).

Obviously (2.111) follows from (2.114). To establish (2.112), note from (2.113) that for any $y \in \mathcal{R}(T)^\perp$, $TT^\dagger y = 0$. Let $v = T^\dagger y$. Then, either $v = 0$ or $Tv = 0$. Since v is in $\mathcal{R}(T^\dagger)$, from (2.115) we have that $v \in \mathcal{N}(T)^\perp$ and therefore $Tv = 0$ only if $v = 0$, which implies that $T^\dagger y = 0$ as required.

We now show that the definition of $T^\dagger$ implies (2.113)–(2.115). Combining (2.111) with the fact that $Tx = 0$ for any $x \in \mathcal{N}(T)$, we have (2.114). To prove (2.113) note that any $y \in \mathcal{R}(T)$ can be written as $y = Tx$ for some $x \in \mathcal{N}(T)^\perp$. Therefore, using (2.114), $TT^\dagger y = T(T^\dagger T)x = Tx = y$. For $y \in \mathcal{R}(T)^\perp$, we have from (2.112) that $T^\dagger y = 0$, verifying (2.113).

It remains to establish (2.115). First, let x be an arbitrary vector in $\mathcal{N}(T)^\perp$. Then, from (2.114), we can write $x = T^\dagger(Tx)$ so that x is in $\mathcal{R}(T^\dagger)$. Next, let x be an arbitrary vector in $\mathcal{R}(T^\dagger)$. We will show that x is also in $\mathcal{N}(T)^\perp$ which proves (2.115). Since $x \in \mathcal{R}(T^\dagger)$, it can be written as $x = T^\dagger y$ for some $y \in \mathcal{S}$. Expressing y as $y = y_1 + y_2$ with $y_1 \in \mathcal{R}(T)$ and $y_2 \in \mathcal{R}(T)^\perp$, and using the fact that from (2.112), $T^\dagger y_2 = 0$, we have $x = T^\dagger y_1$. Now, because $y_1 \in \mathcal{R}(T)$, we can write $y_1 = Tv$ for some $v \in \mathcal{N}(T)^\perp$. It then follows that $x = T^\dagger y_1 = T^\dagger T v = v$, where we used (2.114). Finally, since $v \in \mathcal{N}(T)^\perp$, x is in $\mathcal{N}(T)^\perp$, as required. $\square$

Another well-known definition of the pseudoinverse is given by the *Moore–Penrose conditions* [38, 39]. These conditions state that the pseudoinverse is the unique transformation satisfying

$$TT^\dagger T = T;$$

$$T^\dagger TT^\dagger = T^\dagger;$$

$$(TT^\dagger)^* = TT^\dagger;$$

$$(T^\dagger T)^* = T^\dagger T. \tag{2.116}$$

It can be readily established that these conditions coincide with those of Definition 2.18. One way to prove this is by relying on the equivalent requirements given by Theorem 2.3.

Example 2.29 As we saw in Example 2.17, the range of the transformation T defined by (2.22) is not closed. Therefore, it does not have a pseudoinverse. Let us, then, consider the restriction $\widetilde{T}$ of T to signals defined over $[\varepsilon, 2]$ for some $0 < \varepsilon < 1$. Specifically, for every $x \in L_2([\varepsilon, 2])$, the function $y = \widetilde{T}x$ is defined by

$$y(t) = \begin{cases} tx(t), & \varepsilon \le t \le 1 \\ 0, & 1 < t \le 2. \end{cases} \tag{2.117}$$

By the same arguments as in Example 2.14, the null space $\mathcal{N}(\widetilde{T})$ consists of all signals in $L_2([\varepsilon, 2])$ that vanish almost everywhere on $t \in [\varepsilon, 1]$. Similarly, the range $\mathcal{R}(\widetilde{T})$ comprises all signals that vanish on $t \in (1, 2]$.

Clearly, given a signal $y \in \mathcal{R}(\widetilde{T})$, we cannot determine a unique signal $x \in L_2([\varepsilon, 2])$ such that $y = \widetilde{T}x$. This is because any signal x for which $y = \widetilde{T}x$ can be modified arbitrarily on $t \in (1, 2]$ and still result in the same y. However, among all these valid signals, only one lies in $\mathcal{N}(\widetilde{T})^\perp$, namely the one for which $x(t) = 0$ on $t \in (1, 2]$. Thus, the pseudoinverse $\widetilde{T}^\dagger$ of $\widetilde{T}$ is defined by

$$x(t) = \widetilde{T}^\dagger\{y\}(t) = \begin{cases} \frac{y(t)}{t}, & \varepsilon \le t \le 1 \\ 0, & 1 < t \le 2. \end{cases} \tag{2.118}$$

It is easy to see that for every $x(t)$ that vanishes on $t \in (1, 2]$, we have $\widetilde{T}^\dagger\widetilde{T}\{x\}(t) = x(t)$, and that $\widetilde{T}^\dagger\{y\}(t) = 0$ for all $y(t) \in \mathcal{R}(\widetilde{T})^\perp$.

The following properties are easily verified from any one of the equivalent set of conditions defining the pseudoinverse:

1. The pseudoinverse of the pseudoinverse is the original transform: $(T^\dagger)^\dagger = T$.
2. The pseudoinverse commutes with the adjoint: $(T^*)^\dagger = (T^\dagger)^*$.
3. For any $a \neq 0$, $(aT)^\dagger = a^{-1}T^\dagger$.
4. If T is invertible, then $T^\dagger = T^{-1}$.

Finally, we note that when either T^*T or TT^* is invertible, the pseudoinverse can be expressed more explicitly in terms of T and T^*. In particular, when T^*T is invertible we have that $T^\dagger = (T^*T)^{-1}T^*$. If TT^* is invertible, then $T^\dagger = T^*(TT^*)^{-1}$. These results follow from the definition of the pseudoinverse and are left as an exercise (see Exercise 21).

2.7.2   Matrices

When dealing with matrices and finite linear algebra, there are other well-known characterizations of the pseudoinverse. These, of course, coincide with the more general definitions given in the previous section.

Probably the most popular description of the pseudoinverse in the finite case is that given in terms of the *singular value decomposition (SVD)* (see also Appendix A).

Proposition 2.17. *Let* $\mathbf{T}$ *be an* $n \times m$ *matrix with* $n \ge m$. *Then* $\mathbf{T}$ *can be written as*

$$\mathbf{T} = \mathbf{U}\boldsymbol{\Sigma}\mathbf{V}^H = \sum_{i=1}^{r} \sigma_i \mathbf{u}_i \mathbf{v}_i^H,$$

where r is the rank of $\mathbf{T}$, $\mathbf{U}$ is an $n \times n$ unitary matrix, $\mathbf{\Sigma}$ is an $n \times m$ diagonal matrix with the first r diagonal elements equal to $\sigma_i > 0$ and the remaining diagonal elements equal to 0, and $\mathbf{V}$ is an $m \times m$ unitary matrix. The vectors $\mathbf{u}_i, \mathbf{v}_i$ are the ith columns of $\mathbf{U}, \mathbf{V}$ respectively. The constants σ_i are called the singular values *of $\mathbf{T}$ and the decomposition is referred to as the* SVD.

For an $n \times m$ matrix $\mathbf{T}$ with $n < m$, we can apply the SVD to $\mathbf{T}^H$.

If $\mathbf{T} = \mathbf{U}\mathbf{\Sigma}\mathbf{V}^H$ then $\mathcal{R}(\mathbf{T})$ is spanned by the first r columns of $\mathbf{U}$, and $\mathcal{N}(\mathbf{T})^\perp$ is spanned by the first r columns of $\mathbf{V}$. Note that for any columns $\mathbf{u}_i, \mathbf{v}_i$ of $\mathbf{U}, \mathbf{V}$, we have that $\mathbf{T}\mathbf{v}_i = \sigma_i\mathbf{u}_i$ and $\mathbf{T}^H\mathbf{u}_i = \sigma_i\mathbf{v}_i$. In fact, this property can be used as a definition of the singular values. A positive real number $\sigma > 0$ is a *singular value* of $\mathbf{T}$ if and only if there exist normalized vectors $\mathbf{u}$ and $\mathbf{v}$ such that $\mathbf{T}\mathbf{v} = \sigma\mathbf{u}$ and $\mathbf{T}^H\mathbf{u} = \sigma\mathbf{v}$. The vectors $\mathbf{u}$ and $\mathbf{v}$ are called the left singular and right singular vectors for σ, respectively.

The SVD is closely related to the eigenvalue decompositions of the matrices $\mathbf{T}\mathbf{T}^H$ and $\mathbf{T}^H\mathbf{T}$. To see this, given an SVD of the matrix $\mathbf{T}$ we can immediately verify that

$$\mathbf{T}\mathbf{T}^H = \mathbf{U}(\mathbf{\Sigma}\mathbf{\Sigma}^T)\mathbf{U}^H, \quad \mathbf{T}^H\mathbf{T} = \mathbf{V}(\mathbf{\Sigma}^T\mathbf{\Sigma})\mathbf{V}^H. \tag{2.119}$$

Since $\mathbf{\Sigma}\mathbf{\Sigma}^T$ and $\mathbf{\Sigma}^T\mathbf{\Sigma}$ are diagonal matrices, and $\mathbf{U}$ and $\mathbf{V}$ are unitary, the right-hand sides in both relations above describe the eigenvalue decompositions of the Hermitian matrices on the left-hand sides (see Appendix A). Consequently, the squared singular values are the nonzero eigenvalues of $\mathbf{T}^H\mathbf{T}$ and $\mathbf{T}\mathbf{T}^H$. Furthermore, the columns of $\mathbf{U}$ are eigenvectors of $\mathbf{T}\mathbf{T}^H$ while the columns of $\mathbf{V}$ are eigenvectors of $\mathbf{T}^H\mathbf{T}$.

Given an $n \times m$ matrix $\mathbf{T}$ with SVD $\mathbf{T} = \mathbf{U}\mathbf{\Sigma}\mathbf{V}^H$, the pseudoinverse can be shown to equal $\mathbf{T}^\dagger = \mathbf{V}\mathbf{\Sigma}^\dagger\mathbf{U}^H$, where $\mathbf{\Sigma}^\dagger$ is the $m \times n$ matrix with diagonal elements $1/\sigma_i$ for $\sigma_i > 0$ and 0 otherwise. It is easy to verify that this choice satisfies the pseudoinverse definition. However, it does not offer the insight we obtained by treating general transformations and observing the effect of $\mathbf{T}^\dagger$ on the various subspaces associated with $\mathbf{T}$.

The SVD concept can be extended to a certain class of operators, called *compact operators* (this class is the closure of finite-rank operators). Although every compact operator is bounded, the reverse implication does not hold. The interested reader is referred to [23] for a detailed treatment of compact operators and the SVD.

2.8   Frames

So far we have considered signal expansions onto bases in which the vectors used in the expansion are linearly independent. However, signals can also be expressed in terms of linear combinations of overcomplete vectors that may be linearly dependent. An overcomplete set of vectors in a finite-dimensional space is called a *frame*. In infinite dimensions, we require certain stability bounds along with linear dependence, in order to ensure stable expansions. As shown below, these bounds are similar in spirit to those imposed on Riesz bases.

2.8.1   Definition of frames

Frames, which are generalizations of bases, were introduced in the context of nonharmonic Fourier series by Duffin and Schaeffer [40] (see also [25]). Recently, the theory of frames has been expanded and gained popularity [33, 41, 42, 43], in part owing to the utility of frames in analyzing wavelet expansions. One advantage of using a frame decomposition is that it allows redundancy that may be exploited in the case of corruption or loss of coefficients. Furthermore, since the conditions on frames are not as stringent as those imposed on bases, it allows more flexibility in their design [26, 42, 43, 44, 45, 46, 47]. For example, frame expansions can admit signal representations that are localized in both time and frequency. Frames have played a key role in the development of modern uniform and nonuniform sampling techniques [43, 48]. In this book, our focus is primarily on basis expansions. However, for completeness, in this section we summarize some of the basic definitions and properties of frames. For a more detailed treatment we refer the reader to [26, 33].

In finite dimensions, a set of vectors form a frame if they span the space. In contrast to a basis, the vectors in a frame do not have to be linearly independent. Before providing a formal definition of a frame, which includes the infinite-dimensional setting as well, let us first examine a simple finite-dimensional example.

Example 2.30 Consider the three vectors shown in Fig. 2.11(a): $\mathbf{x}_1 = [1 \; 0]^T$, $\mathbf{x}_2 = [0 \; 1]^T$, and $\mathbf{x}_3 = [-1 \; -1]^T$. These vectors clearly span $\mathbb{R}^2$ and are therefore a frame for $\mathbb{R}^2$. Since we have three vectors in a two-dimensional space, they must be linearly dependent. This means that we can represent every vector $\mathbf{x} \in \mathbb{R}^2$ as a linear combination of only two of them. Note, for example, that $\mathbf{x}_1$ and $\mathbf{x}_2$ form an orthonormal basis so that any $\mathbf{x} \in \mathbb{R}^2$ can be expressed as

$$\mathbf{x} = \langle \mathbf{x}_1, \mathbf{x} \rangle \mathbf{x}_1 + \langle \mathbf{x}_2, \mathbf{x} \rangle \mathbf{x}_2 + \langle \mathbf{0}, \mathbf{x} \rangle \mathbf{x}_3. \tag{2.120}$$

However, this is not the only possible representation of $\mathbf{x}$ using the frame elements. Indeed, we can also write

$$\mathbf{x} = \langle \tilde{\mathbf{x}}_1, \mathbf{x} \rangle \mathbf{x}_1 + \langle \tilde{\mathbf{x}}_2, \mathbf{x} \rangle \mathbf{x}_2 + \langle \tilde{\mathbf{x}}_3, \mathbf{x} \rangle \mathbf{x}_3 \tag{2.121}$$

with $\tilde{\mathbf{x}}_1 = [2/3 \; -1/3]^T$, $\tilde{\mathbf{x}}_2 = [-1/3 \; 2/3]^T$, and $\tilde{\mathbf{x}}_3 = [-1/3 \; -1/3]^T$. The vectors $\tilde{\mathbf{x}}_i$ are referred to as dual vectors and are shown in Fig. 2.11(b). In contrast to (2.120), here all three vectors participate in the representation.

An important observation is that both (2.120) and (2.121) have the form of (2.74). This highlights the fact that, as opposed to bases, in frame expansions the dual vectors are not unique.

In the infinite setting, more restrictions are imposed on the frame vectors in order to ensure stable recovery. In fact, these conditions are the same as those used to define a Riesz basis, without requiring linear independence. Therefore, intuitively, a frame is a complete set of vectors that lead to stable signal expansions, but are not constrained to be linearly independent.

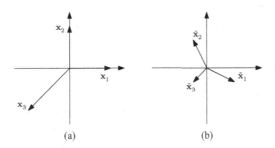

Figure 2.11 (a) A frame in $\mathbb{R}^2$. (b) One choice of a dual for the frame.

Definition 2.19. *Let $\{x_n, n \in \mathcal{I}\}$ denote a set of vectors in a Hilbert space $\mathcal{H}$. The vectors $\{x_n\}$ form a* frame *for $\mathcal{H}$ if there exist constants $\alpha > 0$ and $\beta < \infty$ such that*

$$\alpha\|x\|^2 \le \sum_{n \in \mathcal{I}} |\langle x, x_n \rangle|^2 \le \beta\|x\|^2, \tag{2.122}$$

for all $x \in \mathcal{H}$.

The definition of a frame is equivalent to property (2.55) which we showed holds true for any Riesz basis. Therefore, a Riesz basis is always a frame. In fact, it can be shown that a Riesz basis is a frame with the property that it ceases to be a frame if any of the elements from the set are removed. Although a Riesz basis constitutes a frame, the converse of course is not true. In particular, for frames, (2.50) does not hold in general since the lower bound in that equation necessitates the vectors to be linearly independent. Nonetheless, frame vectors are still guaranteed to be complete. This follows from the lower bound in (2.122). Indeed, suppose that for some $x \in \mathcal{H}$, we have $\langle x, x_n \rangle = 0$ for all n. Then, the lower bound guarantees that $x = 0$ which, by Proposition 2.9, proves completeness.

The frame inequalities in (2.122) imply that if $\|x\|$ is small then so is $\sum_n |\langle x, x_n \rangle|^2$, and vice versa. Therefore, the transformation from x to the coefficients $a = X^*x$ is bounded: $\|a\|^2 \le \beta\|x\|^2$. Similarly, the inequality $\|x\|^2 \le \alpha^{-1}\|a\|^2$ shows that errors in the expansion coefficients a are not amplified in an unbounded manner. A sequence of vectors $\{x_n\}$ that only satisfy the upper bound in (2.122) is called a *Bessel sequence*. Such a set leads to bounded coefficients, but is not necessarily complete.

An important special case is when $\mathcal{H} = \mathbb{C}^m$ and there are a finite number N of frame vectors. To ensure that the vectors $\{x_n\}$ span $\mathcal{H}$ we must have $N \ge m$. For any finite set of vectors, the right-hand inequality of (2.122) is satisfied with $\beta = \sum_{n=1}^N \|x_n\|^2$. This follows from the Cauchy–Schwarz inequality. The left-hand inequality holds when the vectors are complete. Therefore, as we noted at the beginning of this section, any finite set of vectors that spans $\mathcal{H}$ is a frame for $\mathcal{H}$. However, in contrast to basis vectors which are linearly independent, frame vectors with $N > m$ are linearly dependent. The *redundancy* of a finite frame is defined as $r = N/m$, i.e. N vectors in an m-dimensional space. If the bounds $\alpha = \beta$ in (2.122), then we have a *tight frame*. If in addition $\alpha = \beta = 1$, then we have a *normalized tight frame*. When $\|x_n\| = 1$ and the frame is tight, α is the redundancy of the frame. In particular, if $\|x_n\| = 1$ and $\alpha = \beta = 1$, then the frame is an orthonormal basis for $\mathcal{H}$.

2.8.2   Frame expansions

Since a frame is complete, any $x \in \mathcal{H}$ can be written as a linear combination of the frame vectors: $x = \sum_n a[n]x_n$ for some coefficients $a[n]$. Note, however, that in contrast to Riesz basis expansions, here the choice of a is not unique. More specifically, using set transformation notation, for any $x \in \mathcal{H}$ there is some a such that $x = Xa$. But because the vectors $\{x_n\}$ are linearly dependent, the null space of X is not zero (see Proposition 2.10). If b is an arbitrary sequence in $\mathcal{N}(X)$, then $Xb = 0$ and we can also write $x = X(a + \alpha b)$ for any $\alpha \in \mathbb{C}$.

A convenient method for choosing a set of coefficients is via the frame operator. The *frame operator* corresponding to frame vectors $\{x_n\}$ is defined by

$$Sx = XX^*x = \sum_n \langle x_n, x \rangle x_n, \qquad (2.123)$$

for any $x \in \mathcal{H}$. An important property of the frame operator is that it is invertible, as incorporated in the following proposition.

Proposition 2.18.   *Let $\{x_n\}$ be a frame for a Hilbert space $\mathcal{H}$, and let S be the corresponding frame operator, defined by (2.123). Then S is invertible over $\mathcal{H}$.*

Proof: We have seen in the proof of Proposition 2.11 that (2.55) is equivalent to (2.56). Therefore, using the frame operator, we can rewrite (2.122) as

$$\alpha I_{\mathcal{H}} \preceq S \preceq \beta I_{\mathcal{H}}, \qquad (2.124)$$

which shows that S is a Hermitian operator, bounded (from above and below), and consequently invertible [33, p. 58].   □

Using the invertibility of the frame operator, we can now define the *dual frame vectors*, denoted by $\{\tilde{x}_n\}$, as

$$\tilde{x}_n = S^{-1}x_n = (XX^*)^{-1}x_n. \qquad (2.125)$$

Note that we have used the same notation as for the biorthogonal Riesz vectors; this is no coincidence. It can be readily shown that when the vectors $\{x_n\}$ form a Riesz basis, so that X^*X is also invertible, (2.125) coincides with the biorthogonal vectors introduced in Section 2.5.4 (see Exercise 25). However, when the vectors are linearly dependent, X^*X is no longer invertible, while XX^* still is. Therefore, when treating frames, (2.125) is used. Although in this case the biorthogonality property no longer holds, nonetheless the expansion

$$x = \sum_n \langle \tilde{x}_n, x \rangle x_n \qquad (2.126)$$

is still true for any $x \in \mathcal{H}$.

To verify (2.126), we first note that the set transformation corresponding to $\{\tilde{x}_n\}$ is

$$\widetilde{X} = (XX^*)^{-1}X. \qquad (2.127)$$

In terms of $\widetilde{X}$, (2.126) can be written as

$$x = X\widetilde{X}^*x. \qquad (2.128)$$

Substituting (2.127) into (2.128) and noting that

$$X\widetilde{X}^* = XX^*(XX^*)^{-1} = I_{\mathcal{H}} \qquad (2.129)$$

proves (2.126).

From (2.127) it follows that $\widetilde{X}\widetilde{X}^* = (XX^*)^{-1}$. Therefore, the dual frame vectors satisfy

$$\frac{1}{\beta}\|x\|^2 \leq \sum_{i=1}^{n} |\langle x, \tilde{x}_n\rangle|^2 \leq \frac{1}{\alpha}\|x\|^2, \qquad (2.130)$$

for all $x \in \mathcal{H}$, which means that they also constitute a frame for $\mathcal{H}$.

2.8.3 The canonical dual

There are many possibilities to select coefficients $a[n]$ in a frame expansion. The choice $a[n] = \langle \tilde{x}_n, x \rangle$ has some useful properties, rendering this selection popular in many applications. Consequently, the set of dual frame vectors $\{\tilde{x}_n\}$ is often referred to as the *canonical dual frame*.

Properties of the canonical dual

The first property of the canonical dual is that it leads to coefficients with minimal ℓ_2-norm, among all possible choices [33].

Proposition 2.19. *Let $\{x_n\}$ be a set of frame vectors for a Hilbert space $\mathcal{H}$, and let $\{\tilde{x}_n\}$ be the canonical dual frame vectors defined by (2.125). Then, for an arbitrary $x \in \mathcal{H}$, among all coefficients $a[n]$ satisfying $x = \sum_n a[n]x_n$ the choice $a[n] = b[n]$ with $b[n] = \langle \tilde{x}_n, x \rangle$ has minimal ℓ_2-norm.*

Proof: Suppose that $x = \sum_n a[n]x_n = Xa$ for some $a \in \ell_2$. Expressing a as $a = a_1 + a_2$ where $a_1 \in \mathcal{N}(X)^{\perp}$ and $a_2 \in \mathcal{N}(X)$ it follows that $\|a\|^2 = \|a_1\|^2 + \|a_2\|^2 \geq \|a_1\|^2$ with equality if and only if $a_2 = 0$. Therefore, the minimal norm representation must lie in $\mathcal{N}(X)^{\perp}$. Consequently, to prove the proposition we need to show that $b = \widetilde{X}^*x$ is the unique sequence in $\mathcal{N}(X)^{\perp}$ satisfying $x = Xb$.

Uniqueness follows from the fact that we have shown for an arbitrary transformation T that there is a unique z in $\mathcal{N}(T)^{\perp}$ satisfying $y = Tz$ for any y. Therefore, it remains to show that $b \in \mathcal{N}(X)^{\perp}$. By definition, b lies in $\mathcal{R}(\widetilde{X}^*)$. Now, from the properties of the adjoint, $\mathcal{R}(\widetilde{X}^*) = \mathcal{N}(\widetilde{X})^{\perp}$. Since $(XX^*)^{-1}$ is clearly invertible, the null space of $\widetilde{X}$ is equal to that of X which proves the claim. $\qquad \square$

Another useful property of the canonical dual is that in the case of a tight frame it takes on a particularly simple form, leading to an expansion reminiscent of an orthogonal

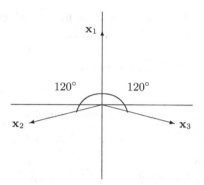

Figure 2.12 Example of a tight frame. The vectors $\mathbf{x}_1, \mathbf{x}_2$, and $\mathbf{x}_3$ span $\mathbb{R}^2$ and therefore form a frame for $\mathbb{R}^2$.

basis expansion. A tight frame is defined by $\alpha = \beta$ in (2.122), or equivalently in (2.124). Therefore, in this case $S = XX^* = \alpha I_{\mathcal{H}}$ and $\tilde{x}_n = (1/\alpha)x_n$ leading to the expansion

$$x = \frac{1}{\alpha} \sum_n \langle x_n, x \rangle x_n. \tag{2.131}$$

With $\alpha = 1$ this formula coincides with the usual reconstruction from an orthonormal basis; however, here the vectors $\{x_n\}$ can be linearly dependent, as we show in the next example.

Example 2.31 A classical example of a tight frame is the frame depicted in Fig. 2.12, in which $\mathbf{x}_1 = [0\ 1]^T$, $\mathbf{x}_2 = [-\sqrt{3}/2\ -1/2]^T$, and $\mathbf{x}_3 = [\sqrt{3}/2\ -1/2]^T$. Since the vectors $\mathbf{x}_1, \mathbf{x}_2, \mathbf{x}_3$ clearly span $\mathbb{R}^2$, they form a frame for $\mathbb{R}^2$. Note that the vectors are linearly dependent and therefore do not constitute a basis for $\mathbb{R}^2$.

We can immediately verify that $\sum_{n=1}^{3} \mathbf{x}_n \mathbf{x}_n^* = \mathbf{XX}^* = (3/2)\mathbf{I}_2$ so that this frame constitutes a tight frame for $\mathbb{R}^2$. In this example, $\|\mathbf{x}_n\| = 1$ for all n rendering $\alpha = 3/2$ the redundancy of the frame: three vectors in a two-dimensional space. The dual frame vectors are given by $\tilde{\mathbf{x}}_n = (2/3)\mathbf{x}_n$ so that any vector $\mathbf{x} \in \mathbb{R}^2$ can be written as

$$\mathbf{x} = \frac{2}{3} \sum_n \langle \mathbf{x}_n, \mathbf{x} \rangle \mathbf{x}_n. \tag{2.132}$$

As we have already seen, the coefficients in (2.132) are not unique. In particular, since $\sum_n \mathbf{x}_n = 0$, adding a constant to the coefficients will still recover $\mathbf{x}$, i.e.

$$\mathbf{x} = \frac{2}{3} \sum_n (\langle \mathbf{x}_n, \mathbf{x} \rangle + a)\mathbf{x}_n, \tag{2.133}$$

for any a. Note however that $\sum_n |\langle \mathbf{x}_n, \mathbf{x} \rangle + a|^2 > \sum_n |\langle \mathbf{x}_n, \mathbf{x} \rangle|^2$ for any $a \neq 0$. Indeed,

$$\sum_n |\langle \mathbf{x}_n, \mathbf{x} \rangle + a|^2 = \|\mathbf{X}^*\mathbf{x} + a\mathbf{e}\|^2, \tag{2.134}$$

where $\mathbf{e} = [1\ 1\ 1]^T$. Now, $\mathbf{e}^*\mathbf{X}^*\mathbf{x} = (\mathbf{X}\mathbf{e})^*\mathbf{x} = 0$ since $\mathbf{X}\mathbf{e} = \sum_n \mathbf{x}_n = 0$.
Therefore,

$$\|\mathbf{X}^*\mathbf{x} + a\mathbf{e}\|^2 = \|\mathbf{X}^*\mathbf{x}\|^2 + a^2\|\mathbf{e}\|^2 \geq \|\mathbf{X}^*\mathbf{x}\|^2 = \sum_n |\langle \mathbf{x}, \mathbf{x}_n\rangle|^2, \tag{2.135}$$

with equality if and only if $a = 0$.

One simple way to create a tight frame with $\alpha = 1$ for a space $\mathcal{H}$ is by considering an orthonormal basis in a larger space $\mathcal{V}$ containing $\mathcal{H}$. The orthogonal projection of this basis onto $\mathcal{H}$ will always lead to a tight frame for $\mathcal{H}$. This result is known in quantum information theory as Naimark's theorem [49]. (It turns out that there is a close relationship between tight frames and quantum measurements; see [31] for more details.)

Computing the canonical dual

In principle, to compute the canonical dual all we need is to invert the frame operator. However, in practice, this computation can be computationally intensive. Instead, a common approach is to approximate the inverse by a power series, as we now show.

Our starting point is the set of inequalities (2.124). If α and β are close to each other, then we can interpret (2.124) as saying that S is "close" to $\frac{\alpha+\beta}{2}I_\mathcal{H}$. This then implies that the inverse, S^{-1}, may be approximated by $\frac{2}{\alpha+\beta}I_\mathcal{H}$. From (2.125), the dual frame vectors are then roughly equal to $\tilde{x}_n \approx \frac{2}{\alpha+\beta}x_n$, which from (2.126) leads to the expansion

$$x \approx \frac{2}{\alpha+\beta}\sum_n \langle x_n, x\rangle x_n. \tag{2.136}$$

To quantify the error in this approximation, note that the right-hand side of (2.136) is equal to $\frac{2}{\alpha+\beta}Sx$. Writing

$$x = \frac{2}{\alpha+\beta}Sx + \left(I_\mathcal{H} - \frac{2}{\alpha+\beta}S\right)x, \tag{2.137}$$

we conclude that

$$x = \frac{2}{\alpha+\beta}\sum_n \langle x_n, x\rangle x_n + Rx, \tag{2.138}$$

with R denoting the error operator:

$$R = I_\mathcal{H} - \frac{2}{\alpha+\beta}S. \tag{2.139}$$

This operator obeys the inequalities

$$-\frac{\beta-\alpha}{\alpha+\beta}I_\mathcal{H} \preceq R \preceq \frac{\beta-\alpha}{\alpha+\beta}I_\mathcal{H}. \tag{2.140}$$

Defining $r = \beta/\alpha - 1$ we can bound R by

$$-\frac{r}{r+2}I_\mathcal{H} \preceq R \preceq \frac{r}{r+2}I_\mathcal{H}, \tag{2.141}$$

so that for any x in $\mathcal{H}$,

$$\|Rx\| \leq \frac{r}{r+2}\|x\|. \tag{2.142}$$

If r is small, meaning $\beta \approx \alpha$, then the norm of the approximation error will also be small, and we can drop the error term in (2.138).

To improve the approximation for large r, we exploit higher orders of R to obtain a power series expansion of S^{-1}. The series follows from writing the frame operator as

$$S = \frac{\alpha + \beta}{2}(I_{\mathcal{H}} - R), \tag{2.143}$$

and relying on the following result [28, Corollary 5.6.16]:

Proposition 2.20. *Suppose that* $\|A\| < 1$ *where* $\| \cdot \|$ *is an arbitrary operator norm (see Appendix A). Then* $(I_{\mathcal{H}} + A)^{-1} = \sum_{k=0}^{\infty}(-A)^k.$

The norm of an operator $T: \mathcal{H} \to \mathcal{S}$ is defined as the smallest value of the constant c such that

$$\|Tx\|_{\mathcal{S}} \leq c\|x\|_{\mathcal{H}} \quad \text{for all } x \in \mathcal{H}. \tag{2.144}$$

Combining Proposition 2.20 with (2.142) and (2.143), we have that

$$S^{-1} = \frac{2}{\alpha + \beta} \sum_{k=0}^{\infty} R^k. \tag{2.145}$$

Retaining only the zero-order term in (2.145) leads to (2.138) with the residual dropped. Keeping the first N terms in (2.145), and using the relation $x = S^{-1}Sx$, leads to the Nth-order approximation

$$x^N = \frac{2}{\alpha + \beta} \sum_{k=0}^{N} R^k Sx. \tag{2.146}$$

The value of x^N can be computed recursively starting with $x^0 = 2Sx/(\alpha+\beta)$ by noting that

$$\begin{aligned}
x^N &= Rx^{N-1} + \frac{2}{\alpha + \beta}Sx \\
&= x^{N-1} + \frac{2}{\alpha + \beta}S(x - x^{N-1}) \\
&= x^{N-1} + \frac{2}{\alpha + \beta}\sum_{n}\langle x_n, x - x^{N-1}\rangle x_n, \tag{2.147}
\end{aligned}$$

where the second equality is a result of substituting $R = I_{\mathcal{H}} - \frac{2}{\alpha+\beta}S$. Equation (2.147) allows us to iteratively approximate x without having to perform an inverse operation. These iterations are proven to converge exponentially to the true value of x [33, p. 62].

An alternative method to invert the frame operator is by using the conjugate gradient algorithm; see [50] for a detailed implementation.

To conclude this section we comment that historically, frames have often been associated with wavelet decompositions. Although defined years before, their popularity was largely spurred by the work of Daubechies and co-authors [41, 42] in the context of wavelets. However, as we have seen throughout this section, frames are more general than wavelet expansions. They do not need to obey any particular structure, as wavelets do, but rather can be used to describe stable redundant expansions in general Hilbert spaces.

2.9   Exercises

1. Consider the space V of antisymmetric functions in $L_2([-1, 1])$, namely finite-energy signals satisfying $v(t) = -v(-t)$ for every $|t| \leq 1$. For each of the following choices of the space W on $L_2([-1, 1])$, determine whether $W \cap V = \{0\}$ and whether $W + V = L_2([-1, 1])$.
   a. W is the set of symmetric functions in $L_2([-1, 1])$, namely finite-energy signals satisfying $w(t) = w(-t)$ for every $|t| \leq 1$.
   b. W is the set of polynomials whose degree does not exceed d, for some given finite integer $d > 0$.
   c. W is the set of polynomials of the form $a_0 + a_2 t^2 + a_4 t^4 + \cdots + a_d t^d$, for some given even integer $d > 0$.
   d. W is the set of finite-energy functions that vanish for every $t < 0$.
   e. W is the set of finite-energy functions that vanish for every $|t| < 1/2$.
   f. W is the set of piecewise constant signals with exactly d discontinuities, for some given finite integer $d > 0$.
2. For each of the choices of W in Exercise 1 for which $W \oplus V = L_2([-1, 1])$, determine the decomposition of $x(t) = e^{-t}$ into a component $x_W \in W$ and a component $x_V \in V$ such that $x(t) = x_W(t) + x_V(t)$.
3. Prove that for any valid inner product over a vector space V:

$$\langle ax + bz, y \rangle = \bar{a}\langle x, z \rangle + \bar{b}\langle z, y \rangle, \tag{2.148}$$

for any $x, z, y \in V$ and $a, b \in \mathbb{C}$.
4. a. Show that the inner product defined in Example 2.10 is a valid inner product.
   b. Show that the inner product defined in Example 2.12 is a valid inner product.
5. Given an arbitrary set $\{v_n\}_{n=1}^N$ of linearly independent vectors which span a space V, the Gram–Schmidt process is a method for constructing an orthonormal set of vectors $\{w_n\}_{n=1}^N$ spanning V. In this method, an auxiliary set $\{u_n\}_{n=1}^N$ is first constructed as follows. We begin with $u_1 = v_1$. For $n \geq 2$, let

$$u_n = v_n - \sum_{k=1}^{n-1} \frac{\langle u_k, v_n \rangle}{\|u_k\|^2} u_k. \tag{2.149}$$

The vectors $\{w_n\}_{n=1}^N$ are then computed as

$$w_n = \frac{u_n}{\|u_n\|}. \tag{2.150}$$

a. Apply the Gram–Schmidt method to the set

$$\mathbf{v}_1 = [1\ 0\ 0]^T, \quad \mathbf{v}_2 = [1\ 1\ 1]^T, \quad \mathbf{v}_3 = [0\ 1\ 0]^T. \tag{2.151}$$

b. Show by induction that the set $\{u_n\}_{n=1}^N$ is orthogonal. Specifically, show that if $\{u_n\}_{n=1}^\ell$ are orthogonal, then the ℓth step of the Gram–Schmidt procedure results in a vector $u_{\ell+1}$ that is orthogonal to $\{u_n\}_{n=1}^\ell$.

c. Show that the set $\{w_n\}_{n=1}^N$ is orthonormal.

6. Let $x(t)$ be an arbitrary function in $L_2(\mathbb{R})$ and define the sequence of functions $x_n(t) = x(t + nT)$ for $n \in \mathbb{Z}$.

a. Show that $\langle x(t + mT), x(t + nT) \rangle = \langle x(t), x(t + (n - m)T) \rangle$ for all $m \in \mathbb{Z}$, where the inner product is the standard inner product on L_2.

b. Use the previous result to show that the functions $\{x_n(t)\}$ form an orthonormal set if and only if $\langle x(t), x(t + nT) \rangle = \delta_{0n}$.

7. Using Definition 2.7, show that for a linear transformation $T : \mathcal{H} \to \mathcal{S}$, the null space $\mathcal{N}(T)$ is a linear subspace of $\mathcal{H}$ and the range space $\mathcal{R}(T)$ is a linear subspace of $\mathcal{S}$.

8. Determine whether each of the following transforms on L_2 is bounded or not and whether its range is closed in L_2 or not. For the unbounded transforms, construct a sequence $\{x_n(t)\}_{n=1}^\infty$ of unit-norm functions for which $\|Tx_n\|^2$ is unbounded from above. For the open-range transforms, construct a sequence $\{x_n(t)\}_{n=1}^\infty$ of unit-norm functions for which $\|Tx_n\|^2$ is unbounded from below.

a. $T\{x\}(t) = x(\alpha t)$ for some $\alpha \neq 0$.

b. $T\{x\}(t) = x(t^3)$.

c. $T\{x\}(t) = x(t^{1/3})$.

d. $T\{x\}(t) = x'(t)$.

e. $T\{x\}(t) = \int_0^t x(\tau)d\tau$.

9. Compute the adjoint of the transform $T : L_2 \to L_2$ defined by $T\{x\}(t) = x(\alpha t)$, where $\alpha \neq 0$ is a given scalar, with respect to the standard inner product on L_2.

10. Define the inner product $\langle \mathbf{x}, \mathbf{y} \rangle = \mathbf{x}^T \mathbf{W} \mathbf{y}$ for all $\mathbf{x}, \mathbf{y} \in \mathbb{R}^m$, where $\mathbf{W}$ is a symmetric positive definite matrix. Let $\mathbf{A}$ be any $m \times m$ matrix. Determine the adjoint $\mathbf{A}^*$ (see Definition 2.11) with respect to the given inner product.

11. Construct an inner product on $\mathbb{R}^2$ under which the vectors $\mathbf{x}_1 = [1\ 0]^T$ and $\mathbf{x}_2 = [1\ 1]^T$ are orthogonal.

12. Let A be a Hermitian operator (see Definition 2.12) on a complex Hilbert space $\mathcal{H}$. Prove that

a. $\langle x, Ax \rangle$ is real for any $x \in \mathcal{H}$;

b. If $\langle x, Ax \rangle = 0$ for all $x \in \mathcal{H}$, then $A = 0$.

13. Let T be a continuous linear transformation. Prove that

a. $\mathcal{N}(T) = \mathcal{R}(T^*)^\perp$.

b. $\mathcal{N}(T)^\perp = \mathcal{R}(T^*)^c$.

c. $\mathcal{N}(T^*) = \mathcal{R}(T)^\perp$.

d. $\mathcal{N}(T^*)^\perp = \mathcal{R}(T)^c$.

14. Let T be a continuous linear transformation. Prove that

    a. $\mathcal{N}(T^*T) = \mathcal{N}(T)$.

    b. $\mathcal{R}(T^*T) = \mathcal{R}(T^*)$.

15. Consider the functions $\{x_n(t)\}_{n\in\mathbb{Z}}$ defined over $[-\pi, \pi]$ as follows:

$$x_n(t) = \begin{cases} t, & n = 0 \\ e^{jnt}, & n \neq 0. \end{cases}$$

Let $X : \ell_2 \rightarrow L_2([-\pi, \pi])$ be the set transformation (see Definition 2.14) corresponding to $\{x_n(t)\}_{n\in\mathbb{Z}}$.

    a. Determine the range space $\mathcal{R}(X)$.

    b. Determine the null space $\mathcal{N}(X)$.

16. Let $\mathbf{x}_1 = [1 \ 0]^T$ and $\mathbf{x}_2 = [\cos(\theta) \ \sin(\theta)]^T$ be two vectors in $\mathbb{R}^2$, where $\theta \in [0, 2\pi)$ is arbitrary.

    a. Determine the values of θ for which $\mathbf{x}_1$ and $\mathbf{x}_2$ form a basis for $\mathbb{R}^2$.

    b. As we have seen, any finite-dimensional basis is a Riesz basis (see Definition 2.16). Compute the bounds (2.50) of the Riesz basis $\{\mathbf{x}_1, \mathbf{x}_2\}$ as a function of θ.

17. Let $\mathcal{A}$ be the space of piecewise constant signals $\mathcal{A} = \{x \in L_2: x(t) = c_n$ for every $t \in [n - 1/2, n + 1/2)\}$, where c_n are arbitrary constants.

    a. Find an orthonormal basis for $\mathcal{A}$.

    b. What is the best approximation of a signal $x(t)$ by a signal in $\mathcal{A}$ in the least squares sense? Provide an intuitive explanation for the result.

18. Consider the set $\mathcal{A}$ of signals in L_2 that vanish for every $|t| > \alpha$, where $\alpha \neq 0$ is some given scalar. Let $T : \mathcal{A} \rightarrow \mathcal{A}$ be defined as follows. For every $x \in \mathcal{A}$, the function $y = Tx$ is obtained as $y(t) = w(t+\beta)$ for some given scalar $\beta \neq 0$, where

$$w(t) = g(x(t - \alpha)) \qquad (2.152)$$

and

$$g(t) = \begin{cases} 1, & |t| \leq \beta \\ 0, & |t| > \beta. \end{cases} \qquad (2.153)$$

    a. Provide a necessary and sufficient condition on α and β such that T is invertible.

    b. For values of α and β for which T is not invertible, determine the pseudoinverse $T^\dagger$ of T.

19. As discussed in Section 2.7.1, given an $n \times m$ matrix $\mathbf{T}$ with SVD $\mathbf{T} = \mathbf{U\Sigma V}^H$, the pseudoinverse can be determined as $\mathbf{T}^\dagger = \mathbf{V\Sigma}^\dagger \mathbf{U}^H$ where $\mathbf{\Sigma}^\dagger$ is the $m \times n$ matrix with diagonal elements $1/\Sigma_{ii}$ wherever $\Sigma_{ii} > 0$ and 0 otherwise. Show that this definition satisfies the conditions of Definition 2.18.

20. Let $T : \mathcal{H} \rightarrow \mathcal{S}$ be a continuous linear transformation with closed range. Prove that the pseudoinverse $T^\dagger$ (see Definition 2.18) of T satisfies the Moore–Penrose conditions (2.116).

21. Let T be a bounded linear transformation with closed range.

    a. Assume that T^*T is invertible. Show that the pseudoinverse of T is given by $T^\dagger = (T^*T)^{-1}T^*$.

b. Assume that TT^* is invertible. Show that the pseudoinverse of T is given by
$T^\dagger = T^*(TT^*)^{-1}$.

22. Let $\mathbf{x}_1 = [1\ 0]^T$, $\mathbf{x}_2 = [0\ 1]^T$, and $\mathbf{x}_3 = [\cos(\theta)\ \sin(\theta)]^T$ be three vectors in $\mathbb{R}^2$, where $\theta \in [0, 2\pi)$ is arbitrary. As we have seen, any finite number of finite-dimensional vectors forms a frame (see Definition 2.19). Compute the frame bounds of $\{\mathbf{x}_1, \mathbf{x}_2, \mathbf{x}_3\}$.

23. In this exercise we show that frames for finite-dimensional spaces can have infinitely many vectors. Let $\mathbf{z}_k, 1 \le k \le N$ be an orthonormal basis for $\mathbb{R}^N$ and define $\mathbf{x}_{k,\ell} = (1/\ell)\mathbf{z}_k, 1 \le k \le N, \ell \ge 1$.
a. Prove that the vectors $\{\mathbf{x}_{k,\ell}\}$ form a frame for $\mathbb{R}^N$.
   Hint: Recall that $\sum_{\ell=1}^{\infty}(1/\ell^2) = \pi^2/6$.
b. Find the canonical dual vectors of $\{\mathbf{x}_{k,\ell}\}$.

24. Let $\mathbf{x}_k, 1 \le k \le N$ be an orthonormal basis for $\mathbb{R}^N$. For each of the sequences of vectors defined below, determine whether or not they constitute a Riesz basis or frame and compute the minimal possible Riesz/frame bounds:
a. $\{\mathbf{x}_1, \mathbf{x}_1, \mathbf{x}_2, \mathbf{x}_2, \mathbf{x}_3, \mathbf{x}_3, \ldots\}$.
b. $\{\mathbf{x}_1, \mathbf{x}_2/2, \mathbf{x}_3/3, \ldots\}$.
c. $\{\mathbf{x}_1, \mathbf{x}_2/\sqrt{2}, \mathbf{x}_2/\sqrt{2}, \mathbf{x}_3/\sqrt{3}, \mathbf{x}_3/\sqrt{3}, \mathbf{x}_3/\sqrt{3}, \ldots\}$.
d. $\{2\mathbf{x}_1, \mathbf{x}_2, \mathbf{x}_3, \mathbf{x}_4, \ldots\}$.

25. Let $\{x_n\}$ be a Riesz basis for a Hilbert space $\mathcal{H}$, and denote by X the corresponding set transformation. As we have seen, the set transformation corresponding to the biorthogonal Riesz basis is given by $\tilde{X} = X(X^*X)^{-1}$. Furthermore, every Riesz basis is also a frame. Therefore, the biorthogonal Riesz basis coincides with the canonical dual frame, which is given by $\tilde{X} = (XX^*)^{-1}X$. Show that the two expressions are equivalent.
Hint: Use (2.99) along with the fact that $\mathcal{R}(X) = \mathcal{H}$.

Chapter 3

Fourier analysis

In the previous chapter we considered expansions of signals in appropriate bases, which can be interpreted as sampling and reconstruction. However, the expansions we discussed were written in abstract form so that it is not immediately obvious how to translate them to concrete sampling methods and practical reconstruction algorithms. From an engineering perspective, our interest is in sampling theories that can be implemented in practice using standard analog components such as filters and modulators. In the next chapter we move away from the abstract formulation and concentrate on concrete signal classes for which the representations we are interested in can be implemented efficiently using common engineering building blocks. The advantage of these settings is that much of the analysis can be carried out efficiently and in a simple manner in the Fourier domain. Therefore, a large part of the tools and methods used in the rest of the book for developing sampling theorems will rely on Fourier analysis.

In this chapter we provide a self-contained presentation of the results needed in the context of linear time-invariant systems and Fourier representations. We review both the continuous-time and discrete-time Fourier representations, and discuss the relationship between the two in the context of sampled signals. There are many excellent textbooks on Fourier analysis; our goal here is to summarize the properties needed in our derivations and the key relations we exploit between continuous-time and discrete-time transforms. For a more thorough treatment of the material in this chapter, we refer the reader to [51, 52, 53, 54].

The first variants of Fourier analysis date back to the mid eighteenth century, in the works of Clairaut, d'Alembert, Euler, Bernoulli, and Lagrange. These were extended in the early nineteenth century by Gauss. However, the real breakthrough in Fourier analysis of arbitrary signals was developed in 1807, when Joseph Fourier showed that any periodic function can be represented as a series of trigonometric functions, a result known today as the Fourier series for periodic (or time-limited) functions. Fourier further demonstrated that representing a function in this way greatly simplifies the study of heat propagation. This is a result of the fact that the exponential functions are eigenfunctions of differentiation, which means that the Fourier representation converts linear differential equations with constant coefficients into algebraic equations, simplifying their analysis considerably. In fact, as we will see in this chapter, Fourier transforms diagonalize any linear time-invariant (LTI) operator, allowing simple frequency domain analysis of filtering and explaining the popularity of Fourier analysis in signal processing.

3.1   Linear time-invariant systems

The Fourier transform is particularly useful in the analysis of LTI systems, which are prevalent in signal processing in general and in sampling theorems in particular. Therefore, we begin our discussion on Fourier transforms with some basic definitions regarding LTI systems.

Strictly speaking, when defining a linear system, we need to specify its domain and range, namely, the input signals on which it operates and the space of outputs. To simplify the discussion, we will typically assume the domain and range are obvious from the context and not define them explicitly. In particular, in examples involving infinite sums and integrals, we assume the input space is such that these operations are well defined.

Throughout the chapter we will be operating in the continuous-time spaces $L_p(\mathbb{R})$ and the discrete-time spaces $\ell_p(\mathbb{Z})$ for $1 \leq p < \infty$. The vector space $L_p(\mathbb{R})$ consists of all functions $x(t)$ for which

$$\|x(t)\|_p^p = \int_{-\infty}^{\infty} |x(t)|^p dt < \infty. \tag{3.1}$$

Similarly, the vector space $\ell_p(\mathbb{Z})$ contains all sequences $x[n]$ such that

$$\|x[n]\|_p^p = \sum_{n \in \mathbb{Z}} |x[n]|^p < \infty. \tag{3.2}$$

For brevity, we will often use the shorthand notation L_p, ℓ_p.

3.1.1   Linearity and time-invariance

A system is said to be *linear* if the relationship between the input $x(t)$ and system output $y(t)$ is linear: if the response to $x_1(t)$ is $y_1(t)$ and the response to $x_2(t)$ is $y_2(t)$, then the input $ax_1(t) + bx_2(t)$ results in an output $ay_1(t) + by_2(t)$ for any scalars a, b. In particular, the output to a zero input is also identically zero.

Time-invariance means that the output of the system does not depend on the input time, up to a simple shift in time. More specifically, if the output due to $x(t)$ is $y(t)$, then the input $x(t - \tau)$ results in an output $y(t - \tau)$.

A system that is both linear and time-invariant is said to be *LTI*. In the following examples we show that some of the popular building blocks of physical and engineering systems are LTI: differential equations, RC circuits, and averaging operators.

Example 3.1 A simple example of an LTI system is a differentiator. Its input–output relationship is given by

$$y(t) = x'(t) = \left(\frac{dx}{dt}\right)(t), \tag{3.3}$$

where we assume implicitly that the input is differentiable. The derivative operator is clearly a linear function, since

$$\frac{d}{dt}(ax_1(t) + bx_2(t)) = a\left(\frac{dx_1}{dt}\right)(t) + b\left(\frac{dx_2}{dt}\right)(t). \tag{3.4}$$

To prove time-invariance, we examine the derivative of $\tilde{x}(t) = x(t - \tau)$. By definition, the output $\tilde{y}(t)$ is

$$\tilde{y}(t) = \left(\frac{d\tilde{x}}{dt}\right)(t) = \left(\frac{dx}{dt}\right)(t - \tau) = y(t - \tau), \tag{3.5}$$

proving that a differentiator is an LTI system.

Example 3.2 Consider a linear differential system in which the relation between the input $x(t)$ and output $y(t)$ is governed by the differential equation

$$y(t) + A_1(t)y^{(1)}(t) + \cdots + A_m(t)y^{(m)}(t) = B_1(t)x(t) + \cdots + B_n(t)x^{(n)}(t), \tag{3.6}$$

for an input $x(t)$ that is sufficiently differentiable. Here m and n are (finite) integers, $A_1(t), \ldots, A_m(t)$ and $B_1(t), \ldots, B_n(t)$ are given functions, and $x^{(n)}(t)$ denotes the nth-order derivative of $x(t)$. We further assume that $x(t)$ and its derivatives are zero for $t < 0$.

To check whether this system is linear, we denote by $y_1(t)$ and $y_2(t)$ the responses of the system to the inputs $x_1(t)$ and $x_2(t)$ respectively, and examine the output $\tilde{y}(t)$ corresponding to the combination $\tilde{x}(t) = ax_1(t) + bx_2(t)$:

$$\tilde{y}(t) + A_1(t)\tilde{y}^{(1)}(t) + \cdots + A_m(t)\tilde{y}^{(m)}(t) = B_1(t)\tilde{x}(t) + \cdots + B_n(t)\tilde{x}^{(n)}(t)$$

$$= a\left(B_1(t)x_1(t) + \cdots + B_n(t)x_1^{(n)}(t)\right) + b\left(B_1(t)x_2(t) + \cdots + B_n(t)x_2^{(n)}(t)\right)$$

$$= a\left(y_1(t) + A_1(t)y_1(t) + \cdots + A_m(t)y_1^{(m)}(t)\right)$$

$$+ b\left(y_2(t) + A_1(t)y_2(t) + \cdots + A_m(t)y_2^{(m)}(t)\right). \tag{3.7}$$

Equating derivatives of the same degree, it is easy to see that this equation is satisfied only if $\tilde{y}(t) = ay_1(t) + by_2(t)$, proving that the system is linear.

To determine whether the system is time-invariant we consider the output $\tilde{y}(t)$ to a shifted input $\tilde{x}(t) = x(t - \tau)$:

$$\tilde{y}(t) + A_1(t)\tilde{y}^{(1)}(t) + \cdots + A_m(t)\tilde{y}^{(m)}(t) = B_1(t)\tilde{x}(t) + \cdots + B_n(t)\tilde{x}^{(n)}(t)$$

$$= B_1(t)x(t - \tau) + \cdots + B_n(t)x^{(n)}(t - \tau). \tag{3.8}$$

In order for $\tilde{y}(t)$ to equal $y(t - \tau)$, the left-hand side needs to equal

$$y(t - \tau) + A_1(t)y^{(1)}(t - \tau) + \cdots + A_m(t)y^{(m)}(t - \tau). \tag{3.9}$$

A sufficient condition for this to hold is that $A_q(t) = A_q(t - \tau)$, $q = 1, \ldots, m$ and $B_q(t) = B_q(t - \tau)$, $q = 1, \ldots, n$, for every t and τ. This implies that $A_1(t) =$

$a_1, \ldots, A_m(t) = a_m$ and $B_1(t) = b_1, \ldots, B_n(t) = b_n$ are constant functions. In this case, (3.9) becomes

$$y(t - \tau) + A_1(t - \tau)y^{(1)}(t - \tau) + \cdots + A_m(t - \tau)y^{(m)}(t - \tau)$$
$$= B_1(t - \tau)x(t - \tau) + \cdots + B_n(t - \tau)x^{(n)}(t - \tau)$$
$$= B_1(t)x(t - \tau) + \cdots + B_n(t)x^{(n)}(t - \tau). \tag{3.10}$$

We conclude that the system (3.6) is LTI if its coefficients are constant.

Example 3.3 The next system we consider is a series RC circuit, in which an input voltage $x(t)$ is imposed on a resistor of R ohms and a capacitor of C farads, which are serially connected. The output $y(t)$ is the voltage measured on the capacitor, as depicted in Fig. 3.1. We assume that $x(t) = 0$ for $t < 0$.

From the current–voltage relation of the capacitor, the current in this circuit is given by $i(t) = Cy^{(1)}(t)$. Using Ohm's law, the input voltage satisfies $x(t) = y(t) + i(t)R$. These two equations imply that

$$y(t) + RCy^{(1)}(t) = x(t). \tag{3.11}$$

As we have seen in the previous example, this differential relation corresponds to an LTI system. The value $\tau = RC$ is often referred to as the time constant of the circuit.

Example 3.4 Consider a system whose output at time t equals the average of the input signal $x(t)$ over a window of width T around t:

$$y(t) = \frac{1}{T} \int_{t - \frac{T}{2}}^{t + \frac{T}{2}} x(\eta)d\eta. \tag{3.12}$$

Clearly this system is linear owing to the linearity of the integral. To check if it is also time-invariant, we apply it on the signal $\tilde{x}(t) = x(t - \tau)$:

$$\tilde{y}(t) = \frac{1}{T} \int_{t - \frac{T}{2}}^{t + \frac{T}{2}} \tilde{x}(\eta)d\eta = \frac{1}{T} \int_{t - \frac{T}{2}}^{t + \frac{T}{2}} x(\eta - \tau)d\eta \tag{3.13}$$

$$= \frac{1}{T} \int_{t - \tau - \frac{T}{2}}^{t - \tau + \frac{T}{2}} x(\eta)d\eta = y(t - \tau),$$

so that the system is LTI.

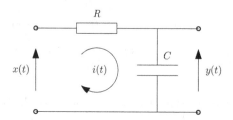

Figure 3.1 A serial RC circuit.

3.1.2   The impulse response

One of the most fundamental results in analysis of LTI systems is that any bounded LTI system can be characterized by a single function:[1] the *impulse response* $h(t)$. This function is the response of the system to a Dirac input. To characterize the impulse response we therefore first define the Dirac delta function.

Dirac functions

The Dirac function, denoted $\delta(t)$, is a generalized function which is defined by its operation under an integral [52]. Specifically,

$$\int_{-\infty}^{\infty} x(t)\delta(t)dt = x(0),\tag{3.14}$$

for any function $x(t)$ belonging to the class of test functions (infinitely differentiable functions with compact support). Since the family of test functions is dense in $L_p(\mathbb{R})$ for $1 \leq p < \infty$, we will use (3.14) for any $x(t) \in L_p(\mathbb{R})$. The derivations below relying on Dirac functions are therefore limited to this class of functions.

The Dirac delta can alternatively be defined by a limit of smooth functions with integral equal to 1. Let $\phi(t)$ be a smooth function with area 1; for example

$$\phi(t) = \frac{1}{2\sqrt{\pi}}e^{-t^2/4}.\tag{3.15}$$

Then,

$$\delta(t) = \lim_{\varepsilon \to 0} \frac{1}{\varepsilon}\phi\left(\frac{t}{\varepsilon}\right).\tag{3.16}$$

Computations using $\delta(t)$ are justified mathematically by the theory of distributions [58]. For our purposes in the book, the definitions here suffice.[2]

From (3.14), we have immediately that

$$\int_{-\infty}^{\infty} \delta(t)dt = 1,\tag{3.17}$$

and

$$\int_{-\infty}^{\infty} x(t)\delta(t - t_0)dt = x(t_0).\tag{3.18}$$

In addition,

$$x(t)\delta(t) = x(0)\delta(t).\tag{3.19}$$

This property will be exploited in the context of sampling in order to represent sampled data in terms of a continuous-time train of delta functions.

---

[1] In an interesting series of papers (see, e.g., [55, 56, 57]), Sandberg showed that not all LTI functions can be written in terms of a convolution; however, loosely speaking, for bounded systems in the sense that bounded inputs result in bounded outputs, such a characterization is possible.

[2] A property that we will use throughout our derivations is that two distributions f_1, f_2 are said to be equal if, for every function $x(t)$ that is infinitely continuously differentiable with compact support, $\int_{-\infty}^{\infty} f_1(t)x(t)dt = \int_{-\infty}^{\infty} f_2(t)x(t)dt$.

Convolution

Equipped with the Dirac delta, we now show that any bounded LTI system is completely characterized by its output $h(t)$ when the input $x(t)$ is equal to $\delta(t)$. The function $h(t)$ is referred to as the *impulse response* of the system. More specifically, the output $y(t)$ to an arbitrary input $x(t)$ is the convolution with $h(t)$: $y(t) = h(t) * x(t)$. The *convolution* between two $L_1(\mathbb{R})$ functions is defined by

$$g(t) = x(t) * y(t) = \int_{-\infty}^{\infty} x(\tau)y(t-\tau)d\tau. \tag{3.20}$$

It is easy to see that the convolution is symmetric, namely, $g(t)$ can also be written as $g(t) = y(t) * x(t)$.

To establish the convolution property of LTI systems note that from (3.18), any function $x(t) \in L_p(\mathbb{R})$ for $1 \le p < \infty$ can be written as

$$x(t) = \int_{-\infty}^{\infty} x(\tau)\delta(t-\tau)d\tau. \tag{3.21}$$

Let $Sx(t)$ denote the operation of the LTI system S. From linearity,

$$Sx(t) = \int_{-\infty}^{\infty} x(\tau)S\delta(t-\tau)d\tau. \tag{3.22}$$

By definition, $S\delta(t) = h(t)$. Using the time-invariance property, $S\delta(t-\tau) = h(t-\tau)$. Therefore, (3.22) becomes

$$Sx(t) = \int_{-\infty}^{\infty} x(\tau)h(t-\tau)d\tau = x(t) * h(t). \tag{3.23}$$

In the examples below, we determine the impulse response of two simple LTI systems which we will encounter frequently throughout the book.

Example 3.5 Consider a delay system, whose output $y(t)$ to an input $x(t)$ is defined by $y(t) = x(t-\eta)$ for some delay η. From the definition of the Dirac delta function and (3.21), $x(t-\eta)$ can be written as

$$x(t-\eta) = \int_{-\infty}^{\infty} x(\tau)\delta(t-\eta-\tau)d\tau. \tag{3.24}$$

Comparing with (3.23), this implies that the system's impulse response is given by

$$h(t) = \delta(t-\eta). \tag{3.25}$$

Example 3.6 Next, we revisit the LTI system of Example 3.4, in which the input–output relation is given by (3.12). We can write this relation as

$$y(t) = \frac{1}{T} \int_{t-\frac{T}{2}}^{t+\frac{T}{2}} x(\tau)d\tau = \int_{-\infty}^{\infty} x(\tau)h(t-\tau)d\tau, \tag{3.26}$$

where $h(t)$ is a square window of width T and height $1/T$:

$$h(t) = \begin{cases} \frac{1}{T}, & |t| \le \frac{T}{2} \\ 0, & |t| > \frac{T}{2}. \end{cases} \tag{3.27}$$

Therefore, the impulse response of the averaging system is a square pulse.

3.1.3 Causality and stability

Many of the characteristics of LTI systems follow from the convolution property (3.23). This relationship allows us to interpret features of LTI systems in terms of properties of the impulse response $h(t)$. Below, we demonstrate this in the context of causality and stability.

An LTI system is *causal* if its output at time t depends only on inputs at times $x(\tau), \tau \le t$. From the convolution property it is easy to see that this is equivalent to $h(t) = 0, t < 0$.

The system is *stable* if the output $y(t) = h(t) * x(t)$ is bounded when $x(t)$ is bounded. This is referred to as bounded-input, bounded-output (BIBO) stability. Using the convolution property once more, we note that

$$|y(t)| \le \int_{-\infty}^{\infty} |x(t - \tau)||h(\tau)|d\tau \le \max_{t \in \mathbb{R}} |x(t)| \int_{-\infty}^{\infty} |h(\tau)|d\tau. \tag{3.28}$$

Therefore, $y(t)$ is bounded if $x(t)$ is bounded and $h(t)$ is absolutely integrable, i.e. $h(t) \in L_1(\mathbb{R})$. It can be shown that this condition is also necessary.

To examine the role of the impulse response in ensuring stability, in the next example we consider the case in which $h(t)$ is not in $L_1(\mathbb{R})$. As we show, we can easily construct an input $x(t)$ that will result in an unbounded output.

Example 3.7 Consider the LTI system whose impulse response is given by $h(t) = \text{sinc}(t/T)$ for some $T > 0$, where

$$\text{sinc}(t) = \begin{cases} \frac{\sin(\pi t)}{\pi t}, & t \ne 0 \\ 1, & t = 0, \end{cases} \tag{3.29}$$

and is shown in Fig. 3.2(a). We will encounter this function quite often in the sequel owing to its role in the Shannon–Nyquist theorem (see Theorem 4.1). We refer to such a system as an ideal low-pass filter (LPF), for reasons that will be explained in Example 3.12. (As we will see, it passes only the low-frequency content of the input.) From a BIBO point of view, this system is not stable, since

$$\int_{-\infty}^{\infty} \left| \text{sinc}\left(\frac{t}{T}\right) \right| dt = \infty. \tag{3.30}$$

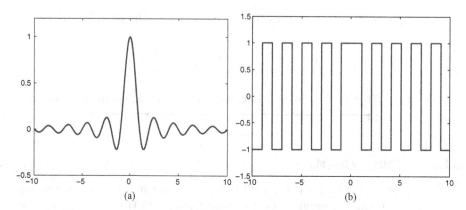

Figure 3.2 (a) The function sinc(t). (b) The function sign(sinc($-t$)).

A simple example of a bounded signal $x(t)$ for which the output of this system explodes is

$$x(t) = \text{sign}\left(\text{sinc}\left(-\frac{t}{T}\right)\right), \tag{3.31}$$

where the signum function is defined by

$$\text{sign}(x) = \begin{cases} 1, & x > 0 \\ 0, & x = 0 \\ -1, & x < 0. \end{cases} \tag{3.32}$$

This input is depicted in Fig. 3.2(b) for $T = 1$. To show that the system is unstable, we compute the system output at $t = 0$:

$$y(0) = \left[\int_{-\infty}^{\infty} \text{sign}\left(\text{sinc}\left(-\frac{\tau}{T}\right)\right) \text{sinc}\left(\frac{t-\tau}{T}\right) d\tau\right]_{t=0}$$

$$= \int_{-\infty}^{\infty} \left|\text{sinc}\left(\frac{\tau}{T}\right)\right| d\tau$$

$$= \infty. \tag{3.33}$$

Evidently, $y(0)$ is infinite, although the magnitude of the input $x(t)$ never exceeds the value 1.

In the previous example we showed that we can construct a specific bounded input for which the output of the LPF is unbounded. This example can be extended to any unstable LTI system whose impulse response $h(t)$ is not absolutely integrable: The output at $t = t_0$ will explode when the system input is the bounded signal $x(t) = \text{sign}(h(t_0 - t))$. Clearly, this is only one particular example; in general, there are many alternative bounded inputs for which the system output is unbounded.

3.1.4   Eigenfunctions of LTI systems

Another important property of LTI systems that follows from the convolution relation is that complex exponentials of the form $e^{j\omega t}$ are their eigenvectors (see Appendix A for a definition of eigenvectors). This is why the Fourier transform, which represents signals by complex exponentials, is so useful in the analysis of LTI operators.

To establish the eigenfunction property note that from (3.23),

$$S e^{j\omega t} = \int_{-\infty}^{\infty} e^{j\omega \tau} h(t - \tau) d\tau = e^{j\omega t} \int_{-\infty}^{\infty} e^{-j\omega \tau} h(\tau) d\tau = e^{j\omega t} H(\omega), \quad (3.34)$$

where $H(\omega) = \int_{-\infty}^{\infty} e^{-j\omega \tau} h(\tau) d\tau$ is a function only of ω and is the eigenvalue associated with $e^{j\omega t}$. In fact, $H(\omega)$ is precisely the continuous-time Fourier transform of $h(t)$ at frequency ω, as we define next.

3.2   The continuous-time Fourier transform

3.2.1   Definition of the CTFT

The *continuous-time Fourier transform (CTFT)* of a signal $x(t)$ in $L_1(\mathbb{R})$ is defined as

$$X(\omega) = \int_{-\infty}^{\infty} x(t) e^{-j\omega t} dt. \quad (3.35)$$

We use the convention that upper-case letters denote Fourier transforms. The inverse Fourier transform is given by

$$x(t) = \frac{1}{2\pi} \int_{-\infty}^{\infty} X(\omega) e^{j\omega t} d\omega. \quad (3.36)$$

The inversion formula (3.36) does not always exactly recover $x(t)$. Conditions under which this formula is exact are discussed in detail in standard Fourier analysis textbooks. One case is when both $x(t)$ and $X(\omega)$ are in $L_1(\mathbb{R})$.

Strictly speaking, the CTFT is defined for absolutely integrable functions $x(t)$, namely functions in $L_1(\mathbb{R})$, in order to avoid convergence issues [52]. For $x(t) \in L_1(\mathbb{R})$, the integral (3.35) always converges and the Fourier transform is bounded:

$$|X(\omega)| \leq \int_{-\infty}^{\infty} |x(t) e^{-j\omega t}| dt = \int_{-\infty}^{\infty} |x(t)| dt < \infty. \quad (3.37)$$

It can also be shown that in this case, $X(\omega)$ is continuous in ω. If $X(\omega)$ is also in $L_1(\mathbb{R})$, then the inverse transform given by (3.36) results in a continuous-time function $x(t)$. However, the formulas (3.35) and (3.36) also hold in an $L_2(\mathbb{R})$ sense when $x(t)$ is in $L_2(\mathbb{R})$. We will therefore freely use these formulas when analyzing $L_2(\mathbb{R})$ signals throughout the text.

The difficulty in analyzing the CTFT of functions in $L_2(\mathbb{R})$ that are not absolutely integrable is that in this case the integrand in (3.35) is also not absolutely integrable. However, since $L_2(\mathbb{R}) \cap L_1(\mathbb{R})$ is dense in $L_2(\mathbb{R})$, any function $x(t)$ in $L_2(\mathbb{R})$ can be approximated arbitrarily closely by functions $x_n(t)$ in $L_2(\mathbb{R}) \cap L_1(\mathbb{R})$. Formally, the CTFT is then defined as the limit point of the CTFTs of $x_n(t)$. Such an extension satisfies all the properties below that are defined for functions in $L_1(\mathbb{R})$.

3.2.2   Properties of the CTFT

The CTFT satisfies several basic and important properties. Below we list some of those properties that will be useful in our developments. For brevity, we denote by $x(t) \leftrightarrow X(\omega)$ a Fourier transform pair.

Linearity
By its definition, the CTFT is linear: $ax_1(t) + bx_2(t) \leftrightarrow aX_1(\omega) + bX_2(\omega)$.

Convolution and multiplication
The convolution property is one of the fundamental results we exploit throughout. If $x(t), y(t)$ are in $L_1(\mathbb{R})$, then their convolution $h(t) = x(t) * y(t)$ is also in $L_1(\mathbb{R})$ and its CTFT is given by

$$H(\omega) = X(\omega)Y(\omega), \tag{3.38}$$

where $X(\omega), Y(\omega)$ are the CTFTs of $x(t), y(t)$ respectively [59, Theorem 2.2]. The dual of the convolution property is the CTFT of multiplication, which is given by the relation

$$x(t)y(t) \leftrightarrow \frac{1}{2\pi}X(\omega) * Y(\omega). \tag{3.39}$$

Symmetry
The CTFT of a real function has conjugate symmetry: $X(\omega) = \overline{X(-\omega)}$.

Shifts, modulation, and scaling
Shifting the signal in time results in modulation in frequency: $x(t - t_0) \leftrightarrow X(\omega)e^{-j\omega t_0}$. The dual property is that modulation in time results in a shift in frequency: $x(t)e^{j\omega_0 t} \leftrightarrow X(\omega - \omega_0)$. Finally, scaling in time results in the reverse scaling in frequency: $x(at) \leftrightarrow (1/|a|)X(\omega/a)$ for a real constant $a \neq 0$. As a special case we have that $x(-t) \leftrightarrow X(-\omega)$.

Cross-correlation
The cross-correlation between two real signals $x(t)$ and $y(t)$ is defined by

$$r_{xy}(t) = \int_{-\infty}^{\infty} x(\tau)y(t + \tau)d\tau. \tag{3.40}$$

Noting that $r_{xy}(t) = x(-t) * y(t)$, and using the scaling and convolution properties, it follows that $r_{xy}(t) \leftrightarrow X(-\omega)Y(\omega)$. From the symmetry property for real functions,

$$r_{xy}(t) \leftrightarrow \overline{X(\omega)}Y(\omega). \tag{3.41}$$

Energy conservation (Parseval's theorem)

An important property of the CTFT is the fact that it preserves energy. This relation is captured by Parseval's theorem

$$\int_{-\infty}^{\infty} \overline{x(t)}y(t)dt = \frac{1}{2\pi} \int_{-\infty}^{\infty} \overline{X(\omega)}Y(\omega)d\omega, \qquad (3.42)$$

which holds for any $x(t), y(t)$ in $L_2(\mathbb{R})$. When $x(t) = y(t)$ we have the energy conservation principle:

$$\int_{-\infty}^{\infty} |x(t)|^2 dt = \frac{1}{2\pi} \int_{-\infty}^{\infty} |X(\omega)|^2 d\omega. \qquad (3.43)$$

Duality

The CTFT and inverse CTFT behave similarly. Specifically, if $y(\omega)$ is the CTFT of the function $x(t)$, then $2\pi x(-\omega)$ is the CTFT of the function $y(t)$.

Regularity

The decay rate of a function's CTFT tells us about its regularity. It can be shown that a function $x(t)$ is bounded and p times continuously differentiable with bounded derivatives if

$$\int_{-\infty}^{\infty} |X(\omega)|(1 + |\omega|^p)d\omega < \infty. \qquad (3.44)$$

Localization

A final important property of the CTFT is that it is impossible to construct a function $x(t) \neq 0$ that has compact support and whose CTFT $X(\omega)$ also has compact support. More specifically, if $x(t)$ has compact support then $X(\omega)$ cannot be zero on an interval. Conversely, if $X(\omega)$ has compact support, then $x(t)$ cannot be zero on an interval [59, Theorem 2.6]. This property is related to the uncertainty principle, which we will discuss in Section 4.4.1.

3.2.3 Examples of the CTFT

We now provide examples of several CTFT pairs that will be used throughout the book.

Example 3.8 The CTFT of $x(t) = \delta(t)$ is given by

$$X(\omega) = \int_{-\infty}^{\infty} \delta(t)e^{-j\omega t}dt = e^{-j\omega \cdot 0} = 1. \qquad (3.45)$$

This formula can also be justified mathematically by Fourier transforms for tempered distributions [58].

Applying the shifting property, the CTFT of $x(t) = \delta(t - t_0)$ is equal to $X(\omega) = e^{-j\omega t_0}$.

Example 3.9 From the duality property and the result of Example 3.8, the CTFT of the constant function $x(t) = 1$ is $X(\omega) = 2\pi\delta(\omega)$. Therefore, a constant function, which is often referred to as a direct-current (DC) function, has energy only at $\omega = 0$.

Using the modulation property we can extend this result to a pure exponent of the form $e^{j\omega_0 t}$. Such a signal has a transform $e^{j\omega_0 t} \leftrightarrow 2\pi\delta(\omega - \omega_0)$, so that it contains energy only at $\omega = \omega_0$.

Example 3.10 A signal which is very important in the context of sampling is the train of Dirac delta functions

$$x(t) = \sum_{n \in \mathbb{Z}} \delta(t - nT), \tag{3.46}$$

also called an *impulse train*. From Example 3.8, the CTFT of $x(t)$ is given by

$$X(\omega) = \sum_{n \in \mathbb{Z}} e^{-j\omega nT}. \tag{3.47}$$

Using the Poisson-sum formula, introduced below in (3.89), we can further show that

$$\sum_{n \in \mathbb{Z}} e^{-j\omega nT} = \frac{2\pi}{T} \sum_{k \in \mathbb{Z}} \delta\left(\omega - \frac{2\pi k}{T}\right). \tag{3.48}$$

This equality holds in the sense of distributions. Therefore, the CTFT of an impulse train in the time domain is an impulse train in the frequency domain. The spacing between adjacent impulses in time and in frequency is inversely proportional, as shown in Fig. 3.3.

Example 3.11 Consider the box signal (3.27) of Example 3.6 with $T = 1$:

$$x(t) = \begin{cases} 1, & |t| \leq \frac{1}{2} \\ 0, & |t| > \frac{1}{2}. \end{cases} \tag{3.49}$$

From the definition of the CTFT, its Fourier transform $X(\omega)$ is given by

$$X(\omega) = \int_{-\frac{1}{2}}^{\frac{1}{2}} e^{-j\omega t} dt = \frac{-1}{j\omega}\left(e^{-j\omega/2} - e^{j\omega/2}\right) = \frac{\sin\left(\frac{\omega}{2}\right)}{\frac{\omega}{2}} = \text{sinc}\left(\frac{\omega}{2\pi}\right), \tag{3.50}$$

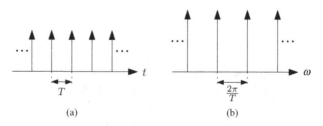

(a)   (b)

Figure 3.3 (a) Impulse train in the time domain. (b) Its CTFT.

where the sinc function was defined in (3.29). Note that since $x(t)$ is discontinuous, its Fourier transform is not integrable. (Otherwise, both $x(t)$ and $X(\omega)$ would be in $L_1(\mathbb{R})$, which implies that $x(t)$ must be continuous.)

Example 3.12 As a final example, we revisit the sinc impulse response of Example 3.7. From the duality property and the result of the previous example, the CTFT of the function $x(t) = \mathrm{sinc}(t)$ is given by

$$X(\omega) = \begin{cases} 1, & |\omega| \le \pi \\ 0, & |\omega| > \pi. \end{cases} \tag{3.51}$$

Using the scaling property we can conclude more generally that the transform of $\mathrm{sinc}(t/T)$ is

$$X(\omega) = \begin{cases} T, & |\omega| \le \frac{\pi}{T} \\ 0, & |\omega| > \frac{\pi}{T}. \end{cases} \tag{3.52}$$

The functions $x(t)$ and $X(\omega)$ are depicted in Fig. 3.4, where it can be seen that the width of the main lobe of $x(t)$ is inversely proportional to the width of $X(\omega)$.

Suppose we are given an LTI system with impulse response $h(t) = \mathrm{sinc}(t/T)$. From the convolution property, the output of the system with $g(t)$ as its input can be written in the Fourier domain as $Y(\omega) = G(\omega)H(\omega)$ where $H(\omega)$ is given by (3.52). Clearly $Y(\omega)$ will contain no frequency components above π/T. Consequently, the LTI system whose impulse response is $\mathrm{sinc}(t/T)$ is referred to as an *ideal LPF*: it only passes the low-frequency components of its input.

3.2.4 Fubini's theorem

When manipulating double integrals (for example when proving the convolution theorem) it is useful to be able to change the order of integration. When the signals are in $L_1(\mathbb{R})$ this is possible owing to Fubini's theorem [60].

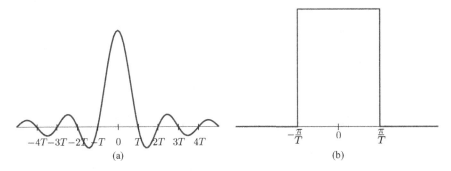

Figure 3.4 (a) The function $x(t) = \mathrm{sinc}(t/T)$. (b) The CTFT $X(\omega)$ of $x(t)$.

Theorem 3.1 (Fubini's theorem). *If either*

$$\int_{-\infty}^{\infty} \left(\int_{-\infty}^{\infty} |x(t,\tau)| dt \right) d\tau < \infty \quad or \quad \int_{-\infty}^{\infty} \left(\int_{-\infty}^{\infty} |x(t,\tau)| d\tau \right) dt < \infty, \quad (3.53)$$

then

$$\int_{-\infty}^{\infty} \left(\int_{-\infty}^{\infty} x(t,\tau) dt \right) d\tau = \int_{-\infty}^{\infty} \left(\int_{-\infty}^{\infty} x(t,\tau) d\tau \right) dt < \infty. \quad (3.54)$$

3.3   Discrete-time systems

Essentially all the results we developed so far for continuous-time systems have a discrete counterpart. Discrete-time systems arise when sampling, or digitizing, continuous-time signals. Very often the sampling is uniform so that the discrete-time signal $x[n]$ is a result of uniformly sampling an analog signal $x(t)$ at times $t = nT$.

Just as in the continuous-time case, many practical discrete-time systems are linear and time-invariant. In the discrete setting, time-invariance is defined with respect to a normalized sampling interval of $T = 1$. Namely, a linear discrete-time operator is *time-invariant* if an input $x[n]$ delayed by an integer m produces a delayed response:

$$S\{x_m\}[n] = S\{x\}[n - m], \quad (3.55)$$

where Sx is the output when $x[n]$ is the input and $x_m[n] = x[n - m]$.

3.3.1   Discrete-time impulse response

Discrete-time LTI systems are defined by their *(discrete-time) impulse response*. In the discrete setting the impulse is the *discrete Dirac*

$$\delta[n] = \begin{cases} 1, & n = 0 \\ 0, & n \neq 0. \end{cases} \quad (3.56)$$

The discrete Dirac is equivalent to the Kronecker delta δ_{n0} defined in (2.18). More generally, $\delta_{nk} = \delta[n - k]$.

The impulse response $h[n]$ is defined by the output of the system when the input is $\delta[n]$. Noting that any signal $x[n] \in \ell_p, 1 \leq p < \infty$ can be written as a sum of shifted Diracs,

$$x[n] = \sum_{m \in \mathbb{Z}} x[m] \delta[n - m], \quad (3.57)$$

we can use the LTI property to express the output of the system as

$$Sx[n] = \sum_{m \in \mathbb{Z}} x[m] S\delta[n - m] = \sum_{m \in \mathbb{Z}} x[m] h[n - m]. \quad (3.58)$$

The relationship given by (3.58) is the *discrete-time convolution* between $x[n]$ and $h[n]$. The impulse response $h[n]$ is also referred to as a *filter*. If the support of $h[n]$ (i.e. the indices of n over which $h[n]$ is not zero) is finite, then the convolution in (3.58) can be computed by a finite number of operations. Such filters are referred to as *finite-impulse response (FIR)*. When $h[n]$ has infinitely many coefficients, it is referred to as an *infinite-impulse response (IIR)* filter.

When a convolution with an IIR filter is written as a sum over past values of $x[n]$ there are infinitely many operations to evaluate. However, it turns out that a large class of IIR filters can be implemented using only a small number of operations per output sample. This is a result of replacing the sum over infinitely many input samples in (3.58) by a recursive equation which involves past values of the output $y[n]$. Systems that can be realized in this form have impulse response functions with rational frequency responses consisting of a small number of zeros and poles. We will discuss such filters in Example 3.15, after introducing the discrete-time Fourier transform. An important observation in this context is that it is sometimes more efficient to implement an IIR filter, rather than an FIR filter with many nonzero values.

As in the continuous-time setting, causality and stability can be characterized by the impulse response $h[n]$. A necessary and sufficient condition for causality is that $h[n] = 0, n < 0$. This guarantees that the output of the system depends only on present and past values of the input $x[n]$. BIBO stability is guaranteed if and only if $h[n]$ is in ℓ_1, i.e. $\sum_{n \in \mathbb{Z}} |h[n]| < \infty$.

3.3.2　Discrete-time Fourier transform

The *discrete-time Fourier transform (DTFT)* of a sequence $x[n]$ in ℓ_1 is defined by

$$X(e^{j\omega}) = \sum_{n \in \mathbb{Z}} x[n] e^{-j\omega n}. \tag{3.59}$$

The DTFT is 2π-periodic; to emphasize this fact we use the notation $X(e^{j\omega})$. The inverse DTFT is given by the equation

$$x[n] = \frac{1}{2\pi} \int_{-\pi}^{\pi} X(e^{j\omega}) e^{j\omega n} d\omega. \tag{3.60}$$

A sufficient condition to guarantee convergence in (3.59) is that $x[n]$ is in ℓ_1, namely, it is absolutely summable. In this case (3.59) converges uniformly to a continuous function. When $x[n]$ is in ℓ_2, (3.59) converges in an L_2 sense. Just as in the CTFT, using distributions, we can define DTFTs for more general sequences that are not in ℓ_2, such as the pair

$$e^{j\omega_0 n} \leftrightarrow 2\pi \sum_{k \in \mathbb{Z}} \delta(\omega - \omega_0 + 2\pi k). \tag{3.61}$$

This relation can be verified by substituting the transform into (3.60), and using the integral representation of the Dirac delta.

Any function that is square integrable over $[-\pi, \pi]$ can be expressed as a DTFT of some sequence $a[n]$ in ℓ_2. This is a result of the fact that the functions $\{e^{j\omega n}\}$ form an orthonormal basis for square-integrable functions over $[-\pi, \pi]$ [59, Theorem 3.2].

We have seen that in continuous time, the continuous complex exponentials are eigenvectors of any LTI system. A similar property holds in the discrete-time setting, where the discrete exponentials $e^{j\omega n}$ are now the eigenvectors. To see this, note that for any LTI system S with impulse response $h[n]$ it follows from (3.58) that

$$Se^{j\omega n} = \sum_{m\in\mathbb{Z}} h[m]e^{j\omega(n-m)} = e^{j\omega n}\sum_{m\in\mathbb{Z}} h[m]e^{-j\omega m} = e^{j\omega n}H(e^{j\omega}), \qquad (3.62)$$

where $H(e^{j\omega})$ is the DTFT of $h[n]$. Therefore, $H(e^{j\omega})$ is the eigenvalue associated with the eigenvector $e^{j\omega n}$, and is referred to as the system's *frequency response*.

3.3.3   Properties of the DTFT

The DTFT satisfies several properties that are reminiscent of those satisfied by the CTFT. In fact, we can view the DTFT as the CTFT of a continuous-time representation $x(t) = \sum_{n\in\mathbb{Z}} x[n]\delta(t-n)$ of the sequence $x[n]$. We will use such continuous-time formulations of discrete-time signals in the next chapter, when discussing sampling theorems. To see the equivalence, we formally compute the CTFT of $x(t)$:

$$\int_{-\infty}^{\infty} x(t)e^{-j\omega t}dt = \sum_{n\in\mathbb{Z}} x[n]\int_{-\infty}^{\infty}\delta(t-n)e^{-j\omega t}dt = \sum_{n\in\mathbb{Z}} x[n]e^{-j\omega n} = X(e^{j\omega}).$$

$$(3.63)$$

Similarly to the CTFT, the DTFT is linear and conjugate symmetric. Time shifts correspond to modulation, and modulation in frequency to shifts in time. The time reversal property implies that $x[-n] \leftrightarrow X(e^{-j\omega})$. Finally, the convolution property and Parseval's theorem also hold for the DTFT. Because of their importance, we write them explicitly:

Convolution
The convolution between two ℓ_1-sequences is defined by

$$y[n] = x[n] * h[n] = \sum_{\ell\in\mathbb{Z}} x[\ell]h[n - \ell], \qquad (3.64)$$

where $y[n]$ is also in ℓ_1. The DTFT of $y[n]$ is related to that of $x[n]$ and $h[n]$ via

$$Y(e^{j\omega}) = X(e^{j\omega})H(e^{j\omega}). \qquad (3.65)$$

Energy conservation (Parseval's theorem)

$$\sum_{n\in\mathbb{Z}} \overline{x[n]}y[n] = \frac{1}{2\pi}\int_{-\pi}^{\pi} \overline{X(e^{j\omega})}Y(e^{j\omega})d\omega. \qquad (3.66)$$

When $x[n] = y[n]$ we have the energy conservation property:

$$\sum_{n \in \mathbb{Z}} |x[n]|^2 = \frac{1}{2\pi} \int_{-\pi}^{\pi} |X(e^{j\omega})|^2 d\omega. \tag{3.67}$$

Discrete correlation sequence

An important sequence we will encounter repeatedly in analyzing sampling theorems is the correlation sequence of real signals:

$$c[n] = \sum_{m \in \mathbb{Z}} d[m]d[n+m]. \tag{3.68}$$

From the definition of convolution $c[n]$ can be represented as $c[n] = d[n] * d[-n]$. Therefore, if $d[n]$ is real, then from the properties of the convolution and time reversal of the DTFT we can express the relationship between $C(e^{j\omega})$ and $D(e^{j\omega})$ in the DTFT domain as

$$C(e^{j\omega}) = |D(e^{j\omega})|^2. \tag{3.69}$$

We now give a few examples of DTFT pairs and the convolution theorem.

Example 3.13 Consider the sequence

$$h[n] = \begin{cases} a^n, & n \geq 0 \\ 0, & n < 0, \end{cases} \tag{3.70}$$

where $|a| < 1$. From the definition of the DTFT, $H(e^{j\omega})$ is given by

$$H(e^{j\omega}) = \sum_{n=0}^{\infty} a^n e^{-j\omega n} = \sum_{n=0}^{\infty} \left(ae^{-j\omega}\right)^n = \frac{1}{1 - ae^{-j\omega}}. \tag{3.71}$$

This example is important in the context of causal IIR filtering. Suppose that $h[n]$ represents an LTI system with frequency response given by (3.71). We will now show that such a system can be implemented recursively, with only one addition and multiplication per output sample. This leads to a far more efficient realization of the IIR filter (3.70) than that obtained by directly computing the convolution sum.

Let $x[n], y[n]$ denote the input and output of the system with impulse response $h[n]$, and let $Y(e^{j\omega}), X(e^{j\omega})$ be the corresponding DTFTs. To derive a recursive formula for computing $y[n]$, we note from (3.71) and the convolution property that

$$Y(e^{j\omega}) = H(e^{j\omega})X(e^{j\omega}) = \frac{1}{1 - ae^{-j\omega}}X(e^{j\omega}), \tag{3.72}$$

or

$$Y(e^{j\omega}) = X(e^{j\omega}) + aY(e^{j\omega})e^{-j\omega}. \tag{3.73}$$

Going back to the time domain by applying the inverse DTFT results in the difference equation

$$y[n] = x[n] + ay[n-1]. \tag{3.74}$$

Consequently, to obtain the output $y[n]$ we combine the current input with the output at time $n - 1$.

We conclude that a frequency response of the form (3.71) corresponds to a first-order difference equation. The value a is called a pole of the system. To understand this choice of terminology, consider the Z-transform $H(z)$ of $h[n]$, which is obtained by replacing the exponent $e^{j\omega}$ in the expression for $H(e^{j\omega})$ by the complex variable z. At the pole $z = a$, the function $H(z)$ is singular since the denominator becomes zero. Therefore, the sequence $h[n]$ is absolutely summable, and thus BIBO stable, only if the pole lies within the complex unit circle, namely $|a| < 1$.

The previous example showed how a system with frequency response given by a single pole can be implemented recursively by relying on past output values. The next example shows that if the pole a lies outside the unit circle $|a| > 1$, then the resulting system still has a recursive implementation, but an anticausal one in which the current output value depends on a future value.

Example 3.14 Consider an LTI system with impulse response

$$
h[n] = \begin{cases} 0, & n > 0 \\ a^n, & n \leq 0, \end{cases} \tag{3.75}
$$

where $|a| > 1$. This anticausal system is a time-reversed version of the system in Example 3.13. From the definition of the DTFT, $H(e^{j\omega})$ is given by

$$
H(e^{j\omega}) = \sum_{n=-\infty}^{0} a^n e^{-j\omega n} = \sum_{n=0}^{\infty} \left(a^{-1} e^{j\omega} \right)^n = \frac{1}{1 - a^{-1} e^{j\omega}}. \tag{3.76}
$$

Note that here the pole a must lie outside the complex unit circle in order to ensure stability.

Following a similar derivation as in the previous example, we can show that the output $y[n]$ of the system with frequency response given by (3.76) and input $x[n]$ can be implemented recursively in the time domain by writing

$$
y[n] = x[n] + \frac{1}{a} y[n + 1]. \tag{3.77}
$$

The recursion expressed by (3.77) is noncausal. It is therefore relevant only for offline applications, in which the entire sequence $x[n]$ is available before the processing begins. In these cases it is possible to use the above recursion, which operates on the elements of $x[n]$ in reverse order. Such situations arise, for example, in image processing tasks, where $x[n]$ corresponds to a row (or column) of pixels.

Example 3.13 (and likewise Example 3.14) can be extended to higher-order systems that include both poles and zeros. As we show, such systems may be implemented recursively by combining both past outputs and past inputs.

Example 3.15 Consider an LTI system corresponding to the difference equation

$$y[n] + a_1 y[n-1] + \cdots + a_p y[n-p] = b_0 x[n] + b_1 x[n-1] + \cdots + b_q x[n-q], \tag{3.78}$$

where p and q are positive integers, and $a_1, \ldots, a_p$ and $b_1 \ldots, b_q$ are constants. When $a_i = 0$ for $i = 1, \ldots, p$, the system is referred to as *moving average (MA)*. On the other hand, if $b_i = 0$ for $i = 1, \ldots, q$, then the system is called *autoregressive (AR)*. The more general system represented by (3.78) is referred to as *ARMA*.

Taking the DTFT of both sides and solving for $Y(e^{j\omega})$,

$$Y(e^{j\omega}) = \frac{b_0 + b_1 e^{-j\omega} + \cdots + b_q e^{-j\omega q}}{1 + a_1 e^{-j\omega} + \cdots + a_p e^{-j\omega p}} X(e^{j\omega}), \tag{3.79}$$

where we used the property that the DTFT of $x[n - n_0]$ is $X(e^{j\omega})e^{-j\omega n_0}$. This shows that (3.78) is an LTI system with frequency response

$$H(e^{j\omega}) = \frac{b_0 + b_1 e^{-j\omega} + \cdots + b_q e^{-j\omega q}}{1 + a_1 e^{-j\omega} + \cdots + a_p e^{-j\omega p}}. \tag{3.80}$$

Owing to its structure, a system of this kind is said to have a rational transfer function. Note that both the denominator and the numerator are polynomials in $e^{-j\omega}$.

To examine when a rational LTI system is BIBO stable, we can express the polynomial in the denominator in terms of its roots

$$\begin{aligned} H(e^{j\omega}) &= \frac{b_0 + b_1 e^{-j\omega} + \cdots + b_q e^{-j\omega q}}{\prod_{i=1}^{p}(1 - \alpha_i e^{-j\omega})} \\ &= \frac{b_0}{\prod_{i=1}^{p}(1 - \alpha_i e^{-j\omega})} + \cdots + \frac{b_q e^{-j\omega q}}{\prod_{i=1}^{p}(1 - \alpha_i e^{-j\omega})}. \end{aligned} \tag{3.81}$$

Therefore, the frequency response consists of $q + 1$ summands which all have the same form up to scaling and modulation (corresponding to a shift in time). Each of these summands, in turn, is a product of p systems of the form (3.71). As discussed in Example 3.13, these subsystems are stable only if their poles lie within the unit circle. Thus, the system (3.78) is stable if and only if $|\alpha_i| < 1$ for every $i = 1, \ldots, p$.

3.4   Continuous–discrete representations

In the context of sampling theorems, we will very often encounter mixed signal representations in which a continuous-time signal $x(t)$ is represented by a discrete-time sequence $d[n]$ via shifts of a given generator $h(t)$:

$$x(t) = \sum_{n \in \mathbb{Z}} d[n] h(t - nT). \tag{3.82}$$

We have already seen such representations in Examples 2.2 and 2.3 of the previous chapter. Equations of the form (3.82) involve both continuous-time signals ($x(t), h(t)$), and discrete-time sequences $(d[n])$. In this section we examine Fourier transforms of such mixed signals.

Proposition 3.1. *Let $x(t) = \sum_{n \in \mathbb{Z}} d[n]h(t - nT)$. Then, in the Fourier domain,*

$$X(\omega) = D(e^{j\omega T})H(\omega), \tag{3.83}$$

where $D(e^{j\omega})$ is the DTFT of the sequence $d[n]$, and $H(\omega)$ is the CTFT of the function $h(t)$.

Proof: By the definition of the CTFT,

$$
\begin{aligned}
X(\omega) &= \int_{-\infty}^{\infty} x(t)e^{-j\omega t}dt \\
&= \sum_{n \in \mathbb{Z}} d[n] \int_{-\infty}^{\infty} h(t - nT)e^{-j\omega t}dt \\
&= \sum_{n \in \mathbb{Z}} d[n]e^{-j\omega nT} \int_{-\infty}^{\infty} h(t)e^{-j\omega t}dt \\
&= D(e^{j\omega T})H(\omega),
\end{aligned}
\tag{3.84}
$$

where in the third equality we used the change of variables $t \to t - nt$. $\qquad\square$

Note that in (3.83), the function $D(e^{j\omega T})$ is $2\pi/T$-periodic and represents the transform of a discrete-time signal. On the other hand, $H(\omega)$ has no periodicity in general, and represents a continuous-time signal. The relationship (3.83) will be used repeatedly throughout our derivations in the rest of the book.

Example 3.16 Consider the signal

$$x(t) = \sum_{n \in \mathbb{Z}} d[n]\operatorname{sinc}(t - n), \tag{3.85}$$

where $d[n]$ is some bounded-norm sequence. From Proposition 3.1, the CTFT $X(\omega)$ of $x(t)$ is given by

$$X(\omega) = D(e^{j\omega})H(\omega), \tag{3.86}$$

where $H(\omega)$ is the CTFT of $\operatorname{sinc}(t)$ (see Example 3.12):

$$H(\omega) = \begin{cases} 1, & |\omega| \leq \pi \\ 0, & |\omega| > \pi. \end{cases} \tag{3.87}$$

Therefore,

$$X(\omega) = \begin{cases} D(e^{j\omega}), & |\omega| \leq \pi \\ 0, & |\omega| > \pi. \end{cases} \tag{3.88}$$

The functions $D(e^{j\omega})$, $H(\omega)$, and $X(\omega)$ are shown in Fig. 3.5.

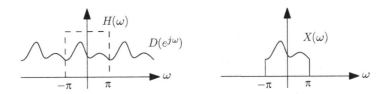

Figure 3.5 The DTFT $D(e^{j\omega})$ of $d[n]$ and the CTFT $H(\omega)$ of sinc(t) (left) are multiplied to yield the CTFT $X(\omega)$ of $x(t) = \sum_n d[n]h(t-n)$ (right).

3.4.1 Poisson-sum formula

We now examine the relation between the Fourier transforms of a continuous-time signal $x(t)$ and its sampled sequence $x(nT)$. A convenient tool in this context is the Poisson-sum formula [58, 62].

Proposition 3.2. *Let $x(t)$ be a continuous-time function in $L_1(\mathbb{R})$, and let $X(\omega)$ denote its CTFT. Then,*

$$\sum_{n\in\mathbb{Z}} x(nT) = \frac{1}{T}\sum_{k\in\mathbb{Z}} X\left(\frac{2\pi k}{T}\right). \tag{3.89}$$

Using Proposition 3.2 we can derive the DTFT of the sampled sequence $x(nT)$, as incorporated in the following theorem. In order to avoid confusion in the theorem statement, we denote the DTFT of $x[n] = x(nT)$ by $B(e^{j\omega})$ and the CTFT of $x(t)$ by $X(\omega)$.

Theorem 3.2 (DTFT of sampled data). *Let $x(t)$ be a continuous-time function in $L_1(\mathbb{R})$ with CTFT $X(\omega)$, let $x[n]$ be the discrete-time sequence obtained by sampling $x(t)$ at points $t = nT$, and let $B(e^{j\omega})$ denote the DTFT of $x[n]$. Then,*

$$B(e^{j\omega}) = \frac{1}{T}\sum_{k\in\mathbb{Z}} X\left(\frac{\omega}{T} - \frac{2\pi k}{T}\right). \tag{3.90}$$

Proof: To prove the theorem, let $g(t) = x(t)e^{-j\omega_0 t/T}$. From the properties of the CTFT, we have that $G(\omega) = X(\omega + \omega_0/T)$. Applying (3.89) to $g(t)$ leads to

$$\sum_{n\in\mathbb{Z}} x(nT)e^{-j\omega_0 n} = \frac{1}{T}\sum_{k\in\mathbb{Z}} X\left(\frac{\omega_0}{T} - \frac{2\pi k}{T}\right). \tag{3.91}$$

The left-hand side of (3.91) is the DTFT of $x[n] = x(nT)$ at frequency ω_0. Using the fact that this equation is true for all ω_0 establishes the result. $\qquad\square$

The connections between the frequency response of a continuous-time signal $x(t)$ and its sequence of samples $x(nT)$ is studied in greater detail in Section 4.2, in the context of the Shannon–Nyquist sampling theorem.

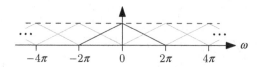

Figure 3.6 Shifted replicas of the triangle function $(G * G)(\omega)$ sum to a constant.

Example 3.17 Consider the signal $x(t) = \text{sinc}^2(t)$. It is seen immediately that its samples at the integers are given by $x[n] = \delta[n]$. We now demonstrate how this relation can be obtained from Theorem 3.2.

Using the multiplication property of the CTFT, we have that $\text{sinc}^2(t) \leftrightarrow (2\pi)^{-1}G(\omega) * G(\omega)$, where $G(\omega)$ is the CTFT of $\text{sinc}(t)$. As we have seen in Example 3.12, $G(\omega)$ is a box function on $[-\pi, \pi]$. Direct computation of the convolution between two box functions leads to

$$M(\omega) = G(\omega) * G(\omega) = \begin{cases} 2\pi - |\omega|, & |\omega| < 2\pi \\ 0, & |\omega| \geq 2\pi. \end{cases} \tag{3.92}$$

Therefore, from Theorem 3.2 with $T = 1$, $X(e^{j\omega})$ is given by

$$X(e^{j\omega}) = \frac{1}{2\pi} \sum_{k \in \mathbb{Z}} M(\omega - 2\pi k) = 1. \tag{3.93}$$

A geometric interpretation of this result is shown in Fig. 3.6. Since $X(e^{j\omega}) = 1$, its inverse DTFT is a discrete impulse: $x[n] = \delta[n]$.

Consider a signal $x(t)$ that can be expressed in the form (3.82). Using Theorem 3.2 and Proposition 3.1 we conclude that the DTFT of its samples $x[n] = x(nT)$ is given by

$$B(e^{j\omega}) = \frac{1}{T} D(e^{j\omega}) \sum_{k \in \mathbb{Z}} H\left(\frac{\omega}{T} - \frac{2\pi k}{T}\right), \tag{3.94}$$

where we used the fact that $D(e^{j\omega T})$ is $2\pi/T$-periodic so that $D(e^{j(\omega/T - 2\pi k/T)T}) = D(e^{j\omega})$. When $h(t)$ has the property that $h(nT) = \delta[n]$, it is easy to see directly from (3.82) that $d[n] = x(nT)$. This result also follows from (3.94) by noting that from Theorem 3.2, $h(nT) = \delta[n]$ implies

$$\frac{1}{T} \sum_{k \in \mathbb{Z}} H\left(\frac{\omega}{T} - \frac{2\pi k}{T}\right) = 1. \tag{3.95}$$

(See Exercise 19.)

3.4.2   Sampled correlation sequences

The relation (3.90) is very useful in determining DTFTs of various sampled sequences. For example, an important sequence encountered in signal recovery problems is the sampled cross-correlation

Figure 3.7 Sampling $y(t)$ after passing through the filter $x(-t)$, to obtain $r_{xy}[n]$.

$$r_{xy}[n] = \langle x(t), y(t+nT) \rangle = \int_{-\infty}^{\infty} x(t)y(t+nT)dt, \qquad (3.96)$$

where the inner product is the usual $L_2(\mathbb{R})$ inner product, and we assume that $x(t)$ is real. This sequence can be viewed as samples at times nT of the continuous-time cross-correlation $r_{xy}(t)$ defined by (3.40). Using the fact that $r_{xy}(t) = x(-t) * y(t)$, the sequence $r_{xy}[n]$ can be obtained by sampling the output of the filter $x(-t)$ with $y(t)$ as its input at times $t = nT$, as depicted in Fig. 3.7. Combining the interpretation of Fig. 3.7 with (3.90), we can express the DTFT of $r_{xy}[n]$ directly in terms of the CTFTs of $x(t)$ and $y(t)$:

$$R_{XY}(e^{j\omega}) = \frac{1}{T}\sum_{k\in\mathbb{Z}} \overline{X\left(\frac{\omega}{T} - \frac{2\pi k}{T}\right)} Y\left(\frac{\omega}{T} - \frac{2\pi k}{T}\right). \qquad (3.97)$$

When $y(t) = x(t)$, the function $r_{xy}(t)$ reduces to the correlation function $r_{xx}(t)$. The sampled correlation, obtained by sampling $r_{xx}(t)$ at times $t = nT$, measures the inner products between $x(t)$ and all of its shifts by nT:

$$r_{xx}[n] = \int_{-\infty}^{\infty} x(t)x(t+nT)dt = \langle x(t), x(t+nT) \rangle. \qquad (3.98)$$

It is easy to see that $\langle x(t+mT), x(t+nT) \rangle = r_{xx}[n-m]$. Therefore, from our definition of orthonormality in the previous chapter, it follows that the functions $\{x(t - nT), n \in \mathbb{Z}\}$ form an orthonormal set if and only if $r_{xx}[n] = \delta[n]$ (see also Exercise 6). Using (3.97), this condition becomes

$$R_{XX}(e^{j\omega}) = \frac{1}{T}\sum_{k\in\mathbb{Z}} \left| X\left(\frac{\omega}{T} - \frac{2\pi k}{T}\right) \right|^2 = 1, \qquad (3.99)$$

so that orthonormality can be determined in the DTFT domain as well.

Example 3.18 In this example we use (3.99) to prove the orthonormality of the set $\{x(t-n)\}_{n\in\mathbb{Z}}$ with

$$x(t) = \text{sinc}\left(\frac{t}{2}\right) \cos\left(\frac{3\pi}{2}t\right). \qquad (3.100)$$

As we have seen in Example 3.12, the CTFT of $\text{sinc}(t/2)$ is a box function of width $\pi/2$ and height 2. Writing $\cos(\alpha) = (e^{j\alpha} + e^{-j\alpha})/2$, it is easy to see that the CTFT of $\cos(3\pi t/2)$ is equal to $\pi(\delta(\omega - 3\pi/2) + \delta(\omega + 3\pi/2))$. Now, the CTFT of the product in (3.100) is given by the convolution of their transforms scaled by 2π:

$$X(\omega) = \begin{cases} 1, & \pi \le |\omega| \le 2\pi \\ 0, & |\omega| < \pi \text{ or } |\omega| > 2\pi. \end{cases} \tag{3.101}$$

Consequently, $\sum_{k \in \mathbb{Z}} |X(\omega - 2\pi k)|^2 = 1$, implying that the set $\{x(t-n)\}_{n \in \mathbb{Z}}$ is orthonormal.

3.5   Exercises

1. Consider two continuous-time signals $x(t)$ and $y(t)$. Based on the definition of the convolution operator (3.20), prove that

$$\frac{d}{dt}(x(t) * y(t)) = x'(t) * y(t) = x(t) * y'(t), \tag{3.102}$$

   where $y'(t)$ denotes the derivative of $y(t)$.
2. Let $x(t)$ be a finite-energy signal and denote its CTFT by $X(\omega)$ (see (3.35)). Show that the CTFT of the derivative $x'(t)$ of $x(t)$ is given by $j\omega X(\omega)$.
3. Let $x(t)$ be a finite-energy complex signal and denote its CTFT by $X(\omega)$. Express the CTFT of $\overline{x(t)}$ in terms of $X(\omega)$.
4. Let ω_0 be an arbitrary constant.
   a. Compute the CTFT of the signal $\sin(\omega_0 t)$.
   b. Compute the CTFT of the signal $\cos(\omega_0 t)$.
   Hint: Use the fact that $\cos(\alpha) = (e^{j\alpha} + e^{-j\alpha})/2$ and $\sin(\alpha) = (e^{j\alpha} - e^{-j\alpha})/(2j)$.
5. Consider a periodic signal $x(t)$ with period T. Prove that its CTFT has the form

$$X(\omega) = \sum_{k \in \mathbb{Z}} a_k \delta(\omega - \omega_0 k). \tag{3.103}$$

   Provide an explicit expression for ω_0 as a function of T, and explain how the constants $\{a_k\}_{k \in \mathbb{Z}}$ are related to the CTFT of one period of $x(t)$.
   Hint: Express $x(t)$ as a linear combination of shifted versions of a single period.
6. Let $x_n(t) = s(t)e^{jnt}$, where $s(t)$ is a given function in $L_2(\mathbb{R})$. Denote by X the set transformation (see Definition 2.14) corresponding to $\{x_n(t)\}_{n \in \mathbb{Z}}$.
   a. Prove that the operator X^*X corresponds to convolution with a sequence $h[n]$. In other words, prove that if $a = X^*Xb$ then $a[m] = \sum_{n \in \mathbb{Z}} h[m-n]b[n]$.
   b. Write an explicit expression for $H(e^{j\omega})$, the DTFT of $h[n]$ (see (3.59)), in terms of $s(t)$.
   Hint: $H(e^{j\omega})$ satisfies $h[n] = (2\pi)^{-1} \int_{-\pi}^{\pi} H(e^{j\omega})e^{-j\omega n} d\omega$.
   c. Derive necessary and sufficient conditions on $H(e^{j\omega})$ such that $\{x_n(t)\}_{n \in \mathbb{Z}}$ is a Riesz basis (see Definition 2.16).
7. Consider a signal $x(t)$ with CTFT $X(\omega)$, whose support is contained in $[\omega_1, \omega_2]$, where $0 < \omega_1 < \omega_2 < \infty$. To transmit $x(t)$ using amplitude modulation (AM), we construct the signal

$$y(t) = (a + x(t)) \cos(\omega_c t), \tag{3.104}$$

where $a > 0$ is a scalar and $\omega_c \gg \omega_2$ is called the carrier frequency. The receiver performs the operation

$$\hat{x}(t) = z(t) * h(t), \tag{3.105}$$

where

$$z(t) = y(t) \cos(\omega_r t). \tag{3.106}$$

Determine a function $h(t)$ and frequency ω_r such that $\hat{x}(t) = x(t) + a$.

8. Assume that $0 < a < 1$ and ω_0 is an arbitrary constant.

a. Compute the CTFT of the signal

$$h(t) = \begin{cases} a^t \cos(\omega_0 t), & t \geq 0 \\ 0, & t < 0. \end{cases} \tag{3.107}$$

b. Compute the DTFT of the sequence

$$h[n] = \begin{cases} a^n \cos(\omega_0 n), & n \geq 0 \\ 0, & n < 0, \end{cases} \tag{3.108}$$

using the definition of the DTFT and verify that the result satisfies Theorem 3.2.

9. Consider a system H_1 whose output $y[n]$ is given by

$$y[n] = \begin{cases} x\left[\frac{n}{L}\right], & \text{if } \frac{n}{L} \text{ is an integer} \\ 0, & \text{else} \end{cases} \tag{3.109}$$

for some integer $L > 1$. The operation performed by this system is called upsampling. Similarly, consider the system H_2, which performs downsampling by an integer factor of $K > 1$:

$$y[n] = x[nK]. \tag{3.110}$$

a. Is the system H_1 LTI?

b. Is the system H_2 LTI?

c. Is the system $H_2 H_1$ LTI?

10. Let ω_0 be an arbitrary constant.

a. Compute the DTFT of the sequence $\sin(\omega_0 n)$.

b. Compute the DTFT of the sequence $\cos(\omega_0 n)$.

11. Consider the frequency response

$$H(e^{j\omega}) = \frac{1}{e^{j\omega} - a e^{-j\omega}}, \tag{3.111}$$

where a is a complex scalar. Determine the values of a for which this system is stable.

12. Consider a discrete-time LTI system whose output is given by

$$y[n] = \begin{cases} 1, & n = 1 \\ -a, & n = 2 \\ 0, & \text{else} \end{cases} \tag{3.112}$$

when fed with the input

$$x[n] = \begin{cases} a^n, & n \geq 0 \\ 0, & n < 0, \end{cases} \tag{3.113}$$

where $a \neq 0$ is some scalar. What is the impulse response $h[n]$ of the system?

13. Let $h[n]$ be the sequence defined by

$$h[n] = \begin{cases} 2, & n = 0 \\ -a, & |n| = 1 \\ 0, & |n| > 1, \end{cases} \tag{3.114}$$

where a is a constant.

a. Compute the DTFT $H(e^{j\omega})$ of $h[n]$.

b. Express the reciprocal $G(e^{j\omega}) = 1/H(e^{j\omega})$ of the frequency response as

$$G(e^{j\omega}) = G_c(e^{j\omega})G_{ac}(e^{j\omega}), \tag{3.115}$$

where $G_c(e^{j\omega})$ is a causal system and $G_{ac}(e^{j\omega})$ is an anticausal system; that is, its impulse response vanishes at nonnegative times.

c. Derive conditions on a such that both systems are stable.

14. As we have seen, the output of a discrete-time LTI system is given by the convolution between the input $x[n]$ and the response $h[n]$ of the system to an impulse $\delta[n]$. We would now like to express this relation in terms of the response of the system to a different sequence. Specifically, let $\tilde{h}[n]$ denote the response of an LTI system to the sequence

$$b[n] = \begin{cases} a^n, & n \geq 0 \\ 0, & n < 0, \end{cases} \tag{3.116}$$

where $|a| < 1$ is some complex constant. Write an explicit expression in the time domain relating the output $y[n]$ of the system to its input $x[n]$, and to $\tilde{h}[n]$.

15. Let $h[n]$ and $g[n]$ be the impulse responses of two causal FIR systems.

a. Determine the impulse response of the system depicted in Fig. 3.8.

b. Provide conditions on $h[n]$ and $g[n]$ such that the system depicted in Fig. 3.8 is stable.

16. Suppose that the signal $x(t)$ can be expressed as

$$x(t) = \sum_{n \in \mathbb{Z}} c[n]h(t - n) \tag{3.117}$$

for some sequence $c[n]$ in ℓ_2 and function $h(t)$ in L_2.

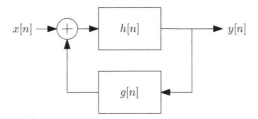

Figure 3.8 A feedback-loop system.

a. Assume that the signal $x(t)$ can also be written as

$$x(t) = \sum_{n \in \mathbb{Z}} d[n]g(t-n) \tag{3.118}$$

for some sequence $d[n]$ in ℓ_2 and function $g(t)$ in L_2. Prove that $h(t)$ can be expressed in terms of $g(t)$ as

$$h(t) = \sum_{n \in \mathbb{Z}} e[n]g(t-n) \tag{3.119}$$

and provide an explicit expression for $E(e^{j\omega})$, the DTFT of $e[n]$, as a function of the DTFTs $C(e^{j\omega})$ and $D(e^{j\omega})$.

b. Assume that $|H(\omega)| > 0$ for all ω and that the set $\{h(t-n)\}_{n \in \mathbb{Z}}$ is not an orthogonal basis. Show how a function $g(t)$ can be constructed from $h(t)$ such that $\{g(t-n)\}_{n \in \mathbb{Z}}$ is an orthogonal basis.

17. Let $s(t)$ and $w(t)$ be given functions in L_2. Consider a discrete-time system whose output $y[n]$ relates to the input $x[n]$ via

$$y[n] = \int_{\infty}^{\infty} s(t-n)a(t)dt \tag{3.120}$$

where

$$a(t) = \sum_{k \in \mathbb{Z}} x[k]w(t-k). \tag{3.121}$$

Determine whether this system is LTI.

18. Let $s(t)$ and $w(t)$ be given functions in L_2. Consider a continuous-time system whose output $y(t)$ relates to the input $x(t)$ by

$$y(t) = \sum_{k \in \mathbb{Z}} d[n]w(t-n), \tag{3.122}$$

where

$$d[n] = \int_{\infty}^{\infty} s(t-n)x(t)dt. \tag{3.123}$$

Determine whether this system is LTI.

19. Assume that $h(t)$ is a function in $L_1(\mathbb{R})$ for which $h(nT) = \delta[n]$. Prove that this implies that

$$\frac{1}{T}\sum_{k\in\mathbb{Z}} H\left(\frac{\omega}{T} - \frac{2\pi k}{T}\right) = 1, \qquad (3.124)$$

where $H(\omega)$ is the CTFT of $h(t)$.

Hint: Use Theorem 3.2 to show that the left-hand side of (3.124) is the DTFT of $h(nT)$.

20. Let $x(t) = \text{sinc}(t)$ where $\text{sinc}(t)$ is defined by (3.29), and let $x_n(t) = x(t - nt)$. Determine for what values T and a the functions $\{ax_n(t)\}$ are orthonormal.

Chapter 4

Signal spaces

We now begin translating the linear algebra results of Chapter 2 into concrete sampling theorems. The Fourier transforms presented in the previous chapter will play an important role in these developments. As we discussed in the Introduction, a key component in any sampling theory is the prior knowledge we have about our signal. Without incorporating such knowledge the problem of recovery from samples is ill-posed; there are always many curves that can pass through a set of points. The challenge in practice is to find the "best" curve in some sense consistent with our prior information.

In this chapter we introduce the classes of signals we will be focusing on throughout the book, along with some of the fundamental mathematical properties associated with these signal sets. Therefore, in some sense this chapter is a continuation of the mathematical introduction we started in the previous chapters, with a focus on function classes that will be relevant when discussing more general sampling theorems. Subsequent chapters consider different recovery strategies given each one of the signal classes discussed herein.

4.1   Structured bases

4.1.1   Sampling and reconstruction spaces

In our introduction to vector spaces (see Theorem 2.1) we have shown that given a Riesz basis $\{w_n\}$ for some space $\mathcal{W}$, any vector in this space can be written as

$$x = \sum_n \langle \tilde{w}_n, x \rangle w_n, \tag{4.1}$$

where $\{\tilde{w}_n\}$ are the vectors biorthogonal to $\{w_n\}$. Since the inner products $c[n] = \langle \tilde{w}_n, x \rangle$ are scalars, we may view them as samples, or generalized samples, of x. Using set transformation notation we can write these samples as $c = \widetilde{W}^* x$ where $\widetilde{W}$ is the set transformation corresponding to the vectors $\{\tilde{w}_n\}$. Equation (4.1) can then be interpreted as a sampling theorem: any signal x in the space $\mathcal{W}$ is recoverable from its generalized samples $c = \widetilde{W}^* x$ via the formula

$$x = \sum_n c[n] w_n = Wc, \tag{4.2}$$

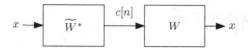

Figure 4.1 Sampling and recovery of a general signal in $\mathcal{W}$.

where W is the set transformation corresponding to the vectors $\{w_n\}$. This interpretation is illustrated in Fig. 4.1. In accordance with this viewpoint, the vectors $\{\tilde{w}_n\}$ are referred to as the *sampling vectors*, since they are used to sample x, while $\{w_n\}$ serve as the *reconstruction vectors*.

An alternative to (4.1) can be obtained by exchanging the role of the sampling and reconstruction vectors:

$$ x = \sum_n \langle w_n, x \rangle \tilde{w}_n. \tag{4.3} $$

In (4.3), $\{w_n\}$ constitute the sampling vectors, while $\{\tilde{w}_n\}$ are now the reconstruction vectors.

Both sets of sampling and reconstruction vectors span subspaces which we refer to as the *sampling* and *reconstruction spaces,* respectively. In (4.1) and (4.3), these subspaces are equal to $\mathcal{W}$. In future chapters we treat more general expansions in which these spaces may differ. Sampling is then accomplished by generalizing the concept of a biorthogonal basis to a broader class that will allow us to work in two different spaces.

4.1.2   Practical sampling theorems

The discussion above highlights the fact that the linear algebra results of Chapter 2 directly lead to sampling theorems over general spaces. However, expansions of the form (4.1) are written in abstract form. It is not immediately obvious how to translate them into concrete sampling methods and practical reconstruction algorithms. From an engineering perspective, it is important to keep in mind that the sampling theorems we develop need to be realized in hardware, as the ultimate goal is to sample real-world analog signals, which cannot be manipulated by software unless they are sampled. Our interest therefore is in sampling theorems that are implementable in practice using standard analog (and digital) components such as filters and modulators. In contrast, the block diagram of Fig. 4.1 contains abstract linear elements which are not immediately implementable in an efficient manner. For example, without structure, sampling x requires computing infinitely many inner products, which involve infinitely many sampling vectors. Our challenge, therefore, is to choose subspaces that contain signals of practical interest, and at the same time allow us to compute the basis and biorthogonal basis vectors efficiently. Consequently, in the rest of the book, we move away from the abstract formulation of Chapter 2 and concentrate our attention on concrete signal classes for which the blocks in Fig. 4.1 have a simple interpretation in terms of standard analog and digital components.

In order to obtain feasible sampling methods, the basis (or frame) expansions we consider need to have structure; that is, the basis vectors should be related to each other

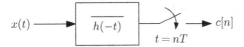

Figure 4.2 Sampling as filtering followed by pointwise sampling.

in some simple way. This will allow us to replace the sampling block in Fig. 4.1 by a more convenient system. For example, suppose that $w_n(t) = h(t - nT), n \in \mathbb{Z}$ where $h(t)$ is a given function in $L_2(\mathbb{R})$ so that the basis vectors are related to each other by a time shift. In this case the samples $c[n] = \langle w_n(t), x(t) \rangle$ of any $x(t) \in L_2(\mathbb{R})$ can be written as

$$c[n] = \langle h(t - nT), x(t) \rangle = \int_{-\infty}^{\infty} \overline{h(\tau - nT)} x(\tau) d\tau = x(t) * \overline{h(-t)}|_{t=nT}. \quad (4.4)$$

The expression in (4.4) amounts to filtering $x(t)$ with a filter whose impulse response is given by $\overline{h(-t)}$, and then sampling the output at uniformly spaced intervals of length T, as depicted in Fig. 4.2.

A special case in which sampling can be modeled as in (4.4) is when W is equal to the space of bandlimited signals. Most of the literature on sampling as well as practical ADCs rely on the Shannon–Nyquist theorem which assumes a bandlimited input. Therefore, we will begin in Section 4.2 by examining this well-known theorem, and recasting it in terms of a basis expansion in an appropriate Hilbert space, in the spirit of our previous discussion. This will then suggest a natural extension to a much broader set of input classes, without the bandlimited constraint, referred to as *shift-invariant (SI) subspaces*. In such spaces, sampling can be implemented as in Fig. 4.2, with different choices of the filter $\overline{h(-t)}$. These spaces include splines, which are very popular in signal and image processing, and various communication signals such as pulse-amplitude modulation.

To enlarge the space of signals that can be sampled efficiently, we may consider a bank of sampling filters $\{h_k(t)\}$, instead of a single filter as in Fig. 4.2. An appealing property of such a structure is that increasing the number of filters allows us to capture a larger set of signals in $L_2(\mathbb{R})$. In fact, it turns out that any signal in $L_2(\mathbb{R})$ is recoverable from filterbank samples, when the number of filters approaches infinity, and the impulse responses are chosen appropriately. The drawback, of course, is that it is infeasible to implement an infinite filterbank unless it possesses further structure. In particular, if all the filters $h_k(t)$ are related to each other in a simple way, then we may be able to exploit this relation to compute the coefficients efficiently. Two popular options are to choose the filters such that they are related by modulation $h_k(t) = e^{j2\pi Wkt}h(t)$ for some frequency W, or by scaling $h_k(t) = 2^{-k/2}h(t/2^k)$. The former leads to *Gabor signal expansions*, while the latter results in *wavelet expansions*. We will very briefly touch upon both of these topics below. However, our main focus throughout this book is on SI subspaces with a single or finite number of generators. These represent strict subspaces of $L_2(\mathbb{R})$ which is the typical setting in sampling theorems. Similar tools to those used for finitely generated SI subspaces can be applied to analyze other structured spaces as well.

4.2   Bandlimited sampling

Without a doubt, the most-studied sampling theorem that has had a major impact on signal processing is the Shannon–Nyquist theorem. This theorem has become a cornerstone in both the mathematical and engineering literature. Below, we provide a formal statement and proof of this result. We then show how it can be implemented simply with modulators and filters, and interpret the theorem as an orthogonal basis expansion. This interpretation will pave the way to extensions of the result to general SI subspaces.

Before stating the theorem, we introduce the notation used throughout this chapter. Since we will be discussing pointwise samples of continuous-time signals, to avoid confusion we use parentheses around continuous-time variables, and square brackets to denote discrete-time variables. Accordingly, samples at times nT of a continuous-time signal $x(t)$ will be denoted by $x[n]$. When we would like to make the sampling time explicit we write $x(nT)$, which is to be interpreted as $x(t = nT)$. This notation will also be used to distinguish between continuous-time and discrete-time deltas (or Dirac functions): $\delta(t)$ denotes the generalized function defined in (3.14), while $\delta[n]$ is the discrete-time sequence defined by (3.56).

4.2.1   The Shannon–Nyquist theorem

The bandlimited sampling theorem was first proved by E. T. Whittaker in 1915 [4] (who referred to it as the cardinal series). Nyquist [2] showed that up to $2B$ independent pulse samples could be sent through a system of bandwidth B; however, he did not explicitly consider the problem of sampling and reconstruction. The theorem was then rediscovered and used by Shannon in 1949 for applications to communication theory [1]. Kotelnikov and J. M. Whittaker published similar results in the 1930s [3, 5]. Earlier work with related ideas date back to Cauchy [6]. For a more detailed account of the sampling theorem history we refer the reader to [7].

Theorem 4.1 (The Shannon–Nyquist theorem).   *Let* $x(t) \in L_2(\mathbb{R})$ *be a signal with CTFT* $X(\omega)$. *If* $X(\omega) = 0$ *for all* $|\omega| \geq \pi/T$, *then* $x(t)$ *can be reconstructed from its samples* $x(nT)$ *using the reconstruction formula*

$$x(t) = \sum_{n \in \mathbb{Z}} x(nT) \operatorname{sinc}((t - nT)/T) \qquad (4.5)$$

where

$$\operatorname{sinc}(t) = \frac{\sin(\pi t)}{\pi t}. \qquad (4.6)$$

Theorem 4.1 asserts that a signal bandlimited to π/T can be recovered from its uniformly spaced samples with period T, or sampling rate $f = 1/T$. This sampling rate is referred to as the *Nyquist rate*. In fact, as the proof will reveal, recovery is possible from uniform samples at any rate equal to or above the Nyquist rate, i.e., the signal can be obtained using (4.5) with T' replacing T for any $T' \leq T$.

Before proving the theorem, we point out that in practice, sampling is performed by an ADC that is often far from ideal. Practical ADCs introduce a variety of different distortions including nonlinearities, noise, and jitter. In addition, the samples $x[n]$ are typically quantized to a certain resolution. The topic of quantization is a very important part of ADC design; however, it is outside the scope of this text. As in many other books on signal processing, we will assume that the samples take on values from a continuous interval.

We now provide a formal proof of the Shannon–Nyquist theorem, by relying on the Poisson-sum formula introduced in Section 3.4.1. In Section 4.2.2 we consider an alternative engineering interpretation of the theorem and its proof, in which the sampling process is modeled as multiplication by an impulse train. Although this viewpoint is slightly problematic from a strict mathematical perspective, it provides nice insight into the frequency domain analysis of the sampling theorem.

Proof: Our goal is to show that the right-hand side of (4.5) is equal to $x(t)$. Denoting $h(t) = \text{sinc}(t/T)$, we can write this expression as

$$y(t) = \sum_{n \in \mathbb{Z}} x[n]h(t - nT), \qquad (4.7)$$

with $x[n] = x(nT)$. In Chapter 3, Eq. (3.82) and (3.83), we considered the CTFT of signals that are given by such a mixed continuous–discrete representation, and established that $Y(\omega) = X(e^{j\omega T})H(\omega)$. In our case, $X(e^{j\omega T})$ represents the DTFT of the sequence of samples $x(nT)$. From Theorem 3.2,

$$X(e^{j\omega T}) = \frac{1}{T} \sum_{k \in \mathbb{Z}} X\left(\omega - \frac{2\pi k}{T}\right). \qquad (4.8)$$

The Fourier transform of $h(t)$ is given by (see Example 3.12)

$$H(\omega) = \begin{cases} T, & |\omega| \leq \pi/T \\ 0, & \text{otherwise,} \end{cases} \qquad (4.9)$$

resulting in

$$Y(\omega) = \begin{cases} \sum_{k \in \mathbb{Z}} X(\omega - 2\pi k/T), & |\omega| \leq \pi/T \\ 0, & \text{otherwise.} \end{cases} \qquad (4.10)$$

Since $X(\omega)$ is zero outside $[-\pi/T, \pi/T]$, the only nonzero element in the sum (4.10) corresponds to $k = 0$. Thus,

$$Y(\omega) = \begin{cases} X(\omega), & |\omega| \leq \pi/T \\ 0, & \text{otherwise,} \end{cases} \qquad (4.11)$$

or $Y(\omega) = X(\omega)$, as required. $\qquad \square$

We point out that the convergence in (4.5) is uniform if $x(t) \in L_2(\mathbb{R})$, that is $\lim_{N \to \infty} \sup_{t \in \mathbb{R}} |x(t) - x_N(t)| = 0$, where

$$x_N(t) = \sum_{n=-N}^{N} x(nT) \text{sinc}((t - nT)/T) \qquad (4.12)$$

is the truncated cardinal series [64]. Uniform convergence also holds when $X(\omega) \in L_1(\mathbb{R})$ [59].

4.2.2   Sampling by modulation

We now consider an engineering interpretation and proof of the Shannon–Nyquist theorem. We begin by noting that the sum in (4.5) can be expressed as a convolution

$$x(t) = x_\delta(t) * \operatorname{sinc}(t/T), \tag{4.13}$$

with

$$x_\delta(t) = \sum_{n\in\mathbb{Z}} x(nT)\delta(t - nT). \tag{4.14}$$

The signal $x_\delta(t)$ is a continuous-time representation of the discrete-time samples, where each sample is replaced by an impulse function with integral equal to the sample value, as demonstrated in Fig. 4.3. This allows us to model sampling as multiplication by a periodic impulse train, or by modulation. An essential difference between the continuous-time representation $x_\delta(t)$ and the sequence $x[n]$ is that the latter is indexed by an integer variable n irrespective of the sampling period T, effectively introducing a time normalization.

Equation (4.13) can be interpreted in terms of the block diagram in Fig. 4.4. The samples are first modulated by an impulse train with period T. The modulated impulse train is then filtered by a filter with impulse response $h(t) = \operatorname{sinc}(t/T)$ and frequency response $H(\omega)$ given by (4.9). To ensure that the output in Fig. 4.4 is indeed equal to

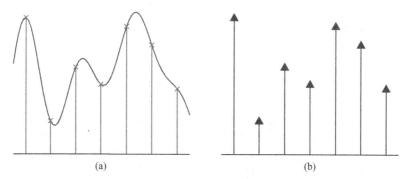

(a)   (b)

Figure 4.3 (a) A bandlimited signal $x(t)$ and its samples $x[n] = x(nT)$, marked by crosses. (b) The continuous-time representation $x_\delta(t)$ of the samples $x[n]$.

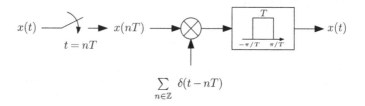

Figure 4.4 Interpretation of the Shannon–Nyquist theorem.

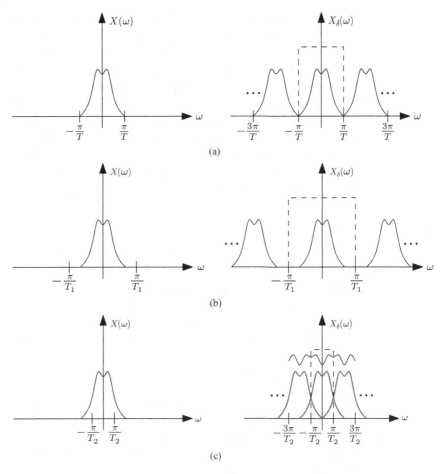

Figure 4.5 (a) The Fourier transform of a signal $x(t)$ bandlimited to π/T and the continuous-time representation of its Nyquist-rate samples $x[n] = x(nT)$. (b) Sampling with period $T_1 < T$. (c) Sampling with period $T_2 > T$.

$x(t)$, we consider the Fourier transform of $x_\delta(t)$. From (3.82) and (3.83), it follows that $X_\delta(\omega) = X(e^{j\omega T})$ with $x[n] = x(nT)$ since the CTFT of $\delta(t)$ is equal to 1. Using (4.8),

$$X_\delta(\omega) = \frac{1}{T} \sum_{k\in\mathbb{Z}} X\left(\omega - \frac{2\pi k}{T}\right). \tag{4.15}$$

The transform in (4.15) describes the effect of sampling in the Fourier domain. The sampling process creates overlaps of $X(\omega)$, at equally spaced frequencies $2\pi/T$, as illustrated in Fig. 4.5. The resulting transform $X_\delta(\omega)$ is periodic with period $2\pi/T$. If the replicas do not overlap, then the signal $X(\omega)$ can be recovered from $X_\delta(\omega)$ by simple low-pass filtering with cutoff π/T. To ensure that no overlapping occurs, the largest frequency of $X(\omega)$ must be smaller than π/T. If $X(\omega)$ is not sufficiently bandlimited, then the sampling process will create overlaps within the low-pass regime, which will remain after filtering. This distortion is referred to as *aliasing*. Figure 4.5 also shows that we can sample with period T_1 smaller than T and still recover the signal.

In the context of Nyquist sampling, aliasing is considered an undesirable distortion which is to be avoided. In later chapters, when we treat sub-Nyquist sampling of certain signal classes, we will see that aliasing can actually be used as a resource. In fact, the methods and hardware prototypes we introduce for sub-Nyquist sampling all rely on aliasing the data prior to low-rate sampling. Thus, when properly controlled, aliasing may be beneficial for certain tasks.

4.2.3 Aliasing

Figure 4.5(c) shows that if the signal $x(t)$ is not originally bandlimited to π/T, then sampling it at a rate $1/T$ will lead to aliasing distortion: this is the error caused by folding of the signal energy from a high frequency, beyond π/T, to a lower one. Mathematically, aliasing can be viewed in the Fourier transform of the sampled sequence $X_\delta(\omega)$ in (4.15) as the contributions to the sum from indices $k \neq 0$ in the low-pass regime $|\omega| < \pi/T$. If $X(\omega)$ has energy out of the area $(-\pi/T, \pi/T)$, then there will be a value of $k \neq 0$ for which $X(\omega - 2\pi k/T)$ is not zero for some $\omega \in (-\pi/T, \pi/T)$. Going back to Fig. 4.5(c), consider values of ω between 0 and π/T_2. For $k = 1$, $X(\omega - 2\pi k/T_2)$ contains the frequency content of $X(\omega)$ in the interval $[-2\pi/T_2, -\pi/T_2]$. This content is aliased (namely, added) to the interval $[0, \pi/T_2]$. A complementary viewpoint is shown in Fig. 4.6: for some $\omega_0 \in (-\pi/T, \pi/T)$, the sum $\sum_{k \in \mathbb{Z}} X(\omega_0 - 2\pi k/T)$ includes contributions from $X(\omega)$ for $\omega \notin (-\pi/T, \pi/T)$ unless the support of $X(\omega)$ is contained in $(-\pi/T, \pi/T)$.

Example 4.1 below is a classical case of aliasing which motivates the choice of terminology: a sinusoid at a high frequency after sampling will appear as a low-frequency tone. The two signals become aliases of each other since they are indistinguishable after sampling. An example of aliasing can be seen in old movies, especially when watching wagon wheels on old Western films. The wheel picks up speed as expected, but then the wheel seems to slow, and as the wagon further accelerates, the wheel appears to turn backwards. This phenomenon occurs as the rate of the wheel's spinning approaches

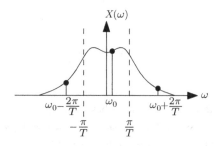

Figure 4.6 Aliasing at a frequency ω_0 corresponds to contributions from $\omega_0 + 2\pi k/T$ for $k \neq 0$.

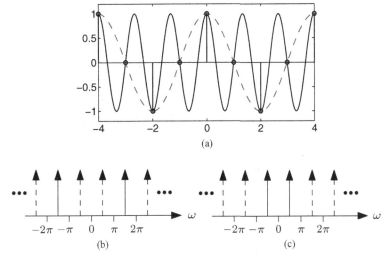

Figure 4.7 (a) The signals $x_1(t) = \cos(1.5\pi t)$ and $x_2(t) = \cos(0.5\pi t)$ and their samples at the integers. (b), (c) The DTFTs $X_1(e^{j\omega})$ and $X_2(e^{j\omega})$ of $x_1(n)$ and $x_2(n)$ respectively. The dashed impulses are 2π-shifted replicas of the solid bold ones.

the rate of the sampler, in this case the camera, operating at a fixed frame rate. When the wagon is first accelerating, the frame rate of the movie camera is much higher than the revolution rate of the wheel. As the spinning rate of the wheel approaches the Nyquist limit, we see only two points on the wheel which are 180 degrees apart. This is perceived as stopping of the wheel. When the wagon continues to accelerate, the spinning rate exceeds the Nyquist rate and the wagon wheels are observed as turning in reverse.

Example 4.1 Consider the two sinusoidal signals $x_1(t) = \cos(1.5\pi t)$ and $x_2(t) = \cos(0.5\pi t)$. Suppose that we sample these signals with period $T = 1$. In this case, as demonstrated in Fig. 4.7(a), the samples coincide:

$$x_1(n) = \cos(1.5\pi n) = \cos(-1.5\pi n + 2\pi n) = \cos(0.5\pi n) = x_2(n). \qquad (4.16)$$

To understand why this happens, we note that the CTFTs of the signals $x_1(t)$ and $x_2(t)$ are given by $X_1(\omega) = \pi(\delta(\omega + 1.5\pi) + \delta(\omega - 1.5\pi))$ and $X_2(\omega) = \pi(\delta(\omega + 0.5\pi) + \delta(\omega - 0.5\pi))$, respectively. Now, the DTFTs of the sample sequences $x_1(n)$ and $x_2(n)$, comprise shifted replicas of $X_1(\omega)$ and $X_2(\omega)$, with a spacing of 2π. As can be seen in Figs. 4.7(b) and 4.7(c), this results in the same impulse train. Therefore, the signal $x_1(t)$ cannot be distinguished from the low-frequency tone $x_2(t)$ based on samples with period $T = 1$.

When aliasing occurs, simply substituting the Nyquist-rate samples into the interpolation formula given by (4.5) will not recover the true signal $x(t)$. In fact, it can result in a large error, especially when the signal contains substantial energy at high

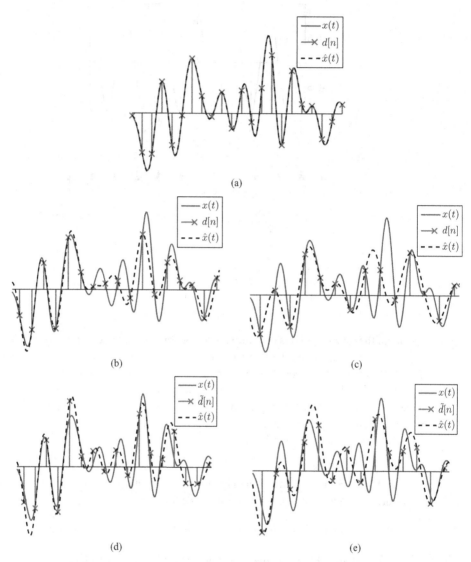

Figure 4.8 (a) Reconstruction of a π/T-bandlimited signal with a sampling period of T. (b), (c) Reconstruction with sampling periods $1.25T$ and $1.5T$ without an anti-aliasing filter leads to signal-to-error ratios of $4.8\,$dB and $0.7\,$dB, respectively. (d), (e) Reconstruction with sampling periods $1.25T$ and $1.5T$ with an anti-aliasing filter leads to signal-to-error ratios of $6.6\,$dB and $2.4\,$dB, respectively.

frequencies, as can be seen in Fig. 4.8. Figures 4.8(a), (b), and (c) demonstrate that for a π/T-bandlimited signal, the reconstruction error becomes higher as the sampling period T_s exceeds T. For $T_s = T$, the recovery $\hat{x}(t)$ is perfect, for $T_s = 1.25T$ it slightly departs from $x(t)$, and for $T_s = 1.5T$ the aliasing error is significant. In the figure, $d[n]$ denotes the samples of $x(t)$ at $t = nT_s$.

To reduce the reconstruction error, an alternative strategy is to force the signal to be bandlimited before sampling. This can be achieved by inserting an LPF with cutoff

π/T_s prior to uniform sampling with period T_s in Fig. 4.4. This filter is referred to as an *anti-aliasing filter* owing to its role in removing aliasing effects. The output of the LPF, denoted $y(t)$, is bandlimited and can therefore be perfectly recovered from its uniform samples, which we denote by $\tilde{d}[n]$, using (4.5) with $T = T_s$. The recovery error is then the error in approximating $x(t)$ by its low-pass version $y(t)$. It turns out that this approach is optimal in the sense that it results in a bandlimited recovery $\hat{x}(t)$ that is as close as possible in L_2-norm to the original signal $x(t)$. We prove this statement after introducing a basis expansion interpretation of the Shannon–Nyquist theorem. Figures 4.8(d) and (e) depict the recovery obtained using this paradigm with sampling periods $1.25T$ and $1.5T$. These reconstructions are closer to the original signal (in an L_2 sense) than the recoveries depicted in Figs. 4.8(b) and (c).

4.2.4 Orthonormal basis interpretation

Now that we have understood the Shannon–Nyquist theorem, our next step is to connect it with the linear algebra viewpoint promoted in Chapter 2. To do this, we need to express the bandlimited sampling theorem using sampling and reconstruction vectors that are biorthogonal to each other:

$$x(t) = \sum_{n \in \mathbb{Z}} \langle \tilde{h}_n(t), x(t) \rangle h_n(t) = \sum_{n \in \mathbb{Z}} a[n] h_n(t) \tag{4.17}$$

for a set of functions $h_n(t)$ that form a basis for the subspace $\mathcal{W}$ of signals bandlimited to π/T, and some coefficients $a[n]$.

The expansion (4.17) has the same form as (4.5) if we choose the reconstruction vectors $h_n(t) = h(t - nT)$ with $h(t) = \mathrm{sinc}(t/T)$ and samples $a[n] = x(nT)$. Therefore, we need to show that $x(nT) = \langle \tilde{h}_n(t), x(t) \rangle$ for the biorthogonal vectors $\tilde{h}_n(t)$ to complete the equivalence with (4.17).

It turns out the basis vectors $\{h(t - nT) = \mathrm{sinc}((t - nT)/T)\}$ are orthogonal. This then implies that the biorthogonal vectors are scaled versions of $\{h(t - nT)\}$. To establish the orthogonality of $\{h_n(t) = h(t - nT)\}$ we compute the inner products between different vectors in the set:

$$\langle h_n(t), h_m(t) \rangle = \int_{-\infty}^{\infty} \mathrm{sinc}((t - nT)/T) \, \mathrm{sinc}((t - mT)/T) dt$$

$$= \frac{1}{2\pi} \int_{-\infty}^{\infty} |H(\omega)|^2 e^{-j(m-n)\omega T} d\omega$$

$$= \frac{T^2}{2\pi} \int_{-\pi/T}^{\pi/T} e^{-j(m-n)\omega T} d\omega. \tag{4.18}$$

The second equality follows from Parseval's formula (3.42) and the shift property of the CTFT; the function $H(\omega)$ is the Fourier transform of $\mathrm{sinc}(t/T)$ given by (4.9). If $m = n$, then $e^{-j(m-n)\omega T} = 1$, and the expression in (4.18) is equal to T. When $m \neq n$,

$$\frac{T^2}{2\pi} \int_{-\pi/T}^{\pi/T} e^{-j(m-n)\omega T} d\omega = j \frac{T}{2\pi(m-n)} e^{-j(m-n)\omega T} \Big|_{-\pi/T}^{\pi/T} = 0. \tag{4.19}$$

Combining these observations with (4.18) we conclude that

$$\langle h_n(t), h_m(t) \rangle = T\delta_{nm}, \tag{4.20}$$

so that the vectors $\{h_n(t) = \mathrm{sinc}(t - nT)\}$ form an orthogonal basis for $\mathcal{W}$. The biorthogonal vectors are therefore particularly simple to compute – they are scaled versions of the original vectors: $\tilde{h}_n(t) = (1/T)h_n(t)$. Indeed, with this choice, $\langle \tilde{h}_n(t), h_m(t) \rangle = \delta_{nm}$.

Equipped with the biorthogonal set, we now evaluate the expansion coefficients. By definition,

$$\begin{aligned} a[n] &= \langle \tilde{h}_n(t), x(t) \rangle \\ &= \frac{1}{T} \int_{-\infty}^{\infty} x(t) \, \mathrm{sinc}((t - nT)/T) dt \\ &= \frac{1}{2\pi} \int_{-\pi/T}^{\pi/T} X(\omega) e^{j\omega nT} d\omega \\ &= x(nT). \end{aligned} \tag{4.21}$$

The third equality is a result of applying Parseval and using the CTFT of the sinc function as in (4.18). The last line follows from the definition of the inverse CTFT. Since $X(\omega)$ is bandlimited to π/T,

$$\frac{1}{2\pi} \int_{-\pi/T}^{\pi/T} X(\omega) e^{j\omega nT} d\omega = \frac{1}{2\pi} \int_{-\infty}^{\infty} X(\omega) e^{j\omega nT} d\omega = x(t = nT). \tag{4.22}$$

We have therefore successfully cast the Shannon–Nyquist theorem as a basis expansion of the form (4.17) with $\tilde{h}_n(t) = (1/T)h(t - nT)$ and $h(t) = \mathrm{sinc}(t/T)$.

To gain more insight that we will shortly extend to other subspaces, it is also useful to interpret the samples $a[n]$ in the spirit of Fig. 4.2. As we have seen, inner products with shifts of a given function can be viewed as samples of the output of an appropriate filter:

$$\begin{aligned} a[n] &= \frac{1}{T} \int_{-\infty}^{\infty} x(t) \, \mathrm{sinc}((t - nT)/T) dt \\ &= \frac{1}{T} \int_{-\infty}^{\infty} x(t) \, \mathrm{sinc}(-(nT - t)/T) dt \\ &= \frac{1}{T} x(t) * \mathrm{sinc}(-t/T)|_{t=nT}. \end{aligned} \tag{4.23}$$

Since the sinc function is symmetric, $a[n] = (1/T)x(t) * \mathrm{sinc}(t/T)|_{t=nT}$. Noting that the CTFT of $(1/T)\mathrm{sinc}(t/T)$ is equal to a LPF with cutoff frequency π/T (namely, $(1/T)H(\omega)$ with $H(\omega)$ given by (4.9)) we may interpret $a[n]$ as samples of a signal $y(t)$ obtained by filtering $x(t)$ by a LPF. This interpretation is illustrated in Fig. 4.9. Since $x(t)$ is itself bandlimited to π/T it is clear that the output of the LPF is equal to its input $x(t)$ and consequently $a[n] = x(nT)$. Figure 4.10 depicts the combined implementation of (4.17).

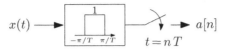

Figure 4.9 Sampling a bandlimited signal.

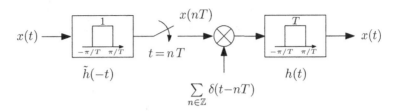

Figure 4.10 Sampling and recovery of a bandlimited signal using the basis expansion interpretation of (4.17).

It is interesting to note the difference between Figs. 4.4 and 4.10. When the input signal $x(t)$ is bandlimited, as we have assumed, they are equivalent. This is because the LPF in Fig. 4.10 has no effect on a bandlimited input. However, clearly they lead to different outputs when $x(t)$ is not bandlimited. Since the Shannon–Nyquist theorem only refers to bandlimited inputs, it leaves open the question of how to interpret it for signals that are not bandlimited.

The method of Fig. 4.4 is equivalent to substituting the samples $x(nT)$ into the recovery formula (4.5) regardless of whether the signal is bandlimited. As we have shown in Fig. 4.8, this leads to aliasing and sometimes large distortion. On the other hand, the approach of Fig. 4.10, which follows from the basis expansion interpretation, is to first filter high-frequency signal components, so that the samples in (4.5) correspond to samples of a bandlimited approximation of the input. Using the results of Section 2.6 we now establish that this intuitive approach is in fact optimal in a squared-error sense.

Recall that given a vector space $\mathcal{W}$, the orthogonal projection of an arbitrary x onto $\mathcal{W}$ can be written as $P_{\mathcal{W}}x = \sum_n \langle \tilde{w}_n, x \rangle w_n$ where the vectors $\{w_n\}$ form a basis for $\mathcal{W}$ and $\{\tilde{w}_n\}$ is the biorthogonal basis. Furthermore, the approximation property of the orthogonal projection given by Proposition 2.15 asserts that $P_{\mathcal{W}}x$ is the signal in $\mathcal{W}$ closest to $x(t)$ (in an L_2 sense). From (4.17) it therefore follows that if $x(t)$ is not bandlimited, then

$$\hat{x}(t) = \sum_{n\in\mathbb{Z}} \langle \tilde{h}_n(t), x(t) \rangle h_n(t), \tag{4.24}$$

with $h_n(t) = \mathrm{sinc}((t - nT)/T)$ is the orthogonal projection of $x(t)$ onto the space of bandlimited signals, as depicted in Fig. 4.11. This signal has the property that it is the closest bandlimited signal to $x(t)$: $\|x(t) - \hat{x}(t)\| \leq \|x(t) - y(t)\|$ for any bandlimited signal $y(t)$.

To summarize, the first LPF in Fig. 4.10 orthogonally projects the input signal onto the space of bandlimited signals. This projection can then be sampled, as in our direct

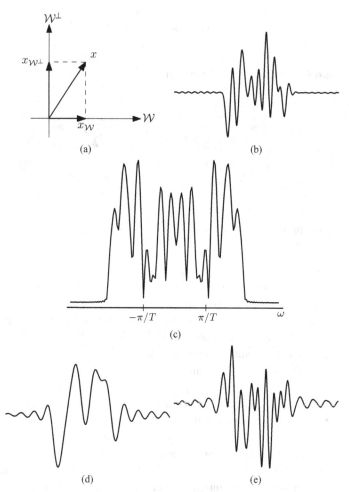

Figure 4.11 (a) Geometric interpretation of the orthogonal projection of a signal x onto a subspace $\mathcal{W}$. (b) A signal $x(t)$ not bandlimited to π/T. (c) The CTFT $X(\omega)$ of $x(t)$. (d) The projection $x_{\mathcal{W}}(t)$ of $x(t)$ onto the space $\mathcal{W}$ of π/T-bandlimited signals. (e) The residual $x_{\mathcal{W}^\perp}(t) = x(t) - x_{\mathcal{W}}(t)$.

interpretation of the Shannon–Nyquist theorem of Fig. 4.4. The recovered output is equal to the projected signal, which is by definition bandlimited. It is also equal to the original signal within the frequency interval $(-\pi/T, \pi/T)$. Obviously, if we used the scheme of Fig. 4.4 directly on the input $x(t)$, then the recovered output would also be bandlimited. However, it is no longer equal to the orthogonal projection of $x(t)$ onto the space $\mathcal{W}$ of bandlimited signals, and consequently, it results in a larger squared error than the output of Fig. 4.10. In addition, because of aliasing, the frequency content in the interval $(-\pi/T, \pi/T)$ is no longer equal to that of the original signal.

4.2.5 Towards more general sampling spaces

So far, we have introduced the Shannon–Nyquist bandlimited sampling theorem, and viewed it as a basis expansion. This interpretation also provided insight into the reconstruction effect on signals that are not bandlimited. Although widely used, this theorem relies on several fundamental assumptions that are often not satisfied in practice. First, natural signals are almost never perfectly bandlimited. Second, the sampling device is usually not ideal, i.e. it does not produce the exact signal values at the sampling locations. A common situation is that the ADC integrates the signal, usually over small neighborhoods around the sampling points. Moreover, nonlinear distortions are often introduced during sampling. Third, the bandlimiting operation tends to result in the Gibbs phenomenon (oscillations near discontinuities) which can be visually disturbing, for example when applied to images. Finally, using the sinc kernel for reconstruction is often impractical owing to its very slow decay.

To design recovery methods adapted to realistic scenarios, there are several issues that need to be properly addressed:

1. The sampling mechanism should be adequately modeled;
2. Prior knowledge on the class of input signals should be accounted for;
3. Limitations should be imposed on the reconstruction algorithm in order to ensure robust and efficient recovery.

In subsequent chapters we treat each of these three essential components of sampling. We focus on several models, which commonly arise in signal processing, image processing and communication systems. The first step we take is to move away from the world of bandlimited signals, to a richer class of inputs. Mathematically, this is equivalent to changing the basis functions. Practically, it allows for simpler sampling and recovery models, which do not rely on the use of ideal filters.

Before proceeding, we point out that one of the limitations listed above is that practically, ideal sampling is difficult to obtain. On the other hand, we illustrated in Fig. 4.2 that inner products with a shifted generator can be viewed as filtering followed by ideal uniform sampling. At first sight, it may seem like the problem remains: we still have ideal samples at the filter output. However, it is important to keep in mind that the interpretation offered by Fig. 4.2 is just a model of the sampling process. In practice, we do not need to use an ideal sampler but rather the filter is part of the sampling stage. For example, consider the scenario in which the sampler averages the signal over intervals of width Δ, so that the nth sample is given by

$$c[n] = \int_0^\Delta x(n\Delta - t)dt. \tag{4.25}$$

Clearly, this process does not involve ideal sampling. Nonetheless, for analysis purposes, it is convenient to model the samples as in Fig. 4.2 with $T = \Delta$ and

$$h(-t) = \begin{cases} 1, & 0 \leq t \leq \Delta \\ 0, & \text{otherwise.} \end{cases} \tag{4.26}$$

This allows a uniform treatment of a variety of different feasible sampling schemes.

4.3   Sampling in shift-invariant spaces

A simple way to generalize the Shannon–Nyquist theorem is to start from the interpretation depicted in Fig. 4.10. Recall that the first LPF is associated with the biorthogonal generator $\tilde{h}(t)$, and the second with the reconstruction generator $h(t)$. All other basis vectors are obtained by appropriate shifts. This viewpoint suggests a natural extension of the previous sampling theorem to a much broader class of theorems that allow the use of more practical filters. We simply replace the LPFs by a general function $h(t)$ and its biorthogonal pair $\tilde{h}(t)$, as depicted in Fig. 4.12. This extension retains the basic, shift-invariant flavor of the classical theory, and as such, will allow Fourier analysis similar to the bandlimited setting. For simplicity, we assume throughout that $h(t)$ and the sampling sequences are real; however, the results easily extend to the complex case.

The advantage of this generalization is that we now have more freedom in the choice of basis functions, so that we can select them to match desired properties. For example, $h(t)$ may be chosen as a compactly supported function if locality is needed. When hardware design is an important factor, it can be selected to have a simple analog implementation.

4.3.1   Shift-invariant spaces

To put the general sampling strategy of Fig. 4.12 into our linear algebra framework as in (4.17), we need to identify the sampling and reconstruction vectors associated with this sampling scheme. The samples $a[n]$ can be expressed as

$$a[n] = \tilde{h}(-t) * x(t)|_{t=nT} = \langle \tilde{h}(t - nT), x(t) \rangle. \qquad (4.27)$$

Therefore the sampling vectors associated with the figure are $\tilde{h}_n(t) = \tilde{h}(t - nT)$. Since all vectors are obtained by shifts of the generator $\tilde{h}(t)$, we refer to the set $\{\tilde{h}_n(t)\}$ as the basis generated by $\tilde{h}(t)$. Our next step is to identify the reconstruction vectors. The relation between $a[n]$ and the output $x(t)$ is given by

$$x(t) = h(t) * \sum_n a[n]\delta(t - nT) = \sum_n a[n]h(t - nT). \qquad (4.28)$$

Consequently, the reconstruction vectors are $\{h(t - nT)\}$, generated by shifts of $h(t)$.

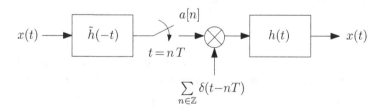

Figure 4.12 Generalizing the basic bandlimited sampling theorem.

We next characterize the class of signals that can be recovered by the system of Fig. 4.12. Regardless of the choice of coefficients $a[n]$, any output will have the form (4.28). For a given generator $h(t)$ in L_2, summing over all possible choices of sequences a leads to a subspace of L_2. This subspace has a shift-invariant property: if $x(t)$ lies in the subspace spanned by shifts of $h(t)$ then so do all of its shifts $x(t - mT)$ by an arbitrary integer m. Indeed,

$$x(t - mT) = \sum_n a[n]h(t - (n+m)T) = \sum_n a[n-m]h(t - nT). \qquad (4.29)$$

Since $x(t - mT)$ can also be expressed in the form $\sum_n b[n]h(t - nT)$ for some sequence b, it lies in the same subspace as $x(t)$. Because of this property, subspaces of the form (4.28) are referred to as *shift-invariant subspaces*. Formally, we have the following definition.

Definition 4.1. *A shift-invariant (SI) subspace (with a single generator) is the space of signals that can be expressed as linear combinations of shifts of a given generator:*

$$\mathcal{W} = \left\{ x(t) \,\middle|\, x(t) = \sum_{n \in \mathbb{Z}} a[n]h(t - nT),\ \textit{for some } a \in \ell_2 \right\}, \qquad (4.30)$$

where $h(t)$ is the SI generator and $a[n]$ are the expansion coefficients.

With a slight abuse of terminology, we will say that $h(t)$ generates $\mathcal{W}$. What we mean more precisely in this case is that the basis vectors are $\{h(t - nT), n \in \mathbb{Z}\}$.

Clearly, any continuous-time function $x(t)$ in $\mathcal{W}$ is defined by a countable set of coefficients $a[n]$. These coefficients provide a discrete representation of the signal that enables discrete-time signal processing of $x(t)$. In general, however, these are not equal to samples of the original signal, as we already saw in Fig. 2.1. Therefore, for a given signal $x(t)$ an important question is how to compute $a[n]$. We consider this problem in the next chapter, which is devoted to the study of SI spaces and their mathematical properties. One of the nice features of sampling in SI subspaces that we will discuss is that sampling and recovery can be obtained by a simple four-stage process: analog prefiltering, uniform sampling, digital correction (filtering), and analog postfiltering. This is precisely the setup in the Shannon–Nyquist theorem with the difference that in this more general setting the analog filters do not have to be ideal. In turn, we compensate by digital processing prior to recovery, if necessary. These steps translate to computation of the biorthogonal functions.

The class of signals that can be represented in the form (4.30) is quite large. Choosing $h(t) = \operatorname{sinc}(t/T)$ leads to the subspace of bandlimited signals, with bandwidth π/T. Other important examples are spline functions, and communication transmissions such as pulse amplitude modulation (PAM) and quadratic amplitude modulation (QAM). In these examples, $h(t)$ can be quite different from the sinc function, and may be simpler to handle numerically. We address each of these examples in the following subsections.

4.3.2   Spline functions

Owing to the importance of splines in signal and image processing and to their pervasive use in sampling and interpolation, here we elaborate more on this class of functions and their properties [65].

Splines were first mathematically described by Schoenberg in 1946 [66], several years before Shannon's landmark paper which promoted the use of bandlimited sampling. A spline is a smooth piecewise polynomial function. The term comes from the flexible spline devices used by shipbuilders and drafters to draw smooth shapes. The use of splines was pervasive in the aircraft and shipbuilding industries long before their formal introduction by Schoenberg. The signal processing and communications community followed in Shannon's footsteps, and to a large extent adopted the use of bandlimited signals, while splines pretty much lay dormant. Only in the 1960s did mathematicians begin to realize the many optimality properties associated with splines, such as the minimum curvature property, which led to research in other branches of applied mathematics including approximation theory and numerical analysis [67, 68]. Splines began impacting the engineering community in the 1980s, particularly in computer graphics [69, 70]. The 1990s witnessed a surge of interesting applications of splines within the signal processing community, due in large part to the fundamental works of Unser, Aldroubi, and Eden [65, 71, 72, 73]. One of their important contributions was to develop simple algorithms for computing the spline representation coefficients; we will discuss these algorithms in Chapter 9.

A spline of order n is a piecewise polynomial of order n, with the constraint that for $n \geq 1$ the segments are connected in such a way as to ensure continuity of the spline and all its derivatives up to order $n - 1$. Although in principle we need $n + 1$ parameters to describe an order-n polynomial, the continuity constraints result in only one degree of freedom per segment. The joining points of the polynomials are called *knots*. Here, we treat only the case of uniform spacing of the knots, and in particular we assume a spacing of $T = 1$. For $n = 0$ we have the class of piecewise constant functions, and for $n = 1$ the class of piecewise linear functions. Figure 4.13 depicts several examples of spline functions.

The remarkable result due to Schoenberg is that a spline of order n (with spacing $T = 1$) is uniquely defined in terms of a B-spline expansion

$$x(t) = \sum_m a[m]\beta^n(t - m), \tag{4.31}$$

where $\beta^n(t)$ is the central *B-spline* of degree n. The term B-spline was coined by Schoenberg and is short for basis spline. B-splines are symmetrical, bell-shaped functions constructed from the $(n + 1)$-fold convolution of a rectangular pulse $\beta^0(t)$:

$$\beta^n(t) = \underbrace{\beta^0(t) * \cdots * \beta^0(t)}_{(n+1) \text{ times}}, \tag{4.32}$$

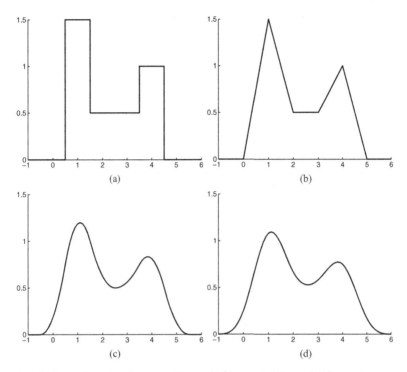

Figure 4.13 Splines of varying degree n. (a) $n = 0$. (b) $n = 1$. (c) $n = 2$. (d) $n = 3$.

where

$$\beta^0(t) = \begin{cases} 1, & -\frac{1}{2} < t < \frac{1}{2} \\ \frac{1}{2}, & |t| = \frac{1}{2} \\ 0, & \text{otherwise.} \end{cases} \tag{4.33}$$

Equation (4.31) establishes that any spline can be characterized by a countable set of coefficients $a[n]$. Therefore, even though a spline is a continuous-time function, it is enough to know its expansion coefficients. Figure 4.14 shows B-spline functions of degree 0 to 3. These functions are computed from (4.32), resulting in

$$\beta^1(t) = \begin{cases} 1 - |t|, & |t| \leq 1 \\ 0, & \text{otherwise,} \end{cases} \tag{4.34}$$

$$\beta^2(t) = \begin{cases} \frac{1}{2}\left(\frac{3}{2} - |t|\right)^2, & \frac{1}{2} \leq |t| \leq \frac{3}{2} \\ \frac{3}{4} - t^2, & |t| \leq \frac{1}{2} \\ 0, & \text{otherwise,} \end{cases} \tag{4.35}$$

$$\beta^3(t) = \begin{cases} \frac{1}{6}\left(2 - |t|\right)^3, & 1 \leq |t| \leq 2 \\ \frac{2}{3} - \frac{1}{2}t^2(2 - |t|), & |t| \leq 1 \\ 0, & \text{otherwise.} \end{cases} \tag{4.36}$$

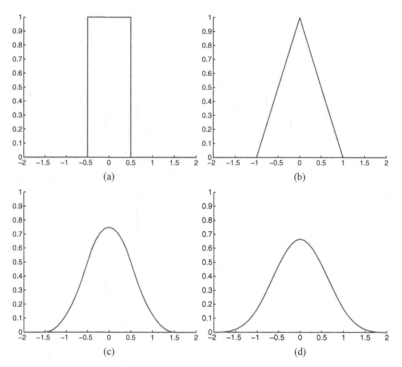

Figure 4.14 B-splines of varying degree n. (a) $n = 0$. (b) $n = 1$. (c) $n = 2$. (d) $n = 3$.

Note that, in sharp contrast to the sinc function, B-splines are compactly supported. Another fundamental difference is that it is no longer true that $a[n] = x(nT)$ in general. It is easy to see that this relation holds for splines of order $n = 0, 1$. However, for $n = 2$ and larger, it is no longer valid. For example, the spline $x(t)$ of degree 3, formed from the expansion coefficients $a[n] = \delta[n]$, is the B-spline function $\beta^3(t)$ of degree 3. As can be seen in Fig. 4.14, the samples of this function at the integers do not coincide with the sequence $\delta[n]$. In particular, $x(0) = 2/3 \neq 1$ and $x(1) = x(-1) = 1/6 \neq 0$. In Chapter 5 we will see more generally how to determine the coefficients in representations of the form (4.31).

One of the more popular choices of splines in applications are cubic splines corresponding to $n = 3$. These splines have a minimal curvature property, namely, they minimize the norm of the second derivative. More generally, it can be shown that splines of order n have the property that from all interpolators of a set of points, they minimize the norm of the $(n-1)$th derivative [67].

4.3.3   Digital communication signals

Digital modulation techniques offer many advantages over analog modulation including cost, power efficiency, spectral usage, error correction codes, speed, improved security and more. In digital communication, information is transferred in the form of bits. This information could be digital to begin with, or analog data that has been sampled and properly quantized. One of the popular modulation strategies is amplitude modulation (others, which we will not discuss, include frequency modulation). These techniques include PAM and QAM.

Generally speaking, in amplitude modulation, the information bit stream is represented by the amplitudes of a transmitted signal. Over a given symbol interval T_s, the bits are encoded onto a generator $h(t)$ which is then modulated to a carrier frequency f_c. In the nth symbol interval, the transmitted signal is

$$g_n(t) = a[n]h(t)\cos(2\pi f_c t). \tag{4.37}$$

The signal $g_n(t)$ is transmitted at time nT_s. Summing over all symbol intervals, the final transmitted signal is given by

$$s(t) = \sum_{n\in\mathbb{Z}} a[n]h(t - nT_s)\cos(2\pi f_c(t - nT_s)) = \sum_{n\in\mathbb{Z}} a[n]g(t - nT_s), \tag{4.38}$$

where $g(t) = h(t)\cos(2\pi f_c t)$. Therefore, PAM signals have the form (4.30).

A similar strategy is used in QAM with the difference that the information bits are encoded in both amplitude and phase of the transmitted signal, so that

$$g_n(t) = a[n]\cos(\phi_n)h(t)\cos(2\pi f_c t) - a[n]\sin(\phi_n)h(t)\sin(2\pi f_c t), \tag{4.39}$$

for some phase ϕ_n.

In the examples above $a[n]$ can typically only obtain values in a finite set, so that strictly speaking PAM and QAM signals do not form a subspace. However, at the receiver, the modified symbols $a[n]$ very often take on continuous values (owing to noise, fading, and other impairments), and therefore it is reasonable to assume an SI model. Furthermore, even when these values are in fact quantized, it is common to first process the signals as if they were continuous-valued in order to obtain a first estimate of $a[n]$. These estimates can then be quantized to their closest values to approximate the true unknown symbols.

Another interesting connection between digital communication and sampling is manifested in the Nyquist criterion for no *intersymbol interference* (ISI), where a symbol interferes with subsequent symbols. In a typical communication setting, the modulated symbol sequence is transmitted over a channel which often can be modeled as LTI. At the receiver, the observed signal has a similar form where the generator $h(t)$ is convolved with the channel impulse response $c(t)$ and a matched filter $\overline{h(-t)}$. The combined response $p(t) = h(t) * \overline{h(-t)} * c(t)$ causes the transmitted symbol to spread in the time domain, which can cause previously transmitted symbols to interfere with the currently received symbol. This interference is referred to as ISI. One way to avoid ISI is to choose $p(t)$ to satisfy what is known as the *Nyquist criterion*:

$$\frac{1}{T_s}\sum_{k\in\mathbb{Z}} P\left(\omega + \frac{2\pi k}{T_s}\right) = C, \tag{4.40}$$

where C is a constant. The left-hand side of (4.40) is the CTFT of an impulse train with spacing T_s and samples $p(nT_s)$. Therefore, this condition implies that $p(nT_s) = C\delta[n]$, so that the effective pulse shape equals zero at sampling points associated with past or future symbols. In Section 5.2.3 we discuss this property in more detail in connection with interpolation: a function $p(t)$ satisfying (4.40) is also called an *interpolation function*. From a sampling viewpoint, these functions have the property that the samples $a[n]$ in (4.30) are pointwise evaluations of the input, i.e., $a[n] = x(nT)$.

We conclude this subsection with several examples of Nyquist pulses (that is, pulses satisfying the Nyquist criterion (4.40)) used commonly in communications; these functions will also be useful in the context of interpolation.

Example 4.2 A simple example of a Nyquist pulse is the scaled B-spline function $p(t) = \beta^1(t/T)$. As can be seen in Fig. 4.14, this function is a triangular pulse which vanishes for every $t \notin (-T, T)$. Therefore, clearly $p(nT) = \delta[n]$, as required from a Nyquist pulse.

Example 4.3 As implied by the Shannon–Nyquist sampling theorem, the function $p(t) = \mathrm{sinc}(t/T)$ is a Nyquist pulse, because $p(nT) = \delta[n]$. The disadvantage of this pulse is that it has very slow decay. One common modification of the sinc kernel, which retains the Nyquist property, is given by

$$p(t) = \mathrm{sinc}\left(\frac{t}{T}\right) \frac{\cos\left(\frac{\pi\beta t}{T}\right)}{1 - \left(\frac{2\beta t}{T}\right)^2} \tag{4.41}$$

where $0 \le \beta \le 1$ is called the *roll-off factor*. Owing to the sinc term, it is immediately clear that this pulse satisfies $p(nT) = \delta[n]$. The parameter β controls the decay of $p(t)$ in the time domain. When $\beta = 0$, $p(t)$ reduces to a sinc pulse, which decays slowly. As β approaches 1, the decay becomes more rapid. This behavior is demonstrated in Fig. 4.15.

The CTFT of $p(t)$ is given by

$$P(\omega) = \begin{cases} T, & |\omega| \le \frac{\pi(1-\beta)}{T} \\ \frac{T}{2}\left(1 - \sin\left(\frac{T}{2\beta}\left(|\omega| - \frac{\pi}{T}\right)\right)\right), & \frac{\pi(1-\beta)}{T} < |\omega| \le \frac{\pi(1+\beta)}{T} \\ 0, & |\omega| > \frac{\pi(1+\beta)}{T}. \end{cases} \tag{4.42}$$

As can be seen from the formula, and is also evident in Fig. 4.15, the larger β is, the larger the bandwidth of the pulse. When $\beta = 1$, $P(\omega)$ reduces to

$$P(\omega) = \begin{cases} \frac{T}{2}\left(1 + \cos\left(\frac{\omega T}{2}\right)\right), & -\frac{2\pi}{T} < \omega < \frac{2\pi}{T} \\ 0, & |\omega| \ge \frac{2\pi}{T}, \end{cases} \tag{4.43}$$

for which reason $p(t)$ is commonly referred to as the *raised cosine pulse*.

The fact that $p(t)$ is a Nyquist pulse can also be easily seen in the frequency domain (see Exercise 10). Specifically, owing to the antisymmetry of $P(\omega)$ around

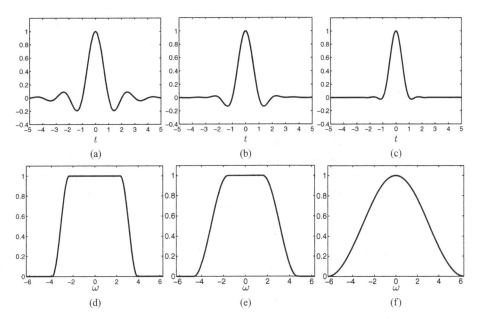

Figure 4.15 Raised cosine pulses $p(t)$ and the corresponding CTFTs $P(\omega)$. (a) $p(t)$ with $\beta = 0.25$. (b) $p(t)$ with $\beta = 0.5$. (c) $p(t)$ with $\beta = 1$. (d) $P(\omega)$ with $\beta = 0.25$. (e) $P(\omega)$ with $\beta = 0.5$. (f) $P(\omega)$ with $\beta = 1$.

$|\omega| = \pi/T$, the function $P(\omega) + P(\omega - 2\pi/T)$ is constant in the transition range $\omega \in [(1-\beta)\pi/T, (1+\beta)\pi/T]$. Therefore, it is easy to verify that $\sum_{k\in\mathbb{Z}} P(\omega + 2\pi k/T)$ is constant everywhere, implying that $p(t)$ is a Nyquist pulse.

4.3.4   Multiple generators

Definition 4.1 can be extended to include subspaces that are defined by more than one generator [45, 74]. The underlying SI space then includes signals of the form

$$x(t) = \sum_{\ell=1}^{N} \sum_{n\in\mathbb{Z}} a_\ell[n] h_\ell(t - nT), \tag{4.44}$$

where $h_\ell(t), 1 \leq \ell \leq N$ are the SI generators, and $a_\ell[n]$ are the corresponding expansion coefficients.

Representations of the form (4.44) allow greater flexibility than subspaces formed by a single generator. For example, such expansions have been used in finite-element methods to facilitate inversion of matrices involved in the computation of differential equations. In this setting the functions $h_\ell(t)$ are chosen to have small support, and no overlap [75]. Using several functions allows for shorter support than a single generator. Multiple generators are also used in the context of multiwavelet constructions. We will briefly discuss these application areas in the following subsection.

Throughout the book, our emphasis is on subspaces of $L_2(\mathbb{R})$ with a finite set of generators; nonetheless, one of the nice properties of the model (4.44) is that if we allow for infinitely many generators $h_\ell(t)$, then the entire space $L_2(\mathbb{R})$ can be represented in terms of their shifts. Implementing the resulting sampling scheme will entail filtering with infinitely many filters, which is obviously impractical. Fortunately, though, if the generators $h_\ell(t)$ are chosen such that they are all related through a simple transformation, then sampling and recovery can be performed efficiently. Two important examples are wavelet and Gabor expansions: both lead to decompositions of an arbitrary signal in $L_2(\mathbb{R})$ into a countable set of coefficients of the form (4.44). In the wavelet setting, the generators are related to one another through scaling, while in the Gabor representation they are related via modulation. These transformations will be discussed in Section 4.4.

We now provide some examples of multiple-generator SI subspaces (4.44).

Example 4.4 An important example of the model (4.44) is the class of multiband signals. We will study such signals in detail in Chapter 14. For convenience, we define a signal as multiband if it consists of at most N frequency bands, each of length no larger than B. In addition, the signal is bandlimited to π/T. We focus here on the positive frequency axis. A representative signal is depicted in Fig. 4.16.

There are several ways to express such signals in the form (4.44). The simplest approach is to associate a generator $h_\ell(t)$ with each one of the frequency bands that exist in the signal. More specifically, we choose $h_\ell(t)$ so that its Fourier transform $H_\ell(\omega)$ is a box function of width B, centered at ω_ℓ:

$$h_\ell(t) = \frac{2\pi}{B} \operatorname{sinc}\left(\frac{Bt}{2\pi}\right) e^{-j\omega_\ell t}. \tag{4.45}$$

The first part of the expression in (4.45) represents an LPF with cutoff $B/2$. The modulation creates the frequency shift.

Using the functions (4.45) to represent $x(t)$ results in an expansion in which the generators depend on the frequencies ω_ℓ. In some applications, these carrier frequencies may not be known. We discuss this scenario at length in Chapter 14. In such cases it is useful to have a representation in which the generators are independent of the unknown frequencies. To this end, we divide the frequency interval $[0, \pi/T)$ into m sections, each of equal length $\pi/(mT)$. It follows that if

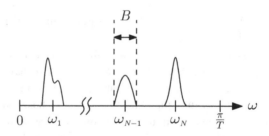

Figure 4.16 A representative multiband signal.

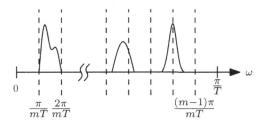

Figure 4.17 Representing multiband signals via slicing of the frequency axis into fixed cells.

$m \leq \pi/(BT)$ then each signal band is contained in no more than two intervals, as demonstrated in Fig. 4.17. Since there are N bands, this implies that at most $2N$ sections contain energy, and this number will generally be much smaller than m. We therefore define the m generators

$$h_\ell(t) = 2mT \operatorname{sinc}\left(\frac{t}{2mT}\right) e^{-j\left(\frac{\ell-1/2}{mT}\right)\pi t} \tag{4.46}$$

for $\ell = 1, \ldots, m$. We then choose among these the $2N$ generators corresponding to the active bands.

Note that the price for using a set of generators on a fixed grid is that the number of generators needed is double that required by (4.45).

Example 4.5 SI spaces with multiple generators are also encountered in filterbank applications. Specifically, assume that a signal $x(t)$ is sampled at times $t = nT$ after passing through the filters $\overline{h_1(-t)}, \ldots, \overline{h_N(-t)}$, as shown in Fig. 4.18. Since the nth sample at the output of the ℓth channel is given by the inner product $\langle h_\ell(t - nT), x(t)\rangle$, the sampling space is the closed linear span of $\{h_\ell(t - nT)\}_{n\in\mathbb{Z},\ell=1,\ldots,N}$, which is a multiple-generator SI space. When using an N-channel sampling system, the average sampling rate is given by N/T. The increase with respect to the sampling rate of a single channel, though, is not obtained by refining the sampling grid, but rather by sampling different properties of the signal on the original low-rate grid. This is a popular method to increase rate, while requiring only low-rate samplers on each channel.

For example, the Shannon–Nyquist theorem asserts that a $\pi N/T$-bandlimited signal can be recovered from its samples taken with period T/N. In 1977, Papoulis [76] showed that alternatively, we can recover such a signal from samples of its value and derivatives up to order $N - 1$ taken with a period of T. This setting corresponds to using the filters

$$H_\ell(\omega) = \begin{cases} (j\omega)^\ell, & |\omega| < \pi N/T \\ 0, & |\omega| \geq \pi N/T, \end{cases} \tag{4.47}$$

in Fig. 4.18. More generally, Papoulis derived conditions on the N filters such that a bandlimited signal can be recovered from the resulting samples. We will discuss

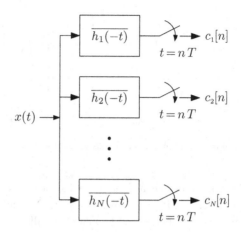

Figure 4.18 A multichannel sampling system.

recovery methods in the context of multiple-generator SI spaces as well as Papoulis' theorem in Section 6.8.

Example 4.6 Another popular application of multiple generators is interleaved ADCs in which several low-rate uniform samplers are combined to achieve an effective higher rate. This approach can be viewed as a special case of Papoulis' generalized sampling, introduced in the previous example.

In applications in which the Nyquist rate N/T is very high, it is often convenient to achieve the same average rate using a series of low-rate ADCs. More specifically, instead of using a single sampler with period T/N, we can apply N samplers, each with period T. If the input to the ℓth sampler is delayed by $\ell T/N$, then the samples at the output of the filterbank will correspond to the original uniform samples at rate N/T. More generally, the ℓth channel input may be delayed by an arbitrary value $0 \leq \tau_\ell < T$, leading to samples

$$c_\ell[n] = x(nT + \tau_\ell), \quad 1 \leq \ell \leq N, n \in \mathbb{Z}. \tag{4.48}$$

These samples can be viewed as the output of a filterbank as in Fig. 4.18, with filters $h_\ell(-t) = \delta(t + \tau_\ell)$.

Such a strategy is commonly referred to as *recurrent nonuniform sampling* or *periodic nonuniform sampling*: the sampling points can be divided into groups of N with a recurrent period of T, where the first group consists of the samples $x(\tau_1), \ldots, x(\tau_N)$ [77]. Recurrent nonuniform sampling often results when using interleaved ADCs that are not perfectly synchronized [78, 79]. As long as the delays are distinct, recovery of the bandlimited input is possible from the samples (4.48). We treat interleaved ADCs in greater detail in Sections 6.8.2 and 14.3.

4.3.5   Refinable functions

Multiple generators have received a lot of attention in the context of wavelet constructions where the goal is to design wavelets with as many desirable properties as possible.
Wavelets are a special case of *refinable functions*, which we briefly describe in this subsection. We consider wavelets in more detail in Section 4.4.2. Refinable functions also
lie at the heart of finite-element methods for solving differential equations. The use of
multiple generators in these contexts is quite popular, owing to the fact that this allows
for functions that are symmetric, orthogonal, and have finite support.

A function $x(t)$ is said to be refinable if there exists a sequence $c[n]$ and a value N
such that

$$x(t) = \sum_{n=0}^{N} c[n]x(2t - n). \tag{4.49}$$

This equation is satisfied, for example, by splines. It is also the starting point for the
construction of wavelets, as we will see below in Section 4.4.2. The properties of $x(t)$
are determined by the coefficients $c[n]$. When using finite elements, a function $f(t)$ is
approximated by combinations of the form

$$f(t) \approx \sum_{n \in \mathbb{Z}} a[n]x(t/h - n), \tag{4.50}$$

for some $h > 0$. The approximation error decreases as h^p, where p is the number of
polynomials $1, t, \ldots, t^{p-1}$ that can be exactly reproduced from combinations of the
translates $x(t - n)$. The value of p may be determined from the properties of $c[n]$ in
(4.49). Specifically, denoting by $C(e^{j\omega})$ the DTFT of $c[n]$, we say that $x(t)$ has a power
of approximation p if

$$C(e^{j\cdot 0}) \neq 0, \quad C(e^{j\omega}) \text{ has a zero of order } p \text{ at } \omega = \pi. \tag{4.51}$$

For example, when $x(t)$ is the box function on $[0, 1]$ (leading to the Haar wavelet in the
context of wavelet theory), $p = 1$. More generally, splines of degree $p-1$ can reproduce
polynomials of degree p and therefore have pth-order accuracy.

To achieve higher orders of approximation than those obtained when using a single function as in (4.49) one may consider vector valued functions: in this case,
$x(t)$ in (4.49) becomes a vector with elements $x_\ell(t)$. The set of functions $\{x_\ell(t)\}$
is said to be refinable if they are combinations of the scaled and translated functions
$\{x_\ell(2t - n)\}$. In the context of wavelets, such functions are referred to as *multiwavelets*
[80]. The advantage of using multiple functions is that the same order of accuracy p
can be achieved as in the scalar case (4.49), but with additional desired properties. For
example, the functions $x_\ell(t)$ may be symmetric with shorter support.

In the context of wavelet theory, a refinable function $x(t)$ is referred to as a *scaling
function* and denoted $\phi(t)$. It can be used to create the corresponding wavelet function $w(t) = \sum_n d[n]\phi(2t - n)$ for appropriate coefficients $d[n]$, chosen such that $w(t)$
and $\phi(t)$ are orthogonal. Dilations and translations of $w(t)$ create an orthogonal basis for
$L_2(\mathbb{R})$. However, designing wavelets with several properties, such as short time support,

orthogonality, and reproducing polynomials of high orders, is difficult. For example, the only orthogonal symmetric wavelet is the Haar wavelet. In contrast, when using multiwavelets, corresponding to vector functions $x(t)$ in (4.49), one can create symmetric orthogonal wavelets with higher orders of approximation. Furthermore, such wavelets may be shorter and have higher p than a single wavelet.

4.4   Gabor and wavelet expansions

In the previous section we considered SI spaces with multiple generators. As we already pointed out, if the sum over ℓ in (4.44) is taken to infinity, then with appropriate choices of $h_\ell(t)$ we can express any signal $x(t)$ in $L_2(\mathbb{R})$ via coefficients $a_\ell[n]$. The fact that any signal in $L_2(\mathbb{R})$ has a discrete representation is not surprising: we already noted that this space is separable which immediately implies that a countable representation exists. However, the nice result is that the basis vectors can be chosen to yield an SI representation with multiple generators. Furthermore, the generators $\{h_\ell(t)\}$ may be constructed to obey a simple relationship leading to efficient computation of the expansion coefficients.

Our primary focus in this book is on representations with finitely many generators. Nonetheless, the ideas we develop in the ensuing chapters are applicable to the infinite setting as well with appropriate modifications. For completeness, in this section we depart from the finite setting, and briefly mention two classes of SI representations that had a profound impact on signal processing: Gabor and wavelet expansions. Both of these transforms have been the subject of extensive research and many excellent textbooks exist that cover these topics; see e.g. [26, 33, 59, 81, 82, 83, 84, 85]. Below, we very briefly highlight some of the main features of these expansions, and their connection to the SI setting treated in subsequent chapters.

Gabor and wavelet expansions are special cases of SI spaces with infinitely many generators. Their importance stems from the fact that the generators are related in a simple way, leading to fast algorithms for computing the expansion coefficients and explicit conditions under which the generators lead to stable expansions. Therefore, in essence, a large part of wavelet and Gabor theory deals with applying the concepts developed in Chapter 2 to the specific structure implied by these settings. Beyond the fact that these expansions are computationally attractive, their main advantage is in their ability to capture important properties of the underlying signals. This results in representation coefficients that are easy to interpret, and carry meaningful signal information.

4.4.1   Gabor spaces

Gabor analysis, named after Dennis Gabor, has become a popular signal processing method for decomposing and reconstructing signals from their time–frequency projections. It has been successfully used in many applications such as noise suppression, blind source separation, echo cancelation, system identification, and pattern and object recognition, among others.

Part of the appeal of finitely generated SI subspaces is their tight relationship with the Fourier transform: the analysis of SI signals is easily carried out in the Fourier domain. However, many realistic signals have time-varying frequency content. Fourier analysis is not well suited to represent local information in time, since the underlying building blocks, the complex exponentials, are not time-localized. If the signal being analyzed changes in time, then typically all the Fourier coefficients will change as well. Gabor analysis attempts to build up local exponential blocks to better represent nonstationary signals, and in such a way studies the frequency content of local sections of a signal as it varies in time. Gabor spaces can also be modeled as SI with generators that are related through modulation. Taking infinitely many shifts and modulations allows one to represent the entire space of $L_2(\mathbb{R})$ by building blocks that are localized in time and frequency. One of the advantages of Gabor analysis is the highly structured system inherited from the uniform time–frequency lattice which allows for efficient computational algorithms. Furthermore, as we have already pointed out, the resulting coefficients capture the local frequency behavior of the signal and therefore can be easily interpreted.

In previous sections, we considered signal expansions by starting from the synthesis equation, namely, by writing $x(t)$ as a linear combination of shifts of a given generator. As we pointed out, SI bases also have SI biorthogonal bases so that the samples, or expansion coefficients, can be computed by taking inner products with shifts of a biorthogonal generator. The same holds true for Gabor spaces. Therefore, in order to follow the traditional exposition of Gabor analysis, we begin here from the expansion coefficients.

The Gabor transform was formulated in Gabor's 1946 paper [86], where he studied methods to locally represent communication signals. His idea was to use a 2D transform, with coordinates given by time and frequency. For an arbitrary signal $x(t)$, the *Gabor transform* is defined as

$$g[k, n] = \int_{-\infty}^{\infty} e^{j2\pi Wkt} x(t) h(t - nT) dt, \tag{4.52}$$

where $W, T > 0$ are typically chosen such that $WT = p/q$ with p and q relatively prime. Here $h(t)$ is a given window, referred to as the *analysis window*. The basic building blocks in this representation are translations by multiples of T, and modulations by multiples of W, of the window $h(t)$, as depicted in Fig. 4.19. The idea behind this transform is that if the window function $h(t)$ is localized in time, and its Fourier transform $H(\omega)$ is localized in frequency, then the function

$$h_{kn}(t) = e^{j2\pi Wkt} h(t - nT) \tag{4.53}$$

is localized in the time–frequency plane around nT, kW. Therefore, the knth coefficient $g[k, n]$ captures information about $x(t)$ over that time–frequency region.

It can be shown that if $\{h_{kn}(t)\}$ constitutes a frame or Riesz basis for $L_2(\mathbb{R})$, then the biorthogonal functions have a similar form. Namely, they are obtained as shifts and modulations of a different generator $v(t)$. In this case, any $x(t) \in L_2(\mathbb{R})$ can be written as

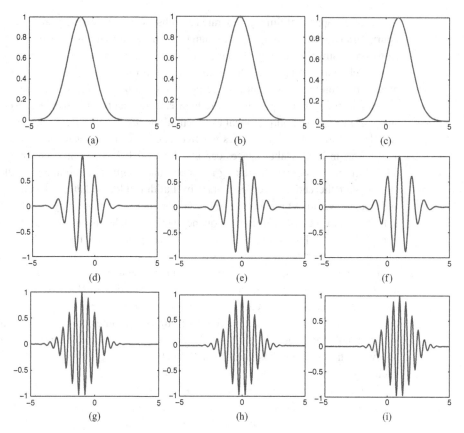

Figure 4.19 Translations and modulations of the function $h(t) = \exp\{-t^2/2\}$ with periods $T = 1$ and $W = 1$ (only real part is shown). (a) $(k, n) = (0, -1)$. (b) $(k, n) = (0, 0)$. (c) $(k, n) = (0, 1)$. (d) $(k, n) = (1, -1)$. (e) $(k, n) = (1, 0)$. (f) $(k, n) = (1, 1)$. (g) $(k, n) = (2, -1)$. (h) $(k, n) = (2, 0)$. (i) $(k, n) = (2, 1)$.

$$x(t) = \sum_{k,n \in \mathbb{Z}} g[k, n] e^{j2\pi Wkt} v(t - nT). \tag{4.54}$$

The synthesis window $v(t)$ is referred to as the *dual* of $h(t)$, and can be found via the Zak transform method [87], or using one of several available iterative algorithms [85]. For given lattice constants W, T the window $v(t)$ generates a dual Gabor frame if and only if

$$\langle v(t), e^{j2\pi kt/T} h(t - n/W) \rangle = WT\delta_{k0}\delta_{n0}, \quad k, n \in \mathbb{Z}. \tag{4.55}$$

This condition is referred to as the *Wexler–Raz condition* [88]. Another interesting relation between the lattices defined by W, T and those defined by $1/T, 1/W$ is given by the Ron–Shen principle [89]. They proved that $h(t)$ generates a frame on the lattice W, T if and only if it generates a Riesz basis for its closed linear span, with lattice constants $1/T, 1/W$.

Uncertainty principle

In his original paper, Gabor chose the Gaussian window $h(t) = e^{-\pi t^2}$, with $WT = 1$. The distinct feature of the Gaussian window is that it is most concentrated in both time and frequency: that is, it satisfies the uncertainty principle on time and frequency resolution with equality.

The *uncertainty principle* states that for any function $x(t)$ in $L_2(\mathbb{R})$, and for all points (t_0, ω_0) in the time–frequency plane,

$$\|x(t)\|^2 \leq 4\pi \|(t - t_0)x(t)\| \|(\omega - \omega_0)X(\omega)\|, \tag{4.56}$$

where $\| \cdot \|$ denotes the usual L_2-norm. Equality is achieved only by functions of the form

$$x(t) = Ce^{-a(t-t_0)^2}e^{j2\pi\omega_0}, \quad a > 0, C \in \mathbb{C}, \tag{4.57}$$

namely, modulated and translated Gaussians. Intuitively, the time–frequency tradeoff can also be seen by noting that the Fourier transform of a scaled function $x(at)$ with $a < 1$ is given by $(1/a)F(\omega/a)$: the Fourier transform is dilated by $1/a$. This means that we lose frequency resolution when trying to localize in time. A formal proof of this result is developed in Exercise 12. The uncertainty principle was first derived by Heisenberg in the context of quantum mechanics, as it poses limits on the ability to determine the position and momentum of a free particle.

Choice of lattice parameters

An important question is how to choose the lattice parameters W, T. In general, the Gabor transform is redundant: that is, it provides more coefficients than needed for recovery of $x(t)$. This redundancy is determined by the product of the sampling intervals in time and frequency, i.e. WT. As W and T are increased, there are fewer coefficients per unit time for any given frequency range. In order for the Gabor coefficients to be sufficient for representing an arbitrary signal $x(t)$ in $L_2(\mathbb{R})$, it is necessary that $WT \leq 1$. When $WT > 1$ it can be shown that the functions $\{h_{kn}(t)\}$ of (4.53) are incomplete in $L_2(\mathbb{R})$ for any window $h(t) \in L_2(\mathbb{R})$. When $WT = 1$, the functions are nonredundant.

When $h(t)$ is chosen as a Gaussian window, the functions $h_{kn}(t)$ are complete in $L_2(\mathbb{R})$ for $WT \leq 1$ [90]. However, if $WT = 1$, then they do not form a Riesz basis [91]. Therefore, in this setting originally proposed by Gabor, there is no numerically stable method to reconstruct $x(t)$ from the samples $\langle h_{kn}(t), x(t) \rangle$. When $WT < 1$ it can be shown that the Gaussian window generates a frame.

Although the Gaussian function does not produce a Riesz basis for $WT = 1$, there are other window functions that do. As an example, it is easy to see that the box function (which is equal to 1 on $0 \leq t \leq 1$ and 0 otherwise), and the sinc function $h(t) = \sin(\pi t)/(\pi t)$, create orthonormal bases functions for $L_2(\mathbb{R})$ when $W = T = 1$. However, both these examples are poorly localized: the box function is badly localized in frequency, while the sinc is poorly localized in time. The fact that for $WT = 1$ it is difficult to generate stable expansions with good localization in both time and frequency is a result of the *Balian–Low theorem*, which is the reason why frame expansions are popular in Gabor analysis [92]. This theorem states that if $h_{kn}(t)$ produce a Riesz basis for the Hilbert space $L_2(\mathbb{R})$, then

$$\int_{\infty}^{\infty} t^2 h(t) dt = \infty \quad \text{or} \quad \int_{\infty}^{\infty} \omega^2 H(\omega) d\omega = \infty. \tag{4.58}$$

Therefore, we cannot have a basis expansion that is localized in both time and frequency.

Shift and modulation (SMI) invariant spaces

If we choose W and T such that $WT > 1$, then perfect recovery is possible only for a subspace of $L_2(\mathbb{R})$. This subspace can be viewed as a generalization of the SI prior to a richer class of subspaces, termed *shift and modulation (SMI) invariant spaces* [93]. This choice of terminology follows from the fact that if $x(t)$ lies in the space, then shifts and modulations of $x(t)$ also lie in the space.

An SMI subspace is the space of signals that can be written as

$$x(t) = \sum_{k,n \in \mathbb{Z}} g[k,n] e^{j2\pi W k t} v(t - nT), \tag{4.59}$$

for some generator $v(t)$ and coefficients $g[k,n]$. This class of functions can be viewed as SI with infinitely many generators

$$g_k(t) = e^{j2\pi W k t} v(t), \quad k \in \mathbb{Z}. \tag{4.60}$$

Evidently, in this setting, all the generators are related to each other. This relationship can be exploited, for example, to obtain explicit conditions under which the generators form a Riesz basis. The interested reader is referred to [93, 94, 95] for more details.

SMI expansions allow us to retain many of the appealing features of Gabor expansions, while reducing the computational load by increasing the time and frequency sampling intervals. For arbitrary signals in $L_2(\mathbb{R})$, this will result in an inevitable recovery error. However, the reduced complexity may outweigh the approximation error, which often can be controlled [93].

4.4.2   Wavelet expansions

Wavelet expansions are used to decompose a signal into a particular set of basis functions called wavelets, which are obtained from scaling and translations of the mother wavelet. The important difference that distinguishes the wavelet transform from Fourier and Gabor analysis is its time and frequency localization properties. In the Gabor transform, the frequency scale is linear, while in contrast, wavelets correspond to logarithmic frequency analysis, or constant relative bandwidth. It is well known, for example, that the human auditory system uses constant relative bandwidth, as is evident in musical notes. Audio compression systems are based on this property.

The basis functions comprising the wavelet transform are related through shifts, and scaling by a factor of 2, resulting in a logarithmic frequency axis. This leads to a very different decomposition of the time–frequency plane, as depicted in the tiling diagrams

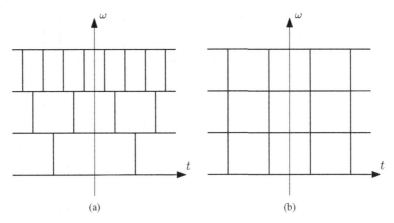

Figure 4.20 Tiling of the time–frequency plane. (a) The wavelet transform. (b) The Gabor transform.

of Fig. 4.20. In these diagrams, the different rectangles represent the energy concentration of a given basis function (Gabor used such tiling diagrams in [86]). When using Gabor analysis, the time–frequency plane is divided into equal rectangles. On the other hand, in wavelet decompositions, lower frequencies have larger time steps, while higher frequencies are sampled more frequently.

Although the wavelet transform has come into prominence during the past two decades, the founding principles behind wavelets can be traced back as far as 1909 when Haar discovered that for any continuous function $x(t)$, the series

$$x(t) = \sum_{k=0}^{\infty} \sum_{n=0}^{2^k - 1} a_k[n]\psi(2^k t - n), \quad 0 \leq t < 1, \tag{4.61}$$

converges uniformly to $x(t)$, where $\psi(t)$ is the *Haar wavelet*. This is the simplest of the orthogonal wavelets: a set of rectangular basis functions depicted in Fig. 4.21. A drawback of the Haar basis function is the fact that it is discontinuous and therefore inefficient in modeling smooth signals.

The final foundations of wavelet analysis were laid in the research of Grossmann and Morlet [96]. The transition from continuous signal processing to DSP was achieved by Mallat [61] and Daubechies [97]. Since then there has been a proliferation of activity, with comprehensive studies expanding on the wavelet transform and its implementation into many fields of science including multiresolution signal processing, image and data compression, telecommunications, numerical analysis, and speech processing.

The wavelet decomposition is based on a multiresolution analysis: that is, a function $x(t)$ is represented at different levels of resolution. To achieve this, we begin by constructing a sequence of embedded spaces $\mathcal{V}_i$ such that

$$0 \cdots \subset \mathcal{V}_{-1} \subset \mathcal{V}_0 \subset \mathcal{V}_1 \cdots \subset L_2(\mathbb{R}), \tag{4.62}$$

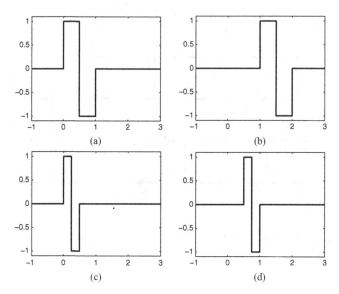

Figure 4.21 Translations and modulations of the Haar mother wavelet function. (a) $(k, n) = (0, 0)$. (b) $(k, n) = (0, 1)$. (c) $(k, n) = (1, 0)$. (d) $(k, n) = (1, 1)$.

with the following properties:

1. $\bigcup_i \mathcal{V}_i$ is dense in $L_2(\mathbb{R})$;
2. $\bigcap_i \mathcal{V}_i = \{0\}$;
3. The embedded subspaces are related by a scaling law:

$$x(t) \in \mathcal{V}_j \text{ only if } x(2t) \in \mathcal{V}_{j+1}; \tag{4.63}$$

4. Each subspace is SI with a single generator.

Once a multiresolution embedding is found, we seek a scaling function $\phi(t)$ such that $\{\phi(t - n)\}$ forms a Riesz basis for $\mathcal{V}_0$. From the properties above, this then implies that $\{\phi(2t - n)\}$ forms a basis for $\mathcal{V}_1$. Since $\mathcal{V}_0 \subset \mathcal{V}_1$, we can express any function in $\mathcal{V}_0$ as a linear combination of $\{\phi(2t - n)\}$:

$$\phi(t) = \sum_{n \in \mathbb{Z}} a[n] \phi(2t - n), \tag{4.64}$$

for some $a \in \ell_2$. The relation (4.64) is referred to as the *dilation equation* or the *scaling relation*. We next define the functions

$$\phi_{kn}(t) = 2^{k/2} \phi(2^k t - n), \tag{4.65}$$

for any integers $k, n \in \mathbb{Z}$. For each k_0, the functions $\{\phi_{k_0 n}\}$ form a basis for $\mathcal{V}_{k_0}$. Clearly any such subspace is SI by shifts of size 2^{-k_0}. Note also that $\phi(t)$ can be used to define a complete set of functions for the Hilbert space L_2 by means of dyadic translations and dilations as in (4.65). The dilation parameter k is referred to as the *scale*.

Example 4.7 In the Haar wavelet transform, the scaling function $\phi(t)$ is given by

$$\phi(t) = \begin{cases} 1, & 0 \le t < 1 \\ 0, & \text{else.} \end{cases} \tag{4.66}$$

This function satisfies the two-scale relation $\phi(t) = \phi(2t) + \phi(2t - 1)$. Therefore, (4.64) holds with $a[0] = a[1] = 1$ and $a[n] = 0$ for $n \ne 0, 1$.

Example 4.8 A B-spline $\phi(t) = \beta^p(t)$ of any odd degree $p \ge 0$ satisfies (4.64). To see this, we use the fact that (4.64) is equivalent to

$$\Phi(2\omega) = \frac{1}{2}A(e^{j\omega})\Phi(\omega), \tag{4.67}$$

where $\Phi(\omega)$ is the CTFT of $\phi(t)$ and $A(e^{j\omega})$ is the DTFT of $a[n]$ (see Exercise 16).

Using the convolution property of B-splines (4.32), the CTFT $\Phi(\omega)$ of a degree-p B-spline is given by

$$\Phi(\omega) = \text{sinc}^{p+1}\left(\frac{\omega}{2\pi}\right). \tag{4.68}$$

Therefore, (4.67) is satisfied if[1]

$$A(e^{j\omega}) = \begin{cases} 2^{-p}\frac{\sin^{p+1}(\omega)}{\sin^{p+1}(\omega/2)}, & \omega/(2\pi) \text{ is not an integer} \\ 2, & \text{else.} \end{cases} \tag{4.69}$$

This function can be written as

$$A(e^{j\omega}) = 2^{-p}\left(\frac{1 - e^{-2j\omega}}{1 - e^{-j\omega}}\right)^{p+1} e^{-j\omega\frac{p+1}{2}}. \tag{4.70}$$

The term inside the parentheses is equal to $1 + e^{-j\omega}$ and is the DTFT of the sequence

$$b[n] = \begin{cases} 1, & n = 0, 1 \\ 0, & \text{else.} \end{cases} \tag{4.71}$$

Therefore, the sequence $a[n]$ of the relation (4.64) can be obtained in this example by convolving $b[n]$ with itself $p + 1$ times, shifting the result by $(p + 1)/2$, and then scaling by 2^{-p}.

We conclude that any scaling space, namely a space spanned by a scaling function, is SI. In fact, as we now show, the corresponding wavelet spaces are also SI. The kth wavelet space $\mathcal{W}_k$ is the orthogonal complement of $\mathcal{V}_k$ in $\mathcal{V}_{k+1}$:

$$\mathcal{V}_{k+1} = \mathcal{V}_k \oplus \mathcal{W}_k, \quad \mathcal{V}_k \perp \mathcal{W}_k. \tag{4.72}$$

---

[1] In fact, the values of $A(e^{j\omega})$ for $\omega/(2\pi)$ an integer can be chosen arbitrarily, as both sides of (4.67) vanish at those locations regardless of $A(e^{j\omega})$. The choice $A(e^{j2\pi k}) = 2$ results in a continuous function $A(e^{j\omega})$.

A nice feature of the wavelet spaces is that they are orthogonal, and their sum is equal to $L_2(\mathbb{R})$.

We next define a wavelet function $\psi(t)$ that generates $\mathcal{W}_0$. Then, similarly to (4.65), we have that

$$\psi_{kn}(t) = 2^{k/2}\psi(2^k t - n),\tag{4.73}$$

is a basis for $\mathcal{W}_k$. Since $\mathcal{W}_0$ is contained in $\mathcal{V}_1$, we can express $\psi(t)$ in terms of the scaling function at the next level:

$$\psi(t) = \sum_{n\in\mathbb{Z}} b[n]\phi(2t - n).\tag{4.74}$$

Example 4.9 As depicted in Fig. 4.21, the Haar mother wavelet function $\psi(t)$ is given by

$$\psi(t) = \begin{cases} 1, & 0 \leq t < 1/2 \\ -1, & 1/2 \leq t < 1 \\ 0, & \text{else.} \end{cases}\tag{4.75}$$

This function can be expressed in terms of the Haar scaling function: $\psi(t) = \phi(2t) - \phi(2t - 1)$. Therefore (4.74) holds with $b[0] = 1$, $b[1] = -1$ and $b[n] = 0$ for every $n \neq 0, 1$.

Since the nested subspaces are dense in $L_2(\mathbb{R})$, any $x(t) \in L_2(\mathbb{R})$ can be expressed in terms of a wavelet expansion as

$$x(t) = \sum_{k,n\in\mathbb{Z}} a_{kn}\psi_{kn}(t),\tag{4.76}$$

where the coefficients a_{kn} are given by inner products of $x(t)$ with the biorthogonal functions of $\psi_{kn}(t)$. The multiresolution structure and the two-scale relationships (4.74) and (4.64) allow efficient computation of the expansion coefficients at each level, in terms of the preceding level. More specifically, the wavelet and scaling function coefficients of a certain level are a weighted sum of the corresponding coefficients from the previous level. It turns out that this weighted sum can be determined efficiently by a digital filterbank, which leads to the discrete wavelet transform.

To see this, suppose we are given a signal $x(t) \in \mathcal{V}_{j+1}$. We can then write

$$x(t) = \sum_{n\in\mathbb{Z}} c_j[n]\phi(2^j t - n) + \sum_{n\in\mathbb{Z}} d_j[n]\psi(2^j t - n).\tag{4.77}$$

If we further assume that $\phi(2^j t - n)$ and $\psi(2^j t - n)$ generate orthonormal bases for $\mathcal{V}_j$ and $\mathcal{W}_j$ respectively, then

$$c_j[n] = \langle \phi(2^j t - n), x(t)\rangle, \quad d_j[n] = \langle \psi(2^j t - n), x(t)\rangle.\tag{4.78}$$

The first component in (4.77) can be further decomposed as

$$\sum_{n\in\mathbb{Z}} c_{\tilde{j}}[n]\phi(2^j t - n) = \sum_{n\in\mathbb{Z}} c_{j-1}[n]\phi(2^{j-1} t - n) + \sum_{n\in\mathbb{Z}} d_{j-1}[n]\psi(2^{j-1} t - n).$$

(4.79)

Using the two-scale relation (4.64) we may express $c_{j-1}[n]$ as

$$c_{j-1}[n] = \sum_{m\in\mathbb{Z}} a[-m]\langle\phi(2^j t - 2n + m), x(t)\rangle = \sum_{m\in\mathbb{Z}} a[-m]c_j[2n - m]. \quad (4.80)$$

Similarly, from (4.74) it follows that

$$d_{j-1}[n] = \sum_{m\in\mathbb{Z}} b[-m]c_j[2n - m]. \quad (4.81)$$

Relations (4.80) and (4.81) lead to a recursive filterbank implementation of the scaling and wavelet coefficients, as depicted in Fig. 4.22.

An important aspect of wavelet theory is designing wavelet functions $\psi_{kn}(t)$ that produce a large number of zero expansion coefficients a_{kn}. This is particularly pertinent in applications such as compression and denoising. The number of nonzero values depends on the regularity of $x(t)$, the number of vanishing moments of $\psi(t)$, and the support size of $\psi(t)$. The Strang–Fix conditions [98] relate the number of vanishing moments of $\psi(t)$ to the properties of the scaling function $\phi(t)$: $\psi(t)$ has p vanishing moments if and only if any polynomial of degree $p - 1$ can be expressed as a linear combination of the functions $\{\phi(t - n)\}$. On the other hand, it can be shown that in an orthogonal wavelet expansion, if $\psi(t)$ has p vanishing moments, then its support is at least of size $2p - 1$. Therefore, in practice, there is a tradeoff between the number of vanishing moments and the support size. Here multiwavelets, discussed briefly in Section 4.3.5, become important. Using several scaling functions (and wavelets) allows increased design flexibility, leading to a better tradeoff between the wavelet support and its vanishing moments [80].

A detailed discussion of wavelets is beyond our scope. However, it is important to point out that many of the analysis tools used in the context of wavelet theory are the same as those used throughout this book: frames, Riesz bases, and biorthogonality. In fact, wavelet theory is one of the driving forces behind the revival of sampling theory in the past two decades.

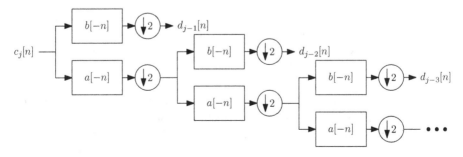

Figure 4.22 Recursive computation of wavelet coefficients.

4.5   Union of subspaces

Until now we have considered signal models in which $x(t)$ lies in a single subspace. In Chapters 6 and 8 we will see that the subspace model is quite powerful, leading to perfect recovery of the signal from its linear and nonlinear samples under very broad conditions. Furthermore, recovery is achieved by digital and analog LTI filtering, effectively generalizing the Shannon–Nyquist theorem to a broader set of input classes.

Despite the simplicity and intuitive appeal of subspace models, in modern applications many signals are characterized by parameters which are not necessarily known to the sampler. As we show now via several examples, we can frequently still describe the signal by a subspace model, but the subspace has to have enough degrees of freedom to capture the uncertainty. Often, in order to include all possible parameter choices, this results in a subspace of very large dimension, leading to extremely high sampling rates. In order to capture the signal representation mathematically without unnecessarily increasing the rate, in this section we introduce the *union of subspaces* (UoS) model which is applicable to many interesting classes of signals. In Chapters 11–15 we will analyze sampling and recovery with these models in detail, and show that although sampling can often still be obtained by LTI filtering, recovery becomes more involved and requires nonlinear algorithms.

Example: Multiband communication

To understand the necessity for the UoS model, consider a typical communication scenario, where several radio devices or stations transmit narrowband signals simultaneously, as depicted in Fig. 4.23. Modern communication signaling operates at very high carrier frequencies, and therefore the maximal possible frequency component at the receiver may be on the order of several GHz. At the receiver, the signal is first sampled and then further processed, either in order to detect the information symbols or for further transmission. Signals of the form depicted in Fig. 4.23 are referred to as *multiband signals*. Besides being very popular in communications, such models have been analyzed in detail within the sampling literature. In Chapter 14 we discuss these signal classes along with the minimal rates at which they can be sampled, and specific recovery algorithms. For now, our focus is on describing a mathematical model that captures such signals.

Multiband signals can be characterized by the set of band locations ω_i, and the width of each band B_i. Let $\mathcal{X}$ be the class of signals with N bands (ω_i, B_i). It is easy to see that $\mathcal{X}$ defines a subspace: any linear combination of two signals in $\mathcal{X}$ also comprises N bands characterized by (ω_i, B_i). Therefore, the results we develop on subspace models are applicable here as well. In fact, any signal in $\mathcal{X}$ can be described by an SI model with N generators of the form (4.44) where $h_i(t)$ represents an LPF of width B_i centered at ω_i. In Example 4.4 we have seen that we can also describe multiband signals by a fixed set of generators that do not depend on the carrier frequencies, at the expense of a possible increase in the number of generators to $2N$. A famous result by Landau [99] shows that the minimal sampling rate achievable in such models is the sum of the bandwidths,

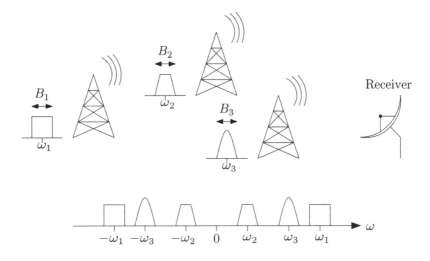

Figure 4.23 The frequency content of a multiband signal.

and not the maximal frequency $\omega_{\max}$ corresponding to the Nyquist rate. A proof of this result along with methods that attain this rate will be discussed in Chapter 14.

Suppose now that the carriers are unknown to the sampler. This situation arises for example in cognitive radios [100], which aim at utilizing unused frequency regions on an opportunistic basis. In such settings the receiver has to deal with many concurrent transmissions whose carriers are unknown, or can be varying in time. In this case, our knowledge may only consist of the maximal number of transmissions N, and their maximal bandwidths B_i, but not the carrier frequencies ω_i. The resulting signal model no longer defines a subspace. Indeed, since the carrier frequencies are arbitrary, adding two signals with N nonoverlapping bands can result in a signal with as many as $2N$ bands, which is no longer in $\mathcal{X}$; this is demonstrated in Fig. 4.24. Here we see an example where the uncertainty associated with the subspace model moves the signal class out of the subspace setting.

Of course, we can still describe such a signal by the subspace of signals bandlimited to $\omega_{\max}$. However, this subspace is much larger than the actual set of signals we are trying to accommodate. In particular, the bandlimited subspace includes signals that occupy the entire bandwidth, which are not included in the set $\mathcal{X}$. Therefore, using a subspace model results in sampling rates much higher than those needed in practice.

4.5.1   Signal model

Evidently, in order to reduce the sampling rate with respect to the high Nyquist rate, we need to better characterize our degrees of freedom. A very convenient representation arises from noting that for each *fixed* set of frequencies $\{\omega_i\}$, we have a subspace model. Therefore, any signal in $\mathcal{X}$ will lie in some subspace $\mathcal{U}_\lambda$ for an appropriate index λ, defined by the frequencies $\{\omega_i\}$. The problem, though, is that we do not know in advance

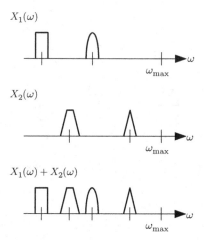

Figure 4.24 Summing two signals in the set $\mathcal{X}$ of functions with two bands can result in a signal having four bands.

which subspace the signal actually resides in. Mathematically, this means that our signals can be represented by a UoS [101]:

$$x \in \bigcup_{\lambda \in \Lambda} \mathcal{U}_\lambda, \tag{4.82}$$

where each subspace $\mathcal{U}_\lambda$ is associated with a specific choice of frequencies, and Λ represents the set of possibilities.

Representations of the form (4.82) are very general and can accommodate many interesting signal priors. Before providing further examples, some of which will also be treated in more detail in the rest of the book, we stress the difference between the union model (4.82) and its subspace counterpart. For simplicity, suppose that Λ is a countable set and λ is an integer. Replacing the union by a sum leads to

$$x \in \mathcal{U}_1 + \mathcal{U}_2 + \cdots + \mathcal{U}_m, \tag{4.83}$$

where m can be infinite. Clearly, the sum of subspaces is itself a subspace, and is the smallest space containing $\mathcal{U}_\lambda, \lambda \in \Lambda$. However, as we now demonstrate, it is much larger than the union.

Example 4.10 Suppose that $\mathcal{U}_1$ and $\mathcal{U}_2$ are the sets of signals whose frequency-domain supports are contained in $\{\omega \colon 0 \leq \omega < u_1\}$ and $\{\omega \colon u_1 \leq \omega < u_2\}$ respectively. Then $\mathcal{U}_1 + \mathcal{U}_2$ is the set of signals with support within $\{\omega \colon 0 \leq \omega < u_2\}$. In contrast, in the union model, the signal lies in one of the spaces. This means that its support is either contained in $\{\omega \colon 0 \leq \omega < u_1\}$ or in $\{\omega \colon u_1 \leq \omega < u_2\}$ but cannot occupy both intervals. The problem is that we do not know the correct support in advance.

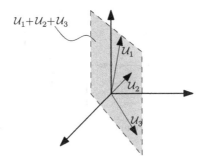

Figure 4.25 The union and the sum of three one-dimensional subspaces in $\mathbb{R}^3$.

Any signal that lies in the union (4.82) will also be in the sum (4.83). However, the latter includes many more signals that do not lie in any of the individual subspaces $\mathcal{U}_\lambda$. Therefore, the signal set described by (4.83) is much larger than that associated with (4.82). Consequently, sampling theories developed for the sum model will require higher sampling rates than those associated with the union, which includes fewer signals.

To visualize the UoS model, consider three one-dimensional subspaces in $\mathbb{R}^3$, shown in Fig. 4.25. Each such subspace corresponds to a ray passing through $(0, 0, 0)$. Therefore, $\mathcal{U}_1 \cup \mathcal{U}_2 \cup \mathcal{U}_3$ is the set of all points in $\mathbb{R}^3$ which lie on one of these rays. On the other hand, the sum $\mathcal{U}_1 + \mathcal{U}_2 + \mathcal{U}_3$ corresponds to the subspace spanned by the three rays. In the example of Fig. 4.25, this is a two-dimensional space, i.e. a plane. It is clear from this example that the sum is much larger than the union of the spaces.

There are many applications in which the signals of interest lie in a UoS. A very simple union model arises in the context of identifying multipath fading channels, as we show in the next example. Despite the apparent simplicity, this model can be used to describe many important application areas such as ultrasound imaging, radar signals, biological processes, and more. In Chapter 15 we demonstrate how properly exploiting the structure in this signal class allows one to greatly reduce the sampling and processing rates in many important examples, leading to a variety of interesting potential benefits such as reduced-size ultrasound machines and sub-Nyquist radar identification. Further examples and applications are provided in Chapter 10.

Example 4.11 Consider a pulse $h(t)$ transmitted through a multipath medium, as schematically depicted in Fig. 4.26. The received signal $x(t)$ comprises a sum of shifted and modulated versions of $h(t)$:

$$x(t) = \sum_{\ell=1}^{k} a_\ell h(t - t_\ell), \quad t \in [0, \tau]. \tag{4.84}$$

Here k is the number of different paths in the medium, $\{t_\ell\}_{\ell=1}^{k}$ are the corresponding time delays, and $\{a_\ell\}_{\ell=1}^{k}$ are associated with the reflectance coefficients of the

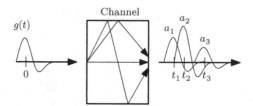

Figure 4.26 The signal at the output of a multipath channel lies in a union of subspaces.

different paths. For fixed time delays, the variability in the set of all possible signals $x(t)$ is due only to the unknown coefficients $\{a_\ell\}_{\ell=1}^k$, forming a subspace of dimension k. When the delays are unknown, the set of possible signals forms a UoS, each corresponding to a different constellation of time shifts. Note that in this example, we have an infinite union of k-dimensional spaces.

4.5.2   Union classes

There are many possibilities for defining a UoS. In contrast to the subspace setting, which, as we will show in Chapter 6, can be treated in a unified way, sampling over unions is much more intricate, due to the fact that the model is no longer linear: the sum of two signals from a union $\mathcal{X}$ is generally no longer in $\mathcal{X}$. This nonlinear behavior of the sampling set complicates the sampling and recovery stages. Therefore, instead of attempting to treat all unions in a unified way, in our development in subsequent chapters we focus attention on some specific classes of union models, in order of complexity.

Finite-dimensional unions

The simplest class of unions results when the number of subspaces comprising the union is finite, and each subspace has finite dimension. This model underlies the field of compressed sensing (CS) [12, 13, 14], which we discuss in detail in Chapter 11. At its core, CS is a mathematical framework that studies accurate recovery of a signal x represented by a vector of length n from a number of measurements $m \ll n$, effectively performing compression during signal acquisition. The underlying assumption enabling CS is that the signal is *sparse*, i.e. in an appropriate basis representation, it consists of only a small number $k \ll n$ of nonzero coefficients and thus lies in a k-dimensional subspace. If we knew the locations of the k nonzero values, then we could recover the signal from k appropriate measurements. Since the location set is unknown, there are $\binom{n}{k}$ possible subspaces in which x may reside, leading to a UoS model.

Ignoring the sparsity structure, x can be recovered from n measurements since it lies in an n-dimensional space. However, the union model implies that only k of its values are nonzero. Thus, intuitively, we would expect on the order of k samples to suffice for recovery. The main question at the heart of CS is how to design sampling and recovery methods that enable reconstruction of a sparse vector x from order-k samples.

In certain applications, the signal may possess structure that is not entirely expressed using sparsity alone. An example is when only certain sparse support patterns are allowable, as is often the case with wavelet transforms in which the dominant coefficients tend to cluster into a connected rooted subtree. When multiple sparse signals are recorded simultaneously, their supports might be correlated according to the properties of the sensing environment. In other cases the nonzero coefficients may appear in blocks. It is possible to leverage such constraints and formulate more concise models. For example, this structure can represent sparse unions of finite-dimensional subspaces in which the signal consists of components in k subspaces, chosen from n possibilities. Such models are treated in Chapter 12.

Union of subspaces for analog signal models

One of the primary motivations for the growing field of CS is to design new sampling systems to acquire continuous-time, analog signals or images. In contrast, the finite-dimensional sparse model described above inherently assumes that the signal is discrete. Furthermore, in the context of sampling, our interest is typically in continuous-time inputs $x(t)$. Using union modeling, the concept of sparsity can be extended to provide UoS models for analog signals. Two of the broader frameworks that treat sub-Nyquist sampling of continuous-time signals are Xampling [21, 102] and finite rate of innovation (FRI) [103], which are discussed in Chapters 13–15.

In general, when treating UoS modeling of analog signals there are three main cases to consider, as elaborated further in Chapter 10:

1. Finite unions of infinite-dimensional spaces;
2. Infinite unions of infinite-dimensional spaces;
3. Infinite unions of finite-dimensional spaces.

In each of the three settings above there is an element that can take on infinite values, which is a result of the fact that we are considering analog signals: either the underlying subspaces are infinite-dimensional, or the number of subspaces is infinite. When dealing with infinite-dimensional subspaces, we will rely on the SI structure, as we do in the context of subspace sampling. We therefore consider both finite and infinite unions of SI subspaces. An example is the multiband model, illustrated in Fig. 4.23, which is a special case of sparse SI unions [104]. Further examples will be given in Chapter 10.

Another signal class that can often be expressed as a UoS is signals with FRI. Depending on the specific structure, this model corresponds to an infinite union of finite- or infinite-dimensional subspaces, and includes many common signals described by a small number of degrees of freedom. In this case, each subspace corresponds to a certain choice of parameter values, with the set of possible values being infinite, and thus the number of subspaces is infinite as well. The signal in Example 4.11 conforms with the FRI model, where the unknown delays t_ℓ dictate the choice of subspace.

More generally, signals of the form

$$x(t) = \sum_{\ell=1}^{k} a_\ell h_\ell(t; \theta_\ell) \tag{4.85}$$

are FRI: for every given set of parameters $\{\theta_\ell\}_{\ell=1}^k$, $x(t)$ lies in the k-dimensional subspace spanned by $\{h_\ell(t;\theta_\ell)\}_{\ell=1}^k$. Since each θ_ℓ can take on a value from a continuous set, (4.85) represents an infinite union of finite-dimensional subspaces. In Example 4.11, $h_\ell(t;\theta_\ell) = h(t - t_\ell)$ with $\theta_\ell = t_\ell$. A similar idea can be used to define infinite unions of infinite-dimensional subspaces, by replacing the finite-dimensional subspace representation by an SI structure:

$$x(t) = \sum_{\ell=1}^k \sum_{n\in\mathbb{Z}} a_\ell[n] h_\ell(t - nT; \theta_\ell). \tag{4.86}$$

One choice we examine in Chapter 15 is when $h_\ell(t - nT; \theta_\ell) = h(t - nT - t_\ell)$ for some known pulse shape $h(t)$. This model allows us to describe more complicated time-varying multipath channels than that captured by Example 4.11.

In each one of the models (4.85) and (4.86), we can sample the underlying signal $x(t)$ at the Nyquist rate of the pulses $\{h_\ell(t)\}$. However, when the generators have a wide bandwidth, this approach seems very wasteful. For example, in the time delay model of (4.84), the pulse shape is known; the uncertainty is a result of the unknown delays. Since there are only $2k$ unknowns (k delays and k amplitudes) we expect the required sampling rate to be proportional to $2k$, and not to the Nyquist rate of $h(t)$ which completely ignores the known pulse shape. Similarly, in the model (4.86), if the parameters θ_ℓ are known, then our signal lies in an SI subspace with k generators. We can then sample and recover $x(t)$ using a filterbank with k branches, each sampled at rate $1/T$. Therefore, even when θ_ℓ are unknown, we expect the rate to be on the order of k/T. These scenarios are treated in detail in Chapters 13 and 15 where we show how the rate can be significantly reduced with respect to the underlying Nyquist rate. This is particulary important in applications in which the Nyquist rate is exceedingly high, challenging current technologies. Large sampling rates also entail high DSP rates so that even when sampling at Nyquist is possible, it is often advantageous to reduce the rate in order to enable low processing rates.

To gain some insight as to why sub-Nyquist sampling should be possible, we note that the Nyquist theorem is a worst-case result: it dictates a rate that can handle all bandlimited signals. It therefore leaves open the possibility of enabling recovery from reduced-rate samples, for signals that possess further structure. As we will show, sampling with a UoS prior can be translated into concrete hardware solutions leading to sampling and recovery at rates far below that specified by Nyquist.

4.6   Stochastic and smoothness priors

Until now our focus has been on input signals $x(t)$ that are deterministic, and lie in a subspace or a UoS. Often, the prior information regarding our signal is much more limited. We may only know that the signal is bounded, or some of its derivatives are bounded. Such priors can be expressed in the form $\|Lx\|_2 \leq U$ for some suitable operator L and upper bound U.

Example 4.12 One common choice of an operator L is a weighted combination of derivatives of various orders, namely

$$L\{x\}(t) = x(t) + \alpha_1 x^{(1)}(t) + \alpha_2 x^{(2)}(t) + \dots, \tag{4.87}$$

for some constants $\alpha_1, \alpha_2, \dots$ In this case, the constraint $\|Lx\|_2 \leq U$ rules out signals which are highly oscillating.

For example, suppose that $\alpha_1 = 1$ and $\alpha_n = 0$ for every $n > 1$, and consider the Gaussian pulse

$$x(t) = \frac{1}{\pi^{1/4}\sigma^{1/2}} \, e^{-\frac{t^2}{2\sigma^2}}. \tag{4.88}$$

The energy of this pulse is given by $\int_{-\infty}^{\infty} x^2(t)dt = 1$, independent of the pulse width, which is determined by σ. The energy of $L\{x\}(t)$, however, is given by

$$
\begin{aligned}
\int_{-\infty}^{\infty} L^2\{x\}(t)dt &= \int_{-\infty}^{\infty} \left(x(t) + \alpha_1 x^{(1)}(t) \right)^2 dt \\
&= \frac{1}{\sqrt{\pi}\sigma} \int_{-\infty}^{\infty} \left(e^{-\frac{t^2}{2\sigma^2}} - \frac{\alpha_1}{\sigma^2} t e^{-\frac{t^2}{2\sigma^2}} \right)^2 dt \\
&= \frac{1}{\sqrt{\pi}\sigma} \int_{-\infty}^{\infty} e^{-\frac{t^2}{\sigma^2}} \left(1 - \frac{\alpha_1}{\sigma^2} t \right)^2 dt \\
&= 1 + \frac{\alpha_1}{2\sigma^2}.
\end{aligned}
\tag{4.89}
$$

As σ becomes smaller, the width of the pulse decreases, and the energy of $L\{x\}(t)$ increases. Therefore, the constraint $\|Lx\|_2 \leq U$ translates into the limitation

$$\sigma^2 \geq \frac{\alpha_1}{2(U^2 - 1)} \tag{4.90}$$

on the pulse width, for $U > 1$. If $U \leq 1$ then there exists no valid value of σ^2.

Example 4.13 A more general class of operators L is the family of LTI systems. In this case, L is associated with a frequency response $L(\omega)$. The squared norm $\|Lx\|_2^2$ then reduces to

$$\|Lx\|_2^2 = \frac{1}{2\pi} \int_{-\infty}^{\infty} |X(\omega)L(\omega)|^2 \, d\omega. \tag{4.91}$$

The function $L(\omega)$ can be chosen such that only smooth signals satisfy the constraint. For instance, if $L(\omega) = \omega^p$ for some $p \geq 1$, then for a signal to be admissible, it must decay faster than $\omega^{-p-1/2}$ as $|\omega| \to \infty$ and it must not tend to infinity faster than $\omega^{-p-1/2}$ as $|\omega| \to 0$.

It can be shown that the rate of decay of the frequency content of a signal is directly related to its smoothness so that (4.91) may be viewed as a smoothness constraint on the signal. To demonstrate this, consider the case $p = 1$ and let

$$x_1(t) = \begin{cases} 1, & |t| \le \frac{1}{2} \\ 0, & |t| > \frac{1}{2}, \end{cases} \qquad x_2(t) = e^{-|t|}. \qquad (4.92)$$

The signal $x_2(t)$ is continuous, whereas $x_1(t)$ is not. The CTFT of $x_1(t)$, which is given by $X_1(\omega) = \text{sinc}(\omega/(2\pi))$, decays like $1/\omega$. This rate is slower than $1/\omega^{p+1/2} = 1/\omega^{3/2}$ and consequently $x_1(t)$ is inadmissible. On the other hand, the CTFT of $x_2(t)$, which is given by $2/(1 + \omega^2)$, decays faster than $1/\omega^{3/2}$, rendering $x_2(t)$ admissible.

Generally, the bounded-norm prior is not enough to guarantee perfect recovery of the signal. Therefore, in such cases we will need to approximate the signal from its samples, subject to this knowledge. In Chapter 7 we discuss several approaches to signal approximation.

Interestingly, the solutions obtained under the bounded-norm constraint are functionally the same as (or similar to) those obtained in a stochastic setting by assuming a certain covariance function of the unknown process. (See Appendix B for a review of stochastic processes.) Specifically, instead of considering $x(t)$ to be deterministic with some prior information, we may treat $x(t)$ as a random function with known second-order statistics. For simplicity, we focus on zero-mean wide-sense stationary (WSS) signals with autocorrelation function

$$R_{xx}(\tau) = E\{x(t)x(t + \tau)\}. \qquad (4.93)$$

The decay of the autocorrelation describes the extent to which distinct points in the process are statistically related. If the autocorrelation decays slowly, then typical realizations of the process will be smooth and nonoscillating. An equivalent characterization of WSS processes is via their spectrum $\Lambda_{xx}(\omega)$, defined as the CTFT of $R_{xx}(\tau)$. The spectrum characterizes the frequency content of a typical realization of the random process $x(t)$.

Figure 4.27 depicts realizations of two Gaussian WSS processes with different spectra. In this example, the spectrum of each signal vanishes for every $\omega > \pi$ while their content within $\omega \in [-\pi, \pi]$ is different. The left column corresponds to a process with narrow autocorrelation function, so that its spectrum contains high frequencies (up to π). As can be seen, the signal realization associated with this process is highly oscillating. The signal in the right column has a wider autocorrelation function, and therefore its spectrum decays rapidly. As a result, the realization associated with this process is smoother.

In Chapter 7 we treat the stochastic setting, and consider methods for interpolating a given set of samples in a minimal mean-squared error (MSE) sense. This implies that the recovered signal $\hat{x}(t)$ is designed to minimize the MSE between itself and the original signal $x(t)$. We will see that in many cases the MSE solutions coincide with those obtained under a minimax criterion subject to a bounded-norm constraint. Therefore, these two separate routes often lead to the same computational solution, while providing complementary insights into the problem as well as on the notion of optimality.

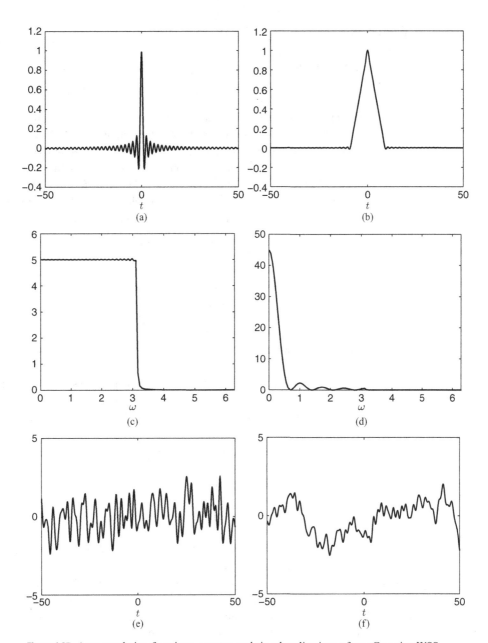

Figure 4.27 Autocorrelation functions, spectra, and signal realizations of two Gaussian WSS processes $x_1(t)$ and $x_2(t)$. (a) The autocorrelation $R_{x_1x_1}(\tau)$. (b) The autocorrelation $R_{x_2x_2}(\tau)$. (c) The spectrum $\Lambda_{x_1x_1}(\omega)$. (d) The spectrum $\Lambda_{x_2x_2}(\omega)$. (e) A signal realization $x_1(t)$. (f) A signal realization $x_2(t)$.

4.7   Exercises

1. Consider the set transformation W associated with the functions

$$w_{nk}(t) = e^{j2\pi k\Omega t} h(t - nT), \quad k = 1, \ldots, K, \; n \in \mathbb{Z}, \tag{4.94}$$

   where $h(t)$ is a given square integrable function and Ω and T are positive constants. We would like to construct a practical system that outputs the coefficients $c = W^* x$ when fed with a signal $x(t)$. Propose a system, comprising K modulators (which multiply their input by a complex exponent e^{jat} with some design parameter a), K filters, and K samplers, which implements the adjoint transformation W^*.

2. Assume that the signal $x(t) = \mathrm{sinc}^2(t/T)$ is sampled at times $t = nT_s$ to obtain the sequence $c_1[n]$. We would like to recover $x(t)$ from its samples using the formula

$$\hat{x}_1(t) = \sum_{n \in \mathbb{Z}} c_1[n] \, \mathrm{sinc}\left(\frac{t - nT_s}{T_s}\right). \tag{4.95}$$

   Write an explicit expression for the recovery error

$$e_1 = \int_{-\infty}^{\infty} (x(t) - \hat{x}_1(t))^2 dt \tag{4.96}$$

   as a function of T and T_s.

   Hint: Use Parseval's relation to compute the error in the frequency domain.

3. Assume that the signal $x(t) = \mathrm{sinc}^2(t/T)$ is sampled at times $t = nT_s$ after being convolved with the anti-aliasing filter $h(t) = \mathrm{sinc}(t/T_s)$ to obtain the sequence $c_2[n]$. We would like to recover $x(t)$ from the samples $c_2[n]$ using the formula

$$\hat{x}_2(t) = \sum_{n \in \mathbb{Z}} c_2[n] \, \mathrm{sinc}\left(\frac{t - nT_s}{T_s}\right). \tag{4.97}$$

   Write an explicit expression for the recovery error

$$e_2 = \int_{-\infty}^{\infty} (x(t) - \hat{x}_2(t))^2 dt \tag{4.98}$$

   as a function of T and T_s. Draw a plot of the errors e_1 of the previous exercise and e_2, as functions of T_s. Explain the results.

4. Let $x(t)$ be a signal whose bandwidth does not exceed π/T and let $g(x)$ be an invertible function on $\mathbb{R}$.

   a. What is the bandwidth of the signal $g(x(t))$, if $g(x) = x^p$ for some $p \geq 1$?

   b. We would like to recover $x(t)$ from its nonlinearly distorted samples $c[n] = g(x(nT))$ using the formula

$$\hat{x}(t) = \sum_{n \in \mathbb{Z}} d[n] \, \mathrm{sinc}\left(\frac{t - nT}{T}\right). \tag{4.99}$$

   Write an expression for the coefficients $d[n]$ as a function of the samples $c[n]$.

5. Let $x(t)$ be a signal whose CTFT vanishes for every $|\omega| \notin [p\pi, (p+1)\pi]$, where $p \geq 1$ is some integer.

a. Plot schematically the DTFT of the sequence $x(n)$, $n \in \mathbb{Z}$.

b. Show how $x(t)$ can be recovered from its samples at the integers.

6. In this exercise we show how the integral of a continuous bandlimited signal can be computed exactly from its samples below the Nyquist rate. Let $x(t)$ be a signal in $L_2(\mathbb{R})$ with $X(\omega) = 0$ for $|\omega| \geq \pi/T$.

   a. Show that $\int_{-\infty}^{\infty} x(t)dt = T \sum_{n \in \mathbb{Z}} x(nT)$.

   Hint: Use the Poisson-sum formula (3.89).

   b. Show that a similar expression holds when $x(t)$ is sampled with a period $T' > T$. How large can T' be chosen?

7. Let $x(t), y(t)$ denote two signals in $L_2(\mathbb{R})$ that are bandlimited to π/T. We are interested in evaluating the inner product $r = \langle x(t), y(t) \rangle$. Show that r can be computed directly from the samples $x(nT), y(nT)$.

8. In this exercise we consider changing the sampling grid to a denser grid, after sampling. Suppose we sampled a bandlimited signal $x(t)$ at its Nyquist rate, resulting in the samples $x[n] = x(nT)$. We would now like to evaluate the samples $y[n] = x(nT/M)$ for $M > 1$ which correspond to sampling at the uniform period T/M, directly from the samples $x[n]$.

   a. Show that $y[n] = z[n] * \operatorname{sinc}(n/M)$ for an appropriate choice of sequence $z[n]$.

   b. Plot schematically the DTFT of the sequences $z[n]$ and $x[n]$.

9. One appealing property of splines is that differential and integral operations on them can be carried out by simple manipulation of the coefficients $a[k]$ in the B-spline expansion (4.31).

   a. Prove that

$$\frac{d\beta^n(t)}{dt} = \beta^{n-1}\left(t + \frac{1}{2}\right) - \beta^{n-1}\left(t - \frac{1}{2}\right). \tag{4.100}$$

   Hint: Use (4.32), the fact that $d\beta^0(t)/dt = \delta(t + 1/2) - \delta(t - 1/2)$ and the property $d(a * b)(t)/dt = (a * (db/dt))(t)$ for any two functions $a(t), b(t)$.

   b. Suppose that $x(t)$ is a spline of degree n whose B-spline coefficients are $a[k]$. Determine the B-spline coefficients of $dx(t)/dt$ as a function of the coefficients $a[k]$.

   c. Prove that

$$\int_{-\infty}^{t} \beta^n(\tau)d\tau = \sum_{k=0}^{\infty} \beta^{n+1}\left(t - \frac{1}{2} - k\right). \tag{4.101}$$

   Hint: Write the relation (4.100) at time $t' = t - 1/2$ with B-spline of degree $n' = n + 1$ and integrate both sides from $-\infty$ to $t - k$.

   d. For a spline $x(t)$ of degree n with B-spline coefficients $a[k]$, determine the B-spline coefficients of $\int_{-\infty}^{t} x(\tau)d\tau$ as a function of $a[k]$.

10. Consider the raised cosine pulse $p(t)$, whose formula is given in (4.41). Prove that $p(t)$ is a Nyquist pulse, by showing that its CTFT $P(\omega)$ satisfies the Nyquist condition (4.40).

11. Suppose that $p(t)$ is a Nyquist pulse, and consider the function

$$h(t) = \sum_{n \in \mathbb{Z}} a[n] p(t - nT) \tag{4.102}$$

for a sequence $a[n]$ with finite support. Show that $h(t)$ is a Nyquist pulse for an appropriate symbol interval $T' = \ell T$ for some ℓ. Provide an explanation both in the time domain and in the frequency domain using (4.40).

12. Our goal in this exercise is to prove the uncertainty principle (4.56). We assume throughout that $x(t)$ is a real function in $L_2(\mathbb{R})$.

a. Using integration by parts, show that

$$\int_{-\infty}^{\infty} t x(t) \frac{dx(t)}{dt} dt = -\frac{1}{2} \|x\|^2. \tag{4.103}$$

b. Using Parseval's theorem and Exercise 2 of Chapter 3, prove that

$$\left| \int_{-\infty}^{\infty} t x(t) \frac{dx(t)}{dt} dt \right|^2 \leq \frac{1}{2\pi} \int_{-\infty}^{\infty} \omega^2 |X(\omega)|^2 d\omega \int_{-\infty}^{\infty} t^2 |x(t)|^2 dt. \tag{4.104}$$

c. Prove (4.56).

13. The (k, n)th coefficient of the Gabor transform (4.52) can be interpreted as the inner product between the signal $x(t)$ and the function $h_{kn}(t)$ of (4.53). This function is obtained by translating and modulating the window $h(t)$. Let $\mathcal{T}_a$ and $\mathcal{M}_b$ denote the translation and modulation operators, given by

$$\mathcal{T}_a\{h(t)\} = h(t - a),$$
$$\mathcal{M}_b\{h(t)\} = e^{j2\pi bt} h(t). \tag{4.105}$$

a. Show that these operators satisfy the relation

$$\mathcal{M}_{kW} \mathcal{T}_{nT} = e^{j2\pi WTkn} \mathcal{T}_{nT} \mathcal{M}_{kW}. \tag{4.106}$$

b. Provide a necessary condition on T and W such that the operators $\mathcal{M}_{kW}$ and $\mathcal{T}_{nT}$ commute for every $n, k \in \mathbb{Z}$.

14. Let $\chi_{[0,1]}$ be the box function on the interval $[0, 1]$. Prove that $\{\mathcal{M}_m \mathcal{T}_{nT}\}$ is a frame for $L_2(\mathbb{R})$ for all $0 < T < 1$ where $\mathcal{T}_a$ and $\mathcal{M}_b$ are defined in Exercise 13.

15. Let $g[k, n]$ denote the Gabor expansion coefficients of a signal $x(t)$ in $L_2(\mathbb{R})$ as defined by (4.52). Express the Gabor expansion of $x(t - t_0)$ and $x(t)e^{j2\pi W_0 t}$ in terms of $g[k, n]$.

16. In this exercise, we study a generalization of the two-scale relation (4.64). Specifically, consider the M-scale relation

$$\phi(t) = \sum_{n \in \mathbb{Z}} a[n] \phi(t/M - n), \tag{4.107}$$

where $\phi \in L_2(\mathbb{R})$, $a \in \ell_2$ and $M > 1$ is some integer. Prove that this relation is equivalent to

$$\Phi(M\omega) = \frac{1}{M} A(e^{j\omega}) \Phi(\omega), \tag{4.108}$$

where $\Phi(\omega)$ is the CTFT of $\phi(t)$ and $A(e^{j\omega})$ is the DTFT of $a[n]$.

17. The wavelet representation of a signal involves the functions $\psi_{kn}(t)$, given by (4.73). These functions are obtained by translating and scaling the mother wavelet function $\psi(t)$. Let $\mathcal{T}_a$ be the translation operator defined in Exercise 13, and let $\mathcal{D}_c$ denote the dilation operator:

$$\mathcal{D}_c\{h(t)\} = |c|^{-1/2} h\left(\frac{t}{|c|}\right). \tag{4.109}$$

Show that these operators satisfy the relation

$$\mathcal{D}_c \mathcal{T}_a = \mathcal{T}_{ac} \mathcal{D}_c. \tag{4.110}$$

18. Let $x(t)$ be defined by

$$x(t) = \begin{cases} 1, & t \in \left[2\left(1 - 2^{-n}\right), 2\left(1 - \frac{3}{4}2^{-n}\right)\right) \\ -1, & t \in \left[2\left(1 - \frac{3}{4}2^{-n}\right), 2\left(1 - 2^{-n-1}\right)\right) \\ 0, & t \notin [0,2). \end{cases} \tag{4.111}$$

Plot $x(t)$ and compute the coefficients a_{kn} of its Haar representation (4.76).

19. Show that $\phi(t) = \mathrm{sinc}(t)$ is a valid scaling function, i.e. that it generates a proper multiresolution embedding, and compute the coefficients $a[n]$ in (4.64).

20. Various classes of matrices arising in engineering problems can be described within the UoS model.

a. Assume that $M \geq N$ and consider the set $\mathcal{X}$ of all $M \times N$ matrices X, which can be factored as

$$X = UD,$$

where U is an $M \times N$ matrix whose columns are orthonormal, and D is a diagonal $N \times N$ matrix. Prove that $\mathcal{X}$ is a UoS. Determine the dimension of each subspace and their number.

b. Prove that the set $\mathcal{Y}$ of all $M \times N$ matrices Y with rank $K \leq \min\{M, N\}$ is a UoS. Determine the dimension of each subspace and their number.

Hint: Use the SVD representation (Proposition 2.17).

21. Consider the set $\mathcal{H}$ of all impulse responses $h[n]$ of first-order IIR filters (3.70). Prove that $\mathcal{H}$ is a UoS. What is the dimension of each subspace?

22. Let $\mathcal{Z}$ be the set of all piecewise constant signals $x(t)$, whose number of discontinuities in every segment of length T is no larger than 1. Prove that $\mathcal{Z}$ is a UoS. What is the dimension of each subspace?

23. Prove that the union of $N < \infty$ subspaces is equal to their sum if and only if one of the subspaces is equal to the sum. Show that this is not necessarily so for an infinite number of spaces.

24. Consider the set of signals, whose CTFT $X(\omega)$ satisfies

$$\int_{-\infty}^{\infty} |\omega^p X(\omega)|^2 \leq U \tag{4.112}$$

for some $p \geq 1$ and $U > 0$. What condition does the degree q of a spline $x(t)$ have to satisfy in order for it to be admissible?

Chapter 5

Shift-invariant spaces

Our primary focus in this book is on signal models involving shift-invariant (SI) spaces. These include subspace priors in which the prior is an SI subspace, and union of subspaces priors in which each one of the individual subspaces is SI. We also treat recovery in SI spaces, for arbitrary input signals. Because of the importance of SI models in the development of many classes of sampling theorems, we devote this chapter to studying some of the mathematical properties associated with these spaces. The material we present here will serve as a basis for many of the developments in the ensuing chapters.

5.1   Riesz basis in SI spaces

We begin by considering SI spaces generated by a single function following Definition 4.1 of the previous chapter. Specifically, we treat the class of signals

$$W = \left\{ x(t) \ \middle| \ x(t) = \sum_{n \in \mathbb{Z}} a[n]h(t - nT), \text{ for some } a \in \ell_2 \right\}, \tag{5.1}$$

where $h(t)$ is the SI generator. In Section 5.5 we extend the derivations to multiple generators. For simplicity we assume that all functions are real-valued. Similar derivations hold in the complex case, with obvious modifications.

Until now we have considered arbitrary generators and have not imposed any constraints on $h(t)$. However, as emphasized in Chapter 2, stability is crucial in the context of sampling. Therefore, it is important to choose $h(t)$ in a way that ensures stable recovery. This can be guaranteed by selecting the vectors $\{h(t - nT)\}$ to form a Riesz basis for the SI subspace W defined by (5.1). Riesz bases were introduced in Section 2.5.2: a Riesz basis consists of a complete set of vectors that ensure stable expansions as defined by (2.50). Our goal now is to translate these requirements to the SI setting, leading to an explicit condition on the generator $h(t)$.

The definition of (5.1) implies that the vectors $\{h(t-nT)\}$ are complete in W. Indeed, every vector in the space is approximated arbitrarily closely by combinations of the basis vectors. Therefore, to form a Riesz basis it remains to verify (2.50). With our choice of basis vectors, this condition becomes

$$\alpha \sum_{n \in \mathbb{Z}} a^2[n] \leq \left\| \sum_{n \in \mathbb{Z}} a[n]h(t - nT) \right\|^2 \leq \beta \sum_{n \in \mathbb{Z}} a^2[n], \tag{5.2}$$

for some $\alpha > 0, \beta < \infty$, and all $a \in \ell_2$. The question is what restrictions need to be imposed on $h(t)$ so that (5.2) holds.

Before discussing conditions on $h(t)$, we show that if $h(t)$ forms a Riesz basis then the sampling process is stable in the energy sense. Specifically, suppose that $x(t) = \sum_n a[n]h(t - nT)$ for some samples $a[n]$. We would like the reconstruction to be stable so that a slight perturbation of the samples $a[n]$ does not result in a large perturbation of the reconstruction $x(t)$. If $a[n]$ is replaced by $a[n] + e[n]$ then the error in the reconstruction is $e(t) = \sum_n e[n]h(t - nT)$. The sampling is stable if

$$\langle e(t), e(t) \rangle = \int_{-\infty}^{\infty} e^2(t)dt \leq C \sum_{n \in \mathbb{Z}} e^2[n]. \tag{5.3}$$

When $h(t)$ generates a Riesz basis, (5.3) follows from the upper bound in (5.2).

The lower bound provides stability in the other direction. That is, if a small error signal is added to $x(t)$ then the lower bound guarantees that the energy in the perturbation of the samples will be small. To see this, suppose that we add an error term $e(t) = \sum_n e[n]h(t - nT)$ to $x(t)$, whose norm is small. When the lower bound holds, the coefficients $e[n]$ must have bounded energy. On the other hand, if the lower bound is not satisfied, then the energy of $e[n]$ can grow arbitrarily large.

5.1.1 Riesz basis condition

We now analyze when $h(t)$ satisfies (5.2). One of the advantages of SI representations is that many properties of the signal as well as the samples can be conveniently described in the Fourier domain. In particular, formulating (5.2) in the Fourier domain leads to an explicit condition on $h(t)$, or on its CTFT $H(\omega)$, such that the inequalities hold.

We begin by explicitly computing the middle expression in (5.2):

$$\left\| \sum_n a[n]h(t - nT) \right\|^2 = \sum_{n,m} a[n]a[m] \int_{-\infty}^{\infty} h(t - nT)h(t - mT)dt$$

$$= \sum_{n,m} a[n]a[m] \int_{-\infty}^{\infty} h(t)h(t - (m - n)T)dt$$

$$= \sum_{n,k} a[n]a[k + n] \int_{-\infty}^{\infty} h(t)h(t - kT)dt, \tag{5.4}$$

where the second line is a result of the change of variables $t \rightarrow t - nT$, and the third line follows from the substitution $k = m - n$. Using Parseval's theorem (3.42),

$$\int_{-\infty}^{\infty} h(t)h(t - kT)dt = \frac{1}{2\pi} \int_{-\infty}^{\infty} |H(\omega)|^2 e^{-j\omega kT} d\omega. \tag{5.5}$$

Substituting into (5.4) leads to

$$\left\|\sum_n a[n]h(t-nT)\right\|^2 = \frac{1}{2\pi}\int_{-\infty}^{\infty} |H(\omega)|^2 \sum_{n,k} a[n]a[k+n]e^{-j\omega kT}d\omega$$

$$= \frac{1}{2\pi}\int_{-\infty}^{\infty} |H(\omega)|^2 G(e^{j\omega T})d\omega, \qquad (5.6)$$

where $G(e^{j\omega})$ is the DTFT of the sequence $g[k] = \sum_n a[n]a[k+n] = a[k]*a[-k]$. From (3.69) we therefore have that $G(e^{j\omega}) = |A(e^{j\omega})|^2$ where $A(e^{j\omega})$ is the DTFT of $a[n]$.

When dealing with expressions involving both discrete- and continuous-time signals, we often encounter mixed integrals of the form (5.6). Since $H(\omega)$ is a CTFT, in principle it exists over the entire real axis. On the other hand, $G(e^{j\omega})$, representing a DTFT, is 2π-periodic, so that $G(e^{j\omega T})$ has period $2\pi/T$. This periodicity can be exploited to conveniently reduce the integral as follows. Treating the integral as a sum over frequency, instead of directly summing over the entire real line, we break up the sum by first fixing a frequency $\omega_0 \in [0, 2\pi/T)$, and then summing over values $\omega_0 + 2\pi k/T$ for all integers k. Clearly this results in coverage of the entire real line. Mathematically, for any function $F(\omega)$, we have

$$\int_{-\infty}^{\infty} F(\omega)d\omega = \int_0^{2\pi/T} \sum_{k\in\mathbb{Z}} F\left(\omega - \frac{2\pi k}{T}\right)d\omega. \qquad (5.7)$$

In (5.6), $F(\omega) = |H(\omega)|^2 G(e^{j\omega T})$. Since $G(e^{j\omega T}) = |A(e^{j\omega T})|^2$ is $2\pi/T$ periodic,

$$\frac{1}{2\pi}\int_{-\infty}^{\infty} |H(\omega)|^2 |A(e^{j\omega T})|^2 d\omega = \frac{1}{2\pi}\int_0^{2\pi/T} |A(e^{j\omega T})|^2 \sum_{k\in\mathbb{Z}} \left|H\left(\omega - \frac{2\pi k}{T}\right)\right|^2 d\omega. \qquad (5.8)$$

Equation (5.8) represents the inner expression of (5.2) in the Fourier domain. We now use the discrete Parseval's theorem (3.66) to do the same for the outer expression:

$$\sum_n a^2[n] = \frac{1}{2\pi}\int_0^{2\pi} |A(e^{j\omega})|^2 d\omega = \frac{T}{2\pi}\int_0^{2\pi/T} |A(e^{j\omega T})|^2 d\omega = \frac{T}{2\pi}I_c, \qquad (5.9)$$

where the last equality is a result of the change of variables $\omega \to \omega/T$, and for brevity we denoted

$$I_c = \int_0^{2\pi/T} |A(e^{j\omega T})|^2 d\omega. \qquad (5.10)$$

Combining (5.9) and (5.8), the function $h(t)$ must satisfy

$$\alpha I_c \le \frac{1}{T}\int_0^{2\pi/T} |A(e^{j\omega T})|^2 \sum_{k\in\mathbb{Z}} \left|H\left(\omega - \frac{2\pi k}{T}\right)\right|^2 d\omega \le \beta I_c, \qquad (5.11)$$

for every choice of $A(e^{j\omega T})$.

Let

$$R_{HH}(e^{j\omega T}) = \frac{1}{T} \sum_{k \in \mathbb{Z}} \left| H\left(\omega - \frac{2\pi k}{T}\right) \right|^2. \tag{5.12}$$

As we have seen in (3.97), $R_{HH}(e^{j\omega})$ is the DTFT of the sampled correlation sequence

$$r_{hh}[n] = \int_{-\infty}^{\infty} h(t)h(t + nT)dt. \tag{5.13}$$

Clearly, if

$$\alpha \le R_{HH}(e^{j\omega T}) \le \beta, \quad \text{almost everywhere on } \omega, \tag{5.14}$$

then (5.11) is satisfied. We can immediately verify that this condition is also necessary. Indeed, if (5.14) is violated over a set of frequencies $\omega \in \mathcal{I}$, then we can always choose $a[n]$ such that $A(e^{j\omega T}) = 1$ over $\mathcal{I}$, and zero everywhere else, which contradicts (5.11).

The Riesz basis condition is summarized in the following theorem:

Theorem 5.1 (Riesz basis condition). *The signals $\{h(t - nT)\}$ form a Riesz basis for their span if and only if there exists $\alpha > 0, \beta < \infty$ such that $\alpha \le R_{HH}(e^{j\omega T}) \le \beta$ almost everywhere, where $R_{HH}(e^{j\omega T})$ is defined by (5.12).*

5.1.2 Examples

We now illustrate the Riesz basis condition with several examples, some of which will be used repeatedly throughout the book.

Example 5.1 As we have seen in (4.18), the functions $\{h_n(t) = \text{sinc}((t - nT)/T)\}$ are orthogonal, and satisfy

$$\langle h_n(t), h_m(t) \rangle = T\delta_{nm}. \tag{5.15}$$

Therefore, they form a Riesz basis with bounds $\alpha = \beta = T$.

To see how this follows from Theorem 5.1, note that the CTFT of $h(t) = \text{sinc}(t/T)$ is equal to $H(\omega) = T$ for $|\omega| < \pi/T$ and $H(\omega) = 0$ otherwise. Consequently, $\sum_{k \in \mathbb{Z}} |H(\omega - 2\pi k/T)|^2 = T^2$ for every $\omega \in \mathbb{R}$. Theorem 5.1 then implies that $\{h(t - nT)\}$ forms a Riesz basis with bounds $\alpha = \beta = T$.

Example 5.2 In this example we show that B-spline functions of arbitrary degree generate Riesz bases. The proof is adapted from [105].

Consider the set of functions $\{h_n(t) = \beta^m(t - n)\}$, where $\beta^m(t)$ is a B-spline of degree $m \ge 0$, as defined by (4.32) and (4.33). Since $\beta^m(t)$ is the convolution of $m + 1$ rectangular windows, its CTFT can be computed by multiplying $m + 1$ sinc functions in the frequency domain:

$$H(\omega) = \text{sinc}^{m+1}\left(\frac{\omega}{2\pi}\right) = \left(\frac{\sin(\omega/2)}{\omega/2}\right)^{m+1}. \tag{5.16}$$

To find the Riesz bounds for $\beta^m(t)$, we need to bound

$$S(e^{j\omega}) = \sum_{k \in \mathbb{Z}} |H(\omega - 2\pi k)|^2 = \sum_{k \in \mathbb{Z}} \left| \frac{\sin(\omega/2 - \pi k)}{\omega/2 - \pi k} \right|^{2m+2} \tag{5.17}$$

from below and above.

Noting that the function $S(e^{j\omega})$ is 2π-periodic and symmetric, it suffices to analyze it on the interval $\omega \in [0, \pi]$. A lower bound can be easily obtained as

$$\sum_{k \in \mathbb{Z}} \left| \frac{\sin(\omega/2 - \pi k)}{\omega/2 - \pi k} \right|^{2m+2} \geq \left| \frac{\sin(\omega/2)}{\omega/2} \right|^{2m+2} \geq \left| \frac{\sin(\pi/2)}{\pi/2} \right|^{2m+2} = \left(\frac{2}{\pi} \right)^{2m+2}$$

$$\tag{5.18}$$

since $\sin(\omega)/\omega$ is strictly decreasing over $[0, \pi]$. Consequently, the lower Riesz bound satisfies $\alpha \geq |2/\pi|^{2m+2}$. To obtain an upper bound, we note that for $k \neq 0$

$$\sup_{|\omega| < \pi} \left| \frac{\sin(\omega/2 - \pi k)}{\omega/2 - \pi k} \right|^{2m+2} \leq \sup_{|\omega| < \pi} |\omega/2 - \pi k|^{-2m-2} \leq |\pi(1/2 - |k|)|^{-2m-2}. \tag{5.19}$$

For $k = 0$, this expression reduces to

$$\sup_{|\omega| < \pi} \left| \frac{\sin(\omega/2)}{\omega/2} \right|^{2m+2} = 1. \tag{5.20}$$

Therefore,

$$\sum_{k \in \mathbb{Z}} \left| \frac{\sin(\omega/2 - \pi k)}{\omega/2 - \pi k} \right|^{2m+2} \leq 1 + \sum_{k \neq 0} \frac{1}{|\pi(1/2 - |k|)|^{2m+2}}$$

$$= 1 + 2 \left(\frac{2}{\pi} \right)^{2m+2} \sum_{k=1}^{\infty} \frac{1}{|1 - 2k|^{2m+2}}$$

$$\leq 1 + 2 \left(\frac{2}{\pi} \right)^{2m+2} \sum_{k=1}^{\infty} \frac{1}{k^{2m+2}}$$

$$= 1 + 2 \left(\frac{2}{\pi} \right)^{2m+2} \zeta(2m + 2), \tag{5.21}$$

where $\zeta(\cdot)$ denotes the Riemann zeta function. It is well known that $\zeta(\ell) < \infty$ for any integer $\ell > 1$. Therefore the upper Riesz bound is finite and satisfies $\beta \leq 1 + 2(2/\pi)^{2m+2}\zeta(2m + 2)$.

Example 5.3 We now examine what happens when a set of functions $\{h_n(t) = h(t - nT)\}$ does not satisfy the lower Riesz bound.

Consider the function

$$h(t) = \frac{1}{\sqrt{2T}} \begin{cases} 1, & |t| \leq T \\ 0, & |t| > T. \end{cases} \tag{5.22}$$

It is easy to verify that the closed linear span of $\{h(t-nT)\}$ is the set of functions that are piecewise constant over intervals $t \in [nT, (n+1)T]$, $n \in \mathbb{Z}$. Note, however, that the support of $h(t)$ has size $2T$, which is twice as wide as the sampling step T, so that $h(t)$ is not equal to the zero-order spline.

To check whether the Riesz condition is satisfied, we consider the CTFT of $h(t)$. Direct computation of the CTFT results in $H(\omega) = \sqrt{2T} \operatorname{sinc}(T\omega/\pi)$. This function vanishes at $\omega = \pi k/T$ for every integer $k \neq 0$. Consequently, $R_{HH}(e^{j\omega T})$ equals 0 at $\omega = \pm\pi/T$. This fact, in itself, does not preclude the possibility of $\{h(t-nT)\}$ constituting a Riesz basis, since $R_{HH}(e^{j\omega T})$ needs to be positive only *almost everywhere*. However, in our case, $R_{HH}(e^{j\omega T})$ is continuous, so that it becomes arbitrarily close to 0 as ω tends to π/T. Therefore, there exists no (positive) lower Riesz bound.

To see the implication of this observation, consider the sequences

$$a_k[n] = \begin{cases} (-1)^n, & |n| \le k \\ 0, & |n| > k \end{cases} \tag{5.23}$$

for $k = 1, 2, \ldots$, and let $x_k(t) = \sum_{n \in \mathbb{Z}} a_k[n]h(t-nT)$. Then, $x_k(t)$ is explicitly given by

$$x_k(t) = \frac{1}{\sqrt{2T}} \begin{cases} (-1)^k, & |t| \in [kT, (k+1)T] \\ 0, & \text{else.} \end{cases} \tag{5.24}$$

The sequences $a_k[n]$ and signals $x_k(t)$ are depicted in Fig. 5.1 for $k = 1, 2, 3$. As can be seen from the figure, the norm of the kth sequence of coefficients is $\|a_k\|_{\ell_2} = \sqrt{2k+1}$, whereas the norm of the kth signal is $\|x_k\|_{L_2} = 1$. Therefore, the fact that there is no lower Riesz bound allows the construction of a set of piecewise constant

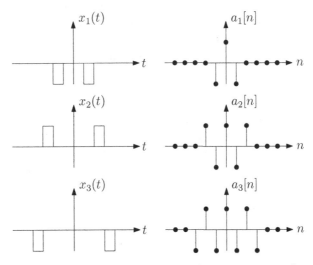

Figure 5.1 A set of unit-norm piecewise constant signals (left). The corresponding representation sequences in the SI basis generated by $h(t)$ of (5.22) (right).

functions, $x_1(t), x_2(t), \ldots$, such that $\|x_k\|_{L_2} = 1$ for every $k \geq 1$, but whose norms of representation coefficients grow indefinitely with k. In other words, there exist bounded-norm signals in the span of $\{h(t - nT)\}$ whose representation as a linear combination of $\{h(t - nT)\}$ requires using a sequence of coefficients with arbitrarily large norms. In cases where there exists a (positive) lower Riesz bound, such a phenomenon cannot occur.

In this example we considered violation of the lower Riesz bound condition. A dual situation arises when the functions $\{h(t - nT)\}$ do not satisfy the *upper* Riesz bound. In this case, there exist signals $x(t)$ with arbitrarily large L_2-norm, whose sequence of representation coefficients $a[n]$ has unit ℓ_2 norm.

5.2   Riesz basis expansions

One of the primary advantages of Riesz bases in Hilbert spaces is that the expansion coefficients can be computed in a stable manner using the biorthogonal vectors. These results are summarized in Theorem 2.1 of Section 2.5.4. Given a Riesz basis corresponding to a set transformation $H\colon \ell_2 \to \mathcal{H}$, the biorthogonal basis corresponds to the set transformation $\widetilde{H} = H(H^*H)^{-1}$. In this section we show that in SI subspaces, these vectors maintain the SI structure so that

$$\tilde{h}_n = \tilde{h}(t - nT) \tag{5.25}$$

for a generator $\tilde{h}(t)$, which can be computed explicitly in the Fourier domain.

5.2.1   Biorthogonal basis

To evaluate $\widetilde{H}$ we consider its action on an arbitrary sequence c in ℓ_2, and decompose the computation into two steps. In the first stage we calculate $b = (H^*H)^{-1}c$; in the second step we compute the $L_2(\mathbb{R})$ signal $x(t) = Hb$. Since H corresponds to an SI basis with generator $h(t)$, this last step can be evaluated as

$$x(t) = \sum_{n \in \mathbb{Z}} b[n]h(t - nT). \tag{5.26}$$

It remains to obtain an explicit expression for $b = (H^*H)^{-1}c$. To this end we show in the next proposition that any inner product operator of the form S^*A, where A and S are set transformations corresponding to SI generators, can be implemented by digital filtering. This proposition is key in developing filtering methods for recovery, and will be used throughout the book.

Proposition 5.1. *Let A and S be set transformations corresponding to the SI bases $\{a(t - nT)\}$ and $\{s(t - nT)\}$, and let $b = S^*Ac$ where c is a given sequence in ℓ_2. Then $b[n]$ can be obtained by filtering the sequence $c[n]$ by the digital filter*

$$r_{sa}[n] = s(-t) * a(t)|_{t=nT} = \int_{-\infty}^{\infty} a(t)s(t - nT)dt = \langle s(t), a(t + nT) \rangle. \tag{5.27}$$

The frequency response of the filter is equal to

$$R_{SA}(e^{j\omega}) = \frac{1}{T}\sum_{k\in\mathbb{Z}}\overline{S\left(\frac{\omega}{T} - \frac{2\pi k}{T}\right)}A\left(\frac{\omega}{T} - \frac{2\pi k}{T}\right). \tag{5.28}$$

Note that $r_{sa}[n]$ is the correlation sequence defined by (3.97).

Proof: Write $b = S^*y$ with $y(t) = Ac = \sum_n c[n]a(t - nT)$. We have seen already that S^*y is equivalent to uniform sampling at the output of the filter $s(-t)$. Thus,

$$\begin{aligned}
b[n] &= \langle s(t - nT), y(t)\rangle \\
&= \sum_{m\in\mathbb{Z}} c[m]\langle s(t - nT), a(t - mT)\rangle \\
&= \sum_{m\in\mathbb{Z}} c[m]\langle s(t), a(t + (n - m)T)\rangle \\
&= \sum_{m\in\mathbb{Z}} c[m]r_{sa}[n - m] \tag{5.29}
\end{aligned}$$

which is a convolution between $r_{sa}[n]$ and $c[n]$. The frequency response of $r_{sa}[n]$ follows from (3.97). $\square$

Proposition 5.1 implies that the operator H^*H is equivalent to a digital filter with impulse response $r_{hh}[n]$ and DTFT $R_{HH}(e^{j\omega})$. Consequently, the inverse $(H^*H)^{-1}$ corresponds to an inverse filter with frequency response $1/R_{HH}(e^{j\omega})$. Note that the Riesz basis condition (5.14) guarantees that the inverse is well defined. Combining this observation with (5.26) allows the operation of $H(H^*H)^{-1}$ to be expressed on a sequence c by a combination of digital filtering followed by impulse modulation and analog filtering, as illustrated in Fig. 5.2.

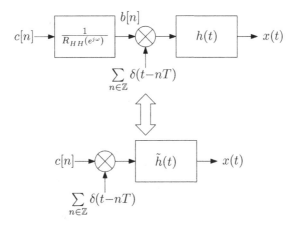

Figure 5.2 Interpretation of the transform $\widetilde{H} = H(H^*H)^{-1}$ in terms of filtering operations. Here $x = \widetilde{H}c = Hb$ with $b = (H^*H)^{-1}c$.

We can recover the mth biorthogonal vector $\tilde{h}_m$ from $\tilde{H}$ by noting that $\tilde{h}_m = \tilde{H}e_m$ where $e_m[n] = \delta[n - m]$. Choosing $c[n] = e_0[n]$ in Fig. 5.2 results in

$$\tilde{H}(\omega) = \tilde{H}_0(\omega) = \frac{1}{R_{HH}(e^{j\omega T})}H(\omega). \tag{5.30}$$

It is easy to see that a shift by m of the input, namely $c[n] = \delta[n - m]$, will result in a shift by mT of the output sequence, establishing that $\tilde{h}_m(t) = \tilde{h}(t - mT)$.

In conclusion, the biorthogonal vectors corresponding to an SI basis generated by $h(t)$ are also SI, with generator $\tilde{h}(t)$ whose CTFT is given by (5.30). In the special case in which the basis is orthonormal, $R_{HH}(e^{j\omega}) = 1$ (see (3.99)), and as we expect $\tilde{h}(t) = h(t)$.

Example 5.4 We demonstrate the computation of the biorthogonal vectors corresponding to the SI basis generated by $\beta^1(t)$, the B-spline function of degree 1.

From the convolution property of splines, we immediately have that

$$\beta^1(t) = \begin{cases} t + 1, & -1 < t \leq 0 \\ 1 - t, & 0 < t < 1 \\ 0, & |t| \geq 1. \end{cases} \tag{5.31}$$

Assume that $T = 1$. We can directly compute the sampled correlation as

$$r_{\beta^1\beta^1}[n] = \int_{-\infty}^{\infty} \beta^1(t)\beta^1(t - n)dt = \begin{cases} 1/6, & |n| = 1 \\ 2/3, & n = 0 \\ 0, & |n| > 1. \end{cases} \tag{5.32}$$

Taking the DTFT of $r_{\beta^1\beta^1}[n]$ we have

$$R_{\beta^1\beta^1}(e^{j\omega}) = \frac{1}{6}e^{j\omega} + \frac{2}{3} + \frac{1}{6}e^{-j\omega} = \frac{2}{3} + \frac{1}{3}\cos(\omega). \tag{5.33}$$

Therefore, the biorthogonal function $\tilde{\beta}^1(t)$ is given by

$$\tilde{\beta}^1(t) = \sum_{n\in\mathbb{Z}} b[n]\beta^1(t - n), \tag{5.34}$$

where $b[n]$ is the inverse DTFT of

$$B(e^{j\omega}) = \frac{3}{2 + \cos(\omega)}. \tag{5.35}$$

The function $\beta^1(t)$ and its biorthogonal function $\tilde{\beta}^1(t)$ are depicted in Fig. 5.3.

For a given set of samples $c[n]$, Figs. 5.3(c)–(d) depict the reconstructions $x_1(t) = \sum_n c[n]\beta^1(t - n)$ and $x_2(t) = \sum_n c[n]\tilde{\beta}^1(t - n)$ respectively. Note that whereas $x_1(t)$ passes through the samples, $x_2(t)$ does not. This is a result of the fact that the function $\beta^1(t)$ satisfies the interpolation property $\beta^1(nT) = \delta[n]$, while $\tilde{\beta}^1(t)$ is not an interpolating function. We will discuss this property in more detail in Section 5.2.3.

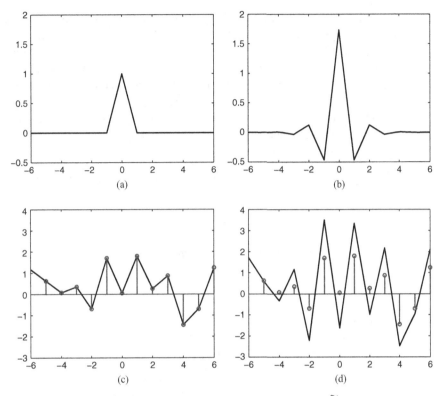

Figure 5.3 (a) B-spline $\beta^1(t)$ of degree 1. (b) The biorthogonal function $\tilde{\beta}^1(t)$. (c) The function $x(t) = \sum_n c[n]\beta^1(t - n)$. (d) The function $x(t) = \sum_n c[n]\tilde{\beta}^1(t - n)$. The values of $c[n]$ are indicated by circles in (c) and (d).

5.2.2 Expansion coefficients

Once we determine the biorthogonal vectors, the expansion coefficients are given by:

$$c[n] = \langle \tilde{h}(t - nT), x(t) \rangle. \tag{5.36}$$

The inner products in (5.36) can be implemented by filtering $x(t)$ with the filter $\tilde{h}(-t)$ followed by uniform sampling at times nT. From Proposition 5.1, these coefficients are expressible in the DTFT domain as

$$C(e^{j\omega}) = R_{\tilde{H}X}(e^{j\omega}), \tag{5.37}$$

where $R_{\tilde{H}X}(e^{j\omega})$ is defined by (5.28). Substituting the expression for $\tilde{H}(e^{j\omega})$ (see (5.30)) into (5.37) leads to

$$C(e^{j\omega}) = \frac{R_{HX}(e^{j\omega})}{R_{HH}(e^{j\omega})}. \tag{5.38}$$

The relation (5.38) can be used to compute the expansion coefficients $c[n]$ in the Riesz basis generated by $h(t)$ directly from $x(t)$.

As a sanity check, we can easily show that (5.38) recovers the correct sequence. Specifically, suppose that $x(t)$ lies in the SI space spanned by $h(t)$, so that $x(t) = \sum_n a[n]h(t - nT)$ for some sequence $a[n]$ with DTFT $A(e^{j\omega})$. We would like to show that $C(e^{j\omega})$ of (5.38) is equal to $A(e^{j\omega})$. To see this, note that from (3.83), $X(\omega) = A(e^{j\omega T})H(\omega)$. Substituting this relation into (5.38), and using the fact that $A(e^{j\omega T})$ is $2\pi/T$-periodic, establishes the result.

The following theorem summarizes our findings regarding the biorthogonal Riesz basis and the corresponding expansion coefficients.

Theorem 5.2 (Riesz basis expansion). *Let $\{h(t - nT)\}$ denote a Riesz basis for a subspace $\mathcal{W}$. Then the biorthogonal Riesz basis is given by $\{\tilde{h}(t - nT)\}$ where in the CTFT domain*

$$\tilde{H}(\omega) = \frac{1}{R_{HH}(e^{j\omega T})}H(\omega),$$

with $R_{HH}(e^{j\omega T})$ defined by (5.12). In the time domain,

$$\tilde{h}(t) = \sum_{n \in \mathbb{Z}} b[n]h(t - nT),$$

where $b[n]$ is the inverse DTFT of $1/R_{HH}(e^{j\omega})$.

Any $x(t)$ in $\mathcal{W}$ can be expressed as $x(t) = \sum_{n \in \mathbb{Z}} c[n]h(t - nT)$ where the expansion coefficients $c[n]$ are given in the DTFT domain by

$$C(e^{j\omega}) = \frac{R_{HX}(e^{j\omega})}{R_{HH}(e^{j\omega})},$$

and $R_{HX}(e^{j\omega})$ is defined by (5.28).

To conclude our discussion on basis expansions in SI subspaces, we summarize two important properties used in our derivations in the following proposition. We will rely on this proposition in future developments.

Proposition 5.2. *Let $x(t) = \sum_n a[n]h(t - nT)$ for a sequence $a \in \ell_2$, and let $c[n] = \langle s(t - nT), x(t) \rangle$ for a function $s(t)$. Then, in the frequency domain we have that:*

1. *The signal $x(t)$ can be written as $X(\omega) = A(e^{j\omega T})H(\omega)$;*
2. *The samples $c[n]$ are given in the DTFT domain by*

$$C(e^{j\omega}) = A(e^{j\omega})R_{SH}\left(e^{j\omega}\right), \tag{5.39}$$

where $R_{SH}(e^{j\omega})$ is defined by (5.28).

5.2.3   Alternative basis expansions

Until now we have focused on expansions of the form $x(t) = \sum_n c[n]h(t - nT)$ where $c[n]$ was computed using the biorthogonal generator $\tilde{h}(t)$. By reversing the roles of $h(t)$ and $\tilde{h}(t)$ we can also expand $x(t)$ as $x(t) = \sum_n c[n]\tilde{h}(t - nT)$ where now $c[n] = \langle h(t - nT), x(t) \rangle$ are the corresponding coefficients.

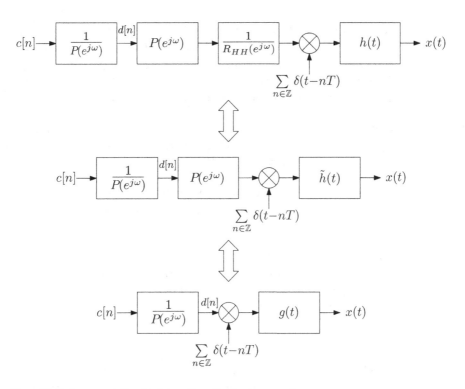

Figure 5.4 A framework for obtaining alternative basis expansions.

For a given subspace $\mathcal{W}$, there are clearly many functions $g(t)$ that will span the space via shifts by nT. We have already seen two examples: the original function $h(t)$, and the biorthogonal function $\tilde{h}(t)$. To see how more examples can be generated, suppose that in the scheme of Fig. 5.2, we insert a stably invertible filter with DTFT $P(e^{j\omega})$ before the digital filter and concatenate it with its inverse $1/P(e^{j\omega})$, as depicted in Fig. 5.4. Obviously, this does not change the output. If we now define

$$g(t) = \sum_{n\in\mathbb{Z}} p[n]\tilde{h}(t - nT), \qquad (5.40)$$

where $p[n]$ is the inverse DTFT of $P(e^{j\omega})$, then we have that

$$x(t) = \sum_{n\in\mathbb{Z}} d[n]g(t - nT), \qquad (5.41)$$

where $d[n] = c[n] * b[n]$ and $b[n]$ is the inverse DTFT of the filter $1/P(e^{j\omega})$. This interpretation is depicted in the lower branch of Fig. 5.4.

Therefore, instead of the basis functions $\{h(t - nT)\}$ or $\{\tilde{h}(t - nT)\}$, we can obtain alternative expansions using any set of functions of the form (5.40). The same of course holds true when we take stable linear combinations of $\{h(t - nT)\}$. We formalize this claim in Proposition 5.3 below, adapted from [106, Proposition 6]. The filter $p[n]$ can be chosen to yield basis expansions with required properties. For example, we may construct $g(t)$ as a function localized in time or frequency, or such that it generates an orthonormal basis. Alternatively, we may want $g(t)$ to possess symmetry properties. Another useful requirement in certain applications is the interpolation property, where $g(t)$ is chosen to exactly interpolate a given set of samples.

Proposition 5.3. *Let $h(t)$ generate a Riesz basis for a subspace $\mathcal{W}$, and let $g(t) = \sum_{n \in \mathbb{Z}} p[n] h(t - nT)$. Then $g(t)$ generates a Riesz basis for $\mathcal{W}$ if and only if*[1] ess sup $|P(e^{j\omega})| < \infty$ and ess inf $|P(e^{j\omega})| > 0$.

We now consider some examples of choices of $p[n]$, leading to generators $g(t)$ with specific properties.

Orthogonal generators

A function $g(t)$ is called an *orthogonal generator* if it generates an orthonormal basis: $\langle g_n(t), g_m(t) \rangle = \delta_{nm}$ where $g_n(t) = g(t - nT)$. Using the structure of $g_n(t)$ this is equivalent to the requirement

$$\langle g(t), g(t - nT) \rangle = \delta[n], \tag{5.42}$$

or in the frequency domain, $R_{GG}(e^{j\omega}) = 1$. An example is the (properly normalized) sinc function. The advantage of an orthogonal generator is that it is equal to its biorthogonal generator: $\tilde{g}(t) = g(t)$.

To obtain an orthogonal generator starting from an arbitrary function $h(t)$ we choose

$$P(e^{j\omega}) = \frac{1}{R_{HH}^{1/2}(e^{j\omega})} = \frac{T^{1/2}}{\left(\sum_{k \in \mathbb{Z}} \left| H \left(\frac{\omega}{T} - \frac{2\pi k}{T} \right) \right|^2 \right)^{1/2}}. \tag{5.43}$$

This function satisfies the conditions of Proposition 5.3 when $h(t)$ generates a Riesz basis.

To show that the resulting generator $g(t) = \sum_{n \in \mathbb{Z}} p[n] h(t - nT)$ is orthogonal, we use Proposition 5.2 to write $G(\omega) = P(e^{j\omega T}) H(\omega)$ so that

$$R_{GG}(e^{j\omega}) = |P(e^{j\omega})|^2 R_{HH}(e^{j\omega}). \tag{5.44}$$

Substituting $P(e^{j\omega})$ of (5.43) results in $R_{GG}(e^{j\omega}) = 1$, as required.

Example 5.5 We demonstrate how to construct an orthonormal basis for the space of piecewise linear functions with knots at the integers.

---

[1] If $a = \text{ess sup}_x f(x)$ then a is the supremum of $f(x)$ except possibly over a set of measure 0. More precisely, ess sup $f(x) = \inf\{b \in \mathbb{R} : \mu(\{x : f(x) > b\}) = 0\}$, where μ denotes the measure of the set. The definition of ess inf follows in a similar fashion.

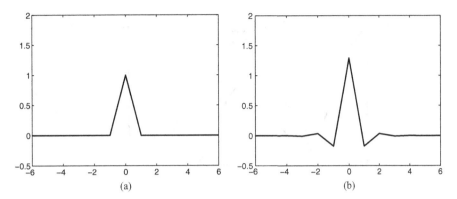

Figure 5.5 (a) B-spline $\beta^1(t)$ of degree 1. (b) The corresponding orthogonal generator $g(t)$.

As we saw in Example 5.4, this space is generated by the B-spline function $h(t) = \beta^1(t)$ of degree 1, for which $\sum_{k \in \mathbb{Z}} |H(\omega - 2\pi k)|^2 = (2 + \cos(\omega))/3$. Therefore, an orthogonal generator for this space is given by

$$g(t) = \sum_{n \in \mathbb{Z}} p[n]\beta^1(t - n) \tag{5.45}$$

where $p[n]$ is the inverse DTFT of

$$P(e^{j\omega}) = \frac{\sqrt{3}}{\sqrt{2 + \cos(\omega)}}. \tag{5.46}$$

The functions $\beta^1(t)$ and $g(t)$ are depicted in Fig. 5.5.

Interpolating generators

In Chapter 4 we encountered interpolating functions when discussing communication signals. A generator $g(t)$ is said to have the *interpolation property* if $g(nT) = \delta[n]$. This means that $g(0) = 1$, and at every other multiple of T, $g(t)$ is equal to 0. In the Fourier domain, this amounts to the requirement that

$$\frac{1}{T} \sum_{k \in \mathbb{Z}} G\left(\omega - \frac{2\pi k}{T}\right) = 1. \tag{5.47}$$

To understand the implications of the interpolation property, suppose that $x(t) = \sum_n a[n]g(t - nT)$ where $g(t)$ satisfies the interpolation condition. Sampling $x(t)$ at points mT we have immediately that $x(mT) = a[m]$. Therefore, at the sampling instances, $x(t)$ is equal to the expansion coefficients. In between these points, $g(t)$ interpolates the samples to a continuous-time function, as shown in Fig. 5.6(a) for $g(t) = \text{sinc}(t)$. This property can therefore be used to interpolate between a given set of samples. When this condition is not satisfied, after fitting a curve $g(t)$ to a set of points, the resulting function $x(t)$ will not generally pass through the given samples, as demonstrated in Fig. 5.6(b). Here we construct $x(t)$ from the samples $a[n]$ using the function

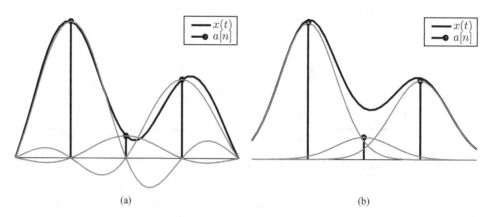

Figure 5.6 (a) Interpolation using the sinc kernel. (b) Reconstruction using a Gaussian kernel, which does not satisfy the interpolation property.

$g(t) = \exp\{-t^2/(2\sigma^2)\}$. Since this kernel does not satisfy the interpolation property, $x(mT)$ is not equal to $a[m]$.

To construct an interpolating function from a given generator $h(t)$, we let

$$P(e^{j\omega}) = \frac{T}{\sum_{k\in\mathbb{Z}} H\left(\frac{\omega}{T} - \frac{2\pi k}{T}\right)} = \frac{1}{B(e^{j\omega})}, \tag{5.48}$$

as long as $B(e^{j\omega}) = (1/T)\sum_{k\in\mathbb{Z}} H(\omega/T - 2\pi k/T)$ is bounded. Note that $B(e^{j\omega})$ is the DTFT of the sampled sequence $h(nT)$. Using the relation $G(\omega) = P(e^{j\omega T})H(\omega)$, it is easy to verify that this choice satisfies (5.47).

Example 5.6 As we saw, a Gaussian window $h(t) = \exp\{-t^2/(2\sigma^2)\}$ does not satisfy the interpolation property. To obtain an interpolating generator for the SI space generated by the Gaussian window, we compute the corresponding function $B(e^{j\omega})$ in (5.48).

The CTFT of $h(t)$ is given by $H(\omega) = \sqrt{2\pi\sigma^2}\exp\{-\sigma^2\omega^2/2\}$. Therefore, assuming for simplicity that $T = 1$, an interpolating kernel can be constructed as

$$g(t) = \sum_{n\in\mathbb{Z}} p[n]\exp\{-(t-n)^2/(2\sigma^2)\} \tag{5.49}$$

where $p[n]$ is the inverse DTFT of

$$P(e^{j\omega}) = \frac{1}{\sqrt{2\pi\sigma^2}\sum_{k\in\mathbb{Z}}\exp\{-\sigma^2(\omega - 2\pi k)^2/2\}}. \tag{5.50}$$

Figure 5.7 depicts a Gaussian generator and its corresponding interpolating generator for the case $T = 1$, $\sigma^2 = 1/4$. Figure 5.8 shows the interpolation obtained with a Gaussian kernel and its interpolating version for the sequence appearing in Fig. 5.6. As can be seen, the interpolating generator produces a recovery that passes through the given samples.

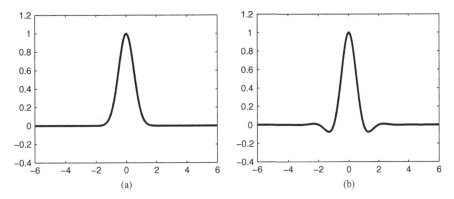

Figure 5.7 (a) Gaussian generator $h(t) = \exp\{-2t^2\}$. (b) The corresponding interpolating generator $g(t)$.

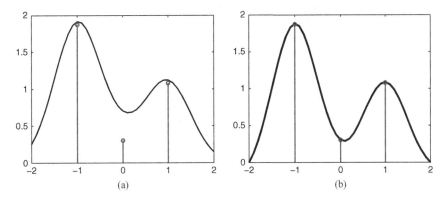

Figure 5.8 (a) Reconstruction using the Gaussian generator. (b) Reconstruction using the interpolating Gaussian generator.

5.3 Partition of unity

Until now we have focused on the Riesz property of the generator $h(t)$, which guarantees stable basis expansions. Another requirement on $h(t)$ that can be advantageous in the context of sampling is that if we increase the sampling rate, namely if T is made smaller, then we hope to be able to approximate any input function as closely as desired. Since the key to controlling the error is the sampling step, we consider the rescaled signal expansion $x(t) = \sum_n a[n]h(t/T-n)$. Here the basis function is dilated by T and shifted by the same amount. When T goes to 0, we expect that choosing $a[n] = x(nT)$ will result in a pretty good approximation to $x(t)$. As we now show, if the generator $h(t)$ has the *partition of unity property*, defined below in (5.52), then the error in approximating $x(t)$ by its samples $x(nT)$ tends to 0 when T is decreased [107].

Proposition 5.4. *Let $x(t)$ be an arbitrary function in L_2, with bounded derivative. Denote by*

$$P_T x(t) = \sum_{n \in \mathbb{Z}} x(nT) h(t/T - n) \tag{5.51}$$

an approximation of $x(t)$ from the samples $x(nT)$, and assume that $h(t)$ is bounded, and compactly supported. If $h(t)$ satisfies the partition of unity property:

$$\sum_{n \in \mathbb{Z}} h(t - n) = 1, \quad \text{for all } t, \tag{5.52}$$

then

$$\max_t |P_T x(t) - x(t)| \to 0, \tag{5.53}$$

when $T \to 0$.

Proof: To prove the proposition we first note that if $h(t)$ has support R, then the sum in (5.51) can include at most R points. In addition, from (5.52),

$$x(t) = x(t) \sum_{n \in \mathbb{Z}} h(t/T - n) = \sum_{n \in \mathbb{Z}} x(t) h(t/T - n). \tag{5.54}$$

Therefore,

$$\sum_{n \in \mathbb{Z}} x(nT) h(t/T - n) - x(t) = \sum_{n \in \mathcal{I}} (x(nT) - x(t)) h(t/T - n), \tag{5.55}$$

where $\mathcal{I}$ is the set of indices n such that $|t - nT| \le RT/2$. From the mean value theorem,

$$|x(nT) - x(t)| \le x'(t_0) RT \le MRT, \tag{5.56}$$

where $x'(t_0)$ is the derivative of $x(t)$ at a point t_0 in the interval defined by nT and t, and M is an upper bound on $x'(t)$. Substituting into (5.55),

$$\left| \sum_{n \in \mathbb{Z}} x(nT) h(t/T - n) - x(t) \right| \le H \sum_{n \in \mathcal{I}} |x(nT) - x(t)| \le HMR^2 T, \tag{5.57}$$

which goes to zero as $T \to 0$. Here H is an upper bound on $h(t)$. $\qquad \square$

The proof above shows sufficiency of the partition of unity property under several assumptions on the generator $h(t)$. A more detailed analysis can be carried out to prove that as long as $h(t)$ is sufficiently smooth (but not necessarily compactly supported) the partition of unity is a necessary and sufficient condition to guarantee that the approximation error tends to 0 as T is made smaller [65, 107].

Condition (5.52) may be easily determined in the Fourier domain using the Poisson-sum formula (3.89). Specifically, letting $g(t) = h(t_0 + t)$, we have from (3.89) that

$$\sum_{n \in \mathbb{Z}} h(t_0 - n) = \sum_{n \in \mathbb{Z}} g(-n) = \sum_{k \in \mathbb{Z}} G(2\pi k) = \sum_{k \in \mathbb{Z}} H(2\pi k) e^{j2\pi k t_0}. \tag{5.58}$$

Therefore, (5.52) becomes

$$\sum_{k \in \mathbb{Z}} H(2\pi k) e^{j2\pi k t_0} = 1 \tag{5.59}$$

for all t_0, which is satisfied only if

$$H(2\pi k) = \delta[k]. \tag{5.60}$$

Example 5.7 The function $h(t) = \operatorname{sinc}(t)$ trivially satisfies (5.60) since

$$H(\omega) = \begin{cases} 1, & |\omega| \leq \pi \\ 0, & |\omega| > \pi \end{cases} \tag{5.61}$$

so that indeed $H(2\pi k) = \delta[k]$.

Example 5.8 Splines of all degrees also satisfy (5.60). This is because, as we have seen in Example 5.2, the CTFT of a B-spline of degree $m \geq 0$ is given by

$$H(\omega) = \operatorname{sinc}^{m+1}\left(\frac{\omega}{2\pi}\right). \tag{5.62}$$

Therefore, here too we have that $H(2\pi k) = \delta[k]$.

Example 5.9 The Gaussian kernel $h(t) = \exp\{-t^2/(2\sigma^2)\}$ does not satisfy (5.60). This can be seen by noting that its CTFT, which is given by $H(\omega) = \sqrt{2\pi\sigma^2} \exp\{-\sigma^2\omega^2/2\}$, does not vanish anywhere. The implication of this is that with the Gaussian kernel, $P_T x(t)$ does not tend to $x(t)$ as T tends to zero. This effect is demonstrated in Fig. 5.9, where the function $x(t) = \sin(t)$ is approximated from its samples at times $t = nT$ using both the sinc kernel and the Gaussian kernel. The former, which satisfies the partition of unity, leads to a recovery $P_T x(t)$ that converges to $x(t)$ as $T \to 0$, whereas the latter does not.

5.4 Redundant sampling in SI spaces

As we noted in Chapter 2, in most of the book we focus on nonredundant representations corresponding to basis expansions. However, for completeness, we now briefly discuss redundant SI representations.

Suppose that we have an SI space of the form (5.1), where the functions $\{h(t-nT)\}$ are not necessarily linearly independent. We would like to determine conditions under which this set forms a frame for $\mathcal{W}$. To this end we express the frame condition (2.122) in the SI setting:

$$\alpha \|x(t)\|^2 \leq \sum_{n \in \mathbb{Z}} |\langle x(t), h(t-nT) \rangle|^2 \leq \beta \|x(t)\|^2 \tag{5.63}$$

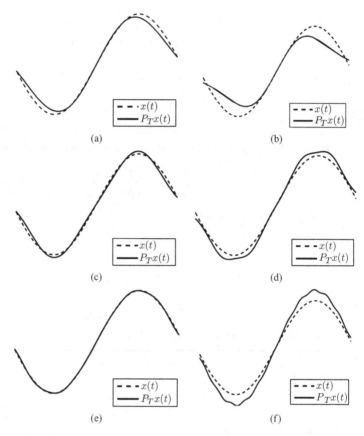

Figure 5.9 Reconstruction of the signal $x(t) = \sin(t)$ from samples at $t = nT$ via (5.51). (a), (c), (e) Using the kernel $h(t) = \mathrm{sinc}(t)$ with $T = 2, 1, 0.5$ respectively. (b), (d), (f) Using the kernel $h(t) = \exp(-2t^2)$ with $T = 2, 1, 0.5$ respectively.

for some $\alpha > 0$, $\beta < \infty$ and for every $x(t) = \sum_n a[n]h(t - nT)$, with some $a \in \ell_2$.

Denote by $c[n]$ the inner product $c[n] = \langle h(t - nT), x(t) \rangle$. From Proposition 5.2, $C(e^{j\omega}) = A(e^{j\omega})R_{HH}(e^{j\omega})$. Therefore,

$$\sum_{n \in \mathbb{Z}} |\langle x(t), h(t - nT) \rangle|^2 = \sum_{n \in \mathbb{Z}} |c[n]|^2$$

$$= \frac{T}{2\pi} \int_0^{2\pi/T} |C(e^{j\omega T})|^2 \, d\omega$$

$$= \frac{1}{2\pi T} \int_0^{2\pi/T} |A(e^{j\omega T})|^2 \left| \sum_{k \in \mathbb{Z}} H\left(\omega - \frac{2\pi k}{T}\right) \right|^2 \, d\omega, \qquad (5.64)$$

where the second equality is a result of Parseval's theorem.

We can express the energy of $x(t)$ in similar form as

$$
\begin{aligned}
\|x(t)\|^2 &= \frac{1}{2\pi} \int_{-\infty}^{\infty} |X(\omega)|^2 d\omega \\
&= \frac{1}{2\pi} \int_{-\infty}^{\infty} |A(e^{j\omega T})|^2 |H(\omega)|^2 d\omega \\
&= \frac{1}{2\pi} \int_{0}^{2\pi/T} |A(e^{j\omega T})|^2 \sum_{k\in\mathbb{Z}} \left| H\left(\omega - \frac{2\pi k}{T}\right) \right|^2 d\omega,
\end{aligned}
\tag{5.65}
$$

where we used Parseval, the relation $X(\omega) = A(e^{j\omega T})H(\omega)$, and (5.7). Thus, if

$$
\alpha < \frac{1}{T} \sum_{k\in\mathbb{Z}} \left| H\left(\omega - \frac{2\pi k}{T}\right) \right|^2 < \beta, \quad \text{almost everywhere on } \omega,
\tag{5.66}
$$

for every frequency ω for which $\sum_{k\in\mathbb{Z}} |H(\omega - 2\pi k/T)|^2 \neq 0$, then from (5.64) and (5.65), the frame condition (5.63) is satisfied. We can also show that this condition is necessary, just as in the Riesz basis case.

It is interesting to compare the Riesz basis condition (5.14) with the frame condition (5.66). These two equations are very similar: the difference is that the latter only has to hold for frequencies ω for which $R_{HH}(e^{j\omega T}) = (1/T)\sum_{k\in\mathbb{Z}} |H(\omega - 2\pi k/T)|^2 \neq 0$. This implies that while for a Riesz basis $R_{HH}(e^{j\omega T})$ cannot be zero over an interval, it can when $h(t)$ generates a frame.

5.4.1 Redundant bandlimited sampling

To illustrate the effect of oversampling, let us consider uniform sampling of bandlimited signals. Suppose that $\mathcal{W}$ is the space of signals bandlimited to π/T where $T > 1$. A basis for this space is $w_n(t) = \text{sinc}((t - nT)/T)$. We now show how to construct a frame for $\mathcal{W}$ so that we have more vectors than necessary to represent a bandlimited signal $x(t)$.

To this end, we begin with an orthonormal basis for a larger space $\mathcal{V}$ containing $\mathcal{W}$ and then orthogonally project it onto $\mathcal{W}$. Since $x(t)$ is bandlimited to π/T and $T > 1$, it is also bandlimited to π. Therefore, we can express $x(t)$ as

$$
x(t) = \sum_{n\in\mathbb{Z}} x(n) \, \text{sinc}(t - n).
\tag{5.67}
$$

The vectors $\text{sinc}(t-n)$ form a basis for the space $\mathcal{V}$ of signals bandlimited to π. We now use these vectors to construct a frame for $\mathcal{W} \subset \mathcal{V}$. Since $x(t) \in \mathcal{W}$, $P_{\mathcal{W}}x(t) = x(t)$ where $P_{\mathcal{W}}$ is the orthogonal projection onto $\mathcal{W}$. Thus,

$$
x(t) = \sum_{n\in\mathbb{Z}} x(n) P_{\mathcal{W}} \, \text{sinc}(t - n).
\tag{5.68}
$$

The orthogonal projection P_W is given by filtering with an LPF with cutoff frequency π/T. Noting that the CTFT of $v(t) = \mathrm{sinc}(t - n)$ is

$$V(\omega) = \begin{cases} e^{-j\omega n}, & |\omega| \leq \pi \\ 0, & \text{otherwise,} \end{cases} \tag{5.69}$$

it follows that the CTFT of $h(t) = P_W \mathrm{sinc}(t - n)$ is given by

$$H(\omega) = \begin{cases} e^{-j\omega n}, & |\omega| \leq \frac{\pi}{T} \\ 0, & \text{otherwise.} \end{cases} \tag{5.70}$$

Transforming back to the time domain,

$$h(t) = P_W \mathrm{sinc}(t - n) = \frac{1}{T}\mathrm{sinc}((t - n)/T). \tag{5.71}$$

We conclude that any $x(t) \in W$ can be expressed as

$$x(t) = \frac{1}{T}\sum_{n \in \mathbb{Z}} x(n)\,\mathrm{sinc}((t - n)/T), \tag{5.72}$$

and therefore the functions $h_n(t) = h(t - n) = (1/T)\mathrm{sinc}((t - n)/T)$ with $h(t) = (1/T)\mathrm{sinc}(t/T)$ span W. The resulting sampling and reconstruction scheme is depicted in Fig. 5.10. For this choice of $h(t)$,

$$\sum_{k \in \mathbb{Z}} |H(\omega - 2\pi k)|^2 = \begin{cases} 1, & |\omega - 2\pi k| \leq \frac{\pi}{T} \text{ for some integer } k \\ 0, & \text{otherwise.} \end{cases} \tag{5.73}$$

Because the sum is identically zero over intervals with measure greater than 0, the vectors $\{h(t - n)\}$ cannot form a basis for W; they do, however, obey the frame condition and therefore constitute a frame for this space.

The block diagram of Fig. 5.10 is different than the conventional oversampling block diagram, depicted in Fig. 5.11, which results from using a smaller sampling period (in this case $T = 1$) in the Shannon–Nyquist theorem. The difference is in the reconstruction LPF. In Fig. 5.10 the LPF has a cutoff frequency of π/T, while in Fig. 5.11 the cutoff frequency is π. This difference may at first seem insignificant, since $x(t)$ is bandlimited to π/T. However, if there is noise added to the samples, then using a frame-based recovery leads to noise reduction due to the lower cutoff frequency which filters out some of the noise.

To generalize the results, suppose that we sample $x(t)$ at times $t = nRT$ where $R \leq 1$ is the redundancy factor. When $R = 1$, we are back at Nyquist-rate samples. Since $x(t)$

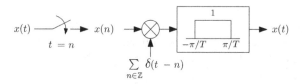

Figure 5.10 Frame expansion of a bandlimited signal.

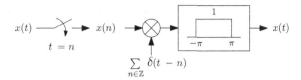

Figure 5.11 Oversampling of a bandlimited signal using the Shannon–Nyquist theorem.

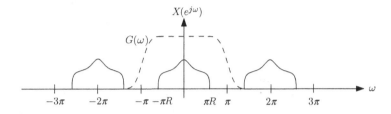

Figure 5.12 Oversampling of a bandlimited signal with a practical reconstruction filter $g(t)$.

is bandlimited to $\pi/(RT)$ for any $R \leq 1$, we can use the Shannon–Nyquist theorem to write

$$x(t) = \sum_{n \in \mathbb{Z}} x(nRT) \operatorname{sinc}((t - nRT)/(RT)). \tag{5.74}$$

Following the same steps as before, we can now project $x(t)$ onto the space of signals bandlimited to π/T, which amounts to filtering $x(t)$ with an LPF with cutoff π/T. The output of the filter is given by

$$x(t) = R \sum_{n \in \mathbb{Z}} x(nRT) \operatorname{sinc}((t - nRT)/T). \tag{5.75}$$

We conclude that a signal $x(t)$ that is oversampled by a factor of R can be recovered from its redundant samples using (5.75).

In the presence of noise, the recovery of (5.75) will preserve the signal while rejecting the largest possible amount of noise energy; this is because it completely filters out any noise above π/T. In practice, it is often difficult to implement an ideal LPF. We will treat interpolation with practical filters in detail in Chapters 6 and 9. However, it is worth pointing out here that when the signal is oversampled we can simplify the design of the recovery filter by noting that in the absence of noise, any filter that is equal to 1 in the region $(-\pi/T, \pi/T)$, and arbitrary for $\pi/T \leq |\omega| \leq (2\pi - \pi R)/(RT)$, can be used. This is a result of the fact that owing to oversampling, the signal content in this regime is equal to 0, as illustrated in Fig. 5.12. In the figure, we plot the DTFT of the sampled sequence $x(nRT)$ for $R < 1$ and $T = 1$. This allows increased design flexibility of the recovery filter, and in particular permits the use of filters with a continuous Fourier transform. One such example is shown in the figure. More specifically, let $g(t)$ denote a continuous-time filter with Fourier transform satisfying

$$G(\omega) = \begin{cases} RT, & |\omega| < \frac{\pi}{T} \\ \text{arbitrary}, & \frac{\pi}{T} \le |\omega| \le \frac{2\pi}{RT} - \frac{\pi}{T} \\ 0, & |\omega| > \frac{2\pi}{RT} - \frac{\pi}{T}. \end{cases} \qquad (5.76)$$

Then, $x(t)$ can be recovered from the samples $x(nRT)$ via

$$x(t) = \sum_{n \in \mathbb{Z}} x(nRT) g(t - nRT). \qquad (5.77)$$

5.4.2   Missing samples

In the previous section we have seen that oversampling is beneficial in the presence of noise and can lead to relaxed constraints on the recovery filter, simplifying its design. Oversampling is also useful in compensating for other distortions, such as missing samples. In this section, we consider the case in which a finite number of samples are lost and show that if the signal is oversampled, then the lost information is recoverable. This result is intuitive since oversampling implies that the samples are dependent in some way, and therefore share information. For concreteness, we focus on the bandlimited setting; however, the results generalize in a straightforward manner to other SI spaces.

Suppose that $x(t)$ is bandlimited to π/T and sampled at times $t = nRT$ with $R < 1$. We assume that a finite number K of samples have been lost, and denote by $\mathcal{K}$ the set of lost samples. Our goal is to recover $x(t)$ from $\{x(nRT), n \notin \mathcal{K}\}$. As we now show, recovery is possible as long as $\mathcal{K}$ is finite and $R < 1$. We note, however, that this approach becomes unstable when K increases and/or R approaches 1.

To develop a recovery strategy we rewrite $x(t)$ of (5.75) as

$$x(t) = R \sum_{n \in \mathcal{K}} x(nRT) \operatorname{sinc}((t - nRT)/T) + y_{\mathcal{K}}(t), \qquad (5.78)$$

where we defined

$$y_{\mathcal{K}}(t) = R \sum_{n \notin \mathcal{K}} x(nRT) \operatorname{sinc}((t - nRT)/T). \qquad (5.79)$$

Note that $y_{\mathcal{K}}(t)$ can be computed from the given samples. Evaluating (5.78) at the missing points $t = mRT, m \in \mathcal{K}$ we have

$$\sum_{n \in \mathcal{K}} x(nRT) \left(\delta[n - m] - R \operatorname{sinc}((m - n)R) \right) = y_{\mathcal{K}}(mRT), \quad m \in \mathcal{K}. \qquad (5.80)$$

The left-hand side of (5.80) depends on the K unknowns $x(nRT), n \in \mathcal{K}$. Therefore, writing out (5.80) for each $m \in \mathcal{K}$ leads to K equations, in K unknowns. Denoting by $\mathbf{x}$ the length-K vector of unknowns $x(nRT), n \in \mathcal{K}$, we can express (5.80) as

$$\mathbf{Hx} = \mathbf{g}, \qquad (5.81)$$

where $\mathbf{g}$ is the vector with mth value $y_{\mathcal{K}}(mRT)$, and $\mathbf{H} = \mathbf{I} - \mathbf{S}$ with $\mathbf{S}$ denoting the matrix of elements $s_{mn} = R \operatorname{sinc}((m - n)R), m, n \in \mathcal{K}$. Thus, as long as the matrix

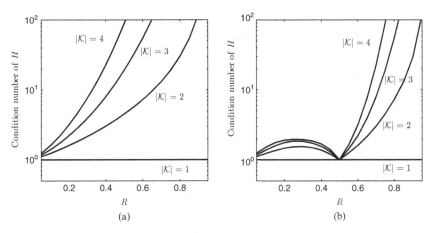

Figure 5.13 Condition number of the matrix **H** as a function of R. (a) Consecutive missing samples. (b) One known sample between every two missing samples.

H is invertible, we can recover the missing samples from (5.81). When $R = 1$ we have immediately that $\mathbf{S} = \mathbf{I}$ and recovery is impossible.

Figure 5.13(a) shows the condition number[2] of **H** as a function of R for the case in which a set of consecutive samples are missing. As can be seen, the condition number increases with the amount $|\mathcal{K}|$ of missing samples and as R approaches 1. Figure 5.13(b) shows the condition number for the case in which the missing samples are separated by one known sample. In this setting, the condition number is lower for every value of R, indicating that the problem is more stable.

5.5   Multiple generators

So far, we have focused on SI spaces $\mathcal{W}$ with a single generator. We now turn to discuss the scenario in which multiple generators are used to define $\mathcal{W}$. As we have seen in Section 4.3.4, in this case any $x(t)$ in the space can be expressed as [45, 74, 80]

$$x(t) = \sum_{\ell=1}^{N} \sum_{n \in \mathbb{Z}} a_\ell[n] h_\ell(t - nT), \qquad (5.82)$$

where $h_\ell(t), 1 \leq \ell \leq N$ are the SI generators, and $a_\ell[n]$ are the corresponding expansion coefficients. In the Fourier domain, (5.82) becomes

$$X(\omega) = \sum_{\ell=1}^{N} A_\ell(e^{j\omega T}) H_\ell(\omega). \qquad (5.83)$$

---

[2] The condition number of a matrix is the ratio of its largest and smallest (nonzero) singular values and is a measure of the stability of the corresponding linear system of equations.

We follow a parallel route to that of the single-generator case, in order to determine under what conditions a set of generators forms a Riesz basis. We then show how to obtain the expansion coefficients $a_\ell[n]$ in (5.82).

5.5.1   Riesz condition

With multiple generators, the Riesz condition (2.50) becomes

$$\alpha \sum_{\ell,n} a_\ell^2[n] \leq \left\| \sum_{\ell,n} a_\ell[n] h_\ell(t - nT) \right\|^2 \leq \beta \sum_{\ell,n} a_\ell^2[n], \qquad (5.84)$$

for some $\alpha > 0, \beta < \infty$, and all $a_\ell \in \ell_2, 1 \leq \ell \leq N$. For brevity, here and in the following, $\sum_{\ell,n}$ is used as shorthand for the double sum $\sum_{\ell=1}^{N} \sum_{n \in \mathbb{Z}}$.

Following similar steps to (5.4) we can write

$$\left\| \sum_{\ell,n} a_\ell[n] h_\ell(t - nT) \right\|^2 = \sum_{\ell,m} \sum_{n,k} a_\ell[n] a_m[k+n] \int_{-\infty}^{\infty} h_\ell(t) h_m(t - kT) dt$$

$$= \frac{1}{2\pi} \int_{-\infty}^{\infty} \sum_{\ell,m} \overline{H_\ell(\omega)} H_m(\omega) \overline{A_\ell(e^{j\omega T})} A_m(e^{j\omega T}) d\omega$$

$$= \frac{T}{2\pi} \int_0^{2\pi/T} \sum_{\ell,m} \overline{A_\ell(e^{j\omega T})} A_m(e^{j\omega T}) R_{H_\ell H_m}(e^{j\omega T}) d\omega. \qquad (5.85)$$

At this point, we depart from the derivations in Section 5.1.1 and express (5.85) in matrix form. Let

$$\mathbf{M}_{HH}(e^{j\omega}) = \begin{bmatrix} R_{H_1 H_1}(e^{j\omega}) & \cdots & R_{H_1 H_N}(e^{j\omega}) \\ \vdots & \vdots & \vdots \\ R_{H_N H_1}(e^{j\omega}) & \cdots & R_{H_N H_N}(e^{j\omega}) \end{bmatrix}, \qquad (5.86)$$

where $R_{H_i H_k}(e^{j\omega})$ is defined by (5.28). We can then write (5.85) as

$$\left\| \sum_{\ell,n} a_\ell[n] h_\ell(t - nT) \right\|^2 = \frac{T}{2\pi} \int_0^{2\pi/T} \mathbf{a}^*(e^{j\omega T}) \mathbf{M}_{HH}(e^{j\omega T}) \mathbf{a}(e^{j\omega T}) d\omega, \qquad (5.87)$$

where $\mathbf{a}(e^{j\omega})$ is the length-N vector with elements $A_\ell(e^{j\omega})$. The left- and right-hand expressions in (5.84) are also expressible in the Fourier domain:

$$\sum_{\ell,n} a_\ell^2[n] = \frac{1}{2\pi} \int_0^{2\pi} \sum_\ell |A_\ell(e^{j\omega})|^2 d\omega = \frac{1}{2\pi} \int_0^{2\pi} \mathbf{a}^*(e^{j\omega}) \mathbf{a}(e^{j\omega}) d\omega = \frac{T}{2\pi} I_\mathbf{a},$$

$$(5.88)$$

where the last equality is a result of the change of variables $\omega \to \omega/T$, and for brevity we denoted

$$I_\mathbf{a} = \int_0^{2\pi/T} \mathbf{a}^*(e^{j\omega T}) \mathbf{a}(e^{j\omega T}) d\omega. \qquad (5.89)$$

Combining (5.88) and (5.87), the functions $\{h_\ell(t)\}$ must satisfy

$$\alpha I_{\mathbf{a}} \leq \int_0^{2\pi/T} \mathbf{a}^*(e^{j\omega T}) \mathbf{M}_{HH}(e^{j\omega T}) \mathbf{a}(e^{j\omega T}) d\omega \leq \beta I_{\mathbf{a}}, \tag{5.90}$$

for every choice of $\mathbf{a}(e^{j\omega T})$. Clearly, if

$$\alpha \mathbf{I} \preceq \mathbf{M}_{HH}(e^{j\omega T}) \preceq \beta \mathbf{I}, \quad \text{almost everywhere on } \omega, \tag{5.91}$$

then (5.90) is satisfied. Indeed, by the definition of matrix inequalities, (5.91) means that

$$\alpha \mathbf{a}^*(e^{j\omega T}) \mathbf{a}(e^{j\omega T}) \leq \mathbf{a}^*(e^{j\omega T}) \mathbf{M}_{HH}(e^{j\omega T}) \mathbf{a}(e^{j\omega T}) \leq \beta \mathbf{a}^*(e^{j\omega T}) \mathbf{a}(e^{j\omega T}), \tag{5.92}$$

for every $\mathbf{a}(e^{j\omega T})$. As in the single-generator case, it is easy to verify that this condition is also necessary: if (5.91) does not hold over a set of frequencies $\omega \in \mathcal{I}$, then we may always choose $\mathbf{a}(e^{j\omega T})$ such that it is nonzero over this set of frequencies, and zero elsewhere, which violates (5.90).

The Riesz basis condition for multiple generators is summarized in the following theorem:

Theorem 5.3 (Riesz basis condition for multiple generators). *The signals $\{h_\ell(t - nT), 1 \leq \ell \leq N\}$ form a Riesz basis for their span if and only if there exists $\alpha > 0, \beta < \infty$ such that $\alpha \mathbf{I} \preceq \mathbf{M}_{HH}(e^{j\omega T}) \preceq \beta \mathbf{I}$, almost everywhere, where $\mathbf{M}_{HH}(e^{j\omega T})$ is defined by (5.86).*

Example 5.10 Consider the set of generators

$$h_\ell(t) = \frac{1}{mT} \operatorname{sinc}\left(\frac{t}{2mT}\right) \cos\left(\frac{\pi t}{mT}\left(\ell + \frac{1}{2}\right)\right) \tag{5.93}$$

for $\ell = 0, 1, \ldots, m - 1$ with shifts of mT. In the frequency domain,

$$H_\ell(\omega) = \begin{cases} 1, & \frac{2\pi\ell}{mT} \leq |\omega| < \frac{2\pi(\ell+1)}{2mT} \\ 0, & \text{else,} \end{cases} \tag{5.94}$$

so that these functions generate the class of multiband signals presented in Example 4.4. Since these functions occupy different frequency ranges, it can immediately be seen that

$$\mathbf{M}_{HH}(e^{j\omega mT}) = \frac{1}{mT} \mathbf{I} \tag{5.95}$$

for every ω. Consequently, the Riesz condition of Theorem 5.3 is satisfied with $\alpha = \beta = 1/(mT)$.

5.5.2   Biorthogonal basis

Our next step is to determine the biorthogonal basis, following a similar path as in the single-generator case.

With multiple generators, the set transformation H is defined on N sequences $a_\ell[n]$. We denote by ℓ_2^N all sets of N sequences in ℓ_2: $\{a_\ell[n]\}_{\ell=1}^N$. We then define H: $\ell_2^N \to L_2$ by

$$Ha = \sum_{\ell=1}^{N} \sum_{n\in\mathbb{Z}} a_\ell[n] h_\ell(t - nT). \tag{5.96}$$

It is easy to see that with an appropriate definition of an inner product on ℓ_2^N, the adjoint H^*: $L_2 \to \ell_2^N$ returns a sequence $b = H^*x(t)$ in ℓ_2^N such that

$$b_\ell[n] = \langle h_\ell(t - nT), x(t)\rangle. \tag{5.97}$$

(See Exercise 13.)

Using these definitions of H and H^*, we are now ready to derive the biorthogonal basis $\widetilde{H} = H(H^*H)^{-1}$. Let $b = H^*Hc$. Then,

$$b_\ell[n] = \langle h_\ell(t - nT), \sum_{k=1}^{N} \sum_{m\in\mathbb{Z}} c_k[m] h_k(t - mT)\rangle$$

$$= \sum_{k=1}^{N} \sum_{m\in\mathbb{Z}} c_k[m]\langle h_\ell(t - nT), h_k(t - mT)\rangle$$

$$= \sum_{k=1}^{N} \sum_{m\in\mathbb{Z}} c_k[m]\langle h_\ell(t), h_k(t + (n - m)T)\rangle$$

$$= \sum_{k=1}^{N} \sum_{m\in\mathbb{Z}} c_k[m] r_{h_\ell h_k}[n - m], \tag{5.98}$$

which is a convolution between $r_{h_\ell h_k}[n]$ and $c_k[n]$. Therefore, in the Fourier domain,

$$B_\ell(e^{j\omega}) = \sum_{k=1}^{N} C_k(e^{j\omega}) R_{H_\ell H_k}(e^{j\omega}). \tag{5.99}$$

Denoting by $\mathbf{b}(e^{j\omega}), \mathbf{c}(e^{j\omega})$ the vectors with ℓth elements $B_\ell(e^{j\omega}), C_\ell(e^{j\omega})$ respectively, we can write (5.99) in matrix form as

$$\mathbf{b}(e^{j\omega}) = \mathbf{M}_{HH}(e^{j\omega})\mathbf{c}(e^{j\omega}), \tag{5.100}$$

where $\mathbf{M}_{HH}(e^{j\omega})$ is defined by (5.86). Therefore, since $c = (H^*H)^{-1}b$,

$$\mathbf{c}(e^{j\omega}) = \mathbf{M}_{HH}^{-1}(e^{j\omega})\mathbf{b}(e^{j\omega}). \tag{5.101}$$

Note that when $\{h_\ell(t-nT)\}$ forms a Riesz basis, $\mathbf{M}_{HH}(e^{j\omega})$ is stably invertible thanks to condition (5.91).

We conclude that the operation of $H(H^*H)^{-1}$ on a sequence $b \in \ell_2^N$ can be described by a combination of digital filtering with the *multiple-input, multiple-output (MIMO) filter* $\mathbf{M}_{HH}^{-1}(e^{j\omega})$, followed by impulse modulation and analog filtering on each branch using the appropriate generator. This interpretation is illustrated in Fig. 5.14. An $N \times N$ MIMO filter $\mathbf{M}(e^{j\omega})$ comprises N^2 scalar filters, as schematically shown in Fig. 5.15.

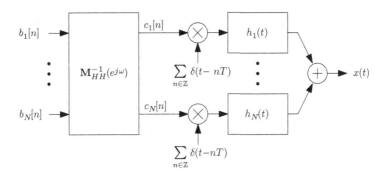

Figure 5.14 Multichannel recovery.

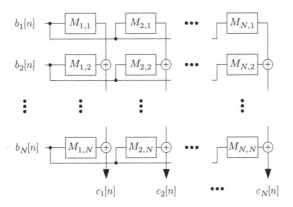

Figure 5.15 A MIMO digital filter $\mathbf{M}(e^{j\omega})$.

In analogy to the single-generator case, it is easy to see that the biorthogonal vectors are SI with multiple generators $\{\tilde{h}_\ell(t), 1 \leq \ell \leq N\}$. Each one of the generators $\tilde{h}_\ell(t)$ can be viewed as the output of the system in Fig. 5.14 when the input sequences are $b_\ell[n] = \delta[n]$ and all other inputs are set to zero. To develop an explicit expression for these functions, consider, for example, $\ell = 1$. Let $b_1[n] = \delta[n]$ and set all other inputs to zero. In this case, from analog filtering, the output on the kth channel is given by $C_k(e^{j\omega T})H_k(\omega)$. Summing over all channels,

$$\widetilde{H}_1(\omega) = \sum_{k=1}^{N} C_k(e^{j\omega T})H_k(\omega) = \mathbf{m}_1^T(e^{j\omega T})\mathbf{h}(\omega), \qquad (5.102)$$

where $\mathbf{m}_1(e^{j\omega})$ is the first column of $\mathbf{M}_{HH}^{-1}(e^{j\omega})$.

Repeating the derivation for all indices ℓ results in

$$\widetilde{\mathbf{h}}(\omega) = \mathbf{M}_{HH}^{-T}(e^{j\omega T})\mathbf{h}(\omega), \qquad (5.103)$$

where $\mathbf{M}_{HH}^{-T}$ denotes the transpose of the inverse $\mathbf{M}_{HH}^{-1}$, and $\widetilde{\mathbf{h}}(\omega)$, $\mathbf{h}(\omega)$ are the vectors with elements $\widetilde{H}_\ell(\omega)$, $H_\ell(\omega)$, respectively.

Once we determine the biorthogonal vectors, the expansion coefficients are given by

$$c_\ell[n] = \langle \tilde{h}_\ell(t - nT), x(t) \rangle, \tag{5.104}$$

or, in the Fourier domain,

$$C_\ell(e^{j\omega}) = R_{\tilde{H}_\ell X}(e^{j\omega}). \tag{5.105}$$

Substituting the expression for $\tilde{H}_\ell(e^{j\omega})$ (see (5.103)) into (5.105) leads to

$$C_\ell(e^{j\omega}) = \sum_{k=1}^{N} [\mathbf{M}_{HH}^{-1}(e^{j\omega})]_{k\ell} R_{H_k X}(e^{j\omega}). \tag{5.106}$$

We conclude this chapter with an example illustrating the computation of the biorthogonal functions via (5.103).

Example 5.11 Consider the SI space with Riesz basis $\{\beta^1(t - 2n), \beta^1(t - 1 - 2n)\}_{n \in \mathbb{Z}}$, where $\beta^1(t)$ is the B-spline of order 1 (see (5.31)). This space is generated by the functions $h_1(t) = \beta^1(t)$ and $h_2(t) = \beta^1(t - 1)$ with spacing $T = 2$. In this example, we compute the biorthogonal generators $\tilde{h}_1(t)$ and $\tilde{h}_2(t)$ using (5.103).

We begin by constructing the matrix $\mathbf{M}_{HH}^{-1}(e^{j\omega})$. It is easy to see that

$$r_{h_1 h_1}[n] = \int_{-\infty}^{\infty} \beta^1(t)\beta^1(t + 2n)dt = \delta[n] \tag{5.107}$$

and

$$r_{h_1 h_2}[n] = \int_{-\infty}^{\infty} \beta^1(t)\beta^1(t - 1 + 2n)dt = \frac{1}{6}\left(\delta[n] + \delta[n - 1]\right). \tag{5.108}$$

Similarly, $r_{h_2 h_2}[n] = \delta[n]$. Therefore, $\mathbf{M}_{HH}(e^{j\omega})$ is given by

$$\mathbf{M}_{HH}(e^{j\omega}) = \begin{pmatrix} 1 & \frac{1}{6}\left(1 + e^{-j\omega}\right) \\ \frac{1}{6}\left(1 + e^{j\omega}\right) & 1 \end{pmatrix} \tag{5.109}$$

and its inverse equals

$$\mathbf{M}_{HH}^{-1}(e^{j\omega}) = \frac{1}{1 - \frac{1}{18}\left(1 + \cos(\omega)\right)} \begin{pmatrix} 1 & -\frac{1}{6}\left(1 + e^{-j\omega}\right) \\ -\frac{1}{6}\left(1 + e^{j\omega}\right) & 1 \end{pmatrix}. \tag{5.110}$$

Substituting $\mathbf{M}_{HH}^{-T}(e^{2j\omega})$ into (5.103) leads to

$$\tilde{H}_1(\omega) = \frac{1}{1 - \frac{1}{18}\left(1 + \cos(2\omega)\right)} \left(H_1(\omega) - \frac{1}{6}\left(1 + e^{2j\omega}\right)H_2(\omega)\right) \tag{5.111}$$

and

$$\tilde{H}_2(\omega) = \frac{1}{1 - \frac{1}{18}\left(1 + \cos(2\omega)\right)} \left(-\frac{1}{6}\left(1 + e^{-2j\omega}\right)H_1(\omega) + H_2(\omega)\right). \tag{5.112}$$

As a sanity check, we can verify for example that $R_{\widetilde{H}_2 H_1}(e^{j\omega}) = 0$. Indeed,

$$R_{\widetilde{H}_2 H_1}(e^{j\omega}) = \frac{1}{1 - \frac{1}{18}(1 + \cos(\omega))}$$
$$\left(-\frac{1}{6}(1 + e^{j\omega}) R_{H_1 H_1}(e^{j\omega}) + R_{H_2 H_1}(e^{j\omega}) \right). \qquad (5.113)$$

Since $R_{H_2 H_1}(e^{j\omega}) = \overline{R_{H_1 H_2}(e^{j\omega})} = (1/6)(1 + e^{j\omega})$, the result follows. In a similar fashion it can readily be seen that $R_{\widetilde{H}_1 H_2}(e^{j\omega}) = 0$, $R_{\widetilde{H}_1 H_1}(e^{j\omega}) = 1$, and $R_{\widetilde{H}_2 H_2}(e^{j\omega}) = 1$.

5.6 Exercises

1. Suppose that the functions $\{h(t - nT)\}$ form a Riesz basis. Determine whether or not the functions $\{g(t - nT)\}$ form a Riesz basis for the following choices of $g(t)$:
   a. $g(t) = h(t - \tau)$ for some arbitrary $\tau \in [0, T]$.
   b. $g(t) = h(t)e^{j\omega_0 t}$ for some arbitrary $\omega_0 \in [-\pi, \pi]$.
   c. $g(t) = h(t + T) + h(t - T)$.
2. Consider the function $h(t) = \alpha \, \mathrm{sinc}(\alpha t)$.
   a. What are the values of α for which the functions $\{h(t - n)\}$ form a Riesz basis?
   b. Compute and plot the lower and upper Riesz bounds as a function of α.
   c. What are the values of α for which the functions $\{h(t - n)\}$ form a frame?
3. Consider the function $h(t) = \alpha \, \mathrm{sinc}^2(\alpha t)$.
   a. What are the values of α for which the functions $\{h(t - n)\}$ form a Riesz basis?
   b. Compute and plot the lower and upper Riesz bounds as a function of α.
   c. What are the values of α for which the functions $\{h(t - n)\}$ form a frame?
4. Consider the SI space generated by the function $h(t) = 2\,\mathrm{sinc}^2(2t)$ with a spacing of $T = 1$. Let

$$a_k[n] = \frac{2k^2}{\pi} \cos\left(\pi n - \frac{1}{k}\right) \mathrm{sinc}\left(\frac{n}{2\pi k}\right), \qquad k = 1, 2, \ldots \qquad (5.114)$$

   denote a set of sequences of expansion coefficients, and denote by

$$x_k(t) = \sum_{n \in \mathbb{Z}} a_k[n] h(t - n), \qquad k = 1, 2, \ldots \qquad (5.115)$$

   the corresponding signals in this space.
   a. Compute the ℓ_2-norm of the sequences $a_k[n]$ and show that it does not depend on k.
   b. Compute the L_2-norm of the signals $x_k(t)$ and show that it grows with k.
   c. Explain how this phenomenon relates to the Riesz bounds associated with $h(t)$.
5. Suppose we wish to shift a signal of the form $x(t) = \sum_{n \in \mathbb{Z}} a[n] h(t - nT)$ by a noninteger number t_0. However, we would like to do so just by modifying the expansion coefficients $a[n]$. Determine the sequence $\tilde{a}[n]$ which results in a recovery $\tilde{x}(t) = \sum_{n \in \mathbb{Z}} \tilde{a}[n] h(t - nT)$ that is closest to $x(t - t_0)$ in an L_2 sense.

6. Let $h(t) = \mathrm{sinc}(t) + \mathrm{sinc}(2t)$ and denote by $\mathcal{H}$ the closed linear span of $\{h(t-n)\}$.
   a. Compute the biorthogonal generator for $\mathcal{H}$.
   b. Compute an orthogonal generator for $\mathcal{H}$.
   c. Compute an interpolating generator for $\mathcal{H}$.
   See Section 5.2 for the definitions of the different expansions.
7. Consider the function

$$h(t) = \frac{\mathrm{sinc}^p(pt)}{\int_{-\infty}^{\infty} \mathrm{sinc}^p(pt)dt}. \tag{5.116}$$

   a. Show that $h(t)$ satisfies the partition of unity property (5.52) for every integer $p \geq 1$.
   b. Show that the functions $\{h(t-n)\}$ do not constitute a Riesz basis or a frame for any integer $p \geq 2$.
8. Let $h(t)$ be any function in L_2. Show that the function $g(t) = (f * h)(t)$ satisfies the partition of unity property (5.52) for any function f of the form $f(t) = \beta^m(t/k)$, where m and k are arbitrary positive integers. For the special case $m = 0$, provide a proof both in the time domain and in the frequency domain.
9. Express (5.74) and (5.75) in the Fourier domain. Explain the difference between these two recovery formulas.
10. Suppose that $x(t)$ is bandlimited to π/T, and sampled at times $t = nRT$ for some $R < 1$.
    a. Suggest two different interpolation kernels $g(t)$ in (5.77) to recover $x(t)$ from its samples.
    b. Suppose now that the samples are contaminated by additive white noise. From all possible choices of $g(t)$, which kernel will lead to the smallest recovery error in a mean-squared error sense? Explain.
11. Suppose that the functions $\{h(t-nT)\}$ are orthogonal. Let N be some finite integer and define the signals

$$h_\ell(t) = h(t - \ell T) \tag{5.117}$$

    for $\ell = 1, \ldots, N$. Show that $\{h_\ell(t - n\tilde{T}), n \in \mathbb{Z}, \ell = 1, \ldots, N\}$ form a Riesz basis with $\tilde{T} = NT$.
    Hint: Express the matrix $\mathbf{M}_{HH}(e^{j\omega})$ of (5.86) in terms of $R_{HH}(e^{j\omega})$, and use Theorem 5.1.
12. Prove that a multiple-generator Riesz basis is orthonormal if and only if the associated matrix $\mathbf{M}_{HH}(e^{j\omega})$ of (5.86) equals the identity matrix $\mathbf{I}$ for all ω.
13. Let ℓ_2^N be the set of all real N sequences in ℓ_2 endowed with the inner product

$$\langle a, b \rangle = \sum_{\ell=1}^{N} \sum_{n \in \mathbb{Z}} a_\ell[n] b_\ell[n]. \tag{5.118}$$

    Prove that the adjoint of the set transformation $H \colon \ell_2^N \to L_2$ defined in (5.96) is given by (5.97).
14. Let $\{h_\ell(t), n \in \mathbb{Z}, 1 \leq \ell \leq N\}$ be a set of functions in L_2 forming a multiple-generator SI Riesz basis. Show that the biorthogonal basis functions also possess

the SI structure and correspond to the generators $\{\tilde{h}_\ell(t), n \in \mathbb{Z}, 1 \leq \ell \leq N\}$ given in (5.103).

15. Consider the signal $x(t) = \sum_{\ell=1}^N \sum_{n \in \mathbb{Z}} c_\ell[n] h_\ell(t - nT)$. Verify that the sequences $c_\ell[n]$ can be written as the inner products $c_\ell[n] = \langle \tilde{h}_\ell(t - nT), x(t) \rangle$, where $\tilde{h}_\ell(t)$ are given by (5.103), in which $\mathbf{M}_{HH}(e^{j\omega})$ is the MIMO filter of (5.86).

16. We have seen in Chapter 2 (see (2.103)) that the orthogonal projection of a vector $x \in \mathcal{H}$ onto a subspace $\mathcal{V}$ can be written as $P_\mathcal{V} x = V\tilde{V}^* x = \sum_n \langle \tilde{v}_n, x \rangle v_n$. In this equation, V is the set transformation associated with the Riesz basis $\{v_n\}$ for $\mathcal{V}$, and $\tilde{V}$ is the set transformation associated with the biorthogonal basis $\{\tilde{v}_n\}$. We would now like to use this formulation to obtain an expression for the best approximation of a signal $x(t)$ by a piecewise linear (affine) signal $\tilde{x}(t)$ with discontinuities at the integers.

    a. Show that the set $\mathcal{V}$ of all piecewise linear signals with discontinuities at $t \in \mathbb{Z}$ is an SI space with generators

$$h_1(t) = \begin{cases} 1, & 0 \leq t < 1 \\ 0, & \text{else,} \end{cases} \qquad h_2(t) = \begin{cases} t, & 0 \leq t < 1 \\ 0, & \text{else.} \end{cases} \qquad (5.119)$$

    b. Prove that the functions $\{h_1(t - n), h_2(t - n)\}_{n \in \mathbb{Z}}$ generate a Riesz basis.

    c. Compute the generators $\tilde{h}_1(t)$ and $\tilde{h}_2(t)$ of the biorthogonal Riesz basis.

    d. Write an explicit expression for the piecewise linear signal $\tilde{x}(t)$ closest to a given signal $x(t)$ in an L_2 sense.

17. Show that the filters (4.47) of Example 4.5 generate a Riesz basis for the subspace of signals bandlimited to $\pi N/T$, and determine the corresponding biorthogonal filter for $N = 1$.

Chapter 6

Subspace priors

We now begin applying the machinery developed in previous chapters to generalized sampling. In this chapter we focus on subspace priors and linear sampling. Our goal is to recover a signal x from its samples, when we know a priori that x lies in some subspace $\mathcal{A}$ of a Hilbert space $\mathcal{H}$. For example, x may be a bandlimited signal, a piecewise polynomial signal, or a pulse amplitude modulation (PAM) signal with known pulse shape. We focus on SI subspaces, which were treated in the previous chapter, and uniform sampling grids. In this setting, we will show that sampling and reconstruction can be performed by filtering operations. However, all the results herein hold in more abstract Hilbert space settings including finite-dimensional spaces and spaces that are not SI, and for sampling grids that are not necessarily uniform.

6.1   Sampling and reconstruction processes

6.1.1   Sampling setups

Although our focus in this chapter is on subspace priors, many of the essential ideas will also be exploited in Chapter 7 when treating smoothness and stochastic priors. Therefore, in this section we provide an overview of the different setups and criteria that will be used in both chapters. We then focus on the subspace setting and defer the discussion on other signal classes to the next chapter. In Chapter 8 we revisit subspace priors and generalize the sampling mechanism to include nonlinear sampling. Subsequent chapters are devoted to nonlinear signal priors taking the form of UoS.

The setups we treat in this and the next chapter are summarized in Table 6.1, and are detailed in the ensuing sections. In the table, unconstrained refers to the scenario in which no constraints are imposed on the recovery process. In contrast, the predefined kernel column refers to the case in which reconstruction is performed by a given analog interpolation kernel. Table 6.2 indicates the design objective used in each scenario. As we discuss further below, different priors and constraints dictate distinct objectives. In some cases perfect recovery of a signal from its given samples is possible, indicated in the table as unique recovery. When constraints are imposed on the reconstruction, we often seek to minimize the squared error between the original signal and the best estimate attainable under the given restrictions. If this is possible, then we refer to the resulting solution as the minimal-error recovery. Unfortunately, there are settings in which the

Table 6.1 Different scenarios treated in Chapters 6 and 7

	Unconstrained	Predefined kernel
Subspace priors	Unique recovery: Sections 6.2–6.4 Nonunique setting: Section 6.5	Section 6.6
Smoothness priors	Section 7.1	Section 7.2
Stochastic priors	Section 7.3.1	Section 7.3.2

Table 6.2 Design objective in each scenario

	Unconstrained	Predefined kernel
Subspace priors	Perfect reconstruction[1]	Minimal error
Smoothness priors	LS/minimax	LS/minimax
Stochastic priors	MSE	MSE

minimal-error solution depends on the unknown input, and therefore cannot be implemented. For example, when the only information we have about the signal is that it is smooth, then the error cannot be minimized uniformly over all signals, and alternative design strategies are needed. For these scenarios, we introduce the least squares (LS) and minimax objectives. When considering stochastic signals, the mean-squared error (MSE) is a natural measure of the recovery error.

6.1.2 Sampling process

In a general Hilbert space setting, sampling is described by taking inner products with a Riesz basis $\{s_n\}$, so that the nth sample is given by $c[n] = \langle s_n, x \rangle$. Denoting by S the set transformation corresponding to $\{s_n\}$ we can write the samples compactly as $c = S^* x$. We refer to the subspace generated by $\{s_n\}$ as the *sampling space*, and denote it throughout by $\mathcal{S}$. The signal x is assumed to lie in a given subspace $\mathcal{A}$, which we refer to as the *prior space*.

We focus most of our attention on the real SI setting, in which $s_n = s(t - nT)$ for a real function $s(t)$ and a given period T. In this case the samples can be modeled as uniform samples, at times nT, of the output of an LTI filter with impulse response $s(-t)$:

$$c[n] = \langle s(t - nT), x(t) \rangle = x(t) * s(-t)|_{t=nT}. \tag{6.1}$$

This interpretation is depicted in Fig. 6.1. The sampling filter $s(t)$ allows one to incorporate imperfections in the ideal sampler, as we explained in detail in Chapter 4 (see also [11, 48, 106, 108, 109, 110]). Throughout the next two chapters, sampling in SI

---

[1] Under a general condition on the subspaces involved perfect recovery is possible. When this condition is violated, we will rely on the LS and minimax criteria. The same holds true in the constrained setting: typically the error can be minimized. Otherwise, we invoke the LS and minimax strategies.

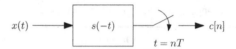

Figure 6.1 LTI or SI sampling.

settings will always have the form (6.1), where $s(t)$ is chosen such that it generates a Riesz basis (cf. Section 5.1). We briefly address multichannel sampling in Section 6.8. In Chapter 8 we treat the more complicated situation in which the sampling process includes nonlinear distortions.

When we refer to the SI setting, we also assume that the prior space $\mathcal{A}$ is SI with a real generator $a(t)$ satisfying the Riesz condition. Any input $x(t)$ can then be written as

$$x(t) = \sum_{n \in \mathbb{Z}} d[n] a(t - nT), \tag{6.2}$$

for some bounded-norm sequence $d[n]$.

Example 6.1 Consider an ADC whose output samples correspond to integrals of the input signal over nonoverlapping intervals, namely

$$c[n] = \int_{nT-T/2}^{nT+T/2} x(t) dt. \tag{6.3}$$

This situation corresponds to using a symmetric rectangular sampling filter $s(t)$ with width T. The sampling space $\mathcal{S}$ is the set of signals $f(t)$ that can be expressed as

$$f(t) = \sum_{n \in \mathbb{Z}} d[n] s(t - nT) \tag{6.4}$$

with some bounded-norm sequence $d[n]$. This space comprises piecewise constant signals with discontinuities at $t = (n - 1/2)T$, $n \in \mathbb{Z}$.

Example 6.2 In magnetic resonance imaging (MRI), each sample corresponds to the two-dimensional (or three-dimensional) CTFT of the signal at a different frequency. Consider for simplicity a one-dimensional signal sampled using this method, resulting in samples

$$c[n] = \int_{-W/2}^{W/2} x(t) e^{-j\omega_n t} dt, \tag{6.5}$$

where $\{\omega_n\}$ is a grid of frequencies and W is the support of $x(t)$ (corresponding to the field of view in MRI). These samples correspond to inner products with the functions $\{e^{j\omega_n t}\}$, which are clearly not shifted versions of one another. Therefore, MRI does not correspond to SI sampling.

6.1.3   Unconstrained recovery

Given samples of a signal, our goal is to develop methods to recover or approximate the input. We begin by treating the setting in which no constraints are imposed on the recovery mechanism. In this case our task is to design interpolation methods that are best adapted to the signal prior according to the objectives in Table 6.2. Often, the fact that we know the signal lies in a subspace $\mathcal{A}$ is enough to ensure perfect reconstruction, as in the Shannon–Nyquist sampling theorem, even when the sampling space $\mathcal{S}$ is different than $\mathcal{A}$. This implies that sampling can be performed with a large class of sampling filters. Mathematically, to enable perfect recovery, we require that a certain direct-sum condition between the spaces involved is satisfied. When this condition is violated, there are several signals that match the given samples. We therefore need additional criteria to choose among all possible solutions. In our development we consider LS and minimax recovery objectives.

The reconstruction methods under the different criteria all have a common structure, depicted in Fig. 6.2. Here $w(t)$ is the impulse response of a continuous-time filter, which serves as the interpolation kernel, while $h[n]$ represents a discrete-time filter used to process the samples prior to reconstruction. Denoting the output of this filter by $d[n]$, the input to the filter $w(t)$ is a modulated impulse train $\sum_n d[n]\delta(t - nT)$. The filter's output is given by

$$\hat{x}(t) = \sum_{n \in \mathbb{Z}} d[n] w(t - nT). \tag{6.6}$$

In the more general Hilbert space setting, the design objectives we consider all lead to recovery of the form $\hat{x} = WHc$, where W is a set transformation corresponding to a basis $\{w_n\}$ for the reconstruction space, and H is a linear transformation operating on the samples c prior to recovery with W.

The optimal interpolation kernels resulting from the various objective functions in the SI setting are typically derived in the frequency domain but very often do not admit a closed form in the time domain. This limits the applicability of these recovery techniques to situations in which the kernel needs to be calculated only on a discrete set of points. The discrete Fourier transform (DFT) can be used in such settings to approximate the desired values. Consequently, these methods seem to have been used, for example in the image processing community, only as a means of enlarging an image by an integer factor [111, 112]. More general geometrical transformations, such as rotation, lens

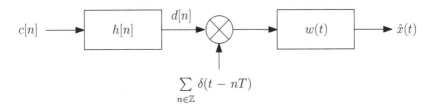

Figure 6.2 Reconstruction using a digital correction filter $h[n]$ and interpolation kernel $w(t)$.

distortion correction, and scaling by an arbitrary factor, are typically not treated using these techniques. One way to resolve this problem is to choose the signal prior so as to yield an efficient interpolation algorithm, for example using B-splines. We will discuss interpolation using B-splines in Chapter 9. Nevertheless, this approach restricts the type of priors that can be handled.

6.1.4   Predefined recovery kernel

To overcome the difficulties in implementing the unconstrained solutions, we may resort to a system that uses a predefined interpolation kernel which is easy to implement. In this setup, the only freedom is in the design of the digital correction filter $h[n]$ in Fig. 6.2, which may be used to compensate for the nonideal behavior of the prespecified kernel $w(t)$ [48, 106, 108, 110, 113, 114, 115]. The filter $h[n]$ is selected to optimize a criterion matched to the signal prior. We treat two such criteria: LS and minimax, which are discussed in the next section.

By restricting the reconstruction to the form (6.6), we are essentially imposing that the recovered signal $\hat{x}(t)$ lie in the SI space generated by the prespecified kernel $w(t)$. We refer to this space as the *reconstruction space*, and denote it throughout by $\mathcal{W}$. Constraining $w(t)$ allows one to choose $\mathcal{W}$ such as to yield highly efficient interpolation methods. For example, B-splines have been used for interpolation in the mathematical literature since the pioneering work of Schoenberg [67]. In signal processing applications the use of B-splines gained popularity due to the work of Unser, Aldroubi, and Eden that showed how B-spline interpolation can be implemented efficiently [72, 73]. Interpolation using splines of degree up to 3 is very common in image processing, owing to their ability to efficiently represent smooth signals and the relatively low computational complexity needed for their evaluation at arbitrary locations [71]. We will review these results and their use for interpolation in Chapter 9.

Other common interpolation approaches are introduced in the following examples.

Example 6.3 One of the most widely used kernels in image processing is Keys' cubic interpolation kernel [116], which is given by

$$w(t) = \begin{cases} \frac{3}{2}|t|^3 - \frac{5}{2}|t|^2 + 1, & |t| < 1 \\ -\frac{1}{2}|t|^3 + \frac{5}{2}|t|^2 - 4|t| + 2, & 1 \le |t| < 2 \\ 0, & 2 \le |t|. \end{cases} \tag{6.7}$$

The appeal of this kernel is that it has small support and satisfies the interpolation property $w(n) = \delta[n]$, resulting in computationally efficient recovery. Note that in contrast, cubic B-splines do not satisfy the interpolation property, requiring preprocessing of the samples, as we discuss in detail in Chapter 9. The Keys kernel is depicted in Fig. 6.3(a) from which we see that it is symmetric, continuous, and has a continuous first-order derivative (see Exercise 1). In fact, these properties can essentially be used to develop the form of the Keys function, as shown in Exercise 2.

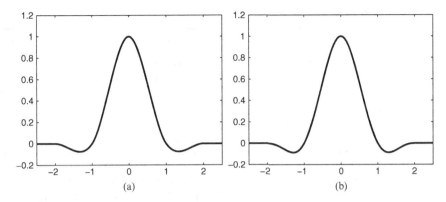

Figure 6.3 (a) Keys' cubic interpolation kernel. (b) Lanczos interpolation kernel.

Example 6.4 Another choice of interpolation kernel, commonly used in image resampling applications, is the Lanczos kernel [117], given by

$$w(t) = \begin{cases} \text{sinc}(t)\,\text{sinc}\left(\frac{t}{\Delta}\right), & |t| < \Delta \\ 0, & |t| \geq \Delta. \end{cases} \tag{6.8}$$

Here $\Delta > 0$ is some integer, typically chosen to be 2 or 3. Similar to the Keys function, the Lanczos kernel is symmetric, continuous, has a small support and satisfies the interpolation property. Moreover, its derivatives up to order 2 are continuous (see Exercise 3). The Lanczos kernel for $\Delta = 2$ is shown in Fig. 6.3(b). As can be seen, it is very similar to the Keys function.

In the more general Hilbert space setting, constrained recovery is obtained by taking linear combinations of a given set of reconstruction vectors $\{w_n\}$ that span a subspace $\mathcal{W}$. The reconstructed signal then has the form $\hat{x} = \sum_n d[n]w_n$ for some coefficients $d[n]$ that are a linear transform of the measurements $c[n]$. We always choose $\{w_n\}$ (or $w(t)$ in the SI case) to form a Riesz basis for $\mathcal{W}$.

The interpolation methods corresponding to the different scenarios outlined so far are summarized in Table 6.3. The numbers in the table indicate the equation numbers containing the reconstruction formulas. Interestingly, we will see that the solutions share a similar structure. Throughout our discussion, the commonalities and equivalence between the different approaches will be emphasized. In particular, we provide a number of different routes (and formulations) that in many cases lead to the same computational solution, while providing several complementary insights into the problem as well as on the notion of optimality.

6.1.5 Design objectives

We now turn to discuss approaches for designing the filter $h[n]$ in Fig. 6.2 so as to yield a good approximation of the original signal. In a general Hilbert space, this corresponds

Table 6.3 Methods for signal recovery

	Unconstrained	Predefined kernel
Subspace priors	Direct-sum: (6.33) LS and minimax: (6.90)	Direct-sum: (6.108) LS: (6.115) Minimax: (6.126)
Smoothness priors	(7.13), (7.14)	LS: (7.53) Minimax: (7.56)
Stochastic priors	(7.82), (7.83)	(7.86)

to choosing a linear transform H such that the corrected measurements $d = Hc$ lead to the desired recovery. In order to present the concepts for the most general case, we focus here on the Hilbert space setting.

In all of the setups described, our recovery problem can be posed as follows: given samples of a signal $c = S^*x$ and some prior knowledge of the form $x \in \mathcal{T}$ for a given set $\mathcal{T}$, produce a reconstruction $\hat{x}$ that is close to x in some sense. The set $\mathcal{T}$ incorporates knowledge about the typical input signals and can be a subspace, or any other type of constraint, for example a bounded-norm restriction (what we refer to in the tables as a *smoothness constraint*).

The samples c together with the set $\mathcal{T}$ determine the set of feasible signals:

$$\mathcal{G} = \{x: x \in \mathcal{T}, S^*x = c\}. \tag{6.9}$$

This is the class of signals that are consistent with the data and our prior information, and therefore could have generated the given samples. When the samples are noisy, we may replace this set by

$$\mathcal{G} = \{x: x \in \mathcal{T}, \|S^*x - \tilde{c}\| \le \alpha\}, \tag{6.10}$$

where $\tilde{c}$ are the noisy samples, and α is an appropriate bound on the noise. For now, we focus on the noiseless setting (6.9). We briefly discuss recovery in the presence of noise in the next chapter.

In some cases, the set $\mathcal{G}$ contains only one signal: these are precisely the settings in which perfect recovery is possible. The problem then is how to determine this solution efficiently from the given samples. More generally, $\mathcal{G}$ can include several signals. This means that there is more than one solution that is consistent with the data and prior information. The question then is how to select one option as a recovery, among all possibilities.

The first strategy we may attempt is to minimize the recovery error $\|\hat{x} - x\|$ over $\mathcal{G}$, where $\hat{x}$ is the reconstructed signal. The difficulty, however, is that this error depends on the actual value of x and therefore in general there is no algorithm that will minimize the error uniformly over all x. Indeed, the minimizer of $\|\hat{x} - x\|$ over $\mathcal{G}$ is given by $\hat{x} = x$, a solution that clearly cannot be implemented without knowledge of x. Nonetheless, there are settings in which the recovery error can be minimized regardless of x. This occurs, for example, in constrained recovery when x lies in an appropriate subspace and the subspaces involved satisfy a direct-sum condition. We treat this scenario in

Section 6.6.1. When the error cannot be minimized uniformly over all feasible x, an alternative objective is needed. Two approaches that are rooted in the statistical signal processing literature [118, 119, 120, 121, 122, 123] are LS and worst-case (minimax) design. These techniques have been used in many different estimation problems in order to obtain good estimates from noisy data. Below, we adapt these objectives to our sampling context.

The LS strategy amounts to replacing the reconstruction error $\|\hat{x} - x\|$ by the error-in-samples objective $\|S^*\hat{x} - c\|$. This approach seeks a signal $\hat{x}$ that produces samples as close as possible to the measured samples c:

$$\hat{x}_{\mathrm{LS}} = \arg\min_{x \in \mathcal{T}} \|S^*x - c\|^2. \tag{6.11}$$

When (6.11) does not have a unique solution, a common alternative is to choose the one with minimal norm:

$$\hat{x}_{\mathrm{LS}} = \arg\min_{x \in \mathcal{G}} \|x\|^2. \tag{6.12}$$

From an optimization perspective, the objective in (6.11) (or (6.12)) is convex (quadratic) in x and therefore, if $\mathcal{T}$ is a convex set, as we assume throughout, then the problem is convex [124]. An advantage of convex problems is that a myriad of efficient algorithms are available to solve them. Furthermore, closed-form solutions can often be obtained by relying on necessary and sufficient optimality conditions that are known for convex problems. Owing to the relative simplicity of the LS objective, this criterion is widely used in inverse problems in general, and in sampling in particular [125, 126]. We will see that it admits a closed-form solution for many interesting priors. However, it is important to note that there are situations where minimization of the error-in-samples leads to a large reconstruction error. This happens, for example, when S is such that large perturbations in x lead to small perturbations in S^*x. Therefore, this method does not guarantee a small recovery error. From a statistical signal processing perspective, it is well known that LS-type objectives often yield unsatisfactory results and can be improved upon. For a detailed discussion on these issues in the broader context of estimation theory see [122, 123] and the many references therein.

Another drawback of the LS approach in our context is that it cannot jointly handle prior constraints on x and recovery constraints on $\hat{x}$. This is because the objectives in (6.11) and (6.12) contain only one variable x. To illustrate this point, suppose, for example, that x lies in a subspace $\mathcal{A}$ and we constraint the recovery to a subspace $\mathcal{W}$. Since the LS objective involves only the variable x we can either perform the minimization over $x \in \mathcal{A}$ or $x \in \mathcal{W}$. Because of this limitation, we show in Section 6.6 that when the recovery kernel is predefined, the LS criterion will generally lead to poor performance.

An alternative to LS is the worst-case (or minimax) design [113, 122, 127, 128, 129, 140]. This method attempts to control the estimation error by minimizing its largest possible value. Specifically, since x is unknown, we seek the reconstruction $\hat{x}$ that minimizes the error for the worst feasible signal:

$$\hat{x}_{MX} = \arg\min_{\hat{x}} \max_{x \in \mathcal{G}} \|\hat{x} - x\|^2. \tag{6.13}$$

In contrast to (6.11), here we attempt to directly control the reconstruction error $\|x - \hat{x}\|$, which is the quantity of interest in many applications. Therefore, from an approximation error perspective, we expect this approach to yield better results than LS in many cases. Furthermore, since the objective in (6.13) has two distinct variables x and $\hat{x}$, both prior and recovery conditions can be incorporated simultaneously. For example, if we know that x is in $\mathcal{A}$ and wish to constraint the recovery to $\mathcal{W}$, then the first constraint can be incorporated into $\mathcal{G}$ while the minimization is performed over $\hat{x} \in \mathcal{W}$. We will show that this property results in superior performance of the minimax technique over LS design in the constrained setting.

A disadvantage of the minimax formulation is that solving (6.13) is more challenging than (6.11). One possible solution method is to establish a lower bound on the objective and then find a vector $\hat{x}$, which is not a function of x, that achieves it. Specifically, suppose that $\max_{x \in \mathcal{G}} \|\hat{x} - x\|^2 \geq \max_{x \in \mathcal{G}} g(x)$ for all $\hat{x}$, where $g(x)$ is some function of x. Then any $\hat{x}$ that achieves the bound and is not a function of x is a solution. Although this approach is not constructive, when applicable, it leads to a closed-form solution to the infinite-dimensional minimax problem (6.13). Luckily, this strategy works in our sampling problems.

When x is a random process, the squared error $\|\hat{x} - x\|^2$ is replaced by the expected squared error or the MSE, which is independent of x. This scenario will be treated in the next chapter.

6.2   Unconstrained reconstruction

6.2.1   Geometric interpretation

We now turn to the focus of this chapter: recovery with subspace priors. Suppose that our signal x is known to lie in a prior subspace $\mathcal{A}$. The generalized samples of x are given by $c = S^*x$ where S is a set transformation corresponding to a basis for the sampling space $\mathcal{S}$. In our discussion below we distinguish between the setting in which the prior subspace and sampling spaces are equal, $\mathcal{A} = \mathcal{S}$, and the case in which they are not. As we will see, even when $\mathcal{A} \neq \mathcal{S}$, perfect recovery of x from its nonideal samples is often possible. Specifically, for any sampling space $\mathcal{S}$ there is a broad class of subspace priors under which x can be perfectly reconstructed. Conversely, for any given subspace $\mathcal{A}$ there are many choices of sampling functions that will allow for perfect recovery. In Chapter 8 we show that these results are valid even when a memoryless, invertible nonlinearity is inserted prior to sampling, as long as the nonlinearity does not vary too fast.

We begin with a simple geometric interpretation of our sampling problem. The key to this interpretation is in noting that knowing the samples $c[n]$ is equivalent to knowing the orthogonal projection of x onto $\mathcal{S}$, which we denote by $x_{\mathcal{S}} = P_{\mathcal{S}}x$. To see this, we

Figure 6.4 Decomposition of the sampling process into two stages.

use the definition of the adjoint together with the fact that P_S is Hermitian to express the samples $c[n]$ in terms of x_S:

$$c[n] = \langle s_n, x \rangle = \langle P_S s_n, x \rangle = \langle s_n, P_S x \rangle = \langle s_n, x_S \rangle. \tag{6.14}$$

The second equality is a result of the fact that $P_S s_n = s_n$, while the third equality follows from the definition of the adjoint. This equation implies that sampling x is tantamount to sampling its orthogonal projection x_S, as illustrated in Fig. 6.4.

The notion of a Riesz basis, studied in Chapter 2, implies that any signal in S can be determined from its samples $\langle s_n, P_S x \rangle$ with respect to a Riesz basis $\{s_n\}$ for S by computing the biorthogonal basis $\{\tilde{s}_n\}$ associated with $\{s_n\}$:

$$x_S = P_S x = \sum_{n \in \mathbb{Z}} \langle s_n, P_S x \rangle \tilde{s}_n = \sum_{n \in \mathbb{Z}} c[n] \tilde{s}_n. \tag{6.15}$$

Combining (6.14) and (6.15) we see that there is a one-to-one correspondence between the samples $c[n]$ and the orthogonal projection $x_S = P_S x$. Therefore, the problem of reconstructing x from $c[n]$ is equivalent to determining x from its orthogonal projection onto S. An immediate consequence is that if x lies in S so that $x = x_S$, then it can be perfectly recovered from the given samples.

At first glance it may seem like signals in S are the only ones that are recoverable since the projection zeros out any component in $S^\perp$. However, closer inspection reveals that if we know in advance that x lies in a space $\mathcal{A}$ with suitable properties (which we will define below), then there is a unique vector in $\mathcal{A}$ with the given orthogonal projection onto S. As depicted in Fig. 6.5, in this case we can draw a vertical line from the projection until we hit the space $\mathcal{A}$ and in such a way obtain the unique vector in $\mathcal{A}$ that is consistent with the given samples. If we take the orthogonal projection $P_S y$ of any other $y \in \mathcal{A}$, then this projection will not equal $P_S x$. Evidently, perfect recovery is possible for a broad class of signals beyond those that lie in S.

Geometrically, therefore, our problem amounts to finding the unique signal in $\mathcal{A}$ with the given projection onto S, when such a unique signal is guaranteed. Although Fig. 6.5 illustrates the recovery, in practice it is not clear how to compute the original signal. Obviously, the biorthogonal vectors in (6.15) will no longer provide a solution since they will always lead to a recovery in S. As we show, when $\mathcal{A} \neq S$, these vectors will be replaced by an *oblique biorthogonal set*. These are vectors that satisfy the biorthogonality property; however, they form a basis for $\mathcal{A}$. Despite the fact that we proved in Proposition 2.12 that the biorthogonal set is unique in S, this does not preclude the possibility of constructing biorthogonal vectors that span a different space $\mathcal{A}$. Oblique biorthogonal expansions will be treated in more detail in Section 6.3.4.

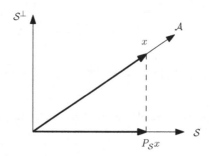

Figure 6.5 A unique vector in $\mathcal{A}$ which is consistent with the samples in $\mathcal{S}$ can be recovered from the given samples.

So far we have assumed that there is a unique vector in $\mathcal{A}$ with the given orthogonal projection onto $\mathcal{S}$. We will show that this is true as long as $\mathcal{A}$ and $\mathcal{S}^{\perp}$ satisfy a direct-sum condition. When this condition does not hold, there can be several signals in $\mathcal{A}$ with the same orthogonal projection. To choose a recovery, we develop LS and minimax strategies. In this specific problem, we show that these two approaches lead to the same solution. However, when the recovery is constrained, this will no longer be the case.

6.2.2   Equal sampling and prior spaces

To develop explicit recovery techniques, we begin with the simple case in which $\mathcal{A} = \mathcal{S}$. In this setting, the sampling functions form a basis for the space that includes the input signal, which is exactly the scenario treated in the more general context of Riesz basis expansions. Therefore, recovery is obtained via (6.15), with the biorthogonal basis $\widetilde{S} = S(S^*S)^{-1}$ (cf. Section 2.5.4).

In Section 5.2 we specialized Riesz basis expansions to the SI setting in which $x = x(t)$ lies in an SI subspace $\mathcal{S}$ generated by $s(t)$, and sampling has the form (6.1). We have seen that the application of $\widetilde{S}$ on a sequence $c[n]$ can be described in terms of the block diagram of Fig. 5.2. This implies that any $x(t)$ in $\mathcal{S}$ may be sampled and recovered using the system of Fig. 6.6 with

$$H(e^{j\omega}) = \frac{1}{R_{SS}(e^{j\omega})}, \tag{6.16}$$

where $R_{SS}(e^{j\omega})$ is defined by (3.97). The Riesz basis condition (5.14) guarantees that the inverse is well defined.

Figure 6.6 represents a whole class of perfect reconstruction sampling theorems in SI spaces. Note that the recovery architecture is a special case of Fig. 6.2 with $w(t) = s(t)$ and $h[n] = r_{ss}^{-1}[n]$, where the inverse of a discrete-time sequence refers to the convolutional inverse: $r_{ss}^{-1}[n] * r_{ss}[n] = \delta[n]$. The partition of unity property proved in Proposition 5.4 implies that if $s(t)$ satisfies $\sum_n s(t-n) = 1$ for all t, then by selecting the sampling period T to be sufficiently small, any input signal that is norm-bounded can be approximated as closely as desired by the scheme of Fig. 6.6 [10].

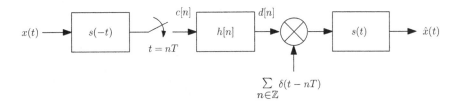

Figure 6.6 Sampling and reconstruction in an SI subspace $\mathcal{S}$.

We now take a closer look at the correction stage reflected by filtering with $h[n]$. The denominator in (6.16) is the DTFT of the sampled correlation function $r_{ss}[n] = \langle s(t), s(t - nT) \rangle$. If the functions $\{s(t-nT)\}$ form an orthonormal basis, then $r_{ss}[n] = \delta[n]$ and $H(e^{j\omega}) = 1$. In this case no preprocessing of the samples is necessary prior to reconstruction. This is precisely the setting of the Shannon–Nyquist sampling theorem: we have seen already in (4.18) that the functions $s(t - nT) = \text{sinc}(t - nT)$ form an orthogonal basis [10, 130]. The correction is then a simple scaling by T so that the block diagram of Fig. 6.6 coincides with that of Fig. 4.10. However, the class of theorems implied by Fig. 6.6 is much more general. When $r_{ss}[n]$ is not concentrated at $n = 0$, the correction stage has a large impact on the recovery error and therefore must be incorporated prior to interpolation by $s(t)$. We illustrate this point by an example.

Example 6.5 Suppose that $x(t)$ is known to be a spline of degree 1 with knots at the integers. Such a signal can be represented as (cf. Section 4.3.2)

$$x(t) = \sum_{n\in\mathbb{Z}} d[n]\beta^1(t - n), \qquad (6.17)$$

where $\beta^1(t)$ is the B-spline function of degree 1 (see (4.32)). In the terminology of this chapter, $x(t)$ lies in the space $\mathcal{A}$ spanned by $a(t) = \beta^1(t)$.

The system of Fig. 6.6 implies that $x(t)$ can be perfectly recovered from samples of its filtered version $x(t) * \beta^1(-t)$ where $s(t) = a(t)$. To recover the signal we need to compute the correction filter $h[n]$, which is the convolutional inverse of

$$r_{ss}[n] = \int_{-\infty}^{\infty} \beta^1(t)\beta^1(t - n)dt = \begin{cases} 1/6, & |n| = 1 \\ 2/3, & n = 0 \\ 0, & |n| > 1. \end{cases} \qquad (6.18)$$

Therefore, in the frequency domain,

$$H(e^{j\omega}) = \frac{1}{R_{SS}(e^{j\omega})} = \frac{1}{\frac{1}{6}e^{j\omega} + \frac{2}{3} + \frac{1}{6}e^{-j\omega}} = \frac{3}{2 + \cos(\omega)}. \qquad (6.19)$$

Figures 6.7(a) and 6.7(b) show respectively the reconstructions obtained by using the scheme of Fig. 6.6 with $H(e^{j\omega})$ of (6.19) and by ignoring the digital correction filter. This example highlights the need for the digital filtering stage.

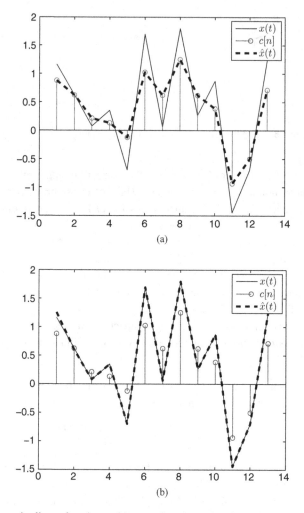

Figure 6.7 A piecewise linear function and its samples after going through the filter $\beta^1(t)$. (a) Reconstruction without using a digital correction filter. (b) Reconstruction using the system of Fig. 6.6.

Arbitrary input

When $x(t)$ lies in $\mathcal{S}$, the output of Fig. 6.6 is equal to $x(t)$. However, what happens when an arbitrary input is injected into the system? Since this block diagram implements a biorthogonal expansion, the output will be equal to the orthogonal projection $P_{\mathcal{S}}x(t)$ onto $\mathcal{S}$. An immediate consequence is that the reconstructed signal $\hat{x}(t)$ is the closest signal in $\mathcal{S}$ to $x(t)$, in an L_2-norm sense.

It is instructive to explicitly verify that the output of Fig. 6.6 equals $P_{\mathcal{S}}x(t)$. We first note that if $x(t)$ is in $\mathcal{S}^\perp$, then $c[n]$ is the zero sequence and the output is equal to zero. This is a result of the fact that the inner product of $x(t)$ with any signal in $\mathcal{S}$ will be zero in this case. On the other hand, if $x(t) \in \mathcal{S}$, then we can write $x(t) = \sum_n b[n]s(t - nT)$

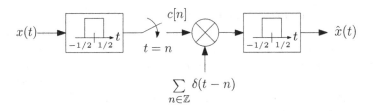

Figure 6.8 Minimal-error piecewise constant approximation.

for some sequence $b[n]$. From Proposition 5.2,

$$C(e^{j\omega}) = B(e^{j\omega})R_{SS}\left(e^{j\omega}\right),\tag{6.20}$$

so that $d[n] = b[n]$ and $\hat{x}(t) = x(t)$.

We conclude that Fig. 6.6 can be used to obtain an explicit expression for the minimal-error approximation of an arbitrary signal in a given subspace S generated by $s(t)$.

Example 6.6 Suppose we want to approximate a signal $x(t)$ by a piecewise constant function over intervals of length one. Namely, we seek a signal of the form

$$\hat{x}(t) = \sum_{n\in\mathbb{Z}} d[n]\beta^0(t - n)\tag{6.21}$$

that is closest to $x(t)$, where $\beta^0(t)$ is the 0th-order B-spline function (rectangular pulse) defined by (4.33).

The minimal-error approximation $\hat{x}(t)$ can be computed via Fig. 6.6 with $s(t) = \beta^0(t)$. Since the functions $\{s(t - n)\}_{n\in\mathbb{Z}}$ are orthonormal, $h[n] = \delta[n]$, leading to the scheme of Fig. 6.8. Determining the resulting samples explicitly:

$$c[n] = \int_{-\infty}^{\infty} s(t - n)x(t)dt = \int_{n-1/2}^{n+1/2} x(t)dt.\tag{6.22}$$

The approximation is then $\hat{x}(t) = \sum_{n\in\mathbb{Z}} c[n]\beta^0(t - n)$ so that the value on each constant interval is equal to the average of $x(t)$ over that segment.

6.3 Sampling in general spaces

To extend recovery beyond the space S, suppose now that x lies in an arbitrary subspace A of a Hilbert space $\mathcal{H}$. Clearly in order to be able to reconstruct any x in A from samples in S we need that A and $S^\perp$ intersect only at zero. Otherwise, any nonzero signal y in the intersection of A and $S^\perp$ will yield zero samples and cannot be recovered. Intuitively, to have a unique solution, we also need A and S to have the same number of degrees of freedom. This requirement can be made precise by assuming a direct-sum condition

$$\mathcal{H} = A \oplus S^\perp.\tag{6.23}$$

Condition (6.23) means that A and $S^\perp$ are disjoint, and together span $\mathcal{H}$.

It is easy to see that for a finite-dimensional Hilbert space $\mathcal{H}$, (6.23) implies that the dimensions of $\mathcal{A}$ and $\mathcal{S}$ are equal. Indeed, we always have that $\mathcal{H} = \mathcal{S} \oplus \mathcal{S}^\perp$. Therefore, if $\dim(\mathcal{S}^\perp) = m$ and $\dim(\mathcal{H}) = n$, then $\dim(\mathcal{S}) = n - m$. But if in addition, $\mathcal{H} = \mathcal{A} \oplus \mathcal{S}^\perp$, then $\dim(\mathcal{A}) = \dim(\mathcal{H}) - \dim(\mathcal{S}^\perp) = n - m$. The fact that the two subspaces have the same dimension implies that there is a one-to-one (bijective) mapping between any vector in $\mathcal{S}$ and that in $\mathcal{A}$. The same intuition follows in infinite-dimensional spaces: under the direct-sum condition it can be shown that there is a bijection between $\mathcal{S}$ and $\mathcal{A}$ [131], so that $\mathcal{S}$ and $\mathcal{A}$ are *isomorphic*. This isomorphism enables recovery of an arbitrary $x \in \mathcal{A}$ from its samples in $\mathcal{S}$.

Proposition 6.1. *Let $\mathcal{A}$ and $\mathcal{S}$ be closed subspaces of a Hilbert space $\mathcal{H}$ with $\mathcal{H} = \mathcal{A} \oplus \mathcal{S}^\perp$. Then the orthogonal projection from $\mathcal{A}$ to $\mathcal{S}$, $P_\mathcal{S}: \mathcal{A} \to \mathcal{S}$, is bijective.*

Proof: We begin by showing that $P_\mathcal{S}$ is injective over $\mathcal{A}$. If $P_\mathcal{S}a = 0$ for some $a \in \mathcal{A}$, then $a \in \mathcal{S}^\perp$. But since $\mathcal{A} \cap \mathcal{S}^\perp = \{0\}$ we conclude that $a = 0$ and $P_\mathcal{S}$ is injective. To show that $P_\mathcal{S}$ is surjective, let $s \in \mathcal{S}$ be arbitrary. Using $\mathcal{H} = \mathcal{A} \oplus \mathcal{S}^\perp$ we can write $s = a + v$ with $a \in \mathcal{A}$ and $v \in \mathcal{S}^\perp$. Since $s \in \mathcal{S}$,

$$s = P_\mathcal{S}s = P_\mathcal{S}a + P_\mathcal{S}v = P_\mathcal{S}a, \tag{6.24}$$

and $P_\mathcal{S}$ is surjective over $\mathcal{A}$. $\square$

Proposition 6.1 suggests that under the direct-sum condition, there is a one-to-one mapping between $\mathcal{A}$ and $\mathcal{S}$. In the following sections we will see how to compute this mapping explicitly.

More generally we may have a situation in which $\mathcal{A}$ and $\mathcal{S}^\perp$ intersect. Clearly in this case there are many possible signals in $\mathcal{A}$ that will fit the samples. Letting e denote a vector in the intersection, if $x \in \mathcal{A}$ is a given recovery, then any signal $y = x + e$ is also in $\mathcal{A}$ and yields the same samples, since the samples of e are zero. Therefore, knowledge of the samples and the prior $\mathcal{A}$ are insufficient to ensure perfect recovery in this case and additional criteria to select an estimate are needed. This scenario will be discussed in Section 6.5.

6.3.1   The direct-sum condition

The direct sum (6.23) is central to our developments in this section. Therefore, we pause to examine this condition and its implications in more detail.

In principle, one can verify whether (6.23) holds by simply attempting to directly check the definition of a direct sum. An alternative approach is to rely on the following proposition (see [46, Appendix A]).

Proposition 6.2. *Let $\mathcal{S} \neq \{0\}$, let $\mathcal{A}$ be a closed subspace of $\mathcal{H}$, and define*

$$\delta(\mathcal{S}, \mathcal{A}) = \|(I - P_\mathcal{A})P_\mathcal{S}\|, \tag{6.25}$$

where $\|\cdot\|$ denotes the spectral norm (see Appendix A). If $\delta(\mathcal{S}, \mathcal{A}) < 1$ and $\delta(\mathcal{A}, \mathcal{S}) < 1$ then $\mathcal{H} = \mathcal{A} \oplus \mathcal{S}^\perp$.

The direct-sum condition is also related to the concept of angles between subspaces. The angle from $\mathcal{A}$ to $\mathcal{S}$ is defined as the unique number in $[0, \pi/2]$ for which

$$\cos(\mathcal{S}, \mathcal{A}) = \inf_{x \in \mathcal{S}, \|x\|=1} \|P_{\mathcal{A}} x\|. \tag{6.26}$$

Similarly we can define the sine of the angle as

$$\sin(\mathcal{S}, \mathcal{A}) = \sup_{x \in \mathcal{S}, \|x\|=1} \|P_{\mathcal{A}^{\perp}} x\|. \tag{6.27}$$

The following proposition [133, Theorem 2.3] establishes that positive angles are equivalent to a direct-sum decomposition:

Proposition 6.3. *Given closed subspaces $\mathcal{A}, \mathcal{S}$ of a separable Hilbert space $\mathcal{H}$, the following are equivalent:*

1. $\mathcal{H} = \mathcal{A} \oplus \mathcal{S}^{\perp}$;
2. $\mathcal{H} = \mathcal{S} \oplus \mathcal{A}^{\perp}$;
3. $\cos(\mathcal{S}, \mathcal{A}) > 0$ and $\cos(\mathcal{A}, \mathcal{S}) > 0$.

In SI spaces, the angle $\cos(\mathcal{S}, \mathcal{A})$ is symmetric in $\mathcal{S}$ and $\mathcal{A}$, i.e. $\cos(\mathcal{S}, \mathcal{A}) = \cos(\mathcal{A}, \mathcal{S})$, and has a particularly simple form [46, 108]:

Proposition 6.4. *Let $\mathcal{A}, \mathcal{S}$ be SI spaces with bases $\{a(t - nT)\}, \{s(t - nT)\}$. Let $A(\omega), S(\omega)$ be the CTFTs of $a(t), s(t)$ respectively. Then,*

$$\cos(\mathcal{S}, \mathcal{A}) = \operatorname{ess\,inf}_{\omega} \frac{|R_{SA}(e^{j\omega})|}{\sqrt{R_{SS}(e^{j\omega}) R_{AA}(e^{j\omega})}}, \tag{6.28}$$

where $R_{SA}\left(e^{j\omega}\right)$ is defined by (5.28).

Proof: See Exercise 5. $\qquad\qquad\square$

Note that from the Cauchy–Schwarz inequality $|R_{SA}(e^{j\omega})|^2 \leq R_{SS}(e^{j\omega}) R_{AA}(e^{j\omega})$, so that the cosine defined by (6.28) is smaller than or equal to 1, as required. Furthermore, since $R_{AS}(e^{j\omega}) = \overline{R_{SA}(e^{j\omega})}$, the angle is symmetric in $\mathcal{S}$ and $\mathcal{A}$.

Propositions 6.3 and 6.4 lead to an easily verifiable condition under which the direct-sum condition is satisfied in SI spaces [46, Proposition 4.8].

Corollary 6.1. *Let $\mathcal{A}, \mathcal{S}$ be SI spaces with bases $\{a(t - nT)\}, \{s(t - nT)\}$. Let $A(\omega), S(\omega)$ be the CTFTs of $a(t), s(t)$ respectively. Then, $L_2(\mathbb{R}) = \mathcal{A} \oplus \mathcal{S}^{\perp}$ if and only if there exists a constant $\alpha > 0$ such that*

$$\left| R_{SA}\left(e^{j\omega}\right) \right| > \alpha, \quad \text{almost everywhere.} \tag{6.29}$$

An important consequence of the direct-sum condition is that any signal $x \in \mathcal{H}$ with $\mathcal{H} = \mathcal{A} \oplus \mathcal{S}^{\perp}$ can be written uniquely as a decomposition

$$x = x_{\mathcal{A}} + x_{\mathcal{S}^{\perp}}, \tag{6.30}$$

where $x_A \in \mathcal{A}$ and $x_{\mathcal{S}^\perp} \in \mathcal{S}^\perp$. These components are given by the oblique projection operator $E_{A\mathcal{S}^\perp}$:

$$x_A = E_{A\mathcal{S}^\perp} x$$
$$x_{\mathcal{S}^\perp} = (I - E_{A\mathcal{S}^\perp})x = E_{\mathcal{S}^\perp A} x. \qquad (6.31)$$

Under (6.23), the oblique projection operator is well defined, and is given by (see Theorem 2.2)

$$E_{A\mathcal{S}^\perp} = A(S^*A)^{-1}S^*. \qquad (6.32)$$

This operator will play an important role in recovery algorithms when $\mathcal{A}$ and $\mathcal{S}$ are not equal.

6.3.2   Unique recovery

A direct consequence of (6.32) is that under the direct-sum condition, we can write any x in $\mathcal{A}$ as $x = E_{A\mathcal{S}^\perp} x = A(S^*A)^{-1}c$ where $c = S^*x$ are the given samples. Therefore, any x in $\mathcal{A}$ can be recovered from its samples c in $\mathcal{S}$ by first multiplying c by the operator $(S^*A)^{-1}$, and then applying the set transformation A. This establishes a simple recovery algorithm as long as (6.23) holds.

Using Proposition 5.1 we can translate these operations in the SI setting into a filtering scheme, as in Fig. 6.2, with $w(t) = a(t)$ and

$$H(e^{j\omega}) = \frac{1}{R_{SA}(e^{j\omega})}. \qquad (6.33)$$

The direct-sum condition ensures that the denominator in (6.33) is invertible (see Corollary 6.1). When $\mathcal{A} = \mathcal{S}$, the filter (6.33) coincides with (6.16).

We can directly verify that (6.33) leads to perfect recovery of signals in $\mathcal{A}$, by using the fact that any $x(t) \in \mathcal{A}$ can be written as $x(t) = \sum_n b[n]a(t - nT)$. From Proposition 5.1, the DTFT of the sequence of samples is then

$$C(e^{j\omega}) = B(e^{j\omega})R_{SA}(e^{j\omega}), \qquad (6.34)$$

from which the result follows.

The overall sampling and reconstruction scheme is illustrated in Fig. 6.9. By construction, this block diagram implements an oblique projection. Therefore, when the input $x(t)$ is arbitrary and does not necessarily lie in $\mathcal{A}$, the output will be equal to $\hat{x}(t) = E_{A\mathcal{S}^\perp} x(t)$.

Example 6.7 Suppose that $x(t)$ is a spline of degree 2 with knots at the integers. Thus, $x(t)$ lies in the space $\mathcal{A}$ generated by $a(t) = \beta^2(t)$ and can be represented as

$$x(t) = \sum_{n \in \mathbb{Z}} d[n]\beta^2(t - n), \qquad (6.35)$$

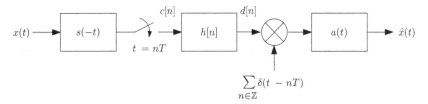

Figure 6.9 Sampling in an SI space $\mathcal{S}$ with reconstruction in the SI space $\mathcal{A}$.

where $\beta^2(t)$ is the B-spline function of degree 2. We wish to recover $x(t)$ from averages of its values over the intervals $[n - 1/2, n + 1/2]$, $n \in \mathbb{Z}$. This setting corresponds to sampling at times $t = n$ at the output of the filter $s(t) = \beta^0(t)$.

Using the system of Fig. 6.9 perfect recovery of $x(t)$ can be achieved by passing the samples through the correction filter $h[n]$, which is the convolutional inverse of

$$r_{sa}[n] = \int_{-\infty}^{\infty} \beta^0(t)\beta^2(t + n)dt = \begin{cases} 1/6, & |n| = 1 \\ 2/3, & n = 0 \\ 0, & |n| > 1. \end{cases} \tag{6.36}$$

Note that this filter is the same as that encountered in Example 6.5, because of the convolution property (4.32) of B-splines. Therefore, the frequency response of the correction filter is given by (6.19). Figures 6.10(a) and 6.10(b) show respectively the reconstructions obtained by using the scheme of Fig. 6.9 with $H(e^{j\omega})$ of (6.19) and without a digital correction filter. Evidently, the digital filtering stage is required in order to perfectly restore the signal from its samples.

Example 6.8 The previous example demonstrated that the effect of an averaging sampling filter can be overcome by proper processing of the samples. To further highlight the possibilities offered by a correction stage, we next show that the signal $x(t)$ of (6.35), which is nonbandlimited, can also be recovered from ideal pointwise samples.

Pointwise samples of $x(t)$ correspond to a sampling filter $s(t) = \delta(t)$, which is not in L_2. However, the main purpose of our requirement that $s(t)$ lie in L_2 was to ensure that the sample sequence $c[n]$ be in ℓ_2. Luckily, in the case of pointwise samples, this can also be guaranteed if $x(t)$ is continuous and decays sufficiently fast [108]. Under this assumption, the system of Fig. 6.9 implies that the required correction filter $h[n]$ is the convolutional inverse of

$$r_{sa}[n] = \int_{-\infty}^{\infty} \delta(t)\beta^2(t + n)dt = \begin{cases} 1/8, & |n| = 1 \\ 3/4, & n = 0 \\ 0, & |n| > 1. \end{cases} \tag{6.37}$$

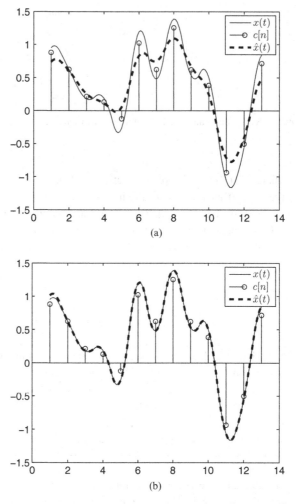

Figure 6.10 A spline of degree 2 and its samples after going through the filter $\beta^0(t)$.
(a) Reconstruction without using a digital correction filter. (b) Reconstruction using the system of Fig. 6.9.

The frequency response of the correction filter is therefore given by

$$H(e^{j\omega}) = \frac{1}{\frac{1}{8}e^{-j\omega} + \frac{3}{4} + \frac{1}{8}e^{j\omega}} = \frac{4}{\cos(\omega) + 3}. \qquad (6.38)$$

Perfect recovery of $x(t)$ can be achieved by first filtering the samples with the correction filter of (6.38) and then reconstructing with $a(t)$ as in Fig. 6.9.

Example 6.9 Perhaps more counter-intuitive is the fact that nonbandlimited signals can be recovered even if the sampling filter completely zeroes out their high frequencies. To demonstrate this, consider a signal $x(t)$ formed by exciting an RC

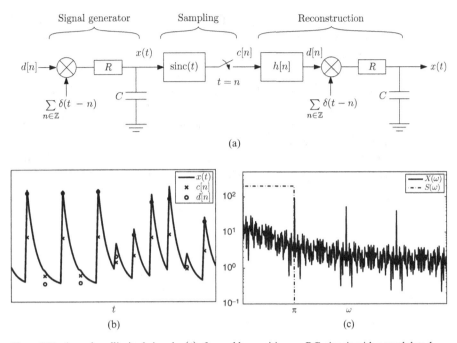

Figure 6.11 A nonbandlimited signal $x(t)$, formed by exciting an RC circuit with a modulated impulse train, is sampled after passing through an ideal LPF and then perfectly reconstructed. (a) Sampling and reconstruction setup. (b) The signal $x(t)$, its samples $c[n]$, and expansion coefficients $d[n]$. (c) The signal $X(\omega)$ and the sampling filter $S(\omega)$.

circuit with a modulated impulse train $\sum_n d[n]\delta(t-n)$, as shown in Fig. 6.11(a). The impulse response of the RC circuit is known to be $a(t) = \tau^{-1}\exp\{-t/\tau\}u(t)$, where $u(t)$ is the unit step function and $\tau = RC$ is the time constant. Therefore

$$x(t) = \frac{1}{\tau}\sum_{n\in\mathbb{Z}} d[n]\exp\{-(t-n)/\tau\}u(t-n). \tag{6.39}$$

Clearly, $x(t)$ is not bandlimited. Now, suppose that $x(t)$ is filtered by an ideal LPF $s(t) = \mathrm{sinc}(t)$ and then sampled at times $t = n$ to obtain the sequence $c[n]$. The signal $x(t)$ and its samples are depicted in Fig. 6.11(b).

Intuitively, there seems to be information loss in the sampling process since the frequency content of $x(t)$ outside $[-\pi, \pi]$ is zeroed out, as shown in Fig. 6.11(c). However, it is easily verified that if $\tau < \pi^{-1}$, then $|R_{SA}(e^{j\omega})| > (1-\pi^2\tau^2)^{-1/2} > 0$ for all $\omega \in [-\pi, \pi]$ so that condition (6.29) is satisfied and perfect recovery is possible. The digital correction filter (6.33) can be calculated as

$$h[n] = \begin{cases} 1, & n = 0 \\ \frac{\tau}{n}(-1)^n, & n \neq 0. \end{cases} \tag{6.40}$$

To reconstruct $x(t)$ we need to excite an identical RC circuit with an impulse train modulated by the sequence $d[n] = h[n] * c[n]$. The entire sampling–reconstruction setup is depicted in Fig. 6.11(a).

The last two examples demonstrate that a signal $x(t) = \sum_{n \in \mathbb{Z}} d[n]a(t - nT)$ lying in an SI space generated by $a(t)$ with period T can be perfectly recovered from its pointwise samples, or its low-pass content, even if the signal is not bandlimited, under appropriate conditions.

Generalizing the derivations in Example 6.8, $x(t)$ can be recovered from its pointwise samples, which are assumed to be in ℓ_2, if the sequence

$$r_{sa}[n] = \int_{-\infty}^{\infty} \delta(t)a(t + nT) = a(nT) \tag{6.41}$$

has a convolutional inverse. Since in this case $r_{sa}[n]$ is equal to pointwise samples of $a(t)$,

$$R_{SA}(e^{j\omega}) = \frac{1}{T} \sum_{k \in \mathbb{Z}} A\left(\frac{\omega}{T} - \frac{2\pi k}{T}\right) = A(e^{j\omega}), \tag{6.42}$$

leading to the following proposition.

Proposition 6.5. *Let $x(t)$ lie in an SI subspace $\mathcal{A}$ generated by $a(t)$ with period T. Assume that the pointwise sample sequence $x(nT)$ is in ℓ_2. Then $x(t)$ can be perfectly recovered from its samples $x(nT)$ using the system of Fig. 6.9 as long as $|A(e^{j\omega})| = |\frac{1}{T} \sum_{k \in \mathbb{Z}} A\left(\frac{\omega}{T} - \frac{2\pi k}{T}\right)|$ is bounded away from zero almost everywhere. Recovery is obtained with $H(e^{j\omega}) = 1/A(e^{j\omega})$.*

Example 6.9 can also be extended to general SI settings. When $s(t)$ is an ideal LPF with cutoff π/T,

$$R_{SA}(e^{j\omega}) = \frac{1}{T} A\left(\frac{\omega}{T}\right), \quad |\omega| \leq \pi. \tag{6.43}$$

Therefore, in this case recovery is possible as long as $|A(\omega)|$ is bounded away from zero almost everywhere on $|\omega| \leq \pi/T$.

Proposition 6.6. *Let $x(t)$ lie in an SI subspace $\mathcal{A}$ generated by $a(t)$ with period T. Then $x(t)$ can be perfectly recovered from pointwise samples with period T of the output of an LPF with cutoff π/T using the system of Fig. 6.9 as long as $|A(\omega)|$ is bounded away from zero almost everywhere on $|\omega| \leq \pi/T$. Recovery is obtained with $H(e^{j\omega}) = T/A(\omega/T)$ for $|\omega| \leq \pi$.*

6.3.3   Computing the oblique projection operator

We have seen in Section 6.3.2 that recovery under the direct-sum condition is equivalent to computing an oblique projection operator. This in turn involves evaluating the inverse operator $(S^*A)^{-1}$, or, in the SI setting, evaluating the filter (6.33). Computing the inverse of a general operator can be computationally demanding. Very often, it is easier to apply S^*A than calculate its inverse. In this section we explore two methods for evaluating $d = (S^*A)^{-1}c$ which involve applications of S^*A alone. These techniques are beneficial when the operator S^*A is simpler to apply than its inverse. The first method is based on the Neumann series of an operator [132], while the second results from viewing d as a solution to a convex optimization problem and applying the well-known steepest-descent iterations [124].

In the SI setting, when $s(t)$ and $a(t)$ are compactly supported, $R_{SA}(e^{j\omega})$ corresponds to filtering with an FIR filter, while the inverse in (6.33) may be a general IIR filter (see Section 3.3 for definitions of FIR and IIR filters). In both strategies below, the inverse operation is replaced by a series of FIR filters, which are typically easier to implement than a general, nonstructured, IIR filter. In Chapter 9 we consider special cases of $s(t)$ and $a(t)$ which lead to IIR filters that can be computed directly, in an efficient manner. Our discussion below, in contrast, treats the general case in which no special structure is available to be exploited.

Filtering via Neumann series

Our first approach is based on the Neumann series representation of an operator Q on a Hilbert space $\mathcal{H}$ [132]. Suppose that $\|Q\| < 1$ where $\|Q\|$ is the operator norm (see Appendix A). Then,

$$(I - Q)^{-1} = \sum_{n=0}^{\infty} Q^n. \tag{6.44}$$

Letting $Q = I - \alpha S^*A$ where $\alpha \neq 0$ is chosen such that $\|Q\| < 1$, we have that

$$(S^*A)^{-1} = \alpha \sum_{n=0}^{\infty} (I - \alpha S^*A)^n. \tag{6.45}$$

Using (6.45) we can implement $(S^*A)^{-1}$ by a sequence of operations involving only S^*A. To obtain a simple recursive implementation, let

$$\hat{d}^m = \alpha \sum_{n=0}^{m} (I - \alpha S^*A)^n c \tag{6.46}$$

denote an approximation of $d = (S^*A)^{-1}c$ by the first $m + 1$ terms of the Neumann series. We can then calculate the next update $\hat{d}^{m+1}$ from $\hat{d}^m$ to be

$$\hat{d}^{m+1} = \alpha \sum_{n=0}^{m+1} (I - \alpha S^*A)^n c$$

$$= \alpha \left(c + \sum_{n=1}^{m+1} (I - \alpha S^*A)^n c \right)$$

$$= \alpha \left(c + (I - \alpha S^*A) \frac{1}{\alpha} \hat{d}^m \right)$$

$$= \hat{d}^m + \alpha \left(c - S^*A \hat{d}^m \right). \tag{6.47}$$

As $\hat{d}^m$ approaches $(S^*A)^{-1}c$, the increment $c - S^*A\hat{d}^m$ converges to zero. In practice the iterations may be terminated when this value is sufficiently small.

In the SI setting, S^*A is equivalent to filtering with $h[n] = h(nT)$ where $h(t) = s(-t) * a(t)$. If both $a(t)$ and $s(t)$ have compact support, then the filter $h(t)$ will also be compactly supported so that $h[n]$ is an FIR filter. In this case, the iterations defined by (6.47) involve only FIR filtering. We can also readily compute an appropriate value of α in this setting: the operator Q is equivalent to filtering with $Q(e^{j\omega}) = 1 - \alpha H(e^{j\omega})$. Therefore, choosing α such that $\|Q\| < 1$ is equivalent to the requirement

$$|1 - \alpha H(e^{j\omega})| < 1, \quad \text{for all } \omega. \tag{6.48}$$

The iterations defined by (6.47) are very simple; however, the convergence rate can be slow because of the constant step size. To improve the convergence rate we may choose α at each iteration to minimize the norm of the error $\varepsilon^{m+1} = c - S^*A\hat{d}^{m+1}$. From (6.47), the error obeys the iterations $\varepsilon^{m+1} = (I - \alpha^m S^*A)\varepsilon^m$. The optimal step size is therefore given by

$$\alpha^m = \arg\min_\alpha \|(I - \alpha S^*A)\varepsilon^m\|^2. \tag{6.49}$$

Since the objective is quadratic in α, the solution can be found by setting the derivative to 0, yielding

$$\alpha^m = \frac{(\varepsilon^m)^* S^*A\varepsilon^m}{\|S^*A\varepsilon^m\|^2}. \tag{6.50}$$

This modification renders each iteration computationally more demanding, since it requires one more filtering operation with S^*A to determine the step size. Nevertheless, it can often lead to faster convergence.

Filtering via steepest-descent iterations

An alternative approach to computing $d = (S^*A)^{-1}c$ is to view d as a solution to a convex optimization problem, and then rely on the vast number of iterative algorithms available to efficiently compute d in practice. As an example of this technique, we consider the steepest-descent method, which is one of the simplest iterative algorithms. Steepest descent is a first-order technique in which a minimum of a function is found by taking steps proportional to the negative of the gradient of the function at the current point. Algorithms that utilize higher-order derivatives can be applied as well in order to improve convergence rate; The interested reader is referred to [133] for an in-depth treatment of iterative methods. The resulting algorithm is computationally more intensive than the varying step-size Neumann approach, but often leads to fast convergence in the first few iterations. Once the iterations approach the true minimum, the method tends to become quite slow.

To develop the steepest-descent iterations, we first view d as the minimizer of the convex objective

$$f(d) = \|S^*Ad - c\|^2. \tag{6.51}$$

Although clearly there is a closed-form solution to (6.51), we can use a steepest-descent algorithm to minimize the objective iteratively. This avoids the need for computing the inverse $(S^*A)^{-1}$ and instead allows us to approximate it by repeated applications of S^*A.

Starting with some initial guess d^0, for example $d^0 = c$, the steepest-descent algorithm updates the current approximation d^m of d by

$$d^{m+1} = d^m - \alpha^m \nabla_d f(d), \tag{6.52}$$

where $\nabla_d f(d)$ is the derivative of the objective with respect to d, and α^m is an appropriate step size. In our case,

$$\nabla_d f(d) = 2(A^*S)(S^*Ad - c). \tag{6.53}$$

The steepest-descent iterations then become

$$\hat{d}^{m+1} = \hat{d}^m + \alpha^m A^*S \left(c - S^*A\hat{d}^m\right). \tag{6.54}$$

It is interesting to note that (6.54) is almost the same as the Neumann update rule (6.47), apart from the factor A^*S.

If α^m is properly chosen at each iteration, then the sequence of approximations $\hat{d}^m$ will converge to d [124]. One option is to choose the value α^m that minimizes (6.51) at each iteration. From (6.54), this is equivalent to

$$\alpha^m = \arg \min_\alpha \|(I - \alpha S^*AA^*S)\varepsilon^m\|^2. \tag{6.55}$$

Setting the derivative to 0 results in

$$\alpha^m = \frac{\|AS^*\varepsilon^m\|^2}{\|S^*AA^*S\varepsilon^m\|^2}. \tag{6.56}$$

A schematic representation of the Neumann and steepest-descent update rules is shown in Fig. 6.12. Each Neumann iteration comprises a single application of S^*A, which corresponds to filtering in SI spaces. The varying step-size Neumann approach requires two filtering operations (or applications of S^*A), while the steepest-descent method requires three filtering operations: one for calculating S^*Ad^m, and two more for computing the expression $(S^*A)(A^*S)(c - S^*Ad^m)$ needed for determining the updates $\hat{d}^m$ and the step size α_m.

Figure 6.13 compares the recoveries produced by all three methods in the setting of Example 6.7. As can be seen, two iterations suffice for all three algorithms to obtain a good approximation of the input signal $x(t)$. Figure 6.14 shows the reconstruction error of the three approaches as a function of the iteration number. The convergence rate of the methods is similar in this case. We note, however, that in situations in which the frequency response $R_{SA}(e^{j\omega})$ of the filter corresponding to S^*A is close to 0 for some ω, convergence of the steepest-descent method becomes much slower. Furthermore, each Neumann iteration is three times as fast as steepest descent.

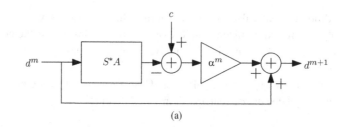

(a)

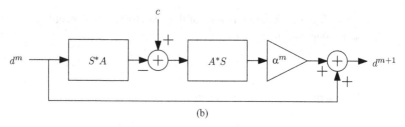

(b)

Figure 6.12 (a) One iteration of the Neumann algorithm. (b) One iteration of the steepest-descent method.

6.3.4   Oblique biorthogonal basis

Just as in the case of equal sampling and recovery spaces, it is interesting to interpret the sampling scheme of Fig. 6.9 in terms of a basis expansion. Since any signal in $\mathcal{A}$ can be recovered from the corrected samples $d[n] = c[n] * h[n]$ via $x(t) = \sum_n d[n]a(t - nT)$, we may view this sequence as the coefficients in a basis expansion. To obtain the corresponding basis we note that by combining the effects of the sampler $s(t)$ and the correction filter $h[n]$ of (6.33), the sequence of samples can be equivalently expressed as $d[n] = \langle v(t - nT), x(t) \rangle$ where

$$v(t) = \sum_{n \in \mathbb{Z}} h[n]s(t - nT). \tag{6.57}$$

In the Fourier domain,

$$V(\omega) = H(e^{j\omega T})S(\omega) = \frac{1}{R_{SA}(e^{j\omega T})}S(\omega). \tag{6.58}$$

We conclude that any $x(t) \in \mathcal{A}$ can be written as

$$x(t) = \sum_{n \in \mathbb{Z}} \langle v(t - nT), x(t) \rangle a(t - nT), \tag{6.59}$$

with $v(t)$ given by (6.57) or (6.58).

From (6.57) it follows immediately that $v(t)$ lies in $\mathcal{S}$. We can further show that the set of functions $\{v(t - nT)\}$ form a Riesz basis by relying on Proposition 5.3 and noting that owing to the direct-sum condition, the function R_{SA} is bounded (Corollary 6.1). These basis functions have an additional appealing property – they are biorthogonal to $\{a(t - mT)\}$ so that $\langle v(t - nT), a(t - mT) \rangle = \delta_{mn}$. This is a result of the fact that $r_{va}[n] = \delta[n]$, as can be seen quite easily in the Fourier domain:

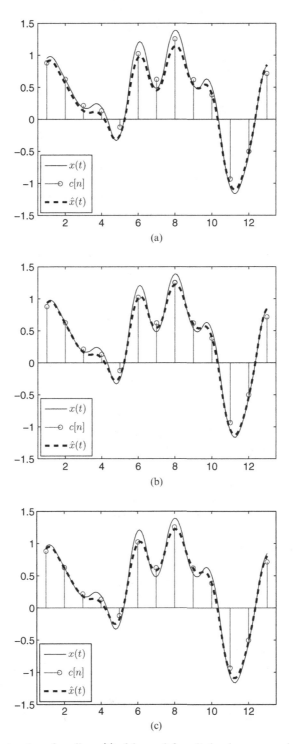

Figure 6.13 Reconstruction of a spline $x(t)$ of degree 2 from its local averages using two iterations of (a) the Neumann method with constant step size. (b) the Neumann method with varying step size. (c) the steepest-descent method.

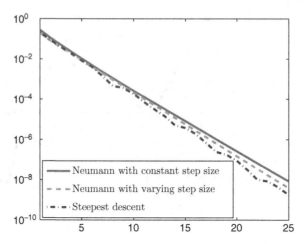

Figure 6.14 Reconstruction error as a function of the iteration number for the Neumann and steepest-descent methods.

$$R_{VA}(e^{j\omega}) = \frac{1}{T}\sum_{k\in\mathbb{Z}}\overline{V\left(\frac{\omega}{T} - \frac{2\pi k}{T}\right)}A\left(\frac{\omega}{T} - \frac{2\pi k}{T}\right)$$

$$= \frac{1}{R_{SA}(e^{j\omega})}\frac{1}{T}\sum_{k\in\mathbb{Z}}\overline{S\left(\frac{\omega}{T} - \frac{2\pi k}{T}\right)}A\left(\frac{\omega}{T} - \frac{2\pi k}{T}\right) = 1. \qquad (6.60)$$

We conclude that the vectors $\{v(t - nT)\}$ form a Riesz basis for $\mathcal{S}$ and are biorthogonal to the vectors $\{a(t - nT)\}$. When $\mathcal{A} = \mathcal{S}$, we recover the conventional biorthogonal basis functions. Equation (6.58) therefore provides a concrete method for constructing a biorthogonal basis of a given basis $\{a(t - nT)\}$ in any subspace $\mathcal{S}$ satisfying the direct-sum condition $L_2 = \mathcal{A} \oplus \mathcal{S}^\perp$. We refer to this more general class of biorthogonal functions as the *oblique biorthogonal vectors*.

When $\mathcal{A}$ and $\mathcal{S}$ are general subspaces in a Hilbert space $\mathcal{H}$ satisfying (6.23), the oblique biorthogonal vectors in $\mathcal{S}$ correspond to the set transformation $V = S(A^*S)^{-1}$. The fact that a biorthogonal Riesz basis exists in $\mathcal{S}$ can actually be used to define the direct-sum condition: it can be shown that $\mathcal{H} = \mathcal{A} \oplus \mathcal{S}^\perp$ if and only if there exist Riesz bases for $\mathcal{A}$ and $\mathcal{S}$ that are biorthogonal [131].

We summarize our results on the oblique biorthogonal basis in the following theorem.

Theorem 6.1 (Oblique biorthogonal basis). *Let $\mathcal{A}$ and $\mathcal{S}$ be subspaces of a Hilbert space $\mathcal{H}$ such that $\mathcal{H} = \mathcal{A} \oplus \mathcal{S}^\perp$, and let the vectors $\{a_n\}$ be a Riesz basis for $\mathcal{A}$ with set transformation A. Then, the oblique biorthogonal vectors $\{v_n\}$ of $\{a_n\}$ in $\mathcal{S}$ correspond to the set transformation $V = S(A^*S)^{-1}$ where S is a set transformation corresponding to an arbitrary Riesz basis for $\mathcal{S}$. When $\mathcal{A}$ is an SI subspace with basis functions $a_n(t) = a(t - nT)$, then $v_n = v(t - nT)$ with*

$$V(\omega) = \frac{1}{R_{SA}(e^{j\omega T})}S(\omega).$$

Note that if instead of (6.57) we define

$$v(t) = \sum_{n \in \mathbb{Z}} h[n]a(t - nT), \tag{6.61}$$

or

$$V(\omega) = H(e^{j\omega T})A(\omega) = \frac{1}{R_{SA}(e^{j\omega T})} A(\omega), \tag{6.62}$$

then we can express any $x(t) \in \mathcal{A}$ as

$$x(t) = \sum_{n \in \mathbb{Z}} \langle s(t - nT), x(t) \rangle v(t - nT) = \sum_{n \in \mathbb{Z}} c[n]v(t - nT). \tag{6.63}$$

The functions $v(t - nT)$ now lie in $\mathcal{A}$, and are biorthogonal to $s(t - mT)$.

Recall that any standard biorthogonal basis in $\mathcal{A}$ can be used to construct an orthogonal projection. Similarly, a biorthogonal basis in $\mathcal{S}$ can be used to define an oblique projection. Indeed, suppose $\{a_n\}$ is a Riesz basis for $\mathcal{A}$, and $\mathcal{S}$ is a subspace such that $\mathcal{H} = \mathcal{A} \oplus \mathcal{S}^{\perp}$. Let $\{v_n\}$ be the biorthogonal basis in $\mathcal{S}$. Then,

$$E_{\mathcal{A}\mathcal{S}^{\perp}}x = \sum_{n \in \mathbb{Z}} \langle v_n, x \rangle a_n \tag{6.64}$$

$$E_{\mathcal{S}\mathcal{A}^{\perp}}x = \sum_{n \in \mathbb{Z}} \langle a_n, x \rangle v_n. \tag{6.65}$$

Both equations can be verified directly from the definition of the oblique projection. We show here how to prove the first equality; the second follows in a similar fashion. To establish (6.64) we need to show that $E_{\mathcal{A}\mathcal{S}^{\perp}}x = 0$ for any $x \in \mathcal{S}^{\perp}$ and $E_{\mathcal{A}\mathcal{S}^{\perp}}x = x$ for $x \in \mathcal{A}$. The first equality is a result of the fact that since $v_n \in \mathcal{S}$, $\langle v_n, x \rangle = 0$ for any $x \in \mathcal{S}^{\perp}$. Now, for $x \in \mathcal{A}$, we can write $x = \sum_m b[m]a_m$ for some coefficients $b[m]$. Since $\langle v_n, a_m \rangle = \delta_{nm}$, we have $\langle v_n, x \rangle = b[n]$, validating (6.64).

Another interesting observation is that if we orthogonally project the biorthogonal basis $\{v_n\}$ onto $\mathcal{A}$, then the resulting functions $w_n = P_{\mathcal{A}}v_n$ are precisely the biorthogonal basis vectors of $\{a_n\}$ in $\mathcal{A}$. The biorthogonality property is preserved since $\langle a_m, v_n \rangle = \langle P_{\mathcal{A}}a_m, v_n \rangle = \langle a_m, P_{\mathcal{A}}v_n \rangle$. The fact that $\{P_{\mathcal{A}}v_n\}$ span $\mathcal{A}$ follows from substituting the equality $\langle v_n, x \rangle = \langle P_{\mathcal{A}}v_n, x \rangle$ into (6.64) for any x in $\mathcal{A}$.

As a final note, similar concepts to those discussed above can be introduced starting from frame expansions, leading to the notion of *oblique dual frames* [46, 47, 110, 114, 115].

6.4 Summary: unique unconstrained recovery

6.4.1 Consistent recovery

Before treating the setting in which the direct-sum condition is violated, we summarize our results so far. We have seen that a signal $x(t)$ in an SI subspace $\mathcal{A}$ generated by $a(t)$, can be reconstructed from its generalized samples in Fig. 6.1 using any choice of $s(t)$ for which (6.23) is satisfied. Thus for a given SI space, there is a broad set of possible sampling filters. Choosing the functions appropriately leads to the formulation of a variety of interesting sampling theorems, such as pointwise sampling of nonbandlimited signals, bandlimited sampling of nonbandlimited functions, and many more. Regardless of the choice of $\mathcal{S}$, as long as the direct-sum condition is satisfied and $x(t)$ lies in $\mathcal{A}$, the output of Fig. 6.9 is always equal to $x(t)$.

The flexibility offered by Fig. 6.9 comes at a cost when the input signal does not lie entirely in $\mathcal{A}$, for example owing to noise or mismodeling. As we have seen, for a general input $x(t) \in L_2(\mathbb{R})$, the output of Fig. 6.9 is equal to $E_{\mathcal{A}\mathcal{S}^\perp}x$. In contrast, when $\mathcal{A} = \mathcal{S}$, the output of Fig. 6.6 equals $P_{\mathcal{A}}x$. The orthogonal projection properties imply that this output is the closest signal in $\mathcal{A}$ to the input signal $x(t)$, in a squared-error sense. Consequently, the error resulting from Fig. 6.9 with an arbitrary choice of $s(t)$, will in general be larger than that of Fig. 6.6, leading to a tradeoff between design simplicity and recovery error. In the next section, we quantify this tradeoff by examining the error for an arbitrary filter $s(t)$. Note, however, that regardless of the choice of $s(t)$, the system of Fig. 6.9 leads to a *consistent* recovery [108]: resampling the output using the given function $s(t)$, namely, reinjecting the output back into the system, results in the same samples $c[n]$. This follows from the fact that $S^* E_{\mathcal{A}\mathcal{S}^\perp} = S^*$ (see Exercise 9).

The concept of a consistent recovery is particulary important when we do not have prior information about the true underlying signal. Suppose, for example, that we are given samples of a signal obtained with a specific sampling filter, and would like to interpolate a continuous-time signal from the samples. Without prior knowledge, many different recoveries are possible; this interpolation problem will be treated in considerable more detail in Chapter 9. For now, assume that we use a specific interpolation kernel $a(t)$, for example a B-spline function which is popular in a variety of different applications. Since no prior about the signal is known, a reasonable approach is to seek a consistent solution, namely, a recovery $\hat{x}$ that has the property that if we resample it using the same sampling filter, then we will obtain the given samples. Our results can be used to determine a consistent recovery by first applying the operator $(S^* A)^{-1}$ to the samples, and then interpolating the corrected samples with the chosen reconstruction kernel. This approach allows us to consider several different kernels $a(t)$, and choose the recovery that best fits our needs. The important aspect is that regardless of the choice of $a(t)$, as long as the samples are properly corrected according to Fig. 6.9, the recovery will be consistent.

Following the discussion in Section 6.3.4 and using the results of Theorem 6.1 we can interpret the correction $H(e^{j\omega}) = 1/R_{SA}(e^{j\omega})$ together with the interpolation by $a(t)$ as computing the oblique biorthogonal dual of $s(t)$ in a space $\mathcal{A}$ that is different from the sampling space $\mathcal{S}$. When $a(t) = s(t)$ this amounts to implementing the conventional biorthogonal basis. These ideas are illustrated in the next example.

Example 6.10 Many practical sampling devices average the signal over non-overlapping segments, which corresponds to using $s(t) = \beta^0(t)$ as a sampling filter. Suppose we are given such local averages of an unknown signal $x(t)$. A possible approach to approximate $x(t)$ from these samples is to use the biorthogonal basis. Since the functions $\{\beta^0(t - n)\}$ are orthonormal, the biorthogonal basis is also spanned by $\beta^0(t)$. Using this filter for reconstruction results in a discontinuous recovery, as illustrated in Fig. 6.16(a), which can be undesirable.

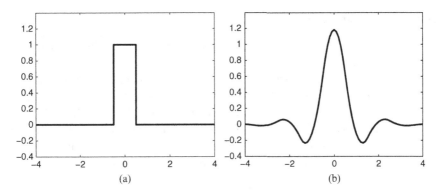

Figure 6.15 (a) The function $\beta^0(t)$. (b) The oblique dual of $\beta^0(t)$ in the space of splines of degree 2.

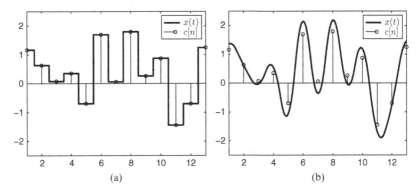

Figure 6.16 (a) A spline of degree 0 with given local averages $c[n] = \int_{n-0.5}^{n+0.5} x(t)dt$. (b) A spline of degree 2 having the same local averages.

As an alternative, we may make use of the oblique biorthogonal basis in a space of smoother functions. For example, Fig. 6.15 depicts the oblique biorthogonal generator of $\beta^0(t)$ in the space of splines of degree 2. This corresponds to choosing a B-spline of degree 2 as the reconstruction filter in Fig. 6.9, and using an appropriate correction filter prior to recovery. The oblique biorthogonal generator is given by

$$v(t) = \sum_{n\in\mathbb{Z}} h[n]\beta^2(t-n) \tag{6.66}$$

where $h[n]$ is the convolutional inverse of $r_{\beta^0\beta^2}[n]$, computed in (6.19), and serves as the correction filter. Fig. 6.16(b) demonstrates that the recovery obtained using $v(t)$ often constitutes a better approximation to signals encountered in applications, owing to its smoothness. Note that both solutions have the same samples and therefore both are consistent reconstructions.

6.4.2   Recovery error

We now turn to analyze the price associated with the flexibility offered by Fig. 6.9 when the input $x(t)$ does not lie entirely in $\mathcal{A}$.

Suppose we design a sampling system assuming that $x(t) \in \mathcal{A}$. We consider two schemes: one in which the sampler is matched to the prior space so that $s(t) = a(t)$, and one in which we choose an arbitrary sampling function $s(t)$ such that the direct-sum condition (6.23) is satisfied. The output of the first system is given by $x_1(t) = P_{\mathcal{A}}x(t)$, while the second approach results in $x_2(t) = E_{\mathcal{A}\mathcal{S}^{\perp}}x(t)$. For an arbitrary input $x(t)$, we associate an error with each reconstructed output:

$$e_1(t) = x(t) - x_1(t) = (I - P_{\mathcal{A}})x(t) = P_{\mathcal{A}^{\perp}}x(t)$$
$$e_2(t) = x(t) - x_2(t) = (I - E_{\mathcal{A}\mathcal{S}^{\perp}})x(t) = E_{\mathcal{S}^{\perp}\mathcal{A}}x(t). \tag{6.67}$$

In the next proposition we quantify the possible increase in squared-norm error when using an arbitrary sampling filter. To make the results more general, we state them for the arbitrary Hilbert space setting.

Proposition 6.7.   *Let $\mathcal{A}$ and $\mathcal{S}$ be two closed subspaces of a Hilbert space $\mathcal{H}$ such that $\mathcal{H} = \mathcal{A} \oplus \mathcal{S}^{\perp}$, and let x be an arbitrary vector in $\mathcal{H}$. Then,*

$$\|x - P_{\mathcal{A}}x\| \leq \|x - E_{\mathcal{A}\mathcal{S}^{\perp}}x\| \leq \frac{1}{\cos(\mathcal{A},\mathcal{S})}\|x - P_{\mathcal{A}}x\|, \tag{6.68}$$

where $\cos(\mathcal{A},\mathcal{S})$ is defined in (6.26).

An immediate consequence of the theorem is that

$$1 \leq \frac{\|e_2\|}{\|e_1\|} \leq \frac{1}{\cos(\mathcal{A},\mathcal{S})}, \tag{6.69}$$

so that in particular, $\|e_2\| \geq \|e_1\|$.

Proof: The lower inequality follows immediately from the projection theorem (Proposition 2.15) since both $P_{\mathcal{A}}x$ and $E_{\mathcal{A}\mathcal{S}^{\perp}}x$ are vectors in $\mathcal{A}$.

To prove the upper bound, denote by e_3 the difference between x_1 and x_2: $e_3 = (P_{\mathcal{A}} - E_{\mathcal{A}\mathcal{S}^{\perp}})x$. It is easy to see that $e_2 = e_1 + e_3$ where $e_1 \in \mathcal{A}^{\perp}$ and $e_3 \in \mathcal{A}$. Therefore, $e_1 = P_{\mathcal{A}^{\perp}}e_2$. In addition, $e_2 \in \mathcal{S}^{\perp}$. Now,

$$\min_{x} \frac{\|x - P_{\mathcal{A}}x\|}{\|x - E_{\mathcal{A}\mathcal{S}^{\perp}}x\|} = \min_{x} \frac{e_1}{e_2} = \min_{x} \frac{P_{\mathcal{A}^{\perp}}e_2}{e_2} = \min_{e_2 \in \mathcal{S}^{\perp}} \frac{P_{\mathcal{A}^{\perp}}e_2}{e_2} = \cos(\mathcal{S}^{\perp},\mathcal{A}^{\perp}). \tag{6.70}$$

It remains to show that

$$\cos(\mathcal{S}^{\perp},\mathcal{A}^{\perp}) = \cos(\mathcal{A},\mathcal{S}). \tag{6.71}$$

From the definition of the cosine:

$$\cos^2(\mathcal{S}^{\perp},\mathcal{A}^{\perp}) = \min_{s \in \mathcal{S}^{\perp}, \|s\|=1} \|P_{\mathcal{A}^{\perp}}s\|^2 = 1 - \max_{s \in \mathcal{S}^{\perp}, \|s\|=1} \|P_{\mathcal{A}}s\|^2. \tag{6.72}$$

Now, from the Cauchy–Schwarz inequality,

$$\|P_A s\|^2 = \max_{\|x\|=1} |\langle x, P_A s\rangle|^2 = \max_{\|x\|=1} |\langle P_A x, s\rangle|^2 = \max_{a\in A, \|a\|=1} |\langle a, s\rangle|^2. \quad (6.73)$$

Similarly, for any $a \in A$,

$$\|P_{S^\perp} a\|^2 = \max_{\|x\|=1} |\langle x, P_{S^\perp} a\rangle|^2 = \max_{\|x\|=1} |\langle P_{S^\perp} x, a\rangle|^2 = \max_{s\in S^\perp, \|s\|=1} |\langle a, s\rangle|^2. \quad (6.74)$$

Combining (6.73) and (6.74) we have that

$$\max_{s\in S^\perp, \|s\|=1} \|P_A s\|^2 = \max_{a\in A, \|a\|=1} \|P_{S^\perp} a\|^2. \quad (6.75)$$

Substituting (6.75) into (6.72) proves (6.71). $\qquad\square$

Proposition 6.7 states that there is a penalty for the flexibility offered by choosing S (almost) arbitrarily: the norm of the reconstruction error for $x \notin A$ is increased. However, in many practical applications this increase may not be very large [10, 134, 135, 136]. When S and A are SI spaces the worst-case norm increase can be computed directly using Proposition 6.4.

Example 6.11 In Example 6.10 we treated a common scenario in which the signal is sampled with a B-spline of degree 0. Rather than interpolating the samples with the same generator, we considered recovery with a higher-order spline, which results in a smoother function. Recovery with an nth-order B-spline involves computing the oblique dual of β^0 in the space A of splines of degree n, as described in Theorem 6.1.

If the input is not an nth-order spline, then the recovery will not be equal to the original signal. For a given nth-order interpolator, the smallest error is obtained when sampling with $s(t) = \beta^n(t)$. However, such sampling is typically more complicated than using $\beta^0(t)$, which is equivalent to local averaging. From Proposition 6.7, the error resulting from using a simple sampling device satisfies $\|x - E_{AS^\perp} x\| \leq \frac{1}{\cos(A,S)} \|x - P_A x\|$ so that the increase in the error over the lowest possible is at most $1/\cos(A, S)$. In Table 6.4 we report the values of $\cos(A, S)$ when S, A are generated by a 0th-order B-spline and an nth-order B-spline, respectively [137]. The second column is an average measure in which rather than computing the minimum value of $|R_{SA}(e^{j\omega})|/\sqrt{R_{SS}(e^{j\omega}) R_{AA}(e^{j\omega})}$ as per the definition (6.28), this expression is integrated over all frequencies.

As can be seen in the table, the values of $\cos(A, S)$ are only slightly smaller than 1, so that the error increase resulting from simple sampling is not too large.

For SI spaces with fixed period T, Proposition 6.7 implies that using an oblique projection can increase the error. However, as the period decreases, the difference between orthogonal and oblique projections becomes negligible. Specifically, it can be shown that in the SI setting, if $s(t)$ has the partition of unity property, i.e. $\sum_n s(t - n) = 1$, then both the orthogonal projection and the oblique projection have the same asymptotic

Table 6.4 Cosine of the angle between spline spaces of degree n and degree 0

n	$\cos(\mathcal{A}, \mathcal{S})$ (worst case)	$\overline{\cos}(\mathcal{A}, \mathcal{S})$ (average over frequency)
0	1	1
1	0.866 025	0.926 420
2	0.872 872	0.930 323
3	0.836 154	0.916 853

behavior as the sampling step goes to zero:

$$\|x(t/T) - P_{\mathcal{A}}x(t/T)\| = \|x(t/T) - E_{\mathcal{AS}^\perp}x(t/T)\| = C\|x^{(-L)}(t)\| + O(T^{-L+1}). \tag{6.76}$$

The integer $L = n + 1$ represents the order of the representation (i.e. the model has the ability to reproduce polynomials of degree n), and $\|x^{(-L)}(t)\|$ is the norm of the Lth derivative of $x(t)$ [10].

Using similar tools to those used in the proof of Proposition 6.7, we can develop an analogous bound on the norm $\|\hat{x}\|$ of the recovered output. When $\mathcal{A} = \mathcal{S}$, $\hat{x} = P_{\mathcal{A}}x$ so that the norm of the output is always bounded by that of the input: $\|P_{\mathcal{A}}x\| \leq \|x\|$. However, if we apply an arbitrary sampling filter, then the norm can actually increase, as incorporated in the following proposition.

Proposition 6.8. *Let $\mathcal{A}$ and $\mathcal{S}$ be closed subspaces of a Hilbert space $\mathcal{H}$ and let $\mathcal{H} = \mathcal{A} \oplus \mathcal{S}^\perp$. Then,*

$$\|E_{\mathcal{AS}^\perp}x\| \leq \frac{1}{\cos(\mathcal{A}, \mathcal{S})}\|x\|, \tag{6.77}$$

where $\cos(\mathcal{A}, \mathcal{S})$ is defined by (6.26).

Proof: To prove the theorem, we note that by definition of $\cos(\mathcal{A}, \mathcal{S})$,

$$\cos(\mathcal{A}, \mathcal{S}) \leq \frac{\|P_{\mathcal{S}}y\|}{\|y\|}, \tag{6.78}$$

for any $y \in \mathcal{A}$. Choosing $y = E_{\mathcal{AS}^\perp}x$ for an arbitrary $x \in \mathcal{H}$ leads to

$$\|E_{\mathcal{AS}^\perp}x\| \leq \frac{\|P_{\mathcal{S}}E_{\mathcal{AS}^\perp}x\|}{\cos(\mathcal{A}, \mathcal{S})}. \tag{6.79}$$

By substituting the algebraic expressions for $P_{\mathcal{S}}$ and $E_{\mathcal{AS}^\perp}$ we can immediately verify that $P_{\mathcal{S}}E_{\mathcal{AS}^\perp} = P_{\mathcal{S}}$. Combining this observation with the fact that $\|P_{\mathcal{S}}x\| \leq \|x\|$ we have

$$\|E_{\mathcal{AS}^\perp}x\| \leq \frac{\|P_{\mathcal{S}}x\|}{\cos(\mathcal{A}, \mathcal{S})} \leq \frac{\|x\|}{\cos(\mathcal{A}, \mathcal{S})}, \tag{6.80}$$

completing the proof. $\square$

6.5   Nonunique recovery

We now turn to treat the case in which $\mathcal{A}$ and $\mathcal{S}^{\perp}$ intersect, implying that there is more than one signal that matches the given samples.

This setting raises a fundamental problem of how to best estimate an unknown signal from a given set of measurements. In the previous settings we treated there was only one signal consistent with the samples, namely the original input. Therefore, estimation techniques were not needed. However, here we have to choose among several possibilities and therefore some criterion for selection is required. In Section 6.1.5 we introduced two design criteria that can be applied here, and in additional problems in which there is no unique recovery from the given samples (such as in noisy settings): LS and minimax. Interestingly, we will see that in the case we treat here, the two solutions coincide. However, in Section 6.6 we show that these strategies lead to quite different reconstruction methods when the recovery process is constrained. In the next chapter we will see that the results also differ under a smoothness prior.

6.5.1   Least squares recovery

We begin by considering the LS formulation (6.11). Since there are now many solutions for which $S^*x = c$, the minimal value of the LS objective is 0. Therefore, from all possible consistent solutions $S^*x = c$ we seek the one with minimal L_2-norm, as suggested in (6.12). Taking into account the prior knowledge that x is in $\mathcal{A}$, the LS recovery problem can be written as

$$\hat{x}_{\text{LS}} = \arg \min_{x \in \mathcal{A}, S^*x = c} \|x\|^2. \tag{6.81}$$

Expressing x in terms of its expansion coefficients in an orthonormal basis $\{a_n\}$ for $\mathcal{A}$ as $x = \sum d[n]a_n = Ad$, the optimal sequence d is the solution to

$$\hat{d}_{\text{LS}} = \arg \min_{S^*Ad = c} \|d\|^2. \tag{6.82}$$

Here we used the fact that since A corresponds to an orthonormal basis, $A^*A = I$, and

$$\|x\|^2 = \langle Ad, Ad \rangle = \langle d, A^*Ad \rangle = \|d\|^2. \tag{6.83}$$

To solve (6.82) we first characterize all possible values of d that are consistent with the samples, namely such that $S^*Ad = c$:

Theorem 6.2 (Consistent solutions).   *The possible sequences d for which $S^*Ad = c$ are given by*

$$\hat{d} = (S^*A)^{\dagger}c + v, \tag{6.84}$$

where v is an arbitrary vector in $\mathcal{N}(S^*A)$.

Before proving the theorem, we verify that the pseudoinverse is well defined (namely, bounded). If $\mathcal{S}$ and $\mathcal{A}$ have finite dimensions, say M and N respectively, then S^*A corresponds to an $M \times N$ matrix and $(S^*A)^{\dagger}$ is trivially a bounded operator. However, this is not necessarily true for infinite-dimensional operators. Fortunately, the fact that S and A are set transformations corresponding to Riesz bases guarantees that $(S^*A)^\dagger$ is bounded, as stated in the next proposition.

Proposition 6.9. *Let S and A be set transformations corresponding to Riesz bases $\{s_n\}$ and $\{a_n\}$ respectively. Then $(S^*A)^\dagger$ is a bounded operator.*

Proof: The proof of the proposition relies on the fact that the pseudoinverse of an operator is bounded if its range is closed. We therefore need to show that the range of $T = S^*A$ is closed. Now, the pseudoinverse is the inverse of T restricted to $\mathcal{N}(T)^\perp$. Since A and S correspond to Riesz bases, T is bounded and therefore has a closed range if it is bounded below (see Proposition 2.6). We now invoke once more the fact that S and A correspond to Riesz bases to conclude that S^* and A are both bounded below on $\mathcal{N}(T)^\perp$, from which it follows that T is bounded below. $\square$

We now prove Theorem 6.2.

Proof of Theorem 6.2:   To see that the set of solutions to $S^*Ad = c$ is given by (6.84), we compute

$$S^*A\hat{d} = (S^*A)((S^*A)^\dagger c + v) = (S^*A)(S^*A)^\dagger c = P_{\mathcal{R}(S^*A)}c = c. \qquad (6.85)$$

The second equality follows from the fact that $v \in \mathcal{N}(S^*A)$, the third equality follows from the properties of the pseudoinverse, and the last equality is a result of $c \in \mathcal{R}(S^*A)$ since $c = S^*x$ for some $x \in \mathcal{A}$. Therefore, $S^*Ad = c$ for any vector described by (6.84). It is also clear from (6.85) that vectors of the form $\hat{d} + w$, where $\hat{d}$ is given by (6.84) and $w \in \mathcal{N}(S^*A)^\perp$, do not satisfy $S^*Ad = c$, completing the proof. $\sqcap$

An immediate corollary to Theorem 6.2 is that the minimal-norm value of d, namely the solution to (6.82), corresponds to $v = 0$.

Corollary 6.2.   *The solution of (6.82) is given by*

$$\hat{d}_{LS} = (S^*A)^\dagger c. \qquad (6.86)$$

Proof: The proof follows by noting that $\mathcal{R}\left((S^*A)^\dagger\right) = \mathcal{N}(S^*A)^\perp$. Since $v \in \mathcal{N}(S^*A)$, for every d of the form (6.84) it holds that

$$\|d\|^2 = \|(S^*A)^\dagger c\|^2 + \|v\|^2 \geq \|(S^*A)^\dagger c\|^2, \qquad (6.87)$$

with equality if and only if $v = 0$. $\square$

We conclude from Corollary 6.2 that LS recovery with minimal-norm amounts to applying

$$H = (S^*A)^\dagger \qquad (6.88)$$

on the samples c, resulting in the expansion coefficients $d = Hc$. This sequence is then used to synthesize $\hat{x}$ via $\hat{x} = \sum_n d[n]a_n$. Thus, $\hat{x}$ is related to x by

$$\hat{x}_{LS} = Ad = AHc = A(S^*A)^\dagger c = A(S^*A)^\dagger S^*x. \qquad (6.89)$$

If the direct-sum condition holds, then $S^* A$ is invertible by Lemma 2.2, and the solution of (6.88) reduces to that obtained in Section 6.3.2.

When $\mathcal{A}$ and $\mathcal{S}$ are SI spaces with generators $a(t)$ and $s(t)$ respectively, the operator $S^* A$ corresponds to convolution with the sequence $r_{sa}[n]$. Therefore, $H = (S^* A)^\dagger$ is a digital filter with frequency response

$$H(e^{j\omega}) = \begin{cases} \frac{1}{R_{SA}(e^{j\omega})}, & R_{SA}\left(e^{j\omega}\right) \neq 0 \\ 0, & R_{SA}\left(e^{j\omega}\right) = 0. \end{cases} \tag{6.90}$$

This solution coincides with (6.33) for frequencies on which $R_{SA}(e^{j\omega})$ is not zero. On frequencies corresponding to $R_{SA}(e^{j\omega}) = 0$, the samples convey no information about the original signal. This follows from the fact that if $x(t) = \sum_n d[n]a(t - nT)$, then the DTFT of the samples is given by $C(e^{j\omega}) = D(e^{j\omega})R_{SA}(e^{j\omega})$, so that $C(e^{j\omega}) = 0$ when $R_{SA}(e^{j\omega}) = 0$.

6.5.2   Minimax recovery

We now treat the recovery of x via a minimax framework:

$$\hat{x}_{MX} = \arg \min_{\hat{x}} \max_{x \in \mathcal{G}} \|\hat{x} - x\|^2, \tag{6.91}$$

where $\mathcal{G}$ is the set of signals $x \in \mathcal{A}$ satisfying $S^* x = c$.

To approach this problem, note that $\hat{x}_{MX}$ must lie in $\mathcal{A}$, as any $\hat{x} \notin \mathcal{A}$ can be improved upon by projecting it onto $\mathcal{A}$: $\|\hat{x} - x\|^2 \geq \|P_A\hat{x} - x\|^2$ for any $\hat{x}$ and $x \in \mathcal{A}$. Therefore, we can express both $\hat{x}$ and x in terms of their expansion coefficients in $\mathcal{A}$, by writing $\hat{x} = A\hat{d}$ and $x = Ad$ for some Riesz basis A. To guarantee that the error in (6.91) is finite, the sequence d must be constrained to a bounded set. We therefore impose the additional requirement that $\|d\| \leq \rho$, for some $\rho > 0$. Problem (6.91) can then be reformulated as

$$\min_{\hat{d}} \max_{d \in \mathcal{D}} \|A\hat{d} - Ad\|^2, \tag{6.92}$$

where $\mathcal{D} = \{d: S^* Ad = c, \|d\| \leq \rho\}$. As we now show, the choice of ρ does not affect the solution, as long as $\mathcal{D}$ is a nonempty set.

Theorem 6.3 (Minimax solution). *The solution of (6.92) is $\hat{d} = (S^* A)^\dagger c$. The minimax recovery is then given by $\hat{x}_{MX} = A(S^* A)^\dagger c$.*

Note that the minimax solution coincides with the LS recovery (6.89).

Proof: As we have seen in Theorem 6.2, when $c \in \mathcal{R}(S^* A)$, a sequence d satisfies $S^* Ad = c$ if and only if it is of the form $d = (S^* A)^\dagger c + v$, where v is in $\mathcal{N}(S^* A)$. Furthermore, $(S^* A)^\dagger c \in \mathcal{N}(S^* A)^\perp$ so that $\|v\|^2 = \|d\|^2 - \|(S^* A)^\dagger c\|^2$. Therefore, the inner maximization in (6.92) becomes

$$\|A(\hat{d} - (S^* A)^\dagger c)\|^2 + \max_{v \in \mathcal{V}}\{\|Av\|^2 - 2v^* A^* A(\hat{d} - (S^* A)^\dagger c)\}, \tag{6.93}$$

where

$$V = \{v : v \in \mathcal{N}(S^*A), \|v\|^2 \le \rho^2 - \|(S^*A)^\dagger c\|^2\}. \qquad (6.94)$$

Since V is a symmetric set, the vector v attaining the maximum in (6.93) must satisfy $v^*A^*A(\hat{d} - (S^*A)^\dagger c) \le 0$; otherwise we can change the sign of v without affecting the constraints. Therefore,

$$\max_{v \in V}\{\|Av\|^2 - 2v^*A^*A(\hat{d} - (S^*A)^\dagger c)\} \ge \max_{v \in V}\|Av\|^2. \qquad (6.95)$$

Combining (6.95) and (6.93) we have that

$$\min_{\hat{d}} \max_{d \in \mathcal{D}} \|A\hat{d} - Ad\|^2 \ge \min_{\hat{d}}\{\|A(\hat{d} - (S^*A)^\dagger c)\|^2 + \max_{v \in V}\|Av\|^2\}$$
$$= \max_{v \in V}\|Av\|^2, \qquad (6.96)$$

where the equality is a result of solving the minimization, which is obtained at $\hat{d} = (S^*A)^\dagger c$.

We now show that the inequality in (6.96) can be achieved with $\hat{d} = (S^*A)^\dagger c$. Indeed, substituting this choice of $\hat{d}$ in (6.93), we have that

$$\max_{d \in \mathcal{D}} \|A\hat{d} - Ad\|^2 = \max_{v \in V}\{\|Av\|^2 - 2v^*A(\hat{d} - (S^*A)^\dagger c)\} = \max_{v \in V}\|Av\|^2. \qquad (6.97)$$

Finally, we note that the solution is unique since the objective function is strictly convex, concluding the proof. $\qquad\square$

We illustrate the minimax and LS recoveries in the following example.

Example 6.12 Consider the signal

$$x(t) = \sum_{n \in \mathbb{Z}} d[n]a(t - n), \qquad (6.98)$$

where the CTFT of $a(t)$ satisfies $A(\omega) > \alpha > 0$ for all $\omega \in [-\pi, \pi]$. Assume that $x(t)$ passes through the filter $s(t) = (1/T_0)\,\mathrm{sinc}(t/T_0)$, with $T_0 > 1$, and is then sampled at the integers. Our goal is to recover $x(t)$ from its samples $c[n]$.

In this example, the frequency content of $x(t)$ outside $[-\pi/T_0, \pi/T_0]$ is zeroed out. Consequently, there are infinitely many sequences $d[n]$ that agree with the samples $c[n]$. Indeed, $C(e^{j\omega}) = D(e^{j\omega})R_{SA}(e^{j\omega})$ where in our setting

$$R_{SA}(e^{j\omega}) = \begin{cases} A(\omega), & |\omega| \le \frac{\pi}{T_0} \\ 0, & \frac{\pi}{T_0} < |\omega| \le \pi. \end{cases} \qquad (6.99)$$

Therefore, the samples are not affected by the content of $D(e^{j\omega})$ in the range $\pi/T_0 < |\omega| \le \pi$.

The LS and minimax recoveries correspond to using the expansion coefficients $\hat{d}[n]$ obtained by passing the samples $c[n]$ through the filter

$$H(e^{j\omega}) = \begin{cases} \frac{1}{A(\omega)}, & |\omega| \le \frac{\pi}{T_0} \\ 0, & \frac{\pi}{T_0} < |\omega| \le \pi. \end{cases} \qquad (6.100)$$

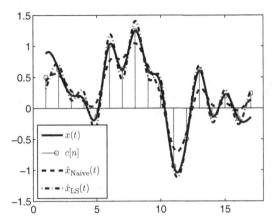

Figure 6.17 A spline of degree 3, its samples $c[n]$ after going through the filter $s(t) = (1/4) \cdot$ sinc$(t/4)$ and the recoveries obtained with the correction filter (6.100) and without it.

Figure 6.17 compares the recovery obtained by this method with the naive approach in which the samples are not processed prior to reconstruction, i.e. $\hat{d}[n] = c[n]$. In this figure, $a(t) = \beta^3(t)$ and $T_0 = 4$.

6.6 Constrained recovery

Until now we have specified the sampling process but did not restrict the reconstruction or interpolation kernel $w(t)$ in Fig. 6.2. For a sampling theorem to be practical, it must also take into account constraints that are imposed on the interpolation method. One aspect of the Shannon–Nyquist sampling theorem, which makes it difficult to implement, is the use of the sinc interpolation kernel. Because of its slow decay, the evaluation of $x(t)$ at a certain time instant t_0 requires using a large number of samples located far away from t_0. In many applications, reduction of computational load is achieved by employing much simpler methods, such as linear interpolation. In image processing applications, kernels with small supports are often used. These include nearest-neighbor, bilinear, bicubic, Lanczos and splines (see Examples 6.3 and 6.4). The kernel $w(t)$ can also represent the pixel shape of an image display. When $w(t)$ is fixed in advance, the filter $h[n]$ should be modified to compensate for the chosen nonideal kernel.

Therefore, given a sampling function $s(-t)$ and a fixed interpolation kernel $w(t)$, we now address the question of how to design the digital filter $h[n]$ in Fig. 6.2 so that the output $\hat{x}(t) = \sum_n d[n]w(t - nT)$ is a good approximation of the input signal $x(t)$ in some sense. In order to obtain a stable reconstruction, we concentrate on cases in which $w(t)$ satisfies the Riesz basis condition (5.14).

Example 6.13 As a motivating toy example, suppose that $x(t)$ is a spline of degree 1. Our goal is to recover $x(t)$ from its pointwise samples using a simple zero-order

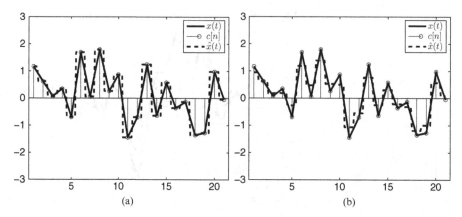

Figure 6.18 Reconstruction of a degree-1 spline from its pointwise samples by zero-order hold. (a) Using the system of Fig. 6.9 with $a(t) = \beta^0(t)$ and $h[n]$ of (6.33). (b) Using the scheme of Fig. 6.20 with $w(t) = \beta^0(t)$ and $a(t) = \beta^1(t)$ (see below).

hold, corresponding to the reconstruction filter $\beta^0(t)$. With this choice, the recovery $\hat{x}(t)$ is a spline of degree 0, and thus cannot equal $x(t)$.

A naive approach to deal with this situation is to use the scheme of Fig. 6.9 with $s(t) = \delta(t)$ and $a(t) = \beta^0(t)$. Since the recovery is in the space spanned by $\beta^0(t)$, we use this function as the generator of our prior space. The resulting approximation with $h[n]$ of (6.33) is depicted in Fig. 6.18(a). The problem with this method, however, is that it disregards our knowledge about $x(t)$. As we will see below, by taking into account that $x(t)$ is a degree-1 spline, we can obtain the reconstruction in Fig. 6.18(b), which is closer to $x(t)$ in terms of squared error.

6.6.1   Minimal-error recovery

In Section 6.3.2 we showed that under a direct-sum condition, perfect recovery of x is possible from its samples in $\mathcal{S}$. However, once we constrain the reconstruction mechanism, we will no longer be able to recover the signal. This is because for every choice of the sequence $d[n]$, $\hat{x}(t) = \sum_n d[n]w(t - nT)$ will lie in the space $\mathcal{W}$, spanned by the generator $w(t)$. If $x(t)$ does not lie in $\mathcal{W}$ to begin with, then $\hat{x}(t)$ cannot be equal to $x(t)$. Instead, we may attempt to choose $h[n]$ so as to yield a recovery $\hat{x}(t)$ that is closest in an L_2-norm sense to $x(t)$; we refer to such a solution as minimal-error recovery.

Since $\hat{x}(t)$ is constrained to lie in $\mathcal{W}$, it follows from the projection theorem (Proposition 2.15) that the minimal-error approximation to $x(t)$ is obtained when $\hat{x}(t) = P_{\mathcal{W}}x(t)$. The question is whether this solution can be generated from the samples $c[n]$. Carrying over the discussion to arbitrary Hilbert spaces, suppose we are given samples $c = S^*x$ of x and we constrain the reconstruction to $\mathcal{W}$. Therefore, $\hat{x} = Wd$ for a set transformation W corresponding to a Riesz basis for $\mathcal{W}$, and $d = Hc$ for some linear transformation H. We would like to choose H to minimize the squared error

$$\min_{H} \|\hat{x} - x\|^2 = \min_{H} \|WHS^*x - x\|^2. \tag{6.101}$$

From the projection theorem, the error is minimized when $WHS^*x = P_{\mathcal{W}}x$. The question is whether there exists an H such that $WHS^*x = P_{\mathcal{W}}x$ for all x in $\mathcal{H}$. In general, the answer is negative without sufficient prior knowledge on the signal, as incorporated in the following proposition.

Proposition 6.10. *Let $H: \ell_2 \to \ell_2$ be an arbitrary linear transformation, and let W and S be bounded transformations with $\mathcal{R}(W) = \mathcal{W}$, $\mathcal{R}(S) = \mathcal{S}$, and $\mathcal{W} \not\subseteq \mathcal{S}$. Then we cannot choose H such that $WHS^*x = P_{\mathcal{W}}x$ for all x.*

Proof: To prove the proposition, suppose to the contrary that there exists an H such that $WHS^*x = P_{\mathcal{W}}x$ for all x. Consider a signal x constructed as $x = x_{\mathcal{S}^\perp} + x_{\mathcal{W}}$ where $x_{\mathcal{S}^\perp}$ is in $\mathcal{S}^\perp$ but not in $\mathcal{W}^\perp$ (such a vector always exists since $\mathcal{W} \not\subseteq \mathcal{S}$) and $x_{\mathcal{W}} \in \mathcal{W}$. For this choice, $S^*x = S^*x_{\mathcal{W}}$ so that

$$WHS^*x = WHS^*x_{\mathcal{W}}. \tag{6.102}$$

On the other hand, due to our assumption, $WHS^*x = P_{\mathcal{W}}x$ and $WHS^*x_{\mathcal{W}} = P_{\mathcal{W}}x_{\mathcal{W}} = x_{\mathcal{W}}$. Substituting $x = x_{\mathcal{S}^\perp} + x_{\mathcal{W}}$ it follows that $P_{\mathcal{W}}x = P_{\mathcal{W}}x_{\mathcal{S}^\perp} + x_{\mathcal{W}}$. Consequently, to satisfy (6.102) we must have $P_{\mathcal{W}}x_{\mathcal{S}^\perp} + x_{\mathcal{W}} = x_{\mathcal{W}}$, or $P_{\mathcal{W}}x_{\mathcal{S}^\perp} = 0$. This implies that $x_{\mathcal{S}^\perp} \in \mathcal{W}^\perp$, contradicting our assumption. $\square$

Note that the proposition considers subspaces $\mathcal{W} \not\subseteq \mathcal{S}$. In the special case in which $\mathcal{W} \subseteq \mathcal{S}$, the minimal-error reconstruction can be achieved with

$$H = (W^*W)^{-1}W^*S(S^*S)^{-1}. \tag{6.103}$$

Indeed, with this choice of H,

$$\hat{x} = W(W^*W)^{-1}W^*S(S^*S)^{-1}S^*x = P_{\mathcal{W}}P_{\mathcal{S}}x = P_{\mathcal{W}}x, \tag{6.104}$$

where the last equality follows from the fact that $\mathcal{W} \subseteq \mathcal{S}$.

Proposition 6.10 establishes that when $\mathcal{W} \not\subseteq \mathcal{S}$ the minimal error cannot be achieved *over all* x. However, as we now show, when x lies in a subspace $\mathcal{A}$ satisfying (6.23), $P_{\mathcal{W}}x$ is attainable. In the next chapter we consider constrained recovery without subspace knowledge on x using the LS and minimax criteria.

Theorem 6.4 (Minimal error recovery). *Consider the problem*

$$\min_{H} \|\hat{x} - x\|^2 = \min_{H} \|WHS^*x - x\|^2, \quad \text{for } x \in \mathcal{A}$$

where $\mathcal{A} \subseteq \mathcal{H}$ is a closed subspace such that $\mathcal{H} = \mathcal{A} \oplus \mathcal{S}^\perp$ and W, S are set transformations corresponding to $\mathcal{W}, \mathcal{S}$ respectively. The solution is given by

$$H_{\mathcal{A}} = (W^*W)^{-1}W^*A(S^*A)^{-1}, \tag{6.105}$$

where A is a set transformation corresponding to a Riesz basis for $\mathcal{A}$. With this choice, $\hat{x}$ is the minimal-error solution $\hat{x} = P_{\mathcal{W}}x$.

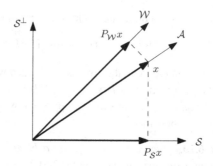

Figure 6.19 A signal $x(t) \in \mathcal{A}$ can be recovered from its samples in $\mathcal{S}$, allowing computation of its orthogonal projection onto $\mathcal{W}$.

Before proving the theorem, we note that from Lemma 2.2 the operator $(S^*A)^{-1}$ is well defined. Furthermore, it is easy to see that $A(S^*A)^{-1}$ is independent of the choice of Riesz basis A, since any other Riesz basis can be expressed as $\tilde{A} = AP$ for some invertible operator $P\colon \ell_2 \to \ell_2$.

Proof: We begin by noting that since $x \in \mathcal{A}$, it can be reconstructed exactly from the samples c as $x = A(S^*A)^{-1}c$. Once we know x, the approximation in $\mathcal{W}$ minimizing the squared error is

$$\hat{x} = P_{\mathcal{W}}x = P_{\mathcal{W}}A(S^*A)^{-1}c = WH_Ac. \tag{6.106}$$

Finally, since $c = S^*x$,

$$\hat{x} = P_{\mathcal{W}}A(S^*A)^{-1}S^*x = P_{\mathcal{W}}E_{\mathcal{AS}^\perp}x = P_{\mathcal{W}}x, \tag{6.107}$$

where we used the fact that for $x \in \mathcal{A}$, we have $E_{\mathcal{AS}^\perp}x = x$. □

To understand Theorem 6.4 geometrically, recall that under the direct-sum condition, any vector $x \in \mathcal{A}$ can be recovered from its samples in $\mathcal{S}$, as illustrated in Fig. 6.5. Here, however, we are constrained to obtain a solution in $\mathcal{W}$. But, once we recover x, we can readily compute $P_{\mathcal{W}}x$, which is the minimal-error approximation in $\mathcal{W}$. This is shown in Fig. 6.19.

When $\mathcal{A}, \mathcal{S}$, and $\mathcal{W}$ are SI subspaces, it follows from (6.105) that $P_{\mathcal{W}}x(t)$ is obtained by filtering the sequence of samples with

$$H_A\left(e^{j\omega}\right) = \frac{R_{WA}\left(e^{j\omega}\right)}{R_{SA}\left(e^{j\omega}\right)R_{WW}\left(e^{j\omega}\right)}, \tag{6.108}$$

where $R_{WA}\left(e^{j\omega}\right)$, $R_{SA}\left(e^{j\omega}\right)$, and $R_{WW}\left(e^{j\omega}\right)$ are defined in (5.28) with the corresponding substitution of the filters $W(\omega)$, $A(\omega)$, and $S(\omega)$. The resulting sampling and recovery scheme is depicted in Fig. 6.20.

Let us now revisit Example 6.13. In this setting $x(t)$ is a degree-1 spline, $s(t) = \delta(t)$, and $w(t) = \beta^0(t)$. Figure 6.18(b) depicts the recovery obtained using the scheme of Fig. 6.20 with $a(t) = \beta^1(t)$. As can be seen, this technique is advantageous over the recovery of Fig. 6.18(a), which does not take into account the subspace prior $x(t) \in \mathcal{A}$.

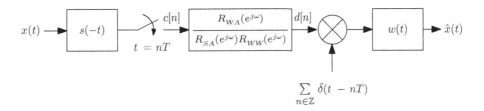

Figure 6.20 Reconstruction in $\mathcal{W}$ of a signal $x(t) \in \mathcal{A}$.

To further demonstrate the constrained recovery procedure, we consider the setting of Example 6.9, and impose a constraint on the reconstruction mechanism.

Example 6.14 Suppose that the signal $x(t)$ of (6.39) is sampled at the integers after passing through the anti-aliasing filter $s(t) = \mathrm{sinc}(t)$, as in Example 6.9. We would now like to recover $x(t)$ from the samples $c[n]$ using zero-order hold. The corresponding reconstruction filter is therefore $w(t) = u(t) - u(t-1)$, where $u(t)$ is the unit step function.

To compute the digital correction filter (6.108), we note that $R_{WW}(e^{j\omega}) = 1$. Furthermore, the filter $1/R_{SA}(e^{j\omega})$ is given by (6.40), as we have already seen in Example 6.9. The remaining term, $R_{WA}(e^{j\omega})$, corresponds to the filter

$$ h_{WA}[n] = \begin{cases} e^{\frac{n}{\tau}}\left(1 - e^{-\frac{1}{\tau}}\right), & n \le 0 \\ 0, & n > 0. \end{cases} \tag{6.109} $$

Therefore, the sequence $d[n]$ feeding the zero-order hold is obtained by convolving the samples $c[n]$ with $h[n]$ of (6.40) and then by $h_{WA}[n]$ of (6.109). Figure 6.21 compares the reconstruction strategy described above with that obtained without using a correction filter. As can be seen, the minimal-error recovery is much more loyal to the input signal in terms of squared error.

We next treat the setting in which the direct-sum condition does not hold. Since the minimal-error solution can no longer be obtained in general, we develop alternatives based on LS and minimax strategies.

6.6.2 Least squares recovery

To obtain a reconstruction in $\mathcal{W}$ within the LS methodology we reformulate (6.11) as

$$ \hat{x}_{\mathrm{CLS}} = \arg \min_{x \in \mathcal{W}} \left\| S^* x - c \right\|^2 \tag{6.110} $$

where CLS stands for constrained least squares. The objective in (6.110) constrains the solution to be in $\mathcal{W}$, but it ignores our prior knowledge that $x \in \mathcal{A}$. A drawback of

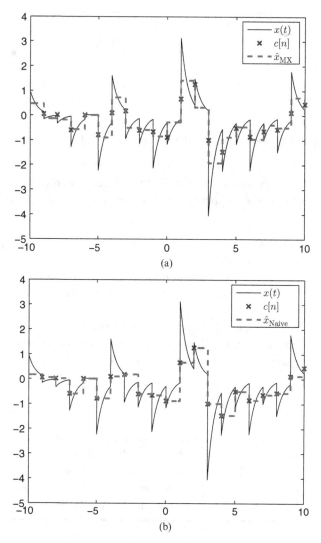

Figure 6.21 Reconstruction of $x(t)$ of (6.39) from its samples $c[n] = (x(t) * \mathrm{sinc}(t))|_{t=n}$ using a zero-order hold filter. (a) Minimal-error recovery. (b) Recovery without digital correction.

the LS criterion is that it cannot simultaneously account for both prior and recovery constraints. As we will see, this results in inferior performance relative to the minimax solution.

An important observation in our setting is that the samples c do not necessarily lie in $\mathcal{R}(S^*W)$, since the true x is in $\mathcal{A}$. Therefore, there may not exist an $x \in \mathcal{W}$ giving rise to the measured samples c, so that the minimal error in (6.110) is generally not equal to zero. The next theorem examines the minimal objective value and the set of signals in $\mathcal{W}$ which attain it.

Theorem 6.5 (CLS solutions). *Let $\hat{x} = W\hat{d}$ be a vector in $\mathcal{W}$ minimizing (6.110). Then $\hat{d}$ has the form*

$$\hat{d} = (S^*W)^\dagger c + v, \tag{6.111}$$

where v is an arbitrary vector in $\mathcal{N}(S^*W)$.

Proof: Write $x \in \mathcal{W}$ as $x = Wd$ and let $\hat{d}$ be a minimizer of (6.110). Denote by $\hat{c} = S^*W\hat{d}$ the samples it produces. Then, (6.110) can be written as

$$\min_{\hat{c} \in \mathcal{R}(S^*W)} \|\hat{c} - c\|^2. \tag{6.112}$$

Clearly, the optimal $\hat{c}$ is the projection of c onto $\mathcal{R}(S^*W)$:

$$\hat{c} = S^*W\hat{d} = P_{\mathcal{R}(S^*W)}c = (S^*W)(S^*W)^\dagger c. \tag{6.113}$$

Therefore, all possible solutions to (6.110) have the form

$$\hat{d} = (S^*W)^\dagger c + v, \tag{6.114}$$

where $v \in \mathcal{N}(S^*W)$. $\qquad\qquad\square$

Noting that $\mathcal{R}((S^*W)^\dagger)^\perp = \mathcal{N}(S^*W)$, for any optimal value of $\hat{d}$ we have $\|\hat{d}\| \geq \|(S^*W)^\dagger c\|$ which leads to the following corollary.

Corollary 6.3. *The minimal-norm solution to (6.110) is $\hat{x}_{CLS} = W(S^*W)^\dagger c$. This solution can be obtained by applying $H = (S^*W)^\dagger$ on the samples c prior to recovery with W.*

The solution of Corollary 6.3 has the same structure as the unconstrained LS reconstruction (6.89) with A replaced by W. We can therefore interpret this recovery as the one that minimizes the LS error assuming that x lies in $\mathcal{W}$.

In the SI setting, the CLS correction $H = (S^*W)^\dagger$ is a digital filter with frequency response

$$H(e^{j\omega}) = \begin{cases} \frac{1}{R_{SW}(e^{j\omega})}, & R_{SW}(e^{j\omega}) \neq 0; \\ 0, & R_{SW}(e^{j\omega}) = 0. \end{cases} \tag{6.115}$$

Therefore, recovery is performed by the block diagram of Fig. 6.2, where the reconstruction kernel is $w(t)$ and the digital correction filter is given by (6.115).

It is interesting to study the relation between the unconstrained and constrained solutions. By construction, $\hat{x}_{LS}$ of (6.89) is consistent, namely $S^*\hat{x}_{LS} = c$. Therefore, $\hat{x}_{CLS}$ can be expressed in terms of $\hat{x}_{LS}$ as:

$$\hat{x}_{CLS} = W(S^*W)^\dagger c = W(S^*W)^\dagger S^*\hat{x}_{LS}. \tag{6.116}$$

The geometric meaning of this relation is best understood when $\mathcal{H} = \mathcal{W} \oplus \mathcal{S}^\perp$. In this case S^*W is invertible (cf. Lemma 2.2) so that $(S^*W)^\dagger = (S^*W)^{-1}$. From Theorem 2.2 it then follows that $\hat{x}_{CLS}$ is the oblique projection of $\hat{x}_{LS}$ onto $\mathcal{W}$ along $\mathcal{S}^\perp$:

$$\hat{x}_{CLS} = E_{\mathcal{W}\mathcal{S}^\perp}\hat{x}_{LS}. \tag{6.117}$$

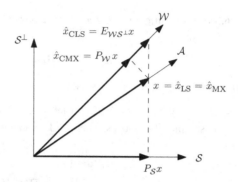

Figure 6.22 Comparison between the CLS and minimax solutions when $\mathcal{H} = \mathcal{A} \oplus \mathcal{S}^\perp$. In this case, $x = \hat{x}_{LS} = \hat{x}_{MX}$ and thus the signal $x \in \mathcal{A}$ can be recovered from its samples $c[n]$, allowing to compute its orthogonal projection onto $\mathcal{W}$. The constrained minimax approach yields $\hat{x}_{CMX} = P_{\mathcal{W}}x$ whereas the CLS criterion leads to $\hat{x}_{CLS} = E_{\mathcal{W}\mathcal{S}^\perp}x$ (assuming in addition that $\mathcal{H} = \mathcal{W} \oplus \mathcal{S}^\perp$).

Figure 6.22 depicts $\hat{x}_{LS}$ and $\hat{x}_{CLS}$ in a situation in which $\mathcal{A}$ and $\mathcal{S}^\perp$ satisfy the direct-sum condition (6.23) so that $\hat{x}_{LS} = \hat{x}_{MX} = x$, and also $\mathcal{H} = \mathcal{W} \oplus \mathcal{S}^\perp$ implying that (6.117) holds. This example highlights the disadvantage of the LS formulation. In this setting we are constrained to yield $\hat{x} \in \mathcal{W}$. But since x can be determined from the samples c, so can its best approximation in $\mathcal{W}$, given by $P_{\mathcal{W}}x = P_{\mathcal{W}}\hat{x}_{LS}$. This alternative is also shown in Fig. 6.22, and is clearly advantageous to $\hat{x}_{CLS}$ in terms of squared error *for every* x. We will see in Section 6.6.3 that orthogonally projecting $\hat{x}_{LS}$ onto the reconstruction space $\mathcal{W}$ can be motivated even when (6.23) does not hold, via a minimax formulation.

6.6.3   Minimax recovery

We now treat the constrained recovery setting via a worst-case design strategy. The constraint $\hat{x} \in \mathcal{W}$ leads to an inherent limitation on the minimal achievable reconstruction error: the best approximation in $\mathcal{W}$ of any signal x is given by $\hat{x} = P_{\mathcal{W}}x$, which in general cannot be computed from the sequence of samples $c[n]$ (see Proposition 6.10). Therefore, we consider here the minimization of the *regret*, which is defined by $\|\hat{x} - P_{\mathcal{W}}x\|^2$ [138, 139, 140]. The regret quantifies how much we lose with respect to the best possible (but unachievable) recovery in $\mathcal{W}$, given by $P_{\mathcal{W}}x$. Since the regret is a function of the unknown signal x, we seek the reconstruction $\hat{x} \in \mathcal{W}$ minimizing the worst-case regret. Our problem is therefore

$$\hat{x}_{CMX} = \min_{\hat{x} \in \mathcal{W}} \max_{x \in \mathcal{G}} \|\hat{x} - P_{\mathcal{W}}x\|^2, \tag{6.118}$$

where $\mathcal{G}$ is the set of signals $x \in \mathcal{A}$ satisfying $S^*x = c$. In contrast to the LS objective, the minimax formulation can take both prior and recovery constraints into account, since the optimization is over two different variables: x and $\hat{x}$. This often results in superior performance of the minimax solution.

To solve (6.118) we express $\hat{x}$ and x in terms of their expansion coefficients in $\mathcal{W}$ and $\mathcal{A}$ respectively, by writing $\hat{x} = W\hat{d}$ and $x = Ad$. As in the unconstrained setting, we require that $\|d\| \leq \rho$ for some $\rho > 0$, in order for the inner maximization to be bounded. Problem (6.118) can then be written as

$$\min_{\hat{d}} \max_{d \in \mathcal{D}} \|W\hat{d} - P_{\mathcal{W}}Ad\|^2, \tag{6.119}$$

where $\mathcal{D} = \{d \colon S^*Ad = c, \|d\| \leq \rho\}$.

Theorem 6.6 (Constrained minimax solution).   *The solution of (6.119) is* $\hat{d} = (W^*W)^{-1}W^*A(S^*A)^\dagger c$. *The constrained minimax recovery is then given by* $\hat{x}_{\text{CMX}} = P_{\mathcal{W}}A(S^*A)^\dagger c$.

Proof: Following the proof of Theorem 6.2, the set $\mathcal{D}$ consists of all sequences of the form $d = (S^*A)^\dagger c + v$, where v is a vector in $\mathcal{N}(S^*A)$, with $\|v\|^2 = \|d\|^2 - \|(S^*A)^\dagger c\|^2$. Therefore, the inner maximization in (6.119) becomes

$$\|W\hat{d} - P_{\mathcal{W}}A(S^*A)^\dagger c\|^2 + \max_{v \in \mathcal{V}}\{\|P_{\mathcal{W}}Av\|^2 - 2(P_{\mathcal{W}}Av)^*(W\hat{d} - P_{\mathcal{W}}A(S^*A)^\dagger c)\}, \tag{6.120}$$

where $\mathcal{V}$ is given by (6.94). Since $\mathcal{V}$ is a symmetric set, the vector v attaining the maximum in (6.120) must satisfy $(P_{\mathcal{W}}Av)^*(W\hat{d} - P_{\mathcal{W}}(S^*A)^\dagger c) \leq 0$, as we can change the sign of v without affecting the constraints. Consequently,

$$\max_{v \in \mathcal{V}}\{\|P_{\mathcal{W}}Av\|^2 - 2(P_{\mathcal{W}}Av)^*(W\hat{d} - P_{\mathcal{W}}A(S^*A)^\dagger c)\} \geq \max_{v \in \mathcal{V}}\|P_{\mathcal{W}}Av\|^2. \tag{6.121}$$

Combining (6.121) and (6.120),

$$\min_{\hat{d}} \max_{d \in \mathcal{D}} \|W\hat{d} - P_{\mathcal{W}}Ad\|^2 \geq \min_{\hat{d}}\{\|W\hat{d} - P_{\mathcal{W}}A(S^*A)^\dagger c\|^2 + \max_{v \in \mathcal{V}}\|P_{\mathcal{W}}Av\|^2\}$$

$$= \max_{v \in \mathcal{V}}\|P_{\mathcal{W}}Av\|^2, \tag{6.122}$$

where the equality is a result of solving the minimization. The inner minimum is obtained at

$$\hat{d} = (W^*W)^{-1}W^*A(S^*A)^\dagger c. \tag{6.123}$$

We now show that the inequality in (6.122) is achieved with $\hat{d}$ given by (6.123). Substituting this solution into (6.120), we have that

$$\max_{d \in \mathcal{D}} \|W\hat{d} - P_{\mathcal{W}}Ad\|^2 = \max_{v \in \mathcal{V}}\{\|P_{\mathcal{W}}Av\|^2 - 2(P_{\mathcal{W}}Av)^*(W\hat{d} - P_{\mathcal{W}}A(S^*A)^\dagger c)\}$$

$$= \max_{v \in \mathcal{V}}\|P_{\mathcal{W}}Av\|^2, \tag{6.124}$$

from which the proof follows. Since the problem (6.119) is strictly convex in $\hat{d}$, the solution is unique.

Finally, the solution to the minimax problem (6.118) is given by

$$\hat{x}_{\text{CMX}} = W\hat{d} = W(W^*W)^{-1}W^*A(S^*A)^\dagger c = P_{\mathcal{W}}A(S^*A)^\dagger c, \tag{6.125}$$

completing the proof. $\square$

In the SI setting, reconstruction can be obtained via Fig. 6.2 with

$$H\left(e^{j\omega}\right) = \begin{cases} \frac{R_{WA}(e^{j\omega})}{R_{SA}(e^{j\omega})R_{WW}(e^{j\omega})}, & R_{SA}(e^{j\omega}) \neq 0 \\ 0, & R_{SA}(e^{j\omega}) = 0. \end{cases} \tag{6.126}$$

In contrast to the CLS reconstruction of Corollary 6.3, the minimax-regret solution of Theorem 6.6 explicitly depends on A. Therefore, the prior knowledge that $x \in \mathcal{A}$ plays a role, as one would expect. It is also readily observed that the relation between the unconstrained and constrained minimax solutions is different than in the LS approach. Identifying in (6.125) the expression $A(S^*A)^\dagger c = \hat{x}_{MX} = \hat{x}_{LS}$, the constrained minimax recovery can be expressed as the orthogonal projection onto $\mathcal{W}$ of the unconstrained reconstruction:

$$\hat{x}_{CMX} = P_{\mathcal{W}}\hat{x}_{MX}. \tag{6.127}$$

In Section 6.6.2 we discussed the superiority of this approach in situations in which the spaces $\mathcal{S}$ and $\mathcal{A}$ satisfy the direct-sum condition (6.23), as shown in Fig. 6.22. We now see that this strategy stems from the minimization of the worst-case regret, for *any two* spaces $\mathcal{S}$ and $\mathcal{A}$.

To summarize, treating the constrained reconstruction scenario within the minimax-regret framework leads to a simple and plausible recovery method. This solution coincides with the minimal-error recovery when $\mathcal{H} = \mathcal{A} \oplus \mathcal{S}^\perp$. In contrast, the CLS approach does not take the prior into account and is thus often inferior in terms of squared error in this setting, as we showed in Fig. 6.18: Fig. 6.18(a) corresponds to the CLS recovery while Fig. 6.18(b) is the constrained minimax solution.

6.7   Unified formulation of recovery techniques

In this chapter, we considered sampling and recovery of a signal $x \in \mathcal{A}$ in a Hilbert space $\mathcal{H}$ from samples $c = S^*x$. We distinguished between unconstrained and constrained recoveries. For each of these settings, we developed solutions under the direct-sum condition $\mathcal{H} = \mathcal{A} \oplus \mathcal{S}^\perp$, and when this condition does not hold.

Interestingly, by examining the various solutions presented in this chapter, it can be seen that all the proposed recovery algorithms can be expressed in the form

$$\hat{x} = W(W^*W)^{-1}W^*P(S^*P)^\dagger c = P_{\mathcal{W}}P(S^*P)^\dagger c, \tag{6.128}$$

where c are the given samples and all transformations correspond to Riesz bases for the appropriate subspaces. Here S is the sampling operator, while W and P depend on the setting and the recovery strategy. In general, W is an appropriate reconstruction basis, and P represents a prior basis. The reconstruction basis W is either chosen in advance to lead to efficient implementation, or is equal to a basis A for the prior space $\mathcal{A}$. The transformation P is typically selected as $P = A$; the only exception is the CLS solution for constrained recovery in which $P = W$.

In the SI setting, recovery has the form depicted in Fig. 6.2 with a digital correction filter $H(e^{j\omega})$ that can be written as

$$H\left(e^{j\omega}\right) = \frac{R_{WP}\left(e^{j\omega}\right)}{R_{SP}\left(e^{j\omega}\right)R_{WW}\left(e^{j\omega}\right)}, \tag{6.129}$$

where $R_{SA}(e^{j\omega})$ is defined by (5.28). Here $S(\omega)$ and $W(\omega)$ are the CTFTs of the sampling and reconstruction filters, and $P(\omega)$ is an appropriately chosen filter corresponding to a Riesz basis for the prior space. The filter in (6.129) is defined for frequencies ω such that $R_{SP}(e^{j\omega}) \neq 0$. At all other frequency values, $H(e^{j\omega}) = 0$.

Table 6.5 summarizes the reconstruction techniques resulting from (6.128) and (6.129) for the different scenarios. The last row in the table indicates the explicit equations when all subspaces involved are SI. Note that, under the direct-sum condition, $R_{SA}(e^{j\omega}) \neq 0$ almost everywhere, as follows from Corollary 6.1. In the case in which no constraint is imposed on $\hat{x}$, shown in columns 1 and 2, the LS and minimax reconstructions coincide. The solutions follow from substituting $W = P = A$ in (6.128) and (6.129). When recovery is constrained to lie in $\mathcal{W}$ (columns 3 and 4), the solutions can be expressed in terms of the unconstrained recoveries (columns 1 and 2). The minimax-regret solutions (column 4) are related to the unconstrained recoveries (column 2) via an orthogonal projection onto the reconstruction space $\mathcal{W}$, and follow from (6.128) and (6.129) by substituting $P = A$, and choosing W equal to the reconstruction basis. The recovery $\hat{x}_{\text{CMX}} = P_{\mathcal{W}}\hat{x}_{\text{MX}}$ is the minimal-error approximation to $\hat{x}_{\text{MX}}$ which implies that $\|\hat{x}_{\text{CMX}}\| \leq \|\hat{x}_{\text{MX}}\|$. The CLS solutions (column 3) correspond to choosing $P = W$. When the sampling and prior spaces satisfy the direct-sum condition, and in addition $\mathcal{H} = \mathcal{W} \oplus \mathcal{S}^{\perp}$, this solution becomes $\hat{x}_{\text{CLS}} = E_{\mathcal{W}\mathcal{S}^{\perp}}x$, which has several implications. First, in contrast to the orthogonal projection, an oblique projection may lead to solutions with arbitrarily large norm, if $\mathcal{W}$ is sufficiently far apart from $\mathcal{S}$. Therefore, the error in the CLS framework is not guaranteed to be bounded, unless a bound on the "distance" between $\mathcal{S}$ and $\mathcal{W}$ is known a priori. Second, the recovery does not depend on the prior, i.e. $\hat{x}_{\text{CLS}}$ is not a function of $\mathcal{A}$.

Finally, we point out that practical evaluation of (6.129) may in general be difficult. One brute-force technique is to truncate the infinite series in (5.28) required for the computation of $R_{SP}(e^{j\omega})$, $R_{WP}(e^{j\omega})$, and $R_{WW}(e^{j\omega})$. Then, any filter design method may be used to approximate this desired response with an FIR or IIR filter. Alternatively,

Table 6.5 Reconstruction from noiseless samples with a subspace prior

Setting	Unconstrained ($\hat{x} \in \mathcal{H}$)		Constrained ($\hat{x} \in \mathcal{W}$)	
	Least squares	Minimax	Least squares	Minimax
$x \in \mathcal{A}$	$\hat{x}_{\text{LS}} = A(S^*A)^{\dagger}c$	$\hat{x}_{\text{MX}} = \hat{x}_{\text{LS}}$	$\hat{x}_{\text{CLS}} = W(S^*W)^{\dagger}S^*\hat{x}_{\text{LS}}$	$\hat{x}_{\text{CMX}} = P_{\mathcal{W}}\hat{x}_{\text{MX}}$
$\mathcal{H} = \mathcal{A} \oplus \mathcal{S}^{\perp}$	$\hat{x}_{\text{LS}} = x$	$\hat{x}_{\text{MX}} = x$	$\hat{x}_{\text{CLS}} = E_{\mathcal{W}\mathcal{S}^{\perp}}x$ $(\mathcal{H} = \mathcal{W} \oplus \mathcal{S}^{\perp})$	$\hat{x}_{\text{CMX}} = P_{\mathcal{W}}x$
SI spaces	(6.90)	(6.90)	(6.115)	(6.126)

we can apply the techniques of Section 6.3.3 to implement the denominator by a series of FIR approximations. There are also cases in which a closed-form expression for $H(e^{j\omega})$ exists. The most important example is when $s(t)$, $w(t)$, and $p(t)$ are all B-splines. The numerator in (6.129) then corresponds to an FIR filter with a simple formula. Each of the terms in the denominator represent a concatenation of causal and anticausal IIR filters. We may therefore first filter the data with a recursive formula running from left to right and then filter the result with the same formula running from right to left [65]. We will discuss this case in detail in Chapter 9.

In the next chapter we turn our attention to deterministic and stochastic smoothness priors. Interestingly, we will see that the recoveries under these priors also take on the form of (6.128) and (6.129) with appropriate choices of P and W.

6.8 Multichannel sampling

6.8.1 Recovery methods

We conclude the chapter by adapting the main results obtained in the single-channel setting (corresponding to a single generator for the SI space) to the multichannel case.

Suppose that the prior space $\mathcal{A}$ is spanned by N generators $\{a_\ell(t), 1 \le \ell \le N\}$ (see Section 5.5). A signal $x(t)$ in this space can then be written as

$$x(t) = \sum_{\ell=1}^{N} \sum_{n \in \mathbb{Z}} d_\ell[n] a_\ell(t - nT). \tag{6.130}$$

We sample $x(t)$ using N sampling filters $s_\ell(t)$, as depicted in Fig. 6.23, and recover using N reconstruction filters $w_\ell(t)$. The ℓth sampling sequence is given by $c_\ell[n] = \langle s_\ell(t - nT), x(t) \rangle$, and the sampling space $\mathcal{S}$ is spanned by $\{s_\ell(t), 1 \le \ell \le N\}$. Many of the results derived for the single-channel case depend only on the underlying spaces $\mathcal{A}$ and $\mathcal{S}$ rather than on a specific basis representation within the spaces. These results therefore carry over in a straightforward manner to the multichannel setting with the appropriate definitions of $\mathcal{A}$ and $\mathcal{S}$.

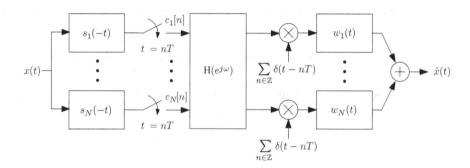

Figure 6.23 Multichannel sampling and reconstruction.

In Section 5.5 we saw that Riesz bases comprising multiple generators behave in a very similar way as ones generated by a single function, with the main difference being that the sampled correlation function $R_{AA}(e^{j\omega})$ is replaced by the correlation matrix $\mathbf{M}_{AA}(e^{j\omega})$ with the $k\ell$th element given by $R_{A_k A_\ell}(e^{j\omega})$. Similarly, in adapting the results of this chapter, summarized in Section 6.7, to the multichannel setting, we can replace expressions involving $R_{SA}(e^{j\omega})$ by the cross-correlation matrix

$$\mathbf{M}_{SA}(e^{j\omega}) = \begin{bmatrix} R_{S_1 A_1}(e^{j\omega}) & \cdots & R_{S_1 A_N}(e^{j\omega}) \\ \vdots & \vdots & \vdots \\ R_{S_N A_1}(e^{j\omega}) & \cdots & R_{S_N A_N}(e^{j\omega}) \end{bmatrix}. \tag{6.131}$$

The general form of (6.129) describing several recovery filters under different scenarios can then be adapted to the multichannel scheme by filtering the samples with a MIMO filter

$$\mathbf{H}\left(e^{j\omega}\right) = \mathbf{M}_{WW}^{-1}(e^{j\omega})\mathbf{M}_{WP}(e^{j\omega})\mathbf{M}_{SP}^{-1}(e^{j\omega}). \tag{6.132}$$

For simplicity, we assume here that $\mathbf{M}_{SP}(e^{j\omega})$ is stably invertible; otherwise, the inverse can be replaced by the pseudoinverse. Recovery is then obtained by modulating the corrected samples onto the reconstruction generators $w_\ell(t)$. As in the previous section, when no recovery constraint is imposed, $W = P = A$, and (6.132) becomes [36, 45, 104, 141, 142]

$$\mathbf{H}\left(e^{j\omega}\right) = \mathbf{M}_{SA}^{-1}(e^{j\omega}). \tag{6.133}$$

If the recovery is constrained to lie in $\mathcal{W}$, then the minimax solution is obtained by choosing $P = A$, and W equal to the reconstruction basis. The CLS approach corresponds to $P = W$.

Extending the discussion of Section 6.3.4 to the multichannel setting with $W = P = A$, we can interpret Fig. 6.23 with $\mathbf{H}(e^{j\omega})$ of (6.133) in terms of a biorthogonal expansion. In particular, generalizing (6.63) leads to the expansion

$$x(t) = \sum_{\ell=1}^{N} \sum_{n \in \mathbb{Z}} c_\ell[n] v_\ell(t - nT), \tag{6.134}$$

where $V_\ell(\omega)$ is the ℓth element of the vector

$$\mathbf{v}(\omega) = \mathbf{M}_{SA}^{-T}\left(e^{j\omega T}\right) \mathbf{a}(\omega), \tag{6.135}$$

and $\mathbf{a}(\omega)$ denotes the vector with elements $A_\ell(\omega)$. Here $\mathbf{M}_{SA}^{-T}\left(e^{j\omega T}\right)$ is the transpose of $\mathbf{M}_{SA}^{-1}\left(e^{j\omega T}\right)$. This interpretation is depicted in Fig. 6.24, and will be the basis for Papoulis' theorem, which we discuss next.

6.8.2 Papoulis' generalized sampling

An interesting special case of Fig. 6.23 when $\mathcal{A}$ is the space of bandlimited signals is Papoulis' generalized sampling theorem which we referred to in Example 4.5 of Section 4.3.4. We now describe this result in detail and show how it follows from (6.134)–(6.135).

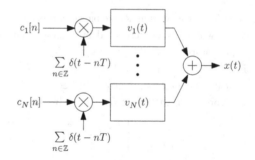

Figure 6.24 Multichannel biorthogonal expansion.

In 1977 Papoulis presented a generalized sampling theorem [76] which unified many special cases of multichannel sampling that were proposed prior to his work. For example, Jagerman and Fogel [143] showed that a bandlimited signal can be sampled at half the Nyquist rate if at each point both the signal and its derivative are given. In fact, this result was stated without proof by Shannon in [1]. Derivative sampling was later extended by Linden and Abramson [144] to allow reconstruction from the signal and its $N - 1$ derivatives, sampled at an Nth of the Nyquist rate. Another example was given by Yen [145] who proved that a signal can be recovered from its recurrent nonuniform samples. Such samples are described by choosing N distinct sampling points in an interval of N times the Nyquist interval T, repeating with period NT. All of these examples are included in Papoulis' generalized sampling theorem, which considers sampling a bandlimited signal using N sampling filters at an Nth of the Nyquist rate. In the examples below we will show specifically how these older results follow from Papoulis' theorem.

In his original paper [76], Papoulis proved his result by developing a recovery procedure involving functions in two variables that solved a linear system of equations. The recovery filters were then obtained by integrating out one of the variables. Brown [146] later showed that the same solution can be expressed in simpler form in terms of a bank of filters, as in Fig. 6.23. The following theorem states Papoulis' result as presented by Brown. For full generality, we treat the case in which the functions may be complex.

Theorem 6.7 (Papoulis' generalized sampling). *Let $x(t)$ be an $L_2(\mathbb{R})$ signal bandlimited to π/T, let $\{\overline{s_\ell(-t)}, 1 \leq \ell \leq N\}$ be a set of N sampling filters, and let $c_\ell[n] = \langle s_\ell(t - nNT), x(t) \rangle$ be the sequence of samples obtained by sampling the output of these N filters with a period of $T' = NT$. Then $x(t)$ can be recovered from $\{c_\ell[n], 1 \leq \ell \leq N\}$ as long as the matrix $\mathbf{B}(\omega)$ with $k\ell$th element*

$$B_{k\ell}(\omega) = \overline{S_\ell\left(\omega + \frac{2\pi(k-1)}{NT}\right)}, \quad 1 \leq \ell, k \leq N \tag{6.136}$$

is stably invertible for all $\omega \in \mathcal{I}_1$ where

$$\mathcal{I}_\ell = \left(-\frac{\pi}{T} + \frac{2\pi(\ell-1)}{NT}, -\frac{\pi}{T} + \frac{2\pi\ell}{NT}\right). \tag{6.137}$$

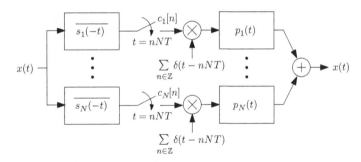

Figure 6.25 Papoulis' sampling and reconstruction scheme.

Recovery is given by

$$x(t) = \sum_{\ell=1}^{N} \sum_{n \in \mathbb{Z}} c_\ell[n] p_\ell(t - nNT), \qquad (6.138)$$

where $P_\ell(\omega)$ is bandlimited to π/T and is given by

$$P_\ell \left(\omega + \frac{2\pi(k-1)}{NT} \right) = NT[\mathbf{B}^{-1}(\omega)]_{\ell k}, \quad 1 \leq \ell, k \leq N, \omega \in \mathcal{I}_1. \qquad (6.139)$$

Papoulis' theorem is illustrated in Fig. 6.25. From (6.136), the ℓth column of the matrix $\mathbf{B}$ contains the CTFT $\overline{S_\ell(\omega)}$ in the interval $(-\pi/T, \pi/T)$, where each row corresponds to a section of width $2\pi/(NT)$. This is demonstrated in Fig. 6.26. The filter $P_\ell(\omega)$ is then constructed from the ℓth row of $\mathbf{B}^{-1}(e^{j\omega})$ by concatenating the columns, as shown in Fig. 6.27.

Establishing the theorem

We now show that Theorem 6.7 is a direct consequence of (6.134)–(6.135). To this end, we first need to choose a set of N functions that span the space of bandlimited signals. Although there are many possibilities, we select

$$A_\ell(\omega) = \begin{cases} 1, & \omega \in \mathcal{I}_\ell \\ 0, & \text{otherwise,} \end{cases} \quad 1 \leq \ell \leq N. \qquad (6.140)$$

From (6.134)–(6.135), recovery has the form (6.138) with

$$P_\ell(\omega) = \sum_{m=1}^{N} \left[\mathbf{M}_{SA}^{-1} \left(e^{j\omega NT} \right) \right]_{m\ell} A_m(\omega), \qquad (6.141)$$

where we incorporated the fact that the sampling period is NT. Below we will see that under the condition of the theorem, $\mathbf{M}_{SA}^{-1} \left(e^{j\omega NT} \right)$ is well defined.

Since $A_k(\omega)$ is nonzero only over the interval $\mathcal{I}_k$, for frequencies ω in $\mathcal{I}_k$ we have

$$P_\ell(\omega) = \left[\mathbf{M}_{SA}^{-1} \left(e^{j\omega NT} \right) \right]_{k\ell}, \quad \omega \in \mathcal{I}_k. \qquad (6.142)$$

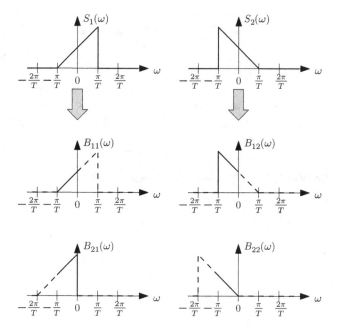

Figure 6.26 Construction of the matrix $\mathbf{B}(\omega)$ in Papoulis' theorem. The relevant values, shown in solid lines, correspond to the range $(-\pi/T, 0)$.

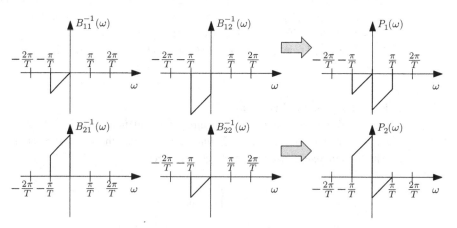

Figure 6.27 Construction of the functions $P_\ell(\omega)$ in Papoulis' theorem from the inverse of the matrix $\mathbf{B}(\omega)$ of Fig. 6.26.

An alternative way to express (6.142) is

$$P_\ell\left(\omega + \frac{2\pi(k-1)}{NT}\right) = \left[\mathbf{M}_{SA}^{-1}\left(e^{j\omega NT}\right)\right]_{k\ell}, \quad \omega \in \mathcal{I}_1. \qquad (6.143)$$

It remains to compute the elements of $\mathbf{M}_{SA}\left(e^{j\omega NT}\right)$ for $\omega \in \mathcal{I}_1$. By definition, the ℓkth element is given by

$$R_{S_\ell A_k}\left(e^{j\omega NT}\right) = \frac{1}{NT} \sum_{m \in \mathbb{Z}} \overline{S_\ell\left(\omega + \frac{2\pi m}{NT}\right)} A_k\left(\omega + \frac{2\pi m}{NT}\right), \quad \omega \in \mathcal{I}_1. \quad (6.144)$$

Since $A_k(\omega)$ is nonzero only over the interval $\mathcal{I}_k$, it follows that for $\omega \in \mathcal{I}_1$, $A_k\left(\omega + \frac{2\pi m}{NT}\right)$ is nonzero only when $m = k - 1$. Thus, over $\mathcal{I}_1$,

$$R_{S_\ell A_k}\left(e^{j\omega NT}\right) = \frac{1}{NT} \overline{S_\ell\left(\omega + \frac{2\pi(k-1)}{NT}\right)} = \frac{1}{NT} B_{k\ell}(\omega), \quad (6.145)$$

so that $\mathbf{M}_{SA}\left(e^{j\omega NT}\right) = 1/(NT)\mathbf{B}^T(e^{j\omega})$. Substituting into (6.143),

$$P_\ell\left(\omega + \frac{2\pi(k-1)}{NT}\right) = NT[\mathbf{B}^{-1}(e^{j\omega})]_{\ell k}, \quad \omega \in \mathcal{I}_1, \quad (6.146)$$

establishing (6.139).

Examples of Papoulis' theorem

We now show how the two examples mentioned at the beginning of this section, derivative sampling and recurrent nonuniform sampling, follow from Theorem 6.7.

Example 6.15 Suppose we wish to recover a π/T-bandlimited signal $x(t)$ from its pointwise samples, and the pointwise samples of its derivative $x'(t)$ at times $t = 2nT$. This corresponds to using sampling filters $S_1(\omega) = 1$ and $S_2(\omega) = -j\omega$ in Fig. 6.25, resulting in

$$\mathbf{B}(\omega) = \begin{bmatrix} 1 & j\omega \\ 1 & j(\omega + \frac{\pi}{T}) \end{bmatrix}, \quad \omega \in [-\pi/T, 0]. \quad (6.147)$$

We now show how to recover $x(t)$ using Papoulis' theorem. We first compute

$$\mathbf{B}^{-1}(\omega) = \frac{T}{\pi} \begin{bmatrix} (\omega + \frac{\pi}{T}) & -\omega \\ j & -j \end{bmatrix}, \quad \omega \in [-\pi/T, 0]. \quad (6.148)$$

From (6.139), we then have that

$$P_1(\omega) = 2T \begin{cases} \frac{T}{\pi}(\omega + \frac{\pi}{T}), & \omega \in [-\pi/T, 0] \\ -\frac{T}{\pi}(\omega - \frac{\pi}{T}), & \omega \in [0, \pi/T] \end{cases}$$

$$= 2T\left(1 - \frac{T}{\pi}|\omega|\right), \quad \omega \in [-\pi/T, \pi/T], \quad (6.149)$$

and

$$P_2(\omega) = 2T \begin{cases} j\frac{T}{\pi}, & \omega \in [-\pi/T, 0] \\ -j\frac{T}{\pi}, & \omega \in [0, \pi/T] \end{cases}$$

$$= -\frac{2jT^2}{\pi} \text{sign}(\omega), \quad \omega \in [-\pi/T, \pi/T], \quad (6.150)$$

where $\text{sign}(x)$ is the signum function defined by (3.32).

Direct computation of the inverse transforms yields

$$p_1(t) = \text{sinc}^2\left(\frac{t}{2T}\right) \tag{6.151}$$

and

$$p_2(t) = -\frac{2T^2}{\pi^2 t}\left(\cos\left(\frac{\pi t}{T}\right) - 1\right) = \frac{4T^2}{\pi^2 t}\sin^2\left(\frac{\pi t}{2T}\right) = t\,\text{sinc}^2\left(\frac{t}{2T}\right). \tag{6.152}$$

The previous example can be extended to sampling of higher-order derivatives. In Exercise 16 we consider recovery from samples of $x(t)$, $x'(t)$, and $x''(t)$ with period $3T$ and show that the reconstruction filters are given by

$$p_1(t) = \text{sinc}^3\left(\frac{t}{3T}\right), \quad p_2(t) = t\,\text{sinc}^3\left(\frac{t}{3T}\right), \quad p_3(t) = \frac{t^2}{2}\text{sinc}^3\left(\frac{t}{3T}\right). \tag{6.153}$$

From this pattern, we conclude more generally that if we sample $x(t)$ and its $N-1$ derivatives with period NT, then it can be recovered from these samples using

$$p_\ell(t) = \frac{t^{\ell-1}}{(\ell-1)!}\text{sinc}^N\left(\frac{t}{NT}\right), \quad 1 \le \ell \le N. \tag{6.154}$$

Substituting (6.154) into (6.138) leads to the following expression for $x(t)$:

$$x(t) = \sum_{\ell=1}^{N}\sum_{n\in\mathbb{Z}}\frac{(t-nNT)^{\ell-1}}{(\ell-1)!}x^{\ell-1}(nNT)\,\text{sinc}^N\left(\frac{t-nNT}{NT}\right), \tag{6.155}$$

where $x^\ell(t)$ denotes the ℓth derivative of $x(t)$.

It is interesting to examine (6.155) when $N \to \infty$. Noting that

$$\lim_{N\to\infty}\text{sinc}^N\left(\frac{t-nNT}{NT}\right) = \delta[n], \tag{6.156}$$

we obtain

$$\lim_{N\to\infty}x(t) = \sum_{\ell=1}^{\infty}\frac{t^{\ell-1}}{(\ell-1)!}x^{\ell-1}(0), \tag{6.157}$$

which is the Taylor series expansion of $x(t)$ about $t = 0$.

Example 6.16 Consider next recovering a π/T-bandlimited signal $x(t)$ from its pointwise samples at $t = 2Tn$ and at $t = 2Tn + \tau$ for some $\tau \in (0, 2T)$. This corresponds to sampling filters $S_1(\omega) = 1$ and $S_2(\omega) = e^{-j\omega\tau}$, resulting in

$$\mathbf{B}(\omega) = \begin{bmatrix} 1 & e^{j\omega\tau} \\ 1 & e^{j(\omega+\frac{\pi}{T})\tau} \end{bmatrix}, \quad \omega \in [-\pi/T, 0]. \tag{6.158}$$

Explicit computation of the inverse gives

$$\mathbf{B}^{-1}(\omega) = \frac{1}{e^{j\omega\tau}(e^{j\frac{\pi}{T}\tau}-1)}\begin{bmatrix} e^{j\frac{\pi}{T}\tau}e^{j\omega\tau} & -e^{j\omega\tau} \\ -1 & 1 \end{bmatrix}, \quad \omega \in [-\pi/T, 0]. \tag{6.159}$$

Substituting into (6.139), the recovery filters are

$$P_1(\omega) = \frac{2T}{e^{j\frac{\pi}{T}\tau} - 1} \begin{cases} e^{j\frac{\pi}{T}\tau}, & \omega \in [-\pi/T, 0) \\ -1, & \omega \in (0, \pi/T], \end{cases} \tag{6.160}$$

and

$$P_2(\omega) = \frac{2Te^{-j\omega\tau}}{e^{j\frac{\pi}{T}\tau} - 1} \begin{cases} -1, & \omega \in [-\pi/T, 0) \\ e^{j\frac{\pi}{T}\tau}, & \omega \in (0, \pi/T]. \end{cases} \tag{6.161}$$

The example above can be extended to more general recurrent nonuniform sampling patterns, where we have N sequences of uniform samples, each with period NT, and shift $\tau_i, 1 \le i \le N - 1$ with $\tau_i \in (0, NT)$. Exact expressions for the recovery filters can be found in [77]. As in the case of $N = 2$ given in the example, these filters turn out to be bandlimited to the same bandwidth as $x(t)$, and each filter is piecewise constant over frequency intervals of length $2\pi/(NT)$ (up to a linear phase factor).

The interpolation identity

The recovery in Fig. 6.25 involves analog filters. Often, it can be easier to interpolate the samples in the digital domain to uniform Nyquist samples, either for further processing in that form or for conversion to continuous time. We now show how this can be achieved using the *interpolation identity* [77].

The interpolation identity is illustrated in Fig. 6.28. The input $x(t) \in L_2(\mathbb{R})$ is bandlimited to π/T, and $H(\omega)$ is an arbitrary filter bandlimited to π/T. The block diagrams in Fig. 6.28 are equivalent for

$$\tilde{H}(e^{j\omega}) = \frac{1}{T}H\left(\frac{\omega}{T}\right), \quad |\omega| \le \pi. \tag{6.162}$$

Note that (6.162) implies that $\tilde{h}[n] = h(nT)$, where $\tilde{h}[n]$ is the inverse DTFT of $\tilde{H}(e^{j\omega})$, and $h(t)$ is the inverse CTFT of $H(\omega)$. A more general form of this theorem can be found in [77].

The block diagram at the bottom of Fig. 6.28 consists of expanding the sequence of samples by a factor of N, via zero padding with $N-1$ zeros between consecutive values, and then filtering by a discrete-time filter with frequency response given by (6.162). The input–output relation of the expander is given by

$$y[n] = \begin{cases} x\left[\frac{n}{N}\right], & n = kN, k \in \mathbb{Z} \\ 0, & \text{otherwise.} \end{cases} \tag{6.163}$$

The filtered output is then used to modulate an impulse train which is subsequently low-pass filtered.

We now apply the interpolation identity to the recovery resulting from Papoulis' theorem of Fig. 6.25. Using the identity on each branch, and moving the impulse modulation

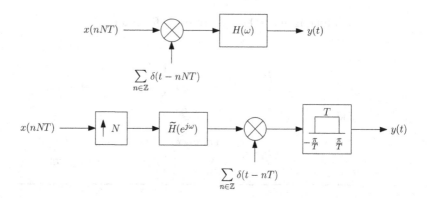

Figure 6.28 The interpolation identity.

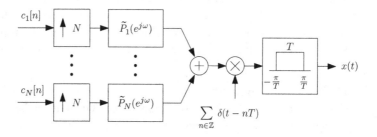

Figure 6.29 Papoulis' scheme via the interpolation identity.

and LPF in each branch outside the summation, we obtain the equivalent implementation in Fig. 6.29 with

$$\widetilde{P}_\ell(e^{j\omega}) = \frac{1}{NT}P_\ell\left(\frac{\omega}{NT}\right), \quad |\omega| \le \pi. \tag{6.164}$$

The overall output of Fig. 6.29 is the original continuous-time signal $x(t)$. Furthermore, since it is reconstructed through low-pass filtering of a uniformly spaced impulse train with period T, the impulse train values must correspond to uniformly spaced samples of $x(t)$ at the Nyquist rate. Thus, the discrete-time filterbank of Fig. 6.29 effectively interpolates the given samples to uniform Nyquist samples. This filterbank can be implemented in practice very efficiently, exploiting the many known results regarding implementation of filterbank structures; see [147] for an excellent reference.

Example 6.17 We now apply the interpolation identity to implement recovery from a signal and its derivative using discrete-time filters.

In Example 6.15 we have seen that a signal $x(t)$ bandlimited to π/T can be reconstructed from samples of $x(t)$ and $x'(t)$ with period $2T$ using the system of Fig. 6.25, with

$$p_1(t) = \text{sinc}^2\left(\frac{t}{2T}\right), \quad p_2(t) = t\,\text{sinc}^2\left(\frac{t}{2T}\right). \tag{6.165}$$

Applying the interpolation identity, we can instead recover with the system of Fig. 6.29 where

$$\tilde{p}_1[n] = \text{sinc}^2\left(\frac{n}{2}\right), \quad \tilde{p}_2[n] = n\,\text{sinc}^2\left(\frac{n}{2}\right). \tag{6.166}$$

This approach also provides a discrete-time mechanism to convert the generalized samples to uniform Nyquist samples.

6.9   Exercises

1. Show that the derivative of the Keys interpolation kernel (6.7) is continuous for every t.

2. In this exercise we develop the general form of the Keys kernel (6.7) by imposing several conditions on the interpolation kernel $w(t)$.
   a. Let $w(t)$ be an arbitrary symmetric interpolation function that is composed of piecewise cubic polynomials on the intervals $[-2, 0), (-1, 0), (0, 1), (1, 2]$. Write down an expression for $w(t)$ as a function of eight unknown variables.
   b. Consider the following conditions on $w(t)$:
      i. $w(t)$ is an interpolation function: $w(n) = \delta[n]$;
      ii. $w(t)$ is continuous;
      iii. $w(t)$ has a continuous derivative.
      Show that any $w(t)$ satisfying these conditions has the form

$$w(t) = \begin{cases} (a+2)|t|^3 - (a+3)|t|^2 + 1, & |t| < 1 \\ a|t|^3 - 5a|t|^2 + 8a|t| - 4a, & 1 \le t < 2 \\ 0, & 2 \le |t|, \end{cases} \tag{6.167}$$

   for some value a.
   c. What value of a leads to the Keys function (6.7)?

3. Show that the first and second derivatives of the Lanczos interpolation kernel (6.8) are continuous for every t.

4. Compute the cosine of the angle between the space of degree-0 splines with knots at the integers and the space of π-bandlimited signals.

5. Our goal in this exercise is to prove that the cosine between two SI spaces $\mathcal{S}$ and $\mathcal{A}$ is given by (6.28).
   a. Show that if $x(t)$ is expressed as $\sum_{n\in\mathbb{Z}} d_1[n]a(t-n)$, where $a(t)$ is an orthonormal generator of $\mathcal{A}$, and $P_\mathcal{S}x(t)$ is expressed as $\sum_{n\in\mathbb{Z}} d_2[n]s(t-n)$, where $s(t)$ is an orthonormal generator of $\mathcal{S}$, then $d_2[n] = d_1[n] * r_{sa}[n]$.
   b. Write explicit expressions for $\|x\|_{L_2}^2$ and $\|P_\mathcal{S}x\|_{L_2}^2$ in terms of the Fourier transforms $D_1(e^{j\omega})$ and $R_{SA}(e^{j\omega})$ of the sequences $d_1[n]$ and $r_{sa}[n]$.

c. Prove that the infimum of $\|P_S x\|_{L_2}^2 / \|x\|_{L_2}^2$ over the set of nonzero functions $D_1(e^{j\omega}) \in L_2([-\pi, \pi])$ is given by $\inf_\omega |R_{SA}(e^{j\omega})|$.

Hint: Show that the infimum is lower-bounded by this quantity, and that there is a sequence of functions $D_1^\ell(e^{j\omega}) \in L_2([-\pi, \pi])$, $\ell = 1, 2, \ldots$ that approach the bound.

d. Extend the result to general (nonorthogonal) generators $s(t)$ and $a(t)$.

6. Derive Equation (6.40) of Example 6.9.

7. Let S and A be SI spaces generated by $\{s(t - n)\}$ and $\{a(t - n)\}$ respectively, such that $\cos(S, A) > 0$. For each of the following choices of functions $\tilde{s}(t)$ and $\tilde{a}(t)$, determine whether the corresponding SI spaces $\tilde{S}, \tilde{A}$ satisfy $\cos(S, A) = \cos(\tilde{S}, \tilde{A})$.

   a. $\tilde{s}(t) = s(t - \tau)$, $\tilde{a}(t) = a(t - \tau)$, for some $\tau \neq 0$.
   b. $\tilde{s}(t) = s(t - \tau_1)$, $\tilde{a}(t) = a(t - \tau_2)$, for some $\tau_1, \tau_2 \in \mathbb{Z}$.
   c. $\tilde{s}(t) = s(t - \tau_1)$, $\tilde{a}(t) = a(t - \tau_2)$, for some $\tau_1, \tau_2 \in \mathbb{R}$.
   d. $\tilde{s}(t) = s(t)e^{j\omega_0 t}$, $\tilde{a}(t) = a(t)e^{j\omega_0 t}$, for some $\omega_0 \in \mathbb{R}$.
   e. $\tilde{s}(t) = s(t)e^{j\omega_1 t}$, $\tilde{a}(t) = a(t)e^{j\omega_2 t}$, for some $\omega_1, \omega_2 \in \mathbb{R}$.
   f. $\tilde{s}(t) = \sum_{n \in \mathbb{Z}} h[n]s(t - n)$, where $0 < \alpha < |H(e^{j\omega})| < \beta < \infty$ for some constants α, β, and $\tilde{a}(t) = a(t)$.

8. For each of the following pairs of SI spaces $S = \text{span}\{s(t - n)\}$ and $A = \text{span}\{a(t - n)\}$, determine whether the direct-sum condition $A \oplus S^\perp = L_2$ is satisfied.

   a. $s(t) = \text{sinc}(t)$, $a(t) = e^{-|t|/\tau}$ for some $\tau > 0$.
   b. $s(t) = \beta^0(t)$, $a(t) = e^{-|t|/\tau}$ for some $\tau > 0$.
   c. $s(t) = \beta^0(t)\cos(\omega_0 t)$ for some $\omega_0 \neq 0$, $a(t) = \beta^0(t)$.

9. Suppose that S and A are two subspaces of $\mathcal{H}$ satisfying $A \oplus S^\perp = \mathcal{H}$ that are generated by Riesz bases with corresponding set transformations S and A. Prove that $S^* E_{AS^\perp} = S^*$, where $E_{AS^\perp}$ is the oblique projection operator onto A along $S^\perp$.

10. Let $\mathbf{X}$ be an $M \times N$ matrix whose left and right singular vectors are known.

    a. Provide a necessary condition on M and N to allow for the recovery of the singular values of $\mathbf{X}$ from knowledge of the sums of the entries of $\mathbf{X}$ along its rows.

    b. Write a sufficient condition on the singular vectors such that $\mathbf{X}$ can be recovered from the sums of its entries along the rows. Show explicitly how to recover $\mathbf{X}$ when the condition is satisfied.

    Hint: Determine a subspace $A \subset \mathbb{R}^{M \times N}$ to which $\mathbf{X}$ is known to belong and a subspace $S \subset \mathbb{R}^{M \times N}$ onto which the projection of $\mathbf{X}$ is known.

11. Consider a degree-n_p spline with knots at the integers that is sampled at $t = n, n \in \mathbb{Z}$, after going through the filter $\beta^{n_s}(t)$ where $\beta^m(t)$ is the B-spline function of degree m defined in (4.32). Write the frequency response $H(e^{j\omega})$ of the digital correction filter required for processing the samples $c[n]$ prior to reconstruction with $\beta^{n_p}(t)$. Show that this filter depends on the spline degrees only through their sum $n_p + n_s$.

12. Compute the oblique dual of the functions $\{(1/\tau_1)e^{(t-nT)/\tau_1}u(nT - t)\}_{n \in \mathbb{Z}}$ in the space spanned by $\{(1/\tau_2)e^{(t-nT)/\tau_2}u(nT - t)\}_{n \in \mathbb{Z}}$, where τ_1, τ_2, and T are arbitrary positive constants and $u(t)$ denotes the unit step function.

13. Suppose that the left and right singular vectors of a square matrix $\mathbf{X}$ are known. Given knowledge of the entries of the first column of $\mathbf{X}$, we would like to approximate $\mathbf{X}$ by a diagonal matrix $\widetilde{\mathbf{X}}$.
    a. Formulate a CLS optimization problem and write an expression for the resulting $\widetilde{\mathbf{X}}$. Explain why this recovery is generally very poor.
    b. Formulate a constrained minimax optimization problem and provide an expression for the resulting $\widetilde{\mathbf{X}}$.

14. Suppose that a degree-n_p spline with knots at the integers is sampled at $t = n$, $n \in \mathbb{Z}$, after going through the filter $s(t) = \mathrm{sinc}(t)$. Our goal is to construct from the samples $c[n]$ a degree-n_r spline $\hat{x}(t)$ that approximates $x(t)$. Write the frequency response $H(e^{j\omega})$ of the digital correction filter required for processing the samples $c[n]$ prior to reconstruction with $\beta^{n_r}(t)$ for the two following cases.
    a. $\hat{x}(t)$ is consistent with the samples $c[n]$, namely $(\hat{x}(t) * s(t))|_{t=n} = c[n]$.
    b. $\hat{x}(t)$ best approximates $x(t)$ (in an L_2-norm sense) within the space of degree-n_r splines.

    Show that both solutions do not depend on the degree n_p of the prior spline space.

15. Derive Equation (6.109) of Example 6.14.

16. Consider recovering a bandlimited signal from samples of the signal and its first two derivatives with period $3T$ where T is the Nyquist period. Show that the recovery filters in Papoulis' theorem are given by (6.153).

17. Example 6.16 treats recovery from recurrent nonuniform samples. Suppose that we chose $\tau = T$ in this example. Derive the expression for the recovery filters in this case. Explain your result.

Chapter 7

Smoothness priors

Until now we have assumed that the signal x being sampled lies in a known subspace. We saw that under an appropriate direct-sum condition, perfect recovery from the samples is possible. We now treat a less restrictive formulation of the sampling problem in which the only prior knowledge on the signal is that it is smooth in some sense. In our development, we consider two models of smoothness. In the first, the weighted norm $\|Lx\|$ of x is bounded where L is an appropriate weighting operator, such as a differential operator. In the SI setting this corresponds to the assumption that the L_2 signal-norm at the output of a continuous-time filter $L(\omega)$ with $x(t)$ as its input is bounded. An alternative formulation is to assume that $x(t)$ is a wide-sense stationary (WSS) process with known spectrum. We will see that under appropriate objective functions, these two priors lead to the same recovery methods with the weighting function L chosen as the whitening filter of the process $x(t)$. Relevant background on WSS signals and whitening is summarized in Appendix B.

Unlike subspace priors, a one-to-one correspondence between smooth signals and their samples does not exist, since smoothness is a far less restrictive constraint than confining the signal to a subspace. Perfect recovery, or even error-norm minimization, is therefore generally impossible. Instead, we focus on LS and minimax approaches for designing the reconstruction system in both the unconstrained and constrained recovery scenarios. Interestingly, we will see that the solutions can be interpreted as recovery methods under a subspace prior, with an appropriate choice of subspace. As a result, the smoothness constraints can effectively be translated into an equivalent subspace prior that depends on the objective to be optimized, the sampling space, and the weighted norm (or signal spectrum in the WSS case).

After comparing the different noiseless sampling and recovery methods developed in this chapter and the previous one, we will turn to treat recovery from noisy samples. We focus on the SI setting, and adapt the LS, minimax, and stochastic approaches to this context.

7.1 Unconstrained recovery

7.1.1 Smoothness prior

Our sampling setup is the same as in the previous chapter: we are given samples $c = S^*x$ of a signal x in an arbitrary Hilbert space $\mathcal{H}$, where S is a set transformation corresponding to a Riesz basis $\{s_n\}$ for the sampling space $\mathcal{S}$. The goal is to recover x from its samples c based on prior knowledge on x. We assume that x is smooth in the sense that $\|Lx\| \leq \rho$ for some invertible operator L and constant ρ, although the results extend to the noninvertible case as well. In the SI setting, $L = L(\omega)$ represents a continuous-time filter; the norm-bound constraint translates to the requirement that the energy at the output of the filter $L(\omega)$ with $x(t)$ as its input is bounded, where we assume that $|L(\omega)| > \alpha > 0$ almost everywhere for some α.

As discussed in Chapter 4, in practice $L(\omega)$ is often chosen to be a first- or second-order derivative in order to constrain the solution to be smooth and nonoscillating, i.e. $L(\omega) = a_0 + a_1 j\omega + a_2(j\omega)^2 + \cdots$ for some constants a_n. For example, choosing $L(\omega) = 1 - aj\omega$ defines the set of signals $x(t)$ for which $\int (1+a^2\omega^2)|X(\omega)|^2 d\omega \leq \rho^2$. This set contains only functions whose CTFT decays at least as fast as $1/\omega^{3/2}$ when $|\omega|$ tends to infinity. In other words, only functions that contain mostly low frequencies are admissible. Another common choice is the filter $L(\omega) = (a_0^2 + \omega^2)^\beta$ with some parameter β that controls the decay rate of the admissible signals.

The smoothness prior is not enough to specify the signal uniquely: there are many signals x for which $S^*x = c$ and their weighted norm $\|Lx\|$ is bounded. This is illustrated in Fig. 7.1. Therefore, some criterion is needed in order to choose among all possibilities. As our objectives, we consider the LS and minimax criteria introduced in the previous chapter. We discuss both the unconstrained setting and the constrained scenario in which recovery is obtained via a given reconstruction basis (or filter). In this section we focus on unconstrained recovery.

7.1.2 Least squares solution

We begin by approximating a smooth signal x via the LS objectives (6.11) and (6.12). To take the smoothness prior into account, we define the set $\mathcal{T}$ of feasible signals as

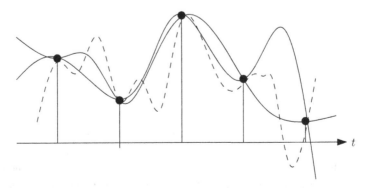

Figure 7.1 Three signals passing through a given set of samples. In solid lines are two smooth signals whose frequency content decays rapidly. In dashed line is a signal whose energy spreads to higher frequencies. We would typically like to avoid the latter as a solution, whereas the former are both valid.

$\mathcal{T} = \{x \colon \|Lx\| \leq \rho\}$. Throughout this chapter, the constant $\rho > 0$ is assumed to be large enough so that $\mathcal{T}$ is nonempty. The LS problem is then

$$\hat{x}_{\mathrm{LS}} = \arg\min_{x \in \mathcal{T}} \|S^*x - c\|^2. \tag{7.1}$$

Since, by assumption, there exists an x in $\mathcal{T}$ giving rise to the measured samples c, the optimal value in (7.1) is 0. However, there may be infinitely many solutions in $\mathcal{T}$ yielding 0 error-in-samples, as demonstrated in Fig. 7.2. In this figure, the solid vertical segment is the set of signals satisfying $S^*x = c$ and $\|Lx\| \leq \rho$. To resolve this ambiguity, we seek the smoothest reconstruction among all possible solutions:

$$\hat{x}_{\mathrm{LS}} = \arg\min_{x \in \mathcal{G}} \|Lx\|^2, \tag{7.2}$$

where now $\mathcal{G} = \{x \colon S^*x = c\}$.

Problem (7.2) is a linearly constrained quadratic program with a convex objective. In finite dimensions there always exists a solution to this class of problems. However, in infinite dimensions this is no longer true [148, Chapter 11]. To guarantee the existence of a solution, we assume that the operator L^*L is bounded from above and below, so that there exist constants $0 < \alpha_L \leq \beta_L < \infty$ such that for all $x \in \mathcal{H}$,

$$\alpha_L \|x\|^2 \leq \|L^*Lx\|^2 \leq \beta_L \|x\|^2. \tag{7.3}$$

This condition implies that $(L^*L)^{-1}$ is bounded and that $\beta_L^{-1}\|x\|^2 \leq \|(L^*L)^{-1}x\|^2 \leq \alpha_L^{-1}\|x\|^2$ for any $x \in \mathcal{H}$ [42]. In the SI setting, (7.3) is equivalent to the requirement that $|L(\omega)|$ is bounded above and below almost everywhere on ω.

Theorem 7.1 (Smooth LS solution). *Assume that the operator L satisfies condition (7.3). Then the solution to (7.2) is given by*

$$\hat{x}_{\mathrm{LS}} = \widetilde{W}(S^*\widetilde{W})^{-1}c, \tag{7.4}$$

where

$$\widetilde{W} = (L^*L)^{-1}S, \tag{7.5}$$

and $S^*\widetilde{W}$ is stably invertible.

The recovery (7.4) can be written as

$$\hat{x}_{\mathrm{LS}} = \widetilde{W}(S^*\widetilde{W})^{-1}S^*x = E_{\widetilde{\mathcal{W}}\mathcal{S}^\perp}x, \tag{7.6}$$

where $\widetilde{\mathcal{W}} = \mathcal{R}(\widetilde{W})$. In other words, $\hat{x}_{\mathrm{LS}}$ is the oblique projection onto $\widetilde{\mathcal{W}}$ along $\mathcal{S}^\perp$ of the original signal x (see Section 2.6.2). Clearly, if x lies in $\widetilde{\mathcal{W}}$ to begin with, then $\hat{x}_{\mathrm{LS}}$ will be equal to x.

In Section 6.3.2, we have seen that (7.4) corresponds to the unique recovery consistent with the samples when we have prior knowledge that x lies in $\widetilde{\mathcal{W}}$. Therefore, the LS solution may be viewed as first determining the optimal reconstruction space given by (7.5), and then computing the unique solution consistent with the data, within this space.

In the special case in which $L = I$, $\widetilde{W} = S$ and $\hat{x}_{\mathrm{LS}} = P_{\mathcal{S}}x$ is the orthogonal projection onto x. Thus, if all we know about the signal is that its norm is bounded, then the minimal-norm solution is equal to the orthogonal projection onto $\mathcal{S}$. This result is very intuitive: we have established in Section 6.2.1 that knowing the samples c is equivalent to knowledge of the orthogonal projection $P_S x$. Therefore, any reconstruction $\hat{x}$ which yields the samples c has the form $\hat{x} = P_S x + v$ with v an arbitrary vector in $S^\perp$. The minimal-norm approximation corresponds to choosing $v = 0$, which is a reasonable choice without further knowledge on x.

Proof: Since $(L^* L)^{-1}$ is upper- and lower-bounded, and S satisfies the Riesz basis condition, $\widetilde{W}$ is a set transformation corresponding to a Riesz basis. In addition,

$$S^* \widetilde{W} = S^* (L^* L)^{-1} S \tag{7.7}$$

is stably invertible.

To solve (7.2), we denote by $E = E_{\widetilde{W} S^\perp}$ the oblique projection (7.6). Any x can then be decomposed as

$$x = Ex + (I - E)x = Ex + v, \tag{7.8}$$

where $v = (I - E)x \in S^\perp$. Direct substitution shows that

$$E^* L^* L (I - E) = 0. \tag{7.9}$$

Consequently,

$$\|Lx\|^2 = \|LEx\|^2 + \|L(I - E)x\|^2 = \|LEx\|^2 + \|Lv\|^2. \tag{7.10}$$

Next, we note that

$$S^* E = S^* (L^* L)^{-1} S (S^* (L^* L)^{-1} S)^{-1} S^* = S^*, \tag{7.11}$$

so that $S^* x = S^* Ex = c$. Therefore, any x for which $S^* x = c$ has $Ex = \hat{x}_{\text{LS}}$ of (7.4): $\hat{x}_{\text{LS}}$ is the unique vector in $\widetilde{W}$ consistent with the samples, and Ex must lie in $\widetilde{W}$. It follows that (7.2) is equivalent to

$$\min_{v \in S^\perp} \{ \|L\hat{x}_{\text{LS}}\|^2 + \|Lv\|^2 \}. \tag{7.12}$$

Clearly, the minimum is obtained at $v = 0$. $\qquad\square$

Figure 7.2(a) provides a geometric interpretation of the LS solution. The set of feasible signals is the subspace $S^* x = c$, which is orthogonal to S (vertical dashed line). Level sets of the objective function $\|Lx\|$ are indicated by dashed ellipsoids. The LS recovery is the intersection between the vertical line and the ellipsoid for which it constitutes a tangent. The reconstruction space $\widetilde{W}$, is the line connecting all possible reconstructions (for all choices of sample sequences c).

In the SI setting, L is an LTI operator corresponding to convolution with the inverse CTFT of $L(\omega)$, so that $(L^* L)^{-1}$ corresponds to filtering with $1/|L(\omega)|^2$. If the sampling space S is SI, then $\widetilde{W}$ is also SI with generator $\tilde{w}(t)$ whose CTFT is

$$\widetilde{W}(\omega) = \frac{S(\omega)}{|L(\omega)|^2}. \tag{7.13}$$

The correction transform H then corresponds to the digital filter

$$H(e^{j\omega}) = \frac{1}{R_{S\widetilde{W}}(e^{j\omega})}. \tag{7.14}$$

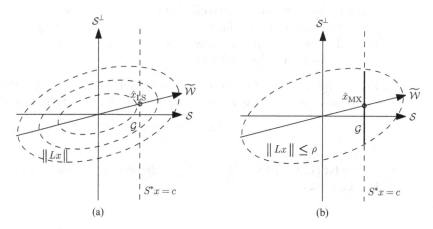

Figure 7.2 Geometric interpretation of (a) the LS and (b) minimax recoveries.

The filter $R_{S\widetilde{W}}(e^{j\omega})$ follows from (5.28) with $A(\omega)$ replaced by $\widetilde{W}(\omega)$. The overall reconstruction scheme is that shown in Fig. 6.2 of the previous chapter, where now the reconstruction kernel is given by (7.13) and the digital correction filter by (7.14). Note that instead of performing digital filtering with (7.14) prior to recovery with (7.13), we can equivalently use the modified continuous-time reconstruction kernel

$$\widehat{W}(\omega) - \frac{\widetilde{W}(\omega)}{R_{S\widetilde{W}}(e^{j\omega T})} \tag{7.15}$$

without digital filtering.

7.1.3   Minimax solution

We now treat the problem of reconstructing a smooth signal from its samples via a worst-case design approach. The prior information we have can be used to construct a set $\mathcal{G}$ of all possible input signals:

$$\mathcal{G} = \{x: S^*x = c, \|Lx\| \le \rho\}. \tag{7.16}$$

This set contains signals that are consistent with the samples and relatively smooth. We now seek the reconstruction that minimizes the worst-case error over $\mathcal{G}$:

$$\hat{x}_{\text{MX}} = \min_{\hat{x}} \max_{x \in \mathcal{G}} \|\hat{x} - x\|^2. \tag{7.17}$$

Theorem 7.2 (Smooth minimax solution).   *The solution to problem (7.17) coincides with the LS approach, namely $\hat{x}_{\text{MX}}$ equals $\hat{x}_{\text{LS}}$ of (7.4).*

Proof: As in (7.8), we write $x = Ex + v$ with $E = E_{\widetilde{W}S^\perp}$, $v \in \mathcal{S}^\perp$ and recall from (7.11) that $S^*E = S^*$.

We have seen in the proof of Theorem 7.1 that the set of signals x satisfying the consistency constraint $S^*x = c$ can be written as $x = \hat{x}_{\text{LS}} + v$ with $v \in \mathcal{S}^\perp$. Furthermore,

$\|Lx\|^2 = \|L\hat{x}_{LS}\|^2 + \|Lv\|^2$. Therefore, we can write the inner maximization in (7.17) as

$$\|\hat{x} - \hat{x}_{LS}\|^2 + \max_{v \in \mathcal{V}} \left\{ \|v\|^2 - 2(\hat{x} - \hat{x}_{LS})^* v \right\}, \tag{7.18}$$

where

$$\mathcal{V} = \left\{ v \colon \|Lv\|^2 \leq \rho^2 - \|L\hat{x}_{LS}\|^2 \right\}. \tag{7.19}$$

Clearly, at the maximum value of v we have that $(\hat{x} - \hat{x}_{LS})^* v \leq 0$ since we can change the sign of v without affecting the constraint. Therefore,

$$\max_{v \in \mathcal{V}} \left\{ \|v\|^2 - 2(\hat{x} - \hat{x}_{LS})^* v \right\} \geq \max_{v \in \mathcal{V}} \|v\|^2. \tag{7.20}$$

Combining (7.20) and (7.18),

$$\min_{\hat{x}} \max_{x \in \mathcal{G}} \|\hat{x} - x\|^2 \geq \min_{\hat{x}} \left\{ \|\hat{x} - \hat{x}_{LS}\|^2 + \max_{v \in \mathcal{V}} \|v\|^2 \right\} = \max_{v \in \mathcal{V}} \|v\|^2, \tag{7.21}$$

where the equality is a result of solving the outer minimization, attained at $\hat{x} = \hat{x}_{LS}$. It is straightforward to see that equality throughout (7.21) is achieved with $\hat{x} = \hat{x}_{LS}$, completing the proof. $\qquad\square$

Figure 7.2(b) illustrates the minimax solution geometrically. The set $\mathcal{G}$ (solid segment) of feasible signals is an intersection of the ellipsoid defined by $\|Lx\| \leq \rho$ and the subspace $S^* x = c$, which is orthogonal to S. Different points on this line correspond to different choices of $v \in S^\perp$. Clearly, for any reconstruction $\hat{x} \in \mathcal{G}$, the worst-case signal x maximizing $\|\hat{x} - x\|^2$, lies on the boundary of $\mathcal{G}$. Therefore, to minimize the worst-case error, $\hat{x}_{MX}$ must be the midpoint of the solid segment, as shown in the figure. The optimal reconstruction space $\widetilde{\mathcal{W}}$ connects the recoveries corresponding to all possible sequences of samples c. This is equivalent to moving the vertical dashed line along the horizontal axis in the figure, and connecting the midpoints of the corresponding feasible sets $\mathcal{G}$.

Although the two approaches we discussed are equivalent in the unrestricted setting, the minimax strategy allows more flexibility in incorporating constraints on the reconstruction, as we show in the next section. In particular, we will demonstrate that it tends to outperform LS when further restrictions are imposed.

7.1.4   Examples

We now consider some examples of the LS and minimax recovery methods (7.13)–(7.15), in which we examine how the smoothness prior affects the solution for several sampling filters.

Example 7.1 Suppose we filter $x(t)$ with an LPF

$$S(\omega) = \begin{cases} 1, & |\omega| \leq \pi \\ 0, & |\omega| > \pi, \end{cases} \tag{7.22}$$

and take pointwise samples of the output at times $t = n$, $n \in \mathbb{Z}$. In this case,

$$\widetilde{W}(\omega) = \frac{S(\omega)}{|L(\omega)|^2} = \begin{cases} \frac{1}{|L(\omega)|^2}, & |\omega| \leq \pi \\ 0, & |\omega| > \pi, \end{cases} \tag{7.23}$$

and $R_{S\widetilde{W}}(e^{j\omega})$ becomes

$$R_{S\widetilde{W}}(e^{j\omega}) = \frac{1}{|L(\omega)|^2}, \quad |\omega| \leq \pi. \tag{7.24}$$

The digital correction filter is then $H(e^{j\omega}) = |L(\omega)|^2$ for $|\omega| \leq \pi$. Finally, the optimal reconstruction filter is given by

$$\widehat{W}(\omega) = \begin{cases} 1, & |\omega| \leq \pi \\ 0, & |\omega| > \pi, \end{cases} \tag{7.25}$$

which does not depend on $L(\omega)$.

The LS recovery is therefore obtained by modulating the samples $c[n]$ onto an impulse train $\sum_n \delta(t - n)$, and filtering by $\widehat{W}(\omega)$, for any choice of $L(\omega)$. Consequently, with LPF sampling, the smoothness prior does not influence the final reconstruction (although $H(e^{j\omega})$ does depend on $L(\omega)$). In this case, the frequency content below π is perfectly recovered, while no frequencies above π are reconstructed. The same conclusion holds true when $S(\omega)$ has an arbitrary nonzero frequency response within the interval $[-\pi, \pi]$.

Example 7.2 We next consider the case in which we take pointwise samples without filtering so that $S(\omega) = 1$ for all ω. We assume that the frequency content of $x(t)$ decays faster than $|\omega|^{-1/2}e^{-\alpha|\omega|}$ for some $\alpha > 0$. This prior can be incorporated into the minimax and LS approaches by noting that

$$\int_{-\infty}^{\infty} |X(\omega)L(\omega)|^2 \, d\omega \leq \rho^2 \tag{7.26}$$

for some $\rho > 0$ with

$$L(\omega) = e^{\alpha|\omega|}. \tag{7.27}$$

In this case, $R_{S\widetilde{W}}(e^{j\omega})$ is given by

$$R_{S\widetilde{W}}(e^{j\omega}) = \sum_{k\in\mathbb{Z}} \frac{1}{|L(\omega + 2\pi k)|^2} = \sum_{k\in\mathbb{Z}} e^{-2\alpha|\omega+2\pi k|}. \tag{7.28}$$

Since this expression is 2π-periodic and symmetric in ω, we may compute it for $\omega \in [0, \pi]$ only, yielding

$$\sum_{k=0}^{\infty} e^{-2\alpha(\omega+2\pi k)} + \sum_{k=-\infty}^{-1} e^{2\alpha(\omega+2\pi k)} = e^{-2\alpha\omega}\frac{1}{1 - e^{-4\pi\alpha}} + e^{2\alpha\omega}\frac{e^{-4\pi\alpha}}{1 - e^{-4\pi\alpha}}. \tag{7.29}$$

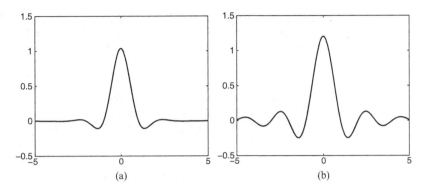

Figure 7.3 Reconstruction filter for the exponential frequency decay prior corresponding to (7.27). (a) $\alpha = 0.5$. (b) $\alpha = 2$.

Consequently, the digital correction filter is given by

$$H(e^{j\omega}) = \frac{1}{R_{S\widetilde{W}}(e^{j\omega})} = \frac{1 - e^{-4\pi\alpha}}{e^{-2\alpha|\omega|} + e^{2\alpha|\omega|}e^{-4\pi\alpha}} \tag{7.30}$$

for $\omega \in [-\pi, \pi]$. The reconstruction filter $\widetilde{W}(\omega)$ satisfies

$$\widetilde{W}(\omega) = \frac{1}{|L(\omega)|^2} = e^{-2\alpha|\omega|}, \tag{7.31}$$

which corresponds to the time-domain filter

$$\tilde{w}(t) = \frac{1}{2\pi} \frac{4\alpha}{4\alpha^2 + t^2}. \tag{7.32}$$

Combining the two leads to the equivalent reconstruction filter $\widehat{W}(\omega)$ of (7.15):

$$\hat{w}(t) = \sum_{n \in \mathbb{Z}} h[n]\tilde{w}(t - n). \tag{7.33}$$

Figure 7.3 shows the shape of $\hat{w}(t)$ for two values of α. As α increases, a larger penalty is placed on higher frequencies, thus causing $\hat{w}(t)$ to widen.

Example 7.3 We now demonstrate the smoothness prior approach in the context of image enlargement.

A digital image $c[m, n]$ is typically the result of sampling a continuous-space scene $x(t, \eta)$, after being convolved with $s(-t, -\eta)$, the point-spread function (PSF) of the lens. Thus,

$$c[m, n] = \iint s(t - m\Delta, \eta - n\Delta)x(t, \eta)dtd\eta, \tag{7.34}$$

where Δ is the sampling spacing. Enlarging a digital image means constructing a more densely spaced sampled version of $x(t, \eta)$. If, for instance, the desired zoom factor is Γ, then the desired (zoomed) digital image is $c_z[k, \ell] = x(k\Delta/\Gamma, \ell\Delta/\Gamma)$. In practice, $x(t, \eta)$ is not available but rather we only have access to its sampled version $c[m, n]$ so that the enlarged image $c_z[k, \ell]$ cannot be computed directly.

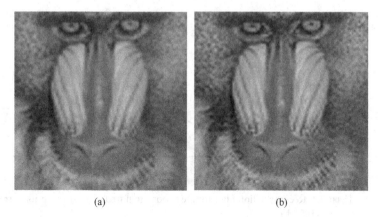

(a)   (b)

Figure 7.4 Mandrill image rescaling: downsampling by a factor of 3 using a rectangular sampling filter followed by upsampling back to the original dimensions using two interpolation methods. (a) The bicubic interpolation kernel leads to a blurry reconstruction with PSNR of 24.18 dB. (b) The minimax method leads to a sharper reconstruction with PSNR of 24.39 dB.

A common alternative is to compute $c_z[k, \ell]$ as $\hat{x}(k\Delta/\Gamma, \ell\Delta/\Gamma)$, where $\hat{x}(t, \eta)$ is a reconstruction of $x(t, \eta)$, obtained from $c[m, n]$ as

$$\hat{x}(t, \eta) = \sum_{m \in \mathbb{Z}} \sum_{n \in \mathbb{Z}} c[m, n]\hat{w}(t - m\Delta, \eta - n\Delta). \tag{7.35}$$

Here, $\hat{w}(t, \eta)$ is some chosen reconstruction kernel.

One common choice for $\hat{w}(t, \eta)$ is the bicubic kernel[1] (6.7). An alternative is to design $\hat{w}(t, \eta)$ according to (7.15) so as to best take into account the PSF $s(t, \eta)$ and an assumed smoothness prior on the original scene $x(t, \eta)$. Natural scenery is often characterized by a polynomial falloff of the frequency content. Therefore, in our example, we used the regularization operator $L(\boldsymbol{\omega}) = \left((0.1\pi)^2 + \|\boldsymbol{\omega}\|^2\right)^{1.3}$, where $\boldsymbol{\omega}$ denotes the 2D frequency vector. The sampling filter $s(t, \eta)$ we assumed was a rectangular kernel supported on $[-\Delta/2, \Delta/2]$ in each axis.

Figure 7.4 compares the minimax (and LS) approach with bicubic interpolation in the task of enlarging an image by a factor of $\Gamma = 3$. To obtain a quantitative comparison, the digital image $c[m, n]$ we started with was obtained by downsampling a high-resolution image. This high-resolution image was compared against the enlarged version $c_z[k, \ell]$ produced by the two approaches.

As can be seen in the figure, in this example minimax recovery is superior to the commonly used bicubic method in terms of peak SNR (PSNR), defined as PSNR $= 10\log_{10}(255^2/\text{MSE})$ with MSE denoting the empirical squared error averaged over all pixel values. In terms of visual quality, the minimax reconstruction is sharper and contains enhanced textures.

---

[1] The 2D bicubic kernel $\hat{w}(t, \eta)$ is given by $w(t)w(\eta)$ with $w(t)$ of (6.7).

7.1.5   Multichannel sampling

Similar to Section 6.8, the LS (and minimax) solution (7.4) can also be applied in the multichannel setting in which $x(t)$ is uniformly sampled at the output of N branches with filters $s_1(-t), \ldots, s_N(-t)$. In this case, the set transformation S corresponds to sampling functions

$$s_{\ell,n}(t) = s_\ell(t - nT), \quad 1 \le \ell \le N, \ n \in \mathbb{Z}. \tag{7.36}$$

Consequently, each of the functions associated with the set transformation $\widetilde{W} = (L^*L)^{-1}S$ is equivalent to filtering the respective sampling function with $1/|L(\omega)|^2$, namely

$$\tilde{w}_{\ell,n}(t) = \tilde{w}_\ell(t - nT), \quad 1 \le \ell \le N, \ n \in \mathbb{Z}, \tag{7.37}$$

where

$$\widetilde{W}_\ell(\omega) = \frac{S_\ell(\omega)}{|L(\omega)|^2}, \quad 1 \le \ell \le N. \tag{7.38}$$

Reconstruction is then achieved using a bank of reconstruction filters:

$$\hat{x}(t) = \sum_{\ell=1}^{N} \sum_{n \in \mathbb{Z}} d_\ell[n] \tilde{w}_\ell(t - nT). \tag{7.39}$$

The sequences $d_\ell[n]$ are the result of applying the operator $(S^*\widetilde{W})^{-1}$ to the sample sequences $c_\ell[n]$. Following the derivations in Section 6.8, this operator corresponds to the MIMO correction filter

$$\mathbf{H}(e^{j\omega}) = \mathbf{M}_{S\widetilde{W}}^{-1}(e^{j\omega}), \tag{7.40}$$

where $\mathbf{M}_{S\widetilde{W}}(e^{j\omega})$ is the cross-correlation matrix between the sampling and reconstruction filters, as defined in (6.131). The overall sampling and reconstruction system is then equivalent to that presented in Fig. 6.23 with reconstruction filters (7.38) and MIMO correction filter (7.40).

In many practical situations, the MIMO digital filter $\mathbf{M}_{S\widetilde{W}}(e^{j\omega})$ is easy to apply, whereas its inverse has an infinite-impulse response with no particularly convenient structure. In these cases, the effect of filtering by $\mathbf{M}_{S\widetilde{W}}^{-1}(e^{j\omega})$ can be approximated by a sequence of filtering operations involving only $\mathbf{M}_{S\widetilde{W}}(e^{j\omega})$, in exactly the same manner as discussed in Section 6.3.3 in the context of single-channel sampling.

Specifically, one option is to employ the Neumann series representation of the operator $S^*\widetilde{W}$, which, as we have seen in (6.47), leads to the following update

$$\hat{d}^{m+1} = \hat{d}^m + \alpha \left(c - S^*\widetilde{W}\hat{d}^m \right). \tag{7.41}$$

Here, c and $\hat{d}^m$ correspond, respectively, to the N sample sequences $\{c_i[n]\}_{n \in \mathbb{Z}, i=1,\ldots,N}$ and to the approximation at the mth iteration of the N corrected sequences

$\{d_i[n]\}_{n\in\mathbb{Z},i=1,\ldots,N}$ at the output of the MIMO filter $\mathbf{M}_{S\widetilde{W}}^{-1}(e^{j\omega})$. The operator $S^*\widetilde{W}$ corresponds to applying $\mathbf{M}_{S\widetilde{W}}(e^{j\omega})$ of (6.131). The constant α should either be chosen to satisfy

$$\|\mathbf{I} - \alpha\,\mathbf{M}_{S\widetilde{W}}(e^{j\omega})\| < 1, \quad \text{for all } \omega \tag{7.42}$$

or it can be modified along the iterations. As we saw in Section 6.3.3, the largest reduction in the approximation error within this framework is attained by using

$$\alpha^m = \frac{(\varepsilon^m)^*\,S^*\widetilde{W}\varepsilon^m}{\|S^*\widetilde{W}\varepsilon^m\|^2} \tag{7.43}$$

at the mth iteration. Here ε^m represents the N error sequences $\{c_i[n] - c_i^m[n]\}_{n\in\mathbb{Z},i=1,\ldots,N}$, where we denoted by $\{c_i^m[n]\}$ the result of applying the MIMO filter $\mathbf{M}_{S\widetilde{W}}(e^{j\omega})$ on the sequences $\{\hat{d}_i^m[n]\}$.

An alternative approach to approximating the effect of filtering by $\mathbf{M}_{S\widetilde{W}}^{-1}(e^{j\omega})$ is to employ steepest-descent iterations which are aimed at minimizing the error $\|S^*\widetilde{W}d - c\|^2$. As shown in (6.54), this approach leads to the update

$$\hat{d}^{m+1} = \hat{d}^m + \alpha^m\widetilde{W}^*S\left(c - S^*\widetilde{W}\hat{d}^m\right) \tag{7.44}$$

for which the optimal step size is

$$\alpha^m = \frac{\|\widetilde{W}S^*\varepsilon^m\|^2}{\|S^*\widetilde{W}\widetilde{W}^*S\varepsilon^m\|^2}. \tag{7.45}$$

Example 7.4 One popular application in which samples from multiple channels are used to recover a signal is super-resolution imaging. In this example, one collects N digital images $c_1[m,n],\ldots,c_N[m,n]$ of a continuous-space (two-dimensional) scene $x(t,\eta)$. Each image is captured with a slightly different translation, say (t_i,η_i) for the ith image. Assuming that the PSF of the lens is $s(-t,-\eta)$, the digital images are given by

$$c_i[m,n] = \iint s(t - \tau_i - m\Delta, \eta - \eta_i - n\Delta)x(t,\eta)dtd\eta, \tag{7.46}$$

where Δ is the sampling spacing. This process can be thought of as using N sampling channels, with the ith sampling filter being $s_i(\tau,\eta) = s(\tau - \tau_i, \eta - \eta_i)$, so that

$$c_i[m,n] = \iint s_i(t - m\Delta, \eta - n\Delta)x(t,\eta)dtd\eta. \tag{7.47}$$

The goal in super-resolution, just like in the image interpolation scenario described in Example 7.3, is to produce a densely sampled version of the original scene. For example, if the desired zoom factor is Γ, then the super-resolved digital image is $c_z[k,\ell] = x(k\Delta/\Gamma, \ell\Delta/\Gamma)$. Since $x(t,\eta)$ is unavailable, $c_z[k,\ell]$ is typically approximated as $\hat{x}(k\Delta/\Gamma, \ell\Delta/\Gamma)$, where $\hat{x}(t,\eta)$ is a reconstruction of $x(t,\eta)$, obtained from the low-resolution images $\{c_i[m,n]\}$.

(a)   (b)

Figure 7.5 Resolution enhancement by a factor of $\Gamma = 4$ from a set of 20 images. (a) One of the low-resolution images. (b) Super-resolved image.

Figure 7.5 shows a super-resolution recovery obtained from the first 20 frames of the sequence Disk taken from [149]. Here, the pixel spacing is set to $\Delta = 1$, the sampling filter was assumed to be the rectangular kernel $s(t, \eta) = \beta^0(t)\beta^0(\eta)$ and the translations (t_i, η_i) were estimated from the low-resolution images $\{c_i[m, n]\}$. The regularization filter $L(\omega)$ was chosen to be $\mathrm{sinc}(\omega/(2\pi))^{-3/2}$. This results in the reconstruction filter $\widetilde{W}(\omega) = S(\omega)/|L(\omega)|^2 = \mathrm{sinc}(\omega/(2\pi))^4$, which in the spatial domain is given by the degree-3 B-spline $\tilde{w}(t, \eta) = \beta^3(t)\beta^3(\eta)$.

7.2   **Constrained recovery**

We next derive LS and minimax approximations of x from its samples c using a prespecified interpolation basis W. In the SI setting this amounts to using a specific interpolation kernel $w(t)$. As we will see, in this setting, the solutions no longer coincide.

7.2.1   Least squares solution

In order to produce a solution $\hat{x} \in \mathcal{W}$, we modify the feasible set $\mathcal{T}$ of (7.1) to include only signals in $\mathcal{W}$:

$$\hat{x}_{\mathrm{CLS}} = \arg\min_{x \in \mathcal{T}} \|S^*x - c\|^2, \tag{7.48}$$

where $\mathcal{T} = \{x \colon \|Lx\| \leq \rho, x \in \mathcal{W}\}$.

We have seen in the proof of Theorem 6.5 that without the constraint $\|Lx\| \leq \rho$, the set of solutions to (7.48) is given by

$$\mathcal{G} = \{x \colon x \in \mathcal{W}, S^*x = P_{\mathcal{R}(S^*W)}c\}. \tag{7.49}$$

To choose among these possibilities, we minimize the smoothness $\|Lx\|^2$ over $\mathcal{G}$:

$$\hat{x}_{\text{CLS}} = \arg\min_{x \in \mathcal{G}} \|Lx\|^2. \tag{7.50}$$

Theorem 7.3 (Smooth CLS solution). *Assume that the operator L satisfies condition (7.3). Then the solution to (7.50) is given by*

$$\hat{x}_{\text{CLS}} = \widehat{W}(S^*\widehat{W})^\dagger c, \tag{7.51}$$

where

$$\widehat{W} = W(W^*L^*LW)^{-1}W^*S. \tag{7.52}$$

Note that W^*L^*LW is invertible, since W corresponds to a Riesz basis and L^*L is bounded.

Proof: The proof of the theorem follows similar steps to those in Section 7.1.2 and utilizes the fact that every signal in $\mathcal{G}$ can be written as $x = \widehat{W}(S^*\widehat{W})^\dagger c + Wv$, where $v \in \mathcal{N}(S^*W)$. This holds true since $\mathcal{R}(S^*\widehat{W}(S^*\widehat{W})^\dagger) = \mathcal{R}(S^*\widehat{W}) = \mathcal{R}(S^*W)$, where the last equality is a result of the fact that W^*L^*LW is invertible and Hermitian. See Exercise 10. □

The CLS solution takes on a particularly simple form when $\mathcal{W} \oplus \mathcal{S}^\perp = \mathcal{H}$. In Section 6.3.2 we have seen that in this case there is a unique $x \in \mathcal{W}$ satisfying $S^*x = c$, which is equal to the oblique projection $E_{\mathcal{WS}^\perp}x$. In addition, S^*W is invertible. Since there is only one signal in the constraint set of (7.50), the smoothness measure in the objective does not play a role and the solution becomes $\hat{x}_{\text{CLS}} = W(S^*W)^{-1}c$. This can also be verified directly from (7.51) and (7.52). Using the fact that the unconstrained solution (7.4) satisfies $S^*x_{\text{LS}} = c$, we can write $\hat{x}_{\text{CLS}} = W(S^*W)^{-1}S^*x_{\text{LS}} = E_{\mathcal{WS}^\perp}\hat{x}_{\text{LS}}$, recovering the relation (6.117) developed for a subspace prior. Consequently, in the LS methodology, the constrained and unconstrained solutions are related through an oblique projection, regardless of the type of prior.

Another interesting scenario where L does not affect the solution is the case in which $\mathcal{W}$ and $\mathcal{S}$ are SI spaces with generators $w(t)$ and $s(t)$ respectively and L is an LTI operator with frequency response $L(\omega)$. The operator $(W^*L^*LW)^{-1}$ then corresponds to a digital filter $1/R_{AA}(e^{j\omega})$ where $A(\omega) = L(\omega)W(\omega)$ and $R_{AA}(e^{j\omega})$ is given by (5.28). Therefore, $\hat{x}_{\text{CLS}}(t)$ of (7.51) can be produced by filtering the sequence of samples $c[n]$ with

$$H(e^{j\omega}) = \begin{cases} \frac{1}{R_{SW}(e^{j\omega})}, & R_{SW}(e^{j\omega}) \neq 0 \\ 0, & \text{else}, \end{cases} \tag{7.53}$$

prior to reconstruction with $W(\omega)$. Evidently, (7.53) does not depend on $L(\omega)$; that is, the smoothness prior does not affect the solution in the SI setting. The resulting scheme is identical to the CLS reconstruction discussed in Section 6.5.1 in the context of subspace priors. Thus, in the SI setting, under the LS objective the constrained recovery is the same independent of the signal prior.

7.2.2 Minimax-regret solution

We next extend the minimax approach of Section 7.1.3 to the setup where $\hat{x}$ is constrained to lie in $\mathcal{W}$. Similar to the case of subspace priors, treated in Section 6.6.3, we minimize the worst-case regret:

$$\hat{x}_{\text{CMX}} = \arg \min_{\hat{x} \in \mathcal{W}} \max_{x \in \mathcal{G}} \|\hat{x} - P_{\mathcal{W}} x\|_2^2, \tag{7.54}$$

where $\mathcal{G} = \{x: S^* x = c, \|Lx\| \leq \rho\}$ and we assume that there is a feasible point.

Theorem 7.4 (Smooth constrained minimax solution). *The solution to (7.54) is given by*

$$\hat{x}_{\text{CMX}} = P_{\mathcal{W}} \widetilde{W} (S^* \widetilde{W})^{-1} c = P_{\mathcal{W}} \hat{x}_{\text{MX}}, \tag{7.55}$$

where $\widetilde{W}$ is given by (7.5) and $\hat{x}_{\text{MX}} = \hat{x}_{\text{LS}}$ is the unconstrained solution of (7.4).

Proof: The proof follows the same steps used to prove Theorem 7.2 in Section 7.1.3. See Exercise 11. □

The result of Theorem 7.4 is intuitive: when the output is constrained to a subspace $\mathcal{W}$, the minimax recovery is the orthogonal projection onto $\mathcal{W}$ of the minimax solution without the restriction. Recall that relation (7.55) is also true for subspace priors, as we have seen in Section 6.5.2.

Figure 7.6 illustrates the minimax-regret solution geometrically. As in the unconstrained scenario of Fig. 7.4, the feasible set of signals $\mathcal{G}$ is the vertical solid segment. Here, however, the reconstruction $\hat{x}$ is constrained to lie in the predefined space $\mathcal{W}$. The regret criterion (7.54) measures the deviation of $\hat{x}$ from $P_{\mathcal{W}} x$. The tilted solid segment is the projection of the feasible set $\mathcal{G}$ onto $\mathcal{W}$. For every reconstruction $\hat{x}$ in $\mathcal{W}$, the signal x leading to the worst regret corresponds to one of the endpoints of this set. Therefore, the minimal regret is attained if we choose $\hat{x}_{\text{CMX}}$ to be the midpoint of this segment.

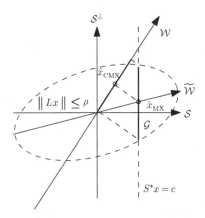

Figure 7.6 Geometric interpretation of minimax-regret recovery in a predefined reconstruction space $\mathcal{W}$.

This solution is also the projection of the midpoint of $\mathcal{G}$ onto $\mathcal{W}$, i.e. the projection of the unconstrained minimax solution (7.4) onto $\mathcal{W}$.

When $\mathcal{S}$ and $\mathcal{W}$ are SI spaces and L is an LTI operator, the correction transform H corresponds to a digital filter $H(e^{j\omega})$ as in Fig. 6.2. This filter can be determined by writing $H = (W^*W)^\dagger W^*\widetilde{W}(S^*\widetilde{W})^\dagger$, where $\widetilde{W} = (L^*L)^{-1}S$ is the set transformation corresponding to the unrestricted minimax solution. The operators W^*W, $W^*\widetilde{W}$, and $S^*\widetilde{W}$ correspond to the digital filters $R_{WW}(e^{j\omega})$, $R_{W\widetilde{W}}(e^{j\omega})$, and $R_{S\widetilde{W}}(e^{j\omega})$ respectively, and

$$H(e^{j\omega}) = \frac{R_{W\widetilde{W}}(e^{j\omega})}{R_{S\widetilde{W}}(e^{j\omega})R_{WW}(e^{j\omega})}. \tag{7.56}$$

In contrast to the CLS solution (7.53), this filter depends on $L(\omega)$ so that the prior affects the solution. The next example demonstrates the effectiveness of this filter in an image processing task.

Example 7.5 In Fig. 7.7 we demonstrate the difference between the LS and minimax-regret methods in an image enlargement task. The setup is the same as that of Fig. 7.4, only now the reconstruction filter is constrained to be a triangular kernel corresponding to linear interpolation, namely $w(t, \eta) = \beta^1(t/\Delta)\beta^1(\eta/\Delta)$. With this interpolation kernel, the direct-sum condition $L_2 = \mathcal{W} \oplus \mathcal{S}^\perp$ is satisfied. The error of the minimax-regret recovery is only 0.7 dB less than the unconstrained minimax solution shown in Fig. 7.4. The CLS approach, on the other hand, is much worse both in terms of PSNR and in terms of visual quality. Its tendency to over-enhance high frequencies stems from the fact that it ignores the smoothness prior.

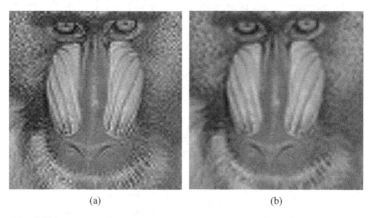

(a)   (b)

Figure 7.7 Mandrill image rescaling: downsampling by a factor of 3 using a rectangular sampling filter followed by upsampling back to the original dimensions using the LS and minimax-regret methods with linear interpolation. (a) The LS approach over-enhances the high frequencies and results in a PSNR of 22.51 dB. (b) The minimax-regret method leads to a smoother reconstruction with PSNR of 23.69 dB.

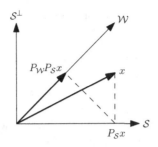

Figure 7.8 Minimax-regret reconstruction with $\|x\| \leq \rho$. The signal is orthogonally projected onto the sampling space, and then onto the reconstruction space.

Norm-bounded error

Many of the interesting properties of the minimax-regret recovery (7.55) can be best understood by examining the case where our only prior on the signal is that it is norm-bounded, that is, when $L = I$ [113]. In this case, $\widetilde{W} = S$ so that $\hat{x}_{\text{CMX}} = P_W P_S x$; this solution is illustrated in Fig. 7.8. We have seen already that knowing the samples $c[n]$ is equivalent to knowing $P_S x$. In addition, our recovery is constrained to lie in W. As illustrated in the figure, the minimax-regret solution is a robust recovery scheme by which the signal is first orthogonally projected onto the sampling space, and then onto the reconstruction space.

In the SI case, setting $L(\omega) = 1$ in (7.56), the correction filter becomes

$$H(e^{j\omega}) = \frac{R_{WS}(e^{j\omega})}{R_{SS}(e^{j\omega}) R_{WW}(e^{j\omega})}, \tag{7.57}$$

since from (7.13), $\tilde{w}(t) = s(t)$. Applying the Cauchy–Schwarz inequality to the numerator of (7.57) and to the denominator of (7.53), it is easy to see that the magnitude of the minimax-regret filter (7.57) is smaller than that of the CLS filter (7.53) at all frequencies. This property renders the minimax-regret approach more resistant to noise in the samples $c[n]$, since perturbations in $\hat{x}(t)$ caused by errors in $c[n]$ are always smaller in the minimax-regret method than in the CLS approach.

Apart from robustness to digital noise, which takes place after sampling, the minimax-regret method is also more resistant to perturbations in the continuous-time signal $x(t)$. To see this, note that since $\hat{x}_{\text{CMX}} = P_W P_S x$, the norm of $\hat{x}_{\text{CMX}}$ is necessarily bounded by that of x. Furthermore, it is easy to show that the resulting reconstruction error is always bounded by twice the norm of x:

$$\|\hat{x}_{\text{CMX}} - x\|^2 = \|(I - P_W P_S)x\|^2 = \|P_W(I - P_S)x\|^2 + \|P_{W^\perp}x\|^2 \leq 2\|x\|^2. \tag{7.58}$$

The last inequality follows from the fact that $I - P_S = P_{S^\perp}$, and $\|P_A P_V x\| \leq \|P_V x\| \leq \|x\|$ for any orthogonal projections P_A, P_V. In contrast, the norm of the error resulting from the CLS solution can, in some cases, grow without bound. To see this, note that in the case of a direct sum $\mathcal{H} = W \oplus S^\perp$, the CLS recovery is given by the oblique projection $\hat{x}_{\text{CLS}} = E_{WS^\perp}x$, which may arbitrarily increase the norm of x.

The following example compares the CLS and minimax-regret solutions with $L = I$.

Example 7.6 When $L = I$, the constrained minimax solution can be expressed as $\hat{x}_{\text{CMX}} = P_{\mathcal{W}} P_{\mathcal{S}} x$, whereas the CLS recovery (assuming $\mathcal{H} = \mathcal{W} \oplus \mathcal{S}^{\perp}$) is $\hat{x}_{\text{CLS}} = E_{\mathcal{W}\mathcal{S}^{\perp}} x$. In Section 7.2.3 we provide a thorough analysis of when each of the approaches is preferable. However, simple intuition can be gained by examining two extreme cases. Specifically, if x lies in the sampling space $\mathcal{S}$, then $\hat{x}_{\text{CMX}}$ becomes $P_{\mathcal{W}} x$, which is the best possible approximation to x in $\mathcal{W}$. If, on the other hand, x lies in $\mathcal{W}$, then $\hat{x}_{\text{CLS}} = x$ which is clearly preferable. We now demonstrate that there is a smooth transition between these two extremes.

Suppose that $s(t) = \beta^0(t)$, $w(t) = \beta^1(t)$ and the sampling interval is $T = 1$. Let us compute the terms $R_{WW}(e^{j\omega})$, $R_{WS}(e^{j\omega})$, and $R_{SS}(e^{j\omega})$, appearing in the CLS and constrained minimax solutions. The function $R_{WW}(e^{j\omega})$ is the DTFT of

$$(w(-t) * w(t))|_{t=n} = (\beta^1(t) * \beta^1(t))|_{t=n} = \beta^3(n) = \begin{cases} \frac{2}{3}, & n = 0 \\ \frac{1}{6}, & |n| = 1 \\ 0, & |n| \geq 2, \end{cases} \tag{7.59}$$

so that

$$R_{WW}(e^{j\omega}) = \frac{1}{6} e^{-j\omega} + \frac{2}{3} + \frac{1}{6} e^{j\omega} = \frac{1}{3} \left(\cos(\omega) + 2 \right). \tag{7.60}$$

Similarly, the term $R_{WS}(e^{j\omega})$ corresponds to the impulse response

$$(w(-t) * s(t))|_{t=n} = (\beta^1(t) * \beta^0(t))_{t=n} = \beta^2(n) = \begin{cases} \frac{3}{4}, & n = 0 \\ \frac{1}{8}, & |n| = 1 \\ 0, & |n| \geq 2, \end{cases} \tag{7.61}$$

whose DTFT is

$$R_{WS}(e^{j\omega}) = \frac{1}{8} e^{-j\omega} + \frac{3}{4} + \frac{1}{8} e^{j\omega} = \frac{1}{4} \left(\cos(\omega) + 3 \right). \tag{7.62}$$

Finally, since $(\beta^0(t) * \beta^0(t))_{t=n} = \beta^1(n) = \delta[n]$, we have that $R_{SS}(e^{j\omega}) = 1$.

From these expressions it follows that the correction filter of the constrained minimax-regret method is given by

$$H(e^{j\omega}) = \frac{3 \left(\cos(\omega) + 3 \right)}{4 \left(\cos(\omega) + 2 \right)}, \tag{7.63}$$

while the CLS correction filter is

$$H(e^{j\omega}) = \frac{4}{\cos(\omega) + 3}. \tag{7.64}$$

To gain intuition as to how these two filters perform, consider the signal

$$x(t) = \sum_{n \in \mathbb{Z}} (-1)^n a(t - n) \tag{7.65}$$

where

$$a(t) = \alpha \beta^0(t) + (1 - \alpha)\beta^1(t) \tag{7.66}$$

with some $\alpha \in [0, 1]$. The value of α controls the amount of energy of $x(t)$ in $\mathcal{S}$ and $\mathcal{W}$. Specifically, when $\alpha = 1$, the signal $x(t)$ lies in the sampling space $\mathcal{S}$, whereas when $\alpha = 0$, $x(t)$ lies in the reconstruction space $\mathcal{W}$. Figures 7.9(a) and 7.9(b) show the recoveries produced by the two approaches for $\alpha = 0.25$ and $\alpha = 0.75$ respectively. In the latter case, constrained minimax is preferable, whereas in the former situation, CLS is advantageous. Figure 7.9(c) depicts the normalized squared-reconstruction error $\|x - \hat{x}\|^2 / \|x\|^2$ of both methods, as a function of α. As expected, the minimax approach is preferable to the CLS method for high values of α, but is worse when α is close to 0.

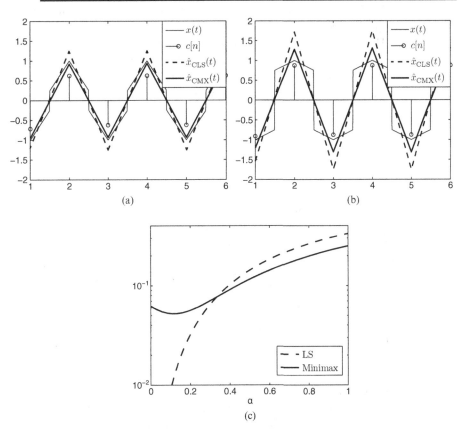

Figure 7.9 Comparison between the CLS and constrained minimax-regret approaches in the task of recovering $x(t)$ of (7.65) when the sampling filter is $s(t) = \beta^0(t)$ and the reconstruction filter is $w(t) = \beta^1(t)$. (a) $\alpha = 0.25$. (b) $\alpha = 0.75$. (c) Normalized squared-reconstruction error $\|x - \hat{x}\|^2 / \|x\|^2$ as a function of α.

7.2.3   Comparison between least squares and minimax

Throughout our derivations, we have pointed out some differences between the LS and minimax solutions. In this section, we take a closer look at the advantages and disadvantages of both methods.

We first note that when recovery is unconstrained, the LS and minimax recoveries coincide. This was also the case under the subspace prior. However, in the constrained setting, the solutions differ. One of the disadvantages of the LS approach in this context is that it often does not depend on the choice of norm, which is dictated by L. This happens when the direct-sum condition $\mathcal{H} = \mathcal{W} \oplus \mathcal{S}^\perp$ holds, and in the SI setting. In the former scenario, the LS recovery is given by the oblique projection $E_{\mathcal{W}\mathcal{S}^\perp} x$. In both cases the solution is the same as the LS recovery assuming that x lies in an arbitrary subspace so that the prior does not play any role. This is in contrast to the minimax solution, which is always affected by the prior and depends explicitly on L. Furthermore, the minimax solution has the appealing structure of an orthogonal projection onto $\mathcal{W}$ of the unconstrained recovery (which is the same under both the LS and minimax objectives). This solution is identical to that obtained when x is known to lie in the subspace spanned by $(L^*L)^{-1}S$.

The discussion above highlights the intuitive appeal of the minimax approach. We next compare directly the reconstruction error resulting from both methods. At the end of the previous section we showed that when $L = I$, the norm of the minimax error is bounded, while the error resulting from the LS solution can, in principle, grow without bound. We now analyze these errors more rigorously and provide further guidelines to select between the LS and minimax methods.

Recovery error

To explicitly compare the recovery error resulting from both schemes we focus our attention on the case in which $L = I$ and $\mathcal{H} = \mathcal{W} \oplus \mathcal{S}^\perp$. The derivations can be readily extended to the more general setting, but the expressions become more involved (see Exercise 12). Under these conditions, $\hat{x}_{\mathrm{CMX}} = P_\mathcal{W} P_\mathcal{S} x$ and $\hat{x}_{\mathrm{CLS}} = E_{\mathcal{W}\mathcal{S}^\perp} x$. Our analysis below shows that if the spaces $\mathcal{S}$ and $\mathcal{W}$ are sufficiently far apart, or if x has enough energy in $\mathcal{S}$, then the minimax-regret method is preferable in a squared-norm error sense to the CLS approach.

Theorem 7.5 provides tight bounds on the error resulting from the minimax and CLS strategies. These error bounds are given with respect to the optimal error that can be achieved when forcing a recovery in $\mathcal{W}$: the best approximation to x in $\mathcal{W}$ is $P_\mathcal{W} x$ resulting in $e_{\mathrm{OPT}}(x) = P_{\mathcal{W}^\perp} x$.

Theorem 7.5 (Error bounds).   *Let* $e_{\mathrm{CMX}}(x) = x - P_\mathcal{W} P_\mathcal{S} x, e_{\mathrm{CLS}}(x) = x - E_{\mathcal{W}\mathcal{S}^\perp} x$
denote the errors resulting from the minimax-regret and CLS reconstructions and let
$e_{\mathrm{OPT}}(x) = P_{\mathcal{W}^\perp} x$ *be the optimal error in the squared-norm sense. Then*

$$\|e_{\mathrm{OPT}}(x)\|_2^2 + \cos^2(\mathcal{S}^\perp, \mathcal{W}) \|P_{\mathcal{S}^\perp} x\|_2^2 \leq \|e_{\mathrm{CMX}}(x)\|_2^2$$
$$\leq \|e_{\mathrm{OPT}}(x)\|_2^2 + \sin^2(\mathcal{W}, \mathcal{S}) \|P_{\mathcal{S}^\perp} x\|_2^2, \tag{7.67}$$

where $\cos(\cdot)$ *and* $\sin(\cdot)$ *are defined in (6.26) and (6.27) respectively, and*

$$\frac{\|e_{\mathrm{OPT}}(x)\|_2^2}{\sin^2(\mathcal{S}^\perp, \mathcal{W})} \leq \|e_{\mathrm{CLS}}(x)\|_2^2 \leq \frac{\|e_{\mathrm{OPT}}(x)\|_2^2}{\cos^2(\mathcal{W}, \mathcal{S})}. \tag{7.68}$$

Proof: Writing $e_{\mathrm{CMX}}(x) = P_{\mathcal{W}^\perp} e_{\mathrm{CMX}}(x) + P_{\mathcal{W}} e_{\mathrm{CMX}}(x)$, we have that

$$\|e_{\mathrm{CMX}}(x)\|_2^2 = \|e_{\mathrm{OPT}}(x)\|_2^2 + \|P_{\mathcal{W}}(I - P_{\mathcal{S}})x\|_2^2 = \|e_{\mathrm{OPT}}(x)\|_2^2 + \|P_{\mathcal{W}} P_{\mathcal{S}^\perp} x\|_2^2. \tag{7.69}$$

Note that for $x \in \mathcal{S}$, $\|e_{\mathrm{CMX}}(x)\|_2^2 = \|e_{\mathrm{OPT}}(x)\|_2^2$, so that the minimax-regret reconstruction is optimal. If $x \notin \mathcal{S}$, then $\|P_{\mathcal{S}^\perp} x\|_2 \neq 0$ and we can rewrite (7.69) as

$$\|e_{\mathrm{CMX}}(x)\|_2^2 = \|e_{\mathrm{OPT}}(x)\|_2^2 + \|P_{\mathcal{W}} v\|_2^2 \|P_{\mathcal{S}^\perp} x\|_2^2, \tag{7.70}$$

where we defined $v = P_{\mathcal{S}^\perp} x / \|P_{\mathcal{S}^\perp} x\|_2$. Since v is a normalized vector in $\mathcal{S}^\perp$ which is orthogonally projected onto $\mathcal{W}$,

$$\cos^2(\mathcal{S}^\perp, \mathcal{W}) \leq \|P_{\mathcal{W}} v\|_2^2 \leq \sin^2(\mathcal{S}^\perp, \mathcal{W}^\perp). \tag{7.71}$$

Combining (7.71) with (7.70) and (6.71), results in (7.67).

If $v \in \mathcal{S}^\perp$ obtains the maximum (minimum)^2 angle with $\mathcal{W}$, then $x = v + s$, where $s = P_{\mathcal{S}} x = S(S^* S)^{-1} c$, achieves the upper (lower) bound of (7.67). Therefore, the bounds of (7.67) are tight.

The upper bound in (7.68) follows from plugging $\mathcal{A} = \mathcal{W}$ into the upper bound in Proposition 6.7. To develop the lower bound, we first note that

$$e_{\mathrm{OPT}}(x) = P_{\mathcal{W}^\perp} e_{\mathrm{CLS}}(x). \tag{7.72}$$

If $e_{\mathrm{CLS}}(x) = e_{\mathrm{OPT}}(x) = 0$, then CLS is optimal. Next suppose that $e_{\mathrm{CLS}}(x) \neq 0$, which occurs if and only if ^{3} $e_{\mathrm{OPT}}(x) \neq 0$. In this case we derive from (7.72):

$$0 < \frac{\|e_{\mathrm{OPT}}(x)\|_2^2}{\|e_{\mathrm{CLS}}(x)\|_2^2} = \frac{\|P_{\mathcal{W}^\perp} e_{\mathrm{CLS}}(x)\|_2^2}{\|e_{\mathrm{CLS}}(x)\|_2^2} \leq \sin^2(\mathcal{S}^\perp, \mathcal{W}), \tag{7.73}$$

obtaining the lower bound. Note that $\sin(\mathcal{S}^\perp, \mathcal{W}) \leq 1$ with equality only if $\mathcal{W} = \mathcal{S}$ (in which case $e_{\mathrm{CLS}}(x) = e_{\mathrm{CMX}}(x) = e_{\mathrm{OPT}}(x)$).

As with the bounds (7.67), it can be shown that the bounds of (7.68) are tight, by taking $v \in \mathcal{S}^\perp$ which achieves the maximum (minimum) angle with respect to $\mathcal{W}^\perp$ and constructing $x = v + w$ where $w \in \mathcal{W}$ satisfies $P_{\mathcal{S}} w = S(S^* S)^{-1} c$ (so that $S^* x = c$). $\qquad\square$

---

^{2} Here we assume that the inf and the sup in the definitions of the angles can be replaced by min and max, respectively. We refer the reader to Theorem 2 of [108] for sufficient conditions for the above to hold, in the case of SI spaces.

^{3} An equivalent claim is $e_{\mathrm{CLS}}(x) = 0$ if and only if $e_{\mathrm{OPT}}(x) = 0$. Indeed, when $e_{\mathrm{CLS}}(x) = 0$, trivially $e_{\mathrm{OPT}}(x) = 0$. On the other hand, assuming $e_{\mathrm{OPT}}(x) = 0$, (7.72) implies that $e_{\mathrm{CLS}}(x) \in \mathcal{W}$. Since $e_{\mathrm{CLS}}(x) \in \mathcal{S}^\perp$, and $\mathcal{W} \cap \mathcal{S}^\perp = \{0\}$, we must have $e_{\mathrm{CLS}}(x) = 0$.

Bound comparison

Using the bounds of Theorem 7.5 we can identify regions of $\|e_{\mathrm{OPT}}(x)\|_2$ for which the regret approach is preferable to the CLS method, for all values of x, and vice versa. Specifically, if the upper bound in (7.67) is smaller than the lower bound in (7.68), then the norm of the error resulting from CLS will be larger than that resulting from the regret approach. Manipulating the equations, it can be shown that this occurs when

$$\|e_{\mathrm{OPT}}(x)\|_2^2 \geq \gamma_1 \|P_{\mathcal{S}^\perp} x\|_2^2, \tag{7.74}$$

where the constant γ_1 is given by

$$\gamma_1 = \frac{\sin^2(\mathcal{S}^\perp, \mathcal{W}) \sin^2(\mathcal{W}, \mathcal{S})}{\cos^2(\mathcal{S}^\perp, \mathcal{W})}. \tag{7.75}$$

Since the numerator of (7.75) is no larger than 1, a sufficient condition to ensure a lower error using the regret reconstruction is

$$\|e_{\mathrm{OPT}}(x)\|_2^2 \geq \frac{1}{\cos^2(\mathcal{S}^\perp, \mathcal{W})} \|P_{\mathcal{S}^\perp} x\|_2^2. \tag{7.76}$$

Evidently, if $\mathcal{W}$ is close to $\mathcal{S}^\perp$ and most of the signal energy is within the sampling space, then the minimax-regret method will result in a lower error than CLS.

Similarly, by comparing the worst-case bound on the CLS reconstruction error with the best-case bound on the regret error, we can show that if

$$\|e_{\mathrm{OPT}}(x)\|_2^2 \leq \gamma_2 \|P_{\mathcal{S}^\perp} x\|_2^2, \tag{7.77}$$

where

$$\gamma_2 = \frac{\cos^2(\mathcal{W}, \mathcal{S}) \cos^2(\mathcal{S}^\perp, \mathcal{W})}{\sin^2(\mathcal{W}, \mathcal{S})}, \tag{7.78}$$

then CLS results in a lower error. A sufficient condition is

$$\|e_{\mathrm{OPT}}(x)\|_2^2 \leq \cos^2(\mathcal{W}, \mathcal{S}) \cos^2(\mathcal{S}^\perp, \mathcal{W}) \|P_{\mathcal{S}^\perp} x\|_2^2. \tag{7.79}$$

These results are illustrated in Fig. 7.10.

As evident from the figure, when $\|e_{\mathrm{OPT}}(x)\|_2^2$ is large (i.e. most of the signal energy is not within the reconstruction space), or the bound $\gamma_1 \|P_{\mathcal{S}^\perp} x\|_2^2$ is small (i.e. most of the signal energy is within the sampling space and $\mathcal{W}$ is "close" to $\mathcal{S}^\perp$), minimax-regret outperforms CLS. Conversely, for small values of $\|e_{\mathrm{OPT}}(x)\|_2^2$, CLS is preferable.

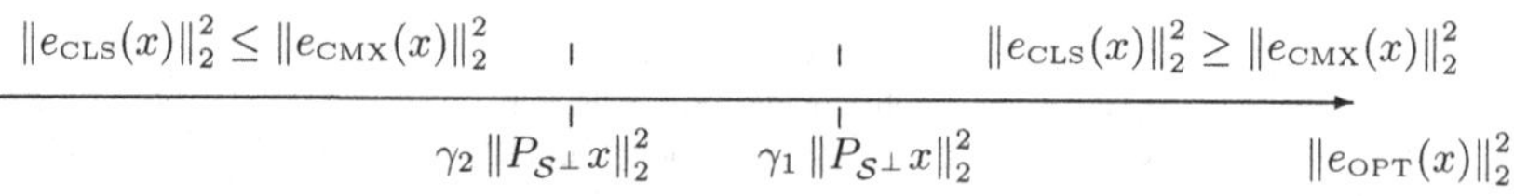

Figure 7.10 Regions of $\|e_{\mathrm{OPT}}(x)\|_2^2$ in which the regret reconstruction leads to a smaller error than CLS recovery, and vice versa.

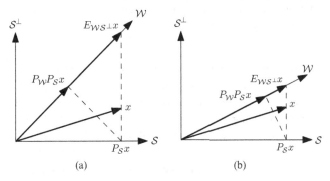

Figure 7.11 Comparison of minimax-regret and CLS reconstructions for two different choices of $\mathcal{W}$ satisfying $\mathcal{H} = \mathcal{W} \oplus \mathcal{S}^{\perp}$. (a) The minimax strategy $(P_{\mathcal{W}} P_{\mathcal{S}} x)$ is preferable to CLS $(E_{\mathcal{W}\mathcal{S}^{\perp}} x)$ when $\mathcal{W}$ is "far" from $\mathcal{S}$. (b) Both methods lead to errors of the same order of magnitude when $\mathcal{W}$ is "close" to $\mathcal{S}$.

These insights are illustrated geometrically in Fig. 7.11. Figure 7.11(a) depicts the CLS and regret reconstructions when $\mathcal{W}$ is far from $\mathcal{S}$. In this case, it is apparent that the error resulting from CLS is large with respect to minimax regret. In Fig. 7.11(b), $\mathcal{W}$ and $\mathcal{S}$ are close, and the errors have roughly the same magnitude.

7.3 Stochastic priors

Until now we have considered deterministic smoothness priors which translated into an L_2-norm bound on the unknown signal. Smoothness can also be interpreted in a stochastic setting where the norm is replaced by the signal's second-order statistics. Loosely speaking, such constraints result from replacing the square norm by its expected value. In contrast to the deterministic setting, in which the prior information is generally not sufficient to minimize the squared error between the recovered signal $\hat{x}$ and the unknown signal x, in the stochastic setting the situation is much more favorable. Replacing the squared error by the average, or projected, MSE leads to a well-defined criterion which can be optimized. This will be made clear in the ensuing derivations. Interestingly, it turns out the optimal solutions resulting from such considerations coincide with the minimax recovery methods treated in the previous sections where the weighting L is replaced by the inverse power spectrum density of the stochastic process.

For simplicity, we focus our attention in this section on the SI setting, although the results of course hold more generally. Specifically, we assume that the unknown signal $x(t)$ is a WSS random process having a power spectral density (PSD) function $\Lambda_x(\omega)$ (see Appendix B for background on random processes). Our goal is to linearly estimate $x(t)$ given the samples $c[n]$. Since our main emphasis in this book is on deterministic settings, we will not consider in detail derivations that relate to manipulating random signals, but rather concentrate on the results and their interpretation. Detailed derivations can be found in [150].

Before delving into the details, we first provide an example of a stochastic signal prior that arises in many applications.

Example 7.7 A common signal prior is associated with the family of Matérn WSS processes [112]. The reason for the popularity of this class of priors is rooted in the fact that the PSD of Matérn processes decays polynomially with frequency, a phenomenon which is shared by numerous types of signals in nature. A few examples are natural images, refractive index variations in biological tissues, and air turbulence.

For 1D signals, the PSD of a Matérn process is of the form

$$\Lambda_x(\omega) = \sigma^2 \prod_{m=1}^{K} \frac{1}{(a_m + \omega^2)^{\gamma_m}}, \tag{7.80}$$

where $a_m > 0$ and $\gamma_m \geq 1$. The autocorrelation function associated with this PSD has a closed-form, though somewhat cumbersome, expression. Here, the parameters γ_m determine the decay rate of the frequency content and σ^2 is proportional to the variance of the process. A few realizations of Gaussian Matérn processes with different parameters are depicted in Fig. 7.12.

A special case of this prior, which is simple to analyze, corresponds to $K = 1$, $\gamma_1 = 1$, resulting in the autocorrelation function

$$r_x(\tau) = \frac{\sigma^2}{2a_1} e^{-a_1|\tau|}. \tag{7.81}$$

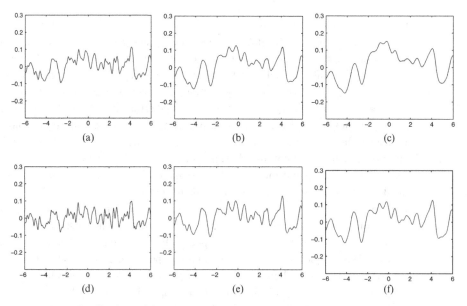

Figure 7.12 Realizations of Gaussian Matérn processes with $K = 1$. (a) $a_1 = 0.25$, $\gamma_1 = 1$. (b) $a_1 = 0.5$, $\gamma_1 = 1$. (c) $a_1 = 0.75$, $\gamma_1 = 1$. (d) $a_1 = 0.25$, $\gamma_1 = 2$. (e) $a_1 = 0.5$, $\gamma_1 = 2$. (f) $a_1 = 0.75$, $\gamma_1 = 2$.

In this case the statistical relation (in the second-order sense) between the variables $x(t_1)$ and $x(t_2)$ decays exponentially with the distance $|t_2 - t_1|$ between the points.

7.3.1 The hybrid Wiener filter

The study of sampling random signals was initiated in the late 1950s by Balakrishnan [151]. His well-known sampling theorem states that a bandlimited WSS random signal $x(t)$ can be perfectly reconstructed in an MSE sense from its uniform samples whenever the sampling rate exceeds twice the signal's bandwidth, where the bandwidth is measured by that of the PSD. Reconstruction is achieved by using the sinc function as an interpolation kernel. This theorem is the analog of the Shannon–Nyquist theorem for the stochastic case, where the transform of the signal is replaced by the transform of its correlation function, and the interpolation formula holds in an MSE sense. A detailed statement of the theorem is given in Appendix B.

Balakrishnan's result was later extended by several authors to account for some of its practical limitations. In [152] a sampling theorem for bandpass and multiband WSS signals was developed. It was shown that under certain conditions on the support of the signal's spectrum $\Lambda_x(\omega)$, perfect reconstruction in an MSE sense is possible using an interpolation filter with the same support. This was a first departure from the bandlimited case to broader classes of random signals. We will consider similar results for deterministic signals in Chapter 14. The more general setting in which nonideal samples of the signal are given was treated in [112, 153, 154]. This is the setting of interest to us here.

We first examine constraint-free reconstruction. In the deterministic setting with a smoothness prior we could not minimize the squared error $\|\hat{x}(t) - x(t)\|^2$ for all smooth $x(t)$, and therefore discussed LS and minimax methods. In contrast, in the stochastic case we can use the PSD $\Lambda_x(\omega)$ of $x(t)$ in order to minimize the MSE $E\{|x(t) - \hat{x}(t)|^2\}$ for every t, which depends only on the statistics of $x(t)$ and not on the signal itself.

We minimize the MSE by linear processing of the samples $c[n]$. As opposed to the common Wiener filtering problem, where both the input and output are either continuous- or discrete-time signals, here we are interested in estimating a continuous-time signal $x(t)$ based on equidistant samples of $y(t) = x(t) * s(-t)$. This problem is therefore referred to as *hybrid Wiener filtering*. The reconstruction $\hat{x}(t)$ minimizing the MSE can be implemented by the block diagram in Fig. 6.2 with an interpolation kernel

$$\widetilde{W}(\omega) = S(\omega)\Lambda_x(\omega), \tag{7.82}$$

and digital correction filter

$$H(e^{j\omega}) = \frac{1}{R_{S\widetilde{W}}(e^{j\omega})}. \tag{7.83}$$

A proof of this result can be found in [112, 153, 154]. It is interesting to observe that (7.82) and (7.83) are identical to (7.13) and (7.14) with $\Lambda_x(\omega) = |L(\omega)|^{-2}$. Therefore, the smoothness operator in the deterministic case corresponds to the whitening filter of the input $x(t)$ in the stochastic setting (see Appendix B for a more detailed discussion on the whitening filter).

This analogy carries over to the multichannel setting as well. Specifically, suppose that samples of $x(t)$ are collected at the output of N sampling filters $s_1(t), \ldots, s_N(t)$. Then the linear system minimizing the MSE is the same as that presented in Section 7.1.5, with $|L(\omega)|$ replaced by $1/\sqrt{\Lambda_x(\omega)}$. This system is referred to as the *vector hybrid Wiener filter*, since the input is a discrete-time vector process $\mathbf{c}[n] = [c_1[n] \ldots c_N[n]]^T$ and the output $\hat{x}(t)$ is a continuous-time signal.

Example 7.8 The hybrid Wiener and deterministic smoothness prior approaches lead to the same solutions. Nevertheless, the stochastic interpretation of these solutions provides further insight into the types of signals to which a certain recovery kernel is best matched. For example, if the samples are pointwise, then $S(\omega) = 1$ and the optimal recovery kernel (7.82) is given by $\widetilde{W}(\omega) = \Lambda_x(\omega)$. This implies that a given kernel $\tilde{w}(t)$ is best suited to recovering random signals whose autocorrelation function is $r_x(\tau) = \tilde{w}(\tau)$.

Figure 7.13 depicts realizations of four Gaussian random processes to which the reconstruction kernels $\mathrm{sinc}(t)$, $\beta^1(t)$, $\beta^3(t)$, and $0.5^2(0.5^2+t^2)^{-1}$ are best matched. The first three are widely used in practical applications in which the sampling interval is $T = 1$. The fourth corresponds to a smoothness prior with exponential frequency decay. As can be seen, whereas the sinc kernel is best matched to signals that vary smoothly on intervals whose length is of order T, linear interpolation (corresponding to the kernel $\beta^1(t)$) is beneficial in circumstances in which the signal rapidly varies in between every two consecutive samples. In this respect, the cubic spline and exponential decay filters are in between these two extremes.

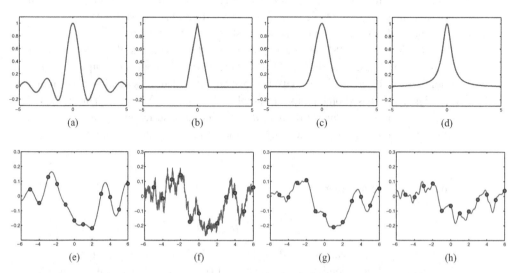

Figure 7.13 Reconstruction kernels and realizations of the Gaussian random signals they are best matched to. (a) $\tilde{w}(t) = \mathrm{sinc}(t)$. (b) $\tilde{w}(t) = \beta^1(t)$. (c) $\tilde{w}(t) = \beta^3(t)$. (d) $\tilde{w}(t) = 0.5^2(0.5^2 + t^2)^{-1}$. (e)–(h) Gaussian random processes to which the kernels (a)–(d) are best suited.

7.3.2   Constrained reconstruction

We now treat a constrained setting, in which the interpolation filter is fixed in advance. Unfortunately, in this case, for a general given interpolation kernel, there is no digital correction filter that minimizes the MSE for every t [129]. In fact, the filter minimizing the MSE at a certain time instant t_0 also minimizes the MSE at times $\{t_0 + nT\}$ for all integer n, but not over the whole continuum. Therefore, error measures other than pointwise MSE must be considered. Before treating the problem of choosing an appropriate criterion, we first discuss how this time-dependence phenomenon is related to artifacts commonly encountered in certain interpolation methods.

The signal $x(t)$ in our setup is assumed to be WSS and, consequently, the sequence of samples $c[n]$ is a discrete WSS random process, as is the output $d[n]$ of the digital correction filter in Fig. 6.2. The reconstruction $\hat{x}(t)$ is formed by modulating the shifts of the kernel $w(t)$ by the WSS discrete-time process $d[n]$. Assuming that the PSD of $d[n]$ is positive everywhere, signals of this type are not stationary unless $w(t)$ is π-bandlimited [150]. Generally, $\hat{x}(t)$ will be a second-order cyclostationary process, i.e. its second-order statistics vary periodically in time. In practice, the interpolation kernels in use have a finite (and usually small) support, and are therefore not bandlimited. In these cases, the periodic correlation in $\hat{x}(t)$ often degrades the reconstruction quality, as subjectively perceived by the visual or auditory system.

Note that although natural signals are rarely stationary to begin with, it is still relevant to study how an interpolation algorithm reacts to stationary signals. In fact, if an interpolation scheme outputs a cyclostationary signal when fed with a stationary input, then it will commonly also produce reconstructions with degraded subjective quality when applied to real-world signals, as demonstrated in Fig. 7.14.

The nonstationary behavior of $\hat{x}(t)$ is the reason why the pointwise MSE cannot be minimized for every t in general. Two alternative error measures are the sampling-period-average MSE and the projected MSE.

The sampling-period-average MSE utilizes the periodicity of the MSE, and integrates it over one period [150]:

$$\text{MSE}_A = \frac{1}{T} \mathsf{E} \left\{ \int_{t_0}^{t_0+T} |x(t) - \hat{x}(t)|^2 dt \right\}, \qquad (7.84)$$

where t_0 is an arbitrary point in time and E denotes the expected value. It turns out that minimization of the average MSE leads to a correction filter independent of t_0 [150]. The second approach makes use of the fact that the best possible approximation to $x(t)$ in $\mathcal{W}$ is $P_{\mathcal{W}}x(t)$. Therefore this method aims at minimizing the projected MSE, defined as the MSE with respect to the optimal approximation in $\mathcal{W}$ [129]:

$$\text{MSE}_P = \mathsf{E} \left\{ |P_{\mathcal{W}}x(t) - \hat{x}(t)|^2 \right\}. \qquad (7.85)$$

This criterion is analogous to the minimax-regret approach in the deterministic setting.

Interestingly, both error measures (7.84) and (7.85) lead to the same digital correction filter, which is given by [129, 150]

(a) Original low-resolution image.

(b) Rectangular kernel.

(c) Bicubic kernel.

(d) Sinc kernel.

Figure 7.14 Periodic structure in an interpolated signal is a phenomenon related to the effective bandwidth of the interpolation kernel. The larger the portion of its energy outside $[-\pi, \pi]$, the stronger the periodic correlation. The three images on the right were obtained by scaling a patch of the original image in (a) by a factor of 5 using three different methods. The portion of energy in the range $[-\pi, \pi]$ of the kernels is: (b) rectangular kernel – 61%. (c) bicubic kernel – 91%. (d) sinc – 100%. Suppressed periodic correlation, however, does not necessarily imply that the reconstruction error is small.

$$H(e^{j\omega}) = \frac{R_{W\widetilde{W}}\left(e^{j\omega}\right)}{R_{S\widetilde{W}}\left(e^{j\omega}\right) R_{WW}\left(e^{j\omega}\right)}, \qquad (7.86)$$

where $\widetilde{W}(\omega) = S(\omega)\Lambda_x(\omega)$ here. This is also the solution obtained by the minimax-regret criterion (see (7.56)) where $|L(\omega)|^{-2}$ replaces the spectrum $\Lambda_x(\omega)$. Therefore, here again, $L(\omega)$ plays the role of the whitening filter of $x(t)$.

The mathematical equivalence between the minimax and Wiener formulas suggests selecting an "optimal" operator $L(\omega)$ in the minimax formulation that "whitens" the signal. In practice, one can either choose $L(\omega)$ in advance to approximately whiten signals typically encountered in a specific application or specify a parametric form for $L(\omega)$ and optimize the parameters based on the samples $c[n]$ [111].

Example 7.9 Suppose that we observe pointwise samples, separated $T = 1$ seconds apart, of a random signal $x(t)$ whose autocorrelation function is $\beta^p(t)$. We would like to recover $x(t)$ from its samples by minimizing the MSE.

We first determine the unconstrained hybrid Wiener filter, given by (7.82) and (7.83). In our setting, $S(\omega) = 1$ since the samples are pointwise, and therefore $\tilde{w}(t) = \beta^p(t)$. Furthermore, $R_{S\widetilde{W}}(e^{j\omega})$ corresponds to a filter with impulse response $\beta^p(n)$. Therefore, (7.83) is simply the convolutional inverse of $\beta^p(n)$.

Next, we constrain the reconstruction kernel to be $w(t) = \beta^0(t)$. The optimal digital correction filter is now given by (7.86). Since $r_{ww}[n] = \delta[n]$ in this situation, $R_{WW}(e^{j\omega}) = 1$ and therefore the denominator of (7.86) is the same as the denominator of the unconstrained filter (7.83). The numerator of (7.86) corresponds to filtering with $r_{\beta^0\beta^p}[n] = \beta^{p+1}(n)$. Thus, in the constrained setting, an additional digital FIR filter is applied. This filter has a smoothing effect, which is required in order to compensate for the nonsmooth behavior of the reconstruction kernel $\beta^0(t)$.

Figure 7.15 depicts the constrained and unconstrained Wiener reconstructions corresponding to $p = 1$ and $p = 3$. Note that the constrained recoveries are not consistent in the sense that they do not pass through the samples.

7.4 Summary of sampling methods

In this section we summarize the recovery methods developed so far in this chapter and the previous one. We highlight commonalities between the various approaches and provide a unified view of the different solutions. Some of the insights we discuss below were also mentioned in Section 6.7, where we reviewed recovery based on subspace priors. Here we show that the unified view and implications drawn in that context are valid for smoothness priors as well.

7.4.1 Summary of methods

Table 7.1 summarizes the reconstruction techniques. We use the superscripts "sub" and "smo" to signify whether a solution corresponds to a subspace or a smoothness prior. The transformations $\widetilde{W}$ and $\widehat{W}$ are given by

$$\widetilde{W} = (L^*L)^{-1}S \tag{7.87}$$

and

$$\widehat{W} = W(W^*L^*LW)^{-1}W^*S \tag{7.88}$$

Table 7.1 Reconstruction from noiseless samples

Prior	Unconstrained ($\hat{x} \in \mathcal{H}$)		Constrained ($\hat{x} \in \mathcal{W}$)	
	Least squares	**Minimax**	**Least squares**	**Minimax**
$x \in \mathcal{A}$	$\hat{x}_{\text{LS}}^{\text{sub}} = A(S^*A)^\dagger c$	$\hat{x}_{\text{MX}}^{\text{sub}} = \hat{x}_{\text{LS}}^{\text{sub}}$	$\hat{x}_{\text{CLS}}^{\text{sub}} = W(S^*W)^\dagger S^*\hat{x}_{\text{LS}}^{\text{sub}}$	$\hat{x}_{\text{CMX}}^{\text{sub}} = P_\mathcal{W}\hat{x}_{\text{MX}}^{\text{sub}}$
$\|Lx\| \le \rho$	$\hat{x}_{\text{LS}}^{\text{smo}} = \widetilde{W}(S^*\widetilde{W})^{-1}c$	$\hat{x}_{\text{MX}}^{\text{smo}} = \hat{x}_{\text{LS}}^{\text{smo}}$	$\hat{x}_{\text{CLS}}^{\text{smo}} = \widehat{W}(S^*\widehat{W})^\dagger S^*\hat{x}_{\text{LS}}^{\text{smo}}$	$\hat{x}_{\text{CMX}}^{\text{smo}} = P_\mathcal{W}\hat{x}_{\text{MX}}^{\text{smo}}$

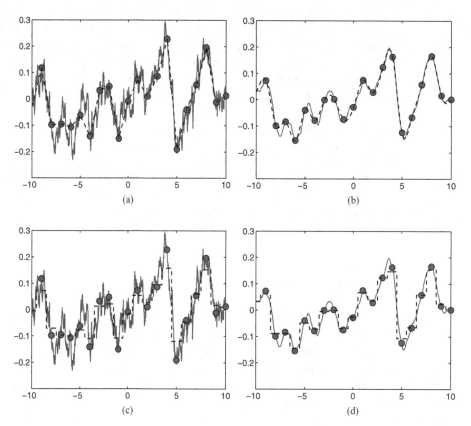

Figure 7.15 Wiener reconstruction $\hat{x}(t)$ (dashed line) of a random signal $x(t)$ (solid line) with autocorrelation $\beta^p(t)$ from pointwise samples (indicated by circles). (a) Unconstrained solution, $p = 1$. (b) Unconstrained solution, $p = 3$. (c) Constrained solution with $w(t) = \beta^0(t)$, $p = 1$. (d) Constrained solution with $w(t) = \beta^0(t)$, $p = 3$.

respectively. Note that in the table we use the pseudoinverse for S^*A since this operator is not necessarily invertible (it is invertible only if $\mathcal{H} = \mathcal{A} \oplus \mathcal{S}^\perp$). On the other hand, $S^*\widetilde{W}$ is always invertible because by our assumption L^*L is bounded and S corresponds to a Riesz basis.

This table highlights the key observations discussed in the previous two chapters. We begin by examining the case in which no constraint is imposed on $\hat{x}$, shown in columns

1 and 2. First, we see that in this situation the LS and minimax reconstructions coincide. This property holds true for both subspace and smoothness priors. Second, smoothness-prior recovery (row 2) has the same structure as subspace-prior recovery (row 1) with $\widetilde{W}$ replacing A. Therefore, we can interpret $\widetilde{\mathcal{W}} = \mathcal{R}(\widetilde{W})$ as the optimal reconstruction space associated with the smoothness prior. Finally, we note that for certain subspace priors, perfect recovery can be achieved, leading to $\hat{x}_{\text{LS}}^{\text{sub}} = x$. Specifically, this happens if the sampling space $\mathcal{S}$ and the prior space $\mathcal{A}$ satisfy the direct-sum condition $\mathcal{H} = \mathcal{A} \oplus \mathcal{S}^{\perp}$. In the smoothness prior case, the direct-sum condition $\mathcal{H} = \widetilde{\mathcal{W}} \oplus \mathcal{S}^{\perp}$ does not imply perfect recovery because the original x does not necessarily lie in $\widetilde{\mathcal{W}}$. Instead, recovery in this setting can be interpreted as an oblique projection of the (unknown) signal x onto the optimal reconstruction space $\widetilde{\mathcal{W}}$, namely $\hat{x}_{\text{LS}}^{\text{smo}} = E_{\widetilde{\mathcal{W}}\mathcal{S}^{\perp}} x$.

We now examine the case in which the recovery is constrained to lie in $\mathcal{W}$ (columns 3 and 4). These solutions are expressed in Table 7.1 in terms of the unconstrained reconstructions (columns 1 and 2). The minimax-regret solutions (column 4) are related to the unconstrained recoveries (column 2) via an orthogonal projection onto the reconstruction space $\mathcal{W}$. This implies that $\|\hat{x}_{\text{CMX}}\| \le \|\hat{x}_{\text{MX}}\|$. When the sampling and reconstruction spaces satisfy the direct-sum $\mathcal{H} = \mathcal{W} \oplus \mathcal{S}^{\perp}$, and in addition $\mathcal{H} = \mathcal{A} \oplus \mathcal{S}^{\perp}$ for the subspace prior, both CLS solutions of column 3 become $\hat{x}_{\text{CLS}} = E_{\mathcal{W}\mathcal{S}^{\perp}} x$. In contrast to an orthogonal projection, an oblique projection may lead to solutions with arbitrary large norm, given that $\mathcal{W}$ is sufficiently far apart from $\mathcal{S}$. Therefore, the error in the constrained CLS framework is not guaranteed to be bounded, unless a bound on the "distance" between $\mathcal{S}$ and $\mathcal{W}$ is known a priori. We further note that recovery in this case does not depend on the prior, i.e. $\hat{x}_{\text{CLS}}^{\text{sub}}$ is not a function of A and $\hat{x}_{\text{CLS}}^{\text{smo}}$ does not depend on L. These properties are clearly undesirable and can lead to unsatisfactory results in practical applications, as demonstrated in Example 7.5.

Table 7.2 summarizes the recovery formulas obtained under the direct-sum assumptions discussed above. The expressions in row 1 are true when $\mathcal{H} = \mathcal{A} \oplus \mathcal{S}^{\perp}$, while the recoveries of column 3 are obtained under the assumption that $\mathcal{H} = \mathcal{W} \oplus \mathcal{S}^{\perp}$.

In the SI setting, namely when $\mathcal{S}$, $\mathcal{A}$, and $\mathcal{W}$ are SI spaces with generators $s(t)$, $a(t)$, and $w(t)$ respectively and L is an LTI operator corresponding to the filter $L(\omega)$, all the reconstruction methods of Table 7.1 can be implemented by digitally filtering the samples $c[n]$ prior to reconstruction, as depicted in Fig. 6.2. The resulting interpolation methods are summarized in Table 7.3 where the numbers indicate the equation numbers containing the reconstruction formulas of the digital correction filter and reconstruction kernel. The optimal kernel corresponding to unconstrained recovery with a subspace

Table 7.2 Reconstruction from noiseless samples under direct-sum assumptions

Prior	Unconstrained ($\hat{x} \in \mathcal{H}$)		Constrained ($\hat{x} \in \mathcal{W}$)	
	Least squares	**Minimax**	**Least squares**	**Minimax**
$x \in \mathcal{A}$	$\hat{x}_{\text{LS}}^{\text{sub}} = x$	$\hat{x}_{\text{MX}}^{\text{sub}} = \hat{x}_{\text{LS}}^{\text{sub}}$	$\hat{x}_{\text{CLS}}^{\text{sub}} = E_{\mathcal{W}\mathcal{S}^{\perp}} x$	$\hat{x}_{\text{CMX}}^{\text{sub}} = P_{\mathcal{W}} x$
$\|Lx\| \le \rho$	$\hat{x}_{\text{LS}}^{\text{smo}} = E_{\widetilde{\mathcal{W}}\mathcal{S}^{\perp}} x$	$\hat{x}_{\text{MX}}^{\text{smo}} = \hat{x}_{\text{LS}}^{\text{smo}}$	$\hat{x}_{\text{CLS}}^{\text{smo}} = E_{\mathcal{W}\mathcal{S}^{\perp}} x$	$\hat{x}_{\text{CMX}}^{\text{smo}} = P_{\mathcal{W}} \hat{x}_{\text{MX}}^{\text{smo}}$

Table 7.3 Reconstruction from noiseless samples in SI spaces

Prior	Unconstrained ($\hat{x} \in \mathcal{H}$)		Constrained ($\hat{x} \in \mathcal{W}$)	
	Least squares	**Minimax**	**Least squares**	**Minimax**
$x \in \mathcal{A}$	(6.90)	(6.90)	(6.115)	(6.126)
$\|Lx\| \leq \rho$	(7.13), (7.14)	(7.13), (7.14)	(7.53)	(7.56)

Table 7.4 Prior filter for different setups

	Subspace prior	Smoothness prior	Stochastic prior
Assumption	$x(t) = \sum d[n]a(t-n)$	$\int \lvert L(\omega)X(\omega)\rvert^2 d\omega \leq \rho^2$	$x(t)$ WSS with PSD $\Lambda_x(\omega)$
Prior filter	$A(\omega)$	$S(\omega)/\lvert L(\omega)\rvert^2$	$S(\omega)\Lambda_x(\omega)$

prior (row 1, columns 1 and 2) is $a(t)$, while with a smoothness prior (row 2, columns 1 and 2) the kernel is $\tilde{w}(t)$. The interpolation kernel in the constrained case (columns 3 and 4) is $w(t)$. The direct-sum conditions, under which Table 7.2 was constructed, can be easily verified in SI spaces using Corollary 6.1.

7.4.2   Unified view

The summary in the previous section highlights the fact that although the recovery methods discussed emerge from different assumptions on the underlying signal, the sampling process, and the reconstruction mechanism, their structure is often similar. Whereas it is generally well understood how to model the sampling process in real-world applications, choosing the signal prior and the reconstruction scheme is typically left to the practitioner. These components affect both the performance and the computational load of the resulting algorithm. Below, we emphasize commonalities and equivalence between the different methods in order to help the user design the most appropriate filter for a particular application. For brevity, the discussion is focused on the SI setting, and under the assumption that $\mathcal{H} = \mathcal{A} \oplus \mathcal{S}^\perp$ in the subspace case. However, similar conclusions hold for the general Hilbert space setting as well.

The linear recovery algorithms corresponding to Table 7.3 share a common structure, also presented in Section 6.7. The digital correction filter $H(e^{j\omega})$ of Fig. 6.2 can be written in all cases in the form

$$H\left(e^{j\omega}\right) = \frac{R_{WP}\left(e^{j\omega}\right)}{R_{SP}\left(e^{j\omega}\right)R_{WW}\left(e^{j\omega}\right)}, \tag{7.89}$$

where $R_{SP}(e^{j\omega})$ is defined by (5.28). Here $S(\omega)$ and $W(\omega)$ are the CTFTs of the sampling and reconstruction filters, and $P(\omega)$, referred to as the prior filter, shapes the spectrum of $\hat{x}(t)$ according to the prior. The different priors together with the corresponding filters are summarized in Table 7.4.

The reconstruction filter $W(\omega)$ can either be chosen in advance so as to lead to efficient implementation, or can be optimized according to the prior. The solutions in the unconstrained case can be recovered from (7.89) with $W(\omega) = P(\omega)$, in which case the filter of (7.89) reduces to

$$H\left(e^{j\omega}\right) = \frac{1}{R_{SP}\left(e^{j\omega}\right)}. \qquad (7.90)$$

Substituting the values of the prior filter $P(e^{j\omega})$ into (7.90) according to Table 7.4 leads to the first two columns of Table 7.3. This filter also guarantees perfect recovery for any signal lying in the SI space spanned by the functions $\{p(t - nT)\}$, offering an additional viewpoint on the prior filter $P(\omega)$: it defines the SI space for which perfect recovery is obtained using (7.90).

When the reconstruction filter $W(\omega)$ is fixed in advance, substituting the values of $P(\omega)$ from Table 7.4 into (7.89) results in the minimax solutions – column 4 of Table 7.3. The CLS solutions follow from choosing $P(\omega) = W(\omega)$.

The unified interpretation of the different interpolation algorithms highlights the importance of choosing the prior filter. This filter should be matched to the typical frequency content of the input signals as best as possible. In addition, we have seen that a general purpose recovery algorithm (i.e. one which can handle resampling at arbitrary points) requires an explicit expression for $w(t)$ in the time domain. This should be taken into consideration when choosing the prior in the unconstrained approach since $w(t) = p(t)$ in this case. Therefore, a kernel $p(t)$ with an analytic formula is beneficial.

Finally, we comment briefly on the reconstruction filter $w(t)$. The key consideration in choosing $w(t)$ is its support, which determines the number of coefficients of the corrected sequence $d[n]$ participating in computing $\hat{x}(t_0) = \sum d[n]w(t_0 - nT)$. Typically, kernels with support up to $4T$ are used, requiring four multiplications per time instant t_0 to compute $\hat{x}(t_0)$ (or 16 in two dimensions). These include B-splines of degree 0 to 3 whose supports are T to $4T$ respectively, the Keys cubic interpolation kernel [116] whose support is $4T$, and the Lanczos kernel [117] with support $4T$. Some of the commonly used kernels, such as Keys and Lanczos, possess the interpolation property, namely $w(nT) = \delta[n]$. This implies that if we are given pointwise samples of the signal $c[n] = x(nT)$, then no correction filter $h[n]$ is needed in order to obtain a consistent reconstruction satisfying $\hat{x}(nT) = c[n]$.

7.5   Sampling with noise

Until now we have considered the noiseless setting, in which the samples were given exactly. In practice, the samples are typically perturbed by noise so that $c = S^*x + u$, where $u[n]$ is an unknown noise sequence. To treat recovery in this context, we can follow similar approaches to those we used in the noiseless setting: LS and minimax-based designs. The LS methodology can be easily adapted to the noisy case by minimizing the error-in-samples $\|S^*x - c\|^2$ over the set of feasible signals. Thus, the optimization problems (6.81), (6.110), (7.1), and (7.48), which correspond to the unconstrained and constrained subspace and smoothness scenarios, remain valid here too. However, note that to solve these problems we assumed that signals x for which $S^*x = c$ (or $S^*x = P_{\mathcal{R}(W^*S)}c$ in the constrained setting) are included in the feasible set. When the samples are noisy, this is not necessarily true so that, for example, the optimal value of the unconstrained problem $\min_{x \in \mathcal{A}} \|S^*x - c\|^2$ is no longer 0. Nevertheless, it can be easily shown that the solutions we obtained under the subspace prior assumption (problems (6.81) and (6.110)) remain the same. Furthermore, under the smoothness prior, the LS and minimax strategies still coincide and the optimal reconstruction space $\widetilde{\mathcal{W}}$ of (7.5) stays the same; only the expansion coefficients of $\hat{x}$ in $\widetilde{\mathcal{W}}$ change. Interestingly, this property holds even when the ℓ_2-norm in the error-in-samples term $\|S^*x - c\|$ is replaced by an ℓ_p-norm with arbitrary $p \in [1, \infty]$ [125].

Since the approaches are similar in principle to the noiseless setting, we do not repeat all the derivations here. Instead, we focus on the smoothness prior and primarily point out the final answers with references to where the detailed derivations can be found. The equivalent results for subspace priors are derived in a similar manner.

We begin by treating the constrained scenario in which the reconstruction filter $w(t)$ is given in advance, and adapt the LS, minimax, and Wiener strategies to our context. As we will show, the solutions for the unconstrained case have the same form with an optimal choice of $w(t)$. The main difference with the noise-free setting lies in the fact that the samples are now random variables, because of the noise. The LS strategy remains unchanged since it is deterministic in nature: the noise properties are not taken into account within this criterion. The Wiener approach can be easily modified to fit our setting by taking the expected value of the squared error over the noise as well as the prior signal. To incorporate noise into the minimax strategy, we consider two possibilities. The first is based on replacing the squared error $\|x(t) - \hat{x}(t)\|^2$ by its expected value $\mathsf{E}\{\|x(t) - \hat{x}(t)\|^2\}$ where the expectation is over the noise. In the second approach, we leave the objective as before, but add a further constraint in the maximization step that $\|S^*x - c\| \leq \alpha$ where α is proportional to the noise standard deviation. It turns out that the first method results in a closed-form solution for the optimal filter [129]. On the other hand, the second approach leads to a more difficult optimization problem which can be solved numerically by relying on ideas from semidefinite and quadratic programming [124, 155]. Below, we focus on the first (MSE-based) objective.

7.5.1   Constrained reconstruction problem

The basic problem we treat is the recovery of a continuous-time signal $x(t)$ given some equally spaced, noisy measurements $c[n]$ of the output of the measurement device $s(-t)$:

$$c[n] = \langle s(t - nT), x(t) \rangle + u[n], \quad n \in \mathbb{Z}, \tag{7.91}$$

where the noise, $u[n]$, is a discrete zero-mean WSS process with known PSD. The measurement model is depicted in Fig. 7.16, in which $c_0[n] = \langle s(t - nT), x(t) \rangle$ denote the noise-free samples. The signal $x(t)$ can either be deterministic, or a zero-mean WSS signal with known PSD.

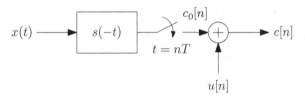

Figure 7.16 Noisy measurement model.

When the sampling is ideal, i.e. when $s(t) = \delta(t)$, we have an interpolation problem with noisy data; otherwise, a deconvolution problem with the distinctive feature that we recover a continuous-time solution. We seek a reconstruction $\hat{x}(t)$ that is included in some SI space $\mathcal{W}$, spanned by the integer shifts of a generating function $w(t)$ so that it has the form

$$\hat{x}(t) = \sum_{n \in \mathbb{Z}} d[n] w(t - nT), \tag{7.92}$$

and is specified in terms of the unknowns $d[n]$. The problem then is to design a correction filter $h[n]$ such that $\hat{x}(t)$ is close to $x(t)$ in some sense, where $\hat{x}(t)$ is specified by (7.92), with expansion coefficients $d[n]$ determined as

$$d[n] = h[n] * c[n]. \tag{7.93}$$

In our derivation in the deterministic setting, we required that $w(t)$ generates a Riesz basis. Since the noisy samples $c[n]$ specified by (7.91) are not necessarily in ℓ_2, we want a slightly stronger L_p stability condition for all $1 \le p \le \infty$, which can be enforced by the additional requirement [48]

$$\sup_{t \in [0,T]} \sum_{n \in \mathbb{Z}} |w(t - nT)| < \infty. \tag{7.94}$$

This condition implies that the continuous-time reconstruction $\hat{x}(t)$ will be bounded whenever its coefficients $d[n]$ are bounded, and vice versa (see Exercise 14). Globally, this ensures BIBO – bounded (discrete) input, bounded (continuous) output – behavior whenever the digital correction filter $h[n]$ is stable (i.e. $h \in \ell_1$).

We now investigate several different design criteria for $h[n]$ and derive the corresponding solutions. We begin with the LS approach which does not assume any prior information. This filter is derived for reference, and is equivalent to that we obtained with a subspace prior. We then move on to the LS objective with a smoothness constraint. This is equivalent to employing a regularization approach which is popular for solving inverse problems in the presence of noise. Since our smoothness prior takes on the form of a quadratic constraint, the resulting LS problem is equivalent to the Tikhonov technique used more generally for solving inverse problems. Departing partially from the deterministic setting, we next assume a random noise process and consider minimax MSE strategies. Finally, we adopt a Wiener-type formulation where both the signal and noise are treated as realizations of stationary processes, and minimize an appropriate MSE measure.

7.5.2   Least squares solution

We begin, for reference, with the LS solution with no prior information. In the noisy setting the LS objective becomes

$$\varepsilon_{\mathrm{LS}} = \sum_{n \in \mathbb{Z}} (\hat{c}[n] - c[n])^2 \tag{7.95}$$

where $c[n]$ is the noisy data and $\hat{c}[n]$ represents the samples derived from the reconstructed signal $\hat{x}(t)$:

$$\hat{c}[n] = \langle s(t - nT), \hat{x}(t) \rangle. \tag{7.96}$$

Using Parseval's theorem (3.66), the error (7.95) can equivalently be written as

$$\varepsilon_{\mathrm{LS}} = \frac{1}{2\pi} \int_0^{2\pi} \left| \widehat{C}(e^{j\omega}) - C(e^{j\omega}) \right|^2 d\omega, \tag{7.97}$$

where $C(e^{j\omega})$ and $\widehat{C}(e^{j\omega})$ are the discrete-time Fourier transforms of $c[n]$ and $\hat{c}[n]$, respectively. More generally, introducing a positive frequency weighting kernel $Q(e^{j\omega})$ leads to the weighted LS criterion

$$\varepsilon_{\mathrm{LS}} = \frac{1}{2\pi} \int_0^{2\pi} Q^{-1}(e^{j\omega}) \left| \widehat{C}(e^{j\omega}) - C(e^{j\omega}) \right|^2 d\omega. \tag{7.98}$$

In practice, the weighting is often chosen to be proportional to the noise spectrum when it is known.

To develop a solution to (7.98) we first note that (cf. (7.93))

$$\widehat{X}(\omega) = D(e^{j\omega T}) W(\omega) = H(e^{j\omega T}) C(e^{j\omega T}) W(\omega), \tag{7.99}$$

where $H(e^{j\omega})$ is the frequency response of the filter $h[n]$. Therefore, from (7.96),

$$\widehat{C}(e^{j\omega T}) = H(e^{j\omega T}) C(e^{j\omega T}) R_{SW}(e^{j\omega T}), \tag{7.100}$$

where we used the fact that $H(e^{j\omega T})$ and $C(e^{j\omega T})$ are $2\pi/T$-periodic. Clearly, $\varepsilon_{\mathrm{LS}}$ is minimized if we can choose $H(e^{j\omega})$ such that $\widehat{C}(e^{j\omega}) = C(e^{j\omega})$ for all $\omega \in \Omega$, where Ω denotes the set of frequencies over which $C(e^{j\omega}) R_{SW}(e^{j\omega}) \neq 0$. This can be achieved with

$$H_{\mathrm{LS}}(e^{j\omega}) = \frac{1}{R_{SW}(e^{j\omega})}, \quad \omega \in \Omega. \tag{7.101}$$

For $\omega \notin \Omega$ the filter $H_{\mathrm{LS}}(e^{j\omega})$ can be chosen arbitrarily. For convenience, throughout this section, we will use subscripts to denote the filters corresponding to the different criteria.

Interestingly, the LS filter does not depend on the frequency weighting kernel. This filter is equivalent to that obtained under a subspace prior in the noise-free setting so that, in effect, it does not take the noise into account. A drawback of the LS solution is that if $R_{SW}(e^{j\omega})$ is close to zero for some frequency ω, then $H_{\mathrm{LS}}(e^{j\omega})$ will be large at that frequency, leading to noise enhancement.

We next improve the performance by adding a smoothness prior and considering both LS and minimax strategies.

7.5.3 Regularized least squares

Suppose now that $\|Lx\| \leq \rho$ where L represents a filter with frequency response $L(\omega)$. In this case the LS criterion can be modified by minimizing the error subject to this constraint. An alternative approach is to add the constraint as a penalty, leading to the well-established Tikhonov regularization method (or regularized LS) for denoising:

$$\varepsilon_{\mathrm{CLS}} = \int_0^{2\pi} Q^{-1}(e^{j\omega}) \left|\widehat{C}(e^{j\omega}) - C(e^{j\omega})\right|^2 d\omega + \lambda \int_{-\infty}^{\infty} |L(\omega)|^2 |\widehat{X}(\omega)|^2 d\omega, \tag{7.102}$$

for some frequency weighting function $|L(\omega)|^2 > 0$ and scalar $\lambda \geq 0$, where $\widehat{C}(e^{j\omega})$ is given by (7.100) and $\widehat{X}(\omega)$ by (7.99).

The error measure $\varepsilon_{\mathrm{CLS}}$ can be viewed as the Lagrangian associated with the problem of minimizing $\varepsilon_{\mathrm{LS}}$ of (7.98) subject to the constraint that the reconstructed signal $\hat{x}(t)$ lies in the class $\mathcal{G}$ defined by

$$\mathcal{G} = \left\{ x(t) \colon \frac{1}{2\pi} \int_{-\infty}^{\infty} |L(\omega)|^2 |X(\omega)|^2 d\omega \leq \rho^2 \right\}. \tag{7.103}$$

In this case, it follows from the Karush–Kuhn–Tucker conditions [124] that at the optimal solution

$$\lambda \left(\|L\widehat{X}\|^2 - \rho^2\right) = 0, \tag{7.104}$$

so that either the inequality constraint is satisfied with equality for $\widehat{X}(\omega)$, or $\lambda = 0$.

The Tikhonov filter can be derived by imposing the condition that all Gateaux differentials are zero; details are given in [129]. The resulting filter is

$$H_{\mathrm{CLS}}(e^{j\omega}) = \frac{R_{WS}(e^{j\omega})}{|R_{WS}(e^{j\omega})|^2 + \lambda Q(e^{j\omega})\frac{1}{T}\sum_{k\in\mathbb{Z}}\left|L(\frac{\omega}{T} + \frac{2\pi k}{T})\right|^2 \left|W(\frac{\omega}{T} + \frac{2\pi k}{T})\right|^2}, \tag{7.105}$$

for frequencies ω such that $C(e^{j\omega}) \neq 0$. Although $H_{\mathrm{CLS}}(e^{j\omega})$ is arbitrary on values of ω for which $C(e^{j\omega}) = 0$, the choice of $H_{\mathrm{CLS}}(e^{j\omega})$ on these frequencies will not affect the reconstructed output $\hat{x}(t)$. For $\lambda = 0$, the Tikhonov filter (7.105) reduces to the LS filter (7.101). If $\lambda > 0$, then $|H_{\mathrm{CLS}}(e^{j\omega})| < |H_{\mathrm{LS}}(e^{j\omega})|$, since $|L(\omega)|^2 > 0$ and $|W(\omega)|^2 > 0$ for some ω because of the Riesz basis condition (5.14).

In the inequality-constrained version of the Tikhonov filter, λ must be chosen to satisfy (7.104). Denoting by $\hat{x}_{\mathrm{CLS},\lambda}$ the reconstructed signal resulting from the Tikhonov filter (7.105) for a fixed value of λ, this implies the following procedure for selecting λ: if $\|L\hat{x}_{\mathrm{CLS},0}\| \leq \rho$, then $\lambda = 0$. The solution in this setting coincides with the LS solution (7.101) which assumes no prior information. Otherwise, $\lambda > 0$ is a parameter that depends on the data $c[n]$, and is chosen such that $\|L\hat{x}_{\mathrm{CLS},\lambda}\| = \rho$.

7.5.4 Minimax MSE filters

Both the LS and constrained LS (Tikhonov) algorithms are based on minimizing a data-error criterion. However, in an estimation context, we typically would like to minimize the estimation error $x(t) - \hat{x}(t)$. In the deterministic setting we considered minimizing $\|\hat{x}(t) - x(t)\|^2$ or $\|\hat{x}(t) - P_{\mathcal{W}} x(t)\|^2$ for the worst-case value of x. Here, however, $\hat{x}(t)$ is a random signal because of the noise. Therefore, instead of minimizing the squared error we may consider the MSE over all noise realizations: $\mathsf{E}\{|\hat{x}(t) - x(t)|^2\}$. (Note that we eliminated the norm since the signal $\hat{x}$ will typically have infinite norm owing to the stationary noise.) Computing the MSE in our setting in which the signal $x(t)$ is deterministic shows that it depends explicitly on $x(t)$, and therefore cannot be minimized. (This is in contrast with the stochastic setting in which $x(t)$ is a stationary random process, as we discuss in Section 7.5.5.) Assuming, as in the previous section, that $x(t)$ belongs to the class $\mathcal{G}$ defined by (7.103), we can obtain a signal-independent error measure by considering the worst-case MSE on $\mathcal{G}$. The generic case in which the only information we have is that $x(t) \in L_2$ can be treated in this framework by choosing $L(\omega) = 1$ and $\rho \to \infty$.

Having combated the signal dependence of the MSE, we are now faced with another problem (which remains also in the stochastic case): it turns out that the filter minimizing the worst-case MSE depends on the time index $t = t_0$ so that, in principle, a different filter $h[n]$ is optimal for each t_0. To obtain a fixed solution $h[n]$ for all t_0 we develop two strategies. The first is based on minimizing the worst-case regret, as we did in the deterministic settings of Sections 6.6.3 and 7.2.2. In the stochastic setting treated here, our problem becomes[4]

$$\min_{h[n]} \max_{x(t) \in \mathcal{G}} \mathsf{E}\{|\hat{x}(t_0) - P_{\mathcal{W}} x(t_0)|^2\}. \tag{7.106}$$

Fortunately, the resulting filter does not depend on t_0. In the second approach, the filter is designed to minimize the time-average worst-case MSE

$$\varepsilon_{\mathrm{AVG}} = \lim_{\tau \to \infty} \frac{1}{2\tau} \int_{-\tau}^{\tau} \max_{x(t) \in \mathcal{G}} \mathsf{E}\{|\hat{x}(t_0) - x(t_0)|^2\} dt_0. \tag{7.107}$$

Interestingly, these two strategies lead to the same reconstruction filter [129].

The derivation of the minimax MSE filters is a bit involved, and includes several tedious calculations as well as manipulations of random signals. Therefore, rather than detailing the derivation we proceed to the final result and refer the interested reader to [129]. The minimax MSE filter is given by

$$H_{\mathrm{MX}}(e^{j\omega}) = \frac{\rho^2 R_{W\widetilde{W}}(e^{j\omega})}{R_{WW}(e^{j\omega}) \left(\Lambda_u(e^{j\omega}) + \rho^2 R_{S\widetilde{W}}(e^{j\omega}) \right)}, \tag{7.108}$$

where $\Lambda_u(e^{j\omega})$ is the PSD of the noise $u[n]$ and

$$\widetilde{W}(\omega) = \frac{S(\omega)}{|L(\omega)|^2}. \tag{7.109}$$

---

[4] Note that strictly speaking the operation $P_{\mathcal{W}} x(t)$ is not well defined since $x(t)$ is not a signal in L_2. However, we can still carry out the operation implied by $P_{\mathcal{W}} x(t)$, namely, sample $x(t)$ with W^*, filter it with $(W^*W)^{-1}$, and then recover with W. Although the result is not a signal in L_2, it will still be a valid (cyclostationary) signal, which with an abuse of notation we call $P_{\mathcal{W}} x(t)$.

When the power of the noise is small with respect to the (weighted) norm of $x(t)$ so that $\Lambda_u(e^{j\omega})/\rho^2$ is small for all ω, the filter of (7.108) becomes

$$H_{\mathrm{MX}}(e^{j\omega}) \approx \frac{R_{W\widetilde{W}}(e^{j\omega})}{R_{WW}(e^{j\omega})R_{S\widetilde{W}}(e^{j\omega})}, \tag{7.110}$$

which is equal to the minimax filter (7.56) in the noise-free setting. Since in this case, $\mathsf{E}\{|\hat{x}(t_0) - P_{\mathcal{W}}x(t_0)|^2\} = |\hat{x}(t_0) - P_{\mathcal{W}}x(t_0)|^2$, besides minimizing the worst-case error energy $\|\hat{x}(t) - P_{\mathcal{W}}x(t)\|^2$ the filter (7.110) also minimizes the worst-case pointwise error.

7.5.5 Hybrid Wiener filter

We may also consider a full stochastic setting where $x(t)$ is a realization of a continuous-time zero-mean WSS random process with PSD $\Lambda_x(\omega)$, and $u[n]$ is a zero-mean WSS noise process with PSD $\Lambda_u(e^{j\omega})$, independent of $x(t)$. Since $x(t)$ is now random, the MSE averages the squared norm also over the signal, leading to a signal-independent expression. In principle, therefore, the filter $h[n]$ can be designed to directly minimize the MSE. Unfortunately, as in the deterministic signal case, the resulting filter depends on the time index t. If instead we minimize the projected MSE $\mathsf{E}\{|\hat{x}(t_0) - P_{\mathcal{W}}x(t_0)|^2\}$, then the optimal solution is independent of time, and given by (7.108), where $\rho^2/|L(\omega)|^2$ in the minimax filter is replaced by the signal PSD $\Lambda_x(\omega)$ in the Wiener filter. We can therefore view the minimax solution as a Wiener filter matched to a power spectrum $\Lambda_x(\omega) = \rho^2/|L(\omega)|^2$.

7.5.6 Summary of the different filters

The filtering algorithms proposed in this section were derived based on the minimization of a suitable cost function. The suggested methods differ in the assumptions that have been made and are summarized in Table 7.5, where $\mathcal{G}$ is the set of signals satisfying $\|Lx\| \le \rho$.

We have seen already that the Wiener and minimax MSE filters have a similar structure, independent of the choice of $s(t)$ and $w(t)$. In Section 7.5.7 we consider bandlimited interpolation, and show that in this case the Tikhonov filter also shares this form.

Table 7.5 Comparison of methods for noisy signal recovery

	Signal model	Noise model	Criterion	Formula
LS	No constraint	Irrelevant	Data term	(7.101)
Tikhonov	Deterministic: $x \in \mathcal{G}$	Not explicit	Data term + regularization	(7.105)
Projected minimax	Deterministic: $x \in \mathcal{G}$	Stationary process	Worst-case projected MSE at $t = t_0$	(7.108)
Average minimax	Deterministic: $x \in \mathcal{G}$	Stationary process	Worst-case MSE averaged over t	(7.108)
Wiener	Stationary process	Stationary process	Projected MSE at $t = t_0$	(7.108)

When $w(t) = s(t)$, it is shown in [129] that the MSE of the minimax MSE filter is smaller than that of LS for *all* $x(t) \in \mathcal{G}$ (not only the worst-case $x(t)$) and is therefore preferable in an MSE sense.

We next consider an example comparing the different recovery techniques.

Example 7.10 To demonstrate the recovery approaches from noisy samples, we now revisit the situation presented in Example 7.6 in which $s(t) = \beta^0(t)$, $w(t) = \beta^1(t)$, $T = 1$, and our only prior knowledge is that the energy of $x(t)$ is bounded so that $L(\omega) = 1$ for all ω. The stochastic equivalent of this prior corresponds to the assumption that $x(t)$ is a random WSS process with flat PSD, meaning that $x(t)$ is a white noise signal. Here, we assume that the samples are perturbed by white Gaussian noise with variance σ^2 and thus set $Q(e^{j\omega}) = 1$.

The constrained LS (Tikhonov) correction filter (7.105) becomes

$$
\begin{aligned}
H_{\mathrm{CLS}}(e^{j\omega}) &= \frac{R_{WS}(e^{j\omega})}{|R_{WS}(e^{j\omega})|^2 + \lambda R_{WW}(e^{j\omega})} \\
&= \frac{\frac{1}{4}(\cos(\omega) + 3)}{\frac{1}{16}(\cos(\omega) + 3)^2 + \frac{1}{3}\lambda(\cos(\omega) + 2)},
\end{aligned}
\tag{7.111}
$$

where we have substituted the expressions (7.60) and (7.62) for $R_{WW}(e^{j\omega})$ and $R_{WS}(e^{j\omega})$. To compute the minimax filter we note that $\widetilde{W}(\omega) = S(\omega)$, and

$$
\begin{aligned}
H_{\mathrm{MX}}(e^{j\omega}) &= \frac{\rho^2 R_{WS}(e^{j\omega})}{R_{WW}(e^{j\omega})(\sigma^2 + \rho^2 R_{SS}(e^{j\omega}))} \\
&= \frac{\rho^2 \frac{1}{4}(\cos(\omega) + 3)}{(\sigma^2 + \rho^2)\frac{1}{3}(\cos(\omega) + 2)},
\end{aligned}
\tag{7.112}
$$

where we have used the fact that here $R_{SS}(e^{j\omega}) = 1$.

Figure 7.17 depicts the recoveries produced by both methods when

$$
x(t) = \cos(\omega_0 t)
\tag{7.113}
$$

with $\omega_0 = 2\pi/\sqrt{200}$. In this case, the noiseless samples are

$$
c[n] = \int_{n-0.5}^{n+0.5} x(t)dt = \frac{1}{\omega_0}\left(\sin(\omega_0(n + 0.5)) - \sin(\omega_0(n - 0.5))\right).
\tag{7.114}
$$

The scalar ρ^2 of the minimax method was chosen as the energy of $x(t)$ over one period, while λ in the LS approach was set to $\lambda = 1/\rho^2$. For illustration purposes, the LS solution developed in Example 7.6, which does not take smoothness into account, is also shown in the figure, and referred to simply as LS. Figures 7.17(a) and 7.17(b) show the reconstructions obtained when the SNR is 1.6 dB and 51.6 dB respectively. As can be seen, the performance of the constrained methods is similar, while the standard LS solution clearly performs worse in the low-SNR regime. A quantitative comparison is provided in Fig. 7.17(c), showing that the minimax approach is preferable at low SNR levels whereas the constrained LS strategy is

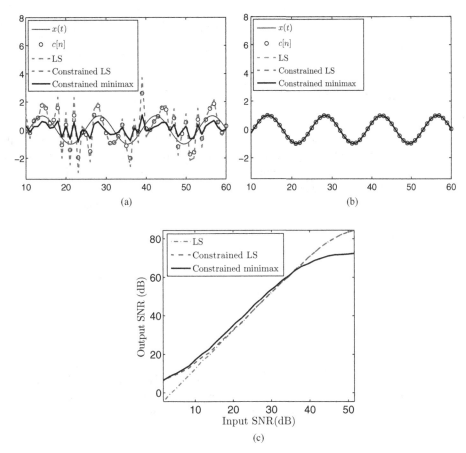

Figure 7.17 LS versus minimax recovery of $x(t) = \cos(\omega_0 t)$ from noisy samples in a constrained setting. (a) 1.6 dB SNR. (b) 51.6 dB SNR. (c) output SNR as a function of input SNR.

advantageous at high SNR. The performance of LS is very similar to that of the constrained LS method at high SNR, but is much worse than both constrained strategies at low SNR.

7.5.7   Bandlimited interpolation

An important special case of constrained recovery is when $w(t) = \text{sinc}(t)$, which corresponds to interpolating the samples to a bandlimited function. It is tempting in this case to replace the continuous–discrete model of Fig. 6.2 by the discrete-time model depicted in Fig. 7.18, and define the different error measures directly on the discrete representation in which $x[n] = x(t)|_{t=nT}$ and $f[n] = s(t) * w(t)|_{t=nT}$ are the samples at times $t = nT$ of the bandlimited version of $s(t)$. As we now show, this equivalence holds under the LS and Tikhonov formulations. In the minimax MSE and Wiener approaches,

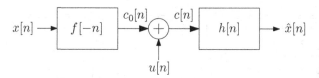

Figure 7.18 Discrete-time deconvolution.

an additional bandlimited constraint on the input is required. However, it is important to stress that in the nonbandlimited scenario, the discrete formulation does not lead to the same solutions as the continuous–discrete one we have been considering in this chapter and the previous one.

In the sequel, we assume that $w(t)$ is an LPF with cutoff frequency π, and that $T = 1$. Starting with the LS method, and using the fact that

$$W(\omega) = \begin{cases} 1, & |\omega| \leq \pi \\ 0, & |\omega| > \pi, \end{cases} \tag{7.115}$$

the LS filter of (7.101) reduces to

$$H_{\mathrm{LS}}(e^{j\omega}) = \frac{1}{S(\omega)}, \quad \omega : S(\omega) \neq 0, |\omega| \leq \pi. \tag{7.116}$$

Noting that the Fourier transform of $f[n]$ is $F(e^{j\omega}) = S(\omega)$ for $|\omega| \leq \pi$, we can express the filter as $H_{\mathrm{LS}}(e^{j\omega}) = 1/F(e^{j\omega})$ which is precisely the LS solution to the problem of minimizing the data error in the discrete-time deconvolution problem of Fig. 7.18.

Similarly, under the model (7.115), the Tikhonov filter becomes

$$H_{\mathrm{CLS}}(e^{j\omega}) = \frac{S(\omega)}{|S(\omega)|^2 + \lambda Q(e^{j\omega})|L(\omega)|^2}, \quad |\omega| \leq \pi. \tag{7.117}$$

This solution can also be interpreted as the Tikhonov filter corresponding to the equivalent discrete-time model, subject to the constraint that $\hat{x}[n]$ lies in the class $\mathcal{G}_d$ defined by

$$\mathcal{G}_d = \left\{ x[n] : \frac{1}{2\pi} \int_{-\pi}^{\pi} |L_d(e^{j\omega})|^2 |X(e^{j\omega})|^2 d\omega \leq \rho^2 \right\}, \tag{7.118}$$

where $L_d(e^{j\omega})$ is the discrete-time filter satisfying $L_d(e^{j\omega}) = L(\omega)$ for $|\omega| \leq \pi$.

Since the LS and Tikhonov criteria depend only on $\hat{x}(t)$, the fact that $w(t)$ is bandlimited reduces the entire problem to the discrete-time version of Fig. 7.18. This is in contrast with the minimax MSE filter which becomes

$$H_{\mathrm{MX}}(e^{j\omega}) = \frac{\rho^2 S(\omega)/|L(\omega)|^2}{\Lambda_u(e^{j\omega}) + \rho^2 \sum_{k \in \mathbb{Z}} |S(\omega + 2\pi k)|^2/|L(\omega + 2\pi k)|^2}, \tag{7.119}$$

for $|\omega| \leq \pi$. A functionally equivalent equation is also obtained for the Wiener filter with $\Lambda_x(\omega)$ playing the role of $\rho^2/|L(\omega)|^2$. The minimax solution (7.119) does not have a corresponding discrete-time interpretation because of the infinite sum in the denominator. To convert the problem into the discrete form associated with Fig. 7.18, we need to add the constraint that the input $x(t)$ is bandlimited, or, equivalently, that the analysis filter is such that $S(\omega) = 0$ for $|\omega| > \pi$. In this case,

$$H_{\text{MX}}(e^{j\omega}) = \frac{\rho^2 S(\omega)/|L(\omega)|^2}{\Lambda_u(e^{j\omega}) + \rho^2|S(\omega)|^2/|L(\omega)|^2}, \quad |\omega| \leq \pi, \qquad (7.120)$$

which is equal to the minimax MSE deconvolution filter that minimizes the worst-case MSE given by $\max_{x[n] \in \mathcal{G}_d} E\{|\hat{x}[n_0] - x[n_0]|^2\}$ [156]. In other words, the continuous–discrete formulation reduces to the discrete one.

If we choose $\Lambda_x(\omega) = \rho^2/|L(\omega)|^2$ in (7.120), then $H_{\text{MX}}(e^{j\omega}) = H_{\text{W}}(e^{j\omega})$ reduces to

$$H_{\text{W}}(e^{j\omega}) = \frac{\Lambda_x(\omega)S(\omega)}{\Lambda_u(e^{j\omega}) + \Lambda_x(\omega)|S(\omega)|^2}. \qquad (7.121)$$

The filter (7.121) is the classical Wiener filter for the problem of estimating a random process $x[n]$ from blurred, noisy observations where the blurring filter is given by $\overline{S(\omega)}$ and the noise PSD is $\Lambda_u(e^{j\omega})$.

Comparing (7.117), (7.120), and (7.121) we see that in the case of bandlimited interpolation the Tikhonov, minimax, and Wiener filters are equivalent provided that $\Lambda_x(\omega) = \rho^2/|L(\omega)|^2$ and $\lambda = 1/\rho^2$. Furthermore, these filters reduce to the classical Wiener solution. However, it is important to note that in the more general, nonbandlimited setting, these conclusions no longer hold.

7.5.8 Unconstrained recovery

We now turn to treat the unconstrained scenario under the LS, minimax, and Wiener strategies. It turns out that the unconstrained solutions in all three cases have the same form, and follow from substituting an optimal choice of $w(t)$ in the corresponding expressions for the constrained setting.

More specifically, the unconstrained scenario under the LS criterion was treated in [157]. The solution has the same form as the constrained filter (7.105) where the given recovery filter $w(t)$ is replaced by the optimal choice $W(\omega) = \widetilde{W}(\omega)$ of (7.109). The resulting filter is then

$$H_{\text{LS}}(e^{j\omega}) = \frac{1}{R_{S\widetilde{W}}(e^{j\omega}) + \lambda Q(e^{j\omega})}, \qquad (7.122)$$

where we have used the fact that the term multiplying $\lambda Q(e^{j\omega})$ is now equal to $R_{S\widetilde{W}}$.

In the noiseless setting, the minimax and LS criteria lead to the same solutions in the unconstrained setting. It can be shown that this property holds also in the presence of noise when we consider the average MSE with $\lambda Q(e^{j\omega})$ replaced by $(1/\rho^2)\Lambda_u(e^{j\omega})$. (Clearly, in the unconstrained setting the projected MSE is no longer a sensible criterion.) This result can be proved by following the derivation in Appendix IV of [129] for the average minimax filter in the constrained setting. The difference is that now, the minimization is both over the filter $h[n]$ and the recovery filter $w(t)$. Thus, the optimal minimax filter in the unconstrained setting is still optimal in the constrained case for a specific choice of $w(t)$. The additional step needed is to optimize the objective over $w(t)$.

Note that the unconstrained solution in the noise-free setting under both LS and minimax criteria can be viewed as a special case of the respective constrained solutions, where we substitute $W(\omega)$ of (7.109) into the expressions for the constrained optimal filters. It is also easy to see that when $Q(\omega) = 0$ or $\Lambda_u(e^{j\omega}) = 0$ the results coincide with those obtained in the noise-free case.

Example 7.11 Suppose that the signal $x(t)$ is known to belong to the set $\mathcal{G}$ of (7.103) with $L(\omega) = \mathrm{sinc}^{-1}(\omega/(2\pi))$. Our goal is to recover $x(t)$ from its noisy samples at the integers. Assume further that the sampling filter is $s(t) = \beta^0(t)$ and that the noise is white with variance σ^2.

The LS and minimax strategies dictate using the reconstruction filter

$$\widetilde{W}(\omega) = \frac{S(\omega)}{|L(\omega)|^2} = \mathrm{sinc}^3\left(\frac{\omega}{2\pi}\right), \tag{7.123}$$

which in the time domain is given by

$$\tilde{w}(t) = \beta^2(t). \tag{7.124}$$

The term $R_{S\widetilde{W}}(e^{j\omega})$ appearing in the expression for the correction filter (7.122) corresponds to the DTFT of the sequence

$$(s(-t) * \tilde{w}(t))_{t=n} = (\beta^0(t) * \beta^2(t))_{t=n} = \beta^3(n) = \begin{cases} \frac{2}{3}, & n = 0 \\ \frac{1}{6}, & |n| = 1 \\ 0, & |n| \geq 2, \end{cases} \tag{7.125}$$

which is given by

$$\frac{1}{6}e^{-j\omega} + \frac{2}{3} + \frac{1}{6}e^{j\omega} = \frac{1}{3}(\cos(\omega) + 2). \tag{7.126}$$

Therefore, the minimax correction filter is equal to

$$H(e^{j\omega}) = \frac{1}{\frac{\sigma^2}{\rho^2} + \frac{1}{3}(\cos(\omega) + 2)}. \tag{7.127}$$

To demonstrate this method, recall that the stochastic equivalent of the deterministic constraint $x \in \mathcal{G}$ is that $x(t)$ is a random signal with spectrum $\Lambda_x(\omega) = \rho^2/|L(\omega)|^2$. In our setting, this corresponds to a random process with autocorrelation $r_x(\tau) = \rho^2\beta^1(\tau)$. Figure 7.19 depicts the recovery produced by the minimax approach in this case. As can be seen, the recovery is not consistent with the samples that is, if we were to sample $\hat{x}(t)$, the samples would not coincide with $c[n]$. This behavior is a result of the fact that the samples are noisy. Indeed, the larger the noise variance, the larger the deviation of $\hat{x}(t)$ from the measured samples.

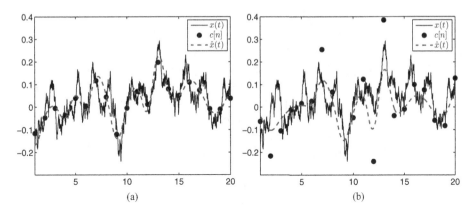

Figure 7.19 Unconstrained LS recovery from noisy samples. (a) Input SNR of $10\log_{10}(\rho^2/\sigma^2)$ = 20 dB. (b) Input SNR of $10\log_{10}(\rho^2/\sigma^2) = 0$ dB.

To summarize, in these past two chapters we have revisited the fundamental problem of reconstructing signals from their linear samples. We considered several models for each of the essential ingredients of the sampling problem: the sampling mechanism (general prefilter, noise), the reconstruction kernel (prespecified, unrestricted), and the signal prior (subspace, smoothness). Our approach was to define an optimization problem that takes into account both the fit of the reconstructed signal to the given samples and the prior knowledge we have about the signal. Each of the settings studied was treated using two optimization strategies: LS and minimax. We showed that when the samples are noise-free, both strategies coincide if the reconstruction mechanism is unrestricted. In this case, perfect recovery is often possible under a subspace prior. In contrast, when the recovery is constrained, the minimax strategy often leads to solutions that are closer to the original signal.

The last part of this chapter was devoted to recovery in the presence of noisy samples, where we limited our attention to the SI setting.

The choice of the most suitable method clearly depends on the context (e.g. deterministic versus stochastic) and on the type of prior information available. Nevertheless, by comparing the various solutions, we have been able to make some interesting links between them, and suggested simple guidelines for selecting the free parameters of the different algorithms.

7.6   Exercises

1. Suppose that the pointwise samples $c[n] = x(nT)$ of a signal are given by $c[n] = \delta[n]$ and assume that our prior knowledge is that $\|Lx\| \le 1$. For each of the following operators, give examples of three signals that are consistent with the samples: two that satisfy the prior constraint and one that does not. Assume that $T = 1$.

a. $L\{x\}(t) = e^{|t|}x(t)$.

b. $L\{x\}(t) = \frac{dx}{dt}(t)$.

c. $L\{x\}(t) = \int_{t-1}^{t+1} x(\tau)d\tau$.

2. Let $s_n(t)$, $n \in \mathbb{Z}$, be a set of sampling functions and let $c[n] = \langle s_n, x \rangle$ be the corresponding samples of $x(t)$. Consider the LS and minimax reconstruction problems (7.2) and (7.17). Explain in words how the parameter α affects the solution for each of the following choices of the operator L.

a. $L\{x\}(t) = x(t - \alpha)$, $\alpha \in \mathbb{R}$.

b. $L\{x\}(t) = \exp\{-\alpha t^2\}x(t)$, $\alpha > 0$.

c. $L\{x\}(t) = t^\alpha x(t)$, $\alpha > 0$.

d. $L\{x\}(t) = (g * x)(t)$, where $g(t) = \exp\{-\alpha t^2\}$, $\alpha > 0$.

e. $L\{x\}(t) = \frac{dx}{dt}(\alpha t)$, $\alpha > 0$.

f. $L\{x\}(t) = x(t^\alpha)$, where α is a positive even integer.

g. $L\{x\}(t) = \alpha \exp\{-|t|\}x(t) + (1 - \alpha \exp\{-|t|\})\frac{dx}{dt}(t)$, $0 < \alpha < 1$.

3. Determine the optimal reconstruction functions $\tilde{w}_n(t)$ corresponding to the operators (a)–(e) of Exercise 2.

4. Consider the LS and minimax reconstruction problems (7.2) and (7.17) with $S(\omega) = 1$. Assume that the sampling period is $T = 1$ and let N be a positive integer. Determine the optimal digital correction filter $H(e^{j\omega})$ corresponding to the smoothness LTI operator

$$L(t) = \sum_{n \in \mathbb{Z}} \mathrm{sinc}(N(t - n)). \tag{7.128}$$

5. Suppose that $x(t)$ is sampled by filtering with $S(\omega)$ such that $|S(\omega)| > \alpha$, $|\omega| \leq \pi$ for a positive α, and is zero otherwise, and then evaluating the output at times $t = n$.

a. Derive the LS and minimax digital correction filters for arbitrary $L(\omega)$.

b. Determine the optimal reconstruction filter $\widehat{W}(\omega)$ of (7.15).

6. Show through an example (possibly in $\mathbb{R}^2$) that the solution to problem (7.54) is not necessarily smooth and not necessarily consistent with the samples. In other words, show that $\hat{x}_{\mathrm{CLS}}$ does not necessarily lie in the constraint set $\mathcal{G}$. Provide geometric intuition through a drawing similar to Fig. 7.6.

7. Determine the matrix whose Frobenius norm is minimal among all 3×3 matrices whose sums of rows are all equal to 1 and whose trace equals 6.

8. For every 3×3 matrix $\mathbf{A}$, denote by $\widetilde{\mathbf{A}}$ the matrix closest to $\mathbf{A}$ in Frobenius norm among all matrices whose elements are equal. Determine the matrix $\hat{\mathbf{A}}$ with all equal entries that minimizes the worst-case Frobenius distance $\|\hat{\mathbf{A}} - \widetilde{\mathbf{A}}\|$ over the set of 3×3 matrices whose sums of rows are equal to 1 and whose trace equals 6.

9. To produce a reconstruction $\hat{x} \in \mathcal{W}$ of $x \in \mathcal{G}$ we proposed solving the minimax-regret problem (7.54). One could argue that it may be preferable to minimize the worst-case error-norm instead, by solving

$$\hat{x}_{\mathrm{CMX}} = \arg \min_{\hat{x} \in \mathcal{W}} \max_{x \in \mathcal{G}} \|\hat{x} - x\|_2^2, \tag{7.129}$$

where $\mathcal{G} = \{x: S^*x = c, \|Lx\| \leq \rho\}$. However, this strategy is over-pessimistic. Prove that when $L = I$, the solution to problem (7.129) is $\hat{x} = 0$ regardless of the value of c.

10. Fill in the details of the proof of Theorem 7.3.

    Hint: Express x as $Wa + Wv$, according to the proof outline, and show that LWa and LWv are orthogonal.

11. Prove Theorem 7.4, following the same steps as in the proof of Theorem 7.2.

12. Assume that $\mathcal{H} = \mathcal{W} \oplus \mathcal{S}^{\perp}$. Show that the error of the constrained minimax-regret solution (7.55) also satisfies condition (7.67) when L is not the identity operator.

13. Assume that a reconstruction $\hat{x}(t)$ is formed as

$$\hat{x}(t) = \sum_{n \in \mathbb{Z}} d[n]w(t - n), \tag{7.130}$$

    where $w(t)$ is a given kernel and $d[n]$ is a zero-mean stationary sequence (namely $E[d[m]d[n]]$ depends only on $m - n$ and is denoted $R_d[m - n]$).

    a. Obtain an expression for $E[\hat{x}^2(t)]$ for a general kernel $w(t)$. Does the expression depend on t?

    b. Show that if $w(t) = \mathrm{sinc}(t)$ then $E[\hat{x}^2(t)]$ is constant as a function of t.

14. Show that when the kernel $w(t)$ satisfies condition (7.94) the signal

$$\hat{x}(t) = \sum_{n \in \mathbb{Z}} d[n]w(t - nT) \tag{7.131}$$

    is bounded if the sequence $d[n]$ is bounded.

15. Suppose that $x[n]$ is a discrete-time stationary random signal with spectrum $\Lambda_x(e^{j\omega})$ and let $y[n] = x[n] + u[n]$, where $u[n]$ is a stationary sequence independent of $x[n]$ with spectrum $\Lambda_u(e^{j\omega})$. It is well known that the linear minimum MSE estimate of $x[n]$ based on $y[n]$ is obtained by feeding $y[n]$ into the discrete-time Wiener filter

$$H(e^{j\omega}) = \frac{\Lambda_x(e^{j\omega})}{\Lambda_x(e^{j\omega}) + \Lambda_u(e^{j\omega})}. \tag{7.132}$$

    If, on the other hand, $y[n]$ corresponds to samples of a continuous-time process $x(t)$, which are contaminated by noise $u[n]$, then the optimal solution is given by (7.108) and (7.109) with $\rho/|L(\omega)|^2$ replaced by $\Lambda_x(\omega)$. Show that if the spectrum $\Lambda_x(\omega)$ of $x(t)$ is bandlimited to π/T and the samples are pointwise (i.e. $S(\omega) = 1$) then the hybrid Wiener solution is equivalent to performing discrete-time Wiener filtering on the samples prior to reconstruction with a sinc kernel.

16. Two of the most commonly used recovery techniques from pointwise samples are nearest-neighbor and linear interpolation. These methods correspond, respectively, to using the B-spline interpolation filters $\beta^0(t)$ and $\beta^1(t)$, with no digital correction. These solutions can also be interpreted as a hybrid Wiener filter corresponding to a certain spectrum $\Lambda_x(\omega)$. Determine $\Lambda_x(\omega)$ for each of the methods. Explain, based on this point of view, which of the methods is expected to yield better results for bandlimited signals.

Chapter 8
Nonlinear sampling

In this chapter we depart from the linear sampling model treated until now and consider recovering signals that lie in a subspace, from their nonlinear samples. Surprisingly, we will see that many types of nonlinearities that are encountered in practice can be completely compensated for without having to increase the sampling rate. Thus, even though typically nonlinearities result in bandwidth expansion, using the algorithms discussed in this chapter we will be able to recover an input from its nonlinear samples at the rate associated with the input space.

Nonlinearities appear in a variety of setups and applications of digital signal processing, including power electronics and wireless communications where an amplifier introduces nonlinear distortions to its input signal, radiometric photography, and charge-coupled device (CCD) image sensors. For example, in radiometric photography, the radiometric response $M(x)$ of a camera relates the radiance x at the input of a sensor to intensity $y = M(x)$ at its output. This mapping is monotonically increasing (which also means that it is invertible), but may exhibit nonlinearity. The intensity signal is sampled by an ADC, and the radiance then needs to be estimated from the samples. Similar nonlinear distortions of amplitude occur in many CCD image sensors, owing to excessive light intensity which causes saturation. More generally, amplifier saturation introduces nonlinear distortion to the input signal, and occurs in many applications such as satellite communications, and power amplifiers in hand-held devices. Operating in the nonlinear region of the amplifier helps to secure high power efficiency. Another application in which nonlinearities are inevitable is in optical fibers used over long distances, which incur chromatic and polarization dispersion.

In some cases, nonlinearity is introduced deliberately in order to increase the possible dynamic range of the signal while avoiding amplitude clipping, or damage to the ADC [158]. For example, in various types of communications channels, a standard method for dealing with limited dynamic range is *companding*. A compander consists of an invertible nonlinearity introduced to the signal prior to transmission that reduces the signal dynamic range; at the receiver, the signal is expanded to the original value. Companding is also sometimes used in digital systems in order to compress the analog input before entering an ADC.

8.1 Sampling with nonlinearities

8.1.1 Nonlinear model

One simple approach to model nonlinearities is to assume that the signal is distorted by a memoryless, nonlinear, invertible mapping prior to sampling by $s(-t)$, as in Fig. 8.1. This rather straightforward model is general enough to capture many systems of practical interest. Therefore, the main focus of this chapter will be on nonlinearities of this form. Extensions to non-SI settings can be found in [159].

Example 8.1 Modern analog circuitry is designed to work at high speeds. One way to achieve this is by incorporating optical components via architectures usually referred to as photonic integrated circuits. A basic building block in this area is the electro-optical amplitude modulator. This component controls the light intensity according to the voltage applied to it.

A common method of implementing an electro-optical amplitude modulator is by using a Mach–Zehnder interferometer with an electro-optical phase modulator in one of its arms. This technique can be designed such that if the light intensity at the input of the device is I, and the voltage applied to it is $x(t)$, then the light intensity at its output is given by

$$y(t) = I \sin^2(\alpha x(t)), \tag{8.1}$$

for some α. To obtain a monotonic behavior, the voltage $x(t)$ is kept in the range $(0, \pi/(2\alpha))$. This nonlinearity is depicted in Fig. 8.2(a) for $\alpha = \pi/2$.

An interface between optical and electrical components in a circuit usually involves light detectors. Suppose that after passing through an amplitude modulator, the light intensity is converted into electrical voltage using a detector, and then sampled. The resulting scheme, shown in Fig. 8.2(b), has the form of Fig. 8.1.

Consider the subspace setting treated in Chapter 6, in which $x(t)$ lies in a subspace $\mathcal{A}$ and is sampled using a filter $s(-t)$ which generates a subspace $\mathcal{S}$, such that $\mathcal{H} = \mathcal{A} \oplus \mathcal{S}^\perp$. However, prior to sampling by $s(-t)$ the signal is distorted by a memoryless, nonlinear, and invertible mapping $M(x(t))$ as in Fig. 8.1. In this chapter we derive conditions under which $x(t)$ can be recovered from the samples $c[n]$, and provide several iterative recovery algorithms.

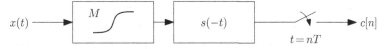

Figure 8.1 Nonlinear and shift-invariant sampling. The signal amplitudes $x(t)$ are distorted by the nonlinear mapping M prior to shift-invariant sampling.

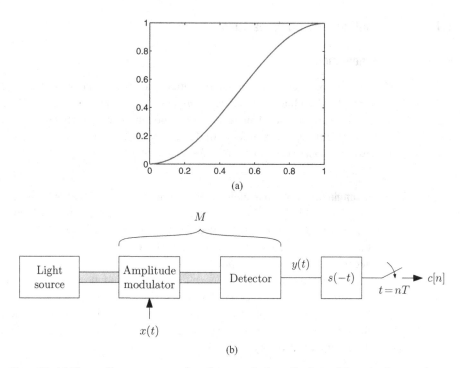

(a)

(b)

Figure 8.2 (a) The nonlinear response of an electro-optical amplitude modulator implemented with a Mach–Zehnder interferometer. (b) A system comprising an electro-optical amplitude modulator, a light detector, an electrical filter, and a sampler.

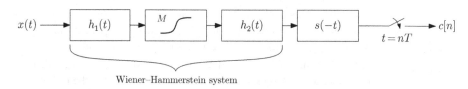

Figure 8.3 Wiener–Hammerstein nonlinearity followed by shift-invariant sampling.

8.1.2   Wiener–Hammerstein systems

The results we develop can also be applied to nonlinearities that are described by the more general *Wiener–Hammerstein model*. A Wiener system is a composition of a linear mapping followed by a memoryless nonlinear distortion, while in a Hammerstein model these blocks are connected in reverse order. Wiener–Hammerstein systems combine the above two models, by enclosing the static nonlinearity with dynamic and linear models from each side, as illustrated in Fig. 8.3. To treat such systems, we note that the first linear mapping can be absorbed into the input constraints $x \in \mathcal{A}$, while the second linear operator can be used to define a modified set of generalized sampling functions. Thus, our derivations extend straightforwardly to Wiener–Hammerstein acquisition devices.

Example 8.2 Suppose that x belongs to an SI space $\mathcal{A}$, so that $x(t) = \sum d[n]a(t - nT)$ for some sequence $d[n]$ and generator $a(t)$. The signal at the output of the first filter in Fig. 8.3 can then be written as

$$\tilde{x}(t) = x(t) * h_1(t) = \sum_{n \in \mathbb{Z}} d[n] \left(a(t - nT) * h_1(t) \right) = \sum_{n \in \mathbb{Z}} d[n]\tilde{a}(t - nT), \quad (8.2)$$

where we denoted $\tilde{a}(t) = a(t) * h_1(t)$. Thus, the signal entering the nonlinearity lies in the SI space $\tilde{\mathcal{A}} = \text{span}\{\tilde{a}(t - nT)\}$. After the nonlinearity, the signal undergoes two filtering operations, which can be represented by a single convolution with impulse response $\tilde{s}(-t) = h_2(t) * s(-t)$.

This shows that the problem of recovering $x \in \mathcal{A}$ from samples at the output of the system in Fig. 8.3 is equivalent to that of recovering a signal $x \in \tilde{\mathcal{A}}$ from samples at the output of the system in Fig. 8.1, with an SI sampling kernel $\tilde{s}(-t)$.

Wiener–Hammerstein systems are a special case of the more general *Volterra model*, which allows time-invariant nonlinearities with memory to be incorporated [160]. The output of a Volterra system of order L with $x(t)$ as its input is given by

$$y(t) = \sum_{n=0}^{L} \int_{-\infty}^{\infty} \cdots \int_{-\infty}^{\infty} k_n(t_1, \ldots, t_n)x(t - t_1) \cdots x(t - t_n)dt_1 \cdots dt_n. \quad (8.3)$$

The function $k_n(t_1, \ldots, t_n)$ is called the nth-order Volterra kernel, and can be regarded as a higher-order impulse response of the system. The Volterra series has the form of a power series and therefore its convergence cannot be guaranteed for arbitrary input signals. Furthermore, it is not general enough to describe all possible nonlinear systems. Nonetheless, Fréchet proved that any continuous, nonlinear system can be uniformly approximated to arbitrary accuracy by a Volterra series of sufficient but finite order if the input signals are restricted to square integrable functions on a finite interval. We will not discuss Volterra systems here, but many of the ideas we develop can be generalized to this setting as well.

Before proceeding, we note that there are many works in the literature that treat the problem of nonlinear system identification based on measurements of its input and output. Although related to our problem, this is a different subject which we will not concentrate on here. The interested reader is referred to [161, 162, 163, 164], to name a few references that deal with this topic. Just as we show here that in many cases the input signal can be recovered without having to increase the sampling rate, the same is true in the context of system identification: many classes of nonlinear systems are identifiable from samples of the input and output at a rate corresponding to the input space.

We begin our discussion by focusing on the bandlimited setting with pointwise sampling, for which it is easiest to build intuition. We will then extend the results to arbitrary input spaces $\mathcal{A}$ with sampling spaces $\mathcal{S} = \mathcal{A}$. Finally, we consider the more general setting of arbitrary sampling and input spaces. In each case we derive conditions on the nonlinearity $M(x)$ such that perfect recovery is possible, and propose iterative algorithms that are guaranteed to recover the input.

8.2 Pointwise sampling

8.2.1 Bandlimited signals

Consider a bandlimited signal $x(t)$ with maximal frequency $\omega_{\max} = \pi/T$ that passes through a memoryless invertible transformation. In general, the nonlinearity will increase the signal bandwidth. For example, suppose that $M(x) = x^3$. Then, at the output of the nonlinearity $y(t) = x^3(t)$ and the bandwidth of $y(t)$ is three times as large as that of $x(t)$. This is demonstrated in Fig. 8.4 for the signal $x(t) = \text{sinc}^2(t)$. A naive attempt to apply the Shannon–Nyquist theorem to $y(t)$ will result in the conclusion that a sampling rate of $6\omega_{\max}$ is necessary in order to be able to recover the signal. This reasoning seems to imply that the nonlinearity inevitably causes an increase in sampling rate. However, as we now show, in many cases such a rate increase is not necessary.

To see how this is possible, we continue to focus on bandlimited signals and pointwise sampling. Thus, after the nonlinear distortion, $y(t)$ is sampled uniformly at times nT, namely the Nyquist rate of the original signal $x(t)$. Although distortion occurs in the frequency domain, it can be easily inverted in time. We illustrate this with an example. Suppose, for instance, that the signal

$$x(t) = \cos(0.4\pi t) \tag{8.4}$$

passes through the nonlinearity $M(x) = x^3$ and is then sampled at times $t = n, n \in \mathbb{Z}$. The Nyquist criterion in this case is satisfied for the original signal but not for its distorted version. More specifically,

$$y(t) = \cos^3(0.4\pi t) = 0.75\cos(0.4\pi t) + 0.25\cos(1.2\pi t), \tag{8.5}$$

so that

$$y[n] = \cos^3(0.4\pi n) = 0.75\cos(0.4\pi n) + 0.25\cos(1.2\pi n). \tag{8.6}$$

The second cosine in this expression can also be written as $\cos(1.2\pi n - 2\pi n) = \cos(-0.8\pi n) = \cos(0.8\pi n)$. This component is therefore aliased, i.e. it cannot be

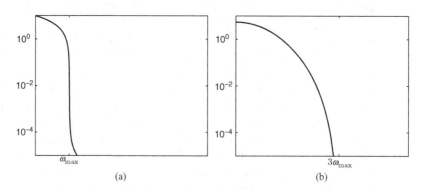

Figure 8.4 The Fourier transform of (a) $x(t) = \text{sinc}^2(t)$ and (b) $y(t) = x^3(t)$.

distinguished from $\cos(0.8\pi t)$. In other words, the samples $y[n]$ we observe could also have been obtained by sampling

$$y(t) = 0.75\cos(0.4\pi t) + 0.25\cos(0.8\pi t), \tag{8.7}$$

at times $t = n$. The important thing to note, however, is that there does not exist another π-bandlimited signal that leads to this $y(t)$ when subjected to the nonlinearity $M(x) = x^3$. Therefore, even though we have aliasing, there is only one π-bandlimited signal that gives rise to the observed samples, so that recovery is possible [165].

To see this, we apply the inverse transform $M^{-1}(x)$ to the samples $y[n]$. The corrected samples $z[n] = M^{-1}(y[n])$ can be expressed in terms of the original signal $x(t)$ as

$$z[n] = M^{-1}(y(nT)) = M^{-1}(M(x(nT))) = x(nT). \tag{8.8}$$

Therefore, we conclude that by applying M^{-1} to the samples $y[n]$ we invert the nonlinearity. The resulting samples are equal to uniformly spaced samples of $x(t)$ from which the signal can be recovered using a simple LPF, according to the Shannon–Nyquist theorem. In our specific example, we have

$$z[n] = y^{\frac{1}{3}}[n] = (\cos^3(0.4\pi n))^{\frac{1}{3}} = \cos(0.4\pi n) = x(n). \tag{8.9}$$

The overall recovery scheme is depicted in Fig. 8.5.

Although until now we have considered the class of bandlimited signals, the same reasoning leads to the conclusion that as long as the nonlinearity is followed by pointwise sampling, it can be removed entirely. In other words, we can easily recover the pointwise samples of the original signal. Therefore, any signal that is determined by its pointwise samples, is also recoverable in the presence of an invertible nonlinearity. We summarize this result in the following theorem.

Theorem 8.1 (Pointwise nonlinear sampling). *Let $x(t)$ be a signal in $L_2(\mathbb{R})$ that is uniquely determined by its samples $x[n] = x(t_n)$ for some sampling points $\{t_n\}$. Suppose we are given samples $y[n] = y(t_n)$ of the signal $y(t) = M(x(t))$ where M is an invertible nonlinearity. Then $x(t)$ can be recovered from the samples $y[n]$ by first applying the inverse mapping M^{-1} to $y[n]$ in order to obtain $x[n]$.*

Since any bandlimited signal can be recovered from its pointwise samples at an appropriate sampling rate, Theorem 8.1 implies that we do not need to increase the sampling rate even in the presence of nonlinearities. Furthermore, this theorem applies to any

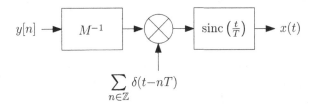

Figure 8.5 Recovery of a π/T-bandlimited signal from pointwise nonlinear samples $y[n] = M(x(nT))$.

signal that is determined from pointwise samples. A broad class of signals with this property are signals lying in a *reproducing kernel Hilbert space*.

8.2.2   Reproducing kernel Hilbert spaces

Stable sampling

In order to reconstruct $x \in \mathcal{A} \subset L_2(\mathbb{R})$ from its ideal samples on a given sampling grid $\{t_n\}$, it is necessary for the function x to be uniquely determined by its values $\{x(t_n)\}$. In particular, this implies that the only function within $\mathcal{A}$ which equals zero for every point on this grid must be the zero function. Such sampling sets are called *uniqueness sets*. Uniqueness, however, is not sufficient to obtain stable reconstruction of x. A reconstruction is *stable* [166] if there exists a positive constant α such that

$$\alpha \|x\|^2 \leq \sum_{n \in \mathbb{Z}} |x(t_n)|^2, \tag{8.10}$$

for all $x \in \mathcal{A}$, i.e. small errors in the samples $x(t_n)$ lead to small errors in the reconstructed signal. Note that (8.10) also implies uniqueness of the sampling set. Indeed, if $x(t_n) = 0$ for all n, then we must have that $\|x\| = 0$, or $x = 0$.

To reconstruct $x(t)$ by processing the samples with a linear and bounded operator, we also require that any finite energy x yields finite-energy samples $\{x(t_n)\}$. This is ensured if there is a constant $\beta < \infty$ such that

$$\sum_{n \in \mathbb{Z}} |x(t_n)|^2 \leq \beta \|x\|^2, \tag{8.11}$$

for all $x \in \mathcal{A}$. Sampling points $\{t_n\}$ satisfying (8.10) and (8.11) are called *stable sets of sampling*, or simply *sampling sets* for $\mathcal{A}$ [48]. Thus, reconstructing $x \in \mathcal{A}$ from pointwise samples of $M(x)$ is possible if the sampling grid $\{t_n\}$ is a stable sampling set for $\mathcal{A}$.

The most studied sampling sets are in the context of bandlimited functions. For example, if $\mathcal{A}$ is the subspace of π/T-bandlimited functions, then it is well known that uniform sampling $\{t_n = nT\}$ is stable. Kadec's $1/4$ theorem states that the (possibly nonuniform) sampling grid $\{t_n\}$ with $\sup_n |t_n - nT| < 1/4$ is also stable. The most general formulation of nonuniform sampling of bandlimited functions is due to Beurling and Landau which, loosely speaking, states that stable sampling is obtained if the average number of samples per unit length exceeds the Nyquist rate. This essential result is incorporated in the following theorem by Yao and Thomas [166] (see also [167]):

Theorem 8.2 (Nonuniform sampling of bandlimited signals).   *Let $x(t)$ be a finite-energy bandlimited signal such that $X(\omega) = 0$ for $|\omega| > W - \varepsilon$ and some $0 < \varepsilon \leq W$. Then $x(t)$ is uniquely determined by its samples $x(t_n)$ if*

$$\left| t_n - n\frac{\pi}{W} \right| < L < \infty, \quad |t_n - t_m| > \delta > 0, \ n \neq m. \tag{8.12}$$

Reconstruction is given by

$$x(t) = \sum_{n \in \mathbb{Z}} x(t_n) \frac{G(t)}{G'(t_n)(t - t_n)}, \qquad (8.13)$$

where

$$G(t) = (t - t_0) \prod_{n \in \mathbb{Z},\ n \neq 0} \left(1 - \frac{t}{t_n}\right), \qquad (8.14)$$

$G'(t_n)$ *is the derivative of* $G(t)$ *evaluated at* $t = t_n$, *and if* $t_n = 0$ *for some* n, *then* $t_0 = 0$.

A special case of this theorem is recurrent nonuniform sampling considered in Example 6.16 (see also [77, 145, 168]). Algorithms for signal recovery from nonuniform samples can be found, for example, in [48, 169, 170, 171].

Reproducing kernels

Stable sampling sets are closely related to the notion of frames [26] and reproducing kernel Hilbert spaces (RKHS) [172, 173]. RKHS are function spaces, namely, spaces whose elements are functions, and are given by the following definition.

Definition 8.1. *A Hilbert space* $\mathcal{H}$ *of functions on an interval* I *is called a reproducing kernel Hilbert space (RKHS), if there is a reproducing kernel function* $k(t, t_0)$ *on* $I \times I$ *such that:*

1. *For all fixed* $t_0 \in I$, $k_{t_0}(t) = k(t, t_0) \in \mathcal{H}$;
2. *The kernel* $k(t, t_0)$ *possesses the reproduction property* $\langle k_{t_0}(t), x(t) \rangle_{\mathcal{H}} = x(t_0)$ *for all* $t_0 \in I$ *and all* $x \in \mathcal{H}$.

Equivalently, $\mathcal{H}$ *is an RKHS if the linear pointwise evaluation operator* $x \to x(t_0)$ *is bounded, for all* $t_0 \in I$ *and all* $x \in \mathcal{H}$. *In this case, for each* $t \in I$ *there exists a unique function* k_t *in* $\mathcal{H}$ *such that* $x(t) = \langle k_t, x \rangle$.

It is easy to see that $k(t, u) = \langle k_t, k_u \rangle$ for $t, u \in I$. The kernel $k(t, u)$ is associated with the operator $K : \mathcal{H} \to \mathcal{H}$ defined by

$$y(t) = \int_I k(t, u) x(u) du. \qquad (8.15)$$

For any reproducing kernel $k(t, u)$, the corresponding operator K is Hermitian and positive semidefinite (see Exercise 6). The Aronszajn–Moore theorem [174] states that, conversely, every positive definite Hermitian kernel $k(t, u)$ on $I \times I$ determines a unique Hilbert space $\mathcal{H}$ for which $k(t, u)$ is the reproducing kernel.

Since pointwise values of a function $x(t)$ in an RKHS can be expressed as an inner product with the reproducing kernel $k(t, u)$, we can write (8.10) and (8.11) as

$$\alpha \|x\|^2 \leq \sum_n |\langle k_{t_n}, x \rangle|^2 \leq \beta \|x\|^2 \qquad (8.16)$$

for all $x \in \mathcal{A}$. Thus, $\{t_n\}$ is a stable sampling set if and only if $\{k_{t_n}\}$ forms a frame sequence for $\mathcal{A}$. This connection to frame theory was established in [175]. Denoting the dual frame of $\{k_{t_n}\}$ by $\{\tilde{k}_{t_n}\}$, reconstruction of x from $\{x(t_n)\}$ is given by

$$x(t) = \sum_n \langle k_{t_n}(t), x(t) \rangle \tilde{k}_{t_n}(t) = \sum_n x(t_n) \tilde{k}_{t_n}(t). \qquad (8.17)$$

We conclude that if $x(t)$ lies in an RKHS whose kernel satisfies (8.16), then $x(t)$ can be recovered from pointwise samples of $y(t) = M(x(t))$ on the grid $\{t_n\}$ for any invertible nonlinearity $M(x)$ by

$$x(t) = \sum_n M^{-1}(y(t_n)) \tilde{k}_{t_n}(t). \qquad (8.18)$$

We now consider some examples of RKHS and related sampling theorems. We begin by noting that the whole space $L_2(\mathbb{R})$ is not an RKHS, since pointwise evaluations of a finite-energy function may not be bounded. However, a closed linear subspace $\mathcal{A}$ of $L_2(\mathbb{R})$ can be an RKHS. An example is the space of π/T-bandlimited signals with $k(t_1, t_2) = \mathrm{sinc}((t_1 - t_2)/T)$ as the reproducing kernel. Another large class of RKHSs are finite-dimensional spaces: if $\mathcal{A}$ is finite dimensional, then it is an RKHS. This is because any linear functional over a finite-dimensional Hilbert space is bounded.

An example of a sampling theorem that can be viewed within the RKHS framework is Kramer's generalization of the sampling theorem [176]. Kramer's result treats applications in which we can model $x(t)$ by an expression of the form

$$x(t) = \int_{\mathcal{I}} X(\omega) \Psi(\omega, t) d\omega, \qquad (8.19)$$

for all $t \in \mathbb{R}$. Here, $X(\omega) \in L_2(\mathcal{I})$ is a generalized "transform" of $x(t)$, $\Psi \in L_2(\mathcal{I})$ defines the transform kernel, and $\mathcal{I}$ is a given interval of interest. An example is the inverse CTFT of bandlimited signals, for which $X(\omega)$ is the CTFT of $x(t)$, $\Psi(\omega, t) = e^{j\omega t}/(2\pi)$, and $\mathcal{I} = [-\omega_{\max}, \omega_{\max}]$ is the frequency band of interest.

Suppose that there exists a countable set of sampling points $\{t_n\}$ such that $\{\Psi(\omega, t_n)\}$ is a complete orthogonal set on $\mathcal{I}$. Then Kramer proved that any such function $x(t)$ can be recovered from its pointwise samples via

$$x(t) = \sum_{n \in \mathbb{Z}} x(t_n) k_n(t), \qquad (8.20)$$

where

$$k_n(t) = \frac{\int_{\mathcal{I}} \Psi(\omega, t) \overline{\Psi(\omega, t_n)} d\omega}{\int_{\mathcal{I}} |\Psi(\omega, t_n)|^2 d\omega}. \qquad (8.21)$$

The corresponding reproducing kernel is

$$k(t, u) = \frac{\int_{\mathcal{I}} \Psi(\omega, u) \overline{\Psi(\omega, t)} d\omega}{\int_{\mathcal{I}} |\Psi(\omega, t)|^2 d\omega}. \qquad (8.22)$$

It is easy to see that $k_n(t) = k(t_n, t)$, $k_n(t_k) = \delta_{nk}$, $\langle k_n(t), k_m(t) \rangle = \delta_{nm}$, and $k(t, u) = \sum_n k_n(t) \overline{k_n(u)}$. In particular, $\tilde{k}_{t_n}(t) = k_{t_n}(t)$.

Kramer also showed some examples of functions $\Psi(\omega, t)$ satisfying the required conditions, such as $\Psi(\omega, t) = J_m(\omega t)$ where $J_m(t)$ is the Bessel function of the first kind of order m, $\Psi(\omega, t) = P_t(\omega)$ where $P_t(\omega)$ is the Legendre function, and the prolate spheroidal functions.

We next present a very simple example of a reproducing kernel together with the associated sampling theorem.

Example 8.3 Consider the space $\mathcal{H}$ of real linear functions on $[0, 1]$. Namely, every $x \in \mathcal{H}$ is of the form $x(t) = at$ for some $a \in \mathbb{R}$. A reproducing kernel for this space is

$$k(t, u) = 3tu. \tag{8.23}$$

To see that this is indeed a valid kernel, note that for all fixed $t_0 \in [0, 1]$, the function $k_{t_0}(t)$ is linear and

$$k_{t_0}(t) = k(t, t_0) = 3t_0 t. \tag{8.24}$$

This implies that $k_{t_0}(t)$ is in $\mathcal{H}$. Furthermore, for any linear function $x(t) = at$,

$$\langle k_{t_0}, x \rangle = \int_0^1 3t_0 t \cdot at\, dt = at_0 = x(t_0). \tag{8.25}$$

We now consider stable sampling sets for x. To determine if a sampling set is stable we need to evaluate (8.10)–(8.11). Noting that $x(t_n) = at_n$ and $\|x\|^2 = a^2/3$, the stability condition becomes

$$\frac{1}{3}\alpha \le \sum_n t_n^2 \le \frac{1}{3}\beta. \tag{8.26}$$

Assuming (8.26) holds, we can recover $x(t)$ from its samples $x(t_n)$ by applying (8.17). To this end we compute the biorthogonal function

$$\tilde{k}_{t_n}(t) = \left(\sum_n k_{t_n}(t) k_{t_n}^*(t) \right)^{-1} k_{t_n}(t) = S^{-1}(t) k_{t_n}(t), \tag{8.27}$$

where $S(t) = \sum_n k_{t_n}(t) k_{t_n}^*(t)$. Now,

$$S(t)x(t) = \sum_n k_{t_n}(t) \langle k_{t_n}(t), x(t) \rangle$$

$$= \sum_n k_{t_n}(t) x(t_n) = 3at \sum_n t_n^2 = 3x(t) \sum_n t_n^2. \tag{8.28}$$

Therefore, $S^{-1}(t) = 1/\left(3 \sum_n t_n^2\right)$, and

$$\tilde{k}_{t_n}(t) = \frac{t_n}{\sum_n t_n^2} t. \tag{8.29}$$

Finally,

$$x(t) = \frac{1}{\sum_n t_n^2} \sum_n t_n x(t_n) t. \tag{8.30}$$

Noting that $x(t_n) = at_n$, it is easy to see that the right-hand side of (8.30) is indeed equal to $x(t) = at$.

8.3   Subspace-preserving nonlinearities

Another class of nonlinearities that are easy to analyze are those that preserve the subspace structure. Specifically, consider the output of the nonlinearity $y(t) = M(x(t))$ for any input $x(t)$. Suppose that the nonlinearity maps the input signals $x(t) \in \mathcal{A}$ to another subspace $\mathcal{W}$. That is, if we consider the outputs $y(t)$ for every possible input $x(t) \in \mathcal{A}$, then we obtain a subspace of signals. In this special case, the class of inputs to the sampler $s(t)$ is a subspace of signals $\mathcal{W}$. If we focus on recovering $y(t)$ from the samples obtained by $s(t)$, then this problem becomes that of recovering an input y from a given subspace $\mathcal{W}$, with sampling filter that spans $\mathcal{S}$. We have seen in Chapter 6 that unique recovery is possible as long as the direct-sum condition $L_2(\mathbb{R}) = \mathcal{W} \oplus \mathcal{S}^\perp$ is satisfied. Once $y(t)$ is determined, we can invert the nonlinearity and recover $x(t)$. This situation is schematically depicted in Fig. 8.6.

Example 8.4 A simple example of a subspace which is unaltered under the operation of a memoryless nonlinearity is the space of piecewise constant signals.

Suppose that

$$x(t) = \sum_{n \in \mathbb{Z}} d[n] \beta_0(t - n), \tag{8.31}$$

or, more explicitly,

$$x(t) = d[n], \quad n - \frac{1}{2} \le t < n + \frac{1}{2}. \tag{8.32}$$

The signal $y(t) = M(x(t))$ is then given by

$$y(t) = M(d[n]), \quad n - \frac{1}{2} \le t < n + \frac{1}{2}, \tag{8.33}$$

which can also be expressed as

$$y(t) = \sum_{n \in \mathbb{Z}} \tilde{d}[n] \beta_0(t - n), \tag{8.34}$$

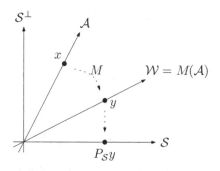

Figure 8.6 A signal $x \in \mathcal{A}$ is mapped to a signal $y \in \mathcal{W}$ by the nonlinearity M and then sampled using sampling kernels that span a subspace $\mathcal{S}$.

where $\tilde{d}[n] = M(d[n])$. This implies that $y(t)$ is also piecewise constant. In other words, memoryless nonlinearities map the space of splines of degree zero to itself.

To demonstrate more complex subspace-preserving scenarios, we depart for a moment from the situation in which M is a memoryless nonlinearity, and consider more general nonlinearities.

Example 8.5 Suppose that $x(t)$ belongs to the subspace $\mathcal{A}$ of signals whose frequency support is contained in some set $\mathcal{I} \in \mathbb{R}$ and consider a nonlinear system M that outputs $y(t) = (x * x * x)(t)$, when fed with $x(t)$. In the frequency domain, we have $Y(\omega) = X^3(\omega)$, so that the frequency support of $Y(\omega)$ is the same as that of $X(\omega)$, namely $\mathcal{I}$. In other words, for every $x \in \mathcal{A}$, the output $y(t)$ of the nonlinearity is also in $\mathcal{A}$.

In this case, if the sampling filter $s(-t)$ is such that every signal in $\mathcal{A}$ can be recovered from its samples, then we can recover $y(t)$. Once we have $y(t)$, we compute $x(t)$ as $X(\omega) = Y^{\frac{1}{3}}(\omega)$.

Example 8.6 Consider again the nonlinearity $y(t) = (x * x * x)(t)$ of the previous example and suppose now that $x(t)$ belongs to the SI space spanned by $\{a(t - n)\}$.

In this case, $x(t)$ can be written as $\sum_n d[n]a(t - n)$ for some sequence $d[n]$, implying that $X(\omega)$ is given by

$$X(\omega) = D(e^{j\omega})A(\omega) \tag{8.35}$$

for some 2π-periodic function $D(e^{j\omega})$. Consequently, the frequency content of the signal $y(t)$ at the output of the nonlinearity is given by

$$Y(\omega) = X^3(\omega) = \tilde{D}(e^{j\omega})\tilde{A}(\omega), \tag{8.36}$$

where we have denoted $\tilde{D}(e^{j\omega}) = D^3(e^{j\omega})$ and $\tilde{A}(\omega) = A^3(\omega)$. Since $\tilde{D}(e^{j\omega})$ is 2π-periodic, $y(t)$ lies in the SI space $\tilde{\mathcal{A}}$ spanned by $\{\tilde{a}(t - n)\}$. Thus, if $s(-t)$ is such that every signal $y \in \tilde{\mathcal{A}}$ can be recovered from its samples, then we can recover $y(t)$ and compute $x(t)$ using the relation $X(\omega) = Y^{\frac{1}{3}}(\omega)$.

8.4 Equal prior and sampling spaces

Until now, we treated two simple cases of nonlinear sampling that lead to direct recovery algorithms: pointwise sampling, and nonlinearities that preserve subspace structure. A more difficult setting is when the sampling is not ideal, and the nonlinearity does not preserve the subspace structure. In this case, the effect of the nonlinearity cannot simply be inverted.

Example 8.7 In many engineering applications, a signal with high dynamic range needs to be transmitted over a noisy channel or a channel with limited dynamic range. In noisy channels, the low-amplitude regions of the signal, which may contain important information, may "drown" in the noise. In channels with limited dynamic range, the high-amplitude regions may be lost owing, for example, to saturation of the analog circuitry.

One way to mitigate these phenomena to some extent is by applying a companding operation (portmanteau for compressing and expanding). Companding refers to the process of compressing the signal values prior to transmission and expanding them back at the receiver. The compression is typically achieved by applying a memoryless monotone function which amplifies low amplitudes more than high amplitudes. Expansion is obtained by applying the inverse of the compression function. A typical compander function is depicted in Fig. 8.7.

Suppose that a π/T-bandlimited signal $x(t)$ is compressed by a function $M(x)$ prior to transmission over a channel that zeros out all frequencies above π/T. This situation can be viewed as filtering the signal $y(t) = M(x(t))$ with an ideal LPF

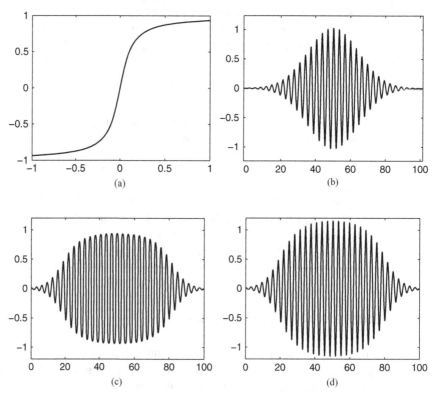

Figure 8.7 (a) Companding function $M(x) = \alpha \arctan(\beta x)$. (b) A π-bandlimited signal $x(t)$. (c) $y(t) = M(x(t))$. (d) A low-pass filtered version of $y(t)$.

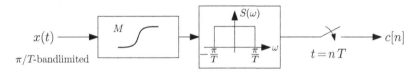

Figure 8.8 Bandlimited sampling with nonlinearities.

$s(t) = \text{sinc}(t/T)$. In this case, we can perfectly recover the resulting received signal from its samples taken at times nT, $n \in \mathbb{Z}$. However, it is not clear whether we can determine $x(t)$. This is because the compressed signal $y(t) = M(x(t))$ is generally not π/T-bandlimited and, actually, often not bandlimited at all. Therefore, the channel's LPF zeros out some of the frequency content of $y(t)$, which, intuitively, may carry valuable information for the purpose of recovering $x(t)$. In this section we will treat such examples and show that surprisingly, recovery of $x(t)$ is indeed possible.

The example above, illustrated in its general form in Fig. 8.8, was first considered by Landau and Miranker in [177]. In this example the signal space is equal to the sampling space, namely the space generated by the sampling filter, which is simply an LPF. Both are the space of bandlimited signals. Here we treat the more general case where $\mathcal{A} = \mathcal{S}$ but is otherwise arbitrary; that is, we are not constrained to the bandlimited setting. In Section 8.5 we discuss scenarios in which $\mathcal{A} \neq \mathcal{S}$. Thus, suppose that $x(t)$ lies in the real space $\mathcal{A}$ generated by $\{a(t - nT)\}$. We sample $y(t) = M(x(t))$ with the sampling filter $a(-t)$ at times $t = nT$. Our goal is to recover $x(t)$ from these samples.

8.4.1 Iterative recovery

Uniqueness

By generalizing the results of [177], it turns out that when $\mathcal{A} = \mathcal{S}$, using a very simple iterative algorithm the original signal can be recovered exactly without having to increase the sampling rate [178]. The fact that recovery is possible stems from the fact that in this setting, invertibility of $M(x)$ guarantees that there is a unique signal in $\mathcal{A}$ consistent with the given measurements.

Theorem 8.3 (Uniqueness of nonlinear sampling). *Let $\mathcal{A}$ be a closed subspace of $L_2(\mathbb{R})$, and let $M(x)$ denote an invertible monotonic function. Then there is a unique signal $x(t) \in \mathcal{A}$ with the given samples $c = A^* M(x)$ where A is a set transformation corresponding to a Riesz basis for $\mathcal{A}$.*

Proof: Suppose that $x_1, x_2 \in \mathcal{A}$ both satisfy $A^* M(x) = c$. This then implies that $P_{\mathcal{A}} M(x_1) = P_{\mathcal{A}} M(x_2)$, where $P_{\mathcal{A}}$ is the orthogonal projection onto $\mathcal{A}$. Now,

$$\langle x_1 - x_2, M(x_1) - M(x_2) \rangle = \langle P_{\mathcal{A}}(x_1 - x_2), M(x_1) - M(x_2) \rangle$$
$$= \langle x_1 - x_2, P_{\mathcal{A}}(M(x_1) - M(x_2)) \rangle = 0. \qquad (8.37)$$

Writing out the inner product explicitly, we have

$$0 = \int_{-\infty}^{\infty} (x_1(t) - x_2(t))(M(x_1) - M(x_2))dt. \tag{8.38}$$

Let $t = t_0$ be such that $x_1(t_0) \neq x_2(t_0)$. Without loss of generality we can assume that $M(x)$ is monotonically increasing, and that $x_1(t_0) > x_2(t_0)$. Since $M(x)$ is strictly monotonic, $M(x_1(t_0)) - M(x_2(t_0)) > 0$, and therefore the integrand in (8.38) is positive. Consequently, the only way for (8.38) to hold is if $x_1(t) = x_2(t)$ almost everywhere, proving the claim. □

Iterative algorithm

Relying on the uniqueness result of Theorem 8.3 we now propose a simple algorithm to recover the true signal based on [178]. Since there is only one signal in $\mathcal{A}$ consistent with the samples, the main idea is to find this signal iteratively by minimizing the error in the samples, while imposing the constraint that the recovery is in $\mathcal{A}$.

Specifically, we know from Chapter 6 that $x \in \mathcal{A}$ can be recovered from its linear samples $c = A^*x$ by first forming the corrected measurements $d = (A^*A)^{-1}c$, and then interpolating d by applying A. Returning to our nonlinear sampling problem, in which $c = A^*M(x)$, a reasonable initial estimate of x is

$$x_0 = A(A^*A)^{-1}c. \tag{8.39}$$

This corresponds to ignoring the nonlinearity and interpolating the samples as though they were sampled by A^* alone.

If M is not equal to the identity, then x_0 will not be equal to the true value of x. In order to test whether or not x_0 is close to x, we resample it and obtain the estimated samples $c_1 = A^*M(x_0)$. If $x = x_0$, then c_1 will be equal to the given samples c, and we have recovered the true underlying signal due to uniqueness. However, in practice this will typically not be the case so that $c_1 - c$ will not be equal to 0. We can use this difference to correct x_0 by forming the interpolated error

$$e_0 = A(A^*A)^{-1}c_1 - x_0 = A(A^*A)^{-1}(c_1 - c), \tag{8.40}$$

and then setting $x_1 = x_0 - \gamma e_0$, where γ is a step size that controls the update size. Continuing these iterations leads to the following update:

$$x_{k+1} = x_k - \gamma A(A^*A)^{-1}(A^*M(x_k) - c). \tag{8.41}$$

Writing x_k in terms of its expansion coefficients in $\mathcal{A}$ as $x_k = Ad_k$, these coefficients are updated by

$$d_{k+1} = d_k - \gamma(A^*A)^{-1}(A^*M(x_k) - c). \tag{8.42}$$

The proposed recovery approach is summarized in Algorithm 8.1.

In the SI setting, the iterations defined by (8.42) take on a particularly simple form, illustrated in Fig. 8.9. Suppose that $x(t) = \sum_n d[n]a(t - nT)$ and sampling is performed by a filter $a(-t)$. In this case the samples are given by $c[n] = \int a(t - nT)M(x(t))dt$, and the estimate $x_k(t)$ at the kth iteration is represented in

---

Algorithm 8.1 Fixed-point iterations for $\mathcal{A} = \mathcal{S}$

Input: Samples $c = A^*M(x)$ of a signal $x \in \mathcal{A}$, step size γ chosen according to
    Theorem 8.4
Output: Recovered signal $\hat{x} \in \mathcal{A}$
Initialize: $x_0 = A(A^*A)^{-1}c$, $k = 0$
while halting criterion false **do**
    $k \leftarrow k + 1$
    $x_k = x_{k-1} - \gamma A(A^*A)^{-1}(A^*M(x_{k-1}) - c)$
end while
return $\hat{x} \leftarrow x_k$

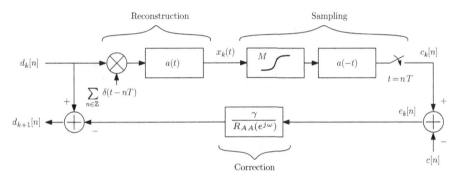

Figure 8.9 Iterative recovery using Algorithm 8.1 of a signal $x(t) \in \text{span}\{a(t - nT)\}$ from samples $c[n]$ of the convolution of $M(x(t))$ and $a(-t)$, at times $t = nT$.

terms of its expansion coefficients $d_k[n]$, such that $x_k(t) = \sum d_k[n]a(t - nT)$. These coefficients are updated from iteration to iteration, as shown in the figure, where we have replaced $(A^*A)^{-1}$ by filtering with $1/R_{AA}(e^{j\omega})$.

Example 8.8 Suppose that $T = 1$ and, as in Example 8.7, $a(t) = \text{sinc}(t)$. In this case both analog filters in Fig. 8.9 are ideal LPFs with cutoff π and $R_{AA}(e^{j\omega}) = 1$ so that the correction stage reduces to a simple gain. The iterations can then be written as $x_{k+1} = x_k - \gamma\text{BL}\{M(x_k) - M(x)\}$, where $\text{BL}\{y\}$ denotes an ideal LPF with cutoff π applied to $y(t)$.

    Figure 8.10 demonstrates the recovery process for the signal $x(t) = \cos(2t)$ $\exp\{-(t - 50)^2/20^2\}$ of Fig. 8.7 and $M(x) = \alpha \arctan(\beta x)$ with $\alpha = 2/\pi$, $\beta = 10$, and $\gamma = 1/Q = 20/\pi$ where $Q = \max M'(x)$. The squared error $\|x_k - x\|^2$, as a function of iteration number, is shown in Fig. 8.11.

Algorithm convergence
Theorem 8.4 below establishes that Algorithm 8.1 converges to the true signal x as long as the nonlinearity is continuous and has a bounded derivative.

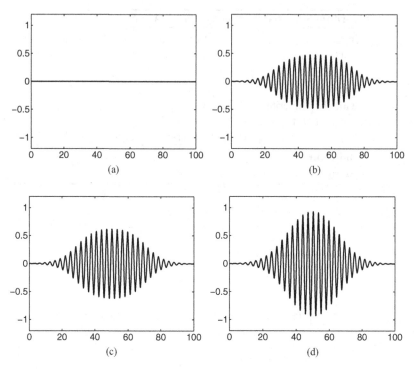

Figure 8.10 The recovery $x_k(t)$ of $x(t)$ from Fig. 8.7 throughout the iterations using the procedure of Fig. 8.9. (a) $k = 0$. (b) $k = 10$. (c) $k = 20$. (d) $k = 100$.

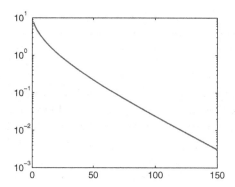

Figure 8.11 Squared error $\|x_k - x\|^2$ as a function of iteration number.

Theorem 8.4 (Convergence of Algorithm 8.1).   *Let $c = A^*M(x)$ be samples of a signal $x \in \mathcal{A}$ where $\mathcal{A}$ corresponds to a Riesz basis for $\mathcal{A}$, and $M(x)$ is a monotonically increasing memoryless nonlinearity that is continuous and differentiable almost everywhere. Let the derivative of M be bounded by Q so that $\sup_x M'(x) = Q$. Then Algorithm 8.1 converges to x as long as $0 < \gamma < 2/Q$.*

We note that the theorem trivially holds if $M(x)$ is monotonically decreasing: we simply apply the result to $\widetilde{M} = -M$.

Proof: It is easy to see that if the iterations converge to x_∞, then $x_\infty = x$: Let $x_{k+1} = x_k = x_\infty$. Then we must have $c = A^* M(x_\infty)$. From Theorem 8.3 we know that there is a unique $x \in \mathcal{A}$ with the given samples. Therefore, $x_\infty = x$, and we only need to establish that the iterations in fact converge.

To prove convergence we rewrite (8.41) as

$$x_{k+1} = T(x_k), \tag{8.43}$$

where $T(x)$ is the mapping on $\mathcal{A}$ defined by

$$T(x) = x + \gamma A (A^* A)^{-1} (c - A^* M(x)). \tag{8.44}$$

Next we note that if x_0 is in $\mathcal{A}$, then x_k will lie in $\mathcal{A}$ for all k since $T(x)$ is in $\mathcal{A}$ for any $x \in \mathcal{A}$. From the fixed-point theorem [179], iterations of the form (8.43) are guaranteed to converge if T is a contraction mapping, namely for any $x, z \in \mathcal{A}$ we have $\|T(x) - T(z)\| < \|x - z\|$.

We now show that for an appropriate step size γ, the mapping T defined by (8.44) is a contraction. We begin by computing the difference $T(x) - T(z)$:

$$\begin{aligned}
T(x) - T(z) &= x - z - \gamma A (A^* A)^{-1} A^* (M(x) - M(z)) \\
&= x - z - \gamma P_{\mathcal{A}} (M(x) - M(z)) \\
&= P_{\mathcal{A}} (x - z - \gamma (M(x) - M(z))), \tag{8.45}
\end{aligned}$$

where in the last equality we used the fact that $x - z \in \mathcal{A}$. Therefore,

$$\|T(x) - T(z)\| \le \|x - z - \gamma (M(x) - M(z))\|, \tag{8.46}$$

since $\|P_{\mathcal{A}} x\| \le \|x\|$ for any x. Consider now a fixed value of t, and assume without loss of generality that $x(t) > z(t)$. Applying the mean value theorem we have that

$$M(x) - M(z) = M'(v)(x - z), \tag{8.47}$$

where for every t, $v(t)$ lies in the interval $[z(t), x(t)]$. Since $M'(x) \le Q$ for all feasible $x(t)$ and owing to the monotonicity of M, $M(x) - M(z) > 0$, we conclude that

$$0 < \frac{M(x) - M(z)}{x - z} \le Q. \tag{8.48}$$

Consequently, for any $0 < \gamma < 2/Q$,

$$\left| 1 - \gamma \frac{M(x) - M(z)}{x - z} \right| < 1. \tag{8.49}$$

Substituting (8.49) into (8.46), we have

$$\begin{aligned}
\|T(x) - T(z)\|^2 &\le \int_{-\infty}^{\infty} (x(t) - z(t))^2 \left(1 - \gamma \frac{M(x(t)) - M(z(t))}{x(t) - z(t)} \right)^2 dt \\
&< \int_{-\infty}^{\infty} (x(t) - z(t))^2 dt = \|x - z\|^2, \tag{8.50}
\end{aligned}$$

completing the proof. $\qquad\square$

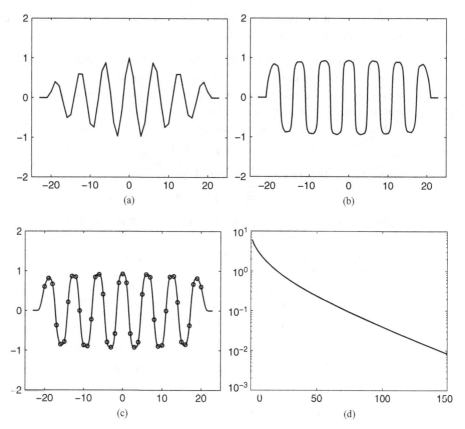

Figure 8.12 (a) A spline $x(t)$ of degree 1. (b) The distorted signal $y(t) = M(x(t))$ with $M(x) = \alpha \arctan(\beta x)$. (c) The filtered signal $(y * \beta^1)(t)$ and its pointwise samples. (d) Squared error as a function of iteration number using the iterations of Fig. 8.9.

Example 8.9 Suppose that a spline $x(t)$ of degree 1 with knots at $n \in \mathbb{Z}$ goes through the nonlinearity $M(x) = \alpha \arctan(\beta x)$ of Fig. 8.7, is filtered with $\beta^1(t)$ and then sampled at $t = n$, $n \in \mathbb{Z}$. An example of such a process is shown in Fig. 8.12. Recovery of $x(t)$ from its samples using the iterative algorithm of Fig. 8.9 involves digital filtering with $1/R_{\beta^1\beta^1}(e^{j\omega})$, which is given by (6.19).

In Fig. 8.12(d) we plot the recovery error as a function of the iteration number using the method of Fig. 8.9 for the same settings as in Example 8.8 where here the true coefficients of $x(t)$ are chosen as $d[n] = \cos(n) \exp\{-n^2/20^2\}$. The iterations converge to the true signal at a rate similar to that encountered in the bandlimited setting of Example 8.8 (though slightly slower). After roughly 150 iterations, the approximation is already very good.

8.4.2   Linearization approach

Algorithm 8.1 is very intuitive and simple to implement; however, as shown in Examples 8.8 and 8.9, its convergence can be slow in certain cases. To speed up convergence

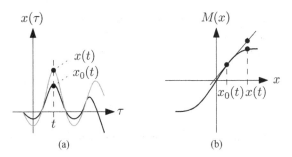

Figure 8.13 (a) The values $x(t)$ and $x_0(t)$ for a specific t. (b) The approximation (8.51) to $M(x(t))$.

(at the expense of a more complicated method) we now present a different approach to the nonlinear recovery problem for $\mathcal{A} = \mathcal{S}$ proposed in [159].

Recall that our measurements are given by $c = A^* M(x(t))$. The difficulty in recovering $x(t)$ arises from the fact that M is a nonlinear mapping. We may therefore consider approximating it by a linear function at each iteration. Clearly, in general, we cannot approximate M over the entire range of input values by a single linear function with small error. However, for a given input $x_k(t)$, we can describe M *locally* as being linear.

Suppose we start with some initial guess of our input signal $x_0(t) \in \mathcal{A}$. We can then evaluate the nonlinearity in the vicinity of $x_0(t)$ using a Taylor expansion:

$$M(x(t)) \approx M(x_0(t)) + M'(x_0(t))(x(t) - x_0(t)). \qquad (8.51)$$

This approximation is valid for every specific t, as demonstrated in Fig. 8.13. Substituting this expansion into the equation $c = A^* M(x(t))$, and assuming equality in (8.51), leads to

$$A^* M'(x_0(t))x(t) = c - A^*(M(x_0(t)) - M'(x_0(t))x_0(t)). \qquad (8.52)$$

The right-hand side of (8.52) describes a sequence in ℓ_2, which we denote temporarily by b. The left-hand side corresponds to sampling a signal $x(t) \in \mathcal{A}$ with sampling functions $S_0 = M'_0 \mathcal{A}$ (the notation M'_k is explained below). We can therefore express (8.52) as $S_0^* x = b$ where $x \in \mathcal{A}$.

The operator $M'_k \colon L_2(\mathbb{R}) \to L_2(\mathbb{R})$ is a linear operator defined by

$$M'_k x(t) = M'(x_k(t))x(t). \qquad (8.53)$$

More generally, we define $M'_h x(t) = M'(h(t))x(t)$ for any $h(t) \in L_2$. Note that M'_h is Hermitian over L_2. We will also use the notation $M'_h(\mathcal{A})$ to denote the subspace obtained by applying M'_h to every $a(t)$ in $\mathcal{A}$. For example, if $M(x) = \arctan(x)$, then $M'(x) = 1/(1 + x^2)$ and $M'_h(\mathcal{A})$ is the space of all signals that are of the form $x(t)/(1 + h^2(t))$, where $x(t)$ is in $\mathcal{A}$.

Assume for now that $S_0^* A$ is invertible. We know from Chapter 6 (cf. Section 6.3.2) that unique recovery of $x(t)$ from b can be obtained via $x = A(S_0^* A)^{-1} b$. This value of x will serve as our next iteration, denoted x_1:

$$x_1(t) = A(A^* M_0' A)^{-1}(c - A^*(M(x_0(t)) - M_0' x_0(t)))$$
$$= x_0(t) + A(A^* M_0' A)^{-1}(c - A^* M(x_0(t))). \tag{8.54}$$

In the last equality we relied on the fact that since $x_0(t)$ is in $\mathcal{A}$,

$$A(A^* M_0' A)^{-1} A^* M_0' x_0(t) = A(S_0^* A)^{-1} S_0^* x_0(t) = E_{\mathcal{A} \mathcal{S}^\perp} x_0(t) = x_0(t), \tag{8.55}$$

where $\mathcal{S} = \mathcal{R}(S_0)$.

We now use $x_1(t)$ as a starting point, approximate the nonlinearity in its vicinity, and reapply the same ideas. Continuing these iterations leads to

$$x_{k+1} = x_k - A(A^* M_k' A)^{-1}(A^* M(x_k) - c). \tag{8.56}$$

In practice, instead of applying (8.56) directly, we add a step size to control the amount of update at each iteration:

$$x_{k+1} = x_k - \gamma A(A^* M_k' A)^{-1}(A^* M(x_k) - c). \tag{8.57}$$

These iterations can be written in terms of the expansion coefficients in a basis for $\mathcal{A}$ as $x_{k+1} = A d_{k+1}$ with

$$d_{k+1} = d_k - \gamma(A^* M_k' A)^{-1}(A^* M(x_k) - c). \tag{8.58}$$

Comparing (8.57) with our previous iterations (8.41) we see that the difference is that the operator $(A^* A)^{-1}$ in the latter is replaced here by $(A^* M_k' A)^{-1}$. As we show in Section 8.4.4, this term can be interpreted in terms of a Newton direction, which is the reason that the iterations (8.57) will converge much more rapidly. We consequently refer to (8.57) as the Newton method. On the other hand, while in the fixed-point approach this operator was constant throughout the iterations, here it depends on the current iterate and therefore needs to be recalculated in each step which can be computationally demanding.

In general, we are not guaranteed that the inverse $(A^* M_k' A)^{-1}$ exists. In the next section (see Corollary 8.1) we address general conditions under which invertibility holds. In particular we will see that as long as the nonlinearity has a bounded derivative, invertibility is guaranteed. Assuming in addition that the derivative is Lipschitz continuous,[1] we show in Theorem 8.7 that the iterations (8.57) converge, with an appropriate choice of step size γ described in Algorithm 8.3 below. The Newton approach is summarized in Algorithm 8.2.

---

[1] A function $f(x)$ is Lipschitz continuous over I with Lipschitz constant $L \geq 0$ if $\|f(x_1) - f(x_2)\| \leq L \|x_1 - x_2\|$ for every x_1, x_2 in I.

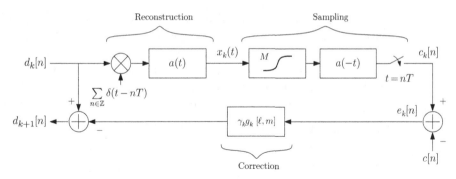

Figure 8.14 Iterative recovery using Algorithm 8.2 of a signal $x(t) \in \mathrm{span}\{a(t - nT)\}$ from samples $c[n]$ of the convolution of $M(x(t))$ and $a(-t)$, at times $t = nT$.

Algorithm 8.2 Newton method for $\mathcal{A} = \mathcal{S}$

Input: Samples $c = A^* M(x)$ of a signal $x \in \mathcal{A}$
Output: Recovered signal $\hat{x} \in \mathcal{A}$
Initialize: $x_0 = A(A^*A)^{-1}c$, $k = 0$
while halting criterion false **do**
  $k \leftarrow k + 1$
  Choose a step size γ_k according to Algorithm 8.3
  $x_k = x_{k-1} - \gamma_k A(A^* M'_{k-1} A)^{-1}(A^* M(x_{k-1}) - c)$
end while
return $\hat{x} \leftarrow x_k$

Figure 8.14 illustrates (8.58) for the SI setting in which $\mathcal{A}$ is an SI space generated by $a(t)$ with period T. In the figure, the block $g_k[\ell, m]$ is a linear system which is the inverse of

$$w_k[\ell, m] = \int_{-\infty}^{\infty} a(t - \ell T) M'(x_k(t)) a(t - mT) dt. \tag{8.59}$$

In general, this is no longer an LTI operation. To compute $g_k[\ell, m]$, suppose that the number N of available samples is finite. In this case, $w_k[\ell, m]$ can be expressed as an $N \times N$ matrix. Consequently, the Newton algorithm requires computation and inversion of an $N \times N$ matrix at each iteration. Thus, although this method converges faster than the fixed-point approach, each iteration involves more computations. We demonstrate this via examples in Section 8.4.5.

8.4.3 Conditions for invertibility

We now study under what conditions $A^* M'_k A$ is invertible. From Lemma 2.2 it follows that $A^* M'_k A$ is invertible over $L_2(\mathbb{R})$ if $L_2(\mathbb{R}) = M'_k(\mathcal{A}) \oplus \mathcal{A}^\perp$. More generally, if S is a set transformation corresponding to a Riesz basis for a subspace $\mathcal{S}$, then $S^* M'_k A$ is invertible if $L_2(\mathbb{R}) = M'_k(\mathcal{A}) \oplus \mathcal{S}^\perp$. Theorem 8.5 below states conditions on $M(x)$

under which the direct-sum condition holds. A detailed proof of this theorem can be found in Section VII of [159]. Since the proof is quite involved we do not include it here. The proof relies on geometrical concepts and frame perturbation theory. The main idea is to view $M'_k(\mathcal{A})$ as a perturbation of the original signal space $\mathcal{A}$. If $L_2(\mathbb{R}) = \mathcal{A} \oplus \mathcal{S}^\perp$ and the perturbation induced by $M'_k(\mathcal{A})$ is not too large, then it can be shown that the direct-sum condition is preserved with respect to the perturbed space $M'_k(\mathcal{A})$.

Theorem 8.5 (Direct-sum condition).   *Let $\mathcal{A}$ and $\mathcal{S}$ be closed subspaces of $L_2(\mathbb{R})$ such that $L_2(\mathbb{R}) = \mathcal{A} \oplus \mathcal{S}^\perp$ and let M be a differentiable, monotonically increasing memoryless nonlinearity. If*

$$|1 - M'(x)| < \frac{1 - \sin(\mathcal{A}, \mathcal{S})}{1 + \sin(\mathcal{A}, \mathcal{S})}, \quad \text{for all } x \in \mathbb{R}, \tag{8.60}$$

where $\sin(\mathcal{A}, \mathcal{S})$ is defined in (6.27), then $L_2(\mathbb{R}) = M'_h(\mathcal{A}) \oplus \mathcal{S}^\perp$ for all $h \in L_2(\mathbb{R})$.

An important feature of the direct-sum condition $L_2(\mathbb{R}) = M'_h(\mathcal{A}) \oplus \mathcal{S}^\perp$ is that it should not be affected by scaling of M. This follows from the obvious fact that the subspace $M'_h(\mathcal{A})$ is equivalent to the subspace $aM'_h(\mathcal{A})$ for any scalar $a \neq 0$. We therefore conclude that it is enough that (8.60) is satisfied with respect to $aM'(x)$ for some $a > 0$, or

$$\frac{2\sin(\mathcal{A}, \mathcal{S})}{1 + \sin(\mathcal{A}, \mathcal{S})} < aM'(x) < \frac{2}{1 + \sin(\mathcal{A}, \mathcal{S})}. \tag{8.61}$$

In order to satisfy this condition we require that (see Exercise 10)

$$\frac{\inf_x M'(x)}{\sup_x M'(x)} > \sin(\mathcal{A}, \mathcal{S}). \tag{8.62}$$

The further apart $\mathcal{A}$ and $\mathcal{S}$ are, the tighter the bound on the ratio between the maximal and minimal derivative value. When $\mathcal{A} = \mathcal{S}$, (8.62) implies that the only requirement on the derivative is that it is bounded above (since $M(x)$ is monotonically increasing we always have that $M'(x) > 0$).

By combining Theorem 8.5 and the discussion above we have the following corollary.

Corollary 8.1.   *Let A and S denote Riesz bases for subspaces $\mathcal{A}$ and $\mathcal{S}$, respectively, such that $L_2(\mathbb{R}) = \mathcal{A} \oplus \mathcal{S}^\perp$. If*

$$\frac{\inf_x M'(x)}{\sup_x M'(x)} > \sin(\mathcal{A}, \mathcal{S}), \tag{8.63}$$

then $L_2(\mathbb{R}) = M'_h(\mathcal{A}) \oplus \mathcal{S}^\perp$ and the operator $S^* M'_h A$ is invertible for all $h \in L_2(\mathbb{R})$. In particular, when $S = A$, $A^* M'_h A$ is invertible, as long as the derivative $M'(x)$ is bounded away from zero, and bounded from above.

8.4.4   Newton algorithm

In order to establish convergence of the iterations (8.57) we first show that these iterations can be formulated as a Newton algorithm, namely, a method for finding stationary points of an appropriate objective function using first- and second-order derivatives. We then rely on known results regarding convergence of Newton methods.

Optimization formulation

Let $\hat{x} \in \mathcal{A}$ be our current estimate of the unknown input. Since $\hat{x}$ is in $\mathcal{A}$, $\hat{x} = Ad$ for some d. If $\hat{x}$ is close to x, then we expect the samples of $\hat{x}$ to be close to the true samples c. It therefore makes sense to try to minimize the error-in-samples

$$f(d) = \frac{1}{2}\|e(d)\|^2 = \frac{1}{2}\|A^*M(Ad) - c\|^2. \tag{8.64}$$

Clearly, the global optimum of $f(d)$ is achieved when d is chosen such that $x = Ad$. From Theorem 8.3 it follows that this is the only global optimum. The proposed criterion is also plausible when the samples c correspond to a perturbation of the true samples by white Gaussian noise. In this case, the minimizer of (8.64) is a maximum-likelihood estimate of d from c.

Unfortunately, since M is nonlinear, $f(d)$ is in general a nonlinear and nonconvex objective function and might possess many local minima. Therefore, in principle, any method designed to minimize $f(d)$ can at best be trapped at a local minimum, namely, a vector d_0 for which the gradient $\nabla f(d_0)$ is zero. Fortunately, we are able to establish the following key result, which shows that if $L_2(\mathbb{R}) = M'_h(\mathcal{A}) \oplus \mathcal{A}^\perp$ for all $h \in L_2$ then any stationary point of f is also a global optimum [159]. In Corollary 8.1 we showed that this direct-sum condition is satisfied as long as $M'(x)$ is bounded. Therefore, under this requirement, any algorithm that is designed to trap a local minimum will in fact find the global optimum, namely, the true underlying vector x.

Theorem 8.6 (Global optimum). *Assume that $M'(x)$ is bounded above and bounded away from zero. Then, any stationary point of $f(d) = \frac{1}{2}\|A^*M(Ad) - c\|^2$ is also its global optimum.*

Proof: From Corollary 8.1, it follows that if $M'(x)$ is bounded then $L_2(\mathbb{R}) = M'_h(\mathcal{A}) \oplus \mathcal{A}^\perp$ for all $h \in L_2(\mathbb{R})$.

Assume that $x_k = Ad_k$ is a stationary point of $f(d)$. This implies that

$$0 = \nabla f(d_k) = A^*M'_k A(A^*M(Ad_k) - c) = (M'_k A)^* Ae(d_k). \tag{8.65}$$

If $e(d_k) = 0$, then x_k is a global optimum. Otherwise, $Ae(d_k)$ is also nonzero, and (8.65) implies that $Ae(d_k)$ lies in $\mathcal{N}((M'_k A)^*) = \mathcal{R}(M'_k A)^\perp$. However, from the direct-sum condition, $\mathcal{A}$ and $\mathcal{R}(M'_k A)^\perp$ intersect only at the zero vector so that $Ae(d_k) = 0$, or $e(d_k) = 0$. □

Optimization algorithms

We now interpret (8.57) as a Newton method aimed at minimizing the cost function f of (8.64). In general, there are many algorithms that can be used to find a stationary point of this objective. Note that our problem is unconstrained, since we incorporated the requirement $x \in \mathcal{A}$ by writing the objective in terms of the sequence d. Many

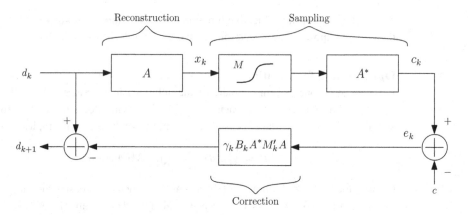

Figure 8.15 Schematic illustration of the iterations (8.69).

unconstrained optimization methods start with an initial guess d and perform iterations of the form

$$d_{k+1} = d_k - \gamma_k B_k \nabla f(d_k), \qquad (8.66)$$

where γ_k is a scalar step size obtained by means of a one-dimensional search and B_k is a positive definite operator such that $B_k \nabla f(d_k)$ is a descent direction:

$$-\langle B_k \nabla f(d_k), \nabla f(d_k) \rangle < \delta < 0. \qquad (8.67)$$

In our case,

$$\nabla f(d_k) = A^* M'_k A(A^* M(Ad_k) - c) = A^* M'_k Ae(d_k), \qquad (8.68)$$

and the iterations become

$$d_{k+1} = d_k - \gamma_k B_k A^* M'_k Ae(d_k). \qquad (8.69)$$

The method of (8.69) is illustrated in Fig. 8.15. At the kth iteration, we construct a signal estimate $x_k = Ad_k$ from the current coefficients d_k and sample this estimate by applying the operator $A^* M(x_k)$. The error sequence $e_k = c_k - c$ is corrected by the operator $\gamma_k B_k A^* M'_k A$ to yield the updated coefficients. Algorithm 8.2 results from choosing

$$B_k = (\nabla^2 f(d_k))^{-1} = (A^* M'_k A)^{-2} \qquad (8.70)$$

in (8.69). This choice leads to Newton's method which is based on the Hessian matrix of $f(d)$. Quasi-Newton algorithms result from approximating the Hessian in order to simplify the computations.

Algorithm convergence
In our setting, the objective function $f(d)$ is bounded from below by zero. In this case, it is shown in [180] that the iterations (8.69) are guaranteed to converge to a stationary point of $f(d)$ if γ_k is chosen to satisfy the Wolfe conditions (a set of inequalities for

performing an inexact line search), $B_k \nabla f(d_k)$ is a descent direction, the gradient $\nabla f(d)$ is Lipschitz continuous, and

$$\cos(\theta_k) = \frac{\langle B_k \nabla f(d_k), \nabla f(d_k) \rangle}{\| B_k \nabla f(d_k) \| \, \| \nabla f(d_k) \|} > \varepsilon \tag{8.71}$$

for some constant $\varepsilon > 0$ independent of k.

A step size satisfying the Wolfe conditions can be found by using the backtracking method [180], as presented in Algorithm 8.3. To show that B_k of (8.70) defines a descent direction, note that

$$-\langle B_k \nabla f(d_k), \nabla f(d_k) \rangle = -\langle (A^* M'_k A)^{-1} e(d_k), A^* M'_k A e(d_k) \rangle$$
$$= -\langle e(d_k), e(d_k) \rangle < 0, \tag{8.72}$$

as long as $A d_k \neq x$. The remaining properties can be proven when $M'(x)$ is Lipschitz continuous [159], as shown in the following theorem.

Algorithm 8.3 Backtracking line search

Input: Function $f(d)$, operator B_k, current iterate d_k, constants $\rho, \eta \in (0, 1)$
Output: Step size γ_k
set $g_k = \nabla f(d_k)$, $z_k = -B_k g_k$, $\delta = 1$
while $f(d_k + \delta z_k) > f(d_k) + \eta \delta \langle z_k, g_k \rangle$ **do**
$\quad \delta \leftarrow \rho \delta$
end while
return $\gamma_k = \delta$

Theorem 8.7 (Convergence of Algorithm 8.2). *Let $c = A^* M(x)$ be samples of a signal $x \in A$ where A corresponds to a Riesz basis for A, and $M(x)$ is a monotonically increasing memoryless nonlinearity such that $M'(x)$ is bounded above and below and Lipschitz continuous. Then Algorithm 8.2 converges to the true input x.*

Proof: To prove the theorem we need to show that (8.71) holds for $B_k = (\nabla^2 f(d_k))^{-1} = (A^* M'_k A)^{-2}$ and that $\nabla f(d)$ is Lipschitz continuous.

We begin by proving that $\cos(\theta_k) > \varepsilon$. Using (8.72) we can write

$$\cos(\theta_k) = \frac{\| e(d_k) \|^2}{\| Q_k^{-1} e(d_k) \| \, \| Q_k e(d_k) \|} \geq \frac{1}{\| Q_k^{-1} \| \| Q_k \|} \tag{8.73}$$

where we denoted $Q_k = A^* M'_k A$, and the norms in the final expression are operator norms (see Appendix A). Since A is a Riesz basis and M'_k is bounded, $\| Q_k \|$ is bounded above. It remains to show that $\| Q_k^{-1} \|$ is upper bounded. By definition, $\| Q_k^{-1} \| = 1/\kappa$, where

$$\kappa = \inf_{\|b\| = 1} \| Q_k b \|. \tag{8.74}$$

From Corollary 8.1 it follows that since $M'(x)$ is bounded, $L_2(\mathbb{R}) = M'_k(\mathcal{A}) \oplus \mathcal{A}^\perp$. This in turn implies that $\cos(M'_k(\mathcal{A}), \mathcal{A}) > \delta$ for some $\delta > 0$ where from (6.26),

$$\cos(M'_k(\mathcal{A}), \mathcal{A}) = \inf_d \frac{\|P_\mathcal{A} M'_k Ad\|}{\|M'_k Ad\|}; \tag{8.75}$$

cf. Proposition 6.3. Since A is a Riesz basis for $\mathcal{A}$, we have that $\|A^* x\| = \|A^* P_\mathcal{A} x\| \geq \alpha \|P_\mathcal{A} x\|$ for some $\alpha > 0$. Furthermore, $\|Ab\| \geq \beta \|b\|$ for all b. Combined with the fact that M'_k is bounded below, we conclude that $\|M'_k Ab\| > U \|b\|$ for $U > 0$. Therefore,

$$\|Q_k b\| = \|A^* P_\mathcal{A} M'_k Ab\| \geq \alpha \|P_\mathcal{A} M'_k Ab\| \geq \alpha \|M'_k Ab\| \cos(M'_k(\mathcal{A}), \mathcal{A}) > \zeta \|b\|, \tag{8.76}$$

where $\zeta = \alpha \delta U$. Consequently, $\kappa \geq \zeta$, which results in $\cos(\theta_k) > \varepsilon$ for an appropriate choice of ε.

The Lipschitz continuity of $\nabla f(d)$ follows from the Lipschitz continuity of $M'(x)$ and the fact that A is a Riesz basis. The proof of this result is straightforward but tedious; we refer the interested reader to Appendix IV of [159]. $\qquad\square$

To conclude, we introduced two iterative methods summarized in Algorithms 8.1 and 8.2 to recover a signal $x(t)$ that is known to lie in a subspace $\mathcal{A}$, from its nonlinear samples. Both iterations are shown to converge to $x(t)$ under appropriate conditions on the derivative of M (cf. Theorems 8.4 and 8.7). In the next section we generalize these techniques to the case of an arbitrary sampling filter.

8.4.5   Comparison between algorithms

We now compare Algorithms 8.1 and 8.2, using the setting of Example 8.9 as a running example to demonstrate the characteristic behavior of the methods.

We begin by studying the convergence rate of the algorithms in terms of the number of iterations they require to produce a good recovery. Figure 8.16 depicts the squared error $\|x - x_k\|^2$ as a function of k for the fixed-point iterations Algorithm 8.1, and for the Newton method Algorithm 8.2 with and without backtracking line search. In the fixed-point iterations and in the constant-step-size Newton algorithm we set $\gamma = 1/Q$ where $Q = \max M'(x)$. As can be seen, the Newton method requires roughly two orders of magnitude fewer iterations to get a good recovery in this example.

Figure 8.16 shows that varying the step size from one iteration to the next may be overly conservative during the first few iterations. In this specific example, during the first 10 steps, the rate of reduction in the reconstruction error with this strategy is in fact very similar to that of the fixed-point iterations and worse than that of the constant-step-size Newton method. Adapting the step size becomes beneficial when the approximation is in the vicinity of the minimum, at which stage the error drops at a much faster rate than with constant step size.

One factor that affects the number of iterations is the accuracy with which the nonlinearity $M(x)$ can be approximated by a linear function over the range of values that $x(t)$ takes. Recall from Fig. 8.7 that our nonlinearity is approximately linear near the origin (for $|x| \ll 1$) and tapers off as $x \to \pm\infty$. Thus, if $|x(t)|$ is small for every $t \in \mathbb{R}$, then

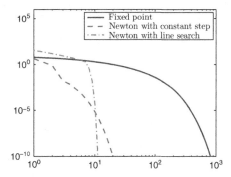

Figure 8.16 Squared error $\|x_k - x\|^2$ as a function of k for the Newton algorithm with a varying and constant step size, and for the fixed-point iterations.

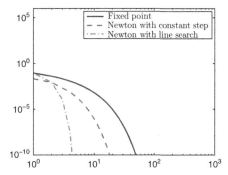

Figure 8.17 Squared error $\|x_k - x\|^2$ as a function of k for a signal with small amplitude.

the nonlinear distortion is not too severe and consequently it is reasonable to expect that fewer iterations suffice. Figure 8.17 depicts the reconstruction error of all three methods as a function of k for an input $\tilde{x}(t) = 0.25x(t)$, where $x(t)$ is the input used in Fig. 8.16. This simulation shows that the fixed-point iterations clearly benefit from this change, requiring an order of magnitude fewer iterations than in the setting of Fig. 8.16. Newton with backtracking line search also improves and converges three times faster than before. The convergence rate of Newton with a constant step size does not improve. Overall, Newton is still preferable in this situation over fixed-point iterations.

It is important to note that each iteration of the Newton method is computationally more demanding than the fixed-point algorithm as it requires the inversion of an $N \times N$ matrix, where N is the number of samples. Furthermore, when using a varying step size, we potentially evaluate the error $\|c_k - c\|^2$ more than once in each iteration (owing to the line search). Therefore, the fact that the Newton algorithm converges in fewer iterations does not necessarily imply that the total running time it requires is smaller. Figure 8.18 depicts the total running time in seconds for all three techniques as a function of the number of samples. In this simulation the expansion coefficients of the signal were randomly generated from a standard Gaussian distribution. As can be seen, for small problem sizes, Newton methods are much faster than the fixed-point iterations. However,

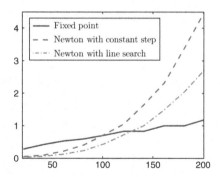

Figure 8.18 Total running time in seconds versus the number of samples.

the running time of the latter grows roughly linearly with the amount of data while the running time of the Newton iterations grow polynomially. Thus, for large problems it becomes advantageous to use the fixed-point algorithm, even though it requires more iterations. Another interesting conclusion from Fig. 8.18 is that varying the step size in the Newton method leads to a smaller total running time, even though the backtracking line search requires several evaluations of the error $\|c_k - c\|^2$.

8.5   Arbitrary sampling filters

Until now we treated the case in which the sampling filter was equal to the generator $a(t)$ of the prior space. We now turn to consider the more complicated setting in which the sampling filter $s(t)$ is chosen arbitrarily such that the direct-sum condition $L_2(\mathbb{R}) = \mathcal{A} \oplus \mathcal{S}^\perp$ is satisfied. Thus, our samples are now given by $c = S^* M(x)$ with $x \in \mathcal{A}$.

8.5.1   Recovery algorithms

To develop recovery methods in this case we follow the same approaches used in the previous section. The difference is that interpolation from linear samples is no longer obtained by $A(A^* A)^{-1} c$ but rather using $A(S^* A)^{-1} c$. Therefore, the fixed-point iterations defined by (8.41) become

$$x_{k+1} = x_k - \gamma A(S^* A)^{-1}(S^* M(x_k) - c). \tag{8.77}$$

Similarly, following the linearization approach, (8.51) stays the same but in (8.52), the transformation A^* is replaced by S^*. This results in the iterations

$$x_{k+1} = x_k - \gamma A(S^* M'_k A)^{-1}(S^* M(x_k) - c). \tag{8.78}$$

Invertibility of $S^* M'_k A$ is guaranteed if the condition of Corollary 8.1 is satisfied. These iterations can be viewed as a Newton method for minimizing the objective

$$f(d) = \frac{1}{2}\|S^* M(Ad) - c\|^2. \tag{8.79}$$

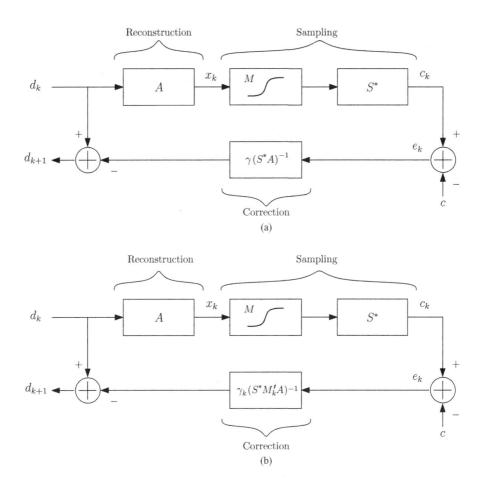

Figure 8.19 Schematic illustration of (a) Algorithm 8.4 and (b) Algorithm 8.5.

The fixed-point and Newton methods for the case in which $\mathcal{A} \neq \mathcal{S}$ are summarized in Algorithms 8.4 and 8.5, respectively. A schematic interpretation of one iteration of each of the two algorithms is depicted in Fig. 8.19. The implementation of these schemes in the SI case is shown in Fig. 8.20. The system $g_k[\ell, m]$ is now the inverse of

$$w_k[\ell, m] = \int_{-\infty}^{\infty} s(t - \ell T) M'(x_k(t)) a(t - mT) dt. \tag{8.80}$$

When $\mathcal{A} = \mathcal{S}$ we showed in Theorem 8.3 that there is a unique vector x with the given nonlinear samples. When sampling in $\mathcal{S}$, this is no longer necessarily true. In the next section we derive conditions under which a unique input with the given samples exists. We will then see under what conditions Algorithms 8.4 and 8.5 converge to the true input. Note that if the iterations above converge, then they will always result in a consistent recovery, namely, a recovery $\hat{x}$ that satisfies $S^* M(\hat{x}) = S^* M(x) = c$.

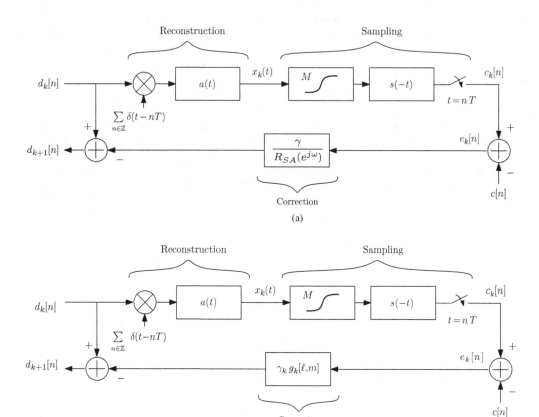

Figure 8.20 Schematic illustration of the SI version of (a) Algorithm 8.4 and (b) Algorithm 8.5.

8.5.2   Uniqueness conditions

To develop uniqueness conditions, we begin by showing that under the direct-sum $M_h'(\mathcal{A}) \oplus \mathcal{S}^\perp = L_2(\mathbb{R})$, there is a unique vector x in $\mathcal{A}$ with the given samples $c = S^*M(x)$. We then rely on Corollary 8.1 to derive conditions on M such that the direct-sum condition is satisfied.

Proposition 8.1. *Let $M(x)$ be a monotonically increasing memoryless nonlinearity that is continuous and differentiable almost everywhere. Suppose that $M_h'(\mathcal{A}) \oplus \mathcal{S}^\perp = L_2(\mathbb{R})$ for any $h \in L_2(\mathbb{R})$. Then there is a unique $x \in \mathcal{A}$ such that $c = S^*M(x)$.*

Proof: Assume that there are two functions $x_1, x_2 \in \mathcal{A}$ which satisfy $c = S^*M(x_1) = S^*M(x_2)$. This implies that $P_\mathcal{S}M(x_1) = P_\mathcal{S}M(x_2)$, where $P_\mathcal{S}$ is the orthogonal projection onto $\mathcal{S}$. Therefore,

$$M(x_1) - M(x_2) \in \mathcal{S}^\perp. \qquad (8.81)$$

Algorithm 8.4 Fixed-point iterations for $\mathcal{A} \neq \mathcal{S}$

Input: Samples $c = S^* M(x)$ of a signal $x \in \mathcal{A}$, step size γ chosen according to
   Theorem 8.9
Output: Recovered signal $\hat{x} \in \mathcal{A}$
Initialize: $x_0 = A(S^* A)^{-1} c$, $k = 0$
while halting criterion false **do**
   $k \leftarrow k + 1$
   $x_k = x_{k-1} - \gamma A(S^* A)^{-1}(S^* M(x_{k-1}) - c)$
end while
return $\hat{x} \leftarrow x_k$

Algorithm 8.5 Newton method for $\mathcal{A} \neq \mathcal{S}$

Input: Samples $c = S^* M(x)$ of a signal $x \in \mathcal{A}$
Output: Recovered signal $\hat{x} \in \mathcal{A}$
Initialize: $x_0 = A(S^* A)^{-1} c$, $k = 0$
while halting criterion false **do**
   $k \leftarrow k + 1$
   Choose a step size γ_k according to Algorithm 8.3
   $x_k = x_{k-1} - \gamma_k A(S^* M'_{k-1} A)^{-1}(S^* M(x_{k-1}) - c)$
end while
return $\hat{x} \leftarrow x_k$

Fix a time instant t and assume without loss of generality that $x_1(t) \leq x_2(t)$. Applying the mean value theorem,

$$M(x_1(t)) - M(x_2(t)) = M'(v(t))(x_1(t) - x_2(t)), \tag{8.82}$$

where $v(t)$ lies in the interval $[x_1(t), x_2(t)]$. Define the function $v(t)$ such that (8.82) is satisfied for all t. We can then use operator notation to write

$$M(x_1) - M(x_2) = M'_v(x_1 - x_2). \tag{8.83}$$

By definition, $M'_v(x_1 - x_2)$ lies in the subspace $M'_h(\mathcal{A})$ with $h = v$ since $x_1 - x_2 \in \mathcal{A}$. On the other hand, (8.83) and (8.81) imply that $M'_v(x_1 - x_2)$ lies in $\mathcal{S}^\perp$. From the direct-sum assumption $M'_h(\mathcal{A}) \oplus \mathcal{S}^\perp = L_2(\mathbb{R})$ it then follows that $M(x_1) - M(x_2) = 0$ or, since M is invertible, $x_1 = x_2$. □

Combining Proposition 8.1 with Corollary 8.1 leads to the following theorem.

Theorem 8.8 (Uniqueness with arbitrary sampling).   *Let $M(x)$ be a monotonically increasing memoryless nonlinearity that is continuous and differentiable almost everywhere. Let $\mathcal{A}$ and $\mathcal{S}$ be closed subspaces of $L_2(\mathbb{R})$ such that $L_2(\mathbb{R}) = \mathcal{A} \oplus \mathcal{S}^\perp$. If*

$$\frac{\inf_x M'(x)}{\sup_x M'(x)} > \sin(\mathcal{A}, \mathcal{S}), \tag{8.84}$$

then there is a unique $x \in \mathcal{A}$ such that $c = S^* M(x)$.

Note that under the conditions of the theorem, we also have from Corollary 8.1 that $S^* M'_k A$ is invertible so that Algorithm 8.5 is well defined.

We conclude that under the assumptions of Theorem 8.8, if (8.77) and (8.78) converge, then they will converge to the true input x. It remains to determine under what conditions these methods converge. To this end we follow similar derivations to those used in the previous section. As we will see, Algorithm 8.4 converges even when (8.84) is not satisfied. In this case, convergence is only guaranteed to a consistent point and not necessarily to the true input x.

Example 8.10 Suppose that a spline of degree 1 with knots at the integers is nonlinearly distorted by a function $M(x)$ and then sampled with the filter $s(t) = \beta^0(t)$ at $t = n$, $n \in \mathbb{Z}$. We would like to determine when uniqueness of the input is guaranteed by using Theorem 8.8.

The term $\sin(\mathcal{A}, \mathcal{S})$ can be computed from (cf. (6.28))

$$\cos(\mathcal{A}, \mathcal{S}) = \inf_{\omega} \frac{|R_{SA}(e^{j\omega})|}{\sqrt{R_{SS}(e^{j\omega}) R_{AA}(e^{j\omega})}} \tag{8.85}$$

using the relation $\sin^2(\mathcal{A}, \mathcal{S}) = 1 - \cos^2(\mathcal{A}, \mathcal{S})$. In our setting,

$$r_{ss}[n] = r_{\beta^0\beta^0}[n] = \beta^1(n) = \delta[n],$$

$$r_{aa}[n] = r_{\beta^1\beta^1}[n] = \beta^3(n) = \frac{1}{6}\delta[n+1] + \frac{2}{3}\delta[n] + \frac{1}{6}\delta[n-1],$$

$$r_{sa}[n] = r_{\beta^0\beta^1}[n] = \beta^2(n) = \frac{1}{8}\delta[n+1] + \frac{3}{4}\delta[n] + \frac{1}{8}\delta[n-1]. \tag{8.86}$$

Consequently,

$$R_{SS}(e^{j\omega}) = 1,$$

$$R_{AA}(e^{j\omega}) = \frac{1}{6}e^{j\omega} + \frac{2}{3} + \frac{1}{6}e^{-j\omega} = \frac{1}{3}\left(2 + \cos(\omega)\right),$$

$$R_{SA}(e^{j\omega}) = \frac{1}{8}e^{j\omega} + \frac{3}{4} + \frac{1}{8}e^{-j\omega} = \frac{1}{4}\left(3 + \cos(\omega)\right). \tag{8.87}$$

Substituting into (8.85):

$$\cos(\mathcal{A}, \mathcal{S}) = \inf_{\omega} \frac{\frac{1}{4}\left(3 + \cos(\omega)\right)}{\sqrt{\frac{1}{3}\left(2 + \cos(\omega)\right)}}$$

$$= \inf_{\omega} \frac{\sqrt{3}}{4}\left(\sqrt{2 + \cos(\omega)} + \frac{1}{\sqrt{2 + \cos(\omega)}}\right)$$

$$= \inf_{x \in [1, \sqrt{3}]} \frac{\sqrt{3}}{4}\left(x + \frac{1}{x}\right)$$

$$= \frac{\sqrt{3}}{2} \tag{8.88}$$

and

$$\sin(\mathcal{A}, \mathcal{S}) = \sqrt{1 - \cos^2(\mathcal{A}, \mathcal{S})} = \frac{1}{2}. \tag{8.89}$$

We conclude that, according to Theorem 8.8, uniqueness is guaranteed if the ratio between the minimal and maximal slopes of $M(x)$ is lower bounded by 0.5. Consider, for illustration, the nonlinearities

$$M_1(x) = \alpha \arctan(\beta x),$$
$$M_2(x) = x + \alpha \arctan(\beta x), \tag{8.90}$$

where α, β are positive scalars. The infimum of the slope of $M_1(x)$ is 0 so that uniqueness is not guaranteed. The slope of the nonlinearity $M_2(x)$ varies between 1 and $1 + \alpha\beta$. Therefore, a unique recovery exists when $\alpha\beta < 1$.

8.5.3 Algorithm convergence

The following theorem studies the convergence of Algorithm 8.4:

Theorem 8.9 (Convergence of Algorithm 8.4). *Let $c = S^* M(x)$ be samples of a signal $x \in \mathcal{A}$ where S corresponds to a Riesz basis for $\mathcal{S}$ such that $L_2(\mathbb{R}) = \mathcal{A} \oplus \mathcal{S}^\perp$, and $M(x)$ is a monotonically increasing memoryless nonlinearity that is continuous and differentiable almost everywhere. Let $Q = \sup_x M'(x) < \infty$ and $q = \inf_x M'(x) > 0$ with*

$$\frac{q}{Q} > \frac{1 - \cos(\mathcal{A}, \mathcal{S})}{1 + \cos(\mathcal{A}, \mathcal{S})}. \tag{8.91}$$

Then Algorithm 8.4 converges as long as

$$\frac{1 - \cos(\mathcal{A}, \mathcal{S})}{q} < \gamma < \frac{1 + \cos(\mathcal{A}, \mathcal{S})}{Q}. \tag{8.92}$$

When $\mathcal{A} = \mathcal{S}$, $\cos(\mathcal{A}, \mathcal{S}) = 1$ and the conditions of Theorem 8.9 reduce to those of Theorem 8.4.

Note that the condition imposed by (8.91) is weaker than that of (8.84). Indeed, for any value θ in $[0, \pi/2)$,

$$\frac{1 - \cos(\theta)}{1 + \cos(\theta)} < \sin(\theta). \tag{8.93}$$

Therefore, the algorithm may converge even when we cannot guarantee uniqueness. As we discussed above, in this case the iterations converge to a consistent solution.

Proof: The proof follows similar lines to Theorem 8.4.

Let $T(x)$ be the mapping on $\mathcal{A}$ defined by

$$T(x) = x + \gamma A(S^* A)^{-1}(c - S^* M(x)), \tag{8.94}$$

so that (8.77) can be written as $x_{k+1} = T(x_k)$. We now prove that $T(x)$ is contracting on $\mathcal{A}$.

Let x, z be arbitrary vectors in $\mathcal{A}$ and denote by $E_{\mathcal{AS}^\perp} = A(S^*A)^{-1}S^*$ the oblique projection onto $\mathcal{A}$ along $\mathcal{S}^\perp$. Since $x - z = E_{\mathcal{AS}^\perp}(x - z)$ we can write

$$\|T(x) - T(z)\| = \|E_{\mathcal{AS}^\perp}(x - z - \gamma(M(x) - M(z)))\|. \tag{8.95}$$

Using the fact that $\|E_{\mathcal{AS}^\perp}\| \leq 1/\cos(\mathcal{A}, \mathcal{S})$ (cf. Proposition 6.8) we have

$$\|T(x) - T(z)\| \leq \frac{1}{\cos(\mathcal{A}, \mathcal{S})}\|x - z - \gamma(M(x) - M(z))\|. \tag{8.96}$$

Consider now a fixed value of t, and assume without loss of generality that $x(t) \geq z(t)$. Applying the mean value theorem,

$$\frac{M(x) - M(z)}{x - z} = M'(v), \tag{8.97}$$

where for every t, $v(t)$ is in the interval $[z(t), x(t)]$. Since $M(x)$ is monotonically increasing,

$$q \leq \frac{M(x) - M(z)}{x - z} \leq Q. \tag{8.98}$$

Consequently, for any γ satisfying (8.92),

$$\left|1 - \gamma\frac{M(x) - M(z)}{x - z}\right| < \cos(\mathcal{A}, \mathcal{S}). \tag{8.99}$$

Substituting into (8.96), we conclude that $\|T(x) - T(z)\| < \|x - z\|$. $\square$

We now turn to discuss convergence of the Newton iterations given by Algorithm 8.5. Our first step is to generalize Theorem 8.6.

Theorem 8.10 (Global optimum with arbitrary sampling). *Let A and S denote Riesz bases for subspaces $\mathcal{A}$ and S respectively such that $L_2(\mathbb{R}) = \mathcal{A} \oplus \mathcal{S}^\perp$. Assume that*

$$\frac{\inf_x M'(x)}{\sup_x M'(x)} > \sin(\mathcal{A}, \mathcal{S}). \tag{8.100}$$

Then, any stationary point of $f(d) = \frac{1}{2}\|S^*M(Ad) - c\|^2$ is also its global optimum.

Proof: From Corollary 8.1, it follows that $L_2(\mathbb{R}) = M'_h(\mathcal{A}) \oplus \mathcal{S}^\perp$ for all $h \in L_2(\mathbb{R})$. Assume that $x_k = Ad_k$ is a stationary point of $f(d)$. This implies that

$$0 = \nabla f(d_k) = (S^*M'_kA)^*(S^*M(Ad_k) - c) = (M'_kA)^*Se(d_k), \tag{8.101}$$

where $e(d_k) = S^*M(Ad_k) - c$. If $e(d_k) = 0$, then from Theorem 8.8, x_k is a global optimum. Otherwise $Se(d_k)$ is also nonzero, and (8.101) implies that $Se(d_k)$ lies in $\mathcal{N}((M'_kA)^*) = \mathcal{R}(M'_kA)^\perp$. However, from the direct-sum condition, S and $\mathcal{R}(M'_kA)^\perp$ intersect only at the zero vector so that $Se(d_k) = 0$, or $e(d_k) = 0$. $\square$

Just as before, (8.78) can be viewed as Newton iterations aimed at minimizing $f(d)$ of (8.79), where now $B_k = (\nabla^2 f(d_k))^{-1} = (S^* M'_k A)^{-1}(S^* M'_k A)^{-*}$. We again choose γ_k to satisfy the Wolfe conditions by using Algorithm 8.3. It is easy to see that the new definition of B_k also leads to a descent direction:

$$-\langle B_k \nabla f(d_k), \nabla f(d_k)\rangle = -\langle (S^* M'_k A)^{-1} e(d_k), (S^* M'_k A)^* e(d_k)\rangle$$
$$= -\langle e(d_k), e(d_k)\rangle < 0, \tag{8.102}$$

as long as $Ad_k \neq x$. To guarantee convergence we generalize Theorem 8.7 to our setting:

Theorem 8.11 (Convergence of Algorithm 8.5). *Let $c = S^* M(x)$ be samples of a signal $x \in \mathcal{A}$ where S corresponds to a Riesz basis for $\mathcal{S}$ such that $L_2(\mathbb{R}) = \mathcal{A} \oplus \mathcal{S}^\perp$, and $M(x)$ is a monotonically increasing memoryless nonlinearity. Suppose that*

$$\frac{\inf_x M'(x)}{\sup_x M'(x)} > \sin(\mathcal{A}, \mathcal{S}), \tag{8.103}$$

and that $M'(x)$ is Lipschitz continuous. Then Algorithm 8.5 converges to the true input x.

Proof: To prove the theorem we need to show that (8.71) holds for $B_k = (\nabla^2 f(d_k))^{-1} = (S^* M'_k A)^{-1}(S^* M'_k A)^{-*}$ and that $\nabla f(x)$ is Lipschitz continuous. The proof follows the same steps as that of Theorem 8.7 where now $Q_k = S^* M'_k A$. Since S, A are Riesz bases and M'_k is bounded, $\|Q_k\|$ is bounded above. It remains to show that $\|Q_k^{-1}\|$ is upper bounded. From Corollary 8.1, $L_2(\mathbb{R}) = M'_k(\mathcal{A}) \oplus \mathcal{S}^\perp$. This in turn implies that $\cos(M'_k(\mathcal{A}), \mathcal{S}) > \delta$ for some $\delta > 0$ where from (6.26),

$$\cos(M'_k(\mathcal{A}), \mathcal{S}) = \inf_d \frac{\|P_{\mathcal{S}} M'_k Ad\|}{\|M'_k Ad\|}; \tag{8.104}$$

cf. Proposition 6.3.

Since S is a Riesz basis for $\mathcal{S}$, we have that $\|S^* x\| = \|S^* P_{\mathcal{S}} x\| \geq \alpha \|P_{\mathcal{S}} x\|$ for some $\alpha > 0$. Combined with the fact that $\|Ab\| \geq \beta \|b\|$ for all b, and M'_k is bounded below, we conclude that $\|M'_k Ab\| > U\|b\|$ for $U > 0$. Therefore,

$$\|Q_k b\| = \|S^* P_{\mathcal{S}} M'_k Ab\| \geq \alpha \|P_{\mathcal{S}} M'_k Ab\| \geq \alpha \|M'_k Ab\| \cos(M'_k(\mathcal{A}), \mathcal{S}) > \zeta \|b\|, \tag{8.105}$$

where $\zeta = \alpha\delta U$. The rest of the proof follows as in Theorem 8.7.

Lipschitz continuity of $\nabla f(x)$ is proven in Appendix IV of [159]. $\qquad\square$

8.5.4 Examples

We end this chapter by demonstrating the behavior of Algorithms 8.4 and 8.5.

Consider a spline $x(t)$ of degree 1 with knots at the integers that is distorted by the nonlinearity

$$M(x) = x + \alpha \arctan(\beta x) \tag{8.106}$$

and then sampled at times $t = n \in \mathbb{Z}$ after passing through a sampling kernel $s(t) = \beta^0(t)$. In this case $Q = 1 + \alpha\beta$, $q = 1$, and, as we have seen in Example 8.10, $\sin(\mathcal{A}, \mathcal{S}) = 1/2$ and $\cos(\mathcal{A}, \mathcal{S}) = \sqrt{3}/2$. From Theorem 8.8, a unique recovery is guaranteed to exist if $q/Q > \sin(\mathcal{A}, \mathcal{S})$, which in our case happens when

$$\alpha\beta < 1. \tag{8.107}$$

According to Theorem 8.11, this condition also guarantees that Algorithm 8.5 converges to the true signal $x(t)$. Theorem 8.9 implies that Algorithm 8.4 converges when $q/Q > (1 - \cos(\mathcal{A}, \mathcal{S}))/(1 + \cos(\mathcal{A}, \mathcal{S}))$ as long as the step size γ is chosen to satisfy (8.92). In our case, this requires that

$$\alpha\beta < 6 + 4\sqrt{3} \approx 12.93. \tag{8.108}$$

Figure 8.21 shows the nonlinear function (8.106) for various values of $\alpha\beta$. This figure illustrates that the limitation $\alpha\beta < 1$, ensuring uniqueness and convergence of Algorithm 8.5, is quite severe, as $M(x)$ is nearly linear for such values. The bound $\alpha\beta \lesssim 13$ is much less strict, but it only ensures convergence of Algorithm 8.4 to a consistent recovery, and not necessarily to the true input $x(t)$.

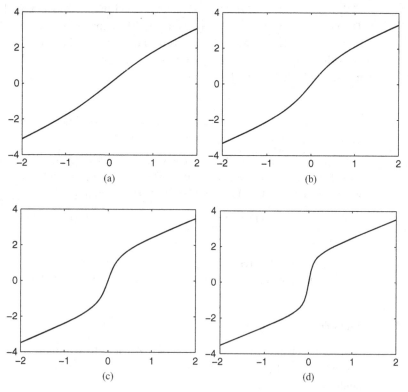

Figure 8.21 The function (8.106) for various values of $\alpha\beta$. (a) $\alpha\beta = 1$. (b) $\alpha\beta = 2$. (c) $\alpha\beta = 6$. (d) $\alpha\beta = 13$.

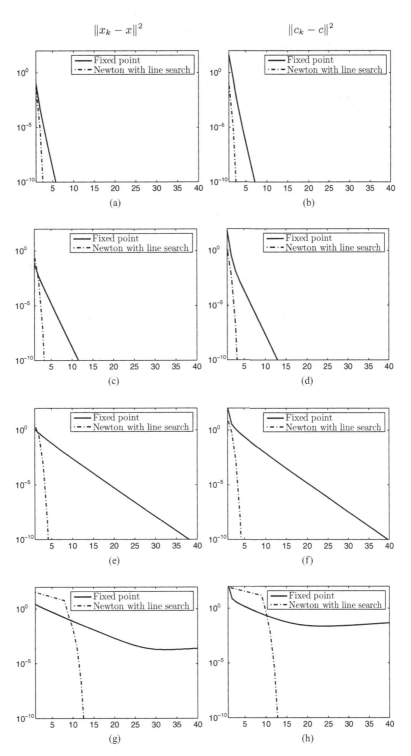

Figure 8.22 The squared error $\|x_k - x\|^2$ (left), and the squared error $\|c_k - c\|^2$ (right), as a function of k, for the fixed-point iterations Algorithm 8.4, and for the Newton iterations with line search Algorithm 8.5. (a), (b) $\alpha\beta = 1$. (c), (d) $\alpha\beta = 2$. (e), (f) $\alpha\beta = 6$. (g), (h) $\alpha\beta = 13$.

Nevertheless, the bounds we developed are generally not tight, so that the performance in practice may be good even beyond the regimes dictated by Theorems 8.9 and 8.11. In fact, this is indeed the case in our setting, when the input signal $x(t)$ is that depicted in Fig. 8.12. Specifically, Fig. 8.22 shows the error in the samples, $\|c_k - c\|$, and the error in recovery, $\|x_k - x\|$, as a function of k using Algorithms 8.4 and 8.5 for various values of $\alpha\beta$. In the fixed-point method we chose γ as the midpoint of the interval given by (8.92): $((1 - \cos(\mathcal{A}, \mathcal{S}))/q, (1 + \cos(\mathcal{A}, \mathcal{S}))/Q)$. Note that both error measures converged in these experiments to zero when using the Newton method. This is despite the fact that we could only guarantee convergence for $\alpha\beta < 1$. The fixed-point iterations, on the other hand, only converged for values of $\alpha\beta$ that were smaller than 13, which is consistent with Theorem 8.9.

Another interesting phenomenon highlighted by Fig. 8.22 is that there do not exist values of $\alpha\beta$ for which the error $\|c_k - c\|$ converges to zero while $\|x_k - x\|$ does not converge to zero. In other words, while for $1 < \alpha\beta < 12.93$ we could only guarantee theoretically that Algorithm 8.4 converges to a consistent solution (namely that $\|c_k - c\| \to 0$), in practice, the solution is unique in this regime (that is, we also have that $\|x_k - x\| \to 0$). In fact, experiments (not shown in Fig. 8.22) imply that in this specific example, we have that $\|x_k - x\| \to 0$ for *every value* of $\alpha\beta$. In other words, there is a unique recovery in this setting for all $\alpha\beta$ and not only for $\alpha\beta < 1$.

8.6   Exercises

1. Show how a general Wiener–Hammerstein system can be represented as a Volterra system.
2. Consider a signal $z(t)$ which is obtained as the product $z(t) = y_1(t)y_2(t)$ of two signals $y_1(t), y_2(t)$, each of which is an output of an LTI system with corresponding impulse responses $h_1(t), h_2(t)$ and the same input $x(t)$. Express the relationship between $x(t)$ and $z(t)$ as a Volterra system and derive the corresponding Volterra kernel.
3. Let $x(t)$ be an SI signal of the form $x(t) = \Sigma d[n]g(t - nT)$ where $g(t)$ satisfies the interpolation property $g(mT) = \delta[m]$. We are given samples $y(nT)$ of $y(t) = \mathcal{M}(x(t))$ where $\mathcal{M}$ is an invertible nonlinearity.
   a. Can $x(t)$ be recovered from $y(nT)$? Explain.
   b. Provide an explicit example of $x(t)$ for which recovery is possible.
4. Consider the missing samples problem in which a finite number of samples are removed from a stable sampling set. Is the remaining set still stable? Explain.
5. In this exercise we use Theorem 8.2 to derive a reconstruction formula for a bandlimited signal from its recurrent nonuniform samples. Consider samples $t_{nm} = mNT + t_n$ for $0 \le n \le N - 1$ and $m \in \mathbb{Z}$, where $t_0 = 0$.
   a. Derive an expression for $G(t)$ of (8.14) by relying on the identity

$$\sin(\pi(t - t_n)/T) = C(t - t_n) \prod_{m \in \mathbb{Z}, m \ne 0} (1 - (t - t_n)/(mT)), \qquad (8.109)$$

   where C is a constant.

b. Use the expression for $G(t)$ to write $x(t)$ explicitly as a function of $x(t_n)$.

6. Let $k(t, u)$ be a reproducing kernel for a Hilbert space $\mathcal{H}$ of functions defined on some interval $I \in \mathbb{R}$.

   a. The kernel $k(t, u)$ defines an operator K through (8.15). Give an explicit characterization of the adjoint operator K^*.

   b. Prove that K is Hermitian (namely that $K^* = K$).

   c. Prove that K is positive semidefinite (namely that $\langle Kx, x \rangle \geq 0$ for all $x \in \mathcal{H}$).

7. Consider an SI subspace $\mathcal{W}$ with period T and generator $w(t)$.

   a. Under what condition on $w(t)$ is $\mathcal{W}$ an RKHS?

   b. Suppose that $\mathcal{W}$ is an RHKS. Write down an expression for the reproducing kernel.

8. Let $\mathcal{A}$ be the subspace of piecewise linear functions and assume that M is a general memoryless nonlinearity. Describe the subspace $\mathcal{W}$ defined by $M(\mathcal{A})$.

9. Suppose we sample an arbitrary signal $x(t) \in L_2$ to obtain the samples $c = A^* M(x)$. We then apply Algorithm 8.1 to the samples.

   a. Will the algorithm converge? Explain.

   b. If the algorithm converges, will the recovery $\hat{x}$ be in $\mathcal{A}$?

10. Prove that (8.62) is equivalent to the existence of a value $a > 0$ satisfying (8.61).
    Hint: Show that the inequality $L < aB(x) < U$ is satisfied for some $a > 0$ if $\frac{\inf_x B(x)}{\sup_x B(x)} > \frac{L}{U}$.

11. Consider a first-order spline $x(t)$ that lies in the space spanned by $\beta^1(t)$. The signal goes through a nonlinear system whose input–output relation is given by

$$y(t) = \big(x(t) * \beta^0(t)\big)^2 * \beta^0(t), \tag{8.110}$$

and is then sampled with a sampling filter $\beta^1(t)$.

   a. Show how this problem can be set up within the framework introduced in the chapter. In particular, identify the subspaces $\mathcal{A}$ and $\mathcal{S}$.

   b. Apply the fixed-point and Newton iterative algorithms to recover $x(t)$ and compare their convergence.

12. Let $x(t)$ be a spline of degree 1 with knots at the integers. Consider the nonlinearity:

$$M(x) = \frac{\alpha}{1 + e^{-\beta x}}, \qquad \alpha, \beta > 0. \tag{8.111}$$

The signal passes through $M(x)$ and is then sampled at times $t = n, n \in \mathbb{Z}$ after going through a filter $s(-t)$. We would like to recover $x(t)$ from these samples.

   a. Suppose that the sampling filter is given by $s(-t) = \beta^1(-t)$. Under what conditions will the fixed-point iterations Algorithm 8.1 and the Newton method Algorithm 8.2 converge to the true signal $x(t)$?

   b. Now suppose that the sampling filter is given by $s(-t) = \beta^0(-t)$ and we use the fixed-point iterations and the Newton method given by Algorithms 8.4 and 8.5, respectively. Will these algorithms converge? If so, then under what conditions?

   c. Repeat the previous part where now

$$M(x) = x + \frac{\alpha}{1 + e^{-\beta x}}, \qquad \alpha, \beta > 0. \tag{8.112}$$

Explain the difference in the results.

13. Suppose that a spline of degree 1 with knots at the integers is nonlinearly distorted by a function $M(x)$ and then sampled with the filter $s(t) = \beta^0(t)$ at $t = n$, $n \in \mathbb{Z}$, as in Example 8.10. We assume that $x(t)$ is in the interval $[-1, 1]$.

    a. Let $M(x) = \sin ax$ for some $a > 0$. What are the conditions on a guaranteeing that there is a unique $x \in \mathcal{A}$ satisfying the measurements?

    b. Now let $M(x) = -x^3/3 + x$ where x is limited to an interval $[-b, b]$. What value of b guarantees that there is a unique $x \in \mathcal{A}$ satisfying the measurements?

Chapter 9

Resampling

In previous chapters we considered the problem of reconstructing a continuous-time signal from its discrete set of samples. Consequently, the recovery systems we designed were of hybrid nature with discrete-time input and continuous-time output. In this chapter we demonstrate that sampling theory also plays a crucial role in the design of fully discrete-time algorithms. In particular, we treat applications in which one would like to obtain a sequence of samples of a signal from a different set of samples of the same signal. In such settings, both the input and the output are digital.

One example is *sampling rate conversion*. Consider, for instance, digital audio files, which nowadays populate personal computers and media devices in enormous amounts. The sources of these files are diverse, and consequently a very common situation is that files are recorded at different sampling rates. When an audio file is played, the samples representing the recording are passed through a DAC. The analog output is then amplified and fed into a speaker. A typical DAC supports only a small number of sampling (and reconstruction) rates. If the audio samples do not correspond to one of these rates, then a preprocessing step is required in order to change their rate. This operation is performed by appropriate digital signal processing and is termed sampling rate conversion. Rate conversion is also required if two audio files with different sampling rates are played simultaneously, even when both sampling rates are supported by the DAC, since a standard DAC operates at a single rate.

Another example, which we already encountered in previous chapters, is *image resampling* or *image interpolation*. These terms refer to changing the resolution of an image in order to obtain an effect of enlargement (zoom-in) or reduction (zoom-out) in dimensions. In resolution conversion, we seek samples of the continuous-space image on a different grid. A denser grid corresponds to zooming in, whereas a more spacious grid induces a zoom-out effect. Since the original continuous-space image is unavailable, image resampling is performed digitally.

From a purely theoretical standpoint, there is no need to develop new tools for resampling. Given samples at some rate, we can reconstruct the underlying continuous-time signal using one of the methods discussed in previous chapters, and then resample the result at the desired rate. In practice, of course, one does not need to go through the analog domain in order to perform conversion, but rather all operations are carried out digitally. Plugging the second sampling stage into the reconstruction formula of the first stage yields an all-digital expression, relating the output sequence to the input set of samples. This paradigm is demonstrated in Fig. 9.1. If both reconstruction and

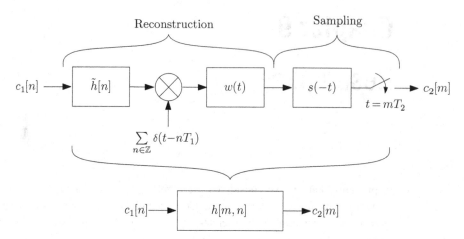

Figure 9.1 Resampling as a two-stage operation. The sequence $c_1[n]$ corresponds to samples of a signal $x(t)$ taken with period T_1. By recovering $x(t)$ and sampling the result at times mT_2, $m \in \mathbb{Z}$, we obtain the rate-converted sequence $c_2[m]$.

sampling are linear, as we assume throughout this chapter, then the overall relation is linear as well. We note, though, that generally rate conversion is not time-invariant, so that the equivalent digital system does not correspond to simple convolution. Rather, it is described by a kernel $h[m, n]$ with two indices: one representing the output rate, and the other corresponding to the input rate.

Why, then, pay special attention to the problem of resampling? The driving force of the developments in the field of resampling is mainly computational. Being such a fundamental building block in many applications, signal resampling often needs to be performed very efficiently. While the naive two-stage approach of Fig. 9.1 is a good solution in terms of recovery error, it is often computationally expensive. Therefore, in practice we may tolerate inferior solutions as long as they require lower computational resources. As we will see in this chapter, such methods can be developed using tools from the previous chapters. We will also consider approaches that require new derivations, which explicitly impose a more convenient structure on the conversion system.

9.1 Bandlimited sampling rate conversion

To understand sampling rate conversion, we begin with the traditional viewpoint, which treats bandlimited signals. We denote the input sampling period by T_1 and the output period by T_2. As a first stage, we examine interpolation by an integer factor I, corresponding to $T_2 = T_1/I$. Since $T_2 < T_1$ the sampling grid at the output is denser than at the input. We then discuss decimation (downsampling) by an integer factor D, namely when $T_2 = T_1 D$ corresponding to $T_2 > T_1$. In both cases we will see that rate conversion can be achieved efficiently using standard building blocks: upsamplers, downsamplers, and filters. Finally, we treat rate conversion by a rational factor I/D.

As long as I and D are small, combining the blocks used for interpolation and decimation leads to efficient conversion. However, for large I or D, this approach requires many computations. Instead, we consider a heuristic which often works quite well: we first interpolate the signal onto a higher sampling grid, and then approximate the signal locally on the desired grid. In Section 9.3 we revisit this approach in a more rigorous setting, and show how to optimally interpolate the signal.

9.1.1 Interpolation by an integer factor I

Suppose that $x(t)$ is a π/T-bandlimited signal and that the sequence $c_1[n]$ corresponds to pointwise samples of $x(t)$ taken at a rate no lower than the Nyquist rate. Thus,

$$c_1[n] = x(nT_1), \tag{9.1}$$

where $T_1 \leq T$. We would like to obtain samples $c_2[n]$ of $x(t)$ spaced T_2 apart, where $T_2 = T_1/I$ for some integer I.

From the Shannon–Nyquist theorem (Theorem 4.1), $x(t)$ can be perfectly reconstructed from $c_1[n]$ using the sinc kernel:

$$x(t) = \sum_{n \in \mathbb{Z}} c_1[n] \operatorname{sinc}\left(\frac{t - nT_1}{T_1}\right). \tag{9.2}$$

Therefore, $c_2[m]$ may be expressed as

$$c_2[m] = x(mT_2) = \sum_{n \in \mathbb{Z}} c_1[n] \operatorname{sinc}\left(\frac{mT_2 - nT_1}{T_1}\right) = \sum_{n \in \mathbb{Z}} c_1[n]h[m, n], \tag{9.3}$$

where we defined

$$h[m, n] = \operatorname{sinc}\left(\frac{m - nI}{I}\right). \tag{9.4}$$

Equation (9.3) represents a simple and direct relationship between the interpolated sequence $c_2[m]$ and the input samples $c_1[n]$. This relation is time-varying, and therefore does not correspond to convolution. Systems of this type are called *multirate systems* because their number of output coefficients in any given time period is different than the number of input samples over the same time frame.

Although the formula relating $c_2[m]$ to $c_1[n]$ is time-varying, it is possible to express it in terms of two very simple building blocks – an *upsampler* and an LTI filter. Upsampling a sequence $c[n]$ by a factor I is an operation that places $I - 1$ zeros between every two consecutive points of $c[n]$, as demonstrated in Fig. 9.2 for $I = 3$. The sequence $d[n]$ at the output of an upsampler fed with $c[n]$ is given by

$$d[n] = \begin{cases} c\left[\frac{n}{I}\right], & n = 0, \pm I, \pm 2I, \ldots \\ 0, & \text{otherwise.} \end{cases} \tag{9.5}$$

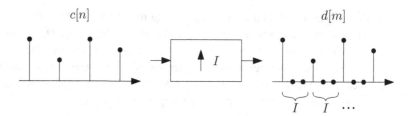

Figure 9.2 Upsampler.

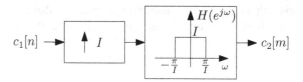

Figure 9.3 Interpolation by an integer factor of I.

Suppose that a sequence $c[n]$ is upsampled by a factor I and the result $d[m]$ is convolved with a filter $h[m]$. The output $e[m]$ is then given by

$$e[m] = \sum_{k \in \mathbb{Z}} d[k]h[m-k] = \sum_{n \in \mathbb{Z}} c[n]h[m-nI]. \qquad (9.6)$$

Comparing this expression with (9.3), we see that increasing the rate of $c_1[n]$ by I can be performed by first upsampling $c_1[n]$ by a factor of I and then filtering the result with

$$h[m] = \operatorname{sinc}\left(\frac{m}{I}\right). \qquad (9.7)$$

This interpretation is illustrated in Fig. 9.3, where $H(e^{j\omega})$ is the DTFT of (9.7).

We can gain intuition into the implementation of Fig. 9.3 by considering the system operation in the frequency domain. To do so, we first treat the interpolation block. The DTFT of the upsampler output $d[n]$ can be expressed in terms of the input $c_1[n]$ as

$$D(e^{j\omega}) = \sum_{m \in \mathbb{Z}} d[m]e^{-j\omega m} = \sum_{n \in \mathbb{Z}} c_1[n]e^{-j\omega nI} = C_1(e^{j\omega I}), \qquad (9.8)$$

where $C_1(e^{j\omega})$ is the DTFT of $c_1[n]$. Therefore, upsampling scales the frequency axis by a factor of I. Since $C_1(e^{j\omega})$ is 2π-periodic, the resulting $D(e^{j\omega})$ comprises I replicas of $C_1(e^{j\omega})$ in the range $[-\pi, \pi]$, shrunk by a factor I, as demonstrated in Fig. 9.4.

Next, we note that the frequency response of the filter (9.7) over $[-\pi, \pi]$ is given by

$$H(e^{j\omega}) = \begin{cases} I, & |\omega| \leq \pi/I \\ 0, & |\omega| > \pi/I. \end{cases} \qquad (9.9)$$

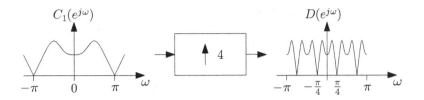

Figure 9.4 Effect of upsampling by a factor of 4 on the frequency content of the input sequence.

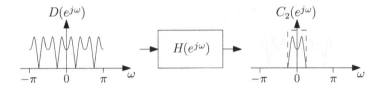

Figure 9.5 Replica suppression at the output of an upsampler.

Therefore, when applied to $D(e^{j\omega})$, this filter suppresses all replicas besides the central one, as seen in Fig. 9.5. Combining (9.8) and (9.9), the relation (9.3) can be expressed in the frequency domain over $[-\pi, \pi]$ as

$$C_2(e^{j\omega}) = \begin{cases} IC_1(e^{j\omega I}), & |\omega| \le \pi/I \\ 0, & |\omega| > \pi/I. \end{cases} \tag{9.10}$$

To see that (9.10) represents pointwise sampling of $x(t)$ with spacing T_2, recall that since $x(t)$ is π/T-bandlimited,

$$C_1(e^{j\omega}) = \frac{1}{T_1} X \left(\frac{\omega}{T_1} \right), \quad |\omega| \le \pi. \tag{9.11}$$

Substituting this expression into (9.10), and using the facts that $T_2 = T_1/I$ and $X(\omega/T_2) = 0$ for $|\omega| > \pi/I$, we have that

$$C_2(e^{j\omega}) = \frac{1}{T_2} X \left(\frac{\omega}{T_2} \right), \quad |\omega| \le \pi, \tag{9.12}$$

as desired.

9.1.2 Decimation by an integer factor D

Next, we consider the problem of converting samples $c_1[n]$ of a π/T-bandlimited signal $x(t)$, obtained with period $T_1 \le T$, into samples $c_2[m]$ taken with period $T_2 = DT_1$ for some integer D. Since we are now decreasing the sampling rate, the sequence $c_2[m]$ may, in general, contain aliased frequencies. Therefore, our goal is that the sequence $c_2[m]$ should correspond to samples of

$$y(t) = (x * s)(t), \tag{9.13}$$

where $s(t) = (1/T_2) \operatorname{sinc}(t/T_2)$ is an anti-aliasing LPF with cutoff π/T_2.

Using (9.2), we have that

$$c_2[m] = y(mT_2)$$

$$= \left(\frac{1}{T_2} \operatorname{sinc}\left(\frac{t}{T_2} \right) * \left(\sum_{n \in \mathbb{Z}} c_1[n] \operatorname{sinc}\left(\frac{t - nT_1}{T_1} \right) \right) \right) \Bigg|_{t = mT_2}$$

$$= \frac{T_1}{T_2} \sum_{n \in \mathbb{Z}} c_1[n] \operatorname{sinc}\left(\frac{mT_2 - nT_1}{T_2} \right)$$

$$= \sum_{n \in \mathbb{Z}} c_1[n] h[m, n], \tag{9.14}$$

where we defined

$$h[m, n] = \frac{1}{D} \operatorname{sinc}\left(\frac{mD - n}{D} \right). \tag{9.15}$$

The third equality follows from the fact that for $T_2 \geq T_1$,

$$\operatorname{sinc}\left(\frac{t}{T_2} \right) * \operatorname{sinc}\left(\frac{t}{T_1} \right) = T_1 \operatorname{sinc}\left(\frac{t}{T_2} \right). \tag{9.16}$$

This relation can be easily established by considering the Fourier transform of both sides of the equation.

As with interpolation, the input–output relation of (9.14) corresponds to a multirate system. This system can be conveniently implemented as a cascade of two basic building blocks – an LTI filter and a *downsampler*. Downsampling a sequence $c[n]$ by a factor D is an operation that retains one out of every D elements of $c[n]$, as demonstrated in Fig. 9.6. Thus, the output $d[m]$ of a factor-D downsampler is given by

$$d[m] = c[mD]. \tag{9.17}$$

If a sequence $c[n]$ is passed through a filter $h[n]$ prior to downsampling by D, then the result is

$$d[m] = \sum_{n \in \mathbb{Z}} c[n] h[mD - n]. \tag{9.18}$$

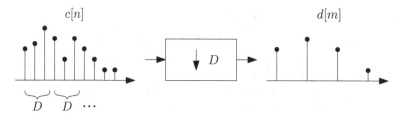

Figure 9.6 Downsampler.

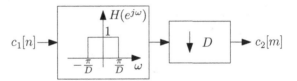

Figure 9.7 Decimation by an integer factor of D.

Comparing this expression with (9.14), we see that our decimation system can be implemented by a cascade of the filter

$$h[n] = \frac{1}{D} \operatorname{sinc}\left(\frac{n}{D}\right) \tag{9.19}$$

and a factor-D downsampler, as shown schematically in Fig. 9.7.

It is insightful to examine the behavior of the system of Fig. 9.7 in the frequency domain. We begin with the downsampler. To this end we write the relation $d[m] = c[mD]$ as $d[m] = \tilde{c}[mD]$, where $\tilde{c}[n] = c[n]p[n]$ and $p[n]$ is the impulse train

$$p[n] = \sum_{m\in\mathbb{Z}} \delta[n - mD]. \tag{9.20}$$

The DTFT of $p[n]$ is given by

$$P(e^{j\omega}) = \sum_{n\in\mathbb{Z}}\left(\sum_{m\in\mathbb{Z}}\delta[n-mD]\right)e^{-j\omega n} = \sum_{m\in\mathbb{Z}} e^{-j\omega mD} = \frac{2\pi}{D}\sum_{k\in\mathbb{Z}}\delta\left(\omega - \frac{2\pi k}{D}\right), \tag{9.21}$$

where the last equality follows from the Poisson-sum formula (3.48). The DTFT of $\tilde{c}[n]$ can be expressed as the convolution between $C(e^{j\omega})$ and $P(e^{j\omega})$ scaled by 2π:

$$\tilde{C}(e^{j\omega}) = \frac{1}{2\pi}\int_{-\pi}^{\pi} C(e^{j\theta})\frac{2\pi}{D}\sum_{k\in\mathbb{Z}}\delta\left(\omega - \frac{2\pi k}{D} - \theta\right)d\theta = \frac{1}{D}\sum_{k=0}^{D-1} C(e^{j(\omega - \frac{2\pi k}{D})}). \tag{9.22}$$

Consequently, the DTFT of $d[m]$ is

$$D(e^{j\omega}) = \sum_{m\in\mathbb{Z}} d[m]e^{-j\omega m} = \sum_{m\in\mathbb{Z}} \tilde{c}[mD]e^{-j\omega m} = \sum_{n\in\mathbb{Z}} \tilde{c}[n]e^{-j\omega n/D}$$

$$= \tilde{C}(e^{j\frac{\omega}{D}}) = \frac{1}{D}\sum_{k=0}^{D-1} C(e^{j(\frac{\omega - 2\pi k}{D})}). \tag{9.23}$$

Thus, whereas upsampling contracts the frequency axis, downsampling expands it, while creating an aliasing effect.

The effect of downsampling in the frequency domain is demonstrated in Fig. 9.8. One way to obtain the output transform from the transform of the input is first to expand the frequency response of the incoming signal in the range $[-\pi, \pi]$ by a factor of D, and then create aliases with spacing π. Alternatively, the same picture can be formed by first creating aliases of the input with spacing π/D, and then expanding the result by D.

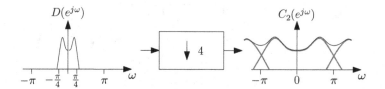

Figure 9.8 Effect of downsampling by a factor of 4 on the frequency content of the input sequence.

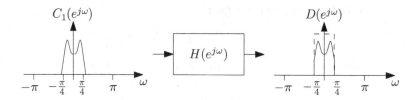

Figure 9.9 Anti-aliasing filtering prior to downsampling.

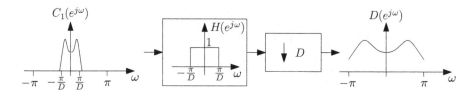

Figure 9.10 The combined effect of low-pass filtering and downsampling by a factor of D.

If the incoming signal is bandlimited to π/D, then no aliasing occurs. In this case, the transform of the output has the same form as that of the input, scaled by D.

The system in Fig. 9.7 contains a prefilter $h[n]$ given by (9.19). The frequency response of $h[n]$ over $[-\pi, \pi]$ is

$$H(e^{j\omega}) = \begin{cases} 1, & |\omega| \leq \pi/D \\ 0, & |\omega| > \pi/D. \end{cases} \tag{9.24}$$

Therefore, it acts as an anti-aliasing filter, as shown in Fig. 9.9. The output of Fig. 9.7 can thus be viewed as downsampling of a properly filtered version of the input. The combined effects of the LPF and downsampler are depicted in Fig. 9.10.

9.1.3   Rate conversion by a rational factor I/D

Finally, consider the problem of converting samples $c_1[n]$ of a π/T-bandlimited signal $x(t)$, taken with period $T_1 \leq T$, into samples $c_2[m]$ with period $T_2 = DT_1/I$, where D and I are integers. Rather than following the derivations of the previous two sections, to obtain a relation between $c_2[m]$ and $c_1[n]$, we can simply concatenate the interpolation

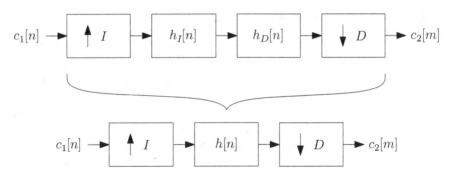

Figure 9.11 Rate conversion by a rational factor of I/D.

and decimation systems we have already derived. Since decimation often introduces errors due to the required anti-aliasing filter, we first perform interpolation and only then decimation (see Exercise 2). The resulting scheme is shown in Fig. 9.11. In this figure, $h_I[n] = \mathrm{sinc}(n/I)$ and $h_D[n] = (1/D)\,\mathrm{sinc}(n/D)$ (see (9.7) and (9.19)), so that $h[n] = (h_I * h_D)[n]$ is given by

$$h[n] = \frac{I}{\max\{I,D\}}\,\mathrm{sinc}\left(\frac{n}{\max\{I,D\}}\right). \tag{9.25}$$

This relation is easy to verify in the DTFT domain (see Exercise 3).

Any real number can be approximated arbitrarily well by a rational number. Therefore, theoretically, the method of Fig. 9.11 can be used to convert between any two rates. However, in practice, this approach may be very inefficient. As a toy example, suppose we wish to increase the size of a digital image by $\sqrt{2}$ in each axis. Using the approximation $\sqrt{2} \approx 577/408$, we can design a rate-conversion system that upsamples the original image by 577, filters it, and then downsamples the result by 408. The intermediate step in this naive implementation requires $577^2 = 332{,}929$ times more memory than that required to store the original image. It is obviously unreasonable to require a system to be able to store nearly half-a-million images just in order to perform a simple factor-$\sqrt{2}$ enlargement. Furthermore, performing digital filtering on the gigantic intermediate image is highly inefficient, and in many systems impractical.

We conclude that if the conversion factor is not rational and cannot be approximated by a rational number I/D, where I is reasonably small, then rate conversion using upsamplers, downsamplers, and filters is computationally expensive. The cost of sampling rate conversion can be reduced by making use of *multirate signal processing* techniques which provide tools to implement multirate systems efficiently. For an excellent tutorial on the topic we refer the reader to [147]. Nonetheless, the basic cost is still determined by the values of I and D. An alternative is to use the relationship (9.2) to compute the samples $x(mT_2)$ at any desired time step T_2. However, since the sinc function has slow decay, it is not clear how to evaluate the resulting expression efficiently. Therefore, in these cases, it is common to resort to inexact solutions, as we discuss next.

9.1.4   Rate conversion by arbitrary factors

To obtain an approximate solution to sampling rate conversion, recall that our goal is to obtain samples of $x(t)$ at times mT_2, $m \in \mathbb{Z}$, where the only available measurements are samples of $x(t)$ at times nT_1, $n \in \mathbb{Z}$. Suppose that one of the desired time instances, say m_0T_2, happens to lie close to one of the original sampling locations, say n_0T_1. It is reasonable to believe that in this case $c_2[m_0] \approx c_1[n_0]$. Consequently, a plausible approximation could be to set $c_2[m_0] = c_1[n_0]$.

Many target sampling locations, however, may not lie "close enough" to available sampling points. Therefore, applying this strategy to all samples can produce large errors. One approach to reduce the approximation error is to increase the sampling rate by some integer factor I as a preliminary step. The resulting sequence corresponds to samples on the denser grid nT_1/I, $n \in \mathbb{Z}$. Now, for any location m_0T_2, there is an available sample located no more than $0.5T_1/I$ time units away, say at n_0T_1/I. Therefore, setting $c_2[m_0] = c_1[n_0]$ in this case is expected to yield smaller errors.

This strategy of integer interpolation followed by a nearest-neighbor fit is sometimes referred to as *first-order approximation*. It can be thought of as a three-stage procedure as follows. First, the original sequence $c_1[n] = x(nT_1)$ is interpolated by an integer factor I to yield $d[\ell] = x(\ell T_1/I)$ as in Fig. 9.3. Second, the resulting sequence $d[\ell]$ is converted into the analog signal

$$\tilde{x}(t) = \sum_{\ell \in \mathbb{Z}} d[\ell] w \left(t - \frac{\ell T_1}{I} \right), \qquad (9.26)$$

where $w(t) = \beta^0(tI/T_1)$ is a rectangular kernel, or a zero-order spline (see (4.32)). Finally, $\tilde{x}(t)$ is sampled at mT_2, $m \in \mathbb{Z}$, to yield the sequence $c_2[m] \approx x(mT_2)$. This scheme is shown in Fig. 9.12, where $h[n]$ is an LPF with cutoff π/I and gain I.

Note that in this approach, the interpolation of $c_1[n]$ onto a higher grid does not take into account the fact that the resampling is performed with a nonoptimal kernel $w(t)$. In Section 9.3.1 we consider dense-grid resampling with subspace priors from a more rigorous perspective and show that the interpolation step can be optimized to minimize the error in the resampling procedure. In our context, this amounts to replacing the LPF in Fig. 9.12 with an optimal filter that takes both the bandlimited prior and the chosen kernel $w(t)$ into account. Here, however, we focus on a simple and intuitive approach that can sometimes yield satisfactory results.

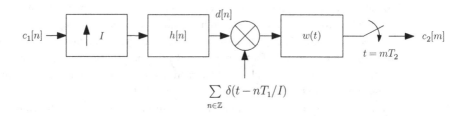

Figure 9.12 Rate conversion by an arbitrary factor.

Example 9.1 An example of applying the approach of Fig. 9.12 on samples of a sinusoidal signal is shown in Fig. 9.13. Here, the frequency of the sinusoid is 1 Hz, $T_1 = 0.174$ s, and $T_2 = T_1/\sqrt{2}$. As can be seen, the samples $c_2[m]$ produced by simple nearest-neighbor interpolation (marked with hollow circles) substantially deviate from the original signal at the desired sampling locations. However, the higher the intermediate interpolation factor we use, the better the approximation becomes.

To further reduce the approximation error, one can replace the nearest-neighbor fit by a linear fit. Specifically, rather than setting $c_2[m_0]$ to equal the value of $d[\ell_0]$, where ℓ_0 is such that $\ell_0 T_1/I$ is closest to mT_2, we can let $c_2[m_0]$ lie on the line connecting $d[\ell_1]$ and $d[\ell_2]$, where ℓ_1 and ℓ_2 are chosen to satisfy $\ell_1 T_1/I \leq m_0 T_2 \leq \ell_2 T_1/I$. It is easily verified that the resulting scheme again takes the form of Fig. 9.12, with the only

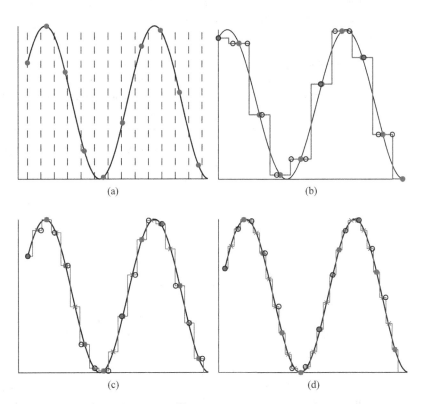

Figure 9.13 Interpolation by a factor of $\sqrt{2}$. (a) Samples $c_1[n]$ at times nT_1, $n \in \mathbb{Z}$, of a sinusoidal signal $x(t)$ and desired sampling locations (marked with vertical dashed lines) at times $nT_1/\sqrt{2}$, $n \in \mathbb{Z}$. (b) Nearest-neighbor interpolation. (c), (d) First-order approximation using an intermediate interpolation factor of 2 and 3, respectively. Intermediate interpolated points are shown in X-marks.

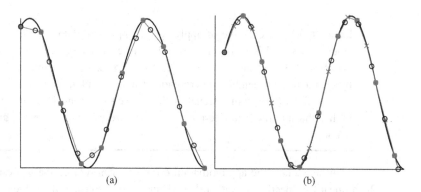

Figure 9.14 Interpolation by a factor of $\sqrt{2}$ in the setting of Fig. 9.13. (a) Linear interpolation. (b) Second-order approximation using an intermediate interpolation factor of 2. Intermediate interpolated points are shown in X-marks.

difference that now $w(t) = \beta^1(tI/T_1)$ is a triangular kernel, or a spline of order 1. This approach is termed *second-order approximation*.

Example 9.2 Figure 9.14 revisits the sinusoidal setting of Example 9.1, this time with second-order approximation. As can be seen, here setting $I = 1$ (that is, performing standard linear interpolation) already produces reasonable results, and $I = 2$ produces an excellent approximation.

The approximation error can be further reduced by considering higher-order splines or other kernels with larger support. For any reconstruction kernel $w(t)$, the samples $c_2[m]$ at the output of the system in Fig. 9.12 are given by

$$c_2[m] = \sum_{n \in \mathbb{Z}} d[n] w \left(mT_2 - \frac{nT_1}{I} \right) = \sum_{n \in \mathbb{Z}} d[n] g[m, n], \qquad (9.27)$$

where we denoted

$$g[m, n] = w \left(mT_2 - \frac{nT_1}{I} \right). \qquad (9.28)$$

If the kernel $w(t)$ is compactly supported, then the multirate system $g[m, n]$ can be applied efficiently since every output sample depends only on a finite number of input coefficients $\{d[n]\}$. This all-digital representation of the scheme of Fig. 9.12 is depicted in Fig. 9.15 with $g[m, n]$ given by (9.28). Note that the number of coefficients $\{d[n]\}$ required to compute each output sample is directly associated with the support of $w(t)$. Thus, while choosing a kernel $w(t)$ with large support may lead to better approximation, it implies a larger computational burden.

The approach outlined in this section assumes that the input signal is bandlimited, and uses splines to approximately interpolate the signal locally onto the required grid. In the next section we discuss an alternative framework, in which we assume to begin

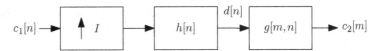

Figure 9.15 Rate conversion by an arbitrary factor with all-digital components. Here $h[n]$ is an LPF with cutoff π/I and gain I, and $g[m,n] = w(mT_2 - nT_1/I)$.

with that the signal to be interpolated is a spline of appropriate degree. Thus, we replace the bandlimited prior with a spline prior. As we will see, this leads to computationally efficient rate-conversion methods. After building intuition based on interpolation with spline priors, in Sections 9.3 and 9.4 we study a broader setup for sampling rate conversion which allows for more general signal priors. In Section 9.3 we focus on increasing the sampling rate, while in Section 9.4 we study rate reduction. The latter is more challenging since reducing the rate often leads to aliasing, or distortion, that needs to be accounted for.

9.2 Spline interpolation

We now depart from the bandlimited assumption and treat spline priors. We focus on standard interpolation, without first interpolating onto a dense grid, in order to build intuition into more general rate-conversion approaches. In Section 9.3 we expand the discussion to include dense-grid interpolation as well as more general signal models. An excellent tutorial on splines and their use in sampling rate conversion can be found in [65].

Suppose that $x(t)$ is known to be a spline of degree p with knots at nT_1, $n \in \mathbb{Z}$. Given pointwise samples $c_1[n] = x(nT_1)$, we would like to determine the samples $c_2[m] = x(mT_2)$ for some $T_2 \neq T_1$. Recall that in the bandlimited case, for sampling rate reduction (namely, when $T_2 > T_1$), we used an anti-aliasing LPF. The analogy of this operation in nonbandlimited SI spaces, such as splines, is not obvious. To keep the discussion focused on the computational aspects of working with splines, we therefore defer the treatment of generalized anti-aliasing operations to Section 9.4.1, and treat here only sampling rate increase, namely $T_2 < T_1$.

9.2.1 Interpolation formula

As we have seen in Chapter 4 (cf. (4.31)), any spline of degree p with knots at nT_1, $n \in \mathbb{Z}$, can be expressed as

$$x(t) = \sum_{n \in \mathbb{Z}} d[n]\beta^p\left(\frac{t - nT_1}{T_1}\right), \qquad (9.29)$$

where β^p is the B-spline of degree p, and $d[n]$ are appropriate coefficients. Once the sequence $d[n]$ is known, we can compute $x(t)$ at any desired time instance t_0 using (9.29). Since the B-spline kernel $\beta^p(t)$ is compactly supported, $x(t_0)$ depends only on a finite number of coefficients from the sequence $d[n]$, allowing for efficient implementation. This is in contrast to bandlimited interpolation, in which the slowly decaying sinc function replaces $\beta^p(t)$.

In practice, though, we are not given the coefficients $d[n]$ but rather the samples $c_1[n] = x(nT_1)$. Therefore, sampling rate conversion requires us to first compute these coefficients. Using (9.29), the samples $c_1[n]$ are related to the coefficients $d[n]$ by

$$c_1[n] = x(nT_1) = \sum_{k \in \mathbb{Z}} d[k]\beta^p(n-k). \tag{9.30}$$

In the frequency domain,

$$D(e^{j\omega}) = C_1(e^{j\omega})\frac{1}{B^p(e^{j\omega})}, \tag{9.31}$$

where $B^p(e^{j\omega})$ is the DTFT of $\beta^p(n)$. It can be shown [65] that $B^p(e^{j\omega})$ is nonzero on $\omega \in [-\pi, \pi]$ for any $p \geq 0$.

The solution (9.31) is equivalent to the filter used for recovering a signal from a known subspace (in this case the appropriate spline space) from given samples. As we have seen in Chapter 6 (cf. (6.33)), the general form of the filter is $H(e^{j\omega}) = 1/R_{SA}(e^{j\omega})$, where $R_{SA}(e^{j\omega})$ is defined by (5.28). In our case, $a(t) = \beta^p(t)$, and $s(t) = \delta(t)$, which results in $H(e^{j\omega}) = 1/B^p(e^{j\omega})$.

It remains to show how to compute (9.31) efficiently. To this end, we note that the sequence $\beta^p(n)$ is supported on $[-\lfloor p/2 \rfloor, \lfloor p/2 \rfloor]$, so that

$$\frac{1}{B^p(e^{j\omega})} = \frac{1}{\sum_{k=-\lfloor \frac{p}{2} \rfloor}^{\lfloor \frac{p}{2} \rfloor} \beta^p(k)e^{-j\omega k}} \tag{9.32}$$

is an infinite-impulse response (IIR) filter with $2\lfloor p/2 \rfloor$ poles. Furthermore, since $\beta^p(k)$ is a symmetric sequence, these poles come in reciprocal pairs. This means that we can write

$$\frac{1}{B^p(e^{j\omega})} = \left(\prod_{k=1}^{\lfloor p/2 \rfloor} \frac{1}{1 - \alpha_k e^{-j\omega}} \right) \left(\prod_{k=1}^{\lfloor p/2 \rfloor} \frac{1}{1 - \alpha_k e^{j\omega}} \right) C, \tag{9.33}$$

where $C \neq 0$ is some constant and $|\alpha_k| < 1$ are the poles that lie within the complex unit circle. The first term in this expression, denoted $H_c(e^{j\omega})$, is a stable causal IIR filter of order $\lfloor \frac{p}{2} \rfloor$, whereas the second term, denoted $H_{ac}(e^{j\omega})$, is a stable anticausal IIR filter of order $\lfloor \frac{p}{2} \rfloor$. Therefore, the coefficients $d[n]$ of the B-spline expansion of $x(t)$ can be obtained by applying two IIR filtering operations on the samples $c_1[n]$. The first casual filter with output $d_1[n] = h_c[n] * c_1[n]$ is implemented using the recursion

$$d_1[n] = \gamma_1 d_1[n-1] + \cdots + \gamma_{\lfloor \frac{p}{2} \rfloor} d_1[n - \lfloor p/2 \rfloor] + c_1[n], \tag{9.34}$$

running from past to future. Here γ_i are the coefficients of the polynomial whose roots are α_i. The second filter yields an output $d_2[n] = h_{ac}[n] * d_1[n]$ which is computed by the backward recursion

$$d_2[n] = \gamma_1 d_2[n+1] + \cdots + \gamma_{\lfloor \frac{p}{2} \rfloor} d_2[n + \lfloor p/2 \rfloor] + d_1[n]. \tag{9.35}$$

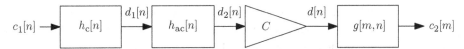

Figure 9.16 Rate conversion by an arbitrary factor with a spline prior using causal and anticausal filtering. Here the filtering and gain operations implement $1/B^p(e^{j\omega})$ and $g[m,n]$ $= \beta^p(mT_2/T_1 - n)$.

The final result is obtained by applying a gain of C:

$$d[n] = Cd_2[n]. \tag{9.36}$$

In practice, we often have a finite sequence of samples $c_1[n]$, $n = 1, \ldots, N$. We can then assume for simplicity that the values outside the range $[1, N]$ are zero, so that the initial conditions in both filtering operations are set to zero. In this case the forward recursion (9.34) is defined for $n = 1, \ldots, N$ and the backward recursion (9.35) for $n = N, \ldots, 1$.

Once we have extracted the expansion coefficients $d[n]$ from the samples $c_1[n]$, we may evaluate $x(t)$ at any desired time instance $t = mT_2$ using (9.29) as illustrated in Fig. 9.15 with $g[m,n] = \beta^p(mT_2/T_1 - n)$. The resulting sampling-rate conversion scheme is depicted in Fig. 9.16.

We now demonstrate computation of the spline coefficients $d[n]$ from pointwise samples $c_1[n] = x(nT_1)$ via several examples.

Example 9.3 Suppose that $x(t)$ is a spline of degree $p = 0$. In this case, $\beta^0(n) = \delta[n]$, namely $B^0(e^{j\omega}) = 1$, so that the expansion coefficients are $d[n] = c_1[n]$, and no digital filtering is needed. The same thing happens for $p = 1$, since $\beta^1(n) = \delta[n]$ as well. These two cases correspond to nearest-neighbor interpolation (a shifted by half version of zero-order hold) and linear interpolation, respectively.

Example 9.4 Consider next a signal $x(t)$ that is a spline of degree $p = 2$. Direct computation shows that

$$\beta^2(n) = \begin{cases} \frac{3}{4}, & n = 0 \\ \frac{1}{8}, & |n| = 1 \\ 0, & |n| \geq 2. \end{cases} \tag{9.37}$$

Therefore,

$$\frac{1}{B^2(e^{j\omega})} = \frac{8}{e^{-j\omega} + 6 + e^{j\omega}}. \tag{9.38}$$

This filter can be expressed as

$$\frac{1}{B^2(e^{j\omega})} = -8\alpha_2 \left(\frac{1}{1 - \alpha_2 e^{-j\omega}} \right) \left(\frac{1}{1 - \alpha_2 e^{j\omega}} \right), \tag{9.39}$$

where $\alpha_2 = -3 + \sqrt{8}$. Therefore, the digital filtering needed to implement (9.31) may be performed by two first-order IIR filters, one causal and one anticausal.

The first filter corresponds to the recursion

$$d_1[n] = c_1[n] + \alpha_2 d_1[n-1] \tag{9.40}$$

and the second filter corresponds to

$$d_2[n] = d_1[n] + \alpha_2 d_2[n+1]. \tag{9.41}$$

The final output is

$$d[n] = -8\alpha_2 d_2[n]. \tag{9.42}$$

Example 9.5 Suppose that $x(t)$ is a spline of degree $p = 3$. In this case,

$$\beta^3(n) = \begin{cases} \frac{2}{3}, & n = 0 \\ \frac{1}{6}, & |n| = 1 \\ 0, & |n| \geq 2, \end{cases} \tag{9.43}$$

and

$$\frac{1}{B^3(e^{j\omega})} = \frac{6}{e^{-j\omega} + 4 + e^{j\omega}}. \tag{9.44}$$

A simple computation shows that

$$\frac{1}{B^3(e^{j\omega})} = -6\alpha_3 \left(\frac{1}{1 - \alpha_3 e^{-j\omega}} \right) \left(\frac{1}{1 - \alpha_3 e^{j\omega}} \right), \tag{9.45}$$

where $\alpha_3 = -2 + \sqrt{3}$. Thus, here as well digital filtering can be performed by two first-order IIR filters. Specifically, we have

$$d_1[n] = c_1[n] + \alpha_3 d_1[n-1] \tag{9.46}$$

and

$$d_2[n] = d_1[n] + \alpha_3 d_2[n+1]. \tag{9.47}$$

The final output is

$$d[n] = -6\alpha_3 d_2[n]. \tag{9.48}$$

9.2.2   Comparison with bandlimited interpolation

The examples above demonstrate that digital processing of the samples required in spline interpolation can be obtained very efficiently when the spline degree is not too high. The reconstruction stage performed with the corrected coefficients $d[n]$ is also quite efficient since the B-spline kernels are compactly supported. To appreciate the difference with respect to (nonapproximate) bandlimited rate conversion by an arbitrary factor, note that if a bandlimited signal $x(t)$ is sampled with a sampling period T_1 smaller than the Nyquist period, then it can be written as

$$x(t) = \sum_{n \in \mathbb{Z}} c_1[n] \operatorname{sinc}\left(\frac{t - nT_1}{T_1}\right). \tag{9.49}$$

Evaluating $x(t)$ at mT_2, for some $T_2 < T_1$, we get

$$c_2[m] = x(mT_2) = \sum_{n \in \mathbb{Z}} c_1[n] \operatorname{sinc}\left(\frac{mT_2 - nT_1}{T_1}\right). \tag{9.50}$$

Thus all samples in the sequence $\{c_1[n]\}$ contribute to the computation of $c_2[m]$, for every m. Suppose that for the sake of efficiency we truncate the sum and use only those coefficients that are multiplied by a factor greater than 0.01 in absolute value. Then, because of the slow decay of the sinc function, we would still be left with roughly 60 samples contributing to each output coefficient.

In contrast, with order-p spline interpolation, the reconstruction stage involves the B-spline kernel $\beta^p(t)$, whose support is of size $2\lfloor p/2 \rfloor$, i.e.

$$c_2[m] = x(mT_2) = \sum_{n \in \mathbb{Z}} d[n] \beta^p\left(\frac{mT_2 - nT_1}{T_1}\right). \tag{9.51}$$

Thus, only $2\lfloor p/2 \rfloor + 1$ coefficients from the corrected sequence $\{d[n]\}$ contribute to the computation of each output sample $c_2[m]$. This difference with respect to the bandlimited case is substantial, especially for splines with a small degree.

9.3   Dense-grid interpolation

The analysis and methods presented in Section 9.1 allowed us to identify the basic building blocks required for practical rate conversion. Namely, we saw that with upsamplers, downsamplers, digital filters, and simple interpolation kernels (such as B-splines of order 0 or 1) we can efficiently convert the rate of a given sequence by any desired factor, assuming a bandlimited prior. In Section 9.2, we expanded the discussion to spline priors. In this case, implementation is convenient since the support of the B-spline kernels is finite. Thus, approximation methods of the type discussed in Section 9.1.4 were unnecessary. In this section, we further extend our framework by allowing for a denser reconstruction grid and treating arbitrary subspace and smoothness priors.

In Chapters 6 and 7 we discussed recovery based on subspace, smoothness, and stochastic priors on $x(t)$. In principle, if we know how to produce a good recovery $\hat{x}(t)$ of $x(t)$ for any $t \in \mathbb{R}$, then we can, in particular, compute $\hat{x}(nT_2)$ for every $n \in \mathbb{Z}$, or, if desired, $(s * \hat{x})(nT_2)$, where $s(t)$ is some anti-aliasing filter (required especially for decimation). Why not use this paradigm, then?

Recall that the unconstrained recovery approaches we developed often led to noncompactly supported reconstruction kernels, rendering them computationally inefficient. Conversely, in the constrained setting, in which we sought an optimal correction filter

$h[n]$ for a fixed recovery kernel $w(t)$, the resulting reconstruction was often poor if $w(t)$ was chosen to have a very small support (such as a B-spline of order 0 or 1). Therefore, to obtain a practical rate-conversion system that takes into account arbitrary priors on $x(t)$, we will extend the reconstruction schemes treated in previous chapters in several ways. First, we consider dense-grid interpolation where, as in Section 9.1.4, we allow for interpolation prior to rate conversion. We also constrain the reconstruction to simple compactly supported kernels. Second, when treating decimation, or sampling rate reduction, we combine standard recovery methods with proper anti-aliasing filtering to minimize the resulting distortion.

As in Section 9.2, we begin by considering sampling rate increase. The treatment of sampling rate decrease requires adaptation of the anti-aliasing operation from the bandlimited scenario to other types of priors, and will be discussed in Section 9.4.1. Motivated by the results of Section 9.1.4, to efficiently increase the sampling rate we focus on the architecture depicted in Fig. 9.12 where we consider both subspace priors and smoothness priors. Namely, we upsample the sequence $c_1[n]$, digitally filter the result with $h[n]$, reconstruct an analog signal $\tilde{x}(t)$ using some simple kernel $w(t)$, and then sample the result at times mT_2, $m \in \mathbb{Z}$ to obtain $c_2[m]$. We constrain ourselves to using a predefined reconstruction kernel $w(t)$ (which we choose to allow for efficient implementation) and a predefined rate-conversion factor I. Our goal is to design the digital correction filter $h[n]$. When $I = 1$ this problem is equivalent to that studied in Chapters 6 and 7; here we extend the discussion to general I and also consider practical methods for implementing $h[n]$. Spline interpolation treated in Section 9.2 is a special case of Fig. 9.12 in which $I = 1$ and $w(t)$ is chosen as an appropriate spline.

9.3.1   Subspace prior

We now formulate our problem in mathematical terms, starting with a subspace prior.

Suppose that the signal $x(t)$ is known to belong to an SI subspace $\mathcal{A}$ spanned by $\{a(t - kT_1)\}_{k \in \mathbb{Z}}$, so that

$$x(t) = \sum_{k \in \mathbb{Z}} b[k]a(t - kT_1), \qquad (9.52)$$

for some sequence $b[n]$. This signal is sampled at times nT_1, $n \in \mathbb{Z}$, after passing through the filter $s(-t)$, to produce the samples

$$c_1[n] = \int_{-\infty}^{\infty} x(t)s(t - nT_1)dt. \qquad (9.53)$$

Our goal is to recover $x(t)$ by using shifts, spaced $\tilde{T}_1 = T_1/I$ time units apart, of a predefined reconstruction kernel $w(t)$, namely

$$\hat{x}(t) = \sum_{n \in \mathbb{Z}} d[n]w\left(t - \frac{nT_1}{I}\right). \qquad (9.54)$$

for some sequence $d[n]$ to be determined. In other words, we would like to produce a recovery $\hat{x}(t)$, which lies in the subspace $\mathcal{W}$ spanned by the functions $\{w(t - nT_1/I)\}_{n \in \mathbb{Z}}$. We will then resample $\hat{x}(t)$ at the desired times mT_2 to perform rate conversion. The important aspect of this approach is that $w(t)$ is chosen such that this final step can be carried out efficiently; that is, the sum

$$c_2[m] = \sum_{n \in \mathbb{Z}} d[n] w \left(mT_2 - \frac{nT_1}{I} \right), \qquad (9.55)$$

may be evaluated using a small number of computations. In particular, we will consider compactly supported kernels $w(t)$.

Recovery formula

In Chapter 6 Theorem 6.4 we developed a general formula for the recovery $\hat{x}$ that minimizes the reconstruction error $\|x - \hat{x}\|$ when x is known to lie in a given subspace spanned by the set transformation A, and $\hat{x}$ is constrained to lie in the range of a set transformation W. Specifically, the samples c_1 are processed by a digital system to obtain the reconstruction coefficients

$$d = (W^*W)^{-1} W^* A (S^* A)^{-1} c_1. \qquad (9.56)$$

In our setting, A, S, and W denote the set transformations associated with $\{a(t - kT_1)\}_{k \in \mathbb{Z}}$, $\{s(t - nT_1)\}_{n \in \mathbb{Z}}$, and $\{w(t - mT_1/I)\}_{m \in \mathbb{Z}}$, respectively. Note that in contrast to the SI discussion in Chapter 6, here this system is not LTI but rather multirate. Indeed, it produces I coefficients for every input sample.

We now explicitly compute the expression given by (9.56). Since both S and A correspond to SI generators with period T_1, the operator $S^* A$ corresponds to convolution with the sequence

$$r_{sa}^{T_1}[n] = \int_{-\infty}^{\infty} a(t) s(t - nT_1) dt \qquad (9.57)$$

whose DTFT is

$$R_{SA}^{T_1}(e^{j\omega}) = \frac{1}{T_1} \sum_{k \in \mathbb{Z}} \overline{S\left(\frac{\omega}{T_1} - \frac{2\pi k}{T_1} \right)} A\left(\frac{\omega}{T_1} - \frac{2\pi k}{T_1} \right). \qquad (9.58)$$

Therefore, applying $(S^* A)^{-1}$ corresponds to filtering with $1/R_{SA}^{T_1}(e^{j\omega})$. Similarly, the operator $W^* W$ is equivalent to convolution with

$$r_{ww}^{\tilde{T}_1}[m] = \int_{-\infty}^{\infty} w(t) w(t - m\tilde{T}_1) dt, \qquad (9.59)$$

where $\tilde{T}_1 = T_1/I$. The DTFT of $r_{ww}^{\tilde{T}_1}[m]$ is

$$R_{WW}^{\tilde{T}_1}(e^{j\omega}) = \frac{1}{\tilde{T}_1} \sum_{k \in \mathbb{Z}} \left| W\left(\frac{\omega}{\tilde{T}_1} - \frac{2\pi k}{\tilde{T}_1} \right) \right|^2. \qquad (9.60)$$

Multiplying by $(W^* W)^{-1}$ is therefore equivalent to filtering with $1/R_{WW}^{\tilde{T}_1}(e^{j\omega})$.

It remains to determine the effect of the operator W^*A. To do so, we note that if $d = W^*Ac$, then

$$
\begin{aligned}
d[m] &= \int_{-\infty}^{\infty} \left(\sum_{n \in \mathbb{Z}} c[n]a(t - nT_1) \right) w(t - m\widetilde{T}_1)dt \\
&= \sum_{n \in \mathbb{Z}} c[n] \int_{-\infty}^{\infty} a(t - nT_1)w(t - m\widetilde{T}_1)dt \\
&= \sum_{n \in \mathbb{Z}} c[n] \int_{-\infty}^{\infty} a(t)w(t + nT_1 - m\widetilde{T}_1)dt \\
&= \sum_{n \in \mathbb{Z}} c[n] \int_{-\infty}^{\infty} a(t)w(t - \widetilde{T}_1(m - nI))dt \\
&= \sum_{n \in \mathbb{Z}} c[n]r_{wa}^{\widetilde{T}_1}[m - nI],
\end{aligned}
\tag{9.61}
$$

where

$$
r_{wa}^{\widetilde{T}_1}[n] = \int_{-\infty}^{\infty} a(t)w(t - n\widetilde{T}_1)dt.
\tag{9.62}
$$

Comparing this expression with (9.6), we see that the operator W^*A corresponds to upsampling the sequence c by a factor I and then filtering the result with $r_{wa}^{\widetilde{T}_1}[n]$.

We conclude that the digital processing of the samples $c_1[n]$ required to produce the coefficients $d[n]$ can be expressed as the concatenation of four primitive building blocks, as shown in Fig. 9.17. The two right-most filters in this system can, of course, be merged into a single filter. Thus, $d[n]$ is obtained by passing $c_1[n]$ through the filter $1/R_{SA}^{T_1}(e^{j\omega})$, upsampling by a factor I, and then filtering the result with the filter $R_{WA}^{\widetilde{T}_1}(e^{j\omega})/R_{WW}^{\widetilde{T}_1}(e^{j\omega})$. The overall rate-conversion system then takes the form shown in Fig. 9.18, in which $g[m, n]$ is as in Fig. 9.15.

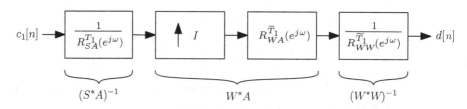

Figure 9.17 Digital processing required for rate conversion via the dense-grid methodology with a subspace prior and $\widetilde{T}_1 = T_1/I$.

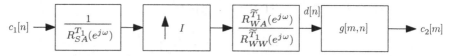

Figure 9.18 Rate conversion using the dense-grid methodology with pre- and postfiltering. Here $g[m, n] = w(mT_2 - nT_1/I)$.

The spline interpolation example studied in Section 9.2 is a special case of Fig. 9.18 in which $I = 1$ and the prior space A is equal to the reconstruction space W. The sampling operator corresponds to pointwise samples which, in the frequency domain, results in $S(\omega) = 1$. With these choices the first three blocks in Fig. 9.18 reduce to the single filter $1/R_{SA}^{T_1}(e^{j\omega})$ which is equal to $1/B^p(e^{j\omega})$.

Equivalent implementation

The scheme of Fig. 9.18 is a bit different than the architecture of Fig. 9.15 we set forth to implement, in that here there is a digital filter both before and after the upsampler. As we now show, the two systems can be made equivalent by removing the first prefilter and replacing the second filter with

$$H(e^{j\omega}) = \frac{R_{WA}^{\widetilde{T_1}}(e^{j\omega})}{R_{WW}^{\widetilde{T_1}}(e^{j\omega})R_{SA}^{T_1}(e^{j\omega I})}. \tag{9.63}$$

In other words, we move the filter $1/R_{SA}^{T_1}(e^{j\omega})$ to after the upsampler and in turn scale the frequency axis so that this filter becomes $1/R_{SA}^{T_1}(e^{j\omega I})$. The resulting system is depicted in Fig. 9.19.

To establish the equivalence, we show more generally that the effect of convolving a sequence $c[n]$ with some impulse response $g[n]$ and upsampling the result can be alternatively achieved by first upsampling $c[n]$ and then convolving the result with a filter $\tilde{g}[n]$ whose frequency response is $\widetilde{G}(e^{j\omega}) = G(e^{j\omega I})$. This correspondence is depicted in Fig. 9.20. Recall from (9.8) that the DTFT of the result of upsampling $(c * g)[n]$ by a factor I is $C(e^{j\omega I})G(e^{j\omega I})$. On the other hand, the DTFT of the convolution of the I-fold upsampled version of $c[n]$ with $\tilde{g}[n]$ is $C(e^{j\omega I})\widetilde{G}(e^{j\omega})$. Therefore, the operations are the same if $\widetilde{G}(e^{j\omega}) = G(e^{j\omega I})$.

Examples

We now consider examples of Fig. 9.18.

Example 9.6 We first revisit the traditional approximation approaches discussed in Section 9.1 for converting the sampling rate of a π/T_1-bandlimited signal $x(t)$.

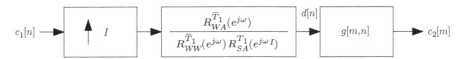

Figure 9.19 Rate conversion using the dense-grid methodology with postfiltering. Here $g[m,n] = w(mT_2 - nT_1/I)$.

Figure 9.20 Equivalence between filtering before and after an upsampler.

Assume that given samples $c_1[n] = x(nT_1)$ of $x(t)$ we would like to compute the samples $c_2[m] = x(mT_2)$ for every $m \in \mathbb{Z}$, where $T_2/T_1 < 1$ is a nonrational factor. Since the samples are pointwise in this scenario, $S(\omega) = 1$. Furthermore, the bandlimited assumption on $x(t)$ corresponds to a subspace prior with

$$A(\omega) = \begin{cases} 1, & |\omega| \leq \frac{\pi}{T_1} \\ 0, & |\omega| > \frac{\pi}{T_1}. \end{cases} \tag{9.64}$$

This implies that

$$R_{SA}^{T_1}(e^{j\omega}) = \frac{1}{T_1} \tag{9.65}$$

for all $\omega \in [-\pi, \pi]$ and

$$\frac{R_{WA}^{\tilde{T}_1}(e^{j\omega})}{R_{WW}^{\tilde{T}_1}(e^{j\omega})} = \begin{cases} \dfrac{\frac{1}{\tilde{T}_1} W\left(\frac{\omega}{\tilde{T}_1}\right)}{R_{WW}^{\tilde{T}_1}(e^{j\omega})}, & |\omega| \leq \frac{\pi}{\tilde{T}} \\ 0, & |\omega| > \frac{\pi}{\tilde{T}}. \end{cases} \tag{9.66}$$

Since the prefilter's frequency response is constant, the resulting scheme can be implemented using an I-fold upsampler followed by a digital correction filter $h[n]$, whose frequency response is

$$H(e^{j\omega}) = \begin{cases} I \dfrac{W\left(\frac{I\omega}{\tilde{T}_1}\right)}{R_{WW}^{\tilde{T}_1}(e^{j\omega})}, & |\omega| \leq \frac{\pi}{\tilde{T}} \\ 0, & |\omega| > \frac{\pi}{\tilde{T}}. \end{cases} \tag{9.67}$$

This approach is very similar to using standard rate conversion by a factor I, except for the fact that, as opposed to (9.9), here the frequency response of $H(e^{j\omega})$ is generally not flat in the pass-band. Instead, its shape is determined according to the chosen reconstruction filter $w(t)$. Therefore, the correction filter $H(e^{j\omega})$ compensates, to some extent, for the effect of the nonideal kernel $w(t)$. This is in contrast to the heuristic approach of Section 9.1.4 in which a standard LPF was used, irrespective of $w(t)$.

For example, in first-order approximation, the kernel $w(t)$ corresponds to

$$W(\omega) = \tilde{T}_1 \, \mathrm{sinc}\left(\frac{\tilde{T}_1 \omega}{2\pi}\right) \tag{9.68}$$

and $R_{WW}^{\tilde{T}_1}(e^{j\omega}) = \tilde{T}_1$. In this case, the correction filter (9.67) becomes

$$H(e^{j\omega}) = \begin{cases} I \, \mathrm{sinc}\left(\frac{\omega}{2\pi}\right), & |\omega| \leq \frac{\pi}{\tilde{T}} \\ 0, & |\omega| > \frac{\pi}{\tilde{T}}. \end{cases} \tag{9.69}$$

A similar analysis can be carried out for second-order approximation (see Exercise 4).

Example 9.7 Suppose that a spline $x(t)$ of degree 2 with knots at nT_1 is sampled at nT_1, $n \in \mathbb{Z}$, after passing through the filter $s(t) = \beta^0(t/T_1)$. Our goal is to produce an approximation of $x(t)$ from the samples with nearest-neighbor fit on the grid mT_1/I, $m \in \mathbb{Z}$, where $I \geq 1$ is some integer. This corresponds to using the kernel $w(t) = \beta^0(t/\widetilde{T}_1)$, with $\widetilde{T}_1 = T_1/I$.

In this case, $a(t) = \beta^2(t/T_1)$ and $s(t) = \beta^0(t/T_1)$, so that $r_{sa}^{T_1}[n] = T_1\beta^3(n)$. This implies that the prefilter in Fig. 9.18, which is given by $1/R_{SA}^{T_1}(e^{j\omega})$, equals $1/(T_1B^3(e^{j\omega}))$. As we have seen in Example 9.5, this filter can be implemented efficiently as a cascade of causal and anticausal filters (see (9.46) and (9.47)). We also have that $r_{ww}^{\widetilde{T}_1}[n] = \widetilde{T}_1\beta^1(n) = \widetilde{T}_1\delta[n]$, so that $R_{WW}^{\widetilde{T}_1} = \widetilde{T}_1$. The impulse response associated with the filter $R_{WA}^{\widetilde{T}_1}(e^{j\omega})$ in Fig. 9.18 is given by

$$r_{wa}^{\widetilde{T}_1}[n] = \int_{-\infty}^{\infty} \beta^2\left(\frac{t}{T_1}\right)\beta^0\left(\frac{t - n\widetilde{T}_1}{\widetilde{T}_1}\right)dt = T_1\int_{\frac{n-0.5}{I}}^{\frac{n+0.5}{I}} \beta^2(t)dt. \tag{9.70}$$

Since $\beta^2(t)$ is compactly supported, this corresponds to an FIR filter.

When $I = 1$, $r_{wa}^{\widetilde{T}_1}[n] = T_1\beta^3(n)$, namely $R_{WA}^{\widetilde{T}_1}(e^{j\omega}) = T_1B^3(e^{j\omega})$. Since no upsampling is performed in this situation, this postfilter is exactly the inverse of the prefilter. Thus, overall, no filtering operation is required. For $I = 2$, direct numerical computation shows that $r_{wa}^{\widetilde{T}_1}[n]$ is given by

$$r_{wa}^{\widetilde{T}_1}[n] = T_1 \begin{cases} 0.365, & n = 0 \\ 0.249, & |n| = 1 \\ 0.069, & |n| = 2 \\ 0.003, & |n| = 3 \\ 0, & |n| \geq 4. \end{cases} \tag{9.71}$$

Figure 9.21 shows reconstruction results obtained for $I = 1$ and $I = 2$. As evident in the plots, the denser the intermediate grid, the better the approximation.

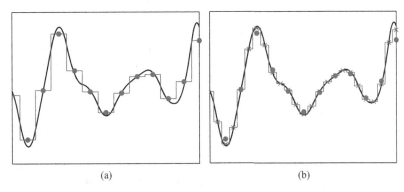

Figure 9.21 Dense-grid recovery of a second-degree spline, sampled with a zero-degree B-spline, using a zero-degree B-spline reconstruction kernel. The samples are marked with circles and the intermediate interpolated points with X-marks. (a) $I = 1$. (b) $I = 2$.

9.3.2   Smoothness prior

Suppose next that our only knowledge about $x(t)$ is that its weighted norm $\|Lx\|$ is bounded, where L is some linear operator, as discussed in Chapter 7.

The sampling and reconstruction mechanisms we assume are identical to (9.53) and (9.54). Therefore, mathematically, our problem is to produce an approximation of the form $\hat{x} = Wd$, with some sequence d, of a signal x, which is known to satisfy $\|Lx\| \leq \rho$ and $c_1 = S^*x$. We have seen in Chapter 7 that it is generally impossible to minimize the reconstruction error $\|\hat{x} - x\|^2$ uniformly over the set of feasible signals in this setting. Instead, we derived an expression for the estimator minimizing the worst-case regret $\|\hat{x} - P_{\mathcal{W}}x\|^2$ over the set of valid signals. The coefficients d of this solution are obtained as

$$d = (W^*W)^{-1}W^*\widetilde{W}(S^*\widetilde{W})^{-1}c_1 \tag{9.72}$$

(see (7.55)), where

$$\widetilde{W} = (L^*L)^{-1}S. \tag{9.73}$$

The expression (9.72) is the same as in the subspace-prior setting (9.56) with A replaced by $\widetilde{W}$. When L represents an LTI filter with frequency response $L(\omega)$, the term $\widetilde{W}$ corresponds to the set transformation associated with $\{\tilde{w}(t - n\widetilde{T}_1)\}_{n \in \mathbb{Z}}$, where

$$\widetilde{W}(\omega) = \frac{S(\omega)}{|L(\omega)|^2} \tag{9.74}$$

and $\widetilde{T}_1 = T_1/I$. Therefore, the dense-grid smoothness-prior minimax-regret solution coincides with the one developed for a subspace prior with $A(\omega)$ replaced by $\widetilde{W}(\omega)$. Specifically, the correction filter of Fig. 9.15 is now given by

$$H(e^{j\omega}) = \frac{R_{W\widetilde{W}}^{\widetilde{T}_1}(e^{j\omega})}{R_{WW}^{\widetilde{T}_1}(e^{j\omega})R_{S\widetilde{W}}^{T_1}(e^{j\omega I})}. \tag{9.75}$$

Example 9.8 We demonstrate dense-grid recovery with a smoothness prior in the context of image enlargement by an irrational factor.

Consider the setting of Example 7.3, in which a continuous-space scene $x(t, \eta)$ is convolved with a rectangular kernel $s(t, \eta) = \beta^0(t/T_1)\beta^0(\eta/T_1)$ and sampled on the grid $\{mT_1, nT_1\}$, $m, n \in \mathbb{Z}$, to produce a digital image $c_1[m, n]$. Our goal is to produce a larger digital image $c_2[k, \ell]$ from $c_1[m, n]$, corresponding to samples of $x(t, \eta)$ on the grid $\{kT_2, \ell T_2\}$, $k, \ell \in \mathbb{Z}$, where $T_2/T_1 < 1$ is some irrational factor.

As discussed in Example 7.3, one common approach to image enlargement is to use Keys' bicubic kernel (6.7). Owing to the compact support of this kernel, reconstruction is very efficient. Nevertheless, this method often produces overly blurry recoveries. We also saw in Example 7.3 that by using a smoothness prior, without any constraints on the recovery mechanism, it is possible to attain a smaller reconstruction error than that achieved using bicubic interpolation. However, this

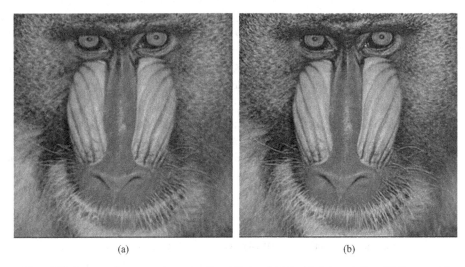

(a)   (b)

Figure 9.22 Image enlargement by an irrational factor. (a) Bicubic interpolation. (b) Dense-grid recovery with a triangular kernel (corresponding to second-order approximation).

requires use of a noncompactly supported kernel, rendering the method highly inefficient. If the system is constrained to operate with a triangular kernel $w(t, \eta) = \beta^1(t/T_1)\beta^1(\eta/T_1)$, then the smoothness-prior solution becomes worse than bicubic interpolation, as shown in Example 7.5.

The dense-grid methodology allows us to benefit from both worlds. By increasing resolution as an intermediate step, we are able to better approximate the unconstrained minimax solution and at the same time still use a compactly supported kernel. In Fig. 9.22 we compare the minimax-regret dense-grid reconstruction approach with that of bicubic interpolation in the task of enlarging an image by the irrational factor $T_2/T_1 = e/\pi$. The regularization operator was chosen to be

$$L(\boldsymbol{\omega}) = \left((0.01\pi)^2 + \|\boldsymbol{\omega}\|^2\right)^{1.3}. \tag{9.76}$$

Filtering with $H(e^{j\omega})$ was performed in the frequency domain by truncating the infinite sums in the expressions for $R_{W\widetilde{W}}^{\widetilde{T}_1}(e^{j\omega})$, $R_{WW}^{\widetilde{T}_1}(e^{j\omega})$, and $R_{S\widetilde{W}}^{T_1}(e^{j\omega I})$. We used an intermediate rate conversion factor of $I = 2$ and a compactly supported kernel $w(t, \eta) = \beta^1(t/\widetilde{T}_1)\beta^1(t/\widetilde{T}_1)$. It is clearly seen that the bicubic method produces a blurry reconstruction whereas in the minimax method the edges are sharp and the textures are better preserved.

9.3.3   Stochastic prior

In Section 7.3 we showed that recovery in the presence of a smoothness constraint is equivalent to a stochastic prior in which the spectral density replaces the weighting function $1/|L(\omega)|^2$. It turns out that the same is true in the context of dense-grid recovery [150]. Specifically, assume that $x(t)$ is a stationary random process with spectral density

$\Lambda_x(\omega)$. Our goal is to reconstruct $x(t)$ from pointwise samples at times nT_1, $n \in \mathbb{Z}$, taken at the output of a filter $s(-t)$. We constrain the reconstruction to be in the span of $\{w(t - nT_1/I)\}$ as shown in Fig. 9.12, where $w(t)$ is predefined.

Following Section 7.3.2 (cf. (7.84)), a plausible error measure in this situation is the sampling-period-average MSE, defined by

$$\text{MSE}_A = \frac{1}{T}\mathsf{E}\left\{\int_{t_0}^{t_0+T_1} |x(t) - \hat{x}(t)|^2 dt\right\}, \tag{9.77}$$

where t_0 is an arbitrary point in time. It turns out that the digital filter $h[n]$ minimizing the average MSE is independent of t_0 and is given by (9.75), where

$$\widetilde{W}(\omega) = \Lambda_x(\omega)S(\omega). \tag{9.78}$$

The detailed derivations can be found in [150]. This solution is the same as in the smoothness scenario (see (9.74)) with $\Lambda_x(\omega)$ replacing $1/|L(\omega)|^2$.

9.4   Projection-based resampling

In the previous section we extended the traditional rate-conversion approach presented in Section 9.1 to general subspace and smoothness priors for the case in which $T_2 < T_1$, i.e. a rate increase. This generalization overlooked one important aspect, which is the aliasing effect that occurs in rate reduction when $T_2 > T_1$. In bandlimited rate reduction we typically are not interested in approximating the pointwise samples of $x(t)$ at times mT_2, $m \in \mathbb{Z}$, since the samples $x(mT_2)$ may not constitute a faithful representation of $x(t)$. Common practice is to apply a digital anti-aliasing filter prior to downsampling, which means that the resulting sequence $c_2[m]$ corresponds to samples of a properly filtered version of $x(t)$.

The use of an anti-aliasing LPF is related to the bandlimited prior assumed in traditional rate conversion. At first sight, it is not obvious how this anti-aliasing operation should be translated to the case in which the bandlimited prior is replaced by an arbitrary subspace prior. When decreasing the sampling rate by an integer factor D, the desired set of samples is a subset of the original sequence $c_1[n]$. Therefore, they can be computed exactly as $c_2[m] = c_1[mD] = x(mT_2)$. However, if we were to try and recover $x(t)$ from $c_2[m]$, then we would incur a large error due to aliasing. This highlights the fact that when designing the filter, it is not the error $|x(mT_2) - c_2[m]|$ in the samples that should interest us but rather the error between the continuous-time signal $x(t)$ and the best possible recovery of $x(t)$ from $c_2[m]$.

In what sense is the choice of a π/T_2-bandlimited anti-aliasing filter optimal? Recall the two-stage interpretation of signal resampling, in which we first recover $x(t)$ and then sample it on the desired grid. This scheme is shown in Fig. 9.23 for the special case of sinc interpolation. Note that when $x(t)$ is assumed to be π/T_1-bandlimited, we can recover it perfectly from $c_1[n]$. At the second stage, we apply a π/T_2-bandlimited anti-aliasing filter prior to sampling at a rate of $1/T_2$. Thus, from the sequence $c_2[m]$ we can recover a signal $\hat{x}(t)$ whose frequency content up to π/T_2 coincides with that

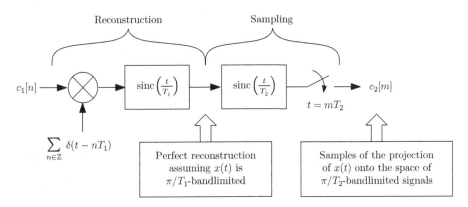

Figure 9.23 Bandlimited resampling as a two-stage operation.

of $x(t)$. In other words, we recover the orthogonal projection of $x(t)$ onto the space of π/T_2-bandlimited functions. We therefore may interpret traditional rate conversion as assuming that the original signal lies in the space spanned by $\{\operatorname{sinc}((t - nT_1)/T_1)\}_{n \in \mathbb{Z}}$, while the output is a set of samples $c_2[m]$ that allow us to recover the projection of $x(t)$ onto the space spanned by $\{\operatorname{sinc}((t - mT_2)/T_2)\}_{m \in \mathbb{Z}}$.

In the next two sections we extend this insight to arbitrary input spaces following [137, 182]. In Section 9.4.1 we focus on rate-conversion methods that result from computing the orthogonal projection onto the appropriate output space, while in Section 9.4.2 we show how in some cases the computations can be simplified by considering an oblique projection. In both sections we are primarily interested in rate reduction so that $T_2 > T_1$. We note, however, that the projection methods are also applicable to the case in which $T_2 < T_1$. Generally, this will lead to a different set of samples than using the direct approach of Fig. 9.12.

In order to keep the discussion simple, we focus on the case in which the input space and reconstruction space are generated by the same kernel $a(t) = w(t)$. We also assume that $I = 1$ so that no interpolation is introduced prior to resampling. The derivations can be easily extended to the more general setting by following the same steps as in the previous section.

9.4.1 Orthogonal projection resampling

Suppose that $x(t)$ is known to lie in a space $\mathcal{A}_{T_1}$ spanned by the functions $\{a((t - nT_1)/T_1)\}_{n \in \mathbb{Z}}$, for some generator $a(t)$. Our goal is to obtain samples of $x(t)$ on the grid mT_2. In the direct method of Fig. 9.12, we performed rate conversion by first interpolating $x(t)$ from the given samples, and then resampling $x(t)$ on the desired grid. Here we take an alternative approach by noting that after resampling we are essentially approximating $x(t)$ in the space $\mathcal{A}_{T_2}$ spanned by $\{a((t - nT_2)/T_2)\}_{n \in \mathbb{Z}}$. From the projection theorem (cf. Proposition 2.15) we know that the signal closest to $x(t)$ in the space $\mathcal{A}_{T_2}$ is the orthogonal projection $P_{\mathcal{A}_{T_2}} x(t)$. Therefore, instead of resampling $x(t)$, we resample $P_{\mathcal{A}_{T_2}} x(t)$. A three-stage representation of this procedure is shown schematically in Fig. 9.24.

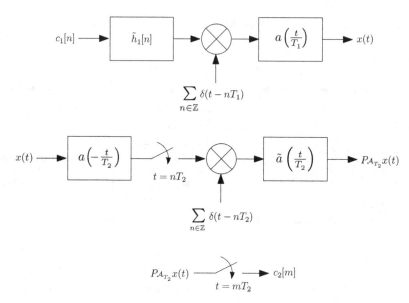

Figure 9.24 Three-stage representation of orthogonal projection resampling: signal recovery (top), computation of the orthogonal projection of $x(t)$ onto the space $\mathcal{A}_{T_2}$ (middle), and sampling of the result at times mT_2 (bottom).

Although the motivation for this strategy is to treat the case in which we downsample the signal, namely $T_2 > T_1$, the proposed resampling approach can also be used when $T_2 < T_1$. In general, this method will differ from direct interpolation since for arbitrary kernels $a(t)$ the space $\mathcal{A}_{T_2}$ spanned by $\{a((t - nT_2)/T_2)\}_{n\in\mathbb{Z}}$ does not contain the space $\mathcal{A}_{T_1}$ spanned by $\{a((t - nT_1)/T_1)\}_{n\in\mathbb{Z}}$ for $T_2 < T_1$. If the spaces are nested so that $\mathcal{A}_{T_1} \subseteq \mathcal{A}_{T_2}$ then orthogonal projection resampling coincides with the direct method. This is the case, for example, in bandlimited sampling where $a(t) = \text{sinc}(t)$. Nonetheless, the advantage of this strategy over the direct approach is most pronounced when $T_2 > T_1$.

Computational steps

We now detail the computational steps in orthogonal projection resampling. The assumption that $x(t)$ lies in $\mathcal{A}_{T_1}$ corresponds to the fact that

$$x(t) = \sum_{n\in\mathbb{Z}} b[n]a\left(\frac{t - nT_1}{T_1}\right) \qquad (9.79)$$

for some expansion coefficients $b[n]$. Following the same steps as in Section 9.2 the sequence $b[n]$ can be obtained from the samples $c_1[n] = x(nT_1)$ using the filter

$$\tilde{H}_1(e^{j\omega}) = \frac{1}{\sum_{k\in\mathbb{Z}} A(\omega - 2\pi k)}. \qquad (9.80)$$

This filter was computed explicitly in Examples 9.3–9.5 for $a(t)$ equal to B-splines of degrees $p = 0, 1, 2, 3$. Once $b[n]$ is known, we may in principle recover $x(t)$, and from it the required orthogonal projection $y(t) = P_{\mathcal{A}_{T_2}} x(t)$. The final reduced-rate samples are then $c_2[m] = y(mT_2)$.

As we have seen in previous chapters, the projection $y(t) = P_{\mathcal{A}_{T_2}} x(t)$ of $x(t)$ onto the space $\mathcal{A}_{T_2}$ spanned by $\{a((t-nT_2)/T_2)\}_{n \in \mathbb{Z}}$ may be computed using the system in the middle row of Fig. 9.24. Specifically, $y(t)$ can be expressed in terms of the biorthogonal functions $\{\tilde{a}((t-nT_2)/T_2)\}_{n \in \mathbb{Z}}$ in the form

$$y(t) = \sum_{n \in \mathbb{Z}} d[n] \tilde{a} \left(\frac{t - nT_2}{T_2} \right). \tag{9.81}$$

The coefficients $d[n]$ correspond to inner products between $x(t)$ and the functions $\{a((t-nT_2)/T_2)\}_{n \in \mathbb{Z}}$:

$$
\begin{aligned}
d[m] &= \left\langle a \left(\frac{t - mT_2}{T_2} \right), x(t) \right\rangle \\
&= \left\langle a \left(\frac{t - mT_2}{T_2} \right), \sum_{n \in \mathbb{Z}} b[n] a \left(\frac{t - nT_1}{T_1} \right) \right\rangle \\
&= \sum_{n \in \mathbb{Z}} b[n] \left\langle a \left(\frac{t - mT_2}{T_2} \right), a \left(\frac{t - nT_1}{T_1} \right) \right\rangle \\
&= \sum_{n \in \mathbb{Z}} b[n] T_1 \int_{-\infty}^{\infty} a \left(\frac{T_1}{T_2} t \right) a \left(t - \left(n - m \frac{T_2}{T_1} \right) \right) dt \\
&= T_1 \sum_{n \in \mathbb{Z}} b[n] v_R(n - mR) \\
&= \sum_{n \in \mathbb{Z}} b[n] g[m, n], \tag{9.82}
\end{aligned}
$$

where we denoted $R = T_2/T_1$,

$$v_R(t) = a \left(\frac{t}{R} \right) * a(-t) \tag{9.83}$$

and

$$g[m, n] = T_1 v_R(n - mR). \tag{9.84}$$

When $a(t)$ is a kernel of compact support, the function $v_R(t)$ is also compactly supported. Therefore, for every m, the expansion coefficient $d[m]$ only depends on a finite (typically small) number of coefficients from the sequence $\{b[n]\}$. Assuming that there is a closed-form expression for the function $v_R(t)$, this renders computation of $d[m]$ efficient.

As a final stage, we need to compute the samples $c_2[m] = y(mT_2)$ of $y(t)$. These are given by

$$c_2[m] = \sum_{n \in \mathbb{Z}} d[n] \tilde{a} \left(\frac{mT_2 - nT_2}{T_2} \right) = \sum_{n \in \mathbb{Z}} d[n] \tilde{a}(m - n) = d[m] * \tilde{a}(m). \tag{9.85}$$

Recall that

$$\tilde{a}(t) = \sum_{n \in \mathbb{Z}} q[n]a(t-n), \tag{9.86}$$

where

$$Q(e^{j\omega}) = \frac{1}{R_{AA}(e^{j\omega})}. \tag{9.87}$$

Consequently,

$$\tilde{a}(m) = \sum_{n \in \mathbb{Z}} q[n]a(m-n) = q[m] * a(m), \tag{9.88}$$

so that

$$c_2[m] = d[m] * q[m] * a(m). \tag{9.89}$$

In other words, $c_2[m]$ is obtained by feeding $d[m]$ into the filter

$$\frac{\sum_{k \in \mathbb{Z}} A(\omega - 2\pi k)}{R_{AA}(e^{j\omega})}. \tag{9.90}$$

The overall rate-conversion scheme is depicted in Fig. 9.25. It comprises a digital prefilter, a multirate system which can be implemented efficiently given that $a(t)$ has small support, and a digital postfilter.

Example 9.9 Consider orthogonal projection resampling with $T_1 = 1$ and $a(t) = \beta^0(t)$. In this case, $a(n) = \delta[n]$, and no prefiltering is required. Furthermore, $r_{aa}[n] = \delta[n]$, so that postfiltering is also not needed. Finally, direct computation of the function $v_R(t)$ shows that it is of trapezoidal shape. For example, if $R < 1$, then (see Exercise 7)

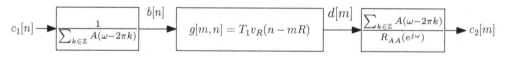

$$c_1[n] \rightarrow \boxed{\frac{1}{\sum_{k \in \mathbb{Z}} A(\omega - 2\pi k)}} \xrightarrow{b[n]} \boxed{g[m,n] = T_1 v_R(n - mR)} \xrightarrow{d[m]} \boxed{\frac{\sum_{k \in \mathbb{Z}} A(\omega - 2\pi k)}{R_{AA}(e^{j\omega})}} \rightarrow c_2[m]$$

Figure 9.25 Orthogonal projection resampling. Here $v_R(t)$ is given by (9.83).

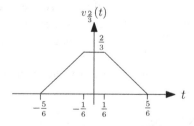

Figure 9.26 The function $v_{\frac{2}{3}}(t)$ of (9.83) corresponding to a rectangular kernel $a(t) = \beta^0(t)$.

$$v_R(t) = \begin{cases} R, & |t| < \frac{1}{2} - \frac{R}{2} \\ \frac{1}{2} + \frac{R}{2} - |t|, & \frac{1}{2} - \frac{R}{2} \le |t| < \frac{1}{2} + \frac{R}{2} \\ 0, & \frac{1}{2} + \frac{R}{2} \le |t|. \end{cases} \tag{9.91}$$

This function is depicted in Fig. 9.26 for the case $R = 2/3$.

B-Spline resampling

We have mentioned already that a very common choice in practice is to use B-splines as reconstruction kernels. When $a(t)$ is chosen to be a B-spline of small degree, the pre- and postfilters possess a simple structure (see, for instance, Examples 5.4, 6.5, 6.7, 6.8). As we now demonstrate, this structure leads to efficient implementation. We also illustrate implementation of the multirate conversion system for B-spline resampling.

Example 9.10 Suppose that $T_1 = 1$ and $a(t) = \beta^1(t)$. In this case $a(n) = \delta[n]$ so that no prefiltering is needed. The postfiltering stage comprises filtering with $a(n) = \delta[n]$ and with $1/R_{AA}(e^{j\omega})$. As we have seen in Example 5.4, $R_{\beta^1\beta^1}(e^{j\omega})$ is the DTFT of

$$r_{\beta^1\beta^1}[n] = \int_{-\infty}^{\infty} \beta^1(t)\beta^1(t-n)dt = \beta^3(n). \tag{9.92}$$

Therefore, from Example 9.5, we have that

$$R_{\beta^1\beta^1}(e^{j\omega}) = \frac{1}{6}e^{j\omega} + \frac{2}{3} + \frac{1}{6}e^{-j\omega}. \tag{9.93}$$

In Section 9.2, we saw that the inverse of (9.93) can be expressed as the concatenation of a stable causal IIR filter of order 1 and a stable anticausal IIR filter of order 1 (see (9.45)). Specifically, this cascade corresponds to the two recursions:

$$d_1[m] = 6d[m] + \alpha_3 d_1[m-1], \tag{9.94}$$

and

$$c_2[m] = -\alpha_3 d_1[m] + \alpha_3 c_2[m+1], \tag{9.95}$$

with $\alpha_3 = -2 + \sqrt{3}$.

Computation of the function $v_R(t)$ in this example involves evaluating the integral of the product of two piecewise linear functions $\beta^1(\tau/R)$ and $\beta^1(\tau - t)$. This leads to a piecewise cubic function (see Exercise 10).

Example 9.11 Consider next the case in which $a(t) = \beta^2(t)$. In this situation, $a(n)$ is given by (9.37) so that the prefilter's transfer function can be expressed in the form (9.39). As explained in Example 9.10, this filter is a concatenation of two stable first-order IIR filters, one causal and one anticausal.

The postfiltering stage comprises filtering with $a(n)$ and with $1/R_{AA}(e^{j\omega})$. The function $R_{AA}(e^{j\omega})$ is the DTFT of the sequence

$$r_{\beta^2\beta^2}[n] = \int_{-\infty}^{\infty} \beta^2(t)\beta^2(t-n)dt = \beta^5(n) = \begin{cases} \frac{11}{20}, & n=0 \\ \frac{13}{60}, & |n|=1 \\ \frac{1}{120}, & |n|=2 \\ 0, & |n| \geq 3. \end{cases} \quad (9.96)$$

Therefore, the frequency response of the postfilter can be expressed as

$$\frac{\frac{1}{8}e^{-j\omega} + \frac{3}{4} + \frac{1}{8}e^{j\omega}}{\frac{1}{120}e^{-2j\omega} + \frac{13}{60}e^{-j\omega} + \frac{11}{20} + \frac{13}{60}e^{j\omega} + \frac{1}{120}e^{2j\omega}}$$

$$= 15\frac{\alpha_5^1\alpha_5^2}{\alpha_2}\left(\frac{1-\alpha_2 e^{-j\omega}}{(1-\alpha_5^1 e^{-j\omega})(1-\alpha_5^2 e^{-j\omega})}\right)\left(\frac{1-\alpha_2 e^{j\omega}}{(1-\alpha_5^1 e^{j\omega})(1-\alpha_5^2 e^{j\omega})}\right), \quad (9.97)$$

where $\alpha_2 = -3 + \sqrt{8}$ (see Example 9.4), $\alpha_5^1 = -0.431$ and $\alpha_5^2 = -0.043$. This filter is also implementable as a cascade of a causal and an anticausal stable IIR filter. The feedback order of these filters is 2 and the feedforward order is 1.

In this example $v_R(t)$ corresponds to the integral of the product of two piecewise quadratic functions $\beta^2(\tau/R)$ and $\beta^2(\tau - t)$, which is a piecewise fifth-order polynomial. We omit this technical and lengthy computation here.

The examples above demonstrate that the larger we choose the B-spline's degree, the more cumbersome the expression for $v_R(t)$ becomes. For example, when $a(t) = \beta^1(t)$, the function $v_R(t)$ is generally a symmetric cubic spline with nine unequally spaced knots (see Exercise 10). To avoid working with such complicated expressions, it is common to approximate $v_R(t)$ by a Gaussian function. As the B-spline's degree increases, this approximation becomes more accurate. Indeed, if $a(t) = \beta^p(t)$, then $a(t)$ corresponds to the $(p+1)$-fold convolution of $\beta^0(t)$. From the central limit theorem, we know that the larger p is, the closer $a(t)$ is to being a Gaussian function. The variance of the Gaussian is $(p+1)$ times the variance of $\beta^0(t)$, which is $1/12$. Similarly, $a(t/R)$ corresponds to a Gaussian with variance $R(p+1)/12$. The convolution of two Gaussian functions is a Gaussian whose variance is the sum of variances of the individual Gaussians. Therefore, $v_R(t) = a(-t) * a(t/R)$ can be approximated by the Gaussian

$$v_R(t) \approx \tilde{v}_R(t) = \begin{cases} \frac{1}{\sqrt{2\pi\sigma^2}}e^{-\frac{t^2}{2\sigma^2}}, & |t| \leq \frac{p+1}{2}(1+R) \\ 0, & |t| > \frac{p+1}{2}(1+R), \end{cases} \quad (9.98)$$

with variance

$$\sigma^2 = (1+R)\frac{p+1}{12}. \quad (9.99)$$

Here, we have truncated the Gaussian to reflect the fact that the true function $v_R(t)$ is of compact support. In practice, this approximation is fairly accurate even for B-splines of degree as low as 1. This is demonstrated in Fig. 9.27 for the special case $a(t) = \beta^1(t)$, $R = 1$, in which $v_R(t) = \beta^3(t)$.

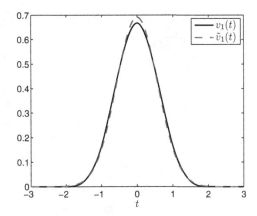

Figure 9.27 The function $v_1(t)$ of (9.83) with $R = 1$ corresponding to a triangular kernel $a(t) = \beta^1(t)$ and its Gaussian approximation $\tilde{v}_1(t)$ of (9.98). In this case $v_1(t) = \beta^3(t)$.

Example 9.12 Orthogonal projection resampling with B-spline functions is useful for enlarging or reducing digital images. This can be performed by applying the method to each dimension separately. Specifically, for an $M \times N$ image, we begin by resampling each row to obtain an intermediate image of size $M \times \lfloor N/R \rfloor$. Then, we resample each column of the intermediate image to obtain the final result, which is an image of size $\lfloor M/R \rfloor \times \lfloor N/R \rfloor$.

Figure 9.28 demonstrates the importance of performing orthogonal projections. In this example, we choose $a(t) = \beta^1(t)$, which is a kernel satisfying the interpolation property. Therefore, the direct solution discussed in Section 9.2 (cf. Fig. 9.16) reduces to performing recovery without a correction filter and then sampling the result without applying an anti-aliasing filter, namely

$$c_2^{\mathrm{direct}}[m] = \sum_{n \in \mathbb{Z}} c_1[n]\beta^1(mR - n). \qquad (9.100)$$

Here we compare this direct method to orthogonal projection resampling (cf. Fig. 9.25). To emphasize the differences between the approaches, we performed 10 consecutive resampling operations by a factor of $R = 1.017$. As can be seen, the direct approach results in a blurry image, whereas the orthogonal projection method manages to preserve the textures and sharp edges of the original image.

9.4.2 Oblique projection resampling

We saw that orthogonal projection resampling, illustrated in Fig. 9.25, can be performed quite conveniently with B-spline kernels of small degree. The disadvantage of this approach, however, is that, as the B-spline degree increases, the technique becomes cumbersome to implement, especially for sampling rate reduction ($R > 1$). This is mainly due to the large support of $v_R(t)$, which determines the number of coefficients $b[n]$ that need to be taken into account in computing $d[m]$ for every m.

(a)

(b)

Figure 9.28 Resampling with the kernel $a(t) = \beta^1(t)$ and $R = 1.017$. (a) Direct resampling. (b) Orthogonal projection resampling.

More specifically, to identify the bottleneck in B-spline-based orthogonal resampling, we note that the prefilter in Fig. 9.25 can be implemented as a cascade of a causal filter and an anticausal filter, each of order $\lfloor p/2 \rfloor$ (see Exercise 9). Similarly, the postfilter can be implemented as a cascade of two IIR filters, each with feedforward order of $\lfloor p/2 \rfloor$ and feedback order of $p+1$, one causal and one anticausal. Finally, the multirate system $g[m, n]$ is of finite-impulse response. For each m, the number of coefficients for which $g[m, n] \neq 0$ is at most $\lceil (p+1)(R+1) \rceil$. We therefore have the following computational requirements:

1. Prefilter: requires $2\lfloor p/2 \rfloor$ multiplications per element of the input sequence $c_1[n]$.
2. Multirate system $g[m,n]$: requires $\lceil (p+1)(R+1) \rceil$ multiplications per element of the output sequence $c_2[m]$. As there are R times fewer coefficients in $c_2[m]$ per time unit than in $c_1[n]$, this corresponds to $\lceil (p+1)(R+1) \rceil / R$ multiplications per coefficient of the input sequence $c_1[n]$.
3. Postfilter: requires $2\lfloor p/2 \rfloor + 2(p+1)$ multiplications per element of $c_2[m]$, or, equivalently, $(2\lfloor p/2 \rfloor + 2(p+1))/R$ multiplications per element of $c_1[n]$.

The above analysis shows that the number of multiplications per time unit required by the first filter is independent of R. The number of multiplications in the last stage decreases like $1/R$ as the rate conversion factor R is increased. This scaling is desirable, especially when decreasing the sampling rate ($R > 1$), since our resampling system only needs to produce $1/R$ output coefficients for each input element. However, the computational load in the second stage does not tend to zero as R is increased. Namely, even for $R \gg 1$, this stage requires no less than $p+1$ multiplications per input sample.

The multirate system $g[m,n]$ is determined by the function $v_R(t)$. The latter implements the last block in the top branch and first block in the middle branch of Fig. 9.24, so that it corresponds to the convolution of $a(t/T_1)$ and $a(-t/T_2)$. The basic idea in oblique projection resampling is to replace the orthogonal projection onto $\mathcal{A}_{T_2} = \mathrm{span}\{a((t - mT_2)/T_2)\}_{m\in\mathbb{Z}}$ by an oblique projection onto $\mathcal{A}_{T_2}$ along $\mathcal{S}_{T_2}^{\perp}$, where $\mathcal{S}_{T_2} = \mathrm{span}\{s((t - mT_2)/T_2)\}_{m\in\mathbb{Z}}$ for some kernel $s(t)$ [137]. This change contributes an additional degree of freedom, which is the choice of $s(t)$. To obtain an efficient implementation, we will choose $s(t) = \beta^0(t)$. The three-stage representation of this approach is schematically depicted in Fig. 9.29. This process involves the convolution $a(t/T_1) * s(-t/T_2)$, whose support is generally much smaller than that of $a(t/T_1) * a(-t/T_2)$. Furthermore, we will see that the resulting system can be implemented by using only one multiplication per output element.

Computational steps
We now develop the computational steps associated with Fig. 9.29. In the first stage, we extract the expansion coefficients $b[n]$ of $x(t)$ in $\mathcal{A}_{T_1}$ from the samples $c_1[n] = x(nT_1)$. This stage comes before the projection, and is therefore the same as in Section 9.4.1 (see (9.79) and (9.80)).

Next, we need to determine from $b[n]$ the oblique projection $y(t)$ of $x(t)$ onto $\mathcal{A}_{T_2}$ along $\mathcal{S}_{T_2}^{\perp}$. To this end, we express $y(t)$ in terms of the oblique biorthogonal vectors $\{\tilde{a}_{\mathcal{S}}((t - nT_2)/T_2)\}_{n\in\mathbb{Z}}$ of $\{s((t - nT_2)/T_2)\}_{n\in\mathbb{Z}}$ in $\mathcal{A}_{T_2}$, which were defined and discussed in Section 6.3.4. Namely, we write

$$y(t) = \sum_{n\in\mathbb{Z}} d[n]\tilde{a}_{\mathcal{S}}\left(\frac{t - nT_2}{T_2}\right), \tag{9.101}$$

where $d[n]$ are the inner products between $x(t)$ and $\{s((t - nT_2)/T_2)\}_{n\in\mathbb{Z}}$:

$$d[m] = \left\langle s\left(\frac{t - mT_2}{T_2}\right), x(t) \right\rangle$$

$$= \sum_{n\in\mathbb{Z}} b[n] \left\langle s\left(\frac{t - mT_2}{T_2}\right), a\left(\frac{t - nT_1}{T_1}\right) \right\rangle$$

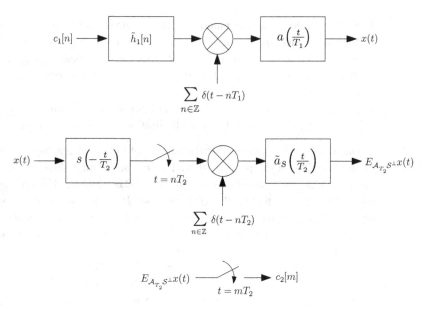

Figure 9.29 Three-stage representation of oblique projection resampling: signal recovery (top), computation of the oblique projection of $x(t)$ onto the space $\mathcal{A}_{T_2}$ along $\mathcal{S}_{T_2}^{\perp}$ (middle), and sampling of the result at times mT_2 (bottom).

$$= \sum_{n\in\mathbb{Z}} b[n]T_1 \int_{-\infty}^{\infty} s\left(\frac{tT_1 + nT_1 - mT_2}{T_2}\right) a(t)dt$$

$$= T_1 \sum_{n\in\mathbb{Z}} b[n] \int_{(m-\frac{1}{2})R-n}^{(m+\frac{1}{2})R-n} a(t)dt$$

$$= T_1 \sum_{n\in\mathbb{Z}} b[n]\tilde{v}_R(n - mR)$$

$$= \sum_{n\in\mathbb{Z}} b[n]g[m,n], \tag{9.102}$$

where we denoted $R = T_2/T_1$,

$$\tilde{v}_R(t) = \int_{-t-\frac{1}{2}R}^{-t+\frac{1}{2}R} a(\tau)d\tau, \tag{9.103}$$

$g[m,n] = T_1\tilde{v}_R(n - mR)$, and used the fact that $s(t) = \beta^0(t)$.

If $a(t)$ is a B-spline of degree larger than 0 then the support of $\tilde{v}_R(t)$ is smaller than that of the function $v_R(t)$ of (9.83) associated with orthogonal projection resampling. This fact, by itself, implies that oblique projection resampling is computationally more efficient. In addition, instead of directly implementing the multirate system $g[m,n]$, we may obtain an even more marked decrease in the computational load by relying on the fact that $a(t)$ is a B-spline function.

Specifically, assume that $a(t) = \beta^p(t)$ for some degree p. Then we can make use of the relation (see Exercise 9 in Chapter 4)

$$\int_{-\infty}^{t} \beta^p(\tau)d\tau = \sum_{\ell=0}^{\infty} \beta^{p+1}\left(t - \frac{1}{2} - \ell\right). \tag{9.104}$$

Using this identity in (9.102) we have that

$$d[m] = T_1 \sum_{n \in \mathbb{Z}} b[n] \left(\sum_{\ell=0}^{\infty} \beta^{p+1}\left(\left(m + \frac{1}{2}\right)R - n - \frac{1}{2} - \ell\right) \right.$$

$$\left. - \sum_{\ell=0}^{\infty} \beta^{p+1}\left(\left(m - \frac{1}{2}\right)R - n - \frac{1}{2} - \ell\right) \right)$$

$$= T_1 \sum_{n \in \mathbb{Z}} \sum_{\ell=n}^{\infty} b[n] \left(\beta^{p+1}\left(\left(m + \frac{1}{2}\right)R - \frac{1}{2} - \ell\right) \right.$$

$$\left. - \beta^{p+1}\left(\left(m - \frac{1}{2}\right)R - \frac{1}{2} - \ell\right) \right)$$

$$= T_1 \sum_{\ell \in \mathbb{Z}} (b * u)[\ell] \left(\beta^{p+1}\left(\left(m + \frac{1}{2}\right)R - \frac{1}{2} - \ell\right) \right.$$

$$\left. - \beta^{p+1}\left(\left(m - \frac{1}{2}\right)R - \frac{1}{2} - \ell\right) \right)$$

$$= T_1 \tilde{g}\left(\left(m + \frac{1}{2}\right)R\right) - T_1 \tilde{g}\left(\left(m - \frac{1}{2}\right)R\right), \tag{9.105}$$

where $u[n]$ denotes the unit step sequence and

$$\tilde{g}(t) = \sum_{\ell \in \mathbb{Z}} e[\ell]\beta^{p+1}\left(t - \ell - \frac{1}{2}\right), \tag{9.106}$$

with

$$e[n] = (b * u)[n] = \sum_{\ell=-\infty}^{n} b[\ell]. \tag{9.107}$$

Note that $e[n]$ can be efficiently computed from $b[n]$ using the recursion

$$e[n] = e[n-1] + b[n]. \tag{9.108}$$

For each m, we need to evaluate $\tilde{g}(t)$ at two distinct points in order to obtain $d[m]$. However, since we need $d[m]$ for all values of m, we can take advantage of previous calculations by noting that

$$d[m+1] = T_1 \tilde{g}\left(\left(m + \frac{3}{2}\right)R\right) - T_1 \tilde{g}\left(\left(m + \frac{1}{2}\right)R\right). \tag{9.109}$$

The second term in this expression is the same as the first term in (9.105). Therefore, for each m we need to evaluate $\tilde{g}(t)$ at a single point. We conclude that this stage requires a single multiplication per output element, so that the amount of computation scales like $1/R$ and tends to zero as R is increased, in contrast to the orthogonal resampling procedure.

The last stage is to compute the samples $c_2[m] = y(mT_2)$ of $y(t)$. These are given by

$$c_2[m] = \sum_{n \in \mathbb{Z}} d[n] \tilde{a}_{\mathcal{S}} \left(\frac{mT_2 - nT_2}{T_2} \right) = \sum_{n \in \mathbb{Z}} d[n] \tilde{a}_{\mathcal{S}}(m - n) = d[m] * \tilde{a}_{\mathcal{S}}(m).$$

(9.110)

From (6.58),

$$\tilde{a}_{\mathcal{S}}(t) = \sum_{n \in \mathbb{Z}} g_{\mathcal{S}}[n] a(t - n), \tag{9.111}$$

where

$$G_{\mathcal{S}}(e^{j\omega}) = \frac{1}{R_{AS}(e^{j\omega})}. \tag{9.112}$$

Consequently,

$$\tilde{a}_{\mathcal{S}}(m) = \sum_{n \in \mathbb{Z}} g_{\mathcal{S}}[n] a(m - n) = g_{\mathcal{S}}[m] * a(m), \tag{9.113}$$

so that

$$c_2[m] = d[m] * g_{\mathcal{S}}[m] * a(m). \tag{9.114}$$

To conclude, $c_2[m]$ is obtained by feeding $d[m]$ into the filter

$$\frac{\sum_{k \in \mathbb{Z}} A(\omega - 2\pi k)}{R_{AS}(e^{j\omega})}. \tag{9.115}$$

The overall rate-conversion scheme based on oblique projections is depicted in Fig. 9.30. The multirate system $g[m, n]$ can be efficiently implemented using (9.105).

Comparing (9.115) with (9.90), we see that $R_{AS}(e^{j\omega})$ replaces $R_{AA}(e^{j\omega})$ in the denominator. When $a(t) = \beta^p(t)$, $R_{AS}(e^{j\omega})$ is the DTFT of $\beta^{p+1}(n)$, while $R_{AA}(e^{j\omega})$ is the DTFT of $\beta^{2p+1}(n)$. Consequently, this stage can be implemented as the cascade of a causal and anticausal filter, requiring only $(2\lfloor p/2 \rfloor + 2\lfloor (p+1)/2 \rfloor)/R$ multiplications per output element in contrast to the $(2\lfloor p/2 \rfloor + 2\lfloor (p+1) \rfloor)/R$ multiplications needed in orthogonal resampling (see Exercise 12). This reduces computation time roughly by a factor of $(2/3)/R$ for large p.

Examples
We now provide several examples comparing oblique and orthogonal resampling.

Example 9.13 We first revisit Example 9.10 in which $a(t) = \beta^1(t)$. The pre-filtering stage in oblique projection resampling is the same as in orthogonal

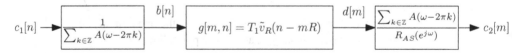

Figure 9.30 Oblique projection resampling. Here $\tilde{v}_R(t)$ is given by (9.103).

projection resampling. As we have seen in Example 9.10, no prefilter is required when $a(t) = \beta^1(t)$. Furthermore, from (9.105)–(9.109), the rate-conversion stage requires one multiplication per output element, regardless of the spline degree. All that remains to be determined, then, is how to implement the postfilter efficiently.

The postfiltering stage involves convolution with $a(n) = \beta^1(n) = \delta[n]$, and with the convolutional inverse of $r_{as}[n] = r_{\beta^1\beta^0}[n] = \beta^2(n)$, which is given by (9.37). Therefore, as discussed in Example 9.10, this filter is a concatenation of a stable causal IIR filter and a stable anticausal IIR filter, each of order 1.

In this setting, then, there is no computational advantage for the oblique method over the orthogonal resampling approach, in terms of the postfilter (see Example 9.10). The only saving in computation time is due to the rate-conversion stage (namely, the system $g[m, n]$). Specifically, as we have seen, we can implement $g[m, n]$ of the oblique approach with only one multiplication per output element. In contrast, the filter $g[m, n]$ of the orthogonal projection method requires $\lceil (p + 1)(R + 1) \rceil$ multiplications per output element, which in this example equals $\lceil 2R + 2 \rceil$. If, for instance, we decrease the rate by $R = 2$, then the multirate stage is 6 times more efficient in the oblique setting than in the orthogonal case.

Example 9.14 We next revisit Example 9.11 in which $a(t) = \beta^2(t)$. The prefiltering stage in the oblique approach is the same as in the orthogonal method, and can be implemented as in Example 9.11.

The postfiltering stage involves convolution with $a(n) = \beta^2(n)$, which is given by (9.37), and with the convolutional inverse of $r_{as}[n] = r_{\beta^2\beta^0}[n] = \beta^3(n) = r_{\beta^1\beta^1}[n]$, computed in (9.93). Therefore, as discussed in Examples 9.10 and 9.11, the frequency response of the filter can be expressed as

$$\frac{\frac{1}{8}e^{-j\omega} + \frac{3}{4} + \frac{1}{8}e^{j\omega}}{\frac{1}{6}e^{-j\omega} + \frac{2}{3} + \frac{1}{6}e^{j\omega}} = \frac{3\alpha_3}{4\alpha_2}\left(\frac{1 - \alpha_2 e^{-j\omega}}{1 - \alpha_3 e^{-j\omega}}\right)\left(\frac{1 - \alpha_2 e^{j\omega}}{1 - \alpha_3 e^{j\omega}}\right) \tag{9.116}$$

with $\alpha_2 = -3 + \sqrt{8}$ and $\alpha_3 = -2 + \sqrt{3}$. This filter is a concatenation of a stable causal IIR filter and a stable anticausal IIR filter, each of order 1. Recall that in the orthogonal projection case discussed in Example 9.11, the postfilter comprised IIR filters of order 2.

The difference in the computational burden associated with $g[m, n]$ is even more pronounced. Namely, as opposed to the single multiplication per output element in the oblique setting, we need to perform $\lceil (p+1)(R+1) \rceil = \lceil 3R + 4 \rceil$ multiplications per output element in the orthogonal case. Therefore, if, for example, we decrease the rate by a factor of $R = 2$, then this stage is 10 times more efficient in the oblique method than when using the orthogonal approach.

Example 9.15 We have seen that the oblique projection technique lends itself to a more efficient implementation than the orthogonal method. We now demonstrate that this advantage does not come at the cost of severe performance degradation. To this end, consider the problem of image resampling with the triangular kernel

(a)

(b)

Figure 9.31 Resampling with the kernel $a(t) = \beta^1(t)$ and $R = 1.017$. (a) Orthogonal projection resampling. (b) Oblique projection resampling.

$a(t) = \beta^1(t)$. Example 9.12 showed that orthogonal projection resampling yields a much better reconstruction than the direct approach of Fig. 9.16, which involves no projection. Figure 9.31 provides a comparison between the orthogonal and oblique strategies following the same experiment as in Example 9.12.

The visual differences between the approaches are very minor. The oblique reconstruction is slightly more blurry in certain textured areas, but not seriously so. As mentioned in Example 9.13, the difference in the number of computations in this setting is only associated with the multirate system $g[m, n]$, which is $\lceil 2R+3 \rceil$ times less efficient in the orthogonal case than in the oblique setting. This corresponds to a factor of 6 in our example.

9.5   Summary of conversion methods

This chapter presented various techniques for sampling rate conversion. Although rate conversion can in principle be treated by reconstructing $\hat{x}(t)$ of $x(t)$ from the available samples using any one of the methods discussed in Chapters 6 and 7, and then evaluating $\hat{x}(t)$ on the desired grid, this direct approach disregards two very important aspects. The first corresponds to computational requirements: how do we convert sampling rate efficiently, and, more importantly, how can we control the tradeoff between performance and computation? The second aspect is associated with the need to perform anti-aliasing filtering, especially when reducing the sampling rate.

9.5.1   Computational aspects

To summarize this chapter, let us first concentrate on the computational aspects and put aside, for the moment, the need for performing anti-aliasing filtering. We showed that with spline priors, resampling can be performed very efficiently. Specifically, it involves applying a digital filter, which is implementable by a cascade of a causal and an anticausal filter of small orders, and a multirate system, which, owing to the compact support of B-splines, does not require many multiplications per output sample.

With a bandlimited prior, resampling can be performed by simple means if the conversion factor is rational, namely of the form I/D, with reasonably small I. In this case, conversion is implemented with upsamplers, downsamplers, and digital filters. It is important to note, though, that the fact that rational bandlimited resampling can be expressed as a combination of simple building blocks does not yet mean that the required amount of computation is small. This is because the digital filters involved in this setting are ideal LPFs. These filters are noncausal and have an infinite-impulse response that is not easily implemented. In practice, filter design techniques are usually employed in order to approximate the desired ideal response with a practical filter. This is in contrast to spline interpolation, in which the needed filters are easy to implement to begin with, and thus require no approximations. If the conversion factor is irrational, then it is common to use suboptimal techniques, such as first- and second-order approximation. These methods are special cases of the general framework of dense-grid recovery.

Although spline resampling is generally more efficient than bandlimited rate conversion, in practical applications, the choice of which of these techniques to use should primarily depend on the type of signals encountered. If the classes of signals we are interested in are closer to being bandlimited than splines, then bandlimited resampling should generally be preferred, and vice versa. In practical situations our signals may be neither bandlimited nor splines. A distinction, to some extent, between the prior and the amount of computation can be obtained by forcing the reconstruction stage to involve a kernel with small support (say, a B-spline of small degree). This, of course, degrades the recovery performance if the reconstruction is not matched to the prior. Dense-grid recovery allows us to control the tradeoff between performance and complexity. With this approach, the prior may be quite general (an arbitrary SI space or a smoothness prior) while computational load is adjusted by choosing the density of the reconstruction grid (the reconstruction kernel is scaled appropriately). As the grid becomes denser, recovery improves, but the number of multiplications per output sample increases.

9.5.2   Anti-aliasing aspects

A key ingredient in sampling rate conversion, which needs to be accounted for especially when reducing sampling rate, is the required anti-aliasing operation. In bandlimited sampling, the anti-aliasing filter is an ideal LPF with cutoff frequency associated with the Nyquist rate of the output sequence. Thus, when increasing the rate, no anti-aliasing operation is required.

In general SI spaces, anti-aliasing can be associated with the operation of orthogonally projecting the signal onto the output SI space (an SI space with shifts corresponding to the output sampling rate). Therefore, generally, anti-aliasing filtering is required both when reducing and when increasing the sampling rate, although it is still more crucial in the former situation. We have seen that performing the orthogonal projection operation in spline spaces is computationally cheap only for splines with very low degree (typically 0 or 1). However, it is possible to replace the orthogonal projection by an oblique projection along an SI space whose generator has a small support, such as a B-spline of degree 0. In this case, implementation becomes much more efficient, while performance is only slightly affected.

To conclude, spline resampling is a very efficient rate-conversion method, especially when the signal can be modeled by a spline, to a good approximation. Even when this is not the case, it is still often advantageous to use a B-spline reconstruction kernel via the dense-grid methodology. Typically, for a large enough density of the grid, the approximation is reasonable, and computations are still affordable. When reducing sampling rate, aliasing should be taken into account. An often desirable way to perform anti-aliasing is by using the oblique projection method, which is extremely efficient and closely approximates the optimal orthogonal projection technique.

9.6   Exercises

1. In this problem we explore the effect of interchanging the filtering and downsampling or upsampling operations.
   a. Suppose that a sequence $c[n]$ is filtered with a filter whose frequency response is $H(e^{j\omega})$, and the output is upsampled by I yielding the sequence $d[n]$.

    i. Derive an expression for $D(e^{j\omega})$, the DTFT of $d[n]$.

    ii. Show that $d[n]$ can equivalently be obtained by first upsampling $c[n]$ by I and then filtering the result with a filter $G(e^{j\omega})$. Derive an expression for $G(e^{j\omega})$ as a function of $H(e^{j\omega})$ and I.

  b. Suppose now that $c[n]$ is downsampled by D and then filtered with a filter whose frequency response is $H(e^{j\omega})$, yielding the sequence $d[n]$.

    i. Derive an expression for $D(e^{j\omega})$.

    ii. Show that $d[n]$ can equivalently be obtained by filtering $c[n]$ with a filter $G(e^{j\omega})$ and then downsampling the result by D. Derive an expression for $G(e^{j\omega})$ as a function of $H(e^{j\omega})$ and D.

2. Consider converting the pointwise samples $c_1[n] = x(nT_1)$ of a π/T_1-bandlimited signal into samples $c_2[m]$ with period $T_2 = DT_1/I$, where D and I are integers. One approach is to first perform interpolation by a factor I with the system of Fig. 9.3 and then decimation by a factor D with the system of Fig. 9.7. An alternative is to first perform decimation and then interpolation. For each of these options, draw schematically the frequency content of the resulting sequence. Explain which method is preferable.

3. Let $h_I[n] = \operatorname{sinc}(n/I)$ and $h_D[n] = (1/D)\operatorname{sinc}(n/D)$, and define $h[n] = (h_I * h_D)[n]$. By considering the DTFTs of $h_I[n]$, $h_D[n]$, show that

$$h[n] = \frac{I}{\max\{I, D\}} \operatorname{sinc}\left(\frac{n}{\max\{I, D\}}\right). \qquad (9.117)$$

4. Let $x(t)$ be a π/T_1-bandlimited signal and assume we are given the pointwise samples $c_1[n] = x(nT_1)$, $n \in \mathbb{Z}$. We would like to approximate the samples $c_2[m] = x(mT_2)$, $m \in \mathbb{Z}$, using the dense-grid approach. For a given intermediate sampling period $\widetilde{T}_1 = T_1/I$ and for the reconstruction kernel $w(t) = \beta^1(t/\widetilde{T}_1)$, what is the optimal correction filter $H(e^{j\omega})$ in Fig. 9.12?

5. Suppose we know that $x \in \mathcal{A} = \operatorname{span}\{a(t - n)\}_{n \in \mathbb{Z}}$. We would like to determine the samples $c_2[m] = x(mT_2)$ from the sequence $c_1[n] = (s(-t) * x(t))|_{t=n}$ using the dense-grid approach with a predefined reconstruction kernel $w(t)$.

  a. For $I = 1$, show that the method calculates $c_2[m]$ without error if $W(\omega) = \alpha(e^{j\omega})A(\omega)$ for some bounded 2π-periodic function $\alpha(e^{j\omega})$.

  b. For arbitrary $I > 1$, show that the method calculates $c_2[m]$ without error if $W(\omega) = \alpha(e^{j\omega/I})A(\omega)$ for some $2\pi I$-periodic function $\alpha(e^{j\omega/I})$.

  Hint: Prove that in these situations $\operatorname{span}\{a(t - n)\}_{n \in \mathbb{Z}} \subseteq \operatorname{span}\{w(t - n/I)\}_{n \in \mathbb{Z}}$.

6. Consider converting the rate of the samples $c_1[n] = (s(-t) * x(t))|_{t=nT_1}$ of a signal $x \in \mathcal{A} = \operatorname{span}\{a(t - nT_1)\}_{n \in \mathbb{Z}}$. Instead of using the dense-grid approach, in which the rate is first increased by a factor I, we would like to use a "spacious-grid" approach, namely first decreasing the rate by a factor D. In other words, we would like to obtain the minimum-norm approximation to $x(t)$ in the space $\mathcal{W} = \operatorname{span}\{w(t - mT_1D)\}_{m \in \mathbb{Z}}$. Determine the digital system that outputs the expansion coefficients of the recovery $\hat{x}(t)$ in $\mathcal{W}$ when fed with the samples $c_1[n]$.

7. Consider orthogonal projection resampling with a rectangular kernel, as in Example 9.9.

a. Derive the expression $v_R(t)$ of (9.91) (where $R < 1$).

b. Derive an expression for $v_R(t)$ for the case $R > 1$.

c. For each of the above cases, determine the maximal number of coefficients $b[n]$ needed for computation of $d[m]$ for every m.

8. Consider orthogonal projection resampling with a B-spline kernel of degree $p = 3$, similar to Examples 9.10 and 9.11. Represent the prefilter as the concatenation of two stable IIR filters, one causal and one anticausal. The postfilter in this case can also be expressed in a similar manner. Determine the order of the causal and anticausal IIR filters associated with the postfilter.

9. The pre- and postfilters in orthogonal projection resampling with a B-spline kernel can always be represented as a concatenation of two stable IIR filters, one causal and one anticausal. Determine the order of these filters for a general B-spline of degree p.

10. Consider orthogonal projection resampling with a B-spline kernel of degree 1. Prove that the function $v_R(t)$ is a symmetric cubic spline. For each of the following cases, determine the number and position of the knots of this spline:

a. $0 < R < 0.5$.

b. $R = 0.5$.

c. $0.5 < R < 1$.

d. $R = 1$. ·

e. $1 < R < 2$.

f. $R = 2$.

g. $R > 2$.

11. In projection-based resampling (both orthogonal and oblique), we assumed that $c_1[n]$ corresponds to pointwise samples of $x(t) \in \mathcal{A}_{T_1}$ and that $c_2[n]$ corresponds to pointwise samples of the projection $y(t)$ of $x(t)$ onto $\mathcal{A}_{T_2}$. In some cases, the samples $c_1[n]$ are not pointwise, but rather given by $c_1[n] = (x(t) * s_1(-t))|_{t=nT_1}$ for some kernel $s(t)$. Similarly, we may desire to compute the samples $c_2[m] = (y(t) * s_2(-t))|_{t=mT_2}$ for some kernel $s_2(t)$. How should the pre- and postfilters be modified, in each of the methods, to account for $s_1(t)$ and $s_2(t)$?

12. Similar to orthogonal projection resampling, the pre- and postfilters in oblique projection resampling with a B-spline kernel can be represented as a concatenation of two stable IIR filters, one causal and one anticausal. Determine the order of these filters for a general B-spline of degree p.

13. The basic idea in oblique projection resampling is to compute the oblique projection of $x(t)$ onto $\mathcal{A}_{T_2}$ along $\mathcal{S}_{T_2}^{\perp}$. This solution, however, is not guaranteed to yield a good approximation to $x(t)$ in $\mathcal{A}_{T_2}$. Assuming that we can perfectly recover the signal $x(t)$ from the samples $c_1[n]$, an alternative approach is to compute the minimax-regret solution $P_{\mathcal{A}_{T_2}} P_{\mathcal{S}_{T_2}} x$. Show that this solution can be implemented by replacing the postfilter of the oblique method by an alternative one. Determine the postfilter corresponding to the minimax approach.

14. We have seen that when working with B-spline kernels, the digital filtering stages in the various resampling methods can be implemented by a cascade of a causal and

an anticausal filter. However, this property holds true for other kernels as well. As an example, let

$$s(-t) = \frac{1}{\tau} e^{-\frac{t}{\tau}} u(t), \qquad (9.118)$$

where $\tau > 0$ is some constant and $u(t)$ is the unit step function.

a. Write an expression for the autocorrelation sequence $r_{ss}^{T_1}[n]$, for some $T_1 > 0$.

b. Express the frequency response of the filter $1/R_{SS}^{T_1}(e^{j\omega})$ as the concatenation of two stable IIR filters, one causal and one anticausal. What is the order of the filters?

Chapter 10

Union of subspaces

Our primary focus in the book thus far has been on sampling and recovery of signals that lie in a single subspace. Subspace models lead to powerful sampling theorems, resulting in perfect recovery of the signal from its linear and nonlinear samples under very broad conditions. An appealing feature of the recovery algorithms is that they can often be implemented using simple digital and analog filtering, generalizing the Shannon–Nyquist theorem to a much broader set of input classes. The subspace viewpoint also leads to nice geometrical interpretations of classic and new sampling theorems in terms of orthogonal and oblique projections.

To a large extent, the notion that any possible vector in a given subspace is a valid signal has driven the explosion in the dimensionality of the data we sample and process. Despite the simplicity and geometric appeal of subspace modeling, there are many classes of signals whose structure is not captured well by a single subspace. This is particularly true in applications in which the sampler does not have full knowledge of the received signal. In response to these challenges, there has been a surge of interest in recent years, across many fields, in a variety of low-dimensional signal representations that quantify the notion that the number of degrees of freedom in high-dimensional signals is often quite small compared with their ambient dimension. Low-dimensional modeling has been used extensively in machine learning, parameter estimation, and detection techniques. These methods generally process the data after it is sampled, typically at the Nyquist rate (or higher). Our goal in the next few chapters is to show that the same models used for dimensionality reduction and parameter estimation can be used to enable sampling a wide class of signals at low rates. This allows direct processing of the signal in its compact representation without first having to sample it at a high rate and only then exploit its structure.

In Chapter 4 we considered several examples in which the signal to be sampled is parameterized by unknown values, such as unknown carrier frequencies in multiband communications, or unknown time delays in multipath environments. In all of these examples, and many similar ones, the signal can still be described by a single low-dimensional subspace; however, this underlying subspace depends on unknown parameters. A naive way to capture this uncertainty is to sum over all possible parameter choices, which typically leads to a subspace of large dimension. This in turn results in sampling methods with extremely high rates. A more compact description can be obtained by noting that any signal of this form still resides within a low-dimensional subspace, even if it is unknown. This structure is conveniently described mathematically by a *union of subspaces (UoS)*

[101, 183]. Using the UoS representation allows us to retain the low dimensionality in the problem and still work within subspaces. The price to pay is an additional detection-like parameter which indicates in which of the possible subspaces the given signal resides.

In this chapter we introduce the union model along with several examples, and explore bounds on the minimal possible sampling rates attainable in such settings. In particular, we derive an explicit condition for a sampling operator to be invertible and stable over a UoS, which relates union sampling to subspace sampling. Much of the material in this chapter relies on [101]. Unfortunately, to date, unlike the single subspace prior, there is no general sampling theory that describes sampling and recovery over arbitrary unions. Instead, in Chapters 11–15 we will focus our attention on certain classes of union models for which low-rate sampling methods have been developed. As we show, although sampling can still be obtained by LTI filtering or modulations, the recovery becomes more involved and requires nonlinear algorithms.

10.1 Motivating examples

Before formally describing the UoS model, we begin by providing some motivating examples.

In Section 4.5 we described in detail several applications in which UoS models arise naturally. Two of these examples, multiband communications and time-delay estimation, are illustrated in Fig. 10.1, and summarized shortly below. We refer the reader to Section 4.5 for a more detailed description.

10.1.1 Multiband sampling

Consider first the scenario of a multiband input $x(t)$, which has sparse spectra, such that its CTFT $X(\omega)$ is supported on N frequency intervals, or bands, with individual

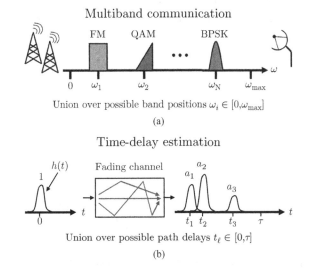

Multiband communication

Union over possible band positions $\omega_i \in [0, \omega_{\max}]$

(a)

Time-delay estimation

Union over possible path delays $t_\ell \in [0, \tau]$

(b)

Figure 10.1 Example applications of UoS modeling.

widths not exceeding B. In addition, the largest frequency component is below ω_{max}. Figure 10.1(a) illustrates a typical multiband spectra. When the band positions are known and fixed, the signal model is linear, since the CTFT of any combination of two inputs is supported on the same frequency bands. This scenario is typical in communication channels, when a receiver intercepts several RF transmissions, each modulated onto a different high carrier frequency. Knowing the carriers allows the receiver to demodulate a transmission of interest to baseband, that is to shift the contents from the relevant RF band to the origin. Subsequent sampling and processing are carried out at a low rate corresponding to the individual band of interest.

When the input consists of a single (double-sided) transmission, an alternative approach to shift contents to baseband is by uniform undersampling at a properly chosen sub-Nyquist rate [20]. We discuss undersampling in detail in Chapter 14, along with its advantages and disadvantages. Example 10.1 below demonstrates undersampling of a bandpass signal by direct uniform sampling.

Example 10.1 Consider a real signal $x(t)$ with CTFT given in Fig. 10.2(a). We refer to such a signal as a bandpass signal. Since $x(t)$ only occupies a small portion of the spectrum, it can be sampled at a rate lower than twice its highest frequency ω_h while still enabling recovery of its content. This is because sampling at a low rate creates aliases, as illustrated in Fig. 10.2(b). As long as the aliases of the positive and negative frequency axes do not overlap, the signal can be retrieved from its replica in baseband.

In Chapter 14 we show that in order to avoid aliasing, the possible sampling rates ω_s are those obeying the following condition:

$$\frac{2\omega_h}{n} \le \omega_s \le \frac{2\omega_\ell}{n-1}$$

where n is an integer in the range $[1, n_{max}]$ with $n_{max} = \left\lfloor \frac{\omega_h}{\omega_h - \omega_\ell} \right\rfloor$. The Nyquist rate, $\omega_N = 2\omega_h$, corresponds to $n = 1$, while the lowest possible rate, is given by

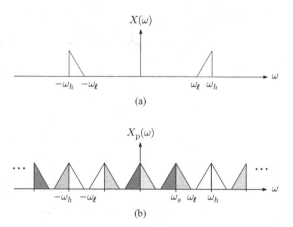

Figure 10.2 (a) The CTFT of $x(t)$; ω_ℓ and ω_h are the lower-frequency and higher-frequency values of the band, respectively. (b) Periodic replicas of the CTFT.

$\omega_{s,\min} = 2\omega_h / n_{\max}$. If ω_h is a multiple of the signal bandwidth $B = \omega_h - \omega_\ell$, a case referred to as *integer positioning*, then $\omega_{s,\min} = 2B$ which is equal to the actual signal occupancy in frequency. Otherwise, $\omega_{s,\min} > 2B$.

Figure 10.2(b) illustrates sampling at the minimal rate when $\omega_{s,\min} > 2B$. In this case, the replicas of $X(\omega)$ do not overlap, implying that we can recover the signal content by appropriate low-pass filtering. Note, however, that the carrier frequency cannot be determined from the samples.

Undersampling of a single band is one way to reduce the sampling rate below Nyquist. However, as we showed in Example 10.1, it achieves the minimal possible rate only for integer band positioning. Furthermore, although the signal content within the band may be determined, the carrier frequency cannot be recovered from the samples. More importantly, this approach provides a valid sub-Nyquist method only when a single band is present. It is not easily extended to treat multiple bands. Nonuniform sampling methods that can accommodate several transmissions were developed in [184, 185, 186], under the assumption that the digital recovery algorithm is provided with knowledge of the spectral support. We will elaborate on these methods in Chapter 14.

When the carrier frequencies are unknown, we are interested in the set of all possible multiband signals that occupy up to NB of the spectrum. In this scenario, the transmissions may lie anywhere below $\omega_{\max}$. At first sight, it may seem that sampling at the Nyquist rate corresponding to $\omega_{\max}$ is necessary, since every frequency interval below $\omega_{\max}$ appears in the support of some multiband $x(t)$. On the other hand, as each specific $x(t)$ in this model fills only a portion of the Nyquist range (only NB), we intuitively expect to be able to reduce the sampling rate below Nyquist. Since the carriers are unknown, standard demodulation cannot be used, which renders this sampling problem quite challenging. As we will see below, the structure in this problem is nicely captured using UoS modeling, paving the way to sub-Nyquist sampling strategies. Practical sampling techniques for multiband sampling will be discussed in detail in Chapter 14.

10.1.2   Time-delay estimation

The second example depicted in Fig. 10.1(b) is that of estimating time delays and amplitudes from an observed signal of the form

$$x(t) = \sum_{\ell=1}^{L} a_\ell \, h(t - t_\ell), \quad t \in [0, \tau], \tag{10.1}$$

where $h(t)$ is a known pulse. We typically assume that the delays t_ℓ are chosen such that $h(t - t_\ell)$ is contained in the interval $[0, \tau]$. For fixed time delays t_ℓ, (10.1) defines a subspace with L degrees of freedom, one for each unknown amplitude a_ℓ. The values of a_ℓ can easily be recovered from L samples $x(t_{k_i}), 1 \le i \le L$ of $x(t)$, as long as t_{k_i} are chosen such that the matrix $\mathbf{H}$ with elements $h(t_{k_i} - t_\ell)$ is invertible. Indeed, denoting by $\mathbf{x}$ the vector containing the samples $x(t_{k_i})$ and by $\mathbf{a}$ the vector containing the unknown amplitudes a_ℓ, we have $\mathbf{x} = \mathbf{Ha}$. An example of this recovery approach is provided in Exercise 2.

In practice, however, there are many interesting situations with unknown t_ℓ. For example, when a communication channel introduces multipath fading, the transmitter can assist the receiver in channel identification by sending a short probing pulse $h(t)$. Since the receiver knows the shape of $h(t)$, it may resolve the delays t_ℓ and use this information to decode the following information messages. Another example is radar, where the delays t_ℓ correspond to target locations, while the amplitudes a_ℓ encode Doppler shifts indicating target speeds. Medical imaging techniques, e.g. ultrasound, use signals of the form (10.1) to probe density changes in human tissues as a vital tool in medical diagnosis. Underwater acoustics also conform with (10.1). We will discuss such applications in detail in Chapter 15. In all these examples, it is desired to use pulses $h(t)$ that are short in time, in order to improve delay resolution. This results in high bandwidth of $h(t)$, so that sampling $x(t)$ according to its Nyquist bandwidth, which is effectively that of $h(t)$, leads to unnecessary large sampling rates.

In contrast, it follows from (10.1) that only $2L$ unknowns determine $x(t)$, namely $t_\ell, a_\ell, 1 \leq \ell \leq L$. Therefore, intuitively, roughly $2L$ samples per time τ should suffice to recover $x(t)$. This rate is typically much smaller than the Nyquist rate, corresponding to the high bandwidth of $h(t)$. Since with unknown delays, (10.1) describes a nonlinear relationship, subspace modeling cannot achieve the optimal sampling rate of $2L/\tau$. Instead, we will rely once more on UoS modeling.

We end this section with an additional example of a union model.

Example 10.2 Many transient signals can be modeled by piecewise polynomials, as in Fig. 10.3. Suppose we define a set $\mathcal{X}$ of all signals consisting of $K \geq 2$ pieces of polynomials supported on $[0, 1]$, where each piece is of degree less than d. Since the points of discontinuity are not specified, the sum $y(t)$ of two signals in $\mathcal{X}$ may have $2K - 1$ polynomial pieces, so that $y(t)$ no longer lies in $\mathcal{X}$, as illustrated in Fig. 10.3(c). This immediately implies that $\mathcal{X}$ is not a subspace. On the other hand, once we fix the locations of the discontinuities, it is easy to see that $\mathcal{X}$ defines a subspace, as shown in Fig. 10.3(d). In this case, the sum remains in $\mathcal{X}$.

For signals given in the above examples, traditional sampling schemes based on a single subspace assumption can be either inapplicable or highly inefficient. This is a result of the fact that the signal obeys a nonlinear model. These applications motivate the need for signal modeling that is more sophisticated than the conventional subspace approach. In order to capture real-world scenarios within a convenient mathematical formulation without unnecessarily increasing the rate, we formally introduce the UoS model in the next section. As we show, this model captures the signal's low dimensionality, while still allowing for uncertainty. We then consider some of the mathematical properties of these sets. In subsequent chapters we analyze sampling strategies for several UoS classes in detail, and show that although sampling can still be obtained by linear filtering, recovery becomes more involved and requires nonlinear algorithms.

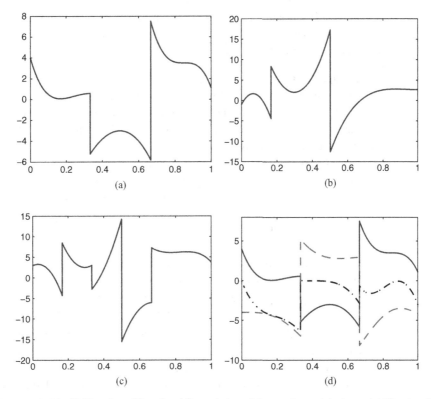

Figure 10.3 (a), (b) Signals $x_1(t)$ and $x_2(t)$ consisting of three polynomial pieces. (c) The signal $y(t) = x_1(t) + x_2(t)$ which consists of five polynomial pieces. (d) When the discontinuities in $x_1(t)$ (solid line) and $x_3(t)$ (dashed line) are at the same locations, the signal $z(t) = x_1(t) + x_3(t)$ (dash-dot line) consists of three polynomial pieces with the same locations of discontinuities.

10.2 Union model

10.2.1 Definition and properties

To define the union model, we denote by $\mathcal{H}$ the ambient Hilbert space which contains our signal x. If $x = x(t)$ is a continuous-time signal with bounded energy then we may choose $\mathcal{H} = L_2(\mathbb{R})$. For the example of piecewise polynomials on $[0, 1]$, we have $\mathcal{H} = L_2([0, 1])$. Multiband signals lie in the space $\mathcal{H}$ of finite-energy signals bandlimited to $\omega_{\max}$. A UoS is defined formally as follows.

Definition 10.1. *A union of subspaces $\mathcal{X}$ in a Hilbert space $\mathcal{H}$ is defined by*

$$\mathcal{X} = \bigcup_{\lambda \in \Lambda} \mathcal{U}_\lambda, \tag{10.2}$$

where Λ is an index set, and each individual $\mathcal{U}_\lambda$ is a subspace of $\mathcal{H}$.

Note that if x lies in the union $\mathcal{X}$, then x resides within $\mathcal{U}_{\lambda^*}$ for some $\lambda^* \in \Lambda$ (there may be more than one choice of λ^* if the subspaces making up the union intersect). The difficulty arises from the fact that a priori the index λ^* is unknown.

Example 10.3 We now revisit Example 10.2 and the set of multiband signals, and show how these signal models can be written in the form (10.2).

The set $\mathcal{X}$ from Example 10.2 is the union of piecewise polynomial subspaces corresponding to all possible discontinuity locations. In this case λ^* indicates the positions of discontinuity. Once λ^* is known, the signal can be described by a subspace of size Kd: each polynomial segment has at most d unknowns and there are K segments. Thus, every $\mathcal{U}_\lambda$ corresponds to a subspace of piecewise polynomial signals with degree less than d and given discontinuity locations. Figure 10.4(a) shows an example of two signals $x_1, x_2 \in \mathcal{U}_\lambda$, where $\mathcal{U}_\lambda$ corresponds to a subspace with fixed discontinuities.

In the multiband communication example, multiband signals form a union over all possible band positions where λ^* denotes the true band locations of a given signal. Each of the subspaces $\mathcal{U}_\lambda$ corresponds to signals with bands at a given location. An example of two multiband signals in a specific $\mathcal{U}_\lambda$ is illustrated in Fig. 10.4(b).

The union (10.2) forms a nonlinear signal set $\mathcal{X}$, where the nonlinearity stems from the fact that the sum (or any linear combination) of $x_1, x_2 \in \mathcal{X}$ does not generally lie in $\mathcal{X}$. More formally, the following proposition shows that a UoS (10.2) is a subspace if and only if the subspaces defining the UoS are contained in each other. We state and prove the proposition for a union of two subspaces.

Proposition 10.1. *Let $\mathcal{X} = \mathcal{U}_1 \cup \mathcal{U}_2$ where $\mathcal{U}_1$ and $\mathcal{U}_2$ are subspaces. Then $\mathcal{X}$ is a subspace if and only if $\mathcal{U}_1 \subseteq \mathcal{U}_2$ or $\mathcal{U}_2 \subseteq \mathcal{U}_1$.*

Proof: If $\mathcal{U}_1 \subseteq \mathcal{U}_2$ then $\mathcal{X} = \mathcal{U}_2$ and $\mathcal{X}$ is a subspace. Similarly, if $\mathcal{U}_2 \subseteq \mathcal{U}_1$ then $\mathcal{X} = \mathcal{U}_1$ and again $\mathcal{X}$ is a subspace.

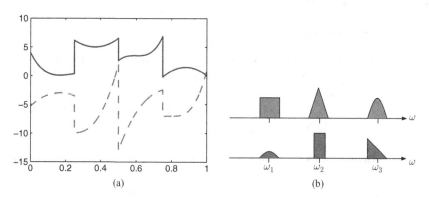

Figure 10.4 (a) An example of two signals, consisting of four polynomial pieces with fixed locations of the discontinuities. (b) An example of two multiband signals with given band locations.

Suppose next that $\mathcal{X}$ is a subspace and that $\mathcal{U}_1$ is not a subspace of $\mathcal{U}_2$ or vice versa. This implies that there exists an element $u \in \mathcal{U}_1$ that is not in $\mathcal{U}_2$. Since $\mathcal{X}$ is a subspace, for any element $v \in \mathcal{U}_2$ the sum $w = u + v$ is in $\mathcal{X}$. This is a result of the fact that both u and v are in $\mathcal{X}$. Now, if $w \in \mathcal{X}$ then w must lie either in $\mathcal{U}_1$ or in $\mathcal{U}_2$. Suppose that $w \in \mathcal{U}_2$. Writing $u = w - v$ and using the fact that both w and v are in $\mathcal{U}_2$ we conclude that $u \in \mathcal{U}_2$, which is a contradiction.

If $w \in \mathcal{U}_1$, then since $v = w - u$ and both w and u are in $\mathcal{U}_1$, we conclude that $v \in \mathcal{U}_1$. Recall that v was an arbitrary element in $\mathcal{U}_2$. We therefore conclude that any $v \in \mathcal{U}_2$ is also in $\mathcal{U}_1$, which is again a contradiction. Consequently, either $\mathcal{U}_1$ is a subspace of $\mathcal{U}_2$ or $\mathcal{U}_2$ is a subspace of $\mathcal{U}_1$. □

An important consequence of Proposition 10.1 is that a union of the form (10.2) is generally a true subset of the space

$$S = \left\{ x = \sum_{\lambda \in \Lambda} a_\lambda x_\lambda \,\middle|\, a_\lambda \in \mathbb{R}, x_\lambda \in \mathcal{U}_\lambda \right\}. \tag{10.3}$$

Since every $x \in \mathcal{X}$ also belongs to S, one can in principle apply conventional sampling strategies with respect to the single subspace S to sample and recover any $x \in \mathcal{X}$. However, this technically correct approach often leads to practically infeasible sampling systems with a tremendous waste of expensive hardware and software resources, because it accommodates a signal set that is typically much larger than $\mathcal{X}$.

Example 10.4 Consider the multiband signal model in which a signal $x(t)$ consists of N bands of width B spread over the frequency range $[0, \omega_{\max}]$. Such a signal can be written in the form (10.2) where each $\mathcal{U}_\lambda$ corresponds to a specific choice of N carrier frequencies in the interval $[0, \omega_{\max}]$. For given carriers, $\mathcal{U}_\lambda$ is the subspace of signals consisting of N bands of width B at those frequencies. The values of λ range over all possible carrier frequencies up to $\omega_{\max}$.

It is easy to see that in this case S is equal to the set of signals supported on $[0, \omega_{\max}]$, for which no rate reduction (with respect to the Nyquist rate) is possible. Since each $\mathcal{U}_\lambda$ is confined to this interval, clearly S cannot contain energy out of this range. On the other hand, any arbitrary signal defined on $[0, \omega_{\max}]$ can be written as a sum $x = x_1 + \cdots + x_n$ where $n = \lceil \omega_{\max}/(NB) \rceil$ and each x_i is supported on N contiguous bands of width B. Thus, x_i is in $\mathcal{U}_\lambda$ for some λ, and $x \in S$.

The UoS model follows the spirit of classic sampling theory by assuming that x belongs to a single underlying subspace $\mathcal{U}_\lambda$. However, in contrast to the traditional paradigm, the union setting permits uncertainty in the exact signal subspace, opening the door to interesting sampling problems. The challenge is to treat this uncertainty at an overall complexity (of hardware and software) that is comparable with a system that knows the exact $\mathcal{U}_\lambda$ rather than being forced to work in the typically high-dimensional S.

10.2.2   Classes of unions

When sampling over UoSs, we will mainly concern ourselves with linear sampling so that the samples are described by $c_n = \langle s_n, x \rangle$ for a set of sampling vectors $\{s_n\}$ (some exceptions are given in Chapters 11 and 15). Given a signal $x \in \mathcal{X}$ and a set of sampling vectors, we would like to be able to answer the following questions:

1. What are the conditions on $\{s_n\}$ such that x is uniquely represented by c_n?
2. What is the minimum sampling requirement for a given union?
3. What are efficient algorithms to reconstruct a signal from its samples?

In the next section, we treat the first two questions, which relate to the sampling mechanism. Recovery over unions is much more involved, so that to date, there is no unifying recovery theory for all choices of subspaces. We will therefore focus on four possible classes and study representatives of each group in the ensuing chapters. The cases we consider differ in two aspects: the number of subspaces in the union (namely, the size of Λ) which can be finite or infinite, and the dimension of each of the subspaces $\mathcal{U}_\lambda$, which again may be finite or infinite, leading to four main categories. We refer to these by the size of Λ and that of the underlying subspace; for example, by a finite–infinite class we mean more explicitly that the number of subspaces is finite, and each subspace has infinite dimension. Ranging over the possible combinations leads to the following settings, which we elaborate on below:

1. Finite–finite: examples include compressed sensing (CS), block sparsity, and structured unions of finite subspaces.
2. Finite–infinite, infinite–infinite: when dealing with infinite-dimensional subspaces we focus on SI subspaces and consider both finite and infinite unions of such subspaces.
3. Infinite–finite: this signal class is related to signals with finite rate of innovation (FRI), examples of which are the pulse streams we have seen in Section 10.1.2.

Finite-dimensional unions

The simplest class of UoS results when the number of subspaces comprising the union is finite, and each subspace has finite dimensions. We call this setup a finite union of subspaces (FUS) model. The problem we are interested in is to recover such signals from as few samples as possible by exploiting the FUS structure.

A popular example within this class is when each subspace is spanned by k columns of the identity matrix of size n. Thus, there are $\binom{n}{k}$ possible subspaces that are aligned with k out of the n coordinate axes of $\mathbb{R}^n$. A vector in one of these spaces has nonzero components only in the chosen k locations, forming a subspace of dimension k. Such a vector is referred to as k-*sparse*. An example of 2-sparse signals embedded in $\mathbb{R}^3$ is illustrated in Fig. 10.5. This model underlies the field of CS, which we review in detail in Chapter 11.

An extension of this class are FUS models with structured sparse supports, namely, sparse vectors that meet additional restrictions on the support (i.e., the set of indices

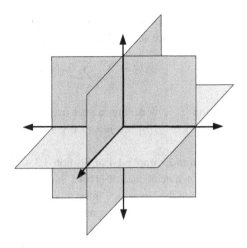

Figure 10.5 Union of subspaces defined by the set of all 2-sparse signals in $\mathbb{R}^3$.

Figure 10.6 A sparse vector with three out of 15 nonzero elements (top) and a block-sparse vector with three nonzero blocks (bottom).

indicating the vector's nonzero entries). This corresponds to only certain choices $\mathcal{U}_i$ out of the $\binom{n}{k}$ subspaces being allowed.

A slightly more complicated scenario is when each subspace is equal to a direct sum of k subspaces chosen out of n possibilities, but now every individual subspace has dimension larger than 1. As we show in Chapter 12, this leads to block sparsity in which certain blocks in a vector are zero, and others are not. Examples of a sparse vector and a block-sparse vector are shown in Fig. 10.6. In the latter, the nonzero values appear in blocks, adding structure that can be exploited to further reduce the number of measurements needed for its representation.

Although these examples do not describe subspaces with infinite dimensions, many of the ideas and results developed in processing and analyzing FUS structures will be essential for extending the results to analog models. Therefore, we will begin our detailed study of unions in Chapters 11 and 12 by considering finite-dimensional settings and exploring conditions that allow such signals to be recovered from a small number of measurements.

Unions of SI subspaces

In the context of sampling, our main goal is to characterize processing of analog signals. Therefore, we again rely on the SI model which has enjoyed widespread use in the context of subspace sampling. To describe unions of SI models, we begin by treating a finite union in which our signals lie in an SI subspace with N generators, where in practice only k of them are active. That is, we can write $x(t)$ as

$$x(t) = \sum_{|\ell|=k} \sum_{n \in \mathbb{Z}} a_\ell[n] h_\ell(t - nT), \qquad (10.4)$$

where the notation $|\ell| = k$ denotes a sum over k elements from $\{1, 2, \ldots, N\}$. Thus, some of the sequences $a_\ell[n]$ are identically zero. This setting is treated in detail in Chapter 13.

As we have seen in Example 4.4 of Section 4.3.4, the multiband model of Fig. 10.1 may be viewed as a special case of (10.4). Suppose, for simplicity, that $\omega_{max} = mB$, so that we can divide the frequency interval $[0, \omega_{max})$ into m sections of equal length B. Then, each band making up the signal is contained in no more than two intervals, as demonstrated in Fig. 4.17. Since there are N bands, this implies that at most $2N$ sections contain energy. We can therefore express $x(t)$ in the form (10.4) with $k = 2N$ and each generator $h(t)$ representing a bandpass filter with energy in one of the m consecutive bands.

A more complicated case is when the union consists of infinitely many subspaces. To model such settings within the SI framework, we first assume, as before, that our signal is described in terms of k generators. However, now, these generators are taken from an infinite set of possibilities:

$$x(t) = \sum_{\ell=1}^{k} \sum_{n \in \mathbb{Z}} a_\ell[n] h_\ell(t - nT; \theta_\ell). \qquad (10.5)$$

Here θ_ℓ is an unknown parameter that takes on values from a continuous set. Therefore, there are infinitely many choices for the generators $h_\ell(t)$.

Example 10.5 The model (10.5) can be used to describe more complicated time-varying multipath channels than that captured by (10.1).

For example, consider a PAM communication system, where the data symbols are transmitted at a symbol rate of $1/T$, and modulated by a known pulse $g(t)$. For this communication system, the transmitted signal $x_T(t)$ is given by

$$x_T(t) = \sum_{n \in \mathbb{Z}} d[n] g(t - nT), \qquad (10.6)$$

where $d[n]$ are the data symbols. The transmitted signal $x_T(t)$ passes through a time-varying multipath channel whose impulse response is modeled as

$$h(\tau, t) = \sum_{\ell=1}^{k} \alpha_\ell(t) \delta(\tau - t_\ell), \qquad (10.7)$$

where $\alpha_\ell(t)$ is the time-varying gain for the ℓth propagation path and t_ℓ is the corresponding time delay. The total number of paths is denoted by k.

We assume that the channel is slowly varying relative to the symbol rate, so that the path gains are considered to be constant over one symbol period:

$$\alpha_\ell(t) = \alpha_\ell[nT] \text{ for } t \in [nT, (n+1)T]. \tag{10.8}$$

In addition, the propagation delays are assumed to be confined to one symbol, i.e., $t_\ell \in [0, T)$. Under these assumptions, the received signal is

$$x(t) = \sum_{\ell=1}^{k} \sum_{n \in \mathbb{Z}} a_\ell[n] g(t - t_\ell - nT), \tag{10.9}$$

where

$$a_\ell[n] = \alpha_\ell[nT] d[n]. \tag{10.10}$$

The resulting signal has the form (10.5) with $h_\ell(t; \theta_\ell) = g(t - t_\ell)$.

Example 10.6 The model (10.5) can also be used to extend the multiband example to situations in which the modulating signals have a known shape, and the carrier frequencies are arbitrary in the range $[0, \omega_{\max}]$.

For example, consider a communication scenario in which k transmissions are received, each with known pulse shape $g_\ell(t)$ and unknown carrier frequency ω_ℓ. The received signal is given by

$$x(t) = \sum_{\ell=1}^{k} \sum_{n \in \mathbb{Z}} a_\ell[n] g_\ell(t - nT) \cos(\omega_\ell(t - nT)), \tag{10.11}$$

which has the form (10.5) with $h_\ell(t; \theta_\ell) = g_\ell(t) \cos(\omega_\ell t)$. Here $a_\ell[n]$ are the symbols of the ℓth transmittance, and $g_\ell(t)$ is an arbitrary pulse shape. This example is more general than the multiband model, in which the generators $g_\ell(t)$ were all equal to an appropriate box function, simplifying the analysis.

Finite rate of innovation signals

An intermediate setting is one in which the number of subspaces in the union is infinite, but each subspace has finite dimension. Such models can arise from (10.5) where the periodic repetition in T is omitted, leading to a class of signals of the form

$$x(t) = \sum_{\ell=1}^{k} a_\ell h_\ell(t; \theta_\ell). \tag{10.12}$$

We have seen a special case of (10.12) in (10.1), in which $h_\ell(t; \theta_\ell) = h(t - t_\ell)$ for appropriate time delays t_ℓ.

More generally, FRI signals are those that can be described by a finite number of parameters per unit time, referred to as their *rate of innovation*. Intuitively, we expect to be able to recover such signals from samples at a rate proportional to the rate of innovation. We note that although many examples of FRI signals fit the UoS model, FRI signals are not restricted to have a UoS structure, as we show in the following example. We treat FRI models in detail in Chapter 15.

Example 10.7 An example of a class of FRI signals that do not conform to the UoS model are continuous-phase modulation (CPM) transmissions. These include continuous-phase frequency shift keying (CPFSK) and minimum shift keying (MSK), tamed frequency modulation (TFM), Gaussian MSK (GMSK), and more. In these settings the transmitted signal is given by

$$x(t) = \cos\left(\omega_0 t + 2\pi h \int_{-\infty}^{t} \sum_{m \in \mathbb{Z}} a_m g(\tau - mT) d\tau\right), \tag{10.13}$$

where ω_0 is a fixed carrier frequency, $a_m \in \pm 1, ..., \pm(Q-1)$ are the message symbols, h is the modulation index, and $g(t)$ is a pulse shape, supported on $[0, LT]$ for some integer $L > 0$, satisfying $\int_0^{LT} g(t) dt = 0.5$.

By evaluating the integral (see Exercise 7), it can be easily seen that such a signal is characterized by a finite number of parameters per unit time; however, it cannot be described by a UoS.

In each one of the models (10.4), (10.5), and (10.12), we can sample the underlying signal $x(t)$ at its associated Nyquist rate. For example, in (10.5) this would imply sampling at the Nyquist rate of the pulses $\{h_\ell(t)\}$. However, when the generators have a wide bandwidth, this approach seems very wasteful. Indeed, in the time-delay model of (10.9), the pulse shape is known; the uncertainty is a result of the time delays. Since there are only k delays over an interval of length T, we expect the rate to be proportional to k/T, and not to the Nyquist rate of $g(t)$ which completely ignores the known pulse shape. Similarly, in the model (10.4), we may sample using a filterbank with N branches, each sampled at rate $1/T$. However, given that only k generators are active, we would again expect to be able to achieve a sampling rate on the order of k/T rather than N/T. We will discuss these scenarios in detail in Chapters 13 and 15 and show how the rate can be reduced significantly with respect to the underlying Nyquist rate by exploiting the union structure.

10.3   Sampling over unions

10.3.1   Unique and stable sampling

In the context of subspace sampling we have seen that any sampling operator corresponding to a Riesz basis or frame for the underlying space leads to unique and stable signal expansions. The situation is more involved when treating union models.

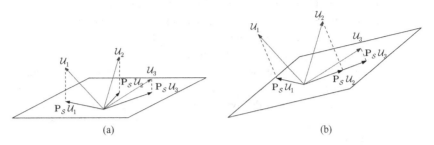

Figure 10.7 An example of a union of subspaces $\mathcal{X} = \bigcup_{\lambda=1}^{3} \mathcal{U}_\lambda$ and its projection onto two sampling spaces. (a) Invertible sampling. (b) Noninvertible sampling.

Suppose we are given measurements $c_n = \langle s_n, x \rangle$, or $c = S^* x$, and let $\mathcal{S}$ denote the sampling space spanned by the sampling functions s_n. We have seen in Section 6.2.1 that knowing the samples is equivalent to knowing the orthogonal projection $P_\mathcal{S} x$ onto $\mathcal{S}$. Consequently, the sampling process is invertible if and only if there is a one-to-one mapping between the union $\mathcal{X}$ and its projection onto $\mathcal{S}$, which we denote by $P_\mathcal{X} = P_\mathcal{S} \mathcal{X}$.

Example 10.8 Consider an example in which the signal space $\mathcal{H}$ is equal to $\mathbb{R}^3$. The signal set $\mathcal{X} = \bigcup_{\lambda=1}^{3} \mathcal{U}_\lambda$ is a union of three one-dimensional subspaces, as depicted in Fig. 10.7. We project the union $\mathcal{X}$ onto two sampling spaces $\mathcal{S}$, yielding $P_\mathcal{X} = \bigcup_{\lambda=1}^{3} P_\mathcal{S} \mathcal{U}_\lambda$. As can be seen in Fig. 10.7, there is an invertible mapping between $\mathcal{X}$ and $P_\mathcal{X}$ as long as there are no two subspaces in $\{\mathcal{U}_\lambda\}_{\lambda=1}^{3}$ that are projected onto the same vector in $\mathcal{S}$. Figure 10.7(a) represents an invertible mapping, so that the input can in principle be recovered from the samples. In contrast, the projection in Fig. 10.7(b) collapses two subspaces onto the same vector, and recovery of the input is impossible. This example trivially extends to vectors in $\mathbb{R}^N$ for any N.

Example 10.8 illustrates that it is possible to project a set of signals onto a lower-dimensional representation while still preserving their information. An important question in this context is how small the dimension of $\mathcal{S}$ can be without losing information on $\mathcal{X}$. This question is tightly connected to the minimal rate requirements for $\mathcal{X}$. In general, there will be many possible choices of $\mathcal{S}$ that lead to invertible transformations. Intuitively, we would like the projected subspaces $P_\mathcal{S} \mathcal{U}_\lambda$ to be such that they are "sufficiently" far apart from each other, so that the sampling does not become too sensitive to noise and numerical errors.

The following definition formalizes uniqueness of a given sampling operator S^* over a union $\mathcal{X}$.

Definition 10.2. *A sampling operator $S^*: \mathcal{H} \to \ell_2$ is invertible over a union of subspaces $\mathcal{X}$ if for every $x_1, x_2 \in \mathcal{X}$,*

$$S^* x_1 = S^* x_2 \iff x_1 = x_2. \tag{10.14}$$

Definition 10.2 implies that S^* is a one-to-one mapping between $\mathcal{X}$ and the set in ℓ_2 defined by $S^*\mathcal{X}$. Thus, for any given sampling sequence $c = S^*x$, there is a unique $x \in \mathcal{X}$ that could have generated those samples.

In practice, it is often useful to define stability beyond uniqueness. Namely, we want to ensure not only that there is a unique recovery from the given samples, but also that the original signal x can be determined in a numerically stable fashion. In the context of subspace sampling, this was ensured by relying on the notions of Riesz bases and frames. To define stable sampling over unions, we extend the concept of a frame to the union model.

Definition 10.3. *A sampling operator S^*: $\mathcal{H} \rightarrow \ell_2$ is stable over a union of subspaces $\mathcal{X}$ if there exist constants $0 < \alpha \leq \beta < \infty$ such that*

$$\alpha\|x_1 - x_2\|^2 \leq \|S^*(x_1 - x_2)\|^2 \leq \beta\|x_1 - x_2\|^2, \tag{10.15}$$

for any $x_1, x_2 \in \mathcal{X}$.

The tightest ratio β/α provides a measure of the stability of the sampling operator. If the lower bound in (10.15) is satisfied, then it is immediate to see that the sampling operator is invertible: in this case, $S^*(x_1 - x_2) = 0$ necessarily implies $x_1 = x_2$.

Note that when defining frames for a subspace $\mathcal{S}$, we required (10.15) to hold for any $x = x_1 - x_2 \in \mathcal{S}$. That is because when $\mathcal{S}$ is a subspace, $x_1 - x_2$ is in $\mathcal{S}$ for any $x_1, x_2 \in \mathcal{S}$. However, over unions it is no longer true that $x_1, x_2 \in \mathcal{X}$ implies $x_1 - x_2 \in \mathcal{X}$, and therefore we write the inequalities specifically in terms of these difference vectors. This crucial difference makes working over unions much more complicated than working over subspaces since the underlying set of signals is nonlinear. Therefore, we cannot directly invoke linear algebra results to study sampling and stability over unions. A similar issue arises when examining the invertibility condition (10.14). In the case of a single subspace $\mathcal{S}$, invertibility reduces to the requirement that for any $x \in \mathcal{S}$, $S^*x = 0$ implies $x = 0$. In contrast, over a union $\mathcal{X}$, we need that for any $x_1, x_2 \in \mathcal{X}$, $S^*(x_1 - x_2) = 0$ implies $x_1 - x_2 = 0$. Again, this is a result of the fact that $x_1 - x_2$ is no longer guaranteed to be in $\mathcal{X}$. Thus, instead of requiring uniqueness and stability over a single subspace, we are forced to extend the definition to a nonlinear set.

To overcome this difficulty, we introduce subspaces constructed from the sums of pairs of subspaces $\mathcal{U}_\lambda$:

$$\mathcal{U}_{\lambda,\gamma} = \mathcal{U}_\lambda + \mathcal{U}_\gamma = \{x|x = x_1 + x_2, \text{ where } x_1 \in \mathcal{U}_\lambda, x_2 \in \mathcal{U}_\gamma\}. \tag{10.16}$$

The sum space $\mathcal{U}_{\lambda,\gamma}$ often has a simple interpretation. For example, for multiband signals, the sum is the set of all signals with $2N$ nonzero bands. It is easy to see that the uniqueness and stability conditions of Definitions 10.2 and 10.3 on the union $\mathcal{X}$ can be stated over single subspaces $\mathcal{U}_{\lambda,\gamma}$.

Proposition 10.2. *A sampling operator S^*: $\mathcal{H} \rightarrow \ell_2$ is invertible over a union of subspaces $\mathcal{X}$ if and only if S^* is invertible for every $\mathcal{U}_{\lambda,\gamma}$ with $\lambda, \gamma \in \Lambda$. In other words, if and only if for any $y \in \mathcal{U}_{\lambda,\gamma}$ with $\lambda, \gamma \in \Lambda$,*

$$S^*y = 0 \quad \Longleftrightarrow \quad y = 0. \tag{10.17}$$

It is a stable sampling operator for $\mathcal{X}$ if and only if there exist constants $0 < \alpha \le \beta < \infty$ such that

$$\alpha\|x\|^2 \le \|S^*x\|^2 \le \beta\|x\|^2, \tag{10.18}$$

for every $x \in \mathcal{U}_{\lambda,\gamma}$ with $\lambda, \gamma \in \Lambda$.

Proposition 10.2 implies that S^* is invertible (stable) over $\mathcal{X}$ if it is invertible (stable) over all choices of subspaces $\mathcal{U}_{\lambda,\gamma}$. The latter can be verified using standard linear algebra results.

Proof: We begin by considering the invertibility condition. Suppose first that S^* is invertible over every possible subspace $\mathcal{U}_{\lambda,\gamma}$, and let x_1, x_2 be vectors in $\mathcal{X}$ for which $S^*x_1 = S^*x_2$. Then, $S^*y = 0$ where $y = x_1 - x_2$ is a vector in $\mathcal{U}_{\lambda,\gamma}$ for some pair λ, γ. Since S^* is invertible over $\mathcal{U}_{\lambda,\gamma}$, we conclude immediately that $y = 0$ and $x_1 = x_2$.

Next, assume that $S^*x_1 = S^*x_2$ implies $x_1 = x_2$ for every $x_1, x_2 \in \mathcal{X}$. Our goal is to show that if y is an arbitrary vector in $\mathcal{U}_{\lambda,\gamma}$ for some pair λ, γ, then $S^*y = 0$ implies $y = 0$. Since $y \in \mathcal{U}_{\lambda,\gamma}$, we can write $y = y_1 - y_2$ for some $y_1 \in \mathcal{U}_\lambda$ and $y_2 \in \mathcal{U}_\gamma$. Suppose that $S^*y = 0$. This implies that $S^*y_1 = S^*y_2$, where $y_1, y_2 \in \mathcal{X}$. From our assumption it then follows that $y_1 = y_2$, concluding the invertibility proof.

The stability condition follows immediately by noting that (10.18) is identical to (10.15). $\square$

Example 10.9 Consider a FUS $\mathcal{X}$ where each subspace is spanned by k columns of the identity matrix of size n. In this case, $\mathbf{x} \in \mathcal{X}$ means that $\mathbf{x}$ is a k-sparse vector of length n. The sampling operator S^* is described by an $m \times n$ matrix $\mathbf{A}$.

For this example, $\mathcal{U}_{\lambda,\gamma}$ with $\lambda, \gamma \in \Lambda$ defines a subspace spanned by at most $2k$ columns of the identity matrix. Ranging over all possible λ, γ implies that (10.17) and (10.18) must be satisfied for any $2k$-sparse vector. Specifically, to ensure invertibility we require that $\mathbf{Ax} = \mathbf{0}$ for any $\mathbf{x}$ that has at most $2k$ nonzero values implies $\mathbf{x} = \mathbf{0}$. This means that every $2k$ columns of $\mathbf{A}$ must be linearly independent. A matrix with this property is said to have a spark that is larger than $2k$. We will introduce the spark formally in Chapter 11.

Stability of the sampling operator is captured by the requirement that

$$\alpha\|\mathbf{x}\|^2 \le \|\mathbf{Ax}\|^2 \le \beta\|\mathbf{x}\|^2, \tag{10.19}$$

for some $0 < \alpha \le \beta < \infty$ and all vectors $\mathbf{x}$ that have at most $2k$ nonzero values. In Chapter 11 we introduce the restricted isometry property (RIP) and show that it is equivalent to (10.19). Note that if a matrix satisfies the RIP, then its spark is larger than $2k$. This is because $\mathbf{Ax}$ cannot be equal to $\mathbf{0}$ for any $2k$-sparse $\mathbf{x}$, since this would violate the lower bound. We therefore conclude that for our choice of $\mathcal{X}$, invertibility and stability are guaranteed when sampling with a matrix $\mathbf{A}$ if it satisfies the RIP.

10.3.2   Rate requirements

We now use the invertibility condition of Proposition 10.2 to bound the minimal number of samples needed to represent an arbitrary $x \in \mathcal{X}$. In our discussion, we distinguish between unions of finite- and infinite-dimensional subspaces.

Finite-dimensional subspaces

Suppose first that each $\mathcal{U}_\lambda$ is a finite-dimensional subspace of dimension d_λ, and let $d_{\lambda,\gamma}$ denote the dimension of $\mathcal{U}_{\lambda,\gamma}$. The size of the set Λ, as well as the dimension of the ambient space $\mathcal{H}$, can be finite or infinite. In this setting, it is clear that for (10.17) to hold over $\mathcal{U}_{\lambda,\gamma}$ the number of samples n must satisfy $n \geq d_{\lambda,\gamma}$. If in addition all subspaces have equal dimension d and they do not intersect, then we obtain the condition $n \geq 2d$.

Proposition 10.3. *Let $S^*: \mathcal{H} \to \mathbb{C}^n$ be an invertible sampling operator over a union $\mathcal{X} = \bigcup_{\lambda \in \Lambda} \mathcal{U}_\lambda$, where each subspace $\mathcal{U}_\lambda$ has finite dimension d_λ. Then $n \geq d$ where $d = \max_{\lambda,\gamma \in \Lambda} \dim(\mathcal{U}_{\lambda,\gamma})$.*

Proof: Let λ^*, γ^* denote the values in Λ for which $\dim(\mathcal{U}_{\lambda,\gamma})$ is maximized, so that $d = \dim(\mathcal{U}_{\lambda^*,\gamma^*})$. If S^* is invertible over $\mathcal{X}$, then from Proposition 10.2, S^* is invertible over any $\mathcal{U}_{\lambda,\gamma}$ and in particular over $\mathcal{U}_{\lambda^*,\gamma^*}$. This in turn implies that $\dim(S^*\mathcal{U}_{\lambda^*,\gamma^*}) = \dim(\mathcal{U}_{\lambda^*,\gamma^*}) = d$. On the other hand, since the range of S^* is contained within $\mathbb{C}^n$, we have immediately that $\dim(S^*\mathcal{U}_{\lambda^*,\gamma^*}) \leq n$ from which we obtain $n \geq d$. $\square$

Example 10.10 Consider an application of Proposition 10.3 to Example 10.2. The union $\mathcal{X}$ is the set of all signals consisting of $K \geq 2$ pieces of polynomials supported on $[0, 1]$, where each piece is of degree less than d_0.

In this case $\mathcal{U}_{\lambda,\gamma}$ are subspaces of piecewise polynomials with at most $2K - 1$ pieces, each of degree less than d_0. According to Proposition 10.3, the number of samples n must satisfy $n \geq d = (2K - 1)d_0$.

It can be easily seen that each signal in $\mathcal{X}$ is completely defined by $Kd_0 + K - 1$ values: Kd_0 parameters represent the coefficients of K polynomial pieces and $K - 1$ parameters specify the locations of the discontinuities. Therefore, contrary to what one would expect, the minimal number of samples, d, is strictly greater than the number of parameters $Kd_0 + K + 1$.

Using Proposition 10.2 we can also obtain a simple condition on stable sampling when the union consists of finite-dimensional spaces. Specifically, let $U_{\lambda,\gamma}: \mathbb{C}^{d_{\lambda,\gamma}} \to \mathcal{H}$ denote an orthonormal basis for $\mathcal{U}_{\lambda,\gamma}$. Then (10.18) may be written as

$$\alpha\|c\|^2 \leq \|S^*U_{\lambda,\gamma}c\|^2 \leq \beta\|c\|^2, \tag{10.20}$$

for any $c \in \mathbb{C}^{d_{\lambda,\gamma}}$. Thus, the operator $S^*U_{\lambda,\gamma}$ must be stably left-invertible for all λ, γ. Assuming that the number of measurements n is finite, this means that the singular values of the matrix $S^*U_{\lambda,\gamma}$ should be bounded below and above for all λ, γ.

Infinite-dimensional subspaces

As we noted earlier, when treating unions over infinite-dimensional subspaces, we restrict our attention to SI subspaces. Thus, let $\mathcal{U}_\lambda$ be an SI subspace, generated by the functions $\{g_k^\lambda(t), 1 \leq k \leq N_\lambda\}$. The sum space $\mathcal{U}_{\lambda,\gamma}$ is then generated by $\{g_k^\lambda(t), g_k^\gamma(t)\}$. Denote by $N_{\lambda,\gamma}$ the minimal number of functions needed to generate $\mathcal{U}_{\lambda,\gamma}$. Clearly, $N_{\lambda,\gamma} \leq N_\lambda + N_\gamma$. In our treatment of SI subspaces we have seen that the sampling rate required to ensure stable recovery over an SI subspace with K generators is at least K/T. Therefore, the minimal rate requirement over unions of SI subspaces is N/T where $N = \sup_{\lambda,\gamma \in \Lambda} N_{\lambda,\gamma}$.

10.3.3   Xampling: compressed sampling methods

Until now, we have focused on theoretical aspects of modeling and sampling over unions. In the ensuing chapters we consider specific acquisition and recovery methods for the different classes of unions surveyed in this chapter. We begin, in Chapter 11, with an overview of the field of CS, which treats recovery of sparse vectors from undetermined systems. In Chapter 12 we show how the basic ideas of CS can be extended to unions of finite-dimensional subspaces. Sampling over unions of continuous-time signals is treated in Chapters 13–15. Unions of SI spaces are considered in Chapter 13, multiband sampling is treated in Chapter 14, and FRI sampling is discussed in Chapter 15 along with applications to problems in radar, ultrasound, and wireless communications.

Besides developing the required theory and sampling techniques, we also describe several hardware prototypes available for sub-Nyquist sampling. Although our focus is mainly on the theory, we feel it is important to highlight some of the issues that need to be accounted for when one wants to translate the theoretical results into concrete hardware devices. It is also important to stress that these recent ideas of sub-Nyquist sampling have led not only to beautiful new concepts and deep mathematical results, but also to practical low-rate ADC prototypes that have the potential for a large impact on the ADC industry.

Even though the sampling schemes and detailed recovery techniques vary from one class of signals to the next, they can all be viewed under a unified framework, referred to as *Xampling* [21, 22], that suggests several guidelines and principles for designing UoS sampling systems. Figure 10.8 shows a high-level Xampling architecture [21, 22] which consists of two main functions: analog compression that narrows down the input bandwidth prior to sampling with commercial devices while retaining all vital signal information, followed by a nonlinear algorithm that detects the input subspace prior to conventional signal processing. The implementation of each of these blocks depends on the specific class of UoS model considered.

The first two blocks in the figure, termed X-ADC, perform the conversion of $x(t)$ to digital. An operator P compresses the high-bandwidth input $x(t)$ into a signal with lower bandwidth, effectively capturing the entire union $\mathcal{X}$ by a subspace $\mathcal{W}$ with substantially lower sampling requirements. Commercial ADC devices (there may be several of them

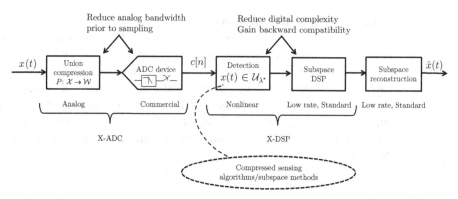

Figure 10.8 Xampling – a pragmatic framework for signal acquisition and processing in union of subspaces.

in a filterbank structure) acquire pointwise samples of the compressed signal, resulting in the sequence of samples $c[n]$. The role of P in Xampling is to narrow down the analog bandwidth, so that low-rate ADC devices can subsequently be used. As in digital compression, the goal is to capture all vital information of the input in the compressed version, though here this functionality is achieved by hardware rather than software. This is accomplished by aliasing the signal prior to sampling in such a way that the samples contain energy from all subspace components. The aliasing is manifested in different domains, depending on the application. For example, the modulated wideband converter (MWC) introduced in Chapter 14 performs aliasing in frequency in order to sample multiband signals at low rates. Aliasing ensures that all signal bands appear in baseband, regardless of the carrier frequencies. We may then sample only the baseband signal and still capture the entire multiband signal energy. In contrast, the sum-of-sincs sampling method (or more generally, Fourier sampling) introduced in Chapter 15 for sensing pulse streams, is based on aliasing in time. In this way, narrow pulses which require high sampling rates are first spread in time, so that the signal energy can be captured from low-rate samples.

In the digital domain, Xampling consists of three computational blocks. A nonlinear step detects the signal subspace $\mathcal{U}_{\lambda^*}$ from the low-rate samples. CS algorithms, considered in the next chapter, or comparable methods for subspace identification, such as MUSIC [15] and ESPRIT [187], which will be studied in Chapter 15, can be used for this purpose. Once the index λ^* is determined, we gain backward compatibility, meaning that standard DSP methods apply and commercial DAC devices may be used for signal reconstruction. The combination of nonlinear detection and standard DSP is referred to as X-DSP. Besides backward compatibility, nonlinear detection decreases computational loads, since the subsequent DSP and DAC stages need to treat only the single subspace $\mathcal{U}_{\lambda^*}$. The important point is that detection is performed efficiently at the low acquisition rate.

Pronounced as CS-Sampling (phonetically /kˈsæmplɪŋ/), the nomenclature Xampling symbolizes the combination between recent developments in CS and the successful machinery of analog sampling theory developed over the past century. It provides a general architecture for low-rate sampling of a large class of analog signals by combining analog compression and standard sampling ideas with the more recent concept of CS. The basic three-step strategy of aliasing (analog projection), subspace identification, and subspace recovery is shared among all examples. Nonetheless, Xampling is a generic template architecture. It does not specify the exact acquisition operator P or nonlinear detection method to be used. These are application-dependent functions which we will study in the ensuing chapters in the context of specific UoS classes.

10.4 Exercises

1. Consider a real signal $x(t)$ with a single band on the positive axis of width B positioned between $\omega_\ell = (n - 1/2)B$ and $\omega_h = (n + 1/2)B$ for some integer $n \geq 1$.
   a. At what rates can the signal be sampled without creating aliasing?
   b. Plot the DTFT of the sampled sequence $x(nT)$ when $T = 2\pi/B$.
2. Consider the signal

$$x(t) = \sum_{\ell=1}^{10} a_\ell h(t - t_\ell), \ t \in [0, 10] \tag{10.21}$$

with known pulse shape

$$h(t) = \exp\left\{-\frac{t^2}{2\sigma^2}\right\}, \ \sigma = 0.45 \tag{10.22}$$

and fixed time delays

$$\{t_\ell\}_{\ell=1}^{10} = \{0.90, 1.30, 2.20, 3.40, 4.50, 5.20, 6.30, 7.10, 8.20, 9.60\}.$$

Find the values of $\{a_\ell\}_{\ell=1}^{10}$ from 10 samples $x(t_{k_i}), 1 \leq i \leq 10$ of $x(t)$ given by:

$$\{t_{k_i}\}_{i=1}^{10} = \{0.50, 1.50, 2.50, 3.50, 4.50, 5.50, 6.50, 7.50, 8.50, 9.50\},$$
$$\{x(t_{k_i})\}_{i=1}^{10} = \{0.14, 0.34, 0.09, 0.25, 0.26, 0.38, 0.35, 0.36, 0.46, 0.11\}.$$

3. Let $x(t)$ be a stream of Diracs $x(t) = \sum_{\ell=1}^{L} \alpha_\ell \delta(t - t_\ell)$ with unknown locations $\{t_\ell\}_{\ell=1}^{L}$ and weights $\{\alpha_\ell\}_{\ell=1}^{L}$. Determine the minimal number of samples required for signal reconstruction.
4. Consider the class of signals $x(t)$ that lie in an SI subspace $x(t) = \sum_{n\in\mathbb{Z}} d[n]h(t - nT)$ for a sequence of numbers $d[n]$. For each of the restrictions on the values $d[n]$ below indicate whether the signals $x(t)$ form a subspace, or a UoS:

a. $d[4] = 5, d[10] = 7$; the remaining values of $d[n]$ are arbitrary.

b. $d[3n] = 0$ for all n; the remaining values of $d[n]$ are arbitrary.

c. $d[3n + \ell] = 0$ for all n and some value of ℓ; the remaining values of $d[n]$ are arbitrary. Here ℓ can be any index from the set $\{0, 1, 2\}$.

5. Consider a pulse-width modulation signal of the form $s(t) = \sum_{n \in \mathbb{Z}} g(t - nT; b_n)$ where

$$g(t; b = 0) = \begin{cases} 1, & 0 \le t \le \frac{T}{2} \\ 0, & \text{otherwise} \end{cases}, \quad g(t; b = 1) = \begin{cases} 1, & 0 \le t \le T \\ 0, & \text{otherwise}. \end{cases} \quad (10.23)$$

a. Is $s(t)$ an FRI signal?

b. Can $s(t)$ be described as a UoS?

6. Let $\mathcal{U}_i$ be the subspace of signals defined by $\mathcal{U}_i = \{x(t)|x(t) = \sum_{n \in \mathbb{Z}} a[n]h_i(t - nT), a[n] \in \mathbb{R}\}$ where $h_i(t) = h(t - iT/L)$ with

$$h(t) = \begin{cases} \frac{L}{T}, & 0 \le t \le \frac{T}{L} \\ 0, & \text{otherwise}. \end{cases} \quad (10.24)$$

We consider sampling with the operator S^* corresponding to shifts by T of the generator

$$s(t) = \begin{cases} 1, & 0 \le t \le T \\ 0, & \text{otherwise}. \end{cases} \quad (10.25)$$

Thus, for an arbitrary signal $x(t)$ the samples are $c[n] = \langle s(t - nT), x(t) \rangle$. Determine whether or not the sampling operator is invertible over the following signal sets $\mathcal{X}$:

a. $\mathcal{X} = \mathcal{U}_i$ for some given $i = 0, 1, \ldots, L - 1$.

b. $\mathcal{X} = \bigcup_{i=0}^{L-1} \mathcal{U}_i$.

c. $\mathcal{X} = \bigcup_{i=0}^{L-1} \mathcal{U}_i$ but now the samples are given by $c[n] = \langle s(t - nT/L), x(t) \rangle$.

7. a. Show that the signal $x(t)$ of (10.13) can be expressed as

$$x(t) = \cos\left(\omega_0 t + \sum_{m \in \mathbb{Z}} b_m f(t - mT)\right), \quad (10.26)$$

for appropriate choices of b_m and $f(t)$.

b. Determine the support of $f(t)$.

c. Compute the rate of innovation of $x(t)$.

8. Consider the signal $x(t)$ of Exercise 7. Explain whether this signal can be described as a UoS.

9. Provide an example of an FRI signal that is not a UoS and vice versa.

10. Propose a sampling method achieving the minimal sampling requirement for piecewise polynomials derived in Example 10.10.

11. Let $\mathcal{X}$ be the set of 2-sparse vectors of length 5 and let the sampling operator be given by a 4×5 matrix $\mathbf{A}$. For the following matrices $\mathbf{A}$, determine whether the sampling operator is invertible and stable:

a.

$$\mathbf{A} = \frac{1}{\sqrt{2}} \begin{bmatrix} 1 & 1 & 0 & 0 & -1 \\ 1 & 0 & 1 & 0 & 0 \\ 0 & -1 & 1 & 1 & 0 \\ 0 & 0 & 0 & -1 & 1 \end{bmatrix}. \tag{10.27}$$

b.

$$\mathbf{A} = \frac{1}{\sqrt{2}} \begin{bmatrix} 1 & 1 & 0 & 0 & -1 \\ 1 & 0 & 1 & 0 & 0 \\ 0 & -1 & 0 & 1 & 0 \\ 0 & 0 & -1 & -1 & 1 \end{bmatrix}. \tag{10.28}$$

12. Consider the union $\mathcal{X}$ of 3-sparse vectors of length 8. We sample a vector $\mathbf{x}$ in $\mathcal{X}$ by multiplying it with an $m \times 8$ matrix $\mathbf{A}$. For each of the following values of m, indicate whether there exists an $\mathbf{A}$ leading to invertible sampling for all possible $\mathbf{x}$, and provide an example of $\mathbf{A}$ when your answer is positive:

a. $m = 3$.
b. $m = 5$.
c. $m = 6$.
d. $m = 8$.

Chapter 11

Compressed sensing

One of the most well-studied examples of a UoS is that of a vector x that is sparse in an appropriate basis. This model underlies the rapidly growing field of compressed sensing (CS), which has attracted considerable attention in signal processing, statistics, and computer science, as well as the broader scientific community. In this chapter, we provide a review of the basic concepts underlying CS. We focus on the theory and algorithms for sparse recovery in finite dimensions. In subsequent chapters, we will see how the fundamentals presented here can be expanded and extended to include richer structures in both analog and discrete-time signals, ultimately leading to sub-Nyquist sampling techniques for a broad class of continuous-time signals.

11.1   Motivation for compressed sensing

The sampling theorems we studied in previous chapters, including the celebrated Shannon–Nyquist theorem, are at the heart of the current digital revolution that is driving the development and deployment of new kinds of sensing systems with ever-increasing fidelity and resolution. Digitization has enabled the creation of sensing and processing systems that are more robust, flexible, cheaper, and, consequently, more widely used than their analog counterparts. As a result of this success, the amount of data generated by sensing systems has grown from a trickle to a torrent. Unfortunately, in many important and emerging applications, the resulting sampling rate is so high that we end up with far too many samples that need to be transmitted, stored, and processed. In addition, in applications involving wideband inputs it is often very costly, and sometimes even physically impossible, to build devices capable of acquiring samples at the necessary rate [188, 189]. Thus, despite extraordinary advances in sampling theory as well as computational power, the acquisition and processing of signals in application areas such as radar, wideband communications, video, medical imaging, remote surveillance, and genomic data analysis continues to pose a tremendous challenge.

To address the challenges involved in dealing with such high-dimensional data, we often depend on compression, which aims at finding the most concise representation of a signal that is able to achieve a target level of acceptable distortion. One of the most popular techniques for signal compression is known as *transform coding*, which typically relies on finding a basis or frame that provides *sparse* or *compressible* descriptions for signals in a class of interest [190]. By a sparse representation of a signal of length n, we mean that it can be represented with $k \ll n$ nonzero coefficients; a compressible representation is one in which the signal is well-approximated by $k < n$ nonzero coefficients. Both sparse and compressible signals can be described with high fidelity by preserving only the values and locations of the largest coefficients of the signal. This process is called *sparse approximation*, and forms the foundation of transform coding schemes that exploit signal sparsity, including the JPEG, JPEG2000, MPEG, and MP3 standards.

Leveraging the concept of transform coding, CS has emerged as a framework for simultaneous sensing and compression of finite-dimensional vectors, which relies on linear dimensionality reduction. CS enables a potentially large reduction in sampling and computation costs for measuring signals that have a sparse or compressible description. While the Shannon–Nyquist sampling theorem states that a certain minimum number of samples is required in order to perfectly capture an arbitrary bandlimited signal, if the signal is sparse in a known basis then we can vastly reduce the number of measurements that need to be stored. Consequently, when sensing sparse signals we might be able to do better than suggested by subspace results. This is the fundamental idea behind CS: rather than first sampling at a high rate and then compressing the sampled data, we would like to find ways to directly measure the data in a compressed form, i.e. at a lower sampling rate. The ideas underlying CS have their origins in abstract results from functional analysis and approximation theory, but were brought to the forefront by the works of Candès, Romberg, and Tao, and of Donoho. These authors showed that a finite-dimensional signal having a sparse or compressible representation can be recovered from a small set of linear measurements [12, 13, 191]. The design of these measurements and their extensions to practical data models and acquisition systems are central challenges in the field of CS. In Chapters 13–15 we treat in detail extensions of CS that consider sub-Nyquist sampling of many interesting classes of continuous-time signals.

The roots of CS date back as far as the eighteenth century. In 1795, Prony proposed an algorithm for estimation of the parameters associated with a small number of complex exponentials sampled in the presence of noise [192]. As we show in Chapter 15, Prony's algorithm and its extensions play a key role in sub-Nyquist sampling and recovery over infinite union of subspaces, and form the basis for sampling in certain FRI models. The next theoretical leap came in the early 1900s, when Carathéodory showed that a positive linear combination of any k sinusoids is uniquely determined by its value at $t = 0$ and at any other $2k$ points in time [193]. This represents far fewer samples than the number of Nyquist-rate samples when k is small and the range of possible frequencies is large. In the 1990s, this work was generalized by Gorodnitsky, Rao, and George, who studied sparsity in biomagnetic imaging and other contexts [194, 195, 196, 197]. These are some of the first few works explicitly exploiting sparsity in signal processing using ideas and techniques that are closely related to those that lie at the heart of CS.

A related problem focuses on recovery of a signal from partial observation of its Fourier transform. Beurling proposed a method for extrapolating these observations to determine the entire Fourier transform [198]. One can show that if the signal consists of a finite number of impulses, then Beurling's technique will correctly recover the entire Fourier transform (of this nonbandlimited signal) from any sufficiently large piece of its Fourier transform. His approach – to find the signal with smallest L_1 norm among all signals agreeing with the acquired Fourier measurements – bears remarkable resemblance to some of the algorithms used in CS.

More recently, Candès, Romberg, and Tao [13, 191], and Donoho [12] showed that a signal having a sparse representation can be recovered exactly from a small set of linear, nonadaptive measurements. This result suggests that it may be possible to sense sparse signals by taking far fewer samples than implied by subspace results. Note, however, that CS differs from classical sampling in three important aspects. First, sampling theory typically considers infinite-length, continuous-time signals. In contrast, CS is a mathematical theory focused on measuring finite-dimensional vectors in $\mathbb{R}^n$. Second, CS systems typically acquire random measurements in the form of inner products between the signal and random test functions. Third, the two frameworks differ in the manner in which they deal with signal recovery, i.e. the problem of determining the original signal from the compressive measurements. In subspace sampling with linear measurements, recovery is achieved through linear processing, while CS relies on highly nonlinear methods.

In the remainder of this chapter, we review some of the important results in CS that will serve as background material for subsequent chapters. We begin in Section 11.2 with a brief overview of the relevant mathematical tools, and then survey several low-dimensional models commonly used in CS. We next focus attention on the theory and algorithms for sparse recovery in finite dimensions. In particular, in Section 11.3 we develop conditions on the sensing matrices to allow recovery of sparse vectors from a small number of measurements. Polynomial-time recovery algorithms are then presented in Section 11.4. We discuss conditions under which these methods are guaranteed to return the true underlying sparse vector in Section 11.5. Finally, we extend the results to the case in which we measure a set of vectors with a joint sparsity pattern in Section 11.6.

11.2   Sparsity models

In CS, a signal $\mathbf{x} \in \mathbb{R}^n$ is acquired by taking $m < n$ linear measurements $\mathbf{y} = \mathbf{A}\mathbf{x}$ using an $m \times n$ *CS matrix* $\mathbf{A}$. We refer to $\mathbf{y}$ as the *measurement vector*. Ideally, the matrix $\mathbf{A}$ is designed to reduce the number of measurements m as much as possible while allowing for recovery of a wide class of signals $\mathbf{x}$ from their measurement vectors $\mathbf{y}$. However, the fact that $m < n$ implies that $\mathbf{A}$ has a nonempty null space; this, in turn, means that for any particular input $\mathbf{x}_0 \in \mathbb{R}^n$, an infinite number of signals $\mathbf{x}$ will yield the same measurements $\mathbf{y}_0 = \mathbf{A}\mathbf{x}_0 = \mathbf{A}\mathbf{x}$ for the chosen CS matrix $\mathbf{A}$.

The motivation behind the design of $\mathbf{A}$ is, therefore, to allow for distinct signals $\mathbf{x}, \mathbf{x}'$ within a class of signals of interest to be uniquely identifiable from their measurements $\mathbf{y} = \mathbf{A}\mathbf{x}, \mathbf{y}' = \mathbf{A}\mathbf{x}'$, even though $m \ll n$. We must therefore choose the set of signals that we aim to recover.

As noted in Chapter 10, much of classical sampling theory is based on modeling signals by vectors in an appropriate subspace. A vector $\mathbf{x}$ in $\mathcal{W}$ can be recovered from measurements $\mathbf{y} = \mathbf{Ax}$ if the null space $\mathcal{N}(\mathbf{A})$ and $\mathcal{W}$ are disjoint. This will ensure that no vector in $\mathcal{W}$ (besides the zero vector) leads to zero measurements. Our goal here, however, is to be able to accommodate signals in higher dimensions that have a low-dimensional parametrization. The UoS class introduced in the previous chapter is one route to describing such signals. In CS we focus on the simplest example of a finite-dimensional UoS in which the signal is a sparse vector in $\mathbb{R}^n$.

Before turning to discuss sparsity in detail, we pause to briefly introduce ℓ_p-norms, which will play a key role in treating sparse vectors.

11.2.1   Normed vector spaces

The ℓ_p-norm for $p \in [1, \infty]$ of a vector $\mathbf{x} \in \mathbb{R}^n$ is defined as

$$\|\mathbf{x}\|_p = \begin{cases} \left(\sum_{i=1}^{n} |x_i|^p\right)^{\frac{1}{p}}, & p \in [1, \infty) \\ \max_{i=1,2,\ldots,n} |x_i|, & p = \infty. \end{cases} \tag{11.1}$$

In some contexts it is useful to extend the notion of ℓ_p-norms to the case where $p < 1$. In this case, the "norm" defined in (11.1) fails to satisfy the triangle inequality, so it is actually a quasinorm. We will also make frequent use of the notation $\|\mathbf{x}\|_0 = |\operatorname{supp}(\mathbf{x})|$, where $\operatorname{supp}(\mathbf{x}) = \{i \colon x_i \neq 0\}$ denotes the support of $\mathbf{x}$ and $|\operatorname{supp}(\mathbf{x})|$ is the cardinality of $\operatorname{supp}(\mathbf{x})$. Thus, $\|\mathbf{x}\|_0$ counts the number of nonzero values in $\mathbf{x}$. Note that $\| \cdot \|_0$ is not even a quasinorm, but one can easily show that $\lim_{p \to 0} \|\mathbf{x}\|_p = |\operatorname{supp}(\mathbf{x})|$, justifying this choice of notation. The ℓ_p-(quasi)norms have notably different properties for different values of p. To illustrate this, in Fig. 11.1 we show the unit sphere, i.e. $\{\mathbf{x} \colon \|\mathbf{x}\|_p = 1\}$, induced by each of these norms in $\mathbb{R}^2$.

The behavior of the different ℓ_p-norms leads to very different approximation properties when using these norms as a measure of a signal approximation error. For example, suppose we are given a vector $\mathbf{x} \in \mathbb{R}^2$ and wish to approximate it using a point in a one-dimensional affine space $\mathcal{A}$, as illustrated in Fig. 11.2. If we measure the approximation error using an ℓ_p-norm, then our task is to find the vector $\hat{\mathbf{x}} \in \mathcal{A}$ that minimizes

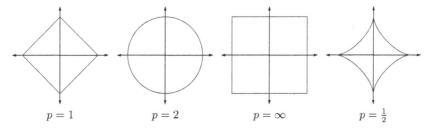

$$p = 1 \qquad p = 2 \qquad p = \infty \qquad p = \tfrac{1}{2}$$

Figure 11.1 Unit spheres in $\mathbb{R}^2$ for the ℓ_p-norms with $p = 1, 2, \infty$, and for the ℓ_p-quasinorm with $p = \tfrac{1}{2}$.

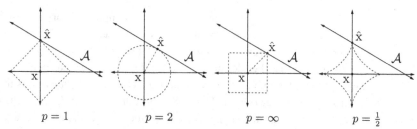

Figure 11.2 Best approximation of the point **x** at the origin by a one-dimensional affine subspace using the ℓ_p-norms for $p = 1, 2, \infty$, and the ℓ_p-quasinorm with $p = \frac{1}{2}$.

$\|\mathbf{x} - \hat{\mathbf{x}}\|_p$. The choice of p will have a significant effect on the properties of the resulting approximation error. To compute the closest point in $\mathcal{A}$ to **x** using each ℓ_p-norm, we can imagine growing an ℓ_p sphere centered on **x** until it intersects with $\mathcal{A}$. This will be the point $\hat{\mathbf{x}} \in \mathcal{A}$ that is closest to **x** in the corresponding ℓ_p-norm. We observe that larger p tends to spread out the error more evenly among the two coefficients, while smaller p leads to an error that is unevenly distributed and tends to be sparse. For $p \leq 1$ the intersection is located on the vertical axis, leading to a sparse solution. This intuition generalizes to higher dimensions, and plays an important role in the development of CS theory.

Three of the norms that are used most frequently in the development of CS are the ℓ_p-norms for $p = 1, 2, \infty$. These norms are related via the following proposition:

Proposition 11.1. *Suppose* **x** *is a vector with at most* k *nonzero values:* $\|\mathbf{x}\|_0 \leq k$. *Then*

$$\frac{\|\mathbf{x}\|_1}{\sqrt{k}} \leq \|\mathbf{x}\|_2 \leq \sqrt{k}\|\mathbf{x}\|_\infty. \tag{11.2}$$

Proof: For any **x**, $\|\mathbf{x}\|_1 = \langle \mathbf{x}\,\mathrm{sign}(\mathbf{x}) \rangle$. By applying the Cauchy–Schwarz inequality we obtain $\|\mathbf{x}\|_1 \leq \|\mathbf{x}\|_2 \|\mathrm{sign}(\mathbf{x})\|_2$, where $\mathrm{sign}(\mathbf{x})$ is the signum function defined by (3.32) for each vector element. The lower bound follows since $\mathrm{sign}(\mathbf{x})$ has at most k nonzero entries all equal to ± 1 and thus $\|\mathrm{sign}(\mathbf{x})\|_2 = \sqrt{k}$. The upper bound is obtained by observing that each of the k nonzero entries of **x** can be upper-bounded by $\|\mathbf{x}\|_\infty$. $\square$

A useful inequality that results from Proposition 11.1 is given in the following proposition.

Proposition 11.2. *Suppose* **u** *and* **v** *are orthogonal vectors. Then*

$$\|\mathbf{u}\|_2 + \|\mathbf{v}\|_2 \leq \sqrt{2}\,\|\mathbf{u} + \mathbf{v}\|_2. \tag{11.3}$$

Proof: Define the 2×1 vector $\mathbf{w} = [\|\mathbf{u}\|_2, \|\mathbf{v}\|_2]^T$. By applying Proposition 11.1 with $k = 2$, we have $\|\mathbf{w}\|_1 \leq \sqrt{2}\,\|\mathbf{w}\|_2$. From this we obtain

$$\|\mathbf{u}\|_2 + \|\mathbf{v}\|_2 \leq \sqrt{2}\sqrt{\|\mathbf{u}\|_2^2 + \|\mathbf{v}\|_2^2}. \tag{11.4}$$

Since **u** and **v** are orthogonal, $\|\mathbf{u}\|_2^2 + \|\mathbf{v}\|_2^2 = \|\mathbf{u} + \mathbf{v}\|_2^2$, which yields the desired result. $\square$

11.2.2 Sparse signal models

Signals can often be well-approximated as a linear combination of just a few elements from a known basis or frame.[1] When this representation is exact we say that the signal is *sparse*. Sparse signal models provide a mathematical framework for capturing the fact that in many cases high-dimensional signals contain relatively little information compared with their ambient dimension. Sparsity can be thought of as one incarnation of *Occam's razor* — when faced with many possible ways to describe a signal, the simplest choice is the best one.

Sparse signals

Mathematically, to define the notion of sparsity, we rely on signal representations in a given basis $\phi_i, 1 \leq i \leq n$ for $\mathbb{R}^n$. Every $\mathbf{x} \in \mathbb{R}^n$ is representable in terms of n coefficients $\theta_i, 1 \leq i \leq n$ as $\mathbf{x} = \sum_{i=1}^{n} \phi_i \theta_i$; arranging the ϕ_i as columns in the $n \times n$ matrix $\mathbf{\Phi}$ and the coefficients θ_i into the $n \times 1$ coefficient vector $\boldsymbol{\theta}$, we can write succinctly that $\mathbf{x} = \mathbf{\Phi}\boldsymbol{\theta}$, with $\boldsymbol{\theta} \in \mathbb{R}^n$. Similarly, if we use a frame $\mathbf{\Phi}$ containing L unit-norm column vectors of length n with $n < L$ (i.e. $\mathbf{\Phi} \in \mathbb{R}^{n \times L}$), then for any vector $\mathbf{x} \in \mathbb{R}^n$ there exist infinitely many decompositions $\boldsymbol{\theta} \in \mathbb{R}^L$ such that $\mathbf{x} = \mathbf{\Phi}\boldsymbol{\theta}$. We say that a signal $\mathbf{x}$ is k-*sparse* in the basis or frame $\mathbf{\Phi}$ if there exists a vector $\boldsymbol{\theta} \in \mathbb{R}^n$ with only $k \ll n$ nonzero entries such that $\mathbf{x} = \mathbf{\Phi}\boldsymbol{\theta}$. We call the set of indices corresponding to the nonzero entries the *support* of $\boldsymbol{\theta}$ and denote it by $\text{supp}(\boldsymbol{\theta})$.

A k-sparse vector can be efficiently compressed by preserving only the values and locations of its nonzero coefficients, using $\mathcal{O}(k \log_2 n)$ bits: coding the locations of each of the k nonzero coefficients requires $\log_2 n$ bits, while coding the magnitudes uses a constant amount of bits that depends on the desired precision, and is independent of n. This process is known as *transform coding*, and relies on the existence of a suitable basis or frame $\mathbf{\Phi}$ that renders signals of interest sparse or approximately sparse.

In the sequel, we typically assume that $\mathbf{\Phi}$ is chosen as the identity basis so that the signal $\mathbf{x} = \boldsymbol{\theta}$ is itself sparse. In certain cases we will explicitly introduce a different basis or frame $\mathbf{\Phi}$ that arises in a specific application of CS. We define the set Σ_k that contains all signals $\mathbf{x}$ that are k-sparse:

$$\Sigma_k = \{\mathbf{x} : \|\mathbf{x}\|_0 \leq k\}. \tag{11.5}$$

When dealing with sparse vectors the following notation will be very useful: let $\Lambda \subset \{1, 2, \ldots, n\}$ be a subset of indices and define $\Lambda^c = \{1, 2, \ldots, n\} \backslash \Lambda$ as the complement of Λ. The notation $\mathbf{x}_\Lambda$ will typically mean the length n vector obtained from setting the entries of $\mathbf{x}$ indexed by Λ^c to zero. Similarly, $\mathbf{A}_\Lambda$ typically denotes the $m \times n$ matrix obtained by setting the columns of $\mathbf{A}$ indexed by Λ^c to zero.[2]

Sparsity is a highly nonlinear model, since the choice of which dictionary elements are used can change from signal to signal. As we saw for general union models, a linear

---

[1] In the sparse approximation literature, it is common for a basis or frame to be referred to as a *dictionary* or *overcomplete dictionary* respectively, with the dictionary elements being called *atoms*.

[2] We note that this notation will occasionally be abused to refer to the length $|\Lambda|$ vector obtained by keeping only the entries corresponding to Λ or the $m \times |\Lambda|$ matrix obtained by keeping the columns corresponding to Λ respectively. The usage should be clear from the context.

combination of two k-sparse vectors is in general no longer k-sparse, since the supports of the individual signals may not coincide. That is, for any $\mathbf{x}, \mathbf{z} \in \Sigma_k$, we do not necessarily have that $\mathbf{x} + \mathbf{z} \in \Sigma_k$ (although it is true that $\mathbf{x} + \mathbf{z} \in \Sigma_{2k}$).

Sparse representations have long been exploited in signal processing and approximation theory for tasks such as compression [190, 199, 200] and denoising [201], and in statistics and learning theory as a method for avoiding overfitting [181]. Sparsity also figures prominently in model selection and statistical estimation theory [202], and has been exploited heavily in image processing tasks, since the multiscale wavelet transform [59] provides nearly sparse representations for many natural images.

Example 11.1 The wavelet transform consists of recursively dividing an input into its low- and high-frequency content, as we have seen in Chapter 4. The lowest-frequency components provide a coarse-scale approximation of the signal, while higher-frequency coefficients fill in the detail and resolve edges.

For a signal $x[n]$, the high-pass and low-pass discrete wavelet transform (DWT) coefficients are given by

$$y_{\text{high}}[k] = \sum_{n} x[n]g[2k - n], \qquad (11.6)$$

$$y_{\text{low}}[k] = \sum_{n} x[n]h[2k - n], \qquad (11.7)$$

where $g[n]$ and $h[n]$ are the impulse responses of the high-pass and low-pass wavelet filters, respectively. The above decomposition can be repeated recursively on the low-pass series for further decomposition.

Most natural images are characterized by large smooth or textured regions and relatively few sharp edges. Signals with this structure are known to be nearly sparse when represented using a multiscale wavelet transform [59]. An example, taken from [203], can be seen in Fig. 11.3, where we depict the wavelet transform of a natural image (for more details on the decomposition see Section 4.4.2). As can be seen, most coefficients are very small.

Sparsity has also been extensively put to use in audio signal processing, especially for lossy compression tasks. In particular, the discrete cosine transform (DCT) [204] is used in many audio/video standards such as the popular MPEG, MP3, and Dolby AAC coding standards, since audio signals are known to be sparsely represented in the DCT domain.

Example 11.2 The DCT transforms a finite discrete-time sequence into a sum of cosine functions oscillating at different frequencies. The DCT-2, proposed in [204] and used in many applications, is given by the following formula:

$$y[k] = w[k] \sum_{n=1}^{N} x[n] \cos\left(\frac{\pi(2n + 1)k}{2N}\right), \quad k = 0, \dots, N - 1, \qquad (11.8)$$

Figure 11.3 Sparse representation of an image via a multiscale wavelet transform. (a) Original image. (b) Wavelet representation. Large coefficients are represented by light pixels, while small coefficients are represented by dark pixels. Observe that most of the wavelet coefficients are close to zero.

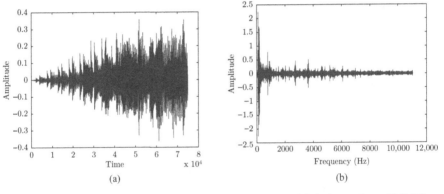

Figure 11.4 Sparse representation of an audio signal via DCT. (a) Original waveform. (b) DCT representation. Most of the DCT coefficients are close to zero.

where

$$w[k] = \begin{cases} \sqrt{\frac{1}{N}}, & k = 0 \\ \sqrt{\frac{2}{N}}, & 1 \leq k \leq N - 1, \end{cases} \tag{11.9}$$

and N is the length of $x(n)$. An example of an audio signal and its DCT transform is shown in Fig. 11.4. Evidently, most of the DCT coefficients are close to zero.

Compressible signals

Few real-world signals are truly sparse; rather they are often *compressible*, meaning they are well-approximated by a sparse signal. This can also be seen in Figs. 11.3 and 11.4

where we showed the wavelet and DCT transforms of an image and audio signal. While there are a relatively small number of dominant coefficients, there are many transform components that are small but not exactly equal to zero.

We can quantify compressibility of $\mathbf{x}$ by calculating the error incurred when approximating it by some $\hat{\mathbf{x}} \in \Sigma_k$:

$$\sigma_k(\mathbf{x})_p = \min_{\hat{\mathbf{x}} \in \Sigma_k} \|\mathbf{x} - \hat{\mathbf{x}}\|_p. \tag{11.10}$$

If $\mathbf{x} \in \Sigma_k$, then clearly $\sigma_k(\mathbf{x})_p = 0$. It is easy to see that thresholding, namely keeping only the k largest coefficients of $\mathbf{x}$, results in the optimal approximation as measured by (11.10) for all ℓ_p-norms (see Exercise 2). A bound on the resulting approximation error is given in the following proposition:

Proposition 11.3. *The approximation error $\sigma_k(\mathbf{x})_p$ is bounded above by $\sigma_k(\mathbf{x})_p \le \|\mathbf{x}\|_q k^{-r}$, for all $q \in [1, \infty)$ where $r = 1/q - 1/p \ge 0$.*

Proof: Let Λ denote the set of indices corresponding to the k largest entries of $\mathbf{x}$, and let ε denote the size of the smallest entry in Λ. Then,

$$\sigma_k(\mathbf{x})_p^p = \sum_{i \notin \Lambda} |x_i|^p = \sum_{i \notin \Lambda} |x_i|^{p-q} |x_i|^q \le \varepsilon^{p-q} \sum_{i \notin \Lambda} |x_i|^q \le \varepsilon^{p-q} \|\mathbf{x}\|_q^q. \tag{11.11}$$

Now,

$$\|\mathbf{x}\|_q^q \ge \sum_{i \in \Lambda} |x_i|^q \ge k\varepsilon^q, \tag{11.12}$$

since $|\Lambda| = k$. Therefore,

$$\varepsilon \le k^{-1/q} \|\mathbf{x}\|_q. \tag{11.13}$$

Substituting (11.13) into (11.11) we have

$$\sigma_k(\mathbf{x})_p \le k^{1/p-1/q} \|\mathbf{x}\|_q^{1-q/p} \|\mathbf{x}\|_q^{q/p} = k^{-r} \|\mathbf{x}\|_q, \tag{11.14}$$

completing the proof. □

It follows from Proposition 11.3 that a signal is well approximated by a sparse representation if its ℓ_q-norm is small. In particular, we would like to ensure that $\|\mathbf{x}\|_q \le C$ for some constant C. One example is when the coefficients obey a power-law decay. Specifically, if we sort the coefficients x_i such that $|x_1| \ge |x_2| \ge \cdots \ge |x_n|$, then we say that the coefficients obey a power-law decay if there exist constants $C_1, q > 0$ such that

$$|x_i| \le C_1 i^{-q}. \tag{11.15}$$

The larger q is, the faster the magnitudes decay, and the more compressible the signal. For such signals there exist constants $C_2, r > 0$ depending only on C_1 and q such that

$$\sigma_k(\mathbf{x})_2 \le C_2 k^{-r}, \tag{11.16}$$

(see Exercise 3). In fact, one can show that $\sigma_k(\mathbf{x})_2$ decays as k^{-r} if and only if the sorted coefficients x_i decay as $i^{-r+1/2}$ [190].

(a)   (b)

Figure 11.5 Sparse approximation of a natural image. (a) Original image. (b) Approximation obtained by keeping only the largest 10% of the wavelet coefficients.

Given a compressible signal, namely a signal whose appropriate ℓ_q-norm is small, we can approximate it by setting the small coefficients to zero, or *thresholding* the coefficients, to obtain a k-sparse representation. It follows from (11.10) that, when measuring the error using an ℓ_p-norm, this procedure yields the best k-term approximation, i.e. the best signal expansion using only k basis elements.[3] To illustrate the effect of thresholding, we revisit Examples 11.1 and 11.2 and recover the original signals after thresholding their respective representations.

Example 11.3 Let us return to Example 11.1. To obtain a sparse representation of the image, we threshold the wavelet coefficients so as to keep only the largest 10% of the values and zero out the rest. We then perform an inverse transform of the result, which leads to the approximation shown in Fig. 11.5. As can be seen, the original and approximate images are very similar, owing to the compressibility of the original image.

We now repeat the thresholding procedure on the audio signal of Example 11.2. Figure 11.6 shows an example of the best k-term approximation of the audio signal in Fig. 11.4 when keeping the largest 10% of the DCT coefficients. The original signal and the approximated one sound exactly the same.

Thresholding of sparse coefficients is the heart of nonlinear approximation [190] – nonlinear because the choice of which coefficients to keep in the approximation depends on the signal itself. This same thresholding operation also serves as an effective

---

[3] Thresholding yields the best k-term approximation with respect to an orthonormal basis. When redundant frames are used, we must rely on sparse approximation algorithms like those described in Section 11.4.2 [59].

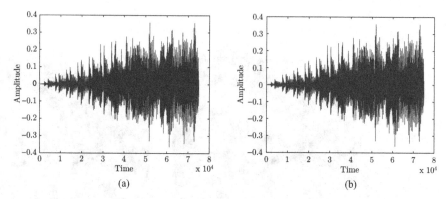

Figure 11.6 Sparse approximation of an audio signal. (a) Original waveform. (b) Approximation obtained by keeping only the largest 10% of the DCT coefficients.

method for rejecting certain common types of noise, which typically do not have sparse transforms [201].

Model-based CS

In many settings, the support of $\mathbf{x}$ obeys additional constraints which can be taken into account. Using structured sparsity models that go beyond simple sparsity may boost the performance of standard sparse recovery algorithms. Broadly speaking, such constraints can be deterministic or probabilistic.

Two typical examples for deterministic models are wavelet trees [205, 206] and block sparsity [183, 207]. The first accounts for the fact that the large wavelet coefficients of piecewise smooth signals and images tend to lie on a rooted, connected tree structure. The second model is based on the assumption that the signal exhibits special structure in the form of the nonzero coefficients occurring in clusters. This is a special case of the more general UoS model which we discuss in detail in Chapter 12. In [206] the authors propose a general framework for structured sparse recovery and demonstrate how both block sparsity and wavelet trees can be merged into standard sparse recovery algorithms.

An alternative route to model-based CS is to place a prior distribution on the support. The most simple prior assumes that the entries of the sparsity pattern are independent and identically distributed (iid) (see e.g. [208]). However, in practice, atoms in the dictionary are often not used with the same frequency. To account for this behavior, we can assign different probabilities to be turned "on" for each entry. In addition, real-life signals often exhibit significant connections between the atoms in the dictionary used for their synthesis. For example, it is well known that when image patches are represented using the DCT or a wavelet transform, the locations of the large coefficients are strongly correlated. In [209] the authors consider general dependency models based on undirected graphs, which are also referred to as Markov random fields, and focus on the special case of a Boltzmann machine. Using this prior, several recovery algorithms are developed based on statistical objectives that take the signal structure into account. Similar ideas are also studied in [210, 211].

11.2.3   Low-rank matrix models

Another model closely related to sparsity is the set of low-rank matrices:

$$\mathcal{L} = \{\mathbf{M} \in \mathbb{R}^{n_1 \times n_2} : \mathrm{rank}(\mathbf{M}) \leq r\}.$$

One way to characterize such matrices is via their SVD. A rank-r matrix can be written as $\mathbf{M} = \sum_{k=1}^{r} \sigma_k \mathbf{u}_k \mathbf{v}_k^H$ where $\sigma_1, \sigma_2, \ldots, \sigma_r > 0$ are the nonzero singular values, and $\mathbf{u}_1, \mathbf{u}_2, \ldots, \mathbf{u}_r \in \mathbb{R}^{n_1}$, $\mathbf{v}_1, \mathbf{v}_2, \ldots, \mathbf{v}_r \in \mathbb{R}^{n_2}$ are the corresponding singular vectors. Rather than constraining the number of elements used to construct the signal, we are limiting the number of nonzero singular values. By counting the number of free parameters in the SVD, and taking into account the orthogonality and normalization constraints on $\mathbf{u}_i$ and $\mathbf{v}_i$, it can be shown that the set $\mathcal{L}$ has $r(n_1 + n_2 - r)$ degrees of freedom (see Exercise 4). For small r this is significantly less than the number of entries $n_1 n_2$ in the matrix.

Low-rank matrices arise in a variety of practical settings. For example, low-rank (Hankel) matrices correspond to low-order LTI systems [212]. In many data-embedding problems, such as sensor geolocation, the matrix of pairwise distances will typically have rank 2 or 3 [213]. Matrices that are well-approximated by a low-rank matrix also arise naturally in the context of collaborative filtering systems such as the now-famous Netflix recommendation system [214] and the related problem of *matrix completion*, where a low-rank matrix is recovered from a small sample of its entries [215, 216]. Another area in which low-rank matrices appear is in the context of semidefinite relaxations of nonconvex optimization problems [217]. Such relaxation techniques can be combined with sparsity priors to treat a variety of interesting problems. As an example, this approach has recently been applied to sparse phase retrieval in optics and image processing [218, 219, 220].

While we do not consider matrix completion or the more general problem of low-rank matrix recovery in this book, we note that many of the concepts and tools developed for CS are highly relevant to this emerging field.

11.3   Sensing matrices

The main design criterion for the CS matrix $\mathbf{A} \in \mathbb{R}^{m \times n}$ is to enable unique identification of a sparse signal $\mathbf{x}$ from its measurements $\mathbf{y} = \mathbf{A}\mathbf{x}$. The matrix $\mathbf{A}$ represents a *linear dimensionality reduction*, i.e. it maps $\mathbb{R}^n$, where n is generally large, into $\mathbb{R}^m$, with m typically much smaller than n. Each sample y_ℓ can be written as $y_\ell = \langle \mathbf{b}_\ell, \mathbf{x} \rangle$ where $\mathbf{b}_\ell$ is the ℓth row of $\mathbf{A}$, or the ℓth column of $\mathbf{B} = \mathbf{A}^*$. Written this way, the measurements have the familiar inner product form we have been discussing throughout the book. Our goal is to push the number of rows m as close as possible to the sparsity level k in order to perform as much signal "compression" during acquisition as possible.

Although the standard CS framework assumes that $\mathbf{x}$ is a finite-length vector with a discrete-valued index (such as time or space), in practice we are often interested in designing measurement systems for acquiring continuously indexed signals such as continuous-time signals or images. It is sometimes possible to extend this model to continuously indexed signals using an intermediate discrete representation. In Chapters 13–15 we will see how the underlying concepts of CS can be generalized to treat certain classes of analog signals. For now we will simply think of $\mathbf{x}$ as a finite-length window of Nyquist-rate samples, and we temporarily ignore the issue of how to directly acquire compressive measurements without first sampling at the Nyquist rate.

Clearly, when we consider the class of k-sparse signals Σ_k, the number of measurements must obey $m \geq k$ for any matrix design, since the identification problem has k unknowns even when the support $\Lambda = \text{supp}(\mathbf{x})$ of $\mathbf{x}$ is provided. Suppose first that Λ is known. In this case we can reduce the measurement equation to:

$$\mathbf{y} = \mathbf{A}\mathbf{x} = \mathbf{A}_\Lambda \mathbf{x}_\Lambda, \tag{11.17}$$

where we restrict the matrix $\mathbf{A}$ to its columns corresponding to the indices in Λ, and the vector $\mathbf{x}$ to the set of indices Λ. Note that $\mathbf{A}_\Lambda \in \mathbb{R}^{m \times k}$ and $\mathbf{x}_\Lambda \in \mathbb{R}^k$. If $\mathbf{A}_\Lambda$ has full column-rank (i.e. all columns are linearly independent), then it is left-invertible and $\mathbf{A}_\Lambda^\dagger \mathbf{A}_\Lambda = \mathbf{I}$ (see Appendix A). In this case,

$$\mathbf{x}_\Lambda = \mathbf{A}_\Lambda^\dagger \mathbf{y}. \tag{11.18}$$

In order for $\mathbf{A}_\Lambda$ to have full column-rank, we must have that $m \geq k$. Note that when Λ is known, the problem reduces to that of recovering a signal that lies in a given subspace: the subspace spanned by the columns of the identity corresponding to these known locations.

The more challenging problem is when the locations are unknown, leading to a non-linear signal model. Two main theoretical questions at the heart of CS are: first, how should we design the sensing matrix $\mathbf{A}$ to ensure that it preserves the information in the signal $\mathbf{x}$ and how many measurements are needed? Second, how can we recover the original signal $\mathbf{x}$ from measurements $\mathbf{y}$? As we will show, in the case where our data are sparse or compressible, we can design matrices $\mathbf{A}$ with $m \ll n$ that enable signal recovery accurately and efficiently using a variety of practical algorithms.

We begin by addressing the question of how to design the sensing matrix $\mathbf{A}$. Rather than directly proposing a design procedure, we instead consider a number of desirable properties that we might wish $\mathbf{A}$ to have. In Section 11.3.5 we discuss how to construct matrices satisfying these properties, with high probability.

11.3.1   Null space conditions

A natural place to begin is by considering the null space of $\mathbf{A}$:

$$\mathcal{N}(\mathbf{A}) = \{\mathbf{z}: \mathbf{A}\mathbf{z} = 0\}. \tag{11.19}$$

If we wish to be able to recover all sparse signals $\mathbf{x}$ from the measurements $\mathbf{A}\mathbf{x}$, then it is immediately clear that for any pair of distinct vectors $\mathbf{x}, \mathbf{x}' \in \Sigma_k$, we must have $\mathbf{A}\mathbf{x} \neq \mathbf{A}\mathbf{x}'$, since otherwise it would be impossible to distinguish $\mathbf{x}$ from $\mathbf{x}'$ based solely on the measurements $\mathbf{y}$. More formally, observe that if $\mathbf{A}\mathbf{x} = \mathbf{A}\mathbf{x}'$ then $\mathbf{A}(\mathbf{x} - \mathbf{x}') = 0$ with $\mathbf{x} - \mathbf{x}' \in \Sigma_{2k}$, since $\mathbf{x} - \mathbf{x}'$ can have at most $2k$ nonzero values, when there is no overlap in the supports of $\mathbf{x}$ and $\mathbf{x}'$. Therefore, $\mathbf{A}$ uniquely represents all $\mathbf{x} \in \Sigma_k$ if and only if $\mathcal{N}(\mathbf{A})$ does not contain any vectors in Σ_{2k}.

Spark condition

While there are many equivalent ways of characterizing the above null space property, one of the most common is known as the *spark*, also referred to in the tensor product literature as the *Kruskal rank* [221].

Definition 11.1. *The spark of a given matrix* $\mathbf{A}$ *is the smallest number of columns of* $\mathbf{A}$ *that are linearly dependent.*

Note that the spark is defined very differently than the rank of $\mathbf{A}$. While the rank measures the maximal number of columns that are linearly independent, the spark measures the minimal number of columns that are linearly dependent. Thus, if $\text{rank}(\mathbf{A}) = r$, then this implies that there exists a set of r columns that are linearly independent. However, the matrix can still have a set of $s \leq r$ columns that are linearly dependent, in which case $\text{spark}(\mathbf{A}) \leq s$.

It is easy to see that for a matrix $\mathbf{A} \in \mathbb{R}^{m \times n}$ with $m \leq n$, $\text{spark}(\mathbf{A}) \in [2, m+1]$. The lower bound follows from the obvious fact that we need two columns to have linear dependence. To obtain the upper bound we note that since the columns of $\mathbf{A}$ are in $\mathbb{R}^m$, any set of $m + 1$ columns will necessarily be linearly dependent.

From the definition of the spark it follows that an invertible $m \times m$ matrix has spark equal to $m + 1$. This is because for an invertible size-m matrix, any set of m or less columns must be linearly independent. However, it is not true in general that an $m \times n$ matrix with full row-rank will have spark equal to $m+1$, as we show in the next example.

Example 11.4 We illustrate the difference between linear independence and spark by considering an example of a full row-rank matrix that does not have full spark, namely, its spark is less than $m + 1$.

Consider the 4×5 matrix:

$$\mathbf{A} = \begin{bmatrix} 1 & 1 & 0 & 0 & 0 \\ 0 & 0 & 1 & 0 & 0 \\ 0 & 0 & 0 & 1 & 0 \\ 0 & 0 & 0 & 0 & 1 \end{bmatrix}. \tag{11.20}$$

It is easy to see that $\text{rank}(\mathbf{A}) = 4$, namely $\mathbf{A}$ has full row-rank. However, the first two columns of $\mathbf{A}$ are identical, and thus linearly dependent. Therefore $\text{spark}(\mathbf{A}) = 2$, and $\mathbf{A}$ does not have full spark.

In contrast to the previous example, random matrices with full row-rank have full spark with probability 1.

Example 11.5 Consider an $m \times n$ real random matrix $\mathbf{A}$ with iid entries. From [222], a real square random matrix with iid elements is invertible with probability 1. Therefore, any $m \times m$ submatrix of $\mathbf{A}$ is invertible with probability 1, and consequently $\text{spark}(\mathbf{A}) = m + 1$ with probability 1.

Another example of a full-spark matrix is given next.

Example 11.6 In this example we consider a Vandermonde matrix, which appears in many different signal processing applications. We will encounter this class of matrices in Chapter 15 when treating sampling of pulse streams.

Consider an $m \times n$ Vandermonde matrix with $m \leq n$ and distinct α_j's:

$$
\mathbf{V} = \begin{bmatrix}
1 & 1 & \cdots & 1 \\
\alpha_1 & \alpha_2 & \cdots & \alpha_n \\
\alpha_1^2 & \alpha_2^2 & \cdots & \alpha_n^2 \\
\vdots & \vdots & \ddots & \vdots \\
\alpha_1^{m-1} & \alpha_2^{m-1} & \cdots & \alpha_n^{m-1}
\end{bmatrix}. \tag{11.21}
$$

We now show that this matrix has full spark $m + 1$. To this end, form an $m \times m$ matrix by choosing any set of m columns of $\mathbf{V}$. By definition, this will result in an $m \times m$ Vandermonde matrix with roots given by a subset of the α_j's. It is well known that a square Vandermonde matrix is invertible as long as its roots are distinct [223]. Therefore, any subset of m columns from $\mathbf{V}$ will lead to an invertible Vandermonde matrix, asserting that any m columns of $\mathbf{V}$ are linearly independent.

Using the spark, we can now pose the following straightforward uniqueness guarantee.

Theorem 11.1 (Uniqueness of sparse recovery).   *For any vector* $\mathbf{y} \in \mathbb{R}^m$, *there exists at most one signal* $\mathbf{x} \in \Sigma_k$ *such that* $\mathbf{y} = \mathbf{Ax}$ *if and only if* $\mathrm{spark}(\mathbf{A}) > 2k$. *In particular, for uniqueness we must have that* $m \geq 2k$.

Proof: We first assume that, for any $\mathbf{y} \in \mathbb{R}^m$, there exists at most one signal $\mathbf{x} \in \Sigma_k$ such that $\mathbf{y} = \mathbf{Ax}$. Now suppose for the sake of contradiction that $\mathrm{spark}(\mathbf{A}) \leq 2k$. This means that there exists some set of at most $2k$ columns that are linearly dependent, which in turn implies that there exists an $\mathbf{h} \in \mathcal{N}(\mathbf{A})$ such that $\mathbf{h} \in \Sigma_{2k}$. Since $\mathbf{h} \in \Sigma_{2k}$ we can write $\mathbf{h} = \mathbf{x} - \mathbf{x}'$ for some $\mathbf{x}, \mathbf{x}' \in \Sigma_k$ with $\mathbf{x} \neq \mathbf{x}'$. Using the fact that $\mathbf{h} \in \mathcal{N}(\mathbf{A})$ we have $\mathbf{A}(\mathbf{x} - \mathbf{x}') = 0$ so that $\mathbf{Ax} = \mathbf{Ax}'$. But this contradicts our assumption that there exists at most one signal $\mathbf{x} \in \Sigma_k$ such that $\mathbf{y} = \mathbf{Ax}$. Therefore, we must have $\mathrm{spark}(\mathbf{A}) > 2k$.

Now suppose that $\mathrm{spark}(\mathbf{A}) > 2k$. Assume that for some $\mathbf{y}$ there exist $\mathbf{x}, \mathbf{x}' \in \Sigma_k$ such that $\mathbf{y} = \mathbf{Ax} = \mathbf{Ax}'$. We therefore have that $\mathbf{A}(\mathbf{x} - \mathbf{x}') = \mathbf{Ah} = 0$, where $\mathbf{h} = \mathbf{x} - \mathbf{x}'$. Since $\mathrm{spark}(\mathbf{A}) > 2k$, all sets of up to $2k$ columns of $\mathbf{A}$ are linearly independent, and therefore $\mathbf{h} = 0$. This in turn implies $\mathbf{x} = \mathbf{x}'$.

Finally we note that since $\mathrm{spark}(\mathbf{A}) \leq m + 1$, the requirement $\mathrm{spark}(\mathbf{A}) > 2k$ immediately leads to $m \geq 2k$. $\square$

Example 11.7 Theorem 11.1 implies that unique recovery of all vectors with sparsity level k is possible as long as $k < \mathrm{spark}(\mathbf{A})/2$. Therefore, in the examples in which $\mathbf{A}$ has full spark, we can recover any signal with sparsity $k < (m + 1)/2$. In Section 11.4 we will describe several concrete recovery algorithms.

In Example 11.4 we considered a matrix with spark($\mathbf{A}$) $= 2$. Therefore, the condition on k becomes $k < 1$, implying that there is no sparsity level at which we can guarantee recovery. Indeed, suppose that $k = 1$ so that the input $\mathbf{x}$ has only one nonzero value. If the nonzero value is at location $i = 1$, then $\mathbf{y}$ is equal to the first column of $\mathbf{A}$. But this column is identical to the second column so that we cannot distinguish between the true input, and a one-sparse vector with a nonzero value in its second element.

Null space property

The spark provides a complete characterization of when recovery of exactly sparse vectors is possible via Theorem 11.1. However, for approximately sparse signals we must consider somewhat more restrictive conditions on the null space of $\mathbf{A}$ [224]. Roughly speaking, we have to ensure that $\mathcal{N}(\mathbf{A})$ does not contain any vectors that are too compressible in addition to vectors that are sparse. This property is captured by the *null space property (NSP)* defined next. In Theorem 11.2 below we show how the NSP is related to recovery of compressible signals.

Definition 11.2. *A matrix $\mathbf{A}$ satisfies the null space property (NSP) of order k if there exists a constant $C > 0$ such that*

$$\|\mathbf{h}_\Lambda\|_2 \leq C \frac{\|\mathbf{h}_{\Lambda^c}\|_1}{\sqrt{k}} \tag{11.22}$$

for all $\mathbf{h} \in \mathcal{N}(\mathbf{A})$ and all Λ with $|\Lambda| \leq k$.

Note that instead of requiring (11.22) to hold for all Λ of size k, it is sufficient that (11.22) is satisfied for Λ corresponding to the k largest values (in magnitude) of $\mathbf{h}$. This follows from the fact that for any other choice of Λ' of size k, $\|\mathbf{h}_{\Lambda'}\|_2 \leq \|\mathbf{h}_\Lambda\|_2$, and $\|\mathbf{h}_{\Lambda^c}\|_1 \leq \|\mathbf{h}_{\Lambda'^c}\|_1$.

Condition (11.22) is also referred to in the literature as the *mixed-norm NSP*. Some references define the (standard) NSP by $\|\mathbf{h}_\Lambda\|_1 \leq C\|\mathbf{h}_{\Lambda^c}\|_1$. Since from Proposition 11.1, $\|\mathbf{x}\|_1 \leq \sqrt{k}\|\mathbf{x}\|_2$ for any $\mathbf{x} \in \mathbb{R}^k$, the standard NSP is implied by (11.22).

Example 11.8 Consider the following matrix:

$$\mathbf{A} = \begin{bmatrix} 1 & 0 & 0 & 1 & 0 \\ 0 & 1 & 0 & 1 & 0 \\ 0 & 0 & 1 & 1 & 0 \\ 0 & 0 & 0 & 0 & 1 \end{bmatrix}. \tag{11.23}$$

We will show that $\mathbf{A}$ satisfies the NSP of order $k \leq 3$ but does not satisfy it for any $k > 3$.

It is easy to see that the null space of $\mathbf{A}$ has dimension equal to 1 and is spanned by the vector $\tilde{\mathbf{h}} = [1\ 1\ 1\ -1\ 0]^T$. Thus, any $\mathbf{h} \in \mathcal{N}(\mathbf{A})$ has the form $\mathbf{h} = [\alpha\ \alpha\ \alpha\ -\alpha\ 0]^T$, $\alpha \in \mathbb{R}$. Since $\|\mathbf{h}\| = |\alpha|\|\tilde{\mathbf{h}}\|$, it is sufficient to show that the

NSP (11.22) is satisfied for $\tilde{\mathbf{h}}$. In addition, we consider only the subset of indices Λ corresponding to the k largest magnitude values of $\tilde{\mathbf{h}}$.

For $k = 1$,

$$\|\tilde{\mathbf{h}}_\Lambda\|_2 = 1 \text{ and } \|\tilde{\mathbf{h}}_{\Lambda^c}\|_1 = 3. \tag{11.24}$$

Therefore, (11.22) holds with $C \geq \frac{1}{3}$. Similarly, it is easy to see that $\tilde{\mathbf{h}}$ satisfies (11.22) for $k = 2$ and $k = 3$, with $C \geq 1$ and $C \geq 3$, respectively. When $k > 3$,

$$\|\tilde{\mathbf{h}}_\Lambda\|_2 > 0 \text{ and } \|\tilde{\mathbf{h}}_{\Lambda^c}\|_1 = 0. \tag{11.25}$$

Therefore, $\tilde{\mathbf{h}}$ does not satisfy the NSP of order k for any $k > 3$.

The NSP quantifies the notion that vectors in the null space of $\mathbf{A}$ should not be too concentrated on a small subset of indices. It is straightforward to see that if a matrix $\mathbf{A}$ satisfies the NSP of order $2k$, then spark$(\mathbf{A}) > 2k$, which from Theorem 11.1 implies that there is at most one k-sparse vector $\mathbf{x}$ such that $\mathbf{y} = \mathbf{Ax}$.

Proposition 11.4. *Suppose that $\mathbf{A}$ satisfies the NSP of order $2k$. Then,* spark$(\mathbf{A}) > 2k$ *and there exists at most one signal $\mathbf{x} \in \Sigma_k$ satisfying $\mathbf{y} = \mathbf{Ax}$.*

Proof: To show that spark$(\mathbf{A}) > 2k$ we need to prove that every $2k$ columns of $\mathbf{A}$ are linearly independent. In other words, if $\mathbf{Ah} = \mathbf{0}$ for some $\mathbf{h} \in \Sigma_{2k}$, then $\mathbf{h} = \mathbf{0}$.

Let $\mathbf{h}$ be an arbitrary vector in Σ_{2k} with $\mathbf{Ah} = \mathbf{0}$, and let $\Lambda = \text{supp}(\mathbf{h})$. Then, $\mathbf{h}_{\Lambda^c} = \mathbf{0}$. Since $\mathbf{A}$ satisfies the NSP of order $2k$, we have from (11.22) that

$$\|\mathbf{h}_\Lambda\|_2 \leq C \frac{\|\mathbf{h}_{\Lambda^c}\|_1}{\sqrt{k}} = 0, \tag{11.26}$$

and $\mathbf{h} = \mathbf{0}$. $\qquad\square$

Example 11.9 We illustrate Proposition 11.4 by returning to Example 11.8. The matrix in that example was shown to satisfy the NSP of order 2. We next compute its spark. It is easy to see that any two or three columns of $\mathbf{A}$ are linearly independent. However, the first four columns of $\mathbf{A}$ are linearly dependent. Therefore spark$(\mathbf{A}) = 4 > 2$, verifying Proposition 11.4.

Now, suppose we obtain the following measurement vector $\mathbf{y} = [1\,1\,1\,0]^T$. One way to recover $\mathbf{x}$ is to check every possible support set and try to invert the resulting system of equations. That is, for increasing k, starting from $k = 1$, we assume $\mathbf{x} \in \Sigma_k$ and consider every possible support Λ of size k. We reduce $\mathbf{x}$ to $\mathbf{x}_\Lambda$, invert the reduced system, and then check whether $\mathbf{y} = \mathbf{A}_\Lambda \mathbf{x}_\Lambda$. In our example, the equation $\mathbf{y} = \mathbf{Ax}$ has a unique solution in Σ_1, $\mathbf{x} = [0\,0\,0\,1\,0]^T$. Here the sparsity of $\mathbf{x}$ ($k = 1$) satisfies the condition of Theorem 11.1.

Consider next the vector $\mathbf{x} = [-1\,0\,0\,1\,0]^T$. Its sparsity ($k = 2$) does not satisfy the requirement of Theorem 11.1. For this choice of $\mathbf{x}$, $\mathbf{y} = \mathbf{Ax} = [0\,1\,1\,0]^T$. It is easy to see that we also have $\mathbf{y} = \mathbf{A}\mathbf{x}'$ where $\mathbf{x}' = [0\,1\,1\,0\,0]^T$, so that the solution is not unique.

Robust signal recovery

To fully illustrate the implications of the NSP in the context of sparse recovery, we now briefly discuss how we measure the performance of sparse recovery algorithms when dealing with general nonsparse $\mathbf{x}$. Towards this end, let $\Delta\colon \mathbb{R}^m \to \mathbb{R}^n$ represent a specific recovery method. We focus primarily on guarantees of the form

$$\|\Delta(\mathbf{A}\mathbf{x}) - \mathbf{x}\|_2 \le C\frac{\sigma_k(\mathbf{x})_1}{\sqrt{k}} \tag{11.27}$$

for all $\mathbf{x}$, where $\sigma_k(\mathbf{x})_1$ is defined in (11.10). This ensures exact recovery of all possible k-sparse signals (since for such signals $\sigma_k(\mathbf{x})_1 = 0$), but also implies a degree of robustness to nonsparse signals that directly depends on how well the signals are approximated by k-sparse vectors. Such guarantees are called *instance-optimal* since they ensure optimal performance for each instance of $\mathbf{x}$ [224]. This distinguishes them from guarantees that only hold for some subset of possible signals, such as sparse or compressible signals. These are also commonly referred to as *uniform guarantees* since they hold uniformly for all $\mathbf{x}$.

Our choice of norms in (11.27) is somewhat arbitrary. We could easily measure the reconstruction error using other ℓ_p-norms. The choice of p, however, limits what kinds of guarantees are possible, and also potentially leads to alternative formulations of the NSP [224]. Moreover, the form of the right-hand side of (11.27) might seem somewhat unusual in that we measure the approximation error by $\sigma_k(\mathbf{x})_1/\sqrt{k}$ rather than simply something like $\sigma_k(\mathbf{x})_2$. However, we will see in Section 11.5.3 that such a guarantee is actually not possible without taking a prohibitively large number of measurements, and that (11.27) represents the best possible guarantee one can hope to obtain. Note, that if (11.27) is satisfied, then from the left-hand inequality in Proposition 11.1 we also have

$$\|\Delta(\mathbf{A}\mathbf{x}) - \mathbf{x}\|_1 \le C\sigma_k(\mathbf{x})_1. \tag{11.28}$$

The following adaptation of a theorem in [224] demonstrates that if there exists any recovery algorithm satisfying (11.27), then $\mathbf{A}$ necessarily satisfies the NSP of order $2k$. In Section 11.5.1 (Theorem 11.12) we show that the NSP of order $2k$ is sufficient to establish a guarantee of the form (11.27) for a practical recovery algorithm (ℓ_1 minimization).

Theorem 11.2 (Stability and the NSP). *Let* $\mathbf{A}\colon \mathbb{R}^n \to \mathbb{R}^m$ *denote a sensing matrix and* $\Delta\colon \mathbb{R}^m \to \mathbb{R}^n$ *an arbitrary recovery algorithm. If the pair* $(\mathbf{A}, \Delta)$ *satisfies (11.27) then* $\mathbf{A}$ *has the NSP of order* $2k$. *Conversely, if*

$$\|\mathbf{h}\|_2 \le \frac{C}{2}\sigma_{2k}(\mathbf{h})_1 \tag{11.29}$$

for all $\mathbf{h} \in \mathcal{N}(\mathbf{A})$, *then there exists an algorithm* Δ *satisfying (11.27).*

Proof: We first assume that $\mathbf{A}$ and Δ satisfy (11.27). Suppose $\mathbf{h} \in \mathcal{N}(\mathbf{A})$ and let Λ be the indices corresponding to the $2k$ largest entries of $\mathbf{h}$. Split Λ into Λ_0 and Λ_1, where $|\Lambda_0| = |\Lambda_1| = k$. Set $\mathbf{x} = \mathbf{h}_{\Lambda_1} + \mathbf{h}_{\Lambda^c}$ and $\mathbf{x}' = -\mathbf{h}_{\Lambda_0}$, so that $\mathbf{h} = \mathbf{x} - \mathbf{x}'$. By construction $\mathbf{x}' \in \Sigma_k$, and therefore $\sigma_k(\mathbf{x}')_1 = 0$. Applying (11.27) we obtain $\mathbf{x}' = \Delta(\mathbf{A}\mathbf{x}')$. Moreover, since $\mathbf{h} \in \mathcal{N}(\mathbf{A})$, we have

$$\mathbf{A}\mathbf{h} = \mathbf{A}(\mathbf{x} - \mathbf{x}') = 0$$

so that $\mathbf{A}\mathbf{x}' = \mathbf{A}\mathbf{x}$ and $\mathbf{x}' = \Delta(\mathbf{A}\mathbf{x})$. Finally,

$$\|\mathbf{h}_\Lambda\|_2 \le \|\mathbf{h}\|_2 = \|\mathbf{x} - \Delta(\mathbf{A}\mathbf{x})\|_2 \le C\frac{\sigma_k(\mathbf{x})_1}{\sqrt{k}} = \sqrt{2}C\frac{\|\mathbf{h}_{\Lambda^c}\|_1}{\sqrt{2k}}, \tag{11.30}$$

where the second inequality follows from (11.27), and the last equality from the definition of $\mathbf{x}$. From (11.30) it follows that $\mathbf{A}$ satisfies the NSP of order $2k$.

To prove sufficiency of (11.29) we define the algorithm Δ as:

$$\Delta(\mathbf{y}) = \arg\min_{\mathbf{y}=\mathbf{A}\mathbf{z}} \min_{\mathbf{v}\in\Sigma_k} \|\mathbf{z} - \mathbf{v}\|_1 = \arg\min_{\mathbf{y}=\mathbf{A}\mathbf{z}} \sigma_k(\mathbf{z})_1. \tag{11.31}$$

Denoting $\hat{\mathbf{z}} = \Delta(\mathbf{y})$ it follows from the definition of $\Delta(\mathbf{y})$ that $\mathbf{A}\hat{\mathbf{z}} = \mathbf{y}$. Thus, for any $\mathbf{y}$, we have $\mathbf{A}\Delta(\mathbf{y}) = \mathbf{y}$. Choosing $\mathbf{y} = \mathbf{A}\mathbf{x}$ for an arbitrary $\mathbf{x}$, we conclude that $\mathbf{A}\mathbf{w} = 0$ with $\mathbf{w} = \mathbf{x} - \Delta(\mathbf{A}\mathbf{x})$. From (11.29),

$$\|\mathbf{w}\|_2 = \|\mathbf{x} - \Delta(\mathbf{A}\mathbf{x})\|_2 \le \frac{C}{2}\sigma_{2k}(\mathbf{x} - \Delta(\mathbf{A}\mathbf{x}))_1. \tag{11.32}$$

Next, we show that for any two vectors $\mathbf{x}, \mathbf{z}$, we have $\sigma_{2k}(\mathbf{x} \pm \mathbf{z})_1 \le \sigma_k(\mathbf{x})_1 + \sigma_k(\mathbf{z})_1$. Indeed,

$$\begin{aligned}
\sigma_{2k}(\mathbf{x} \pm \mathbf{z})_1 &= \min_{\mathbf{v}\in\Sigma_{2k}} \|\mathbf{x} \pm \mathbf{z} - \mathbf{v}\|_1 \\
&= \min_{\mathbf{v}_1,\mathbf{v}_2\in\Sigma_k} \|\mathbf{x} \pm \mathbf{z} - \mathbf{v}_1 - \mathbf{v}_2\|_1 \\
&\le \min_{\mathbf{v}_1\in\Sigma_k} \|\mathbf{x} - \mathbf{v}_1\| + \min_{\mathbf{v}_2\in\Sigma_k} \|\mathbf{z} \mp \mathbf{v}_2\|_1 \\
&= \sigma_k(\mathbf{x})_1 + \sigma_k(\mathbf{z})_1.
\end{aligned} \tag{11.33}$$

Substituting into (11.32),

$$\|\mathbf{w}\|_2 \le \frac{C}{2}\left[\sigma_k(\mathbf{x})_1 + \sigma_k(\Delta(\mathbf{A}\mathbf{x}))_1\right] \le C\sigma_k(\mathbf{x})_1, \tag{11.34}$$

since by definition of Δ, $\sigma_k(\Delta(\mathbf{A}\mathbf{x}))_1 \le \sigma_k(\mathbf{x})_1$. $\qquad\square$

11.3.2   The restricted isometry property

While the NSP is both necessary and sufficient for establishing guarantees of the form (11.27), these guarantees do not account for noise. If the measurements are contaminated with noise or have been corrupted by some error such as quantization, then it will be useful to consider somewhat stronger conditions.

When dealing with subspace sampling, the notion of frames was shown to be useful in determining robustness of sampling operators. In [225], Candès and Tao introduce the *restricted isometry property* (RIP), which is an extension of the notion of frames to the subspaces Σ_k:

Definition 11.3. *A matrix $\mathbf{A}$ satisfies the restricted isometry property (RIP) of order k if there exists a $\delta_k \in (0, 1)$ such that*

$$(1 - \delta_k)\|\mathbf{x}\|_2^2 \leq \|\mathbf{A}\mathbf{x}\|_2^2 \leq (1 + \delta_k)\|\mathbf{x}\|_2^2, \tag{11.35}$$

for all $\mathbf{x} \in \Sigma_k$.

If a matrix $\mathbf{A}$ satisfies the RIP of order $2k$, then we can interpret (11.35) as saying that $\mathbf{A}$ approximately preserves the distance between any pair of k-sparse vectors. To see this, choose $\mathbf{x} = \mathbf{x}_1 - \mathbf{x}_2$ where $\mathbf{x}_1, \mathbf{x}_2$ are k-sparse, so that $\mathbf{x}$ is at most $2k$-sparse. This will clearly have fundamental implications concerning robustness to noise.

It is important to note that while in our definition of the RIP we assume bounds that are symmetric about 1, this is merely for notational convenience. In practice, we could instead consider arbitrary bounds

$$\alpha\|\mathbf{x}\|_2^2 \leq \|\mathbf{A}\mathbf{x}\|_2^2 \leq \beta\|\mathbf{x}\|_2^2 \tag{11.36}$$

with $0 < \alpha \leq \beta < \infty$. This equation has the familiar form defining frames and Riesz bases. Given any such bounds, one can always scale $\mathbf{A}$ so that it satisfies the symmetric bound about one in (11.35). Specifically, multiplying $\mathbf{A}$ by $\sqrt{2/(\beta + \alpha)}$ will result in an $\widetilde{\mathbf{A}}$ for which (11.35) holds with constant $\delta_k = (\beta - \alpha)/(\beta + \alpha)$. While we will not explicitly show this, it can be verified that all of the results in this chapter based on the assumption that $\mathbf{A}$ satisfies the RIP are valid as long as there exists some scaling of $\mathbf{A}$ for which the RIP holds. We therefore lose nothing by restricting our attention to the definition in (11.35).

In general, computing RIP constants is NP-hard. Below we present some simple examples in which the RIP for small size matrices can be computed.

Example 11.10 Consider the case where $k = 1$. Let $\mathbf{x} \in \Sigma_1$ and denote by x_i the only nonzero entry of $\mathbf{x}$. The RIP property of order one becomes:

$$(1 - \delta_1)x_i^2 \leq \|\mathbf{a}_i\|_2^2 x_i^2 \leq (1 + \delta_1)x_i^2, \quad 1 \leq i \leq n, \tag{11.37}$$

where $\mathbf{a}_i$ denotes the ith column of $\mathbf{A}$. Taking $\mathbf{x} \neq \mathbf{0}$ we obtain

$$(1 - \delta_1) \leq \|\mathbf{a}_i\|_2^2 \leq (1 + \delta_1). \tag{11.38}$$

Therefore, a matrix $\mathbf{A}$ satisfies the RIP of order 1 if the norm of each of its columns is approximately equal to 1.

Example 11.11 We next treat a higher sparsity level $k = 2$, and let

$$\mathbf{A} = \begin{bmatrix} \frac{1}{\sqrt{2}} & 0 & -1 \\ -\frac{1}{\sqrt{2}} & 1 & 0 \end{bmatrix}. \tag{11.39}$$

We show below that $\mathbf{A}$ satisfies the RIP of order 2 for $\delta_2 = 1/\sqrt{2}$.

First note, that from the previous example, $\mathbf{A}$ satisfies the RIP of order 1 for every $\delta_1 \in (0, 1)$ since the columns of $\mathbf{A}$ are normalized to 1. Next, let $\mathbf{x} = [x_1 \; x_2 \; x_3]^T$ be an arbitrary vector with two nonzero values. Consider first the case where $x_3 = 0$, and define $\alpha = x_2/x_1$ ($x_1 \neq 0$, otherwise $\mathbf{x}$ would have only one nonzero value). The RIP property for $\mathbf{x}$ can then be written as

$$(1 - \delta_2)(1 + \alpha^2)x_1^2 \leq (\alpha^2 - \sqrt{2}\alpha + 1)x_1^2 \leq (1 + \delta_2)(1 + \alpha^2)x_1^2. \tag{11.40}$$

Condition (11.40) is satisfied for every $\alpha \in \mathbb{R}$ if

$$\delta_2 \geq \max_{\alpha \in \mathbb{R}} \frac{\sqrt{2}|\alpha|}{\alpha^2 + 1} = \frac{1}{\sqrt{2}}. \tag{11.41}$$

We obtain the same equation when we assume that $x_2 = 0$. Finally, suppose that $x_1 = 0$. In this case, $\mathbf{A}\mathbf{x} = [-x_3 \; x_2]^T$ and $\|\mathbf{A}\mathbf{x}\|_2^2 = \|\mathbf{x}\|_2^2$ so that (11.35) holds for any $\delta_2 \in (0, 1)$. We conclude that the RIP is satisfied for all $\mathbf{x} \in \Sigma_2$ with $\delta_2 = \frac{1}{\sqrt{2}}$.

Properties of RIP

It is easy to see that if $\mathbf{A}$ satisfies the RIP of order k with constant δ_k, then for any $k' < k$ we automatically have that $\mathbf{A}$ satisfies the RIP of order k' with constant $\delta_{k'} \leq \delta_k$.

It is also immediate that if $\mathbf{A}$ satisfies the RIP of order $2k$ for any $\delta \in (0, 1)$, then spark$(\mathbf{A}) > 2k$. This follows from the lower bound in (11.35). Indeed, let $\mathbf{x} \neq \mathbf{0}$ be an arbitrary vector in Σ_{2k}. From the RIP property we have that $\|\mathbf{A}\mathbf{x}\| > 0$ so that $\mathbf{x}$ is not in the null space of $\mathbf{A}$. This in turn implies that every $2k$ columns of $\mathbf{A}$ are linearly independent, and spark$(\mathbf{A}) > 2k$.

Another useful property that results from the RIP is given in the following proposition.

Proposition 11.5. *If $\mathbf{A}$ satisfies the RIP of order $2k$, then for any pair of vectors $\mathbf{u}, \mathbf{v} \in \Sigma_k$ with disjoint support,*

$$|\langle \mathbf{A}\mathbf{u}, \mathbf{A}\mathbf{v} \rangle| \leq \delta_{2k} \|\mathbf{u}\|_2 \|\mathbf{v}\|_2. \tag{11.42}$$

Proof: Suppose $\mathbf{u}, \mathbf{v} \in \Sigma_k$ with disjoint support and that $\|\mathbf{u}\|_2 = \|\mathbf{v}\|_2 = 1$. Then, $\mathbf{u} \pm \mathbf{v} \in \Sigma_{2k}$ and $\|\mathbf{u} \pm \mathbf{v}\|_2^2 = 2$. Using the RIP we have

$$2(1 - \delta_{2k}) \leq \|\mathbf{A}\mathbf{u} \pm \mathbf{A}\mathbf{v}\|_2^2 \leq 2(1 + \delta_{2k}). \tag{11.43}$$

Finally, applying the polarization identity (see (2.13))

$$|\langle \mathbf{A}\mathbf{u}, \mathbf{A}\mathbf{v} \rangle| \leq \frac{1}{4} \left| \|\mathbf{A}\mathbf{u} + \mathbf{A}\mathbf{v}\|_2^2 - \|\mathbf{A}\mathbf{u} - \mathbf{A}\mathbf{v}\|_2^2 \right| \leq \delta_{2k}. \tag{11.44}$$

For arbitrary $\mathbf{u}, \mathbf{v}$, define $\mathbf{u}' = \mathbf{u}/\|\mathbf{u}\|_2$, $\mathbf{v}' = \mathbf{v}/\|\mathbf{v}\|_2$. Since $\mathbf{u}', \mathbf{v}'$ have norm equal to 1, we can apply (11.44) to conclude that

$$|\langle \mathbf{A}\mathbf{u}, \mathbf{A}\mathbf{v} \rangle| = \|\mathbf{u}\|\|\mathbf{v}\| |\langle \mathbf{A}\mathbf{u}', \mathbf{A}\mathbf{v}' \rangle| \leq \delta_{2k} \|\mathbf{u}\|_2 \|\mathbf{v}\|_2, \tag{11.45}$$

establishing the proposition. $\square$

Further properties of the RIP are explored in the Exercises.

The RIP and stability

We will see in Sections 11.4.2 and 11.5.1 that the RIP is sufficient for a variety of algorithms to successfully recover a sparse signal from noisy measurements. First, however, we take a closer look at whether the RIP is actually necessary. It should be clear that the lower bound in the RIP is a necessary condition to be able to recover all sparse signals $\mathbf{x}$ from the measurements $\mathbf{Ax}$ for the same reason that the NSP is necessary. We can say even more about the necessity of the RIP by considering the following notion of stability [226].

Definition 11.4. *Let* $\mathbf{A}: \mathbb{R}^n \to \mathbb{R}^m$ *denote a sensing matrix and* $\Delta: \mathbb{R}^m \to \mathbb{R}^n$ *a recovery algorithm. We say that the pair* $(\mathbf{A}, \Delta)$ *is* C-*stable if for any* $\mathbf{x} \in \Sigma_k$ *and any* $\mathbf{e} \in \mathbb{R}^m$ *we have that*

$$\|\Delta(\mathbf{Ax} + \mathbf{e}) - \mathbf{x}\|_2 \leq C\|\mathbf{e}\|_2. \tag{11.46}$$

This definition implies that if we add a small amount of noise to the measurements, then the impact on the recovered signal is not arbitrarily large. Theorem 11.3 [226] below establishes that the existence of a stable decoding algorithm requires that $\mathbf{A}$ satisfy the lower bound of (11.35) with a constant determined by C.

Theorem 11.3 (Stability and the RIP). *If the pair* $(\mathbf{A}, \Delta)$ *is* C-*stable, then*

$$\frac{1}{C}\|\mathbf{x}\|_2 \leq \|\mathbf{Ax}\|_2 \tag{11.47}$$

for all $\mathbf{x} \in \Sigma_{2k}$.

Proof: For any $\mathbf{x} \in \Sigma_{2k}$ write $\mathbf{x} = \mathbf{v} - \mathbf{z}$ with $\mathbf{v}, \mathbf{z} \in \Sigma_k$. Define

$$\mathbf{e}_v = \frac{\mathbf{A}(\mathbf{z} - \mathbf{v})}{2} \quad \text{and} \quad \mathbf{e}_z = \frac{\mathbf{A}(\mathbf{v} - \mathbf{z})}{2}, \tag{11.48}$$

and note that

$$\mathbf{Av} + \mathbf{e}_v = \mathbf{Az} + \mathbf{e}_z = \frac{\mathbf{A}(\mathbf{v} + \mathbf{z})}{2}. \tag{11.49}$$

Let $\hat{\mathbf{x}} = \Delta(\mathbf{Av} + \mathbf{e}_v) = \Delta(\mathbf{Az} + \mathbf{e}_z)$. From the triangle inequality and the definition of C-stability, we have that

$$\begin{aligned}
\|\mathbf{v} - \mathbf{z}\|_2 &= \|\mathbf{v} - \hat{\mathbf{x}} + \hat{\mathbf{x}} - \mathbf{z}\|_2 \\
&\leq \|\mathbf{v} - \hat{\mathbf{x}}\|_2 + \|\hat{\mathbf{x}} - \mathbf{z}\|_2 \\
&\leq C\|\mathbf{e}_v\|_2 + C\|\mathbf{e}_z\|_2 \\
&= C\|\mathbf{Av} - \mathbf{Az}\|_2.
\end{aligned} \tag{11.50}$$

The last equality follows from $\mathbf{Av} - \mathbf{Az} = \mathbf{e}_z - \mathbf{e}_v$ and the fact that $\mathbf{e}_z = -\mathbf{e}_v$. Noting that $\mathbf{x} = \mathbf{v} - \mathbf{z}$ completes the proof.   □

As $C \to 1$, $\mathbf{A}$ must satisfy the lower bound of (11.35) with $\delta_{2k} = 1 - 1/C^2 \to 0$. Thus, if we desire to reduce the impact of noise in our recovered signal then we must adjust $\mathbf{A}$ so that the lower bound of (11.35) holds with a tighter constant.

One might respond to this result by arguing that since the upper bound is not necessary, we can rescale $\mathbf{A}$ so that as long as $\mathbf{A}$ satisfies the RIP with $\delta_{2k} < 1$, the rescaled version $\alpha\mathbf{A}$ will satisfy (11.47) for any constant C. In settings where the noise power is independent of our choice of $\mathbf{A}$, this is a valid point: by scaling $\mathbf{A}$ we are essentially adjusting the gain on the signal part of our measurements, and if increasing this gain does not impact the noise, then we can achieve arbitrarily high SNRs, so that eventually the noise is negligible compared to the signal. However, in practice we will typically not be able to rescale $\mathbf{A}$ to be arbitrarily large. Moreover, in many practical settings the noise is not independent of $\mathbf{A}$. For example, consider the case where e represents quantization noise produced by a quantizer with finite dynamic range. Suppose the measurements lie in the interval $[-T, T]$, and we have adjusted the quantizer to capture this range. If we rescale $\mathbf{A}$ by α, then the measurements now lie between $[-\alpha T, \alpha T]$, and we must scale the dynamic range of our quantizer by α. In this case the resulting quantization error is αe, and we have not reduced the reconstruction error.

The relationship between the RIP and the NSP

The RIP is strictly stronger than the NSP in the sense that if a matrix satisfies the RIP, then it also satisfies the NSP, as incorporated in Theorem 11.4 below. Historically, many of the proofs of results in CS have relied on the RIP rather than on the NSP, which explains its importance in the theory of CS, despite being more difficult to satisfy than the NSP. Both measures are NP-hard to compute.

Theorem 11.4 (The RIP and the NSP).   *Suppose that* $\mathbf{A}$ *satisfies the RIP of order* $2k$ *with* $\delta_{2k} < \sqrt{2} - 1$. *Then* $\mathbf{A}$ *satisfies the NSP of order* $2k$ *with constant*

$$C = \frac{2\delta_{2k}}{1 - (1 + \sqrt{2})\delta_{2k}}.$$

The proof of this theorem involves a useful proposition, stated below. This result holds for arbitrary h, not just vectors $\mathbf{h} \in \mathcal{N}(\mathbf{A})$. When $\mathbf{h} \in \mathcal{N}(\mathbf{A})$, the argument could be simplified considerably. However, this proposition will prove useful when we turn to the problem of sparse recovery from noisy measurements in Section 11.5.1, and thus we establish it now in its full generality.

Proposition 11.6.   *Suppose that* $\mathbf{A}$ *satisfies the RIP of order* $2k$, *and let* $\mathbf{h} \in \mathbb{R}^n$, $\mathbf{h} \neq \mathbf{0}$ *be arbitrary. Let* Λ_0 *be any subset of* $\{1, 2, \ldots, n\}$ *such that* $|\Lambda_0| \leq k$. *Define* Λ_1 *as the index set corresponding to the* k *entries of* $\mathbf{h}_{\Lambda_0^c}$ *with largest magnitude, and let* $\Lambda = \Lambda_0 \cup \Lambda_1$. *Then*

$$\|\mathbf{h}_\Lambda\|_2 \leq \alpha\frac{\|\mathbf{h}_{\Lambda_0^c}\|_1}{\sqrt{k}} + \beta\frac{|\langle \mathbf{Ah}_\Lambda, \mathbf{Ah}\rangle|}{\|\mathbf{h}_\Lambda\|_2}, \tag{11.51}$$

where

$$\alpha = \frac{\sqrt{2}\delta_{2k}}{1 - \delta_{2k}}, \quad \beta = \frac{1}{1 - \delta_{2k}}.$$

Proof: The proof of the proposition is based on the following lemma.

Lemma 11.1. *Let Λ_0 be an arbitrary subset of $\{1, 2, \ldots, n\}$ such that $|\Lambda_0| \leq k$. For any vector $\mathbf{u} \in \mathbb{R}^n$, define Λ_1 as the index set corresponding to the k entries of $\mathbf{u}_{\Lambda_0^c}$ with largest magnitude, Λ_2 as the index set corresponding to the next k largest entries, and so on. Then*

$$\sum_{j \geq 2} \|\mathbf{u}_{\Lambda_j}\|_2 \leq \frac{\|\mathbf{u}_{\Lambda_0^c}\|_1}{\sqrt{k}}. \tag{11.52}$$

Proof: We begin by observing that for $j \geq 2$,

$$\|\mathbf{u}_{\Lambda_j}\|_\infty \leq \frac{\|\mathbf{u}_{\Lambda_{j-1}}\|_1}{k} \tag{11.53}$$

since the Λ_j sort $\mathbf{u}$ to have decreasing magnitude. Applying Proposition 11.1 we have

$$\sum_{j \geq 2} \|\mathbf{u}_{\Lambda_j}\|_2 \leq \sqrt{k} \sum_{j \geq 2} \|\mathbf{u}_{\Lambda_j}\|_\infty \leq \frac{1}{\sqrt{k}} \sum_{j \geq 1} \|\mathbf{u}_{\Lambda_j}\|_1 = \frac{\|\mathbf{u}_{\Lambda_0^c}\|_1}{\sqrt{k}}, \tag{11.54}$$

proving the lemma. $\qquad \square$

We now prove Proposition 11.6. The key ideas in this proof follow from [227]. Since $\mathbf{h}_\Lambda \in \Sigma_{2k}$, the lower bound on the RIP immediately yields

$$(1 - \delta_{2k}) \|\mathbf{h}_\Lambda\|_2^2 \leq \|\mathbf{A}\mathbf{h}_\Lambda\|_2^2. \tag{11.55}$$

Define Λ_j as in Lemma 11.1. Since $\mathbf{A}\mathbf{h}_\Lambda = \mathbf{A}\mathbf{h} - \sum_{j \geq 2} \mathbf{A}\mathbf{h}_{\Lambda_j}$, we can rewrite (11.55) as

$$(1 - \delta_{2k}) \|\mathbf{h}_\Lambda\|_2^2 \leq \langle \mathbf{A}\mathbf{h}_\Lambda, \mathbf{A}\mathbf{h} \rangle - \langle \mathbf{A}\mathbf{h}_\Lambda, \sum_{j \geq 2} \mathbf{A}\mathbf{h}_{\Lambda_j} \rangle. \tag{11.56}$$

In order to bound the second term in (11.56), we use Proposition 11.5, which implies that

$$\left| \langle \mathbf{A}\mathbf{h}_{\Lambda_i}, \mathbf{A}\mathbf{h}_{\Lambda_j} \rangle \right| \leq \delta_{2k} \|\mathbf{h}_{\Lambda_i}\|_2 \|\mathbf{h}_{\Lambda_j}\|_2, \tag{11.57}$$

for any $i \neq j$. Furthermore, Proposition 11.2 yields $\|\mathbf{h}_{\Lambda_0}\|_2 + \|\mathbf{h}_{\Lambda_1}\|_2 \leq \sqrt{2} \|\mathbf{h}_\Lambda\|_2$. Substituting into (11.57) we obtain

$$\left| \langle \mathbf{A}\mathbf{h}_\Lambda, \sum_{j \geq 2} \mathbf{A}\mathbf{h}_{\Lambda_j} \rangle \right| = \left| \sum_{j \geq 2} \langle \mathbf{A}\mathbf{h}_{\Lambda_0}, \mathbf{A}\mathbf{h}_{\Lambda_j} \rangle + \sum_{j \geq 2} \langle \mathbf{A}\mathbf{h}_{\Lambda_1}, \mathbf{A}\mathbf{h}_{\Lambda_j} \rangle \right|$$

$$\leq \sum_{j \geq 2} \left| \langle \mathbf{A}\mathbf{h}_{\Lambda_0}, \mathbf{A}\mathbf{h}_{\Lambda_j} \rangle \right| + \sum_{j \geq 2} \left| \langle \mathbf{A}\mathbf{h}_{\Lambda_1}, \mathbf{A}\mathbf{h}_{\Lambda_j} \rangle \right|$$

$$\leq \delta_{2k} \|\mathbf{h}_{\Lambda_0}\|_2 \sum_{j \geq 2} \|\mathbf{h}_{\Lambda_j}\|_2 + \delta_{2k} \|\mathbf{h}_{\Lambda_1}\|_2 \sum_{j \geq 2} \|\mathbf{h}_{\Lambda_j}\|_2$$

$$\leq \sqrt{2}\delta_{2k} \|\mathbf{h}_\Lambda\|_2 \sum_{j \geq 2} \|\mathbf{h}_{\Lambda_j}\|_2. \tag{11.58}$$

From Lemma 11.1, this reduces to

$$\left| \langle \mathbf{Ah}_\Lambda, \sum_{j\geq 2} \mathbf{Ah}_{\Lambda_j} \rangle \right| \leq \sqrt{2}\delta_{2k} \|\mathbf{h}_\Lambda\|_2 \frac{\|\mathbf{h}_{\Lambda_0^c}\|_1}{\sqrt{k}}. \tag{11.59}$$

Combining (11.59) with (11.56) we have

$$(1 - \delta_{2k}) \|\mathbf{h}_\Lambda\|_2^2 \leq \left| \langle \mathbf{Ah}_\Lambda, \mathbf{Ah} \rangle - \langle \mathbf{Ah}_\Lambda, \sum_{j\geq 2} \mathbf{Ah}_{\Lambda_j} \rangle \right|$$

$$\leq |\langle \mathbf{Ah}_\Lambda, \mathbf{Ah} \rangle| + \left| \langle \mathbf{Ah}_\Lambda, \sum_{j\geq 2} \mathbf{Ah}_{\Lambda_j} \rangle \right|$$

$$\leq |\langle \mathbf{Ah}_\Lambda, \mathbf{Ah} \rangle| + \sqrt{2}\delta_{2k} \|\mathbf{h}_\Lambda\|_2 \frac{\|\mathbf{h}_{\Lambda_0^c}\|_1}{\sqrt{k}}, \tag{11.60}$$

which yields the desired result upon rearranging. □

In order to prove Theorem 11.4, all we need is to apply Proposition 11.6 to the case where $\mathbf{h} \in \mathcal{N}(\mathbf{A})$.

Proof of Theorem 11.4: Suppose that $\mathbf{h} \in \mathcal{N}(\mathbf{A})$. It is sufficient to show that

$$\|\mathbf{h}_\Lambda\|_2 \leq C \frac{\|\mathbf{h}_{\Lambda^c}\|_1}{\sqrt{2k}} \tag{11.61}$$

holds for the case where Λ is the index set corresponding to the $2k$ largest entries of $\mathbf{h}$. Thus, we can take Λ_0 to be the index set corresponding to the k largest entries of $\mathbf{h}$ and apply Proposition 11.6.

The second term in Proposition 11.6 vanishes since $\mathbf{Ah} = 0$, and thus we have

$$\|\mathbf{h}_\Lambda\|_2 \leq \alpha \frac{\|\mathbf{h}_{\Lambda_0^c}\|_1}{\sqrt{k}}, \tag{11.62}$$

where $\alpha = \sqrt{2}\delta_{2k}/(1 - \delta_{2k})$. Using Proposition 11.1,

$$\|\mathbf{h}_{\Lambda_0^c}\|_1 = \|\mathbf{h}_{\Lambda_1}\|_1 + \|\mathbf{h}_{\Lambda^c}\|_1 \leq \sqrt{k} \|\mathbf{h}_{\Lambda_1}\|_2 + \|\mathbf{h}_{\Lambda^c}\|_1 \tag{11.63}$$

resulting in

$$\|\mathbf{h}_\Lambda\|_2 \leq \alpha \left(\|\mathbf{h}_{\Lambda_1}\|_2 + \frac{\|\mathbf{h}_{\Lambda^c}\|_1}{\sqrt{k}} \right). \tag{11.64}$$

Since $\|\mathbf{h}_{\Lambda_1}\|_2 \leq \|\mathbf{h}_\Lambda\|_2$, we have that

$$(1 - \alpha) \|\mathbf{h}_\Lambda\|_2 \leq \alpha \frac{\|\mathbf{h}_{\Lambda^c}\|_1}{\sqrt{k}}. \tag{11.65}$$

The assumption $\delta_{2k} < \sqrt{2} - 1$ ensures that $\alpha < 1$. We may therefore divide by $1 - \alpha$ without changing the direction of the inequality to establish (11.61) with constant

$$C = \frac{\sqrt{2}\alpha}{1 - \alpha} = \frac{2\delta_{2k}}{1 - (1 + \sqrt{2})\delta_{2k}}, \tag{11.66}$$

completing the proof. □

11.3.3   Coherence

While the spark, NSP, and RIP all provide guarantees for the recovery of k-sparse signals, verifying that a general matrix $\mathbf{A}$ satisfies any of these properties has combinatorial computational complexity, since in each case one must essentially consider $\binom{n}{k}$ submatrices. In many cases it is preferable to use properties of $\mathbf{A}$ that are easily computable to provide more concrete recovery conditions. The *coherence* of a matrix is one such property [221, 228].

Definition 11.5.   *The coherence of a matrix $\mathbf{A}$, denoted $\mu(\mathbf{A})$, is the largest absolute inner product between any two columns $\mathbf{a}_i$, $\mathbf{a}_j$ of $\mathbf{A}$:*

$$\mu(\mathbf{A}) = \max_{1 \leq i < j \leq n} \frac{|\langle \mathbf{a}_i, \mathbf{a}_j \rangle|}{\|\mathbf{a}_i\|_2 \|\mathbf{a}_j\|_2}. \tag{11.67}$$

Properties of the coherence
The following proposition bounds the value of the coherence for an arbitrary matrix $\mathbf{A}$.

Proposition 11.7.   *Let $\mathbf{A}$ be a matrix of size $m \times n$ with $m \leq n$, $(n \geq 2)$ whose columns are normalized so that $\|\mathbf{a}_i\| = 1$ for all i. Then the coherence of $\mathbf{A}$ satisfies*

$$\sqrt{\frac{n - m}{m(n - 1)}} \leq \mu(\mathbf{A}) \leq 1. \tag{11.68}$$

The lower bound in (11.68) is known as the *Welch bound* [229]. Note that when $n \gg m$, the lower bound is approximately $\mu(\mathbf{A}) \geq 1/\sqrt{m}$. When $m \geq n$, we can always choose $\mathbf{A}$ to have orthonormal columns, in which case $\mu(\mathbf{A}) = 0$. The lower bound is consequently of interest only when $m < n$.

Proof: The upper bound is a direct consequence of the Cauchy–Schwarz inequality.
To prove the lower bound, consider the $n \times n$ Gram matrix of inner products $\mathbf{G} = \mathbf{A}^* \mathbf{A}$ so that the ijth element of $\mathbf{G}$ is $G_{ij} = \langle \mathbf{a}_i, \mathbf{a}_j \rangle$. Let $\lambda_i, 1 \leq i \leq r$ denote the nonzero eigenvalues of $\mathbf{G}$. Since the rank of $\mathbf{G}$ is upper bounded by m, we have that $r \leq m$. Because the columns of $\mathbf{A}$ are normalized, $\mathrm{Tr}(\mathbf{G}) = n$. Now,

$$n^2 = \mathrm{Tr}^2(\mathbf{G}) = \left(\sum_{i=1}^{r} \lambda_i \right)^2 \leq r \sum_{i=1}^{r} \lambda_i^2 \leq m \sum_{i=1}^{r} \lambda_i^2, \tag{11.69}$$

where the first inequality is a result of the Cauchy–Schwarz inequality. We also have that

$$\mathrm{Tr}(\mathbf{G}^2) = \sum_{i=1}^{n} \sum_{j=1}^{n} |\langle \mathbf{a}_i, \mathbf{a}_j \rangle|^2 = \sum_{i=1}^{r} \lambda_i^2. \tag{11.70}$$

From (11.69) and (11.70) we conclude that

$$\sum_{i=1}^{n} \sum_{j=1}^{n} |\langle \mathbf{a}_i, \mathbf{a}_j \rangle|^2 = n + \sum_{i \neq j} |\langle \mathbf{a}_i, \mathbf{a}_j \rangle|^2 \geq \frac{n^2}{m}, \tag{11.71}$$

or

$$\sum_{i \neq j} |\langle \mathbf{a}_i, \mathbf{a}_j \rangle|^2 \geq \frac{n(n-m)}{m}. \tag{11.72}$$

To complete the proof we note that the sum in (11.72) consists of $n(n-1)$ elements, and each one of them is no larger than $\mu^2(\mathbf{A})$. Thus,

$$\mu^2(\mathbf{A}) \geq \frac{1}{n(n-1)} \sum_{i \neq j} |\langle \mathbf{a}_i, \mathbf{a}_j \rangle|^2 \geq \frac{(n-m)}{m(n-1)}, \tag{11.73}$$

from which the result follows.   □

The proof of Proposition 11.7 also reveals when the lower bound in (11.68) is achieved: this requires that $\mathbf{G}$ has rank m, that all the nonzero eigenvalues of $\mathbf{G}$ are equal, and that $|\langle \mathbf{a}_i, \mathbf{a}_j \rangle| = c$ with $i \neq j$ for some constant c. A set of vectors $\{\mathbf{a}_i\}$ achieving this inequality is referred to as a *Welch-bound equality set* or a *Grassmanian frame* [230] (see Example 11.12). Such vector sets exist only if $n \leq m(m+1)/2$ over the reals, and $n \leq m^2$ over the complex plane [231]. An iterative algorithm to construct such frames is given in [232].

Example 11.12 Consider the vectors $\{\mathbf{a}_i = [\cos(i\pi/3)\,\sin(i\pi/3)]^T,\ i = 1, 2, 3\}$. We will show that this set forms a Grassmanian frame. Since $m = 2$ and $n = 3$, the Welch bound is

$$\sqrt{\frac{n-m}{m(n-1)}} = \frac{1}{2}. \tag{11.74}$$

It is straightforward to see that

$$\frac{|\langle \mathbf{a}_i, \mathbf{a}_j \rangle|}{\|\mathbf{a}_i\|_2 \|\mathbf{a}_j\|_2} = \frac{1}{2}, \quad 1 \leq i < j \leq 3, \tag{11.75}$$

establishing the result.

Note that in this example the matrix $\mathbf{G}$ is given by

$$\mathbf{G} = \begin{bmatrix} 1 & 0.5 & -0.5 \\ 0.5 & 1 & 0.5 \\ -0.5 & 0.5 & 1 \end{bmatrix}. \tag{11.76}$$

Its eigenvalues are equal to $0, 0.5, 0.5$ so that $\mathbf{G}$ has rank 2 and equal nonzero eigenvalues, as required.

If we now modify the set to be equal to $\{\mathbf{a}_i = [\cos(2i\pi/3)\,\sin(2i\pi/3)]^T,\ i = 1, 2, 3\}$, then $\mathbf{G}$ will change to

$$\mathbf{G} = \begin{bmatrix} 1 & -0.5 & -0.5 \\ -0.5 & 1 & -0.5 \\ -0.5 & -0.5 & 1 \end{bmatrix}. \tag{11.77}$$

Thus, (11.75) still holds. It can also be verified that the eigenvalues of $\mathbf{G}$ are equal to $0, 1.5, 1.5$ so that this set is also a Grassmanian frame.

On the other hand, if we modify $\mathbf{G}$ to

$$\mathbf{G} = \begin{bmatrix} 1 & 0.5 & 0.5 \\ 0.5 & 1 & 0.5 \\ 0.5 & 0.5 & 1 \end{bmatrix}, \tag{11.78}$$

then its eigenvalues are equal to $0.5, 0.5, 2$. This matrix therefore no longer corresponds to the Gram matrix of a Grassmanian frame.

Example 11.13 Consider the following 3×4 matrix:

$$\mathbf{A} = \begin{bmatrix} 1 & -1 & -1 & -1 \\ 1 & 1 & -1 & 1 \\ 1 & 1 & 1 & -1 \end{bmatrix}. \tag{11.79}$$

The Welch bound for this setting is

$$\sqrt{\frac{n-m}{m(n-1)}} = \frac{1}{3}. \tag{11.80}$$

We can easily check that

$$\frac{|\langle \mathbf{a}_i, \mathbf{a}_j \rangle|}{\|\mathbf{a}_i\|_2 \|\mathbf{a}_j\|_2} = \frac{1}{3}, \qquad 1 \le i < j \le 4, \tag{11.81}$$

and the columns of $\mathbf{A}$ constitute a Grassmanian frame. The eigenvalues of the Gram matrix $\mathbf{G} = \mathbf{A}^*\mathbf{A}$ are equal to $0, 4, 4, 4$, so that all the nonzero eigenvalues are equal, as we expect.

Example 11.14 We now show an example of a set that does not satisfy the Welch bound. Consider the columns of the matrix:

$$\mathbf{A} = \begin{bmatrix} 1 & 0 & 0 & \frac{1}{\sqrt{2}} \\ 0 & 1 & 0 & \frac{1}{\sqrt{2}} \\ 0 & 0 & 1 & 0 \end{bmatrix}. \tag{11.82}$$

Since the dimensions of this matrix are equal to those of the previous example, the Welch bound is given as before by $1/3$. On the other hand,

$$\mu(\mathbf{A}) = \max_{1 \le i < j \le 4} \frac{|\langle \mathbf{a}_i, \mathbf{a}_j \rangle|}{\|\mathbf{a}_i\|_2 \|\mathbf{a}_j\|_2} = \frac{1}{\sqrt{2}}, \tag{11.83}$$

and the columns of $\mathbf{A}$ do not satisfy the Welch-bound equality.

Note, that in this example, $\langle \mathbf{a}_i, \mathbf{a}_j \rangle = 0$ for $i, j = 1, 2, 3$ and $i \ne j$, while $\langle \mathbf{a}_i, \mathbf{a}_j \rangle = 1/\sqrt{2}$ for $i = 4$ and $j = 1, 2$ so that the values of $|\langle \mathbf{a}_i, \mathbf{a}_j \rangle|$ are not all equal.

Relation to spark and RIP

The spark and RIP can both be related to the coherence. To connect the coherence and the spark we employ the Geršgorin disk theorem [233] (see Appendix A).

Theorem 11.5 (Eigenvalue spread).   *The eigenvalues of an $n \times n$ matrix $\mathbf{M}$ with entries M_{ij}, $1 \leq i, j \leq n$, lie in the union of n disks $d_i = d_i(c_i, r_i)$, $1 \leq i \leq n$, centered at $c_i = M_{ii}$ and with radius $r_i = \sum_{j \neq i} |M_{ij}|$.*

The following corollary is an immediate consequence of Theorem 11.5:

Corollary 11.1.   *Let $\mathbf{M}$ be a symmetric matrix. If for any i,*

$$\sum_{j \neq i} |M_{ij}| < M_{ii}, \tag{11.84}$$

then $\mathbf{M}$ is positive definite.

Proof: If (11.84) holds, then it follows from Theorem 11.5 that all the eigenvalues of $\mathbf{M}$ must be positive. The corollary then follows from the fact that a symmetric matrix is positive definite if and only if all its eigenvalues are positive. $\square$

Applying Theorem 11.5 on the Gram matrix $\mathbf{G} = \mathbf{A}_\Lambda^* \mathbf{A}_\Lambda$ leads to the following straightforward result.

Lemma 11.2.   *For any matrix $\mathbf{A}$,*

$$\text{spark}(\mathbf{A}) \geq 1 + \frac{1}{\mu(\mathbf{A})}. \tag{11.85}$$

Proof: Since $\text{spark}(\mathbf{A})$ does not depend on the scaling of the columns, we can assume without loss of generality that $\mathbf{A}$ has unit-norm columns. Let $\Lambda \subseteq \{1, \ldots, n\}$ with $|\Lambda| = p$ determine a set of indices. We consider the restricted Gram matrix $\mathbf{G} = \mathbf{A}_\Lambda^* \mathbf{A}_\Lambda$, which satisfies the following properties:

1. $G_{ii} = 1, 1 \leq i \leq p$;
2. $|G_{ij}| \leq \mu(\mathbf{A}), 1 \leq i, j \leq p, i \neq j$.

From Corollary 11.1, if $\sum_{j \neq i} |G_{ij}| < G_{ii}$ then the matrix $\mathbf{G}$ is positive definite, so that the columns of $\mathbf{A}_\Lambda$ are linearly independent. Thus, the spark condition implies $(p - 1)\mu(\mathbf{A}) < 1$ or, equivalently, $p < 1 + 1/\mu(\mathbf{A})$ for all $p < \text{spark}(\mathbf{A})$, yielding $\text{spark}(\mathbf{A}) \geq 1 + 1/\mu(\mathbf{A})$. $\square$

By merging Theorem 11.1 with Lemma 11.2, we can pose the following condition on $\mathbf{A}$ that guarantees uniqueness [221, Theorem 12].

Theorem 11.6 (Uniqueness via coherence).   *If*

$$k < \frac{1}{2} \left(1 + \frac{1}{\mu(\mathbf{A})} \right), \tag{11.86}$$

then for each measurement vector $\mathbf{y} \in \mathbb{R}^m$ there exists at most one signal $\mathbf{x} \in \Sigma_k$ such that $\mathbf{y} = \mathbf{A}\mathbf{x}$.

Theorem 11.6, together with the Welch bound of Proposition 11.7, provides an upper bound on the level of sparsity k that guarantees uniqueness using coherence: $k = O(\sqrt{m})$. Note, on the other hand, that the spark of a matrix can be of order m so that Theorem 11.1 allows for a sparsity of order m. Thus, Theorem 11.6 is a pessimistic result, and often recovery is possible for larger values of k than those predicted by the theorem.

Example 11.15 Consider the matrix

$$\mathbf{A} = \begin{bmatrix} 1 & -1 & -1 & 1 \\ -1 & 1 & -1 & 1 \\ 1 & 1 & -1 & -1 \end{bmatrix}. \tag{11.87}$$

It is easy to see that $\mu(\mathbf{A}) = 1/3$.

Suppose we obtain the measurement vector $\mathbf{y} = [1\,-1\,1]^T$. By checking every possibility, just as in Example 11.9, we can see that the equation $\mathbf{y} = \mathbf{A}\mathbf{x}$ has a unique solution in Σ_1, $\mathbf{x} = [1\,0\,0\,0]^T$. Here the sparsity of $\mathbf{x}$ ($k = 1$) satisfies the condition of Theorem 11.6.

Consider next the input $\mathbf{x} = [1\,1\,0\,0]^T$ whose sparsity ($k = 2$) does not satisfy the conditions of the theorem. In this case, $\mathbf{A}\mathbf{x} = [0\,0\,2]^T$. We can obtain the same output by choosing $\mathbf{x}' = [0.5\,0.5\,-0.5\,-0.5]^T$; therefore, there is no unique solution to the equation $\mathbf{y} = \mathbf{A}\mathbf{x}$ with $\mathbf{x} \in \Sigma_2$.

Another straightforward application of the Geršgorin disk theorem (Theorem 11.5) connects the RIP to the coherence property.

Lemma 11.3. *If $\mathbf{A}$ has unit-norm columns and coherence $\mu = \mu(\mathbf{A})$, then $\mathbf{A}$ satisfies the RIP of order k with $\delta_k \leq (k-1)\mu$.*

Proof: Let $\mathbf{x}$ be an arbitrary vector with $\mathrm{supp}(\mathbf{x}) \leq k$. Without loss of generality, we will assume that the k first values are nonzero. Then,

$$\|\mathbf{A}\mathbf{x}\|_2^2 = \sum_{i,j=1}^{k} \langle \mathbf{a}_i, \mathbf{a}_j \rangle x_i x_j = \|\mathbf{x}\|_2^2 + \sum_{i \neq j} \langle \mathbf{a}_i, \mathbf{a}_j \rangle x_i x_j. \tag{11.88}$$

We now bound the second term:

$$\left| \sum_{i \neq j} \langle \mathbf{a}_i, \mathbf{a}_j \rangle x_i x_j \right| \leq \mu \sum_{i \neq j} |x_i x_j| \leq \mu \left(\sum_{i,j=1}^{k} |x_i x_j| - \|\mathbf{x}\|_2^2 \right). \tag{11.89}$$

From the Cauchy–Schwarz inequality,

$$\sum_{i,j=1}^{k} |x_i x_j| = \left(\sum_{i=1}^{k} |x_i| \right)^2 \leq k\|\mathbf{x}\|_2^2. \tag{11.90}$$

Substituting into (11.89),

$$\left| \sum_{i \neq j} \langle \mathbf{a}_i, \mathbf{a}_j \rangle x_i x_j \right| \leq \mu(k-1)\|\mathbf{x}\|_2^2. \tag{11.91}$$

Combining with (11.88) we have

$$(1 - (k - 1)\mu)\|\mathbf{x}\|_2^2 \leq \|\mathbf{A}\mathbf{x}\|_2^2 \leq (1 + (k - 1)\mu)\|\mathbf{x}\|_2^2, \qquad (11.92)$$

so that $\delta_k \leq (k - 1)\mu$. $\qquad\qquad\qquad\qquad\qquad\qquad\qquad\qquad\qquad\qquad\square$

11.3.4   Uncertainty relations

The coherence introduced in the previous subsection plays an important role in uncertainty relations involving signal decompositions in orthonormal bases [234, 235].

The uncertainty principle was first derived by Heisenberg in the context of quantum mechanics, as it poses limits on the ability to determine the position and momentum of a free particle. The classical formulation of the uncertainty principle states that two conjugate variables such as position and momentum, or any other pair coupled by the Fourier transform, cannot both be determined simultaneously to arbitrary precision. We have seen a mathematical statement of this relation in (4.56). Here we discuss uncertainty relations for the minimum number of nonzero entries required to represent a signal in a pair of bases. Uncertainty relations for signal representations were first studied by Donoho and Stark in [236] with the bases chosen as the DFT and identity matrices. In [234, 235] it was shown that uncertainty relations can be used to develop thresholds on the sparsity level required for uniqueness of signal representations. We will discuss these results below after introducing an uncertainty relation for sparse decompositions in pairs of orthonormal bases.

The uncertainty principle for sparse vectors
Uncertainty relations for sparse signals are concerned with representations of a vector $\mathbf{x} \in \mathbb{C}^n$ in two different orthonormal bases for $\mathbb{C}^n$: $\{\phi_\ell, 1 \leq \ell \leq n\}$ and $\{\psi_\ell, 1 \leq \ell \leq n\}$. Any vector $\mathbf{x} \in \mathbb{C}^n$ can be expanded uniquely in terms of each one of these bases according to:

$$\mathbf{x} = \sum_{\ell=1}^{n} a_\ell \phi_\ell = \sum_{\ell=1}^{n} b_\ell \psi_\ell, \qquad (11.93)$$

with appropriate coefficients a_ℓ, b_ℓ. Uncertainty relations set limits on the sparsity of the decompositions (11.93) for any $\mathbf{x} \in \mathbb{C}^n$. Specifically, denoting $A = \|\mathbf{a}\|_0$ and $B = \|\mathbf{b}\|_0$, where $\mathbf{a}$ and $\mathbf{b}$ represent the vectors with elements a_ℓ and b_ℓ respectively, Theorem 11.7 below shows that A and B cannot both be small simultaneously. Thus, a signal cannot be sparsely decomposed into two orthonormal bases. We denote by $\mathbf{\Phi}$ and $\mathbf{\Psi}$ the matrices whose ℓth columns are ϕ_ℓ and ψ_ℓ, respectively.

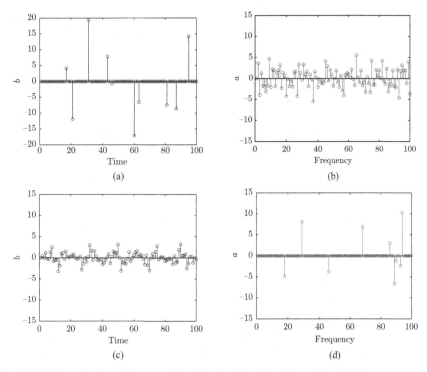

Figure 11.7 Representations of a sparse signal in the spike (identity) and Fourier bases. (a) Time-domain representation of a sparse signal in the identity basis. (b) Frequency-domain representation of a sparse signal in the identity basis. We observe that the signal is not sparse in the frequency domain. (c) Time-domain representation of a sparse signal in the Fourier basis. Now observe that the signal is not sparse in the time domain. (d) Frequency-domain representation of a sparse signal in the Fourier basis.

Example 11.16 Let $\boldsymbol{\Phi}$ be the Fourier basis and $\boldsymbol{\Psi}$ be the spike (identity) basis. Both of these bases are orthonormal. Figure 11.7 shows that a signal whose time-domain representation is sparse has a frequency-domain representation which is not sparse and vice versa.

The level of sparsity achievable should clearly depend on some measure of distance between $\boldsymbol{\Phi}$ and $\boldsymbol{\Psi}$. In particular, if $\boldsymbol{\Phi} = \boldsymbol{\Psi}$ then we can choose $\mathbf{x}$ to be equal to one of the columns of $\boldsymbol{\Phi}$ and obtain $\|\mathbf{a}\|_0 = \|\mathbf{b}\|_0 = 1$. In order to introduce an appropriate uncertainty relation, we generalize the definition of coherence to two orthonormal bases leading to the *mutual coherence*:

$$\mu(\boldsymbol{\Phi}, \boldsymbol{\Psi}) = \max_{\ell,r} |\boldsymbol{\phi}_\ell^* \boldsymbol{\psi}_r|. \tag{11.94}$$

Note that since the columns of $\boldsymbol{\Phi}$ and $\boldsymbol{\Psi}$ are each orthonormal, $\mu(\boldsymbol{\Phi}, \boldsymbol{\Psi})$ is equal to $\mu(\mathbf{A})$ where $\mathbf{A} = [\boldsymbol{\Phi} \ \boldsymbol{\Psi}]$ is defined as the concatenation of $\boldsymbol{\Phi}$ and $\boldsymbol{\Psi}$.

Theorem 11.7 (Uncertainty relation for sparse decompositions).   *Let* $\boldsymbol{\Phi}, \boldsymbol{\Psi}$ *be two unitary* $n \times n$ *matrices with columns* $\{\boldsymbol{\phi}_\ell, \boldsymbol{\psi}_\ell, 1 \leq \ell \leq n\}$ *and let* $\mathbf{x} \in \mathbb{C}^n$ *satisfy (11.93).*
Then

$$\frac{1}{2}(A + B) \geq \sqrt{AB} \geq \frac{1}{\mu(\boldsymbol{\Phi}, \boldsymbol{\Psi})} \tag{11.95}$$

where $A = \|\mathbf{a}\|_0$ *and* $B = \|\mathbf{b}\|_0$.

Proof: Without loss of generality, we assume that $\|\mathbf{x}\|_2^2 = 1$ and denote for brevity
$\mu = \mu(\boldsymbol{\Phi}, \boldsymbol{\Psi})$. Writing $\mathbf{x}$ once as $\boldsymbol{\Phi}\mathbf{a}$ and once as $\boldsymbol{\Psi}\mathbf{b}$ we have

$$1 = \|\mathbf{x}\|_2^2 = \mathbf{a}^* \mathbf{G} \mathbf{b} = \sum_{i,j=1}^{n} \overline{a_i} G_{ij} b_j \tag{11.96}$$

where we denoted $\mathbf{G} = \boldsymbol{\Phi}^* \boldsymbol{\Psi}$. Since by definition (11.94), $|G_{ij}| \leq \mu$,

$$1 \leq \sum_{i,j=1}^{n} |a_i G_{ij} b_j| \leq \mu \sum_{i=1}^{n} |a_i| \sum_{i=1}^{n} |b_j| = \mu \|\mathbf{a}\|_1 \|\mathbf{b}\|_1. \tag{11.97}$$

From Proposition 11.1,

$$\|\mathbf{a}\|_1 \leq \sqrt{A} \|\mathbf{a}\|_2 = \sqrt{A}, \tag{11.98}$$

where we used the fact that $\|\mathbf{a}\|_2 = \|\mathbf{x}\|_2 = 1$ since $\boldsymbol{\Phi}$ is unitary. Similarly, we have
that $\|\mathbf{b}\|_1 \leq \sqrt{B}$. Substituting into (11.97) and using the inequality of arithmetic and
geometric means (for every $x, y > 0$ we have $\sqrt{xy} \leq (x+y)/2$) completes the proof. $\square$

Example 11.17 Let $\boldsymbol{\Phi}$ be the spike (identity) basis and $\boldsymbol{\Psi}$ be the Haar basis. The
family of Haar functions $h_j(t), j \geq 0$ are defined on the interval $0 \leq t \leq 1$, and
labeled by two parameters p and q such that $p = \lfloor \log_2 j \rfloor + 1$ and $q = j - 2^{p+1} + 1$.
The normalized functions are defined as follows [237]:

$$h_j(t) = \begin{cases} \frac{1}{\sqrt{n}} 2^{(p-1)/2}, & (q-1)/2^{p-1} < t < (q - \frac{1}{2})/2^{p-1} \\ -\frac{1}{\sqrt{n}} 2^{(p-1)/2}, & (q - \frac{1}{2})/2^{p-1} < t < q/2^{p-1} \\ 0, & \text{otherwise.} \end{cases} \tag{11.99}$$

To form an $n \times n$ discrete Haar transform matrix, the n Haar functions $h_j(t), 0 \leq
j \leq n - 1$ are sampled at $t = i/n$, where $0 \leq i \leq n - 1$. The function $h_j(i), 0 \leq
i \leq n - 1$ then constitutes the jth row of $\boldsymbol{\Psi}$. As an example, for $n = 4$:

$$\mathbf{H}_4 = \frac{1}{2} \begin{bmatrix} 1 & 1 & 1 & 1 \\ 1 & 1 & -1 & -1 \\ \sqrt{2} & -\sqrt{2} & 0 & 0 \\ 0 & 0 & \sqrt{2} & -\sqrt{2} \end{bmatrix}. \tag{11.100}$$

Figure 11.8 shows the representation of a sparse signal $\mathbf{x}$ of length $n = 128$ in
these two orthonormal bases. In this case, $A = 8$ and $B = 28$. The coherence of the
two bases can be computed as

$$\mu(\boldsymbol{\Phi}, \boldsymbol{\Psi}) = \max_{0 \leq j, i \leq n-1} |h_j(i)| = \frac{2^{(p_{max}-1)/2}}{\sqrt{n}} = \frac{8}{\sqrt{128}} = \frac{1}{\sqrt{2}},$$

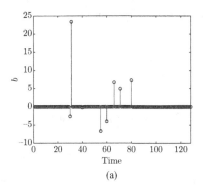 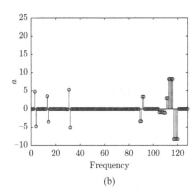

Figure 11.8 Representation of a sparse signal in the spike (identity) and Haar bases.
(a) Representation in the identity basis. (b) Representation in the Haar basis.

where $p_{\max} = \lfloor \log_2(n-1) \rfloor + 1$. Since $\sqrt{AB} = 4\sqrt{14}$ and $1/\mu(\boldsymbol{\Phi}, \boldsymbol{\Psi}) = \sqrt{2}$, (11.95) is satisfied for this choice of $\mathbf{x}$.

Example 11.18 We now return to Example 11.16 and consider the uncertainty principle for the signal depicted in Fig. 11.7(a) and (b).

In this example, $A = 10$ and $B = 100$. To compute the coherence of the identity and Fourier bases we note that the magnitude of the elements of the $n \times n$ Fourier basis are all equal to $1/\sqrt{n}$. Therefore, $\mu(\boldsymbol{\Phi}, \boldsymbol{\Psi}) = 1/\sqrt{n}$. In this case, $\sqrt{AB} = \sqrt{1000}$ which is larger than $1/\mu(\boldsymbol{\Phi}, \boldsymbol{\Psi}) = \sqrt{128}$.

Bounds on the mutual coherence

In Proposition 11.7 we developed a lower bound on the coherence of an arbitrary matrix $\mathbf{A}$. Applying this bound to our setting in which $\mathbf{A}$ is a concatenation of two orthonormal bases of size $n \times n$ corresponds to choosing $m' = n$ and $n' = 2n$. The lower bound then becomes

$$\mu(\mathbf{A}) \geq \sqrt{\frac{n' - m'}{m'(n' - 1)}} = \sqrt{\frac{1}{2n - 1}}. \tag{11.101}$$

In the next proposition we show that a tighter bound can be obtained on the mutual coherence between two orthonormal bases by exploiting the special structure of $\mathbf{A}$ in this case:

Proposition 11.8. *Let* $\boldsymbol{\Phi}, \boldsymbol{\Psi}$ *be two unitary* $n \times n$ *matrices. Then the mutual coherence* $\mu(\boldsymbol{\Phi}, \boldsymbol{\Psi})$ *satisfies*

$$\frac{1}{\sqrt{n}} \leq \mu(\boldsymbol{\Phi}, \boldsymbol{\Psi}) \leq 1. \tag{11.102}$$

Proof: The upper bound follows from the Cauchy–Schwarz inequality and the fact that the basis elements have norm 1.

To obtain the lower bound, denote $\mathbf{G} = \mathbf{\Phi}^*\mathbf{\Psi}$. Since $\mathbf{G}$ is a unitary matrix,

$$\sum_{\ell=1}^{n}\sum_{j=1}^{n}|\phi_\ell^*\psi_j|^2 = \mathrm{Tr}(\mathbf{G}^*\mathbf{G}) = \mathrm{Tr}(\mathbf{I}_n) = n. \tag{11.103}$$

By definition $|\phi_\ell^*\psi_j| \le \mu(\mathbf{\Phi}, \mathbf{\Psi})$. Therefore,

$$n = \sum_{\ell=1}^{n}\sum_{j=1}^{n}|\phi_\ell^*\psi_j|^2 \le n^2\mu^2(\mathbf{\Phi}, \mathbf{\Psi}), \tag{11.104}$$

from which the result follows. $\qquad\square$

From the proof of the proposition it is easy to see that the lower bound is achieved when $|\phi_\ell^*\psi_j| = \mu$ for all ℓ, j. With this choice, the uncertainty relation (11.95) becomes

$$A + B \ge 2\sqrt{AB} \ge 2\sqrt{n}. \tag{11.105}$$

We already saw in Example 11.18 that when $\mathbf{\Phi}$ and $\mathbf{\Psi}$ are chosen as the spike (identity) and Fourier bases, $\mu(\mathbf{\Phi}, \mathbf{\Psi}) = 1/\sqrt{n}$. Another example in which the lower bound in (11.102) is satisfied is given next.

Example 11.19 Let $\mathbf{\Phi}$ be the $n \times n$ orthonormal Walsh–Hadamard basis and $\mathbf{\Psi}$ be the $n \times n$ spike (identity) basis. The Hadamard matrix of order $m = 2^n$ is given recursively by [238]:

$$\mathbf{H}_m = \frac{1}{\sqrt{2}}\begin{bmatrix} \mathbf{H}_{m-1} & \mathbf{H}_{m-1} \\ \mathbf{H}_{m-1} & -\mathbf{H}_{m-1} \end{bmatrix}, \tag{11.106}$$

with

$$\mathbf{H}_2 = \begin{bmatrix} 1 & 1 \\ 1 & -1 \end{bmatrix}. \tag{11.107}$$

Since $\mathbf{\Phi}_{i,j} = \frac{1}{\sqrt{n}}(-1)^{ij}$,

$$|\phi_\ell^*\psi_j| = \frac{1}{\sqrt{n}}, \qquad 0 \le \ell, j \le n, \tag{11.108}$$

and the lower bound of (11.102) is satisfied.

In the next example we consider an input signal that is maximally sparse in two orthonormal bases as given by equality in the uncertainty principle (11.105).

Example 11.20 We have seen already that if $\mathbf{\Phi}$ is the $n \times n$ Fourier basis and $\mathbf{\Psi}$ the $n \times n$ identity basis, then the mutual coherence is $1/\sqrt{n}$. We now provide an example of a signal $\mathbf{x}$ whose representation in both bases satisfies (11.105).

Let n be such that $\sqrt{n}$ is an integer, and choose $\mathbf{x}$ as an impulse train $\delta_{\sqrt{n}}$ with spacing $\sqrt{n}$, as shown in Fig. 11.9(a) for $n = 100$. Since $\delta_{\sqrt{n}}$ has $\sqrt{n}$ nonzero values, $A = \sqrt{n}$. Now, it is well known that the Fourier transform of $\delta_{\sqrt{n}}$ is also

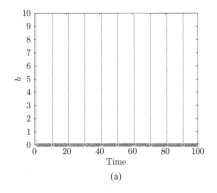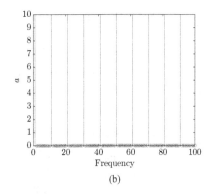

Figure 11.9 Representation of an impulse train ($n = 100$) in the spike (identity) and Fourier bases. (a) Time-domain representation. (b) Frequency domain.

$\delta_{\sqrt{n}}$ (see Fig. 11.9(b)). Therefore, $B = \sqrt{n}$ as well and the relations in (11.95) are all satisfied with equality:

$$\frac{1}{2}(A + B) = \sqrt{AB} = \frac{1}{\mu(\boldsymbol{\Phi}, \boldsymbol{\Psi})} = \sqrt{n}. \tag{11.109}$$

Uniqueness conditions

The uncertainty relation of Theorem 11.7 can be used to develop a condition on the uniqueness of a sparse representation $\mathbf{y} = \mathbf{Ax}$ when $\mathbf{A}$ is the concatenation of $\boldsymbol{\Phi}$ and $\boldsymbol{\Psi}$. In Theorem 11.6 we developed a condition on the sparsity level k and the coherence $\mu(\mathbf{A})$ such that $\mathbf{x}$ is unique. As we now show, in the special case in which $\mathbf{A} = [\boldsymbol{\Phi} \ \boldsymbol{\Psi}]$ is a concatenation of two orthonormal bases, this condition can be relaxed, allowing for larger sparsity levels.

Theorem 11.8 (Uniqueness via mutual coherence).   *Let $\mathbf{A} = [\boldsymbol{\Phi} \ \boldsymbol{\Psi}]$ be a concatenation of two orthonormal bases $\boldsymbol{\Phi}$ and $\boldsymbol{\Psi}$ where both matrices are unitary and of size $m \times m$. If*

$$k < \frac{1}{\mu(\boldsymbol{\Phi}, \boldsymbol{\Psi})} = \frac{1}{\mu(\mathbf{A})}, \tag{11.110}$$

then for each measurement vector $\mathbf{y} \in \mathbb{R}^m$ there exists at most one signal $\mathbf{x} \in \Sigma_k$ such that $\mathbf{y} = \mathbf{Ax}$.

Since $1/\mu(\mathbf{A})$ may be as large as $\sqrt{n}$ in this case, it follows that signals with sparsity level on the order of $\sqrt{n}$ are recoverable.

Proof: To prove the theorem, suppose that there exist two k-sparse vectors $\mathbf{x}_1, \mathbf{x}_2$ such that $\mathbf{y} = \mathbf{Ax}_1 = \mathbf{Ax}_2$. We will then show that we must have $k \geq 1/\mu(\mathbf{A})$.

Let $\mathbf{e} = \mathbf{x}_1 - \mathbf{x}_2$. Then $\mathbf{0} = \mathbf{Ae}$ and $\|\mathbf{e}\|_0 \leq 2k$. We now write the length-$2m$ vector $\mathbf{e}$ as $\mathbf{e} = [\mathbf{e}_1^T \ \mathbf{e}_2^T]^T$ where $\mathbf{e}_1, \mathbf{e}_2$ are length-m vectors. With these definitions, we have

$$\boldsymbol{\Phi}\mathbf{e}_1 = -\boldsymbol{\Psi}\mathbf{e}_2. \tag{11.111}$$

From Theorem 11.7 it follows that

$$\|e_1\|_0 + \|e_2\|_0 \geq \frac{2}{\mu(\mathbf{A})}. \tag{11.112}$$

In addition, by definition of **e**, $\|e\|_0 = \|e_1\|_0 + \|e_2\|_0$. Thus, we must have that

$$2k \geq \|e\|_0 = \|e_1\|_0 + \|e_2\|_0 \geq \frac{2}{\mu(\mathbf{A})}, \tag{11.113}$$

from which the result follows.   □

The uncertainty relations and corresponding recovery thresholds developed in this section can be generalized to concatenations of two arbitrary bases, namely, bases $\mathbf{\Phi}$ and $\mathbf{\Psi}$ that are not necessarily orthonormal, as incorporated in the following theorem [239].

Theorem 11.9 (Uncertainty relation for arbitrary bases).   *Let* $\mathbf{\Phi}, \mathbf{\Psi}$ *be two matrices with coherence* $\mu(\mathbf{\Phi}), \mu(\mathbf{\Psi})$, *and let* $\mathbf{x} \in \mathbb{C}^n$ *satisfy* $\mathbf{x} = \mathbf{\Phi}\mathbf{a} = \mathbf{\Psi}\mathbf{b}$. *Then*

$$\frac{1}{2}(A + B) \geq \sqrt{AB} \geq \frac{\sqrt{[1 - \mu(\mathbf{\Phi})(A - 1)]_+ [1 - \mu(\mathbf{\Psi})(B - 1)]_+}}{\mu(\mathbf{\Phi}, \mathbf{\Psi})} \tag{11.114}$$

where $A = \|a\|_0$, $B = \|b\|_0$, *and* $[c]_+ = \max(c, 0)$.

Note that when $\mathbf{\Phi}$ and $\mathbf{\Psi}$ are unitary, $\mu(\mathbf{\Phi}) = \mu(\mathbf{\Psi}) = 0$ and (11.114) reduces to the uncertainty relation (11.95). Theorem 11.9 allows for arbitrary matrices $\mathbf{\Phi}$ and $\mathbf{\Psi}$. In particular, they do not have to have the same number of columns, or form a basis. This result can be used to develop tighter recovery guarantees then those implied by Theorem 11.8; see [239] for further details.

Finally, we point out that the proposed uncertainty relations have also been extended to other structured signal models such as block-sparse signals [207], and analog signals in SI spaces [240].

11.3.5   Sensing matrix constructions

We have seen in the previous subsections that if the coherence of $\mathbf{A}$ is low enough, or if $\mathbf{A}$ satisfies the RIP with small enough constant, then we can recover a k-sparse vector $\mathbf{x}$ from measurements $\mathbf{y} = \mathbf{A}\mathbf{x}$. We now address the construction of matrices $\mathbf{A}$ that satisfy these properties.

Before proceeding with the details, we note that constructing matrices satisfying the coherence and RIP relies on a variety of tools from probability theory and geometric functional analysis. Therefore, we do not attempt to prove any of the results here, but rather refer the reader to the excellent tutorial [241] from which many of the results below are quoted.

Measurement bounds

To begin with, it is interesting to obtain insight into how many measurements are necessary to achieve the RIP. If we ignore the impact of δ and focus only on the dimensions of the problem (n, m, and k) then we can establish a simple lower bound [226, Theorem 3.5].

Theorem 11.10 (RIP and the number of measurements). *Let* $\mathbf{A}$ *be an* $m \times n$ *matrix that satisfies the RIP of order* $2k$ *with constant* $\delta \in (0, \frac{1}{2}]$. *Then*

$$m \geq Ck \log \left(\frac{n}{k}\right) \tag{11.115}$$

where $C = 1/2 \log(\sqrt{24} + 1) \approx 0.28$.

Similar results can be established by examining the *Gelfand width* of the ℓ_1 ball [242], or by exploiting results concerning the *Johnson–Lindenstrauss lemma*, which relates to embeddings of finite sets of points in low-dimensional spaces [243]. Specifically, it is shown in [244] that if we are given a point cloud with p points and wish to embed these points in $\mathbb{R}^m$ such that the squared ℓ_2 distance between any pair of points is preserved up to a factor of $1 \pm \varepsilon$, then we must have that

$$m \geq \frac{c_0 \log p}{\varepsilon^2}, \tag{11.116}$$

where $c_0 > 0$ is a constant. The Johnson–Lindenstrauss lemma is closely related to the RIP. In [245] it is shown that any procedure for generating a linear, distance-preserving embedding of a point cloud can be used to create a matrix satisfying the RIP. Furthermore, if a matrix $\mathbf{A}$ satisfies the RIP of order $k = c_1 \log p$ with constant δ, then $\mathbf{A}$ can be used to construct a distance-preserving embedding for p points with $\varepsilon = \delta/4$ [246].

Matrix construction

The lower bounds above imply that we need at least on the order of $k \log(n/k)$ measurements in order to recover a k-sparse vector $\mathbf{x}$. The question, though, is how to build an appropriate measurement matrix of these dimensions.

It is straightforward to show that an $m \times n$ Vandermonde matrix $\mathbf{V}$ constructed from m distinct scalars has $\mathrm{spark}(\mathbf{V}) = m+1$ [224] (see Example 11.6). Unfortunately, these matrices are poorly conditioned for large values of n, rendering the recovery problem numerically unstable. Similarly, there are known matrices $\mathbf{A}$ of size $m \times m^2$ that achieve the coherence lower bound $\mu(\mathbf{A}) = 1/\sqrt{m}$, such as a Gabor frame generated from the Alltop sequence [247] and more general equiangular tight frames [248]. These constructions restrict the number of measurements needed to recover a k-sparse signal to be $m = O(k^2 \log n)$. It is also possible to generate deterministic matrices of size $m \times n$ that satisfy the RIP of order k, but the resulting values of m are relatively large [249, 250]. For example, the construction in [250] requires $m = O(k^2 \log n)$, while the matrices in [249] have $m = O(kn^\alpha)$ for some constant α.

Fortunately, these limitations can be overcome by randomizing the matrix construction. For example, random matrices $\mathbf{A}$ of size $m \times n$ whose entries are iid with continuous distributions have $\mathrm{spark}(\mathbf{A}) = m + 1$ with probability 1 (see Example 11.5). More significantly, random matrices satisfy the RIP with high probability if the entries are chosen according to a Gaussian, Bernoulli, or more generally any sub-Gaussian distribution. To state these results more formally, we introduce the following definition.

Definition 11.6. *A random variable* x *is called a sub-Gaussian random variable if* $P(|x| \geq t) \leq 2 \exp(-t^2/C^2)$ *for some constant* $C > 0$. *A random vector* $\mathbf{x}$ *is called a* sub-Gaussian random vector *if any linear combination of the form* $\mathbf{x}^T\mathbf{a}$ *for an arbitrary vector* $\mathbf{a}$ *is sub-Gaussian. Finally, a random vector* $\mathbf{x}$ *is isotropic if its covariance matrix is equal to the identity.*

Examples of sub-Gaussian random variables are Gaussian, Rademacher (namely $P(x = 1) = P(x = -1) = 1/2$), and all bounded random variables (random variables x such that $|x| \leq C$ almost surely for some C). A vector $\mathbf{x}$ whose elements are iid sub-Gaussian random variables is a sub-Gaussian random vector.

Consider one of the following two models for $\mathbf{A}$:

Row-independent model: the rows of $\mathbf{A}$ are independent sub-Gaussian isotropic random vectors in $\mathbb{R}^n$;

Column-independent model: the columns $\mathbf{a}_i$ of $\mathbf{A}$ are independent sub-Gaussian isotropic random vectors in $\mathbb{R}^m$ with $\|\mathbf{a}_i\|_2 = \sqrt{m}$ almost surely.

We can then state the following theorem ([241], Theorem 5.65).

Theorem 11.11 (RIP of sub-Gaussian matrices). *Let* $\mathbf{A}$ *be an* $m \times n$ *sub-Gaussian random matrix with independent rows or columns, which follows either of the two models above. Then the normalized matrix* $\bar{\mathbf{A}} = \frac{1}{\sqrt{m}}\mathbf{A}$ *satisfies the following for every sparsity level* $1 \leq k \leq n$ *and every* $\delta \in (0,1)$:

$$\text{if} \quad m \geq C\delta^{-2}k\log(en/k) \quad \text{then} \quad \delta_k(\bar{\mathbf{A}}) \leq \delta$$

with probability at least $1 - 2\exp(-c\delta^2 m)$. *The constants* $C, c > 0$ *depend only on the distribution of the rows or columns of* $\mathbf{A}$.

Theorem 11.11 states that if a matrix $\mathbf{A}$ is chosen according to a sub-Gaussian distribution with $m = O\left(k\log(n/k)/\delta_{2k}^2\right)$, then $\mathbf{A}$ will satisfy the RIP of order $2k$ with probability at least $1 - 2\exp(-c\delta_{2k}^2 m)$. Note that in light of the measurement bound of Theorem 11.10, this achieves the optimal number of measurements up to a constant. It also follows from Theorem 11.4 that these random constructions provide matrices satisfying the NSP. Similarly, it can be shown that a partial Fourier matrix with $m = O\left(k\log^4 n/\delta_{2k}^2\right)$ rows, namely a matrix formed from an $n \times n$ Fourier matrix by taking m of its rows uniformly at random, satisfies the RIP of order $2k$ with high probability. A similar result holds true for random submatrices of bounded orthogonal matrices. Random matrices that do not necessarily satisfy the RIP can also be used to guarantee unique recovery with high probability; see [251] for a detailed discussion.

Random matrices also result in low coherence: when the distribution used has zero mean and finite variance, then in the asymptotic regime (as m and n grow) the coherence converges to $\mu(\mathbf{A}) = \sqrt{(2\log n)/m}$ [12, 252, 253].

Recently, there has been growing interest in the literature in random matrices that possess further structure. As an example, partial random circulant and Toeplitz matrices are studied in detail in [254]. Random Gabor systems are considered in [255].

Using random matrices to build $\mathbf{A}$ has a number of additional benefits. To illustrate these, we will focus on the RIP. In practice we are often more interested in the setting where $\mathbf{x}$ is sparse with respect to some basis $\mathbf{\Phi}$. In this case, what we actually require is that the product $\mathbf{A}\mathbf{\Phi}$ satisfies the RIP. If we were to use a deterministic construction then we would need to explicitly take $\mathbf{\Phi}$ into account. However, when $\mathbf{A}$ is chosen randomly we can avoid this consideration. For example, if $\mathbf{A}$ is chosen according to a Gaussian distribution and $\mathbf{\Phi}$ is an orthonormal basis then it is easy to see that $\mathbf{A}\mathbf{\Phi}$ also has a Gaussian distribution. Therefore, provided that m is sufficiently high $\mathbf{A}\mathbf{\Phi}$ will satisfy the RIP with high probability, just as before. Although less obvious, similar results hold for sub-Gaussian distributions [245]. This property, sometimes referred to as *universality*, constitutes a significant advantage of using random $\mathbf{A}$ matrices.

Finally, we note that since the fully random matrix approach is sometimes impractical to build in hardware, several architectures have been implemented and/or proposed that enable random measurements to be acquired in practical settings. We will consider examples in Chapters 14 and 15 in the context of sub-Nyquist ADCs. Such architectures typically use a reduced amount of randomness and are modeled via matrices $\mathbf{A}$ that have significantly more structure than a fully random matrix. Perhaps somewhat surprisingly, while it is typically not quite as easy as in the fully random case, it can be shown that many of these constructions also satisfy the RIP and/or have low coherence. Examples include circulant matrices corresponding to filtering with a random filter [254]. Furthermore, one can analyze the effect of inaccuracies in the matrix $\mathbf{A}$ implemented by the system [256]; in the simplest cases, such sensing matrix errors may be addressed through system calibration.

11.4   Recovery algorithms

We now focus on solving the CS recovery problem: given $\mathbf{y}$ and $\mathbf{A}$, find a sparse signal $\mathbf{x}$ such that $\mathbf{y} = \mathbf{A}\mathbf{x}$ exactly or approximately. There are a variety of algorithms that have been used in applications such as sparse approximation, statistics, geophysics, and theoretical computer science that were developed to exploit sparsity in other contexts and can be brought to bear on the CS recovery problem. We briefly review some of these in this section, and refer the reader to the overview in [257] for further details.

We begin by considering a natural first approach to the problem. Given measurements $\mathbf{y}$ and the knowledge that our original signal $\mathbf{x}$ is sparse or compressible, it is natural to attempt to recover $\mathbf{x}$ by solving an optimization problem of the form

$$\hat{\mathbf{x}} = \arg\min_{\mathbf{x}} \|\mathbf{x}\|_0 \quad \text{s.t. } \mathbf{x} \in \mathcal{B}(\mathbf{y}), \tag{11.117}$$

where $\mathcal{B}(\mathbf{y})$ ensures that $\hat{\mathbf{x}}$ is consistent with the measurements $\mathbf{y}$ (here s.t. indicates "subject to"). For example, when the measurements are exact and noise-free, we can set $\mathcal{B}(\mathbf{y}) = \{\mathbf{x}: \mathbf{A}\mathbf{x} = \mathbf{y}\}$. If the measurements are contaminated with a small amount of bounded noise, then we may instead consider $\mathcal{B}(\mathbf{y}) = \{\mathbf{x}: \|\mathbf{A}\mathbf{x} - \mathbf{y}\|_2 \leq \varepsilon\}$ for some $\varepsilon > 0$. In both cases, (11.117) finds the sparsest $\mathbf{x}$ that is consistent with $\mathbf{y}$.

The formulation (11.117) inherently assumes that $\mathbf{x}$ itself is sparse. When $\mathbf{x} = \mathbf{\Phi}\theta$ with θ sparse, we can easily modify the approach to

$$\hat{\theta} = \arg\min_{\theta} \|\theta\|_0 \quad \text{s.t. } \theta \in \mathcal{B}(\mathbf{y}) \tag{11.118}$$

where $\mathcal{B}(\mathbf{y}) = \{\theta: \mathbf{A}\mathbf{\Phi}\theta = \mathbf{y}\}$ or $\mathcal{B}(\mathbf{y}) = \{\theta: \|\mathbf{A}\mathbf{\Phi}\theta - \mathbf{y}\|_2 \leq \varepsilon\}$. By defining $\widetilde{\mathbf{A}} = \mathbf{A}\mathbf{\Phi}$ we see that (11.117) and (11.118) are essentially identical. Moreover, as noted in Section 11.3.5, in many cases the introduction of $\mathbf{\Phi}$ does not significantly complicate the construction of matrices $\mathbf{A}$ such that $\widetilde{\mathbf{A}}$ satisfies the desired properties. Thus, we restrict attention to the case where $\mathbf{\Phi} = \mathbf{I}$. It is important to note, however, that this restriction does impose certain limits in our analysis when $\mathbf{\Phi}$ is a general dictionary and not an orthonormal basis. For example, in this case $\|\hat{\mathbf{x}} - \mathbf{x}\|_2 = \|\mathbf{\Phi}\widehat{\boldsymbol{\theta}} - \mathbf{\Phi}\boldsymbol{\theta}\|_2 \neq \|\widehat{\boldsymbol{\theta}} - \boldsymbol{\theta}\|_2$, and therefore a bound on $\|\widehat{\boldsymbol{\theta}} - \boldsymbol{\theta}\|_2$ cannot directly be translated into a bound on $\|\hat{\mathbf{x}} - \mathbf{x}\|_2$, which is often the metric of interest. For further discussion of these and related issues see [258].

Solving (11.117) relies on an exhaustive search over all possible support sets of size k (as we discussed, for instance, in Example 11.9). For increasing k, starting from $k = 1$ up to $k = \lfloor m/2 \rfloor$, we assume $\mathbf{x} \in \Sigma_k$ and consider every possible support Λ of size k. Given a choice of Λ, we replace $\mathbf{x}$ by $\mathbf{x}_\Lambda$ and solve the reduced system as in (11.18). In the noise-free setting with $\mathcal{B}(\mathbf{y}) = \{\mathbf{x}: \mathbf{A}\mathbf{x} = \mathbf{y}\}$, the solution coincides with the true signal $\mathbf{x}$ as long as $\mathbf{A}$ has the sparse solution uniqueness property (i.e. for m as small as $2k$; see Theorems 11.1 and 11.6). However, this algorithm has combinatorial computational complexity, since we must check whether the measurement vector $\mathbf{y}$ belongs to the span of each set of k columns of $\mathbf{A}$. In fact, one can show that for a general matrix $\mathbf{A}$, even finding a solution that approximates the true minimum is NP-hard [259, Sec. 0.8.2]. Our goal, therefore, is to find computationally feasible algorithms that can successfully recover a sparse vector $\mathbf{x}$ from a measurement vector $\mathbf{y}$ for the smallest possible number of measurements m.

11.4.1 ℓ_1 recovery

One avenue for translating (11.117) into a tractable problem is to replace $\|\cdot\|_0$ with its convex approximation $\|\cdot\|_1$. Specifically, we consider

$$\hat{\mathbf{x}} = \arg\min_{\mathbf{x}} \|\mathbf{x}\|_1 \quad \text{s.t. } \mathbf{x} \in \mathcal{B}(\mathbf{y}). \tag{11.119}$$

Provided that $\mathcal{B}(\mathbf{y})$ is convex, (11.119) is computationally feasible. When $\mathcal{B}(\mathbf{y}) = \{\mathbf{x}: \mathbf{A}\mathbf{x} = \mathbf{y}\}$, (11.119) reduces to

$$\hat{\mathbf{x}} = \arg\min_{\mathbf{x}} \|\mathbf{x}\|_1 \quad \text{s.t. } \mathbf{y} = \mathbf{A}\mathbf{x}. \tag{11.120}$$

This problem can be posed as a linear program [260] making its computational complexity polynomial in the signal length. The resulting algorithm is referred to in the literature as *basis pursuit (BP)*.[4]

In the presence of noise, we choose $\mathcal{B}(\mathbf{y}) = \{\mathbf{x}: \|\mathbf{A}\mathbf{x} - \mathbf{y}\|_2 \leq \varepsilon\}$ in (11.119), resulting in a quadratic optimization problem, which again is solvable using known polynomial-time methods [124]. The Lagrangian relaxation of this quadratic program is written as

$$\hat{\mathbf{x}} = \arg\min_{\mathbf{x}} \|\mathbf{x}\|_1 + \lambda\|\mathbf{y} - \mathbf{A}\mathbf{x}\|_2, \tag{11.121}$$

for some $\lambda > 0$, and is known as *basis pursuit denoising (BPDN)*. There exist many efficient algorithms and solvers to find BPDN solutions; see e.g. [262, 263, 264, 265,

---

[4] A similar set of recovery algorithms, known as total variation minimizers, operate on the gradient of $\mathbf{x}$. When $\mathbf{x}$ represents an image, for example, the gradient often exhibits sparsity, especially when the image is piecewise smooth [261].

266, 267, 268, 269, 270, 271]. In the examples in this chapter we use a popular software package called CVX [272] which is a Matlab-based modeling system for convex optimization. As we show, it is very simple to use and can be applied to solving many of the optimization problems encountered in the context of CS.

Oftentimes, a bounded-norm noise model is overly pessimistic, and it may be reasonable instead to assume that the noise is random. For example, additive white Gaussian noise $\mathbf{w} \sim \mathcal{N}(0, \sigma^2 \mathbf{I})$ is a common choice. The BPDN algorithm can be used to treat such problems where λ is adjusted to take the noise level into account [273]. Bayesian approaches can also be designed to address stochastic noise by further assuming a prior on the the set of observable signals [274]. Other optimization-based approaches have also been formulated in this case; one of the most popular techniques is the Dantzig selector [225]:

$$\hat{\mathbf{x}} = \arg \min_{\mathbf{x}} \|\mathbf{x}\|_1 \quad \text{s.t.} \ \|\mathbf{A}^T(\mathbf{y} - \mathbf{Ax})\|_\infty \leq \lambda \sqrt{\log n} \sigma, \tag{11.122}$$

where λ is a parameter that controls the probability of successful recovery.

Example 11.21 In this example, we explore the influence of λ in BPDN (11.121) in a simple setting. Consider the matrix

$$\mathbf{A} = \frac{1}{\sqrt{4.44}} \begin{bmatrix} 1.2 & -1 & -1.2 & 1 & -1 \\ -1 & 1 & -1 & 1.2 & 1.2 \\ 1 & 1.2 & -1 & -1 & 1 \\ 1 & -1 & 1 & 1 & 1 \end{bmatrix}. \tag{11.123}$$

We first treat the noise-free setting and let

$$\mathbf{b} = \mathbf{Ax} = \begin{bmatrix} 0.2 & 0 & 2.2 & 0 \end{bmatrix}^T. \tag{11.124}$$

Using ℓ_1 minimization, we obtain perfect recovery (see implementation in CVX below):

$$\mathbf{x} = \begin{bmatrix} \sqrt{4.44} & \sqrt{4.44} & 0 & 0 & 0 \end{bmatrix}^T. \tag{11.125}$$

Note, that in this example, $\mu(\mathbf{A}) = 0.5405$ so that the bound in Theorem 11.6 implies that recovery is possible for $k = 1$. Here we show recovery of a signal with sparsity $k = 2$ demonstrating that Theorem 11.6 is often pessimistic.

Next, consider additive white Gaussian noise $\mathbf{w} \sim \mathcal{N}(0, \sigma^2 \mathbf{I})$ and denote $\mathbf{y} = \mathbf{b} + \mathbf{w}$. We numerically explore the influence of λ in BPDN. To this end, we generate a white Gaussian noise vector and solve (11.121) for various λ. The experiment is repeated 1000 times for each λ. Figure 11.10 shows the normalized recovery error, namely the norm of the error between the true vector and the recovered vector $\mathbf{x}$, divided by the norm of the true vector as a function of λ, for $\sigma^2 = 0.0001$ and $\sigma^2 = 1$. BPDN can be implemented in CVX as shown below.

As can be seen in the figure, when the variance of the noise is small, the solution of BPDN is equal to that of BP for λ high enough. The recovery error is then zero. In noisier settings, BPDN outperforms BP for some λ. (The BP solution is obtained when λ is large.)

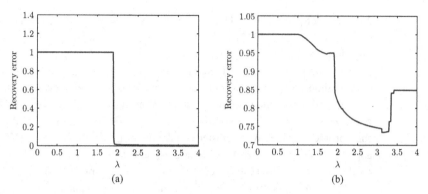

Figure 11.10 Influence of λ on the (normalized) recovery error in BPDN with additive white Gaussian noise with (a) $\sigma^2 = 0.0001$ and (b) $\sigma^2 = 1$.

CVX code for BP without noise

```
m = 4; n =5;
A=1/sqrt(4.44)*[1.2 -1 1 1; -1 1 1.2 -1; -1.2 -1 -1 1;
                              1 1.2 -1 1; -1 -0.8 1 1]';
b=[0.2; 0; 2.2; 0];
cvx_begin
  variable x(n)
  minimize(norm(x,1))
  subject to
    A*x == b
cvx_end
```

CVX code for BPDN with additive white Gaussian noise

```
m = 4; n =5;
A=1/sqrt(4.44)*[1.2 -1 1 1; -1 1 1.2 -1; -1.2 -1 -1 1;
                              1 1.2 -1 1; -1 -0.8 1 1]';
b=[0.2; 0; 2.2; 0];
sigma=0.1;
w=sigma*randn(m,1);
y=b+w;
lambda=1;
cvx_begin
  variable x(n)
  minimize(norm(x,1)+lambda*norm(y-A*x,2))
cvx_end
```

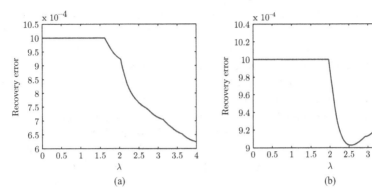

Figure 11.11 Influence of λ on the (normalized) recovery error in BPDN using a Gaussian matrix in the presence of additive white Gaussian noise with (a) $\sigma^2 = 0.01$ and (b) $\sigma^2 = 1$.

Example 11.22 We now examine the influence of λ in BPDN in a problem of larger scale. We choose $m = 32$, $n = 64$ and consider a real-valued random matrix $\mathbf{A}$ whose entries are drawn independently from a Gaussian distribution with zero mean and variance 1. We construct a 10-sparse vector $\mathbf{x}$ of size 64×1. The nonzero locations are drawn uniformly at random, and the nonzero values are drawn from a Gaussian distribution as described above. The experiment is repeated 1000 times for each λ. Figure 11.11 shows the normalized recovery error, namely the norm of the error between the true and recovered vectors, divided by the norm of the true vector, as a function of λ, for $\sigma^2 = 0.01$ and $\sigma^2 = 1$.

While it is clear that replacing (11.117) with (11.119) transforms a computationally intractable problem into a tractable one, it may not be immediately obvious that the solution to (11.119) will be at all similar to the solution to (11.117). However, there are certainly intuitive reasons to expect that the use of ℓ_1 minimization will indeed promote sparsity. As an example, recall that in Fig. 11.2, the solution to the ℓ_1 minimization problem coincided exactly with the solution to the ℓ_p minimization problem for any $p < 1$, and notably, was sparse. Moreover, the use of ℓ_1 minimization to promote or exploit sparsity has a long history, dating back at least to the work of Beurling on Fourier transform extrapolation from partial observations [198]. In Example 11.21, we showed a case in which ℓ_1 perfectly recovers the signal when there is no noise, and provides a reasonable approximation in the presence of noise.

Additionally, in a somewhat different context, in 1965 Logan [275, 276] showed that a bandlimited signal can be perfectly recovered in the presence of arbitrary corruptions on a small interval (see also extensions of these conditions in [276]). Again, the recovery method consists of searching for the bandlimited signal that is closest to the observed signal in the L_1-norm. This can be viewed as further validation of the intuition gained from Fig. 11.2 – the ℓ_1-norm is well-suited to sparse errors.

Historically, the use of ℓ_1 minimization on large problems finally became practical with the explosion of computing power in the late 1970s and early 1980s. In one of its first applications, it was demonstrated that geophysical signals consisting of spike trains could be recovered from only the high-frequency components of these signals by exploiting ℓ_1 minimization [277, 278, 279]. Finally, in the 1990s there was renewed interest in these approaches within the signal processing community for the purpose of finding sparse approximations to signals and images when represented in overcomplete dictionaries or unions of bases [59, 260]. Separately, ℓ_1 minimization received significant attention in the statistics literature as a method for variable selection in regression, known as the Lasso [280]. In Section 11.5.1 we prove that under appropriate conditions on $\mathbf{A}$, ℓ_1 can indeed recover a sparse vector from a small number of measurements and provide a good approximation of the unknown signal in the presence of noise.

11.4.2   Greedy algorithms

An alternative to optimization-based approaches are *greedy algorithms* for sparse signal recovery [281, 282, 283, 284, 285, 286, 287, 288, 289, 290, 291]. These methods are iterative in nature and select columns of $\mathbf{A}$ according to their correlation with the measurements $\mathbf{y}$ determined by an appropriate inner product. Some greedy methods can actually be shown to have performance guarantees that match those obtained for convex optimization approaches, as we discuss in Section 11.5.

Broadly speaking, greedy algorithms can be divided into two categories:

1. Greedy pursuits, which build up an estimate of $\mathbf{x}$ by starting from $\mathbf{0}$ and iteratively adding new components. At each iteration, the values of the nonzero elements are optimized according to an appropriate criterion. Representatives of this class are the matching pursuit (MP) and orthogonal matching pursuit (OMP) algorithms [281, 292].
2. Thresholding algorithms, which in each stage both select sets of nonzero elements, and remove unwanted ones. Examples include compressive sampling matching pursuit (CoSaMP) [293] and iterative hard thresholding (IHT) [294].

Greedy methods in general have a long history in various disciplines. For example, the MP method in signal processing is very similar to the forward stepwise regression in statistics [295], the pure greedy algorithm in nonlinear approximation [296], and the CLEAN algorithm in radio astronomy [297], although they have all been discovered independently.

Matching pursuit

One of the simplest pursuit algorithms is MP [281]. This method was inspired by the concept of projection pursuit [298], in which a given data set is projected onto some direction and then tested for deviation from Gaussianity. The notion of projecting data onto different directions was adopted by Mallat and Zhang [281] in MP for signal approximation. Both MP and its extension, OMP, proceed by finding the column of $\mathbf{A}$ most correlated to the signal residual, which is obtained by subtracting the contribution of a partial estimate of the signal from $\mathbf{y}$. Once the support is determined, the coefficients over the support set are updated. The difference between the two techniques is in the coefficient update stage. Whereas in OMP all nonzero elements are chosen so as

---

Algorithm 11.1 Orthogonal matching pursuit

Input: CS matrix $\mathbf{A}$, measurement vector $\mathbf{y}$
Output: Sparse representation $\hat{\mathbf{x}}$
Initialize: $\hat{\mathbf{x}}_0 = \mathbf{0}, \mathbf{r} = \mathbf{y}, \Lambda = \emptyset, \ell = 0$
while halting criterion false **do**
    $\ell \leftarrow \ell + 1$
    $\mathbf{b} \leftarrow \mathbf{A}^*\mathbf{r}$ {form residual signal estimate}
    $\Lambda \leftarrow \Lambda \cup \operatorname{supp}(\mathcal{T}(\mathbf{b}, 1))$ {update support with residual}
    $\hat{\mathbf{x}}_\ell|_\Lambda \leftarrow \mathbf{A}_\Lambda^\dagger \mathbf{y}, \hat{\mathbf{x}}_\ell|_{\Lambda^c} \leftarrow \mathbf{0}$ {update signal estimate}
    $\mathbf{r} \leftarrow \mathbf{y} - \mathbf{A}\hat{\mathbf{x}}_\ell$ {update measurement residual}
end while
return $\hat{\mathbf{x}} \leftarrow \hat{\mathbf{x}}_\ell$

---

to minimize the residual error, in MP only the component associated with the currently selected column is updated.

To develop MP/OMP, we begin by seeking a one-sparse vector that best explains the data in a squared-error sense and assume for simplicity that all variables are real valued. In this case, $\mathbf{Ax} = \mathbf{a}_i x$ for some index i and scalar x so that our problem becomes

$$\min_{i,x} \|\mathbf{y} - \mathbf{a}_i x\|^2. \tag{11.126}$$

By setting the derivative to zero it is easy to see that the solution is given by

$$i = \arg\max_j \frac{|\mathbf{a}_j^*\mathbf{y}|^2}{\|\mathbf{a}_j\|^2}, \quad x = \frac{\mathbf{a}_i^*\mathbf{y}}{\|\mathbf{a}_i\|^2}. \tag{11.127}$$

After finding the first index, we update the residual by subtracting from $\mathbf{y}$ the contribution of the ith column, and continue iterating. MP and OMP remove the effect of the selected column in different ways.

In MP the residual is updated by subtracting from it the contribution of $\mathbf{a}_i$ assuming it is the only active vector, as given by (11.127). Thus, $\mathbf{r}_0 = \mathbf{y}$ and at each iteration ℓ (see Exercise 18),

$$\mathbf{r}_\ell = \mathbf{r}_{\ell-1} - \frac{\mathbf{a}_i^*\mathbf{r}_{\ell-1}}{\|\mathbf{a}_i\|_2^2}\mathbf{a}_i. \tag{11.128}$$

The signal estimate at the ℓth stage is correspondingly given by

$$\hat{\mathbf{x}}_\ell \leftarrow \hat{\mathbf{x}}_{\ell-1}, \quad \hat{\mathbf{x}}_\ell|_i \leftarrow \frac{\mathbf{a}_i^*\mathbf{r}_{\ell-1}}{\|\mathbf{a}_i\|_2^2}. \tag{11.129}$$

On the other hand, in the OMP algorithm, the signal update step is the solution to

$$\min_{\mathbf{x}_\Lambda} \|\mathbf{y} - \mathbf{A}_\Lambda \mathbf{x}_\Lambda\|^2, \tag{11.130}$$

which is given by $\mathbf{x}_\Lambda = \mathbf{A}_\Lambda^\dagger \mathbf{y}$. Here Λ is the current estimated support set. Thus, all elements of $\mathbf{x}_\Lambda$ are updated at each iteration, and not just the current support index as in

MP. Computationally, this step can be performed efficiently by maintaining a QR factorization of $\mathbf{A}_\Lambda$. Since the objective in (11.130) is convex, we find its optimal solution by setting the derivative to zero which results in

$$\mathbf{A}_\Lambda^*(\mathbf{y} - \mathbf{A}_\Lambda \mathbf{x}_\Lambda) = \mathbf{A}_\Lambda^*\mathbf{r} = 0, \tag{11.131}$$

where $\mathbf{r}$ here denotes the current residual. Thus, in each iteration, the residual is orthogonal to the columns of $\mathbf{A}$ that are included in the current estimated support, giving OMP its name. This orthogonality also ensures that these columns will not be chosen again in future iterations.

The OMP method (assuming that the columns of $\mathbf{A}$ are normalized for simplicity) is formally defined by Algorithm 11.1, where $\mathcal{T}(\mathbf{x}, k)$ denotes a *thresholding* operator on $\mathbf{x}$ that sets all but the k entries of $\mathbf{x}$ with the largest magnitudes to zero, and $\mathbf{x}|_\Lambda$ denotes the restriction of $\mathbf{x}$ to the entries indexed by Λ.

Similarly to MP, the residual in the OMP algorithm can also be written in a recursive form [292] (see Exercise 19). Let

$$\beta_j = \mathbf{a}_j - \sum_{i \in \Lambda} \langle \mathbf{a}_j, \mathbf{a}_i \rangle \mathbf{a}_i, \tag{11.132}$$

where Λ is the support of $\hat{\mathbf{x}}_{\ell-1}$ as defined in Algorithm 11.1 and j is the new support index. Then

$$\mathbf{r}_\ell = \mathbf{r}_{\ell-1} - \frac{\mathbf{a}_j^*\mathbf{r}_{\ell-1}}{\|\beta_j\|_2^2}\beta_j. \tag{11.133}$$

The convergence criterion used to find sparse representations consists of checking whether $\mathbf{y} = \mathbf{A}\hat{\mathbf{x}}$ exactly or approximately, or a limit on the number of iterations, which also limits the number of nonzeros in $\hat{\mathbf{x}}$. Alternative approaches consist of measuring the norm of the residual error and stopping when the error does not decrease significantly, or stopping when the norm of the difference between consecutive estimates is small.

There are several variations of MP/OMP in which instead of selecting a single element to add to the support at each iteration, a group of elements are chosen simultaneously, a process referred to as stagewise selection [284]. A natural approach is to select all support elements corresponding to $|b_i| > \alpha$ with $\mathbf{b} = \mathbf{A}^*\mathbf{r}$ and α an appropriate threshold value (which may change between iterations). Another modification is to allow for support adaptation once the full support is found, by swapping between indices in the support and out of it. This allows correction for possible erroneous support selection throughout the iterations and often improves performance [299].

Thresholding algorithms

Other greedy techniques that have a similar flavor to OMP include CoSaMP [293], detailed as Algorithm 11.2, and subspace pursuit (SP) [282] (we assume here once more that the columns of $\mathbf{A}$ are normalized). These two algorithms are very similar, and therefore treated jointly. Both CoSaMP and SP keep track of an active set Λ and add as well as remove elements in each iteration. Specifically, at each iteration, a k-sparse approximation is used to compute the current error. The $2k$ columns of $\mathbf{A}$ that correlate best with this error are then selected and added to the support set (in SP only the k best columns are used). An LS estimate is then found over the current support. The k largest elements are identified, and their corresponding locations are chosen as the new support set. CoSaMP and SP differ in the last step of residual update: while CoSaMP uses the output of the thresholding step, SP optimizes the error over the new support.

Algorithm 11.2 CoSaMP

Input: CS matrix $\mathbf{A}$, measurement vector $\mathbf{y}$, sparsity k
Output: k-sparse representation $\hat{\mathbf{x}}$
Initialize: $\hat{\mathbf{x}}_0 = \mathbf{0}, \mathbf{r} = \mathbf{y}, \ell = 0$
while halting criterion false **do**
$\quad \ell \leftarrow \ell + 1$
$\quad \mathbf{e} \leftarrow \mathbf{A}^*\mathbf{r}$ {form residual signal estimate}
$\quad \Lambda \leftarrow \mathrm{supp}(\mathcal{T}(\mathbf{e}, 2k))$ {prune residual}
$\quad T \leftarrow \Lambda \cup \mathrm{supp}(\hat{\mathbf{x}}_{\ell-1})$ {merge supports}
$\quad \mathbf{b}|_T \leftarrow \mathbf{A}_T^\dagger \mathbf{y}, \mathbf{b}|_{T^C} \leftarrow \mathbf{0}$ {form signal estimate}
$\quad \hat{\mathbf{x}}_\ell \leftarrow \mathcal{T}(\mathbf{b}, k)$ {prune signal using model}
$\quad \mathbf{r} \leftarrow \mathbf{y} - \mathbf{A}\hat{\mathbf{x}}_\ell$ {update measurement residual}
end while
return $\hat{\mathbf{x}} \leftarrow \hat{\mathbf{x}}_\ell$

A simpler variant of CoSaMP and SP that is based only on thresholding steps is known as IHT [294]: starting from an initial estimate $\hat{\mathbf{x}}_0 = 0$, the algorithm iterates a gradient descent step followed by hard thresholding, i.e.

$$\hat{\mathbf{x}}_i = \mathcal{T}(\hat{\mathbf{x}}_{i-1} + \mathbf{A}^*(\mathbf{y} - \mathbf{A}\hat{\mathbf{x}}_{i-1}), k) \tag{11.134}$$

until a convergence criterion is met.

Finally, we mention that there also exist iterative sparse recovery techniques based on message passing schemes for sparse graphical models [300].

Example 11.23 In this example, we compare the performance of five different recovery algorithms: ℓ_1 minimization, OMP, MP, IHT, and CoSaMP as a function of the sparsity level k. We use a residual-based halting criterion for the OMP. The IHT algorithm is provided with the true sparsity.

We consider two different dictionaries: a real-valued random matrix $\mathbf{A}$ whose entries are drawn independently from a Gaussian distribution with zero mean and variance one and a random partial Fourier matrix. For each value of the sparsity $1 \leq k \leq m$, we construct a k-sparse vector $\mathbf{x}$ of size $n \times 1$. The nonzero locations are drawn uniformly at random, and the nonzero values are drawn from a Gaussian distribution as described above. The experiment is repeated 5000 times for each sparsity value. We choose $m = 512$ and $n = 1024$. Figure 11.12 shows the normalized recovery error of the different reconstruction algorithms.

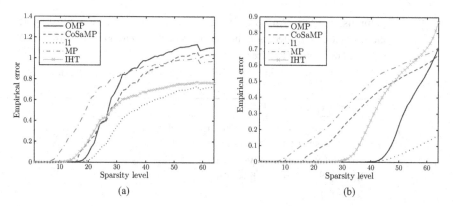

Figure 11.12 Performance comparison of ℓ_1 minimization, OMP, MP, IHT, and CoSaMP as a function of the sparsity level for two choices of **A**. (a) Gaussian matrix. (b) Random partial Fourier matrix.

11.4.3   Combinatorial algorithms

In addition to ℓ_1 minimization and greedy approaches, there is another important class of sparse recovery methods referred to as *combinatorial algorithms*, which have been mostly developed by the theoretical computer science community.

The historically oldest of these algorithms were developed in the context of *combinatorial group testing* [301]. In this problem we suppose that there are n total items and k anomalous elements that we wish to find. For example, we might wish to identify defective products in an industrial setting, or identify a subset of diseased tissue samples in a medical context. In both cases the vector **x** indicates which elements are anomalous, i.e. $x_i \neq 0$ for the k anomalous elements and $x_i = 0$ otherwise. Our goal is to design a collection of tests that allow us to identify the support (and possibly the values of the nonzeros) of **x** while also minimizing the number of tests performed. In the simplest practical setting these tests are represented by a binary matrix **A** whose entries a_{ij} are equal to one if and only if the jth item is used in the ith test. If the output of the test is linear with respect to the inputs, then the problem of recovering the vector **x** is essentially the same as the standard sparse recovery problem in CS.

Another application area in which combinatorial algorithms have proven useful is computation on *data streams* [259]. As an example of a typical data streaming problem, suppose that x_i represents the number of packets passing through a network router with destination i. Simply storing the vector **x** is typically infeasible since the total number of possible destinations (represented by a 32-bit IP address) is $n = 2^{32}$. Thus, instead of attempting to store **x** directly, one can store **y** = **Ax** where **A** is an $m \times n$ matrix with $m \ll n$. In this context the vector **y** is often called a *sketch*. Note that in this problem **y** is computed in a different manner than in the CS context. Specifically, in the network traffic example we never observe x_i directly; rather, we view increments to x_i (when a packet with destination i passes through the router). Thus we construct **y** iteratively by adding the ith column to **y** each time we observe an increment to x_i, which we can do since **y** = **Ax** is linear. When the network traffic is dominated by traffic to a small number of destinations, the vector $\mathbf{x}$ is compressible, and thus the problem of recovering $\mathbf{x}$ from the sketch $\mathbf{Ax}$ is again essentially the same as the sparse recovery problem in CS.

Despite the fact that in both of these settings we ultimately wish to recover a sparse signal from a small number of linear measurements, there are also some important differences with CS. First, in CS and in data streams, it is natural to assume that the designer of the reconstruction algorithm also has full control over $\mathbf{A}$, and is thus free to choose $\mathbf{A}$ in a manner that reduces the amount of computation required to perform recovery. For example, it is often useful to choose $\mathbf{A}$ so that it has very few nonzeros [302]. This additional optimization can lead to significantly faster algorithms [303, 304, 305].

Second, note that the computational complexity of the convex and greedy algorithms described above is at least linear in n, since in order to recover $\mathbf{x}$ we must at least incur the computational cost of reading out all n entries of $\mathbf{x}$. While this may be acceptable in most typical CS applications, it becomes impractical when n is extremely large, as in the network monitoring example. In this context, one may seek to develop algorithms whose complexity is linear only in k. Such algorithms do not return a complete reconstruction of $\mathbf{x}$ but instead determine only its k largest elements (and their indices); see [303, 304, 305] for examples.

11.4.4   Analysis versus synthesis methods

Until now we have focused our attention on signals that are sparse in some basis so that $\mathbf{x} = \mathbf{\Phi}\theta$ for a sparse vector θ. Our goal was to seek a sparse θ that matches the measurements $\mathbf{y} = \mathbf{A}\mathbf{\Phi}\theta$:

$$\hat{\theta} = \arg\min_{\theta} \|\theta\|_0 \quad \text{s.t. } \mathbf{y} = \mathbf{A}\mathbf{\Phi}\theta. \tag{11.135}$$

In this setting, we are in effect synthesizing $\mathbf{x}$ using a sparse representation. This approach therefore is referred to as the *synthesis method*.

An alterative viewpoint is the *analysis method* in which we assume that there is an analysis operator $\mathbf{D}$ such that $\mathbf{D}^*\mathbf{x}$ is sparse. Examples include the finite-difference operator, the wavelet transform, and more. In this case, (11.135) is replaced by

$$\hat{\theta} = \arg\min_{\mathbf{x}} \|\mathbf{D}^*\mathbf{x}\|_0 \quad \text{s.t. } \mathbf{y} = \mathbf{A}\mathbf{x}. \tag{11.136}$$

If $\mathbf{D}$ and $\mathbf{\Phi}$ are square and invertible, then both approaches coincide by choosing $\mathbf{D}^* = \mathbf{\Phi}^{-1}$ and $\mathbf{x} = \mathbf{\Phi}\theta$. In particular, when $\mathbf{\Phi}$ is an orthonormal basis, $\mathbf{D} = \mathbf{\Phi}$. However, for general dictionaries, the methods differ, and lead to distinct solutions.

The theory and algorithms developed for the synthesis model may be adapted to treat the analysis setting as well. For example, the parallel of BP results from replacing the ℓ_0-norm in (11.136) by the ℓ_1-norm [258]. Greedy methods can also be extended to this formulation. The interested reader is referred to [306] for further details.

To develop recovery guarantees for the analysis setting, a generalization of the RIP (see Section 11.3.2) was suggested in [258] and is termed D-RIP.

Definition 11.7. *A matrix* $\mathbf{A}$ *satisfies the* restricted isometry property adapted to $\mathbf{D}$ (D-RIP) *of order* k *if there exists a* $\delta_k \in (0, 1)$ *such that*

$$(1 - \delta_k)\|\mathbf{D}\mathbf{x}\|_2^2 \leq \|\mathbf{A}\mathbf{D}\mathbf{x}\|_2^2 \leq (1 + \delta_k)\|\mathbf{D}\mathbf{x}\|_2^2, \tag{11.137}$$

for all $\mathbf{x} \in \Sigma_k$.

Note that (11.137) can be viewed as requiring the standard RIP to hold for vectors of the form $\mathbf{v} = \mathbf{D}\mathbf{x}$ where $\mathbf{x}$ is sparse. As an example of the results available using the analysis approach, it is shown in [258, 307] that if $\mathbf{A}$ satisfies the D-RIP of order $2k$ with $\delta_{2k} < 0.493$, then the solution $\mathbf{x}'$ of (11.136) satisfies

$$\|\mathbf{x}' - \mathbf{x}\|_2 \leq \frac{C}{\sqrt{k}}\|\mathbf{D}^*\mathbf{x} - \sigma_k(\mathbf{D}^*\mathbf{x})_1\|_1, \tag{11.138}$$

where C is a constant. In particular, if $\mathbf{D}^*\mathbf{x}$ is k-sparse, then (11.136) will recover it exactly. This result can be extended to the noisy setting as well.

The analysis approach may be useful, for example, when $\mathbf{x} = \mathbf{D}\theta$ and $\mathbf{D}$ is a highly coherent dictionary. In this case, the RIP will generally not hold, since the columns of $\mathbf{A}\mathbf{D}$ will typically be highly correlated. However, the D-RIP can still be satisfied. Therefore, it may be beneficial in this scenario to recover $\mathbf{x}$ by solving (11.136) rather than using (11.135).

11.5 Recovery guarantees

In the previous section we described a variety of methods for determining a sparse vector $\mathbf{x}$ from measurements $\mathbf{y} = \mathbf{A}\mathbf{x}$ where the number of measurements m is much smaller than the dimension n of $\mathbf{x}$. We also saw some performance examples for random input signals. We now study systematically under what conditions we can guarantee that sparse recovery is possible using these computationally efficient approaches. As we will see, many of the CS algorithms come with guarantees on their performance. However, it is important to note at the outset that all the results we present are pessimistic in nature. In other words, they provide worst-case bounds under which recovery is possible. In practice, as can be seen by examining the various examples, recovery is often ensured for much more relaxed conditions than those stated by the results below.

In our discussion, we distinguish between recovery guarantees for ℓ_1 optimization, and for greedy methods. For each class of algorithms, we further group the results according to the matrix metric used to obtain the guarantee: RIP-based results, and coherence-based results. We first state results based on RIP and then turn to treat coherence guarantees.

Proving the various recovery results is typically quite involved. Since the focus of the book is not on CS, but rather more on its use in the context of analog signal sampling and recovery, we will not detail the proofs of all methods. Nonetheless, in order to give a flavor of the developments, we provide full proofs for the first class of recovery guarantees: ℓ_1 recovery when $\mathbf{A}$ satisfies the RIP. For the remaining cases (ℓ_1 recovery based on coherence and recovery guarantees for greedy algorithms) we state the results and refer the interested reader to where the proofs can be found.

11.5.1 ℓ_1 recovery: RIP-based results

We begin by studying the properties of ℓ_1 recovery when $\mathbf{A}$ satisfies the RIP. We first consider the case in which the measurements are noise-free and then turn to treat the noisy setting.

Noise-free signal recovery
In order to analyze ℓ_1 minimization algorithms using the RIP for various specific choices of $\mathcal{B}(\mathbf{y})$, we require the following general result which builds on Proposition 11.6.

Proposition 11.9. *Suppose that $\mathbf{A}$ satisfies the RIP of order $2k$ with $\delta_{2k} < \sqrt{2}-1$. Let $\mathbf{x}, \hat{\mathbf{x}} \in \mathbb{R}^n$ be given, and define $\mathbf{h} = \hat{\mathbf{x}} - \mathbf{x}$. Let Λ_0 denote the index set corresponding to the k entries of $\mathbf{x}$ with largest magnitude and Λ_1 the index set corresponding to the k entries of $\mathbf{h}_{\Lambda_0^c}$ with largest magnitude. Set $\Lambda = \Lambda_0 \cup \Lambda_1$. If $\|\hat{\mathbf{x}}\|_1 \le \|\mathbf{x}\|_1$, then*

$$\|\mathbf{h}\|_2 \le C_0 \frac{\sigma_k(\mathbf{x})_1}{\sqrt{k}} + C_1 \frac{|\langle \mathbf{A}\mathbf{h}_\Lambda, \mathbf{A}\mathbf{h}\rangle|}{\|\mathbf{h}_\Lambda\|_2}, \tag{11.139}$$

where

$$C_0 = 2\frac{1-(1-\sqrt{2})\delta_{2k}}{1-(1+\sqrt{2})\delta_{2k}}, \quad C_1 = \frac{2}{1-(1+\sqrt{2})\delta_{2k}}. \tag{11.140}$$

Proof: The key ideas in this proof follow from [227].

We begin by observing that $\mathbf{h} = \mathbf{h}_\Lambda + \mathbf{h}_{\Lambda^c}$, so that from the triangle inequality

$$\|\mathbf{h}\|_2 \le \|\mathbf{h}_\Lambda\|_2 + \|\mathbf{h}_{\Lambda^c}\|_2. \tag{11.141}$$

We first aim to bound $\|\mathbf{h}_{\Lambda^c}\|_2$. From Lemma 11.1 we have

$$\|\mathbf{h}_{\Lambda^c}\|_2 = \left\|\sum_{j \ge 2} \mathbf{h}_{\Lambda_j}\right\|_2 \le \sum_{j \ge 2} \|\mathbf{h}_{\Lambda_j}\|_2 \le \frac{\|\mathbf{h}_{\Lambda_0^c}\|_1}{\sqrt{k}}, \tag{11.142}$$

where the Λ_j are defined as in Lemma 11.1, i.e. Λ_1 is the index set corresponding to the k largest entries of $\mathbf{h}_{\Lambda_0^c}$ (in absolute value), Λ_2 is the index set corresponding to the next k largest entries, and so on.

We now wish to bound $\|\mathbf{h}_{\Lambda_0^c}\|_1$. Since $\|\mathbf{x}\|_1 \ge \|\hat{\mathbf{x}}\|_1$, by applying the triangle inequality we obtain

$$\begin{aligned}
\|\mathbf{x}\|_1 \ge \|\mathbf{x} + \mathbf{h}\|_1 &= \|\mathbf{x}_{\Lambda_0} + \mathbf{h}_{\Lambda_0}\|_1 + \|\mathbf{x}_{\Lambda_0^c} + \mathbf{h}_{\Lambda_0^c}\|_1 \\
&\ge \|\mathbf{x}_{\Lambda_0}\|_1 - \|\mathbf{h}_{\Lambda_0}\|_1 + \|\mathbf{h}_{\Lambda_0^c}\|_1 - \|\mathbf{x}_{\Lambda_0^c}\|_1.
\end{aligned} \tag{11.143}$$

Rearranging and again applying the triangle inequality,

$$\left\|\mathbf{h}_{\Lambda_0^c}\right\|_1 \leq \|\mathbf{x}\|_1 - \|\mathbf{x}_{\Lambda_0}\|_1 + \|\mathbf{h}_{\Lambda_0}\|_1 + \left\|\mathbf{x}_{\Lambda_0^c}\right\|_1$$

$$\leq \|\mathbf{x} - \mathbf{x}_{\Lambda_0}\|_1 + \|\mathbf{h}_{\Lambda_0}\|_1 + \left\|\mathbf{x}_{\Lambda_0^c}\right\|_1. \tag{11.144}$$

Recalling that $\sigma_k(\mathbf{x})_1 = \left\|\mathbf{x}_{\Lambda_0^c}\right\|_1 = \|\mathbf{x} - \mathbf{x}_{\Lambda_0}\|_1$, we can write

$$\left\|\mathbf{h}_{\Lambda_0^c}\right\|_1 \leq \|\mathbf{h}_{\Lambda_0}\|_1 + 2\sigma_k(\mathbf{x})_1. \tag{11.145}$$

Substituting (11.145) into (11.142) leads to

$$\|\mathbf{h}_{\Lambda^c}\|_2 \leq \frac{\|\mathbf{h}_{\Lambda_0}\|_1 + 2\sigma_k(\mathbf{x})_1}{\sqrt{k}} \leq \|\mathbf{h}_{\Lambda_0}\|_2 + 2\frac{\sigma_k(\mathbf{x})_1}{\sqrt{k}} \tag{11.146}$$

where the last inequality follows from Proposition 11.1. Combining (11.146) with (11.141) and observing that $\|\mathbf{h}_{\Lambda_0}\|_2 \leq \|\mathbf{h}_\Lambda\|_2$ yields

$$\|\mathbf{h}\|_2 \leq 2\|\mathbf{h}_\Lambda\|_2 + 2\frac{\sigma_k(\mathbf{x})_1}{\sqrt{k}}. \tag{11.147}$$

We now bound the norm $\|\mathbf{h}_\Lambda\|_2$. Combining Proposition 11.6 with (11.145) and applying Proposition 11.1 leads to

$$\|\mathbf{h}_\Lambda\|_2 \leq \alpha\frac{\left\|\mathbf{h}_{\Lambda_0^c}\right\|_1}{\sqrt{k}} + \beta\frac{|\langle \mathbf{Ah}_\Lambda, \mathbf{Ah}\rangle|}{\|\mathbf{h}_\Lambda\|_2}$$

$$\leq \alpha\frac{\|\mathbf{h}_{\Lambda_0}\|_1 + 2\sigma_k(\mathbf{x})_1}{\sqrt{k}} + \beta\frac{|\langle \mathbf{Ah}_\Lambda, \mathbf{Ah}\rangle|}{\|\mathbf{h}_\Lambda\|_2}$$

$$\leq \alpha\|\mathbf{h}_{\Lambda_0}\|_2 + 2\alpha\frac{\sigma_k(\mathbf{x})_1}{\sqrt{k}} + \beta\frac{|\langle \mathbf{Ah}_\Lambda, \mathbf{Ah}\rangle|}{\|\mathbf{h}_\Lambda\|_2}. \tag{11.148}$$

Since $\|\mathbf{h}_{\Lambda_0}\|_2 \leq \|\mathbf{h}_\Lambda\|_2$,

$$(1 - \alpha)\|\mathbf{h}_\Lambda\|_2 \leq 2\alpha\frac{\sigma_k(\mathbf{x})_1}{\sqrt{k}} + \beta\frac{|\langle \mathbf{Ah}_\Lambda, \mathbf{Ah}\rangle|}{\|\mathbf{h}_\Lambda\|_2}. \tag{11.149}$$

The assumption $\delta_{2k} < \sqrt{2} - 1$ ensures that $\alpha < 1$. Dividing by $(1 - \alpha)$ and combining with (11.147) results in

$$\|\mathbf{h}\|_2 \leq \left(\frac{4\alpha}{1 - \alpha} + 2\right)\frac{\sigma_k(\mathbf{x})_1}{\sqrt{k}} + \frac{2\beta}{1 - \alpha}\frac{|\langle \mathbf{Ah}_\Lambda, \mathbf{Ah}\rangle|}{\|\mathbf{h}_\Lambda\|_2}. \tag{11.150}$$

Plugging in the values of α and β yields the desired constants. □

Proposition 11.9 establishes an error bound for the class of ℓ_1 minimization algorithms described by (11.119) when combined with a measurement matrix $\mathbf{A}$ satisfying the RIP. In order to obtain specific bounds for concrete examples of $\mathcal{B}(\mathbf{y})$, we must examine how requiring $\hat{\mathbf{x}} \in \mathcal{B}(\mathbf{y})$ affects $|\langle \mathbf{Ah}_\Lambda, \mathbf{Ah}\rangle|$. As an example, in the case of noise-free measurements we obtain the following theorem [227, Theorem 1.1].

Theorem 11.12 (RIP-based noise-free ℓ_1 recovery).   *Suppose that* $\mathbf{A}$ *satisfies the RIP of order* $2k$ *with* $\delta_{2k} < \sqrt{2} - 1$ *and we obtain measurements of the form* $\mathbf{y} = \mathbf{A}\mathbf{x}$. *Then when* $\mathcal{B}(\mathbf{y}) = \{\mathbf{z}: \mathbf{A}\mathbf{z} = \mathbf{y}\}$, *the solution* $\hat{\mathbf{x}}$ *to (11.119) obeys*

$$\|\hat{\mathbf{x}} - \mathbf{x}\|_2 \leq C_0 \frac{\sigma_k(\mathbf{x})_1}{\sqrt{k}}, \tag{11.151}$$

where C_0 *is given by* (11.140).

Proof: Since $\mathbf{x} \in \mathcal{B}(\mathbf{y})$, $\|\hat{\mathbf{x}}\|_1 \leq \|\mathbf{x}\|_1$. We can therefore apply Proposition 11.9 to obtain that for $\mathbf{h} = \hat{\mathbf{x}} - \mathbf{x}$,

$$\|\mathbf{h}\|_2 \leq C_0 \frac{\sigma_k(\mathbf{x})_1}{\sqrt{k}} + C_1 \frac{|\langle \mathbf{A}\mathbf{h}_\Lambda, \mathbf{A}\mathbf{h} \rangle|}{\|\mathbf{h}_\Lambda\|_2}. \tag{11.152}$$

Furthermore, since $\mathbf{x}, \hat{\mathbf{x}} \in \mathcal{B}(\mathbf{y})$ we also have $\mathbf{y} = \mathbf{A}\mathbf{x} = \mathbf{A}\hat{\mathbf{x}}$ so that $\mathbf{A}\mathbf{h} = \mathbf{0}$. Therefore the second term vanishes, leading to the desired result.   □

Theorem 11.12 is rather remarkable. By considering the case where $\mathbf{x} \in \Sigma_k$, we see that provided $\mathbf{A}$ satisfies the RIP – which as shown in Section 11.3.5 allows for as few as $O(k \log(n/k))$ measurements – we can recover any k-sparse $\mathbf{x}$ exactly. Here, O denotes the worst-case complexity, that is the upper bound on the resources required by an algorithm. This result seems rather surprising, and so one might expect that the procedure would be highly sensitive to noise. However, we will see below in Theorem 11.13 that Proposition 11.9 can also be used to demonstrate that this approach is actually stable.

Note that Theorem 11.12 assumes that $\mathbf{A}$ satisfies the RIP. One could easily modify the argument to replace this with the assumption that $\mathbf{A}$ satisfies the NSP instead. Specifically, if we are only interested in the noiseless setting, in which case $\mathbf{h}$ lies in the null space of $\mathbf{A}$, then Proposition 11.9 simplifies and its proof could essentially be broken into two steps: (i) show that if $\mathbf{A}$ satisfies the RIP then it satisfies the NSP (as shown in Theorem 11.4), and (ii) show that the NSP implies the simplified version of Proposition 11.9. This proof directly mirrors that of Proposition 11.9. Thus, by the same argument as in the proof of Theorem 11.12, if $\mathbf{A}$ satisfies the NSP then it will obey the same error bound.

Example 11.24 In this example, we demonstrate the recovery rate of ℓ_1 minimization (11.119) as a function of the sparsity level k. We consider two different dictionaries: a real-valued random matrix $\mathbf{A}$ whose entries are drawn independently from a Gaussian distribution with zero mean and variance one and a random partial Fourier matrix. For each value of the sparsity $1 \leq k \leq m$, we construct a k-sparse vector $\mathbf{x}$ of size $n \times 1$. The nonzero locations are drawn uniformly at random and the nonzero values are drawn from a Gaussian distribution as described above. The experiment is repeated 5000 times for each value of the sparsity. Figure 11.13 shows the empirical recovery rate of the reconstruction algorithm for $m = 32$, $n = 64$ and $m = 64$, $n = 128$, respectively. The empirical recovery rate is calculated as the percentage of correct solutions, for which the normalized recovery error is less than 10^{-1}.

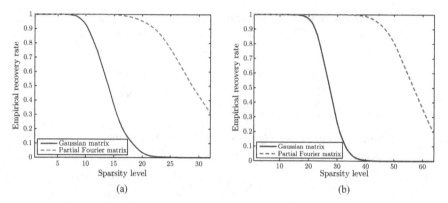

Figure 11.13 Empirical recovery rate of ℓ_1 minimization. (a) $m = 32$ and $n = 64$. (b) $m = 64$ and $n = 128$.

As can be seen in the figure, for small enough values of k, we obtain perfect recovery. It is also interesting to note the threshold effect: as long as k is below a certain value, recovery is possible. Beyond this value the performance deteriorates rapidly, dropping to zero recovery rate. The exact value of this threshold does not follow from our general results since they are all specified up to a constant. However, the figure provides the overall trend.

Signal recovery in noise

The ability to perfectly reconstruct a sparse signal from noise-free measurements represents a very promising result. However, in most real-world systems the measurements are likely to be contaminated by some form of noise. For instance, in order to process data in a computer we must be able to represent it using a finite number of bits, so that the measurements will typically be subject to quantization error. Moreover, systems which are implemented in physical hardware will be subject to a variety of different types of noise depending on the setting. Another important noise source is on the signal itself. In many settings the signal $\mathbf{x}$ to be estimated is contaminated by some form of random noise. The implications of this type of noise on the achievable sampling rates has been recently analyzed in [308, 309, 310]. Here we focus on measurement noise, which has received much more attention in the literature. Furthermore, as shown explicitly in [310], this analysis can be used for the case in which $\mathbf{x}$ is contaminated directly as well.

Perhaps somewhat surprisingly, one can show that it is possible to stably recover sparse signals under a variety of common noise models [191, 311, 312]. As might be expected, both the RIP and coherence are useful in establishing performance guarantees in noise. We begin our discussion below with robustness guarantees for matrices satisfying the RIP. We then turn to results for matrices with low coherence.

We first provide a bound on the worst-case performance for uniformly bounded noise [191, Theorem 1.2].

Theorem 11.13 (RIP-based noisy ℓ_1 recovery). *Suppose that $\mathbf{A}$ satisfies the RIP of order $2k$ with $\delta_{2k} < \sqrt{2} - 1$, and let $\mathbf{y} = \mathbf{A}\mathbf{x} + \mathbf{e}$ where $\|\mathbf{e}\|_2 \leq \varepsilon$. Then, when*

$\mathcal{B}(\mathbf{y}) = \{\mathbf{z}: \|\mathbf{A}\mathbf{z} - \mathbf{y}\|_2 \leq \varepsilon\}$, *the solution $\hat{\mathbf{x}}$ to (11.119) obeys*

$$\|\hat{\mathbf{x}} - \mathbf{x}\|_2 \leq C_0 \frac{\sigma_k(\mathbf{x})_1}{\sqrt{k}} + C_2 \varepsilon, \tag{11.153}$$

where

$$C_0 = 2\frac{1 - (1 - \sqrt{2})\delta_{2k}}{1 - (1 + \sqrt{2})\delta_{2k}}, \quad C_2 = 4\frac{\sqrt{1 + \delta_{2k}}}{1 - (1 + \sqrt{2})\delta_{2k}}.$$

Proof: We are interested in bounding $\|\mathbf{h}\|_2 = \|\hat{\mathbf{x}} - \mathbf{x}\|_2$ with $\mathbf{h} = \hat{\mathbf{x}} - \mathbf{x}$. Since $\|\mathbf{e}\|_2 \leq \varepsilon$, $\mathbf{x} \in \mathcal{B}(\mathbf{y})$, and therefore we know that $\|\hat{\mathbf{x}}\|_1 \leq \|\mathbf{x}\|_1$. Thus we may apply Proposition 11.9 by bounding $|\langle \mathbf{A}\mathbf{h}_\Lambda, \mathbf{A}\mathbf{h} \rangle|$. To do this, we observe that

$$\|\mathbf{A}\mathbf{h}\|_2 = \|\mathbf{A}\hat{\mathbf{x}} - \mathbf{y} + \mathbf{y} - \mathbf{A}\mathbf{x}\|_2 \leq \|\mathbf{A}\hat{\mathbf{x}} - \mathbf{y}\|_2 + \|\mathbf{y} - \mathbf{A}\mathbf{x}\|_2 \leq 2\varepsilon \tag{11.154}$$

where the last inequality follows since $\mathbf{x}, \hat{\mathbf{x}} \in \mathcal{B}(\mathbf{y})$. Combining this with the RIP and the Cauchy–Schwarz inequality we obtain

$$|\langle \mathbf{A}\mathbf{h}_\Lambda, \mathbf{A}\mathbf{h} \rangle| \leq \|\mathbf{A}\mathbf{h}_\Lambda\|_2 \|\mathbf{A}\mathbf{h}\|_2 \leq 2\varepsilon\sqrt{1 + \delta_{2k}} \|\mathbf{h}_\Lambda\|_2. \tag{11.155}$$

Thus,

$$\|\mathbf{h}\|_2 \leq C_0 \frac{\sigma_k(\mathbf{x})_1}{\sqrt{k}} + C_1 2\varepsilon\sqrt{1 + \delta_{2k}} = C_0 \frac{\sigma_k(\mathbf{x})_1}{\sqrt{k}} + C_2 \varepsilon, \tag{11.156}$$

completing the proof. ∎

Example 11.25 We now illustrate the bounds of Theorems 11.12 and 11.13, for both the noise-free and noisy settings. Consider the following 3×4 matrix:

$$\mathbf{A} = \frac{1}{\sqrt{3}} \begin{bmatrix} 1 & -1 & -1 & 1 \\ -1 & 1 & -1 & 1 \\ 1 & 1 & -1 & -1 \end{bmatrix}. \tag{11.157}$$

It is easy to see that $\mu(\mathbf{A}) = \frac{1}{3}$. Thus, from Lemma 11.3, $\mathbf{A}$ satisfies the RIP of order 2 with $\delta_2 \leq \frac{1}{3}$.

Let us first consider noise-free recovery. In Example 11.15, we saw that we can perfectly recover any vector from Σ_1. For any such $\mathbf{x}$, $\sigma_1(\mathbf{x})_1 = 0$. Since $\delta_2 < \sqrt{2} - 1$, Theorem 11.12 also implies that perfect recovery is possible using ℓ_1 optimization.

Now, let $\mathbf{x} = [1\,1\,0\,0]^T$. In Example 11.15 we showed that $\mathbf{x}$ is not recoverable. The solution $\hat{\mathbf{x}}$ to (11.119), obtained using CVX, is $\hat{\mathbf{x}} = 1/2\,[1\,1\,-1\,-1]^T$. For $k = 1$, we have $\sigma_1(\mathbf{x})_1 = 1$ and $C_0 = 2\frac{1-(1-\sqrt{2})\delta_{2k}}{1-(1+\sqrt{2})\delta_{2k}} \approx 11.66$ with $\delta_2 = \frac{1}{3}$. Thus,

$$\|\hat{\mathbf{x}} - \mathbf{x}\|_2 = 1 \leq C_0 \frac{\sigma_1(\mathbf{x})_1}{\sqrt{1}}, \tag{11.158}$$

verifying Theorem 11.12.

We next treat the noisy setting. Assume that $\mathbf{y} = \mathbf{A}\mathbf{x} + \mathbf{e}$ where $\mathbf{e} = [0.1\,0.1\,0.1]^T$ and $\mathbf{x} = [1\,0\,0\,0]^T$. Now, $C_2 = 4\frac{\sqrt{1+\delta_{2k}}}{1-(1+\sqrt{2})\delta_{2k}} \approx 23.65$ (from Theorem 11.13) for $\delta_2 = \frac{1}{3}$. The solution $\hat{\mathbf{x}}$ to (11.119), obtained using CVX, is $\hat{\mathbf{x}} = [0.8939\,0\,-0.0671\,0]^T$ for which

$$\|\hat{\mathbf{x}} - \mathbf{x}\|_2 = 0.1255 \le C_2\varepsilon, \tag{11.159}$$

where $\varepsilon = \|\mathbf{e}\|_2 = \sqrt{0.03}$, so that Theorem 11.13 holds.

Finally, let $\mathbf{x} = [1\,1\,0\,0]^T$. Using CVX now leads to the recovery $\hat{\mathbf{x}} = [0.4620\,0.4620\,-0.5612\,-0.3880]^T$. In this case,

$$\|\hat{\mathbf{x}} - \mathbf{x}\|_2 = 1.0220 \le C_0\frac{\sigma_1(\mathbf{x})_1}{\sqrt{1}} + C_2\varepsilon, \tag{11.160}$$

and Theorem 11.13 is again satisfied.

Example 11.26 We next demonstrate the recovery rate of ℓ_1 minimization (11.121) in the presence of noise, as a function of the SNR, for different values of the sparsity level k. The setting is the same as in Example 11.24. In particular, we consider two choices of dictionaries: a real-valued random matrix $\mathbf{A}$ whose entries are drawn independently from a Gaussian distribution with zero mean and variance one, and a random partial Fourier matrix.

Figure 11.14 shows the empirical recovery rate of the reconstruction algorithm for each one of the two dictionaries and for two different sizes: $m = 32, n = 64$ and $m = 64, n = 128$. The parameter λ was set to 4.75. The empirical recovery rate is calculated as the percentage of correct solutions for which the normalized recovery error is less than 10^{-1}. Similarly to Fig. 11.13, here also we have a threshold effect: perfect recovery is possible for high enough SNR values up to a threshold value beyond which the recovery rate quickly drops to zero.

In Theorem 11.13, the reconstruction error is proportional to the noise magnitude ε. This is because the only assumption on the noise is its magnitude, so that $\mathbf{e}$ might be aligned to maximally harm the estimation process. If, instead, we assume that the noise is Gaussian, then with high probability we can guarantee an error that is proportional to $\sqrt{k}\sigma$ where σ is the noise level. This is much lower than the expected noise norm $\sqrt{m}\sigma$.

More specifically, suppose that the coefficients of $\mathbf{e} \in \mathbb{R}^m$ are iid according to a Gaussian distribution with zero mean and variance σ^2. In the random noise case, bounds on $\|\hat{\mathbf{x}} - \mathbf{x}\|$ can only be stated with high probability, since there is always a small probability that the noise will be very large and completely overpower the signal. On the other hand, the random noise model allows us to obtain stronger guarantees. For example, it is shown in [312, 313] that for $\mathbf{x} \in \Sigma_k$, with an appropriate choice of regularization parameter the error $\|\hat{\mathbf{x}} - \mathbf{x}\|_2^2$ using BPDN is bounded by a constant times $k\sigma^2 \log n$, with high probability. Note that since we typically require $m > k \log n$, this can be substantially lower than the expected noise power $E\{\|\mathbf{e}\|_2^2\} = m\sigma^2$, illustrating the fact that sparsity-based techniques are highly successful in reducing the noise level. Such a result cannot be obtained in the case of bounded noise, wherein the error in $\mathbf{x}$ is on the order of the noise power ε.

The value $k\sigma^2 \log n$ is nearly optimal in several respects. First, an "oracle" estimator which knows the locations of the nonzero components and uses a LS technique to estimate their values achieves an estimation error on the order of $k\sigma^2$. The Cramér–Rao

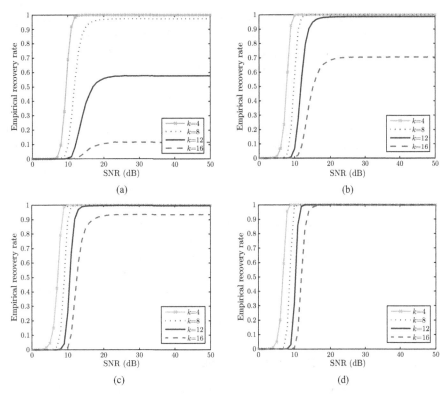

Figure 11.14 Empirical recovery rate of ℓ_1 minimization as a function of SNR for different values of k and two choices of **A**. (a) iid Gaussian matrix with $m = 32$ and $n = 64$. (b) iid Gaussian matrix with $m = 64$ and $n = 128$. (c) random partial Fourier matrix with $m = 32$ and $n = 64$. (d) random partial Fourier matrix with $m = 64$ and $n = 128$.

bound (CRB) for estimating $\mathbf{x}$, which is derived in [314], is also on the order of $k\sigma^2$. This is of practical interest since the CRB is achieved by the maximum likelihood estimator at high SNR, implying that for low-noise settings, an error of $k\sigma^2$ is achievable. However, the maximum likelihood estimator is NP-hard to compute, so that near-oracle results are still of interest.

In order to place this result in context, recall that if we know the support of the k nonzero coefficients, which we denote by Λ_0, then a natural approach is to use the LS estimate given by[5] (11.18):

$$\hat{\mathbf{x}}_{\Lambda_0} = \mathbf{A}_{\Lambda_0}^\dagger \mathbf{y} = (\mathbf{A}_{\Lambda_0}^* \mathbf{A}_{\Lambda_0})^{-1} \mathbf{A}_{\Lambda_0}^* \mathbf{y}. \tag{11.161}$$

The implicit assumption in (11.161) is that $\mathbf{A}_{\Lambda_0}$ has full column-rank. With this choice, the recovery error is given by

$$\|\hat{\mathbf{x}} - \mathbf{x}\|_2 = \left\|(\mathbf{A}_{\Lambda_0}^* \mathbf{A}_{\Lambda_0})^{-1} \mathbf{A}_{\Lambda_0}^* (\mathbf{A}\mathbf{x} + \mathbf{e}) - \mathbf{x}\right\|_2 = \left\|(\mathbf{A}_{\Lambda_0}^* \mathbf{A}_{\Lambda_0})^{-1} \mathbf{A}_{\Lambda_0}^* \mathbf{e}\right\|_2. \tag{11.162}$$

---

[5] Note that while this approach can be improved upon in a statistical setting (in terms of MSE) by considering alternative biased estimators [315, 316, 317], this does not fundamentally change our conclusions.

We now consider the worst-case bound for this error. Using standard properties of the SVD, it is straightforward to show that if $\mathbf{A}$ satisfies the RIP of order $2k$ (with constant δ_{2k}), then the largest singular value of $\mathbf{A}_{\Lambda_0}^{\dagger}$ lies in the range $[1/\sqrt{1 + \delta_{2k}}, 1/\sqrt{1 - \delta_{2k}}]$ (see Exercise 8). Thus, for any e such that $\|\mathbf{e}\|_2 \leq \varepsilon$,

$$\frac{\varepsilon}{\sqrt{1 + \delta_{2k}}} \leq \|\hat{\mathbf{x}} - \mathbf{x}\|_2 \leq \frac{\varepsilon}{\sqrt{1 - \delta_{2k}}}. \tag{11.163}$$

Consequently, the guarantee on the LS estimate, which is given perfect knowledge of the support of $\mathbf{x}$, cannot improve upon the bound in Theorem 11.13 by more than a constant.

Example 11.27 Let us return to Example 11.25 with $\mathbf{x} = [1\,1\,0\,0]^T$ and $\mathbf{e} = [0.1\,0.1\,0.1]^T$. Assuming that the locations of the nonzero coefficients of $\mathbf{x}$ are known, (11.161) results in the estimate $\hat{\mathbf{x}} = [1.0866\,1.0866\,0\,0]^T$, whose error is given by $\|\hat{\mathbf{x}} - \mathbf{x}\|_2 = 0.1225$. This error is smaller than in the case in which the support is unknown (see Example 11.25). Surprisingly, though, the errors are quite similar. However, as shown in Example 11.25, ℓ_1 optimization was not able to recover the true support.

11.5.2 ℓ_1 recovery: coherence-based results

Thus far, we have examined performance guarantees based on the RIP. We have seen that $m = O(k)$ measurements are sufficient to guarantee recovery in the noise-free setting, and to ensure recovery with small error in the noisy case, with high probability. In practice, however, it is typically very difficult to verify that a matrix $\mathbf{A}$ satisfies the RIP or calculate the corresponding RIP constant for large dimensions. In this respect, results based on coherence are appealing, since they can be used with arbitrary dictionaries. The main difference between the guarantees that rely solely on coherence and those that rely on RIP is the scaling of the number of measurements m needed for successful recovery of k-sparse signals. The drawback of coherence-based results is that they typically suffer from the so-called square-root bottleneck: they require $m = O(k^2)$ measurements, or $k = O(\sqrt{m})$, to ensure good recovery results. This provides an additional reason for the popularity of random CS matrices.

In the noise-free case, a simple result can be obtained under the same condition for uniqueness given in Theorem 11.6. Specifically, if

$$k < \frac{1}{2}\left(1 + \frac{1}{\mu(\mathbf{A})}\right), \tag{11.164}$$

then ℓ_1 optimization will recover the true underlying vector $\mathbf{x}$ [289]. Coherence results have also been extended to the noisy setting. One quick route to coherence-based performance guarantees is to combine RIP results with coherence bounds such as Lemma 11.3. However, this technique typically leads to guarantees that are overly pessimistic. Tighter results are derived by directly exploiting coherence [253, 311, 312]. Representative examples for both bounded and Gaussian noise are given in the following theorems [311, Theorem 3.1].

Theorem 11.14 (Coherence-based ℓ_1 recovery with bounded noise). *Suppose that* $\mathbf{A}$ *has coherence* μ *and that* $\mathbf{x} \in \Sigma_k$ *with* $k < (1/\mu + 1)/4$*. Furthermore, suppose that we obtain measurements of the form* $\mathbf{y} = \mathbf{Ax} + \mathbf{e}$ *with* $\gamma = \|\mathbf{e}\|_2$*. Then when* $\mathcal{B}(\mathbf{y}) = \{\mathbf{z}: \|\mathbf{Az} - \mathbf{y}\|_2 \le \varepsilon\}$ *with* $\varepsilon > \gamma$*, the solution* $\hat{\mathbf{x}}$ *to (11.119) obeys*

$$\|\mathbf{x} - \hat{\mathbf{x}}\|_2 \le \frac{\gamma + \varepsilon}{\sqrt{1 - \mu(4k - 1)}}. \tag{11.165}$$

Note that the ℓ_1 optimization method must be aware of the noise magnitude γ to choose $\varepsilon \ge \gamma$. Additionally, the error is proportional to the noise magnitude γ rather than the sparsity level.

An additional type of coherence-based guarantee for BPDN (11.121) is given below. This result goes beyond what we have seen so far by providing explicit guarantees concerning the recovery of the support of $\mathbf{x}$ [312, Corollary 1].

Theorem 11.15 (Coherence-based ℓ_1 recovery with random noise). *Suppose that* $\mathbf{A}$ *has coherence* μ *and that* $\mathbf{x} \in \Sigma_k$ *with* $k \le 1/(3\mu)$*. Furthermore, suppose that we obtain measurements of the form* $\mathbf{y} = \mathbf{Ax} + \mathbf{e}$ *where the entries of* $\mathbf{e}$ *are iid* $\mathcal{N}(0, \sigma^2)$*. Set*

$$\lambda = \sqrt{8\sigma^2(1 + \alpha)\log(n - k)} \tag{11.166}$$

for some fairly small value $\alpha > 0$*. Then with probability exceeding*

$$\left(1 - \frac{1}{(n - k)^\alpha}\right)(1 - \exp(-k/7)),$$

the solution $\hat{\mathbf{x}}$ *to (11.121) is unique,* $\mathrm{supp}(\hat{\mathbf{x}}) \subset \mathrm{supp}(\mathbf{x})$*, and*

$$\|\hat{\mathbf{x}} - \mathbf{x}\|_2^2 \le \left(\sqrt{3} + 3\sqrt{2(1 + \alpha)\log(n - k)}\right)^2 k\sigma^2. \tag{11.167}$$

In this case we are guaranteed that any nonzero of $\hat{\mathbf{x}}$ corresponds to a true nonzero of $\mathbf{x}$. This analysis allows for the worst-case signal $\mathbf{x}$. It is possible to improve upon this result by instead assuming that $\mathbf{x}$ has a limited amount of randomness. Specifically, in [253] it is shown that if $\mathrm{supp}(\mathbf{x})$ is chosen uniformly at random and the signs of the nonzero entries of $\mathbf{x}$ are independent and equally likely to be ± 1, then it is possible to significantly relax the assumption on μ. Moreover, by requiring the nonzeros of $\mathbf{x}$ to exceed some minimum magnitude, perfect recovery of the true support is guaranteed.

11.5.3   Instance-optimal guarantees

We now briefly return to the noise-free setting to take a closer look at instance-optimal guarantees for recovering nonsparse signals. To begin, recall that in Theorem 11.12 we bounded the ℓ_2-norm of the reconstruction error $\|\hat{\mathbf{x}} - \mathbf{x}\|_2$ by a constant C_0 times $\sigma_k(\mathbf{x})_1/\sqrt{k}$. One can generalize this result to measure the recovery error using the ℓ_p-norm for any $p \in [1, 2]$. For example, by a slight modification of these arguments, it can be shown that $\|\hat{\mathbf{x}} - \mathbf{x}\|_1 \le C_0\sigma_k(\mathbf{x})_1$ (see [227] and Proposition 11.1). This leads to the question of whether we might replace the bound for the ℓ_2 error with a result of the form $\|\hat{\mathbf{x}} - \mathbf{x}\|_2 \leq C\sigma_k(\mathbf{x})_2$. Unfortunately, this would require an unreasonably large number of measurements, as quantified by the following theorem [224, Theorem 5.1].

Theorem 11.16 (Instance-optimal recovery). *Suppose that* $\mathbf{A}$ *is an* $m \times n$ *matrix and that* $\Delta \colon \mathbb{R}^m \to \mathbb{R}^n$ *is a recovery algorithm that satisfies*

$$\|\mathbf{x} - \Delta(\mathbf{A}\mathbf{x})\|_2 \leq C\sigma_k(\mathbf{x})_2 \tag{11.168}$$

for some $k \geq 1$. *Then* $m > \left(1 - \sqrt{1 - 1/C^2}\right)n$.

Thus, if we want a bound of the form (11.168) that holds for all signals $\mathbf{x}$ with a constant $C \approx 1$, then regardless of the recovery algorithm we need to take $m \approx n$ measurements. Using the results from Section 11.5.1 we now show that we can overcome this limitation by essentially treating the approximation error as noise.

Towards this end, notice that all the results concerning ℓ_1 minimization stated thus far are deterministic instance-optimal guarantees that apply simultaneously to all $\mathbf{x}$ given any matrix that satisfies the RIP. This is an important theoretical property, but as noted in Section 11.3.5, in practice it is very difficult to obtain a deterministic guarantee that the matrix $\mathbf{A}$ satisfies the RIP. In particular, constructions that rely on randomness are only known to satisfy the RIP with high probability.

Even within the class of probabilistic results, there are two distinct flavors. The typical approach is to combine a probabilistic construction of a matrix that will satisfy the RIP with high probability with the previous results in this chapter. This yields a procedure that, with high probability, satisfies a deterministic guarantee applying to all possible signals $\mathbf{x}$. A weaker result is one that states that given a signal $\mathbf{x}$, we can draw a random matrix $\mathbf{A}$ and with high probability expect certain performance for that $\mathbf{x}$. This type of guarantee is sometimes called *instance-optimal in probability*. The distinction is essentially whether or not we need to draw a new random $\mathbf{A}$ for each $\mathbf{x}$. This may be an important distinction in practice, but if we assume for the moment that it is permissible to draw a new matrix $\mathbf{A}$ for each $\mathbf{x}$, then Theorem 11.16 may be somewhat pessimistic, as exhibited by the following result [203, Theorem 1.14].

Theorem 11.17 (Instance-optimal in probability). *Let* $\mathbf{x} \in \mathbb{R}^n$ *be fixed. Set* $\delta_{2k} < \sqrt{2} - 1$ *and let* $\mathbf{A}$ *be an* $m \times n$ *sub-Gaussian random matrix with* $m = O\left(k \log(n/k)/\delta_{2k}^2\right)$. *Suppose we obtain measurements of the form* $\mathbf{y} = \mathbf{A}\mathbf{x}$. *Then with probability exceeding* $1 - 2\exp(-c_1\delta^2 m) - \exp(-c_0 m)$, *when* $\mathcal{B}(\mathbf{y}) = \{\mathbf{z} \colon \|\mathbf{A}\mathbf{z} - \mathbf{y}\|_2 \leq \varepsilon\}$ *with* $\varepsilon = 2\sigma_k(\mathbf{x})_2$, *the solution* $\hat{\mathbf{x}}$ *to* (11.119) *obeys*

$$\|\hat{\mathbf{x}} - \mathbf{x}\|_2 \leq \frac{8\sqrt{1 + \delta_{2k}} - (1 + \sqrt{2})\delta_{2k}}{1 - (1 + \sqrt{2})\delta_{2k}}\sigma_k(\mathbf{x})_2. \tag{11.169}$$

Thus, while it is not possible to achieve a deterministic guarantee of the form (11.168) without taking a prohibitively large number of measurements, such results may hold with high probability using far fewer measurements than suggested by Theorem 11.16. This observation requires the parameter ε to be selected correctly, which implies some limited knowledge of $\mathbf{x}$, namely $\sigma_k(\mathbf{x})_2$. In practice a parameter selection technique may be used such as cross-validation [318]. There also exists more intricate analysis of ℓ_1 minimization that shows it is possible to obtain similar performance without requiring an oracle for parameter selection [319]. Theorem 11.17 can be generalized to handle other measurement matrices and to the case where $\mathbf{x}$ is compressible rather than sparse.

11.5.4 The cross-polytope and phase transitions

While RIP-based results on ℓ_1 minimization allow a variety of guarantees to be established under different noise settings, one drawback is that the analysis of how many measurements are actually required for a matrix to satisfy the RIP is relatively loose. An alternative approach is to examine ℓ_1 algorithms from a geometric perspective. Towards this end, we define the closed ℓ_1 ball, also known as the *cross-polytope*:

$$C^n = \{\mathbf{x} \in \mathbb{R}^n \colon \|\mathbf{x}\|_1 \le 1\}. \tag{11.170}$$

Note that C^n is the convex hull of $2n$ points $\{p_i\}_{i=1}^{2n}$. Let $\mathbf{A}C^n \subseteq \mathbb{R}^m$ denote the convex polytope defined as either the convex hull of $\{\mathbf{A}p_i\}_{i=1}^{2n}$ or equivalently as

$$\mathbf{A}C^n = \{\mathbf{y} \in \mathbb{R}^m \colon \mathbf{y} = \mathbf{A}\mathbf{x}, \mathbf{x} \in C^n\}. \tag{11.171}$$

To any $\mathbf{x} \in \Sigma_k$, we associate a k-face of C^n with the support and sign pattern of $\mathbf{x}$. The number of k-faces of $\mathbf{A}C^n$ is precisely the number of index sets of size k for which signals supported on them can be recovered by (11.119) with $\mathcal{B}(\mathbf{y}) = \{\mathbf{z} \colon \mathbf{A}\mathbf{z} = \mathbf{y}\}$. Thus, ℓ_1 minimization yields the same solution as ℓ_0 minimization for all $\mathbf{x} \in \Sigma_k$ if and only if the number of k-faces of $\mathbf{A}C^n$ is identical to the number of k-faces of C^n. Moreover, by counting the number of k-faces of $\mathbf{A}C^n$, we can quantify exactly what fraction of sparse vectors are recoverable using ℓ_1 minimization with $\mathbf{A}$ as the sensing matrix. See [12, 320, 321] for more details. Note also that by replacing the cross-polytope with certain other polytopes (the simplex and the hypercube), the same techniques may be used to obtain results concerning the recovery of more limited signal classes, such as sparse signals with nonnegative or bounded entries [321].

These geometric considerations lead to probabilistic bounds on the number of k-faces of $\mathbf{A}C^n$ with $\mathbf{A}$ generated at random. Under the assumption that $k = \rho m$ and $m = \gamma n$, one can obtain asymptotic results as $n \to \infty$. This analysis reveals the *phase transition* phenomenon, where for very large problem sizes there are sharp thresholds dictating that the fraction of k-faces preserved will tend to either 1 or 0 with very high probability, depending on ρ and γ [321].

Example 11.28 This example is taken from [321]. Our goal is to assess the success rate for recovery of a k-sparse vector of length n from a measurement vector of length $m < n$. We consider the two following experiments. In the first, we generate an $m \times n$ submatrix $\mathbf{A}$ of the $n \times n$ Fourier matrix, and draw a random vector $\mathbf{x}$ with k entries equal to ± 1 while all other entries are 0. The problem is solved via ℓ_1 minimization and we record a success if the solution is equal to $\mathbf{x}$ up to a relative error of 10^{-6}. The parameters that were chosen are: $n = 1600$, m varies

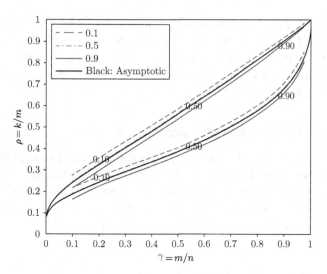

Figure 11.15 Empirical phase transitions (taken from [321]). Lower/higher set of curves: level curves of success probabilities for Experiment 1/2 at values 0.1, 0.5, and 0.9. The heavy black curve is the asymptotic theoretical result.

from $m = 160$ to $m = 1440$, and k varies from $k = 1$ to $k = m$. In the second experiment, the Fourier matrix is replaced by a random Bernoulli matrix with entries independently chosen as 0 or 1 with equal probability, and the k nonzero coefficients of $\mathbf{x}$ are set to 1.

Figure 11.15 shows three levels of success rates for both the experiments, namely 90%, 50%, and 10%, as a function of the undersampling and sparsity coefficients, $\gamma = m/n$ and $\rho = k/m$ respectively. We observe two clean phases: one where the fraction of success is essentially one and another where it is essentially zero. In between these phases is a narrow transition zone. The experimental data illustrates the phenomenon of phase transition. A fourth curve added to each set, derived from combinatorial geometry [321], shows the transition boundary. This curve plots the sparsity below which we can recover the original vector with probability 1, and above which it can be recovered with probability 0. The phase transition curve is derived in [320, 321] and is given by

$$\rho(\gamma) \sim [2\log(1/\gamma)]^{-1}, \quad \gamma \to 0.$$

The cross-polytope method provides sharp bounds on the minimum number of measurements required in the noiseless case. In general, these bounds are significantly stronger than the corresponding results obtained within the RIP framework, which tend to be extremely loose in terms of the constants involved. However, these sharper results require somewhat more intricate analysis and typically more restrictive assumptions on $\mathbf{A}$. Thus, one of the main strengths of the RIP-based analysis presented in this chapter is that it gives results for a very broad class of matrices that can also be extended to noisy settings.

11.5.5   Guarantees on greedy methods

Many of the results that were derived for ℓ_1 optimization have counterparts for greedy methods as well. However, there are also some important differences.

Starting with pursuit algorithms, in general, we can no longer obtain uniform recovery guarantees that hold for all $\mathbf{x}$ when $m = O(k \log n)$. Such results are available only with $m = O(k^2 \log n)$ as incorporated in the following theorem [290].

Theorem 11.18 (RIP-based OMP recovery).   *Suppose that $\mathbf{A}$ satisfies the RIP of order $k + 1$ with $\delta_{k+1} < 1/(3\sqrt{k})$ and let $\mathbf{y} = \mathbf{Ax}$. Then OMP can recover a k-sparse signal exactly in k iterations.*

We have seen in Theorem 11.11 that if we draw a random measurement $\mathbf{A}$ with iid entries from a sub-Gaussian distribution, then for $m = O(k/\delta^2 \log(n/k))$, $\mathbf{A}$ will satisfy the RIP of order k. Since Theorem 11.11 requires $\delta^2 = 1/(9k)$ we conclude that $m = O(k^2 \log n)$ measurements are needed. Guarantees also exist for noisy measurements of the same flavor; see [322] for details. A similar analysis to that of Theorem 11.18 holds assuming a bounded coherence condition of the form (11.164) [289]. Here again the results apply when $m = O(k^2 \log n)$.

There have been many efforts to improve upon these basic results. For example, in [323] the number of measurements is reduced to $m = O(k^{1.6} \log n)$ by allowing OMP to run for more than k iterations. This result is improved in [324], requiring the familiar $m = O(k \log n)$ measurements, and establishing stability of OMP for sufficiently many iterations. Both of these analyses exploit the RIP. Note, however, that since more than k steps are needed, the true support set is not recovered even when the error $\|\mathbf{y} - \mathbf{Ax}\|$ is minimized. In fact, it is shown in [325] that using $O(k \log n)$ measurements it is impossible to uniformly recover the support set (in k iterations) with OMP. Thus, to minimize the error with a small number of measurements, one needs to run OMP for more than k steps.

All of the above efforts have aimed at establishing uniform guarantees. In light of our discussion on probabilistic results in Section 11.5.3, one might expect to see improvements by considering less restrictive guarantees. As an example, it has been shown that by considering random $\mathbf{A}$ matrices, OMP can recover k-sparse signals in k iterations with high probability using only $m = O(k \log n)$ measurements [228]. Similar results are also possible by placing restrictions on the smallest nonzero value of the signal, as in [311]. Furthermore, such conditions enable near-optimal recovery guarantees when the measurements are corrupted by Gaussian noise [312].

The bounds above can be improved significantly when using greedy methods of the thresholding type. This class of algorithms has performance guarantees that are on a par with those available for convex optimization. Instead of detailing the result for each method separately, we collect a set of independent statements in a single theorem. For details see the analysis in [322].

Theorem 11.19 (RIP-based thresholding recovery).   *Suppose that $\mathbf{A}$ satisfies the RIP of order ck with constant δ and let $\mathbf{y} = \mathbf{Ax} + \mathbf{e}$ where $\|\mathbf{e}\|_2 \leq \varepsilon$. Then the outputs $\hat{\mathbf{x}}$ of the CoSaMP, subspace pursuit, and IHT algorithms obey*

$$\|\hat{\mathbf{x}} - \mathbf{x}\|_2 \le C_1 \sigma_k(\mathbf{x})_2 + C_2 \frac{\sigma_k(\mathbf{x})_1}{\sqrt{k}} + C_3 \varepsilon. \tag{11.172}$$

The requirements on the parameters c, δ of the RIP and the values of C_1, C_2, and C_3 are specific to each algorithm.

Theorem 11.19 can be adapted to recovery of exactly sparse signals from noiseless measurements by noting that in this case $\varepsilon = \sigma_k(\mathbf{x})_2 = \sigma_k(\mathbf{x})_1 = 0$.

Results based on coherence also apply to greedy methods. In the noise-free case, under the same condition (11.164) for ℓ_1 optimization, OMP is guaranteed to recover the true underlying vector $\mathbf{x}$ [289]. In the noisy setting, we have the following theorem [311].

Theorem 11.20 (Coherence-based OMP recovery with bounded noise). *Suppose that $\mathbf{A}$ has coherence μ and that $\mathbf{x} \in \Sigma_k$ with $k < (1/\mu + 1)/4$. Furthermore, suppose that we obtain measurements $\mathbf{y} = \mathbf{Ax} + \mathbf{e}$ with $\gamma = \|\mathbf{e}\|_2$. Then the output of OMP with halting criterion $\|\mathbf{r}\| \le \gamma$ obeys*

$$\|\mathbf{x} - \hat{\mathbf{x}}\|_2 \le \frac{\gamma}{\sqrt{1 - \mu(k-1)}}, \tag{11.173}$$

provided that $\gamma \le \alpha(1 - \mu(2k - 1))/2$ with α being a positive lower bound on the magnitude of the nonzero entries of $\mathbf{x}$.

Note that OMP must be aware of the minimum signal nonzero magnitude to set an appropriate convergence criterion. Furthermore, the error is proportional to the noise magnitude γ rather than the sparsity level.

The bounds on the number of measurements can be improved by considering Gaussian noise [312, Theorem 4].

Theorem 11.21 (Coherence-based OMP recovery with random noise). *Suppose that $\mathbf{A}$ has coherence μ and that $\mathbf{x} \in \Sigma_k$. Furthermore, suppose that we obtain measurements $\mathbf{y} = \mathbf{Ax} + \mathbf{e}$ where the entries of $\mathbf{e}$ are iid $\mathcal{N}(0, \sigma^2)$ and that*

$$(1 - (2k - 1)\mu)|x_{\min}| \ge 2\sigma\sqrt{2(1+\alpha)\log n},$$

for some fairly small value $\alpha > 0$. Then with probability exceeding

$$\left(1 - \frac{1}{n^\alpha \sqrt{\pi(1+\alpha)\log n}}\right)$$

the OMP estimate $\hat{\mathbf{x}}$ identifies the correct support of $\mathbf{x}$, and

$$\|\hat{\mathbf{x}} - \mathbf{x}\|_2^2 \le 8(1+\alpha)k\sigma^2 \log n. \tag{11.174}$$

Comparing Theorems 11.15 and 11.21, we see that OMP requires a bound on the magnitude of the entries of $\mathbf{x}$ which is not needed for BPDN. Thus, greedy approaches may outperform optimization-based methods when the entries of $\mathbf{x}$ are large compared with the noise, but they will generally deteriorate when the noise level increases.

Example 11.29 We return to Example 11.23 and compare the performance of five different recovery algorithms: ℓ_1 minimization, OMP, MP, IHT, and CoSaMP as a function of the sparsity level k in the presence of noise. We use the same settings as

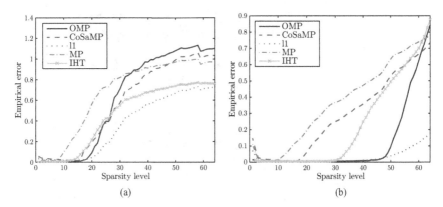

Figure 11.16 Performance comparison of ℓ_1 minimization, OMP, MP, IHT, and CoSaMP as a function of the sparsity level in the presence of noise with variance $\sigma^2 = 0.000\,025$ for two choices of $\mathbf{A}$. (a) Gaussian matrix. (b) Random partial Fourier matrix.

in Example 11.23 but contaminate the measurements with iid zero-mean Gaussian noise with variance $\sigma^2 = 0.000\,025$. The experiment is repeated 5000 times for each sparsity value. Figure 11.16 shows the normalized recovery error of the reconstruction algorithms.

11.6   Multiple measurement vectors

11.6.1   Signal model

Many applications that match the properties of CS involve distributed acquisition of multiple correlated signals. The multiple signal case where all L signals involved are sparse and exhibit the same indices for their nonzero locations is referred to in the CS literature as the *multiple measurement vector (MMV) problem* [196, 197, 291, 326, 327, 328, 329]. In the MMV setting, we are given L measurements $\mathbf{y}_i = \mathbf{A}\mathbf{x}_i$ where the vectors $\mathbf{x}_i, 1 \leq i \leq L$ are jointly sparse. Stacking these vectors into the columns of a matrix $\mathbf{X}$, there will be at most k nonzero rows in $\mathbf{X}$. That is, not only is each vector k-sparse, but the nonzero values occur on a common location set. We therefore say that $\mathbf{X}$ is *row-sparse* and use the notation $\Lambda = \mathrm{supp}(\mathbf{X})$ to denote the index set corresponding to nonzero rows. An example of such a matrix $\mathbf{X}$ is depicted in Fig. 11.17. Rather than trying to recover each single sparse vector $\mathbf{x}_i$ independently, the goal is to jointly recover the set of vectors by exploiting their common sparse support.

MMV problems appear quite naturally in many different application areas. Early work on MMV algorithms focused on magnetoencephalography, which is a modality for imaging the brain [197]. Similar ideas were also developed in the context of array processing [330], equalization of sparse communication channels [331, 332], and more recently cognitive radio and multiband communications [102, 333, 334, 335, 336]. We will discuss these later examples in detail in Chapter 14.

As in standard CS, we assume that we are given measurements $\{\mathbf{y}_i\}_{i=1}^{L}$ where each vector is of length $m < n$. Letting $\mathbf{Y}$ be the $m \times L$ matrix with columns $\mathbf{y}_i$, our problem

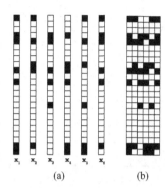

$$x_1 \quad x_2 \quad x_3 \quad x_4 \quad x_5 \quad x_6$$

$$\text{(a)} \qquad\qquad \text{(b)}$$

Figure 11.17 (a) Example of sparse vectors ($L = 6$) sharing a common support. Each square corresponds to a vector entry, with the filled squares representing nonzero elements and blank squares indicating zeros. (b) The corresponding matrix $\mathbf{X} = [\mathbf{x}_1 \, \mathbf{x}_2 \, \ldots \, \mathbf{x}_L]$ has a small number of nonzero rows.

is to recover $\mathbf{X}$ from $\mathbf{Y} = \mathbf{AX}$. Clearly, we can apply any CS method to recover $\mathbf{x}_i$ from $\mathbf{y}_i$ as before. However, since the vectors $\{\mathbf{x}_i\}$ all have a common support, we expect intuitively to improve the recovery ability by exploiting this joint information. In other words, we should in general be able to reduce the number of measurements mL needed to represent $\mathbf{X}$ below sL, where s is the number of measurements required to recover one vector $\mathbf{x}_i$ for a given matrix $\mathbf{A}$.

Since $|\Lambda| = k$, the rank of $\mathbf{X}$ satisfies $\text{rank}(\mathbf{X}) \leq k$. When $\text{rank}(\mathbf{X}) = 1$, all the sparse vectors $\mathbf{x}_i$ are multiples of each other, so that there is no advantage to their joint processing. However, when $\text{rank}(\mathbf{X})$ is large, the diversity in its columns should enable us to benefit from joint recovery. This essential result is captured by the following necessary and sufficient uniqueness condition [337].

Theorem 11.22 (Uniqueness in MMV). *A necessary and sufficient condition for the measurements* $\mathbf{Y} = \mathbf{AX}$ *to uniquely determine the jointly sparse matrix* $\mathbf{X}$ *is that*

$$|\operatorname{supp}(\mathbf{X})| < \frac{\operatorname{spark}(\mathbf{A}) - 1 + \operatorname{rank}(\mathbf{X})}{2}. \tag{11.175}$$

As shown in [327, 337], we can replace $\text{rank}(\mathbf{X})$ by $\text{rank}(\mathbf{Y})$ in (11.175). The sufficient part of this condition was shown in [329] to hold even in the case where there are infinitely many vectors $\mathbf{x}_i$. A direct consequence of Theorem 11.22 is that matrices $\mathbf{X}$ with larger rank can be recovered from fewer measurements. When $\text{rank}(\mathbf{X}) = k$ and $\text{spark}(\mathbf{A})$ takes on its largest possible value equal to $m+1$, condition (11.175) becomes $m \geq k + 1$. Therefore, in this best-case scenario, only $k + 1$ measurements per signal are needed to ensure uniqueness. This is much lower than the value of $2k$ obtained in standard CS via the spark (cf. Theorem 11.1), which we refer to here as the single measurement vector (SMV) setting. Furthermore, as we show in Subsection 11.6.2, in the noiseless setting $\mathbf{X}$ can be recovered by a simple algorithm, in contrast to the combinatorial complexity needed to solve the SMV problem from $2k$ measurements for general matrices $\mathbf{A}$.

Example 11.30 Consider the following 4×7 Vandermonde matrix:

$$\mathbf{A} = \begin{bmatrix} 1 & 1 & 1 & 1 & 1 & 1 & 1 \\ 1 & 2 & 3 & 4 & 5 & 6 & 7 \\ 1 & 4 & 9 & 16 & 25 & 36 & 49 \\ 1 & 8 & 27 & 64 & 125 & 216 & 343 \end{bmatrix}. \tag{11.176}$$

From Example 11.6, we know that $\mathrm{spark}(\mathbf{A}) = 5$. Now, let

$$\mathbf{X} = \begin{bmatrix} 0 & 0 & 0 \\ 0 & 1 & 1 \\ 0 & 0 & 0 \\ 0 & 0 & 0 \\ 1 & 1 & 0 \\ 0 & 1 & 0 \\ 0 & 0 & 0 \end{bmatrix}, \tag{11.177}$$

resulting in the measurement matrix:

$$\mathbf{Y} = \mathbf{AX} = \begin{bmatrix} 1 & 3 & 1 \\ 5 & 13 & 2 \\ 25 & 65 & 4 \\ 125 & 349 & 8 \end{bmatrix}. \tag{11.178}$$

If we recover $\mathbf{X}$ column by column, that is, we do not exploit the joint sparsity, then from Theorem 11.1, we require $\mathrm{spark}(\mathbf{A}) > 6$, since the least sparse column of $\mathbf{X}$ is in Σ_3. That is, we need at least $m = 6$ measurements per signal in order to recover all columns of $\mathbf{X}$. Since $\mathbf{A}$ has only four rows, the spark requirement is not met, and there is no unique solution. For example, choosing $\mathbf{x}_2 = [0\,0\,4\,-6\,5\,0\,0]^T$ leads to the same $\mathbf{y}_2$. With this choice, the joint sparsity is $k = 4$.

Taking the joint sparsity into account, Theorem 11.22 allows for less measurements. In our case $\mathrm{rank}(\mathbf{X}) = k = 3$. Therefore, we only need $m = k + 1 = 4$ measurements per signal in order to recover $\mathbf{X}$. We will show below in (11.184) and Example 11.33 how to actually perform the recovery from four measurements.

11.6.2   Recovery algorithms

A variety of algorithms have been proposed that exploit the joint sparsity in different ways. As in the SMV setting, two main approaches to solving MMV problems are based on convex optimization and greedy methods. An exception is the ReMBo algorithm, introduced in [329], which reduces the MMV problem to an SMV counterpart, and then utilizes standard SMV techniques. We will discuss this algorithm below.

The analog of (11.117) in the MMV case is

$$\widehat{\mathbf{X}} = \arg\min_{\mathbf{X}} \|\mathbf{X}\|_{0,q} \quad \text{s.t. } \mathbf{Y} = \mathbf{AX}, \tag{11.179}$$

where we define the $\ell_{p,q}$-norms for matrices as

$$\|\mathbf{X}\|_{p,q} = \left(\sum_i \|\mathbf{x}^i\|_p^q \right)^{1/q} \tag{11.180}$$

with $\mathbf{x}^i$ denoting the ith row of $\mathbf{X}$. With a slight abuse of notation, we also consider the quasinorms with $p = 0$ such that $\|\mathbf{X}\|_{0,q} = |\,\text{supp}(\mathbf{X})|$ for any q. Optimization-based algorithms relax the ℓ_0-norm in (11.179) and attempt to recover $\mathbf{X}$ by mixed-norm minimization [183, 291, 327, 328, 338, 339, 340]:

$$\widehat{\mathbf{X}} = \arg\min_{\mathbf{X}} \|\mathbf{X}\|_{p,q} \quad \text{s.t.} \quad \mathbf{Y} = \mathbf{AX} \tag{11.181}$$

for some $p, q \geq 1$; values of $p, q = 1, 2$, and ∞ have been advocated.

The standard greedy approaches in the SMV setting have also been extended to the MMV case [207, 291, 337, 341]. The basic idea is to replace the residual vector $\mathbf{r}$ by a residual matrix $\mathbf{R}$, which contains the residuals with respect to each of the measurements, and to replace the surrogate vector $\mathbf{A}^*\mathbf{r}$ by the q-norms of the rows of $\mathbf{A}^*\mathbf{R}$. For example, making these changes to OMP (Algorithm 11.1) yields a variant known as *simultaneous orthogonal matching pursuit*, shown as Algorithm 11.3, where $\mathbf{X}|_\Lambda$ denotes the restriction of $\mathbf{X}$ to the rows indexed by Λ. A similar approach can be used to extend thresholding to the MMV setting [341].

Algorithm 11.3 Simultaneous orthogonal matching pursuit

Input: CS matrix $\mathbf{A}$, MMV matrix $\mathbf{Y}$
Output: Row-sparse matrix $\widehat{\mathbf{X}}$
Initialize: $\widehat{\mathbf{X}}_0 = 0, \mathbf{R} = \mathbf{Y}, \Lambda = \emptyset, \ell = 0$
while halting criterion false **do**
  $\ell \leftarrow \ell + 1$
  $\mathbf{b}(i) \leftarrow \|\mathbf{a}_i^*\mathbf{R}\|_q, 1 \leq i \leq n$ {form residual matrix ℓ_q-norm vector}
  $\Lambda \leftarrow \Lambda \cup \text{supp}(\mathcal{T}(\mathbf{b}, 1))$ {update row support with index of residual's row with largest magnitude}
  $\widehat{\mathbf{X}}_\ell|_\Lambda \leftarrow \mathbf{A}_\Lambda^\dagger \mathbf{Y}, \widehat{\mathbf{X}}_\ell|_{\Lambda^c} \leftarrow 0$ {update signal estimate}
  $\mathbf{R} \leftarrow \mathbf{Y} - \mathbf{A}\widehat{\mathbf{X}}_\ell$ {update measurement residual}
end while
return $\widehat{\mathbf{X}} \leftarrow \widehat{\mathbf{X}}_\ell$

Example 11.31 In this example, we show empirically that simultaneous OMP performs better than standard OMP. We choose $m = 32$, $n = 64$, and $L = 5$ for the MMV system. We consider a real-valued random matrix $\mathbf{A}$ whose entries are drawn independently from a Gaussian distribution with zero mean and variance 1. For each value of the sparsity $1 \leq k \leq m$, we construct a k-sparse matrix $\mathbf{X}$ of size 30×5. The nonzero locations are drawn uniformly at random and the nonzero values are drawn from a Gaussian distribution as described above. The experiment

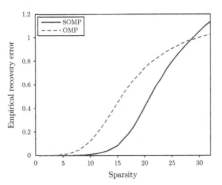

Figure 11.18 Recovery rate of OMP and simultaneous OMP (SOMP).

is repeated 5000 times for each value of k. Figure 11.18 shows the empirical recovery error of the OMP algorithm that reconstructs each channel independently, and simultaneous OMP which performs multichannel reconstruction. The stopping criteria are based on the residual. Evidently, the multichannel approach outperforms SMV reconstruction.

The ReMBo algorithm

An alternative MMV strategy is the ReMBo (reduce MMV and boost) algorithm [329]. ReMBo first reduces the problem to an SMV that preserves the sparsity pattern, and then recovers the signal support set; given the support, the measurements are inverted to recover the input. The reduction is performed by merging the measurement columns $\mathbf{Y}$ with random coefficients $\mathbf{a}$ to form $\mathbf{y} = \mathbf{Ya}$. Using random coefficients ensures that with probability one, nonzero rows in $\mathbf{X}$ are transformed into a nonzero element in the reduced vector $\mathbf{Xa}$. The details of the approach together with a recovery guarantee are given in the following theorem.

Theorem 11.23 (ReMBo recovery). *Let $\overline{\mathbf{X}}$ be the unique k-sparse solution matrix of $\mathbf{Y} = \mathbf{A}\mathbf{X}$ and let $\mathbf{A}$ have spark greater than $2k$. Let $\mathbf{a}$ be a random vector with an absolutely continuous distribution[6] and define the random vectors $\mathbf{y} = \mathbf{Ya}$ and $\overline{\mathbf{x}} = \overline{\mathbf{X}}\mathbf{a}$. Consider the random SMV system $\mathbf{y} = \mathbf{A}\mathbf{x}$. Then:*
1. For every realization of $\mathbf{a}$, the vector $\overline{\mathbf{x}}$ is the unique k-sparse solution of the SMV;
2. $\Lambda(\overline{\mathbf{x}}) = \Lambda(\overline{\mathbf{X}})$ with probability 1.

According to Theorem 11.23, the MMV problem is first reduced to an SMV counterpart $\mathbf{y} = \mathbf{A}\mathbf{x}$ with $\mathbf{y} = \mathbf{Ya}$ for an appropriate random vector $\mathbf{a}$, for which the optimal solution $\overline{\mathbf{x}}$ is found. We then choose the support of $\overline{\mathbf{X}}$ to be equal to that of $\overline{\mathbf{x}}$, and invert the measurement vectors $\mathbf{Y}$ over this support. This reduction can be extremely beneficial in large-scale problems, such as those resulting from analog sampling. In practice,

---

[6] A probability distribution is absolutely continuous if every event of measure 0 occurs with probability 0. Examples include Gaussian and uniform distributions.

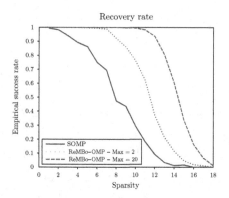

Figure 11.19 Comparison of OMP and ReMBo-OMP.

computationally efficient methods are used to solve the SMV counterpart, which may result in recovery errors in the presence of noise, or when insufficient measurements are taken. By repeating the procedure with different choices of $\mathbf{a}$, the empirical recovery rate is boosted significantly [329]. Denoting by $\mathbf{a}_i$ the ith choice of $\mathbf{a}$, and by $\overline{\mathbf{x}}_i$ the corresponding recovery, we can choose the value of $\overline{\mathbf{x}}_j$ with the smallest support as our final estimate. Alternatively, for each support set $\mathcal{S}_i$ we can compute the corresponding $\overline{\mathbf{X}}_i$ and select the index j for which the error $\|\mathbf{Y} - \mathbf{A}\overline{\mathbf{X}}_j\|$ is minimized.

Example 11.32 Consider an MMV system where $\mathbf{A}$ is a real-valued 20×30 Gaussian matrix. Let $L = 5$ be the number of jointly sparse vectors we wish to recover. We compare the recovery rate of simultaneous OMP and ReMBo-OMP as a function of the sparsity k. We consider two values of the maximal number of iterations for the latter: 2 and 20. At each iteration, a vector $\mathbf{a}$ is generated. When the sparsity of the reconstructed vector is satisfying, namely below k, the algorithm terminates and the final support is chosen equal to the last solution. Success is recorded if $\mathbf{X}$ is recovered exactly up to machine precision. The experiment is carried out 500 times for each technique. Figure 11.19 shows the empirical recovery rate of both methods. We observe that the recovery rate of ReMBo-OMP improves upon simultaneous OMP owing to the boosting effect.

Rank-aware methods

The MMV techniques discussed so far are *rank-blind*: that is, they do not explicitly take the rank of $\mathbf{X}$, or that of $\mathbf{Y}$, into account. Theorem 11.22 highlights the role of the rank of $\mathbf{X}$ in the recovery guarantees. If $\operatorname{rank}(\mathbf{X}) = k$ where k is the size of the support of $\mathbf{X}$, and (11.175) is satisfied, then every k columns of $\mathbf{A}$ are linearly independent (since $\operatorname{spark}(\mathbf{A}) > k + 1$). This in turn can be shown to imply that $\mathcal{R}(\mathbf{Y}) = \mathcal{R}(\mathbf{A}_\Lambda)$ via the following proposition [335].

Proposition 11.10. *For every two matrices $\mathbf{A}, \mathbf{P}$, if the size of the support of $\mathbf{P}$ is less than $\operatorname{spark}(\mathbf{A})$, then $\operatorname{rank}(\mathbf{P}) = \operatorname{rank}(\mathbf{AP})$.*

Proof: Let $r = \text{rank}(\mathbf{P})$. Reorder the columns of $\mathbf{P}$ so that the first r columns are linearly independent. This operation does not change the rank of $\mathbf{P}$ or the rank of $\mathbf{AP}$. Define

$$\mathbf{P} = [\mathbf{P}^{(1)} \ \ \mathbf{P}^{(2)}], \tag{11.182}$$

where $\mathbf{P}^{(1)}$ contains the first r columns of $\mathbf{P}$ and the rest are contained in $\mathbf{P}^{(2)}$. Then,

$$r \geq \text{rank}(\mathbf{AP}) = \text{rank}(\mathbf{A}[\mathbf{P}^{(1)} \ \ \mathbf{P}^{(2)}]) \geq \text{rank}(\mathbf{AP}^{(1)}). \tag{11.183}$$

The inequalities result from the properties of the rank of concatenation and of multiplication of matrices. It is therefore sufficient to prove that $\mathbf{AP}^{(1)}$ has full column-rank.

Let $\mathbf{b}$ be a vector of coefficients such that $\mathbf{AP}^{(1)}\mathbf{b} = 0$. Denote by k the size of the support of $\mathbf{P}$. Since the support of $\mathbf{P}^{(1)}$ is included in that of $\mathbf{P}$, the vector $\mathbf{P}^{(1)}\mathbf{b}$ is k-sparse. From $\text{spark}(\mathbf{A}) > k$, it follows that the null space of $\mathbf{A}$ cannot contain a k-sparse vector unless it is the zero vector. Finally, since $\mathbf{P}^{(1)}$ consists of linearly independent columns this implies that $\mathbf{b} = 0$. $\qquad\square$

Since $\mathbf{Y} = \mathbf{A}_\Lambda \mathbf{X}_\Lambda$, from Proposition 11.10 it is clear that $\text{rank}(\mathbf{Y}) = \text{rank}(\mathbf{X}_\Lambda) = k$. In addition, the k columns of $\mathbf{A}_\Lambda$ are linearly independent so that $\mathcal{R}(\mathbf{Y}) = \mathcal{R}(\mathbf{A}_\Lambda)$. We can therefore identify the support of $\mathbf{X}$ by determining the columns $\mathbf{a}_i$ that lie in $\mathcal{R}(\mathbf{Y})$. One way to accomplish this is by minimizing the norm of the projections onto the orthogonal complement of $\mathcal{R}(\mathbf{Y})$:

$$\min_i \|(\mathbf{I} - P_{\mathcal{R}}(\mathbf{Y}))\mathbf{a}_i\|_2, \tag{11.184}$$

where $P_{\mathcal{R}}(\mathbf{Y})$ is the orthogonal projection onto the range of $\mathbf{Y}$. The objective in (11.184) is equal to zero if and only if $i \in \Lambda$. Since, by assumption, the columns of $\mathbf{A}_\Lambda$ are linearly independent, once we find the support we can determine $\mathbf{X}$ as $\mathbf{X}_\Lambda = \mathbf{A}_\Lambda^\dagger \mathbf{Y}$. We therefore have the following guarantee [337].

Theorem 11.24 (MMV recovery). *If $\text{rank}(\mathbf{X}) = k$ and (11.175) holds, then the algorithm (11.184) is guaranteed to recover $\mathbf{X}$ from $\mathbf{Y}$ exactly.*

Example 11.33 We now return to Example 11.30 and show how to use (11.184) in order to find the sparse input.

For our choice of $\mathbf{Y}$,

$$\mathbf{I} - P_{\mathcal{R}}(\mathbf{Y}) = \mathbf{I} - \mathbf{Y}(\mathbf{Y}^*\mathbf{Y})^{-1}\mathbf{Y}^* = \begin{bmatrix} 0.5561 & -0.4819 & 0.1205 & -0.0093 \\ -0.4819 & 0.4177 & -0.1044 & 0.0080 \\ 0.1205 & -0.1044 & 0.0261 & -0.0020 \\ -0.0093 & 0.0080 & -0.0020 & 0.0002 \end{bmatrix}. \tag{11.185}$$

Computing $\|(\mathbf{I} - P_{\mathcal{R}}(\mathbf{Y}))\mathbf{a}_i\|_2$ for every $1 \leq i \leq 7$ we see that this expression is equal to 0 for $i = 2, 5, 6$. We therefore conclude that $\Lambda = \{2, 5, 6\}$. Finally,

$$\mathbf{X}_\Lambda = \mathbf{A}_\Lambda^\dagger \mathbf{Y} = \begin{bmatrix} 0 & 1 & 1 \\ 1 & 1 & 0 \\ 0 & 1 & 0 \end{bmatrix}. \tag{11.186}$$

By adding zero entries at the corresponding indices, we recover $\mathbf{X}$.

The criterion in (11.184) is similar in spirit to the MUSIC (MUltiple SIgnal Classification) algorithm [15], popular in array signal processing, which also exploits the signal subspace properties. As we will see in Chapter 15, array processing methods are useful in treating a variety of structured analog sampling problems. In particular, we discuss MUSIC in more detail in Section 15.2.6.

In the presence of noise, (11.184) can be adapted by choosing the k values for which the expression is minimized. Since (11.184) leverages the rank to achieve recovery, we say that this method is *rank-aware*. More generally, any method whose performance improves with increasing rank will be termed rank-aware. While MUSIC provides guarantees for the MMV problem in the maximal rank case, there are no performance guarantees for rank$(\mathbf{X}) < k$, and empirically MUSIC does not perform well in this scenario. Therefore, it is of interest to extend classical greedy algorithms to be rank-aware.

It turns out that this is surprisingly simple: for example, to modify simultaneous OMP and thresholding, in each iteration, instead of taking inner products with respect to the current residual $\mathbf{R}$, the inner products are computed with respect to an orthonormal basis $\mathbf{U}$ for the range of $\mathbf{R}$ [337]. Another modification that is crucial to avoid rank degeneration of the residual matrix is to properly normalize the columns of $\mathbf{A}$ in the selection rule. Denote by $\mathbf{P} = \mathbf{I} - \mathbf{A}_\Lambda \mathbf{A}_\Lambda^\dagger$ the orthogonal projection onto $\mathcal{R}(\mathbf{A}_\Lambda)^\perp$. Then the current index is chosen as

$$j = \arg\max_i \frac{\|\mathbf{a}_i^* \mathbf{U}\|_2}{\|\mathbf{P}\mathbf{a}_i\|_2}. \tag{11.187}$$

The resulting algorithm is summarized in Algorithm 11.4. Like MUSIC, Algorithm 11.4 is guaranteed to correctly identify $\mathbf{X}$ when its rank is equal to k and the proper conditions on $\mathbf{A}$ are satisfied [337].

Algorithm 11.4 Rank-aware orthogonal matching pursuit

Input: CS matrix $\mathbf{A}$, MMV matrix $\mathbf{Y}$
Output: Row-sparse matrix $\widehat{\mathbf{X}}$
Initialize: $\widehat{\mathbf{X}}_0 = \mathbf{0}$, $\mathbf{R} = \mathbf{Y}$, $\Lambda = \emptyset$, $\ell = 0$
while halting criterion false **do**
$\quad \ell \leftarrow \ell + 1$
$\quad$ Compute an orthonormal basis $\mathbf{U}$ for $\mathcal{R}(\mathbf{R})$
$\quad$ Compute the orthogonal projection $\mathbf{P}$ onto $\mathcal{R}(\mathbf{A}_\Lambda)^\perp$: $\mathbf{P} = \mathbf{I} - \mathbf{A}_\Lambda \mathbf{A}_\Lambda^\dagger$
$\quad \mathbf{b}(i) \leftarrow \|\mathbf{a}_i^* \mathbf{U}\|_2 / \|\mathbf{P}\mathbf{a}_i\|_2$, $1 \le i \le n$ {form residual matrix vector}
$\quad \Lambda \leftarrow \Lambda \cup \mathrm{supp}(\mathcal{T}(\mathbf{b}, 1))$ {update row support with index of residual's row with
$\quad\quad$ largest magnitude}
$\quad \widehat{\mathbf{X}}_\ell|_\Lambda \leftarrow \mathbf{A}_\Lambda^\dagger \mathbf{Y}$, $\widehat{\mathbf{X}}_\ell|_{\Lambda^c} \leftarrow 0$ {update signal estimate}
$\quad \mathbf{R} \leftarrow \mathbf{Y} - \mathbf{A}\widehat{\mathbf{X}}_\ell$ {update measurement residual}
end while
return $\widehat{\mathbf{X}} \leftarrow \widehat{\mathbf{X}}_\ell$

11.6.3   Performance guarantees

In terms of theoretical guarantees, it can be shown that MMV extensions of SMV algorithms will recover $\mathbf{X}$ under similar conditions to the SMV setting in the worst-case scenario [183, 327, 340]. Thus, theoretical equivalence results for arbitrary values of $\mathbf{X}$ do not predict any performance gain with joint sparsity. In practice, however, multichannel reconstruction techniques perform much better than recovering each channel individually. The reason for this discrepancy is that these results apply to all possible input signals, and are therefore worst-case measures. Clearly, if we input the same signal to each channel, namely when $\mathrm{rank}(\mathbf{X}) = 1$, no additional information on the joint support is provided from multiple measurements. However, as we have seen in Theorem 11.22, higher ranks of the input $\mathbf{X}$ improve the recovery ability. In particular, when $\mathrm{rank}(\mathbf{X}) = k$, rank-aware algorithms such as (11.184) and Algorithm 11.4 recover the true value of $\mathbf{X}$ from the minimal number of measurements given in Theorem 11.22. This property is not shared by the other MMV methods.

Another way to improve performance guarantees is by posing a probability distribution on $\mathbf{X}$ and developing conditions under which $\mathbf{X}$ is recovered with high probability [340, 341, 342]. Average-case analysis can be used to show that fewer measurements are needed in order to recover $\mathbf{X}$ exactly [340].

Theorem 11.25 (Average-case MMV recovery).   *Let $\mathbf{X} \in \mathbb{R}^{n \times L}$ be drawn from a probabilistic model in which the indices for its k nonzero rows are drawn uniformly at random from the $\binom{n}{k}$ possibilities and its nonzero rows (when concatenated) are given by $\Sigma\Delta$, where Σ is an arbitrary diagonal matrix and each entry of Δ is an iid standard Gaussian random variable. If $k \leq \min(C_1/\mu^2(\mathbf{A}), C_2 n/\|\mathbf{A}\|^2)$ where C_1 and C_2 are constants, then $\mathbf{X}$ can be recovered exactly from $\mathbf{Y} = \mathbf{AX}$ via (11.181), with $p = 2$ and $q = 1$, with high probability.*

While worst-case results limit the sparsity to $k = \mathcal{O}(\sqrt{m})$, average-case analysis shows that sparsity up to order $k = \mathcal{O}(m)$ enables recovery with high probability. Moreover, under a mild condition on the sparsity and on the matrix $\mathbf{A}$, the failure probability decays exponentially in the number of channels L [340].

Example 11.34 In this example, we show the recovery rate of (11.181), with $p = 2$ and $q = 1$ for different values of the number of channels L. We choose $m = 32$, $n = 64$. We consider a real-valued random matrix $\mathbf{A}$ whose entries are drawn independently from a Gaussian distribution with zero mean and variance 1. For each value of the sparsity $1 \leq k \leq 32$, we construct a k-sparse matrix $\mathbf{X}$ of size $64 \times L$. The nonzero locations are drawn uniformly at random and the nonzero values are drawn from a Gaussian distribution as described above. The experiment is repeated 1000 times for each sparsity level and number of channels.

Figure 11.20 shows the empirical recovery rate. We consider a success to be when the infinity norm of the error is below 10^{-2}. The algorithm is able to recover the original signals with high probability for sparsity up to $k \approx 15$, which is on the order of $m = 32$. The improvement with increasing L is clearly evident.

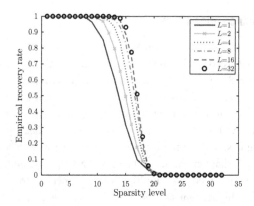

Figure 11.20 Multichannel recovery rate.

11.6.4   Infinite measurement vectors

The MMV model can be further extended to include the case in which there are possibly infinitely many measurement vectors

$$\mathbf{y}(\lambda) = \mathbf{A}\mathbf{x}(\lambda), \quad \lambda \in \Omega, \tag{11.188}$$

where Ω indicates an appropriate index set (which can be countable or uncountable). Here again the assumption is that the set of vectors $\{\mathbf{x}(\lambda), \lambda \in \Omega\}$ all have common support Λ of size k. Such an infinite set of equations is referred to as an *infinite measurement vector (IMV)* system. In Chapters 13–14 we will see that IMV systems arise naturally when considering sparse analog signals. Therefore, developing methods to treat such systems is very important in the context of analog sampling.

In order to treat IMV problems, we begin by showing that the common support can be found by solving a single MMV problem, as incorporated in the following theorem [336].

Theorem 11.26 (IMV recovery). *Suppose that the system of equations (11.188) has a k-sparse solution set with support Λ, and that the matrix $\mathbf{A}$ satisfies*

$$|\Lambda| < \frac{\text{spark}(\mathbf{A}) - 1 + \dim(\text{span}(\mathbf{y}(\lambda)))}{2}. \tag{11.189}$$

Let $\mathbf{V}$ be a matrix whose column span is equal to the span of $\{\mathbf{y}(\lambda), \lambda \in \Omega\}$. Then, the linear system $\mathbf{V} = \mathbf{A}\mathbf{U}$ has a unique k-sparse solution with row support equal to Λ.

Note that since $\dim(\text{span}(\mathbf{y}(\lambda))) \leq k = |\Lambda|$, the condition in Theorem 11.26 implies that $k < \text{spark}(\mathbf{A}) - 1$.

Once Λ is found by seeking a matrix $\mathbf{U}$ that satisfies $\mathbf{V} = \mathbf{A}\mathbf{U}$ with minimal row support, the IMV system reduces to $\mathbf{y}(\lambda) = \mathbf{A}_\Lambda \mathbf{x}_\Lambda(\lambda)$, which can be solved simply by computing the pseudoinverse $\mathbf{x}_\Lambda(\lambda) = \mathbf{A}_\Lambda^\dagger \mathbf{y}(\lambda)$.

Proof: Let $r = \dim(\text{span}(\mathbf{y}(\Lambda)))$ and let $\mathbf{A}$ be an $m \times n$ matrix. Construct an $m \times r$ matrix $\mathbf{Y}$ by taking some set of r linearly independent vectors from $\mathbf{y}(\Lambda)$. Similarly, construct the matrix $\overline{\mathbf{X}}$ of size $n \times r$ by taking the corresponding r vectors from the true input set $\overline{\mathbf{x}}(\Lambda)$. The proof is based on observing the linear system

$$\mathbf{Y} = \mathbf{A}\overline{\mathbf{X}}. \tag{11.190}$$

We first prove that $\overline{\mathbf{X}}$ is the unique k-sparse solution matrix of (11.190) and that its support, denoted here by S, is equal to Λ. Based on this result, the matrix $\mathbf{U}$ is constructed, proving the theorem.

It is easy to see that $S \subseteq \Lambda$, since the columns of $\overline{\mathbf{X}}$ are a subset of the true inputs $\overline{\mathbf{x}}(\Lambda)$. This means that $\overline{\mathbf{X}}$ is a k-sparse solution set of (11.190). Moreover, $\overline{\mathbf{X}}$ is also the unique k-sparse solution of (11.190) according to Theorem 11.22. To conclude the claim on $\overline{\mathbf{X}}$ it remains to prove that $i \in \Lambda$ implies $i \in S$. If $i \in \Lambda$, then for some $\lambda_0 \in \Lambda$ the ith entry of the vector $\mathbf{x}(\lambda_0)$ is nonzero. Now, if $\mathbf{x}(\lambda_0)$ is one of the columns of $\overline{\mathbf{X}}$, then the claim follows trivially. Therefore, assume that $\mathbf{x}(\lambda_0)$ is not a column of $\overline{\mathbf{X}}$. From Proposition 11.10, $\text{rank}(\overline{\mathbf{X}}) = r$. In addition it follows from the same proposition that $\dim\{\text{span}(\overline{\mathbf{x}}(\Lambda))\} = r$. Thus, $\mathbf{x}(\lambda_0)$ must be a (nontrivial) linear combination of the columns of $\overline{\mathbf{X}}$. Since the ith entry of $\mathbf{x}(\lambda_0)$ is nonzero, it implies that at least one column of $\overline{\mathbf{X}}$ has a nonzero value in its ith entry, which means $i \in S$.

Summarizing the first step of the proof, we have that every r linearly independent columns of $\mathbf{y}(\Lambda)$ form an MMV model (11.190) having a unique k-sparse solution matrix $\overline{\mathbf{X}}$, whose support is equal to Λ. As the column span of $\mathbf{V}$ is equal to the column span of $\mathbf{Y}$ we have that $\text{rank}(\mathbf{V}) = r$. Since $\mathbf{V}$ and $\mathbf{Y}$ have the same rank, and $\mathbf{Y}$ also has full column-rank, we get that $\mathbf{V} = \mathbf{Y}\mathbf{R}$ for a unique matrix $\mathbf{R}$ of r linearly independent rows. This immediately implies that $\mathbf{U} = \overline{\mathbf{X}}\mathbf{R}$ is a solution of $\mathbf{V} = \mathbf{A}\mathbf{U}$. Moreover, $\mathbf{U}$ is k-sparse, as each of its columns is a linear combination of the columns of $\overline{\mathbf{X}}$. Theorem 11.22 implies the uniqueness of $\mathbf{U}$ among the k-sparse solution matrices of $\mathbf{V} = \mathbf{A}\mathbf{U}$.

It remains to prove that the support of $\mathbf{U}$ is equal to that of $\overline{\mathbf{X}}$. To simplify notation, we write $\overline{\mathbf{X}}^i$ for the ith row of $\overline{\mathbf{X}}$. Now, $\mathbf{U}^i = \overline{\mathbf{X}}^i\mathbf{R}$, for every $1 \leq i \leq n$. Thus, if $\overline{\mathbf{X}}^i$ is a zero row, then so is $\mathbf{U}^i$. However, for a nonzero row $\overline{\mathbf{X}}^i$, the corresponding row $\mathbf{U}^i$ cannot be zero since the rows of $\mathbf{R}$ are linearly independent. $\quad\square$

The CTF block

Theorem 11.26 implies that in order to solve (11.188) we first find the common support by reducing the problem to an MMV. This reduction is performed by what is referred to as the *continuous to finite (CTF) block*. The CTF builds a frame (or a basis) from the measurements in order to determine the matrix $\mathbf{V}$ in Theorem 11.26. Clearly, there are many possible ways to select $\mathbf{V}$. The theorem ensures that for any such choice, the recovered sparse matrix $\mathbf{U}$ will have the same row support. The matrix $\mathbf{U}$ can be found by using any of the algorithms for solving MMV systems.

One way of constructing $\mathbf{V}$ is by using an eigendecomposition. Specifically, we first form the correlation matrix $\mathbf{Q} = \sum_{\lambda \in \Omega} \mathbf{y}(\lambda)\mathbf{y}^*(\lambda)$. We then perform an eigendecomposition of $\mathbf{Q}$ and choose $\mathbf{V}$ as the matrix of eigenvectors corresponding to the nonzero eigenvalues. Note, that since $\{\mathbf{x}(\lambda)\}$ are jointly k-sparse, the dimension of the span of $\{\mathbf{y}(\lambda)\}$, and consequently the rank of $\mathbf{Q}$, will be no larger than k. When the measurements are noisy, this approach can also be used to eliminate part of the noise, by choosing as $\mathbf{V}$ the eigenvectors corresponding to eigenvalues that are above a threshold. Thus, in this case, besides reducing the computational cost, the CTF process helps improve robustness to noise. In practice, rather than forming the correlation matrix by summing over all vectors $\mathbf{y}(\lambda)$, one can rely on the fact that if the input vectors are random then with high probability summing over order-k vectors will lead to the correct signal space. An alternative choice is to select $\mathbf{V}$ as a sufficiently large set of vectors $\mathbf{y}(\lambda)$. Once again if the vectors are random, then with high probability the range of $\mathbf{V}$ will be equal to the range of the entire signal set $\{\mathbf{y}(\lambda), \lambda \in \Omega\}$, which has dimension no larger than k.

Example 11.35 We consider the CTF operation via an example. Let $\mathbf{A}$ be a real-valued random 16×32 matrix whose entries are drawn independently from a Gaussian distribution with zero mean and variance 1. We construct a k-sparse matrix $\mathbf{X}$ of size 16×10^6 in order to simulate an infinite number of vectors and choose $k = 5$. The nonzero locations are drawn uniformly at random and the nonzero values are drawn from a Gaussian distribution as described above. In our experiment, the support of $\mathbf{X}$ is set to $\{1, 2, 21, 28, 31\}$.

To determine the support from the measurement vectors $\mathbf{y}_i = \mathbf{A}\mathbf{x}_i$, we use the CTF which requires choosing a matrix $\mathbf{V}$ whose columns span the range space of $\mathbf{Y}$. We compare two methods for constructing $\mathbf{V}$: The first is by forming the correlation matrix $\mathbf{Q} = \sum_{i=1}^{10^6} \mathbf{y}_i \mathbf{y}_i^*$ and then using an eigendecomposition, and the second by selecting k columns of $\mathbf{Y}$ uniformly at random. In our example, these techniques lead to different matrices $\mathbf{U}$, denoted $\mathbf{U}^1$ and $\mathbf{U}^2$ for the eigendecomposition and random selection, respectively. Specifically, the nonzero rows are given by

$$\mathbf{U}_\lambda^1 = 1000 \begin{bmatrix} 1.6665 & 6.3115 & -0.3662 & 7.5249 & 0.5905 \\ -3.1875 & -4.7760 & -6.9128 & 4.3368 & 0.7463 \\ -6.4037 & 0.5349 & 2.8612 & 1.6601 & -6.9097 \\ 6.4048 & -1.4024 & -2.6207 & 0.1911 & -7.0702 \\ -2.2773 & 5.9023 & -6.0901 & -4.6591 & -1.0929 \end{bmatrix}, \quad (11.191)$$

$$\mathbf{U}_\lambda^2 = \begin{bmatrix} -7.2314 & -2.1224 & 5.8637 & -4.4914 & 5.9805 \\ 6.6744 & 3.3027 & -1.6938 & -0.7727 & 0.9958 \\ 1.4692 & 0.2108 & 12.9384 & 3.1924 & 18.8706 \\ 12.8984 & 8.3227 & -14.5056 & -9.1430 & -5.6768 \\ 11.6154 & -1.7872 & -3.9152 & -10.3459 & -14.9185 \end{bmatrix}. \quad (11.192)$$

However, the matrices have the same row support, and in both cases the correct support is recovered.

Example 11.36 We next demonstrate the performance of the approaches described in Example 11.35, namely the eigendecomposition and random selection based CTFs,

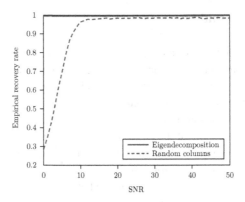

Figure 11.21 Performance of the eigendecomposition method versus the random columns selection method.

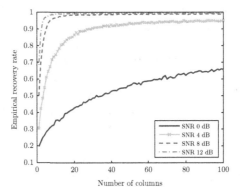

Figure 11.22 Performance of the random columns selection method as a function of the number of columns.

in the presence of noise. We use the same setting as Example 11.35 but add white Gaussian noise to the measurement vectors. The experiment is repeated 1000 times for each SNR value. Figure 11.21 shows the recovered support as a function of SNR. Evidently, the eigendecomposition method yields more robust solutions than the random columns selection approach.

Example 11.37 As a final illustration of the CTF, we focus on the random columns selection method and demonstrate the performance as a function of the number of columns selected from $\mathbf{Y}$ in noisy settings. The experiment is repeated 1000 times for every value of the number of columns and each SNR level. Figure 11.22 shows the recovered support as a function of the number of chosen columns for different SNRs. As we expect, the performance increases as more columns are used beyond k.

11.7   Summary and extensions

To conclude this chapter, CS is an exciting, rapidly growing field that has attracted considerable attention in signal processing, statistics, and computer science, as well as the broader scientific community. Since its initial development only a few years ago, thousands of papers have appeared in this area, and hundreds of conferences, workshops, and special sessions have been dedicated to this growing research field. In this chapter, we reviewed some of the basics of the theory underlying CS. In subsequent chapters of the book, we will see how the fundamentals presented here are expanded and extended to include new models for describing structure in both analog and discrete-time signals, and new sensing design techniques that allow us to sample a wide class of analog signals at sub-Nyquist rates using practical hardware devices.

A more extensive treatment of CS and some of its applications can be found in [14]. Various extensions of the basic CS paradigm are continuously developing, which we do not have space to expand on here. One example of a recent extension of CS is to problems in which we are interested in recovering signal statistics, rather than the signal itself. A special case is power spectrum estimation from compressive measurements, which has applications to array processing and cognitive radio [343, 344, 345, 346, 347].

Another recent extension of the theory that we mention upon closing, and which has been receiving growing interest, is that of recovery of sparse vectors from *nonlinear* measurements. An example arises in *phase retrieval* problems, in which the measurements have the form $y_i = |\langle \mathbf{a}_i, \mathbf{x} \rangle|^2$ for a set of vectors $\mathbf{a}_i$. Note that here only the magnitude of $\langle \mathbf{a}_i, \mathbf{x} \rangle$ is measured, and not the phase. Phase retrieval problems arise in many areas of optics, where the detector can only measure the magnitude of the received optical wave. Several important applications of phase retrieval include X-ray crystallography, transmission electron microscopy, and coherent diffractive imaging [348, 349, 350, 351].

Many methods have been developed for phase recovery [349] which often rely on prior information about the signal, such as positivity or support constraints. One of the most popular techniques is based on alternating projections, where the current signal estimate is transformed back and forth between the object and Fourier domains. The prior information and observations are used in each domain in order to form the next estimate. Two of the main approaches of this type are Gerchberg–Saxton [352] and Fienup [353]. In general, these algorithms are not guaranteed to converge, and often require careful parameter selection and sufficient prior information. To circumvent the difficulties associated with alternating projections, more recently, phase retrieval problems have been treated by assuming that the input signal $\mathbf{x}$ is sparse, thus extending CS ideas to nonlinear measurements.

One of the first works to consider sparse phase retrieval was [354]. Recent interest in this problem has led to several alternative approaches. In [218], sparse recovery from general quadratic measurements was considered based on semidefinite relaxation combined with a row-sparsity constraint on the resulting matrix. An iterative thresholding algorithm was then proposed that approximates the solution. Similar techniques were used in [219, 220, 355]. An alternative algorithm was recently designed in [299, 356] using a greedy search method which is far more efficient than the semidefinite relaxation, and often yields more accurate solutions. Conditions under which a sparse vector $\mathbf{x}$ can be recovered from quadratic measurements are studied in [357, 358]. In particular, it is shown in [358] that stable recovery of a k-sparse vector of length n is possible from $O(k \log(n/k))$ magnitude measurements, where the measurement vectors $\mathbf{a}_i$ are chosen at random. One can also guarantee stable recovery from $O(k \log(n/k)(\log^2 k + \log^2 \log(n/k)))$ noisy measurements.

11.8   Exercises

1. Show that for any $\mathbf{x} \in \mathbb{C}^n$ and $p > q > 0$ we have

$$\|\mathbf{x}\|_p \le \|\mathbf{x}\|_q \le n^{1/q - 1/p} \|\mathbf{x}\|_p. \tag{11.193}$$

   Hint: Use Hölder's inequality.

2. Consider the problem of approximating a signal $\mathbf{x}$ by some $\hat{\mathbf{x}} \in \Sigma_k$, where Σ_k contains all signals $\mathbf{x}$ that are k-sparse (see (11.5)). Prove that thresholding, namely keeping only the k largest coefficients of $\mathbf{x}$, results in the optimal approximation as measured by $\min_{\hat{\mathbf{x}} \in \Sigma_k} \|\mathbf{x} - \hat{\mathbf{x}}\|_p$ for all ℓ_p-norms.

3. Consider a signal $\mathbf{x}$ whose coefficients obey a power-law decay, as defined in (11.15). Show that there exist constants $C_2, r > 0$ depending only on C_1 and q such that $\sigma_k(\mathbf{x})_2 \le C_2 k^{-r}$.

4. In this exercise we will compute the number of degrees of freedom in an $n_1 \times n_2$ matrix $\mathbf{M}$ of rank r.
   a. Consider the SVD $\mathbf{M} = \sum_{k=1}^{r} \sigma_k \mathbf{u}_k \mathbf{v}_k^H$ where $\sigma_1, \sigma_2, \ldots, \sigma_r > 0$ are the nonzero singular values of $\mathbf{M}$, and $\mathbf{u}_1, \mathbf{u}_2, \ldots, \mathbf{u}_r \in \mathbb{R}^{n_1}$, $\mathbf{v}_1, \mathbf{v}_2, \ldots, \mathbf{v}_r \in \mathbb{R}^{n_2}$ are the corresponding singular vectors. Compute the number of variables in the SVD decomposition.
   b. Compute the number of constraints on $\mathbf{u}_i$ and $\mathbf{v}_i$.
   c. Use the previous two results to calculate the number of degrees of freedom in $\mathbf{M}$.

5. Let $\mathbf{A}$ be an $m \times n$ matrix. Prove the rank–nullity theorem, namely

$$\text{rank}(\mathbf{A}) + \text{nullity}(\mathbf{A}) = n, \tag{11.194}$$

   where nullity$(\mathbf{A})$ denotes the dimension of the null space of $\mathbf{A}$.

6. Compute the spark of the following matrices:
   a.

$$\mathbf{A} = \begin{bmatrix} 1 & -1 & 0 & -1 \\ -1 & 1 & 1 & -1 \end{bmatrix}. \tag{11.195}$$

   b.

$$\mathbf{B} = \begin{bmatrix} 1 & 2 & 3 & 4 \\ 4 & 1 & 2 & 3 \\ 3 & 4 & 1 & 2 \end{bmatrix}. \tag{11.196}$$

7. Let $\mathbf{A}, \mathbf{B} \in \mathbb{R}^{m \times n}$. Prove or disprove:

a. $\mathrm{spark}(\mathbf{A}) \leq \mathrm{rank}(\mathbf{A})$.

b. $\mathrm{spark}(\mathbf{A}) \leq \mathrm{rank}(\mathbf{A}) + 1$.

c. $\mathrm{spark}(\mathbf{A}\mathbf{A}^T) \leq \mathrm{spark}(\mathbf{A})$.

d. $\mathrm{spark}(\mathbf{A} + \mathbf{B}) \leq \mathrm{spark}(\mathbf{A}) + \mathrm{spark}(\mathbf{B})$.

e. $\mathrm{spark}(\mathbf{A}\mathbf{B}) = \mathrm{spark}(\mathbf{A}) + \mathrm{spark}(\mathbf{B})$.

8. Let $\mathbf{A}$ be a matrix satisfying the RIP of order k with constant δ_k and let Λ be a set of size k. Show that:

a. The restricted isometry constants are ordered: $\delta_1 \leq \delta_2 \leq \dots$

b. $\|\mathbf{A}_\Lambda^T \mathbf{A}_\Lambda - \mathbf{I}\| \leq \delta_k$.

c. $\|\mathbf{A}_\Lambda^T \mathbf{x}\|_2^2 \leq (1 + \delta_k)\|\mathbf{x}\|_2^2$.

d. $\frac{1}{1+\delta_k}\|\mathbf{x}\|_2^2 \leq \|\mathbf{A}_\Lambda^\dagger \mathbf{x}\|_2^2 \leq \frac{1}{1-\delta_k}\|\mathbf{x}\|_2^2$.

e. $(1 - \delta_k)^2\|\mathbf{x}\|_2^2 \leq \|\mathbf{A}_\Lambda^T \mathbf{A}_\Lambda \mathbf{x}\|_2^2 \leq (1 + \delta_k)^2\|\mathbf{x}\|_2^2$.

f. $\frac{1}{(1+\delta_k)^2}\|\mathbf{x}\|_2^2 \leq \|(\mathbf{A}_\Lambda^T \mathbf{A}_\Lambda)^{-1}\mathbf{x}\|_2^2 \leq \frac{1}{(1-\delta_k)^2}\|\mathbf{x}\|_2^2$.

9. Consider the following 6×7 matrix

$$\mathbf{A} = \begin{bmatrix} 1 & 0 & 0 & 0 & 0 & \frac{1}{\sqrt{6}} & 0 \\ 0 & 1 & 0 & 0 & 0 & \frac{1}{\sqrt{6}} & 0 \\ 0 & 0 & 1 & 0 & 0 & \frac{1}{\sqrt{6}} & 0 \\ 0 & 0 & 0 & 1 & 0 & \frac{1}{\sqrt{6}} & 0 \\ 0 & 0 & 0 & 0 & 1 & \frac{1}{\sqrt{6}} & 0 \\ 0 & 0 & 0 & 0 & 0 & \frac{1}{\sqrt{6}} & 1 \end{bmatrix}. \tag{11.197}$$

a. Show that $\mathbf{A}$ satisfies the RIP of order 2 with $\delta_2 = 1/\sqrt{6}$.

b. Show that $\mathbf{A}$ satisfies Theorem 11.4.

10. Consider the matrix defined in Exercise 9.

a. Compute the coherence $\mu(\mathbf{A})$ of $\mathbf{A}$. Show that $\mu(\mathbf{A})$ satisfies Proposition 11.7.

b. Compute the spark of $\mathbf{A}$. Show that $\mathrm{spark}(\mathbf{A})$ satisfies Lemma 11.2.

c. Show that $\mu(\mathbf{A})$ satisfies Lemma 11.3.

11. Consider the $m \times n$ partial Fourier matrix $\mathbf{F}$ with elements

$$F_{r\ell} = \frac{1}{\sqrt{n}} e^{-j\frac{2\pi}{n} r\ell}, \tag{11.198}$$

where $r = -M, \dots, M$ (so that $m = 2M + 1$) for some M such that $m \leq n$ and $\ell = 0, \dots, n - 1$.

a. Compute the coherence of $\mathbf{F}$.

Hint: Your answer can be a function of the Dirichlet kernel

$$D_n(x) = \sum_{k=-n}^{n} e^{jkx} = \frac{\sin((n + 1/2)x)}{\sin(x/2)}. \tag{11.199}$$

b. Let $n = 256$. Plot the coherence of $\mathbf{F}$ as a function of odd m. Explain the results.

c. Now let $n \to \infty$ with m fixed. Determine the coherence in this setting. Use this result to explain why super-resolution is difficult.

d. Assume now that $\mathbf{y} = \mathbf{F}\mathbf{x}$ and that the nonzero values in $\mathbf{x}$ are known to be separated by a value Δ. Compute the coherence of $\mathbf{F}$ as a function of Δ when restricted to the possible inputs $\mathbf{x}$.

12. Let $\mathbf{H}$ be the 4×4 Haar matrix given by (11.100), and $\mathbf{F}$ be the 4×4 normalized Fourier matrix defined in Exercise 11.

a. Compute the mutual coherence $\mu(\mathbf{H}, \mathbf{F})$.

b. Show that $\mu(\mathbf{H}, \mathbf{F})$ satisfies Proposition 11.8.

13. Suppose we are given three vectors in $\mathbb{R}^2$.

a. What is the smallest coherence possible?

b. Give an example of a choice of vectors that attain this coherence.

14. Consider the matrices defined in Exercise 12. Let

$$\mathbf{x} = \begin{bmatrix} 1 & 1 & 1 & 1 \end{bmatrix}^T. \tag{11.200}$$

a. Compute $\mathbf{a}$ and $\mathbf{b}$ such that $\mathbf{x} = \mathbf{H}\mathbf{a}$ and $\mathbf{x} = \mathbf{F}\mathbf{b}$.

b. Show that $\mu(\mathbf{H}, \mathbf{F})$ satisfies Theorem 11.7.

15. Let $\mathbf{y} = \mathbf{A}\mathbf{x}$ where $\mathbf{x}$ is k-sparse. Show that $\mathbf{x}$ is the unique solution of $\min_{\mathbf{z}} \|\mathbf{z}\|_1$ subject to $\mathbf{y} = \mathbf{A}\mathbf{z}$ if and only if $\mathbf{A}$ satisfies the modified null-space property of order k such that $\|\mathbf{h}_\Lambda\|_1 \leq \|\mathbf{h}_{\Lambda^c}\|_1$ holds for all $\mathbf{h} \in \mathcal{N}(\mathbf{A})$ and for all Λ for which $|\Lambda| \leq k$.

16. Solve the optimization problem (11.117) with $\mathcal{B}(\mathbf{y}) = \{\mathbf{z} : \mathbf{A}\mathbf{z} = \mathbf{y}\}$ where $\mathbf{A}$ is the following 4×8 Vandermonde matrix:

$$\mathbf{A} = \begin{bmatrix} 1 & 1 & 1 & 1 & 1 & 1 & 1 & 1 \\ 1 & 2 & 3 & 4 & 5 & 6 & 7 & 8 \\ 1 & 4 & 9 & 16 & 25 & 36 & 49 & 64 \\ 1 & 8 & 27 & 64 & 125 & 216 & 343 & 512 \end{bmatrix}, \tag{11.201}$$

and

$$\mathbf{y} = \begin{bmatrix} 2 & 6 & 26 & 126 \end{bmatrix}^T. \tag{11.202}$$

17. Repeat Exercise 16 for the ℓ_1 formulation (11.119). Compare your result with the one in Exercise 16.

18. Show that the residual in the MP algorithm can be written in the recursive form:

$$\mathbf{r}_\ell = \mathbf{r}_{\ell-1} - \frac{\mathbf{a}_i^* \mathbf{r}_{\ell-1}}{\|\mathbf{a}_i\|_2^2} \mathbf{a}_i, \tag{11.203}$$

where $\mathbf{a}_i$ is given by (11.127).

19. Show that the residual in the OMP algorithm can be written in the recursive form:

$$\mathbf{r}_\ell = \mathbf{r}_{\ell-1} - \frac{\mathbf{a}_j^* \mathbf{r}_{\ell-1}}{\|\beta_j\|_2^2} \beta_j, \tag{11.204}$$

where $\beta_j = \mathbf{a}_j - \sum_{i \in \Lambda} \langle \mathbf{a}_j, \mathbf{a}_i \rangle \mathbf{a}_i$, Λ is the support of $\hat{\mathbf{x}}_{\ell-1}$ as defined in Algorithm 11.1, and j is the new support index.

20. Let $\mathbf{A}$ be a matrix with iid elements drawn from a zero-mean Gaussian distribution with unit variance. Suppose $\mathbf{A}$ satisfies the RIP with constant δ with some probability p. Let $\mathbf{D}$ be a unitary matrix. Find a value of δ' and p' such that $\mathbf{A}$ satisfies the D-RIP with constant δ' and probability p'. Explain your result.

21. Suppose that $\mathbf{A}$ is a Vandermonde matrix of size 6×20 with distinct roots.
    a. If $\mathbf{x}$ is an arbitrary k-sparse vector of length 20, how large can k be so that $\mathbf{x}$ can be recovered from $\mathbf{y} = \mathbf{Ax}$?
    b. Let $\mathbf{X}$ be a matrix with k nonzero rows of size $20 \times L$. How large can k be so that $\mathbf{X}$ can be recovered from $\mathbf{Y} = \mathbf{AX}$?

22. We are given an MMV system $\mathbf{Y} = \mathbf{AX}$ where $\mathbf{X}$ is k row-sparse.
    a. Define $\mathbf{y}$ as the vector obtained by stacking the columns of $\mathbf{Y}$ one after another, and define $\mathbf{x}$ as the vector obtained by stacking the rows of $\mathbf{X}$. Find a matrix $\mathbf{B}$ such that $\mathbf{y} = \mathbf{Bx}$.
    b. Compute the coherence of $\mathbf{B}$ and compare it with the coherence of $\mathbf{A}$. Explain your result.
    c. What is the sparsity pattern of $\mathbf{x}$?

Chapter 12

Sampling over finite unions

In the previous chapter we considered sampling of sparse finite-dimensional vectors. As we have seen, this setting can be viewed as a special case of a union model, where the subspaces $\mathcal{U}_i$ comprising the union are a direct sum of k one-dimensional subspaces, spanned by columns of the identity matrix. Thus, for each $\mathcal{U}_i = \mathcal{U}$, we have that

$$\mathcal{U} = \mathcal{E}_{i_1} \oplus \ldots \oplus \mathcal{E}_{i_k}, \tag{12.1}$$

where $\mathcal{E}_j, 1 \leq j \leq N$ is the subspace spanned by the jth column of the $N \times N$ identity matrix, and $i_j, 1 \leq j \leq k$ are indices between 1 and N.

Viewing the sparsity model in the form of a UoS such as (12.1) leads to an immediate extension which allows for much more general signal classes. Specifically, we may replace each of the one-dimensional subspaces $\mathcal{E}_j$ of $\mathbb{R}^N$ by a subspace $\mathcal{A}_j$. These subspaces have arbitrary dimension, and are described over an arbitrary Hilbert space $\mathcal{H}$. In particular, they can represent subspaces of analog signals. In this chapter we focus on finite-dimensional unions of this form, on methods for recovering signals that lie in the union from few measurements, and on performance guarantees. We also consider learning the subspaces from training data, when the possible subspaces are not known in advance. Finally, we treat the more difficult scenario of subspace learning from compressed data, leading to an extension of compressed sensing referred to as *blind compressed sensing*.

12.1 Finite unions

12.1.1 Signal model

The model we treat in this chapter is one in which $x \in \mathcal{X}$ where $\mathcal{X}$ is a finite union of subspaces (FUS) $\mathcal{X} = \bigcup_i \mathcal{U}_i$ with each $\mathcal{U}_i$ having the additional structure

$$\mathcal{U}_i = \bigoplus_{|j|=k} \mathcal{A}_j. \tag{12.2}$$

Here $\{\mathcal{A}_j, 1 \leq j \leq m\}$ are a given set of disjoint subspaces, and $|j| = k$ denotes a sum over k indices. Thus, each subspace $\mathcal{U}_i$ corresponds to a different choice of k subspaces $\mathcal{A}_j$ that comprise the sum. We assume that m and the dimensions $d_i = \dim(\mathcal{A}_i)$ of the subspaces $\mathcal{A}_i$ are finite. An advantage of the structure (12.2) is that it allows us to convert the problem into one of block-sparse recovery. Thus, exploiting the structure is necessary in order to develop general purpose recovery algorithms. In fact, as we

475

will see in the broader context of UoS modeling, requiring the subspaces making up the union to have some underlying structure is key to enabling signal recovery in a computationally efficient way. The differences between the models we treat in the ensuing chapters are in the underlying structure, the number of subspaces comprising the union, and the dimensionality of the individual subspaces.

Given n samples

$$\mathbf{y} = S^*x, \tag{12.3}$$

where S is a general sampling operator, and the knowledge that x lies in exactly one of the subspaces $\mathcal{U}_i$, we would like to recover the unknown signal x. In this setting, there are $\binom{m}{k}$ possible subspaces constituting the union. An alternative interpretation of our model is that of a signal x which can be written as

$$x = \sum_{i=1}^{k} x_i, \tag{12.4}$$

where each x_i lies in $\mathcal{A}_j$ for some index j.

In the previous chapter we considered recovery of sparse vectors from compressed samples. Several of the scenarios we treated are in fact special cases of (12.2), where $\mathcal{A}_j$ are subspaces of $\mathbb{R}^N$ for some N. For example, the standard sparsity model in which $x = \mathbf{x}$ is a vector of length N that has a k-sparse representation in a given basis defined by an invertible matrix $\mathbf{W}$ fits (12.2) by choosing $\mathcal{A}_i$ as the space spanned by the ith column of $\mathbf{W}$. In this setting $m = N$, and there are $\binom{N}{k}$ subspaces that make up the union.

Another important example over $\mathbb{R}^N$, is the block-sparsity model [183, 207, 359, 360] in which $\mathbf{x}$ is divided into equal-length blocks of size d, and at most k blocks are nonzero. Such a vector fits our setting if we choose $\mathcal{A}_i$ to be the space spanned by the corresponding d columns of the identity matrix. Here $m = N/d$ and there are $\binom{N/d}{k}$ subspaces in the union. An example of such a vector is depicted in Fig. 12.1. In this example, $\mathbf{x}$ is a vector of length $N = 20$ which is comprised of blocks of size 4, with three nonzero blocks.

Block sparsity appears naturally in a variety of problems such as DNA microarray analysis [361, 362], equalization of sparse communication channels [331], source localization [330], and source separation [363]. In the context of sampling of analog signals, which is our focus here, their importance stems from the fact that union models of the form (12.2) can be converted into block-sparse recovery problems. This equivalence will be developed in detail in the next section.

A special case of block sparsity is the MMV model discussed in Section 11.6, in which the goal is to recover a matrix $\mathbf{X}$ from measurements $\mathbf{Y} = \mathbf{AX}$, for a given sampling

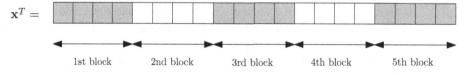

Figure 12.1 A block-sparse vector $\mathbf{x}$, of dimension 20, with equal block size ($d = 4$). The grey areas represent 12 nonzero entries which occupy three blocks.

matrix $\mathbf{A}$. The matrix $\mathbf{X}$ is assumed to have at most k nonzero rows. This problem can be transformed into that of recovering a k-block sparse signal by stacking the rows of $\mathbf{X}$ and $\mathbf{Y}$, leading to the relationship

$$\mathrm{vec}(\mathbf{Y}^T) = (\mathbf{A} \otimes \mathbf{I})\, \mathrm{vec}(\mathbf{X}^T). \qquad (12.5)$$

Here $\otimes$ denotes the Kronecker product, and $\mathrm{vec}(\mathbf{B})$ is the vector formed by stacking the columns of $\mathbf{B}$. Relation (12.5) follows from the properties of the Kronecker product and the vec operation, which are discussed in Appendix A. Since $\mathbf{X}$ is k-row sparse, $\mathbf{X}^T$ is k-column sparse, namely, only k of its columns are nonzero. This in turn implies that $\mathrm{vec}(\mathbf{X}^T)$ is k-block sparse.

Example 12.1 In this example, we demonstrate block-sparse modeling of a music piece.

Consider a musical signal which we divide into time frames of length N corresponding to 100 msec. Each time interval is modeled as a superposition of at most 10 musical notes (i.e. Do, Re, Mi, ...), corresponding to the number of fingers. In addition, a note is itself a superposition of single tones: the fundamental frequency and its harmonics. This can be seen in Fig. 12.2, where we depict the power spectrum of two realizations of the note Do. These examples are formed by playing the same key Do, consecutively. From the figure we see that each of the notes is well approximated by a weighted superposition of several tones, where the weights are not necessarily equal. In this example, there are about eight tones. Assuming more generally that a note is modeled by a superposition of at most 10 single tones, we may represent a note by a matrix $\mathbf{A}$ of size $N \times 20$, where the factor 2 is due to phase ambiguity. The two first columns are a single tone (the fundamental frequency) and its shifted version (by 25 percent of its period),

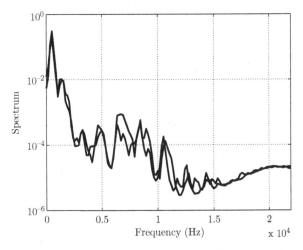

Figure 12.2 Power spectrum of two realizations of the note Do. Each note can be approximated by a weighted superposition of several tones ($\sim$8 here).

while the other columns represent its harmonics. These fundamental frequencies are obtained by a set of training data.

A typical piano has eight octaves where every octave is divided into 12 notes (seven white and five black keys) resulting in a total of 96 notes. Representing each note i by an $N \times 20$ matrix $\mathbf{A}_i$ as described above we can write a segment of a music piece as

$$\mathbf{y} = \mathbf{Ax} = [\mathbf{A}_1 \ \mathbf{A}_2 \ \dots \ \mathbf{A}_{96}] \, \mathbf{x}, \tag{12.6}$$

where $\mathbf{x}$ is a block-sparse vector with at most $k = 10$ active blocks. Within a block, the values of $\mathbf{x}$ indicate which tones are dominant within the specific notes chosen.

12.1.2   Problem formulation

Given k and the subspaces $\mathcal{A}_i$, we would like to address the following questions:

1. What are the conditions on the sampling vectors $s_i, 1 \leq i \leq n$ in order to guarantee that the sampling is stable and invertible?
2. How can we recover the unique x (regardless of computational complexity)?
3. How can we recover the unique x in an efficient and stable manner?

In the case in which x lies in a single subspace $\mathcal{A} \subseteq \mathcal{H}$, we have seen in Chapter 6 that it can be recovered from its samples S^*x as long as $\mathcal{H} = \mathcal{A} \oplus S^\perp$. In particular, when $\mathcal{A}, \mathcal{S}$ are generated by $\{a(t - nT)\}$ and $\{s(t - nT)\}$ respectively, recovery is possible when $R_{SA}(e^{j\omega})$ defined by (5.28) is stably invertible. Reconstruction is obtained by filtering the samples with the filter $H(e^{j\omega}) = 1/R_{SA}(e^{j\omega})$ and then interpolating the corrected samples with $a(t)$. Geometrically, we showed that knowing the samples is equivalent to knowing the orthogonal projection $P_S x$ onto $\mathcal{S}$. If the direct-sum condition is satisfied, then there is a unique vector in $\mathcal{A}$ with the given projection (see Fig. 6.5).

When x lies in one of a set of possible subspaces, sampling and recovery are much more involved. Assuming that sampling is linear, this process can still be thought of as a projection onto the space $\mathcal{S}$ spanned by the sampling vectors. However, we can no longer simply draw a vertical line to match the prior space, since there are many possibilities, as depicted in Fig. 12.3. Therefore, in contrast to previous priors treated in earlier chapters, union priors necessitate nonlinear recovery techniques to determine the correct subspace and reconstruct the signal.

In order to answer the questions above, we rely on an equivalence between the UoS problem assuming the model (12.2) and that of recovering block-sparse vectors. This relationship is established in the next section. In the remainder of the chapter we therefore focus on the k-block sparse model and develop our results in that context. In particular, we introduce a block RIP condition that ensures uniqueness and stability of our sampling problem. We then present an efficient recovery algorithm based on convex optimization (referred to as block BP [207]) which approximates an unknown block-sparse vector $\mathbf{c}$, as well as several greedy block-sparse recovery methods. Based on block RIP and block coherence, we prove that $\mathbf{c}$ can be reconstructed exactly using these techniques.

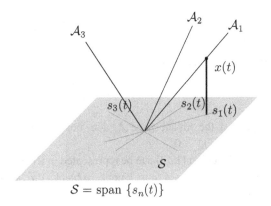

$$\mathcal{S} = \text{span } \{s_n(t)\}$$

Figure 12.3 Sampling and reconstruction of $x(t)$. The union consists of three subspaces $\mathcal{A}_1, \mathcal{A}_2$, and $\mathcal{A}_3$ where $x(t)$ lies in $\mathcal{A}_1$. The sampling vectors are represented by $s_1(t)$, $s_2(t)$, and $s_3(t)$. If $x(t)$ is known to be in $\mathcal{A}_1$, then it can be reconstructed from its projection in the direction of $s_1(t)$. Here, however, we are interested in the case in which it is unknown a priori in which direction the signal lies.

The block-sparse model we present here has also been studied in the statistical literature, where often the objective is quite different. For example, group selection consistency is treated in [364, 365], asymptotic prediction properties are considered in [365, 359], and block sparsity for logistic regression is introduced in [366].

12.1.3 Connection with block sparsity

Consider the model of a signal x in the union of k out of m subspaces $\mathcal{A}_i$, with $d_i = \dim(\mathcal{A}_i)$ as in (12.2). To write x explicitly, we choose a basis for each $\mathcal{A}_i$. Denoting by $A_i \colon \mathbb{R}^{d_i} \to \mathcal{H}$ the set transformation corresponding to a basis for $\mathcal{A}_i$, any such x can be written as

$$x = \sum_{|i|=k} A_i \mathbf{c}_i, \tag{12.7}$$

where $\mathbf{c}_i \in \mathbb{R}^{d_i}$ are the representation coefficients in $\mathcal{A}_i$, and $|i| = k$ denotes a sum over a set of k indices. The choice of indices depends on the signal x and is unknown in advance.

To develop the equivalence with block sparsity, it is useful to introduce some further notation. First, we define $A \colon \mathbb{R}^N \to \mathcal{H}$ as the set transformation that is a result of (column-wise) concatenating the different A_i, with $N = \sum_{i=1}^m d_i$. When $\mathcal{H} = \mathbb{R}^M$ for some M, $A_i = \mathbf{A}_i$ is a matrix of size $M \times d_i$, and $A = \mathbf{A}$ is the $M \times N$ matrix obtained by column-wise concatenating $\mathbf{A}_i$. Next, we define the ith sub-block $\mathbf{c}[i]$ of a length-N vector $\mathbf{c}$ over $\mathcal{I} = \{d_1, \ldots, d_m\}$. The ith sub-block is of length d_i, and the blocks are formed sequentially so that

$$\mathbf{c}^T = [\underbrace{c_1 \cdots c_{d_1}}_{\mathbf{c}[1]} \cdots \underbrace{c_{N-d_m+1} \cdots c_N}_{\mathbf{c}[m]}]^T. \tag{12.8}$$

With these definitions we have that

$$A\mathbf{c} = \sum_{i=1}^{m} A_i \mathbf{c}[i]. \tag{12.9}$$

If for a given x the jth subspace $\mathcal{A}_j$ does not appear in the sum (12.2), or equivalently in (12.7), then $\mathbf{c}[j] = \mathbf{0}$.

Any x in the union (12.2) can be represented in terms of k of the bases A_i. Therefore, $x = A\mathbf{c}$ where there are at most k nonzero blocks $\mathbf{c}[i]$, so that x is described by a sparse vector $\mathbf{c}$ in an appropriate basis. The sparsity pattern here has a unique form which we will exploit in our conditions and algorithms: the nonzero elements appear in blocks.

Definition 12.1. *A vector* $\mathbf{c} \in \mathbb{R}^N$ *is called* k-*block sparse over* $\mathcal{I} = \{d_1, \ldots, d_m\}$ *if* $\mathbf{c}[i]$ *is nonzero for at most* k *indices* i *where* $N = \sum_i d_i$.

An example of a block-sparse vector with $k = 2$ is depicted in Fig. 12.4. When $d_i = 1$ for each i, block sparsity reduces to the conventional definition of a sparse vector. Denoting

$$\|\mathbf{c}\|_{0,\mathcal{I}} = \sum_{i=1}^{m} I(\|\mathbf{c}[i]\|_2 > 0), \tag{12.10}$$

where $I(\|\mathbf{c}[i]\|_2 > 0)$ is an indicator function that obtains the value one if $\|\mathbf{c}[i]\|_2 > 0$, and zero otherwise, a k-block sparse vector $\mathbf{c}$ is defined by $\|\mathbf{c}\|_{0,\mathcal{I}} \leq k$.

Evidently, there is a one-to-one correspondence between any vector x in the union (12.2), and a block-sparse vector $\mathbf{c}$. The measurements (12.3) can also be represented explicitly in terms of $\mathbf{c}$ as

$$\mathbf{y} = S^* x = S^* A\mathbf{c} = \mathbf{D}\mathbf{c}, \tag{12.11}$$

where $\mathbf{D}$ is the $n \times N$ matrix defined by

$$\mathbf{D} = S^* A. \tag{12.12}$$

We may therefore phrase our problem in terms of $\mathbf{D}$ and $\mathbf{c}$ as that of recovering a k-block sparse vector $\mathbf{c}$ over $\mathcal{I}$ from the measurements (12.11).

Note that the choice of basis A_i for each subspace does not affect our sparsity model. Indeed, choosing alternative bases will lead to $x = A\mathbf{W}\mathbf{c}$ where $\mathbf{W}$ is a block diagonal

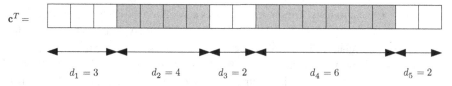

Figure 12.4 A block-sparse vector $\mathbf{c}$ over $\mathcal{I} = \{d_1, \ldots, d_5\}$. The grey areas represent 10 nonzero entries which occupy two blocks.

matrix with blocks of size d_i. Defining $\tilde{\mathbf{c}} = \mathbf{W}\mathbf{c}$, the block sparsity pattern of $\tilde{\mathbf{c}}$ is equal to that of $\mathbf{c}$.

Since our problem is equivalent to that of recovering a block-sparse vector over $\mathcal{I}$ from linear measurements $\mathbf{y} = \mathbf{D}\mathbf{c}$, in the remainder of the chapter we focus our attention on this problem. Our goal is to explicitly take the block structure of $\mathbf{c}$ into account, both in terms of the recovery algorithms and the measures that are used to characterize their performance. As we show, making explicit use of block sparsity yields provably better reconstruction properties than treating the signal as being sparse in the conventional sense, thereby ignoring the additional structure in the problem. We will also consider recovery when $\mathbf{D}$, or equivalently the underlying subspaces $\mathcal{A}_i$, are unknown, but are rather estimated from data.

Example 12.2 Consider the following subspaces:

$$\mathcal{A}_i = \text{span}\{\sin(\omega_i t), \cos(\omega_i t)\}, \qquad i = 1, 2, 3, \qquad (12.13)$$

where $\omega_i \neq \omega_j$ for $i \neq j$. Let $\mathcal{X} = \bigcup_i \mathcal{U}_i$ be the union of subspaces where

$$\mathcal{U}_i = \bigoplus_{|j|=2} \mathcal{A}_j, \qquad i = 1, 2, 3. \qquad (12.14)$$

Thus, any $x(t) \in \mathcal{X}$ is a linear combination of sines and cosines with at most two different frequencies.

We now show how to represent $x(t)$ in $\mathcal{X}$ in terms of a block-sparse vector. We begin by constructing the set transformations A_i corresponding to the subspaces $\mathcal{A}_i$ as $A_i\mathbf{a} = a_1 \sin(\omega_i t) + a_2 \cos(\omega_i t)$ where $\mathbf{a} \in \mathbb{R}^2$. Let

$$x(t) = \sin(\omega_1 t) + 2\cos(\omega_1 t) + 3\sin(\omega_3 t) + 4\cos(\omega_3 t) \qquad (12.15)$$

be a given signal in $\mathcal{X}$. It is then easy to see that $x = A\tilde{\mathbf{c}}$ where A is the set transformation that results from concatenating A_1, A_2, A_3, and $\tilde{\mathbf{c}}$ is the 2-block sparse vector with blocks of size 2 given by

$$\tilde{\mathbf{c}} = \begin{bmatrix} 1 & 2 & 0 & 0 & 3 & 4 \end{bmatrix}^T. \qquad (12.16)$$

Suppose now that we choose a different basis for each $\mathcal{A}_i$:

$$\mathcal{A}_i = \text{span}\{e^{-j\omega_i t}, e^{j\omega_i t}\}, \qquad i = 1, 2, 3. \qquad (12.17)$$

With this choice of basis, x can be written as

$$x = \left(1 + \frac{1}{2j}\right)e^{j\omega_1 t} + \left(1 - \frac{1}{2j}\right)e^{-j\omega_1 t} + \left(2 + \frac{3}{2j}\right)e^{j\omega_3 t} + \left(2 - \frac{3}{2j}\right)e^{-j\omega_3 t}. \qquad (12.18)$$

Defining

$$\mathbf{W} = \text{diag}\left(\begin{bmatrix} j & -j \\ 1 & 1 \end{bmatrix}, \begin{bmatrix} j & -j \\ 1 & 1 \end{bmatrix}, \begin{bmatrix} j & -j \\ 1 & 1 \end{bmatrix}\right), \qquad (12.19)$$

we can express x as $x = A\mathbf{Wc}$ with

$$\mathbf{c} = \left[\left(1+\frac{1}{2j}\right) \ \left(1-\frac{1}{2j}\right) \ \ 0 \ \ 0 \ \ \left(2+\frac{3}{2j}\right) \ \left(2-\frac{3}{2j}\right)\right]^T. \tag{12.20}$$

Although the values of $\mathbf{c}$ and $\tilde{\mathbf{c}}$ are different, their block sparsity pattern remains the same: in both vectors, the first and third blocks have nonzero energy.

Finally, suppose we sample $x(t)$ by observing the signal values at times $t = 0, \pi/6, \pi/4, \pi/3, \pi/2, 2\pi/3$. The sampling matrix $\mathbf{D}$ is obtained by sampling the set transformation A at times t_i. Assuming for simplicity that $\omega_i = i$, we have

$$\mathbf{D} = S^*A = \begin{bmatrix} 0 & 1 & 0 & 1 & 0 & 1 \\ \frac{1}{2} & \frac{\sqrt{3}}{2} & \frac{\sqrt{3}}{2} & \frac{1}{2} & 1 & 0 \\ \frac{\sqrt{2}}{2} & \frac{\sqrt{2}}{2} & 1 & 0 & \frac{\sqrt{2}}{2} & -\frac{\sqrt{2}}{2} \\ \frac{\sqrt{3}}{2} & \frac{1}{2} & \frac{\sqrt{3}}{2} & -\frac{1}{2} & 0 & -1 \\ 1 & 0 & 0 & -1 & -1 & 0 \\ \frac{\sqrt{3}}{2} & -\frac{1}{2} & -\frac{\sqrt{3}}{2} & -\frac{1}{2} & 0 & 1 \end{bmatrix}. \tag{12.21}$$

The measurements $\mathbf{y}$ then correspond to $\mathbf{y} = \mathbf{D}\tilde{\mathbf{c}}$.

12.2   Uniqueness and stability

We begin by studying uniqueness and stability of sampling over the union (12.2). These properties are intimately related to the RIP, which we generalize here to the block-sparse setting.

The first question we address is that of uniqueness, namely conditions under which a block-sparse vector $\mathbf{c}$ is uniquely determined by the measurement vector $\mathbf{y} = \mathbf{Dc}$.

Proposition 12.1. *Let* $\mathbf{y} = \mathbf{Dc}$ *for an arbitrary* k-*block sparse vector* $\mathbf{c}$. *Then,* $\mathbf{c}$ *is unique if and only if* $\mathbf{Dz} \neq 0$ *for every nonzero* $2k$-*block sparse* $\mathbf{z}$.

Proof: We first assume that $\mathbf{Dz} \neq 0$ for every nonzero $2k$-block sparse $\mathbf{z}$. Let $\mathbf{y}$ be arbitrary and suppose that there are two k-block sparse vectors $\mathbf{c}_1$ and $\mathbf{c}_2$ such that $\mathbf{y} = \mathbf{Dc}_1 = \mathbf{Dc}_2$. Then,

$$0 = \mathbf{D}(\mathbf{c}_1 - \mathbf{c}_2) = \mathbf{Dz}, \tag{12.22}$$

where we denoted $\mathbf{z} = \mathbf{c}_1 - \mathbf{c}_2$. Since $\mathbf{c}_1, \mathbf{c}_2$ are each $2k$-block sparse, $\mathbf{z}$ is at most $2k$-block sparse. If $\mathbf{Dz} \neq 0$ for every nonzero $2k$-block sparse $\mathbf{z}$, then (12.22) holds only for $\mathbf{z} = 0$, or $\mathbf{c}_1 = \mathbf{c}_2$, so that $\mathbf{c}$ is unique.

Assume next that $\mathbf{c}$ is unique, and suppose for the sake of contradiction that there exists a $2k$-block sparse $\mathbf{z}$ such that $\mathbf{Dz} = 0$. Since $\mathbf{z}$ has $2k$ nonzero blocks, we can express it as $\mathbf{z} = \mathbf{c}_1 - \mathbf{c}_2$ for some k-block sparse vectors $\mathbf{c}_1, \mathbf{c}_2$ with $\mathbf{c}_1 \neq \mathbf{c}_2$. We then have that $\mathbf{Dz} = \mathbf{D}(\mathbf{c}_1 - \mathbf{c}_2) = 0$ which implies that $\mathbf{Dc}_1 = \mathbf{Dc}_2$. But this contradicts our assumption that there exists at most one k-block sparse signal $\mathbf{c}$ such that $\mathbf{y} = \mathbf{D}\mathbf{c}$. Therefore, $\mathbf{D}\mathbf{z}$ must be equal to $\mathbf{0}$ for every $2k$-block sparse $\mathbf{z}$, proving the proposition.   □

We next address the issue of stability. In Chapter 2, we saw that a sampling operator S for an arbitrary set $\mathcal{T} \in \mathcal{H}$ is stable if and only if there exist constants $0 < \alpha \le \beta < \infty$ such that

$$\alpha\|x_1 - x_2\|_{\mathcal{H}}^2 \le \|S^*x_1 - S^*x_2\|_2^2 \le \beta\|x_1 - x_2\|_{\mathcal{H}}^2, \qquad (12.23)$$

for every x_1, x_2 in $\mathcal{T}$. This condition ensures that small differences in elements from the set do not imply large differences in the samples. The ratio $\kappa = \beta/\alpha$ provides a measure for stability of the sampling operator: the operator is maximally stable when $\kappa = 1$. When $\mathcal{T}$ is a subspace (e.g. $\mathcal{V} \subset \mathcal{H}$), (12.23) can alternatively be expressed as

$$\alpha\|x\|_{\mathcal{H}}^2 \le \|S^*x\|_2^2 \le \beta\|x\|_{\mathcal{H}}^2, \qquad (12.24)$$

for every $x \in \mathcal{V}$. This is due to the fact that if x_1, x_2 are in $\mathcal{V}$, then $x_1 - x_2$ is also in $\mathcal{V}$.

In our setting, S^* is replaced by $\mathbf{D}$, and the set $\mathcal{T}$ contains k-block sparse vectors. Note that $\mathcal{T}$ is not a subspace. Proposition 12.2 below follows immediately from (12.23) by noting that given two k-block sparse vectors $\mathbf{c}_1, \mathbf{c}_2$ their difference $\mathbf{c}_1 - \mathbf{c}_2$ is $2k$-block sparse (see the proof of Proposition 12.1).

Proposition 12.2. *Let* $\mathbf{y} = \mathbf{D}\mathbf{c}$ *for an arbitrary k-block sparse vector* $\mathbf{c}$*. The measurement matrix* $\mathbf{D}$ *is stable for every k-block sparse vector* $\mathbf{c}$ *if and only if there exist* $0 < C_1 \le C_2 < \infty$ *such that*

$$C_1\|\mathbf{v}\|_2^2 \le \|\mathbf{D}\mathbf{v}\|_2^2 \le C_2\|\mathbf{v}\|_2^2, \qquad (12.25)$$

for every $\mathbf{v}$ *that is $2k$-block sparse.*

It is easy to see that if $\mathbf{D}$ satisfies (12.25) then $\mathbf{D}\mathbf{c} \ne \mathbf{0}$ for all $2k$-block sparse vectors $\mathbf{c}$. Therefore, this condition implies both uniqueness and stability.

12.2.1   Block RIP

Property (12.25) is related to the RIP, which we defined in Chapter 11 (cf. Definition 11.3). Recall that a matrix $\mathbf{D}$ of size $n \times N$ is said to have the RIP if there exists a constant $\delta_k \in [0, 1)$ such that for every k-sparse $\mathbf{c} \in \mathbb{R}^N$,

$$(1 - \delta_k)\|\mathbf{c}\|_2^2 \le \|\mathbf{D}\mathbf{c}\|_2^2 \le (1 + \delta_k)\|\mathbf{c}\|_2^2. \qquad (12.26)$$

Extending this property to block-sparse vectors leads to the following definition:

Definition 12.2. *Let* $\mathbf{D}: \mathbb{R}^N \to \mathbb{R}^n$ *be a given matrix. Then* $\mathbf{D}$ *has the block RIP over* $\mathcal{I} = \{d_1, \ldots, d_m\}$ *with parameter* $\delta_{k|\mathcal{I}} \in [0, 1)$ *if for every* $\mathbf{c} \in \mathbb{R}^N$ *that is k-block sparse over* $\mathcal{I}$ *we have that*

$$(1 - \delta_{k|\mathcal{I}})\|\mathbf{c}\|_2^2 \le \|\mathbf{D}\mathbf{c}\|_2^2 \le (1 + \delta_{k|\mathcal{I}})\|\mathbf{c}\|_2^2. \qquad (12.27)$$

By abuse of notation, we use δ_k for the block RIP constant $\delta_{k|\mathcal{I}}$ when it is clear from the context that we refer to blocks. Block RIP is a special case of the $\mathcal{A}$-restricted isometry defined in [285]. This definition extends the standard RIP by requiring that the inequalities in (12.26) must be satisfied for $\mathbf{c}$ in a given set $\mathcal{A}$. From Proposition 12.1 it follows that if $\mathbf{D}$ satisfies the RIP (12.27) with $\delta_{2k} < 1$, then there is a unique block-sparse vector $\mathbf{c}$ consistent with (12.11).

Note that a k-block sparse vector over $\mathcal{I}$ is M-sparse in the conventional sense where M is the sum of the k largest values in $\mathcal{I}$, since it has at most M nonzero elements. If $\mathbf{D}$ satisfies the RIP for all M-sparse vectors, then (12.27) must hold for any M-sparse vector $\mathbf{c}$. Since we only require the RIP for block sparse signals, (12.27) has to be satisfied only for a subset of M-sparse signals. As a result, the block RIP constant $\delta_{k|\mathcal{I}}$ is typically smaller than δ_M (where M depends on k; for blocks with equal size d, $M = kd$).

In Section 12.4, we will see that the ability to recover $\mathbf{c}$ in a computationally efficient way depends on the constant $\delta_{2k|\mathcal{I}}$ in the block RIP (12.27). The smaller the value of $\delta_{2k|\mathcal{I}}$, the fewer samples are needed in order to guarantee stable recovery. Both standard and block RIP constants $\delta_k, \delta_{k|\mathcal{I}}$ are by definition increasing with k. Therefore, it is common to normalize each of the columns of $\mathbf{D}$ to 1, so as to start with $\delta_1 = 0$. In the same spirit, we choose the bases for $\mathcal{A}_i$ such that $\mathbf{D} = S^* A$ has unit-norm columns.

The next example emphasizes the advantage of block RIP over standard RIP.

Example 12.3 Consider the following matrix, separated into three blocks of two columns each:

$$\mathbf{D} = \begin{bmatrix} -1 & 1 & 0 & 0 & 0 & 1 \\ 0 & 2 & -1 & 0 & 0 & 3 \\ 0 & 3 & 0 & -1 & 0 & 1 \\ 0 & 1 & 0 & 0 & -1 & 1 \end{bmatrix}. \tag{12.28}$$

Here, $m = 3$ and $\mathcal{I} = \{d_1 = 2, d_2 = 2, d_3 = 2\}$. We would like to compute the block RIP of $\mathbf{D}$ for block sparsity levels $k = 1$ and $k = 2$.

First, we note that in accordance with our normalization convention discussed before the example, we compute the RIP values for the normalized matrix

$$\mathbf{D}' = \begin{bmatrix} -1 & 1 & 0 & 0 & 0 & 1 \\ 0 & 2 & -1 & 0 & 0 & 3 \\ 0 & 3 & 0 & -1 & 0 & 1 \\ 0 & 1 & 0 & 0 & -1 & 1 \end{bmatrix} \mathbf{B} = \begin{bmatrix} -1 & 0.258 & 0 & 0 & 0 & 0.289 \\ 0 & 0.516 & -1 & 0 & 0 & 0.866 \\ 0 & 0.775 & 0 & -1 & 0 & 0.289 \\ 0 & 0.258 & 0 & 0 & -1 & 0.289 \end{bmatrix},$$

$$\tag{12.29}$$

where $\mathbf{B}$ is a diagonal matrix that results in unit-norm columns of $\mathbf{D}'$, i.e. $\mathbf{B} = \text{diag}^{-1/2}(1, 15, 1, 1, 1, 12)$. Next, suppose that $\mathbf{c}$ is 1-block sparse, which corresponds to at most two nonzero values. Brute-force calculations show that the smallest value of δ_2 satisfying the standard RIP (12.26) for $\mathbf{D}'$ is $\delta_2 = 0.866$ which results from choosing the second and sixth columns in $\mathbf{D}'$. On the other hand, the block RIP (12.27) corresponding to the case in which the two nonzero elements are restricted to occur in one block is satisfied with $\delta_{1|\mathcal{I}} = 0.289$. This result is attained when choosing the last block.

Increasing the number of nonzero elements to $k = 4$, we can verify that the standard RIP (12.26) does not hold for any $\delta_4 \in [0, 1)$. Indeed, there exist two 4-sparse vectors that result in the same measurements $\mathbf{y} = \mathbf{D}'\mathbf{c}_1 = \mathbf{D}'\mathbf{c}_2$. An example is

$$\mathbf{c}_1 = [0\ 0\ 1\ -1\ -1\ 0.1]^T, \quad \mathbf{c}_2 = [-0.0289\ 0\ 0.9134\ -1.0289\ -1.0289\ 0]^T. \tag{12.30}$$

We will show this explicitly in Example 12.6. In contrast, $\delta_{2|\mathcal{I}} = 0.966$ satisfies the lower bound in (12.27) when restricting the four nonzero values to two blocks. Consequently, the measurements $\mathbf{y} = \mathbf{D}'\mathbf{c}$ uniquely specify a single 2-block sparse $\mathbf{c}$.

12.2.2 Block coherence and subcoherence

Determining the RIP constants of a given matrix is in general an NP-hard problem. A simpler and more convenient way to characterize recovery properties of a dictionary is via the coherence measure defined in (11.67) [234, 235, 289]. We have seen in Chapter 11 that appropriate conditions on the coherence guarantee that BP, MP, OMP and other sparse recovery techniques correctly find a sparse vector from underdetermined measurements. We also showed in Section 11.3.4 that the coherence plays an important role in uncertainty relations for sparse signals.

In this section we generalize the coherence to the block setting [207], by defining two separate notions of coherence: coherence within a block, referred to as *subcoherence* and capturing the local behavior of the dictionary, and *block coherence*, describing global dictionary properties. We then show, in Section 12.5, that when the block coherence is low enough, a variety of different computationally efficient algorithms are guaranteed to recover the true unknown vector.

Definition of block coherence
To simplify the exposition we assume without loss of generality that the block sizes all have equal length d so that the length of $\mathbf{c}$ is equal to $N = md$. We furthermore assume that the number of measurements $n = Rd$ with R an integer. Similarly to (12.8), we represent $\mathbf{D}$ as a concatenation of column-blocks $\mathbf{D}[\ell]$ of size $n \times d$:

$$\mathbf{D} = [\underbrace{\mathbf{d}_1 \ \ldots \ \mathbf{d}_d}_{\mathbf{D}[1]} \ \underbrace{\mathbf{d}_{d+1} \ \ldots \ \mathbf{d}_{2d}}_{\mathbf{D}[2]} \ \ldots \ \underbrace{\mathbf{d}_{N-d+1} \ \ldots \ \mathbf{d}_N}_{\mathbf{D}[m]}]. \tag{12.31}$$

In order to have a unique k-block sparse $\mathbf{c}$ satisfying (12.11), we need $R > k$ and the columns within each block $\mathbf{D}[\ell], 1 \leq \ell \leq m$, must be linearly independent. We assume throughout the derivations that the dictionaries we consider satisfy the condition of Proposition 12.1, and, furthermore, $\|\mathbf{d}_r\|_2 = 1, 1 \leq r \leq N$.

We now introduce a generalization of the coherence defined in Chapter 11 to the block-sparse setting. We then show, in Section 12.5, that this coherence measure occurs naturally in recovery thresholds for the block-sparse case.

Definition 12.3. *The block coherence of a matrix* $\mathbf{D}$ *is defined as*

$$\mu_B = \max_{\ell,\, r \neq \ell} \frac{1}{d} \rho(\mathbf{M}[\ell, r]) \tag{12.32}$$

where $\rho(\mathbf{A})$ *denotes the spectral norm (i.e. the largest singular value) of* $\mathbf{A}$, *and*

$$\mathbf{M}[\ell, r] = \mathbf{D}^*[\ell]\mathbf{D}[r]. \tag{12.33}$$

The normalization by d ensures that $\mu_B \leq \mu$, as we show below in Proposition 12.3. Note that $\mathbf{M}[\ell, r]$ is the (ℓ, r)th $d \times d$ block of the $N \times N$ matrix $\mathbf{M} = \mathbf{D}^*\mathbf{D}$. When $d = 1$, as expected, μ_B coincides with the coherence μ of (11.67). For more details on the properties of the spectral norm, see Appendix A.

While μ_B quantifies global properties (i.e. the coherence between blocks) of the dictionary $\mathbf{D}$, local properties (i.e. the coherence between elements of each block) are characterized by the subcoherence.

Definition 12.4. *The subcoherence of* $\mathbf{D}$ *is defined as*

$$\nu = \max_{\ell} \max_{i,\, j \neq i} |\mathbf{d}_i^* \mathbf{d}_j|, \quad \mathbf{d}_i, \mathbf{d}_j \in \mathbf{D}[\ell]. \tag{12.34}$$

When $d = 1$ *we set* $\nu = 0$.

If the columns of $\mathbf{D}[\ell]$ are orthogonal for each ℓ, then $\nu = 0$.

Properties and examples

Since the columns of $\mathbf{D}$ have unit norm, the coherence μ defined in (11.67) satisfies $\mu \in [0, 1]$ and therefore, as a consequence of $\nu \in [0, \mu]$, we have $\nu \in [0, 1]$. The following proposition establishes the same limits for the block coherence μ_B, which explains the choice of normalization by $1/d$ in its definition (12.32).

Proposition 12.3. *The block coherence* μ_B *satisfies* $0 \leq \mu_B \leq \mu$.

Proof: Since the spectral norm is nonnegative, clearly $\mu_B \geq 0$. To prove that $\mu_B \leq \mu$, note that the entries of $\mathbf{M}[\ell, r]$ for $\ell \neq r$ have absolute value smaller than or equal to μ. Now, for any ℓ, r:

$$\rho^2(\mathbf{M}[\ell, r]) = \lambda_{\max}(\mathbf{M}^*[\ell, r]\mathbf{M}[\ell, r]) \leq \max_i \sum_{j=1}^{d} |(\mathbf{M}^*[\ell, r]\mathbf{M}[\ell, r])_{i,j}| \tag{12.35}$$

$$\leq \max_i \sum_{j=1}^{d} d\mu^2 = d^2\mu^2, \tag{12.36}$$

where the inequality in (12.35) is a consequence of the Geršgorin disk theorem (see Appendix A). Therefore, $\mu_B = \max_{\ell,\, r \neq \ell} \frac{1}{d}\rho(\mathbf{M}[\ell, r]) \leq \mu$. $\qquad \square$

From $\mu \leq 1$, with Proposition 12.3, it now follows trivially that $\mu_B \leq 1$. When the columns of $\mathbf{D}[\ell]$ are orthonormal for each ℓ, we can further bound μ_B.

Proposition 12.4. *If* $\mathbf{D}$ *consists of orthonormal blocks, i.e.,* $\mathbf{D}^*[\ell]\mathbf{D}[\ell] = \mathbf{I}_d$ *for all* ℓ*, then* $\mu_B \leq 1/d$.

Proof: Using the submultiplicativity property of the spectral norm (see Appendix A), we have

$$\mu_B = \max_{\ell, r \neq \ell} \frac{1}{d}\rho(\mathbf{D}^*[\ell]\mathbf{D}[r]) \leq \max_{\ell, r \neq \ell} \frac{1}{d}\rho(\mathbf{D}^*[\ell])\rho(\mathbf{D}[r]) = \frac{1}{d}. \tag{12.37}$$

The second equality in (12.37) follows from the fact that

$$\rho^2(\mathbf{D}^*[\ell]) = \lambda_{\max}(\mathbf{D}[\ell]\mathbf{D}^*[\ell]) = \lambda_{\max}(\mathbf{D}^*[\ell]\mathbf{D}[\ell]). \tag{12.38}$$

Since, $\mathbf{D}^*[\ell]\mathbf{D}[\ell] = \mathbf{I}_d$ for all ℓ, and $\lambda_{\max}(\mathbf{I}_d) = 1$, the result follows. $\square$

Example 12.4 In this example we compare the coherence and block coherence for a specific choice of dictionary $\mathbf{D}$.

Let $\mathbf{F}$ be the DFT matrix of size $R = L/d$ with normalized columns. Define $\mathbf{\Phi} = \mathbf{I}_L$ and $\mathbf{\Psi} = \mathbf{F} \otimes \mathbf{U}$ where $\mathbf{U}$ is an arbitrary $d \times d$ unitary matrix and $\otimes$ is the Kronecker product operator. Thus, $\mathbf{\Psi}$ consists of blocks of the form $\mathbf{F}_{\ell,r}\mathbf{U}$, where $\mathbf{F}_{\ell,r}$ is the ℓrth element of $\mathbf{F}$. We choose $\mathbf{D} = [\mathbf{\Phi} \ \mathbf{\Psi}]$ as the $L \times 2L$ dictionary with blocks of size $L \times d$. We now compute the block coherence μ_B and coherence μ of $\mathbf{D}$.

To obtain μ_B note that since $\mathbf{\Phi}$ and $\mathbf{\Psi}$ are unitary matrices, $\mathbf{\Phi}^*[\ell]\mathbf{\Phi}[r] = 0$ and $\mathbf{\Psi}^*[\ell]\mathbf{\Psi}[r] = 0$ for $\ell \neq r$. In addition, $\mathbf{\Phi}^*[\ell]\mathbf{\Psi}[r] = \mathbf{F}_{\ell,r}\mathbf{U}$. Since $|\mathbf{F}_{\ell,r}| = \frac{1}{\sqrt{R}}$, it follows that

$$\mu_B = \frac{1}{d}\rho(\mathbf{F}_{\ell,r}\mathbf{U}) = \frac{1}{d\sqrt{R}}\rho(\mathbf{U}) = \frac{1}{d\sqrt{R}}, \tag{12.39}$$

where we have used the fact that for any unitary matrix $\mathbf{U}$, $\rho(\mathbf{U}) = 1$ (see Appendix A).

We next compute the coherence μ. Denoting the columns of $\mathbf{\Phi}$ and $\mathbf{\Psi}$ by ϕ_i and ψ_i respectively, it follows that for $i \neq j$, $\phi_i^*\phi_j = 0$ and $\psi_i^*\psi_j = 0$ owing to the fact that $\mathbf{\Phi}$ and $\mathbf{\Psi}$ are unitary matrices. Furthermore, $\phi_i^*\psi_j = \mathbf{\Psi}_{ij}$ is the ijth component of $\mathbf{\Psi}$. The elements of $\mathbf{\Psi}$ are the product of a component of $\mathbf{F}$ with a component of $\mathbf{U}$. Since $|\mathbf{F}_{ij}| = \frac{1}{\sqrt{R}}$,

$$\mu = \max_{\ell, r \neq \ell} |\mathbf{F}_{\ell,r}||\mathbf{U}_{\ell,r}| = \frac{\max_{\ell, r \neq \ell}|\mathbf{U}_{\ell,r}|}{\sqrt{R}} = \frac{\|\text{vec}(\mathbf{U})\|_\infty}{\sqrt{R}}. \tag{12.40}$$

In order to compare μ and μ_B we note that since $\mathbf{U}$ is a unitary matrix, each of its columns $\mathbf{u}_i$ satisfies $\|\mathbf{u}_i\|_2 = 1$. This in turn implies that

$$\|\mathbf{u}_i\|_\infty \geq \frac{1}{\sqrt{d}}. \tag{12.41}$$

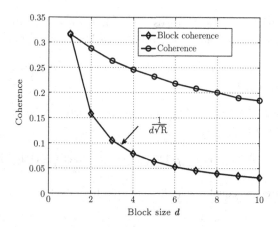

Figure 12.5 Block coherence and coherence for $R = 10$ as a function of d.

Indeed, if all elements of $\mathbf{u}_i$ were smaller in magnitude than $1/\sqrt{d}$, then we would have $\|\mathbf{u}_i\| < 1$. Therefore,

$$\mu \geq \frac{1}{\sqrt{dR}} \geq \frac{1}{d\sqrt{R}} = \mu_{\mathrm{B}}, \tag{12.42}$$

with equality possible only when $d = 1$.

In Fig. 12.5 we show that μ can be much larger than μ_{B} for random choices of $\mathbf{U}$. Specifically, we plot the coherence and block coherence averaged over 1000 randomly chosen unitary matrices $\mathbf{U}$ as a function of the block size d. Here, $\mathbf{U}$ is constructed as $\mathbf{U} = \exp(j\mathbf{A}\mathbf{A}^*)$ where $\mathbf{A}$ is an iid Gaussian matrix.

We conclude that the recovery thresholds for dictionaries of the form $\mathbf{D} = [\mathbf{I}_L \ \mathbf{F} \otimes \mathbf{U}]$ obtained by taking block sparsity into account can be significantly higher than those for conventional sparsity.

12.3   Signal recovery algorithms

12.3.1   Exponential recovery algorithm

We have shown that if $\mathbf{D}$ satisfies the RIP (12.27) with $\delta_{2k} < 1$, then there is a unique block-sparse vector $\mathbf{c}$ consistent with (12.11). The question is how to find $\mathbf{c}$ in practice. Below we present an algorithm that will in principle find the unique $\mathbf{c}$ from samples $\mathbf{y}$, though unfortunately it has exponential complexity. We then turn to develop several efficient methods that can recover $\mathbf{c}$ as long as certain conditions on μ_{B}, ν, or δ_{2k} are satisfied.

Proposition 12.5. *Let* $\mathbf{y} = \mathbf{D}\mathbf{c}_0$ *be measurements of a k-block sparse vector $\mathbf{c}_0$ and suppose that $\mathbf{D}$ satisfies the block RIP (12.27) with $\delta_{2k} < 1$. Then $\mathbf{c}_0$ is the unique solution to the problem*

$$\min_{\mathbf{c}} \|\mathbf{c}\|_{0,\mathcal{I}} \quad \text{s.t. } \mathbf{y} = \mathbf{Dc}. \tag{12.43}$$

Proof: Suppose that there exists a $\mathbf{c}'$ such that $\mathbf{Dc}' = \mathbf{y}$ and $\|\mathbf{c}'\|_{0,\mathcal{I}} \leq \|\mathbf{c}_0\|_{0,\mathcal{I}} \leq k$. Since both $\mathbf{c}_0, \mathbf{c}'$ are consistent with the measurements,

$$0 = \mathbf{D}(\mathbf{c}_0 - \mathbf{c}') = \mathbf{Dd}, \tag{12.44}$$

where $\|\mathbf{d}\|_{0,\mathcal{I}} \leq 2k$ so that $\mathbf{d}$ is a $2k$-block sparse vector. But if $\mathbf{D}$ satisfies (12.27) with $\delta_{2k} < 1$, then we must have $\mathbf{d} = \mathbf{0}$ or $\mathbf{c}' = \mathbf{c}_0$. □

In principle, (12.43) can be solved by searching over all possible sets of k blocks to see whether there exists a $\mathbf{c}$ that is consistent with the measurements. The uniqueness condition (12.27) ensures that there is only one such value. However, clearly this approach is inefficient.

Example 12.5 In Example 12.1, we illustrated how a piece of piano music can be modeled as block sparse. Suppose that we are given an interval of music and we want to decide which notes were played. Since at most 10 notes are played simultaneously (corresponding to the number of fingers), there exist $\binom{96}{10}$ different options. If checking each of the candidates takes one millisecond, then it will take about 358 years to examine all possibilities!

We next turn to consider several efficient alternatives based on convex optimization and extensions of MP to the block-sparse setting. We then show that under appropriate conditions on the block RIP and block coherence, these techniques are guaranteed to recover the true underlying signal.

12.3.2 Convex recovery algorithm

To develop a relaxation of (12.43), we replace the indicator functions in the sum by block norms. Although, in principle, any norm can be used, we focus on the 2-norm for concreteness. To write down the problem explicitly, we define the mixed ℓ_2/ℓ_1-norm over the index set $\mathcal{I} = \{d_1, \ldots, d_m\}$ as

$$\|\mathbf{c}\|_{2,\mathcal{I}} = \sum_{i=1}^{m} \|\mathbf{c}[i]\|_2. \tag{12.45}$$

The proposed algorithm is then [183, 364, 365, 367]

$$\min_{\mathbf{c}} \|\mathbf{c}\|_{2,\mathcal{I}} \quad \text{s.t. } \mathbf{y} = \mathbf{Dc}, \tag{12.46}$$

which we refer to as *block BP* [207]. When the blocks have length 1, (12.46) reduces to the standard BP algorithm (11.120) for sparse recovery. Using similar ideas to those that have been applied to BP, efficient iterative algorithms can be developed to solve (12.46).

In order to use standard off-the-shelf solvers, it is convenient to formulate block BP as a *second-order cone program (SOCP)* [124]. In a general SOCP, a linear function is minimized over the intersection of an affine set and the product of second-order (quadratic) cones [368]. SOCPs are nonlinear convex problems that include linear and (convex) quadratic programs as special cases, and arise in many engineering problems. To formulate (12.46) as an SOCP we define $t_i = \|\mathbf{c}[i]\|_2$, which results in

$$
\min_{\mathbf{c}, t_i} \quad \sum_{i=1}^{m} t_i
$$

$$
\text{s.t.} \quad \mathbf{y} = \mathbf{Dc}
$$

$$
t_i \geq \|\mathbf{c}[i]\|_2, \quad 1 \leq i \leq m. \tag{12.47}
$$

The inequalities in (12.47) are second-order cone constraints.

A variety of software packages have been developed to solve SOCPs efficiently, and any one of them is applicable to (12.47). In Chapter 11 we used CVX to solve BP problems. In the next example we illustrate its use for block-sparse recovery.

Example 12.6 To demonstrate the advantage of block BP over standard BP, consider the matrix $\mathbf{D}$ of (12.29). In Example 12.3, the standard and block RIP constants of $\mathbf{D}$ were calculated, and it was shown that block RIP constants are smaller. This suggests that there are input vectors $\mathbf{x}$ for which block BP will be able to recover them exactly from measurements $\mathbf{y} = \mathbf{Dc}$ while standard ℓ_1 minimization will fail.

To illustrate this behavior, let $\mathbf{c} = [0\ 0\ 1\ -1\ -1\ 0.1]^T$ be a 4-sparse vector, in which the nonzero elements are known to appear in blocks of length 2. The prior knowledge that $\mathbf{c}$ is 4-sparse is not sufficient to determine $\mathbf{c}$ from $\mathbf{y}$. In contrast, there is a unique block-sparse vector consistent with $\mathbf{y}$. Furthermore, block BP finds the correct $\mathbf{c}$ while standard ℓ_1 minimization fails in this case; its output is $\hat{\mathbf{c}} = [-0.0289\ 0\ 0.9134\ -1.0289\ -1.0289\ 0]^T$.

The output of block BP was found by solving the SOCP given by (12.47) using the following code in CVX:

```
cvx_begin
variable c(6)
variable t(3)
minimize(sum(t))
subject to
    D*c == y
    t(1)>=norm(c(1:2));
    t(2)>=norm(c(3:4));
    t(3)>=norm(c(5:6));
cvx_end
```

12.3.3   Greedy algorithms

As an alternative to block BP, we introduce several greedy algorithms, similar in spirit to those used for sparse recovery. We begin by extending MP-type methods to the block-sparse case.

Algorithm 12.1 Block orthogonal matching pursuit

Input: Measurement matrix $\mathbf{D}$, measurement vector $\mathbf{y}$
Output: Block-sparse representation $\hat{\mathbf{c}}$
Initialize: $\hat{\mathbf{c}}_0 = \mathbf{0}, \mathbf{r} = \mathbf{y}, \Lambda = \emptyset, \ell = 0$
while halting criterion false **do**
    $\ell \leftarrow \ell + 1$
    $\mathbf{b} \leftarrow \mathbf{D}^*\mathbf{r}$ {form residual signal estimate}
    $\Lambda \leftarrow \Lambda \cup \mathrm{supp}(\mathcal{T}(\mathbf{b}, 1))$ {update support with residual}
    $\hat{\mathbf{c}}_\ell|_\Lambda \leftarrow \mathbf{D}_\Lambda^\dagger \mathbf{y}, \hat{\mathbf{c}}_\ell|_{\Lambda^c} \leftarrow \mathbf{0}$ {update signal estimate}
    $\mathbf{r} \leftarrow \mathbf{y} - \mathbf{D}\hat{\mathbf{c}}_\ell$ {update measurement residual}
end while
return $\hat{\mathbf{c}} \leftarrow \hat{\mathbf{c}}_\ell$

The block OMP (BOMP) algorithm, summarized as Algorithm 12.1, generalizes OMP (Algorithm 11.1) by matching the residual to blocks rather than individual columns. More specifically, BOMP begins by initializing the residual as $\mathbf{r}_0 = \mathbf{y}$. At the ℓth stage ($\ell \geq 1$) we choose the block that is best matched to $\mathbf{r}_{\ell-1}$ according to:

$$i_\ell = \arg\max_i \|\mathbf{D}^*[i]\mathbf{r}_{\ell-1}\|_2. \tag{12.48}$$

Define

$$\mathcal{T}(\mathbf{b}, k) = \arg_{\mathbf{c} \in \mathcal{M}} \min \|\mathbf{b} - \mathbf{c}\|_2, \tag{12.49}$$

where $\mathcal{M}$ is the set of k-block sparse signals. This function selects the k blocks in $\mathbf{b}$ with the largest ℓ_2-norm. Using this notation, we have that $i_\ell = \mathcal{T}(\mathbf{b}, 1)$ with $\mathbf{b} = \mathbf{D}^*\mathbf{r}_{\ell-1}$.

Once the index i_ℓ is chosen, we find $\mathbf{c}_\ell[i]$ as the solution to

$$\arg\min_{\{\tilde{\mathbf{c}}_\ell[i]\}_{i \in \mathcal{I}}} \left\| \mathbf{y} - \sum_{i \in \mathcal{I}} \mathbf{D}[i]\tilde{\mathbf{c}}_\ell[i] \right\|_2^2 \tag{12.50}$$

with $\mathcal{I}$ denoting the set of chosen indices $i_j, 1 \leq j \leq \ell$. The solution is given by $\mathbf{D}_\Lambda^\dagger \mathbf{y}$ where $\mathbf{D}_\Lambda$ is the restriction of $\mathbf{D}$ to the support Λ dictated by the indices $\mathcal{I}$. The residual is then updated as

$$\mathbf{r}_\ell = \mathbf{y} - \sum_{i \in \mathcal{I}} \mathbf{D}[i]\mathbf{c}_\ell[i]. \tag{12.51}$$

When the columns of $\mathbf{D}[\ell]$ are orthonormal for each ℓ (the elements across different blocks do not have to be orthonormal), we can consider an extension of the MP algorithm to the block case. The resulting algorithm, termed BMP, starts by initializing the residual as $\mathbf{r}_0 = \mathbf{y}$, and at the ℓth stage ($\ell \geq 1$) chooses the block that is best matched to $\mathbf{r}_{\ell-1}$ according to (12.48). Then, however, the algorithm does not perform an LS minimization over the blocks that have already been selected, but directly updates the residual according to

$$\mathbf{r}_\ell = \mathbf{r}_{\ell-1} - \mathbf{D}[i_\ell]\mathbf{D}^*[i_\ell]\mathbf{r}_{\ell-1}. \tag{12.52}$$

Both the BOMP and BMP algorithms are faster than block BP and are easier to implement.

Example 12.7 We illustrate the improvement in performance of OMP and BP obtained by taking block sparsity explicitly into account and performing recovery using BOMP and block BP, respectively.

In our simulations we choose two types of dictionaries: a random dictionary, and the dictionary used in Example 12.4. In both cases, $L = 20, N = 40$, and the block length is chosen as $d = 2$. The random dictionaries are selected by drawing from iid zero-mean, unit-variance, Gaussian matrices and normalizing the resulting columns to have unit norm. The dictionary is divided into consecutive blocks of length $d = 2$. The locations of the nonzero blocks in the block-sparse vector to be recovered are chosen uniformly at random among all $\binom{M}{k}$ possible locations, where $M = N/d$ is the number of blocks. The nonzero entries are generated as iid Gaussian with zero mean and unit variance.

In Fig. 12.6 we plot the recovery success rate as a function of the block-sparsity level for both choices. Success is declared if the recovered vector is within a small Euclidean distance (we chose 0.01) of the original vector. For each block-sparsity level we average over 200 pairs of realizations of the dictionary and the block-sparse signal. Evidently, BOMP outperforms OMP and block BP outperforms BP significantly.

Although our focus here is on MP-type methods, we note that other greedy approaches originally developed for sparse approximation can also be adapted to the block-sparse setting. For example, a simple thresholding algorithm results from choosing the k blocks $\mathbf{b}[i]$ of $\mathbf{b} = \mathbf{D}^*\mathbf{y}$ with the largest norm, namely $\mathcal{T}(\mathbf{b}, k)$. Iterating this process leads to an extension of IHT to the block-sparse setting. We refer to these iterations, summarized in Algorithm 12.2, as block IHT [369].

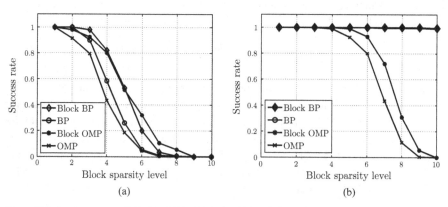

Figure 12.6 Performance of BP, block BP, OMP, and BOMP for: (a) Random dictionary. (b) Dictionary defined in Example 12.4, with $L = 20, N = 40$, and $d = 2$.

Algorithm 12.2 Block iterative hard thresholding

Input: Measurement matrix $\mathbf{D}$, measurement vector $\mathbf{y}$
Output: Block-sparse representation $\hat{\mathbf{c}}$
Initialize: $\hat{\mathbf{c}}_0 = \mathbf{0}$, $\mathbf{r} = \mathbf{y}$, $\ell = 0$
while halting criterion false **do**
    $\ell \leftarrow \ell + 1$
    $\mathbf{b} = \hat{\mathbf{c}}_{\ell-1} + \mathbf{D}^*\mathbf{r}$ {form residual signal estimate}
    $\hat{\mathbf{c}}_\ell \leftarrow \mathcal{T}(\mathbf{b}, k)$ {prune estimate}
    $\mathbf{r} \leftarrow \mathbf{y} - \mathbf{D}\hat{\mathbf{c}}_\ell$ {update measurement residual}
end while
return $\hat{\mathbf{c}} \leftarrow \hat{\mathbf{c}}_\ell$

12.4 RIP-based recovery results

We now turn to discuss recovery results of block BP when the RIP constant is sufficiently small. In the next section we extend the discussion to BMP and BOMP, based on block coherence.

12.4.1 Block basis pursuit recovery

The following theorem establishes that the solution to (12.46) is the true $\mathbf{c}$ as long as δ_{2k} is small enough [183].

Theorem 12.1 (Block BP recovery). *Let* $\mathbf{y} = \mathbf{D}\mathbf{c}_0$ *be measurements of a* k-*block sparse vector* $\mathbf{c}_0$. *If* $\mathbf{D}$ *satisfies the block RIP (12.27) with* $\delta_{2k} < \sqrt{2} - 1$ *then*

1. *there is a unique* k-*block sparse vector* $\mathbf{c}$ *consistent with* $\mathbf{y}$;
2. *the SOCP (12.47) has a unique solution;*
3. *the solution to the SOCP is equal to* $\mathbf{c}_0$.

Proof: We first note that $\delta_{2k} < 1$ guarantees uniqueness of $\mathbf{c}_0$ from Proposition 12.1.

To prove parts (2) and (3) we show that any solution to (12.46) has to be equal to $\mathbf{c}_0$. To this end let $\mathbf{c}' = \mathbf{c}_0 + \mathbf{h}$ be a solution of (12.46). The true value $\mathbf{c}_0$ is nonzero over at most k blocks. We denote by $\mathcal{I}_0$ the block indices for which $\mathbf{c}_0$ is nonzero, and by $\mathbf{h}_{\mathcal{I}_0}$ the restriction of $\mathbf{h}$ to these blocks. Next we decompose $\mathbf{h}$ as

$$\mathbf{h} = \mathbf{h}_{\mathcal{I}_0} + \mathbf{h}_{\mathcal{I}_0^c} = \sum_{i=0}^{\ell-1} \mathbf{h}_{\mathcal{I}_i}, \tag{12.53}$$

where $\mathbf{h}_{\mathcal{I}_i}$ is the restriction of $\mathbf{h}$ to the set $\mathcal{I}_i$ which consists of k blocks, chosen such that the norm of $\mathbf{h}_{\mathcal{I}_0^c}$ over $\mathcal{I}_1$ is largest, the norm over $\mathcal{I}_2$ is second largest, and so on. Our goal is to show that $\mathbf{h} = \mathbf{0}$. We prove this by noting that

$$\|\mathbf{h}\|_2 = \|\mathbf{h}_{\mathcal{I}_0 \cup \mathcal{I}_1} + \mathbf{h}_{(\mathcal{I}_0 \cup \mathcal{I}_1)^c}\|_2 \leq \|\mathbf{h}_{\mathcal{I}_0 \cup \mathcal{I}_1}\|_2 + \|\mathbf{h}_{(\mathcal{I}_0 \cup \mathcal{I}_1)^c}\|_2. \tag{12.54}$$

In the first part of the proof we show that $\|\mathbf{h}_{(\mathcal{I}_0 \cup \mathcal{I}_1)^c}\|_2 \leq \|\mathbf{h}_{\mathcal{I}_0 \cup \mathcal{I}_1}\|_2$. In the second part we establish that $\|\mathbf{h}_{\mathcal{I}_0 \cup \mathcal{I}_1}\|_2 = 0$, which completes the proof.

Part I: $\|\mathbf{h}_{(\mathcal{I}_0 \cup \mathcal{I}_1)^c}\|_2 \leq \|\mathbf{h}_{\mathcal{I}_0 \cup \mathcal{I}_1}\|_2$

We begin by noting that

$$\|\mathbf{h}_{(\mathcal{I}_0 \cup \mathcal{I}_1)^c}\|_2 = \left\|\sum_{i=2}^{\ell-1} \mathbf{h}_{\mathcal{I}_i}\right\|_2 \leq \sum_{i=2}^{\ell-1} \|\mathbf{h}_{\mathcal{I}_i}\|_2. \tag{12.55}$$

Therefore, it is sufficient to bound $\|\mathbf{h}_{\mathcal{I}_i}\|_2$ for $i \geq 2$. Now,

$$\|\mathbf{h}_{\mathcal{I}_i}\|_2 \leq k^{1/2} \|\mathbf{h}_{\mathcal{I}_i}\|_{\infty,\mathcal{I}} \leq k^{-1/2} \|\mathbf{h}_{\mathcal{I}_{i-1}}\|_{2,\mathcal{I}}, \tag{12.56}$$

where we defined $\|\mathbf{a}\|_{\infty,\mathcal{I}} = \max_i \|\mathbf{a}[i]\|_2$. The first inequality follows from the fact that for any k-block sparse $\mathbf{c}$,

$$\|\mathbf{c}\|_2^2 = \sum_{|i|=k} \|\mathbf{c}[i]\|_2^2 \leq k\|\mathbf{c}\|_{\infty,\mathcal{I}}^2. \tag{12.57}$$

The second inequality in (12.56) is a result of the fact that the norm of each block in $\mathbf{h}_{\mathcal{I}_i}$ is by definition smaller than or equal to the norm of each block in $\mathbf{h}_{\mathcal{I}_{i-1}}$. Since there are at most k nonzero blocks, $k\|\mathbf{h}_{\mathcal{I}_i}\|_{\infty,\mathcal{I}} \leq \|\mathbf{h}_{\mathcal{I}_{i-1}}\|_{2,\mathcal{I}}$. Substituting (12.56) into (12.55),

$$\|\mathbf{h}_{(\mathcal{I}_0 \cup \mathcal{I}_1)^c}\|_2 \leq k^{-1/2} \sum_{i=1}^{\ell-2} \|\mathbf{h}_{\mathcal{I}_i}\|_{2,\mathcal{I}} \leq k^{-1/2} \sum_{i=1}^{\ell-1} \|\mathbf{h}_{\mathcal{I}_i}\|_{2,\mathcal{I}} = k^{-1/2} \|\mathbf{h}_{\mathcal{I}_0^c}\|_{2,\mathcal{I}}, \tag{12.58}$$

where the equality is a result of the fact that $\|\mathbf{c}_1 + \mathbf{c}_2\|_{2,\mathcal{I}} = \|\mathbf{c}_1\|_{2,\mathcal{I}} + \|\mathbf{c}_2\|_{2,\mathcal{I}}$ if $\mathbf{c}_1$ and $\mathbf{c}_2$ are nonzero on disjoint blocks.

To develop a bound on $\|\mathbf{h}_{\mathcal{I}_0^c}\|_{2,\mathcal{I}}$ note that since $\mathbf{c}'$ is a solution to (12.46), $\|\mathbf{c}_0\|_{2,\mathcal{I}} \geq \|\mathbf{c}'\|_{2,\mathcal{I}}$. Using the fact that $\mathbf{c}' = \mathbf{c}_0 + \mathbf{h}_{\mathcal{I}_0} + \mathbf{h}_{\mathcal{I}_0^c}$ and $\mathbf{c}_0$ is supported on $\mathcal{I}_0$ we have

$$\|\mathbf{c}_0\|_{2,\mathcal{I}} \geq \|\mathbf{c}_0 + \mathbf{h}_{\mathcal{I}_0}\|_{2,\mathcal{I}} + \|\mathbf{h}_{\mathcal{I}_0^c}\|_{2,\mathcal{I}} \geq \|\mathbf{c}_0\|_{2,\mathcal{I}} - \|\mathbf{h}_{\mathcal{I}_0}\|_{2,\mathcal{I}} + \|\mathbf{h}_{\mathcal{I}_0^c}\|_{2,\mathcal{I}}, \quad (12.59)$$

from which we conclude that

$$\|\mathbf{h}_{\mathcal{I}_0^c}\|_{2,\mathcal{I}} \leq \|\mathbf{h}_{\mathcal{I}_0}\|_{2,\mathcal{I}} \leq k^{1/2} \|\mathbf{h}_{\mathcal{I}_0}\|_2. \tag{12.60}$$

The last inequality follows from applying Cauchy–Schwarz to any k-block sparse vector $\mathbf{c}$:

$$\|\mathbf{c}\|_{2,\mathcal{I}} = \sum_{|i|=k} \|\mathbf{c}[i]\|_2 \cdot 1 \leq k^{1/2} \|\mathbf{c}\|_2. \tag{12.61}$$

Substituting (12.60) into (12.58):

$$\|\mathbf{h}_{(\mathcal{I}_0 \cup \mathcal{I}_1)^c}\|_2 \leq \|\mathbf{h}_{\mathcal{I}_0}\|_2 \leq \|\mathbf{h}_{\mathcal{I}_0 \cup \mathcal{I}_1}\|_2, \tag{12.62}$$

which completes the first part of the proof.

Part II: $\|\mathbf{h}_{\mathcal{I}_0 \cup \mathcal{I}_1}\|_2 = 0$

We next show that $\mathbf{h}_{\mathcal{I}_0 \cup \mathcal{I}_1}$ must be equal to 0. In this part we invoke the RIP.

Since $\mathbf{Dc}_0 = \mathbf{Dc}' = \mathbf{y}$, we have $\mathbf{Dh} = 0$. Using the fact that $\mathbf{h} = \mathbf{h}_{\mathcal{I}_0 \cup \mathcal{I}_1} + \sum_{i \geq 2} \mathbf{h}_{\mathcal{I}_i}$,

$$\|\mathbf{Dh}_{\mathcal{I}_0 \cup \mathcal{I}_1}\|_2^2 = \langle \mathbf{D}(\mathbf{h}_{\mathcal{I}_0} + \mathbf{h}_{\mathcal{I}_1}), \mathbf{Dh}_{\mathcal{I}_0 \cup \mathcal{I}_1} \rangle = -\sum_{i=2}^{\ell-1} \langle \mathbf{D}(\mathbf{h}_{\mathcal{I}_0} + \mathbf{h}_{\mathcal{I}_1}), \mathbf{Dh}_{\mathcal{I}_i} \rangle, \tag{12.63}$$

where in the first equality we used the fact that $\mathbf{h}_{\mathcal{I}_0}$ and $\mathbf{h}_{\mathcal{I}_1}$ have disjoint supports so that $\mathbf{h}_{\mathcal{I}_0 \cup \mathcal{I}_1} = \mathbf{h}_{\mathcal{I}_0} + \mathbf{h}_{\mathcal{I}_1}$. From the parallelogram law (2.12) and the block RIP it can be shown that (see Proposition 11.5)

$$|\langle \mathbf{Dc}_1, \mathbf{Dc}_2 \rangle| \leq \delta_{2k} \|\mathbf{c}_1\|_2 \|\mathbf{c}_2\|_2, \tag{12.64}$$

for any two k-block sparse vectors with disjoint support. Therefore,

$$|\langle \mathbf{Dh}_{\mathcal{I}_0}, \mathbf{Dh}_{\mathcal{I}_i} \rangle| \leq \delta_{2k} \|\mathbf{h}_{\mathcal{I}_0}\|_2 \|\mathbf{h}_{\mathcal{I}_i}\|_2, \tag{12.65}$$

and similarly for $\langle \mathbf{Dh}_{\mathcal{I}_1}, \mathbf{Dh}_{\mathcal{I}_i} \rangle$. Substituting into (12.63),

$$\|\mathbf{Dh}_{\mathcal{I}_0 \cup \mathcal{I}_1}\|_2^2 = \left| \sum_{i=2}^{\ell-1} \langle \mathbf{D}(\mathbf{h}_{\mathcal{I}_0} + \mathbf{h}_{\mathcal{I}_1}), \mathbf{Dh}_{\mathcal{I}_i} \rangle \right|$$

$$\leq \sum_{i=2}^{\ell-1} (|\langle \mathbf{Dh}_{\mathcal{I}_0}, \mathbf{Dh}_{\mathcal{I}_i} \rangle| + |\langle \mathbf{Dh}_{\mathcal{I}_1}, \mathbf{Dh}_{\mathcal{I}_i} \rangle|)$$

$$\leq \delta_{2k}(\|\mathbf{h}_{\mathcal{I}_0}\|_2 + \|\mathbf{h}_{\mathcal{I}_1}\|_2) \sum_{i=2}^{\ell-1} \|\mathbf{h}_{\mathcal{I}_i}\|_2. \tag{12.66}$$

From the Cauchy–Schwarz inequality, any length-2 vector $\mathbf{a}$ satisfies $|a_1 + a_2| \leq \sqrt{2}\|\mathbf{a}\|_2$. Therefore,

$$\|\mathbf{h}_{\mathcal{I}_0}\|_2 + \|\mathbf{h}_{\mathcal{I}_1}\|_2 \leq \sqrt{2}\sqrt{\|\mathbf{h}_{\mathcal{I}_0}\|_2^2 + \|\mathbf{h}_{\mathcal{I}_1}\|_2^2} = \sqrt{2}\|\mathbf{h}_{\mathcal{I}_0 \cup \mathcal{I}_1}\|_2, \tag{12.67}$$

where the last equality is a result of the fact that $\mathbf{h}_{\mathcal{I}_0}$ and $\mathbf{h}_{\mathcal{I}_1}$ have disjoint support. Substituting into (12.66) and using (12.56), (12.58), and (12.60),

$$\|\mathbf{Dh}_{\mathcal{I}_0 \cup \mathcal{I}_1}\|_2^2 \leq \sqrt{2}k^{-1/2}\delta_{2k}\|\mathbf{h}_{\mathcal{I}_0 \cup \mathcal{I}_1}\|_2 \|\mathbf{h}_{\mathcal{I}_0^c}\|_{2,\mathcal{I}}$$

$$\leq \sqrt{2}\delta_{2k}\|\mathbf{h}_{\mathcal{I}_0 \cup \mathcal{I}_1}\|_2 \|\mathbf{h}_{\mathcal{I}_0}\|_2$$

$$\leq \sqrt{2}\delta_{2k}\|\mathbf{h}_{\mathcal{I}_0 \cup \mathcal{I}_1}\|_2^2, \tag{12.68}$$

since $\|\mathbf{h}_{\mathcal{I}_0}\|_2 \leq \|\mathbf{h}_{\mathcal{I}_0 \cup \mathcal{I}_1}\|_2$. Combining (12.68) with the RIP (12.27) we have

$$(1 - \delta_{2k})\|\mathbf{h}_{\mathcal{I}_0 \cup \mathcal{I}_1}\|_2^2 \leq \|\mathbf{Dh}_{\mathcal{I}_0 \cup \mathcal{I}_1}\|_2^2 \leq \sqrt{2}\delta_{2k}\|\mathbf{h}_{\mathcal{I}_0 \cup \mathcal{I}_1}\|_2^2. \tag{12.69}$$

Recalling that $\delta_{2k} < \sqrt{2} - 1$, (12.69) can hold only if $\|\mathbf{h}_{\mathcal{I}_0 \cup \mathcal{I}_1}\|_2 = 0$, completing the proof. $\qquad\square$

It is interesting to compare Theorem 12.1 with Theorem 11.12 in Chapter 11, which discusses recovery guarantees of standard sparse vectors using BP (11.120). The latter implies that if c is k-sparse and the measurement matrix $\mathbf{D}$ satisfies the standard RIP with $\delta_{2k} < \sqrt{2} - 1$, then c can be recovered exactly from the measurements $\mathbf{y} = \mathbf{D}\mathbf{c}$ via ℓ_1 optimization. Since any k-block sparse vector is also M-sparse with $M = kd$, we can find $\mathbf{c}_0$ of Theorem 12.1 using BP if δ_{2M} is small enough. However, this standard CS approach does not exploit the fact that the nonzero values appear in blocks, and not in arbitrary locations within the vector $\mathbf{c}_0$. On the other hand, the SOCP (12.47) explicitly takes the block structure of $\mathbf{c}_0$ into account. Therefore, the condition of Theorem 12.1 is not as stringent as that of Theorem 11.12. Indeed, the value of δ_{2k} in (12.27) may be lower than that obtained from (12.26) with $k = 2M$, as we illustrated in Example 12.3.

Theorem 12.1 can also be extended to the case in which the observations are noisy, and the vector $\mathbf{c}_0$ is not exactly k-block sparse. Specifically, suppose that the measurements (12.11) are corrupted by bounded noise so that

$$\mathbf{y} = \mathbf{D}\mathbf{c} + \mathbf{z}, \tag{12.70}$$

where $\|\mathbf{z}\|_2 \leq \varepsilon$. In order to recover $\mathbf{c}$ we use the modified SOCP:

$$\min_{\mathbf{c}} \|\mathbf{c}\|_{2,\mathcal{I}} \quad \text{s.t.} \quad \|\mathbf{y} - \mathbf{D}\mathbf{c}\|_2 \leq \varepsilon. \tag{12.71}$$

In addition, given a vector $\mathbf{c} \in \mathbb{R}^N$, we denote by $\mathbf{c}^k$ the best approximation of $\mathbf{c}$ by a vector with k nonzero blocks, so that $\mathbf{c}^k$ minimizes $\|\mathbf{c} - \mathbf{d}\|_{2,\mathcal{I}}$ over all k-block sparse vectors $\mathbf{d}$. Theorem 12.2 shows that even when $\mathbf{c}$ is not k-block sparse and the measurements are noisy, $\mathbf{c}^k$ is well approximated using (12.71) [183].

Theorem 12.2 (Noisy block BP recovery). *Let $\mathbf{y} = \mathbf{D}\mathbf{c}_0 + \mathbf{z}$ be noisy measurements of a vector $\mathbf{c}_0$ with $\|\mathbf{z}\|_2 \leq \varepsilon$. Let $\mathbf{c}^k$ denote the best k-block sparse approximation of $\mathbf{c}_0$, such that $\mathbf{c}^k$ is k-block sparse and minimizes $\|\mathbf{c}_0 - \mathbf{d}\|_{2,\mathcal{I}}$ over all k-block sparse vectors $\mathbf{d}$, and let $\mathbf{c}'$ be a solution to (12.71). If $\mathbf{D}$ satisfies the block RIP (12.27) with $\delta_{2k} < \sqrt{2} - 1$ then*

$$\|\mathbf{c}_0 - \mathbf{c}'\|_2 \leq \frac{2(1 - \delta_{2k})}{1 - (1 + \sqrt{2})\delta_{2k}} k^{-1/2} \|\mathbf{c}_0 - \mathbf{c}^k\|_{2,\mathcal{I}} + \frac{4\sqrt{1 + \delta_{2k}}}{1 - (1 + \sqrt{2})\delta_{2k}} \varepsilon. \tag{12.72}$$

The first term in (12.72) is a result of the fact that $\mathbf{c}_0$ is not exactly k-block sparse. The second expression quantifies the recovery error due to the noise.

Proof: The proof is very similar to that of Theorem 12.1 with a few differences which we indicate.

Denote by $\mathbf{c}' = \mathbf{c}_0 + \mathbf{h}$ the solution to (12.71). Owing to the noise and the fact that $\mathbf{c}_0$ is not k-block sparse, we no longer obtain $\mathbf{h} = 0$. However, we will show that $\|\mathbf{h}\|_2$ is bounded. To this end, we begin as in the proof of Theorem 12.1 by using (12.54). In the first part we establish that $\|\mathbf{h}_{(\mathcal{I}_0 \cup \mathcal{I}_1)^c}\|_2 \leq \|\mathbf{h}_{\mathcal{I}_0 \cup \mathcal{I}_1}\|_2 + 2e_0$ where $e_0 = k^{-1/2}\|\mathbf{c}_0 - \mathbf{c}_{\mathcal{I}_0}\|_{2,\mathcal{I}}$ and $\mathbf{c}_{\mathcal{I}_0}$ is the restriction of $\mathbf{c}_0$ onto the k blocks corresponding to the largest ℓ_2-norm. Note that $\mathbf{c}_{\mathcal{I}_0} = \mathbf{c}^k$. In the second part, we develop a bound on $\|\mathbf{h}_{\mathcal{I}_0 \cup \mathcal{I}_1}\|_2$.

Part I: Bound on $\|\mathbf{h}_{(\mathcal{I}_0 \cup \mathcal{I}_1)^c}\|_2$
We begin by decomposing $\mathbf{h}$ as in the proof of Theorem 12.1. The inequalities until (12.59) hold here as well. Instead of (12.59) we have

$$\|\mathbf{c}_0\|_{2,\mathcal{I}} \geq \|\mathbf{c}_{\mathcal{I}_0} + \mathbf{h}_{\mathcal{I}_0}\|_{2,\mathcal{I}} + \|\mathbf{c}_{\mathcal{I}_0^c} + \mathbf{h}_{\mathcal{I}_0^c}\|_{2,\mathcal{I}}$$
$$\geq \|\mathbf{c}_{\mathcal{I}_0}\|_{2,\mathcal{I}} - \|\mathbf{h}_{\mathcal{I}_0}\|_{2,\mathcal{I}} + \|\mathbf{h}_{\mathcal{I}_0^c}\|_{2,\mathcal{I}} - \|\mathbf{c}_{\mathcal{I}_0^c}\|_{2,\mathcal{I}}. \tag{12.73}$$

Therefore,

$$\|\mathbf{h}_{\mathcal{I}_0^c}\|_{2,\mathcal{I}} \leq 2\|\mathbf{c}_{\mathcal{I}_0^c}\|_{2,\mathcal{I}} + \|\mathbf{h}_{\mathcal{I}_0}\|_{2,\mathcal{I}}, \tag{12.74}$$

where we used the fact that $\|\mathbf{c}_0\|_{2,\mathcal{I}} - \|\mathbf{c}_{\mathcal{I}_0}\|_{2,\mathcal{I}} = \|\mathbf{c}_{\mathcal{I}_0^c}\|_{2,\mathcal{I}}$. Combining (12.58), (12.61), and (12.74) leads to

$$\|\mathbf{h}_{(\mathcal{I}_0 \cup \mathcal{I}_1)^c}\|_2 \leq \|\mathbf{h}_{\mathcal{I}_0}\|_2 + 2e_0 \leq \|\mathbf{h}_{\mathcal{I}_0 \cup \mathcal{I}_1}\|_2 + 2e_0, \tag{12.75}$$

where $e_0 = k^{-1/2}\|\mathbf{c}_0 - \mathbf{c}_{\mathcal{I}_0}\|_{2,\mathcal{I}}$.

Part II: Bound on $\|\mathbf{h}_{\mathcal{I}_0 \cup \mathcal{I}_1}\|_2$
Using the fact that $\mathbf{h} = \mathbf{h}_{\mathcal{I}_0 \cup \mathcal{I}_1} + \sum_{i \geq 2} \mathbf{h}_{\mathcal{I}_i}$ we have

$$\|\mathbf{D}\mathbf{h}_{\mathcal{I}_0 \cup \mathcal{I}_1}\|_2^2 = \langle \mathbf{D}\mathbf{h}_{\mathcal{I}_0 \cup \mathcal{I}_1}, \mathbf{D}\mathbf{h} \rangle - \sum_{i=2}^{\ell-1} \langle \mathbf{D}(\mathbf{h}_{\mathcal{I}_0} + \mathbf{h}_{\mathcal{I}_1}), \mathbf{D}\mathbf{h}_{\mathcal{I}_i} \rangle. \tag{12.76}$$

From (12.27),

$$|\langle \mathbf{D}\mathbf{h}_{\mathcal{I}_0 \cup \mathcal{I}_1}, \mathbf{D}\mathbf{h} \rangle| \leq \|\mathbf{D}\mathbf{h}_{\mathcal{I}_0 \cup \mathcal{I}_1}\|_2 \|\mathbf{D}\mathbf{h}\|_2 \leq \sqrt{1+\delta_{2k}}\|\mathbf{h}_{\mathcal{I}_0 \cup \mathcal{I}_1}\|_2 \|\mathbf{D}\mathbf{h}\|_2. \tag{12.77}$$

Since both $\mathbf{c}'$ and $\mathbf{c}_0$ are feasible

$$\|\mathbf{D}\mathbf{h}\|_2 = \|\mathbf{D}(\mathbf{c}_0 - \mathbf{c}')\|_2 \leq \|\mathbf{D}\mathbf{c}_0 - \mathbf{y}\|_2 + \|\mathbf{D}\mathbf{c}' - \mathbf{y}\|_2 \leq 2\varepsilon, \tag{12.78}$$

and (12.77) becomes

$$|\langle \mathbf{D}\mathbf{h}_{\mathcal{I}_0 \cup \mathcal{I}_1}, \mathbf{D}\mathbf{h} \rangle| \leq 2\varepsilon \sqrt{1+\delta_{2k}}\|\mathbf{h}_{\mathcal{I}_0 \cup \mathcal{I}_1}\|_2. \tag{12.79}$$

Substituting into (12.76),

$$\|\mathbf{D}\mathbf{h}_{\mathcal{I}_0 \cup \mathcal{I}_1}\|_2^2 \leq |\langle \mathbf{D}\mathbf{h}_{\mathcal{I}_0 \cup \mathcal{I}_1}, \mathbf{D}\mathbf{h} \rangle| + \sum_{i=2}^{\ell-1} |\langle \mathbf{D}(\mathbf{h}_{\mathcal{I}_0} + \mathbf{h}_{\mathcal{I}_1}), \mathbf{D}\mathbf{h}_{\mathcal{I}_i} \rangle| \tag{12.80}$$

$$\leq 2\varepsilon \sqrt{1+\delta_{2k}}\|\mathbf{h}_{\mathcal{I}_0 \cup \mathcal{I}_1}\|_2 + \sum_{i=2}^{\ell-1} |\langle \mathbf{D}(\mathbf{h}_{\mathcal{I}_0} + \mathbf{h}_{\mathcal{I}_1}), \mathbf{D}\mathbf{h}_{\mathcal{I}_i} \rangle|.$$

Combining with (12.66) and (12.68),

$$\|\mathbf{D}\mathbf{h}_{\mathcal{I}_0 \cup \mathcal{I}_1}\|_2^2 \leq \left(2\varepsilon \sqrt{1+\delta_{2k}} + \sqrt{2}\delta_{2k}k^{-1/2}\|\mathbf{h}_{\mathcal{I}_0^c}\|_{2,\mathcal{I}}\right)\|\mathbf{h}_{\mathcal{I}_0 \cup \mathcal{I}_1}\|_2. \tag{12.81}$$

Using (12.61) and (12.74) results in the upper bound

$$\|\mathbf{D}\mathbf{h}_{\mathcal{I}_0 \cup \mathcal{I}_1}\|_2^2 \leq \left(2\varepsilon \sqrt{1+\delta_{2k}} + \sqrt{2}\delta_{2k}(\|\mathbf{h}_{\mathcal{I}_0}\| + 2e_0)\right)\|\mathbf{h}_{\mathcal{I}_0 \cup \mathcal{I}_1}\|_2. \tag{12.82}$$

On the other hand, the RIP implies the lower bound

$$\|\mathbf{Dh}_{\mathcal{I}_0 \cup \mathcal{I}_1}\|_2^2 \geq (1 - \delta_{2k})\|\mathbf{h}_{\mathcal{I}_0 \cup \mathcal{I}_1}\|_2^2. \tag{12.83}$$

From (12.82) and (12.83),

$$(1 - \delta_{2k})\|\mathbf{h}_{\mathcal{I}_0 \cup \mathcal{I}_1}\|_2 \leq 2\varepsilon\sqrt{1 + \delta_{2k}} + \sqrt{2}\delta_{2k}(\|\mathbf{h}_{\mathcal{I}_0 \cup \mathcal{I}_1}\| + 2e_0), \tag{12.84}$$

or

$$\|\mathbf{h}_{\mathcal{I}_0 \cup \mathcal{I}_1}\|_2 \leq \frac{2\sqrt{1 + \delta_{2k}}}{1 - (1 + \sqrt{2})\delta_{2k}}\varepsilon + \frac{2\sqrt{2}\delta_{2k}}{1 - (1 + \sqrt{2})\delta_{2k}}e_0. \tag{12.85}$$

The condition $\delta_{2k} < \sqrt{2} - 1$ ensures that the denominator in (12.85) is positive. Substituting (12.85) results in

$$\|\mathbf{h}\|_2 \leq \|\mathbf{h}_{\mathcal{I}_0 \cup \mathcal{I}_1}\|_2 + \|\mathbf{h}_{(\mathcal{I}_0 \cup \mathcal{I}_1)^c}\|_2 \leq 2\|\mathbf{h}_{\mathcal{I}_0 \cup \mathcal{I}_1}\|_2 + 2e_0, \tag{12.86}$$

which completes the proof of the theorem. □

To summarize our discussion so far, we have seen that as long as $\mathbf{D}$ satisfies the block RIP (12.27) with a suitable constant, any k-block sparse vector can be perfectly recovered from its samples $\mathbf{y} = \mathbf{Dc}$ using the convex SOCP (12.46). This algorithm is stable in the sense that by slightly modifying it as in (12.71) it can tolerate noise in a way that ensures that the norm of the recovery error is bounded by the noise level. Furthermore, if $\mathbf{c}$ is not k-block sparse, then its best block-sparse approximation can be approached by solving the SOCP. These results are summarized in Table 12.1, in which δ_{2k} denotes the block RIP constant.

Example 12.8 We now demonstrate how to apply the CVX toolbox for solving (12.71) in order to recover a signal from noisy observations using the block BP algorithm. We consider the dictionary $\mathbf{D}$ used in Example 12.3 and assume that the signal is 1-block sparse. We generate 100 random 1-block sparse signals for each ε. The sparse vectors to be recovered have iid Gaussian entries on a randomly chosen support set according to a uniform prior. The additive noise is distributed uniformly in the interval $[-\varepsilon, \varepsilon]$.

In Fig. 12.7, we plot the maximum recovery error as a function of ε. We also depict the upper bound of Theorem 12.2. As can be seen, the bound is very pessimistic: the maximum error is much lower than the upper bound.

Table 12.1 Comparison of algorithms for signal recovery from $\mathbf{y} = \mathbf{Dc}_0 + \mathbf{z}$

	Algorithm (12.46)	Algorithm (12.71)
$\mathbf{c}_0$	k-block sparse	Arbitrary
Noise z	None ($\mathbf{z} = \mathbf{0}$)	Bounded $\|\mathbf{z}\|_2 \leq \varepsilon$
Condition on D	$\delta_{2k} \leq \sqrt{2} - 1$	$\delta_{2k} \leq \sqrt{2} - 1$
Recovery c'	$\mathbf{c}' = \mathbf{c}_0$	$\|\mathbf{c}_0 - \mathbf{c}'\|_2$ small; see (12.72)

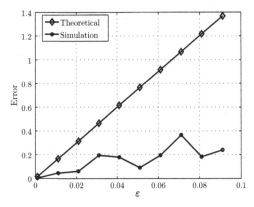

Figure 12.7 Maximum recovery error as a function of the noise bound ε together with the upper bound given in Theorem 12.2.

```
cvx_begin
variable c(6)
variable t(3)
minimize(sum(t))
subject to
    norm(D*c-y)<epsilon
    t(1)>=norm(c(1:2));
    t(2)>=norm(c(3:4));
    t(3)>=norm(c(5:6));
cvx_end
```

12.4.2 Random matrices and block RIP

Theorems 12.1 and 12.2 establish that a sufficiently small block RIP constant ensures exact recovery of the coefficient vector **c**. Using results on random matrices it can be shown that such matrices are likely to satisfy this requirement. Furthermore, fewer measurements are needed to ensure recovery than with standard sparsity [245, Theorem 3.3].

Proposition 12.6. *Consider a matrix* **D** *of size* $n \times N$ *with entries drawn independently from a sub-Gaussian distribution, and block-sparse signals over* $\mathcal{I} = \{d_1 = d, \dots, d_m = d\}$, *where* $N = md$ *for some integer* m. *Let* $t > 0$ *and* $0 < \delta < 1$ *be constant numbers. If*

$$n \geq \frac{36}{7\delta}\left(\ln(L) + kd\ln\left(\frac{12}{\delta}\right) + t\right), \tag{12.87}$$

where $L = \binom{m}{k}$, *then* **D** *satisfies the block RIP (12.27) with restricted isometry constant* $\delta_{k|\mathcal{I}} = \delta$, *with probability at least* $1 - e^{-t}$.

The first term in (12.87) has the dominant impact on the required number of measurements in an asymptotic sense. This term quantifies the amount of measurements that are needed to code the exact subspace in which the sparse signal resides. Specifically, for block-sparse signals

$$\left(\frac{m}{k}\right)^k \le L = \binom{m}{k} \le \left(\frac{e\,m}{k}\right)^k. \tag{12.88}$$

Thus, for a given fraction of nonzeros $r = kd/N$, roughly $n \approx k\log(m/k) = -k\log r$ measurements are needed. For comparison, to satisfy the standard RIP a larger number $n \approx -kd\log r$ is required. Block sparsity reduces the total number of subspaces and therefore requires d times less measurements to code the signal subspace. The second term in (12.87) has a smaller contribution to the number of measurements. This term is proportional to kd, which is the number of nonzero values. Since the number of nonzeros is the same regardless of the sparsity structure, this term is not reduced in the block setting.

Example 12.9 In Fig. 12.8 we plot the RIP and block RIP constants for several instances of $\mathbf{D}$ from a Gaussian ensemble. The values were computed brute-force. The figures qualitatively affirm that block-RIP constants are more "likely" to be smaller than their standard RIP counterparts, even when the dimensions n, N are relatively small.

12.5   Coherence-based recovery results

In the previous section we showed that if the block RIP constant of a dictionary is sufficiently small, then the corresponding block-sparse representation can be recovered from appropriate measurements using block BP. In this section, we extend the discussion to greedy methods by analyzing recovery based on block coherence. More specifically, we present a sufficient condition using block coherence, under which a block-sparse vector can be recovered using either the BOMP or block BP algorithms.

12.5.1   Recovery conditions

Theorems 12.3 and 12.4 below show that any k-block sparse vector $\mathbf{c}$ may be recovered from measurements $\mathbf{y} = \mathbf{D}\mathbf{c}$ using either the BOMP or block BP techniques if the block coherence satisfies $kd < (\mu_{\mathrm{B}}^{-1} + d - (d-1)\nu\mu_{\mathrm{B}}^{-1})/2$. When the columns of $\mathbf{D}[\ell]$ are orthonormal for each ℓ, we have $\nu = 0$ and therefore the recovery condition becomes $kd < (\mu_{\mathrm{B}}^{-1} + d)/2$. In this setting BMP exhibits exponential convergence rate (see Theorem 12.5). If the block-sparse vector $\mathbf{c}$ was treated as a (conventional) kd-sparse vector without exploiting knowledge of the block sparsity structure, then a sufficient condition for perfect recovery using OMP or (12.46) for $d = 1$ (namely, BP) is $kd < (\mu^{-1} + 1)/2$ (see Theorem 11.6). Comparing with $kd < (\mu_{\mathrm{B}}^{-1} + d)/2$, we see that, thanks to $\mu_{\mathrm{B}} \le \mu$, making explicit use of block sparsity leads to guaranteed recovery for a potentially higher sparsity level.

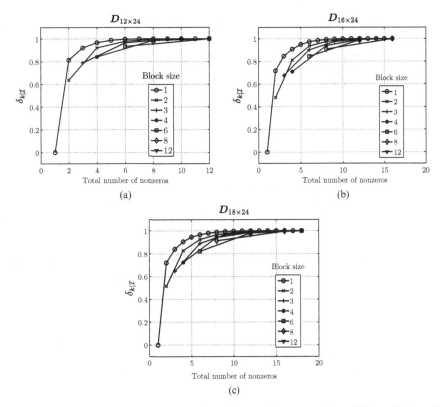

Figure 12.8 The standard and block RIP constants $\delta_{k|\mathcal{I}}$ for $N = 24$ and three different choices of the number of measurements n: (a) $n = 12$. (b) $n = 16$. (c) $n = 18$. Each graph represents an average over 10 instances of the random matrix $\mathbf{D}$. The values of $\mathbf{D}$ are scaled such that (12.23) is satisfied with $\alpha + \beta = 2$.

To formally state the recovery results, suppose that $\mathbf{c}_0$ is a length-N k-block sparse vector, and let $\mathbf{y} = \mathbf{D}\mathbf{c}_0$. Let $\mathbf{D}_0$ denote the $n \times (kd)$ matrix whose blocks correspond to the nonzero blocks of $\mathbf{c}_0$, and let $\overline{\mathbf{D}}_0$ be the matrix of size $n \times (N - kd)$ which contains the $n \times d$ blocks of $\mathbf{D}$ that are not in $\mathbf{D}_0$. We then have the following theorem. The proofs of the theorems in this section are concentrated in Section 12.5.3.

Theorem 12.3 (Block coherence based recovery). *Let $\mathbf{c}_0 \in \mathbb{C}^N$ be a k-block sparse vector with blocks of length d, and let $\mathbf{y} = \mathbf{D}\mathbf{c}_0$ for a given $n \times N$ matrix $\mathbf{D}$. A sufficient condition for the BOMP and block BP algorithms to recover $\mathbf{c}_0$ is that*

$$\rho_c(\mathbf{D}_0^\dagger \overline{\mathbf{D}}_0) < 1 \tag{12.89}$$

where

$$\rho_c(\mathbf{A}) = \max_r \sum_\ell \rho(\mathbf{A}[\ell, r]), \tag{12.90}$$

$\rho(\mathbf{A})$ *is the spectral norm of* $\mathbf{A}$, *and* $\mathbf{A}[\ell, r]$ *is the* (ℓ, r)*th* $d \times d$ *block of* $\mathbf{A}$. *In this case, BOMP picks up a correct new block in each step, and consequently converges in at most* k *steps.*

Note that

$$\rho_c(\mathbf{D}_0^\dagger \overline{\mathbf{D}}_0) = \max_r \rho_c(\mathbf{D}_0^\dagger \overline{\mathbf{D}}_0[r]). \tag{12.91}$$

Therefore, (12.89) implies that for all r,

$$\rho_c(\mathbf{D}_0^\dagger \overline{\mathbf{D}}_0[r]) < 1. \tag{12.92}$$

The sufficient condition (12.89) depends on $\mathbf{D}_0$ and therefore on the location of the nonzero blocks in $\mathbf{c}_0$, which, of course, is not known in advance. Nonetheless, as the following theorem shows, (12.89) holds universally under certain conditions on μ_B and ν associated with the dictionary $\mathbf{D}$.

Theorem 12.4 (Coherence condition). *Let* μ_B *be the block coherence and* ν *the subcoherence of the dictionary* $\mathbf{D}$. *Then* (12.89) *is satisfied if*

$$kd < \frac{1}{2}\left(\mu_\mathrm{B}^{-1} + d - (d-1)\frac{\nu}{\mu_\mathrm{B}}\right). \tag{12.93}$$

For $d = 1$, and therefore $\nu = 0$, we recover the corresponding condition $k < (\mu^{-1} + 1)/2$ of Theorem 11.6 to allow recovery for k-sparse vectors. In the special case in which the columns of $\mathbf{D}[\ell]$ are orthonormal for each ℓ, we have $\nu = 0$ and (12.93) becomes

$$kd < \frac{1}{2}(\mu_\mathrm{B}^{-1} + d). \tag{12.94}$$

Since $\mu_\mathrm{B} < \mu$ this bound is higher than that resulting from sparsity alone.

The next theorem shows that under condition (12.94), BMP exhibits an exponential convergence rate when each block $\mathbf{D}[\ell]$ consists of orthonormal columns.

Theorem 12.5 (MP recovery). *If* $\mathbf{D}^*[\ell]\mathbf{D}[\ell] = \mathbf{I}_d$ *for all* ℓ, *and* $kd < (\mu_\mathrm{B}^{-1} + d)/2$, *then:*

1. *BMP picks up a correct block in each step.*
2. *The energy of the residual decays exponentially, i.e.* $\|\mathbf{r}_\ell\|_2^2 \leq \beta^\ell \|\mathbf{r}_0\|_2^2$ *with*

$$\beta = 1 - \frac{1 - (k-1)d\mu_\mathrm{B}}{k}. \tag{12.95}$$

Theorem 12.4 indicates under which conditions exploiting block sparsity leads to higher recovery thresholds than treating the block-sparse signal as a conventionally sparse signal. One example is when the blocks $\mathbf{D}[\ell]$ consist of orthonormal columns. If the individual blocks $\mathbf{D}[\ell]$ are not orthonormal, then $\nu > 0$, and (12.93) shows that ν has to be small for block-sparse recovery to result in higher recovery thresholds. In this case, it seems natural to orthogonalize the individual blocks $\mathbf{D}[\ell]$ to enforce $\nu = 0$. The comparison that is meaningful here is between the recovery threshold of the original dictionary $\mathbf{D}$ without exploiting block sparsity and the recovery threshold of the orthogonalized dictionary taking block sparsity into account. Unfortunately, it seems difficult to derive general results on the relation between μ before and μ_B after orthogonalization.

Nonetheless, it can be shown (see [207] and Exercise 7) that for a dictionary of size $L \times N$ with $L = Rd$ and $N = Md$ with integers R, M, if $d > RM/(M - R)$, then the recovery threshold obtained from taking block sparsity into account in the orthogonalized dictionary is higher than the threshold corresponding to conventional sparsity in the original dictionary. This is true irrespective of the dictionary we start from as long as it satisfies the conditions of Proposition 12.1.

Example 12.10 In this example, we compare performance of BOMP and block BP with and without orthogonalization on the same dictionary.

We randomly generate dictionaries of size $L = 20$ by $N = 40$, by drawing from iid Gaussian matrices and normalizing the resulting columns to have unit norm. The dictionary is divided into consecutive blocks of length $d = 2$. The sparse vector to be recovered has iid Gaussian entries on a randomly chosen support set according to a uniform prior, i.e. the locations of the nonzero blocks of the block-sparse vector are chosen uniformly at random among all $\binom{M}{k}$ possible locations, where $M = N/d$ is the number of blocks.

In Fig. 12.9 we plot the success rate as a function of the block sparsity level where success is declared if the recovered vector is within a distance of 0.01 from the original vector. For each block-sparsity level we average over 200 realizations of the dictionary and the block-sparse signal. Figure 12.9(a) compares the performance of OMP, BOMP, and BOMP with an orthogonal dictionary. Fig. 12.9(b) compares BP, block BP, and the block BP orthogonal methods. It can be seen that using an orthogonal dictionary yields slightly better performance.

We conclude this section by noting that as in Chapter 11, in general, the theoretical thresholds derived based on both RIP and coherence are quite pessimistic. Simulation results suggest that the performance of recovery algorithms is far better than that predicted by the theoretical guarantees. We illustrate this point by the following example.

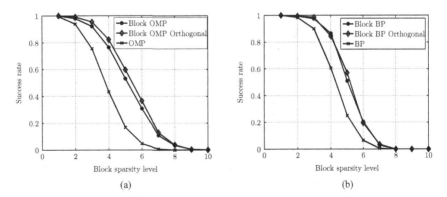

Figure 12.9 Performance of BOMP and block BP with and without orthogonalization on the same dictionary. (a) Comparison between OMP, BOMP, and BOMP with an orthogonal dictionary. (b) Comparison between BP, block BP, and block BP with an orthogonal dictionary.

Example 12.11 To demonstrate the fact that the theoretical bounds may be overly conservative, we compare the performance of several recovery algorithms to the analytical thresholds based on coherence obtained in (12.93) and (12.94) and to the threshold for conventional sparsity of Theorem 11.6.

We randomly generate dictionaries of size $L = 80$ by $N = 160$ and corresponding block-sparse inputs using the same method as in Example 12.10, with the only difference being that here the block size is chosen as $d = 8$. Thus, the dictionaries are drawn from iid Gaussian matrices and their columns are normalized. The block-sparse vectors to be recovered have iid Gaussian entries on a support set chosen uniformly at random. For each block-sparsity level we average over 200 realizations of the dictionary and the block-sparse signal. We also compute the corresponding analytical thresholds and find the following average analytical thresholds (over all realizations of the dictionary):

- BOMP and block BP: $kd < 3.2 \Rightarrow k = 0$
- BOMP orthogonal and block BP orthogonal: $kd < 17.7 \Rightarrow k \leq 2$
- OMP and BP: $kd < 3 \Rightarrow k = 0$.

The success rate as a function of the block-sparse level is plotted in Fig. 12.10 for the various methods.

Evidently, the analytical thresholds are more pessimistic than the numerical thresholds seen in the figure. The latter indicate success rates close to 100% for BOMP, BOMP orthogonal, block BP, and block BP orthogonal up to a block sparsity level of $k = 4$ and for OMP and BP up to $k = 2$.

12.5.2   Extensions

Redundant blocks

Until now we have addressed the basic problem of block-sparse recovery, where the dictionary consists of blocks containing linearly independent vectors. In [370], the authors

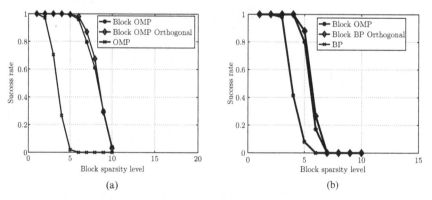

Figure 12.10 Performance of BOMP and block BP with and without orthogonalization on the same dictionary, with blocks of size $d = 8$.

extend the notion of block coherence to subspace coherence, in order to address the case in which the dictionary blocks $\mathbf{D}[\ell]$ may be overcomplete, meaning that the number of columns in each block is larger than the dimension of the subspace they span.

The *subspace coherence* is given by (12.32) where $\mathbf{M}[\ell, r]$ is replaced by

$$\mathbf{M}[\ell, r] = \mathbf{A}^*[\ell]\mathbf{A}[r], \tag{12.96}$$

and $\mathbf{A}[\ell]$ is a matrix whose columns form an orthonormal basis for the subspace spanned by the columns of $\mathbf{D}[\ell]$. Clearly, when the blocks $\mathbf{D}[\ell]$ are orthonormal, subspace coherence reduces to block coherence. Using this definition they prove recovery results for the redundant case. They also suggest a modification of (12.46) that accounts for the redundant blocks, in which the norm of the coefficients is replaced by the norm of the signal part of the corresponding dictionary:

$$\min_{\mathbf{c}} \|\mathbf{D}\mathbf{c}\|_{2,\mathcal{I}} \quad \text{s.t. } \mathbf{y} = \mathbf{D}\mathbf{c}. \tag{12.97}$$

As they show, (12.97) often leads to superior performance.

Hierarchical sparsity

Another extension of the basic block-sparse recovery problem was developed in [363], in which the blocks possess internal sparsity. This corresponds to allowing for internal structure within the subspaces making up the union. Such scenarios can be accounted for by adding an ℓ_1 penalty in (12.46) on the individual blocks, an approach that is dubbed *HiLasso* in [363]. HiLasso allows one to combine the sparsity-inducing property of ℓ_1 optimization at the individual feature level with the block sparsity property of (12.46) on the group level, obtaining a hierarchically structured sparsity pattern. The resulting optimization problem becomes

$$\min_{\mathbf{c}} \|\mathbf{c}\|_{2,\mathcal{I}} + \lambda_1 \|\mathbf{c}\|_1 \quad \text{s.t. } \mathbf{y} = \mathbf{D}\mathbf{c}, \tag{12.98}$$

where λ_1 is a suitable regularization parameter.

In the presence of noise, we can replace (12.98) by

$$\min_{\mathbf{c}} \|\mathbf{y} - \mathbf{D}\mathbf{c}\|_2 + \lambda_2 \|\mathbf{c}\|_{2,\mathcal{I}} + \lambda_1 \|\mathbf{c}\|_1. \tag{12.99}$$

The selection of λ_1 and λ_2 has an important influence on the sparsity of the obtained solution. Intuitively, as λ_2/λ_1 increases, the group constraint becomes dominant and the solution tends to be more sparse at a group level but less sparse within groups. The hierarchical sparsity pattern produced by the solutions of (12.98) is depicted in Fig. 12.11(a).

Combining these ideas within the MMV setting, where multiple vectors are obtained that share the same group sparsity but not necessarily the same internal sparsity, allows for a general modeling framework well suited for applications such as source identification and separation. This approach is referred to as *C-HiLasso (collaborative HiLasso)*. The sparsity patterns obtained in this case are depicted in Fig. 12.11(b). The model in C-HiLasso encourages all signals to share the same groups, while allowing the active sets inside each group to vary between signals. Efficient algorithms for solving the

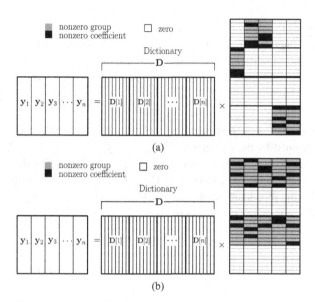

Figure 12.11 Sparsity patterns induced by (a) HiLasso and (b) C-HiLasso. C-HiLasso imposes the same group sparsity pattern in all samples, whereas the in-group sparsity patterns are allowed to vary between samples.

C-HiLasso problem are developed in [363], along with recovery conditions that ensure correct sparse recovery.

A simple example of where this approach can be useful is a piece of music, in which only a few instruments are active at a time (each instrument is a group), and the sound produced by an instrument at each instant is efficiently represented by a few atoms of the subdictionary/group corresponding to it. Using HiLasso in this case permits efficient source identification and separation, where the individual sources (classes/groups) that generated the signal are identified at the same time as their representation is reconstructed. If in addition we know that the same few instruments will be playing simultaneously during different passages of the piece, then we can assume that the active groups at each instant, within the same passage, will be the same. C-HiLasso exploits this information by enforcing that the same groups are active at all instants within a passage, while still allowing each group to have its own unique internal sparsity pattern. This accounts for the fact that the actual sound produced by a given instrument varies from sample to sample.

Example 12.12 An example featuring sparse sums of subspaces that exploits additional internal structure is shown in Fig. 12.12. This example is based on identification of digits from a given image and is taken from [363].

The images are chosen from the US Postal Service digits data set [371], which contains 9298 handwritten single digits between 0 and 9, each of which consists of 16×16 pixel images. The signals are constructed from the images by

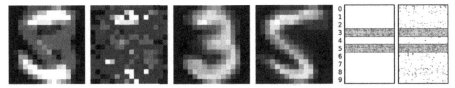

Figure 12.12 Example of recovered digits (3 and 5) from a mixture with 60% of missing components. From left to right: noiseless mixture; observed mixture with missing pixels shown in dark grey (black represents background); recovered digits 3 and 5; and active sets recovered for 180 different mixtures using C-HiLasso and standard BPDN respectively. The last two figures show the active sets of the recovered coefficients as a matrix (one column per mixture), where black dots indicate nonzero coefficients. The coefficients corresponding to the subspace bases for digits 3 and 5 are marked as grey bands.

obtaining vectors containing the unwrapped grey intensities. For each digit, a subspace is learned such that a given image of the digit can be represented sparsely in that subspace. Training is performed using the subspace learning methods we discuss in Section 12.6. The bases that span each of the 10 subspaces $\mathcal{A}_0, \ldots, \mathcal{A}_9$ are collected into a dictionary $\mathbf{D}$.

We consider the task of separating a mixture of digits from subsampled information. An average of 60% of the pixels per mixed image are randomly discarded. To recover the images, we apply the C-HiLasso algorithm on 180 different mixtures. The last two figures illustrate the active sets where the coefficients corresponding to digits 3 and 5 are marked by grey bands. As seen in Fig. 12.12, this approach allows for successful source identification and separation. Specifically, C-HiLasso succeeds in recovering the correct active groups in all the samples. In contrast, standard BPDN, which ignores the block sparsity, is not capable of doing so, and active sets spread all over the groups.

12.5.3 Proofs of theorems

We now prove Theorems 12.3, 12.4, and 12.5 (the proofs are based on [207]). The reader not interested in these proofs can safely skip this section. Note that when substituting $d = 1$ into the developments below, the derivations prove the coherence-based recovery results stated without proof in Chapter 11.

Background results

Before proceeding with the proofs, we start with some definitions and basic results.

For $\mathbf{c} \in \mathbb{C}^N$, we define the general mixed ℓ_2/ℓ_p-norm ($p = 1, 2, \infty$ here and in the following):

$$\|\mathbf{c}\|_{2,p} = \|\mathbf{v}\|_p, \quad \text{where } v_\ell = \|\mathbf{c}[\ell]\|_2 \tag{12.100}$$

and the $\mathbf{c}[\ell]$ are consecutive length-d blocks. For an $n \times N$ matrix $\mathbf{A}$ with $n = Rd$ and $N = md$, where R and m are integers, we define the mixed matrix norm (with block size d) as

$$\|\mathbf{A}\|_{2,p} = \max_{\mathbf{c} \neq 0} \frac{\|\mathbf{Ac}\|_{2,p}}{\|\mathbf{c}\|_{2,p}}. \tag{12.101}$$

The following lemma provides bounds on $\|\mathbf{A}\|_{2,p}$, which will be used in the sequel.

Lemma 12.1. *Let* $\mathbf{A}$ *be an* $n \times N$ *matrix with* $n = Rd$ *and* $N = md$. *Denote by* $\mathbf{A}[\ell, r]$ *the* (ℓ, r)*th* $d \times d$ *block of* $\mathbf{A}$. *Then,*

$$\|\mathbf{A}\|_{2,\infty} \leq \max_{\ell} \sum_{r} \rho(\mathbf{A}[\ell, r]) \triangleq \rho_r(\mathbf{A}) \tag{12.102}$$

$$\|\mathbf{A}\|_{2,1} \leq \max_{r} \sum_{\ell} \rho(\mathbf{A}[\ell, r]) \triangleq \rho_c(\mathbf{A}). \tag{12.103}$$

In particular, $\rho_r(\mathbf{A}) = \rho_c(\mathbf{A}^*)$.

Proof: We first prove (12.102):

$$\|\mathbf{Ax}\|_{2,\infty} = \max_{j} \left\| \sum_{i} \mathbf{A}[j, i]\mathbf{x}[i] \right\|_2 \leq \max_{j} \sum_{i} \|\mathbf{A}[j, i]\mathbf{x}[i]\|_2$$

$$\leq \max_{j} \sum_{i} \|\mathbf{x}[i]\|_2 \, \rho(\mathbf{A}[j, i]) \leq \|\mathbf{x}\|_{2,\infty} \max_{j} \sum_{i} \rho(\mathbf{A}[j, i]). \tag{12.104}$$

Therefore, for any $\mathbf{x} \in \mathbb{C}^N$ with $\mathbf{x} \neq 0$, we have

$$\frac{\|\mathbf{Ax}\|_{2,\infty}}{\|\mathbf{x}\|_{2,\infty}} \leq \rho_r(\mathbf{A}) \tag{12.105}$$

which establishes (12.102). The proof of (12.103) is similar:

$$\|\mathbf{Ax}\|_{2,1} = \sum_{j} \left\| \sum_{i} \mathbf{A}[j, i]\mathbf{x}[i] \right\|_2 \leq \sum_{j} \sum_{i} \|\mathbf{A}[j, i]\mathbf{x}[i]\|_2$$

$$\leq \sum_{i} \|\mathbf{x}[i]\|_2 \sum_{j} \rho(\mathbf{A}[j, i]) \leq \max_{i} \sum_{j} \rho(\mathbf{A}[j, i]) \sum_{i} \|\mathbf{x}[i]\|_2$$

$$= \rho_c(\mathbf{A})\|\mathbf{x}\|_{2,1} \tag{12.106}$$

from which the result follows. Finally, we have $\rho_c(\mathbf{A}^*) = \max_r \sum_{\ell} \rho(\mathbf{A}^*[\ell, r]) = \max_r \sum_{\ell} \rho(\mathbf{A}[r, \ell]) = \rho_r(\mathbf{A})$. □

Lemma 12.2.   *The norm* $\rho_c(\mathbf{A})$ *as defined in (12.90) is a matrix norm and as such satisfies the following properties:*

- *Nonnegative:* $\rho_c(\mathbf{A}) \geq 0$;
- *Positive:* $\rho_c(\mathbf{A}) = 0$ *if and only if* $\mathbf{A} = 0$;
- *Homogeneous:* $\rho_c(\alpha\mathbf{A}) = |\alpha|\rho_c(\mathbf{A})$ *for all* $\alpha \in \mathbb{C}$;
- *Triangle inequality:* $\rho_c(\mathbf{A} + \mathbf{B}) \leq \rho_c(\mathbf{A}) + \rho_c(\mathbf{B})$;
- *Submultiplicative:* $\rho_c(\mathbf{AB}) \leq \rho_c(\mathbf{A})\rho_c(\mathbf{B})$.

Proof: Nonnegativity and positivity follow immediately from the fact that the spectral norm is a matrix norm [28, p. 295]. Homogeneity follows by noting that

$$\rho_c(\alpha \mathbf{A}) = \max_r \sum_\ell \rho(\alpha \mathbf{A}[\ell, r]) = \max_r \sum_\ell |\alpha| \rho(\mathbf{A}[\ell, r]) = |\alpha| \rho_c(\mathbf{A}). \quad (12.107)$$

The triangle inequality is obtained as follows:

$$\rho_c(\mathbf{A} + \mathbf{B}) = \max_r \sum_\ell \rho(\mathbf{A}[\ell, r] + \mathbf{B}[\ell, r])$$

$$\leq \max_r \left(\sum_\ell \rho(\mathbf{A}[\ell, r]) + \sum_\ell \rho(\mathbf{B}[\ell, r]) \right) \quad (12.108)$$

$$\leq \max_r \sum_\ell \rho(\mathbf{A}[\ell, r]) + \max_r \sum_\ell \rho(\mathbf{B}[\ell, r])$$

$$= \rho_c(\mathbf{A}) + \rho_c(\mathbf{B}) \quad (12.109)$$

where (12.108) is a consequence of the spectral norm satisfying the triangle inequality. Finally, to verify submultiplicativity, we use the fact that

$$\rho_c(\mathbf{AB}) = \max_\ell \rho_c(\mathbf{AB}[\ell]). \quad (12.110)$$

Therefore, if we prove the inequality

$$\rho_c(\mathbf{AB}[\ell]) \leq \rho_c(\mathbf{A})\rho_c(\mathbf{B}[\ell]) \quad (12.111)$$

then the result follows from (12.110) and the fact that $\max_\ell \rho_c(\mathbf{B}[\ell]) = \rho_c(\mathbf{B})$. To prove (12.111), note that

$$\rho_c(\mathbf{AB}[\ell]) = \sum_i \rho \left(\sum_j \mathbf{A}[i, j] \mathbf{B}[j, \ell] \right)$$

$$\leq \sum_i \sum_j \rho(\mathbf{A}[i, j] \mathbf{B}[j, \ell]) \leq \sum_j \rho(\mathbf{B}[j, \ell]) \sum_i \rho(\mathbf{A}[i, j]) \quad (12.112)$$

where we used the triangle inequality for, and the submultiplicativity of, the spectral norm. Now, we have

$$\sum_i \rho(\mathbf{A}[i, j]) \leq \max_\ell \sum_i \rho(\mathbf{A}[i, \ell]) = \rho_c(\mathbf{A}). \quad (12.113)$$

Substituting into (12.112) yields

$$\rho_c(\mathbf{AB}[\ell]) \leq \rho_c(\mathbf{A}) \sum_j \rho(\mathbf{B}[j, \ell]) = \rho_c(\mathbf{A})\rho_c(\mathbf{B}[\ell]), \quad (12.114)$$

completing the proof. $\square$

Proof of Theorem 12.3 for BOMP

We begin by proving that (12.89) is sufficient to ensure recovery using the BOMP algorithm. We first show that if $r_{\ell-1}$ is in $\mathcal{R}(\mathbf{D}_0)$, then the next chosen index i_ℓ will correspond to a block in $\mathbf{D}_0$. Assuming that this is true, it follows immediately that i_1 is correct since clearly $\mathbf{r}_0 = \mathbf{y}$ lies in $\mathcal{R}(\mathbf{D}_0)$. Noting that $\mathbf{r}_\ell$ lies in the space spanned by $\mathbf{y}$ and $\mathbf{D}[i], i \in \mathcal{I}_\ell$, where $\mathcal{I}_\ell$ denotes the indices chosen up to stage ℓ, it follows that if $\mathcal{I}_\ell$ corresponds to correct indices, i.e. $\mathbf{D}[i]$ is a block of $\mathbf{D}_0$ for all $i \in \mathcal{I}_\ell$, then $\mathbf{r}_\ell$ also lies in $\mathcal{R}(\mathbf{D}_0)$ and the next index will be correct as well. Thus, at every step a correct $n \times d$ block of $\mathbf{D}$ is selected. As we will show below, no index will be chosen twice since the new residual is orthogonal to all the previously chosen subspaces spanned by the columns of the blocks $\mathbf{D}[i], i \in \mathcal{I}_\ell$; consequently the correct $\mathbf{c}_0$ will be recovered in k steps.

Suppose that $\mathbf{r}_{\ell-1} \in \mathcal{R}(\mathbf{D}_0)$. Then, under (12.89), the next chosen index corresponds to a block in $\mathbf{D}_0$. This is equivalent to requiring that

$$z(\mathbf{r}_{\ell-1}) = \frac{\|\overline{\mathbf{D}}_0^*\mathbf{r}_{\ell-1}\|_{2,\infty}}{\|\mathbf{D}_0^*\mathbf{r}_{\ell-1}\|_{2,\infty}} < 1. \tag{12.115}$$

From the properties of the pseudoinverse, it follows that $\mathbf{D}_0\mathbf{D}_0^\dagger$ is the orthogonal projection onto $\mathcal{R}(\mathbf{D}_0)$. Therefore, $\mathbf{D}_0\mathbf{D}_0^\dagger\mathbf{r}_{\ell-1} = \mathbf{r}_{\ell-1}$. Since $\mathbf{D}_0\mathbf{D}_0^\dagger$ is Hermitian, we have

$$(\mathbf{D}_0^\dagger)^*\mathbf{D}_0^*\mathbf{r}_{\ell-1} = \mathbf{r}_{\ell-1}. \tag{12.116}$$

Substituting (12.116) into (12.115) yields

$$z(\mathbf{r}_{\ell-1}) = \frac{\|\overline{\mathbf{D}}_0^*(\mathbf{D}_0^\dagger)^*\mathbf{D}_0^*\mathbf{r}_{\ell-1}\|_{2,\infty}}{\|\mathbf{D}_0^*\mathbf{r}_{\ell-1}\|_{2,\infty}} \le \rho_r(\overline{\mathbf{D}}_0^*(\mathbf{D}_0^\dagger)^*) = \rho_c(\mathbf{D}_0^\dagger\overline{\mathbf{D}}_0) \tag{12.117}$$

where we used Lemma 12.1.

It remains to show that in each step BOMP chooses a new block participating in the (unique) representation $\mathbf{y} = \mathbf{D}\mathbf{c}$. We start by defining $\mathbf{D}_\ell = [\mathbf{D}[i_1] \cdots \mathbf{D}[i_\ell]]$ where $i_j \in \mathcal{I}_\ell, 1 \le j \le \ell$. It follows that the solution of the minimization problem in (12.50) is given by

$$\hat{\mathbf{c}} = (\mathbf{D}_\ell^*\mathbf{D}_\ell)^{-1}\mathbf{D}_\ell^*\mathbf{y} \tag{12.118}$$

which upon inserting into (12.51) yields

$$\mathbf{r}_\ell = (\mathbf{I} - \mathbf{D}_\ell(\mathbf{D}_\ell^*\mathbf{D}_\ell)^{-1}\mathbf{D}_\ell^*)\mathbf{y}. \tag{12.119}$$

Now, we note that $\mathbf{D}_\ell(\mathbf{D}_\ell^*\mathbf{D}_\ell)^{-1}\mathbf{D}_\ell^*$ is the orthogonal projection onto the range space of $\mathbf{D}_\ell$. Therefore $\|\mathbf{D}^*[i]\mathbf{r}_\ell\|_2 = 0$ for all blocks $\mathbf{D}[i]$ that lie in the span of the matrix $\mathbf{D}_\ell$. By the assumption in Proposition 12.1 we are guaranteed that as long as $\ell < k$ there exists at least one block in $\mathbf{D}_0$ that does not lie in the span of $\mathbf{D}_\ell$. Since this block will lead to strictly positive $\|\mathbf{D}^*[i]\mathbf{r}_\ell\|_2$, the result is established, concluding the proof.

Proof of Theorem 12.3 for block BP

We next show that (12.89) is also sufficient to ensure recovery using block BP. To this end we rely on the following lemma:

Lemma 12.3. *Suppose that* $\mathbf{v} \in \mathbb{C}^{kd}$ *with* $\|\mathbf{v}[\ell]\|_2 > 0$, *for all* ℓ, *and that* $\mathbf{A}$ *is a matrix of size* $n \times (kd)$, *with* $n = Rd$ *and* $d \times d$ *blocks* $\mathbf{A}[\ell, r]$. *Then*, $\|\mathbf{A}\mathbf{v}\|_{2,1} \le \rho_c(\mathbf{A})\|\mathbf{v}\|_{2,1}$.

If in addition the values of $\rho_c(\mathbf{AJ}_\ell)$ are not all equal, then the inequality is strict. Here, $\mathbf{J}_\ell$ is a $(kd) \times d$ matrix that is all zero except for the ℓth $d \times d$ block which equals $\mathbf{I}_d$.

Proof: The proof of the statement $\|\mathbf{Av}\|_{2,1} \leq \rho_c(\mathbf{A})\|\mathbf{v}\|_{2,1}$ follows directly from (12.106) by replacing $\mathbf{A}$ by an $n \times (kd)$ matrix and $\mathbf{x} \in \mathbb{C}^N$ by $\mathbf{v} \in \mathbb{C}^{kd}$ with $\|\mathbf{v}[\ell]\|_2 > 0$, for all ℓ. If the $a_i = \sum_j \rho(\mathbf{A}[j,i])$ are not all equal, then the last inequality in (12.106) is strict. Since $a_i = \rho_c(\mathbf{AJ}_i)$ the result follows. $\qquad\square$

To prove that block BP recovers the correct vector $\mathbf{c}_0$, let $\mathbf{c}' \neq \mathbf{c}_0$ be another length-N k-block sparse vector for which $\mathbf{y} = \mathbf{Dc}'$. Denote by $\mathbf{c}_0$ and $\mathbf{c}'$ the length-kd vectors consisting of the nonzero elements of $\mathbf{c}_0$ and $\mathbf{c}'$, respectively. Let $\mathbf{D}_0$ and $\mathbf{D}'$ denote the corresponding columns of $\mathbf{D}$ so that $\mathbf{y} = \mathbf{D}_0\mathbf{c}_0 = \mathbf{D}'\mathbf{c}'$. From the assumption in Proposition 12.1, it follows that there cannot be two different representations using the same blocks $\mathbf{D}_0$. Therefore, $\mathbf{D}'$ must contain at least one block, $\mathbf{Z}$, that is not included in $\mathbf{D}_0$. From (12.92), we get $\rho_c(\mathbf{D}_0^\dagger \mathbf{Z}) < 1$. For any other block $\mathbf{U}$ in $\mathbf{D}$, we must have that

$$\rho_c(\mathbf{D}_0^\dagger \mathbf{U}) \leq 1. \tag{12.120}$$

Indeed, if $\mathbf{U} \in \mathbf{D}_0$, then $\mathbf{U} = \mathbf{D}_0[\ell] = \mathbf{D}_0\mathbf{J}_\ell$ for some ℓ where $\mathbf{J}_\ell$ was defined in Lemma 12.3. In this case, $\mathbf{D}_0^\dagger \mathbf{D}_0[\ell] = \mathbf{J}_\ell$ and therefore $\rho_c(\mathbf{D}_0^\dagger \mathbf{U}) = \rho_c(\mathbf{J}_\ell) = 1$. If, on the other hand, $\mathbf{U} = \overline{\mathbf{D}}_0[\ell]$ for some ℓ, then it follows from (12.92) that $\rho_c(\mathbf{D}_0^\dagger \mathbf{U}) < 1$.

Now, suppose first that the $(kd) \times d$ blocks in $\mathbf{D}_0^\dagger \mathbf{D}'$ do not all have the same[1] ρ_c. Then,

$$\|\mathbf{c}_0\|_{2,1} = \|\mathbf{D}_0^\dagger \mathbf{D}_0\mathbf{c}_0\|_{2,1} = \|\mathbf{D}_0^\dagger \mathbf{D}'\mathbf{c}'\|_{2,1} < \rho_c(\mathbf{D}_0^\dagger \mathbf{D}')\|\mathbf{c}'\|_{2,1} \leq \|\mathbf{c}'\|_{2,1}, \tag{12.121}$$

where the first equality is a consequence of the columns of $\mathbf{D}_0$ being linearly independent (a result of the assumption in Proposition 12.1), the first inequality follows from Lemma 12.3 since $\|\mathbf{c}'[\ell]\|_2 > 0$ for all ℓ, and the last inequality follows from (12.120). If all the $(kd) \times d$ blocks in $\mathbf{D}_0^\dagger \mathbf{D}'$ have identical ρ_c, then the first inequality is no longer strict, but the second inequality becomes strict instead as a consequence of $\rho_c(\mathbf{D}_0^\dagger \mathbf{Z}) < 1$; therefore $\|\mathbf{c}_0\|_{2,1} < \|\mathbf{c}'\|_{2,1}$ still holds.

Since $\|\mathbf{c}_0\|_{2,1} = \|\mathbf{c}_0\|_{2,1}$ and $\|\mathbf{c}'\|_{2,1} = \|\mathbf{c}'\|_{2,1}$, we conclude that under (12.92), any set of coefficients used to represent the original signal that is not equal to $\mathbf{c}_0$ will result in a larger ℓ_2/ℓ_1-norm.

Proof of Theorem 12.4

We start by deriving an upper bound on $\rho_c(\mathbf{D}_0^\dagger \overline{\mathbf{D}}_0)$ in terms of μ_B and ν. Writing out $\mathbf{D}_0^\dagger$, we have that

$$\rho_c(\mathbf{D}_0^\dagger \overline{\mathbf{D}}_0) = \rho_c((\mathbf{D}_0^* \mathbf{D}_0)^{-1}\mathbf{D}_0^* \overline{\mathbf{D}}_0). \tag{12.122}$$

---

[1] Note that for an $(sd) \times d$ matrix $\mathbf{A}$, $\rho_c(\mathbf{A}) = \sum_\ell \rho(\mathbf{A}[\ell])$, where $\mathbf{A}[\ell]$, $1 \leq \ell \leq s$, denotes the $d \times d$ block of $\mathbf{A}$ made up of the rows $\{(\ell - 1)d + 1, \ldots, \ell d\}$.

Submultiplicativity of $\rho_c(\mathbf{A})$ (Lemma 12.2) implies that

$$\rho_c(\mathbf{D}_0^\dagger \overline{\mathbf{D}}_0) \leq \rho_c((\mathbf{D}_0^*\mathbf{D}_0)^{-1})\rho_c(\mathbf{D}_0^*\overline{\mathbf{D}}_0)$$
$$= \rho_c((\mathbf{D}_0^*\mathbf{D}_0)^{-1}) \max_{j \notin \Lambda_0} \sum_{i \in \Lambda_0} \rho(\mathbf{D}^*[i]\mathbf{D}[j]) \qquad (12.123)$$

where Λ_0 is the set of indices ℓ for which $\mathbf{D}[\ell]$ is in $\mathbf{D}_0$. Since Λ_0 contains k indices, the last term in (12.123) is bounded above by $kd\mu_B$, which allows us to conclude that

$$\rho_c(\mathbf{D}_0^\dagger \overline{\mathbf{D}}_0) \leq \rho_c((\mathbf{D}_0^*\mathbf{D}_0)^{-1})kd\mu_B. \qquad (12.124)$$

It remains to develop a bound on $\rho_c((\mathbf{D}_0^*\mathbf{D}_0)^{-1})$. To this end, we express $\mathbf{D}_0^*\mathbf{D}_0$ as $\mathbf{D}_0^*\mathbf{D}_0 = \mathbf{I} + \mathbf{A}$, where $\mathbf{A}$ is a $(kd) \times (kd)$ matrix with blocks $\mathbf{A}[\ell, r]$ of size $d \times d$ such that $\mathbf{A}_{i,i} = 0$, for all i. This follows from the fact that the columns of $\mathbf{A}$ are normalized. Since $\mathbf{A}[\ell, r] = \mathbf{D}_0^*[\ell]\mathbf{D}_0[r]$, for all $\ell \neq r$, and $\mathbf{A}[r, r] = \mathbf{D}_0^*[r]\mathbf{D}_0[r] - \mathbf{I}_d$, we have

$$\rho_c(\mathbf{A}) = \max_r \sum_\ell \rho(\mathbf{A}[\ell, r])$$
$$\leq \max_r \rho(\mathbf{A}[r, r]) + \max_r \sum_{\ell \neq r} \rho(\mathbf{A}[\ell, r]) \qquad (12.125)$$
$$\leq (d - 1)\nu + (k - 1)d\mu_B \qquad (12.126)$$

where the first term in (12.126) is obtained by applying Geršgorin's disk theorem (see Appendix A) together with the definition of ν; the second term in (12.126) follows from the fact that the summation in the second term of (12.125) is over $k - 1$ elements and $\rho(\mathbf{A}[\ell, r])$, for all $\ell \neq r$, can be upper-bounded by $d\mu_B$. Assumption (12.93) now implies that $(d-1)\nu+(k-1)d\mu_B < 1$ and therefore, from (12.126), we have $\rho_c(\mathbf{A}) < 1$.
We next use the following result.

Lemma 12.4. *Suppose that $\rho_c(\mathbf{A}) < 1$. Then $(\mathbf{I} + \mathbf{A})^{-1} = \sum_{k=0}^{\infty}(-\mathbf{A})^k$.*

Proof: Follows immediately by using the fact that $\rho_c(\mathbf{A})$ is a matrix norm (cf. Lemma 12.2) and applying Corollary 5.6.16 of [28]. □

Thanks to Lemma 12.4, we have that

$$\rho_c((\mathbf{D}_0^*\mathbf{D}_0)^{-1}) = \rho_c\left(\sum_{k=0}^{\infty}(-\mathbf{A})^k\right) \leq \sum_{k=0}^{\infty}(\rho_c(\mathbf{A}))^k \qquad (12.127)$$
$$= \frac{1}{1 - \rho_c(\mathbf{A})} \leq \frac{1}{1 - (d - 1)\nu - (k - 1)d\mu_B}. \qquad (12.128)$$

Here, (12.127) is a consequence of $\rho_c(\mathbf{A})$ satisfying the triangle inequality and being submultiplicative, and (12.128) follows by using (12.126).
Combining (12.128) with (12.124), we get

$$\rho_c(\mathbf{D}_0^\dagger \overline{\mathbf{D}}_0) \leq \frac{kd\mu_B}{1 - (d - 1)\nu - (k - 1)d\mu_B} < 1 \qquad (12.129)$$

where the last inequality is a consequence of (12.93).

Proof of Theorem 12.5

The proof of the first part of Theorem 12.5 follows from the arguments in the proofs of Theorems 12.3 and 12.4 for $\nu = 0$. As a consequence of the first statement of Theorem 12.5, we get that the residual $\mathbf{r}_\ell$ in each step of the algorithm will be in $\mathcal{R}(\mathbf{D}_0)$. The proof of the second statement in Theorem 12.5 mimics the corresponding proof in [372]. Since the derivations are a bit lengthy, we refer the interested reader to the detailed proof in [207].

12.6 Dictionary and subspace learning

Until now, we have assumed that the possible subspaces $\mathcal{A}_j$ comprising the union in (12.2) are known. The uncertainty is a result of the fact that the subspaces actually used are unspecified. In practice, the possibilities $\mathcal{A}_j$ may not be given a priori. To treat this scenario, we will assume that we are given a set of training data $x_n, 1 \leq n \leq L$ from which we learn the subspaces $\mathcal{A}_j$. The simplest case, which we treat first, is when the signals are not sampled, so that each x_n is an element in the union. In other words, x_n can be represented as a sum of a small number of vectors, and each such vector belongs to one of the unknown subspaces $\mathcal{A}_j, 1 \leq j \leq m$. More difficult is the setting in which we are only given sampled data $\mathbf{z}_n, 1 \leq n \leq L$ of a set of vectors $x_n, 1 \leq n \leq L$ belonging to the union, and our goal is to learn the possible subspaces along with the representations of each of the individual sampled vectors.

To formulate the subspace learning problem, suppose we are given training data $x_n, 1 \leq n \leq L$ where each individual signal x_j lies in a union of the form

$$\mathcal{U}_i = \bigoplus_{|j|=k} \mathcal{A}_j, \tag{12.130}$$

in some finite-dimensional subspace $\mathcal{H}$. When $x = x(t)$ represents a continuous-time signal, $\mathcal{H}$ is a subspace of $L_2(\mathbb{R})$. Our goal is to learn the subspaces $\mathcal{A}_j$ from the training data. More specifically, we seek to find bases A_j that span each of the subspaces $\mathcal{A}_j$. As we have seen, by choosing appropriate orthonormal bases A_j for $\mathcal{A}_j$, each x_j can be written as $x_j = A\mathbf{c}_j$ where A represents the concatenation of the A_js, and $\mathbf{c}_j$ is $2k$-block sparse. Letting X denote the set transformation corresponding to the signals x_j, and denoting by $\mathbf{C}$ the matrix with columns $\mathbf{c}_j$, our problem becomes that of finding a set transformation A consisting of orthonormal blocks A_j such that

$$X = A\mathbf{C} = \begin{bmatrix} A_1 | & \cdots & | A_m \end{bmatrix} \mathbf{C}, \tag{12.131}$$

for some matrix $\mathbf{C}$ consisting of k-block sparse columns. In the presence of noise and mismodeling, we may replace the equality by minimizing the squared error:

$$\min_{A,\mathbf{C}} \|X - A\mathbf{C}\|_F^2 \quad \text{s.t. } \|\mathbf{c}_j\|_{0,\mathcal{I}} \leq k. \tag{12.132}$$

Here $\|T\|_F^2 = \mathrm{Tr}(T^*T)$ denotes the Frobenius norm of the finite-dimensional transformation T.

Since $\mathcal{H}$ is finite-dimensional, we can represent x_j in terms of an orthonormal basis expansion in $\mathcal{H}$. To be concrete, suppose that the dimension of $\mathcal{H}$ is equal to r, and let $H: \mathbb{C}^r \rightarrow \mathcal{H}$ be a set transformation that represents an orthonormal basis for $\mathcal{H}$. Then, we can express x_j as $x_j = H\mathbf{y}_j$ for some length-r vector $\mathbf{y}$. Similarly, each of the bases A_j can also be represented in terms of H as $A_j = H\mathbf{D}[j]$ where $\mathbf{D}[j]$ is a coefficient matrix of size $r \times d_j$ consisting of orthonormal columns, and $d_j = \dim(\mathcal{A}_j)$. Problem (12.132) is then equivalent to

$$\min_{\mathbf{D},\mathbf{C}} \|\mathbf{Y} - \mathbf{D}\mathbf{C}\|_F^2 \quad \text{s.t.} \quad \|\mathbf{c}_j\|_{0,\mathcal{I}} \leq k, \tag{12.133}$$

where $\mathbf{Y}$ is the matrix with columns $\mathbf{y}_j$, $\mathbf{D} = [\mathbf{D}[1]| \cdots |\mathbf{D}[m]]$ is a matrix with orthonormal blocks $\mathbf{D}[j]$, and $\|\mathbf{c}_j\|_{0,\mathcal{I}}$ is defined in (12.10).

Clearly (12.133) is a nonconvex optimization problem and therefore finding its minimum is in general difficult. It is also obvious that there are certain invariances that cannot be resolved, such as rotation of $\mathbf{D}$ and $\mathbf{C}$ by a block-diagonal unitary matrix, and permutation of the blocks within $\mathbf{D}$ and $\mathbf{C}$. However, even beyond these issues it is unclear whether (12.133) possesses a unique solution. In the special case in which the columns $\mathbf{c}_j$ are sparse in the conventional sense (that is $d_j = 1$ for all j), problems of the form (12.133) are termed *dictionary learning (DL)*, and have been studied extensively in the CS literature, following the pioneering work of Olshausen and Field [373, 374, 375]. In the next section we review some of the main results and algorithms for DL, which is a special case of subspace learning in which every subspace has dimension one. We then extend these ideas, in Section 12.6.2, to subspace learning. In Section 12.7, we consider the more challenging problem in which the subspaces are learned from sampled data. As we discuss, this problem can be viewed as CS with an unknown sparsifying basis. We therefore refer to this setting as *blind CS (BCS)*.

We note that both DL and BCS were originally formulated for standard sparse recovery. Nonetheless, we review them here in the context of block sparsity and subspace learning since they can be viewed as special cases. Furthermore, they form the basis for the extensions to the subspace setting.

12.6.1 Dictionary learning

Following the work of Olshausen and Field, a variety of methods to learn a dictionary from training data have been developed [376, 377, 378, 379]. These works demonstrate that using a carefully designed dictionary instead of a predefined one, can improve the sparsity in jointly representing a class of signals. Below we will focus on two examples of DL algorithms: the method of directions (MOD) [378] and the K-SVD [379]. Since (12.132) is nonconvex, there is no guarantee that a method attempting to solve it will converge to the correct minima. Nonetheless, it can be shown that if $\mathbf{D}$ satisfies the conditions for uniqueness in sparse recovery, and there is a sufficiently rich set of examples $\mathbf{C}$ which are diverse with respect to both the locations and the values of the nonzero elements, then there is a unique dictionary solving this problem up to scaling and permutation [380]. However, there is currently no general algorithm for finding this unique dictionary.

Algorithm 12.3 Method of directions for dictionary learning

Input: Measurement matrix $\mathbf{Y}$, sparsity level k
Output: Dictionary $\widehat{\mathbf{D}}$
Initialize: $\widehat{\mathbf{D}}_0$ can be initialized by a random matrix, or a random sample of $\mathbf{Y}$, $\ell = 0$
while halting criterion false **do**
  $\ell \leftarrow \ell + 1$
  for each $1 \le i \le L$ approximate the solution $\mathbf{c}_\ell$ of
    $\min_{\mathbf{c}_i} \|\mathbf{y}_i - \widehat{\mathbf{D}}_{\ell-1}\mathbf{c}_i\|^2$ s.t. $\|\mathbf{c}_i\|_0 \le k$ {update sparse coefficients}
  $\widehat{\mathbf{D}}_\ell = \mathbf{Y}\mathbf{C}_\ell^*(\mathbf{C}_\ell\mathbf{C}_\ell^*)^{-1}$ {update dictionary}
end while
return $\widehat{\mathbf{D}} \leftarrow \widehat{\mathbf{D}}_\ell$

To approach a local minimum of (12.133), Engan, Aase, and Husoy [378] suggested the use of alternating minimization, where we first fix the dictionary $\mathbf{D}$ and optimize the sparse representation $\mathbf{C}$, and then reverse roles. More specifically, at the ℓth step we use the dictionary $\mathbf{D}_{\ell-1}$ from the previous iteration to optimize $\mathbf{C}$ by finding a sparse approximation that best fits the columns $\mathbf{y}_i$, using any of the known sparse recovery techniques. Once $\mathbf{C}_\ell$ is found, we proceed to determine a dictionary that best fits $\mathbf{Y}$ and $\mathbf{C}_\ell$. This corresponds to solving

$$\mathbf{D}_\ell = \arg\min_{\mathbf{D}} \|\mathbf{Y} - \mathbf{D}\mathbf{C}_\ell\|_F^2 = \mathbf{Y}\mathbf{C}_\ell^*(\mathbf{C}_\ell\mathbf{C}_\ell^*)^{-1}, \qquad (12.134)$$

assuming $\mathbf{C}_\ell\mathbf{C}_\ell^*$ is invertible; otherwise we replace the inverse by the pseudoinverse. This approach is summarized as Algorithm 12.3. Note that after the dictionary update it is common to scale the columns of $\mathbf{D}_\ell$ to have unit norm. The stopping criterion can be a fixed number of iterations, or a small update in the dictionary.

The K-SVD algorithm follows a similar route, where the difference is in the dictionary update stage. Instead of updating the entire dictionary at once, as done in the dictionary update of Algorithm 12.3, the K-SVD iteration updates the columns sequentially. Denoting by $\mathbf{v}_j^*$ the jth row of the matrix $\mathbf{C}$, we can write the matrix product $\mathbf{D}\mathbf{C}$ in the form

$$\mathbf{D}\mathbf{C} = \sum_{i=1}^{N} \mathbf{d}_i\mathbf{v}_i^*, \qquad (12.135)$$

where N is the number of columns in $\mathbf{D}$. Keeping all columns fixed besides the jth column, the objective in (12.133) may be expressed as

$$\|\mathbf{Y} - \mathbf{D}\mathbf{C}\|_F^2 = \|\mathbf{E}_j - \mathbf{d}_j\mathbf{v}_j^*\|_2^2, \qquad (12.136)$$

with

$$\mathbf{E}_j = \mathbf{Y} - \sum_{i=1,\, i\neq j}^{N} \mathbf{d}_i\mathbf{v}_i^*. \qquad (12.137)$$

Note that in the jth iteration, $\mathbf{E}_j$ is fixed.

Algorithm 12.4 K-SVD algorithm for dictionary learning

Input: Measurement matrix $\mathbf{Y}$, sparsity level k

Output: Dictionary $\widehat{\mathbf{D}}$

Initialize: $\widehat{\mathbf{D}}_0$ can be initialized by a random matrix, or a random sample of $\mathbf{Y}$, $\ell = 0$

while halting criterion false **do**

$\quad \ell \leftarrow \ell + 1$

$\quad$ for each $1 \leq i \leq L$ approximate the solution $\mathbf{c}_\ell$ of

$\quad\quad \min_{\mathbf{c}_i} \|\mathbf{y}_i - \widehat{\mathbf{D}}_{\ell-1}\mathbf{c}_i\|^2 \quad$ s.t. $\|\mathbf{c}_i\|_0 \leq k$ {update sparse coefficients}

$\quad$ {update dictionary}

$\quad$ **for** $j = 1, 2, \ldots, N$ **do**

$\quad\quad$ Compute the residual matrix

$\quad\quad\quad \mathbf{E}_j = \mathbf{Y} - \sum_{i \neq j} \mathbf{d}_i \mathbf{v}_i^*$

$\quad\quad$ where $\mathbf{v}_i^*$ is the ith row of the matrix $\mathbf{C}_\ell$

$\quad\quad$ Obtain $\mathbf{E}_j^\Lambda$ by choosing the columns of $\mathbf{E}_j$ corresponding to the support of $\mathbf{v}_j$

$\quad\quad$ Apply the SVD decomposition $\mathbf{E}_j^\Lambda = \mathbf{U}\boldsymbol{\Sigma}\mathbf{Q}^*$

$\quad\quad$ Update the dictionary atom $\mathbf{d}_j = \mathbf{u}_1$ and the representation by $\mathbf{v}_j^\Lambda = \sigma_1 \mathbf{q}_1$

$\quad$ **end for**

end while

return $\widehat{\mathbf{D}} \leftarrow \widehat{\mathbf{D}}_\ell$

Our goal therefore is to choose $\mathbf{d}_j$ to minimize (12.136). Along with updating $\mathbf{d}_j$ we also update the nonzero values of $\mathbf{v}_j$, while keeping the sparsity pattern determined by the previous iteration intact. Omitting the index j for convenience, and letting Λ denote the support of $\mathbf{v}$, our problem becomes

$$\min_{\mathbf{d},\mathbf{v}} \|\mathbf{E}_\Lambda - \mathbf{d}\mathbf{v}_\Lambda^*\|_F^2, \tag{12.138}$$

where $\mathbf{E}_\Lambda$ is the matrix $\mathbf{E}$ restricted to the columns indicated by the support set Λ. Problem (12.138) can be interpreted as that of finding the best rank-1 approximation to a given matrix $\mathbf{E}_\Lambda$. The optimal solution is given by the outer product of the left- and right singular vectors of $\mathbf{E}_\Lambda$ denoted $\mathbf{u}_1, \mathbf{q}_1$ corresponding to the largest singular value σ_1 multiplied by the largest singular value (see Theorem A.4 in Appendix A). We therefore update $\mathbf{d}_j$ and $\mathbf{v}_j$ by these values as $\mathbf{d}_j = \mathbf{u}_1$ and $\mathbf{v}_j^\Lambda = \sigma_1 \mathbf{q}_1$ and continue to the next column. The K-SVD method is summarized in Algorithm 12.4.

Example 12.13 We compare the performance of the MOD and K-SVD algorithms for DL using a synthetic experiment.

First, we generate a random dictionary consisting of iid Gaussian entries with normalized columns of size 40×80. Then, we generate $N = 3000$ signals each by a random weighted combination of three atoms of the dictionary. The weights are

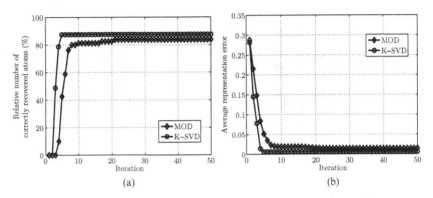

Figure 12.13 K-SVD versus MOD for dictionary learning. (a) Relative number (in %) of correctly recovered atoms. (b) Average representation error.

drawn from an iid Gaussian distribution of zero mean and unit variance. Performing 50 iterations of MOD and K-SVD on this set of signals, we attempt to recover the original dictionary. Both methods are used with a fixed cardinality $k = 3$, and initialized by choosing the first 80 examples as the dictionary atoms.

Figure 12.13(a) depicts the relative number of correctly recovered atoms, while Fig. 12.13(b) shows the average representation error which is computed as $(1/N) \sum_{i=1}^{N} \|\mathbf{y}_i - \mathbf{D}\mathbf{c}_i\|^2$. In this example, K-SVD converges more rapidly than the MOD method.

12.6.2   Subspace learning

A natural approach to solve the subspace learning problem of (12.133), is to extend the DL algorithms presented in the previous section to treat block-sparse representations. A complication that is present here is that this requires sorting the dictionary columns into blocks, as a priori we do not know which columns belong together. In [381], the authors suggest a subspace learning algorithm that builds upon K-SVD, and does not require prior knowledge on the association of the training signals into groups (subspaces), but rather automatically detects the underlying block structure. This structure is inferred by agglomerative clustering of dictionary atoms that induce similar sparsity patterns. In other words, after finding the sparse representations of the training signals individually, the atoms are progressively merged according to the similarity of the sets of signals they represent.

Before explaining the idea behind the algorithm in more detail, we note that besides its importance in the context of sampling signals from a finite-dimensional union, subspace learning arises in several other applications such as face recognition [382, 383] and motion segmentation [384]. In these settings it is typically assumed that each data point belongs to a single subspace, and the goal is to learn the underlying collection of subspaces based on the assumption that each of the samples lies close to one of them

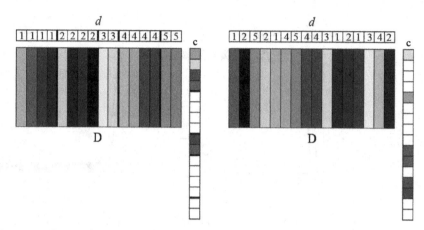

Figure 12.14 Two equivalent examples of dictionaries $\mathbf{D}$ and block structures d with five blocks, together with 2-block sparse representations $\mathbf{c}$. Both examples represent the same signal, since the atoms in $\mathbf{D}$ and the entries of d and $\mathbf{c}$ are permuted in the same manner (in both cases $\|\mathbf{c}\|_{0,d} = 2$).

[385, 386, 387]. Here, however, we assume that each data point belongs to a small sum of subspaces, rather than to a single one.

Dictionary learning for block sparsity

To develop the subspace learning algorithm, we assume that the maximal block size, denoted by s, is known. We further assume that the columns of the matrix $\mathbf{D}$ can be arranged into blocks, where each block spans a corresponding subspace. Every column $\mathbf{d}_i$ of $\mathbf{D}$ is then assigned an index d_i which indicates to which subspace (or block) it belongs, as illustrated in Fig. 12.14. The representation $\mathbf{c}$ is k-block sparse over $\mathbf{D}$ if its nonzero values are concentrated in at most k blocks. Figure 12.14 presents examples of two different block structures and their corresponding block-sparse vectors and dictionaries. Our goal is to find a dictionary $\mathbf{D}$, a block structure d with maximal block size s, and a corresponding block-sparse representation $\mathbf{C}$, that minimizes (12.133).

The algorithm proceeds as follows: we first update the dictionary using the K-SVD method (or MOD) for DL. In other words, we ignore the block structure. This corresponds to initializing the index vector d as N blocks of size 1 (where N is the number of columns in $\mathbf{D}$). To initialize $\mathbf{C}$ we seek a (standard) sparse representation best matched to $\mathbf{Y}$ given $\mathbf{D}$, just as in the sparse representation update of Algorithm 12.3. Then, at each iteration ℓ, we perform the following three steps. Based on the obtained $\mathbf{C}_\ell$, and keeping the dictionary fixed, we recover the block structure described by the vector d using *sparse agglomerative clustering (SAC)*. Agglomerative clustering builds blocks by progressively merging the closest atoms according to some distance metric [388]; SAC uses the ℓ_0-norm for this purpose. We then keep the block structure fixed and update each column of $\mathbf{C}_\ell$ by seeking the best block-sparse representation using any of the algorithms for block-sparse recovery. Finally, we optimize the dictionary using an extension of K-SVD, referred to as *block K-SVD (BK-SVD)*. BK-SVD is similar to

---

Algorithm 12.5 Block-sparse dictionary design

Input: Measurement matrix $\mathbf{Y}$, block sparsity level k, maximal block size s
Output: Dictionary $\widehat{\mathbf{D}}$ and block structure $\hat{d}$
Initialize: $\widehat{\mathbf{D}}_0$ can be initialized by the outcome of K-SVD, $\mathbf{C}_0$ can be initialized by
    any standard sparse recovery method
while halting criterion false **do**
    $\ell \leftarrow \ell + 1$
    fix $\widehat{\mathbf{D}}_{\ell-1}$, $\mathbf{C}_{\ell-1}$ and update $\hat{d}_\ell$ by SAC
    fix $\hat{d}_\ell$ and for each $1 \leq i \leq L$ approximate the solution $\mathbf{c}_\ell$ of
        $\min_{\mathbf{c}_i} \|\mathbf{y}_i - \widehat{\mathbf{D}}_{\ell-1}\mathbf{c}_i\|^2$ s.t. $\|\mathbf{c}_i\|_{0,\mathcal{I}} \leq k$ {update block-sparse coefficients}
    update dictionary by applying BK-SVD {update dictionary}
end while
return $\widehat{\mathbf{D}} \leftarrow \widehat{\mathbf{D}}_\ell$

K-SVD where instead of updating one dictionary column at a time using the best rank-1 approximation to the corresponding error matrix, BK-SVD updates one block at a time using the best rank-r_i approximation, where r_i is the rank of block i. For a more detailed discussion see [381]. The main steps of this subspace learning approach are summarized in Algorithm 12.5, and are elaborated on next.

Sparse agglomerative clustering

To update the block structure d with $\mathbf{D}$ and $\mathbf{C}$ fixed, we wish to assign the columns of $\mathbf{D}$ to blocks so as to minimize the induced block sparsity of $\mathbf{C}$:

$$\min_d \sum_{i=1}^{L} \|\mathbf{c}_i\|_{0,d} \quad \text{s.t. } |d_j| \leq s, \tag{12.139}$$

where $d_j = \{i | d[i] = j\}$ is the set of indices belonging to block j (i.e., the list of columns of $\mathbf{D}$ in block j), and $|d_j|$ denotes the size of this set. The objective seeks to minimize the number of nonzero blocks, which encourages merging of active blocks subject to an upper bound on the maximal block size. SAC provides an approach to approximating (12.139) by merging the pair of blocks that have the most similar pattern of nonzeros in $\mathbf{C}$, leading to the steepest descent in the objective. We allow merging of blocks as long as the maximum block size s is not exceeded. More specifically, let $\omega_j(\mathbf{C}, d)$ denote the list of columns in $\mathbf{C}$ that have nonzero values in rows corresponding to block d_j, i.e., $\omega_j(\mathbf{C}, d) = \{i \in 1, \ldots, L | \|\mathbf{c}_i^{d_j}\|_2 > 0\}$ where $\mathbf{c}_i^{d_j}$ is the vector $\mathbf{c}_i$ restricted to the block indicated by d_j. Problem (12.139) can now be rewritten as

$$\min_d \sum_j |\omega_j(\mathbf{C}, d)| \quad \text{s.t. } |d_j| \leq s, \tag{12.140}$$

where $|\omega_j|$ denotes the size of the list $|\omega|$.

Next, using a suboptimal tractable agglomerative clustering algorithm, the objective in (12.140) is minimized. At each step the pair of blocks that have the most similar pattern of nonzeros in $\mathbf{C}$ are merged, as long as the maximum block size s is not exceeded.

While this aims at obtaining blocks of size s, some of the blocks could be smaller since, for example, two blocks of size $s - 1$ will not be further merged since their joint dimension exceeds s. In detail, at each step we seek the pair of blocks (j_1^*, j_2^*) such that

$$[j_1^*, j_2^*] = \arg \max_{j_1 \neq j_2} |\omega_{j_1} \cap \omega_{j_2}| \quad \text{s.t.} \ |d_{j_1}| + |d_{j_2}| \leq s. \tag{12.141}$$

We then merge j_1^* and j_2^* by setting $d[i] \leftarrow j_1, \omega_{j_1} \leftarrow \{\omega_{j_1} \cup \omega_{j_2}\}$ for all i in d_{j_2}, and $\omega_{j_2} \leftarrow \emptyset$. This is repeated until no blocks can be merged without breaking the constraint on the block size. We do not limit the intersection size for merging blocks from below, since merging is always beneficial. Merging blocks that have nothing in common may not reduce the objective of (12.139); however, this can still lower the representation error at the next BK-SVD iteration. Indeed, while the number of blocks stays fixed, the number of atoms that can be used to reduce the error increases.

Example 12.14 Table 12.2 illustrates the steps of SAC for a specific example of a matrix C with columns $c_i, 1 \leq i \leq 4$. Here s is the maximal block size set to $s = 2$ and d represents the unknown block structure. The goal is to choose d so as to minimize the block sparsity $\sum_{i=1}^{L} \|c_i\|_{0,d}$. This example is taken from [381].

The block structure is initialized to $d = [1, 2, 3, 4]$, which implies that the objective of (12.139) is $\sum_{i=1}^{L} \|c_i\|_{0,d} = 2 + 1 + 2 + 2 = 7$. This is just the sum of the number of nonzero values in each column. In the first iteration, ω_1 and ω_3, which represent the supports of rows 1 and 3, have the largest intersection. Consequently, blocks 1 and 3 are merged. In the second iteration, blocks 2 and 3 are merged. This results in the block structure $d = [1, 2, 1, 2]$ where no blocks can be merged without surpassing the maximal block size. The objective of (12.139) is reduced to $\sum_{i=1}^{L} \|c_i\|_{0,d} = 4$, since all four columns in C are 1-block sparse. Note that since every column contains nonzero values, this is the global minimum, and therefore the algorithm succeeded in solving (12.139).

Once the block assignments have been updated, we determine the corresponding C_ℓ by solving a block-sparse recovery problem on each column. We then turn to update the dictionary D assuming the block assignment of the previous step.

Block K-SVD

Inspired by the K-SVD algorithm, the blocks in D are updated sequentially, along with the corresponding nonzero coefficients in C. Denoting by $\Theta^*[j]$ the jth row-block of the matrix C, we have

$$DC = \sum_{i=1}^{m} D[i]\Theta^*[i], \tag{12.142}$$

where m is the number of blocks in D. The objective in (12.133) can then be written in terms of the jth block as

$$\|Y - DC\|_F^2 = \|E_j - D[j]\Theta^*[j]\|_2^2, \tag{12.143}$$

Table 12.2 A detailed example of the decision making process in SAC

$$
\mathbf{C}_{4\times4} =
\begin{bmatrix}
1.2 & 0 & 0 & -3.5 \\
0 & 0 & 0.2 & 0 \\
1.9 & 0 & 0 & 2.3 \\
0 & 4.7 & -0.8 & 0
\end{bmatrix}
$$
$$s = 2$$

$$\Downarrow$$

$$
d = \begin{bmatrix} 1 & 2 & 3 & 4 \end{bmatrix} \Rightarrow
\begin{cases}
\omega_1 = \{1, 4\} \\
\omega_2 = \{3\} \\
\omega_3 = \{1, 4\} \\
\omega_4 = \{2, 3\}
\end{cases}
\Rightarrow \sum_{i=1}^{4} \|\mathbf{c}_i\|_{0,d} = 7
$$
Step 1: $|\omega_1 \cap \omega_3| = 2 \Rightarrow d[3] = 1$

$$\Downarrow$$

$$
d = \begin{bmatrix} 1 & 2 & 1 & 4 \end{bmatrix} \Rightarrow
\begin{cases}
\omega_1 = \{1, 4\} \\
\omega_2 = \{3\} \\
\omega_4 = \{2, 3\}
\end{cases}
\Rightarrow \sum_{i=1}^{4} \|\mathbf{c}_i\|_{0,d} = 5
$$
Step 2: $|\omega_2 \cap \omega_4| = 1 \Rightarrow d[4] = 2$

$$\Downarrow$$

$$
d = \begin{bmatrix} 1 & 2 & 1 & 2 \end{bmatrix} \Rightarrow
\begin{cases}
\omega_1 = \{1, 4\} \\
\omega_2 = \{2, 3\}
\end{cases}
\Rightarrow \sum_{i=1}^{4} \|\mathbf{c}_i\|_{0,d} = 4
$$

with

$$
\mathbf{E}_j = \mathbf{Y} - \sum_{i=1,\ i\neq j}^{m} \mathbf{D}[i]\mathbf{\Theta}^*[i]. \tag{12.144}
$$

As in K-SVD, we proceed by keeping the support set of $\mathbf{\Theta}$ intact, but updating the values on the support. Let Λ be the set of indices corresponding to rows of $\mathbf{\Theta}$ that are not identically 0, and let $\mathbf{E}_\Lambda$ denote the matrix $\mathbf{E}_j$ restricted to this support set. Omitting the index j for convenience, the objective (12.143) can be written as

$$
\min_{\mathbf{D},\mathbf{\Theta}} \|\mathbf{E}_\Lambda - \mathbf{D}\mathbf{\Theta}_\Lambda^*\|_F^2. \tag{12.145}
$$

Let $r_i = |d_i|$ be the rank of $\mathbf{D}$. We can then interpret problem (12.145) as that of finding the best rank-r_i approximation to $\mathbf{E}_\Lambda$. The optimal solution is given by the outer product of the left and right singular vectors of $\mathbf{E}_\Lambda$ corresponding to the r_i largest singular values, weighted by the singular values (see Theorem A.4 in Appendix A). More specifically, let $\mathbf{E}_\Lambda = \mathbf{U}\mathbf{\Sigma}\mathbf{V}^*$. We then update $\mathbf{D}[j]$ and $\mathbf{\Theta}[j]$ by

$$D[j] = [u_1 \ u_2 \ \dots \ u_{r_i}]$$
$$\Theta[j] = [\sigma_1 v_1 \ \sigma_2 v_2 \ \dots \ \sigma_{r_i} v_{r_i}], \qquad (12.146)$$

and continue to the next block. Note that this update results in unitary blocks $D[j]$.

Example 12.15 In this example, we illustrate the advantage of the BK-SVD dictionary update step, compared with K-SVD.

Let d_1 and d_2 be the atoms in a block of size 2. Suppose that we use K-SVD where the first update of D is $d_1 \leftarrow u_1$ and $\Theta_1[j] \leftarrow \sigma_1 v_1^*$. A possible scenario is that in this specific iteration, $d_2 = u_1$ and $\Theta_2^*[j] = -\sigma_1 v_1^*$. In this case, the second update would leave d_2 and $\Theta_2^*[j]$ unchanged. As a consequence, only the highest-rank component of E_j is removed. In contrast, in the proposed BK-SVD algorithm, the atoms d_1 and d_2 are updated simultaneously, resulting in the two highest-rank components of E_j being removed.

12.7   Blind compressed sensing

Consider now the scenario in which we only have sampled data, projected onto a lower-dimensional subspace. Thus, instead of having access to the data matrix $Y = DC$, we are given samples $Z = ADC$ for some $n \times r$ measurement matrix A, where $n < r$. Our goal is to learn the dictionary D and the representation C given only the low-dimensional data matrix Z. This setup is referred to in the literature as blind CS (BCS) [389, 390]: the source of the name is the fact that to recover C we are facing a CS problem, where in addition the sparsifying dictionary is unknown. BCS combines elements from both CS and DL. On the one hand, as in CS and in contrast to DL, we obtain only low-dimensional measurements of the signals we are trying to recover. On the other hand, we do not require prior knowledge of the sparsity basis, and in this sense it is similar to the DL problem.

To simplify the discussion of BCS we focus on the standard sparsity setting, in which c_i is k-sparse. The ideas we develop here can easily be extended to the block-sparse setting following the steps of the previous section.

12.7.1   BCS problem formulation

Let Z be an $n \times L$ matrix of measurement vectors obtained as $Z = ADC$, where $Y = DC$ is an $r \times L$ matrix containing the original vectors to be recovered. The vectors y_i are all sparse under some unknown dictionary D, so that each column c_i of C is k-sparse. Our goal is to recover Y from Z, if possible.

Unfortunately, as we now show, the BCS problem is not unique for any choice of measurement matrix A, for any number of signals L, and for any sparsity level k. Note that as in DL, uniqueness here is defined up to an equivalence class involving permutations and rotations of the dictionary atoms. One way to guarantee uniqueness is to enforce constraints on the dictionary D [389]. Another possibility is to measure the signals Y

with different matrices $\mathbf{A}$ [390]; in other words, instead of using the same matrix $\mathbf{A}$ to sense all columns $\mathbf{y}_i$, we use different measurement matrices $\mathbf{A}_j$ for different groups of columns. In Section 12.7.2 we consider several different constraints on the dictionary which allow for BCS. The use of multiple measurement matrices is considered in Section 12.7.3.

BCS can be viewed as a DL problem with a dictionary $\mathbf{T} = \mathbf{AD}$. However, there is an important difference in the output of DL and BCS. DL provides the dictionary $\mathbf{T} = \mathbf{AD}$ and the sparse matrix $\mathbf{C}$. On the other hand, in BCS we are interested in recovering the unknown signals $\mathbf{Y} = \mathbf{DC}$. Therefore, after performing DL some postprocessing is needed to retrieve $\mathbf{D}$ from $\mathbf{T}$. This is an important distinction which makes it hard to directly apply DL algorithms.

In particular, even under conditions that guarantee uniqueness of DL, namely that a unique $\mathbf{T}$ can be found up to scaling and permutations, the BCS problem is still generally nonunique. To see this, note that in order to solve the BCS problem we need to recover $\mathbf{D}$ from $\mathbf{T} = \mathbf{AD}$. Since $\mathbf{A}$ has a null space, without further constraints on $\mathbf{D}$ there may be multiple solutions that result in the same $\mathbf{T}$. This is true even if we constrain $\mathbf{D}$ to be a unitary matrix, corresponding to an orthonormal sparsifying basis for $\mathbf{Y}$, as incorporated in the following proposition.

Proposition 12.7. *Let* $\mathbf{T}$ *and* $\mathbf{A}$ *be given matrices of size* $n \times r$ *with* $n < r$ *and assume that there exists a unitary* $\mathbf{D}$ *such that* $\mathbf{T} = \mathbf{AD}$. *Then there are multiple unitary matrices* $\widetilde{\mathbf{D}}$ *that solve the equation* $\mathbf{T} = \mathbf{A}\widetilde{\mathbf{D}}$.

Proof: Assume that $\mathbf{D}_1$ is a unitary matrix solving $\mathbf{T} = \mathbf{AD}$. Decompose $\mathbf{D}_1$ as $\mathbf{D}_1 = \mathbf{D}_{N^\perp} + \mathbf{D}_N$ where the columns of $\mathbf{D}_N$ are in the null space $\mathcal{N}(\mathbf{A})$, of $\mathbf{A}$, and those of $\mathbf{D}_{N^\perp}$ are in its orthogonal complement $\mathcal{N}(\mathbf{A})^\perp$. Note that necessarily $\mathbf{D}_N \neq 0$, otherwise the matrix $\mathbf{D}_1 = \mathbf{D}_{N^\perp}$ is in $\mathcal{N}(\mathbf{A})^\perp$ and has full rank. However, since the dimension of $\mathcal{N}(\mathbf{A})^\perp$ is at most $n < r$, it contains at most n linearly independent vectors. Therefore, there is no $r \times r$ full-rank matrix whose columns are all in $\mathcal{N}(\mathbf{A})^\perp$.

Next define the matrix $\mathbf{D}_2 = \mathbf{D}_{N^\perp} - \mathbf{D}_N \neq \mathbf{D}_1$. It is easy to see that $\mathbf{T} = \mathbf{AD}_2$. In addition, since the columns of $\mathbf{D}_N$ are orthogonal to those of $\mathbf{D}_{N^\perp}$,

$$\mathbf{D}_1^*\mathbf{D}_1 = \mathbf{D}_2^*\mathbf{D}_2 = \mathbf{D}_{N^\perp}^*\mathbf{D}_{N^\perp} + \mathbf{D}_N^*\mathbf{D}_N. \tag{12.147}$$

Since $\mathbf{D}_1^*\mathbf{D}_1 = \mathbf{I}$, it follows that we also have $\mathbf{D}_2^*\mathbf{D}_2 = \mathbf{I}$ so that $\mathbf{D}_2$ is unitary as well. Alternative solutions can be created similarly, for example by changing the signs of only a part of the columns of $\mathbf{D}_N$. $\qquad\square$

Proposition 12.7 shows that the BCS problem does not generally have a unique solution. Two ways to ensure uniqueness are to impose constraints on the basis $\mathbf{D}$, or to allow for multiple measurement matrices $\mathbf{A}_j$.

12.7.2   BCS with a constrained dictionary

We begin by discussing BCS with constraints on $\mathbf{D}$. Although there are many possible restrictions one can impose, we focus below on each of the following:

1. $\mathbf{D}$ is one of a finite and known set of bases.

2. $\mathbf{D}$ is sparse under some known dictionary.
3. $\mathbf{D}$ is unitary and has a block diagonal structure.

It is shown in [389] that under these constraints and appropriate conditions on $\mathbf{A}$, the BCS problem has a unique solution. In particular, random Gaussian matrices $\mathbf{A}$ satisfy these conditions with high probability. We refer the interested reader to the original derivations in order to find the exact statement of the results. Here, we focus on algorithms to retrieve the solutions assuming the conditions hold.

Finite set of bases

Over the years a variety of bases, such as wavelet [59] and DCT [204], have been proven to lead to sparse representations of many natural signals. These bases have fast implementations and are known to fit many types of signals. Therefore, when the dictionary is unknown, it is natural to try one of these choices. Motivated by this intuition, we constrain $\mathbf{D}$ to a finite and known set of dictionaries $\{\mathbf{D}^i\}$.

To recover $\{\mathbf{D}^i\}$, we consider the subproblem of finding the sparsest $\mathbf{C}^i$ that fits the data $\mathbf{Z} = \mathbf{A}\mathbf{D}^i\mathbf{C}^i$. Then, for each data vector $\mathbf{z}_j$, we choose the dictionary $\mathbf{D}^i$ that resulted in the sparsest solution. Note that for each value j we may obtain a different dictionary $\mathbf{D}^i$. We select the final dictionary to be the one that is chosen by the largest number of signals. When the data is noisy we can replace the first step by selecting, for each $\mathbf{z}_j$, the basis that leads to the smallest representation error. For example, we may first solve

$$\min_{\mathbf{c}} \left\{ \|\mathbf{z}_j - \mathbf{A}\mathbf{D}^i\mathbf{c}\|_2 + \lambda\|\mathbf{c}\|_1 \right\} \tag{12.148}$$

for some parameter λ, and each value i, and then choose the value of i for which $\|\mathbf{z}_j - \mathbf{A}\mathbf{D}^i\mathbf{c}\|$ is minimized.

Example 12.16 We demonstrate the performance of BCS when $\mathbf{D}$ is constrained to a finite and known set of dictionaries $\{\mathbf{D}^i\}$. The measurement matrix $\mathbf{A}$ is chosen as an iid Gaussian matrix of size 32×64. We selected five bases $\{\mathbf{D}^i\}$ of size 64×64: the identity, DCT, Haar wavelet, and two random iid Gaussian dictionaries. The dictionaries $\mathbf{D}^i$ are scaled so that $\mathbf{A}\mathbf{D}^i$ has normalized columns. We created 500 signals of length 64 by generating random sparse vectors and multiplying them by the DCT basis, which we denote by $\mathbf{\Psi}$. Each sparse vector $\mathbf{c}$ contained six nonzero elements in uniformly random locations, and values from an iid Gaussian distribution.

To test the BCS algorithm we generated noisy measurements $\mathbf{z}_i = \mathbf{A}\mathbf{\Psi}\mathbf{c}_i + \mathbf{w}_i$, where $\mathbf{w}$ is iid Gaussian noise with variance set to match varying SNR values from 30 dB to 5 dB. For each noise level, we implemented the BCS method with a finite set of bases using OMP as the CS algorithm. Table 12.3 summarizes the results. The misdetected column in the table contains the percentage of signals that indicated a false basis. The average error is the average value of E_i where

$$E_i = \frac{\|\mathbf{z}_i - \hat{\mathbf{z}}_i\|_2}{\|\mathbf{z}_i\|_2} \cdot 100. \tag{12.149}$$

Table 12.3 Simulation results for BCS with a finite set of bases

SNR (dB)	Misdetected %	Average error %
∞	0	5
30	0	9
25	0	13
20	0	20
15	3	31
10	15	47
5	53	64

Here z_i and $\hat{z}_i$ are the columns of the real signal matrix $\mathbf{Z}$ and the reconstructed signal matrix $\widehat{\mathbf{Z}}$ respectively. The average is performed only over the signals that indicated the correct basis. For all noise levels, selecting the basis according to the majority was correct.

As can be seen from Table 12.3, the error grows with the noise level. For high SNR there are no false reconstructions, but as the SNR is decreased beyond 15dB the percentage of false selections increased. In these cases, it is important to use several signals for recovery, so that even if some of the signals indicate a wrong dictionary, there will be enough signals that lead to correct detection.

Sparse basis

A different constraint that may be added to the BCS problem is sparsity of $\mathbf{D}$. That is, the columns of $\mathbf{D}$ are assumed to be sparse under some known dictionary $\boldsymbol{\Phi}$, so that there exists an unknown sparse matrix $\mathbf{W}$ such that $\mathbf{D} = \boldsymbol{\Phi}\mathbf{W}$. We denote the sparsity of the columns of $\mathbf{W}$ by k_d.

The motivation for using a sparse dictionary is to overcome the disadvantage of the previously discussed constraint in which the dictionaries are fixed. Specifically, we may choose a dictionary $\boldsymbol{\Phi}$ with a fast implementation, but enhance its adaptivity to different signals by allowing any sparse enough combination of the columns of $\boldsymbol{\Phi}$. Furthermore, to combine the benefits of this approach with the previous method, we can solve the problem separately for several different dictionaries $\boldsymbol{\Phi}$, and choose the best solution.

The resulting BCS problem is restated as follows. Given measurements $\mathbf{Z}$, the measurement matrix $\mathbf{A}$ and a dictionary $\boldsymbol{\Phi}$, which we assume has full row-rank, find a signal matrix $\mathbf{Y}$ such that $\mathbf{Z} = \mathbf{A}\mathbf{Y}$ where $\mathbf{Y} = \boldsymbol{\Phi}\mathbf{W}\mathbf{C}$ for some k-sparse matrix $\mathbf{C}$ and k_d-sparse and full column-rank matrix $\mathbf{W}$.

Under appropriate conditions this problem has a unique solution even when there is only one signal. Therefore, instead of matrices $\mathbf{Y}$, $\mathbf{C}$, $\mathbf{Z}$ we treat vectors $\mathbf{y}$, $\mathbf{c}$, $\mathbf{z}$ respectively. Since $\|\mathbf{c}\|_0 \leq k$ and $\mathbf{W}$ is k_d-sparse, the vector $\mathbf{b} = \mathbf{W}\mathbf{c}$ necessarily satisfies $\|\mathbf{b}\|_0 \leq k_d k$. Therefore, our problem can be written as:

$$\hat{\mathbf{b}} = \arg\min_{\mathbf{b}} \|\mathbf{b}\|_0 \quad \text{s.t. } \mathbf{z} = \mathbf{A}\boldsymbol{\Phi}\mathbf{b}, \tag{12.150}$$

or,

$$\hat{\mathbf{b}} = \arg \min_{\mathbf{b}} \|\mathbf{z} - \mathbf{A}\boldsymbol{\Phi}\mathbf{b}\|_2^2 \quad \text{s.t.} \quad \|\mathbf{b}\|_0 \leq k_d k, \tag{12.151}$$

where the recovery is $\mathbf{y} = \boldsymbol{\Phi}\hat{\mathbf{b}}$. The solutions to (12.150) and (12.151) are unique if $\sigma(\mathbf{A}\boldsymbol{\Phi}) > 2k_d k$ where $\sigma(\mathbf{A})$ is the spark of $\mathbf{A}$. If there is more than one signal, then we solve (12.150) and (12.151) for each signal separately. These two optimization problems can be approximated by standard CS algorithms.

An alternative approach to solve this BCS problem is to use the sparse K-SVD algorithm [391], which is an extension of K-SVD to the case in which one seeks a sparse dictionary. That is, given measurements $\mathbf{Z}$ and a fixed dictionary $\mathbf{D}$, sparse K-SVD finds a k_d-sparse $\mathbf{W}$ and k-sparse $\mathbf{C}$, such that $\mathbf{Z} = \mathbf{DWC}$. Here, when we say a matrix is k-sparse we mean that each of its columns is k-sparse. In our setting we can apply sparse K-SVD to $\mathbf{Z}$ with $\mathbf{D} = \mathbf{A}\boldsymbol{\Phi}$ in order to find $\mathbf{W}$ and $\mathbf{C}$, and then recover the signals by $\mathbf{Y} = \boldsymbol{\Phi}\mathbf{WC}$.

As in any DL algorithm, for sparse K-SVD to perform well, it requires many diverse signals. In contrast, the uniqueness conditions developed in [389] for the sparse BCS problem do not require multiple signals. Furthermore, the sparse K-SVD algorithm is much more computationally demanding than the direct method (i.e. using a standard CS algorithm). Nonetheless, sparse K-SVD has an advantage when $k_d k$ is large relative to n. Specifically, in order for the direct method to succeed, the value of $k_d k$ must be small. On the other hand, sparse K-SVD attempts to separately find a k-sparse $\mathbf{C}$ and k_d-sparse $\mathbf{W}$, and therefore requires each of k and k_d to be small, rather than their product $k_d k$. Therefore, we conclude that when there are few signals and $k_d k$ is small, the direct method is preferable. However, if $k_d k$ is large, but still satisfies $\sigma(\mathbf{A}\boldsymbol{\Phi}) > 2k_d k$, and there are enough diverse signals, then sparse K-SVD is beneficial.

Example 12.17 We now present simulation results for the direct method. Results for sparse K-SVD can be found in [392].

First, we investigate the influence of the sparsity level of the basis. We generated a random sparse matrix $\mathbf{W}$ of size 256×256 with up to $k_d = 6$ nonzero elements in each column. The value of k (i.e. the number of nonzero elements in $\mathbf{C}$), was gradually increased from 1 to 20. For each k we generated $\mathbf{C}$ as a random k-sparse matrix of size 256×100, and created the signal matrix $\mathbf{Y} = \boldsymbol{\Phi}\mathbf{WC}$, where $\boldsymbol{\Phi}$ was chosen as the DCT basis. We measured $\mathbf{Y}$ using a random Gaussian matrix $\mathbf{A}$ of size 128×256, resulting in $\mathbf{Z} = \mathbf{AY}$. We solved BCS given $\mathbf{A}$ and $\mathbf{Z}$ using the direct method (12.150), where again OMP was used as the CS algorithm. For comparison we also performed OMP with the real basis (i.e. $\mathbf{D} = \boldsymbol{\Phi}\mathbf{W}$, which is unknown in practice). Figure 12.15 summarizes the results. For every value of k, the error is an average over all signals, calculated as in (12.149). Both errors are similar for $k \leq 8$, but for larger k, the error of BCS is much higher. Therefore, as long as the signal is sufficiently sparse, we can learn the basis along with the signal representation.

Since $\mathbf{A}$ is an iid Gaussian matrix and the DCT matrix is orthogonal, $\sigma(\mathbf{A}\boldsymbol{\Phi}) = 129$ with probability 1. Therefore, sparse BCS is unique (with probability 1) for

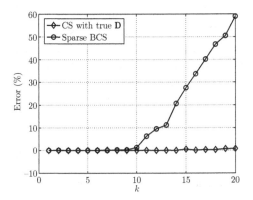

Figure 12.15 Reconstruction error using the direct method as a function of the sparsity level for BCS and standard CS with known dictionary.

$k_d k \leq 64$, or $k \leq 10$. The error began to grow before this sparsity level because OMP is a suboptimal algorithm that is not guaranteed to find the solution even when it is unique, but works well on sparse enough signals. The reconstruction error of the OMP which used the true $\mathbf{D}$ grows much less rapidly for the same values of k. This is because when $\mathbf{D}$ is known, k itself, instead of $k_d k$, needs to be small relative to n for OMP to succeed.

Sparse K-SVD can improve the results for high values of k, assuming of course it is small enough for the solution to be unique. However, in this simulation the number of signals is even less than the length of the vectors, and sparse K-SVD does not work well with such a small number of signals. In the sparse K-SVD simulations presented in [392], the number of signals is at least 100 times the signal length.

Structural constraint

The final constraint we discuss on $\mathbf{D}$ is that of a block diagonal structure. This constraint is motivated by multichannel systems, where the signals from each channel are sparse under separate bases. In such systems we can construct the set of signals $\mathbf{Z}$ by concatenating signals from several different channels. The sparsity dictionary is then block diagonal, where the number of blocks equals the number of channels, and each block is the sparsity dictionary of the corresponding channel.

For example, in microphone arrays or antenna arrays, we can divide the samples from each microphone/antenna into time intervals in order to obtain the ensemble of signals $\mathbf{Y}$. Each column of $\mathbf{Y}$ is a concatenation of the signals from all the microphones/antennas over the same time interval. Alternatively, consider large images that are divided into patches such that each patch is sparse under a different basis. In this case, every column of $\mathbf{Y}$ is a concatenation of the patches in the same locations in different images. This partition into patches is used, for example, in JPEG compression [199].

The advantage of the block structure of $\mathbf{D}$ is that with the right choice of $\mathbf{A}$ the problem can be decomposed into a set of separate simple problems, as we now show.

Furthermore, there is a unique solution to the resulting BCS problem, as long as the signals are sufficiently diverse, and $\mathbf{A}$ satisfies certain conditions detailed in [389]. An example is when $\mathbf{A}$ consists of orthonormal blocks, an assumption we will make below.

Consider the case in which $\mathbf{D}$ consists of orthonormal blocks, so that

$$\mathbf{D} = \begin{bmatrix} \mathbf{D}^1 & \mathbf{0} & \cdots & \mathbf{0} \\ \mathbf{0} & \mathbf{D}^2 & \cdots & \mathbf{0} \\ & & \ddots & \\ \mathbf{0} & \cdots & \mathbf{0} & \mathbf{D}^R \end{bmatrix}, \tag{12.152}$$

where each $\mathbf{D}^i$ is a matrix of size $d \times d$. Corresponding to the structure of $\mathbf{D}$, we represent $\mathbf{A}$ as

$$\mathbf{A} = \begin{bmatrix} \mathbf{A}^1 & \mathbf{A}^2 & \cdots & \mathbf{A}^R \end{bmatrix}, \tag{12.153}$$

where each $\mathbf{A}^i$ is of size $d \times d$. We assume that these blocks consist of orthonormal columns so that $(\mathbf{A}^i)^*\mathbf{A}^i = \mathbf{I}$.

To learn the dictionary and the sparse representation in this setting, we consider the *orthonormal block diagonal BCS (OBD-BCS)* algorithm, proposed in [389]. Given $\mathbf{Z}$, we aim to solve

$$\min_{\mathbf{D},\mathbf{C}} \|\mathbf{Z} - \mathbf{A}\mathbf{D}\mathbf{C}\|_F^2$$

s.t. $\mathbf{C}$ is k-sparse and $\mathbf{D}$ consists of unitary blocks. $\tag{12.154}$

This solution is approximated by alternating minimization. In the first step, $\mathbf{D}$ is fixed and $\mathbf{C}$ is updated using sparse approximation. The second step updates the basis $\mathbf{D}$ while $\mathbf{C}$ is held fixed.

To update $\mathbf{C}$, we note that the objective (12.154) with $\mathbf{D}$ fixed is separable in the columns of $\mathbf{C}$. For each column $\mathbf{z}$ of $\mathbf{Z}$ and $\mathbf{c}$ of $\mathbf{C}$, (12.154) becomes

$$\min_{\mathbf{c}} \|\mathbf{z} - \mathbf{A}\mathbf{D}\mathbf{c}\|_2^2 \quad \text{s.t. } \|\mathbf{c}\|_0 \le k, \tag{12.155}$$

which can be solved using standard CS algorithms. In the next step, $\mathbf{C}$ is fixed and we update $\mathbf{D}$ by exploiting its block structure. Divide the $r \times L$ matrix $\mathbf{C}$, with $r = Rd$, into R submatrices of size $d \times L$ such that:

$$\mathbf{C} = \begin{bmatrix} \mathbf{C}^1 \\ \vdots \\ \mathbf{C}^R \end{bmatrix}. \tag{12.156}$$

Then (12.154) with fixed $\mathbf{C}$ becomes:

$$\min_{\mathbf{D}^1,\ldots,\mathbf{D}^R} \left\| \mathbf{Z} - \sum_{j=1}^R \mathbf{A}^j \mathbf{D}^j \mathbf{C}^j \right\|_F^2 \quad \text{s.t. } (\mathbf{D}^i)^*\mathbf{D}^i = \mathbf{I}. \tag{12.157}$$

Algorithm 12.6 OBD-BCS algorithm

Input: Measurement matrix $\mathbf{Z}$, sparsity level k

Output: Dictionary $\widehat{\mathbf{D}}$

Initialize: $\widehat{\mathbf{D}}_0 = \mathbf{I}$, $\ell = 0$

while halting criterion false **do**

   $\ell \leftarrow \ell + 1$

   for each $1 \leq i \leq L$ approximate the solution $\mathbf{c}_\ell$ of
   $$\min_{\mathbf{c}_i} \|\mathbf{z}_i - \mathbf{A}\widehat{\mathbf{D}}_{\ell-1}\mathbf{c}_i\|^2 \quad \text{s.t.} \quad \|\mathbf{c}_i\|_0 \leq k \quad \{\text{update sparse coefficients}\}$$
   $\{\text{update dictionary}\}$

   update each block $\mathbf{D}^i$ of $\mathbf{D}_{\ell-1}$ by:

      Calculate $\mathbf{Z}^i = \mathbf{Z} - \sum_{j \neq i} \mathbf{A}^j \mathbf{D}^j \mathbf{C}^j$

      Compute the SVD $\mathbf{C}^i(\mathbf{Z}^i)^* \mathbf{A}^i = \mathbf{U}\mathbf{\Sigma}\mathbf{V}^*$

      Update $\mathbf{D}^i = \mathbf{V}\mathbf{U}^*$

end while

return $\widehat{\mathbf{D}} \leftarrow \widehat{\mathbf{D}}_\ell$

To minimize (12.157), we iteratively fix all the blocks $\mathbf{D}^j$ except one, denoted by $\mathbf{D}^i$, and solve

$$\min_{\mathbf{D}^i} \left\|\mathbf{Z}^i - \mathbf{A}^i\mathbf{D}^i\mathbf{C}^i\right\|_F^2 \quad \text{s.t.} \quad (\mathbf{D}^i)^*\mathbf{D}^i = \mathbf{I}, \tag{12.158}$$

where $\mathbf{Z}^i = \mathbf{Z} - \sum_{j \neq i} \mathbf{A}^j \mathbf{D}^j \mathbf{C}^j$. Since $(\mathbf{D}^i)^*\mathbf{D}^i = (\mathbf{A}^i)^*\mathbf{A}^i = \mathbf{I}$, for any block i, $\|\mathbf{A}^i\mathbf{D}^i\mathbf{C}^i\|_F^2 = \|\mathbf{C}^i\|_F^2$. Abandoning the index i, ignoring constant terms, and using the fact that $\|\mathbf{A}\|_F^2 = \mathrm{Tr}(\mathbf{A}^*\mathbf{A})$ for every matrix $\mathbf{A}$, (12.158) reduces to

$$\max_{\mathbf{D}} \Re\{\mathrm{Tr}(\mathbf{Z}^*\mathbf{A}\mathbf{D}\mathbf{C})\} \quad \text{s.t.} \quad \mathbf{D}^*\mathbf{D} = \mathbf{I}. \tag{12.159}$$

Let the SVD of the matrix $\mathbf{R} = \mathbf{C}\mathbf{Z}^*\mathbf{A}$ be $\mathbf{R} = \mathbf{U}\mathbf{\Sigma}\mathbf{V}^*$, where $\mathbf{U}, \mathbf{V}$ are unitary matrices and $\mathbf{\Sigma}$ is a diagonal matrix. Using the fact that $\mathrm{Tr}(\mathbf{X}\mathbf{Y}\mathbf{Z}) = \mathrm{Tr}(\mathbf{Z}\mathbf{X}\mathbf{Y})$, we can manipulate the trace in (12.159) as follows:

$$\mathrm{Tr}(\mathbf{Z}^*\mathbf{A}\mathbf{D}\mathbf{C}) = \mathrm{Tr}(\mathbf{D}\mathbf{R}) = \mathrm{Tr}(\mathbf{\Sigma}\mathbf{V}^*\mathbf{D}\mathbf{U}) = \mathrm{Tr}(\mathbf{\Sigma}\mathbf{Q}) = \sum_i \sigma_i q_{ii}, \tag{12.160}$$

where $\mathbf{Q} = \mathbf{V}^*\mathbf{D}\mathbf{U}$ and q_{ii} denotes the ith diagonal element of $\mathbf{Q}$. Noting that $\mathbf{V}, \mathbf{U}, \mathbf{D}$ are all unitary matrices, $\mathbf{Q}^*\mathbf{Q} = \mathbf{I}$, and consequently $|q_{ii}| \leq 1$ with equality for all i if and only if $\mathbf{Q} = \mathbf{I}$. Therefore, $\sigma_i \Re\{q_{ii}\} \leq \sigma_i |q_{ii}| \leq \sigma_i$, with equality when $\mathbf{Q} = \mathbf{I}$. Thus, $\mathbf{D} = \mathbf{V}\mathbf{U}^*$ achieves a minimum of (12.158). Note that when σ_i is zero for some value i, $\mathbf{Q} = \mathbf{I}$ is not the only optimum; nonetheless this choice still achieves a minimum even if it is not unique.

The resulting OBD-BCS algorithm is summarized in Algorithm 12.6. Note that the initiation can be any block diagonal matrix, not necessarily the identity. Each iteration of OBD-BCS employs a standard CS algorithm and R SVDs.

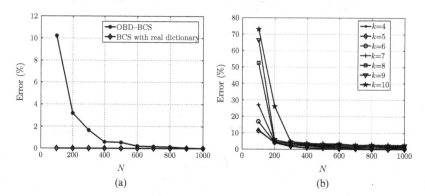

Figure 12.16 Reconstruction error of OBD-BCS as a function of the number of signals. (a) For sparsity level $k = 4$. (b) For different sparsity levels.

Example 12.18 We evaluate the performance of the OBD-BCS algorithm using synthetic data.

Consider a signal matrix $\mathbf{Z}$ constructed as $\mathbf{Z} = \mathbf{ADC}$ where $\mathbf{C}$ is a random sparse matrix with the nonzero elements chosen as iid Gaussian, and $\mathbf{D}$ is an orthogonal four-block diagonal matrix consisting of the identity matrix, Haar wavelet, DCT, and a random matrix generated from a Gaussian distribution followed by Gram–Schmidt orthogonalization. The measurement matrix $\mathbf{A}$ is a concatenation of two random 64×64 orthogonal matrices, generated from a Gaussian distribution followed by Gram–Schmidt orthogonalization. The algorithm terminated when the change in the matrices $\mathbf{C}$ and $\mathbf{D}$ was small, which typically occurred after about 30 iterations.

The number of signals N and the sparsity level k were gradually changed in order to investigate their influence. For each value of N from 150 to 1000, the error is averaged over 20 simulations of the OBD-BCS algorithm. In each simulation the sparse vectors and the orthogonal matrix where generated independently, but the measurement matrix was not changed. The error of each signal was calculated according to (12.149). For comparison, in Fig. 12.16(a), we plot the average error of OMP using the true basis $\mathbf{D}$, which is unknown in practice. As expected, the results of OMP are independent of the number of signals, since it is performed separately on each signal. The average error of OMP is 0.08%, indicating that the algorithm fails for a small number of signals.

In order to examine the influence of k we performed the same experiment as before but for different values of $k \leq 10$. The results are presented in Fig. 12.16(b). It can be seen that for all values of k the graph has the same basic shape: the error decreases with N until a critical value, after which the error is almost constant. As k grows this critical N increases and so does the value of the constant error. The graphs for $k = 1$, $k = 2$, $k = 3$ follow the same pattern; they are not plotted since they are not visible on the same scale as the rest.

12.7.3 BCS with multiple measurement matrices

An alternative way to ensure uniqueness of BCS without imposing constraints on the dictionary is to use multiple measurement matrices, as suggested in [390]. In this setting, the compressed measurements become

$$\mathbf{z}_i = \mathbf{A}_i \mathbf{D} \mathbf{c}_i, \quad 1 \leq i \leq L, \tag{12.161}$$

where $\mathbf{A}_i$ are a set of measurement matrices. If these matrices are sufficiently diverse, then recovery of $\mathbf{D}$ and $\mathbf{c}_i$ is possible. Exact conditions are developed in [390].

One application area in which (12.161) naturally arises is in image inpainting, where we observe an incomplete image, i.e. we only know the intensity values at a subset of pixel locations. Additionally, the image is processed in (often overlapping) patches, which can be converted to vectors $\mathbf{y}_i$. In each patch, only n_i pixels are observed, selected at random. Therefore, the locations of the missing pixels are in general different for each patch. This corresponds to $\mathbf{A}_i$ being chosen as random rows of the identity matrix. We further assume that the image patches are all sparse under a shared dictionary $\mathbf{D}$, which is learned together with image interpolation.

BCS with sparse vectors

To recover $\mathbf{D}$ and $\mathbf{c}_i$ in this case we can use an algorithm that is a combination of MOD (Algorithm 12.3) and the K-SVD (Algorithm 12.4). Specifically, our goal is to solve

$$\min_{\mathbf{D}, \mathbf{C}} \sum_{i=1}^{L} \|\mathbf{z}_i - \mathbf{A}_i \mathbf{D} \mathbf{c}_i\|^2. \tag{12.162}$$

We begin with a coefficient update stage in which $\mathbf{D}$ is held fixed and we approximate the solution to

$$\min_{\mathbf{c}_i} \|\mathbf{z}_i - \mathbf{A}_i \mathbf{D} \mathbf{c}_i\|^2 \quad \text{s.t.} \ \|\mathbf{c}_i\|_0 \leq k, \tag{12.163}$$

using any standard CS algorithm. Once the optimal $\mathbf{c}_i$ are found, just as in K-SVD, we keep the sparsity pattern, namely, the indices of the nonzero values. We then update $\mathbf{D}$ one column at a time together with the coefficients.

Consider the jth update. We first look for all vectors $\mathbf{c}_i$ whose support includes the jth element, and define by Λ the corresponding index set. Denoting by $\mathbf{c}_i(k)$ the kth entry of $\mathbf{c}_i$, and by $\mathbf{d}_k$ the kth column of $\mathbf{D}$, we next define the error vector

$$\mathbf{e}_i = \mathbf{z}_i - \mathbf{A}_i \sum_{k \neq j} \mathbf{d}_k \mathbf{c}_i(k). \tag{12.164}$$

Our problem then becomes

$$\min_{\mathbf{d}} \sum_{i \in \Lambda} \|\mathbf{e}_i - \mathbf{c}_i(j) \mathbf{A}_i \mathbf{d}\|^2. \tag{12.165}$$

A solution to (12.165) is given by the least squares update

$$\mathbf{d} = \left(\sum_{i \in \Lambda} |\mathbf{c}_i(j)|^2 \mathbf{A}_i^* \mathbf{A}_i \right)^{\dagger} \sum_{i \in \Lambda} (\mathbf{c}_i(j) \mathbf{A}_i)^* \mathbf{e}_i. \tag{12.166}$$

Algorithm 12.7 BCS with multiple measurement matrices

Input: CS matrices $\mathbf{A}_i$, measurement matrix $\mathbf{Z}$, sparsity level k
Output: Dictionary $\widehat{\mathbf{D}}$
Initialize: $\widehat{\mathbf{D}}_0$ can be initialized by a random matrix, or a random sample of $\mathbf{Z}$, $\ell = 0$
while halting criterion false **do**
$\quad \ell \leftarrow \ell + 1$
$\quad$ for each $1 \leq i \leq L$ approximate the solution $\mathbf{c}_\ell$ of
$\qquad \min_{\mathbf{c}_i} \|\mathbf{z}_i - \mathbf{A}_i \widehat{\mathbf{D}}_{\ell-1} \mathbf{c}_i\|^2$ s.t. $\|\mathbf{c}_i\|_0 \leq k$ {update sparse coefficients}
$\quad$ {update dictionary}
$\quad$ update each column $\mathbf{d}_j$ of $\mathbf{D}_{\ell-1}$ by:
$\qquad$ Let Λ be the indices i for which $\mathbf{c}_i(j) \neq 0$
$\qquad$ Update $\mathbf{d}_j$ according to (12.166) where $\mathbf{e}_i$ is defined by (12.164)
$\qquad$ Update $\mathbf{c}_i(j)$ for $i \in \Lambda$ according to (12.167)
end while
return $\widehat{\mathbf{D}} \leftarrow \widehat{\mathbf{D}}_\ell$

Once the jth column is updated, we modify the corresponding coefficients $\mathbf{c}_i(j)$ for $i \in \Lambda$ by minimizing the objective in (12.165), which leads to

$$\mathbf{c}_i(j) = \frac{1}{\mathbf{d}^*\mathbf{A}_i^*\mathbf{A}_i\mathbf{d}} \mathbf{d}^* \mathbf{A}_i^* \mathbf{e}_i. \tag{12.167}$$

In a similar fashion we update all columns of $\mathbf{D}$ and the corresponding coefficients $\mathbf{c}_i$. The resulting algorithm is summarized in Algorithm 12.7.

BCS with block-sparse vectors

Algorithm 12.7 is easily extended to include block-sparse vectors. In this case, we first determine the block assignment using SAC, just as in Algorithm 12.5. Step 1 is then replaced by a block-sparse recovery algorithm. In step 2, instead of cycling through the columns of $\mathbf{D}$, we treat blocks of $\mathbf{D}$.

Consider updating the jth block $\mathbf{D}[j]$. We first look for all vectors $\mathbf{c}_i$ whose support includes the jth block, and denote the corresponding index set by Λ. We next define the error vector

$$\mathbf{e}_i = \mathbf{z}_i - \mathbf{A}_i \sum_{k \neq j} \mathbf{D}[k]\mathbf{c}_i[k]. \tag{12.168}$$

Our problem then becomes

$$\min_{\mathbf{D}[j]} \sum_{i \in \Lambda} \|\mathbf{e}_i - \mathbf{A}_i\mathbf{D}[j]\mathbf{c}_i[j]\|^2. \tag{12.169}$$

Using the relation

$$\mathbf{A}_i\mathbf{D}[j]\mathbf{c}_i[j] = (\mathbf{c}_i^*[j] \otimes \mathbf{A}_i) \, \text{vec}(\mathbf{D}[j]), \tag{12.170}$$

we can express (12.169) as a standard least squares problem with the variable $\mathbf{d} = \text{vec}(\mathbf{D}[j])$:

$$\min_{\mathbf{d}} \sum_{i \in \Lambda} \|\mathbf{e}_i - (\mathbf{c}_i^*[j] \otimes \mathbf{A}_i)\mathbf{d}\|^2 . \tag{12.171}$$

(See Appendix A for some properties of the Kronecker operator.) A solution to (12.171) is given by

$$\mathbf{d} = \left(\sum_{i \in \Lambda} \mathbf{G}_i^* \mathbf{G}_i\right)^{\dagger} \sum_{i \in \Lambda} \mathbf{G}_i^* \mathbf{e}_i, \tag{12.172}$$

where we denoted $\mathbf{G}_i = \mathbf{c}_i^*[j] \otimes \mathbf{A}_i$. The corresponding block $\mathbf{D}[j]$ is obtained by rearranging the vector $\mathbf{d}$ into appropriate matrix form. The coefficient update is then given by

$$\mathbf{c}_i[j] = (\mathbf{D}^*[j]\mathbf{A}_i^*\mathbf{A}_i\mathbf{D}[j])^{\dagger}\mathbf{D}^*[j]\mathbf{A}_i^*\mathbf{e}_i. \tag{12.173}$$

Example 12.19 We demonstrate the BCS algorithm with multiple measurement matrices using an example from [390]. This example consists of inpainting the well-known "Barbara" (512×512 pixels) and "House" (256×256 pixels) images, when only 50% of the pixels are observed. The algorithm used in this simulation is the extension of Algorithm 12.7 to the block-sparse model. The images are processed in 8×8 overlapping patches, treated as vectors of dimension $n = 64$.

The original images are presented in Figs. 12.17(a) and 12.18(a). In Figs. 12.17(b) and 12.18(b), we show the test versions, with only 50% of the pixel values observed (selected uniformly at random). In all experiments, the total number of dictionary elements is set to $r = 256$. The inpainting results achieved by BCS using maximum block size of $k = 8$ are depicted in Figs. 12.17(c) and 12.18(c). The peak SNRs for "Barbara" and "House" are 27.93 and 31.80 dB, respectively.

To conclude, we have shown in this chapter that finite union of subspaces can be used to describe a wide variety of signal models. We also demonstrated that recovery under this model is often possible using simple methods that generalize the

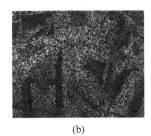

| (c) | (b) | (a) |

Figure 12.17 (a) Original 512×512 Barbara image. (b) Test version with 50% observed pixel values (the remainder are removed). (c) Inpainted 512×512 Barbara image. The peak SNR of this estimate is 27.93 dB.

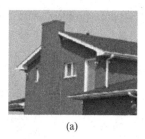 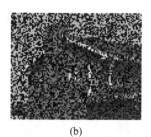 

(a)   (b)   (c)

Figure 12.18 (a) Original 256×256 House image. (b) Test version with 50% observed pixel values (the remainder are removed). (c) Inpainted 256×256 House image. The peak SNR of this estimate is 31.80 dB.

recovery techniques of sparse vectors. Furthermore, when the underlying subspaces are not known a priori, they can be learned directly from training data, or from appropriately designed measured data. The latter problem, referred to as blind CS, may be viewed as CS without prior knowledge of the sparsity basis of the signals.

12.8   Exercises

1. Consider a signal $x(t)$ defined on $[0, 1]$. On this interval $x(t)$ can be one of three options:

   (i) A polynomial of degree 4;

   (ii) A sinusoid with known frequency ω_0;

   (iii) A sum of two pulses of the form $x(t) = a_0 h_0(t) + a_1 h_1(t)$ where $h_0(t), h_1(t)$ are known but the values of a_0, a_1 are arbitrary.

   a. Show that $x(t)$ lies in a union of subspaces and define the individual subspaces. Indicate the dimension of each subspace.

   b. Show how to represent $x(t)$ in terms of a block-sparse vector.

2. Let $Y = AX$ be an MMV system, where X is assumed to have at most k nonzero rows.

   a. Define vectors y and x and a matrix D such that this problem can be transformed into that of recovering a k-block sparse vector x from measurements y. Indicate the dimensions of y and D.

   b. Consider the stability condition (12.25) for the resulting D. Express this condition directly in terms of A.

   c. Relate the block RIP constant of D to the RIP constant of A.

3. Consider the following matrix, separated into three blocks of two columns each:

$$D = \begin{bmatrix} 1 & 1 & 0 & 1 & 1 & 1 \\ 0 & 3 & -1 & 0 & 0 & 3 \\ 1 & 4 & 1 & -1 & 0 & 1 \\ 0 & 1 & 0 & 0 & -1 & 2 \end{bmatrix}. \tag{12.174}$$

   a. Compute the block RIP of D for block sparsity level $k = 1$.

b. Which block must be chosen to attain the result in part (a)?

c. Using brute-force calculations, find the smallest value of δ_2 satisfying the standard RIP (12.26).

d. Which two elements must be chosen to attain the result in part (c)?

4. Compute the block RIP of order 2 of the following 7×8 matrix:

$$\mathbf{D} = \begin{bmatrix} 1 & 0 & 0 & 0 & 0 & 0 & 0 & \frac{1}{\sqrt{7}} \\ 0 & 1 & 0 & 0 & 0 & 0 & 0 & \frac{1}{\sqrt{7}} \\ 0 & 0 & 1 & 0 & 0 & 0 & 0 & \frac{1}{\sqrt{7}} \\ 0 & 0 & 0 & 1 & 0 & 0 & 0 & \frac{1}{\sqrt{7}} \\ 0 & 0 & 0 & 0 & 1 & 0 & 0 & \frac{1}{\sqrt{7}} \\ 0 & 0 & 0 & 0 & 0 & 1 & 0 & \frac{1}{\sqrt{7}} \\ 0 & 0 & 0 & 0 & 0 & 0 & 1 & \frac{1}{\sqrt{7}} \end{bmatrix}. \tag{12.175}$$

5. Consider the matrix $\mathbf{D}$ defined in Exercise 4. Compute the block coherence and the standard coherence of $\mathbf{D}$, and show that $\mu_B(\mathbf{D})$ and $\mu(\mathbf{D})$ satisfy Proposition 12.3.

6. Prove that for a dictionary $\mathbf{D}$ consisting of N elements in a vector space of dimension $n < N$, the coherence μ can be lower-bounded as

$$\mu \geq \sqrt{\frac{N-n}{n(N-1)}} \stackrel{N \gg 1}{\approx} \sqrt{\frac{N-n}{nN}}. \tag{12.176}$$

Hint: Let $\mathbf{G} = \mathbf{D}^*\mathbf{D}$ be the $N \times N$ Gram matrix and let λ_i denote its eigenvalues. Show that $\sum_{i=1}^{N} \lambda_i = N$ and $\sum_{i=1}^{N} \lambda_i^2 \geq N^2/n$. The bound then follows by bounding $\mathrm{Tr}(\mathbf{G}^2)$ and the elements of $\mathbf{G}$.

7. In this exercise, we show that the recovery threshold obtained from taking block sparsity into account in the orthogonalized dictionary is higher than the recovery threshold corresponding to conventional sparsity in the original dictionary.

Let $\mathbf{D}[\ell]$ denote a block of size $L \times d$ in a general dictionary $\mathbf{D}$ of size $L \times N$. We assume that $N = Md$ and $L = Rd$ for integers M, R.

a. Argue that since the columns of $\mathbf{D}[\ell]$ are linearly independent, $\mathbf{D}[\ell]$ can be written as $\mathbf{D}[\ell] = \mathbf{A}[\ell]\mathbf{W}_\ell$ where $\mathbf{A}[\ell]$ consists of orthonormal columns that span $\mathcal{R}(\mathbf{D}[\ell])$ and $\mathbf{W}_\ell$ is invertible.

b. Show that orthogonalization preserves the block sparsity level and that the block coherence is invariant to the choice of orthonormal basis $\mathbf{A}[\ell]$ for $\mathcal{R}(\mathbf{A}[\ell]) = \mathcal{R}(\mathbf{D}[\ell])$.

c. Prove that if $d > RM/(M - R)$, then the recovery threshold obtained from taking block sparsity into account in the orthogonalized dictionary is higher than the recovery threshold corresponding to conventional sparsity in the original dictionary.

Hint: Use the lower bound in Exercise 6 together with Proposition 12.4 and the fact that after orthogonalization $\nu = 0$.

8. Denote by $\mathbf{F}$ the $m \times n$ (n is even) partial Fourier matrix with elements

$$F_{r\ell} = \frac{1}{\sqrt{n}} e^{-j\frac{2\pi}{n}r\ell}, \tag{12.177}$$

where $r = -M, \ldots, M$ (so that $m = 2M + 1, m < n$) and $\ell = 0, \ldots, n - 1$. Suppose that the matrix $\mathbf{F}$ is divided into $n/2$ blocks of size 2 where each block consists of two consecutive columns of $\mathbf{F}$.

a. Compute the block coherence of $\mathbf{F}$.

b. Let $n = 256$. Plot the block coherence of $\mathbf{F}$ as a function of odd m.

c. Now let $n \to \infty$ with m fixed. Determine the block coherence in this setting.

d. Compare the results with that of Exercise 11 of Chapter 11.

e. Suppose now that we divide $\mathbf{F}$ into consecutive blocks of size $L \geq 2$. Let $n = 256, m = 99$. Plot the block coherence of $\mathbf{F}$ as a function of L. Explain the results.

9. Let $\mathbf{F}$ be the DFT matrix of size $R = L/d$ with normalized columns. Define $\boldsymbol{\Phi} = \mathbf{I}_L$ and $\boldsymbol{\Psi} = \mathbf{F} \otimes \mathbf{U}$ where $\mathbf{U}$ is a $d \times d$ Haar matrix, as defined in Example 11.17, and $\otimes$ denotes the Kronecker product operator. We choose $\mathbf{D} = \begin{bmatrix} \boldsymbol{\Phi} & \boldsymbol{\Psi} \end{bmatrix}$ as the $L \times 2L$ dictionary with blocks of size $L \times d$. Compute the block coherence μ_B and coherence μ of $\mathbf{D}$ for $d = 4, 8$.

10. a. Let $\mathbf{D}$ be an $L \times N$ dictionary with normalized columns and define by $d_{\min}^2 = \min_{i \neq j} \|\mathbf{d}_i - \mathbf{d}_j\|_2^2$ the minimum distance between two dictionary columns. Derive an expression for $d_{\min}^2$ as a function of the coherence μ of (11.67).

b. Let $\mathbf{D}$ be an $L \times N$ dictionary with blocks $\mathbf{D}[\ell]$ of size $L \times d$ and $N = Md$. Assume that the matrices $\mathbf{D}[\ell]$ are unitary so that $\mathbf{D}^*[\ell]\mathbf{D}[\ell] = \mathbf{I}$. Let $d_{\min,B}^2 = \min_{i \neq j} \|\mathbf{D}[i]\mathbf{c}_i - \mathbf{D}[j]\mathbf{c}_j\|_2^2$ with $\|\mathbf{c}_i\|_2^2 = \|\mathbf{c}_j\|_2^2 = 1$. Derive an expression for $d_{\min,B}^2$ as a function of the block coherence μ_B of (12.32).

11. Consider the following matrix, separated into three blocks of two columns each:

$$\mathbf{D} = \begin{bmatrix} 1 & 1 & 0 & 1 & 1 & 1 \\ 0 & 3 & -1 & 0 & 0 & 3 \\ 1 & 4 & 1 & -1 & 0 & 1 \\ 0 & 1 & 0 & 0 & -1 & 2 \end{bmatrix}. \tag{12.178}$$

Find an input vector $\mathbf{c}$ for which block BP recovers it exactly from measurements $\mathbf{y} = \mathbf{D}\mathbf{c}$ while standard BP fails.

12. The K-SVD and block K-SVD algorithms are based on the SVD. In this exercise we explore some of its properties.

a. What are the singular values of a unitary matrix?

b. Give an example of a matrix whose SVD is not unique. What remains unique even in this example?

c. Provide a simple method for computing the SVD of the matrix

$$\mathbf{A} = \begin{bmatrix} 3 & -3 \\ -1 & 1 \\ 2 & -2 \end{bmatrix}. \tag{12.179}$$

d. Prove that the rank of a matrix is equal to the number of its nonzero singular values.

13. In this exercise we will develop an online algorithm for dictionary learning. Consider the problem (12.133) where $\mathbf{c}$ is a standard sparse vector. To approximate its solution we consider an ℓ_1 regularized problem:

$$\min_{\mathbf{D}, \mathbf{c}_i} \sum_{i=1}^{L} \left\{ \|\mathbf{y}_i - \mathbf{D}\mathbf{c}_i\|_2^2 + \lambda \|\mathbf{c}_i\|_1 \right\}. \tag{12.180}$$

For given $\mathbf{C}$, denote the objective by $F(\mathbf{D})$ and assume that all variables are real valued.

a. Compute the gradient of $F(\mathbf{D})$. Show that it can be expressed in terms of the matrices $\mathbf{A} = \sum_{i=1}^{L} \mathbf{c}_i \mathbf{c}_i^T$ and $\mathbf{B} = \sum_{i=1}^{L} \mathbf{y}_i \mathbf{c}_i^T$.

b. For fixed $\mathbf{c}_i$, suggest an update rule for $\mathbf{D}$ based on the gradient.

c. Suggest a dictionary learning algorithm that also incorporates optimization over $\mathbf{c}_i$.

d. Suppose now that a new data point is added $\mathbf{y}_{L+1}$ with its corresponding sparse coefficients $\mathbf{c}_{L+1}$. Suggest a method to update the gradient $F(\mathbf{D})$ based on your previous calculations.

14. We illustrate in this exercise the use of dictionary learning for image denoising. Choose an image that you can process in Matlab. From this image we will create training data $\mathbf{Y}$ with columns $\mathbf{y}_i$ by breaking the image into patches of size 8×8. Create a vector $\mathbf{y}_i$ of length 64 out of each such patch by stacking the columns of the patch.

a. Learn a dictionary $\mathbf{D}$ from the training data $\mathbf{Y}$ using K-SVD.

b. Given the dictionary $\mathbf{D}$, find a sparse representation for each one of the patches $\mathbf{y}_i$ such that $\mathbf{y}_i \approx \mathbf{D}\mathbf{c}_i$ for a sparse vector $\mathbf{c}_i$.

c. Compute the approximation error $\sum_i \|\mathbf{y}_i - \mathbf{D}\mathbf{c}_i\|^2$ obtained with the learned dictionary.

d. Create a noisy image $\widetilde{\mathbf{Y}} = \mathbf{Y} + \mathbf{N}$ where $\mathbf{N}$ is a noise matrix with iid random variables. Repeat part b with respect to the columns $\tilde{\mathbf{y}}_i$ of $\widetilde{\mathbf{Y}}$ using the dictionary $\mathbf{D}$ learned in a and determine the sparse coefficients $\hat{\mathbf{c}}_i$.

e. Create a denoised image by first forming the clean patches $\mathbf{D}\hat{\mathbf{c}}_i$ and then concatenating them into an image.

f. Repeat the computation in c with $\hat{\mathbf{c}}_i$ and compare with the error obtained using the clean image.

15. Write out the steps of the sparse agglomerative clustering algorithm for the matrix

$$\mathbf{C} = \begin{bmatrix} -1.1 & 1.8 & 0 & 0 & 0 & 2 \\ 0 & 0 & 0 & 3.4 & 4 & 0 \\ 0 & 0 & 0 & -1 & 0 & 1 \\ -3 & 1 & 0 & 0 & -2.3 & 0 \\ 0 & 0 & 3.4 & 0 & 5 & -2 \\ 1 & 0 & 0 & -6.7 & 0 & 0 \end{bmatrix}. \tag{12.181}$$

Assume that $s = 3$ and that the initial block structure is $d = [1, 2, 3, 4, 5, 6]$.

16. Suppose that the output of the sparse agglomerative clustering algorithm is a matrix $\mathbf{C}$ and block structure vector $d = [1, 2, 1, 1, 2, 3, 2, 3]$.

    a. Give an example of such a matrix C with 10 columns.

    b. Rearrange C so that its columns are block-sparse according to the above pattern.

17. In this exercise we will extend Algorithm 12.7 to the block-sparse setting.

    a. Suggest a modification to the update sparse coefficients step for the case in which each c_i is block sparse (consider Algorithm 12.5).

    b. Assuming a block assignment and given sparse coefficients, suggest an extension of (12.166) to the block setting in order to update the jth dictionary block.

    c. Suggest an extension of (12.167) to the block setting in order to update the jth block in the sparse coefficients.

18. Show that, in the 1-block sparse setting in which only one block per signal is nonzero, the extension of the BCS problem stated in (13.162) to the block-sparse case is equivalent to low-rank matrix completion. Specifically, construct matrices Y, X, C such that Y has low rank and can be decomposed as $Y = XC$ where X and C are unknown, and only a subset of the entries of Y are observed.

Hint: Rearrange the observations z_i to constitute Y.

Chapter 13

Sampling over shift-invariant unions

We now turn to discuss unions of shift-invariant (SI) subspaces. As we have seen throughout the book, this class of subspaces plays a key role in the development of sampling theory. It is therefore natural to incorporate the SI structure within the UoS framework. To do so, we distinguish between two classes of unions: finite unions of SI subspaces, and infinite unions. In this chapter we focus on finite SI unions. A special case is that of multiband signals which are studied in detail in the next chapter. Infinite SI unions will be treated in Chapter 15 in the context of FRI sampling.

13.1 Union model

13.1.1 Sparse union of SI subspaces

Recall from Chapters 5 and 6 that SI signals are characterized by a set of generators $\{h_\ell(t), 1 \le \ell \le N\}$ where in principle N may be finite or infinite (for example when considering Gabor or wavelet expansions of L_2). Here we focus on the case in which N is finite. Any signal in such an SI space can be written as

$$x(t) = \sum_{\ell=1}^{N} \sum_{n \in \mathbb{Z}} d_\ell[n] h_\ell(t - nT), \tag{13.1}$$

for some set of sequences $\{d_\ell[n] \in \ell_2, 1 \le \ell \le N\}$ and period T. As we have seen, this model encompasses many signals used in communication and signal processing, including bandlimited functions, splines, multiband signals (with known carrier positions), and pulse amplitude modulation signals.

The subspace of signals described by (13.1) has infinite dimension, since every signal is associated with infinitely many coefficients $\{d_\ell[n], 1 \le \ell \le N\}$. In Section 6.8 we treated such signal models and showed that $x(t)$ can be recovered from samples at a rate of N/T. Before turning to treat structured SI models, we briefly review the main ideas here.

One possible sampling paradigm at the minimal rate is given in Fig. 13.1. In this approach, $x(t)$ is filtered with a bank of N filters, each with impulse response $\overline{a_\ell(-t)}$ which can be almost arbitrary. The outputs are then uniformly sampled with period T, resulting in the sample sequences $c_\ell[n]$. Denote by $\mathbf{c}(e^{j\omega})$ the vector collecting the

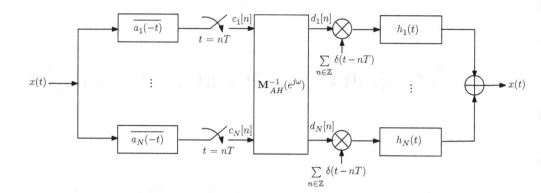

Figure 13.1 Sampling and reconstruction in shift-invariant spaces.

DTFTs of $c_\ell[n]$, $1 \le \ell \le N$, and similarly by $\mathbf{d}(e^{j\omega})$ the vector containing the DTFTs of $d_\ell[n]$, $1 \le \ell \le N$. Then it is shown in Section 6.8 that

$$\mathbf{c}(e^{j\omega}) = \mathbf{M}_{AH}(e^{j\omega})\mathbf{d}(e^{j\omega}), \tag{13.2}$$

where $\mathbf{M}_{AH}(e^{j\omega})$ is an $N \times N$ matrix, with entries

$$\left[\mathbf{M}_{AH}(e^{j\omega})\right]_{i\ell} = \frac{1}{T}\sum_{k\in\mathbb{Z}} \overline{A_i\left(\frac{\omega}{T} - \frac{2\pi k}{T}\right)} H_\ell\left(\frac{\omega}{T} - \frac{2\pi k}{T}\right). \tag{13.3}$$

Here $A_i(\omega)$ and $H_\ell(\omega)$ are the CTFTs of $a_i(t)$ and $h_\ell(t)$, respectively. We will use the result of (13.2) repeatedly throughout the chapter (see Exercise 1 for an explicit derivation).

To recover $x(t)$ from the samples represented by $\mathbf{c}(e^{j\omega})$, we filter the samples using a filterbank with frequency response $\mathbf{M}_{AH}^{-1}(e^{j\omega})$, obtaining the vectors

$$\mathbf{z}(e^{j\omega}) = \mathbf{M}_{AH}^{-1}(e^{j\omega})\mathbf{c}(e^{j\omega}). \tag{13.4}$$

Using (13.2) we have

$$\mathbf{z}(e^{j\omega}) = \mathbf{M}_{AH}^{-1}(e^{j\omega})\mathbf{c}(e^{j\omega}) = \mathbf{M}_{AH}^{-1}(e^{j\omega})\mathbf{M}_{AH}(e^{j\omega})\mathbf{d}(e^{j\omega}) = \mathbf{d}(e^{j\omega}), \tag{13.5}$$

thus recovering the coefficients $\mathbf{d}(e^{j\omega})$. Each output sequence $d_\ell[n]$ is then modulated by a periodic impulse train $\sum_{n\in\mathbb{Z}} \delta(t-nT)$ with period T, followed by filtering with the corresponding analog filter $h_\ell(t)$. In practice, interpolation with finitely many samples results in sufficiently accurate reconstruction, provided that $h_\ell(t)$ decay fast enough [393], similarly to finite interpolation in the Shannon–Nyquist theorem.

In order to incorporate further structure into the generic SI model (13.1), we treat signals that can be represented by a small number k of generators, chosen from a finite set of N functions. Specifically, we consider the input model

$$x(t) = \sum_{|\ell|=k}\sum_{n\in\mathbb{Z}} d_\ell[n]h_\ell(t - nT), \tag{13.6}$$

where $|\ell| = k$ means a sum over at most k elements. We assume throughout that the generators form linearly independent functions. If the k active generators are known, then it suffices to sample at a rate of k/T corresponding to uniform samples with period T at the output of k appropriate filters as in Fig. 13.1. A more difficult question is whether the rate can be reduced if we know that only k of the generators are active, but do not know in advance which ones. In terms of (13.6) this means that only k of the sequences $d_\ell[n]$ have nonzero energy. Consequently, for each value n, $\|\mathbf{d}[n]\|_0 \leq k$, where $\mathbf{d}[n] = [d_1[n], \ldots, d_N[n]]^T$ collects the unknown generator coefficients at time n.

We have seen several examples of this model in Chapters 4 and 10. In particular, Example 4.4 shows how a multiband signal sparsely populated on $[0, \pi/T)$, consisting of N bands of width at most B with $NB \ll \pi/T$, can be viewed as a special case of (13.6). To this end we split the available bandwidth into m fixed bands of width $\pi/(mT)$, where each interval is at least as large as B: $B \leq \pi/(mT)$. We then define m generators that correspond to these bands:

$$h_\ell(t) = 2mT \, \text{sinc}\left(\frac{t}{2mT}\right) e^{-j\left(\frac{\ell-1/2}{mT}\right)\pi t}, \quad \ell = 1, \ldots, m. \tag{13.7}$$

The CTFT $H_\ell(\omega)$ of $h_\ell(t)$ is a box function on $\mathcal{I}_\ell = [\pi(\ell-1)/(mT), \pi\ell/(mT))$. Since the exact locations of the active signal bands are unknown, each band can split into at most two adjacent intervals $\mathcal{I}_\ell$ of size π/mT. Therefore, any multiband $x(t)$ may be written in the form (13.6) using no more than $2N$ generators.

From Proposition 10.2 and the discussion following it, the minimal rate required to recover a signal $x(t)$ of the form (13.6) is at least $2k/T$. Thus, the fact that we do not know the exact subspace leads to an increase of at least a factor 2 in the minimal rate. Our goal is to show how this rate can be achieved in practice by combining ideas from analog sampling and CS.

13.1.2 Sub-Nyquist sampling

Our problem is similar in spirit to finite CS: we would like to sense a sparse signal using fewer measurements than required without the sparsity assumption. However, a fundamental difference between the two stems from the fact that the problem here is defined over an infinite-dimensional space of continuous functions. As we now show, trying to represent it in the same form as CS, by replacing the finite matrices by appropriate operators, raises several difficulties that preclude direct application of CS-type results.

To see this, suppose we represent $x(t)$ in terms of a sparse expansion $x(t) = \Phi(t)\alpha$ which resembles the finite expansion $\mathbf{x} = \Phi\alpha$. Here $\Phi(t)$ is an infinite-dimensional operator corresponding to the concatenation of the functions $h_\ell(t - nT)$, and $\alpha \in \ell_2$ is an infinite sequence consisting of the concatenation of the sequences $d_\ell[n]$. Since $d_\ell[n]$ is identically zero for several values of ℓ, α will contain many zero elements. Next, we can define a measurement operator $M(t)$ so that the measurements are given by $y = A\alpha$ where $A = M(t)\Phi(t)$.

In analogy to the finite setting, the recovery properties of α should depend on A. However, it is not clear how to apply CS ideas to this operator equation. As we have seen, the ability to recover $\boldsymbol{\alpha}$ in the finite setting depends on its sparsity. In our case, the sparsity of α is always infinite. Furthermore, a practical way to ensure stable recovery with high probability in conventional CS is to draw the elements of $\mathbf{A}$ at random, with the number of rows in $\mathbf{A}$ proportional to the sparsity. In the operator setting, we cannot clearly define the dimensions of A or draw its elements at random. Furthermore, even if we are able to develop conditions on A such that the measurement sequence $y = A\alpha$ uniquely determines α, it is still not clear how to recover α from y. For example, the immediate extension of BP (11.120) to this context would be:

$$\min_{\alpha} \|\alpha\|_1 \quad \text{s.t.} \ y = A\alpha. \tag{13.8}$$

Although (13.8) is a convex problem, it is defined over infinitely many variables, with infinitely many constraints. Therefore, it cannot be solved directly using standard optimization packages as in finite-dimensional CS.

This discussion raises three important questions we need to address in order to adapt CS results to the analog setting:

1. How do we choose a compressive analog sampling operator?
2. Can we introduce structure into the sampling operator and still preserve stability?
3. How do we solve the resulting infinite-dimensional recovery problem?

To answer these questions and develop a general CS framework for SI analog signals, we capitalize on two key elements:

1. Fourier domain analysis of the sequences of samples.
2. Choosing the sampling functions such that we obtain an IMV model, as defined in Section 11.6.4.

Once we arrive at an IMV formulation of the problem, we can use the techniques of Section 11.6.4 to recover the desired expansion coefficients via finite-dimensional CS methods.

The sampling approach we present below, developed in [104], consists of a multi-channel filterbank with $p < N$ filters $s_i(t)$. Their design relies on two ingredients:

1. A $p \times N$ matrix $\mathbf{A}$ corresponding to p measurements chosen such that it solves a finite-dimensional CS problem in the dimensions N (vector length) and k (sparsity).
2. A set of functions $a_i(t), 1 \le i \le N$ for which $\mathbf{M}_{AH}(e^{j\omega})$ of (13.3) is stably invertible almost everywhere.

The value of p is chosen either to guarantee exact recovery (in the noiseless setting), in which case p may be as low as[1] $2k$, or to enable efficient and robust recovery (possibly only with high probability) requiring $p > 2k$. The functions $\{a_i(t)\}$ are selected such that they allow recovery of any $x(t)$ of the form (13.1), that is, such that $\mathbf{M}_{AH}(e^{j\omega})$ is stably invertible. This choice does not take the sparsity into account, namely the fact that only k out of the N generators are active, and therefore leads to more measurements than actually

---

[1] In fact, in certain cases we can choose $p = k + 1$ since we will be using MMV recovery techniques; however, for simplicity, we focus on recovery results for standard CS that do not rely on multiple measurements.

needed. We will see how linear combinations of these functions, determined by the matrix $\mathbf{A}$, can be used to reduce the number of filters and consequently the sampling rate.

We derive the proposed sampling scheme in three steps. First, we consider the problem of compressively measuring the vector sequence $\mathbf{d}[n]$, whose ℓth component is given by $d_\ell[n]$ in (13.6), where only k out of the N sequences $d_\ell[n]$ are nonzero. We show that this can be accomplished by using the matrix $\mathbf{A}$ above and IMV recovery theory. More specifically, we target a compressive sampling system that produces a vector of low-rate samples $\mathbf{y}[n] = [y_1[n], \ldots, y_p[n]]^T$ satisfying the relation

$$\mathbf{y}[n] = \mathbf{A}\mathbf{d}[n], \quad \|\mathbf{d}[n]\|_0 \leq k, \tag{13.9}$$

with a sensing matrix $\mathbf{A}$ that allows recovery of k-sparse vectors. The choice $p < N$ reduces the sampling rate below Nyquist.

Since in practice the vector sequence $\mathbf{d}[n]$ is not available to us, in the second step we use the system of Fig. 13.1 to obtain $\mathbf{d}[n]$ from the input signal $x(t)$ using an appropriate filterbank of N analog filters, and sampling their outputs. Finally, we merge the first two steps to arrive at a system of $p < N$ analog filters that compressively sample $x(t)$ directly. These steps are detailed in the next section.

13.2 Compressed sensing in sparse unions

13.2.1 Union of discrete sequences

We begin by treating the problem of sampling and recovering the vector sequence $\mathbf{d}[n]$. This is accomplished using the IMV model introduced in Section 11.6.4. Indeed, suppose we measure $N \times 1$ vectors $\mathbf{d}[n]$ with a measurement matrix $\mathbf{A}$ of size $p \times N$, which allows CS of k-sparse vectors of length N. Then, for each n,

$$\mathbf{y}[n] = \mathbf{A}\mathbf{d}[n], \quad n \in \mathbb{Z}. \tag{13.10}$$

The system of (13.10) is an IMV model – for every n the vector $\mathbf{d}[n]$ is k-sparse. Furthermore, the infinite set of vectors $\{\mathbf{d}[n], n \in \mathbb{Z}\}$ shares a joint sparsity pattern: at most k of the sequences $d_\ell[n]$ are nonzero. As we described in Section 11.6.4, such a set of equations can be solved by transforming it into an equivalent MMV system, whose recovery properties are determined by those of $\mathbf{A}$. Since $\mathbf{A}$ was designed to enable CS techniques, we are guaranteed that $\mathbf{d}[n]$ can be perfectly recovered for each n (or recovered with high probability).

For completeness, we briefly review the IMV setup. In an IMV problem our goal is to recover a set of unknown vectors $\mathbf{x}(\lambda)$ with joint support S of size k from measurement vectors

$$\mathbf{y}(\lambda) = \mathbf{A}\mathbf{x}(\lambda), \quad \lambda \in \Lambda, \tag{13.11}$$

where Λ is a set whose cardinality may be infinite countable or uncountable. Recovery is performed by first finding the joint support S, and then inverting the system of equations over this support. Inversion is possible due to the assumption that $\mathbf{A}$ is a valid CS matrix and is therefore invertible over any support set of size k.

The key observation in solving an IMV system is that the joint support S can be determined by solving a single MMV problem. The essential idea is that every finite collection of vectors spanning the subspace span($\mathbf{y}(\lambda)$), contains sufficient information to recover S. Since $\mathbf{y}(\lambda)$ is a length-p vector, span($\mathbf{y}(\lambda)$) has dimension at most equal to p and therefore p vectors are sufficient to span the space. Stacking these vectors into a matrix $\mathbf{V}$, we consider the MMV problem $\mathbf{V} = \mathbf{AU}$ where $\mathbf{U}$ is a row-sparse matrix; that is, it has at most k nonzero rows. In Section 11.6.4 we proved that if we seek the sparsest row-sparse matrix $\mathbf{U}$ solving $\mathbf{V} = \mathbf{AU}$, then the support of $\mathbf{U}$ is equal to S. The steps used to formulate the MMV recovery of S are grouped under a block referred to as the continuous to finite (CTF) block: these operations enable solving an IMV system using only finite-dimensional CS results.

There are many possibilities for choosing $\mathbf{V}$ that are all guaranteed to lead to the same sparsity pattern S. One choice is to evaluate the integral

$$Q = \int_{\lambda \in \Lambda} \mathbf{y}(\lambda)\mathbf{y}^*(\lambda)d\lambda. \tag{13.12}$$

If it exists, then the column space of every matrix $\mathbf{V}$ satisfying $\mathbf{Q} = \mathbf{VV}^*$ is equal to span($\mathbf{y}(\lambda)$). Such a matrix can be found, for example, by relying on the eigendecomposition of $\mathbf{Q}$. If Λ is a discrete set, then the integral can be replaced by the sum

$$Q = \sum_n \mathbf{y}[n]\mathbf{y}^*[n]. \tag{13.13}$$

Any other choice of $\mathbf{V}$ with the required span is equally valid. The overall procedure is depicted in Fig. 13.2 (further details are provided in Section 11.6.4).

Going back to our problem, we can apply the CTF to the time-domain representation (13.10), or alternatively consider the frequency-domain set of equations:

$$\mathbf{y}(e^{j\omega}) = \mathbf{Ad}(e^{j\omega}), \quad 0 \le \omega < 2\pi, \tag{13.14}$$

where $\mathbf{y}(e^{j\omega}), \mathbf{d}(e^{j\omega})$ are the vectors whose components are the DTFTs $Y_\ell(e^{j\omega}), D_\ell(e^{j\omega})$.

When designing the measurements (13.10) or (13.14), the only freedom we have is in choosing $\mathbf{A}$. To generalize the class of sensing operators we note that $\mathbf{d}[n]$ may also be recovered from

$$\mathbf{y}(e^{j\omega}) = \mathbf{W}(e^{j\omega})\mathbf{Ad}(e^{j\omega}), \quad 0 \le \omega < 2\pi, \tag{13.15}$$

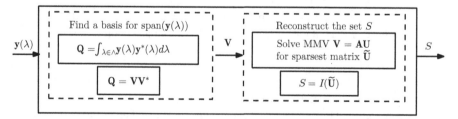

Figure 13.2 Fundamental stages for the recovery of the nonzero location set S in an IMV model by solving only one finite-dimensional problem. Here $I(\widetilde{\mathbf{U}})$ represents the support of $\widetilde{\mathbf{U}}$.

for any invertible $p \times p$ matrix $\mathbf{W}(e^{j\omega})$ with elements $W_{i\ell}(e^{j\omega})$. The extra freedom offered by choosing an arbitrary invertible matrix $\mathbf{W}(e^{j\omega})$ may be useful when designing the corresponding analog sampling filters.

The measurements of (13.15) can be obtained directly in the time domain as

$$y_i[n] = \sum_{\ell=1}^{p} w_{i\ell}[n] * \left(\sum_{r=1}^{N} \mathbf{A}_{\ell r} d_r[n] \right), \quad 1 \le i \le p, \tag{13.16}$$

where $w_{i\ell}[n]$ is the inverse DTFT of $W_{i\ell}(e^{j\omega})$, and $*$ denotes the convolution operator. To recover $d_\ell[n]$ from $\mathbf{y}(e^{j\omega})$, we note that the modified measurements $\tilde{\mathbf{y}}(e^{j\omega}) = \mathbf{W}^{-1}(e^{j\omega})\mathbf{y}(e^{j\omega})$ obey an IMV model:

$$\tilde{\mathbf{y}}(e^{j\omega}) = \mathbf{A}\mathbf{d}(e^{j\omega}), \quad 0 \le \omega < 2\pi. \tag{13.17}$$

Therefore, the CTF block can be applied to $\tilde{\mathbf{y}}(e^{j\omega})$. As in (13.16), we may use the CTF in the time domain by noting that

$$\tilde{y}_i[n] = \sum_{\ell=1}^{p} b_{i\ell}[n] * y_\ell[n], \tag{13.18}$$

where $b_{i\ell}[n]$ is the inverse DTFT of $B_{i\ell}(e^{j\omega})$, with $\mathbf{B}(e^{j\omega}) = \mathbf{W}^{-1}(e^{j\omega})$.

13.2.2 Reduced-rate sampling

The previous section established that given the ability to sample the N sequences $d_\ell[n]$, we can recover them exactly from $p < N$ discrete-time sequences obtained via (13.15) or (13.16). Reconstruction is performed by applying the CTF block to the modified measurements either in the frequency domain (13.17) or in the time domain (13.18). The drawback of this approach is that we do not have access to $d_\ell[n]$ but rather we are given $x(t)$.

Analog front end

In Section 13.1 we have seen that the sequences $d_\ell[n]$ can be obtained by sampling $x(t)$ with a set of functions $a_\ell(t)$ for which $\mathbf{M}_{AH}(e^{j\omega})$ of (13.3) is stably invertible, and then filtering the sampled sequences with a multichannel discrete-time filter $\mathbf{M}_{AH}^{-1}(e^{j\omega})$. Thus, we first apply this front end to $x(t)$, in order to produce the sequence of vectors $\mathbf{d}[n]$. We then use the technique of the previous subsection in order to sense these sequences efficiently. The resulting measurement sequences $y_\ell[n]$ are depicted in Fig. 13.3, where $\mathbf{A}$ is a $p \times N$ matrix satisfying the requirements of CS in the appropriate dimensions, and $\mathbf{W}(e^{j\omega})$ is a size $p \times p$ filterbank that is invertible almost everywhere.

Combining the analog filters $a_\ell(t)$ with the discrete-time multichannel filter $\mathbf{M}_{AH}^{-1}(e^{j\omega})$, we can express $d_\ell[n]$ as

$$d_\ell[n] = \langle v_\ell(t - nT), x(t) \rangle, \quad 1 \le \ell \le N, \; n \in \mathbb{Z}, \tag{13.19}$$

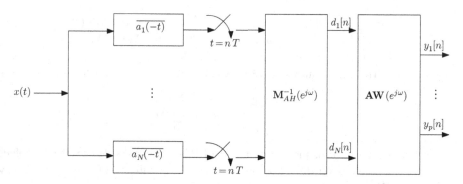

Figure 13.3 Analog compressed sampling with arbitrary filters $a_i(t)$.

where

$$\mathbf{v}(\omega) = \overline{\mathbf{M}_{AH}^{-1}(e^{j\omega T})}\mathbf{a}(\omega). \tag{13.20}$$

Here $\mathbf{v}(\omega), \mathbf{a}(\omega)$ are the vectors with ℓth elements $V_\ell(\omega), A_\ell(\omega)$. The inner products in (13.19) can be obtained by filtering $x(t)$ with a bank of filters $\{v_\ell(-t)\}$, and uniformly sampling their outputs at times nT.

To see that (13.19) and (13.20) hold, let $c_\ell[n]$ be the samples resulting from filtering $x(t)$ with N filters $\overline{v_\ell(-t)}$ and uniformly sampling their outputs at rate $1/T$. Then, from (13.2),

$$\mathbf{c}(e^{j\omega}) = \mathbf{M}_{VH}(e^{j\omega})\mathbf{d}(e^{j\omega}). \tag{13.21}$$

It is easy to see that (see Exercise 5) $\mathbf{M}_{VH}(e^{j\omega}) = \mathbf{I}$ and therefore $c_\ell[n] = d_\ell[n]$, establishing (13.19). This also means that the functions $\{v_\ell(t - nT)\}$ are biorthogonal to $\{h_\ell(t - nT)\}$.

Note that there are many other choices of $v_\ell(t)$ for which $\mathbf{M}_{VH}(e^{j\omega}) = \mathbf{I}$. However, the functions of (13.20) corespond to a front end of filters $a_\ell(-t)$, as required here.

Although the sampling scheme of Fig. 13.3 results in compressed measurements $\{y_\ell[n]\}$, they are still acquired by an analog front end that operates at the high rate N/T. Our final goal is to reduce the rate at the analog front end. This can be easily accomplished by moving the discrete filters $\mathbf{M}_{AH}^{-1}(e^{j\omega})$, $\mathbf{AW}(e^{j\omega})$ back to the analog domain. In this way, the compressed measurement sequences $y_\ell[n]$ are obtained directly from $x(t)$, by filtering $x(t)$ with p filters and uniformly sampling their outputs at times nT, leading to a system with sampling rate p/T. An explicit expression for the corresponding sampling functions is given in the following theorem.

Theorem 13.1 (SI compressed sampling). *Let the compressed measurements $y_\ell[n]$, $1 \le \ell \le p$ be the output of the hybrid filterbank in Fig. 13.3. Then $\{y_\ell[n]\}$ can be obtained by filtering $x(t)$ of (13.6) with p filters $\{s_\ell(-t)\}$ and sampling the outputs at rate $1/T$, where*

$$\mathbf{s}(\omega) = \overline{\mathbf{W}(e^{j\omega T})\mathbf{A}}\mathbf{v}(\omega) = \overline{\mathbf{W}(e^{j\omega T})\mathbf{A}\mathbf{M}_{AH}^{-1}(e^{j\omega T})}\mathbf{a}(\omega). \tag{13.22}$$

Here $\mathbf{s}(\omega), \mathbf{h}(\omega)$ *are the vectors with* ℓ*th elements* $S_\ell(\omega), H_\ell(\omega)$ *respectively, and the components* $V_\ell(\omega)$ *of* $\mathbf{v}(\omega) = \overline{\mathbf{M}_{AH}^{-1}(e^{j\omega T})}\mathbf{a}(\omega)$ *are Fourier transforms of generators* $v_\ell(t)$ *such that* $\{v_\ell(t - nT)\}$ *are biorthogonal to* $\{h_\ell(t - nT)\}$. *In the time domain,*

$$s_i(t) = \sum_{\ell=1}^{N} \sum_{r=1}^{p} \sum_{n\in\mathbb{Z}} \overline{w_{ir}[-n]A_{r\ell}}v_\ell(t - nT), \qquad (13.23)$$

where $w_{ir}[n]$ *is the inverse DTFT of* $W_{ir}(e^{j\omega})$ *and*

$$v_i(t) = \sum_{\ell=1}^{N} \sum_{n\in\mathbb{Z}} \overline{\psi_{i\ell}[-n]}a_\ell(t - nT), \qquad (13.24)$$

where $\psi_{i\ell}[n]$ *is the inverse DTFT of* $[\mathbf{M}_{AH}^{-1}(e^{j\omega})]_{i\ell}$. *When* $\mathbf{W}(e^{j\omega}) = \mathbf{I}$,

$$s_i(t) = \sum_{\ell=1}^{N} \overline{A_{i\ell}}v_\ell(t). \qquad (13.25)$$

Proof: Suppose that $x(t)$ is filtered by the p filters $s_i(t)$ and then uniformly sampled at nT. From (13.2), the samples can be expressed in the Fourier domain as

$$\mathbf{c}(e^{j\omega}) = \mathbf{M}_{SH}(e^{j\omega})\mathbf{d}(e^{j\omega}). \qquad (13.26)$$

In order to prove the theorem we need to show that for the choice of filters given by (13.22), $\mathbf{M}_{SH}(e^{j\omega}) = \mathbf{W}(e^{j\omega})\mathbf{A}$.

Let

$$\mathbf{B}(e^{j\omega}) = \overline{\mathbf{W}(e^{j\omega})\mathbf{A}}, \qquad (13.27)$$

so that $\mathbf{s}(\omega) = \mathbf{B}(e^{j\omega T})\mathbf{v}(\omega)$. Then,

$$
\begin{aligned}
[\mathbf{M}_{SH}(e^{j\omega})]_{i\ell} &= \frac{1}{T}\sum_{k\in\mathbb{Z}} \overline{S_i\left(\frac{\omega}{T} - \frac{2\pi}{T}k\right)} H_\ell\left(\frac{\omega}{T} - \frac{2\pi}{T}k\right) \\
&= \frac{1}{T}\sum_{r=1}^{N} \overline{B_{ir}(e^{j\omega})} \sum_{k\in\mathbb{Z}} \overline{V_r\left(\frac{\omega}{T} - \frac{2\pi}{T}k\right)} H_\ell\left(\frac{\omega}{T} - \frac{2\pi}{T}k\right) \\
&= [\overline{\mathbf{B}(e^{j\omega})}]^i [\mathbf{M}_{VH}(e^{j\omega})]_\ell,
\end{aligned}
\qquad (13.28)
$$

where $[\mathbf{Q}]^i, [\mathbf{Q}]_i$ are the ith row and column respectively of the matrix $\mathbf{Q}$. The first equality follows from the fact that $\mathbf{B}(e^{j\omega})$ is 2π periodic. From (13.28),

$$\mathbf{M}_{SH}(e^{j\omega}) = \overline{\mathbf{B}(e^{j\omega})}\mathbf{M}_{VH}(e^{j\omega}) = \mathbf{W}(e^{j\omega})\mathbf{A}, \qquad (13.29)$$

where we used the fact that $\mathbf{M}_{VH}(e^{j\omega}) = \mathbf{I}$ owing to the biorthogonality property (see Exercise 5).

Finally, if $\mathbf{s}(\omega) = \mathbf{B}(e^{j\omega T})\mathbf{v}(\omega)$, then

$$s_i(t) = \sum_{\ell=1}^{N} \sum_{n\in\mathbb{Z}} b_{i\ell}[n]v_\ell(t - nT), \qquad (13.30)$$

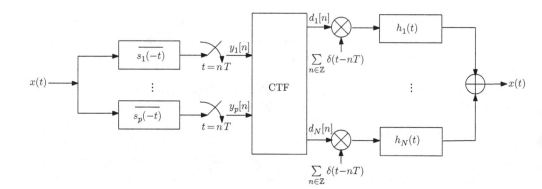

Figure 13.4 Compressed sensing of analog signals. The sampling functions $s_i(t)$ are obtained by combining the blocks in Fig. 13.3 and are given in Theorem 13.1.

where $b_{i\ell}[n]$ is the inverse DTFT of $B_{i\ell}(e^{j\omega})$. Using (13.27) together with the fact that the inverse DTFT of $\overline{Q_{i\ell}(e^{j\omega})}$ is $\overline{q_{i\ell}[-n]}$, results in (13.23). The relation (13.24) follows from the same considerations. $\qquad\square$

Theorem 13.1 is the main result which allows compressive sampling of analog signals. Specifically, starting from any matrix $\mathbf{A}$ that satisfies the CS requirements of finite vectors, and a set of sampling functions $a_i(t)$ for which $\mathbf{M}_{AH}(e^{j\omega})$ is stably invertible, we can create a multitude of sampling functions $s_i(t)$ to compressively sample the underlying analog signal $x(t)$. Sensing is performed by filtering $x(t)$ with the $p < N$ corresponding filters (13.22), and sampling their outputs at rate $1/T$. Reconstruction from the compressed measurements $y_i[n], 1 \leq i \leq p$ is obtained by applying the CTF block in order to recover the sequences $d_i[n]$. The original signal $x(t)$ is then constructed by modulating appropriate impulse trains and filtering with $h_i(t)$, as depicted in Fig. 13.4.

Examples

We now consider two examples illustrating the details of Fig. 13.4.

Example 13.1 In this example, we compressively sample a signal with a simple periodic sparsity model.

Consider sampling a signal $x(t)$ that lies in an SI subspace with period 1, generated by

$$h(t) = \begin{cases} 1, & 0 \leq t \leq 1 \\ 0, & \text{otherwise.} \end{cases} \qquad (13.31)$$

Thus, $x(t) = \sum_{n\in\mathbb{Z}} d[n]h(t-n)$ for a sequence of numbers $d[n]$. The coefficients $d[n]$ have a periodic sparsity pattern: out of each consecutive group of seven coefficients, there are at most two nonzero values, in a given pattern. For example, suppose that the sparsity profile is $S = \{2, 5\}$. Then $d[n]$ can be nonzero only at indices $n = 1 + 7\ell$ or $n = 4 + 7\ell$ for some integer ℓ. An example of such a sequence $d[n]$

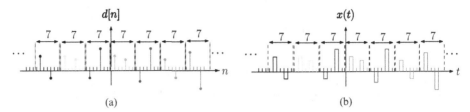

Figure 13.5 An example of a signal with periodic sparsity. (a) The coefficients $d[n]$. (b) The resulting analog signal $x(t)$.

with the corresponding analog signal $x(t)$ is depicted in Fig. 13.5, where each period is marked with a different shade.

A simple way to sample $x(t)$ is to filter it with $h(-t)$ and uniformly sample the output at times $t = n$ resulting in one sample per time step. With this choice, it is easy to see that the samples are given by $c[n] = d[n]$, and clearly $x(t)$ can be recovered from these samples. However, this approach does not exploit the sparsity of $d[n]$. In order to reduce the sampling rate, we now show how to formulate our problem within the sparse SI framework. Applying Fig. 13.4 results in a sampling rate of $4/7$, namely, four samples per period of seven time steps.

The signal $x(t)$ may be viewed as a special case of the general model (13.6) with $T = 7$ and

$$h_\ell(t) = \begin{cases} 1, & \ell - 1 \le t \le \ell \\ 0, & \text{otherwise}, \end{cases} \qquad 1 \le \ell \le 7. \qquad (13.32)$$

The sequences $d_\ell[n]$ are defined by

$$d_\ell[n] = d[\ell - 1 + 7n], \qquad 1 \le \ell \le 7. \qquad (13.33)$$

If we now decompose $d[n]$ into blocks $\mathbf{d}[n]$ of length 7 with elements $d_\ell[n]$, then the sparsity pattern of $x(t)$ implies that the vectors $\mathbf{d}[n]$ are jointly 2-sparse. In our example, the nonzero rows correspond to $\ell = 2$ and $\ell = 5$.

Since there are two nonzero rows, we require at least four sampling filters according to the results of standard CS, which can be determined from Theorem 13.1. To apply the theorem, we first need to construct a set of functions $\{v_\ell(t - n)\}$ that are biorthogonal to $\{h_\ell(t - n)\}$. Since the functions $\{h_\ell(t - n)\}$ are orthonormal, we can choose $v_\ell(t) = h_\ell(t)$. Next, we construct a 4×7 matrix $\mathbf{A}$ that has full spark, so that every four of its columns are linearly independent. To this end we generated a random matrix whose elements where chosen uniformly at random over the integers $[1, 2, \ldots, 10]$, and checked that the spark condition is satisfied. The resulting matrix is given by

$$\mathbf{A} = \begin{bmatrix} 8 & 6 & 10 & 10 & 4 & 7 & 7 \\ 9 & 1 & 10 & 5 & 9 & 0 & 8 \\ 1 & 3 & 2 & 8 & 8 & 8 & 7 \\ 9 & 5 & 10 & 1 & 10 & 9 & 4 \end{bmatrix}. \qquad (13.34)$$

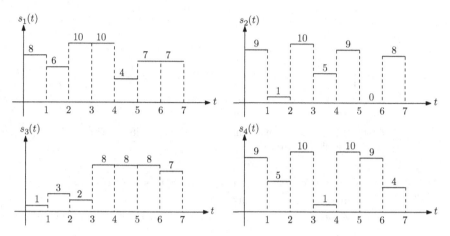

Figure 13.6 Sampling functions $s_1(t)$ through $s_4(t)$.

With this choice of $\mathbf{A}$ and $\mathbf{W} = \mathbf{I}$, the sampling functions follow from (13.25):

$$s_i(t) = \sum_{\ell=1}^{7} \overline{\mathbf{A}_{i\ell}} h_\ell(t) = \sum_{\ell=1}^{7} \overline{\mathbf{A}_{i\ell}} h(t - (\ell - 1)), \qquad 1 \leq i \leq 4. \qquad (13.35)$$

These functions are depicted in Fig. 13.6.

Let $y_i[n]$ denote the ith sampling sequence, resulting from sampling $x(t)$ with $s_i(t)$. Each sample is now equal to a linear combination of seven consecutive values of $d[n]$. This is in contrast to the high-rate method in which each sample is exactly equal to $d[n]$. For example, $y_i[0] = \sum_{\ell=1}^{7} A_{i\ell} d[\ell]$. More generally, $\mathbf{y}[n] = \mathbf{A} d[n]$ where $\mathbf{d}[n]$ contains seven consecutive values of $d[n]$. Since only two values out of each block $\mathbf{d}[n]$ are nonzero, we can compute them from the four linear combinations in $\mathbf{y}[n]$.

To determine $\mathbf{d}[n]$ we begin by finding the support S via the CTF. This requires choosing a basis for the span of $\mathbf{y}[n]$. We do this by evaluating the 4×4 correlation matrix $\mathbf{Q} = (1/L) \sum_{n=1}^{L} \mathbf{y}[n] \mathbf{y}^*[n]$ where $d[n]$ where chosen as iid zero-mean Gaussian random variables with unit variance and L is the number of time frames set to $L = 1000$. We then computed the square root of $\mathbf{Q}$ using the eigendecomposition. Namely, if $\mathbf{Q}$ has an eigendecomposition $\mathbf{Q} = \mathbf{M}\Sigma\mathbf{M}^*$ with $\mathbf{M}$ unitary and Σ diagonal, then we choose $\mathbf{V} = \mathbf{M}\Sigma^{1/2}\mathbf{M}^*$, resulting in

$$\mathbf{V} = \begin{bmatrix} 5.14 & 0.77 & 2.51 & 4.22 \\ 0.77 & 5.98 & 4.50 & 5.21 \\ 2.51 & 4.50 & 4.19 & 5.35 \\ 4.22 & 5.21 & 5.35 & 7.12 \end{bmatrix}. \qquad (13.36)$$

We next solve the MMV problem $\mathbf{V} = \mathbf{A}\mathbf{U}$ where we seek the matrix $\mathbf{U}$ with as few nonzero rows as possible. MMV techniques where discussed at length in Section 11.6. Since in our problem there is no noise, and $\{\mathbf{d}[n]\}$ span the full two-dimensional space, we can use the simple test given by (11.184). This test states that the value i is in the support if

$$\|(\mathbf{I} - \mathbf{V}\mathbf{V}^\dagger)\mathbf{a}_i\|_2 = 0, \tag{13.37}$$

and results in the correct support S (i.e. $\{2, 5\}$). In fact, this test yields the correct support when using only $k+1$ measurement sequences, or three in our case. See the discussion in Section 11.6 for more details. In the presence of noise we may instead use any of the well-known MMV algorithms to determine the support. Finally, the signal $\mathbf{d}[n]$ is obtained by

$$\mathbf{d}_S[n] = \mathbf{A}_S^\dagger \mathbf{y}[n], \quad \mathbf{d}_{S^c}[n] = \mathbf{0}. \tag{13.38}$$

Example 13.2 Our next example treats a simplified version of the multiband problem and suggests a sampling scheme based on Theorem 13.1. We point out that Chapter 14 is devoted entirely to multiband sampling. There, we introduce practical sampling methods that can be implemented using standard hardware elements. Our purpose here is not to offer a practical sampling approach for multiband signals, but rather to illustrate the use of Theorem 13.1 in a simple setting.

Suppose that $x(t)$ is a complex signal that consists of at most two frequency bands, each of width no larger than B. In addition, the signal is bandlimited to $5B$. For simplicity, we will further assume that the bands are limited to one of the intervals $[(m-1)B, mB)$ for $m = 1, \ldots, 5$. As we have seen earlier in the chapter, to treat the multiband problem within the sparse SI setting we do not need to assume that the bands are aligned with a grid of size B. This is done in this example merely for convenience. The more general setting will be treated in Chapter 14.

A typical example of a signal $x(t)$ is depicted in Fig. 13.7. If the band locations are known then we can recover such a signal from samples at a rate of $2B/(2\pi)$, which is smaller than the Nyquist rate $5B/(2\pi)$. One simple way to achieve this rate is to modulate each of the nonzero bands down to baseband, filter it, and then sample it using the standard Shannon–Nyquist theorem.

We now use Theorem 13.1 to sample $x(t)$ at a rate lower than Nyquist even when the band locations are unknown. Clearly we can describe any $x(t)$ in the form (13.6) where $T = 2\pi/B$ and

$$H_\ell(\omega) = \begin{cases} 1, & (\ell-1)B \leq \omega \leq \ell B \\ 0, & \text{otherwise}, \end{cases} \quad 1 \leq \ell \leq 5. \tag{13.39}$$

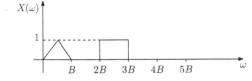

Figure 13.7 A multiband signal, consisting of two (aligned) frequency bands, each of width no larger than B.

(See also (13.7) and the discussion surrounding it, but note that here the occupied bands are assumed to be fixed to a grid.) To sample $x(t)$ we begin by finding a set of biorthogonal functions. It is easy to see that a possible choice is $v_\ell(t) = (2\pi/B)h_\ell(t)$. We next set $\mathbf{W} = \mathbf{I}$ and construct a 4×5 matrix $\mathbf{A}$ at random that has full spark, as in the previous example, resulting in

$$\mathbf{A} = \begin{bmatrix} 9 & 7 & 6 & 2 & 7 \\ 5 & 5 & 3 & 4 & 4 \\ 4 & 3 & 0 & 7 & 7 \\ 7 & 1 & 8 & 4 & 7 \end{bmatrix}. \tag{13.40}$$

Finally, the sampling functions follow from (13.25) as

$$s_i(t) = \sum_{\ell=1}^{5} \overline{\mathbf{A}_{i\ell}} v_\ell(t), \qquad 1 \le i \le 4. \tag{13.41}$$

The CTFTs of these functions are depicted in Fig. 13.8. In the Fourier domain, $S_i(\omega)$ is bandlimited to $5B$ and piecewise constant with values $\overline{\mathbf{A}_{i\ell}}$ over intervals of length B. The DTFTs of the resulting sequences of samples can be easily computed, and are depicted in Fig. 13.9.

To recover the input from the samples, we proceed as in the previous example. First, we compute

$$\mathbf{Q} = \int_0^{2\pi} \mathbf{y}(e^{j\omega})\mathbf{y}^*(e^{j\omega}) d\omega = \begin{bmatrix} 735.1327 & 386.4159 & 150.7964 & 791.6813 \\ 386.4159 & 203.1563 & 79.5870 & 415.7374 \\ 150.7964 & 79.5870 & 33.5103 & 159.1740 \\ 791.6813 & 415.7374 & 159.1740 & 856.6076 \end{bmatrix}, \tag{13.42}$$

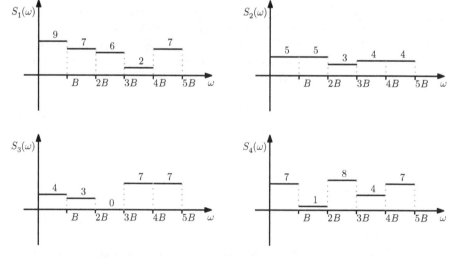

Figure 13.8 The CTFT of the sampling functions $s_1(t)$ through $s_4(t)$.

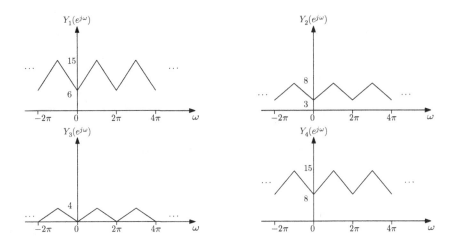

Figure 13.9 Sequences of samples in the Fourier domain.

where we used the sequences in Fig. 13.9. Next, we find the matrix square root which is given by

$$
\mathbf{V} = \begin{bmatrix} 17.41 & 9.21 & 4.02 & 18.18 \\ 9.21 & 4.88 & 2.26 & 9.44 \\ 4.02 & 2.26 & 2.02 & 2.83 \\ 18.18 & 9.44 & 2.83 & 20.70 \end{bmatrix}.
$$

Finally, we determine the support S by choosing the indices i for which $\| (\mathbf{I} - \mathbf{V}\mathbf{V}^\dagger) \mathbf{a}_i \|_2 = 0$, which results in the correct support: $i = 1, 3$. The sequences $d_1[n]$ and $d_3[n]$ are then the elements of

$$
\mathbf{d}_S[n] = \mathbf{A}_S^\dagger \mathbf{y}[n] = \begin{bmatrix} 9 & 6 \\ 5 & 3 \\ 4 & 0 \\ 7 & 8 \end{bmatrix}^\dagger \mathbf{y}[n] = \begin{bmatrix} 0.077 & 0.056 & 0.145 & -0.079 \\ -0.033 & -0.037 & -0.166 & 0.164 \end{bmatrix} \mathbf{y}[n].
$$

$$(13.43)$$

Their DTFTs are depicted in Fig. 13.10.

As in the previous example, the number of sampling filters can be reduced to three by exploiting the joint sparsity pattern.

13.3 Application to detection

In this section, we illustrate how the ideas of analog CS can be applied to standard detection scenarios treated in communication systems, in order to reduce the receiver complexity. Alternatively, with the same receiver, analog CS may be used to convey

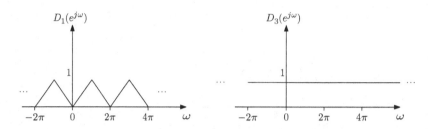

Figure 13.10 The outputs of the CTF block $D_1(e^{j\omega})$ and $D_3(e^{j\omega})$. Note that $D_2(e^{j\omega}) = D_4(e^{j\omega}) = D_5(e^{j\omega}) = 0$.

more information over the channel. We begin by providing some basic background on detection over noisy channels and review the maximum likelihood criterion and the matched-filter receiver. We then show how to simplify the receiver structure by exploiting analog CS principles.

13.3.1   Matched-filter receiver

Consider a basic communication system in which a transmitter conveys digital data to a receiver by sending one of a set of N known signals $\{h_i(t), 1 \le i \le N\}$ over a symbol duration of T. Below, we treat processing over a duration of T, so that for simplicity, we omit the time interval T. To further simplify the exposition, we assume that the waveforms $\{h_i(t)\}$ are linearly independent, and denote by $\mathcal{S}$ the N-dimensional subspace spanned by these signals. We also make the assumption that the waveforms have unit energy: $\|h_i(t)\| = 1$ for all i, and that they are transmitted with equal probability. The channel corrupts the signal by zero-mean white Gaussian noise $n(t)$ with variance σ^2, so that the received signal is given by

$$y(t) = h_\ell(t) + n(t), \tag{13.44}$$

for some index ℓ. Our goal is to design a receiver to determine the index ℓ in order to decode the transmitted symbol.

To describe the receiver operation, it is convenient to decompose it into two parts: a demodulator and a detector. The demodulator projects the received signal $y(t)$ onto an M-dimensional subspace by filtering $y(t)$ with M filters $\overline{s_i(-t)}$ and sampling their outputs at $t = 0$. The demodulator output is a vector $\mathbf{y}$ of length M with elements $y_i = \langle s_i(t), y(t)\rangle$, as depicted in Fig. 13.11. Typically, this subspace is chosen to be equal to the range space $\mathcal{S}$ of all possible signals $\{h_i(t)\}$ resulting in $M = N$. The detector then uses $\mathbf{y}$ to decide which signal was actually transmitted.

An important question is how to design the filters $s_i(t)$ in the demodulator. A standard approach is to choose $s_i(t) = h_i(t)$. The corresponding system is referred to as the *matched-filter (MF) receiver*. A well-known fact is that the MF maximizes the output SNR which is a simple consequence of the Cauchy–Schwarz inequality (see Exercise 9 for a derivation of this result). More importantly, the MF receiver is a *sufficient statistic*

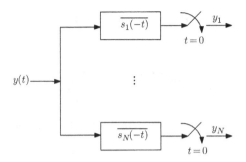

Figure 13.11 A typical matched-filter receiver.

for detection; that is, the optimal detector that minimizes the probability of a detection error assuming Gaussian noise can be computed based on the MF outputs. Furthermore, any other statistic computed from the received signal will not provide additional information about the transmitted signal.

To see this, we first write the MF outputs y_i in terms of their signal and noise components: $y_i = s_i + n_i$ where $s_i = \langle h_i(t), h_\ell(t) \rangle$ and $n_i = \langle h_i(t), n(t) \rangle$. Let $\{v_i(t)\}$ be the biorthogonal functions associated with $\{h_i(t)\}$. Then, for any $x(t)$ in the space $\mathcal{S}$ spanned by the signals $\{h_i(t)\}$ we have

$$x(t) = \sum_{i=1}^{N} \langle h_i(t), x(t) \rangle v_i(t). \tag{13.45}$$

Using (13.45) with $x(t) = h_\ell(t)$, the received signal can be expressed as

$$y(t) = h_\ell(t) + n(t) = \sum_{i=1}^{N} (s_i + n_i) v_i(t) + w(t) = g(t) + w(t) \tag{13.46}$$

where

$$g(t) = \sum_{i=1}^{N} (s_i + n_i) v_i(t) = \sum_{i=1}^{N} y_i v_i(t) \tag{13.47}$$

is the part of the received signal that depends on the MF output, and

$$w(t) = n(t) - \sum_{i=1}^{N} n_i v_i(t), \tag{13.48}$$

is a function of the noise.

We now show that $w(t)$ is statistically independent of $g(t)$ so that $w(t)$ does not contain any information in it that is relevant to the decision as to which signal was transmitted. We emphasize that since the noise is Gaussian, to show statistical independence it is sufficient to show that $g(t)$ and $w(t)$ are uncorrelated.

First, owing to the fact that the noise has zero mean, $E[w(t)] = 0$. We next compute the correlation between $w(t)$ and $g(t)$ given that $h_\ell(t)$ has been transmitted, where we assume for simplicity that the noise is real (the same conclusion holds for complex noise as well). From (13.47),

$$E[w(t)g(t)] = \sum_{i=1}^{N} s_i E[w(t)]v_i(t) + \sum_{i=1}^{N} E[n_i w(t)]v_i(t) = \sum_{i=1}^{N} E[n_i w(t)]v_i(t).$$
(13.49)

Recalling that $n_i = \langle h_i(t), n(t) \rangle$ and that the noise is white we have

$$E[n_i w(t)] = \int h_i(\tau) E[n(\tau)n(t)]d\tau - \sum_{j=1}^{N} v_j(t) \int \int h_i(\tau)h_j(s) E[n(\tau)n(s)]d\tau ds$$

$$= \sigma^2 \int h_i(\tau)\delta(t-\tau)d\tau - \sigma^2 \sum_{j=1}^{N} v_j(t) \int \int h_i(\tau)h_j(s)\delta(s-\tau)d\tau ds$$

$$= \sigma^2 h_i(t) - \sigma^2 \sum_{j=1}^{N} v_j(t)\langle h_j(t), h_i(t)\rangle = 0,$$
(13.50)

where in the first equality we used (13.48) and the last equality is a result of (13.45) applied to $x(t) = h_i(t)$. We conclude that all the relevant information for detection is in the MF output $\mathbf{y}$.

13.3.2 Maximum-likelihood detector

Suppose we wish to design a receiver that maximizes the probability of correct detection. Namely, given the measurements $\mathbf{y}$, we would like to maximize $P(\ell|\mathbf{y}) = P(\text{signal } h_\ell(t) \text{ was transmitted}|\mathbf{y})$. This decision criterion is called the *maximum a posteriori (MAP) criterion* [394]. Using Bayes' rule, we can write these probabilities in terms of the *likelihood functions* $P(\mathbf{y}|\ell)$:

$$P(\ell|\mathbf{y}) = \frac{P(\mathbf{y}|\ell)P(\ell)}{P(\mathbf{y})}.$$
(13.51)

Here $P(\ell)$ is the probability of transmitting the ℓth signal, $P(\mathbf{y})$ is the probability of observing the measurement vector $\mathbf{y}$, and $P(\mathbf{y}|\ell)$ is the probability of observing $\mathbf{y}$ assuming the ℓth signal was transmitted. In our setting, we assume that all signals are transmitted with equal probability so that $P(\ell) = 1/N$. Since $P(\mathbf{y})$ is independent of ℓ it follows that maximizing $P(\ell|\mathbf{y})$ is equivalent to maximizing the likelihood $P(\mathbf{y}|\ell)$. Performing detection by maximizing the likelihood is referred to as the *maximum-likelihood (ML) criterion*. When the prior probabilities are equal, MAP and ML decision rules coincide.

We now explicitly compute the likelihood function. For convenience we write $\mathbf{y}$ as $\mathbf{y} = \mathbf{s} + \mathbf{n}$. Here $\mathbf{s} = H^* h_\ell(t)$ is a vector containing the signal components $s_i = \langle h_i(t), h_\ell(t) \rangle$ and $\mathbf{n} = H^* n(t)$ consists of the noise elements $n_i = \langle h_i(t), n(t) \rangle$, where $H: \mathbb{R}^N \to L_2$ is the set transformation corresponding to $\{h_i(t)\}$. The signal component $\mathbf{s}$ is dependent on our choice of transmitted signal $h_\ell(t)$, and can be expressed as $\mathbf{s} = H^* H \mathbf{e}_\ell$ with $\mathbf{e}_\ell$ denoting the ℓth unit vector. Clearly $\mathbf{y}$ is a Gaussian vector with mean

$\mathbf{s} = \mathbf{s}(\ell)$ where here we explicitly introduce the dependence on ℓ. Its covariance is given by the covariance of the noise, which can be computed as

$$E[n_i n_j] = \int \int h_i(t) h_j(\tau) E[n(t)n(\tau)] dt d\tau = \sigma^2 \int h_i(t) h_j(t) dt. \qquad (13.52)$$

In matrix form, $E[\mathbf{nn}^*] = \sigma^2 H^* H$. The log likelihood function can then be expressed as

$$\ln P(\mathbf{y}|\ell) = -K(\mathbf{y} - \mathbf{s}(\ell))^* (H^* H)^{-1} (\mathbf{y} - \mathbf{s}(\ell)) \qquad (13.53)$$

where K is a constant, independent of ℓ.

Since the ln function is monotonic, maximizing $P(\mathbf{y}|\ell)$ is achieved by choosing the value ℓ that minimizes the weighted distance between $\mathbf{y}$ and $\mathbf{s}(\ell)$:

$$D(\ell) = (\mathbf{y} - \mathbf{s}(\ell))^* (H^* H)^{-1} (\mathbf{y} - \mathbf{s}(\ell)). \qquad (13.54)$$

Using the fact that $\mathbf{s}(\ell) = H^* H \mathbf{e}_\ell$,

$$\mathbf{s}^*(\ell) (H^* H)^{-1} \mathbf{s}(\ell) = \mathbf{e}_\ell^* H^* H \mathbf{e}_\ell = \|h_\ell(t)\|^2 = 1, \qquad (13.55)$$

independent of ℓ. Furthermore,

$$\mathbf{y}^* (H^* H)^{-1} \mathbf{s}(\ell) = \mathbf{y}^* \mathbf{e}_\ell = y_\ell. \qquad (13.56)$$

Finally, $\mathbf{y}^* (H^* H)^{-1} \mathbf{y}$ is independent of ℓ. Therefore the likelihood is maximized by choosing the index ℓ for which $\mathcal{R}\{y_i\}$ achieves its largest value:

$$\ell = \arg\max_i \mathcal{R}\{y_i\} = \arg\max_i \mathcal{R}\{\langle h_i(t), y(t)\rangle\}. \qquad (13.57)$$

13.3.3 Compressed-sensing receiver

We now show how to exploit the ideas of analog CS to reduce the complexity of the MF receiver. Namely, instead of using a demodulator consisting of N filters, as in Fig. 13.11, we reduce the number of filters, and in return modify the detector (13.57). We begin by showing how to reformulate the detection problem as a CS recovery problem.

Using the set transformation H, we can write any signal $h_\ell(t)$ in the form $h_\ell(t) = H\mathbf{x}$ where $\mathbf{x}$ is a sparse vector that contains a 1 in the ℓth position, and is 0 otherwise. Thus, the transmitted signal is sparse under the basis defined by the transformation H. This scenario may be viewed as a special case of (13.6) where we consider only one symbol interval. With this interpretation, we can apply analog CS in order to recover $\mathbf{x}$ using a front-end receiver that consists of $p < N$ correlators, chosen according to Theorem 13.1. Specifically, let $\mathbf{A}$ be an arbitrary $p \times N$ matrix that is designed for CS of a 1-sparse vector. The proposed demodulator then consists of filters $\{s_\ell(-t), 1 \le \ell \le p\}$, where

$$s_\ell(t) = \sum_{m=1}^{N} \overline{\mathbf{A}_{\ell m}} v_m(t), \qquad (13.58)$$

and $\{v_m(t)\}$ are the biorthogonal functions defined by

$$v_m(t) = \sum_{i=1}^{N} \phi_{mi} h_i(t),\tag{13.59}$$

with $\mathbf{\Phi} = (H^*H)^{-1}$. In operator notation, $S = V\mathbf{A}^* = H(H^*H)^{-1}\mathbf{A}^*$ and $V^*H = I$. Note that in the simple case in which all signals $\{h_\ell(t)\}$ are orthogonal, $v_\ell(t) = h_\ell(t)$.

Now that we have reduced the number of filters in the demodulator, it remains to modify the detector. Clearly, it is no longer appropriate to simply choose the maximum output value. To develop a meaningful detector, we begin by considering the case in which there is no noise added to the channel. This will allow us to gain intuition into the detector structure. We then generalize the results to account for noise.

Noise-free case

Suppose that $y(t) = h_\ell(t) = H\mathbf{x}$ for some index ℓ. After applying the proposed demodulator, the output vector, denoted by $\mathbf{c}$, is equal to

$$\mathbf{c} = S^* y(t) = \mathbf{A}(H^*H)^{-1}H^* y(t) = \mathbf{Ax}.\tag{13.60}$$

Thus, our problem reduces to that of recovering a 1-sparse vector $\mathbf{x}$ from compressed measurements $\mathbf{c} = \mathbf{Ax}$. We can now go ahead and apply any of the standard CS algorithms described in Chapter 11 to obtain an estimate $\hat{\mathbf{x}}$ of $\mathbf{x}$. The transmitted signal is then chosen as the index ℓ for which $\mathcal{R}\{\hat{x}_\ell\}$ is maximized.

Since the vector $\mathbf{x}$ is known to be 1-sparse, the CS algorithms and appropriate conditions on $\mathbf{A}$ to ensure recovery can be greatly simplified. In particular, the requirement that every $2k = 2$ columns of $\mathbf{A}$ are linearly independent boils down to the condition that $\mathbf{A}$ is a $2 \times N$ matrix with the property that no two columns are multiples of each other. In addition, we construct $\mathbf{A}$ to have normalized columns. We can then recover the support of $\mathbf{x}$ simply by choosing

$$\ell = \arg\max_i \mathcal{R}\{\langle \mathbf{a}_i, \mathbf{c}\rangle\}.\tag{13.61}$$

This follows from noting that $\mathbf{c} = \mathbf{Ax} = \mathbf{a}_\ell$, and from a straightforward application of the Cauchy–Schwarz inequality:

$$\mathcal{R}\{\langle \mathbf{a}_i, \mathbf{c}\rangle\} = \mathcal{R}\{\langle \mathbf{a}_i, \mathbf{a}_\ell\rangle\} \le 1,\tag{13.62}$$

with equality if and only if $\mathbf{a}_i$ is a multiple of $\mathbf{a}_\ell$. From our assumption on $\mathbf{A}$, this happens only when $i = \ell$. Note that this criterion is similar in spirit to the MF decision rule (13.57): the received signal is projected onto each of the possible columns, and the transmitted signal is chosen as the one for which this projection is maximized.

To conclude, in the absence of noise we can recover the transmitted signal exactly by using a demodulator consisting of only two filters, regardless of the number of signals N. These filters are chosen as a linear combination of all biorthogonal signals $\{v_i(t), 1 \le i \le N\}$, where the combinations are given by the rows of a $2 \times N$ matrix $\mathbf{A}$. The only condition on $\mathbf{A}$ is that no two columns are the same (up to a constant). Recovery is obtained by computing the inner products $\mathcal{R}\{\langle \mathbf{a}_i, \mathbf{c}\rangle\}$ between each column of $\mathbf{A}$ and the demodulator output, and choosing the maximum. The overall receiver is depicted in Fig. 13.12.

We illustrate the proposed CS receiver in the context of a concrete example.

Example 13.3 Consider a transmitter sending one of the four signals depicted in Fig. 13.13. A standard MF detector requires correlating the received signal with these four waveforms and then choosing the maximum value. Relying on analog CS, we now demonstrate perfect detection of the input using only two correlators. For simplicity and without loss of generality we assume that $T = 1$.

To implement the receiver of Fig. 13.12 we first note that since the functions $h_i(t)$ are orthonormal, $v_i(t) = h_i(t)$. Next, we choose a matrix $\mathbf{A}$ with normalized columns that satisfies the requirements of CS for $k = 1$. As we have seen, this reduces to the condition that no two columns are multiples of each other. In this example, we choose

$$\mathbf{A} = \begin{bmatrix} 0 & \frac{\sqrt{2}}{2} & 1 & \frac{\sqrt{2}}{2} \\ 1 & \frac{\sqrt{2}}{2} & 0 & -\frac{\sqrt{2}}{2} \end{bmatrix}. \tag{13.63}$$

Using (13.58), the new correlating signals are given by

$$s_1(t) = \sum_{m=1}^{4} \mathbf{A}_{1m} v_m(t) = \frac{\sqrt{2}}{2} h_2(t) + h_3(t) + \frac{\sqrt{2}}{2} h_4(t)$$

$$s_2(t) = \sum_{m=1}^{4} \mathbf{A}_{2m} v_m(t) = h_1(t) + \frac{\sqrt{2}}{2} h_2(t) - \frac{\sqrt{2}}{2} h_4(t). \tag{13.64}$$

These signals are depicted in Fig. 13.14.

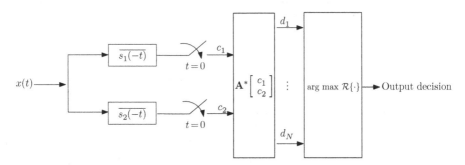

Figure 13.12 A noise-free digital detector based on analog compressed sensing. The functions $s_\ell(t)$ are given by (13.58) and (13.59). Only two branches are needed regardless of the number N of transmitted signals.

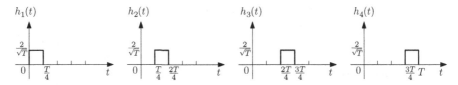

Figure 13.13 Signal waveforms used for transmission.

Table 13.1 The signal obtained at different points in the receiver of Fig. 13.12

Transmitted signal	$[c_1$	$c_2]$		$A^* \begin{bmatrix} c_1 \\ c_2 \end{bmatrix}$			Detected signal
$h_1(t)$	$[0$	$1]$	$\begin{bmatrix} 1$	$\frac{\sqrt{2}}{2}$	0	$-\frac{\sqrt{2}}{2} \end{bmatrix}$	$h_1(t)$
$h_2(t)$	$\begin{bmatrix} \frac{\sqrt{2}}{2}$	$\frac{\sqrt{2}}{2} \end{bmatrix}$	$\begin{bmatrix} \frac{\sqrt{2}}{2}$	1	$\frac{\sqrt{2}}{2}$	$0 \end{bmatrix}$	$h_2(t)$
$h_3(t)$	$[1$	$0]$	$\begin{bmatrix} 0$	$\frac{\sqrt{2}}{2}$	1	$\frac{\sqrt{2}}{2} \end{bmatrix}$	$h_3(t)$
$h_4(t)$	$\begin{bmatrix} \frac{\sqrt{2}}{2}$	$-\frac{\sqrt{2}}{2} \end{bmatrix}$	$\begin{bmatrix} -\frac{\sqrt{2}}{2}$	0	$\frac{\sqrt{2}}{2}$	$1 \end{bmatrix}$	$h_4(t)$

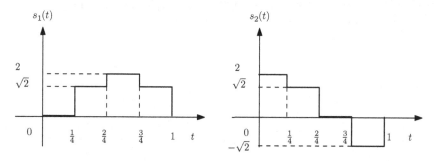

Figure 13.14 The sampling functions in the CS detector.

The CS detector results from inserting these signals into Fig. 13.12. Table 13.1 indicates the values obtained at different points in the receiver when the input is each of the possible transmitted signals (i.e. $h_1(t)$ through $h_4(t)$). As can be seen, the receiver successfully decodes the input in all cases.

Noisy setting

Until now, we have shown that only two correlators are required to recover the transmitted waveform exactly. On the other hand, we have seen that the MF, which consists of N correlators, maximizes the probability of correct detection. To understand this apparent paradox, note that the scheme of Fig. 13.12 is guaranteed to correctly detect the transmitted signal only in the noise-free setting. When noise is present, the number of branches must increase in order to achieve good performance. In this case, strictly maximizing the probability of correct detection will require N filters, as in the MF. However, we can get pretty good performance by using $p < N$ correlators, as we show below. This allows us to simplify the receiver at the expense of only a small impact on performance, as in standard CS. The difference is that here we are interested in detection, while in CS we focused on estimation.

The question we need to address is how to modify the detector in Fig. 13.12 when using $p > 2$ correlators, in order to account for noise. To answer this question, we follow the same steps used to develop the MF receiver and seek the MAP detector that maximizes the probability of correct detection $P(\ell|\mathbf{c})$ given the output $\mathbf{c} = S^* y$, where the correlating functions corresponding to S are given by (13.58) and (13.59). To compute the required probabilities, we first determine the statistics of $\mathbf{c}$.

Using the fact that $S = V\mathbf{A}^*$ and $V^*H = I$ we have

$$\mathbf{c} = S^*y(t) = \mathbf{A}V^*h_\ell(t) + \mathbf{A}V^*n(t) = \mathbf{A}V^*H\mathbf{x} + \mathbf{A}V^*n(t) = \mathbf{A}\mathbf{x} + \mathbf{w}, \quad (13.65)$$

where $\mathbf{w} = \mathbf{A}V^*n(t)$ is the noise component. Following similar computations as in (13.52), it can be shown that

$$E[\langle v_j(t), n(t)\rangle \langle n(t), v_i(t)\rangle] = \sigma^2 \langle v_j(t), v_i(t)\rangle. \quad (13.66)$$

Therefore,

$$\mathbf{R}_w = E[\mathbf{w}\mathbf{w}^*] = \sigma^2 \mathbf{A}V^*V\mathbf{A}^* = \sigma^2 \mathbf{A}(H^*H)^{-1}\mathbf{A}^*. \quad (13.67)$$

If the signals $h_\ell(t)$ are orthogonal, and in addition $\mathbf{A}$ is chosen such that its rows are orthonormal with squared-norm c, then $\mathbf{R}_w = c\sigma^2\mathbf{I}$ and $\mathbf{w}$ is a white noise vector. However, in general, the noise will not be white.

Our problem is to recover $\mathbf{x}$ from noisy measurements $\mathbf{c} = \mathbf{A}\mathbf{x} + \mathbf{w}$ where we know that $\mathbf{x}$ is 1-sparse. Although in principle we can use standard CS algorithms for this task, since we are interested in detection we proceed to develop the MAP detector. Denote by $P(\ell|\mathbf{c})$ the probability that x_ℓ is nonzero, given $\mathbf{c}$. Our goal is to choose the ℓ that maximizes this probability. By applying the Bayes rule (cf. (13.51)), and using the fact that $P(\ell) = 1/N$, our problem becomes that of maximizing $P(\mathbf{c}|\ell)$. From (13.65) and (13.67), $\mathbf{c}$ is a Gaussian vector with mean $\mathbf{A}\mathbf{x}$ and covariance $\mathbf{R}_w = \sigma^2\mathbf{A}(H^*H)^{-1}\mathbf{A}^*$. We assume that $\mathbf{A}$ has full row-rank, as is always the assumption in CS, so that $\mathbf{R}_w$ is invertible. In this case,

$$\ln P(\mathbf{c}|\ell) = -K(\mathbf{c} - \mathbf{a}_\ell)^*(\mathbf{A}(H^*H)^{-1}\mathbf{A}^*)^{-1}(\mathbf{c} - \mathbf{a}_\ell) \quad (13.68)$$

where K is a constant, independent of ℓ. Maximizing $P(\mathbf{c}|\ell)$ reduces to choosing the value of ℓ that minimizes the weighted distance between $\mathbf{c}$ and the ℓth column of $\mathbf{A}$:

$$D(\ell) = (\mathbf{c} - \mathbf{a}_\ell)^*(\mathbf{A}(H^*H)^{-1}\mathbf{A}^*)^{-1}(\mathbf{c} - \mathbf{a}_\ell). \quad (13.69)$$

In the special case in which the signals $h_\ell(t)$ are orthonormal, and the rows of $\mathbf{A}$ are orthogonal with equal norm, the matrix $\mathbf{A}(H^*H)^{-1}\mathbf{A}^*$ is proportional to the identity. If in addition the columns of $\mathbf{A}$ are chosen to have equal norm, then minimizing $D(\ell)$ is equivalent to the selection rule

$$\ell = \arg\max_i \mathcal{R}\{\langle \mathbf{a}_i, \mathbf{c}\rangle\}, \quad (13.70)$$

which is the same as the noise-free decision rule given by (13.61). Note that in the real case, this is also the first step in the OMP algorithm for CS recovery, or the thresholding method. However, in general, the covariance matrix of the noise should be taken into account by minimizing (13.69).

A more explicit form for the decision rule (13.69) can be obtained by whitening the received vector. Suppose that we whiten the data $\mathbf{c}$, multiplying it by $\mathbf{R}_w^{-1/2}$

(see Appendix A for a definition of the matrix square root) to create the new measurement vector

$$\mathbf{z} = \mathbf{R}_w^{-1/2}\mathbf{c} = \widetilde{\mathbf{A}}\mathbf{x} + \tilde{\mathbf{w}}. \tag{13.71}$$

Here $\tilde{\mathbf{w}} = \mathbf{R}_w^{-1/2}\mathbf{w}$ is a white noise vector, and the new measurement matrix is $\widetilde{\mathbf{A}} = \mathbf{R}_w^{-1/2}\mathbf{A}$. We next consider maximizing the likelihood $P(\mathbf{z}|\ell)$. Following the same steps as above, it is easy to see that the optimal value of ℓ minimizes

$$\widetilde{D}(\ell) = (\mathbf{z} - \tilde{\mathbf{a}}_\ell)^*(\mathbf{z} - \tilde{\mathbf{a}}_\ell). \tag{13.72}$$

This in turn is equivalent to the decision rule

$$\ell = \arg\max_i\{2\mathcal{R}\{\langle\tilde{\mathbf{a}}_i, \mathbf{z}\rangle\} - \|\tilde{\mathbf{a}}_i\|^2\}. \tag{13.73}$$

Clearly $\widetilde{D}(\ell)$ and $D(\ell)$ are equal. However, (13.73) allows one to interpret the decision rule in a simple form. We begin by whitening the data. We then project the new measurement vector onto each of the columns of $\widetilde{\mathbf{A}}$, and account for the possibly different column norms. Viewed in this way, the MAP receiver is closely connected to CS algorithms, since it can be interpreted as the first step in standard iterative techniques (such as OMP – see Algorithm 11.1) after accounting for the channel noise.

Example 13.4 We demonstrate the effect of additive noise on the proposed CS detector via an example. Consider a receiver like the one introduced in Example 13.3, with $N = 100$ different transmitted signals given by

$$h_\ell(t) = \begin{cases} 1, & (\ell - 1) \leq t \leq \ell \\ 0, & \text{otherwise,} \end{cases} \qquad \ell = 1, 2, \ldots, N, \tag{13.74}$$

and p correlators. The sensing matrix $\mathbf{A}$ is chosen to be equal to p random rows of the $N \times N$ Fourier matrix, with columns normalized to have unit norm.

In Fig. 13.15 we plot the probability of correct detection as a function of the number of correlators p for different values of the SNR. The probability is estimated by running 1000 Monte Carlo simulations. In every iteration, we choose the transmitted signal, the noise, and the p rows of the Fourier matrix at random. While the MF receiver consists of 100 correlators, we see that perfect detection is achieved using a much smaller number of filters in the analog CS-based receiver, so long as the SNR is high enough. As expected, the number of correlators required increases when the SNR decreases.

Noise folding

Example 13.4 demonstrates that the MF performance can be approached using an analog CS receiver with fewer branches. Nonetheless, with a small number of correlators the performance deteriorates quite rapidly as a function of SNR. To understand this behavior, note that the use of the CS matrix $\mathbf{A}$ reduces the SNR. This can be seen by examining the statistics of the noise vector $\mathbf{w}$ in (13.65).

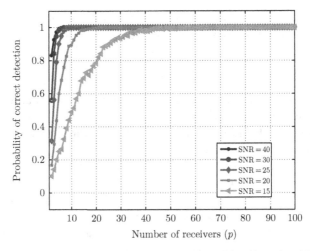

Figure 13.15 Probability of correct detection using the analog CS-based receiver (13.73) as a function of the number of correlators for different SNR values.

Suppose, for simplicity, that the signals $\{h_i(t)\}$ are orthonormal, resulting in $H^*H = I$. Then, from (13.67), $\mathbf{R}_w = \sigma^2 \mathbf{A}\mathbf{A}^*$. Consider the case in which $\mathbf{A}$ is chosen as random rows of a Fourier matrix, like in Example 13.4. Since the columns of $\mathbf{A}$ are assumed to be normalized, this means that $\mathbf{A}_{i\ell} = (1/\sqrt{p}) \exp\{-j2\pi s_i \ell/N\}$ where s_i is the ith row chosen. In this case, $\mathbf{A}\mathbf{A}^* = (N/p)\mathbf{I}$. Therefore, the noise variance in $\mathbf{w}$ is increased by a factor of N/p. Similar behavior can be demonstrated for more general CS matrices; see [310]. This effect is referred to as *noise folding*: the CS matrix $\mathbf{A}$ aliases, or combines, all the noise components, even those corresponding to zero elements in $\mathbf{x}$, thus leading to a noise increase in the compressed measurements. To compensate for this effect, a sufficient number of filters must be employed. In [395], it is shown that around $\log N$ correlators are needed to counterbalance this noise increase.

13.4 Multiuser detection

We now extend the detection ideas from the single-user setting described in the previous section, to a multiuser setting, following the work of [395].

Multiuser detection (MUD) is a classical problem in multiuser communications and signal processing [396]. In multiuser systems, users communicate simultaneously with a given receiver by modulating information symbols onto their unique signature waveforms (also called the spreading code), which we denote by $\{h_i(t), 1 \leq i \leq N\}$. Here N is the number of users supported by the system. We will assume that the signatures are linearly independent and have unit norm, and that the transmitter and receiver are synchronized. The data are chosen from a given constellation, depending on the specific modulation scheme. For simplicity, we consider binary phase shift keying (BPSK), so that the ith user's bit takes on values $b_i = \pm 1$. Each user modulates its data onto the corresponding signature waveform over a symbol duration period of T. The received signal consists of a noisy version of the superposition of the transmitted waveforms:

$$y(t) = \sum_{|i|=k} r_i b_i h_i(t) + n(t), \quad 0 \leq t \leq T, \tag{13.75}$$

where $n(t)$ is zero-mean Gaussian white noise with variance σ^2, and r_i is the channel gain for user i which includes the transmitted signal power. The values of r_i are assumed to be real (but may be negative) and known at the receiver. The sum in (13.75) is only over the active users in the system. That is, in any given symbol time, only k out of the N users are actually transmitting. The goal of MUD is to detect the symbols $\{b_i\}$ of all active users simultaneously.

Based on the ideas of [395], we show below how to exploit the fact that typically the number of active users k is much smaller than the number of users N supported by the system, in order to reduce the complexity of standard MUDs. The complexity reduction is achieved by using analog CS in order to decrease the number of correlators at the receiver front end, similar to our approach in single-user detection.

In the next subsection we discuss conventional MUDs that do not assume sparsity in the number of users. We then show how these approaches can be modified to take sparsity into account.

13.4.1   Conventional multiuser detectors

Conventional MUD structures typically consist of an MF front end, matched to the signature waveforms, followed by a linear or nonlinear digital MUD. As in the single-user case, the MF obtains a set of sufficient statistics when the receiver noise is Gaussian. However, if the signature waveforms are nonorthogonal, then there is mutual interference among users, which degrades system performance. MUD schemes aim at trying to recover the data in the presence of this interference.

Let $\mathbf{y}$ denote the output of the MF with filters $\overline{h_i(-t)}$. Then,

$$\mathbf{y} = H^* y(t) = \mathbf{G}\mathbf{R}\mathbf{b} + \mathbf{n}, \tag{13.76}$$

where we denoted by $\mathbf{G} = H^* H$ the Gram matrix of inner products, by $\mathbf{R}$ the diagonal matrix with diagonal elements r_i, and by $\mathbf{b} = [b_1, \ldots, b_N]^T$ the symbol vector where we set $b_i = 0$ for inactive users. The vector $\mathbf{n} = H^* n(t)$ represents the noise at the MF output which is zero-mean Gaussian, and has covariance $\mathbf{R}_n = \sigma^2 \mathbf{G}$ (cf. (13.52)). To recover user data from the MF output, various digital detectors have been developed. These can be roughly categorized into linear and nonlinear detectors.

Linear detectors apply a linear transform $\mathbf{T}$ to the MF output and then detect the symbol for each user using a sign detector:

$$\hat{b}_i = \text{sign}(r_i[\mathbf{T}\mathbf{y}]_i). \tag{13.77}$$

Note that the standard MUD model assumes that all users are active and therefore $\hat{b}_i \in \pm 1$. Several commonly used detectors are the single-user detector which ignores interference between users and corresponds to $\mathbf{T} = \mathbf{I}$, the decorrelator, and the minimum mean-squared error (MMSE) detector. The decorrelator aims at

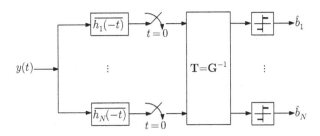

Figure 13.16 Decorrelating detector for multiuser detection.

removing the interference between users by applying $\mathbf{T} = \mathbf{G}^{-1}$, as depicted in Fig. 13.16. This choice allows the symbols to be recovered perfectly in the absence of noise. Indeed, without noise, $\mathbf{Ty} = \mathbf{Rb}$ and interference is eliminated. However, the decorrelator will generally amplify the noise when $\mathbf{G} \neq \mathbf{I}$. Note that if we consider the signals resulting from the combination of the MF front end and $\mathbf{T} = \mathbf{G}^{-1}$, then we obtain the biorthogonal set to $\{h_i(t)\}$ (see Exercise 13). Thus, the decorrelator can be equivalently implemented by correlating the incoming signal with the biorthogonal functions, and then using a single-user detector on each of the outputs. When the noise variance is known, a compromise between the decorrelator and the single-user detector is the MMSE receiver, which minimizes the MSE between the symbols and $\mathbf{Ty}$. This approach results in $\mathbf{T} = (\mathbf{G} + \sigma^2 \mathbf{R}^{-2})^{-1}$. (See Exercise 12 for a derivation of the MMSE solution.)

The decorrelator has received considerable attention in the MUD literature and is one of the most common linear detectors. Therefore, below, we will focus on this choice as an example of a linear detector in the sparse setting.

Nonlinear detectors detect symbols jointly and/or iteratively. The optimal MUD detector minimizing the probability of symbol error is the maximum likelihood sequence estimator (MLSE) [396]. It basically calculates the probability assuming each one of the possible input sequences was transmitted. The complexity per bit of the MLSE detector is exponential in the number of users when the signature waveforms are nonorthogonal, since it has to check all possible input options. For orthogonal signatures, the MLSE detector reduces to the single-user detector. An alternative prominent nonlinear detector is successive interference cancelation (SIC) [397]. This detector decodes symbols iteratively by subtracting the detected symbols of strong users in order to facilitate detection of weak users. More specifically, SIC first finds the active user with the largest gain, detects its symbol, subtracts its effect from the received signal, and iterates the process using the residual. After k iterations, all active users are determined. This approach is very similar in spirit to the OMP algorithm used for CS. Below, we will see how we can exploit these ideas in a reduced-dimensional setting in order to perform MUD using a small number of correlators.

13.4.2   Reduced-dimension MUD (RD-MUD)

As we have seen, conventional receivers require correlating the received signal with N correlators, which can be costly in systems that employ thousands of users. We now show how the ideas of analog CS can be used to reduce the number of correlators by exploiting the fact that at any given time the number of active users, k, is typically much smaller than the total number of users, N. The front-end output is then processed using algorithms that combine ideas from CS and conventional MUD. The resulting structure is referred to as a *reduced-dimension multiuser detector (RD-MUD)* [395], and achieves performance similar to that of conventional MUDs even though it uses far fewer correlators. We will consider two RD-MUD detectors: the reduced-dimension decorrelating (RDD) detector, a linear detector that combines subspace projection and thresholding to determine active users with a sign detector for data recovery, and the reduced-dimension decision-feedback (RDDF) detector, a nonlinear detector that combines decision-feedback for active user detection with sign detection for data recovery in an iterative manner.

Front-end receiver

The RD-MUD front end is equivalent to that of the CS receiver introduced in Section 13.3.3, and given by the filters described in (13.58) and (13.59), with $h_i(t)$ being the signature waveforms. Using this front end, and the fact that $y(t) = H\mathbf{R}\mathbf{b} + n(t)$, the output of the filterbank in Fig. 13.12 can be expressed as

$$\mathbf{c} = S^* y(t) = \mathbf{A}V^* y(t) = \mathbf{A}\mathbf{R}\mathbf{b} + \mathbf{w}, \tag{13.78}$$

where $\mathbf{w}$ is a zero-mean Gaussian noise vector with covariance $\mathbf{R}_w$ given by (13.67), and we used the fact that $V^*H = I$. The vector $\mathbf{c}$ can be viewed as a linear projection of the MF front-end output onto a lower-dimensional subspace referred to as the detection subspace [395]. Since there are at most k active users, $\mathbf{b}$ has at most k nonzero entries. The idea of RD-MUD is that with proper choice of $\mathbf{A}$, the detection performance based on $\mathbf{c}$ can often be made similar to the performance based on the MF output (13.76).

We next discuss how to recover $\mathbf{b}$ from the RD-MUD output $\mathbf{c}$ of (13.78) using digital detectors. In principle, (13.78) has a similar form to the observation model in standard CS. Two main differences are the fact that $\mathbf{b}$ is binary, and that the noise $\mathbf{w}$ is colored. As we have seen in Section 13.3.3, the latter is a result of the fact that the noise is present in the received signal prior to sampling, while in the standard CS literature noise is added to the measurements [310].

One approach to detect $\mathbf{b}$ is to ignore the fact that the noise is colored and use conventional CS algorithms on $\mathbf{c}$ adapted to a binary input. Alternatively, we may first whiten the data via multiplying $\mathbf{c}$ by $\mathbf{R}_w^{-1/2}$ as in (13.71) to create the new measurement vector

$$\mathbf{z} = \mathbf{R}_w^{-1/2}\mathbf{c} = \widetilde{\mathbf{A}}\mathbf{R}\mathbf{b} + \widetilde{\mathbf{w}}, \tag{13.79}$$

where now $\widetilde{\mathbf{w}}$ is a white noise vector, and the new measurement matrix is given by $\widetilde{\mathbf{A}} = \mathbf{R}_w^{-1/2}\mathbf{A}$. Although this premultiplication whitens the noise, at the same time it modifies the measurement matrix $\mathbf{A}$. Recall from Chapter 11 that in order for CS methods to work properly, this matrix must obey certain properties. In particular, we would like its coherence, or RIP constant, to be low. By whitening the noise, we also modify the properties of $\mathbf{A}$. Thus, in practice, careful consideration is needed in order to evaluate whether the whitening step improves performance, or deteriorates it. This will depend on

---

Algorithm 13.1 RD decorrelating detector

Input: CS matrix $\mathbf{A}$, measurement vector $\mathbf{c}$, and number of active users k

Output: Detected bits $\hat{\mathbf{b}}$

Detect active users: find Λ that contains indices of the k largest $|\mathcal{R}\{\langle \mathbf{a}_i, \mathbf{c} \rangle\}|$

Detect symbols: $\hat{b}_n = \text{sign}(r_i \mathcal{R}\{\langle \mathbf{a}_i, \mathbf{c} \rangle\}), i \in \Lambda$ and $\hat{b}_n = 0, i \notin \Lambda$

---

the specific choice of signatures and matrix $\mathbf{A}$ – a more detailed discussion on this issue can be found in [310]. In general, when the signatures are not highly correlated, noise whitening does not tend to improve performance [395]. Therefore, in our development below we do not assume pre-whitening. Nonetheless, the proposed algorithms can be applied in a straightforward manner when whitening is performed by simply replacing $\mathbf{A}$ with $\widetilde{\mathbf{A}}$.

Reduced-dimension decorrelating (RDD) detector

The simplest detection approach at the output of the RD-MUD front end is to ignore the correlation in the noise vector, and apply a thresholding rule, similar in spirit to (13.70) adapted to the multiuser setting. Specifically, we look for the k columns of $\mathbf{A}$ that are most correlated with $\mathbf{c}$ by choosing the largest k values of $|\mathcal{R}\{\langle \mathbf{a}_i, \mathbf{c} \rangle\}|$, or alternatively, declare the active set of users to consist of the indices for which $|\mathcal{R}\{\langle \mathbf{a}_i, \mathbf{c} \rangle\}|$ exceeds a given threshold. Once the active set of users is determined, the actual bits are estimated by sign detection as in (13.77). Thus, the output of the RDD detector is given by

$$\hat{b}_i = \begin{cases} \text{sign}(r_i \mathcal{R}\{\langle \mathbf{a}_i, \mathbf{c} \rangle\}), & i \in \Lambda \\ 0, & i \notin \Lambda, \end{cases} \tag{13.80}$$

where Λ is the estimated active user set. This algorithm is summarized in Algorithm 13.1. Note that instead of using (13.80), once we detect the active set we can use any of the conventional linear MUD detectors on this set. More details on alternative RD-MUD structures are given in [395].

In detecting active users and their symbols, we take the real parts of the inner products because the imaginary part of $\langle \mathbf{a}_i, \mathbf{c} \rangle$ contains only noise. To see this, suppose we are interested in detecting the ith symbol by considering $\langle \mathbf{a}_i, \mathbf{c} \rangle$. We can then write

$$\langle \mathbf{a}_i, \mathbf{c} \rangle = r_i b_i + \sum_{\ell \neq i} r_\ell b_\ell \langle \mathbf{a}_i, \mathbf{a}_\ell \rangle + \langle \mathbf{a}_i, \mathbf{w} \rangle, \tag{13.81}$$

where $b_i \in \{\pm 1, 0\}$ and we assumed that the columns of $\mathbf{A}$ are normalized. Since r_i is real, the term that contains the transmitted symbol is real, while the rest of the expression consisting of interference and noise can be complex. Therefore, the desired symbol does not contribute to the imaginary part of $\langle \mathbf{a}_i, \mathbf{c} \rangle$.

Reduced-dimension decision-feedback (RDDF) detector

The RDDF detector determines active users and their corresponding symbols iteratively in an OMP-type approach summarized in Algorithm 13.2. It starts by setting all estimated bits to 0, and uses the front-end output $\mathbf{c}$ as the residual vector. Subsequently,

---

Algorithm 13.2 RD decision-feedback detector

Input: CS matrix $\mathbf{A}$, measurement vector $\mathbf{c}$, and number of active users k

Output: Detected bits $\hat{\mathbf{b}}$

Initialize: $\hat{\mathbf{b}}_0 = \mathbf{0}, \mathbf{r} = \mathbf{c}, \ell = 0$

while halting criterion false **do**

    $\ell \leftarrow \ell + 1$

    $m = \arg\max |\mathcal{R}\{\langle \mathbf{a}_i, \mathbf{r} \rangle\}|$ {detect active user}

    $\hat{b}_\ell(i) = \begin{cases} \text{sign}(r_i \mathcal{R}\{\langle \mathbf{a}_i, \mathbf{c} \rangle\}), & i = m \\ \hat{b}_{\ell-1}(i), & i \neq m \end{cases}$ {update bit estimate}

    $\mathbf{r} \leftarrow \mathbf{y} - \mathbf{A}\mathbf{R}\hat{\mathbf{b}}_\ell$ {update measurement residual}

end while

return $\hat{\mathbf{b}} \leftarrow \hat{\mathbf{b}}_\ell$

---

in each iteration, the algorithm selects the index m whose corresponding column $\mathbf{a}_m$ is most highly correlated with the residual as the detected active user in the ℓth iteration. The symbol of user m is then detected, with the other detected symbols staying the same as in the previous iteration, and the residual is updated. Note that in each iteration $\mathbf{r}_\ell$ can be expressed as $\mathbf{r}_\ell = \mathbf{r}_{\ell-1} - \mathbf{A}\mathbf{R}\mathbf{z}_\ell$, where $\mathbf{z}_\ell$ is a vector that is all zeros except in the mth position, and m is the detected active user at iteration ℓ. In this position, the value of $\mathbf{z}_\ell$ is given by the difference $\mathbf{b}_\ell(m) - \mathbf{b}_{\ell-1}(m)$.

13.4.3   Performance of RD-MUD

We now consider the performance of the RD-MUD detector. We begin by briefly addressing the choice of $\mathbf{A}$.

In principle one can select any of the matrices $\mathbf{A}$ that are used in CS. One difference is that here the choice of $\mathbf{A}$ also affects the statistics of the noise and not only the signal. This is because, as we explained before, the noise is added by the channel, prior to the front-end receiver. Therefore, we would like to choose a class of matrices $\mathbf{A}$ that have the property that when the number of correlators $p = N$, the detection performance based on RD-MUD will be the same as that using a conventional decorrelator receiver. This will be the case when $\mathbf{A}$ is constructed as rows of a unitary matrix, such as the Fourier matrix. With this choice, the statistics $\{\langle \mathbf{a}_i, \mathbf{c} \rangle\}$ on which the RDD and RDDF detectors are based have the same distribution as the decorrelator output.

To see this, note that using (13.78) and the fact that now $\mathbf{a}_i^* \mathbf{a}_j = \delta_{ij}$ (since $p = N$),

$$\langle \mathbf{a}_i, \mathbf{c} \rangle = \mathbf{a}_i^*(\mathbf{A}\mathbf{R}\mathbf{b} + \mathbf{w}) = r_i b_i + \langle \mathbf{a}_i, \mathbf{w} \rangle. \tag{13.82}$$

The random variables $\{z_i = \langle \mathbf{a}_i, \mathbf{w} \rangle\}$ are zero-mean, Gaussian, with covariance

$$E[z_i \overline{z_j}] = \mathbf{a}_i^* \mathbf{R}_w \mathbf{a}_j = \sigma^2 \mathbf{G}_{ij}^{-1}, \tag{13.83}$$

where we used (13.67) and $\mathbf{G} = H^*H$. On the other hand, the decorrelator output is given by

$$\mathbf{G}^{-1}\mathbf{y} = \mathbf{Rb} + \tilde{\mathbf{n}}, \qquad (13.84)$$

where $\mathbf{y}$ is the MF output of (13.76), and $\tilde{\mathbf{n}} = \mathbf{G}^{-1}\mathbf{n}$ with $\mathbf{n}$ being the noise at the MF output. Since $\mathbf{R}_n = \sigma^2\mathbf{G}$, the covariance of $\tilde{\mathbf{n}}$ is equal to $\sigma^2\mathbf{G}^{-1}$. Therefore the variables defined by (13.82) and (13.84) have the same statistics.

We now turn to analyze the performance of the RD-MUD detectors. The performance measure we use is the probability of error, which includes both the case in which the set of active users is detected incorrectly, and the possibility that any of their symbols are erroneously recovered. Theorem 13.2 bounds the probability of error for both the RDD and RDDF algorithms. The detailed development of these results is quite involved. Therefore, we refer the interested reader to [395], and only summarize the main conclusions.

Theorem 13.2 (RD-MUD performance). *Let $\mathbf{b}$ be a vector of bits $b_i \in \{\pm 1\}, i \in \Lambda$ and $b_i = 0$ otherwise, where Λ is the set of active users, and let k denote the number of active users. Denote by N the number of possible users in the sysem, and by*

$$|r_{\max}| = \max |r_i|, \quad |r_{\min}| = \min |r_i|. \qquad (13.85)$$

Let the output of the RD-MUD receiver be given by $\mathbf{y} = \mathbf{ARb} + \mathbf{w}$ where $\mathbf{A}$ is a $p \times N$ matrix with normalized columns and $\mathbf{w}$ is a Gaussian zero-mean noise vector with covariance $\mathbf{R}_w = \sigma^2\mathbf{AG}^{-1}\mathbf{A}^*$ with $\mathbf{G} = H^*H$ denoting the Gram matrix of the signature waveforms. Let $\mu = \max_{i \neq j} |\mathbf{a}_i^*\mathbf{a}_j|$ denote the coherence of $\mathbf{A}$. If

$$|r_{\min}| - (2k-1)\mu|r_{\max}| \geq 2\sigma\sqrt{2(1+\alpha)\log N}\sqrt{\lambda_{\max}(\mathbf{G}^{-1})}\sqrt{\max_i(\mathbf{a}_i^*\mathbf{AA}^*\mathbf{a}_i)},$$
$$(13.86)$$

for some constant $\alpha > 0$, then the probability of error for the RDD detector is upper-bounded by

$$P_e \leq N^{-\alpha}[\pi(1+\alpha)\log N]^{-1/2}. \qquad (13.87)$$

If the weaker condition

$$|r_{\min}| - (2k-1)\mu|r_{\min}| \geq 2\sigma\sqrt{2(1+\alpha)\log N}\sqrt{\lambda_{\max}(\mathbf{G}^{-1})}\sqrt{\max_i(\mathbf{a}_i^*\mathbf{AA}^*\mathbf{a}_i)},$$
$$(13.88)$$

holds, then the probability of error for the RDDF detector satisfies (13.87).

Note in Theorem 13.2 that the condition of having a small probability of error for the RDDF detector is weaker than for the RDD detector. Intuitively, the iterative approach removes the effect of the largest element in $\mathbf{Rb}$ iteratively, which helps the detection of weaker users.

A special case of the theorem is when the signatures are orthogonal so that $\mathbf{G} = \mathbf{I}$, and $\mathbf{AA}^* = c\mathbf{I}$ for some factor c, chosen such that the columns of $\mathbf{A}$ have equal norm. This happens, for example, when $\mathbf{A}$ is a partial Fourier matrix, with normalized columns

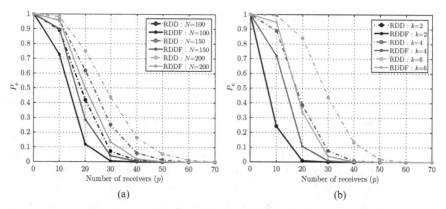

Figure 13.17 Probability of error using the RDD and RDDF detectors as a function of p. (a) The number of users is $k = 2$; shown for different values of N. (b) The number of potential users is $N = 100$; shown for different values of k.

and $c = N/p$. In this case, the noise at the RD-MUD output is white, $a_i^* \mathbf{A}\mathbf{A}^* a_i = c$ and the right-hand sides of (13.86) and (13.88) become $2\sigma\sqrt{c}\sqrt{2(1+\alpha)\log N}$. If we scale σ^2 by c, then this expression is identical to the corresponding quantity in Theorem 11.21, which describes the performance of the OMP algorithm for sparse recovery in standard CS. This demonstrates the noise folding effect we mentioned previously: the variance is increased by c. Note, however, that Theorem 11.21 only guarantees support recovery, namely, detecting the correct active user set, while our theorem here ensures correct detection of not only the active users but their symbols as well. This is because it can be shown that when the conditions for active user detection are satisfied, we are automatically guaranteed to also detect the bits correctly [395].

Example 13.5 In this example, we present numerical experiments illustrating the performance of RD-MUD.

For the matrix $\mathbf{A}$ we choose p random rows of an $N \times N$ DCT matrix and average the results over 5×10^5 Monte Carlo trials. We assume that the signature waveforms are orthonormal. In each trial, we generate a Gaussian random noise vector and random bits: $b_i \in \{-1, 1\}$ with probability $1/2$. The amplitudes are set at $r_i = 1$ for all i, the noise variance is $\sigma^2 = 0.005$, and $\mathbf{G} = \mathbf{I}$.

Figure 13.17(a) shows the probability of error P_e of the RDD and RDDF detectors as a function of the number of correlators p, for fixed sparsity $k = 2$, and different values of N. Clearly the RDDF detector has better performance than the RDD detector. As N increases, the number of receivers p needs to increase as well to ensure acceptable performance. Figure 13.17(b) demonstrates the P_e of the RDD and RDDF detectors as a function of p, for a fixed $N = 100$, and different values of k. Here again the number of correlators must increase as more users join the system.

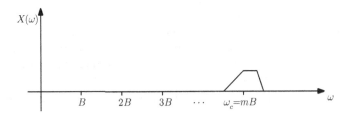

Figure 13.18 An example of a multiband signal with one band.

To summarize, we illustrated how to decrease the number of correlators at the front end of an MUD receiver by exploiting the fact that the number of active users is typically much smaller than the total number of users in the system and applying ideas of analog CS. In the next chapter we use the concept of analog CS to sample multiband signals at sub-Nyquist rates.

13.5   Exercises

1. Let $x(t)$ be an arbitrary signal in L_2, and let $\{\overline{a_\ell(-t)}, 1 \leq \ell \leq N\}$ be a set of sampling filters as depicted in Fig. 13.1.
   a. Write an expression for the DTFT $c(e^{j\omega})$ of the vector of samples $c_\ell[n] = \langle a_\ell(t - nT), x(t) \rangle$ in terms of $X(\omega)$ and $A_\ell(\omega)$.
   b. Assume now that $x(t)$ has the form (13.1). Write down an expression for the Fourier transform $X(\omega)$ of $x(t)$ in terms of $H_\ell(\omega)$ and $D_\ell(e^{j\omega})$.
   c. Use your previous answers to derive (13.2).
2. Let $x(t)$ be a signal bandlimited to π. Consider the sampling framework of Fig. 13.1 with $a_i(t) = \delta(t - t_i), 1 \leq i \leq N$ where the values of t_i are distinct, and $0 \leq t_i < 1$.
   a. Suggest a set of reconstruction filters $h(t)$ such that $x(t)$ can be recovered from the corresponding samples.
   b. Derive an explicit expression for the reconstruction filter $\mathbf{M}_{AH}^{-1}(e^{j\omega})$.
3. Consider the complex signal depicted in Fig. 13.18 where m is unknown. Assume that its bandwidth is no larger than B and that $\omega_c + B/2 \leq \pi/T$ for some T. Show how this signal can be written in the form (13.6). Determine $d_\ell[n], h_\ell(t)$, and N explicitly.
4. Let $\{\mathbf{y}[n] = \mathbf{A}\mathbf{x}[n], n \in \mathbb{Z}\}$ be a given set of vectors.
   a. Show that the span of $\{\mathbf{y}[n]\}$ is equal to the range space of $\mathbf{Q}_y = \sum_{n \in \mathbb{Z}} \mathbf{y}[n]\mathbf{y}^*[n]$.
   b. Define $\mathbf{Q}_x = \sum_{n \in \mathbb{Z}} \mathbf{x}[n]\mathbf{x}^*[n]$. Determine the relationship between $\mathbf{Q}_y$ and $\mathbf{Q}_x$.
   c. Suppose that the vectors $\mathbf{x}[n]$ are jointly k-sparse, and that the matrix $\mathbf{A}$ has spark equal to $2k + 1$. Develop an upper bound on the rank of $\mathbf{Q}_y$.
5. Let $\mathbf{v}(\omega) = \overline{\mathbf{M}_{AH}^{-1}(e^{j\omega T})}\mathbf{a}(\omega)$ be the functions defined in (13.20).
   a. Write down an explicit expression for $V_\ell\left(\frac{\omega}{T} - \frac{2\pi}{T}k\right)$ where $V_\ell(\omega)$ is the ℓth element of $\mathbf{v}(\omega)$ and k is arbitrary.
   b. Use this expression to show that $\mathbf{M}_{VH}(e^{j\omega}) = \mathbf{I}$.

c. Propose another set of functions $\{v_\ell(t), 1 \le \ell \le N\}$ for which $\mathbf{M}_{VH}(e^{j\omega}) = \mathbf{I}$.

6. In Example 13.1 it was mentioned right after (13.37) that three measurements per period are sufficient to recover $x(t)$. Provide an explicit example of sampling filters $s_i(t), 1 \le i \le 3$ such that $x(t)$ can be determined from the corresponding samples. Explain the result.

7. In this exercise, we revisit Example 13.1 considering an arbitrary generator $h(t)$ in (13.31).

   a. Suggest a sampling scheme with rate equal to 1 leading to perfect recovery for arbitrary $h(t)$.

   b. Suggest a sub-Nyquist scheme operating at a rate of 4/7 by generalizing the approach in the example to allow for an arbitrary generator.

8. We now revisit Example 13.2 using different sampling functions. Consider the functions

$$s_i(t) = \delta(t - c_i T), \qquad 1 \le i \le 4 \qquad (13.89)$$

where $T = \frac{2\pi}{B}$ and $\{c_i\}$ are four distinct integer values in the range $1 \le c_i \le 5$.

   a. Show that these filters can be obtained from our general framework incorporated in Theorem 13.1 with an appropriate choice of $\mathbf{W}(e^{j\omega})$ and $\mathbf{A}$.

   b. Develop a recovery algorithm from these samples.

   c. Discuss under what conditions it is sufficient to use only three of the sampling filters.

9. In this exercise we prove that if a signal $s(t), 0 \le t \le T$ is corrupted by additive white Gaussian noise $n(t)$ with variance N_0, then the filter with impulse response matched to $s(t)$ maximizes the output SNR among all possible filters.

   a. Let $y(t) = s(t) + n(t)$. Suppose that $y(t)$ is passed through a filter with impulse response $h(t), 0 \le t \le T$ and its output is sampled at $t = T$. Compute the signal and noise components at $t = T$.

   b. Define the output SNR as the ratio of the power of the signal component to that of the noise component. Show that the output SNR can be written as

$$\text{SNR} = \frac{\left(\int_0^T h(\tau)s(T - \tau)d\tau \right)^2}{N_0 \int_0^T h^2(T - t)dt}. \qquad (13.90)$$

   c. Use the Cauchy–Schwarz inequality to find $h(t)$ which maximizes the numerator of (13.90) and conclude that $h(t)$ is matched to $s(t)$.

10. Suppose that a transmitter uses the signals depicted in Fig. 13.19 to transfer information over a channel. Note that the signals are not orthonormal. We would like to derive the CS receiver as in Fig. 13.12 for this case.

    a. Find the set $\{v_i(t)\}$ biorthogonal to $\{h_i(t)\}$.

    b. Choose a matrix $\mathbf{A}$ appropriate for this setting.

    c. Determine the resulting compressed correlating functions $s_i(t)$.

    d. Construct a table like Table 13.1 for the resulting CS receiver.

11. Show that when $p = N$, the signals are orthonormal, and the rows of $\mathbf{A}$ are orthonormal, the receiver of (13.70) is identical to the MF receiver.

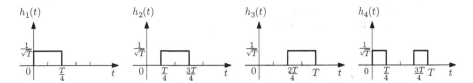

Figure 13.19 Signals for Exercise 10.

12. In this exercise, we derive the MMSE receiver that is the solution to $\min_{\mathbf{T}} E\left\{\|\mathbf{Ty} - \mathbf{b}\|^2\right\}$ with $\mathbf{y} = H^*y(t) = \mathbf{GRb} + \mathbf{n}$ where $\mathbf{n} = H^*y(t)$ as given in (13.76).

   a. Show that the vector $\mathbf{n}$ is zero-mean Gaussian with covariance $\mathbf{R}_n = \sigma^2 \mathbf{G}$.

   b. Derive the MMSE estimator by relying on the fact that the noise and data are uncorrelated.

13. Let $\mathbf{y} = \mathbf{G}^{-1}H^*y(t)$ be the output of the decorrelator receiver of Fig. 13.16.

   a. Determine functions $\{v_i(t), 1 \leq i \leq N\}$ such that we can write $\mathbf{y}$ as $\mathbf{y} = V^*y(t)$.

   b. Show that the functions $\{v_i(t)\}$ are biorthogonal to $\{h_i(t)\}$.

Chapter 14

Multiband sampling

In this chapter we consider sampling of multiband signals, namely, signals whose Fourier transform comprises a small number of bands, spread over a possibly wide frequency range. We begin by discussing the case in which the carrier frequencies, or band positions, are known. We review both classical methods and techniques based on interleaved ADC structures to sample such signals at a rate proportional to the actual band occupancy, and not to the high Nyquist rate associated with the highest frequency. The more challenging scenario in which the carrier frequencies are unknown, resulting in a UoS model, is treated next. Combining sampling approaches for the case in which the carriers are known, with methods for compressed sampling of analog signals developed in the previous chapter, leads to several sub-Nyquist sampling techniques for multiband signals with unknown band positions. Along with developing the theoretical concepts, we address practical constraints and present an example of a sub-Nyquist hardware realization for this class of signals.

14.1 Sampling of multiband signals

Consider the scenario of a multiband input $x(t)$, which has sparse spectra, such that its CTFT $X(\omega)$ is supported on N frequency intervals, or bands, with individual widths not exceeding B. In addition, the largest frequency component is below $\omega_{\max}$. Figure 14.1 illustrates a typical multiband spectrum. Multiband models are common in communication channels, when a receiver intercepts several RF transmissions, each modulated onto a different (possibly high) carrier frequency.

From a mathematical and engineering perspective, there is a fundamental difference between the case in which the carrier frequencies are given and the one in which they are not. Knowing the carriers allows the receiver to demodulate a transmission of interest to baseband: that is, to shift the contents from the relevant RF band to the origin, by multiplying the input with an appropriate sinusoid. Filtering may then be used to remove high-frequency images from the modulated output. Subsequent sampling and processing are carried out at a low rate corresponding to the individual band of interest, as detailed in Section 14.2.1.

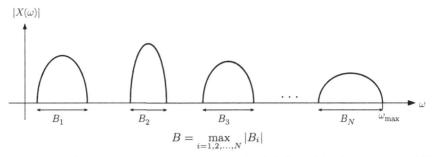

Figure 14.1 A typical multiband spectrum. The width of each individual band does not exceed B.

Cognitive radio

Unknown carriers occur, for example, in cognitive radio (CR) receivers [100], which are designed to exploit unused frequency regions on an opportunistic basis. Commercialization of CR technology necessitates a fast and accurate *spectrum sensing* mechanism to support real-time cognitive decisions, where spectrum sensing refers to the task of identifying the frequency support of a given input signal. CRs are motivated by the fact that over the years, government agencies allocated the majority of the spectrum to legacy users, reserving a particular frequency interval for each owner. This resource allocation methodology has led to spectrum congestion, to such a point that, today, the increasing demand for transmission bands is difficult to satisfy by a permanent allocation. Studies conducted by the Federal Communications Commission (FCC) in the United States and by similar agencies in other countries indicate that the spectrum is underutilized; in a given geographical location, only a small number of legacy users transmit concurrently. The low spectrum utilization, as illustrated in Fig. 14.2, is what drives CR technology.

The idea underlying CR is to exploit temporarily available spectrum holes, namely those frequency intervals whose primary user is inactive. Spectrum sensing therefore takes place whenever a CR seeks available holes for transmission. Furthermore, the CR has to monitor the spectrum continuously, even after a certain frequency band is chosen. Once the primary user becomes active, the CR has to choose another working band, or tailor its transmission to reduce in-band power. Real-time knowledge of available holes makes this transition smooth. Quick and efficient spectrum sensing is therefore an essential function in CR [399, 400]. This sensing task reduces to sampling a wideband spectrum that is sparsely populated, leading to a multiband model.

Chapter outline

Mathematically, when the band positions are known and fixed, the resulting class of multiband signals defines a linear subspace since the CTFT of any combination of two inputs is supported on the same frequency bands. Linear multiband models, which we treat in Sections 14.2 and 14.3, have been studied extensively in the sampling literature. The simplest approach to sample multiband signals is through simultaneous demodulation of all frequency bands followed by low-rate sampling of each band individually. Although simple and intuitive, this approach requires a tunable bank of demodulators,

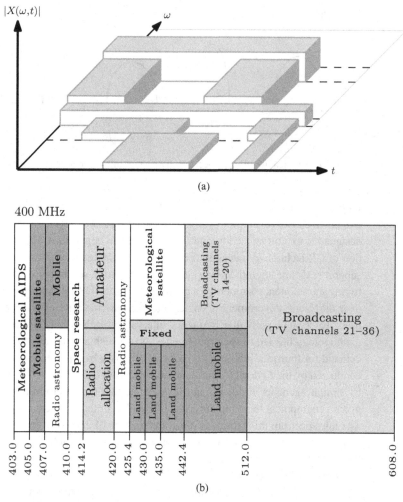

Figure 14.2 Low spectrum utilization. (a) The concept of frequency holes. For each time slot, only a small portion of the spectrum is utilized. (b) A section (the 400 MHz zone) of the FCC allocation map. Despite the fact that the entire frequency band is allocated, in practice, frequency holes as in (a) are common.

which can be expensive to implement in hardware and also depends on the input signal through its carrier frequencies. An alternative technique to convert multiband signals to digital is to directly undersample the incoming signal, without the need for prior analog hardware. The aliases created by undersampling effectively shift the frequency content to baseband, as we detail in Section 14.2.3. In this context, it is interesting to study the minimal sampling rate that can be obtained by direct sampling of a multiband input $x(t)$. This question was addressed by Landau in the early 1960s, leading to what is now known as the *Landau rate* [99], discussed in Section 14.2.2.

Interleaved ADCs, studied in Section 14.3, generalize the concept of direct undersampling to signals with more than one band. A drawback of such systems is that they necessitate high analog bandwidth, which in some cases can be difficult to achieve. To circumvent the analog bandwidth requirement, we introduce the modulated wideband converter (MWC) in Section 14.4 which also allows sampling at the Landau rate but uses only low bandwidth ADCs and does not rely on accurate delay components as do interleaved ADCs. In Section 14.5 we combine the ideas of analog compressed sensing and multiband sampling with known carriers to develop sub-Nyquist methods for the case in which the carrier frequencies are unknown. A hardware realization of the MWC is presented in Section 14.6. We end the chapter with a brief discussion on noise and other imperfections, as well as several simulations illustrating the system's behavior.

14.2   Multiband signals with known carriers

14.2.1   *I/Q* demodulation

We begin by treating real multiband models with known carrier frequencies. The simplest approach to sample such a signal is by using demodulation to first isolate each band and shift it down to baseband, followed by independent low-rate sampling of each band.

To understand demodulation of real signals more precisely, we consider a typical communication system in which two real information signals $I(t), Q(t)$, each with bandwidth B, are modulated onto a carrier frequency ω_i with a relative phase shift of $90°$. The output is then given by [394]

$$r_i(t) = I(t)\cos(\omega_i t) + Q(t)\sin(\omega_i t). \tag{14.1}$$

The signal $I(t)$ is referred to as the *in-phase component* while $Q(t)$ is called the *quadrature component*. Each signal $r_i(t)$ has a total bandwidth of $2B$ with a bandwidth of B on the positive frequency axis and B on the negative axis. These two bands are conjugates of each other, as shown in Fig. 14.3.

Some examples that appear frequently in communications are given below:

- In amplitude modulation (AM), the information of interest is the amplitude of $I(t)$, while $Q(t) = 0$. Pulse AM (PAM) has a similar form with $I(t) = A_m g(t)$ where $g(t)$ is a given pulse shape and A_m is one of M possible digital amplitudes. Typically,

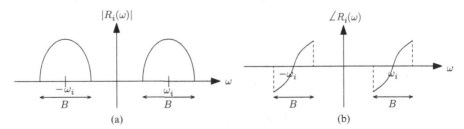

Figure 14.3 A typical conjugate frequency response of a real bandpass signal $r_i(t)$ with carrier ω_i and bands of width B. (a) Symmetric magnitude bands. (b) Antisymmetric phase.

$A_m = (2m - 1 - M)d$ for $1 \leq m \leq M$ where $2d$ is the distance between adjacent signal amplitudes. Digital PAM is also referred to as *amplitude-shift keying*.

- Quadrature AM (QAM) results when both $I(t)$ and $Q(t)$ are used to convey information bits by choosing $I(t) = A_m g(t)$ and $Q(t) = B_m g(t)$. Here both A_m and B_m contain information about the signal.

- Phase- and frequency-modulation (PM/FM) obey (14.1) such that the analog message is $g(t) = \arctan(Q(t)/I(t))$. In digital communication, e.g. phase- or frequency-shift keying (PSK/FSK), $I(t), Q(t)$ carry symbols. Each symbol encodes one, two, or more bits. In digital PM, for each value $1 \leq m \leq M$, we have $I(t) = g(t) \cos(\theta_m)$ and $Q(t) = -g(t) \sin(\theta_m)$ where $g(t)$ is the pulse shape and $\theta_m = 2\pi(m - 1)/M$ are the M possible phases of the carrier that convey the transmitted information. Digital PM is also referred to as *phase-shift keying (PSK)*. FSK (in its simple form) corresponds to choosing $I(t) = g(t) \cos((m-1)\Delta\omega t)$ and $Q(t) = -g(t) \sin((m-1)\Delta\omega t)$ where $\Delta\omega$ is an appropriate frequency step.

The information bearing signals $I(t)$ and $Q(t)$ can be recovered from $r_i(t)$ using an *I/Q*-demodulator, depicted in Fig. 14.4. To understand the system operation, consider multiplying the received signal $r_i(t)$ by $\cos(\omega_i t)$:

$$
\begin{aligned}
s(t) &= r_i(t) \cos(\omega_i t) \\
&= I(t) \cos^2(\omega_i t) + Q(t) \sin(\omega_i t) \cos(\omega_i t) \\
&= \frac{1}{2}I(t) + \frac{1}{2}[I(t) \cos(2\omega_i t) + Q(t) \sin(2\omega_i t)].
\end{aligned}
\tag{14.2}
$$

Low-pass filtering the outcome $s(t)$ with a filter whose cutoff is smaller than $2\omega_i - \frac{B}{2}$ removes the high-frequency terms containing $2\omega_i t$ leaving only the $I(t)$ term. This filtered signal is unaffected by $Q(t)$, showing that the in-phase component can be received independently of the quadrature component. Similarly, multiplying $r_i(t)$ by $\sin(\omega_i t)$ followed by low-pass filtering allows one to extract $Q(t)$. Once $I(t)$ and $Q(t)$ are obtained by the hardware, a pair of low-rate ADC devices acquire uniform samples at rate B,

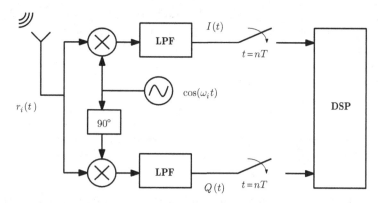

Figure 14.4 A block diagram of a typical *I/Q* demodulator.

where B is the pulse width of $I(t)$ and $Q(t)$. The subsequent DSP block can infer the analog message or decode the bits from the received symbols.

Reconstruction of each $r_i(t)$, and consequently recovery of the multiband input $x(t)$, is performed by remodulating the information on their carrier frequencies ω_i, according to (14.1). This option is used in relay stations or regenerative repeaters which decode the information $I(t), Q(t)$, apply digital error correction algorithms, and then remodulate the signal back to high frequencies for the next transmission section [394].

The I/Q demodulator uses two ADCs each with rate B, leading to a total sampling rate of $2B$, which matches the bandwidth of $r_i(t)$. Thus, if we have N signals $r_i(t)$ with total bandwidth $2NB$, then the required sampling rate using I/Q demodulation is $2NB$, which is clearly the minimal sampling rate possible.

I/Q demodulation has different names in the literature: zero-IF receiver, direct conversion, or homodyne; cf. [19] for various demodulation topologies. Each band of interest requires two hardware channels to extract the relevant $I(t), Q(t)$ signals. A similar principle is used in low-IF, or heterodyne, receivers, which demodulate a band of interest to low frequencies but not around the origin. Heterodyne receivers are widely used in radios, TV tuners, wireless communications, and more. One disadvantage associated with homodyne receivers is DC bias which results from leakage of the local oscillators into the RF port of the mixer, where it is mixed down to DC. This offset can be considerable with respect to the measured signals, leading to a narrower dynamic range. On the other hand, a homodyne receiver allows conversion to baseband in a single frequency conversion, while low-IF receivers require two (or more) frequency conversions, adding hardware complexity.

Although simple and intuitive, in the context of CR demodulation can be difficult to implement in hardware as it necessitates exact knowledge of the carrier frequencies and a bank of tunable modulators. In the next sections we study alternative techniques to convert multiband signals to the digital domain which also ensure sampling at the minimal possible rate, namely the Landau rate. Before introducing these methods, we discuss the Landau sampling theorem for multiband signals, which provides a limit on the sampling rate associated with multiband inputs.

14.2.2 Landau rate

Consider the space of bandlimited functions $\mathcal{B}_{\mathcal{T}}$ whose Fourier transform is restricted to a known support $\mathcal{T}$ which is a subset of $\mathcal{F} = [-\omega_{\max}/2, \omega_{\max}/2]$:

$$\mathcal{B}_{\mathcal{T}} = \{x(t) \in L_2(\mathbb{R}) \,|\, \mathrm{supp}\, X(\omega) \subseteq \mathcal{T}\}. \tag{14.3}$$

In other words, $X(\omega) = 0$ for $\omega \notin \mathcal{T}$. A set $R = \{r_n\}$ is called a *sampling set* for $\mathcal{B}_{\mathcal{T}}$ if the points are separated by at least some distance $d > 0$, and the sequence of samples $x_R[n] = x(r_n)$ is stable; that is, there exist constants $\alpha > 0$ and $\beta < \infty$ such that

$$\alpha \|x - y\|^2 \le \|x_R - y_R\|^2 \le \beta \|x - y\|^2, \quad \forall x, y \in \mathcal{B}_{\mathcal{T}}. \tag{14.4}$$

Landau proved[1] [99] that if R is a sampling set for $\mathcal{B}_{\mathcal{T}}$ then it must have density $D^-(R) \geq \lambda(\mathcal{T})$, where

$$D^-(R) = \lim_{r \to \infty} \inf_{y \in \mathbb{R}} \frac{|R \cap [y, y+r]|}{r} \tag{14.5}$$

is the lower *Beurling density*, and $\lambda(\mathcal{T})$ is the Lebesgue measure of $\mathcal{T}$. The numerator in (14.5) counts the number of points from R in every interval of width r of the real axis.[2] The Beurling density (14.5) reduces to the usual concept of average sampling rate for uniform and periodic nonuniform sampling, as we show in the examples below. The Landau rate, $\lambda(\mathcal{T})$, is then the minimal average sampling rate for $\mathcal{B}_{\mathcal{T}}$.

Example 14.1 In this example we calculate the Beurling density for uniform sampling.

Let $R_u = \{nd_0\}_{n \in \mathbb{Z}}$ be a uniform sampling set with period d_0. As shown in Fig. 14.5, $R_u \cap [y, y+r] = \{n_1 d_0, n_2 d_0, \ldots, n_N d_0\}$, with $n_1 = \lceil \frac{y}{d_0} \rceil$ and $n_N = \lfloor \frac{y+r}{d_0} \rfloor$. Since $|R_u \cap [y, y+r]| = n_N - n_1 + 1$,

$$C_{R_u}(y) = |R_u \cap [y, y+r]| = \left\lfloor \frac{y+r}{d_0} \right\rfloor - \left\lceil \frac{y}{d_0} \right\rceil + 1. \tag{14.6}$$

It is easy to see that for $r > 0$, $C_{R_u}(y)$ is a periodic function with period d_0. Therefore, to compute $D^-(R)$ of (14.5) it is sufficient to consider y in the interval $[0, d_0)$.

For any $r > 0$, we have that $\lceil \frac{y}{d_0} \rceil \leq 1$, $\lfloor \frac{y+r}{d_0} \rfloor \geq \lfloor \frac{r}{d_0} \rfloor$ and both inequalities are achievable for a sufficiently small y. Therefore,

$$\inf_{y \in \mathbb{R}} C_{R_u}(y) = \min_{0 \leq y < d_0} C_{R_u}(y) = \left\lfloor \frac{r}{d_0} \right\rfloor. \tag{14.7}$$

Since every $x > 0$ can be written as $x = \lfloor x \rfloor + s(x)$ where $0 \leq s(x) < 1$ we can compute $D^-(R)$ as

$$D^-(R_u) = \lim_{r \to \infty} \frac{\left\lfloor \frac{r}{d_0} \right\rfloor}{r} = \lim_{r \to \infty} \frac{\frac{r}{d_0} - s(\frac{r}{d_0})}{r} = \frac{1}{d_0}, \tag{14.8}$$

where the last equation is a result of the fact that $0 \leq s(x) < 1$.

Landau's result then implies that for uniform sampling with period d_0, the rate $1/d_0$ must be greater than or equal to the Lebesgue measure of our set $\mathcal{T}$. In particular, for a bandlimited signal with bandwidth B, we have that $1/d_0 \geq B$, which is precisely the limit obtained from the Shannon–Nyquist theorem.

---

[1] A simpler proof of this result can be found in [401].
[2] The numerator is not necessarily finite but as the sampling set is countable the infimum takes on a finite value.

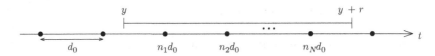

Figure 14.5 Intersection of a uniform sampling set and an interval of length r.

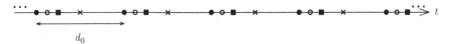

Figure 14.6 Periodic sampling set with $N = 4$ and $\tau_1 = 0, \tau_2 = \frac{1}{2}, \tau_3 = \frac{1}{4}, \tau_4 = \frac{1}{8}$.

Example 14.2 We next compute the Beurling density for periodic sampling sets. Let $R_p = \{d_0(n+\tau_1), d_0(n+\tau_2), \ldots, d_0(n+\tau_N)\}_{n\in\mathbb{Z}}$ be a periodic sampling set with period d_0, where $0 \leq \tau_i < 1$ for $1 \leq i \leq N$ and $\tau_i \neq \tau_j$ for $i \neq j$. Figure 14.6 illustrates an example of such a sampling set.

To compute the Beurling density we note that R_p can be expressed as

$$R_p = R_{u_1} \cup R_{u_2} \cup \cdots \cup R_{u_N} \tag{14.9}$$

where for every i, $R_{u_i} = \{d_0(n + \tau_i)\}_{n\in\mathbb{Z}}$ is a uniform sampling set. Combining (14.6) with (14.9) and the fact that $\{R_{u_i}\}_{i\in 1,\ldots,N}$ are disjoint sets leads to

$$
\begin{aligned}
C_{R_p}(y) = |R_p| &= \sum_{i=1}^{N} |R_{u_i}| \\
&= \sum_{i=1}^{N} \left(\left\lfloor \frac{y - \tau_i d_0 + r}{d_0} \right\rfloor - \left\lceil \frac{y - \tau_i d_0}{d_0} \right\rceil + 1 \right).
\end{aligned} \tag{14.10}
$$

Since, in general, $\inf\,[\,f\,(x) + g(x)\,] \neq \inf f\,(x) + \inf g(x)$ we can only conclude that

$$N \left\lfloor \frac{r}{d_0} \right\rfloor \leq \inf_{y\in\mathbb{R}} C_{R_p}(y) \leq N \left(\left\lfloor \frac{r}{d_0} \right\rfloor + 1 \right). \tag{14.11}$$

Dividing by r and taking the limit as $r \to \infty$ leads to

$$D^-(R_p) = \lim_{r\to\infty} \inf_{y\in\mathbb{R}} \frac{|R_p \cap [y, y+r]|}{r} = \frac{N}{d_0}. \tag{14.12}$$

The limit calculation is the same as in Example 14.1.

As we expect, the Beurling density of a periodic sampling set consisting of N uniform sequences with period d_0 is N times the density of a single uniform sampling set with the same period, treated in Example 14.1. This then implies that the use of such a sampling set allows us to increase the period, or reduce the rate, of each individual sequence, by a factor of N.

Example 14.3 Our last example considers the minimal sampling rate required for multiband signals of the form depicted in Fig. 14.1, using uniform and periodic sampling. More specifically, we treat signals consisting of N bands with individual widths B.

The Lebesgue measure[3] $\lambda(\text{supp}\,|X(\omega)|)$ of such a multiband signal equals NB. From Landau's theorem and Example 14.1 we conclude that the minimal sampling rate needed for perfect reconstruction from uniform samples is $\frac{1}{d_0} \geq NB$. This means that with a sampling rate below NB it is impossible to reconstruct the signal perfectly. Similarly, from Example 14.2, the sampling rate for periodic sampling with M uniform sets has to satisfy $\frac{M}{d_0} \geq NB$. Thus, if we choose $M = N$, then each individual sampler has to operate at a rate greater than or equal to B.

From Example 14.3 we see that Landau's theorem supports the intuitive expectation that a multiband signal $x(t)$ with N bands of individual widths B requires a sampling rate no lower than the sum of the bandwidths, i.e. NB.

The Landau rate constitutes a lower bound on the minimal sampling rate possible. However, Landau's theorem does not provide a method for approaching this minimal rate. Later work in the sampling literature derived methods to achieve this bound in the multiband setting by using direct uniform undersampling and time-interleaved ADC structures [78, 402, 403], which result in periodic (or recurrent) sampling. Undersampling, which we discuss in the next subsection, is often used when there is a single real band (resulting in two frequency bands – one in the positive frequency range and one in the negative frequency range). Such a signal is referred to as a *bandpass signal*. For the single-band (bandpass) setting undersampling provides a simple alternative to demodulation. Nonetheless, as we show, it will often necessitate sampling above the minimal rate and is difficult to extend to a larger number of bands. In the context of generalized sampling theorems for bandlimited signals we introduced a popular alternative to uniform sampling in Section 6.8.2, using interleaved ADCs. Such sampling architectures utilize Papoulis' theorem with a bank of time-delay elements. In Section 14.3 we will see that interleaved ADCs can approach the Landau rate for multiband signals. In fact, we prove a stronger result, namely that appropriately designed interleaved ADC architectures can approach the minimal rate even when the carriers are unknown.

Despite the advantages of interleaved sampling systems, they also suffer from several limitations including the need for high analog bandwidth. As an alternative, we introduce the modulated wideband converter (MWC) in Section 14.4, and show that it too achieves minimal-rate sampling in both the case of known and unknown carrier frequencies. The MWC creates the desired aliasing affect by modulating the incoming signal with a set of periodic functions, rather than by direct undersampling, as done in interleaved structures.

14.2.3   Direct undersampling of bandpass signals

Consider a real signal $x(t)$ of the form depicted in Fig. 14.7. Since the input is real, its CTFT is conjugate symmetric and contains energy in both the positive and negative frequency range. Suppose we would like to sample this signal uniformly without any

---

[3] Any closed interval $[a, b]$ of real numbers is Lebesgue measurable, and its Lebesgue measure is its length $b - a$. In addition, a disjoint union of countable Lebesgue measurable sets is also Lebesgue measurable, and its Lebesgue measure is equal to the sum of the measures of the individual sets.

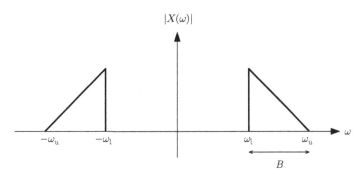

Figure 14.7 The magnitude of a typical single symmetric band, with ω_u and ω_l denoting the upper and lower band limits, such that $B = \omega_u - \omega_l$.

prior analog hardware. We can always treat the signal as bandlimited to ω_u, and sample it at the corresponding Nyquist rate $2\omega_u$. However, since most of the CTFT is zero, this approach is clearly wasteful. Indeed, by using a pre-modulator, the signal can be sampled at rate $2B$ where $B = \omega_u - \omega_l$. We would like to try and approach this rate with a simple sampling strategy. To this end, we consider *direct undersampling*, which corresponds to uniform sampling of the signal at a sub-Nyquist rate [20]. As we now show, this technique also shifts contents to baseband without requiring additional hardware.

Lower bound on sampling rate

Suppose we sample $x(t)$ with rate ω_s where $\omega_s \geq 2B$. Such sampling will create aliasing of the spectrum with shifts of ω_s as depicted in Fig. 14.8. This undersampling has the affect of shifting the signal contents down to baseband (and over the entire spectrum). In order to be able to recover the signal from the samples we need to ensure that the aliases of the bands from the positive and negative frequency axes caused by undersampling do not overlap. Under this condition, recovery is possible by filtering the desired frequency content, as illustrated in Fig. 14.9.

The minimal sampling rate achievable using undersampling depends on the band position, which is the fractional number of bandwidths from the origin at which the upper band edge resides:

$$n_B = \left\lfloor \frac{\omega_u}{B} \right\rfloor. \tag{14.13}$$

A special case is integer band positioning, $\omega_u = n_B B$, in which the band is located at an integral number of bandwidths from the origin. The classical bandpass theorem for uniform sampling states that the sampling rate ω_s must satisfy:

$$\omega_s \geq 2\frac{\omega_u}{n_B} \geq 2B. \tag{14.14}$$

Clearly we can achieve $\omega_s = 2B$ only for integer band positioning. The minimal sampling rate for a bandpass signal with bandwidth B as a function of the upper frequency ω_u is plotted in Fig. 14.10.

To prove (14.14) we illustrate that sampling at a rate lower than $2\omega_u/n_B$ will result in images of the spectrum from the left- and right-hand side overlapping at baseband,

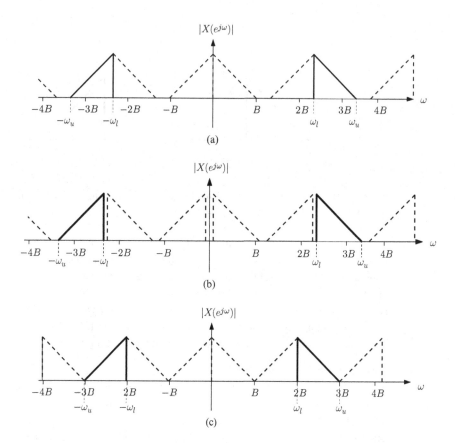

Figure 14.8 Spectrum aliases of a bandpass signal sampled at rate ω_s. The aliases appear at shifts of ω_s. (a) Noninteger positioning with $\omega_u = \frac{10}{3}B$, $\omega_s = \frac{7}{3}B$. (b) Noninteger positioning with $\omega_u = \frac{10}{3}B$, $\omega_s = \frac{9}{4}B$. (c) Integer positioning with $\omega_u = 3B$, $\omega_s = 2B$.

creating aliasing. Consider first sampling the signal in Fig. 14.11 with rate $\omega_s < 2B$. Owing to symmetry, it is sufficient to treat the images of the positive band and ensure that they do not overlap the original negative band. As shown in the figure, if $\omega_s < 2B$, then the spacing between images is smaller than B. As a result, one of the images of the positive band will fold onto the original negative band. The lower bound on the sampling rate $\omega_s \geq 2B$ is independent of the band position ω_u, and easy to establish, but not tight in general.

Finding a tight bound on the sampling rate requires that the right edge of the first image of the positive band that exceeds the original negative band will not overlap it, namely,

$$\omega_u - n\omega_s \leq -\omega_u \tag{14.15}$$

for a chosen n. Figure 14.12 illustrates this condition, corresponding to $\omega_s \geq 2\frac{\omega_u}{n}$. When $n = 1$, this bound reduces to the familiar Nyquist rate condition. The lower bound

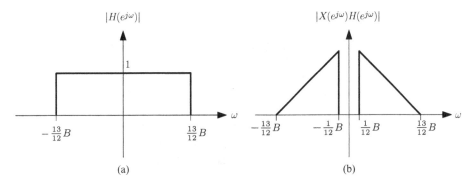

Figure 14.9 Bandpass recovery of the signal in Fig. 14.8(b). (a) The desired low-pass filter with cutoff $\frac{13}{12}B$. (b) Reconstructed bands in baseband.

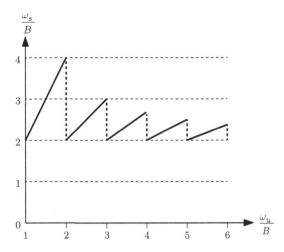

Figure 14.10 The minimal required sampling rate ω_s for a bandpass signal with bandwidth B as a function of the upper frequency ω_u (both normalized by B).

(14.14) follows from finding the largest possible n. Since we must also have $\omega_s \geq 2B$, the largest value of n is the integer satisfying $2\frac{\omega_u}{n} \geq 2B$, or,

$$n_{\max} = \left\lfloor \frac{\omega_u}{B} \right\rfloor = n_B, \tag{14.16}$$

leading to (14.14). For integer band positioning, $n_{\max} = \frac{\omega_u}{B}$, and the bound reduces to $\omega_s \geq 2B$.

Upper bound on sampling rate

When uniformly sampling a low-pass signal, any sampling rate above the minimal rate of $2B$ allows for perfect signal recovery. However, this is not the case when sampling a bandpass signal. This is a result of the fact that the sampling rate has to be chosen to ensure that the images of the positive and negative frequency contents do not overlap. Until now we have treated the lower bound by examining the first image of the positive

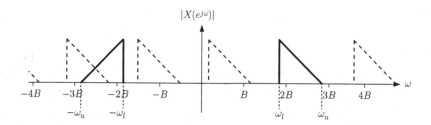

Figure 14.11 Images of the positive band separated by spaces with length smaller than B when $\omega_s < 2B$. This results in images overlapping. The negative images are omitted for clarity.

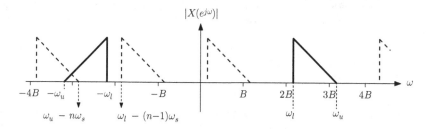

Figure 14.12 Illustration of the condition on the band edges for nonoverlapping images: $\omega_u - n\omega_s \le -\omega_u$ and $\omega_l - (n-1)\omega_s \ge -\omega_l$. In this case $n = 3$.

side to exceed the negative band. We can similarly derive an upper bound, as illustrated in Fig. 14.12, by requiring that the positive band images do not fold into the original negative band, not on its right, $\omega_u - n\omega_s \le -\omega_u$, and not on its left, $\omega_l - (n-1)\omega_s \ge -\omega_l$. As a result, the sampling rate ω_s has to satisfy

$$2\frac{\omega_u}{n} \le \omega_s \le 2\frac{\omega_l}{n-1}, \tag{14.17}$$

where n is any integer in the range $1 \le n \le n_B$. These conditions are shown graphically in Fig. 14.13.

The figure and (14.17) show that $\omega_s = 2B$ is achieved only with integer band positioning of $x(t)$. Furthermore, as the rate reduction factor n increases, the valid region of sampling rates becomes narrower. For a given ω_u, the region corresponding to the maximal $n \le \omega_u/B$ is the most sensitive to slight deviations in the exact values of $\omega_s, \omega_l, \omega_u$ [20]. Consequently, besides the fact that $\omega_s = 2B$ cannot be achieved in general (even in ideal noiseless settings), a significantly higher rate is likely to be required in practice in order to cope with design imperfections.

To conclude, undersampling can be used to directly sample a bandpass signal at sub-Nyquist rates without the need for analog preprocessing. This creates aliases of the signal with spacing equal to the sampling frequency ω_s. However, this approach has several drawbacks. First, the resulting reduced rate is generally significantly higher than the minimal possible, as evident from Figs. 14.10 and 14.13. The minimal rate of $2B$ is achievable only for integer band positioning. Furthermore, although we can recover the signal content within the band, the carrier frequency cannot be determined from the

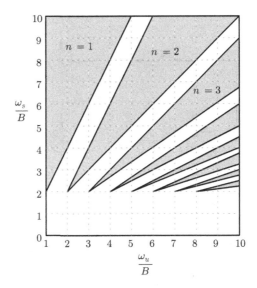

Figure 14.13 Allowed (grey) and forbidden (white) undersampling rates of a bandpass signal as a function of its band position.

samples. Second, and more importantly, undersampling is difficult to adapt to multiband inputs. In this scenario, each individual band defines a range of valid values for ω_s according to (14.17). The sampling rate must be chosen in the intersection of these conditions, and also such that aliases due to different bands do not interfere. As noted in [404], satisfying all these constraints simultaneously, if possible, is likely to require a considerable rate increase.

14.3   Interleaved ADCs

An alternative approach to sample multiband signals without analog preprocessing is based on using interleaved ADCs, which we discuss next.

14.3.1   Bandpass sampling

Kohlenberg [405] appears to be the first who treated interleaved ADCs for bandpass sampling. More specifically, he suggested using two interleaved samplers each with rate B, where the second sampler has a time delay τ with respect to the first sampler, as seen in Fig. 14.14. This approach allows sampling any bandpass input at the minimal rate of $2B$, regardless of the band position. Furthermore, the carrier frequency can be recovered from the resulting samples. Interleaved ADCs are also well suited for sampling multiband inputs by increasing the number of samplers, with each sampler introducing a different delay [184, 185, 186].

We begin by analyzing the case of bandpass sampling of a single real input, following the presentation in [184]. We then treat the more challenging problem of multiband

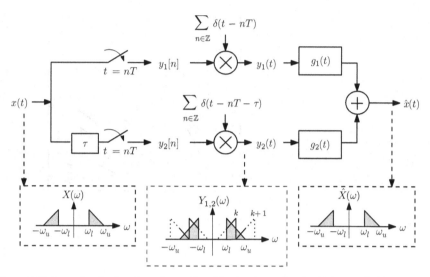

Figure 14.14 Second-order interleaved sampling. The bandpass signal $x(t)$ is sampled by two rate-B uniform samplers with relative time delay τ. The interpolation filters $g_1(t), g_2(t)$ cancel out the contribution of the undesired aliases.

sampling. For the latter, we introduce an alternative approach to analyze the sampling mechanism based on [186, 335], which will turn out to be useful when considering the setting in which the carriers are unknown.

The sampling process

Suppose $x(t)$ is supported on $\mathcal{I} = (\omega_l, \omega_u) \cup (-\omega_u, -\omega_l)$ and $B = \omega_u - \omega_l$ as illustrated in Fig. 14.14. We sample $x(t)$ by using the filterbank structure depicted in Fig. 14.14 with a sampling period of $T = 2\pi/B$ in each channel. Due to the undersampling in the individual channels, aliases of the band contents tile the spectrum, so that the positive and negative images fold on each other, as visualized in the figure.

If we focus on the positive frequencies in the support of the original signal, namely (ω_l, ω_u), then after sampling at rate B we see the original signal $x(t)$, and its aliases that originate from the negative frequencies, shifted by multiples of B. The positions of the shifts depend on the band position, as defined in (14.13). In particular, it is easy to see that moving the image on the negative axis by $k = \left\lceil \frac{2\omega_l}{B} \right\rceil$ shifts of B will result in an image that overlaps the original in the interval (ω_l, ω_m) with $\omega_m = kB - \omega_l$. Moving $k + 1$ shifts results in an image overlapping the original in (ω_m, ω_u). Defining $\beta(\omega)$ as the folding index equal to the number of shifts k needed to create the alias at ω, it follows that

$$\beta(\omega) = \begin{cases} k, & \omega_l \leq \omega \leq \omega_m \\ k+1, & \omega_m < \omega \leq \omega_u, \end{cases} \qquad k = \left\lceil \frac{2\omega_l}{B} \right\rceil, \qquad (14.18)$$

and $\beta(\omega) = -\beta(-\omega)$.

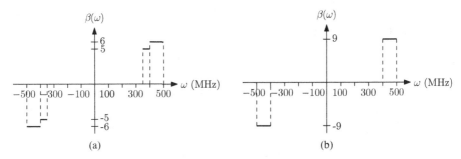

Figure 14.15 The folding index $\beta(\omega)$ for Example 14.4. (a) Noninteger band positioning: $\omega_l = 350\,\text{MHz}$, $\omega_u = 500\,\text{MHz}$. (b) Integer band positioning: $\omega_l = 400\,\text{MHz}$, $\omega_u = 500\,\text{MHz}$.

Example 14.4 We illustrate the computation of the folding index $\beta(\omega)$ for two different bandpass signals.

(1) Consider a signal with $\omega_l = 350\,\text{MHz}$ and $\omega_u = 500\,\text{MHz}$. We can easily see that $B = 150\,\text{MHz}$, $k = 5$, and $\omega_m = 400\,\text{MHz}$. Therefore,

$$\beta(\omega) = \begin{cases} 5, & 350 \leq \omega \leq 400 \\ 6, & 400 < \omega \leq 500, \end{cases} \tag{14.19}$$

and $\beta(\omega) = -\beta(-\omega)$, as shown in Fig. 14.15(a).

(2) Next, suppose that $\omega_l = 400\,\text{MHz}$ and $\omega_u = 500\,\text{MHz}$, so that the signal is bandpass with integer band positioning. In this case $B = 100\,\text{MHz}$, $k = 8$, and $\omega_m = 400\,\text{MHz}$. Thus,

$$\beta(\omega) = 9, \ 400 \leq \omega \leq 500, \tag{14.20}$$

as shown in Fig. 14.15(b). More generally, for integer band positioning, $k = \left\lceil \frac{2\omega_l}{B} \right\rceil = 2\frac{\omega_l}{B}$ is an even number and $\omega_m = \frac{2\omega_l}{B} B - \omega_l = \omega_l$. Thus, $\beta(\omega)$ takes on only two opposite values, $k + 1$ and $-(k + 1)$, which are odd numbers.

The recovery process

To recover $x(t)$ from the interleaved samples, we modulate each of the individual sampling sequences onto an appropriate impulse train which represents the sampling times, and then filter the modulated sequences with filters $G_i(\omega)$, as depicted in Fig. 14.14. In practice, the analog filters can be realized digitally using the interpolation identity introduced in Fig. 6.28 of Section 6.8.2. In this case the output is the Nyquist-rate sequence $x(nT')$, with $T' = 2\pi/(2\omega_u)$ equal to the Nyquist interval. Subsequently, a DAC device may interpolate the continuous signal $x(t)$.

To determine the recovery filters, denote by $y_i(t)$ the modulated signal in the ith branch, as shown in Fig. 14.14. Then,

$$y_1(t) = \sum_{n \in \mathbb{Z}} x(nT)\delta(t - nT) = \sum_{n \in \mathbb{Z}} x(t)\delta(t - nT),$$

$$y_2(t) = \sum_{n \in \mathbb{Z}} x(nT + \tau)\delta(t - nT - \tau) = \sum_{n \in \mathbb{Z}} x(t)\delta(t - nT - \tau). \tag{14.21}$$

In the frequency domain,

$$T\,Y_1(\omega) = X(\omega) + X(\omega - \beta(\omega)B),$$

$$T\,Y_2(\omega) = X(\omega) + X(\omega - \beta(\omega)B)e^{-j\beta(\omega)\tau B}, \tag{14.22}$$

for $\omega \in \mathcal{I}$ (the support of $x(t)$), where $\beta(\omega)$ is defined by (14.18). After filtering with $G_i(\omega)$ and summing the outputs, the reconstructed signal is given by

$$\widehat{X}(\omega) = Y_1(\omega)G_1(\omega) + Y_2(\omega)G_2(\omega). \tag{14.23}$$

Our goal is to choose $G_i(\omega), i = 1, 2$ such that $\widehat{X}(\omega) = X(\omega)$.

Clearly $G_i(\omega) = 0$ for $\omega \notin \mathcal{I}$. Substituting (14.22) into (14.23),

$$T\widehat{X}(\omega) = X(\omega)[G_1(\omega) + G_2(\omega)] + X(\omega - \beta(\omega)B)[G_1(\omega) + e^{-j\beta(\omega)\tau B}G_2(\omega)], \tag{14.24}$$

for $\omega \in \mathcal{I}$. Therefore, to ensure that $\widehat{X}(\omega) = X(\omega)$ we must have for $\omega \in \mathcal{I}$,

$$G_1(\omega) + G_2(\omega) = T,$$

$$G_1(\omega) + e^{-j\beta(\omega)\tau B}G_2(\omega) = 0. \tag{14.25}$$

For each $\omega \in \mathcal{I}$, the relations given in (14.25) represent a system of linear equations in the two unknowns $G_i(\omega), i = 1, 2$. Denoting $\mathbf{g}(\omega) = [G_1(\omega)\,G_2(\omega)]^T$ and $\mathbf{x} = [T0]^T$, we can write (14.25) in matrix form as $\mathbf{x} = \mathbf{B}(\omega)\mathbf{g}(\omega)$ with

$$\mathbf{B}(\omega) = \begin{bmatrix} 1 & 1 \\ 1 & e^{-j\beta(\omega)\tau B} \end{bmatrix}, \quad \omega \in \mathcal{I}. \tag{14.26}$$

The matrix $\mathbf{B}(\omega)$ is invertible over $\omega \in \mathcal{I}$ as long as τ obeys

$$e^{-j\beta(\omega)\tau B} \neq 1. \tag{14.27}$$

Since $\beta(\omega)$ can take on only four distinct values within $\omega \in \mathcal{I}$, there are many possible selections of τ satisfying (14.27). Under this condition $\mathbf{g}(\omega) = \mathbf{B}^{-1}(\omega)\mathbf{x}$, resulting in

$$G_1(\omega) = \frac{-e^{-j\beta(\omega)\tau B}T}{1 - e^{-j\beta(\omega)\tau B}}, \quad G_2(\omega) = \frac{T}{1 - e^{-j\beta(\omega)\tau B}} \tag{14.28}$$

for $\omega \in \mathcal{I}$.

It is interesting to note that the filters (14.28) are piecewise constant; the only dependence on ω is through the function $\beta(\omega)$ which takes on at most four values for $\omega \in \mathcal{I}$. We have seen in Example 6.16 of Section 6.8.2 that filters used to interpolate a bandlimited signal from an interleaved ADC output also have this property.

Example 14.5 Consider interleaved sampling with $\tau = (2\pi)/(mB)$ for an integer $m \geq 2$.

In the special case of $m = 2$, $\tau = \pi/B$, interleaved sampling reduces to uniform sampling of the original signal at a rate of $2B$, namely the Landau rate of the signal. As we have seen in Section 14.2.3, direct undersampling of a bandpass signal at the Landau rate is possible only in the case of integer band positioning. Therefore, we expect condition (14.27) to hold only in this case.

From Example 14.4, we know that with integer band positioning, $\beta(\omega)$ takes on two opposite odd values, $k + 1$ and $-(k + 1)$, which means that $e^{-j\beta(\omega)\pi} \neq 1$ for every $\omega \in \mathcal{I}$ and recovery is possible, as expected. On the other hand, for any other value of the band position, $\beta(\omega)$ takes on two distinct even values within $\omega \in \mathcal{I}$ for which $e^{-j\beta(\omega)\pi} = 1$. Thus, (14.27) is satisfied when $\tau = \pi/B$ only for signals with integer band positioning.

For $m > 2$ we have to satisfy $e^{-j2\pi\beta(\omega)/m} \neq 1$, which is equivalent to the condition that $\beta(\omega)/m$ is not an integer, namely $\beta(\omega) \neq m\ell$ for all $\ell \in \mathbb{Z}$. As m increases, the set of valid band positions leading to perfect recovery grows.

We conclude that for a given τ, the band positions determine whether a bandpass input can be recovered. On the other hand, for a given signal, there always exists a delay τ which enables perfect recovery.

Noise robustness

Although in principle any τ satisfying (14.27) can be used, the choice of τ affects the recovery behavior in the presence of noise.

Suppose that independent white noise is added to each of the sequences of samples prior to recovery with the filters $G_i(\omega)$. The output noise power is then proportional to $\int(|G_1(\omega)|^2 + |G_2(\omega)|^2)d\omega$, or $\int \|\mathbf{g}(\omega)\|^2 d\omega$. (See Appendix B for a discussion on the noise power at the output of an LTI system.) Since $\mathbf{g}(\omega) = \mathbf{B}^{-1}(\omega)\mathbf{x}$, it follows that $\|\mathbf{g}(\omega)\|^2 = T^2\|\mathbf{b}(\omega)\|^2$ where $\mathbf{b}(\omega)$ is the first column of $\mathbf{B}^{-1}(\omega)$. Denote by $\Delta = e^{-j\beta(\omega)\tau B} - 1$ the determinant of $\mathbf{B}(\omega)$. It is then easy to see that $\mathbf{b}(\omega) = (1/\Delta)[e^{-j\beta(\omega)\tau B} - 1]^T$ and

$$\|\mathbf{b}(\omega)\|^2 = \frac{2}{|\Delta|^2} = \frac{1}{2\sin^2\left(\frac{\beta(\omega)\tau B}{2}\right)} \geq \frac{1}{2}. \tag{14.29}$$

Thus, the smaller the value of $|\Delta|^2$, the larger the noise enhancement of the system.

Example 14.6 In this example we explore the noise robustness of an interleaved system as a function of τ. In general, owing to the folding index $\beta(\omega)$, $\|\mathbf{g}(\omega)\|^2$ depends on ω and can take on two distinct values within $\omega \in (\omega_l, \omega_u)$. We consider two simple cases in which it is easy to determine the value of Δ:

(1) $\tau = \pi/B$. As we saw in Example 14.5, the choice of $\tau = \pi/B$ is equivalent to uniform undersampling, and recovery is possible only if the signal has an integer band positioning. In this case, $\beta(\omega)$ takes on a single odd value within

$\omega \in (\omega_l, \omega_u)$ so that $|\Delta|^2 = 4$, which is the largest value $|\Delta|^2$ can have. Thus, uniform sampling, when valid, leads to the most robust system.

(2) $\tau \to 0$. When a small value of τ is chosen, $|\Delta|^2 \to 0$ for any given $\beta(\omega)$. Thus, as the delay between the ADCs becomes smaller, the system robustness decreases.

14.3.2   Multiband sampling

An advantage of interleaved ADCs is that they can be used to sample multiband inputs at the Landau rate. In principle, higher-order interleaved structures may be analyzed by taking a similar approach to that described above for bandpass sampling. However, this requires defining the folding index (14.18) for each of the bands, which becomes quite cumbersome. Instead, we follow the developments in [186, 335], and present an alternative way of viewing the sampling and recovery. To further simplify the analysis we consider sampling of complex inputs so that we only need to treat the positive frequency axis; the extension to real signals is straightforward. Since there are several occupied bands, aliasing will result from shifts of the different bands. Note that in the case of bandpass sampling, if we have a complex signal with a single positive band, then as long as we sample at a rate of a single band or higher, no aliasing occurs, which explains why in that setting we treated real signals.

Consider a complex multiband signal $x(t) \in L_2$ consisting of N frequency bands, each with width no larger than B on the interval $[0, 2\pi/T_{\text{Nyq}})$ where T_{Nyq} is the Nyquist period. Note that bands of different widths can be treated by viewing the bands as consisting of narrower intervals that are integer multiples of a common length. For simplicity we assume that $2\pi/T_{\text{Nyq}} = LB$ for some integer L; this assumption is not essential and is made only for convenience. Figure 14.16 depicts a typical spectral support of a multiband signal.

The Landau rate associated with this signal is NB (see Example 14.3). To sample the signal we propose using an interleaved ADC of the form depicted in Fig. 14.17 with $p \geq N$ samplers, where each ADC samples at a rate of B.

Multicoset sampling
Interleaved ADCs were discussed in Section 6.8.2 in the context of Nyquist rate sampling. An interleaved sampling architecture results in recurrent nonuniform samples

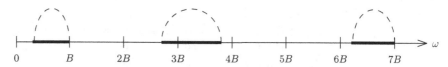

Figure 14.16 A typical support of a complex multiband signal. The signal consists of N frequency bands, each with width no larger than B on the interval $[0, 2\pi/T_{\text{Nyq}})$, where $2\pi/T_{\text{Nyq}} = LB$. In this example: $N = 3$, $L = 7$.

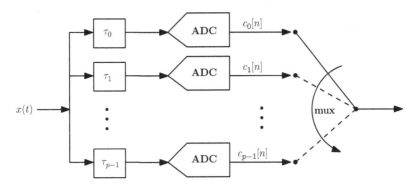

Figure 14.17 An interleaved ADC structure with p delay components followed by p samplers and a multiplexer (mux).

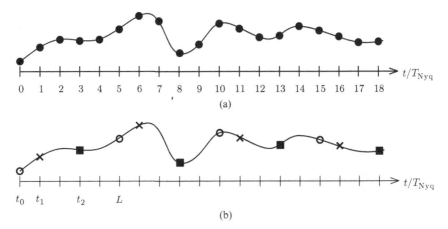

Figure 14.18 Multicoset samples viewed as a subset of Nyquist-rate samples. (a) A Nyquist-rate sampling set. (b) A multicoset sampling set with $L = 5, p = 3, t_0 = 0, t_1 = 1, t_2 = 3$.

consisting of p undersampled sequences with relative time shifts τ_i:

$$c_i[n] = x(nT + \tau_i), \quad 0 \leq i \leq p - 1 \tag{14.30}$$

where T is the sampling period. Whereas in standard interleaved ADCs, T is chosen such that the total sampling rate $2\pi p/T$ is greater than or equal to the Nyquist rate LB, here we exploit the sparsity of the spectrum and choose p so that $2\pi p/T$ is greater than or equal to the Landau rate NB, with $N < L$. Recall that $LB = 2\pi/T_{\mathrm{Nyq}}$.

A special case of recurrent nonuniform sampling is *multicoset sampling* [186]. For this choice, T is a multiple of the Nyquist period T_{Nyq} so that $T = LT_{\mathrm{Nyq}} = 2\pi/B$ and $\tau_i = t_i T_{\mathrm{Nyq}}$ where t_i is an integer in the interval $[0, L-1]$. The resulting samples are a subset of the Nyquist-rate samples, as illustrated in Fig. 14.18.

Taking into account that each sampler has rate B, the Landau rate corresponds to choosing $p = N$ samplers. Nonetheless, our analysis below is presented for general p since we will require $p > N$ when the carriers are unknown. Using $p > N$ branches is

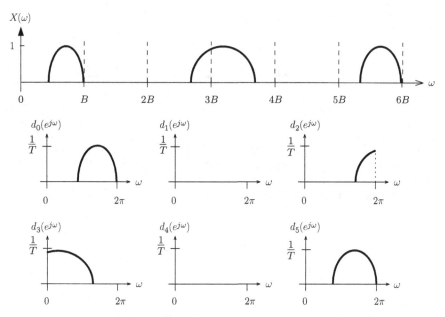

Figure 14.19 A typical multiband signal $X(\omega)$ and its vector-valued function $\mathbf{d}(e^{j\omega})$. In this example $N = 3$, $L = 6$.

also beneficial in the presence of noise, and helps in simplifying the recovery process when the carrier frequencies are arbitrary. To ensure that the total sampling rate $2\pi p/T$ is still below the Nyquist rate $2\pi/T_{\mathrm{Nyq}}$, we need to ensure that $p < L$.

In some applications, sampling at the Nyquist rate is feasible, but processing at the Nyquist rate is difficult. In such cases one can first sample at the Nyquist rate, and then use the techniques we develop here to downsample the sequence to a lower rate for more efficient processing. Multicoset sampling is particularly beneficial in this context: rate reduction can be achieved by simply selecting a subset of the Nyquist-rate samples defined by the integer delays t_i. More specifically, the Nyquist sampling points are divided into groups of size L, and from each subset only the points at locations $t_i, 0 \leq i \leq p - 1$ are retained.

To conveniently analyze multicoset sampling, we introduce the vector-valued function $\mathbf{d}(e^{j\omega})$ of length L, whose kth element is given by

$$d_k(e^{j\omega}) = \frac{1}{T}X\left(\frac{\omega}{T} + \frac{2\pi k}{T}\right), \quad 0 \leq \omega \leq 2\pi. \tag{14.31}$$

As shown in Fig. 14.19, $d_k(e^{j\omega})$ corresponds to the frequency response of samples at rate $B = 2\pi/T$ of the kth slice of $X(\omega)$, where we divide $X(\omega)$ into slices of width B. Since $X(\omega)$ is bandlimited to $2\pi/T_{\mathrm{Nyq}}$ with $T = LT_{\mathrm{Nyq}}$, there are L slices in $X(\omega)$. Each band is then sampled at its Nyquist rate to yield $d_k(e^{j\omega})$. Clearly, recovery of $x(t)$ is equivalent to recovering the samples of each of the slices. Our goal therefore is to determine these spectrum slices, or to recover $\mathbf{d}(e^{j\omega})$.

Sparsity structure

Since $x(t)$ has a multiband structure, many of the signals $d_k(e^{j\omega})$ are identically zero. More specifically, each band of $x(t)$ can contribute to at most two entries of $\mathbf{d}(e^{j\omega})$. Note, however, that if a band of $x(t)$ is split between two entries of $\mathbf{d}(e^{j\omega})$, then these entries do not overlap in frequency. Thus, for a specific ω, the complex vector $\mathbf{d}(e^{j\omega})$ has at most N nonzero values. On the other hand, as a vector of complex functions, $\mathbf{d}(e^{j\omega})$ contains at most $2N$ nonzero functions. Figure 14.20 illustrates the two different sparsity types of $\mathbf{d}(e^{j\omega})$. Evidently, when viewed as a vector, the support can depend in general on ω. We will exploit this sparsity below in the recovery process.

Example 14.7 We illustrate two examples of the support of $\mathbf{d}(e^{j\omega})$: one in which it is fixed and equal to N, and an example in which the support varies over ω. In both cases the support over $2N$ values is fixed.

Figure 14.21 depicts two multiband signals with $N = 3$ bands and $L = 9$ entries in $\mathbf{d}(e^{j\omega})$. The first signal $X_1(\omega)$ has fixed support of $N = 3$ where the nonzero entries are $\{1, 5, 8\}$. For each ω the support is equal to 3 (or less). The second signal $X_2(\omega)$ occupies five different entries $\{0, 1, 3, 4, 8\}$ and therefore can not have a fixed support of $N = 3$ for all ω. Although for each ω at most $N = 3$ entries are nonzero, when considered as a vector of functions the (fixed) support is equal to 5. In both cases the support is no larger than $2N = 6$.

We summarize our observations on the support of $\mathbf{d}(e^{j\omega})$ in the following proposition.

Proposition 14.1. *Consider a multiband signal with N bands of maximal bandwidth B, and Nyquist frequency $2\pi/T_{\mathrm{Nyq}} = LB$ for some integer L. Let $\mathbf{d}(e^{j\omega})$ be the vector defined by (14.31). If $T \leq 2\pi/B$, then for each value ω, $\mathbf{d}(e^{j\omega})$ is N-sparse. In addition, the set of vectors $\{\mathbf{d}(e^{j\omega})\}$ for $\omega \in [0, 2\pi)$ is $2N$-jointly sparse. Similarly, the vectors $\{\mathbf{d}[n]\}$ for $n \in \mathbb{Z}$ are $2N$-jointly sparse, where $\mathbf{d}[n]$ is the inverse DTFT of $\mathbf{d}(e^{j\omega})$.*

Proof: The bands of $x(t)$ are continuous intervals with width upper-bounded by B. The vector $\mathbf{d}(e^{j\omega})$ is constructed by dividing the interval $[0, 2\pi/T_{\mathrm{Nyq}})$ into equal intervals of length $2\pi/T$. Therefore, if $T \leq 2\pi/B$, then each band can either be fully contained in one of these intervals or it can be split between two consecutive intervals such that for each ω only one of the intervals contains a signal value. Since the number of bands is no larger than N it follows that for each ω, the support set is bounded above by N. Furthermore, the joint support set of $\{\mathbf{d}(e^{j\omega})\}$ is no larger than $2N$.

Finally, for any value i for which $d_i(e^{j\omega})$ is zero for all ω, we also have that $d_i[n]$ is zero for all n. Consequently, $\{\mathbf{d}[n]\}$ is at most $2N$-jointly sparse. $\qquad\square$

Recovery process

To recover the signal from its multicoset samples, we first express the DTFT of the samples $c_i[n]$ as a function of $\mathbf{d}(e^{j\omega})$. To this end we write $c_i[n]$ as $c_i[n] = y_i(nT)$ with $y_i(t) = x(t + t_i T_{\mathrm{Nyq}})$. Then,

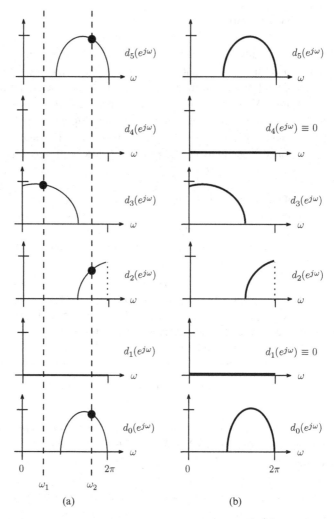

Figure 14.20 Illustrating the support of the vector-valued function $\mathbf{d}(e^{j\omega})$ from Fig. 14.19. In this example $N = 3$ and $L = 6$. (a) For each specific ω, $\mathbf{d}(e^{j\omega})$ has at most N nonzero values. In particular, $\mathbf{d}(e^{j\omega_1}) = [0\ 0\ 0\ c_0\ 0\ 0]^T$, $\mathbf{d}(e^{j\omega_2}) = [c_1\ 0\ c_2\ 0\ 0\ c_3]^T$, where $c_i \neq 0$, $i = 0, 1, 2, 3$. (b) For all ω, $\mathbf{d}(e^{j\omega})$, as a vector of functions, has at most $2N$ nonzero functions: $\mathbf{d}(e^{j\omega}) = [C_0(\omega)\ 0\ C_1(\omega)\ C_2(\omega)\ 0\ C_3(\omega)]^T$ where $C_i(\omega) \not\equiv 0$, $i = 0, 1, 2, 3$.

$$
\begin{aligned}
C_i(e^{j\omega}) &= \frac{1}{T} \sum_{k \in \mathbb{Z}} Y_i\left(\frac{\omega}{T} + \frac{2\pi k}{T}\right) \\
&= \frac{1}{T} \sum_{k=0}^{L-1} Y_i\left(\frac{\omega}{T} + \frac{2\pi k}{T}\right) \\
&= \frac{1}{T} e^{j\omega t_i/L} \sum_{k=0}^{L-1} X\left(\frac{\omega}{T} + \frac{2\pi k}{T}\right) e^{j2\pi k t_i/L},
\end{aligned} \tag{14.32}
$$

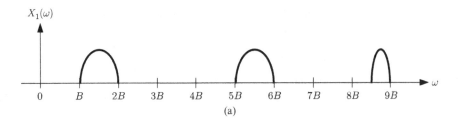

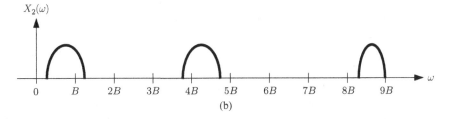

Figure 14.21 Illustrating the support of $\mathbf{d}(e^{j\omega})$. (a) An example of a signal with fixed support $N = 3$. The bands of $X_1(\omega)$ occupy entries $\{1, 5, 8\}$ for all ω: the vector $\mathbf{d}(e^{j\omega})$ has the form $\mathbf{d}(e^{j\omega}) = [0\ d_1(e^{j\omega})\ 0\ 0\ 0\ d_5(e^{j\omega})\ 0\ 0\ d_8(e^{j\omega})]^T$ where $d_i(e^{j\omega}) \not\equiv 0$, $i = 1, 5, 8$, for all $\omega \in [0, 2\pi)$. (b) An example of a signal whose fixed support is equal to 5. The bands of $X_2(\omega)$ occupy five of the nine entries, and $\mathbf{d}(e^{j\omega}) = [d_0(e^{j\omega})\ d_1(e^{j\omega})\ 0\ d_3(e^{j\omega})\ d_4(e^{j\omega})\ 0\ 0\ 0\ d_8(e^{j\omega})]^T$.

where in the second equality we used the fact that $Y(\omega)$ is bandlimited to $2\pi/T_{\mathrm{Nyq}}$ and $T = LT_{\mathrm{Nyq}}$. Let $\mathbf{c}(e^{j\omega})$ be the length-p vector with elements $C_i(e^{j\omega})$. Then from (14.32) it follows that we can write the DTFT of the samples as

$$\mathbf{c}(e^{j\omega}) = \mathbf{W}(\omega)\mathbf{A}\mathbf{d}(e^{j\omega}), \quad 0 \le \omega \le 2\pi, \tag{14.33}$$

where $\mathbf{A}$ is the $p \times L$ matrix with elements

$$A_{ik} = e^{\frac{j2\pi k t_i}{L}}, \tag{14.34}$$

and $\mathbf{W}(\omega)$ is a $p \times p$ invertible diagonal matrix with diagonal elements $w_i = e^{j\omega t_i/L}$.

Recall that our goal is to recover $\mathbf{d}(e^{j\omega})$. The set of equations defined by (14.33) is in general underdetermined since we have p equations but $L > p$ unknowns for each ω. However, we can now exploit the fact that $\mathbf{d}(e^{j\omega})$ is sparse, so that many of its entries are actually zero. As we have seen in Proposition 14.1, for each value of ω, the vector $\mathbf{d}(e^{j\omega})$ has at most N values. Let $\mathcal{S} = \mathcal{S}(\omega)$ be the support of $\mathbf{d}(e^{j\omega})$ for a given ω. We may then express (14.33) as

$$\mathbf{c} = \mathbf{W}\mathbf{A}_{\mathcal{S}}\mathbf{d}_{\mathcal{S}}, \tag{14.35}$$

where $\mathbf{d}_{\mathcal{S}}$ is the vector $\mathbf{d}(e^{j\omega})$ on the support $\mathcal{S}$, namely $\mathbf{d}(e^{j\omega})$ without its zeros, and $\mathbf{A}_{\mathcal{S}}$ is $\mathbf{A}$ with the columns corresponding to zero entries removed. For simplicity, we omitted the index ω in (14.35).

In order to recover $\mathbf{d}_{\mathcal{S}}$ for any choice of support, we must ensure that $\mathbf{A}_{\mathcal{S}}$ is left-invertible for every $\mathcal{S}$ (see Appendix A for more details on solving linear equations). In other words, the columns of $\mathbf{A}_{\mathcal{S}}$ should be linearly independent. In this case, we can use the pseudoinverse to invert $\mathbf{A}_{\mathcal{S}}$ by noting that $\mathbf{A}_{\mathcal{S}}^{\dagger}\mathbf{A}_{\mathcal{S}} = \mathbf{I}$. Therefore, $\mathbf{d}_{\mathcal{S}}$ can be recovered from (14.35) as

$$\mathbf{d}_{\mathcal{S}}(e^{j\omega}) = \mathbf{A}_{\mathcal{S}}^{\dagger}\mathbf{W}^{-1}(\omega)\mathbf{c}(e^{j\omega}). \tag{14.36}$$

To complete our analysis of the recovery process we first consider how to implement the operation $\mathbf{y}(e^{j\omega}) = \mathbf{W}^{-1}(\omega)\mathbf{c}(e^{j\omega})$, and then discuss conditions under which $\mathbf{A}_{\mathcal{S}}$ is left-invertible.

Recall that $\mathbf{W}(\omega)$ is a diagonal matrix with diagonal elements $w_i = e^{j\omega t_i/L}$. Therefore, $\mathbf{W}^{-1}(\omega)\mathbf{c}(e^{j\omega})$ corresponds to filtering each sequence $c_i[n]$ with a filter whose frequency response is equal to $e^{-j\omega t_i/L}$. Since $t_i/L < 1$, this filter is referred to as a *fractional delay filter*. A variety of approaches to implement such filters are described in the literature; see e.g. [406] and the references therein. A straightforward method in our setting is to note that w_i represents an integer delay on a finer grid: if we first interpolate the input to a grid denser by a factor of L, then w_i can be implemented as a delay of t_i samples. We then downsample the result to obtain the output sequence at the original rate. This interpretation is shown in Fig. 14.22. Finally, to obtain $\mathbf{d}_{\mathcal{S}}(e^{j\omega})$, we multiply the output $\mathbf{y}(e^{j\omega}) = \mathbf{W}^{-1}(\omega)\mathbf{c}(e^{j\omega})$ by $\mathbf{A}_{\mathcal{S}}^{\dagger}$. Since $\mathbf{A}_{\mathcal{S}}^{\dagger}$ is a constant matrix whose elements do not depend on ω, this multiplication is tantamount to taking linear combinations of the output sequences $Y_i(e^{j\omega})$, and no further filtering is required.

In principle, the set $\mathcal{S}$ depends on ω, so that the support may be different for varying frequency values ω. Solving (14.36) then becomes more involved as we need to distinguish between the different support sets. Instead, we can always exploit the joint sparsity by choosing $\mathcal{S}$ to be of size $2N$ and independent of ω. In this case we require $p \geq 2N$ samplers to ensure left-invertibility of $\mathbf{A}_{\mathcal{S}}$. An advantage of this approach is that once the support is fixed (i.e. independent of ω), $\mathbf{A}_{\mathcal{S}}$ is fixed as well. This means that we can convert the equation $\mathbf{d}_{\mathcal{S}}(e^{j\omega}) = \mathbf{A}_{\mathcal{S}}^{\dagger}\mathbf{y}(e^{j\omega})$ to the time domain resulting in:

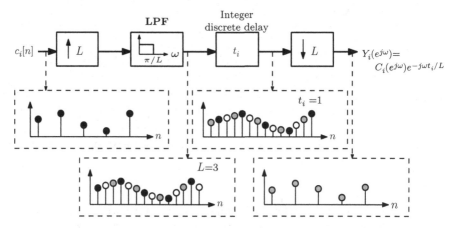

Figure 14.22 Implementation of a fractional delay filter. The input is interpolated to a denser grid by upsampling by L and filtering by a LPF with cutoff frequency π/L. The output is delayed by t_i samples, an integer number, and then downsampled by L. Below the block diagram is an illustration of a signal passing through the system with $L = 3$ and $t_i = 1$.

$$d_{\mathcal{S}}[n] = \mathbf{A}_{\mathcal{S}}^{\dagger} \mathbf{y}[n], \tag{14.37}$$

where $d_{\mathcal{S}}[n], \mathbf{y}[n]$ are the inverse DTFT vectors of $d_{\mathcal{S}}(e^{j\omega}), \mathbf{y}(e^{j\omega})$ and $\mathbf{y}(e^{j\omega}) = \mathbf{W}^{-1}(\omega)\mathbf{c}(e^{j\omega})$. This relation allows us to solve for the nonzero slices $d_{\mathcal{S}}[n]$ directly, without the need to first transform the samples into the frequency domain. A disadvantage of this strategy is an increase in the number of channels required for recovery where instead of $p \geq N$ we now have $p \geq 2N$.

Example 14.8 We demonstrate the recovery of a multiband signal with known carriers from its samples in two different ways: using the varying support of size no larger than N, or a fixed support of size at most $2N$.

Let $x(t)$ be a signal with CTFT:

$$X(\omega) = \begin{cases} X_1(\omega - \omega_1), & \frac{1}{4}B \leq \omega \leq \frac{5}{4}B \\ X_2(\omega - \omega_2), & \frac{15}{4}B < \omega \leq \frac{19}{4}B \\ X_3(\omega - \omega_3), & \frac{21}{4}B < \omega \leq \frac{25}{4}B \\ X_4(\omega - \omega_4), & \frac{33}{4}B < \omega \leq 9B \\ 0, & \text{otherwise}, \end{cases} \tag{14.38}$$

where $X_i(\omega)$ are arbitrary transforms defined on $0 \leq \omega \leq B$ and $\omega_1 = \frac{1}{4}B$, $\omega_2 = \frac{15}{4}B$, $\omega_3 = \frac{21}{4}B$, $\omega_4 = \frac{33}{4}B$. For this signal $N = 4$ and $L = 9$. The corresponding vector $\mathbf{d}(e^{j\omega})$ is depicted in Fig. 14.23. After sampling with p ADCs at rate B we want to recover $x(t)$ from its samples $\{c_i[n], 0 \leq i \leq p - 1\}$.

(1) Recovery using the fact that for each value ω, $\mathbf{d}(e^{j\omega})$ is N-sparse: from Proposition 14.1 it follows that $\mathbf{d}(e^{j\omega})$ is (at most) N-sparse. Indeed, as seen in Fig. 14.23, $\mathbf{d}(e^{j\omega})$ has three different supports for $\omega \in [0, 2\pi)$:

$$\mathcal{S} = \begin{cases} \{1, 4, 6\}, & \omega \in [0, \pi/2) \\ \{0, 4, 5, 8\}, & \omega \in [\pi/2, 3\pi/2) \\ \{0, 3, 5, 8\}, & \omega \in [3\pi/2, 2\pi). \end{cases} \tag{14.39}$$

Thus, recovery of $\mathbf{d}(e^{j\omega})$ by using its N-sparsity requires splitting the range $[0, 2\pi)$ into three different sections and solving $\mathbf{c} = \mathbf{W}\mathbf{A}_{\mathcal{S}}d_{\mathcal{S}}$ in frequency, each time for a different $d_{\mathcal{S}}$ and $\mathbf{A}_{\mathcal{S}}$. To do this, we first interpolate the samples $c_i[n]$ as illustrated in Fig. 14.22 to obtain the sequences $y_i[n]$ with DTFTs $Y_i(e^{j\omega})$ corresponding to $\mathbf{y}(e^{j\omega}) = \mathbf{W}^{-1}(\omega)\mathbf{c}(e^{j\omega}) = \mathbf{A}_{\mathcal{S}}d_{\mathcal{S}}(e^{j\omega})$. We then set up the appropriate system of equations over the different frequency intervals.

For example, for $\omega \in [0, \pi/2)$, we get:

$$\mathbf{A}_{\mathcal{S}}d_{\mathcal{S}} = \begin{bmatrix} 1 & 1 & 1 \\ \lambda_1 & \lambda_4 & \lambda_6 \\ \lambda_1^2 & \lambda_4^2 & \lambda_6^2 \\ \vdots & \vdots & \vdots \\ \lambda_1^{p-1} & \lambda_4^{p-1} & \lambda_6^{p-1} \end{bmatrix} \begin{bmatrix} d_1(e^{j\omega}) \\ d_4(e^{j\omega}) \\ d_6(e^{j\omega}) \end{bmatrix}, \tag{14.40}$$

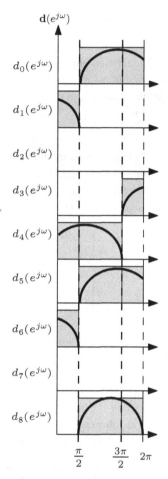

Figure 14.23 Support of the vector-valued function $\mathbf{d}(e^{j\omega})$ for different values of ω. In this example the support varies over the sections $[0, \pi/2), [\pi/2, 3\pi/2), [3\pi/2, 2\pi)$.

and for $\omega \in [3\pi/2, 2\pi)$:

$$
\mathbf{A}_\mathcal{S}\mathbf{d}_\mathcal{S} = \begin{bmatrix} 1 & 1 & 1 & 1 \\ 1 & \lambda_3 & \lambda_5 & \lambda_8 \\ 1 & \lambda_3^2 & \lambda_5^2 & \lambda_8^2 \\ \vdots & \vdots & \vdots & \vdots \\ 1 & \lambda_3^{p-1} & \lambda_5^{p-1} & \lambda_8^{p-1} \end{bmatrix} \begin{bmatrix} d_0(e^{j\omega}) \\ d_3(e^{j\omega}) \\ d_5(e^{j\omega}) \\ d_8(e^{j\omega}) \end{bmatrix}
\tag{14.41}
$$

where $\lambda_k = e^{j2\pi k/L}$, $k = 0, 1, \ldots, 8$. The support choice for $\omega = 5\pi/3$ and the corresponding system $\mathbf{A}_\mathcal{S}\mathbf{d}_\mathcal{S}$ are illustrated in Fig. 14.24.

Solving $\mathbf{y} = \mathbf{A}_\mathcal{S}\mathbf{d}_\mathcal{S}$ for all ω requires that $\mathbf{A}_\mathcal{S}$ is left-invertible in each of the relevant intervals. In particular, this implies that the number of samplers p has to be at least as large as the greatest support size, namely $p \geq 4$.

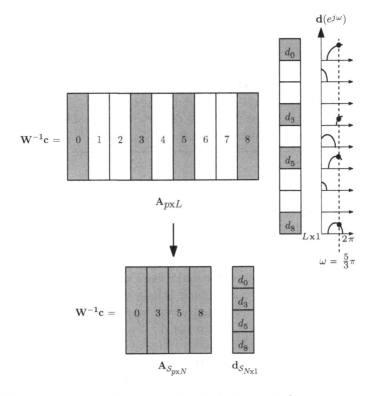

Figure 14.24 Support choice and the corresponding $\mathbf{A}_\mathcal{S}\mathbf{d}_\mathcal{S}$ for $\omega = 5\pi/3$.

(2) Recovery using the fact that $\mathbf{d}(e^{j\omega})$ is $2N$-jointly sparse for all ω: In order to choose a fixed support for all ω, so that we do not need to compute several pseudoinverses and can process the samples in time, we now exploit the joint sparsity of $\mathbf{d}(e^{j\omega})$. From Proposition 14.1, $\mathbf{d}(e^{j\omega})$ is at most $2N$-jointly sparse. In our example, the bands of $x(t)$ occupy seven of its nine entries and thus $\mathbf{d}(e^{j\omega})$ has a fixed support of size 7 for all ω:

$$\mathcal{S} = \{0, 1, 3, 4, 5, 6, 8\}, \quad \omega \in [0, 2\pi). \tag{14.42}$$

As a result, we can recover $\mathbf{d}(e^{j\omega})$ by interpolating the sequences as before, and then considering the equations:

$$\mathbf{y}(e^{j\omega}) = \begin{bmatrix} 1 & 1 & 1 & 1 & 1 & 1 & 1 \\ 1 & \lambda_1 & \lambda_3 & \lambda_4 & \lambda_5 & \lambda_6 & \lambda_8 \\ 1 & \lambda_1^2 & \lambda_3^2 & \lambda_4^2 & \lambda_5^2 & \lambda_6^2 & \lambda_8^2 \\ \vdots & \vdots & \vdots & \vdots & \vdots & \vdots & \vdots \\ 1 & \lambda_1^{p-1} & \lambda_3^{p-1} & \lambda_4^{p-1} & \lambda_5^{p-1} & \lambda_6^{p-1} & \lambda_8^{p-1} \end{bmatrix} \begin{bmatrix} d_0(e^{j\omega}) \\ d_1(e^{j\omega}) \\ d_3(e^{j\omega}) \\ d_4(e^{j\omega}) \\ d_5(e^{j\omega}) \\ d_6(e^{j\omega}) \\ d_8(e^{j\omega}) \end{bmatrix}. \tag{14.43}$$

Solving this system of equations for all ω requires that $\mathbf{A}_S$, which is fixed within $\omega \in [0, 2\pi)$, is left-invertible and that the number of samplers p is greater or equal than the size of the joint sparsity of $\mathbf{d}(e^{j\omega})$, in our case $p \geq 7$. Note that now there is only a single pseudoinverse to compute.

Once the pseudoinverse is calculated, it can be applied directly on the interpolated sequences $y_i[n]$ to determine the nonzero slices $d_i[n]$ using (14.37), without the need to first compute the DTFTs $C_i(e^{j\omega})$. Thus, all of the recovery operations may be performed in the time domain.

14.3.3   Universal sampling patterns

We now turn to discuss (left-)invertibility of $\mathbf{A}_S$. In our analysis below we assume sparsity of size N; however, all the results adapt to the case in which we exploit joint sparsity by replacing N with $2N$.

Invertibility depends on the number of rows p, and on the selected columns, which in turn are a function of the carrier frequencies. In practice, we would like to be able to design a sampling system that is independent of the exact choice of carriers. That way the same system can be used even when the carriers vary. This means that we want to select the values of t_i such that $\mathbf{A}_S$ is left-invertible for *any* choice of N columns, independent of which columns are selected. Such a set of sampling instances is referred to as *universal*. Note that signal recovery will still depend on the carrier values, since we need to know the set S in order to solve for $\mathbf{d}_S$ in (14.36). We will relax this requirement in Section 14.5, by using a larger value of p and utilizing analog CS recovery techniques. Since recovery is performed in DSP it is easier to adapt to the carrier values in the recovery stage. On the other hand, the sampling is performed in analog hardware and therefore there is a benefit to having a fixed hardware structure corresponding to constant sampling times, irrespective of the carrier frequencies.

Definition 14.1. *Let $\mathbf{A}$ be the $p \times L$ matrix with elements $A_{ik} = \exp\{j2\pi kt_i/L\}$ where $t_i, 0 \leq i \leq p-1$ is a given sampling pattern. Each t_i can take on a distinct integer value in the interval $[0, L-1]$. The set $\{t_i\}$ is called a universal pattern if any N columns of $\mathbf{A}$ are linearly independent.*

By the definition of a universal pattern, we must have that $p \geq N$. When $p = N$ the overall sampling rate is given by NB, which is equal to the Landau rate. Therefore, in order to achieve the Landau rate with interleaved ADCs for any multiband input we need to ensure that $\{t_i\}$ forms a universal sampling pattern.

It turns out that for $p \geq N$, there always exist universal patterns. In particular, we have the following proposition:

Proposition 14.2. *The choice $t_i = i$ for $i = 0, 1, \ldots, p-1$ with $p \geq N$ is a universal sampling pattern.*

The choice $t_i = i$ is referred to as *bunched sampling* since the corresponding nonuniform sampling pattern results in samples that are bunched at the beginning of the period, as illustrated in Fig. 14.25.

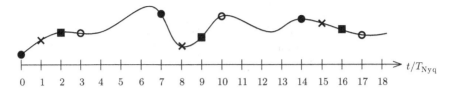

Figure 14.25 A bunched sampling pattern with $L = 7$ and $p = 4$. The samples are all bunched at the beginning of the period; that is, $t_i = i$ for $i = 0, 1, 2, 3$.

Proof: When $t_i = i$, we can write $\mathbf{A}$ as

$$
\mathbf{A} = \begin{bmatrix}
1 & 1 & \cdots & 1 \\
1 & \lambda_1 & \cdots & \lambda_{L-1} \\
1 & \lambda_1^2 & \cdots & \lambda_{L-1}^2 \\
\vdots & & & \vdots \\
1 & \lambda_1^{p-1} & \cdots & \lambda_{L-1}^{p-1}
\end{bmatrix} \tag{14.44}
$$

where $\lambda_k = e^{j2\pi k/L}$, $k = 0, 1, \ldots, L-1$. The matrix $\mathbf{A}$ is a Vandermonde matrix with roots λ_k. Clearly, any choice of N columns from $\mathbf{A}$ also forms a Vandermonde matrix. It is well known that a $p \times N$ Vandermonde matrix is left-invertible if and only if $p \geq N$ and its roots are distinct (see Appendix A). Thus, choosing a subset of N columns of $\mathbf{A}$ will always result in a left-invertible matrix. $\qquad \square$

A less restrictive condition is given by the following proposition [407].

Proposition 14.3. *Let $\{t_i\}$ be an arithmetic progression with difference $d \neq 0$:*

$$
t_i = (t_0 + id) \bmod L, \quad i = 0, 1, \ldots p - 1. \tag{14.45}
$$

If d and L are relatively prime and $p \geq N$, then $\{t_i\}$ is a universal sampling pattern.

Note that Proposition 14.2 is a special case of Proposition 14.3 with $d = 1$ and $t_0 = 0$. We further remark that the difference d can have arbitrary sign, and can be larger than L. Consequently, the resulting values t_i need not be in any particular order, nor will the final values in the interval $[0, L-1]$ necessarily have equal differences.

Proof: With t_i given by (14.45) we can write $\mathbf{A}$ of (14.34) as

$$
\mathbf{A} = \begin{bmatrix}
1 & 1 & \cdots & 1 \\
1 & \lambda_1 & \cdots & \lambda_{L-1} \\
1 & \lambda_1^2 & \cdots & \lambda_{L-1}^2 \\
\vdots & & & \vdots \\
1 & \lambda_1^{p-1} & \cdots & \lambda_{L-1}^{p-1}
\end{bmatrix} \mathbf{D} = \mathbf{A}_1 \mathbf{D}, \tag{14.46}
$$

where $\lambda_k = e^{j2\pi kd/L}$, $k = 0, 1, \ldots, L-1$ and $\mathbf{D}$ is a diagonal matrix with diagonal elements $e^{j2\pi kt_0/L}$. Clearly $\mathbf{D}$ is invertible. The matrix $\mathbf{A}_1$ is a Vandermonde matrix with roots λ_k. Therefore, choosing a subset of N columns of $\mathbf{A}_1$ will result in a left-invertible matrix as long as λ_k are distinct, and $p \geq N$.

Suppose that there exist two integers $0 \leq m < n < L$ such that $\lambda_m = \lambda_n$. This then implies that $(m - n)d = kL$ for some integer k. However, if d and L are relatively prime then we must have $m = n$ and $k = 0$. Thus, $\lambda_m \neq \lambda_n$ and $\mathbf{A}_1$ is left-invertible. Finally, $\mathbf{A}^\dagger = \mathbf{D}^{-1}\mathbf{A}_1^\dagger$. □

Example 14.9 Consider an example in which $L = 12$ and $p = 4$. To create a universal pattern we choose a value of d that is coprime with L and select t_0, for example $d = 5$ and $t_0 = 0$. The resulting arithmetic progression leads to the set

$$\mathcal{K} = \{0, 5, 10, 3\}. \tag{14.47}$$

According to Proposition 14.3, if we generate the sampling matrix $\mathbf{A}$ from this set, then any choice of $N \leq p$ columns from $\mathbf{A}$ is linearly independent.

If we further develop this progression then we obtain a periodic sequence with period L. Using the same reasoning, every four consecutive terms in the resulting sequence form a universal pattern.

Another important case of a universal pattern is when L is prime, as incorporated in the following proposition [408].

Proposition 14.4. *If L is prime then any pattern is universal.*

Proof: The proof relies on the following lemma, proved by Chebotarëv [414, Theorem 6].

Lemma 14.1. *Let L be a prime number, and let $\mathcal{I}$ and $\mathcal{K}$ denote sets of N integers between 0 and $L - 1$. Construct a square $N \times N$ matrix whose elements are given by $e^{\frac{j2\pi ik}{L}}$ where $i \in \mathcal{I}$ and $k \in \mathcal{K}$. Then the matrix is invertible for any choice of $\mathcal{I}$ and $\mathcal{K}$.*

Choose N columns of the matrix $\mathbf{A}$. Then, when L is prime, the resulting submatrix satisfies the conditions of Lemma 14.1 and is consequently invertible. Since this holds for all N rows and columns, we conclude that any set $\{t_i\}$ is universal. □

It is also easy to see that if L is not prime then one can always find a pattern that is not universal. Indeed, let $L = nk$ for some $1 < n, k < L$. The pattern $t_0 = 0, t_1 = n$ is not universal since choosing the support set $\mathcal{S} = \{0, k\}$ results in the matrix

$$\mathbf{A}_\mathcal{S} = \begin{bmatrix} w^{0 \cdot 0} & w^{0 \cdot k} \\ w^{n \cdot 0} & w^{n \cdot k} \end{bmatrix} = \begin{bmatrix} 1 & 1 \\ 1 & 1 \end{bmatrix}, \quad w = e^{j\frac{2\pi}{L}} \tag{14.48}$$

which is rank-deficient.

Example 14.10 Once we find a universal set, we can generate alternative sets from it by using the fact that any cyclic shift of a universal set is also universal. More specifically, suppose that $\{t_i\}$ is a universal set. Then the set $\{\tilde{t}_i\}$ where $\tilde{t}_i = (t_i + c) \bmod L$ for $c \in \mathbb{Z}$, is also universal. The proof is very similar to that of Proposition 14.3 with $t_0 = c$ and where id is replaced by t_i. The matrix $\mathbf{A}_1$ is then replaced by a matrix corresponding to $\{t_i\}$ which is left-invertible since $\{t_i\}$ is a universal set.

As a result, we can extract from every universal set another $L - 1$ universal sets. For instance, from the universal set of Example 14.9 $\{0, 5, 10, 3\}$ with $L = 12$ we can create the universal sets $\{2, 7, 0, 5\}$ $(c = 2)$ or $\{10, 3, 8, 1\}$ $(c = -2)$ as well as another nine distinct sets.

For general choices of L and t_i it is difficult to determine whether or not a pattern is universal. Nevertheless, a random pattern drawn uniformly among all $\binom{L}{p}$ combinations is universal with high probability; see [13] for details.

The propositions above show that there are many possible choices of universal patterns for any N and L. However, it is important to keep in mind that different universal patterns will lead to varying behavior in the presence of noise. Specifically, although a universal pattern guarantees that the relevant submatrix $\mathbf{A}_S$ is invertible, it does not ensure any stability properties, so that the pseudoinverse $\mathbf{A}_S^\dagger$ may have a large norm. In Appendix A we show that for a general matrix $\mathbf{B}$, the maximum amplification $\|\mathbf{Bx}\|_2 / \|\mathbf{x}\|_2$ of an input vector $\mathbf{x}$ is given by its largest singular value $\sigma_{\max} = \|\mathbf{B}\|_2$. Thus, if noise is added to the measurements, then it can be amplified by as much as $\|\mathbf{A}_S\|_2$ in the recovery process. We may therefore use $\|\mathbf{A}_S\|_2$ as a stability measure of the sampling pattern. The next example demonstrates that this value is highly affected by the choice of sampling times.

Example 14.11 We demonstrate the difference in behavior of two distinct sampling patterns in the presence of noise.

Let $\mathbf{A}_1$ and $\mathbf{A}_2$ be two 5×11 matrices generated from the patterns $\{0, 1, 2, 3, 4\}$ and $\{0, 2, 3, 4, 8\}$ $(p = 5, L = 11)$, respectively, and assume that the multiband input consists of three bands, i.e. $|S| = 3$. Since L is prime, Proposition 14.4 ensures that these patterns are universal.

To evaluate the systems' behavior in the presence of noise, we consider $\|\mathbf{A}_S\|_2$ for each of the matrices on the active support S. In general, this value will depend on the chosen support. For each of the systems, we compute the worst-case value, which results in $\max\limits_{S \in S} \|\mathbf{A}_{1_S}^\dagger\|_2 \approx 1.8728$ and $\max\limits_{S \in S} \|\mathbf{A}_{2_S}^\dagger\|_2 \approx 0.7361$. Here S denotes the set of all $\binom{11}{3}$ possible support choices. As can be seen, the second pattern, which has a larger spread, is more stable than the first, whose values are bunched. This claim is true more generally: bunched universal patterns are less stable than patterns that are further spread out.

We summarize our results on recovery of multiband inputs in the following theorem.

Theorem 14.1 (Multicoset recovery). *Let $x(t)$ be a multiband signal on $[0, 2\pi/T_{\mathrm{Nyq}})$ consisting of N bands each with width no larger than B, and let $2\pi/T_{\mathrm{Nyq}} = LB$ for some integer L. Consider multicoset sampling of $x(t)$ with p samplers at rate T, and time shifts $\{t_i\}$. The vector $\mathbf{d}(e^{j\omega})$ defined by (14.31) is the unique N-sparse solution of (14.33) if*

1. $T \leq 2\pi/B$;
2. $p \geq N$;
3. $\{t_i\}$ is a universal pattern.

The vector $\mathbf{d}(e^{j\omega})$ *can be recovered via* $\mathbf{d}_S(e^{j\omega}) = \mathbf{A}_S^\dagger \mathbf{y}(e^{j\omega}) = \mathbf{A}_S^\dagger \mathbf{W}^{-1}(\omega)\mathbf{c}(e^{j\omega})$, *where* $\mathbf{y}(e^{j\omega})$ *is the output of the fractional delay filter illustrated in Fig. 14.22,* $\mathbf{c}(e^{j\omega})$ *is the DTFT of the sequence of samples, and* $S = S(\omega)$ *is the support set at* ω.

14.3.4   Hardware considerations

Both direct undersampling and interleaved ADCs reduce the sampling rate below Nyquist by direct sampling of the analog input. That is, the ADC is applied directly to $x(t)$ with no preceding analog preprocessing components, in contrast to the RF hardware used in I/Q demodulation. However, this comes at a price: not every ADC device fits interleaved or undersampling systems. Only convertors whose front-end analog bandwidth exceeds $\omega_{\max} = 2\pi/T_{\mathrm{Nyq}}$ are viable. To understand this important observation, we briefly discuss the ADC operation.

An ADC device, in its most basic form, repeatedly alternates between two states: track-and-hold (T/H) and quantization. During T/H, the ADC tracks the signal variation. When an accurate track is accomplished, the value is held steady by the ADC allowing the quantizer to convert the amplitude into a finite representation. Both operations must end before the next point is acquired. In signal processing, it is common to model the ADC as an ideal pointwise sampler that captures values of $x(t)$ at a constant rate of r samples per second. However, as with any analog circuitry, the T/H function is limited in the range of frequencies it can accept: it cannot track arbitrarily fast inputs. In practice, an LPF with cutoff b can be used to model the T/H capability [21, 410], as illustrated in Fig. 14.26(a).

In most off-the-shelf ADCs, the analog bandwidth parameter b is specified higher than the maximal sampling rate r of the device. Figure 14.26(b) lists example convertors. In our discussion, we assume that the LPF is ideal, meaning it is a brick-wall filter, so that b represents its true cutoff frequency. Otherwise, our conclusions are still valid with respect to an effective b which can be made a bit higher, capturing the fact that some signal passes through the LPF at frequencies above b, but is attenuated by the filter's

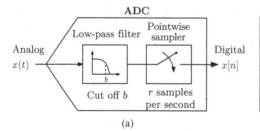

Device	Max. rate (Megasamples/sec)	Analog BW (MHz)
ADS5474	400	1440
AD12401	400	480
AD1020	105	250
AD9433	105	750

(a)   (b)

Figure 14.26 ADC devices. (a) A practical ADC model consisting of a low-pass filter with cutoff b and a pointwise sampler acquiring r samples per second. (b) Off-the-shelf ADC devices and their parameters. Here BW indicates bandwidth.

frequency response. When using an ADC at the Nyquist rate of the input, the filter can be omitted from the model, since the signal is bandlimited to $r/2 \leq b$. In contrast, for sub-Nyquist purposes, the analog bandwidth b becomes an important factor in accurate modeling and actual selection of the ADC, as it defines the maximal input frequency that can be undersampled. Typically, b specifies the -3 dB point of the T/H frequency response. Thus, if flat response in the passband is of interest, $\omega_{\max}$ cannot approach b too closely.

Example 14.12 Let $x(t)$ be a bandpass signal in the range $[600, 625]$MHz. Undersampling at rate $\omega_s = 50$ MHz satisfies condition (14.17), so we can sample the signal at a sub-Nyquist rate. While both AD9433 and AD1020 from the table in Fig. 14.26(b) are capable of sampling at a rate $r \geq 50$ MHz, only the former, with $b = 750$, is applicable to $x(t)$ owing to the analog bandwidth constraint (assuming an ideal LPF in the sampler).

Undersampling ADCs have a wider spacing between consecutive samples. This advantage is translated into simplifying design constraints, especially in the duration allowed for quantization. However, regardless of the sampling rate r, the T/H stage must still hold a pointwise value of a fast-varying signal. In terms of analog bandwidth there is no substantial difference between Nyquist and undersampling ADCs; both have to accommodate the Nyquist rate of the input.

To conclude, undersampling or interleaved ADCs can be used to directly sample a multiband signal at sub-Nyquist rates without the need for analog preprocessing used in demodulation. Furthermore, as we show in the next section, interleaved ADCs can be adapted to the case of unknown carriers, while demodulation requires knowledge of the carrier frequencies. However, undersampling systems also have several drawbacks. First, the uniform low-rate sampler used in undersampling must have sufficient analog bandwidth in order to sample the incoming signal positioned at a high carrier frequency. Second, undersampling methods alias the noise from the entire bandwidth down to baseband which entails a loss in SNR. In contrast, when using demodulation, an LPF is applied prior to sampling which rejects out of band noise. In principle, we can apply a bandpass filter to the bandpass signal prior to undersampling. However, this filter will need to be tunable, depending on the input spectrum, and thus can be complicated to implement. Third, interleaved ADCs require exact knowledge of the sampling times t_i in the recovery stage. In practice, time offsets are hard to avoid so that the actual sampling times are difficult to obtain. Other degradations that result from having a parallel set of samplers such as gain mismatches among the ADCs as well as timing skew of the clocks distributed to them degrade overall system performance [402, 403, 411]. A variety of algorithms have been proposed in the literature in order to compensate for these distortions, but they tend to add substantial complexity to the receiver.

In [336] an alternative front end was proposed, mainly to overcome analog bandwidth issues. We will discuss this sampling system, referred to as the *modulated wideband converter (MWC)*, in the next section. As with interleaved ADCs, the MWC can also be used when the carriers are unknown. Although the MWC, like any sub-Nyquist scheme, suffers from noise aliasing, it alleviates the analog bandwidth issue, as well as the need for different sampling times in each channel. Furthermore, we will see that the MWC can be realized by a single sampling channel, which further reduces impairments in interleaved structures that are a result of mismatches between the different channels, as well as the hardware requirements.

14.4   Modulated wideband converter

Similar to multicoset sampling, the MWC aliases the high-bandwidth spectrum, by combining the spectrum slices $d_k(e^{j\omega})$ of (14.31). In order to achieve the desired aliasing, the system exploits spread-spectrum techniques from communication theory [412, 413] and modulates the input by a periodic function with period T_p, according to the scheme depicted in Fig. 14.27. A particular choice of periodic function is shown in the figure, and will be studied below along with its exact parameter values. The analog mixing

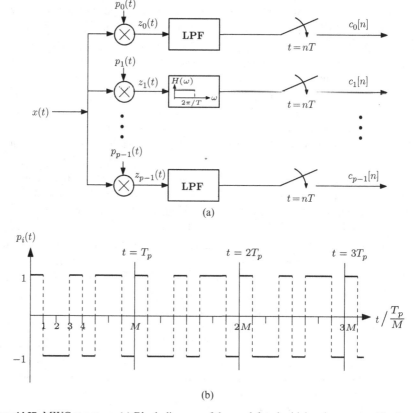

Figure 14.27 MWC structure. (a) Block diagram of the modulated wideband converter. The input passes through p parallel branches, is mixed with a set of periodic functions $p_i(t)$, low-pass filtered, and then sampled at a low rate. (b) A possible choice of periodic function.

front-end aliases the spectrum, such that a spectrum portion from each band appears in baseband. Each channel implements a different mixture by using a distinct modulating sequence $p_i(t)$, so that, in principle, a sufficiently large number of mixtures allows to recover a relatively sparse multiband signal. After mixing, the output is filtered by an LPF and sampled at a low rate $2\pi/T$. The total sampling rate of the MWC is $2\pi p/T$.

The main difference between multicoset and MWC sampling is in the method in which aliasing is performed: while in the former aliasing is achieved by undersampling and the different mixtures are a result of the delays in each channel, in the latter aliasing results from modulation with periodic sequences which precede the sampling operation. Since aliasing is performed prior to sampling, the input to the ADC is no longer a high-bandwidth signal, but rather a signal with bandwidth limited to $2\pi/T$. Such a signal can be sampled using a standard low-rate ADC which is not required to have high analog bandwidth. Another advantage of this architecture over multicoset sampling is that the number of branches can be reduced even to a single sampling channel, by increasing the sampling rate on the channel. We discuss this tradeoff in Section 14.4.3, after explaining the MWC operation. In contrast, multicoset sampling with general sampling times requires a separate branch per delay which can result in a large number of branches that need to be implemented simultaneously.

The front-end preprocessing of the MWC must be carried out by analog means, since both the mixer and the analog filter operate on wideband signals. The purely analog front end is the key to overcoming the bandwidth limitation of ADCs. This is also the difficult part of the sampling architecture: while the sampling operator now only has to deal with low-bandwidth inputs, the mixers must handle high bandwidth. We note, however, that standard radios also face the problem of mixing signals at high frequencies.

Before detailing the operation of the MWC we point out some of its potential advantages:

1. Analog mixers are a proven technology in the wideband regime [414, 415]. In fact, since transmitters use mixers to modulate the information by a high carrier frequency, the mixer bandwidth defines the input bandwidth.

2. Sign alternating functions can be implemented by standard (high-rate) digital circuits. Today's technology allows alternation rates of 23 GHz [416] and even 80 GHz [417] to be reached.

3. The sampling rate $2\pi/T$ matches the LPF cutoff. Therefore, an ADC with conversion rate $r = 2\pi/T$, and any bandwidth $b \geq 0.5r$ (in the real case) can be used to implement this block, where the LPF serves as a preceding anti-aliasing filter. In the sequel, we choose $2\pi/T$ on the order of B, which is the width of a single band of $x(t)$. In practice, this sampling rate allows flexible choice of an ADC from a variety of commercial devices in the low-rate regime.

4. Sampling is synchronized in all channels; that is, there are no time shifts. This is beneficial since the trigger for all ADCs can be generated accurately (e.g. with a zero-delay synchronization device [418]). The same clock can be used for a subsequent digital processor which receives the sample sets at rate $2\pi/T$.

14.4.1   MWC operation

For simplicity of exposition, we follow the same route as in the presentation of multi-coset sampling, and describe the MWC assuming a complex input and restricting attention to the positive frequency axis.

The MWC consists of an analog front end with p channels. In the ith channel, $x(t)$ is multiplied by a mixing function $p_i(t)$, which is T_p-periodic. After mixing, the signal spectrum is truncated by a (complex) low-pass filter $H(\omega)$ with cutoff $2\pi/T$ and the filtered signal is sampled at rate $2\pi/T$ resulting in the sequence of samples $c_i[n]$, $0 \leq i \leq p-1$. The sampling rate on each channel is sufficiently low to allow for the use of existing commercial ADCs. The design parameters are therefore the number of channels p, the period T_p, the sampling rate $2\pi/T$, and the mixing functions $p_i(t)$ for $0 \leq i \leq p-1$.

The mixing operation aliases the spectrum of $x(t)$, such that a portion of the energy of all bands appears in baseband. Specifically, since $p_i(t)$ is periodic with period T_p, it has a Fourier expansion

$$p_i(t) = \sum_{\ell=-\infty}^{\infty} c_{i\ell} e^{j\frac{2\pi}{T_p}\ell t}, \tag{14.49}$$

where

$$c_{i\ell} = \frac{1}{T_p} \int_0^{T_p} p_i(t) e^{-j\frac{2\pi}{T_p}\ell t} dt. \tag{14.50}$$

The Fourier transform of $p_i(t)$ is therefore given by

$$P_i(\omega) = 2\pi \sum_{\ell=-\infty}^{\infty} c_{i\ell} \delta\left(\omega - \frac{2\pi\ell}{T_p}\right). \tag{14.51}$$

In the frequency domain, mixing by $p_i(t)$ is tantamount to convolution between $X(\omega)$ and $(1/2\pi)P_i(\omega)$. Therefore, the CTFT of the mixer output $z_i(t)$ is given by

$$Z_i(\omega) = \sum_{\ell=-\infty}^{\infty} c_{i\ell} X\left(\omega - \frac{2\pi\ell}{T_p}\right). \tag{14.52}$$

The aliasing effect is evident in $z_i(t)$: the input to the filter $H(\omega)$ is a linear combination of shifted copies of $X(\omega)$. Since $X(\omega) = 0$ for $\omega \notin [0, 2\pi/T_{\mathrm{Nyq}})$, for each specific ω, the sum in (14.52) contains (at most) $\lceil T_{\mathrm{Nyq}}/T_p \rceil$ nonzero terms. The filter $H(\omega)$ has a frequency response which is an ideal rectangular function, as depicted in Fig. 14.27. Nonideal filters can also be accommodated; see [419] for details. Consequently, only frequencies in the interval $\mathcal{F}_s = [0, 2\pi/T)$ are present in the uniform sequence $c_i[n]$. These frequencies contain components from all different segments of $X(\omega)$ owing to the aliasing affect. As in multicoset sampling, this aliasing is helpful in order to capture the energy of $x(t)$ from low-rate samples. Repeating this process in each of the branches, using different modulating sequences $p_i(t)$, leads to a linear system of equations in the digital domain that can be solved once the signal structure is exploited. In Section 14.4.4, we discuss how to choose $p_i(t)$ as time-shifts of a basic sequence $p(t)$, which simplifies the hardware design. An additional hardware simplification considered in Section 14.4.3 is to collapse the p-channels in Fig. 14.27 to a single channel with sampling rate increased by a factor of p. The price to pay for the hardware reduction is an increase in digital processing and reduced design flexibility.

An important consequence of periodicity is robustness to time-domain variability. As long as the waveform $p_i(t)$ is periodic, the coefficients $c_{i\ell}$ can be computed, or calibrated. The exact shape of the waveform does not play a role in the recovery process. In particular, suppose we are using a periodic sign wave, as illustrated in Fig. 14.27(b). A sign wave whose alternations do not occur exactly on the Nyquist grid, and whose levels are not accurate ± 1 levels is fine, as long as the same pattern repeats every T_p seconds. This greatly simplifies the design as it is much easier to maintain periodicity than accurate amplitude levels.

14.4.2 MWC signal recovery

We now consider signal recovery from the MWC samples. To ease the analysis we make some assumptions on the system parameters. These restrictions are not necessary for the validity of the sampling method, and are made only for convenience.

Similar to multicoset sampling, we assume that $T_p = LT_{\mathrm{Nyq}}$ and $B = (2\pi)/(LT_{\mathrm{Nyq}}) = 2\pi/T_p$, and choose a sampling period $T = T_p$. In this case, there are at most L aliases of $X(\omega)$ in $Z_i(\omega)$ of (14.52), so that

$$Z_i(\omega) = \sum_{\ell=0}^{L-1} c_{i,-\ell} X\left(\omega + \frac{2\pi\ell}{T}\right), \quad 0 \le \omega \le \frac{2\pi}{T}. \tag{14.53}$$

The DTFT of the ith sample sequence $c_i[n]$ is then given by

$$C_i(e^{j\omega}) = \frac{1}{T} \sum_{\ell=0}^{L-1} c_{i,-\ell} X\left(\frac{\omega}{T} + \frac{2\pi\ell}{T}\right). \tag{14.54}$$

The MWC sampling process is illustrated in Fig. 14.28.

Note that the mixer output $z_i(t)$ is not bandlimited. Therefore, depending on the coefficients $c_{i\ell}$, the Fourier transform (14.52) may not be well defined. This technicality, however, is resolved in (14.54) since the filter output involves only a finite number of aliases of $x(t)$.

Relation (14.54) ties the known DTFTs of $c_i[n]$ to the unknown $X(\omega)$. This equation is the key to recovery of $x(t)$, and is very similar to (14.32) which results from multicoset sampling. In the latter, linear combinations of the spectrum slices of $X(\omega)$ are determined by the delays in each channel while in (14.54), they result from the Fourier coefficients of the periodic functions.

Defining the spectrum slices $d_i(e^{j\omega})$ as in (14.31), we can write (14.54) in matrix form:

$$c(e^{j\omega}) = \mathbf{A}d(e^{j\omega}), \quad 0 \le \omega \le 2\pi, \tag{14.55}$$

where $c(\omega)$ is the length-p vector with ith element $C_i(e^{j\omega})$. The $p \times L$ matrix $\mathbf{A}$ contains the coefficients $c_{i,-\ell}$:

$$A_{i\ell} = c_{i,-\ell}. \tag{14.56}$$

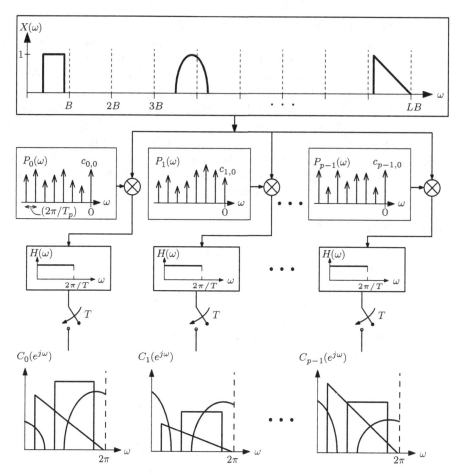

Figure 14.28 MWC operation. A multiband signal with $N = 3, L = 9$ passes through p mixers, each mixing the signal using a periodic function $p_i(t)$ with period $T_p = 2\pi/B$. After mixing, the output is filtered by an LPF with cutoff frequency $2\pi/T$ ($T = T_p$) and sampled at a low rate $2\pi/T$. The total sampling rate of the MWC is given by $2\pi p/T$.

The role of the mixing functions appears implicitly in (14.55) through the coefficients $c_{i\ell}$ defining the matrix in (14.56). Each $p_i(t)$ provides a single row in $\mathbf{A}$. Roughly speaking, $p_i(t)$ should have enough transients within the time period T_p so that its Fourier expansion (14.49) contains at least L dominant terms. In this case, the channel output $c_i[n]$ is a mixture of all (nonidentically zero) spectrum slices in $\mathbf{d}(e^{j\omega})$.

The problem of recovering $\mathbf{d}(e^{j\omega})$ from the MWC samples (14.55) is similar to that of recovering $\mathbf{d}(e^{j\omega})$ from multicoset samples (14.33). An important difference is that here there is no fractional delay filter (represented by the matrix $\mathbf{W}$ in (14.33)), so that upsampling is not necessary. Carrying over the results from the multicoset setting, we conclude that recovery is possible as long as every N columns of $\mathbf{A}$ are linearly independent. This implies that every submatrix $\mathbf{A}_S$ formed by taking the N columns of $\mathbf{A}$ corresponding to the support S of $\mathbf{d}(e^{j\omega})$ for a particular ω is left-invertible. In the terminology of CS studied in Chapter 11 this means that $\mathbf{A}$ must have spark at least $N + 1$. In analogy with multicoset sampling, we will say that the periodic sequences in the MWC are *universal* if $\mathbf{A}_S$ is left-invertible for every choice of S. Under this condition, $\mathbf{d}_S(e^{j\omega}) = \mathbf{A}_S^{\dagger}\mathbf{c}(e^{j\omega})$.

Designing an appropriate matrix $\mathbf{A}$ is easier here than in multicoset sampling, since in principle the values of $c_{i\ell}$ can be chosen as desired, while in multicoset sampling the elements of $\mathbf{A}$ are a function of the sampling times t_i. Recall from Chapter 11, that if the elements $c_{i\ell}$ are generated at random, then with high probability $\mathbf{A}$ will have spark equal to $p + 1$. Thus, as long as $p \geq N$, the sequences will be universal. Chapter 11 also provides alternative choices of $\mathbf{A}$ with this property. For example, if $\mathbf{A}$ consists of $p \geq N$ random rows of an $L \times L$ unitary matrix, then with high probability its spark will equal $p + 1$. In Section 14.4.4 we study periodic sign waveforms and the setting in which all sequences are generated as shifts of a single function.

Example 14.13 Suppose we choose p sequences

$$p_i(t) = T_p \sum_{n \in \mathbb{Z}} \delta(t - t_i T_{\mathrm{Nyq}} - nT_p), \tag{14.57}$$

with $T_p = LT_{\mathrm{Nyq}}$. For this choice, the Fourier coefficients are equal to

$$c_{i\ell} = \exp\{-j2\pi\ell t_i/L\}. \tag{14.58}$$

Thus, the matrix $\mathbf{A}$ of (14.55) coincides with the matrix $\mathbf{A}$ obtained by multicoset sampling with cosets defined by $\{t_i\}$. An important difference however is that in the MWC the actual sampling is performed after aliasing with $p_i(t)$, while in multicoset sampling the aliasing is a result of the sampling itself, thus necessitating a high analog-bandwidth sampler. Furthermore, in multicoset sampling, the samples have to be preprocessed by the matrix $\mathbf{W}^{-1}$ before recovery. This involves implementing a fractional delay filter, which is avoided here.

Example 14.14 We next consider an MWC system with $p = 3$ channels and sequences with period $T_p = 1$, where

$$p_0(t) = \sum_{n \in \mathbb{Z}} \delta(t - n)$$

$$p_1(t) = \sum_{k=0}^{L-1} (k+1)e^{-j2\pi kt}$$

$$p_2(t) = \sum_{k=0}^{L-1} 2(k+1)^2 \cos(2\pi kt) - 1. \tag{14.59}$$

After mixing, the result is filtered by an ideal LPF with cutoff 2π and sampled by an ADC with sampling rate $2\pi/T_p = 2\pi/T = 2\pi$. Direct calculation of the Fourier coefficients of $p_i(t)$ results in

$$c_{0\ell} = 1, \quad c_{1\ell} = \begin{cases} |\ell| + 1, & -(L-1) \le \ell \le 0 \\ 0, & \text{otherwise,} \end{cases} \quad c_{2\ell} = \begin{cases} (|\ell| + 1)^2, & |\ell| \le L - 1 \\ 0, & \text{otherwise.} \end{cases}$$

(14.60)

This system enables sampling of multiband signals with $N = 3, B = 2\pi$ and at the minimum possible rate, $3 \cdot 2\pi/T$, while guaranteeing perfect reconstruction.

As an example, consider sampling a multiband input with $L = 6$. In this case the coefficients form the 3×6 matrix

$$\mathbf{A} = \begin{bmatrix} 1 & 1 & 1 & 1 & 1 & 1 \\ 1 & 2 & 3 & 4 & 5 & 6 \\ 1 & 4 & 9 & 16 & 25 & 36 \end{bmatrix}.$$

(14.61)

More generally, $A_{ij} = \lambda_j^i$ with $\lambda_j = j + 1$ and $0 \le i \le p - 1, 0 \le j \le L - 1$. By selecting any three columns of $\mathbf{A}$ we get a 3×3 Vandermonde matrix with distinct roots which is therefore invertible.

We summarize our discussion in the following theorem, which is an adaptation of Theorem 14.1 to the MWC.

Theorem 14.2 (MWC recovery).   *Let $x(t)$ be a multiband signal on $[0, 2\pi/T_{\text{Nyq}})$ consisting of N bands each with width no larger than B, and let $2\pi/T_{\text{Nyq}} = LB$ for some integer L. Consider the MWC system with mixing functions $p_i(t)$ that are T-periodic. The vector $\mathbf{d}(e^{j\omega})$ defined by (14.31) is the unique N-sparse solution of (14.55) if*

1. $T \le 2\pi/B$;
2. $p \ge N$;
3. The Fourier series coefficients $\{c_{i\ell}\}$ form a universal pattern.

The vector $\mathbf{d}(e^{j\omega})$ can be recovered via $\mathbf{d}_S(e^{j\omega}) = \mathbf{A}_S^\dagger \mathbf{c}(e^{j\omega})$ where $\mathbf{c}(e^{j\omega})$ is the DTFT of the sequence of samples, and $S = S(\omega)$ is the support set at ω.

14.4.3   Collapsing channels

For simplicity we have assumed until now that $T_p = T$ where T is the sampling period. We now briefly comment on more general parameter choices.

The period T_p determines the aliasing of $X(\omega)$ by setting the shift intervals in (14.52) to $\omega_p = 2\pi/T_p$. We choose $2\pi/T_p \ge B$ so that each band contributes only a single nonzero element to $\mathbf{d}(e^{j\omega})$, and consequently $\mathbf{d}(e^{j\omega})$ has at most N nonzeros. In practice ω_p is chosen slightly more than B to avoid edge effects.

The sampling rate of a single channel $\omega_s = 2\pi/T$, and the number of channels p, determines the overall sampling rate $p\omega_s$ of the system. The simplest choice $\omega_s = \omega_p \simeq B$, which we treated until now, allows us to control the sampling rate at a resolution of ω_p. The burden on the hardware implementation is highly affected by the total number of hardware devices, which includes the mixers, LPFs, and ADCs. We now examine a method which reduces the number of channels at the expense of a higher sampling rate in each channel and additional digital processing.

Sampling approach

Suppose that the sampling rate in each channel is increased by a factor of q, leading to $\omega_s = q\omega_p$ and $T = T_p/q$. In this case, the LPF width becomes $2\pi q/T_p$. Its output in the ith branch of Fig. 14.27 is equal to $Z_i(\omega)$ of (14.52) over the interval $[0, 2\pi q/T_p]$. For frequencies in $[0, 2\pi/T_p]$, $Z_i(\omega)$ is given as before by (14.53). However, since the LPF is now wider, we have an additional $q - 1$ bands to consider $[0, 2\pi/T_p] + 2\pi k/T_p$ for $k = 1, \ldots, q - 1$:

$$Z_i(\omega) = \sum_{\ell=-k}^{L-1-k} c_{i,-\ell} X\left(\omega + \frac{2\pi\ell}{T_p}\right), \quad \frac{2\pi k}{T_p} \leq \omega \leq \frac{2\pi(k+1)}{T_p}. \tag{14.62}$$

Alternatively, we can write

$$Z_i(\omega) = \sum_{\ell=0}^{L-1} c_{i,-\ell+k} X\left(\omega + \frac{2\pi(\ell-k)}{T_p}\right), \quad \frac{2\pi k}{T_p} \leq \omega \leq \frac{2\pi(k+1)}{T_p}. \tag{14.63}$$

Denoting by $Z_{ik}(\omega)$ the frequency response in the kth interval $[2\pi k/T_p, 2\pi(k+1)/T_p]$ shifted down to baseband, namely to $[0, 2\pi/T_p]$, it follows that

$$Z_{ik}(\omega) = \sum_{\ell=0}^{L-1} c_{i,-\ell+k} X\left(\omega + \frac{2\pi\ell}{T_p}\right), \quad 0 \leq \omega \leq \frac{2\pi}{T_p}. \tag{14.64}$$

Thus, each band provides an additional equation on $X(\omega)$, so that from one channel we now have q equations. This enables acquisition with fewer channels.

Let us now see how we obtain these additional equations in the digital domain, from the sequences $c_i[n]$. Sampling $Z_i(\omega)$ at rate $2\pi q/T_p$, results in the sampled sequences

$$C_i(e^{j\omega}) = \frac{q}{T_p} \sum_{\ell=0}^{L-1} c_{i,-\ell+k} X\left(\frac{q\omega}{T_p} + \frac{2\pi(\ell-k)}{T_p}\right), \quad \frac{2\pi k}{q} \leq \omega \leq \frac{2\pi(k+1)}{q}. \tag{14.65}$$

As before, we denote by $C_{ik}(e^{j\omega})$ the DTFT on the kth interval $[2\pi k/q, 2\pi(k+1)/q]$ shifted down to $[0, 2\pi/q]$:

$$C_{ik}(e^{j\omega}) = \frac{q}{T_p} \sum_{\ell=0}^{L-1} c_{i,-\ell+k} X\left(\frac{q\omega}{T_p} + \frac{2\pi\ell}{T_p}\right), \quad 0 \leq \omega \leq \frac{2\pi}{q}. \tag{14.66}$$

Denote by $c_{ik}[n]$ the inverse DTFT of $C_{ik}(e^{j\omega})$. Since the DTFT of this sequence is defined over an interval of length $2\pi/q$, we can filter it with an LPF with cutoff $2\pi/q$, and downsample it by q. The output is then a sequence $q_{ik}[n]$ at rate $\omega_p = 2\pi/T_p$ with frequency response

$$Q_{ik}(e^{j\omega}) = \frac{1}{T_p} \sum_{\ell=0}^{L-1} c_{i,-\ell+k} X\left(\frac{\omega}{T_p} + \frac{2\pi\ell}{T_p}\right), \quad 0 \leq \omega \leq 2\pi. \tag{14.67}$$

This sequence is related to the samples $c_i[n]$ through

$$q_{ik}[\tilde{n}] = \left(c_i[n] e^{-j\frac{2\pi}{q} kn}\right) * h[n]\Big|_{n=\tilde{n}q}. \tag{14.68}$$

The brackets describe modulation by $e^{-j\frac{2\pi}{q}kn}$ which performs a frequency shift of $2\pi k/q$, followed by filtering with an LPF with cutoff $2\pi/q$, which we denote by $h[n]$, and decimation by q. We refer to the set of operations which obtain the sequences $q_{ik}[n]$ from $c_i[n]$ as the *expander*. Each sampled sequence is expanded into q digital sequences at $1/q$ of the rate.

Figure 14.29 illustrates the relationship between the input multiband signal $X(\omega)$ and the output discrete-time sequences $Q_{ik}(e^{j\omega})$. After mixing with periodic functions and sampling at rate $2\pi q/T_p$, we form q discrete-time signals from each sampled sequence, by modulations, low-pass filtering, and downsampling. Evidently, each channel requires q digital filters to reduce the rate back to ω_p, which increases the computational load in the digital domain. It is also important to take into account that as q grows, approximating a digital filter with cutoff $2\pi/q$ requires many taps. Thus, the saving in analog channels is at the expense of increased digital computations. It also results in a higher sampling rate and more signal power in each channel, which increases the noise floor and affects the dynamic range of the samplers.

Choosing a collapse factor of $q = p$ reduces a system with p channels to a single channel with sampling rate $\omega_s = p\omega_p$. The resulting sampling architecture is shown in Fig. 14.30. This allows sampling of an arbitrary multiband input at twice the Landau rate using a single analog channel.

Recovery process

Having obtained the sequences $Q_{ik}(e^{j\omega})$ via the expander, we now use them to recover $\mathbf{d}(e^{j\omega})$. Defining $\mathbf{q}_i(e^{j\omega})$ as the vector of length q with elements $Q_{ik}(e^{j\omega})$, we can write (14.67) in matrix form as

$$\mathbf{q}_i(e^{j\omega}) = \mathbf{C}_i\mathbf{d}(e^{j\omega}), \tag{14.69}$$

where

$$\mathbf{C}_i = \begin{bmatrix} c_{i,0} & c_{i,-1} & \cdots & c_{i,-L+1} \\ c_{i,1} & c_{i,0} & \cdots & c_{i,-L+2} \\ \vdots & & & \vdots \\ c_{i,q-1} & c_{i,q-2} & \cdots & c_{i,-L+q} \end{bmatrix}, \quad 0 \le i \le \lceil p/q \rceil - 1. \tag{14.70}$$

This relation shows that from each channel we obtain q equations in the unknown vector $\mathbf{d}(e^{j\omega})$ via shifts of the Fourier coefficient sequence. Concatenating the vectors $\mathbf{q}_i(e^{j\omega})$ to one vector $\mathbf{q}(e^{j\omega})$ leads to a relationship of the form (14.55) where $\mathbf{q}(e^{j\omega})$ replaces $\mathbf{c}(e^{j\omega})$, and the matrix $\mathbf{A}$ is given by the (vertical) concatenation of the matrices $\mathbf{C}_i$ for all values i. When $q = 1$, $\mathbf{C}_1$ becomes the row vector (14.56). It is therefore sufficient to reduce the number of channels by a factor of q, as long as we choose the sequences $c_{i,\ell}$ such that the combined matrix $\mathbf{A}$ has the desired invertibility properties; i.e. every N columns of $\mathbf{A}$ must be linearly independent, or $\mathbf{A}$ should have spark equal to $N + 1$.

Evidently, collapsing the channels leads to a structured matrix $\mathbf{A}$, since each $\mathbf{C}_i$ is constructed from shifts of a single set of Fourier coefficients. The larger the value of q the more restricted $\mathbf{A}$ becomes. Therefore, care must be taken in order to ensure that $\mathbf{A}$ still has the required spark properties.

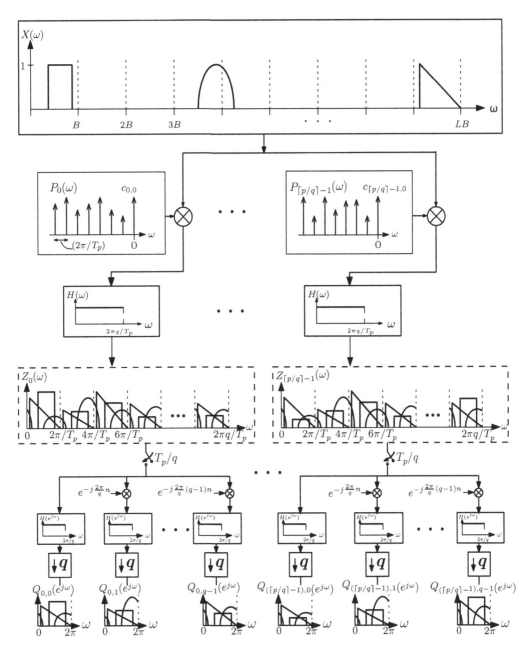

Figure 14.29 Collapsing channels. An MWC system with $\lceil p/q \rceil$ sampling channels. A multiband signal ($N = 3$, $L = 9$) passes through $\lceil p/q \rceil$ analog channels, each mixing it by $p_i(t)$, $0 \le i \le \lceil p/q \rceil - 1$. After filtering by an LPF with cutoff $2\pi q/T_p$ the resulting signal at each channel is sampled at rate $2\pi q/T_p$. The sampled signal then passes through q digital channels, where each shifts the relevant content to baseband by time modulation, filters it by a digital LPF with cutoff $2\pi/q$, and decimates it by q. This approach allows reduction in the number of analog channels at the expense of a higher sampling rate on each channel and additional digital processing. The total sampling rate stays (roughly) the same.

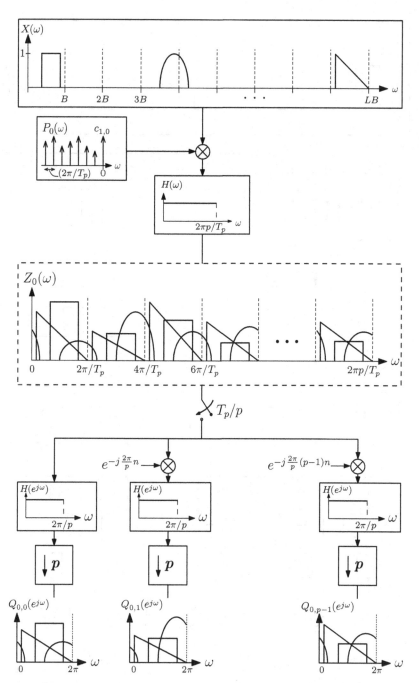

Figure 14.30 An MWC system with a single sampling channel for the special case where $q = p$. A multiband signal ($N = 3$, $L = 9$) passes through a single channel with mixing function $p_0(t)$. After filtering by an LPF with cutoff $2\pi p/T_p$ the signal is sampled at rate $2\pi p/T_p$. The sampled signal passes through p digital channels. Each channel shifts the relevant content to baseband using time modulation, filters by a digital LPF with cutoff $2\pi/p$, and decimates by p.

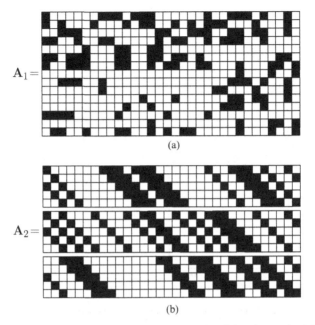

Figure 14.31 Reducing the number of analog channels. For simplicity, the matrix **A** is represented by two values, black and white. (a) No channel collapse: 15 independent analog channels, $q = 1$. (b) Channel collapse: three independent channels with collapse factor $q = 5$. The rows of $\mathbf{A}_2$ correspond to the first three rows of $\mathbf{A}_1$, and their shifts.

Example 14.15 We demonstrate the effect of channel collapse on the matrix **A**.

Consider a MWC with $p = 15$ channels. The resulting matrix **A**, denoted by $\mathbf{A}_1$, is depicted in Fig. 14.31(a). It consists of 15 rows, each corresponding to a different Fourier sequence. In order to visually illustrate the structure of the matrix after channel collapse, we simplify the discussion by representing **A** using two values, black and white.

Suppose now we collapse the channels by a factor of $q = 5$, resulting in only three analog channels. For these channels, we use the first three sequences of the full system. In this case, the new sensing matrix $\mathbf{A}_2$ consists of three blocks, each having five rows, containing the first three rows of $\mathbf{A}_1$ and their shifts. The new matrix is illustrated in Fig. 14.31(b).

Example 14.16 Next, we illustrate the expander operation on a signal comprising three bands with randomly chosen carrier frequencies.

Let

$$x(t) = \sqrt{E_1 B} \operatorname{sinc}\left(\frac{B}{5}(t - \tau_1)\right) \operatorname{sinc}\left(\frac{4B}{5}(t - \tau_1)\right) e^{j2\pi f_1(t - \tau_1)}$$

$$+ \sqrt{E_2 B} \operatorname{sinc}^2\left(\frac{B}{2}(t - \tau_2)\right) e^{j2\pi f_2(t - \tau_2)}$$

$$+ \sqrt{E_3 B} \operatorname{sinc}^3\left(\frac{B}{3}(t - \tau_3)\right) e^{j2\pi f_3(t - \tau_3)},$$

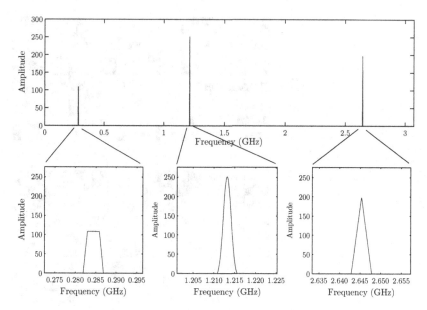

Figure 14.32 Fourier transform of the signal $x(t)$ in Example 14.16 comprising three bands.

with energy $E_i = \{3, 4, 5\}$, time offsets $\tau_i = \{4, 5, 6\}\,\mu s$, and carriers $f_i = \{11.85, 110.22, 50.55\} \cdot f_p$. We choose $B = 5\,\text{MHz}$ and $f_p = 1/T_p = 24\,\text{MHz}$. The signal's Fourier transform is shown in Fig. 14.32.

We assume that the MWC has four channels with a collapse factor of $q = 5$. In each channel the signal is mixed with a sign-alternating sequence of length $M = 263$ (see next subsection) and period T_p. The sampling rate is set to $f_s = qf_p = 120\,\text{MHz}$. In Fig. 14.33 we show the signal in one channel at the output of an LPF with width f_s. We next extract five bands in the digital domain using the expander, as shown in the bottom part of the figure.

14.4.4   Sign-alternating sequences

Mathematical derivation

We now treat a specific choice of periodic modulating signals depicted in Fig. 14.27: sign-alternating functions. For this choice, $p_i(t)$ is a piecewise constant function that alternates between levels ± 1 over M intervals of length T_p/M. In order to ensure sufficient transitions we choose $M \geq L$. Formally,

$$p_i(t) = \alpha_{ik}, \quad k\frac{T_p}{M} \leq t \leq (k+1)\frac{T_p}{M}, \quad 0 \leq k \leq M - 1, \tag{14.71}$$

with $\alpha_{ik} \in \{+1, -1\}$, and $p_i(t + nT_p) = p_i(t)$ for every $n \in \mathbb{Z}$. An advantage of digital sequences is that they are easy to implement in hardware and are simple to manipulate. Furthermore, it is straightforward to introduce delays in the sequences $p_i(t)$ when they are generated in digital form. This allows to create all $p_i(t)$ from delays of a basic

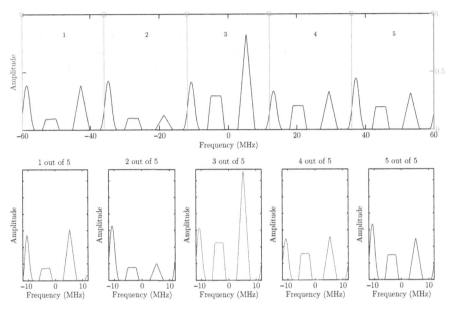

Figure 14.33 The signal of Fig. 14.32 after mixing and low-pass filtering (top) and after the expander (bottom).

sign-alternating function $p(t)$, as we discuss below. A sign-alternating function can be implemented by a shift register, where M determines the number of flip-flops, and $\{\alpha_{ik}\}$ initializes the shift register. The clock rate of the register $(T_p/M)^{-1}$ is also dictated by M.

Calculating the Fourier coefficients $c_{i\ell}$ in this setting gives

$$c_{i\ell} = \frac{1}{T_p}\int_0^{\frac{T_p}{M}}\sum_{k=0}^{M-1}\alpha_{ik}e^{-j\frac{2\pi}{T_p}\ell\left(t+k\frac{T_p}{M}\right)}dt = \frac{1}{T_p}\sum_{k=0}^{M-1}\alpha_{ik}e^{-j\frac{2\pi}{M}\ell k}\int_0^{\frac{T_p}{M}}e^{-j\frac{2\pi}{T_p}\ell t}dt.$$

Evaluating the integral we have

$$d_\ell = \frac{1}{T_p}\int_0^{\frac{T_p}{M}}e^{-j\frac{2\pi}{T_p}\ell t}dt = \begin{cases} \frac{1}{M}, & \ell = 0 \\ \frac{1-\theta^{-\ell}}{j2\pi\ell}, & \ell \neq 0, \end{cases} \tag{14.72}$$

where $\theta = e^{j2\pi/M}$, and thus

$$c_{i\ell} = d_\ell \sum_{k=0}^{M-1}\alpha_{ik}\theta^{-k\ell}. \tag{14.73}$$

Let $\mathbf{F}$ be the $M \times M$ matrix with elements $(1/\sqrt{M})\theta^{\ell k}, 0 \leq \ell, k \leq M-1$, and let $\mathbf{F}_L$ denote the first L columns of $\mathbf{F}$. Then, (14.55) can be written as

$$\mathbf{c}(e^{j\omega}) = \mathbf{SF}_L\mathbf{D}\mathbf{d}(e^{j\omega}), \tag{14.74}$$

where $\mathbf{S}$ is a $p \times M$ sign matrix with $S_{ik} = \alpha_{ik}$, and $\mathbf{D} = \sqrt{M}\,\mathrm{diag}\,(d_0,\ldots,d_{-L+1})$ is an $L \times L$ diagonal matrix with d_ℓ defined by (14.72). For simplicity, we choose $M = L$ in which case $\mathbf{F}_L = \mathbf{F}$ becomes a unitary matrix.

Note that the magnitude of d_ℓ decays as ℓ moves away from $\ell = 0$. This is a consequence of the specific choice of sign-alternating waveforms for the mixing functions $p_i(t)$. Under this selection, spectrum regions of $X(\omega)$ are weighted according to their proximity to the origin. In the presence of noise, the SNR will depend on the band locations owing to this asymmetry.

In order to guarantee recovery we need to design the alternating sequences such that $\mathbf{A} = \mathbf{SF}$ has spark equal to $N + 1$. We ignore the diagonal matrix $\mathbf{D}$, since it does not change the support of $\mathbf{d}(e^{j\omega})$, and can therefore be absorbed in it. In Chapter 11 we introduced the RIP measure, which is a stronger notion than the spark as it ensures a stable inverse for every support choice of size N. It is known that a random sign matrix $\mathbf{S}$ of size $p \times L$, whose entries are drawn independently with equal probability, obeys the RIP of order N if $p \geq CN \log(M/N)$, where C is a positive constant [241]. Furthermore, the RIP of matrices with random signs remains unchanged under any fixed unitary transform of the rows so that multiplying $\mathbf{S}$ by $\mathbf{F}$ preserves the RIP [245]. In [420] it is shown that deterministic sequences such as maximum length sequences and Gold codes can be used as the rows of $\mathbf{S}$ in the MWC, yielding near-optimal performance.

Example 14.17 Let $\mathbf{S}_1$ be a 7×8 matrix, constructed from the first seven rows of the 8×8 Hadamard[4] matrix:

$$\mathbf{S}_1 = \begin{bmatrix} 1 & 1 & 1 & 1 & 1 & 1 & 1 & 1 \\ 1 & -1 & 1 & -1 & 1 & -1 & 1 & -1 \\ 1 & 1 & -1 & -1 & 1 & 1 & -1 & -1 \\ 1 & -1 & -1 & 1 & 1 & -1 & -1 & 1 \\ 1 & 1 & 1 & 1 & -1 & -1 & -1 & -1 \\ 1 & -1 & 1 & -1 & -1 & 1 & -1 & 1 \\ 1 & 1 & -1 & -1 & -1 & -1 & 1 & 1 \end{bmatrix}. \tag{14.75}$$

It is easy to check by brute-force calculations that $\mathbf{S}_1$ has full spark, i.e. any seven of its columns are linearly independent. However, in this example, $\mathbf{S}_1\mathbf{F}$ does not have full spark.

Next we generate a random sign matrix $\mathbf{S}_2$ which results in

$$\mathbf{S}_2 = \begin{bmatrix} 1 & -1 & -1 & 1 & 1 & 1 & -1 & 1 \\ -1 & -1 & 1 & 1 & 1 & 1 & -1 & 1 \\ -1 & 1 & -1 & 1 & -1 & -1 & 1 & 1 \\ -1 & -1 & -1 & -1 & -1 & 1 & -1 & -1 \\ -1 & 1 & 1 & 1 & -1 & 1 & 1 & -1 \\ -1 & 1 & -1 & -1 & 1 & -1 & 1 & 1 \\ -1 & -1 & 1 & 1 & 1 & 1 & 1 & -1 \end{bmatrix}. \tag{14.76}$$

It again can be verified that $\mathbf{S}_2$ has full spark. Moreover, in this case, $\mathbf{S}_2\mathbf{F}$ also satisfies this requirement and is therefore suitable to use in the MWC.

---

[4] The Hadamard matrix is a square matrix whose entries are ± 1 and whose rows are mutually orthogonal.

Reducing shift registers

In the basic MWC design, each channel is associated with a different mixing function $p_i(t), 0 \leq i \leq p - 1$, which requires a shift register of M flip-flops. One of the advantages of the choice of sign-alternating functions is that several sequences $p_i(t)$ can be generated from a single function using shifts, thus reducing the number of flip-flops in the design. In this approach we generate r rows randomly, and use their cyclic shifts to fill out the matrix $\mathbf{S}$, thus reducing the number of flip-flops from pM to rM. Note, that this strategy results in fewer degrees of freedom in $\mathbf{S}$. Therefore, the sequences have to be carefully designed to ensure that performance is not degraded.

Example 14.18 Consider two 8×16 sign matrices $\mathbf{S}_1$ and $\mathbf{S}_2$. The matrix $\mathbf{S}_1$ is created from a single sequence where we use its cyclic shifts to fill the remaining rows of $\mathbf{S}_1$. This requires only 16 flip-flops to form the entire matrix. To construct $\mathbf{S}_2$ we rely on $r = 4$ shift registers corresponding to 64 flip-flops, i.e. we generate four random sign sequences and use their shifts to fill the other four rows. The structure of $\mathbf{S}_1$ and $\mathbf{S}_2$ is shown in Fig. 14.34.

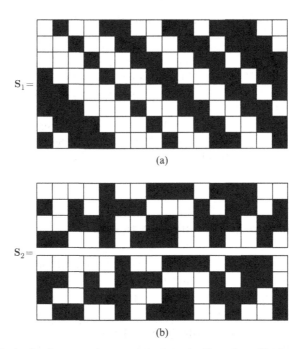

(a)

(b)

Figure 14.34 Reducing hardware requirements by using a small number of basic sequences and their cyclic shifts. White squares denote 1, while black squares denote -1. In this example, $p = 8$, $M = 16$. (a) Using a single shift register corresponding to $r = 1$. (b) Using four shift registers $r = 4$ and a single shift of each register.

In [421] the authors analyze a particularly simple choice of sequences which result from a single shift register and its cyclic shifts, and yields almost optimal performance. That is, the performance of the MWC using these sequences can be proven to be essentially equivalent to an MWC using p random sign functions. This allows the MWC to be implemented with only M flip-flops.

The basic idea is to first create an $M \times M$ circulant matrix

$$
\mathbf{C} = \begin{bmatrix} c_0 & c_1 & \cdots & c_{M-1} \\ c_{M-1} & c_0 & \cdots & c_1 \\ \vdots & \vdots & \ddots & \vdots \\ c_1 & c_2 & \cdots & c_0 \end{bmatrix}, \tag{14.77}
$$

in which $\mathbf{c} = [c_0, c_1, \ldots, c_{M-1}]$ is a properly chosen deterministic sequence. The matrix $\mathbf{S}$ is then constructed by choosing p rows of $\mathbf{C}$ uniformly at random. Since all rows are formed from the same sequence, only one shift register is needed to create $\mathbf{S}$. The resulting matrix $\mathbf{S}$ is shown to have the RIP of order N if $p \geq CN \log^4 M$, where C is a positive constant, and $\mathbf{c}$ is such that its Fourier transform has approximately constant magnitude.

Two possible choices of $\mathbf{c}$ are maximum-length sequences, and Legendre sequences [422]. The former exist for $M = 2^\beta - 1$ where β is a positive integer. The latter are defined for prime values of M and are given by

$$
c_0 = 1,
$$
$$
c_i = \begin{cases} 1, & \text{if } i \text{ is a square (mod } M) \\ -1, & \text{if } i \text{ is a non-square (mod } M), \end{cases} \quad i \geq 1. \tag{14.78}
$$

Other values of $\mathbf{c}$ for even M are given in [421]. Simulations provided in [421] show that using these sequences results in essentially the same performance as a full random binary matrix $\mathbf{S}$, while only requiring a single shift register.

14.5   Blind sampling of multiband signals

We now turn to consider the blind scenario in which the carrier frequencies are unknown. We will see that combining the multicoset and MWC architectures with the analog CS ideas discussed in Chapter 13 leads to sub-Nyquist strategies that achieve the minimal possible rate in this setting.

When the carrier frequencies are unknown, we are interested in the set of all possible multiband signals that occupy up to NB of the spectrum. In this scenario, the transmissions can lie anywhere below $\omega_{\max} = 2\pi/T_{\mathrm{Nyq}}$. Before proceeding, we formally define the set of signals we will be considering throughout this section, which we denote by $\mathcal{M}$ (for multiband).

Definition 14.2. *The set $\mathcal{M}$ contains all signals $x(t)$, such that the support of the Fourier transform $X(\omega)$ is contained within a union of N disjoint intervals (bands) in $[0, 2\pi/T_{\mathrm{Nyq}})$, and each of the bandwidths does not exceed B.*

14.5.1 Minimal sampling rate

An important question prior to discussing sampling strategies is what rate we can hope to achieve when sampling the set $\mathcal{M}$. To answer this question we consider a wider set $\mathcal{N}_\Omega$ of signals bandlimited to $[0, 2\pi/T_{\mathrm{Nyq}})$ with bandwidth occupation no more than $0 < \Omega < 1$, so that

$$\lambda\left(\mathrm{supp}\, X(\omega)\right) \leq \frac{2\pi\Omega}{T_{\mathrm{Nyq}}}, \quad \forall x(t) \in \mathcal{N}_\Omega, \tag{14.79}$$

where $\lambda\left(\mathrm{supp}\, X(\omega)\right)$ is the Lebesgue measure of the support of $X(\omega)$. Clearly, $\mathcal{M}$ is a subset of $\mathcal{N}_\Omega$ with $2\pi\Omega = NBT_{\mathrm{Nyq}}$.

The Nyquist rate for $\mathcal{N}_\Omega$ is $2\pi/T_{\mathrm{Nyq}}$. Note that $\mathcal{N}_\Omega$ is not a subspace so that the Landau theorem does not apply here. Nevertheless, it is intuitive to argue that the minimal sampling rate for $\mathcal{N}_\Omega$ cannot be below $2\pi\Omega/T_{\mathrm{Nyq}}$ as this value is the Landau rate had the spectrum support been known.

A *blind sampling set* R for $\mathcal{N}_\Omega$ is a stable sampling set whose design does not assume knowledge of $\mathrm{supp}\, X(\omega)$. The stability of R requires the existence of $\alpha > 0$ and $\beta < \infty$ such that (14.4) is satisfied for all $x, y \in \mathcal{N}_\Omega$. With this definition, we have the following result [335].

Theorem 14.3 (Blind sampling rate). *Let R be a blind sampling set for $\mathcal{N}_\Omega$. Then,*

$$D^-(R) \geq \min\left\{\frac{2 \cdot 2\pi\Omega}{T_{\mathrm{Nyq}}}, \frac{2\pi}{T_{\mathrm{Nyq}}}\right\}, \tag{14.80}$$

where $D^-(R)$ is defined in (14.5).

Since the Landau rate is equal to $2\pi\Omega/T_{\mathrm{Nyq}}$, the price to pay for not knowing the support is at most only a constant factor of 2.

Proof: The set $\mathcal{N}_\Omega$ is of the form

$$\mathcal{N}_\Omega = \bigcup_{\mathcal{T} \in \Gamma} \mathcal{B}_{\mathcal{T}}, \tag{14.81}$$

where

$$\mathcal{B}_{\mathcal{T}} = \{x(t) \in L_2(\mathbb{R}) | \, \mathrm{supp}\, X(\omega) \subseteq \mathcal{T}\}, \tag{14.82}$$

and

$$\Gamma = \{\mathcal{T} | \, \mathcal{T} \subseteq [0, 2\pi/T_{\mathrm{Nyq}}), \lambda(\mathcal{T}) \leq 2\pi\Omega/T_{\mathrm{Nyq}}\}. \tag{14.83}$$

This representation implies that $\mathcal{N}_\Omega$ is an uncountable union of subspaces.

For every $\gamma, \theta \in \Gamma$ define the subspaces

$$\mathcal{B}_{\gamma,\theta} = \mathcal{B}_\gamma + \mathcal{B}_\theta = \{x + y \mid x \in \mathcal{B}_\gamma, y \in \mathcal{B}_\theta\}. \tag{14.84}$$

Since R is a sampling set for $\mathcal{N}_\Omega$, (14.4) holds for some constants $\alpha > 0, \beta < \infty$. We have seen in Proposition 10.2 that (14.4) is valid if and only if

$$\alpha\|x - y\|^2 \leq \|x_R - y_R\|^2 \leq \beta\|x - y\|^2, \quad \forall x, y \in \mathcal{B}_{\gamma,\theta} \tag{14.85}$$

holds for every $\gamma, \theta \in \Gamma$. In particular, R is a sampling set for every $\mathcal{B}_{\gamma,\theta}$ with $\gamma, \theta \in \Gamma$.

Now, the space $\mathcal{B}_{\gamma,\theta}$ is of the form (14.82) with $\mathcal{T} = \gamma \cup \theta$. Applying Landau's theorem for each $\gamma, \theta \in \Gamma$ results in

$$D^-(R) \geq \lambda(\gamma \cup \theta), \quad \forall \gamma, \theta \in \Gamma. \tag{14.86}$$

Choosing

$$\gamma = \left[0, \frac{2\pi\Omega}{T_{\mathrm{Nyq}}}\right], \quad \theta = \left[\frac{2\pi(1-\Omega)}{T_{\mathrm{Nyq}}}, \frac{2\pi}{T_{\mathrm{Nyq}}}\right], \tag{14.87}$$

we have that for $\Omega \leq 0.5$,

$$D^-(R) \geq \lambda(\gamma \cup \theta) = \lambda(\gamma) + \lambda(\theta) = \frac{2 \cdot 2\pi\Omega}{T_{\mathrm{Nyq}}}. \tag{14.88}$$

If $\Omega \geq 0.5$, then $\gamma \cup \theta = [0, 2\pi/T_{\mathrm{Nyq}})$ and

$$D^-(R) \geq \lambda(\gamma \cup \theta) = \frac{2\pi}{T_{\mathrm{Nyq}}}. \tag{14.89}$$

Combining (14.88) and (14.89) completes the proof.   □

An immediate corollary of Theorem 14.3 is that if $\Omega \geq 0.5$ then uniform sampling at the Nyquist rate and reconstruction with an ideal LPF achieves the minimal possible rate, and provides perfect reconstruction for every $x(t) \in \mathcal{N}_\Omega$. Therefore, in the sequel we assume that $\Omega < 0.5$ so that the minimal sampling rate is twice the Landau rate.

The proof of Theorem 14.3 can be adapted to show that for the set $\mathcal{M}$, sampling at an average rate of $2NB$ is necessary to allow blind perfect reconstruction (for $NBT_{\mathrm{Nyq}} < \pi$). Since for known spectral support of $x(t) \in \mathcal{M}$, the Landau rate is NB, there is a penalty of 2 for not knowing the sampling rate.

Both Landau's theorem and Theorem 14.3 state a lower bound but do not provide a method to achieve the bound. An important distinction between the two theorems is that for known spectrum the set of possible multiband signals is linear (that is, closed under linear combinations) and the sampling and recovery methods such as interleaved ADCs and the MWC are linear as well. Consequently, the stability condition (14.4) suffices to ensure stability of the reconstruction. In contrast, $\mathcal{N}_\Omega$ is a nonlinear set (as is $\mathcal{M}$), and the proposed reconstruction schemes, which we discuss in the next section, are also nonlinear. Therefore, while Theorem 14.3 states the necessary density for a blind stable sampling set, a higher density may be required to ensure stable reconstruction.

14.5.2   Blind recovery

We now turn to discuss reconstruction methods that approach the minimal sampling rate of Theorem 14.3 while ensuring perfect reconstruction.

Multicoset and MWC sampling do not require knowledge of the carrier frequencies. Therefore, these approaches can be used in the blind scenario as well. The recovery stage, however, is more involved: we now need to solve (14.33) and (14.55) where the support $\mathcal{S}$ of $\mathbf{d}(e^{j\omega})$ is unknown. As in the known carrier setting (cf. Proposition 14.1), it is still true that $\mathbf{d}(e^{j\omega})$ is N-sparse as long as $L \leq 2\pi/(BT_{\mathrm{Nyq}})$. For every value of ω we therefore have a sparse recovery problem where we need to recover an N-sparse vector from p measurements. From the results of Chapter 11 we know that $2N$ samples are sufficient to guarantee uniqueness as long as the sensing matrix has spark $2N + 1$. In our setting this is equivalent to requiring that the multicoset sampling or the periodic sequences in the MWC form universal patterns for $2N$. We therefore have the following uniqueness condition.

Theorem 14.4 (Uniqueness in blind recovery).   *Let $x(t) \in \mathcal{M}$ be a multiband signal. The vector $\mathbf{d}(e^{j\omega})$ defined by (14.31) is the unique N-sparse solution of (14.33) and (14.55) if*

1. *$T \leq 2\pi/B$;*
2. *$p \geq 2N$;*
3. *$\{t_i\}$ and $\{c_{i\ell}\}$ are universal patterns for $2N$.*

Under the conditions of Theorem 14.4 we can in principle recover $\mathbf{d}(e^{j\omega})$ from the measurements by seeking the sparsest vector $\mathbf{d}(e^{j\omega})$ that solves (14.33) and (14.55) for each ω. However, in theory, this requires solving infinitely many CS problems. A straightforward approach to approximate the solution is to find the sparsest vector over a dense grid. However, this discretization strategy cannot guarantee perfect reconstruction. Furthermore, it necessitates working in the Fourier domain, thus precluding online computations, since we need to first acquire a large batch of measurements and only then compute their DTFT. The complexity in solving many CS problems is also large. Finally, in noisy environments discretization is inherently problematic since the noise amplitude can be locally higher than the signal.

To simplify the calculations and obtain a more accurate recovery, instead of solving the equations assuming an input that is sparse for each ω, we target the joint sparsity of $\{\mathbf{d}(e^{j\omega})\}$ in the spirit of analog CS introduced in Chapter 13. As we have seen in Proposition 14.1, the vectors $\{\mathbf{d}(e^{j\omega})\}$ and $\{\mathbf{d}[n]\}$ are $2N$-jointly sparse. Algorithms based on joint sparsity are computationally more efficient, and are more effective in noisy environments since they aggregate energy of the signal in each spectrum slice. Furthermore, exploiting joint sparsity allows the problem to be solved in the time domain, thus alleviating the need to Fourier transform the data.

Choosing a universal pattern or appropriate periodic functions, we can guarantee that every $4N$ columns of the respective $\mathbf{A}$ matrices are linearly independent as long as $p \geq 4N$. If we consider (14.33) and (14.55) for all $\omega \in [0, 2\pi)$ then the resulting set of equations form an IMV system and can be solved using the IMV techniques developed in

Section 11.6.4. Specifically, to present the recovery for multicoset and MWC in a unified way, let $\tilde{\mathbf{c}}(e^{j\omega}) = \mathbf{W}^{-1}(\omega)\mathbf{c}(e^{j\omega})$ for multicoset sampling and $\tilde{\mathbf{c}}(e^{j\omega}) = \mathbf{c}(e^{j\omega})$ for the MWC. The vector $\tilde{\mathbf{c}}(e^{j\omega})$ corresponding to multicoset sampling may be computed using the method of Fig. 14.22. The relations (14.33) and (14.55) can then both be written as

$$\mathbf{c}(e^{j\omega}) = \mathbf{A}\mathbf{d}(e^{j\omega}), \quad 0 \le \omega \le 2\pi. \tag{14.90}$$

Since $\mathbf{A}$ is independent of frequency, taking the inverse DTFT leads to

$$\mathbf{c}[n] = \mathbf{A}\mathbf{d}[n], \quad n \in \mathbb{Z}, \tag{14.91}$$

where the vectors $\{\mathbf{d}[n]\}$ are jointly $2N$-sparse. We may therefore find the support $\mathcal{S}$ by using the IMV methods of Section 11.6.4 directly in the time domain. Once $\mathcal{S}$ is determined, recovery of $\mathbf{d}[n]$ is possible for every n by $\mathbf{d}[n] = \mathbf{A}_{\mathcal{S}}^{\dagger}\mathbf{c}[n]$.

As detailed in Section 11.6.4, the first step in solving the IMV system (14.91) is to find the joint support of $\{\mathbf{d}[n]\}$ using the CTF. To implement the CTF we first build a frame (or basis) for the range of $\{\mathbf{c}[n]\}$, and stack these frame elements into a matrix $\mathbf{V}$. One option is to form the $p \times p$ correlation matrix $\mathbf{Q} = \sum_{n \in \mathbb{Z}} \mathbf{c}[n]\mathbf{c}^*[n]$ and then decompose it using, for example, the eigendecomposition, into $\mathbf{Q} = \mathbf{V}\mathbf{V}^*$. Other options, as well as their expected performance, are discussed in Section 11.6.4. We next solve an MMV system $\mathbf{V} = \mathbf{A}\mathbf{U}$ to find the sparse matrix $\mathbf{U}_0$ with the largest number of zero rows that matches $\mathbf{V}$. The support of $\mathbf{U}_0$, namely the rows of $\mathbf{U}_0$ which are nonidentically zero, is then the desired support of $\{\mathbf{d}[n]\}$. The CTF block diagram is illustrated in Fig. 14.35.

Support recovery estimates the frequency support of $x(t)$ at a coarse resolution of slice width $\omega_p = 2\pi/T_p$. Continuous reconstruction is then obtained by standard low-pass interpolation of the active sequences $d_\ell[n]$ and modulation to the corresponding positions in the spectrum.

This overall procedure is termed SBR4 in [335], where 4 designates that under the choice of $p \ge 4N$ sampling sequences, and when $\{t_i\}$ and $p_i(t)$ are chosen to be universal, this algorithm guarantees perfect reconstruction of a multiband $x(t)$. Sampling at a rate of B on each channel, the average sampling rate is $4NB$.

The rate can be further reduced by a factor of 2 exploiting the way a multiband spectrum is arranged in spectrum slices. Using bisection in the frequency domain and iterations of the CTF, an algorithm reducing the required rate was developed in [335] under the name SBR2, requiring $p \ge 2N$ sampling branches, so that the sampling rate approaches $2NB$. This is essentially the provable optimal rate as stated in Theorem 14.3.

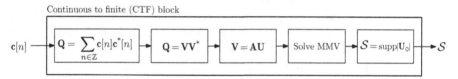

Figure 14.35 Block diagram of the CTF. The first step is to form the correlation matrix $\mathbf{Q}$, decompose it into $\mathbf{Q} = \mathbf{V}\mathbf{V}^*$, and then find the sparsest solution $\mathbf{U}_0$ that solves $\mathbf{V} = \mathbf{A}\mathbf{U}$. The support $\mathcal{S}$ of $\mathbf{U}_0$ is the desired support of $\{\mathbf{d}[n]\}$.

An alternative method to approach the rate $2NB$ is to choose frequency slices of length B/m rather than slices of length B. In this case, the sampling rate on each channel is equal to B/m, and every signal band occupies at most $m+1$ slices. The vector $\mathbf{d}(e^{j\omega})$ in (14.55) is then $N(m+1)$ jointly sparse. Therefore, in order to recover $\mathbf{d}(e^{j\omega})$ we need $N(m+1)$ channels in the case of known band positions, and $2N(m+1)$ in the blind setting. Focusing on the later, and multiplying the number of channels by the rate we end up with a total sampling rate of $2N(1+1/m)B$. As m increases, this rate approaches the minimal value of $2NB$.

Following the same ideas as in Section 14.4.3, the techniques discussed in this section can be combined with collapsing of channels in order to reduce the number of hardware branches.

In the context of real-time processing, we comment that the CTF is executed only when the spectral support changes. In a real-time environment, about $2N$ consecutive input vectors $\mathbf{c}[n]$ should be stored in memory, so that in case of a support change the CTF has enough time to provide a new support estimate before the recovery of $\mathbf{d}[n]$ reaches the point that this information is needed. In order to detect the support changes once they occur, we can either relay an indication from the application layer, or automatically identify the spectral variation in the sequences $\mathbf{d}[n]$. To implement the latter option, let S be the last support estimate of the CTF, and define $\tilde{S} = S \bigcup \{i\}$ for some entry $i \notin S$. We then monitor the value of the sequence $d_i[n]$. As long as the support S does not change, the sparsity of $\mathbf{d}[n]$ implies that $d_i[n] = 0$ or contains only small values due to noise. Whenever this sequence crosses a threshold (over a certain number of consecutive time instances) we trigger the CTF to obtain a new support estimate. Note that the recovery of $d_i[n]$ requires to implement only one row from $\mathbf{A}_{\tilde{S}}^\dagger$. Since the values are not important for detection purposes, this multiplication can be carried out at low resolution.

14.5.3 Multicoset sampling and the sparse SI framework

In the case of multicoset sampling, we can cast the resulting blind sampling and recovery scheme into the sparse SI framework introduced in Chapter 13. To this end, we first view the multiband model as a finite union of bandpass subspaces, termed spectrum slices [336]. To obtain the finite union viewpoint, the Nyquist range $[0, 2\pi/T_{\mathrm{Nyq}})$ is conceptually divided into L consecutive, nonoverlapping slices of individual widths $\omega_p = 2\pi/(LT_{\mathrm{Nyq}})$.

Each spectrum slice represents a SI subspace corresponding to a single band. By choosing $\omega_p \geq B$, we ensure that no more than $2N$ of the L spectrum intervals are active (that is, contain signal energy). We may then define an SI union over the active bandpass subspaces, which can be written in the form (13.6), or in the Fourier domain as

$$X(\omega) = \sum_{i=1}^{L} A_i(\omega) D_i(e^{j\omega LT}). \tag{14.92}$$

Here $A_i(\omega) = \sqrt{LT} H(\omega - \frac{2\pi(i-1)}{LT})$, $H(\omega)$ is an LPF on $[0, 2\pi/(LT))$, and $D_i(e^{j\omega LT}) \triangleq \sum_{n \in \mathbb{Z}} d_i[n] e^{-j\omega nLT}$, $1 \leq i \leq L$. This representation is illustrated in Fig. 14.36. Note

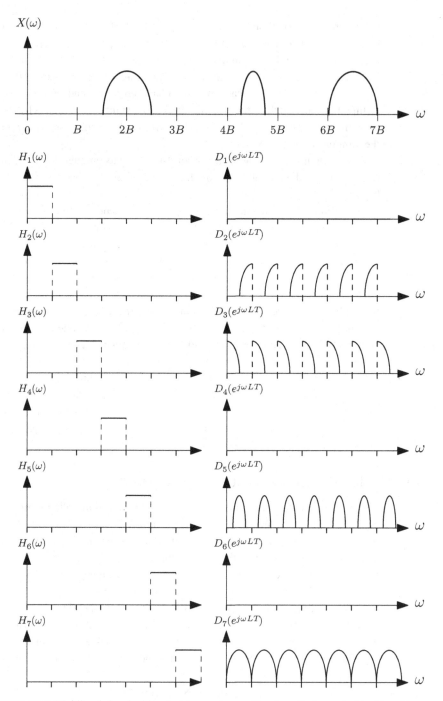

Figure 14.36 Multiband signal with unknown carriers as SI union of subspaces. The Nyquist range $[0, 2\pi/T_{\mathrm{Nyq}})$ is divided into L consecutive, nonoverlapping, slices of individual widths $\omega_p = 2\pi/(LT_{\mathrm{Nyq}})$ where at most $2N$ out of the L slices are active. In this example, $N = 3, L = 7$, and the active slices are $\{2, 3, 5, 7\}$.

that the conceptual division to spectrum slices does not restrict the band positions; a single band may split between adjacent slices. Formulating the multiband model with unknown carriers as a sparse SI problem, we can now apply the analog CS sub-Nyquist sampling scheme of Fig. 13.4.

To realize multicoset sampling, we choose

$$s_i(t) = \delta(t - t_i L T_{\text{Nyq}}), \quad 0 \le i \le p - 1, \tag{14.93}$$

and use a sampling period of $T = LT_{\text{Nyq}}$. Note that these filters are simply causal delay filters. It can then be seen that the general analog CS method reduces to the recovery results presented in this section.

14.5.4   Sub-Nyquist baseband processing

Software packages for DSP typically expect baseband inputs, namely the information signals $I(t), Q(t)$ of (14.1), or equivalently their uniform samples at the narrowband rate. These inputs are obtained by classic demodulation when the carrier frequencies are known. A digital algorithm developed in [21] translates the sequences $\mathbf{d}[n]$ to the desired DSP format with only low-rate computations, enabling smooth interfacing with existing DSP software packages. The algorithm is referred to as Back-DSP, and is illustrated in Fig. 14.37.

The Back-DSP algorithm translates the sequences $\mathbf{d}[n]$ to the narrowband signals $I_i(t), Q_i(t)$ that make up the ith band of interest:

$$s_i(t) = I_i(t)\cos(2\pi f_i t) + Q_i(t)\sin(2\pi f_i t). \tag{14.94}$$

The input to the algorithm are the sequences $\mathbf{d}[n]$ corresponding to the spectrum slices of $x(t)$. In general, as depicted in Fig. 14.37, a spectrum slice may contain more than a single information band. The energy of a band of interest may also split between adjacent slices. To correct for these two effects, the algorithm performs the following actions:

1. Refine the coarse support estimate $\mathcal{S}$ to the actual band edges, using power spectral density estimation;
2. Separate bands occupying the same slice to distinct sequences $s_i[n]$. Stitch together energy that was split between adjacent slices; and

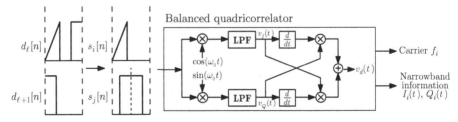

Figure 14.37 The flow of information extraction begins with detecting the band edges. The slices are filtered, aligned, and stitched appropriately to construct distinct quadrature sequences $s_i[n]$ per information band. The balanced quadricorrelator finds the carrier f_i and extracts the narrowband information signals.

3. Apply a common carrier recovery technique, the balanced quadricorrelator [423], on $s_i[n]$. This step estimates the carrier frequencies f_i and outputs uniform samples of the narrowband signals $I_i(t), Q_i(t)$.

A detailed description of each step can be found in [21].

14.5.5   Noise folding

Noise is inevitable in practical measurement devices. A common property of many existing sub-Nyquist methods, including multicoset sampling and the MWC, is that they aggregate wideband noise from the entire Nyquist range, as a consequence of treating all possible spectral supports. The digital reconstruction algorithm we presented based on the IMV framework partially compensates for this noise enhancement by digital denoising. One option to solve the IMV equations is to form the $p \times p$ correlation matrix $\mathbf{Q} = \sum_{n \in \mathbb{Z}} \mathbf{c}[n]\mathbf{c}^*[n]$ (the first level in the CTF) and then decompose it using the eigen-decomposition to $\mathbf{Q} = \mathbf{V}\mathbf{V}^*$. This decomposition allows removal of some of the noise space by retaining only the eigenvectors corresponding to eigenvalues that are above a threshold. In practice, we implement the CTF in real time so that we need to choose the number of samples $\mathbf{c}[n]$ to accumulate in forming $\mathbf{Q}$. Note that the dimension of $\mathbf{Q}$ is always $p \times p$, independent of the number of samples. The more samples used, the better the denoising that can be achieved. This effect is simulated in the following section. Alternatively, we may use a sufficiently large set of vectors $\{\mathbf{c}[n]\}$ as the basis set. As we have seen in the examples in Section 11.6.4, although this approach is computationally simpler, it is less effective in removing noise.

Another route to noise reduction is to carefully design the sequences $p_i(t)$. However, noise aggregation remains a practical limitation of all current sub-Nyquist techniques. One case in which this approach is effective in reducing noise is when the carrier frequencies are known, and we have the ability to modify the modulating sequences when the carriers change. We can then design the sequences such that they have energy only within the nonzero bands. In this way we still alias all the signal energy down to baseband, but do not alias noise from vacant bands.

To better understand the effect of noise on the MWC, denote the wideband analog noise by $e_{\text{analog}}(t)$ and the measurement noise by $\mathbf{e}_{\text{meas}}[n]$. The MWC measurements then become

$$\mathbf{c}[n] = \mathbf{A}(\mathbf{d}[n] + \mathbf{e}_{\text{analog}}[n]) + \mathbf{e}_{\text{meas}}[n] = \mathbf{A}\mathbf{d}[n] + \mathbf{e}_{\text{eff}}[n], \qquad (14.95)$$

with an effective error term $\mathbf{e}_{\text{eff}}[n]$. This means that noise has the same effect in analog CS as it has in the standard CS framework with an increase in variance due to the term $\mathbf{A}\mathbf{e}_{\text{analog}}[n]$. To see that the variance is increased, suppose for simplicity that all the noise vectors are white with variance σ^2, and that $\mathbf{A}$ is a $p \times L$ partial unitary matrix with unit-norm columns such that $\mathbf{A}\mathbf{A}^* = (L/p)\mathbf{I}$. This is the case, for example, when $\mathbf{A}$ consists of p rows from the Fourier matrix, properly normalized. The covariance of $\mathbf{e}_{\text{eff}}[n]$ is then given by $\sigma^2(\mathbf{I} + \mathbf{A}\mathbf{A}^*) = \sigma^2(1 + L/p)\mathbf{I}$. The increase in noise due to subsampling is proportional to L/p, which is referred to in [310] as *noise folding*. It is shown in [310] that noise folding by L/p holds more generally even when $\mathbf{A}\mathbf{A}^*$ is not proportional to the identity.

We conclude from (14.95) that our recovery problem becomes that of a standard IMV model with larger variance. Therefore, existing recovery algorithms can be used to try and combat the noise. In particular, the fact that we exploit multiple measurements in the support recovery helps robustify the proposed sub-Nyquist techniques to a certain extent. This will be evident in the simulations in Section 14.7. Furthermore, we may translate known results and error guarantees developed in the context of CS to handle noisy analog environments.

14.6   Hardware prototype of sub-Nyquist multiband sensing

A board-level hardware prototype of the MWC is reported in [102]. The MWC board appears to be the first reported wideband hardware example borrowing ideas from CS to realize a sub-Nyquist sampling system for wideband multiband signals, where the sampling and processing rates are directly proportional to the actual bandwidth occupation and not the highest frequency.

The hardware specifications cover 2 GHz Nyquist-rate inputs with spectral occupation up to $NB = 120$ MHz with $N = 6$ and $B = 20$ MHz. The sub-Nyquist rate is 280 MHz. Since $N = 6$, the minimal sampling rate corresponding to Theorem 14.3 is $2NB = 240$ MHz. Thus, the sampling rate is very close to the minimum rate required. The slight oversampling helps overcome noise and distortion from the nonideal hardware components. Photos of the hardware appear in Fig. 14.38. In order to reduce the number of analog components, the hardware realization incorporates the advanced MWC configuration in which several channels are collapsed into one. We also use a single shift register to generate all modulating sequences. More specifically, the following parameters are chosen:

- A collapsing factor $q = 3$. Since $N = 6$, the minimal value of channels needed is 12. With a collapse of 3 we end up with $p = 4$ hardware branches. The sampling rate on each channel is set to $1/T = 70$ MHz.

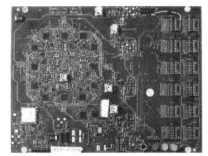

Figure 14.38 A hardware realization of the MWC consisting of two circuit boards. The left pane implements $p = 4$ sampling channels, whereas the right pane provides four sign-alternating periodic waveforms of length $M = 108$, derived from a single shift register.

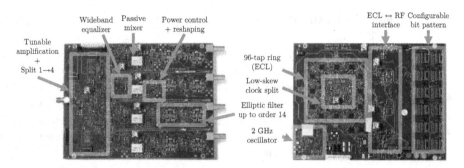

Figure 14.39 A detailed description of the MWC boards.

- The p periodic waveforms are derived using delays of a single shift register output, that is, by tapping p different locations of the register.

Further technical details on this representative hardware exceed the level of practice we are interested in here, although we emphasize below a few conclusions that connect back to the theory.

The Nyquist burden always manifests itself in some part of the design. For example, in pointwise methods, implementation requires ADC devices with Nyquist-rate front-end bandwidth. The MWC shifts the Nyquist burden to an analog RF preprocessing stage that precedes the ADC devices. The motivation behind this choice is to enable capture of the largest possible range of input signals, since, in principle, when the same technology is used by the source and sampler, this range is maximal. In particular, as wideband multiband signals are often generated by RF sources, the MWC framework can treat an input range that scales with advances in RF technology. While this explains the choice of RF preprocessing, the actual sub-Nyquist circuit design can be quite challenging and call for nonordinary solutions. Several RF blocks in the MWC prototype are highlighted in Fig. 14.39. These circuitries stem from the unique application of sub-Nyquist sampling as described in detail in [102]. To give a taste of circuit challenges, we briefly consider two design problems that are studied in [102].

Low-cost analog mixers are typically specified for a pure sinusoid in their oscillator port, whereas the periodic mixing requires simultaneous mixing with the many sinusoids making up $p_i(t)$. This creates nonlinear distortions and complicates the gain selections along the RF path. To address this challenge, power control, special equalizers, and local adjustments on data sheet specifications were used in [102] in order to design the analog acquisition, taking into account the nonordinary mixer behavior due to the periodic mixing.

Another circuit challenge pertains to generating $p_i(t)$ with 2 GHz alternation rates. The waveforms can be generated either by analog or by digital means. Analog waveforms, such as sinusoid, square, or sawtooth waveforms, are smooth within the period, and therefore do not have enough transients at high frequencies which is necessary to ensure sufficient aliasing. On the other hand, digital waveforms can be programmed to any desired number of alternations within the period, but require meeting timing constraints on the order of the clock period. For 2 GHz transients, the clock interval $1/f_{\mathrm{Nyq}} = 480$ picosecs leads to tight timing constraints that are difficult to satisfy with existing digital devices. These timing requirements were overcome by operating commercial devices beyond their datasheet specifications. The reader is referred to [102] for further technical details.

Going back to a high-level practical viewpoint, besides matching the source and sampler technology and addressing circuit challenges, an important point is to verify that the recovery algorithms do not limit the input range through constraints that they impose on the hardware. In the MWC case, periodicity of the waveforms $p_i(t)$ is important since it creates the aliasing effect with the Fourier coefficients $c_{i\ell}$ in (14.49). The hardware implementation and experiments in [102] demonstrate that the appearance in time of $p_i(t)$ does not have a large impact on the system as long as periodicity is maintained. This property is crucial, since precise sign alternations at speeds of 2 GHz are difficult to maintain, whereas relatively simple hardware wirings ensure that $p_i(t) = p_i(t + T_p)$ for every $t \in \mathbb{R}$. Figure 14.40(a) depicts the spectrum of $p_i(t)$, which consists of equally spaced Diracs, i.e. highly concentrated energy peaks, as expected for periodic waveforms. The Dirac spectral lines appear steady, ensuring the periodicity of $p_i(t)$. In Fig. 14.40(b) we observe the time-domain appearance of $p_i(t)$; as can be seen, the waveform is far from nice rectangular transitions on the Nyquist grid. However, since periodicity is the only essential requirement of the MWC, the nonideal time-domain appearance has little effect in practice.

Correct support detection and signal reconstruction in the presence of three narrow-band transmissions using the MWC board is verified in [102]. Figure 14.41 depicts the setup of three signal generators that were combined at the input terminal of the MWC prototype: an amplitude-modulated (AM) signal at 807.8 MHz with 100 kHz envelope, a frequency-modulation (FM) source at 631.2 MHz with 1.5 MHz frequency deviation and 10 kHz modulation rate, and a pure sine waveform at 981.9 MHz. Signal powers were set to about 35 dB SNR with respect to the wideband noise that folded to baseband. The carrier positions were chosen such that their aliases overlay at baseband, as

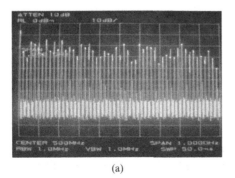
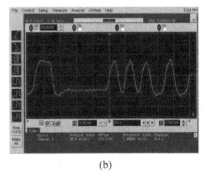

(a)   (b)

Figure 14.40 Nonideal time-domain appearance. (a) The spectrum of $p_i(t)$. Highly concentrated energy in equally spaced Diracs. (b) Time-domain appearance of $p_i(t)$. The waveform is far from rectangular transitions on the Nyquist grid.

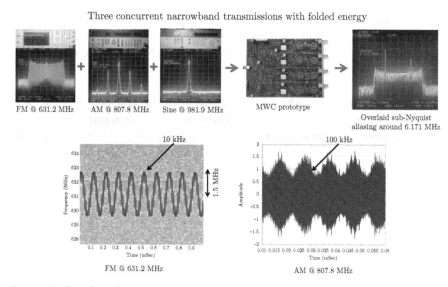

Figure 14.41 Top: three signal generators are combined to the system input terminal. The spectrum of the low-rate samples (first channel) reveals overlapped aliasing at baseband. Bottom: the recovery algorithm finds the correct carriers and reconstructs the original individual signals.

the photos in Fig. 14.41 demonstrate. The CTF was executed and detected the correct support. The unknown carrier frequencies were correctly estimated within 10 kHz. In addition, the figure demonstrates correct reconstruction of the AM and FM signal contents. Laboratory experiments also indicate an average of 10 millisecond duration for the digital computations, including CTF support detection and carrier estimation. The small dimensions of $\mathbf{A}$ (12 × 100 in the prototype configuration) are what renders the MWC practically feasible from a computational perspective.

14.7   Simulations

In this section we explore the performance of practical blind multiband sampling and reconstruction via simulations.

14.7.1   MWC designs

As we saw earlier, $4N$ samplers (when there is no channel collapse) are sufficient to recover a multiband signal with N bands, using the CTF block. However, in the presence of noise, oversampling is needed to ensure good performance. We begin by examining the performance of two different MWC systems using SBR4, as a function of the SNR and the number of sampling channels. Performance is measured by the average recovery rate, namely the percentage of successful support recovery. More precisely, 500 random multiband signals were generated as

$$x(t) = \sum_{i=1}^{3} \sqrt{E_i B}\, \text{sinc}(B(t - \tau_i)) \cos(2\pi f_i (t - \tau_i)), \tag{14.96}$$

Table 14.1 Simulated MWC design parameters for a multiband model with $N = 6$, $B = 50\,\text{MHz}$, and $f_{\text{Nyq}} = 10\,\text{GHz}$

	Design A	**Design B**
Mixing functions frequency	$f_p = \frac{f_{\text{Nyq}}}{M} \approx 51.3\,\text{MHz}$	$f_p = \frac{f_{\text{Nyq}}}{M} \approx 51.3\,\text{MHz}$
ADCs sampling rate	$f_s = f_p \approx 51.3\,\text{MHz}$	$f_s = 5 f_p \approx 256.4\,\text{MHz}$
Analog channels number	$p \geq 4N = 24$	$p \geq \lceil \frac{4N}{5} \rceil = 5$
Sign-alternations within a period	$M = 195$	$M = 195$
Average sampling rate	$p f_s \geq 1.23\,\text{GHz}$	$p f_s \geq 1.54\,\text{GHz}$

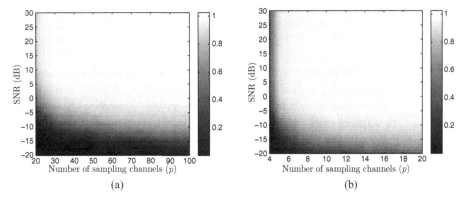

Figure 14.42 MWC simulation results. Recovery rate as a function of SNR and number of sampling channels. The design parameters are detailed in Table 14.1. (a) Design A: no channel collapse. (b) Design B: channel collapse with factor $q = 5$.

with energy coefficients $E_i = \{1, 2, 3\}$ and time offsets $\tau_i = \{0.2, 0.4, 0.7\}\mu\text{s}$. For every signal the carriers f_i were chosen uniformly at random in $[-f_{\text{Nyq}}/2, f_{\text{Nyq}}/2]$ with $f_{\text{Nyq}} = 10$ GHz. These signals all have $N = 6$ bands, each with width $B = 50$ MHz. To implement the MWC we used p sign-alternating sequences $p_i(t)$ with M sign-alternations, generated randomly to form the matrix **S**. The signals are contaminated by white Gaussian noise, $w(t)$, with varying variance to achieve the desired SNR level defined by $10 \log(\|x\|^2 / \|w\|^2)$ with the standard L_2-norm. The channels are collapsed in the second system. The design parameters of both systems are detailed in Table 14.1. Each signal was sampled by both systems and reconstructed using SBR4. For applying the CTF we averaged **Q** over 48 samples in design A, and 163 samples in design B.

The simulation results are shown in Fig. 14.42. An obvious trend is the improvement in performance as a function of SNR and as a function of the number of sampling functions. When the number of channels p increases, the spark requirement of **A** is fulfilled with higher probability. The results for design A show that in the high-SNR regime, correct recovery is achieved when using $p \geq 4N = 24$ channels, which amounts to less than 12% of the Nyquist rate. The reason that recovery is possible even for values lower than 24 is that not all of the multiband signals occupy $2N$ entries; such signals can be

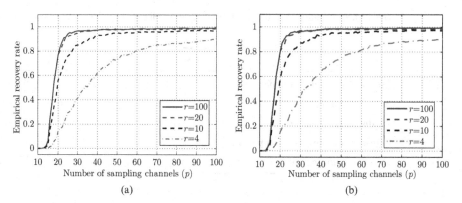

Figure 14.43 Percentage of correct support recovery, when drawing the sign patterns randomly only for the first r channels. (a) SNR = 10 dB. (b) SNR = 25 dB.

recovered from fewer than $4N$ sampling channels. Furthermore, the joint sparsity can be exploited to reduce the rate requirements as shown in Theorem 11.22. The same conclusions are relevant to design B: here, every $q = 5$ sampling channels are collapsed to a single channel with a higher sampling rate. In particular, $24/q \approx 5$ channels achieve acceptable recovery. This implies a significant saving in hardware components.

14.7.2   Sign-alternating sequences

We next examine several aspects related to sign-alternating sequences. We first consider the effect of using sign-alternating sequences generated from a small number of shift registers, as discussed in Section 14.4.4.

To qualitatively evaluate this approach, we use design A from the previous section, and generate sign matrices whose first r rows are drawn randomly as before. The ith row, $r < i \le p$, is chosen to be five cyclic shifts (to the right) of the $(i - r)$th row (uniform delay of five taps). Figure 14.43 depicts the success recovery for several choices of r and two SNR levels. As evident, this strategy enables savings of 80% of the total number of flip-flops, with no empirical degradation in performance.

Next, we fix the SNR at 10 dB and evaluate the recovery with several different types of alternating sequences. Specifically, we compare Gold sequences, commonly used in wireless communications (CDMA) and satellite navigation (GPS), and random sequences. For both choices we consider an MWC with collapse factor $q = 5$, and $p = 4$ channels. The value of B is chosen as 20 MHz, $f_s = qf_p$ with $f_p = 24$ MHz, and the Nyquist rate is set to 6.144 MHz. We use sequences with $M = 263$ alternations. In our simulations, we consider two options. In the first, four different sequences are generated either randomly, or by using four rows of an $M \times M$ Gold matrix. In the second scenario, we construct only one sequence of each type, and form the other three rows of **S** by taking its cyclic shifts. The correlation between the original and recovered signals for the different choices is plotted in Fig. 14.44. The Gold sequences exhibit the best behavior.

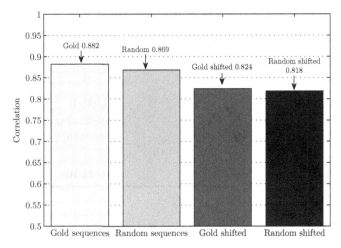

Figure 14.44 Correlation between original and recovered signals for different choices of sequences.

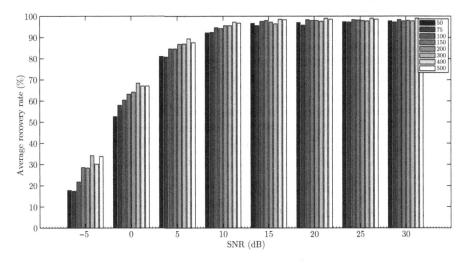

Figure 14.45 Recovery rate as a function of SNR and CTF block length.

14.7.3 Effect of CTF length

We now consider the effect of the CTF block length on the recovery rate. As mentioned in Section 14.5.5, to construct $\mathbf{Q}$ we choose a finite number of samples $\mathbf{c}[n]$. Taking a larger number will lead to more averaging of the measurement noise, and consequently improved recovery rate. However, increasing the block length causes higher delay in the system.

We again use the MWC system of design A from Table 14.1, with $p = 24$. Figure 14.45 shows the recovery rate as a function of SNR for different values of block length. It can be seen that the longer the block length, the better the recovery rate.

Table 14.2 Design parameters for testing f_p

Parameter	Value
Number of channels (p)	4
q	5
B	23.5 MHz
$f_p = 1/T_p$	{20, 22, 24, 26}MHz
$f_s = qf_p$	{100, 110, 120, 130}MHz
M	313
f_{Nyq}	6.144 GHz

However, when the SNR level is sufficient to achieve reasonable performance, the effect of the CTF length is less prominent.

14.7.4   Parameter limits

Finally, we investigate the limits on the MWC parameters. Throughout the simulations, we choose $E_i = \{3, 4, 5\}$ and $\tau_i = \{4, 5, 6\}$μs in (14.96). For the mixing sequence, we use alternating function values $\{\pm1\}$ chosen as rows of a Gold sequence matrix. Gold sequences have small cross-correlations and were shown to yield good results when used within the MWC system. To incorporate distortions that result from practical filters, we use two LPFs throughout the simulations. The first is an analog filter prior to sampling. The second is a digital LPF used in the expander to extract the collapsed channels. We model the analog filter by a digital FIR Type-I with 2000 coefficients. The digital filter is chosen as an FIR Type-I with 100 coefficients. The CTF uses 100 data samples in all simulations.

Limit on the sequence period
We begin by examining the limit $f_p = 1/T_p \geq B$ on the frequency of the periodic sequence. For constant B, we choose $f_p = B \cdot \{0.85, 0.94, 1.02, 1.1\}$. We consider an MWC with four channels and a collapse factor $q = 5$, so that the sampling rate in each channel is $f_s = qf_p$. The parameters used in the simulation are summarized in Table 14.2.

In Fig. 14.46 we plot the (normalized) correlation between the original and recovered signals as a function of SNR for different choices of f_p. Clearly, the results for $f_p \geq B$ are much better than those for $f_p < B$.

Limit on the sequence length
The next simulation is aimed at demonstrating that the length of the alternating sequence M should satisfy $M \geq L$ to ensure good recovery performance, where for real signals

$$L = 2 \left\lceil \frac{f_{\text{Nyq}} + f_s}{2f_p} \right\rceil - 1. \tag{14.97}$$

Table 14.3 Design parameters for testing M

Parameter	Value
Number of channels (p)	4
q	5
B	20 MHz
$f_p = 1/T_p$	24 MHz
f_s	120 MHz
M	$\{155, 191, 263, 299\}$
f_{Nyq}	6.144 GHz

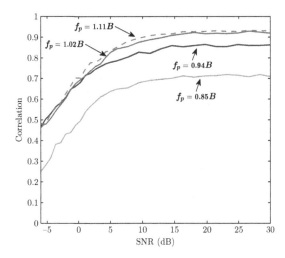

Figure 14.46 MWC performance for different values of f_p (as a fraction of B).

The simulation parameters chosen are shown in Table 14.3. For these parameters, $L = 261$. The correlation between the original and recovered inputs is plotted in Fig. 14.47, from which it is evident that the results with $M > L$ are far better than those obtained with $M < L$.

Limit on the sampling rate

In the next simulation we demonstrate that $m \geq 2N$ is necessary for blind recovery, where $m \equiv qp$ is the number of effective channels of the MWC. Here $N = 6$ is the number of bands, p is the number of hardware channels, and q is the collapse factor. Keeping all other parameters fixed, as detailed in Table 14.4, we chose several values for p and q: $p = \{1, 2, 3, 4, 6, 10, 20\}$ and $q = \{1, 3, 5, 9, 15, 21\}$. The correlation versus p and q is depicted in Fig. 14.48.

Table 14.4 Design parameters for testing the number of effective channels

Parameter	Value
Number of channels (p)	$\{1, 2, 3, 4, 6, 10, 20\}$
q	$\{1, 3, 5, 9, 15, 21\}$
B	20 MHz
$f_p = 1/T_p$	24 MHz
$f_s = qf_p$	$\{24, 72, 120, 216, 360, 504\}$ MHz
M	263
f_{Nyq}	6.144 GHz
SNR	30 dB

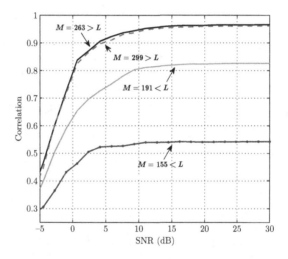

Figure 14.47 MWC performance for different values of the sequence length M.

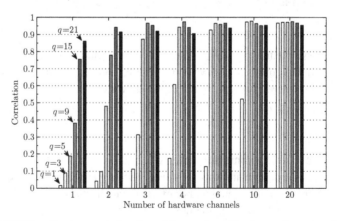

Figure 14.48 MWC performance as a function of the number of hardware channels p for different values of the collapse factor q.

Table 14.5 Design parameters for testing the effect of p and q

Parameter	$m = 105$	$m = 15$
Number of channels (p)	$\{1, 15, 21, 35, 105\}$	$\{1, 3, 5, 15\}$
q	$\{105, 7, 5, 3, 1\}$	$\{15, 5, 3, 1\}$
B	20 MHz	20 MHz
$f_p = 1/T_p$	24 MHz	24 MHz
$f_s = qf_p$	$\{2520, 168, 120, 72, 24\}$MHz	$\{360, 120, 72, 24\}$MHz
M	399	271
f_{Nyq}	6.144 GHz	6.144 GHz

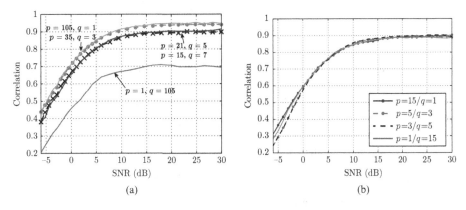

Figure 14.49 MWC performance as a function of m for different values of p and q. (a) $m = 105$. (b) $m = 15$.

Once $m > 2N$, the correlation approaches 1. However, for $m < 2N$ much lower correlation values are obtained.

Collapsed versus non-collapsed channels

As a final simulation, we examine the performance of the MWC with the same number of effective channels $m = pq$ but with different values of the number of analog channels, p, and the collapse factor, q. We consider the results for two choices of m: $m = 105$ and $m = 15$. The remaining parameter values are given in Table 14.5. The correlation versus SNR is plotted in Fig. 14.49.

In the case of $m = 105$, collapse to a single channel leads to poor results. This is because the same sequence is used in all channels so that it is hard to maintain good properties of the matrix $\mathbf{A}$ after the expander. For other parameter values, results improve significantly. When $m = 15$, all factors q lead to almost identical correlation. This implies that the performance is almost unaffected by the reduction in the number of analog channels.

14.8   Exercises

1. Consider the signal

$$x(t) = A_1 \cos(\omega_1 t) + A_2 \cos(\omega_2 t + \phi) \qquad (14.98)$$

with $\omega_1 < \omega_2 - \Delta$ for a given $0 < \Delta < \omega_1$. The signal is multiplied by $\cos(\omega_0 t)$ and then filtered with an LPF with cutoff Δ. Plot the CTFT of the output for:

a. $\omega_0 = \omega_1$.

b. $\omega_0 = \omega_2$.

c. $\omega_0 < \omega_0 < \omega_2$.

2. Calculate the lower Beurling density of the following nonuniform, nonperiodic sampling set:

$$\{t_1 + d_1 n, t_2 + d_2 n\}_{n \in \mathbb{Z}}, \qquad (14.99)$$

where $t_1 \in \mathbb{Q}$, $t_2 \notin \mathbb{Q}$, and $d_1, d_2 \in \mathbb{Z}$ such that $d_1 \neq d_2$. Here $\mathbb{Q}$ denotes the set of rational numbers.

3. Let $R = \{r_n\}$ be a sampling set with lower Beurling density $D^-(R)$. Compute the lower Beurling densities of the following sets:

a. The set R with a finite number of points removed.

b. The set R with half the number of points removed.

c. The set consisting of points $\{2r_n\}$.

4. Calculate the Landau rate of the following multiband signal:

$$x(t) = 3\,\mathrm{sinc}(3t)\cos(\omega_0 t) + \mathrm{sinc}(t)\cos(\omega_1 t) + 5\,\mathrm{sinc}(2t)\cos(\omega_2 t), \quad (14.100)$$

assuming that $\omega_0 \ll \omega_1 \ll \omega_2$.

5. Let $x(t)$ be a real bandpass signal with support $\mathcal{I}$ and folding index $\beta(\omega)$. Prove that $e^{j\beta(\omega)\pi} \neq 1$ for all $\omega \in \mathcal{I}$ if and only if $x(t)$ has integer band positioning.

6. Consider a complex signal $x(t)$ consisting of two bands: one band of width B with carrier $\omega_p > B/2$ and one band of width $2B$ with carrier $\omega_n < -B$. We would like to sample $x(t)$ with direct undersampling at rate ω_s. Determine the valid values of ω_s.

7. Suppose we want to sample a complex bandpass signal $x(t)$ supported on $\mathcal{I} = (\omega_\ell, \omega_u) \cup (-\omega_u, -\omega_\ell)$ with $B = \omega_u - \omega_\ell$, using an interleaved ADC with two branches.

a. Determine the possible values of the delay τ and the sampling rate $1/T$ such that x(t) can be recovered from the samples.

b. Derive an expression for the reconstruction filters.

8. Consider bandpass sampling with an interleaved ADC as in Fig. 14.14 of an input signal with $\omega_\ell = 375$ MHz and $\omega_u = 475$ MHz.

a. Compute the folding index $\beta(\omega)$ of (14.18).

b. Plot the value of $\|\mathbf{b}(\omega)\|^2$ given by (14.29) as a function of τ. Explain the results.

9. Throughout the chapter we have considered complex multiband signals. In this exercise we extend our results to real-valued multiband signals, for which $X(\omega)$ is conjugate symmetric.

Let $x(t)$ be a real-valued multiband signal with $2N$ bands (N on the positive axis and N on the negative axis), each with width not exceeding B. Let L be the number of spectrum slices, and denote the Nyquist rate by T_{Nyq}.

a. Determine the intervals of the spectrum slices as a function of T_{Nyq} and L. Distinguish between odd and even values of L.

b. What is the Landau rate of $x(t)$?

c. Suggest a method to reduce the sampling rate to NB, by preprocessing the signal and exploiting the symmetry of $X(\omega)$.

d. Explain why the communication rate is not reduced even if the sampling rate is reduced in this case.

10. Let $\{t_i\}_{i=0}^{p-1}$ be a universal pattern for a given $L \geq p$ as in Definition 14.1. Prove or disprove:

a. Any cyclic shift $\{\tilde{t}_i\}$, where $\tilde{t}_i = (t_i + c) \mod L$ for $c \in \mathbb{Z}$, is a universal pattern.

b. Any mirroring operation $\{\tilde{t}_i\}$ where $\tilde{t}_i = L - 1 - t_i$ is a universal pattern.

c. Any one-to-one operation $\{\tilde{t}_i\} = \{Tt_i\}$ where $T: \{0, 1, \ldots, L-1\} \longrightarrow \{0, 1, \ldots, L-1\}$ is a one-to-one operator, is a universal pattern.

11. Determine whether the signals below can be sampled with each of the ADC devices listed in Fig. 14.26(b) assuming an ideal LPF. Explain your answers.

a. A bandpass signal consisting of bands in the range $[750, 800]$ MHz, with direct undersampling at $\omega_s = 100$ MHz.

b. A multiband signal, $x(t)$, with the following CTFT:

$$X(\omega) = \begin{cases} X_1(\omega - \omega_1), & 100 \leq \omega \leq 200 \text{ MHz} \\ X_2(\omega - \omega_2), & 320 < \omega \leq 350 \text{ MHz} \\ X_3(\omega - \omega_3), & 450 < \omega \leq 500 \text{ MHz} \\ 0, & \text{otherwise,} \end{cases} \quad (14.101)$$

sampled with interleaved ADCs at the Landau rate ($B = 100$ MHz).

c. $X(\omega)$, sampled with a MWC at the Landau rate.

12. Determine which of the following sensing matrices enables reconstruction of the multiband signal of (14.101) ($p = 3$, $B = 100$ MHz, $f_{\text{Nyq}} = 500$ MHz). Explain your answers.

a. $\mathbf{A}_1 = \begin{bmatrix} -3 & 2 & 4 & 1 & 3 \\ 3 & 3 & 4 & 1 & 4 \\ 3 & 4 & 4 & -1 & 3 \end{bmatrix}$.

b. $\mathbf{A}_2 = \begin{bmatrix} 1 & 1 & 1 & 1 & 1 \\ 1 & \lambda_1 & \lambda_2 & \lambda_3 & \lambda_4 \\ 1 & \lambda_1^2 & \lambda_2^2 & \lambda_3^2 & \lambda_4^2 \end{bmatrix}$, where $\lambda_k = e^{j2\pi k/5}$.

c. $\mathbf{A}_3 = \begin{bmatrix} -1 & 1 & 1 & 1 & 1 \\ 1 & 1 & -1 & 1 & -1 \\ 1 & 1 & -1 & -1 & 1 \end{bmatrix}$.

13. Consider an MWC system with sign-alternating periodic functions, $p_i(t)$. We have seen that in order to ensure enough transitions, the number of sign-alternations, M, within a period, should satisfy $M \geq L$. Give an example of a multiband signal $x(t)$

that we cannot reconstruct perfectly by an MWC system which has sign-alternating functions with $M < L$.

14. Let **S** be the 7×8 matrix given by

$$\mathbf{S} = \begin{bmatrix} 1 & -1 & -1 & 1 & 1 & 1 & -1 & 1 \\ -1 & -1 & 1 & 1 & 1 & 1 & -1 & 1 \\ -1 & 1 & -1 & 1 & -1 & -1 & 1 & 1 \\ -1 & -1 & -1 & -1 & -1 & 1 & -1 & -1 \\ -1 & 1 & 1 & 1 & -1 & 1 & 1 & -1 \\ -1 & 1 & -1 & -1 & 1 & -1 & 1 & 1 \\ -1 & -1 & 1 & 1 & 1 & 1 & 1 & -1 \end{bmatrix}. \tag{14.102}$$

Suppose that **S** is a sensing matrix of an MWC system with no channel collapse, $p = 7$ channels, and $M = 8$ sign-alternations within a period of $p_i(t)$.

a. Assuming we implement the sequences using shift registers and flip-flops, what is the minimal required number of shift registers and flip-flops in such a system?

b. Using SBR4, what is the maximal number of bands, N, which can be reconstructed assuming that the signal is arbitrary?

15. Consider an MWC system with three channels and modulating sequences

$$p_0(t) = \sum_{n \in \mathbb{Z}} \delta(t - n)$$

$$p_1(t) = \sum_{k=0}^{L-1} (k+1) e^{-j 2\pi kt}$$

$$p_2(t) = \sum_{k=0}^{L-1} 2(k+1)^2 \cos(2\pi kt) - 1. \tag{14.103}$$

We want to blindly sample and reconstruct a multiband signal with $N = 3$ bands, each not exceeding $B = 100$ MHz, and $f_{\mathrm{Nyq}} = 900$ MHz.

a. Assuming no channel collapse, how many samplers with sampling rate 100 MHz are needed to satisfy the requirement?

b. Consider now collapsing the channels. What is the minimal collapsing factor q possible? Write down the resulting sensing matrices $\mathbf{C}_i$ of (14.70) explicitly.

16. An electrical engineer ordered a simple MWC board (no channel collapse) and an interleaved ADC board in order to blindly sample multiband signals with $N = 3$ bands each not exceeding $B = 100$ MHz, and to recover them using a CTF block. An appropriate design of the boards requires ADCs with sampling rate $r = 100$ MHz. Unfortunately, these devices were out of stock. As an alternative, the engineer decided to order boards which have ADCs with sampling rate $r = 50$ MHz. The engineer sampled some instances of this model. We assume that these ADCs have sufficient analog bandwidth to handle $x(t)$ in both of the techniques.

a. What is the sparsity of the vector-valued function $\mathbf{d}(e^{j\omega})$? What is its joint sparsity?

b. Following your previous answer, what is the minimal number of ADCs, p, required for perfect support recovery using a CTF block? Compare the average resulting rate to the rate required by SBR4 and explain the results.

c. The $r = 100$ MHz ADCs are now back in stock. The engineer decided to order the ADCs and to replace the current $r = 50$ MHz samplers. To compensate for the higher sampling rate he canceled half the channel outputs. On which of the new boards will he be able to reconstruct the signal using SBR2?

17. Consider the following real signal:

$$x(t) = \text{sinc}\left(\frac{B}{5}(t - \tau_1)\right)\text{sinc}\left(\frac{4B}{5}(t - \tau_1)\right)\cos\left(2\pi f_1(t - \tau_1)\right)$$
$$+ \text{sinc}^2\left(\frac{B}{2}(t - \tau_2)\right)\cos\left(2\pi f_2(t - \tau_2)\right) + \text{sinc}^3\left(\frac{B}{3}t\right)\cos\left(2\pi f_3 t\right),$$

$$(14.104)$$

with parameters $B = 10$MHz, $f_1 = 10$MHz, $f_2 = 20$MHz, $f_3 = 60$MHz and $\tau_k = k\mu$sec, $k = 1, 2$.

a. Plot the magnitude of the signal's CTFT $|X(f)|$ in the frequency range $\mathcal{F} \in [-100, 100]$MHz.

b. What is the Landau rate for sampling this signal?

c. We would like to sample $x(t)$ using an MWC system with no collapse so that the sampling rate on each channel f_s is equal to the mixing function frequency f_p. Suggest a design of such a system. Specify explicitly the number of hardware channels used, and the rate f_s.

d. Repeat (c) where now we use an expander with $q = 3$ and $f_s = qf_p$.

18. In this exercise we consider the same signal as in the previous question where now the carriers f_1, f_2, f_3 are unknown and can take on any value on the interval $[0, 90]$MHz.

a. What is the minimal sampling rate at which the signal can be sampled?

b. Suggest an MWC system to sample $x(t)$ with no collapse so that the sampling rate on each channel f_s is equal to the mixing function frequency f_p. Specify explicitly the number of hardware channels used, and the rate f_s.

c. Repeat (b) where now we use an expander with $q = 3$ and $f_s = qf_p$.

19. Consider an MWC system with a single channel and mixing series:

$$p(t) = \cos^2\left(2\pi f_c t\right), \quad f_c = 30\text{MHz}. \qquad (14.105)$$

The output of the mixer $y(t) = p(t)x(t)$ is filtered with a real LPF with cutoff 30MHz and then sampled uniformly at a rate $f_s = 60$MHz. Specify conditions on a multiband input signal $x(t)$ consisting of N bands of width B such that it can be recovered from these samples assuming that:

a. Its carrier frequencies are known.

b. Its carrier frequencies are unknown.

In both cases make sure to specify B, N and the maximal possible carrier frequency, and describe the complete recovery process.

20. Consider sampling a complex multiband signal with one band of width B.

    a. Suggest an MWC system for sampling such a signal consisting of a single channel with periodic mixing function $p(t)$, a complex LPF, and a uniform sampler. Specify the period of $p(t)$, the cutoff frequency of the LPF and the rate of the sampler.

    b. Detail the recovery procedure.

    c. Show that in the case of a single band the parameters can be chosen such that the CTF block is not needed. Explain this result.

Chapter 15

Finite rate of innovation sampling

In previous chapters we have seen that the UoS model can pave the way to sub-Nyquist sampling of certain categories of structured analog signals. In this chapter we consider an alternative model that relies on parametric representations: *finite rate of innovation (FRI) signals* [103]. This class corresponds to families of functions defined by a finite number of parameters per unit time, a quantity referred to as the *rate of innovation.* More specifically, an FRI signal $x(t)$ is characterized by the fact that any finite duration segment of length r is completely determined by no more than k parameters. In this case, the function $x(t)$ is said to have a local rate of innovation equal to k/r. The FRI viewpoint complements the UoS framework: a signal may lie in a UoS and have FRI; however, not all FRI signals can be described by a UoS model and vice versa, as we will show in examples below.

Interest in this class of signals emerges from the observation that several commonly encountered FRI signals can be perfectly recovered from samples taken at their rate of innovation. The advantage of this result is self-evident: FRI signals need not be bandlimited, and even if they are, the Nyquist frequency may be much higher than their rate of innovation. Thus, by using FRI techniques, the sampling rate required for perfect reconstruction can be reduced substantially. However, exploiting these capabilities requires careful design of the sampling mechanism and of the digital postprocessing. One of the most popular families of functions studied within this framework is streams of pulses which appear in many applications including bio-imaging, radar, and spread-spectrum communication.

This chapter focuses on the theory, recovery techniques, and several applications of the FRI model. We mainly concentrate on pulse streams, and consider in particular the cases of periodic, finite, infinite, and semiperiodic pulse streams. Towards the end of the chapter we will also discuss more general FRI settings.

15.1 Finite rate of innovation signals

The signals we treat in this chapter are those that are determined by a finite number of parameters per time unit. FRI signals were introduced in [103] and are given formally by the following definition:

Definition 15.1. *Let $N_r(t)$ denote a counting function that is equal to the number of parameters defining the segment of a signal $x(t)$ over the time interval $[t, t + r]$. The r-local rate of innovation of $x(t)$, denoted ρ_r, is defined as*

$$\rho_r = \max_{t \in \mathbb{R}} \frac{N_r(t)}{r}. \tag{15.1}$$

The rate of innovation, ρ, is given by

$$\rho = \lim_{r \to \infty} \rho_r. \tag{15.2}$$

A signal is said to have a finite rate of innovation (FRI) *if ρ is finite.*

By definition, ρ_r measures the maximal number of parameters over an interval of length r. The rate of innovation is the limit of this value as r increases to infinity. Given an FRI signal with rate of innovation ρ, we expect to be able to recover $x(t)$ from a number of samples per unit time proportional to ρ. As we will see, for many types of FRI signals, recovery methods exist that operate at the rate of innovation. This rate turns out to have another interesting interpretation in the presence of noise: it is a lower bound on the ratio between the MSE achievable by any unbiased estimator of $x(t)$ and the noise variance, regardless of the sampling method [309].

Before discussing sampling and recovery of FRI signals, we begin by providing some examples of this class of functions.

15.1.1   Shift-invariant spaces

Perhaps the simplest example of an FRI signal corresponds to an SI function that can be expressed as

$$x(t) = \sum_{n \in \mathbb{Z}} a[n]h(t - n\tau) \tag{15.3}$$

for some sequence $a[n] \in \ell_2$, where $h(t)$ is a known pulse in L_2 and $\tau > 0$ is a given scalar.[1] Intuitively, every signal lying in an SI space with spacing τ has one degree of freedom per τ seconds (corresponding to one coefficient from the sequence $a[n]$). It is thus tempting to regard the rate of innovation of such signals as $1/\tau$. As we now show, this is indeed the (asymptotic) rate of innovation for compactly supported pulses $h(t)$. For any finite window size r, the r-local rate of innovation ρ_r is generally larger.

Specifically, suppose that the support of $h(t)$ is contained in $[t_a, t_b]$ and consider intervals of the form $[t, t + r]$, where $r > 0$. Owing to the overlap of the pulses, for any such interval we can only assure that there are no more than $\lceil (t_b - t_a + r)/\tau \rceil$ coefficients $a[n]$ affecting the value of $x(t)$. This is demonstrated in Fig. 15.1(a). The r-local rate of innovation of signals of the form (15.3) is therefore

---

[1] We note that in earlier chapters the period of SI signals was denoted by T. In the context of FRI we will see that the sampling period is typically different than the underlying signal period. We therefore modify the notation so that T is used to denote the sampling period, while the signal period is represented by τ.

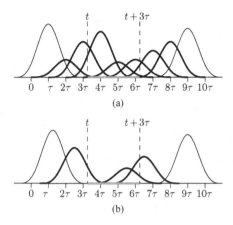

Figure 15.1 Streams of shifted versions of a pulse $h(t)$, supported on $[-2\tau, 2\tau]$. Bold pulses are those that affect the observation segment $[t, t + 3\tau]$. (a) Fixed pulse positions, spaced τ seconds apart. Here, the segment $[t, t + 3\tau]$ is affected by seven pulses so that $\rho_{3\tau} = 7/(3\tau)$. (b) Unknown pulse positions with minimal separation τ. Here, the rate of innovation is $\rho_{3\tau} = 2 \cdot 7/(3\tau) = 14/(3\tau)$. Note that the specific segment $[t, t + 3\tau]$ is affected only by three pulses so that there are $(2 \cdot 3)/(3\tau) = 2/\tau$ parameters per time unit at that location.

$$\rho_r = \frac{1}{r} \left\lceil \frac{t_b - t_a + r}{\tau} \right\rceil. \tag{15.4}$$

The rate of innovation, which is determined by taking r to infinity, is equal $1/\tau$.

We note that if $h(t)$ is not compactly supported, then the rate of innovation is infinite. Thus, for example, bandlimited signals (which correspond to $h(t) = \text{sinc}(t/\tau)$) are not FRI by Definition 15.1. This reflects the fact that it is impossible to recover any finite-duration segment $[t_a, t_b]$ of such signals from a finite number of measurements.

We have seen in earlier chapters that SI signals of the form (15.3) can be recovered from samples at rate $1/\tau$ irrespective of the bandwidth of $h(t)$. The recovery techniques we studied were based on simple filtering operations. Thus, in this case, sampling at the rate of innovation is sufficient to recover the signal using linear filtering methods. The SI model corresponds to a subspace prior. In contrast, many FRI families of signals form nonlinear signal models, rather than a subspace. Nonetheless, they can often still be sampled and perfectly reconstructed at the rate of innovation. In this case, though, more elaborate recovery techniques are needed, as shown later in the chapter.

15.1.2   Channel sounding

Single-burst channel sounding

A more complicated model than (15.3) results when the location of the pulses are unknown a priori, which often happens in channel sounding, ultrasound, and radar applications. Specifically, we have seen in Example 4.11 that in certain medium identification and channel sounding scenarios, as well as in radar and ultrasound, the echoes of a transmitted pulse $h(t)$ are analyzed to identify the positions and reflectance coefficients of scatterers in the medium [394, 424, 425, 426, 427]. In these cases, the received signal has the form

$$x(t) = \sum_{\ell=1}^{L} a_\ell h(t - t_\ell),$$ (15.5)

where L is the number of scatterers. The amplitudes $\{a_\ell\}_{\ell=1}^{L}$ and time delays $\{t_\ell\}_{\ell=1}^{L}$ correspond to the reflectance and location of the scatterers. Such signals can be thought of as belonging to a UoS, where the parameters $\{t_\ell\}_{\ell=1}^{L}$ determine an L-dimensional subspace, and the coefficients $\{a_\ell\}_{\ell=1}^{L}$ describe the position within the subspace. Since there are infinitely many possible values for the parameters $t_1, \ldots, t_L$ we have a union of an infinite number of subspaces.

For any window of size $r > \max_\ell \{t_\ell\} - \min_\ell \{t_\ell\}$, the r-local rate of innovation is given by

$$\rho_r = \frac{2L}{r}.$$ (15.6)

Therefore, if the signal is defined on the interval $[0, \tau]$, then the local rate of innovation over τ is $2L/\tau$. This is consistent with the fact that the signal has $2L$ degrees of freedom over its domain.

Periodic channel sounding

Occasionally, channel sounding techniques consist of repeatedly probing the medium [428]. Assuming the medium does not change throughout the experiment, the result is a periodic signal

$$x(t) = \sum_{\ell=1}^{L} \sum_{n \in \mathbb{Z}} a_\ell h(t - t_\ell - n\tau).$$ (15.7)

As before, the set of feasible signals is an infinite union of finite-dimensional subspaces in which $\{t_\ell\}_{\ell=1}^{L}$ determine the subspace and $\{a_\ell\}_{\ell=1}^{L}$ define the position within the subspace. The r-local rate of innovation coincides with (15.6).

Semiperiodic channel sounding

Another interesting setting is when a channel consists of L paths whose amplitudes change, but the time delays can be assumed constant throughout the duration of the experiment [428, 429, 430]. One example where this occurs is in the context of radar applications in which the targets are moving at a constant velocity, as we discuss in Section 15.7.1. In these cases the received signal has the form:

$$x(t) = \sum_{\ell=1}^{L} \sum_{n \in \mathbb{Z}} a_\ell[n] h(t - t_\ell - n\tau),$$ (15.8)

where $a_\ell[n]$ is the amplitude of the ℓth path at the nth probing experiment. This is, once again, a UoS, but here each subspace is infinite-dimensional, as it is determined by the infinite set of parameters $\{a_\ell[n]\}$.

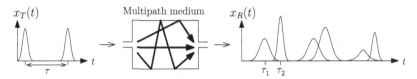

Figure 15.2 A multipath medium. Pulses with known shape are transmitted over the channel. The output signal is composed of delayed and weighted replicas of the transmitted pulses.

The model (15.8) can also describe the case in which the signal consists of a finite number of periods. In this setting $n \in \mathbb{Z}$ is replaced by a finite sum over $n = 1, \ldots, P$, as in the following example.

Example 15.1 We show how the model (15.8) can be applied to time-varying channel estimation in wireless communications [429].

In this setting, the aim of the receiver is to estimate the channel's parameters from samples of the received signal [431]. In a typical scenario, pulses with known shape are transmitted through a multipath medium, which consists of several propagation paths. As a result of the multiple paths, the received signal is composed of delayed and weighted replicas of the transmitted pulses, as illustrated in Fig. 15.2. In order to identify the medium, the time delay and gain coefficient of each multipath component has to be estimated from the received signal.

Consider a baseband communication system working with pulse amplitude modulation (PAM). The transmitted signal has the form

$$x_T(t) = \sum_{n=1}^{N} d[n]h(t - n\tau), \qquad (15.9)$$

where $d[n]$ are the data symbols taken from a finite alphabet, and N is the total number of transmitted symbols. The signal $x_T(t)$ passes through a baseband time-varying multipath channel whose impulse response at time s to an impulse at time t is modeled as [394] $h(s, t) = \sum_{\ell=1}^{L} \alpha_\ell(t)\delta(s - \tau_\ell)$, where $\alpha_\ell(t)$ is the time-varying complex gain for the ℓth multipath propagation path and τ_ℓ is the corresponding time delay. The total number of paths is denoted by L. We assume that the channel is slowly varying relative to the symbol rate, so that the path gains are considered to be constant over one symbol period:

$$\alpha_\ell(t) = \alpha_\ell[n\tau], \quad \text{for } t \in [n\tau, (n+1)\tau].$$

In addition, we confine the propagation delays to one symbol, i.e. $\tau_\ell \in [0, \tau)$. Under these assumptions, the received signal (in a noise-free environment) is given by

$$x_R(t) = \sum_{\ell=1}^{L} \sum_{n=1}^{N} a_\ell[n]h(t - \tau_\ell - n\tau), \qquad (15.10)$$

where $a_\ell[n] = \alpha_\ell[n\tau]d[n]$.

This signal can be seen as a special case of (15.8). Thus, the techniques we study in this chapter can be used to sample such multipath signals at rates below Nyquist. The resulting sampling rate will depend only on the number of multipath components and the transmission rate, but not on the bandwidth of the transmitted pulse. This can lead to significant sampling rate reduction when only a small number of propagation paths exists, or when the bandwidth of the transmitted pulse is relatively high. This application is explored in more detail in [429].

Assume as in the SI setting that the support of $h(t)$ is contained in $[t_a, t_b]$, and consider intervals of the form $[t, t + r]$, where $r > 0$. Then, similar to the SI case, there are no more than $L\lceil (t_b - t_a + r)/\tau \rceil$ coefficients $a[n]$ affecting the value of $x(t)$ for every such interval. In addition, there are at most L unknown delays. Thus,

$$\rho_r = \frac{L}{r} \left(\left\lceil \frac{t_b - t_a + r}{\tau} \right\rceil + 1 \right). \tag{15.11}$$

Taking r to infinity, the rate of innovation is $\rho = L/\tau$, which is half the rate of the single-burst case. Here ρ is not affected by the fact that we do not know the L delays but rather is determined only by the L unknown amplitudes. This is because when we increase the observation period, the effect of the delays becomes negligible.

15.1.3   Other examples

Multi-carrier signals
The model (15.5) and its variants have received the largest amount of attention in the FRI literature. However, other interesting FRI signal classes exist. As an example, suppose that L transmissions of the form (15.3) are modulated, each with a different carrier frequency, to yield

$$x(t) = \sum_{\ell=1}^{L} \sum_{n \in \mathbb{Z}} a_\ell[n] h(t - n\tau) \sin(\omega_\ell t). \tag{15.12}$$

Here, $a_\ell[n]$ is the data transmitted by the ℓth user on the carrier frequency ω_ℓ. This setting is analogous in many respects to the semiperiodic channel sounding case; in particular, the r-local rate of innovation is the same.

Continuous-phase modulation
The examples we have discussed so far all correspond to special cases of a UoS. Clearly, not all UoS models are FRI: in particular, as we have seen, when $h(t)$ of (15.3) is not compactly supported then the corresponding model does not have a finite number of parameters per unit time. Conversely, although the UoS representation is currently the most common setting treated within the FRI literature, FRI signals do not have to conform to the UoS model. An example is continuous-phase modulation (CPM) transmissions. These include continuous-phase frequency shift keying (CPFSK) and minimum shift keying (MSK), tamed frequency modulation (TFM), Gaussian MSK (GMSK), and more. Here, the transmitted signal has the form

$$x(t) = \cos\left(\omega_0 t + 2\pi h \int_{-\infty}^{t} \sum_{n\in\mathbb{Z}} a[n]h(r - n\tau)dr\right), \qquad (15.13)$$

where ω_0 is a fixed carrier frequency, $a[n] \in \{\pm 1, \pm 3, \ldots, \pm(Q - 1)\}$ are the message symbols, h is the modulation index, and $h(t)$ is a pulse shape that is supported on $[0, L\tau]$ for some integer $L > 0$ and satisfies $\int_0^{L\tau} h(t)dt = 0.5$.

The rate of innovation of CPM signals can be determined by expressing (15.13) as (see Exercise 3)

$$x(t) = \cos\left(\omega_0 t + \sum_{m\in\mathbb{Z}} \tilde{a}[m]\tilde{h}(t - m\tau)\right), \qquad (15.14)$$

where $\tilde{a}[m] = \sum_{n=-\infty}^{m} a[n]$ and

$$\tilde{h}(t) = 2\pi h \int_{-\infty}^{t} (h(r) - h(r - \tau))\, dr. \qquad (15.15)$$

Since knowing $a[n]$ is equivalent to knowing $\tilde{a}[n]$ (up to initial boundary conditions) and $\tilde{h}(t)$ is supported on $[0, (L + 1)\tau]$, the number of coefficients affecting $x(t)$ on any interval $[t, t + r]$ is the same as in (15.3) with $t_a = 0$ and $t_b = (L + 1)\tau$. Consequently, the r-local rate of innovation of CPM signals is

$$\rho_r = \frac{1}{r}\left(\left\lceil\frac{r + (L + 1)\tau}{\tau}\right\rceil\right) = \frac{1}{r}\left(\left\lceil\frac{r}{\tau}\right\rceil + L + 1\right) \qquad (15.16)$$

and their asymptotic rate is $1/\tau$.

Nonlinearly distorted shift-invariant spaces
Another example of an FRI model that is not a UoS is when a signal belonging to a UoS is distorted by a nonlinear operation. In Chapter 8 we considered in detail sampling in the presence of nonlinear distortions such as companding methods used to avoid clipping in various communication settings. If the original signal lies in an SI space with a compactly supported generator $h(t)$, then the resulting transmission takes the form

$$x(t) = M\left(\sum_{n\in\mathbb{Z}} a[n]h(t - n\tau)\right), \qquad (15.17)$$

where $M(\cdot)$ is a nonlinear, invertible function. Clearly, the r-local rate of innovation ρ_r of this type of signal is the same as that of the underlying SI function, and is thus given by (15.4). As we have seen in Chapter 8, under appropriate conditions on the nonlinear distortion, $x(t)$ can be recovered from samples at the rate of innovation $1/\tau$.

Chapter outline
Having presented several examples of FRI signals, we next study sampling theorems for this model. We begin by focusing on streams of pulses. Periodic FRI signals turn out to be particularly convenient for analysis, and will be discussed in depth in Section 15.2. After introducing a matched-filtering approach to signal recovery, we will see that it is beneficial to analyze the problem in the frequency domain. In this domain, FRI recovery

is equivalent to estimating a sum of sinusoids with unknown frequencies – a problem which has been well-studied in the array processing literature. The techniques used in this setting can then be extended to treat more general pulse streams including the infinite and semi-infinite cases.

We focus on several key approaches to treating the sum-of-sinusoids problem, including Prony's method, Cadzow denoising, matrix pencil, MUSIC, ESPRIT, and algorithms based on compressed sensing. These techniques all assume that we are given Fourier coefficients of the periodic pulse stream from which the signal parameters are estimated. Approaches for obtaining such coefficients from low-rate samples of the signal are discussed in Sections 15.3 and 15.4. In Section 15.3 we treat two sub-Nyquist sampling methods based on single channel filters: coset sampling and sampling with a sum-of-sincs prefilter. Multichannel structures for sub-Nyquist sampling, including modulation banks and filterbanks, are introduced in Section 15.4.

In Section 15.5 we discuss ultimate bounds that can be achieved when estimating FRI signals in the presence of noise. Iterative recovery techniques for general FRI models are treated in Section 15.6. We end the chapter in Section 15.7 with several applications of the pulse stream model to radar and ultrasound. These examples demonstrate that FRI methods can pave the way to sub-Nyquist sampling and processing in a variety of application domains.

15.2   Periodic pulse streams

Consider a τ-periodic stream of pulses, defined as

$$x(t) = \sum_{\ell=1}^{L} \sum_{n \in \mathbb{Z}} a_\ell h(t - t_\ell - n\tau), \tag{15.18}$$

where $h(t)$ is a known pulse shape, τ is the known period, and $\{t_\ell, a_\ell\}_{\ell=1}^{L}$, $t_\ell \in [0, \tau)$, $a_\ell \in \mathbb{C}$, $\ell = 1, \ldots, L$ are the unknown delays and amplitudes. Our goal is to sample $x(t)$ and reconstruct it, from a minimal number of samples. Since the signal has $2L$ degrees of freedom per period, we expect the minimal number of samples over one period to be $2L$. We are primarily interested in pulses which have large bandwidth corresponding to small time support. In this case the Nyquist rate of the signal $x(t)$ of (15.18) is high so that sub-Nyquist methods can be beneficial. Note that the bandwidth of $x(t)$ is no larger than that of $h(t)$ since delays and summations do not increase bandwidth.

When the number of degrees of freedom in $x(t)$ is less than its bandwidth, we expect to be able to recover $x(t)$ from a small number of measurements. The question, though, is how to choose these low-rate samples. Owing to the short time support, direct uniform sampling of $2L$ samples will typically result in many zeros, since the probability of a sample to hit a pulse is very low, as illustrated in Fig. 15.3(a). Therefore, more sophisticated sampling schemes must be designed.

In the previous chapter we developed techniques for low-rate sampling of multiband signals which were based on aliasing the signal in frequency prior to sampling. This aliasing allowed us to collapse the high bandwidth signal content onto a low-dimensional subspace which was then sampled at a low rate. We will follow a similar approach here,

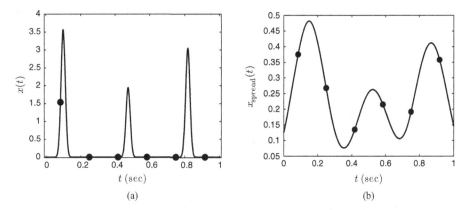

Figure 15.3 Direct uniform sampling of the pulse stream $x(t)$ of (15.18) and its time-aliased (spread) version. Here $h(t)$ is a Gaussian pulse with small variance $\sigma^2 = 3 \times 10^{-4}$, and the amplitudes and delays were chosen randomly. (a) Direct uniform sampling at a low rate. (b) Time-aliased version of $x(t)$ generated using a low-pass filter, and its (nonzero) samples.

where now the aliasing is performed in time rather than in frequency. More specifically, suppose that we spread each of the pulses in time prior to sampling, as illustrated in Fig. 15.3(b). This now enables sampling at a low rate while still acquiring information about the signal. Once the signal is sampled, we can gain back the seemingly lost resolution by employing appropriate recovery methods in the digital domain. These techniques exploit ideas from spectral analysis and compressed sensing by analyzing the stream of pulses in the frequency domain. Recovering the delays and amplitudes from the low-rate samples will then correspond to determining the frequencies in a sum of complex exponentials from a small number of Fourier coefficients.

15.2.1 Time-domain formulation

We begin by considering standard time-domain methods for estimating the delays and amplitudes which are based on Nyquist-rate sampling of $x(t)$ in (15.18). The most popular (and simplest) approach is through a process known in the communication literature as a *rake receiver* [394]. This receiver was proposed back in the 1950s, and is designed to counter the effects of multipath fading.

To develop the rake receiver, suppose that $L = 1$ so that there is a single unknown delay. In the presence of noise, we then have that $x(t) \approx a_1 h(t - t_1)$ for $t \in [0, \tau)$ where a_1 and t_1 are the unknown amplitude and delay, respectively. To simplify the derivations, we assume that the pulse is normalized to unit energy so that $\int_0^\tau |h(t)|^2 dt = 1$ and that $h(t - t')$ is confined to $[0, \tau)$ for all possible delays t'. A reasonable approach to estimate a_1, t_1 is to seek the values that minimize the error:

$$\min_{a_1, t_1} \int_0^\tau |x(t) - a_1 h(t - t_1)|^2 dt. \tag{15.19}$$

This is also the maximum-likelihood estimate when the noise is Gaussian.

By differentiating the objective in (15.19) and equating it to zero, the optimal value of a_1 is given by

$$a_1 = \frac{\int_0^\tau \overline{h(t - t_1)} x(t) dt}{\int_0^\tau |h(t - t_1)|^2 dt}. \tag{15.20}$$

Substituting this value back into the objective (15.19), the optimal value of t_1 can be computed as

$$t_1 = \arg\max_{t'} \frac{\left| \int_0^\tau \overline{h(t - t')} x(t) dt \right|^2}{\int_0^\tau |h(t - t')|^2 dt} = \arg\max_{t'} \left| \int_0^\tau \overline{h(t - t')} x(t) dt \right|^2, \tag{15.21}$$

where in the last equality we relied on the assumption that $h(t - t')$ is confined to $[0, \tau)$ for all feasible t'. To implement (15.21), we note that

$$y(t) = \int_0^\tau \overline{h(\alpha - t)} x(\alpha) d\alpha = \overline{h(-t)} * x(t). \tag{15.22}$$

Thus we can find t_1 by first creating the function $y(t)$ which is the convolution between $x(t)$ and $\overline{h(-t)}$, and then choosing t_1 as the value at which $|y(t)|$ obtains its maximum. This process is known as *matched filtering* [432].

When there are multiple delays, a popular approach is to follow the same procedure, where now the estimated delays are chosen as the arguments corresponding to the L largest values of $|y(t)|$. This method will no longer be optimal in general owing to the interference between overlapping pulses. If L is not known in advance, then it is common to choose all peaks that are beyond a given threshold, where the bound is often set as a percentage of the largest peak or as a function of the background noise level. Model order selection methods can also be used [17].

In practice, the matched filter (MF) is usually implemented digitally. Since the bandwidth of $y(t)$ is no larger than that of $h(t)$, the MF output $y(t)$ can be determined from its samples at the Nyquist rate, corresponding to the bandwidth of $h(t)$. Let $1/T$ denote the Nyquist rate of $y(t)$ and let $y[n] = y(nT)$. Then $y[n] = \overline{h[-n]} * x[n]$ where $h[n] = h(nT)$ and $x[n]$ are samples at rate $1/T$ of a bandlimited version of $x(t)$ (see Exercise 4). Therefore, to estimate t_1 one can first compute the digital MF output $y[n]$, as illustrated in Fig. 15.4, and then construct $y(t)$ by bandlimited interpolation of $y[n]$. Note that in the figure, instead of performing the MF operation and then sampling its output, we can equivalently sample $x(t)$ and then compute $y[n]$ digitally.

To retain computational efficiency, the sinc interpolation step is typically omitted entirely, or approximated by local interpolation. In the former case the delays are found on the discrete (Nyquist) grid; that is, we seek the peaks in $|y[n]|$ and estimate the delays as the values of n at which the peaks are attained. The performance of this simple MF

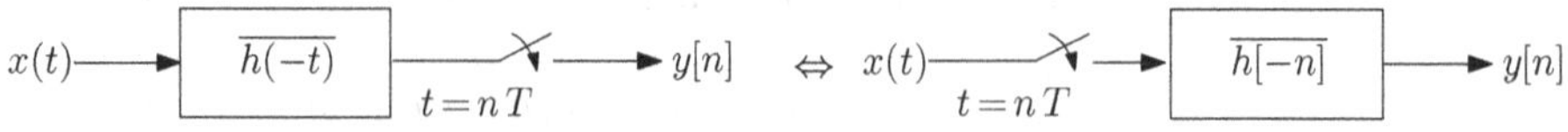

Figure 15.4 Matched filtering in the analog and digital domains.

Algorithm 15.1 Matched filtering

Input: Samples $x[n], h[n]$ of the received signal $x(t)$ and the pulse shape $h(t)$ respectively, with period T

Output: Time delays $t_\ell, \ell = 1, \ldots, L$

Create the matched-filter output $y[n] = \overline{h[-n]} * x[n]$

Optional: Apply local interpolation

Find the L largest peaks of $|y[n]|$

Determine t_ℓ as the values at which the maximum is obtained

---

strategy can be improved by first interpolating the discrete sequence $y[n]$ locally, for example using cubic interpolation, and then determining the peaks.

Evidently, performing the MF operation in the digital domain requires sampling the data at the Nyquist rate or higher, in order to approximate the continuous MF output. If $h(t)$ is time-limited and has N nonzero Fourier coefficients over the interval $[0, \tau)$, then at least N samples are needed over the period τ. Equivalently, if $h(t)$ is bandlimited to $2\pi f_c$ and we consider its τ-periodic extension $h_\tau(t) = \sum_{k \in \mathbb{Z}} h(t + k\tau)$, then at least $f_c \tau$ samples per period are required. We summarize the MF approach to time-delay estimation in Algorithm 15.1.

The MF procedure is illustrated by the following example.

Example 15.2 Consider a signal $x(t) = \sum_{\ell=1}^{L} \sum_{n \in \mathbb{Z}} a_\ell h(t - t_\ell - n\tau)$ consisting of $L = 4$ pulses along the period $\tau = 1$. The waveform $h(t)$ is a sinc function with bandwidth equal to 81 Hz (this bandwidth is chosen in order to enable a comparison with the methods demonstrated in Examples 15.3, 15.4, and 15.8). The rate of innovation of $x(t)$ is equal to $2L$ (see Exercise 2). The delays and amplitudes are set to $a_\ell = 1, t_\ell = \ell\Delta$, where Δ is a parameter. The signal is sampled at its Nyquist rate resulting in 81 samples of $x(t)$ along the period $\tau = 1$. The sampling phase is chosen randomly so that the delays are not aligned to a grid. We consider the case in which the signal $x(t)$ is contaminated by noise so that we are given noisy versions of the samples $x[n]$. The added noise is white Gaussian with variance $1/\text{SNR}$. The digital MF output $y[n]$ is computed by filtering the received sampled signal $x[n]$ with the sampled MF $h[n]$.

We estimate the delays and amplitudes using (15.21) and (15.20) in two ways. In the first, we approximate the values of $\{t_\ell\}$ on the Nyquist grid by seeking the maximum of the absolute MF output over the sampled data. In the second, we use local interpolation to improve the result. Figure 15.5 depicts the MSE (in a log scale) in estimating the delays and amplitudes as a function of the SNR averaged over 500 simulations, for two choices of Δ: distant delays, $\Delta = 0.2$, and close delays, $\Delta = 0.025$. The MF approach is compared with the performance of the matrix pencil method, a frequency-based technique described in Section 15.2.5. Observing the figures it can be seen that estimation using the MF method is limited by the sampling rate, particularly at high SNR. The superior performance of the frequency-based matrix pencil approach is clearly evident.

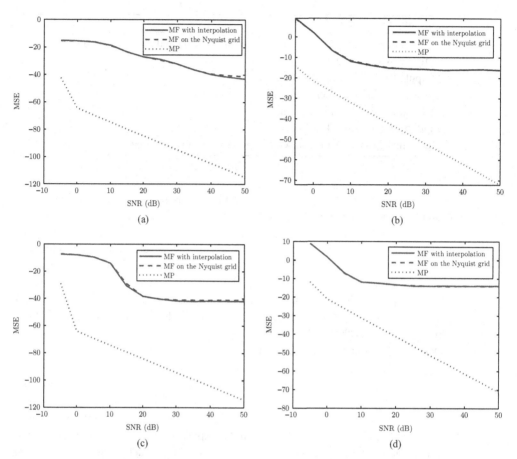

Figure 15.5 Delay and amplitude estimation using the MF and matrix pencil (MP) methods for $L = 4$. (a) MSE in estimating the delays $t_\ell = \ell\Delta$ for $\Delta = 0.2$. (b) MSE in estimating the amplitudes for $\Delta = 0.2$. (c) MSE in estimating the delays for $\Delta = 0.025$. (d) MSE in estimating the amplitudes for $\Delta = 0.025$.

The previous example illustrates that estimating the delays using an MF may lead to poor performance, especially for close delays. Furthermore, this approach requires sampling the data at the Nyquist rate which can be quite high when the underlying pulse $h(t)$ is very narrow in time, as is the case in many applications, particularly those involving localization (such as radar). In the next subsection we introduce a frequency-based framework for time-delay estimation which leads to several alternative recovery methods. These techniques can be adapted to operate on low-rate samples and tend to lead to superior performance when multiple targets are present.

15.2.2   Frequency-domain formulation

Instead of considering time-domain methods for delay estimation, we now analyze the signal in the Fourier domain. This allows us to connect our problem with that of spectral estimation and exploit powerful methods developed in that context, resulting in a convenient route to sampling-rate reduction. These approaches tend to lead to better resolution in high enough SNR, namely, they improve the ability to distinguish between close delays. CS techniques are also readily applicable in frequency, once again suggesting ways to recover the delays and amplitudes from low-rate samples in a robust fashion, particularly in low-SNR regimes.

Fourier series representation

Consider the signal $x(t)$ of (15.18). Since $x(t)$ is periodic, we can represent it in terms of its Fourier series. To this end, define the periodic continuation of $h(t)$ as $f(t) = \sum_{m \in \mathbb{Z}} h(t + m\tau)$. Using Poisson's sum formula (3.89) with $x(t') = h(t' + t)$, $f(t)$ may be written as

$$f(t) = \frac{1}{\tau} \sum_{k \in \mathbb{Z}} H\left(\frac{2\pi k}{\tau}\right) e^{j2\pi kt/\tau}, \qquad (15.23)$$

where $H(\omega)$ denotes the CTFT of the pulse $h(t)$. Substituting (15.23) into (15.18) we obtain

$$x(t) = \sum_{\ell=1}^{L} a_\ell f(t - t_\ell)$$

$$= \sum_{k \in \mathbb{Z}} \left(\frac{1}{\tau} H\left(\frac{2\pi k}{\tau}\right) \sum_{\ell=1}^{L} a_\ell e^{-j2\pi kt_\ell/\tau}\right) e^{j2\pi kt/\tau}$$

$$= \sum_{k \in \mathbb{Z}} X[k] e^{j2\pi kt/\tau}, \qquad (15.24)$$

where we denoted

$$X[k] = \frac{1}{\tau} H\left(\frac{2\pi k}{\tau}\right) \sum_{\ell=1}^{L} a_\ell e^{-j2\pi kt_\ell/\tau}. \qquad (15.25)$$

The expansion in (15.24) is the Fourier series representation of the τ-periodic signal $x(t)$ with Fourier coefficients given by (15.25).

We next show that if $2L$ or more consecutive Fourier coefficients of $x(t)$ are given, then $\{t_\ell, a_\ell\}_{\ell=1}^{L}$ can be determined from (15.25) using the fact that $h(t)$ is known. This number of coefficients is typically much smaller than the number N of nonzero values of $H\left(\frac{2\pi k}{\tau}\right)$ which is the number of samples obtained over τ when sampling at the Nyquist rate. Furthermore, if we are willing to recover the delays on a grid, then we can formulate the recovery in the frequency domain using tools of CS (see Chapter 11). This will allow for robust recovery in the presence of strong noise. In Section 15.3 we discuss methods to obtain the required Fourier coefficients from low-rate samples of $x(t)$ in the time domain. Intuitively, since only on the order of $2L$ Fourier measurements are needed, these can be acquired from low-rate samples in time. Thus, the ability to recover $x(t)$ from a small number of Fourier coefficients immediately implies that low-rate sampling is possible, where the number of samples is equal to the number of Fourier coefficients used in the recovery process.

Sum-of-sinusoids problem

Suppose we are given $m \geq 2L$ coefficients $X[k], k \in \mathcal{K}$ where $\mathcal{K}$ is a consecutive index set of size $|\mathcal{K}| = m$. We assume that $\mathcal{K}$ can be chosen such that $H\left(\frac{2\pi k}{\tau}\right) \neq 0$ for $k \in \mathcal{K}$. Let $Y[k] = \tau X[k]/H\left(\frac{2\pi k}{\tau}\right)$. Then,

$$Y[k] = \sum_{\ell=1}^{L} a_\ell e^{-j\omega_\ell k}, \quad k \in \mathcal{K}, \tag{15.26}$$

where $\omega_\ell = 2\pi t_\ell/\tau$. The problem (15.26) has been studied extensively in the array processing literature and is referred to as the *sum-of-sinusoids problem*. The goal is to recover the unknown frequencies ω_ℓ and unknown amplitudes a_ℓ from the sums $Y[k]$. Many methods have been proposed to determine the unknown frequencies, several of which we outline below. Further details can be found, for example, in [433]. These techniques allow exact recovery when there is no noise, and work well in small to moderate noise levels. However, when the noise increases, it is very difficult to determine ω_ℓ. An alternative is to discretize the possible frequency values so that we assume $\omega_\ell = (2\pi/\tau)\Delta s_\ell$ for some integer s_ℓ where Δ is the chosen time resolution. By discretizing the delays, we can formulate (15.26) within the framework of CS and use any of the methods for sparse vector recovery to find the values of s_ℓ. This approach, discussed in Section 15.2.8, is typically more robust in the presence of noise, although it entails a loss of resolution due to the discretization.

Once the values of ω_ℓ are found, the amplitudes a_ℓ can be determined by solving a simple least squares problem:

$$\min_{a_\ell} \sum_{k \in \mathcal{K}} \left| Y[k] - \sum_{\ell=1}^{L} a_\ell e^{-j\omega_\ell k} \right|^2. \tag{15.27}$$

To solve (15.27) let $\mathcal{K}$ consist of the values $k_1, k_2, \ldots, k_m$ with $k_m = k_1 + (m-1)$ for some value k_1. We can then express the objective in matrix form as

$$\min_{\mathbf{a}} \|\mathbf{y} - \mathbf{V}(\{\omega_\ell\}, k_1)_{m \times L} \mathbf{a}\|^2, \tag{15.28}$$

where $\mathbf{a} = [a_1, \ldots, a_L]^T$ is the vector containing the unknown amplitudes, $\mathbf{y}$ is the length-m vector of measurements $Y[k], k \in \mathcal{K}$, and

$$\mathbf{V}(\{\omega_\ell\}, k_1)_{m \times L} = \begin{bmatrix} e^{-j\omega_1 k_1} & e^{-j\omega_2 k_1} & \cdots & e^{-j\omega_L k_1} \\ e^{-j\omega_1 k_2} & e^{-j\omega_2 k_2} & \cdots & e^{-j\omega_L k_2} \\ \vdots & & & \vdots \\ e^{-j\omega_1 k_m} & e^{-j\omega_2 k_m} & \cdots & e^{-j\omega_L k_m} \end{bmatrix}. \tag{15.29}$$

When $k_1 = 0$ we use the notation

$$\mathbf{V}(\{\omega_\ell\})_{m \times L} = \begin{bmatrix} 1 & 1 & \cdots & 1 \\ e^{-j\omega_1} & e^{-j\omega_2} & \cdots & e^{-j\omega_L} \\ \vdots & & & \vdots \\ e^{-j\omega_1(m-1)} & e^{-j\omega_2(m-1)} & \cdots & e^{-j\omega_L(m-1)} \end{bmatrix}, \tag{15.30}$$

which is an $m \times L$ Vandermonde matrix with roots $\lambda_i = e^{-j\omega_i}$. Note that $\mathbf{V}(\{\omega_\ell\}, k_1)_{m \times L}$ can be expressed as $\mathbf{V}(\{\omega_\ell\}, k_1) = \mathbf{V}(\{\omega_\ell\}) \operatorname{diag}(e^{-j\omega_1 k_1}, e^{-j\omega_2 k_1}, \ldots, e^{-j\omega_L k_1})$.

Throughout the chapter we will rely on several useful properties of Vandermonde matrices. In particular, an important result which we use often is the following (see Appendix A):

Proposition 15.1. *Let* $\mathbf{V}$ *be an* $m \times n$ *Vandermonde matrix with* $m \geq n$ *and roots* λ_i *so that* $V_{i\ell} = \lambda_\ell^i$. *Then* $\mathbf{V}$ *has full column-rank if and only if the roots are distinct. In this case it holds in addition that any* n *rows of* $\mathbf{V}$ *are linearly independent.*

An immediate consequence of the proposition is that $\mathbf{V}(\{\omega_\ell\}, k_1)$ also has full column-rank if the values ω_ℓ are different modulo 2π.

Assuming that $e^{-j\omega_i}$ are distinct, the solution to (15.28) is given by (see Appendix A)

$$\hat{\mathbf{a}} = (\mathbf{V}^* \mathbf{V})^{-1} \mathbf{V}^* \mathbf{y}, \tag{15.31}$$

where for brevity we denoted $\mathbf{V} = \mathbf{V}(\{\omega_\ell\}, k_1)_{m \times L}$. Since $m \geq L$, $\mathbf{V}^* \mathbf{V}$ is invertible.

In the next subsections we outline several methods for recovering the frequencies $\{\omega_\ell\}$ in (15.26):

1. Prony's method and its extensions: total least squares Prony, Cadzow denoising;
2. Matrix pencil;
3. Subspace methods: Pisarenko and MUSIC;
4. Covariance methods: stochastic MUSIC and ESPRIT;
5. Compressed sensing formulation.

For simplicity, we assume throughout that $k_1 = 0$. The first four methods are based on spectral analysis techniques and recover the frequencies exactly when the number of measurements $m \geq 2L$ and there is no noise. The last approach is based on discretizing the frequencies (or delays), but is more robust in the presence of moderate to large noise. To describe the various techniques we rewrite (15.26) as a power series

$$Y[k] = \sum_{\ell=1}^{L} a_\ell u_\ell^k, \tag{15.32}$$

where $u_\ell = e^{-j\omega_\ell}$. Our goal then is to recover $\{u_\ell\}$ from a small number $m \geq 2L$ of Fourier coefficients $Y[k]$.

Before concluding this section we note that our derivations below are based on the model (15.26) (or (15.32)). To arrive at this formulation, we divided the Fourier coefficients of the signal $X[k]$ by those of the pulse shape $H(2\pi k/\tau)$. When the pulse is relatively flat, this is a reasonable route to take. However, if the pulse decays in frequency, then dividing by $H(2\pi k/\tau)$ can lead to noise enhancement. In such cases we may prefer an MF approach where we multiply $X[k]$ by $\overline{H(2\pi k/\tau)}$, or if the noise variance σ^2 is known, then we can use a Wiener prefilter. This corresponds to multiplying $X[k]$ by

$$W[k] = \frac{\overline{H(2\pi k/\tau)}}{|H(2\pi k/\tau)|^2 + \sigma^2}. \tag{15.33}$$

Note that when $\sigma \to 0$, namely, the noise is low, $W[k]$ approaches $1/H(2\pi k/\tau)$. On the other hand, in low SNR corresponding to large σ^2, $W[k]$ is proportional to the MF $\overline{H(2\pi k/\tau)}$. For the remainder of this chapter, we continue to assume the model (15.26) (or (15.32)).

15.2.3   Prony's method

Prony's method [192] for estimating $\{u_\ell\}$ in (15.32) begins by defining a filter $G(z)$ whose roots equal the values u_ℓ to be found:

$$G(z) = \sum_{\ell=0}^{L} g_\ell z^{-\ell} = \prod_{\ell=1}^{L} \left(1 - u_\ell z^{-1}\right), \tag{15.34}$$

where $\{g_\ell\}_{\ell=0}^{L}$ are the filter coefficients. The important property of $G(z)$ is that it *annihilates* the measurements $Y[k]$. Specifically, define the sequence $q[k] = g_k * Y[k]$ as the convolution of g_k and $Y[k]$. Then, for $L \leq k \leq m-1$ we have

$$q[k] = \sum_{i=0}^{L} g_i Y[k-i] = \sum_{i=0}^{L}\sum_{\ell=1}^{L} a_\ell g_i u_\ell^{k-i} = \sum_{\ell=1}^{L} a_\ell u_\ell^{k} \underbrace{\sum_{i=0}^{L} g_i u_\ell^{-i}}_{=0} = 0 \tag{15.35}$$

where the last equality is due to the fact that $G(u_\ell) = 0$. Note that $m-1 \geq L$ since $m \geq 2L$. The filter $\{g_\ell\}$ is therefore called an *annihilating filter*. Its roots uniquely define $\{u_\ell\}$, provided that these values are distinct. We can thus recover $\{u_\ell\}$ once we determine the filter coefficients $\{g_\ell\}$ (up to a scaling factor which does not affect the resulting zeros).

The identity in (15.35) can be written in matrix/vector form as

$$\mathbf{Y}_{(m-L)\times(L+1)}\mathbf{g} = \mathbf{0}, \tag{15.36}$$

where $\mathbf{g} = [g_L, \ldots, g_1, g_0]^T$ and $\mathbf{Y}_{s\times n}$ is the data matrix defined by

$$\mathbf{Y}_{s\times n} = \begin{bmatrix} Y[0] & Y[1] & \cdots & Y[n-1] \\ Y[1] & Y[2] & \cdots & Y[n] \\ \vdots & \vdots & \ddots & \vdots \\ Y[s-1] & Y[s] & \cdots & Y[s+n-2] \end{bmatrix}. \tag{15.37}$$

When the dimensions are clear from the context, we use the shorthand notation $\mathbf{Y} = \mathbf{Y}_{s\times n}$. Note that the $k\ell$th element of $\mathbf{Y}$ is given by

$$\mathbf{Y}_{k\ell} = Y[k+\ell], \quad 0 \leq k \leq s-1 \text{ and } 0 \leq \ell \leq n-1, \tag{15.38}$$

thus it depends only on the sum $k + \ell$. This results in a matrix whose skew-diagonals are all equal and is called a *Hankel matrix*. The Cadzow technique below exploits this property to improve recovery in the presence of noise.

In order to be able to uniquely identify the vector of unknowns $\mathbf{g}$ (up to scaling) we need to ensure that the null space of $\mathbf{Y}_{(m-L)\times(L+1)}$ has dimension equal to 1. For this to happen, we must have that the number of rows $s = m - L$ satisfies $s \geq n - 1$ where $n = L + 1$ denotes the number of columns. This is because if $s \leq n - 2$ then the null space will always have dimension at least 2. Substituting in the values of s and n, we arrive at the condition $m \geq 2L$. Thus, we need at least $2L$ consecutive values of $Y[k]$ to solve (15.36). In Proposition 15.2 (and Corollary 15.1) below we show that this condition is also sufficient, namely, if $m \geq 2L$ then there is a unique solution to (15.36) (up to scaling). Once the filter has been found, the locations $\{t_\ell\}$ are retrieved from the zeros $\{u_\ell\}$ of (15.34) via $u_\ell = e^{-j\omega_\ell}$ and $\omega_\ell = 2\pi t_\ell/\tau$. Given the locations, the weights $\{a_\ell\}$ can be obtained by (15.31).

Proposition 15.2. *Suppose that the values of $Y[k]$ in (15.36) are not all equal to zero and let $\mathbf{Y} = \mathbf{Y}_{s \times n}$ be the matrix defined by (15.37). Then we can express $\mathbf{Y}$ as*

$$\mathbf{Y} = \mathbf{V}(\{\omega_\ell\})_{s \times L} \, \mathrm{diag}\,(\mathbf{a})\mathbf{V}^T(\{\omega_\ell\})_{n \times L}, \tag{15.39}$$

where $\mathbf{a}$ is the length-L vector of coefficients a_ℓ. In addition, if $s, n \geq L$ then the rank of $\mathbf{Y}$ satisfies $\mathrm{rank}(\mathbf{Y}) = L$.

Proposition 15.2 is related to the well-known Carathéodory–Toeplitz theorem [193, 434] which states that any nonnegative definite Toeplitz matrix can be represented in a similar form as (15.39) with $a_\ell \geq 0$ and $\mathbf{V}^T$ replaced by $\mathbf{V}^*$.

Proof: To prove the proposition, we note that for any two matrices $\mathbf{A}, \mathbf{B}$ with L columns $\mathbf{a}_\ell, \mathbf{b}_\ell$ and a diagonal matrix $\mathbf{D} = \mathrm{diag}\,(\mathbf{d})$ with diagonal elements d_ℓ we have

$$\mathbf{A} \, \mathrm{diag}\,(\mathbf{d})\mathbf{B}^T = \sum_{\ell=1}^{L} d_\ell \mathbf{a}_\ell \mathbf{b}_\ell^T. \tag{15.40}$$

Let $\mathbf{C} = \mathbf{V}(\{\omega_\ell\})_{s \times L} \, \mathrm{diag}\,(\mathbf{a})\mathbf{V}^T(\{\omega_\ell\})_{n \times L}$. Using (15.40), we can write

$$\mathbf{C} = \sum_{\ell=1}^{L} a_\ell \mathbf{e}(\omega_\ell)_s \mathbf{e}^T(\omega_\ell)_n, \tag{15.41}$$

where $\mathbf{e}(\omega_\ell)_s$ is the length-s vector defined by

$$\mathbf{e}(\omega_\ell)_s = \begin{bmatrix} 1 \\ e^{-j\omega_\ell} \\ \vdots \\ e^{-j\omega_\ell(s-1)} \end{bmatrix} = \begin{bmatrix} 1 \\ u_\ell \\ \vdots \\ u_\ell^{(s-1)} \end{bmatrix}. \tag{15.42}$$

Algorithm 15.2 Prony's method (combined with total least squares)

Input: $m \geq 2L$ measurements $Y[k]$, $k = 0, \ldots, m - 1$, number of delays L
Output: Time delays t_ℓ, $\ell = 1, \ldots, L$
Build the measurement matrix $\mathbf{Y}_{(m-L) \times (L+1)}$ of (15.37)
Choose $\mathbf{g} = [g_L, \ldots, g_1, g_0]^T$ as a right singular vector of $\mathbf{Y}_{(m-L) \times (L+1)}$ corresponding to its smallest singular value
Find the L roots $\{u_\ell\}$ of $G(z) = \sum_{\ell=0}^{L} g_\ell z^{-\ell}$
Determine t_ℓ via $u_\ell = e^{-j\omega_\ell}$ and $\omega_\ell = 2\pi t_\ell / \tau$

Therefore, the kmth value of $\mathbf{C}$ for $k = 0, \ldots, s - 1$ and $m = 0, \ldots, n - 1$ is equal to

$$c_{km} = \sum_{\ell=1}^{L} a_\ell u_\ell^k u_\ell^m = \sum_{\ell=1}^{L} a_\ell u_\ell^{k+m} = Y[k + m], \tag{15.43}$$

where we used (15.32). From (15.38) it follows that $\mathbf{C} = \mathbf{Y}$.

We next exploit (15.39) to prove that $\text{rank}(\mathbf{Y}) = L$. From Proposition 15.1, $\mathbf{V}(\{\omega_\ell\})_{s \times L}$ has full column-rank equal to L and $\mathbf{V}^T(\{\omega_\ell\})_{n \times L}$ has full row-rank equal to L. Finally, since $a_\ell \neq 0$, $\text{diag}(\mathbf{a})$ is invertible so that its rank is also equal to L. We now rely on the following lemma, whose proof is straightforward (see Exercise 7).

Lemma 15.1. *Let $\mathbf{A}_{n \times m}$ with $n \geq m$ be a full column-rank matrix, let $\mathbf{C}_{m \times k}$ with $m \leq k$ be a full row-rank matrix, and let $\mathbf{B}$ be a matrix with rank r. Then the rank of $\mathbf{ABC}$ is equal to r.*

From Lemma 15.1 and (15.39) it follows immediately that the rank of $\mathbf{Y}$ equals L, as long as $n, s \geq L$. $\qquad \square$

Corollary 15.1. *Suppose that the values of $Y[k]$ in (15.36) are not equal to zero and let $m \geq 2L$. Then there is a unique (nonzero) solution to (15.36) (up to scaling).*

Proof: From Proposition 15.2, the rank of $\mathbf{Y}_{(m-L) \times (L+1)}$ is equal to L as long as $m - L \geq L$, or $m \geq 2L$. The dimension of the null space is therefore equal to $L + 1 - L = 1$. $\qquad \square$

Prony's method is summarized in Algorithm 15.2. In the literature, this approach is also referred to as the *annihilating filter method*. In line 4 of the algorithm description we included the total least squares step which will be described in the next subsection. In the noise-free setting in which there exists a $\mathbf{g}$ solving $\mathbf{Y}_{(m-L) \times (L+1)} \mathbf{g} = \mathbf{0}$, this step is identical to finding $\mathbf{g}$ in the null space of $\mathbf{Y}$.

The annihilating filter technique can be improved by noticing that in addition to (15.35) we also have the relation:

$$\sum_{i=0}^{L} g_i \overline{Y[k + i]} = \sum_{i=0}^{L} \sum_{\ell=1}^{L} \overline{a_\ell} g_i u_\ell^{-(k+i)} = \sum_{\ell=1}^{L} \overline{a_\ell} u_\ell^{-k} \sum_{i=0}^{L} g_i u_\ell^{-i} = 0, \tag{15.44}$$

for $0 \le k \le m - L - 1$. This equality can be written in matrix/vector form as

$$\overline{\mathbf{Y}}_{(m-L)\times(L+1)}\mathbf{J}\mathbf{g} = \mathbf{0}, \tag{15.45}$$

where

$$\mathbf{J} = \begin{bmatrix} 0 & \cdots & 0 & 1 \\ 0 & \cdots & 1 & 0 \\ \vdots & & & \vdots \\ 1 & \cdots & 0 & 0 \end{bmatrix}. \tag{15.46}$$

The matrix $\mathbf{J}$ flips the elements in the vector $\mathbf{g}$ so that $\mathbf{J}\mathbf{g} = [g_0, g_1, \ldots, g_L]^T$. Alternatively, $\overline{\mathbf{Y}}\mathbf{J}$ flips the order of the columns in $\overline{\mathbf{Y}}$. Combining (15.36) and (15.45) leads to more equations, which may be useful in order to render the solution more robust in the presence of noise.

15.2.4 Noisy samples

When the samples are noisy, several possible modifications to Prony's method have been suggested in order to improve the robustness of the estimates.

Total least squares approach

One possibility is to combine Prony's algorithm with the total least squares (TLS) [435] approach. Specifically, in the presence of noise, the measurements $\{Y[k]\}$ are not known precisely, and consequently we only have access to a noisy version of $\mathbf{Y}$, which we denote here by $\widetilde{\mathbf{Y}}$. For notational brevity, we have omitted the indices. Thus, instead of having $\mathbf{Y}\mathbf{g} = \mathbf{0}$ as in (15.36) we have $\widetilde{\mathbf{Y}}\mathbf{g} \approx \mathbf{0}$. The TLS technique seeks the vector $\mathbf{g}$ that minimizes the squared-norm of $\widetilde{\mathbf{Y}}\mathbf{g}$. To preclude the trivial solution $\mathbf{g} = \mathbf{0}$, without loss of generality, we constrain the norm of $\mathbf{g}$ to be equal to 1, resulting in

$$\min_{\mathbf{g}} \|\widetilde{\mathbf{Y}}\mathbf{g}\|^2 \quad \text{s.t. } \|\mathbf{g}\|^2 = 1. \tag{15.47}$$

Clearly, if $\widetilde{\mathbf{Y}} = \mathbf{Y}$ so that there is no noise, then the minimum value is 0 and $\mathbf{g}$ is a normalized vector in the null space of $\mathbf{Y}$.

From the properties of the SVD (see Appendix A) it follows that the vector solving (15.47) is given by the right singular vector corresponding to the smallest singular value of $\widetilde{\mathbf{Y}}$. Once the filter $\mathbf{g}$ is found, one can determine its roots and identify the time delays, as in Prony's method.

Although Prony's algorithm is simple, when the measurements $Y[k]$ are noisy, the error in estimating $\{t_\ell\}$ can be quite large even when the TLS approach is used, as we show below in Example 15.3. Furthermore, it is known that the resulting estimates are not consistent [436], namely, the estimates do not necessarily converge in probability to their true values as the number of samples increases.

Algorithm 15.3 Cadzow denoising

Input: Noisy measurement matrix $\widetilde{\mathbf{Y}}$
Output: Denoised matrix $\widehat{\mathbf{Y}}$ (which can then be used in Prony's method)
Initialize: $\widehat{\mathbf{Y}}_0 = \widetilde{\mathbf{Y}}, \ell = 0$
while halting criterion false **do**
   $\ell \leftarrow \ell + 1$
   Compute the SVD $\widehat{\mathbf{Y}}_\ell = \mathbf{USV}^*$, where $\mathbf{U}$ and $\mathbf{V}$ are unitary and $\mathbf{S}$ is diagonal
   Build the diagonal matrix $\mathbf{S}'$ consisting of the L largest elements in $\mathbf{S}$, and zero
     elsewhere
   $\mathbf{B} = \mathbf{US}'\mathbf{V}^*$ {construct best rank-L approximation}
   $\mathbf{A} = \text{Hankel}(\mathbf{B})$ {construct best Hankel approximation}
   $\widehat{\mathbf{Y}}_\ell \leftarrow \mathbf{A}$
end while
return $\widehat{\mathbf{Y}} \leftarrow \widehat{\mathbf{Y}}_\ell$

Cadzow iterative denoising algorithm

One way to improve the performance of TLS is to reduce the noise prior to applying (15.47). Tufts and Kumaresan [437], and Rahman and Yu [438], proposed exploiting the fact that the noise-free matrix $\mathbf{Y}_{(m-L)\times(L+1)}$ of (15.37) has rank L (for $m \geq 2L$), as we showed in Proposition 15.2 (and Corollary 15.1). This fact can be used to first approximate the noisy matrix $\widetilde{\mathbf{Y}}$ by its best rank-L approximation using the SVD (see Theorem A.4 in Appendix A), and then apply the TLS method to the resulting rank-L matrix. Since $\widetilde{\mathbf{Y}}$ is of size $(m - L) \times (L + 1)$ its rank is upper-bounded by $\text{rank}(\widetilde{\mathbf{Y}}) \leq \min(m-L, L+1)$. If $m = 2L$, then $\text{rank}(\widetilde{\mathbf{Y}}) \leq L$ so that the best rank-L approximation is the matrix itself. Therefore, this approach is beneficial only when $m > 2L$.

A further improvement was suggested by Cadzow [439]; in addition to the rank, Cadzow proposed incorporating the fact that $\mathbf{Y}$ is a Hankel matrix as we saw in (15.38). Cadzow denoising consists of first finding a rank-L Hankel matrix that is as close as possible to $\widetilde{\mathbf{Y}}$ by solving

$$\min_{\widehat{\mathbf{Y}}} \|\widetilde{\mathbf{Y}} - \widehat{\mathbf{Y}}\|_F^2 \quad \text{s.t. } \text{rank}(\widehat{\mathbf{Y}}) \leq L \text{ and } \widehat{\mathbf{Y}} \text{ is Hankel}, \quad (15.48)$$

and then using the resulting estimate with TLS. In the objective, $\|\mathbf{A}\|_F^2 = \text{Tr}(\mathbf{A}^*\mathbf{A})$ denotes the Frobenius norm. This approximation is meaningful even when m is equal to its minimal value $m = 2L$.

An approximate solution to (15.48) can be computed by alternating between finding the best rank-L estimate, and determining the optimal Hankel approximation. Thus, for a given target matrix $\mathbf{C}$, we independently solve the two optimization problems

$$\min_{\mathbf{B}} \|\mathbf{C} - \mathbf{B}\|_F^2 \quad \text{s.t. } \text{rank}(\mathbf{B}) = L \quad (15.49)$$

and

$$\min_{\mathbf{A}} \|\mathbf{B} - \mathbf{A}\|_F^2 \quad \text{s.t. } \mathbf{A} \text{ is Hankel}. \quad (15.50)$$

To solve (15.49), we compute the SVD $\mathbf{C} = \mathbf{U}\mathbf{S}\mathbf{V}^*$ of $\mathbf{C}$, where $\mathbf{U}$ and $\mathbf{V}$ are unitary and $\mathbf{S}$ is a diagonal matrix whose diagonal entries are the singular values of $\mathbf{C}$. We then discard all but the L largest singular values in $\mathbf{S}$. In other words, we construct a diagonal matrix $\mathbf{S}'$ whose diagonal contains the L largest entries in $\mathbf{S}$, and is zero elsewhere. The rank-L matrix closest to $\mathbf{C}$ is then given by $\mathbf{B} = \mathbf{U}\mathbf{S}'\mathbf{V}^*$ (see Appendix A). The solution to (15.50) is easily obtained by averaging the skew-diagonals of $\mathbf{B}$ (see Exercise 8). The resulting iterative algorithm is summarized in Algorithm 15.3. In the algorithm description, Hankel($\mathbf{B}$) denotes the operation of replacing $\mathbf{B}$ by a Hankel matrix obtained by averaging the skew-diagonals.

Applying even a small number of iterations of Cadzow's algorithm tends to yield a matrix whose error with respect to the true unknown data matrix is much lower than the error of the original measurement matrix $\widetilde{\mathbf{Y}}$. The denoised matrix $\widehat{\mathbf{Y}}$ can then be used in conjunction with the TLS technique.

We note that from Proposition 15.2, the noise-free matrix $\mathbf{Y}_{s \times n}$ of (15.37) has rank equal to L, as long as $n, s \geq L$. Therefore, we can apply Cadzow iterations to the matrix $\widetilde{\mathbf{Y}}_{s \times n}$ with any $n, s \geq L$. In order to sum over a large number of values when averaging the diagonals, in the simulations below we choose $\widetilde{\mathbf{Y}}$ to be a square matrix of dimensions $n = s = \lfloor m/2 \rfloor$. After denoising, we reshape the data matrix to have dimensions $(m - L) \times (L+1)$ and then use TLS and Prony's method. As the number of samples m and/or delays L increase, the superiority of choosing $\widetilde{\mathbf{Y}}$ to be square for denoising is more noticeable.

Example 15.3 In this example we examine Prony's method combined with the TLS approach and Cadzow denoising. In particular, we demonstrate the benefits of using Cadzow denoising and its effects on estimation accuracy. We will also highlight some drawbacks of this approach.

The signal we consider is the same as in Example 15.2: $x(t) = \sum_{\ell=1}^{L} \sum_{n \in \mathbb{Z}} a_\ell$ $h(t - t_\ell - n\tau)$ along the period $\tau = 1$ where $a_\ell = 1$, $t_\ell = \ell\Delta$, and $\Delta = 0.025$. The samples are corrupted by noise (in the frequency domain) so that we are given $m \geq 2L$ noisy versions of the coefficients $Y[k], k \in \mathcal{K}$ where $\mathcal{K}$ is a consecutive index set of size m. The additive noise is complex white Gaussian with variance chosen to match the required SNR. The noisy matrix coefficients (which we earlier denoted by $\widetilde{\mathbf{Y}}$) are denoised using Cadzow's algorithm. The delays are then estimated using Prony's method combined with the TLS approach. Finally the amplitudes are given by (15.31).

To evaluate the performance we computed the MSE in estimating the delays and amplitudes by performing 1000 simulations for each SNR value and for different numbers of Cadzow iterations, denoted by J. One of the drawbacks of Cadzow's algorithm is that in some cases it leads to missed detections: two of the zeros become identical so that one (or more) of the delays is completely missed. In calculating the error, missed detections were not taken into account. Thus, we chose 1000 simulations in which no missed detections occurred.

We begin by considering $L = 6$ delays and $m = 81$ samples. Note that as explained in Example 15.2, this corresponds to taking samples at the Nyquist rate

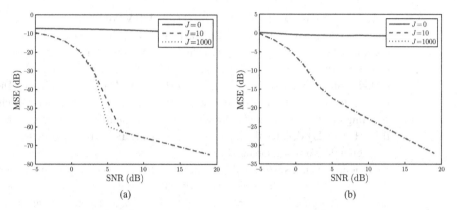

Figure 15.6 Delay and amplitude estimation using Cazdow denoising for $L = 6$ delays and $m = 81$ samples. (a) MSE in estimating the delays. (b) MSE in estimating the amplitudes.

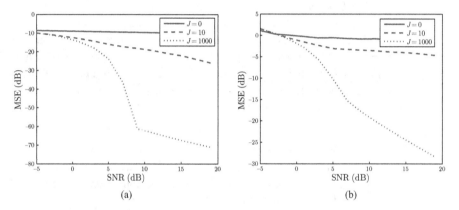

Figure 15.7 Delay and amplitude estimation using Cadzow denoising for $L = 12$ delays and $m = 81$ samples. (a) MSE in estimating the delays. (b) MSE in estimating the amplitudes.

of $x(t)$. In Fig. 15.6 we plot the MSE (in log scale) in estimating the delays and amplitudes as a function of the SNR for different numbers of Cadzow iterations: $J = 0, 10, 1000$. Choosing $J = 0$ results in Prony's method. The figure illustrates that even a small number of Cadzow iterations produces a significant improvement in MSE. It can also be observed that for high enough SNR, proper estimation of the delays and amplitudes is attainable. On the other hand, Prony's method combined with TLS without performing Cadzow denoising is unreliable. In comparison with Fig. 15.5(c) and (d), we see that the Cadzow algorithm is more effective than the MF, even at the Nyquist rate.

When we increase the number of delays, the MSE deteriorates. Figure 15.7 shows the performance with $L = 12$. Although good performance is still achievable, it requires increasing the number of Cadzow iterations markedly. In Fig. 15.8 we plot the performance for different values of L, using 1000 Cadzow iterations. As the number of delays increases, the performance of Cadzow denoising worsens.

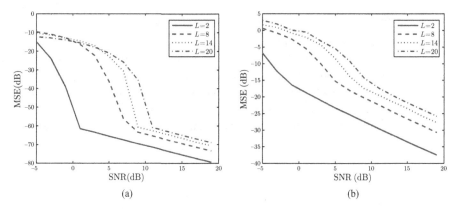

Figure 15.8 Delay and amplitude estimation using 1000 Cadzow iterations for varying values of L and $m = 81$ samples. (a) MSE in estimating the delays. (b) MSE in estimating the amplitudes.

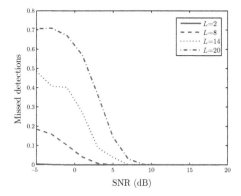

Figure 15.9 Missed detection rates for different number of delays.

The main drawback of using a large number of Cadzow iterations is the missed detection rate. To illustrate this, we calculated the number of missed detections in the simulations performed above, and plotted the missed detection rate in Fig. 15.9 as a function of SNR. The rate was calculated as the number of iterations with misses divided by the total number of iterations.

To conclude, we introduced Prony's method and its extensions to the noisy setting (TLS and Cadzow denoising). These are two-step approaches for evaluating the frequencies ω_ℓ in (15.26), or zeros u_ℓ in (15.32): first, we seek the coefficients of the annihilating filter, and then we find roots of a polynomial. These techniques are therefore also referred to as *polynomial methods*. A drawback of this approach is that finding the roots of a polynomial becomes difficult for large values of L and tends to be sensitive to noise. Alternative algorithms that do not require polynomial rooting include matrix pencil and subspace methods which we discuss next.

15.2.5 Matrix pencil

Matrix pencil algorithm

The matrix pencil [440, 441] algorithm is a one-step process of solving a generalized eigenvalue problem that directly yields the values u_ℓ. This approach is not only more computationally efficient, but it can be shown to lead to better statistical properties than polynomial-based techniques.

We begin by constructing a data matrix $\mathbf{Y}_{(m-M)\times(M+1)}$ from the m measurements where

$$L \leq M \leq m - L \tag{15.51}$$

is the *pencil parameter* and is very useful in eliminating noise in the data as we show below. (In Prony's method and its extensions M was set equal to the number of exponents L.) The condition on M will be made clear in Proposition 15.4. In particular, it implies that we must have $m \geq 2L$. We next create two matrices $\mathbf{Y}_1$ and $\mathbf{Y}_2$ of size $m - M \times M$ by choosing the first and last M columns of $\mathbf{Y}_{(m-M)\times(M+1)}$ respectively. Writing out the matrices explicitly, we have

$$\mathbf{Y}_1 = \begin{bmatrix} Y[0] & Y[1] & \cdots & Y[M-1] \\ Y[1] & Y[2] & \cdots & Y[M] \\ \vdots & \vdots & \ddots & \vdots \\ Y[m-M-1] & Y[m-M] & \cdots & Y[m-2] \end{bmatrix},$$

$$\mathbf{Y}_2 = \begin{bmatrix} Y[1] & Y[2] & \cdots & Y[M] \\ Y[2] & Y[3] & \cdots & Y[M+1] \\ \vdots & \vdots & \ddots & \vdots \\ Y[m-M] & Y[m-M+1] & \cdots & Y[m-1] \end{bmatrix}. \tag{15.52}$$

Note that the knth element of $\mathbf{Y}_i$ is given by

$$\mathbf{Y}_i(k,n) = Y[k+n+i-1], \quad 0 \leq k \leq m-M-1 \text{ and } 0 \leq n \leq M-1. \tag{15.53}$$

The matrix pencil approach exploits the fact that the data matrices $\mathbf{Y}_i$ have a simple decomposition in terms of the Vandermonde matrices $\mathbf{V}(\{\omega_\ell\})_{m-M\times L}$ of (15.30):

Proposition 15.3. *Let $Y[k]$ satisfy (15.26). Then the matrices $\mathbf{Y}_1, \mathbf{Y}_2$ defined by (15.52) can be written as*

$$\mathbf{Y}_1 = \mathbf{V}(\{\omega_\ell\})_{m-M\times L} \operatorname{diag}(\mathbf{a}) \mathbf{V}^T(\{\omega_\ell\})_{M\times L}$$
$$\mathbf{Y}_2 = \mathbf{V}(\{\omega_\ell\})_{m-M\times L} \operatorname{diag}(\mathbf{a}) \operatorname{diag}(\mathbf{u}) \mathbf{V}^T(\{\omega_\ell\})_{M\times L}, \tag{15.54}$$

where $\mathbf{a}$ is the length-L vector of coefficients a_ℓ, and $\mathbf{u}$ is the length-L vector with elements $u_\ell = e^{-j\omega_\ell}$.

Proof: The result for $\mathbf{Y}_1$ follows immediately from Proposition 15.2. To prove the result for $\mathbf{Y}_2$, let $\mathbf{C} = \mathbf{V}(\{\omega_\ell\})_{m-M\times L} \operatorname{diag}(\mathbf{a}) \operatorname{diag}(\mathbf{u}) \mathbf{V}^T(\{\omega_\ell\})_{M\times L}$. Using

(15.40) we can write

$$C = \sum_{\ell=1}^{L} a_\ell u_\ell e(\omega_\ell)_{m-M} e^T(\omega_\ell)_M, \tag{15.55}$$

where $e(\omega_\ell)_s$ is defined by (15.42). Therefore, the knth value of C for $0 \le k \le m - M - 1$ and $0 \le n \le M - 1$ is equal to

$$c_{kn} = \sum_{\ell=1}^{L} a_\ell u_\ell u_\ell^k u_\ell^n = \sum_{\ell=1}^{L} a_\ell u_\ell^{k+n+1} = Y[k+n+1], \tag{15.56}$$

where we applied (15.26). From (15.53), $C = Y_2$. □

Based on Proposition 15.3 we consider the matrix pencil

$$Q(\lambda) = Y_2 - \lambda Y_1 = V(\{\omega_\ell\})_{m-M \times L} \operatorname{diag}(a)(\operatorname{diag}(u) - \lambda I)V^T(\{\omega_\ell\})_{M \times L}. \tag{15.57}$$

In Proposition 15.4 below we show that when λ is equal to one of the roots u_ℓ, $Q(\lambda)$ loses rank. This property can be used to efficiently find the values of u_ℓ by a generalized eigendecomposition, as shown in Corollary 15.2.

Proposition 15.4. *Let $Q(\lambda)$ be the matrix pencil defined by (15.57) with Y_1, Y_2 given by (15.52), and let $L \le M \le m - L$. Define $r = \operatorname{rank}(Q(\lambda))$. Then*

- *If $\lambda \ne u_\ell$ then $r = L$;*
- *If $\lambda = u_\ell$ for some ℓ then $r = L - 1$.*

In addition, if $Q(\lambda)x = 0$ for a nonzero vector x in $\mathcal{N}(Y_2)^\perp$, then $\lambda = u_\ell$ for some ℓ.

Proof: The proof of the proposition is similar to that of Proposition 15.2. Specifically, using Proposition 15.1 we conclude that $V(\{\omega_\ell\})_{m-M \times L}$ has rank L (since $m - M \ge L$) and $V^T(\{\omega_\ell\})_{M \times L}$ has rank L (since $M \ge L$).

Denote $Z(\lambda) = \operatorname{diag}(a)(\operatorname{diag}(u) - \lambda I)$. Since $Z(\lambda)$ is an $L \times L$ diagonal matrix and $a_\ell \ne 0$, its rank is equal to L for $\lambda \ne u_\ell$, and $L - 1$ for $\lambda = u_\ell$. From Lemma 15.1 (see the proof of Proposition 15.2) it follows immediately that the rank of $Q(\lambda)$ is equal to that of $Z(\lambda)$.

Finally, from the decomposition (15.54) of Y_2 and the fact that $V(\{\omega_\ell\})_{m-M \times L}$ has full column-rank, we conclude that $\mathcal{N}(Y_2) = \mathcal{N}(V^T(\{\omega_\ell\})_{M \times L})$. Suppose that x is a nonzero vector in $\mathcal{N}(Y_2)^\perp$. Then, $z = V^T(\{\omega_\ell\})_{M \times L}x \ne 0$ so that $Q(\lambda)x = 0$ only if $Z(\lambda)z = 0$. This can only happen if Z does not have full rank, so that $\lambda = u_\ell$ for some ℓ. □

Note that from the proof of the proposition condition (15.51) on M is needed in order to ensure that $Q(\lambda)$ has rank L for all $\lambda \ne u_\ell$.

Proposition 15.4 can be used to establish that the values of u_ℓ are simply the nonzero eigenvalues of $C = Y_1^\dagger Y_2$.

Corollary 15.2. *Let C be the $M \times M$ matrix defined by $C = Y_1^\dagger Y_2$. Then the M eigenvalues of C are equal to $\lambda_\ell = u_\ell$ for $\ell = 1, \ldots, L$ and $\lambda_\ell = 0$ for $\ell = L + 1, \ldots, M$.*

Algorithm 15.4 Matrix pencil

Input: $m \geq 2L$ measurements $Y[k], k = 0, \ldots, m - 1$, number of delays L
Output: Time delays $t_\ell, \ell = 1, \ldots, L$
Build the measurement matrix $\mathbf{Y}_{(m-M) \times (M+1)}$ of (15.37) for $L \leq M \leq m - L$
Construct the matrices $\mathbf{Y}_1$ and $\mathbf{Y}_2$ as the first and last M columns of $\mathbf{Y}$ respectively
Compute the L nonzero eigenvalues $\{u_\ell\}$ of $\mathbf{C} = \mathbf{Y}_1^\dagger \mathbf{Y}_2$
Determine t_ℓ via $u_\ell = e^{-j\omega_\ell}$ and $\omega_\ell = 2\pi t_\ell / \tau$

Proof: From (15.54), $\mathcal{R}(\mathbf{Y}_1) = \mathcal{R}(\mathbf{Y}_2)$ and $\dim(\mathcal{N}(\mathbf{Y}_2)) = M - L$. Since $\mathcal{N}(\mathbf{Y}_1^\dagger) = \mathcal{R}(\mathbf{Y}_1)^\perp = \mathcal{R}(\mathbf{Y}_2)^\perp$, we conclude that $\mathcal{N}(\mathbf{C}) = \mathcal{N}(\mathbf{Y}_2)$. Indeed, for any $\mathbf{x} \in \mathcal{N}(\mathbf{Y}_2)^\perp$, the nonzero vector $\mathbf{Y}_2 \mathbf{x}$ is in $\mathcal{R}(\mathbf{Y}_2) = \mathcal{N}(\mathbf{Y}_1^\dagger)^\perp$ and consequently $\mathbf{C}\mathbf{x} \neq \mathbf{0}$. We conclude that $\mathcal{N}(\mathbf{C})$ has dimension $M - L$ so that $\mathbf{C}$ has $M - L$ eigenvalues equal to 0.

Consider now an eigenvector of $\mathbf{C}$ corresponding to a nonzero eigenvalue λ such that $\mathbf{C}\mathbf{x} = \lambda \mathbf{x}$ with $\mathbf{x} \in \mathcal{N}(\mathbf{C})^\perp = \mathcal{N}(\mathbf{Y}_2)^\perp$. This then implies that

$$\mathbf{Y}_1^\dagger \mathbf{Y}_2 \mathbf{x} = \lambda \mathbf{x}. \tag{15.58}$$

Multiplying the equation on both sides by $\mathbf{Y}_1$ we have

$$\mathbf{Y}_1 \mathbf{Y}_1^\dagger \mathbf{Y}_2 \mathbf{x} = \lambda \mathbf{Y}_1 \mathbf{x}. \tag{15.59}$$

Noting that $\mathbf{Y}_1 \mathbf{Y}_1^\dagger = P_{\mathcal{R}(\mathbf{Y}_1)} = P_{\mathcal{R}(\mathbf{Y}_2)}$ it follows that $\mathbf{Y}_1 \mathbf{Y}_1^\dagger \mathbf{Y}_2 = \mathbf{Y}_2$. Therefore, (15.59) can be written as

$$(\mathbf{Y}_2 - \lambda \mathbf{Y}_1)\mathbf{x} = \mathbf{Q}(\lambda)\mathbf{x} = \mathbf{0}. \tag{15.60}$$

Since $\mathbf{x} \in \mathcal{N}(\mathbf{Y}_2)^\perp$ it follows from Proposition 15.4 that (15.60) has a nonzero solution $\mathbf{x}$ only if $\lambda = u_\ell$. $\square$

The matrix pencil method is summarized in Algorithm 15.4. When the data is noisy we can first apply the Cadzow denoising approach to $\mathbf{Y}_{(m-M) \times (M+1)}$ of (15.37) and then use the denoised matrix to construct $\mathbf{Y}_1$ and $\mathbf{Y}_2$ by taking the appropriate columns. For efficient noise filtering, it has been observed empirically that a good choice of M is $m/3 \leq M \leq m/2$.

Example 15.4 We now evaluate the performance of matrix pencil and compare it with Prony's method using the same setting as Example 15.3.

In Fig. 15.10, we plot the MSE (in log scale) as a function of SNR for different choices of the pencil parameter M, where the results are averaged over 4000 simulations. We choose $L = 6$ delays and $m = 81$ samples. In each simulation the matrices $\mathbf{Y}_1$ and $\mathbf{Y}_2$ are calculated and denoised by taking only the L largest entries in their SVD. The values of $\{u_\ell\}_{\ell=1}^L$ are determined by the L largest eigenvalues of $\mathbf{C} = \mathbf{Y}_1^\dagger \mathbf{Y}_2$. Finally the amplitudes are given by (15.31). The figure illustrates the importance of choosing M correctly. A popular rule of thumb is to select M

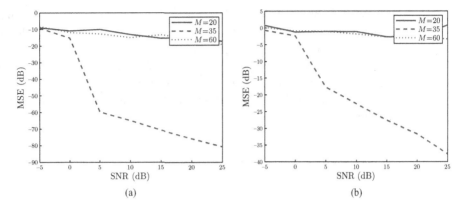

Figure 15.10 Delay and amplitude estimation using matrix pencil for $L = 6$ delays and $m = 81$ samples for various values of the pencil parameter M. (a) MSE in estimating the delays. (b) MSE in estimating the amplitudes.

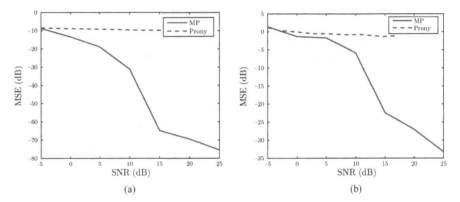

Figure 15.11 Delay and amplitude estimation using Prony's method and matrix pencil (MP) with pencil parameter $M = 35$ for $L = 12$ delays and $m = 81$ samples. (a) MSE in estimating the delays. (b) MSE in estimating the amplitudes.

approximately in the range $m/3 \leq M \leq m/2$, which is consistent with the results in the figure.

We next compare matrix pencil to Prony's method (where in order to prevent missed detections no Cadzow iterations are performed) for $M = 35$, $L = 12$ and $m = 81$. The results can be seen in Fig. 15.11 from which it is clear that the matrix pencil approach achieves much lower MSE in comparison with Prony's algorithm. Comparing with Fig. 15.7, at high SNR matrix pencil outperforms Cadzow denoising followed by Prony, and does not result in missed detections.

Unknown number of delays

Until now we have assumed that the number of delays L is known. When L is unknown in advance, its value can be estimated using the matrix pencil approach by considering the singular values of $\mathbf{Y} = \mathbf{Y}_{(m-M)\times(M+1)}$ or equivalently by examining the eigenvalues of $\mathbf{C} = \mathbf{Y}_1^\dagger \mathbf{Y}_2$. As we have seen in Corollary 15.2, $\mathbf{C}$ should consist of L nonzero eigenvalues. In the presence of noise, these L values will be noisy, and in addition the other $M - L$ zero eigenvalues will contain noise as well. Assuming that the noise is smaller than the signal, we choose L as the number of singular values that are beyond a threshold. The remaining singular values are assumed to be due to noise and are therefore set to zero.

Example 15.5 Suppose that number of delays L is unknown. We illustrate some considerations in choosing a proper threshold in order to estimate its value from the eigenvalues of $\mathbf{C}$ defined in Corollary 15.2.

We again choose $x(t)$ to be the same as in Example 15.2: $x(t) = \sum_{\ell=1}^{L} \sum_{n\in\mathbb{Z}} a_\ell$ $h(t - t_\ell - n\tau)$ along the period $\tau = 1$ where $a_\ell = 1, t_\ell = \ell\Delta$ for a chosen Δ. The samples are corrupted by noise (in the frequency domain) so that we are given $m \geq 2L$ noisy measurements of the coefficients $Y[k], k \in \mathcal{K}$ where $\mathcal{K}$ is a consecutive index set of size m. The noise is complex white Gaussian with variance chosen to match the required SNR. Assume for simplicity that m is even. When no information about L is available, the only choice of pencil parameter M satisfying (15.51) for any $L \leq m/2$ is $M = m/2$. We therefore use this value throughout the simulations.

In each simulation the eigenvalues of $\mathbf{C} = \mathbf{Y}_1^\dagger \mathbf{Y}_2$ are sorted according to their absolute values, and then normalized by the largest eigenvalue. We denote the normalized and sorted absolute eigenvalues by $\lambda_\ell \geq 0$. From Corollary 15.2, in the absence of noise, $\mathbf{C}$ should have L nonzero eigenvalues and $M - L$ zero eigenvalues. When noise is present in the measurements we expect to have L larger eigenvalues due to the delayed pulses and $M - L$ smaller eigenvalues corresponding to noise.

In order to set a proper threshold we examine the probability density function (pdf) of the smallest absolute eigenvalue due to the delayed pulses, λ_L, and of the largest absolute eigenvalue caused by the noise, λ_{L+1}. We approximate the pdfs empirically from 100,000 simulations for different numbers of delays, choices of Δ, and SNR levels. The pdfs of λ_L and λ_{L+1} are depicted in Fig. 15.12 (in the figure we see the sum of the pdfs) for $L = 2$, $\Delta = 0.5$, and two different levels of SNR. These results can be used to set an appropriate threshold, together with any prior information on the desired missed detection or false alarm rates. From the figure, it can be seen that as the SNR decreases the peaks widen so that separating them using a threshold becomes harder.

In Fig. 15.13 we plot the pdfs for $L = 4$, SNR = 10 dB, and two different values of Δ. Decreasing the distance Δ between pulses results in the same effect of peak widening. Similar behavior occurs when the number of delays increases, as demonstrated in Fig. 15.14. Here we plot the pdfs for SNR = 10 dB, $\Delta = 0.1$, and three different numbers of delays. When more delays are added, the peaks widen and become closer, leading to a more noticeable tradeoff between false alarms and missed detections.

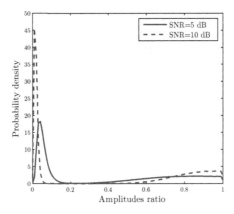

Figure 15.12 Probability density functions of λ_2 and λ_3 for two different SNR levels, with $L = 2$ and $\Delta = 0.5$. The left peak corresponds to the pdf of λ_3, while the right peak is the pdf of λ_2.

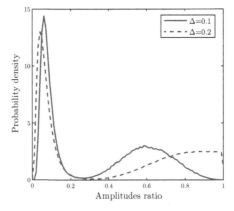

Figure 15.13 Probability density functions of λ_4 and λ_5 for two different values of Δ, with $L = 4$ and SNR $= 10$ dB. The left peak corresponds to the pdf of λ_5, while the right peak is the pdf of λ_4.

15.2.6 Subspace methods

Pisarenko algorithm

Another class of techniques for finding the frequencies in (15.26) is *subspace algorithms*. The first example of this approach is the Pisarenko method [442] which was later generalized to the popular MUltiple SIgnal Classification (MUSIC) algorithm. This class of techniques is based on the fact that any vector in the null space of an appropriate data matrix is orthogonal to the frequency vectors $\mathbf{e}(-\omega_\ell)$ defined in (15.42). The desired frequencies can therefore be found by exploiting this subspace orthogonality.

As in the Prony and TLS methods, we first find a vector in the null space of $\mathbf{Y}_{(m-L)\times(L+1)}$. When the data is noisy, this vector, denoted $\mathbf{v}_{\min}$, is chosen according to the TLS approach as the right singular vector corresponding to the smallest singular value of $\mathbf{Y}_{(m-L)\times(L+1)}$. But then, instead of using $\mathbf{v}_{\min}$ to construct the annihilating

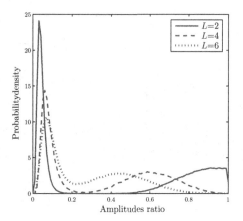

Figure 15.14 Probability density functions of λ_L and λ_{L+1} for three different numbers of delays, with $\Delta = 0.1$ and SNR $= 10$ dB. The left peak corresponds to the pdf of λ_{L+1}, while the right peak is the pdf of λ_L.

filter (15.34) and finding its zeros, we build a *pseudospectrum*

$$S(e^{j\omega}) = \frac{1}{|\mathbf{e}^*(-\omega)_{L+1}\mathbf{v}_{\min}|^2} \tag{15.61}$$

where $\mathbf{e}(-\omega)_{L+1}$ is defined in (15.42). The values of ω_ℓ are then chosen by looking for the peaks in $S(e^{j\omega})$.

The motivation behind the Pisarenko algorithm is that any vector in the null space of $\mathbf{Y} = \mathbf{Y}_{(m-L)\times(L+1)}$ is orthogonal to any vector in the range space of $\mathbf{Y}^*$ (since $\mathcal{N}(\mathbf{Y}) = \mathcal{R}(\mathbf{Y}^*)^\perp$). Furthermore, any vector of the form $\mathbf{e}(-\omega_\ell)_{L+1}$ is in the range of $\mathbf{Y}^*$, as incorporated in the following proposition.

Proposition 15.5. *Let $\mathbf{e}(-\omega)_{M+1}$ be defined by (15.42) and let $\mathbf{Y} = \mathbf{Y}_{(m-M)\times(M+1)}$ be defined by (15.37) for $L \leq M \leq m - L$. Then any vector of the form $\mathbf{e}(-\omega_\ell)_{M+1}$ is in $\mathcal{R}(\mathbf{Y}^*)$ for $m - M \geq L$.*

Proof: From (15.39),

$$\mathbf{Y}^*_{(m-M)\times(M+1)} = \overline{\mathbf{V}(\{\omega_\ell\})}_{M+1\times L}\,\mathrm{diag}\,(\mathbf{a})\mathbf{V}^*(\{\omega_\ell\})_{m-M\times L}. \tag{15.62}$$

Since $m-M \geq L$, the range of $\mathbf{V}^*(\{\omega_\ell\})_{m-M\times L}$ is equal to $\mathbb{C}^L$ (i.e. $\mathbf{V}^*(\{\omega_\ell\})_{m-M\times L}$ is of full row-rank). Together with the fact that $\overline{\mathrm{diag}\,(\mathbf{a})}$ is invertible, this implies that $\mathcal{R}(\mathbf{Y}^*_{(m-M)\times(M+1)})$ is equal to $\mathcal{R}(\overline{\mathbf{V}(\{\omega_\ell\})}_{M+1\times L})$. The result then follows from noting that the columns of $\overline{\mathbf{V}(\{\omega_\ell\})}_{M+1\times L}$ are equal to $\mathbf{e}(-\omega_\ell)_{M+1}$. $\qquad\square$

For noise free data, Proposition 15.5 implies that at $\omega = \omega_\ell$ we will see very strong peaks in the pseudospectrum (15.61). When the data is noisy, the peaks will be less evident, as seen in the following example.

Example 15.6 Consider the pseudospectrum (15.61) using the same settings as in Example 15.2 with delays $t_\ell = \ell\Delta$ and $a_\ell = 1$.

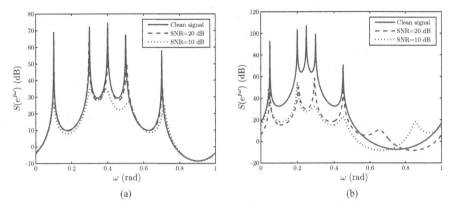

Figure 15.15 Pseudospectrum using Pisarenko's algorithm for different delays. (a) $\Delta = 0.1$. (b) $\Delta = 0.05$.

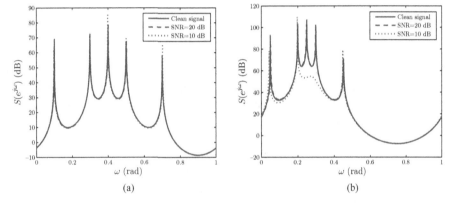

Figure 15.16 Pseudospectrum using Pisarenko's algorithm combined with Cadzow denoising for different delays. (a) $\Delta = 0.1$. (b) $\Delta = 0.05$.

In Fig. 15.15 we plot $S(e^{j\omega})$ for different choices of Δ and varying SNR with $L = 5$. The figure illustrates that decreasing Δ results in erroneous peak locations as well as missed detections. This phenomenon is more pronounced as the SNR decreases. Robustness to noise can be achieved by applying Cadzow denoising on $\mathbf{Y}_{(m-L)\times(L+1)}$ prior to computing its required singular vector, as can be seen in Fig. 15.16. In this example, 100 iterations of Cadzow were taken although in practice, after about 25 iterations, no significant improvement was observed.

MUSIC algorithm
Pisarenko's algorithm can be extended by noting that from Proposition 15.5, $\mathbf{e}(-\omega_\ell)_{M+1}$ is orthogonal to any vector in the null space of $\mathbf{Y} = \mathbf{Y}_{(m-M)\times(M+1)}$ where $L \leq M \leq m - L$. When $M = L$, the null space has dimension equal to 1 leading to Pisarenko's method. However, for larger values of M, the null space dimension grows, and is equal

---

Algorithm 15.5 MUSIC algorithm

Input: $m \geq 2L$ measurements $Y[k], k = 0, \ldots, m - 1$, number of delays L
Output: Time delays $t_\ell, \ell = 1, \ldots, L$
Build the measurement matrix $\mathbf{Y}_{(m-M) \times (M+1)}$ of (15.37) for $L \leq M \leq m - L$
Compute the SVD of $\mathbf{Y}$ and denote by $\{\mathbf{v}_i\}$ the right singular vectors corresponding
  to the $M - L + 1$ smallest singular values
Search for the L peaks ω_ℓ in the pseudospectrum $S(e^{j\omega})$ of (15.64) where $\mathbf{e}(\omega)$ is
  defined by (15.42)
Determine t_ℓ via $\omega_\ell = 2\pi t_\ell / \tau$

---

to $M - L + 1$. Therefore, there are $M - L + 1$ orthonormal vectors in the null space
that can be used.

Let $\{\mathbf{v}_i, 1 \leq i \leq M - L + 1\}$ be an orthonormal basis for $\mathcal{N}(\mathbf{Y})$. Then, in the absence
of noise,

$$\sum_{i=1}^{M-L+1} |\mathbf{e}^*(-\omega_\ell)\mathbf{v}_i|^2 = 0, \quad \ell = 1, \ldots, L. \tag{15.63}$$

For noisy data, (15.63) will not be satisfied with equality; however, we still expect the
left-hand side to be small. This leads to the *multiple signal classification (MUSIC)* [15]
approach in which the values of ω_ℓ are found by searching for the peaks in the pseudospectrum

$$S(e^{j\omega}) = \frac{1}{\sum_{i=1}^{M-L+1} |\mathbf{e}^*(-\omega)_{M+1}\mathbf{v}_i|^2}. \tag{15.64}$$

The vectors $\{\mathbf{v}_i\}$ can be determined by computing the SVD of $\mathbf{Y}$. Let $\mathbf{Y}$ have an SVD
$\mathbf{Y} = \mathbf{U}\boldsymbol{\Sigma}\mathbf{V}^*$, and let $\sigma_i, 1 \leq i \leq M + 1$ denote its sorted singular values. Then the
vectors $\{\mathbf{v}_i\}$ are the right singular vectors corresponding to the $M - L + 1$ smallest
singular values. When L is unknown, it can be estimated, e.g., as the number of singular
values above a threshold.

We summarize the MUSIC approach in Algorithm 15.5. For $M = L$, MUSIC coincides with Pisarenko's method. However, in MUSIC, we average the denominator over
all basis vectors in the null space and in that way improve the estimator's performance.

Example 15.7 We compare the MUSIC pseudospectrum (15.64) with that of (15.61)
by repeating the setting in Example 15.6.

In Fig. 15.17 we examine the pseudospectrum for different choices of the parameter M and the delay separation Δ with an SNR of 15 dB and $L = 5$. The number
of samples is equal to 81. As expected, averaging the denominator over all basis
vectors in the null space improves the estimator's robustness to noise. Choosing
$M = 5$ is equivalent to Pisarenko's method. When M is increased, the performance
improves. However, a decrease in Δ still results in erroneous locations and missed
detections.

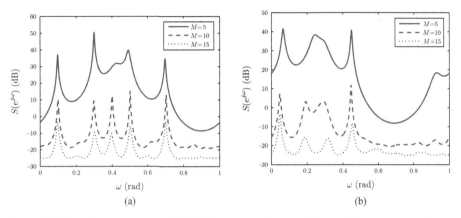

Figure 15.17 Pseudospectrum using MUSIC for varying choices of M and different delays. (a) $\Delta = 0.1$. (b) $\Delta = 0.05$.

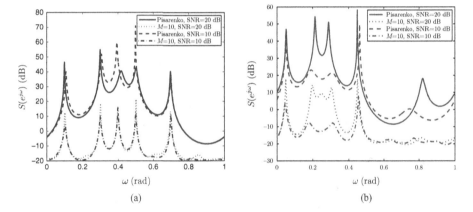

Figure 15.18 Pseudospectrum using Pisarenko and MUSIC for varying choices of SNR and different delays with $M = 10$ in MUSIC. (a) $\Delta = 0.1$. (b) $\Delta = 0.05$.

In Fig. 15.18 the pseudospectrum using the Pisarenko and MUSIC algorithms is depicted for different choices of Δ and varying SNR. We set $M = 10$ in computing the MUSIC spectrum. The superiority of MUSIC over Pisarenko's method in low SNR is clearly evident.

Finally, in Fig. 15.19 we repeat the previous simulations where we first apply Cadzow denoising to the data matrix $\mathbf{Y}_{(m-L)\times(L+1)}$. We performed 100 Cadzow iterations, although after the 30th iteration there was no significant improvement. The differences between the Pisarenko and MUSIC spectra are now much less pronounced.

Root-MUSIC algorithm

A popular variation on MUSIC is the *root-MUSIC algorithm* [443]. This method begins as the standard MUSIC technique by constructing the denominator in (15.64). However, instead of looking for the peaks of its inverse, we express the denominator as

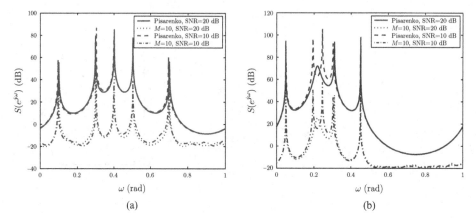

Figure 15.19 Pseudospectrum combined with Cadzow denoising using Pisarenko and MUSIC for varying choices of SNR and different delays with $M = 10$ in MUSIC. (a) $\Delta = 0.1$. (b) $\Delta = 0.05$.

a polynomial $D(z)$ of the form:

$$D(z) = \sum_{k=-L}^{L} b_k z^k,$$

(15.65)

and then search for its zeros, which leads to the desired values ω_ℓ.

Note that when $M = L$ root-MUSIC is the same as TLS Prony.

15.2.7 Covariance-based methods

The methods we have reviewed until now directly used the data matrix $\mathbf{Y}$ defined in (15.37) with appropriate dimensions. The same techniques can be applied where we replace the data matrix by a correlation matrix

$$\mathbf{R} = \begin{bmatrix} r[0] & r[1] & \cdots & r[M] \\ r[1] & r[2] & \cdots & r[M+1] \\ \vdots & \vdots & \ddots & \vdots \\ r[m-M-1] & r[m-M] & \cdots & r[m-1] \end{bmatrix}.$$

(15.66)

Covariance-based approaches can be derived by assuming that in (15.26), the amplitudes a_ℓ are zero-mean iid random variables with variance σ_ℓ^2. In this case, the correlation sequence $r[k]$ satisfies a sum-of-sinusoids equation as well. Indeed,

$$r[p] = E\left\{Y[k]\overline{Y[k-p]}\right\} = \sum_{i=1}^{L}\sum_{\ell=1}^{L} E\{a_\ell \overline{a_i}\} e^{-j\omega_\ell k} e^{j\omega_i(k-p)} = \sum_{\ell=1}^{L} \sigma_\ell^2 e^{-j\omega_\ell p},$$

(15.67)

which has the same form as (15.26) with new amplitudes σ_ℓ^2. Here we used the fact that $E\{a_\ell \overline{a_i}\} = \sigma_\ell^2 \delta_{\ell i}$. Therefore, instead of applying the methods detailed in the previous subsections to $\mathbf{Y}$ of (15.37), we can apply them to the correlation matrix $\mathbf{R}$ with the same dimensions. For example, using the matrix pencil approach with the correlation matrix leads to the well-known *estimation of signal parameters by rotational invariance (ESPRIT)* algorithm [187]. We discuss this method in more detail below.

Stochastic MUSIC

Both MUSIC and ESPRIT can be formulated for more general stochastic models, in which the coefficients $\{a_\ell\}$ do not have to be iid. Specifically, let

$$\mathbf{y} = \mathbf{V}(\{\omega_\ell\})_{m \times L}\mathbf{a} + \mathbf{w}, \tag{15.68}$$

where $\mathbf{V}(\{\omega_\ell\})_{m \times L}$ is the Vandermonde matrix defined by (15.30), $\mathbf{a}$ is a random vector representing the desired signal with zero mean and covariance $\mathbf{R}_a = E\{\mathbf{a}\mathbf{a}^*\}$, and $\mathbf{w}$ is a zero-mean iid noise vector with covariance $\sigma^2\mathbf{I}$, independent of $\mathbf{a}$. From (15.68), the covariance of $\mathbf{y}$ is given by:

$$\mathbf{R}_y = \mathbf{V}(\{\omega_\ell\})\mathbf{R}_a\mathbf{V}^*(\{\omega_\ell\}) + \sigma^2\mathbf{I}, \tag{15.69}$$

where we have omitted the indices in $\mathbf{V}(\{\omega_\ell\})$. Assuming that $m \geq L$ and that $\mathbf{R}_a$ is invertible, the matrix $\mathbf{T} = \mathbf{V}(\{\omega_\ell\})\mathbf{R}_a\mathbf{V}^*(\{\omega_\ell\})$ has rank equal to L and its range space is equal to that of $\mathbf{V}(\{\omega_\ell\})$. Therefore, the sorted eigenvalues of $\mathbf{T}$ satisfy $\lambda_\ell > 0, \ell = 1, \ldots, L$ and $\lambda_\ell = 0, \ell = L+1, \ldots, m$. The sorted eigenvalues of $\mathbf{R}_y$ are then equal to

$$\lambda'_\ell = \begin{cases} \lambda_\ell + \sigma^2, & \ell = 1, \ldots, L \\ \sigma^2, & \ell = L+1, \ldots, m. \end{cases} \tag{15.70}$$

Consider now a matrix $\mathbf{U}$ of size $m \times m - L$ that consists of the eigenvectors of $\mathbf{R}_y$ corresponding to the $m - L$ smallest eigenvalues, with $m \geq L+1$. These eigenvectors span the *noise space* since they are a result of noise only. Furthermore, they are orthogonal to the *signal space* determined by the range of $\mathbf{V}(\{\omega_\ell\})$. To see this, note that since $\mathbf{U}$ consists of eigenvectors with eigenvalues $\lambda'_\ell, \ell = L+1, \ldots, m$:

$$\mathbf{R}_y\mathbf{U} = \mathbf{U}\,\text{diag}\,(\lambda'_{L+1}, \ldots, \lambda'_m) = \sigma^2\mathbf{U}. \tag{15.71}$$

On the other hand,

$$\mathbf{R}_y\mathbf{U} = \mathbf{V}(\{\omega_\ell\})\mathbf{R}_a\mathbf{V}^*(\{\omega_\ell\})\mathbf{U} + \sigma^2\mathbf{U}, \tag{15.72}$$

from which we conclude that

$$\mathbf{V}(\{\omega_\ell\})\mathbf{R}_a\mathbf{V}^*(\{\omega_\ell\})\mathbf{U} = 0. \tag{15.73}$$

Since $\mathbf{R}_a$ is invertible and $\mathbf{V}(\{\omega_\ell\})$ has full column-rank, (15.73) implies that

$$\mathbf{V}^*(\{\omega_\ell\})\mathbf{U} = 0, \tag{15.74}$$

and $\mathbf{U}$ is orthogonal to the columns of $\mathbf{V}(\{\omega_\ell\})$. Note that (15.74) is identical to (15.63), where now we are considering the eigenvectors of the data covariance rather than the data itself.

Algorithm 15.6 Stochastic MUSIC algorithm

Input: N vectors $\mathbf{y}_i = \mathbf{V}(\{\omega_\ell\})_{m \times L} \mathbf{a}_i + \mathbf{w}_i$ with $m \geq L+1$, number of delays L

Output: Time delays $t_\ell, \ell = 1, \dots, L$

Build the correlation matrix $\mathbf{R}_y = \frac{1}{N} \sum_{i=0}^{N-1} \mathbf{y}_i \mathbf{y}_i^*$

Perform an eigendecomposition of $\mathbf{R}_y$ and construct the matrix $\mathbf{U}$ consisting of the
  $m - L$ eigenvectors associated with the smallest eigenvalues in its columns

Search for the L peaks ω_ℓ in the pseudospectrum $S(e^{j\omega})$ of (15.75) where $\mathbf{e}(\omega)$ is
  defined by (15.42)

Determine t_ℓ via $\omega_\ell = 2\pi t_\ell / \tau$

As before, we can determine the frequencies ω_ℓ by searching for peaks in the pseudospectrum

$$S(e^{j\omega}) = \frac{1}{\mathbf{e}^*(\omega)_m \mathbf{U} \mathbf{U}^* \mathbf{e}(\omega)_m} \tag{15.75}$$

where $\mathbf{e}(\omega)_m$ is defined in (15.42). Alternatively, we may use root-MUSIC and search for roots of the polynomial $\mathbf{e}^*(\omega)_m \mathbf{U} \mathbf{U}^* \mathbf{e}(\omega)_m$.

In practice, the correlation matrix $\mathbf{R}_y$ is estimated from the data, for example by using the sample correlation matrix

$$\mathbf{R}_y = \frac{1}{N} \sum_{i-0}^{N-1} \mathbf{y}_i \mathbf{y}_i^*, \tag{15.76}$$

where N is the number of snapshots available and $\mathbf{y}_i = \mathbf{V} \mathbf{a}_i + \mathbf{w}_i$ with $\mathbf{a}_i$ and $\mathbf{w}_i$ denoting iid realizations of the signal and noise respectively. This approach is summarized in Algorithm 15.6.

ESPRIT algorithm

The ESPRIT algorithm [187] exploits the special structure of the correlation matrix (15.69).

From (15.69) it follows that in the absence of noise, the range space of $\mathbf{R}_y$ is identical to that of $\mathbf{V} = \mathbf{V}(\{\omega_\ell\})$. As we noted earlier, this space is referred to as the *signal subspace* and can be determined by choosing the eigenvectors of $\mathbf{R}_y$ corresponding to its L largest eigenvalues. We now show that we can use the structure of $\mathbf{V}$ to extract the desired frequencies from the eigendecomposition of an appropriate matrix constructed from these eigenvectors.

Proposition 15.6. *Let $\mathbf{R}_y$ be the $m \times m$ covariance matrix given by (15.69), and let $\mathbf{E}$ denote the matrix of size $m \times L$ consisting of the eigenvectors corresponding to the L largest eigenvalues of $\mathbf{R}_y$. Let $\mathbf{E}_1$ be equal to the first $m-1$ rows of $\mathbf{E}$, and let $\mathbf{E}_2$ be equal to the last $m-1$ rows of $\mathbf{E}$. We assume that $m \geq L+1$. Then the eigenvalues of $\mathbf{E}_1^\dagger \mathbf{E}_2$ are equal to $\lambda_\ell = u_\ell = e^{-j\omega_\ell}$.*

Proof: Let $\mathbf{V}_1$ be the $m-1 \times L$ matrix equal to the first $m-1$ rows of $\mathbf{V} = \mathbf{V}(\{\omega_\ell\})$, and let $\mathbf{V}_2$ be the $m - 1 \times L$ matrix equal to the last $m - 1$ rows of $\mathbf{V}$. From the

---

Algorithm 15.7 ESPRIT algorithm

Input: N vectors $\mathbf{y}_i = \mathbf{V}(\{\omega_\ell\})_{m \times L} \mathbf{a}_i + \mathbf{w}_i$ with $m \geq L + 1$, number of delays L

Output: Time delays t_ℓ, $\ell = 1, \ldots, L$

Build the correlation matrix $\mathbf{R}_y = \frac{1}{N} \sum_{i=0}^{N-1} \mathbf{y}_i \mathbf{y}_i^*$

Perform an eigendecomposition of $\mathbf{R}_y$ and construct the matrix $\mathbf{E}$ consisting of the L eigenvectors associated with the largest eigenvalues in its columns

Compute the matrix $\mathbf{C} = \mathbf{E}_1^\dagger \mathbf{E}_2$ where $\mathbf{E}_1$ and $\mathbf{E}_2$ consist of the first and last $m - 1$ rows of $\mathbf{E}$ respectively

Compute the L eigenvalues $\{\lambda_\ell = u_\ell\}$ of $\mathbf{C}$

Determine t_ℓ via $u_\ell = e^{-j\omega_\ell}$ and $\omega_\ell = 2\pi t_\ell / \tau$

---

structure of $\mathbf{V}$,

$$\mathbf{V}_2 = \mathbf{V}_1 \operatorname{diag}(\mathbf{u}), \tag{15.77}$$

where $\mathbf{u}$ is the length-L vector with elements $u_\ell = e^{-j\omega_\ell}$. Since the matrices $\mathbf{V}$ and $\mathbf{E}$ span the same space, there exists an invertible $L \times L$ matrix $\mathbf{T}$ such that

$$\mathbf{V} = \mathbf{ET}. \tag{15.78}$$

By deleting the last row in (15.78) we have

$$\mathbf{V}_1 = \mathbf{E}_1 \mathbf{T}. \tag{15.79}$$

Similarly, deleting the first row in (15.78) and using the rotational invariance property (15.77), we have

$$\mathbf{V}_1 \operatorname{diag}(\mathbf{u}) = \mathbf{E}_2 \mathbf{T}. \tag{15.80}$$

Combining (15.79) and (15.80) leads to the following relation between the matrices $\mathbf{E}_1$ and $\mathbf{E}_2$:

$$\mathbf{E}_2 = \mathbf{E}_1 \mathbf{T} \operatorname{diag}(\mathbf{u}) \mathbf{T}^{-1}. \tag{15.81}$$

The matrix $\mathbf{E}_1$ is an $m - 1 \times L$ matrix with full column-rank ($m - 1 \geq L$ by our assumption on m). Therefore, $\mathbf{E}_1^\dagger \mathbf{E}_1 = \mathbf{I}$. Multiplying (15.81) on the left by $\mathbf{E}_1^\dagger$ leads to

$$\mathbf{E}_1^\dagger \mathbf{E}_2 = \mathbf{T} \operatorname{diag}(\mathbf{u}) \mathbf{T}^{-1}. \tag{15.82}$$

From (15.82), the eigenvalues of $\mathbf{E}_1^\dagger \mathbf{E}_2$ are $\lambda_\ell = u_\ell$, completing the proof. □

In practice, $\mathbf{R}_y$ is estimated from the given measurements, as in the MUSIC method. The resulting ESPRIT algorithm is summarized in Algorithm 15.7.

We note that in both MUSIC and ESPRIT we assumed that the correlation matrix $\mathbf{R}_a$ is invertible. In practice, this requirement can be relaxed by performing an additional spatial smoothing stage before applying the algorithms. For details, the reader is referred to [444].

In summary, MUSIC and ESPRIT are subspace methods, which are based on separating the space containing the measurements into signal and noise subspaces. Estimating the unknown set of parameters using MUSIC involves a continuous one-dimensional search over the parameter range. The ESPRIT approach estimates the parameters by solving an eigendecomposition problem and exploiting the structure of the correlation matrix.

15.2.8   Compressed sensing formulation

The deterministic methods we have reviewed until now assumed we are given consecutive values of $Y[k]$ in (15.26). In turn, they allowed for recovery of the frequencies ω_ℓ on a continuous grid (in the noise-free setting). As the noise increases, estimating ω_ℓ becomes more difficult so that in practice it can only be found up to a certain error, limiting the resulting resolution. CS offers an alternative approach to recovery of ω_ℓ. On the one hand, it requires discretizing the possible frequency values. On the other hand, CS-based methods do not require consecutive Fourier measurements, and are often more robust to noise especially when the noise is large.

To formulate (15.26) within the CS framework, we begin by quantizing the analog time axis with a resolution step of Δ, so that $t_\ell = s_\ell \Delta$ for some integer value s_ℓ. We then approximate (15.26) as

$$Y[k] \approx \sum_{\ell=1}^{L} a_\ell e^{-j\frac{2\pi}{\tau}k s_\ell \Delta}, \quad 0 \leq s_\ell \leq N-1, \tag{15.83}$$

where we recall that $\omega_\ell = 2\pi t_\ell / \tau$, and $N = \tau/\Delta$ is the number of possible time steps in the period τ. For simplicity, we assume that N is an integer. Selecting a finite subset of m measurements, namely, choosing k from $\mathcal{K} = \{k_1, k_2, \ldots, k_m\} \subset \mathbb{Z}$, (15.83) may be written as

$$\mathbf{y} = \mathbf{A}\mathbf{x}, \tag{15.84}$$

where $\mathbf{A}$ is an $m \times N$ matrix, formed by taking the set $\mathcal{K}$ of rows from a scaled $N \times N$ Fourier matrix, and $\mathbf{x}$ is an L-sparse vector with nonzero entries $\{a_\ell\}$ at indices $\{s_\ell\}$. CS-based techniques allow flexibility in choosing the Fourier coefficients, or the set $\mathcal{K}$.

Our goal is to find the nonzero entries of $\mathbf{x}$ from the measurements $\mathbf{y}$. This is a standard CS problem, with $\mathbf{A}$ being a partial Fourier matrix. Therefore, any of the methods for CS recovery detailed in Chapter 11 can be used to determine $\mathbf{x}$. Note that the properties of the matrix $\mathbf{A}$ are affected by two main variables: the grid resolution manifested in the number of columns N, and the choice of frequencies which affects which rows of the Fourier matrix are used.

From the results of Chapter 11 we know that high recovery performance of CS algorithms is guaranteed, provided that the sensing matrix $\mathbf{A}$ satisfies desired properties such as the coherence or RIP. These values depend both on the number of columns N, and on the choice of frequency set $\mathcal{K}$. Selecting the frequency samples uniformly at random, it is known that if $m \geq CL \log^4 N$, for some positive constant C, then $\mathbf{A}$ obeys the RIP with high probability (see Section 11.3.5). Since N is determined by the chosen grid resolution, trying to obtain higher resolution by using a fine grid leads to an increase in

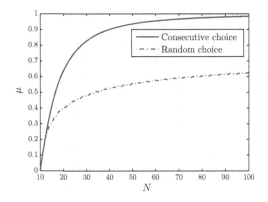

Figure 15.20 Coherence of the matrix $\mathbf{A}$ in (15.84) as a function of N.

the number of measurements needed to obtain good recovery, as judged by this condition. For consecutive frequency selection the RIP is not generally satisfied, unless the cardinality of $\mathcal{K}$ is significantly increased.

These tradeoffs can also be seen through the coherence of the columns of $\mathbf{A}$. Recall from Chapter 11 that low coherence ensures high recovery probability. In Fig. 15.20 we plot the coherence of $\mathbf{A}$ as a function of the length N of $\mathbf{x}$ (which determines the resolution), where the coherence is defined as $\mu(\mathbf{A}) = \max_{i \neq j} |\langle \mathbf{a}_i, \mathbf{a}_j \rangle| / (\|\mathbf{a}_i\| \|\mathbf{a}_j\|)$. The coherence is plotted for two selections of the set $\mathcal{K}$: a consecutive set of $m = 10$ values, and a random selection of m frequency values from $[0, N-1]$. In the latter case we plot the average coherence over 1000 random choices. The figure illustrates two important aspects of CS recovery in our context. The first is that spreading out the frequency values leads to lower coherence and therefore to improved recovery performance. The second is that as the time resolution increases for a fixed number of samples, the coherence increases.

Unfortunately, applying random frequency sampling is not always realistic from a hardware perspective. In Section 15.4 we discuss practical methods for choosing the desired frequency values in combination with CS recovery.

In Examples 15.8 and 15.9 below we examine the performance of CS in several settings, and compare the MSE in delay estimation with that resulting from matrix pencil. As expected, we will see that CS with spread-out frequency values is more robust to noise, and is therefore preferable in the low-SNR regime. This will also be demonstrated in Example 15.15. However, owing to discretization and the coherence of the matrix columns, obtaining high resolution requires many samples. Thus, in the high-SNR regime, matrix pencil is preferable to CS, especially when the number of measurements is small.

There are several more recent approaches to frequency estimation based on convex optimization techniques that do not require discretization and offer improved performance over both CS and matrix pencil type algorithms. The tradeoff is that they tend to be more computationally intensive. The interested reader is referred to [445, 446] and the references therein.

15.2.9   Sub-Nyquist sampling

Until now in all the examples shown we have assumed that the number of samples in time, or the number of frequency values in the set $\mathcal{K}$, was chosen equal to that obtained when sampling at the Nyquist rate. In the following examples we illustrate the use of MF, matrix pencil, and CS with a number of Fourier coefficients m that is smaller than that corresponding to the Nyquist rate. As we will see, while the performance of the MF deteriorates significantly with small values of m, the behavior of matrix pencil is not heavily influenced by the number of measurements, as long as $m \geq 2L$. CS also leads to good performance when m is large enough and the frequency values in $\mathcal{K}$ are properly spread. This demonstrates the value of frequency-based methods in the presence of sub-sampling. In Sections 15.3 and 15.4 we will discuss how to obtain the desired Fourier coefficients directly from low-rate samples of $x(t)$.

Example 15.8 Consider the setting of Example 15.2, used in previous examples as well. The waveform $h(t)$ has bandwidth equal to 81 Hz so that sampling at the Nyquist rate results in $m = 81$ samples. We compare the performance achieved using $m = 36$ and 81 samples where in the MF we use local interpolation. In matrix pencil, the pencil parameter is chosen at the middle of the recommended range $m/3 \leq M \leq m/2$ so that $M = 15, 35$ for $m = 36, 81$ respectively. We used the OMP algorithm (Algorithm 11.1) for CS recovery, and consider two choices of the frequency set $\mathcal{K}$: a consecutive choice and a random choice where the m frequencies are chosen uniformly at random in the range $[0, 10m]$. The grid in time was chosen to have a resolution of $1/10,000$, i.e. we used 10,000 points over the interval $[0, 1)$. The results were averaged over 1000 simulations (for fixed frequency set in the case of random $\mathcal{K}$).

In Fig. 15.21 we plot the MSE (in log scale) as a function of SNR using matrix pencil and MF for $L = 6$. We compare matrix pencil and CS under the

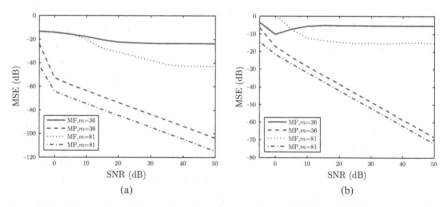

Figure 15.21 Delay and amplitude estimation using matrix pencil (MP) and MF for $L = 6$ with $m = 36$ or 81 samples. (a) MSE in estimating the delays. (b) MSE in estimating the amplitudes.

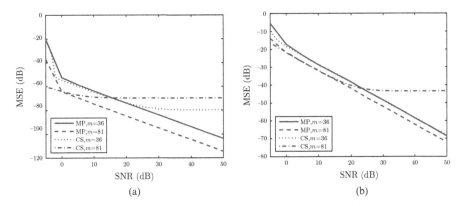

Figure 15.22 Delay and amplitude estimation using matrix pencil (MP) and CS with consecutive frequency values for $L = 6$ using $m = 36$ or 81 samples. (a) MSE in estimating the delays. (b) MSE in estimating the amplitudes.

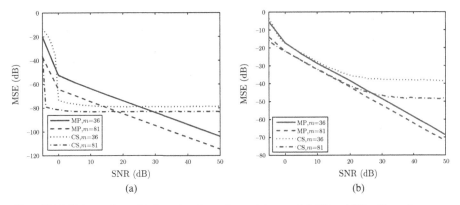

Figure 15.23 Delay and amplitude estimation using matrix pencil (MP) and CS with random frequency values spread over a wide aperture, for $L = 6$ using $m = 36$ or 81 samples. (a) MSE in estimating the delays. (b) MSE in estimating the amplitudes.

same settings in Figs. 15.22 and 15.23. In Fig. 15.22 we sample consecutive frequency values, while the CS results in Fig. 15.23 are based on random frequency locations.

These simulations show that when sampling below Nyquist, matrix pencil and CS achieve much lower MSE in comparison with MF. It is also evident that when using CS methods, spreading out the frequency values is important to ensure good recovery, and leads to superior performance over consecutive frequency measurements. Comparing CS and matrix pencil shows that the former is advantageous in the low-SNR regime, where the performance is noise-limited. However, at high SNR values, matrix pencil leads to improved performance and allows for finer resolution.

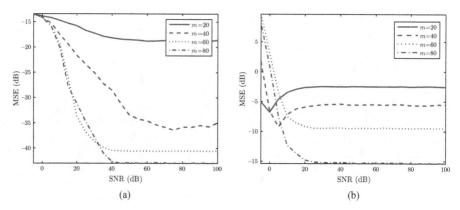

Figure 15.24 Delay and amplitude estimation using the MF with $m = 20, 40, 60, 80$ samples for $L = 6$. (a) MSE in estimating the delays. (b) MSE in estimating the amplitudes.

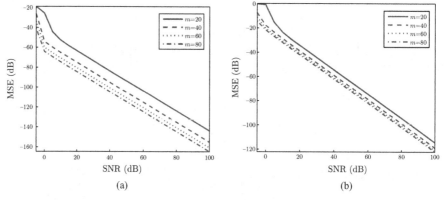

Figure 15.25 Delay and amplitude estimation using matrix pencil (MP) with $m = 20, 40, 60, 80$ samples for $L = 6$. (a) MSE in estimating the delays. (b) MSE in estimating the amplitudes.

Example 15.9 We next consider the behavior of matrix pencil, CS, and MF as a function of the number of samples m. In particular, we demonstrate the degradation caused by sub-Nyquist sampling when using the MF even at very high SNR. In contrast, we show that the performance of matrix pencil and CS is not heavily influenced by the value of m.

We use the same settings as in the previous example for $m = 20, 40, 60, 80$ samples. The pencil parameter for each choice of m is again set as the middle of the recommended range resulting in $M = 8, 17, 25, 33$. Figure 15.24 shows the MSE (in log scale) using the MF for $L = 6$ pulses. The performance of matrix pencil is illustrated in Fig. 15.25. In Fig. 15.26 we plot the results using CS with consecutive frequency values, while a random frequency set is chosen in Fig. 15.27.

Evidently, the performance of MF degrades quite rapidly. When $m < 60$ the MSE is very poor even at high SNR. On the other hand, matrix pencil leads to

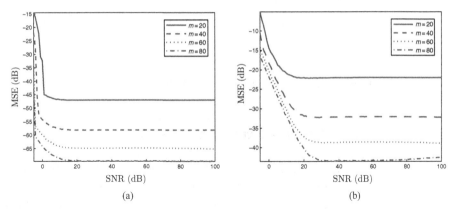

Figure 15.26 Delay and amplitude estimation using CS with consecutive frequency values with $m = 20, 40, 60, 80$ samples for $L = 6$. (a) MSE in estimating the delays. (b) MSE in estimating the amplitudes.

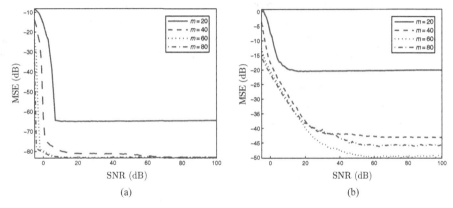

Figure 15.27 Delay and amplitude estimation using CS with random (spread) frequency values with $m = 20, 40, 60, 80$ samples for $L = 6$. (a) MSE in estimating the delays. (b) MSE in estimating the amplitudes.

reliable results even when $m = 20$ and is only slightly affected by the reduction in the number of samples. CS also provides good performance as long as $m > 20$ and random frequency values are chosen. The figures demonstrate once again the superiority of CS over matrix pencil in the low-SNR regime, and the advantage of matrix pencil in the high-SNR region. The importance of using random spread-out frequencies in conjunction with CS is also nicely illustrated.

Example 15.9 demonstrates the potential of sub-Nyquist methods. However, to allow for low-rate samples we must replace the MF by a more sophisticated recovery process. In addition, the sampling scheme needs to be designed such that the required frequency samples can be determined from the resulting low-rate samples. We discuss this topic next.

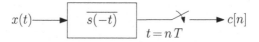

Figure 15.28 Single-channel sampling scheme with $T = \tau/m$.

15.3   Sub-Nyquist sampling with a single channel

In the previous section we introduced a variety of methods for recovering the delays and amplitudes in a pulse stream of the form (15.7), given a set $m \geq 2L$ of Fourier coefficients $X[k], k \in \mathcal{K}$. In practice, however, the signal is sampled in the time domain, and therefore we do not have direct access to samples of $X[k]$. We now show how these coefficients can be obtained conveniently by sampling $x(t)$ at a low rate. In this section we consider the simplest form of sampling: a single filter, followed by uniform low-rate sampling, which leads to the notion of *coset sampling*. In Section 15.4 we treat multi-channel structures that consist of a bank of modulators and integrators, and filterbank systems that contain multiple sampling filters. Although multichannel schemes require more hardware, they often result in simpler filters and lower average sampling rates (in the nonperiodic setting).

We begin by focusing on periodic pulse streams of the form (15.18), which is the easiest case to analyze. We will then see how the results generalize to the finite and infinite pulse stream settings.

15.3.1   Coset sampling

The most straightforward approach to reduce the sampling rate is to filter the periodic pulse stream $x(t)$ with a filter $\overline{s(-t)}$, and then sample the output with period T, as illustrated in Fig. 15.28. In order to obtain m samples per period τ, we choose $T = \tau/m$. Below we derive conditions on the filter $s(t)$ and the set of Fourier coefficients $\mathcal{K}$ such that the required values $X[k], k \in \mathcal{K}$ can be determined from the samples. As we will see, the constraints on $\mathcal{K}$ lead to sampling over cosets, where $\mathcal{K}$ contains only one value from each coset. The cosets consist of frequencies that are aliased together when sampling at the given reduced rate. Requiring that $\mathcal{K}$ contains only one value from each coset ensures that no aliasing occurs, as explained further below. For clarity, we confine ourselves to uniform sampling of the filter output, although the results extend to nonuniform sampling as well.

Condition on sampling filter
Using the general scheme of Fig. 15.28, the samples are given by

$$c[n] = \int_{-\infty}^{\infty} x(t)\overline{s(t - nT)}dt = \langle s(t - nT), x(t)\rangle. \tag{15.85}$$

Substituting (15.24) into (15.85) we have

$$c[n] = \sum_{k \in \mathbb{Z}} X[k] \int_{-\infty}^{\infty} e^{j\frac{2\pi k}{\tau}t}\overline{s(t-nT)}dt$$

$$= \sum_{k \in \mathbb{Z}} X[k] e^{j\frac{2\pi k}{\tau}nT} \int_{-\infty}^{\infty} e^{j\frac{2\pi k}{\tau}t}\overline{s(t)}dt$$

$$= \sum_{k \in \mathbb{Z}} X[k] e^{j\frac{2\pi k}{\tau}nT} \overline{S(2\pi k/\tau)}, \qquad (15.86)$$

where $S(\omega)$ is the CTFT of $s(t)$. Choosing any filter $s(t)$ which satisfies

$$S(\omega) = \begin{cases} 0, & \omega = 2\pi k/\tau, \ k \notin \mathcal{K} \\ \text{nonzero}, & \omega = 2\pi k/\tau, \ k \in \mathcal{K} \\ \text{arbitrary}, & \text{otherwise}, \end{cases} \qquad (15.87)$$

we can rewrite (15.86) as

$$c[n] = \sum_{k \in \mathcal{K}} X[k] e^{j\frac{2\pi k}{\tau}nT} \overline{S(2\pi k/\tau)} = \sum_{k \in \mathcal{K}} X[k] e^{j\frac{2\pi k}{m}n} \overline{S(2\pi k/\tau)}, \qquad (15.88)$$

where we used the fact that $T = \tau/m$. In contrast to (15.86), the sum in (15.88) is finite. Note that (15.87) implies that any real filter meeting this condition will satisfy $k \in \mathcal{K} \Rightarrow -k \in \mathcal{K}$, and in addition $S(2\pi k/\tau) = \overline{S(-2\pi k/\tau)}$, owing to the conjugate symmetry of real filters.

Define the $m \times m$ diagonal matrix $\mathbf{S}$ with kth entry $\overline{S(2\pi k/\tau)}$ for all $k \in \mathcal{K}$, and the length-m vector $\mathbf{c}$ with nth element $c[n]$. We may then write (15.88) as

$$\mathbf{c} = \mathbf{V}(\{-2\pi\ell/m\})_{m \times m}\mathbf{S}\mathbf{x}, \qquad (15.89)$$

where $\mathbf{V}$ is defined by (15.30) with $\omega_\ell = -2\pi\ell/m$ for $\ell \in \mathcal{K}$ and $\mathbf{x}$ is the length-m vector with elements $X[k]$, $k \in \mathcal{K}$. Thus, we need to choose $\mathcal{K}$ and $S(\omega)$ such that $\mathbf{V}(\{-2\pi\ell/m\})_{m \times m}\mathbf{S}$ is left-invertible, so that $\mathbf{x}$ can be determined from (15.89).

Condition on $\mathcal{K}$

The matrix $\mathbf{S}$ is invertible by construction. To ensure invertibility of $\mathbf{V}$, the values of $\ell \in \mathcal{K}$ must be distinct modulo m. Otherwise, we will have two columns that are identical. This condition on the frequencies in $\mathcal{K}$ can be understood as follows. Let $\mathcal{I}$ denote the set of frequencies $\omega_k = 2\pi k/\tau$ for which $X[k] \neq 0$. We divide $\mathcal{I}$ into m subsets where the ℓth group, $0 \leq \ell \leq m - 1$, contains the frequencies $\Omega_\ell = \{2\pi(\ell + km)/\tau\}$ for all values of k such that Ω_ℓ is in $\mathcal{I}$. These sets, which we refer to as *cosets*, are illustrated in Fig. 15.29. In order to construct a valid $\mathcal{K}$ we must choose one element from each group Ω_ℓ. This type of sampling is therefore referred to as coset sampling. Under this requirement on $\mathcal{K}$, and assuming that $S(\omega)$ satisfies (15.87), the coefficients $\mathbf{x}$ can be found from (15.89).

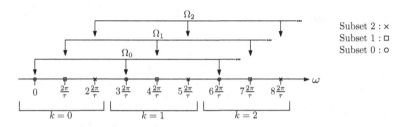

Figure 15.29 Coset sampling for $m = 3$.

Example 15.10 Consider the case illustrated in Fig. 15.29, corresponding to $m = 3$. In this example the three subsets (circles, squares, and stars, respectively) are:

$$\Omega_0 = \{\ldots, -6\tfrac{2\pi}{\tau}, -3\tfrac{2\pi}{\tau}, 0, 3\tfrac{2\pi}{\tau}, 6\tfrac{2\pi}{\tau}, \ldots\}$$
$$\Omega_1 = \{\ldots, -5\tfrac{2\pi}{\tau}, -2\tfrac{2\pi}{\tau}, \tfrac{2\pi}{\tau}, 4\tfrac{2\pi}{\tau}, 7\tfrac{2\pi}{\tau}, \ldots\}$$
$$\Omega_2 = \{\ldots, -4\tfrac{2\pi}{\tau}, -1\tfrac{2\pi}{\tau}, 2\tfrac{2\pi}{\tau}, 5\tfrac{2\pi}{\tau}, 8\tfrac{2\pi}{\tau}, \ldots\}.$$

A valid sampling scheme will include one element from each subset. As an example, one possible choice is $\mathcal{K} = \{0, 1, 2\}$. Another possibility is $\mathcal{K} = \{-6, -5, 8\}$. Both sets will lead to the same matrix $\mathbf{V}$ since they are identical modulo 3.

Once $\mathcal{K}$ is selected, we need to ensure that $S(\omega)$ is nonzero over the frequency values $2\pi/\tau \cdot \mathcal{K}$ and zero for all other values on the grid $2\pi k/\tau$.

When the frequencies are chosen as described above, the values of $\mathcal{K}$ modulo m are always equal to $0, 1, \ldots, m-1$. Thus, with appropriate rearrangement of the columns of $\mathbf{V}$ (and correspondingly those of $\mathbf{S}$ and $\mathbf{x}$), the matrix $\mathbf{V}$ is equal to $\mathbf{V} = \sqrt{m}\mathbf{F}^*$ where $\mathbf{F}$ is the $m \times m$ Fourier matrix (see Appendix A). We can then write

$$\mathbf{x} = \frac{1}{\sqrt{m}}\mathbf{S}^{-1}\mathbf{F}\mathbf{c}. \tag{15.90}$$

The vector $\mathbf{x}$ is therefore obtained by applying the discrete Fourier transform (DFT) on the (possibly reordered) vector of samples, followed by a correction matrix related to the sampling filter. Once $X[k]$, $k \in \mathcal{K}$ are determined, we can apply the techniques in the earlier subsections in order to recover the time delays.

We summarize coset sampling in the following theorem.

Theorem 15.1 (Coset sampling). *Consider the τ-periodic stream of pulses of order L:*

$$x(t) = \sum_{\ell=1}^{L} \sum_{n \in \mathbb{Z}} a_\ell h(t - t_\ell - n\tau).$$

Choose a set $\mathcal{K}$ of indices of size $m = |\mathcal{K}|$ for which $H(2\pi k/\tau) \neq 0$ for all $k \in \mathcal{K}$ and such that the values in $\mathcal{K}$ are distinct modulo m. Set $T = \tau/m$. Then the samples

$$c[n] = \langle s(t - nT), x(t) \rangle, \quad n = 0, \ldots, m - 1,$$

uniquely determine the signal $x(t)$ *for any* $s(t)$ *satisfying condition (15.87), as long as* $m \geq 2L$. *The Fourier coefficients* $X[k]$, $k \in \mathcal{K}$ *can be determined from the samples* $c[n]$ *via (15.90)*.

Example 15.11 A simple example of a filter satisfying the conditions of Theorem 15.1 is $s(t) = (1/T)\,\text{sinc}(t/T)$, with $T = \tau/m$ and $m \geq 2L$ [103]. In this case $s(t)$ is an ideal LPF with bandwidth π/T (see Example 3.12) so that $S(\omega) = 1$ for $|\omega| \leq \pi/T$. The condition in (15.87) is therefore satisfied with $\mathcal{K} = \{-\lfloor m/2 \rfloor, \ldots, \lfloor m/2 \rfloor\}$. Note that because this filter is real-valued, $k \in \mathcal{K}$ implies $-k \in \mathcal{K}$, i.e., the indices come in pairs except for $k = 0$. Since $k = 0$ is part of the set $\mathcal{K}$, the cardinality $m = |\mathcal{K}|$ must be odd-valued leading to $m \geq 2L + 1$ samples, rather than the minimal rate $m \geq 2L$.

The previous example showed that we can sample pulse streams at a low rate by filtering the input with an LPF and uniformly sampling the output. This scheme is very simple to implement in practice. However, it has several drawbacks. First, the ideal LPF has infinite time-support, so that it cannot be extended to finite and nonperiodic infinite streams of pulses. Second, in many cases we can gain robustness to noise by spreading out the Fourier samples over the support of $X(\omega)$, rather than choosing consecutive values in $\mathcal{K}$.

We next propose a class of nonbandlimited sampling kernels, which exploit the additional degrees of freedom in condition (15.87). These filters have compact support in the time domain, and can be designed to pass any choice of frequencies $\mathcal{K}$. The compact support allows this class to be extended to finite and infinite streams, as we show in Section 15.3.4.

15.3.2   Sum-of-sincs filter

One possibility to obtain a filter satisfying (15.87) is by positioning a sinc function $\text{sinc}(\omega/(2\pi/\tau) - k)$ at every value $k \in \mathcal{K}$. Each such sinc will contribute a value of one at k, and zero at other multiples of $2\pi/\tau$. The resulting filter is referred to as a *sum-of-sincs (SoS) filter* [424], and can be written as

$$G(\omega) = \tau \sum_{k \in \mathcal{K}} b_k \, \text{sinc}\left(\frac{\omega}{2\pi/\tau} - k\right) \tag{15.91}$$

where $b_k \neq 0$, $k \in \mathcal{K}$ are arbitrary coefficients.

Since for each sinc in the sum

$$\text{sinc}\left(\frac{\omega}{2\pi/\tau} - k\right) = \begin{cases} 1, & \omega = 2\pi k'/\tau, \; k' = k \\ 0, & \omega = 2\pi k'/\tau, \; k' \neq k, \end{cases} \tag{15.92}$$

the filter $G(\omega)$ satisfies (15.87) by construction. Switching to the time domain

$$g(t) = \text{rect}\left(\frac{t}{\tau}\right) \sum_{k \in \mathcal{K}} b_k e^{j2\pi kt/\tau}, \tag{15.93}$$

which is a time compact filter with support τ. Here

$$\text{rect}(t) = \begin{cases} 1, & |t| \leq \tau/2 \\ 0, & |t| > \tau/2. \end{cases} \tag{15.94}$$

Example 15.12 A simple example of an SoS filter $g(t)$ is when we choose $\mathcal{K} = \{-p, \ldots, p\}$ and set all coefficients $\{b_k\}$ to one. This leads to

$$g(t) = \text{rect}\left(\frac{t}{\tau}\right) \sum_{k=-p}^{p} e^{j2\pi kt/\tau} = \text{rect}\left(\frac{t}{\tau}\right) D_p(2\pi t/\tau), \tag{15.95}$$

where $D_p(t)$ is the Dirichlet kernel defined by

$$D_p(t) = \sum_{k=-p}^{p} e^{jkt} = \frac{\sin\left(\left(p + \frac{1}{2}\right)t\right)}{\sin(t/2)}. \tag{15.96}$$

The resulting filter for $p = 10$ and $\tau = 1$, is depicted in Fig. 15.30. This filter is also optimal in an MSE sense for the case in which $H(\omega)$ is flat over its support, e.g., $h(t) = \delta(t)$, as we show in Theorem 15.2 below.

Example 15.13 As another example, suppose we choose $b_k, 1 \leq k \leq m$ to be a length-m symmetric Hamming window:

$$b_k = 0.54 - 0.46 \cos\left(2\pi \frac{k + m/2}{m}\right), \quad k \in \mathcal{K}, \tag{15.97}$$

with $\mathcal{K} = \{-10, -9, \ldots, 9, 10\}$ and $m = |\mathcal{K}| = 21$. The resulting filter is depicted in Fig. 15.31.

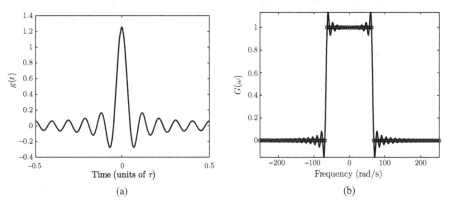

(a)   (b)

Figure 15.30 The filter $g(t)$ of (15.95) in the (a) time and (b) frequency domains. The values of $G(\omega)$ in $\{2\pi k/\tau\}$ for $k \in \mathbb{Z}$ are marked by circles, and are equal to one for $k \in \mathcal{K}$ and zero for $k \notin \mathcal{K}$.

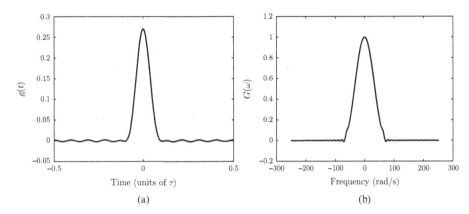

Figure 15.31 The filter $g(t)$ of (15.97) in the (a) time and (b) frequency domains.

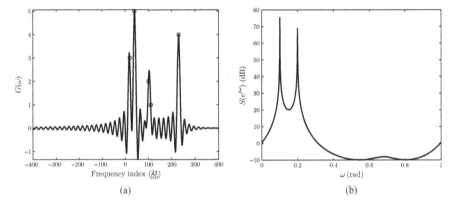

Figure 15.32 (a) The filter $g(t)$ of Example 15.14 in the frequency domain. The circles mark the values of $G(\omega)$ for $k \in \mathcal{K}$, which are equal to b_k in appropriate order. (b) The MUSIC pseudospectrum.

Example 15.14 Our final example shows a case in which the $m = 5$ frequencies in $\mathcal{K}$ are not consecutive. Specially, let $\mathcal{K} = \{11, 10, 2, 23, 4\}$, and choose the values b_k to be $\{1, 2, 3, 4, 5\}$. Modulo $m = 5$ the set $\mathcal{K}$ becomes $\{1, 0, 2, 3, 4\}$ so that the values are all distinct, as required. The resulting SoS filter $G(\omega)$ is depicted in Fig. 15.32. Note that for $k \in \mathcal{K}$, the values of $G(2\pi k/\tau)$ are equal to b_k.

We now consider using this filter for sampling and recovery of $x(t)$, where $x(t) = \delta(t-0.1) + \delta(t-0.2)$ consists of $L = 2$ delayed delta pulses along the period $\tau = 1$. To sample $x(t)$ we filter it with $s(-t)$ where $s(t) = g(t)$, and then sample the output uniformly $m = 5$ times with period $T = 1/5$. Given the samples $\mathbf{c}$, we use (15.89) to determine the vector $\mathbf{x}$ with elements $X[k]$ for $k \in \mathcal{K}$.

First, we reorder the samples $\mathbf{c}$ so that $\mathbf{x}$ corresponds to $X[k]$ for values k mod $m = \{0, 1, 2, 3, 4\}$, namely $k = \{10, 11, 2, 23, 4\}$. This means that we swap the first two entries in $\mathbf{c}$ and $\mathbf{x}$. In return, we also need to exchange these two values in $\mathbf{S}$, and the corresponding columns of $\mathbf{V}$. This leads to the matrix $\mathbf{S} = \mathrm{diag}\,(2,\,1,\,3,\,4,\,5)$. We can then determine $\mathbf{x}$ from (15.90). Given $\mathbf{x}$, we apply MUSIC to determine the delays. The resulting pseudospectrum is depicted in Fig. 15.32. As can be seen, there are two peaks at the correct values of the delays.

In this example we have recovered the signal $x(t)$, which is not bandlimited, from only five samples.

The SoS class in (15.93) may be extended to

$$G(\omega) = \tau \sum_{k \in \mathcal{K}} b_k \phi \left(\frac{\omega}{2\pi/\tau} - k \right) \tag{15.98}$$

where $b_k \neq 0$, $k \in \mathcal{K}$, and $\phi(\omega)$ is any function satisfying:

$$\phi(\omega) = \begin{cases} 1, & \omega = 0 \\ 0, & |\omega| \in \mathbb{Z} \\ \text{arbitrary}, & \text{otherwise.} \end{cases} \tag{15.99}$$

This more general structure allows for smooth versions of the rectangular function, which can be important when practically implementing analog filters.

The function $G(\omega)$ of (15.98) represents a class of filters determined by the parameters $\{b_k\}_{k \in \mathcal{K}}$. These degrees of freedom offer a filter design tool where the values $\{b_k\}_{k \in \mathcal{K}}$ may be optimized for different goals, e.g. parameters which will result in a feasible analog filter. In Theorem 15.2 below, we show how to choose $\{b_k\}$ to minimize the MSE in the presence of noise.

15.3.3 Noise effects

In the presence of noise, the choice of $\{b_k\}_{k \in \mathcal{K}}$ will affect the performance. Consider the case in which digital noise is added to the samples $\mathbf{c}$, so that $\mathbf{y} = \mathbf{c} + \mathbf{w}$, with $\mathbf{w}$ denoting a white Gaussian noise vector with variance σ^2. Using (15.89),

$$\mathbf{y} = \sqrt{m}\mathbf{F}^*\mathbf{B}\mathbf{x} + \mathbf{w} \tag{15.100}$$

where $\mathbf{B}$ is a diagonal matrix, having $\{\tau b_k\}$ on its diagonal. Note here that b_k represents a nonzero coefficient which can be located at any frequency $2\pi(k + rm)/\tau$ for an arbitrary integer r. To choose the optimal $\mathbf{B}$ we assume that the amplitudes $\{a_\ell\}$ are uncorrelated with variance σ_a^2, independent of $\{t_\ell\}$, and that $\{t_\ell\}$ are uniformly distributed in $[0, \tau)$. Since the noise is added to the samples after filtering, increasing the filter's amplification will always reduce the MSE. Therefore, the filter's energy must be normalized, which we ensure by adding the constraint $\mathrm{Tr}(\mathbf{B}^*\mathbf{B}) = 1$. Under these assumptions, the following theorem was derived in [424]:

Theorem 15.2 (MMSE sum-of-sincs filter). *The minimal MSE of a linear estimator of* $\mathbf{x}$ *from the noisy samples* $\mathbf{y}$ *in (15.100) is achieved by choosing the coefficients*

$$b_i = \begin{cases} \frac{\sigma^2}{m\tau^2} \left(\sqrt{\frac{m}{\lambda \sigma^2}} - \frac{1}{|\tilde{h}_i|^2} \right), & \lambda \leq |\tilde{h}_i|^4 m/\sigma^2 \\ 0, & \lambda > |\tilde{h}_i|^4 m/\sigma^2, \end{cases} \tag{15.101}$$

where $\tilde{h}_k = H(2\pi(k + mr)/\tau)\sigma_a\sqrt{L}/\tau$ for the integer value of r that leads to the largest absolute value, and are arranged in an increasing order of $|\tilde{h}_k|$,

$$\sqrt{\lambda} = \frac{(|\mathcal{K}| - M)\sqrt{m/\sigma^2}}{m/\sigma^2 + \displaystyle\sum_{i=M+1}^{|\mathcal{K}|} 1/|\tilde{h}_i|^2}, \qquad (15.102)$$

and M is the smallest index for which $\lambda \le |\tilde{h}_{M+1}|^4 m/\sigma^2$.

Note that the theorem derives the optimal filter coefficients under the assumption of linear recovery. In practice, however, all the methods we outlined for estimating the values of a_ℓ and t_ℓ are nonlinear. Nonetheless, this allows us to obtain some insight into the optimal choice of filter values. In particular, the coefficients b_k of (15.101) have the intuitive property that they are larger for higher values of h_k: we expect to give more emphasis to Fourier coefficients corresponding to high signal values.

In the case in which $\tilde{h}_k$ are equal, we have the following corollary:

Corollary 15.3. *If $\tilde{h}_k$ are all equal then the optimal coefficients are $\beta_k = 1/(m\tau)$ for all $k \in \mathcal{K}$.*

When $h(t) = \delta(t)$, or more generally, $H(\omega)$ is flat over its support, Corollary 15.3 implies that the filter coefficients should all be chosen equally. Following this result, we will set $b_k = b_j$ for all $k, j \in \mathcal{K}$ in noisy simulations in which $h(t) = \delta(t)$. Note, however, that this still leaves open the question of how to choose $\mathcal{K}$. Some practical guidelines for selecting the frequencies are suggested in [427, 447]. Roughly speaking, these recommendations amount to choosing at least two Fourier coefficients consecutively, but at the same time spreading out the frequencies in order to cover a wide aperture. The motivation behind these rules is that widening the frequency aperture results in finer resolution, while choosing consecutive coefficients helps avoid ambiguities. As an example, the sub-Nyquist radar prototype reported in [427] uses a constellation consisting of four groups of consecutive coefficients, where the individual bands are spread randomly over the signal bandwidth. In practice, this type of sampling is achieved by using four bandpass filters and low-rate sampling of each filtered output. See Fig. 15.40 later for an image of the prototype, as well as the discussion in Section 15.4.1 for more details. On the other hand, in the context of ultrasound imaging, the authors of [448] consider a single bandpass filter centered around the high-energy portion of the spectrum, in accordance with Theorem 15.2.

Example 15.15 We now examine the importance of choosing the coefficients b_k optimally via Theorem 15.2.

Consider an input $x(t) = \delta(t - 0.2) + \delta(t - 0.4)$ consisting of $L = 2$ delayed delta pulses along the period $\tau = 1$. Note that this signal has infinite bandwidth. Nonetheless, we will use only 11 samples to recover it by choosing the frequency values corresponding to indices $\mathcal{K} = \{-5, \ldots, 5\}$. To obtain the desired coefficients, we filter $x(t)$ with $g(t)$ of (15.91). The filter output is sampled uniformly

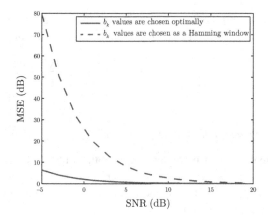

Figure 15.33 MSE in estimating the Fourier coefficients using the optimal values of b_k and using a Hamming window.

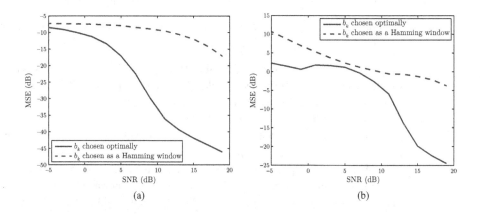

Figure 15.34 Delay and amplitude estimation using matrix pencil (MP) with the optimal values of b_k and with a Hamming window. (a) MSE in estimating the delays. (b) MSE in estimating the amplitudes.

$m = 11$ times, with sampling period $T = 1/m$. White Gaussian noise with variance σ^2 to achieve the desired SNR is added to the samples.

In Fig. 15.33 we examine the MSE between the frequency coefficients resulting from (15.90) and the real frequency coefficients (15.26) for two different choices of the sampling filter $s(t)$. The first is the optimal filter according to Corollary 15.3, which is depicted in Fig. 15.30. The second is the length-m symmetric Hamming window of Fig. 15.31. Clearly the optimal choice of b_k results in improved MSE.

The performance of matrix pencil using the estimated frequency coefficients is demonstrated in Fig. 15.34. We next consider the performance of CS using the estimated frequency coefficients with a grid of 1000 points and consecutive frequency values (as used in matrix pencil). The results are depicted in Fig. 15.35. Here again the advantage of the optimal choice is evident.

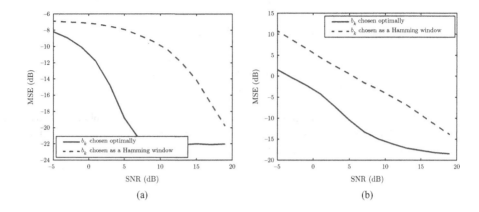

Figure 15.35 Delay and amplitude estimation using CS with consecutive frequencies with optimal values of b_k and with a Hamming window. (a) MSE in estimating the delays. (b) MSE in estimating the amplitudes.

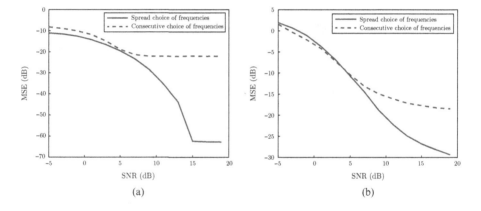

Figure 15.36 Delay and amplitude estimation using CS with consecutive frequencies and with a random choice of frequencies, and optimal values of b_k. (a) MSE in estimating the delays. (b) MSE in estimating the amplitudes.

Finally, it is interesting to compare different optimal choices of b_k. While Theorem 15.2 implies that the values should be set to $b_k = 1$, it leaves open the question of how to select the frequencies $k \in \mathcal{K}$. In Fig. 15.36 we plot the results using a random choice of frequencies from the range $[0,110]$. As we have seen in Examples 15.8 and 15.9, spreading out the frequency values improves the performance considerably.

15.3.4   Finite and infinite pulse streams

Relying on the compact support property of the SoS kernels, we now show how they can be used for sampling of finite and infinite pulse streams.

Finite streams

Consider a finite stream of pulses, defined by

$$x(t) = \sum_{\ell=1}^{L} a_\ell h(t - t_\ell), \quad t_\ell \in [0, \tau), \tag{15.103}$$

where $h(t)$ is a known pulse shape, and $\{t_\ell, a_\ell\}_{\ell=1}^{L}$ are the unknown delays and amplitudes. The time delays $\{t_\ell\}_{\ell=1}^{L}$ are restricted to lie in a finite time interval $[0, \tau)$, and we assume that the pulse $h(t)$ has finite support R, namely

$$h(t) = 0, \quad |t| \geq R/2. \tag{15.104}$$

Recall that our primary interest is in very short pulses which have wide, or even infinite, frequency support, and therefore cannot be sampled efficiently using classical sampling results for bandlimited signals.

In the case of a periodic pulse stream, the samples $c[n]$ in Fig. 15.28 are given by

$$c[n] = \sum_{i \in \mathbb{Z}} \sum_{\ell=1}^{L} a_\ell \int_{-\infty}^{\infty} h(t - t_\ell - i\tau)\overline{s(t - nT)}dt = \sum_{i \in \mathbb{Z}} \sum_{\ell=1}^{L} a_\ell \varphi(nT - t_\ell - i\tau), \tag{15.105}$$

where we defined

$$\varphi(\vartheta) = \langle s(t - \vartheta), h(t) \rangle. \tag{15.106}$$

Choosing $s(t)$ as a compactly supported filter that vanishes for all $|t| > \tau/2$, the support of $\varphi(t)$ is $R + \tau$:

$$\varphi(t) = 0, \quad |t| \geq (R + \tau)/2. \tag{15.107}$$

Using this property, the summation in (15.105) will be over nonzero values for indices i satisfying

$$|nT - t_\ell - i\tau| < (R + \tau)/2. \tag{15.108}$$

Sampling within the window $[0, \tau)$, and noting that the time delays lie in the interval $t_\ell \in [0, \tau)$, $\ell = 1, \ldots, L$, (15.108) implies that

$$(R + \tau)/2 > |nT - t_\ell - i\tau| \geq |i|\tau - |nT - t_\ell| > (|i| - 1)\tau. \tag{15.109}$$

Here we used the triangle inequality and the fact that $|nT - t_\ell| < \tau$ in our setting. Therefore,

$$|i| < \frac{R/\tau + 3}{2} \Rightarrow |i| \leq \left\lceil \frac{R/\tau + 3}{2} \right\rceil - 1 \overset{\triangle}{=} r, \tag{15.110}$$

i.e. the elements of the sum in (15.105) vanish for all i but the values in (15.110). Consequently, the infinite sum in (15.105) reduces to a finite sum over $i \leq |r|$ leading to

$$c[n] = \sum_{i=-r}^{r} \sum_{\ell=1}^{L} a_\ell \int_{-\infty}^{\infty} h(t - t_\ell) \overline{s(t - nT + i\tau)} dt$$

$$= \left\langle \sum_{i=-r}^{r} s(t - nT + i\tau), \sum_{\ell=1}^{L} a_\ell h(t - t_\ell) \right\rangle. \qquad (15.111)$$

Defining a function which consists of $2r + 1$ periods of $s(t)$:

$$s_r(t) = \sum_{i=-r}^{r} s(t + i\tau), \qquad (15.112)$$

we conclude that

$$c[n] = \langle s_r(t - nT), x(t) \rangle. \qquad (15.113)$$

Therefore, the samples $c[n]$ can be obtained by filtering $x(t)$ with the filter $\overline{s_r(-t)}$ prior to sampling. This filter has compact support equal to $(2r + 1)\tau$.

Suppose, for example, that the support R of $h(t)$ satisfies $R \leq \tau$. We then obtain from (15.110) that $r = 1$. Therefore, the filter $s_r(t)$ consists of three periods of $s(t)$:

$$s_1(t) = s(t - \tau) + s(t) + s(t + \tau). \qquad (15.114)$$

Example 15.16 Consider sampling the finite signal $x(t) = \sum_{\ell=1}^{L} a_\ell \delta(t - t_\ell)$ for $t \in [0, 1)$, where $a_\ell = 1, t_\ell = \ell\Delta$, and $\Delta = 0.025$. The samples are corrupted by zero-mean white Gaussian noise with variance chosen to meet the required SNR. We choose the frequency set $\mathcal{K} = \{-L, \ldots, L\}$, so that $m = |\mathcal{K}| = 2L + 1$. Thus, our sampling rate is very close to its minimal rate of $2L$ samples.

Since the support of $h(t) = \delta(t)$ satisfies $R \leq \tau = 1$ the parameter r in (15.110) equals one, and therefore we filter $x(t)$ with $s_1(t)$ of (15.114). For $s(t)$, we use an SoS filter with coefficients b_k all set to 1. The output of the filter is sampled uniformly m times, with sampling period $T = 1/m$. The frequency coefficients are obtained from (15.90).

In Fig. 15.37 we plot the MSE as a function of SNR using the matrix pencil method. Evidently, the results are quite robust even for large values of L.

Infinite streams

A similar technique can also be used to sample and recover infinite-length FRI pulse streams of the form

$$x(t) = \sum_{\ell \in \mathbb{Z}} a_\ell h(t - t_\ell). \qquad (15.115)$$

We assume that the infinite signal has a "bursty" character, i.e. the signal has two distinct phases: (a) bursts of maximal duration τ containing at most L pulses, and (b) quiet

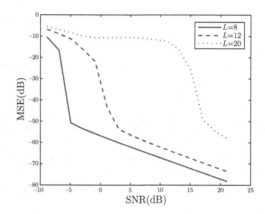

Figure 15.37 The MSE in estimating L delays in a finite pulse stream, for different choices of L, as a function of the SNR.

phases between bursts, in which there is no signal. For simplicity, we begin with the case $h(t) = \delta(t)$. For this choice the filter $\overline{s_r(-t)}$ in (15.112) reduces to $\overline{s_1(-t)}$ of (15.114).

Since the filter $\overline{s_1(-t)}$ has compact support 3τ we are assured that the current burst cannot influence samples taken $3\tau/2$ seconds before or after it. In the finite setting, we confined ourselves to sampling within the interval $[0, \tau)$. Similarly, here, we assume that the samples are taken during the burst duration. Therefore, if the minimal spacing between any two consecutive bursts is $3\tau/2$, then we are guaranteed that each sample is influenced by one burst only. The infinite problem then reduces to sequential local distinct finite-order problems.

Extending this result to a general pulse $h(t)$ is straightforward, as long as $h(t)$ is compactly supported with support R, and we filter with $\overline{s_r(-t)}$ as defined in (15.112) with the appropriate r from (15.110). In this case we require that the minimal spacing between two adjacent bursts is greater than $((2r + 1)\tau + R)/2$.

An alternative approach to sample finite and infinite pulse streams without having to assume separation between bursts is by using a multichannel system, which we discuss in Section 15.4. This allows us to avoid forming a delayed pulse as in $s_r(t)$, which may be difficult to implement in hardware.

Sampling with exponential reproducing kernels

Another class of compact support kernels that can be used to sample finite-length FRI signals is given by the family of exponential reproducing kernels [449, 450].

An exponential reproducing kernel is any function $\varphi(t)$ that, together with its shifted versions, can generate complex exponentials of the form $e^{\alpha_m t}$. Specifically,

$$\sum_{n \in \mathbb{Z}} c_{kn} \varphi(t - n) = e^{\alpha_k t} \tag{15.116}$$

where $k = 0, 1, \ldots, m - 1$. The coefficients are given by $c_{kn} = \langle e^{\alpha_k t}, \tilde{\varphi}(t - n) \rangle$, where $\tilde{\varphi}(t)$ is the biorthogonal function of $\varphi(t)$, that is, $\langle \varphi(t - n), \tilde{\varphi}(t - k) \rangle = \delta_{nk}$. When sampling pulse streams, it is suggested in [449] to choose $\alpha_k = \alpha_0 + k\lambda$ for some α_0, $\lambda \in \mathbb{C}$.

To see how we can use $\varphi(t)$ to sample finite pulse streams, suppose for simplicity that $h(t) = \delta(t)$ so that

$$x(t) = \sum_{\ell=1}^{L} a_\ell \delta(t - t_\ell). \qquad (15.117)$$

We filter $x(t)$ with $(1/T)\varphi(-t/T)$ where $T = \tau/m$, and sample the output at times nT. This results in measurements

$$y[n] = \sum_{\ell=1}^{L} a_\ell \varphi\left(\frac{t_\ell}{T} - n\right). \qquad (15.118)$$

We now linearly combine the samples with the coefficients c_{kn} of (15.116) to obtain the new measurements

$$s[k] = \sum_{n} c_{kn} y[n], \quad 0 \le k \le m - 1. \qquad (15.119)$$

Using (15.118),

$$s[k] = \sum_{\ell=1}^{L} a_\ell \sum_{n} c_{kn} \varphi\left(\frac{t_\ell}{T} - n\right) = \sum_{\ell=1}^{L} a_\ell e^{\frac{\alpha_0 t_\ell}{T}} e^{\frac{\lambda t_\ell k}{T}} = \sum_{\ell=1}^{L} \tilde{a}_\ell u_\ell^k, \qquad (15.120)$$

where $\tilde{a}_\ell = a_\ell e^{\frac{\alpha_0 t_\ell}{T}}$ and $u_\ell = e^{\frac{\lambda t_\ell k}{T}}$. These measurements have the form of a power series as we studied in (15.32), and therefore can be solved using the same class of methods.

15.4 Multichannel sampling

The techniques discussed so far were based on uniform sampling of the signal $x(t)$ after filtering with a single filter $\overline{s(-t)}$ (see Fig. 15.28). Lower sampling rates for the finite and infinite settings and more practical hardware devices can often be achieved at the cost of using several channels. In this section we consider multichannel sampling architectures for sub-Nyquist sampling of pulse streams [427, 429, 430, 451, 452]. In particular, we focus on two different systems: modulation and integration channels, and filterbanks.

The first architecture we treat consists of p channels of modulators and integrators. The output of each branch is given by

$$c_\ell[n] = \int_{(n-1)T}^{nT} x(t) s_\ell(t) dt, \quad 1 \le \ell \le p, \quad n \in \mathbb{Z}, \qquad (15.121)$$

where $s_\ell(t)$ is the modulating function on the ℓth branch. Thus, in every period T we obtain p outputs, resulting in a total sampling rate of p/T. This scheme is particularly simple, and as we show below, can be used to treat all classes of FRI signals: periodic, finite, infinite, and semiperiodic, under the assumption that the pulse $h(t)$ is compactly supported.

An alternative approach is to use filterbanks, where the signal $x(t)$ is convolved with p kernels $\overline{s_1(-t)}, \ldots, \overline{s_p(-t)}$, and the output of each channel is sampled at a rate $1/T$. The set of samples in this case is given by

$$c_\ell[n] = \langle s_\ell(t - nT), x(t) \rangle, \quad 1 \le \ell \le p, \quad n \in \mathbb{Z}. \tag{15.122}$$

Since there are p channels, the total sampling rate is again p/T. We illustrate the filterbank approach for the semiperiodic setting and show that it can accommodate arbitrary pulse shapes $h(t)$, including infinite length. In certain cases, this structure can be implemented as a single sampling channel followed by a serial to parallel converter, leading to hardware savings while still retaining the benefits of multichannel structures.

15.4.1   Modulation-based multichannel systems

We begin by discussing the modulation-based multichannel structure and focus first on periodic pulse streams.

Periodic pulse streams
Consider a τ-periodic stream of L pulses, as in (15.18). Recall from Section 15.2 that if the Fourier coefficients of this signal are available, then standard techniques of spectral analysis or CS can be used to recover the unknown pulse shifts and amplitudes. The multichannel setup of Fig. 15.38 provides a simple and intuitive method for obtaining these Fourier coefficients by correlating the signal $x(t)$ with the Fourier basis functions

$$s_k(t) = \begin{cases} e^{-j\frac{2\pi}{\tau}kt}, & t \in [0, \tau] \\ 0, & \text{elsewhere,} \end{cases} \tag{15.123}$$

for $k \in \mathcal{K}$, where $\mathcal{K}$ are the desired Fourier coefficients and $|\mathcal{K}| = m \ge 2L$. We set the sampling interval T to be equal to the signal period τ, resulting in a total sampling rate of m/τ for all channels. We thus have a sampling system functioning at the rate of innovation when $m = 2L$, and yielding the desired Fourier coefficients of $x(t)$. An additional advantage of this approach is that the kernels have compact support – their support corresponds to one period of the FRI signal, which is smaller than the support of the kernel proposed in Section 15.3.2. This property will facilitate the extension of this scheme to infinite FRI signals.

In principle, we can use the system of Fig. 15.38 to obtain any desired set of frequencies. However, when m is large, this requires implementing many modulators in parallel, which tends to be difficult. A practical alternative, suggested in [427], is to select the frequencies to consist of several bands of consecutive values, spread over a wide aperture. This offers a compromise between the bandwidth of the chosen values (which we often would like to be large), the minimal separation between the points (which we want to be small enough to avoid ambiguities), and the number of required channels. In this case, instead of acquiring each frequency separately as in Fig. 15.38, we aggregate consecutive values and obtain them jointly by filtering the input with a bandpass

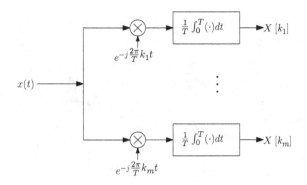

Figure 15.38 Modulation-based multichannel sampling for periodic FRI signals. The resulting samples are the Fourier series coefficients of $x(t)$ at the required frequencies. Note that we only sample once every period, thus $T = \tau$.

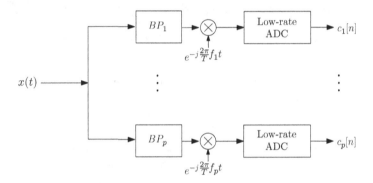

Figure 15.39 Multichannel receiver choosing bands of Fourier coefficients. Here BP_i indicates a bandpass filter centered at f_i.

filter that encompasses the required values, followed by demodulation to baseband and lowrate sampling of the desired band. This idea is illustrated schematically in Fig. 15.39 and forms the basis for the sampling prototype we discuss next.

Hardware design

A board-level hardware prototype of the multichannel receiver of Fig. 15.39 was developed in [427] for radar detection, and is shown in Fig. 15.40. The board consists of four parallel channels which sample distinct bands of the signal spectrum. Each channel comprises a bandpass crystal filter with an effective random carrier frequency that filters the desired band, demodulates it to baseband, and then samples the band at its Nyquist rate. In this scheme, instead of sampling isolated Fourier coefficients, we acquire four sets of consecutive values. This allows to trade off between the theoretical algorithmic requirements, which may benefit from a fully distributed selection, and the constraints of practical analog filters.

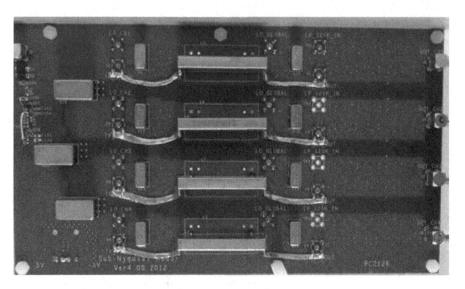

Figure 15.40 A four-channel sub-Nyquist board for pulse streams.

The advantage of crystal filters is that they have an extremely narrow transition band which allows low-rate sampling to be achieved, while extracting a sufficient number of Fourier coefficients. Since crystal filters are standard, off-the-shelf components, the channel design must be adapted to their properties, in order to maximize their efficiency. The prototype developed in [427] therefore includes appropriate filtering and modulation stages. A detailed description of the board and its properties is given in [427].

In the context of radar, it has been shown that this prototype can lead to a 30-fold reduction in sampling rate while maintaining reasonable target location and velocity estimation even at very low SNR values [426, 427]. The application to radar along with simulation results using the prototype are discussed in more detail in Section 15.7.1.

Periodic waveforms

Instead of functions of the form (15.123), one can use sampling kernels which are a linear combination of these sinusoids, as in Fig. 15.41. In each branch, we modulate the signal using a weighted sum of exponentials given by

$$s_\ell(t) = \sum_{k \in \mathcal{K}} s_{\ell k} e^{-j\frac{2\pi}{T}kt}, \tag{15.124}$$

where the weights $s_{\ell k}$ vary in every channel. The samples on the ℓth branch are then

$$c_\ell = \frac{1}{T} \int_0^T x(t) \sum_{k \in \mathcal{K}} s_{\ell k} e^{-j\frac{2\pi}{T}kt} dt = \sum_{k \in \mathcal{K}} s_{\ell k} X[k]. \tag{15.125}$$

This choice can be advantageous from a hardware point of view. If we appropriately choose the linear combinations, then the resulting modulating functions $s_\ell(t)$ may have a simple form, such as low-pass versions of binary sequences [452]. In fact, we can

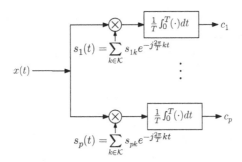

Figure 15.41 Mixing the Fourier coefficients differently in each channel.

use similar sequences to those suggested in the context of the MWC in Chapter 14. In particular, this allows us to reduce hardware complexity by using a small number of binary sequences and their shifts. In addition, in real-life scenarios one or more channels might fail, owing to malfunction or noise corruption, leading to loss of information stored in that branch. By mixing the coefficients we distribute the information about each Fourier coefficient among several sampling channels. Consequently, when one or more branches fail, the required Fourier coefficients may still be recovered from the remaining samples [452].

To relate the samples and the Fourier coefficients, we define the $p \times m$ matrix $\mathbf{S}$ with $s_{\ell k}$ as its ℓkth element, and by $\mathbf{c}$ the length-p sample vector with ℓth element c_ℓ. We then write (15.125) in matrix form as $\mathbf{c} = \mathbf{S}\mathbf{x}$. As long as $\mathbf{S}$ has full column-rank, where $p \geq m$ is a necessary condition, we can recover the length-m vector $\mathbf{x}$ from the samples via $\mathbf{x} = \mathbf{S}^\dagger \mathbf{c}$.

We next discuss in more detail the use of periodic waveforms. Suppose that $p_i(t)$ is a periodic waveform with period T. We can expand $p_i(t)$ using a Fourier series as

$$p_i(t) = \sum_{k \in \mathbb{Z}} d_i[k] e^{j \frac{2\pi}{T} kt}, \qquad (15.126)$$

where $d_i[k]$ are the Fourier series coefficients. The sum in (15.126) is generally infinite, in contrast to the finite sum in (15.124). Therefore, we filter $p_i(t)$ with a filter $g(t)$ which rejects the unwanted elements in (15.126). The filtered waveforms $\tilde{p}_i(t) = p_i(t) * g(t)$ are also periodic. It can be readily shown that their Fourier series coefficients are given by (see Exercise 11)

$$\tilde{d}_i[k] = d_i[k] G\left(\frac{2\pi}{T} k\right), \qquad (15.127)$$

where $G(\omega)$ is the CTFT of $g(t)$. From (15.127), the shaping filter $g(t)$ has to satisfy

$$G(\omega) = \begin{cases} \text{nonzero}, & \omega = \frac{2\pi}{T} k, \ k \in \mathcal{K} \\ 0, & \omega = \frac{2\pi}{T} k, \ k \notin \mathcal{K} \\ \text{arbitrary}, & \text{otherwise}, \end{cases} \qquad (15.128)$$

so that $\tilde{d}_i[k] = 0$ for $k \notin \mathcal{K}$. This condition is similar to (15.87) obtained for single-channel sampling.

An important special case is when $p_i(t)$ consists of $N \geq m$ values, flipping at rate N/T:

$$s_i(t) = \sum_{\ell \in \mathbb{Z}} \sum_{n=0}^{N-1} \alpha_i[n] p(t - nT/N - \ell T), \tag{15.129}$$

where $p(t)$ is a unit pulse of length T/N, and $\alpha_i[n]$ takes on the values ± 1. These are precisely the sequences used in the implementation of the MWC discussed in Section 14.4.4. We can often choose the sequence such that one periodic stream is sufficient for all channels, where each channel uses a delayed version of this common waveform. Therefore, the requirement for multiple oscillators and the need for accurate multiples of the basic frequency are both removed. In addition, periodic streams are easily designed and implemented digitally.

Suppose that we use the filtered version of (15.129) as modulating waveforms. To compute the mixing matrix $\mathbf{S}$ in this case we first note that the Fourier coefficients $d_i[k]$ of $s_i(t)$ are given by

$$d_i[k] = \frac{1}{T} \sum_{n=0}^{N-1} \alpha_i[n] \sum_{\ell \in \mathbb{Z}} \int_{-\ell T}^{-(\ell-1)T} p(t - nT/N) e^{-j\frac{2\pi}{T}kt} dt$$

$$= \frac{1}{T} \sum_{n=0}^{N-1} \alpha_i[n] P\left(\frac{2\pi}{T}k\right) e^{-j\frac{2\pi}{N}kn}, \tag{15.130}$$

where $P(\omega)$ denotes the CTFT of $p(t)$. After filtering with $g(t)$, the resulting matrix $\mathbf{S}$ may be decomposed as

$$\mathbf{S} = \mathbf{A}\mathbf{W}\mathbf{\Phi}, \tag{15.131}$$

where $\mathbf{A}$ is a $p \times N$ matrix with inth element equal to $\alpha_i[n]$, $\mathbf{W}$ is an $N \times m$ matrix with nkth element equal to $e^{-j\frac{2\pi}{N}kn}$, and $\mathbf{\Phi}$ is an $m \times m$ diagonal matrix with kth diagonal element

$$\Phi_{kk} = \frac{1}{T} P\left(\frac{2\pi}{T}k\right) G\left(\frac{2\pi}{T}k\right). \tag{15.132}$$

To guarantee left-invertibility of $\mathbf{\Phi}$ we must have that $P\left(\frac{2\pi}{T}k\right) \neq 0$ for $k \in \mathcal{K}$. Left-invertibility of the $p \times m$ matrix $\mathbf{A}\mathbf{W}$ can be ensured by proper selection of the sequences $\alpha_i[n]$, where a necessary condition is that $p \geq m$.

Example 15.17 As an example consider creating all sequences by cyclic shifts of one common sequence $\alpha[n]$ so that

$$\alpha_i[n] = \alpha[n - i + 1 \bmod N], \tag{15.133}$$

where we assume $p = N = m$. This suggests that in contrast to the direct scheme in Fig. 15.38, which requires multiple frequency sources, here only one pulse generator is needed, simplifying the hardware design.

It is easy to see that with this choice, $\mathbf{A}$ will be a circulant matrix:

$$\mathbf{A} = \begin{bmatrix} \alpha\,[0] & \alpha\,[1] & \cdots & \alpha\,[N-1] \\ \alpha\,[N-1] & \alpha\,[0] & \cdots & \alpha\,[N-2] \\ \vdots & \ddots & \ddots & \vdots \\ \alpha\,[1] & \alpha\,[2] & \cdots & \alpha\,[0] \end{bmatrix}. \tag{15.134}$$

Such a matrix can be decomposed as [453] (see also Appendix A)

$$\mathbf{A} = \mathbf{F}^* \operatorname{diag}\,(\mathbf{F}\alpha)\,\mathbf{F}, \tag{15.135}$$

where $\mathbf{F}$ is the $N \times N$ Fourier matrix, and α is a length-N vector containing the elements of the sequence $\alpha[n]$. Therefore, for $\mathbf{A}$ to be invertible the DFT of the sequence $\alpha[n]$ cannot take on the value 0. As the pulse $p(t)$ we choose

$$p(t) = \begin{cases} 1, & t \in [0, \frac{T}{N}] \\ 0, & t \notin [0, \frac{T}{N}]. \end{cases} \tag{15.136}$$

The frequency response of this pulse satisfies

$$P(\omega) = \frac{T}{N} e^{-j\frac{T}{2N}\omega} \operatorname{sinc}\left(\frac{T}{2\pi N}\omega\right). \tag{15.137}$$

In Fig. 15.42, we show a modulating waveform in the time and frequency domains, for $p = N = m = 7$. The original time-domain waveform comprises rectangular pulses, whereas low-pass filtering results in a smooth modulating waveform. Switching to the frequency domain, the Fourier series coefficients are shaped by $P(\omega)$, the CTFT of the pulse shape. The shaping filter frequency response, $G(\omega)$, is designed to transfer only the Fourier coefficients whose index is a member of the set $\mathcal{K} = \{-3, \dots, 3\}$, suppressing all other coefficients.

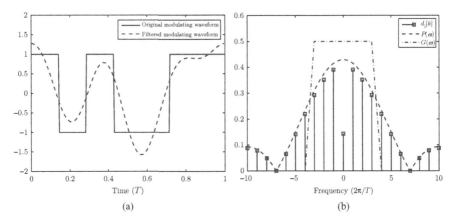

(a) (b)

Figure 15.42 Modulating waveform before and after filtering. (a) Time domain. (b) Frequency domain.

Although the concept of using modulation waveforms is based on the MWC, there are some differences between the methods. First, following the mixing stage, here we use an integrator in contrast to the LPF used in the MWC. This is a result of the different signal quantities measured: Fourier coefficients here as opposed to the frequency band's content in the MWC. The second distinction is in the purpose of the mixing procedure. In the MWC mixing is performed in order to reduce the sampling rate relative to the Nyquist rate. In our setting, mixing is used to simplify hardware implementation and to improve robustness to failure in any of the sampling channels.

Infinite FRI signals

We now turn to treat infinite-duration FRI signals of the form (15.115). Suppose that the τ-local rate of innovation is $2L/\tau$, for some value τ. Thus, there are no more than L pulses in any interval $I_n = [(n-1)\tau, n\tau]$. Assume further that the pulses do not overlap interval boundaries, i.e., if $t_k \in I_n$ then $h(t - t_k) = 0$ for all $t \notin I_n$. Such a requirement automatically holds if $h(t)$ is a Dirac delta, and will hold as long as the support of $h(t)$ is substantially smaller than τ.

Under these assumptions, we may treat the signal parameters in each interval separately. Specifically, consider the τ-periodic signal obtained by periodic continuation of the values of $x(t)$ within a particular interval I_n. This periodic signal can be recovered by obtaining $2L$ of its Fourier coefficients using sampling kernels of the form (15.123). Since the support of these kernels is limited to the interval I_n (rather than its periodic continuation), this technique can be applied on the nonperiodic signal $x(t)$. This requires obtaining a sample from each of the channels once every τ seconds, and using $p \geq 2L$ branches, as illustrated in Fig. 15.43. The success of this approach hinges on the availability of sampling kernels whose support is limited to a single period of the periodic waveform.

Denoting by $\mathbf{c}[n]$ the vector containing the samples $c_1[n], \ldots, c_p[n]$ at time n, we have

$$\mathbf{c}[n] = \mathbf{S}\mathbf{x}[n] \qquad (15.138)$$

where $\mathbf{S}$ is the matrix of elements $s_{\ell k}$, and $\mathbf{x}[n]$ are the desired Fourier coefficients of $x(t)$ over the interval I_n ($\mathbf{x}[n]$ is a length-m vector with $m \geq 2L$). Inverting $\mathbf{S}$ we

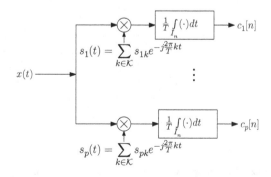

Figure 15.43 Multichannel modulation-based sampling scheme for infinite FRI signals, with $T = \tau$.

obtain the Fourier coefficients over each interval, from which we can find the delays and amplitudes in the nth period. Indeed from (15.25),

$$\mathbf{x}[n] = \mathbf{H}\mathbf{V}(\{\omega_\ell[n]\})\mathbf{a}[n], \tag{15.139}$$

where $\omega_\ell[n]$ are the frequencies defining the nth interval, $\mathbf{H}$ is a diagonal matrix containing the Fourier coefficients of the pulse shape, and $\mathbf{a}[n]$ is the vector of amplitudes α_ℓ over the nth interval. Therefore, for each value of n, we may apply the methods studied in this chapter (such as Prony's method, matrix pencil, etc.) to (15.139) in order to recover $\omega_\ell[n]$ and $\mathbf{a}[n]$.

Semiperiodic FRI signals
Finally, we consider semiperiodic FRI signals (15.8), which consist of L pulses occurring at repeated intervals τ, with amplitudes $a_\ell[n]$ that vary from one period to the next. The modulator approach of Fig. 15.43 can be used as in the infinite case, with the difference that now the samples from different periods may be jointly processed to improve performance.

Specifically, as before, we recover $\mathbf{x}[n]$ from the output of the modulator bank with $T = \tau$ using (15.138). Since the delays are constant, in the frequency domain (after normalizing the Fourier coefficients by the Fourier coefficients of the pulse if necessary) we have the relation

$$\mathbf{x}[n] = \mathbf{V}(\{\omega_\ell\})_{p \times L}\mathbf{a}[n], \quad n \in \mathbb{Z}, \tag{15.140}$$

where $\mathbf{a}[n]$ is the vector of amplitudes $a_k[n]$, and $\omega_\ell = 2\pi t_\ell/\tau$. When there are only a finite number of periods, the same reasoning holds where the set $\mathbb{Z}$ is replaced by a finite set. The derivation is exactly the same as in (15.139) with the exception that here $\omega_\ell[n]$ are independent of n. When only one time instant n is available, we can solve (15.140) using the methods described in Section 15.2. However, now we have many vectors $\mathbf{x}[n]$ that share the same delays – that is, they are related to the Fourier coefficients via a shared matrix $\mathbf{V}$. This allows the use of robust methods that recover the delays more reliably, by jointly processing the samples for all n.

For example, we can use stochastic subspace algorithms, such as MUSIC and ESPRIT, described in Section 15.2.7. These approaches are based on computing the correlation matrix $\sum_{n \in \mathbb{Z}} \mathbf{x}[n]\mathbf{x}^*[n]$. An alternative is to use the CS formulation of Section 15.2.8, where we extend it to our scenario by solving an IMV (or MMV) problem (see Section 11.6) instead of a single measurement formulation.

Clearly, the condition for the general infinite model $p \geq 2L$ is a sufficient condition here as well in order to ensure recovery of $x(t)$. However, the additional prior on the signal's structure can help to reduce the number of sampling channels. In fact, it turns out that it is sufficient to use

$$p \geq 2L - \eta + 1 \tag{15.141}$$

channels, where η is the dimension of the minimal subspace containing the vector set $\{\mathbf{a}[n], n \in \mathbb{Z}\}$ [429]. Since each vector is of length L, the dimension of the subspace spanned by all $\mathbf{a}[n]$ satisfies $\eta \leq L$. When there are a finite number of vectors, this result coincides with Theorem 11.22 which considers necessary and sufficient conditions for MMV recovery. Noting that $\eta \geq 1$, the lower bound is bounded above by $2L$, resulting in a worst-case sampling rate of $2L/T$. However, when $\eta > 1$, the number of channels p can be reduced beyond the lower limit $2L$ for the periodic model. In Chapter 10 we saw that the minimum sampling rate required for perfect recovery of a union of SI subspaces with L generators is $2L/T$, which is consistent with our results here.

Note that standard MUSIC and ESPRIT methods require that $\eta = L$, namely that the coefficient vectors $\mathbf{a}[n]$ vary enough from period to period so that they span the entire space $\mathbb{C}^L$. This requirement ensures that the empirical correlation matrix is invertible. In this case, (15.141) implies that the delays can be recovered using only $L + 1$ sampling channels. When $\eta < L$, an additional smoothing stage is needed prior to applying subspace methods, as we explained in Section 15.2.7. Alternatively, rank-aware MMV methods, discussed in Section 11.6, can be used.

To conclude, when the pulse amplitudes vary sufficiently from period to period, which is expressed by the condition $\eta = L$, the common information about the delays can be utilized to reduce the sampling rate to $(L+1)/T$ with $T = \tau$. Moreover, joint processing improves the estimation of delays in the presence of noise since it uses the mutual information between periods, rather than recovering the delays for each period separately.

Example 15.18 We now demonstrate the advantage of joint processing of semi-periodic FRI signals by comparing the MSE in time delay estimation when recovering the delays from each period separately versus joint recovery.

Consider the signal $x(t) = a_1\delta(t - 0.213) + a_2\delta(t - 0.452) + a_3\delta(t - 0.664) + a_4\delta(t - 0.7453)$, where a_i are Gaussian random variables with mean $\mu = 1$ and standard deviation $\sigma = 1$ which are drawn independently over several periods. The signal is sampled by directly obtaining $m = 9$ consecutive Fourier coefficients symmetric around zero using pure tones. Joint processing is performed using the ESPRIT algorithm (Algorithm 15.7) while recovering the delays from each period separately is obtained via matrix pencil (Algorithm 15.4) with pencil parameter $M = 4$.

The time delay MSE versus SNR for a varying number of periods is depicted in Fig. 15.44. The results are averaged over 1000 simulations. As expected, the use of the mutual information on the delays between periods improves the estimation performance significantly.

15.4.2   Filterbank sampling

An alternative scheme that can be used in the semiperiodic setting is a filterbank system [429]. The advantage of this technique is that it does not require the assumption of distinct pulse intervals, nor is it necessary for the pulse shape to have compact support. Here as well we exploit the semiperiodicity to jointly process the samples by using subspace or MMV methods.

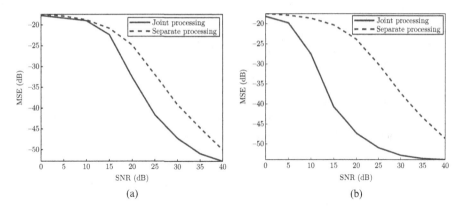

Figure 15.44 MSE in estimating the delays by joint processing versus individual processing. (a) Joint processing over 10 periods. (b) Joint processing over 50 periods.

When the pulse shape $h(t)$ is arbitrary, the derivation departs somewhat from the modulation bank technique. This is a result of the fact that the signal is not periodic and cannot be divided into distinct intervals, so that one can no longer speak of its Fourier series. Instead, in parallel to how we derived the Fourier series approach, we first consider recovery given an appropriate transform of $x(t)$. We will then discuss how the required transform can be obtained from low-rate samples of the signal.

We begin by computing the Fourier transform of the semiperiodic signal $x(t)$ of (15.8):

$$
X(\omega) = \sum_{\ell=1}^{L} \sum_{n \in \mathbb{Z}} a_\ell[n] \int_{-\infty}^{\infty} h(t - t_\ell - n\tau) e^{-j\omega t} dt
$$

$$
= H(\omega) \sum_{\ell=1}^{L} e^{-j\omega t_\ell} \sum_{n \in \mathbb{Z}} a_\ell[n] e^{-j\omega n\tau}
$$

$$
= H(\omega) \sum_{\ell=1}^{L} A_\ell \left(e^{j\omega\tau} \right) e^{-j\omega t_\ell}. \tag{15.142}
$$

To recover t_ℓ we consider slices of $X(\omega)$ of length $2\pi/\tau$, similarly to Chapter 14. Let the kth slice be defined by

$$
X_k(e^{j\omega\tau}) = X\left(\omega + \frac{2\pi k}{\tau}\right), \qquad 0 \leq \omega \leq \frac{2\pi}{\tau}. \tag{15.143}
$$

Namely, it is equal to the frequency content of $X(\omega)$ over $[2\pi k/\tau, 2\pi(k+1)/\tau]$. Using the fact that $A\left(e^{j\omega\tau}\right)$ is $2\pi/\tau$-periodic,

$$
X_k(e^{j\omega\tau}) = H\left(\omega + \frac{2\pi k}{\tau}\right) \sum_{\ell=1}^{L} A_\ell \left(e^{j\omega\tau}\right) e^{-j\omega t_\ell} e^{-j2\pi k t_\ell/\tau}. \tag{15.144}
$$

We now show that we can recover the unknown variables from a small number of slices so that we do not need to know the entire CTFT $X(\omega)$. This, in turn, translates to a reduction in sampling rate. Let $\mathcal{K}$ be a set of indices k such that $H(\omega + \frac{2\pi k}{\tau})$

is nonzero almost everywhere on $[0, 2\pi/\tau)$, and introduce the vector-valued function $\mathbf{x}(e^{j\omega\tau})$, whose elements are given by $X_k(e^{j\omega\tau})$ for $k \in \mathcal{K}$. We can then express $\mathbf{x}(e^{j\omega\tau})$ in matrix form as

$$\mathbf{x}(e^{j\omega\tau}) = \mathbf{H}(e^{j\omega\tau})\mathbf{V}(\{\omega_\ell\})\mathbf{a}(e^{j\omega\tau}), \tag{15.145}$$

where $\omega_\ell = 2\pi t_\ell/\tau$, $\mathbf{H}(e^{j\omega\tau})$ is a diagonal matrix with diagonal elements $H\left(\omega + \frac{2\pi k}{\tau}\right)$ for $k \in \mathcal{K}$, and $\mathbf{a}(e^{j\omega\tau})$ is the length-L vector of elements $e^{-j\omega t_\ell}A_\ell\left(e^{j\omega t_\ell}\right)$. Since by definition $\mathbf{H}(e^{j\omega\tau})$ is invertible, we multiply (15.145) by $\mathbf{H}^{-1}(e^{j\omega\tau})$ to obtain

$$\mathbf{y}(e^{j\omega\tau}) = \mathbf{V}(\{\omega_\ell\})\mathbf{a}(e^{j\omega\tau}), \tag{15.146}$$

where $\mathbf{y}(e^{j\omega\tau}) = \mathbf{H}^{-1}(e^{j\omega\tau})\mathbf{x}(e^{j\omega\tau})$. Taking the inverse DTFT of both sides of the equation results in

$$\mathbf{y}[n] = \mathbf{V}(\{\omega_\ell\})\mathbf{a}[n], \quad n \in \mathbb{Z}, \tag{15.147}$$

which has the same structure as (15.140) and can therefore be treated in a similar fashion. Thus, given the set of vectors $\{\mathbf{y}[n]\}$ we can recover $\{t_\ell\}$ and $\{\mathbf{a}[n]\}$ by using, for example, MUSIC, ESPRIT, or MMV methods. The latter requires discretizing the delays, as described in Section 15.2.8.

It remains to discuss how to obtain $\mathbf{y}[n]$ from low-rate samples of $x(t)$. Since we do not need the entire $X(\omega)$, the rate will be lower than the Nyquist rate. Suppose we sample $x(t)$ with $p = |\mathcal{K}|$ sampling filters $\overline{s_\ell(-t)}$. In this case the DTFT of the sample sequence (15.122) is given by

$$C_\ell\left(e^{j\omega T}\right) = \frac{1}{T}\sum_{m\in\mathbb{Z}}\overline{S_\ell\left(\omega + \frac{2\pi}{T}m\right)}X\left(\omega + \frac{2\pi}{T}m\right), \quad 0 \leq \omega \leq \frac{2\pi}{T}. \tag{15.148}$$

To proceed, we choose $T = \tau$. Our goal then is to be able to recover $X_k(e^{j\omega T})$, $k \in \mathcal{K}$ from $C_\ell(e^{j\omega T})$, $1 \leq \ell \leq p$. To this end we choose $S_\ell(\omega)$ such that for every $0 \leq \omega \leq 2\pi/T$ we have

$$S_\ell\left(\omega + \frac{2\pi}{T}k\right) = \begin{cases} 0, & k \notin \mathcal{K} \\ \text{nonzero,} & \text{otherwise.} \end{cases} \tag{15.149}$$

We can then express (15.148) as

$$\mathbf{c}(e^{j\omega T}) = \mathbf{S}(e^{j\omega T})\mathbf{x}(e^{j\omega T}), \tag{15.150}$$

where $\mathbf{S}(e^{j\omega T})$ is a matrix whose ℓkth element is given by $\frac{1}{T}\overline{S_\ell\left(\omega + \frac{2\pi}{T}k\right)}$, $k \in \mathcal{K}$. Thus, if $\mathbf{S}(e^{j\omega T})$ is invertible, then $\mathbf{x}(e^{j\omega T})$ can be recovered from $\mathbf{c}(e^{j\omega T})$.

Example 15.19 A simple example of a set of sampling kernels satisfying the requirements is the ideal bandpass filterbank given by

$$S_\ell(\omega) = \begin{cases} T, & \omega \in [(\ell-1)\frac{2\pi}{T}, \ell\frac{2\pi}{T}], \\ 0, & \text{otherwise,} \end{cases} \quad 1 \leq \ell \leq p, \tag{15.151}$$

where $\mathcal{K} = \{0, \dots, p-1\}$. The resulting $\mathbf{S}(e^{j\omega T})$ is equal to the identity.

We can also consider sampling using a single filter followed by uniform sampling with rate p/T with $T = \tau$. In this case the sample sequence is given by

$$C\left(e^{j\omega T/p}\right) = \frac{1}{T} \sum_{k \in \mathbb{Z}} \overline{S\left(\omega + \frac{2\pi}{T}kp\right)} X\left(\omega + \frac{2\pi}{T}kp\right). \qquad (15.152)$$

In order to recover $X_k(e^{j\omega T})$, $k \in \mathcal{K}$ we divide the bandwidth of $X(\omega)$ into intervals of length $2\pi/T$. We then define the ℓth subset as the intervals starting at frequencies $\Omega_\ell = \{2\pi(\ell + km)/T\}$ for all values of k such that Ω_ℓ is still in the bandwidth of $X(\omega)$. To construct a valid set $\mathcal{K}$ we must choose one element from each subset Ω_ℓ, as we did in coset sampling (see Section 15.3). The filter $S(\omega)$ is then chosen such that for every $0 \le \omega \le 2\pi/T$,

$$S\left(\omega + \frac{2\pi}{T}k\right) = \begin{cases} 0, & k \notin \mathcal{K} \\ \text{nonzero}, & \text{otherwise}. \end{cases} \qquad (15.153)$$

Next we note that $C\left(e^{j\omega T/p}\right)$ is defined over $[0, 2\pi p/T)$. We may therefore represent $C\left(e^{j\omega T/p}\right)$ in terms of p consecutive bands $C_\ell(e^{j\omega T})$ of size $2\pi/T$. Formally, let $C_\ell(e^{j\omega T}) = C(\omega + 2\pi\ell/T)$ for $\ell = 0, \ldots, p-1$ and $\omega \in [0, 2\pi/T)$ and construct a vector $\mathbf{c}(e^{j\omega T})$ whose elements are $C_\ell(e^{j\omega T})$. With these choices we can express (15.152) as

$$\mathbf{c}(e^{j\omega T}) = \mathbf{S}(e^{j\omega T})\mathbf{x}(e^{j\omega T}), \qquad (15.154)$$

where $\mathbf{S}(e^{j\omega T})$ is a diagonal matrix with diagonal elements $\frac{1}{T}S\left(\omega + \frac{2\pi}{T}k\right)$, $k \in \mathcal{K}$. Thus, $\mathbf{x}(e^{j\omega T})$ can be recovered from $\mathbf{c}(e^{j\omega T})$ as $\mathbf{x}(e^{j\omega T}) = \mathbf{S}^{-1}(e^{j\omega T})\mathbf{c}(e^{j\omega T})$.

15.5 Noisy FRI recovery

Real-world signals are often contaminated by noise and thus do not conform precisely to the FRI assumptions. Furthermore, like any mathematical model, the FRI framework is an approximation which does not precisely hold in practical scenarios, an effect known as mismodeling error. We have seen in Section 15.2 how some of the techniques for FRI recovery can be adapted to account for noise. Noise may arise both in the analog and digital domains, namely before and after sampling, as illustrated in Fig. 15.45. When noise is present, it is no longer possible to recover the original signal perfectly from its samples. However, one can sometimes mitigate the effects of noise by oversampling, i.e. by increasing the sampling rate beyond the rate of innovation.

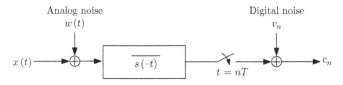

Figure 15.45 The continuous signal $x(t)$ can be corrupted in either the analog or digital paths.

In this section, we describe some results that allow us to analyze the effect of noise on the accuracy with which FRI signals can be recovered. A standard tool for accomplishing this is the *Cramér–Rao bound (CRB)*, which is a lower bound on the MSE achievable by any unbiased estimator [455]. As such, it provides a measure of the difficulty of a given estimation problem, and indicates whether or not existing techniques come close to optimal. It is also useful in evaluating the relative merit of different types of measurements. In particular, we will see that for a large class of FRI signals, sampling with complex exponentials is in fact optimal in a certain MSE sense, justifying the approach we have been following in previous sections.

As our focus in this book is not on performance analysis which involves tools of random processes, we will not go into detailed derivations, but rather concentrate on the final results and conclusions that are relevant to our discussion. The results in this section are based on [309, 456].

15.5.1   MSE bounds

Consider a finite-duration FRI signal $x(t) = h(t, \boldsymbol{\theta})$, which is determined by a finite number k of parameters represented by the vector $\boldsymbol{\theta}$, over some time interval τ. Suppose we sample a noisy version of the signal

$$y(t) = x(t) + w(t), \quad t \in [0, \tau], \tag{15.155}$$

where $w(t)$ is continuous-time white Gaussian noise with variance σ^2. The MSE of an estimator $\hat{x}(t)$ of $x(t)$ is defined as

$$\text{MSE}(\hat{x}, x) \triangleq E\{\|\hat{x} - x\|_2^2\} = E\left\{ \int_0^\tau |\hat{x}(t) - x(t)|^2 dt \right\}. \tag{15.156}$$

An estimator $\hat{x}(t)$ is said to be unbiased if

$$E\{\hat{x}(t)\} = x(t) \text{ for all possible signals } x(t) \text{ and almost all } t \in [0, \tau]. \tag{15.157}$$

Sampling-indifferent bound
To find a bound on the MSE achievable by any sampling method, it is of interest to derive the CRB for estimating $x(t)$ directly from the continuous-time process $y(t)$. Clearly, no sampling mechanism can do better than exhausting all of the information contained in $y(t)$.

This bound turns out to have a particularly simple closed-form expression which depends on the rate of innovation, but not on the class of FRI signals being estimated. Indeed, under suitable regularity conditions (detailed in [309]) the MSE of any unbiased, finite-variance estimator $\hat{x}(t)$ of a signal $x(t)$ of duration τ, satisfies

$$\frac{1}{\tau}\text{MSE}(\hat{x}, x) \geq \rho_\tau \sigma^2 \tag{15.158}$$

where $\rho_\tau = k/\tau$ is the τ-local rate of innovation. Note that when $x(t)$ does not have structure, no finite-MSE unbiased estimators exist.

The bound in (15.158) offers a new interpretation of the rate of innovation in the noisy setting, as the ratio between the best achievable MSE and the noise variance σ^2. This is to be contrasted with the characterization of the rate of innovation in the noise-free case as the lowest sampling rate allowing for perfect recovery of the signal; indeed, when noise is present, perfect recovery is no longer possible.

Bound for sampled measurements

We next present a lower bound on estimating $x(t)$ from samples of the signal $y(t)$ of (15.155). To keep the discussion general, we consider samples of the form

$$c_n = \langle \varphi_n(t), y(t) \rangle + v_n, \quad n = 0, \ldots, N - 1 \tag{15.159}$$

where $\{\varphi_n(t)\}$ is a set of sampling kernels and v_n is measurement noise, which we assume is a white-noise process with variance σ_d^2. For example, pointwise sampling at the output of an anti-aliasing filter $\varphi(-t)$ corresponds to sampling kernels $\varphi_n(t) = \varphi(t - nT)$. We denote by Φ the subspace spanned by the sampling kernels. In this setting, the samples inherit the noise $w(t)$ embedded in the signal $y(t)$, and may suffer from additional discrete-time noise, for example due to quantization. Note that unless the sampling kernels $\{\varphi_n(t)\}$ happen to be orthogonal, the resulting measurements will not be statistically independent.

We assume that there exists a Fréchet derivative $\frac{\partial x}{\partial \theta}$ which quantifies the sensitivity of $x(t)$ to changes in θ. Informally, $\frac{\partial x}{\partial \theta}$ is an operator from $\mathbb{R}^k$ to the space of square-integrable functions L_2 such that

$$x(t)|_{\theta+\delta} \approx x(t)|_\theta + \frac{\partial x}{\partial \theta}\delta. \tag{15.160}$$

More formally,

$$\lim_{\delta \to 0} \frac{\left\| x(\theta + \delta) - x(\theta) - \frac{\partial x}{\partial \theta}\delta \right\|_{\mathbb{R}^N}}{\|\delta\|_{\mathbb{R}^k}} = 0. \tag{15.161}$$

Suppose for a moment that there exist elements in the range space of $\frac{\partial x}{\partial \theta}$ which are orthogonal to Φ. This implies that one can perturb $x(t)$ without changing the distribution of the measurements $c_0, \ldots, c_{N-1}$. This situation occurs, for example, when the number of measurements N is smaller than the number k of parameters defining $x(t)$. While it may still be possible to reconstruct some of the information concerning $x(t)$ from these measurements, this is an undesirable situation from an estimation point of view. Thus we will assume that

$$\frac{\partial x}{\partial \theta} \cap \Phi^\perp = \{0\}. \tag{15.162}$$

Under these assumptions, it can be shown that any unbiased, finite-variance estimator $\hat{x}(t)$ of $x(t)$ from the samples (15.159) satisfies [309]

$$\mathrm{MSE}(\hat{x}, x) \geq \mathrm{Tr}\left[\left(\frac{\partial x}{\partial \theta}\right)^* \left(\frac{\partial x}{\partial \theta}\right) \left(\left(\frac{\partial x}{\partial \theta}\right)^* \mathbf{H}_\Phi \left(\frac{\partial x}{\partial \theta}\right) \right)^{-1} \right], \tag{15.163}$$

where $\mathbf{H}_\Phi$ is given by

$$\mathbf{H}_\Phi = \Phi \left(\sigma^2 \Phi^* \Phi + \sigma_d^2 \mathbf{I}_N \right)^{-1} \Phi^* \tag{15.164}$$

and Φ is the set transformation corresponding to the functions $\{\varphi_n(t)\}_{n=1}^N$. If $\sigma_d = 0$ so that there is only analog noise, then $\mathbf{H}_\Phi$ is proportional to the orthogonal projection onto the range space of Φ.

Note that despite the involvement of continuous-time operators, the expression within the trace in (15.163) is a $k \times k$ matrix and can therefore be computed numerically. Also observe that, in contrast to the continuous-time bound of (15.158), the sampled bound depends on the value of $\boldsymbol{\theta}$. Thus, for a specific sampling scheme, some signals can potentially be more difficult to estimate than others.

We now draw several conclusions from (15.163).

Discrete-time noise

Suppose first that $\sigma^2 = 0$, so that only digital noise is present. The effects of this noise can be surmounted either by increasing the gain of the sampling kernels, or by increasing the number of measurements. These intuitive conclusions are easily verified from (15.163). Consider the modified kernels $\tilde{\varphi}_n(t) = 2\varphi_n(t)$. The set transformation $\tilde{\Phi}$ corresponding to the modified kernels is $\tilde{\Phi} = 2\Phi$, and since $\sigma^2 = 0$, this implies that the expression within the trace is reduced by a factor of 4. Thus, a sufficient increase in sampling gain reduces the bound (15.163) arbitrarily close to zero. Similarly, it is possible to increase the number of samples, for example, by repeating each measurement twice. Let Φ and $\tilde{\Phi}$ denote the transformations corresponding to the original and doubled sets of measurements. It is then readily seen from the definition of the set transformation that $\tilde{\Phi}\tilde{\Phi}^* = 4\Phi\Phi^*$. Consequently, the same argument leads to the conclusion that in the absence of continuous-time noise one can achieve arbitrarily low error by repeated measurements. In practice, rather than repeating each measurement, an increase in sampling rate is often obtained by sampling on a denser grid.

Continuous-time noise

Consider next the situation in which $\sigma_d^2 = 0$, meaning that only continuous-time noise affects the samples. In this case, it is generally impossible to achieve arbitrarily low reconstruction error, regardless of the sampling kernels used. Indeed, it is never possible to outperform the continuous-time CRB of (15.158), which is typically nonzero.

The two bounds (15.163) and (15.158) may sometimes coincide. If this occurs, then at least in terms of the performance bounds, estimators based on the samples (15.159) will suffer no degradation compared with an estimator based on the entire continuous-time function. Such a situation occurs if $x(t) \in \Phi$ for any feasible value of $x(t)$, a situation which we refer to as "Nyquist-equivalent" sampling. In this case, $\mathbf{H}_\Phi \frac{\partial x}{\partial \theta} = \frac{1}{\sigma^2} \frac{\partial x}{\partial \theta}$, so that (15.163) reduces to

$$\frac{1}{\tau} \mathrm{MSE}(\hat{x}, x) \geq \frac{\sigma^2}{\tau} \mathrm{Tr}(\mathbf{I}_{k \times k}) = \sigma^2 \rho_\tau. \tag{15.165}$$

Many practical FRI signal models are not contained in any finite-dimensional subspace, and in these cases, any increase in the sampling rate can improve estimation performance. Even if there exists a subspace containing the entire family of FRI signals, its dimension is often much larger than the number of parameters k defining the signal. Consequently, fully exploiting the information in the signal requires sampling at the Nyquist-equivalent rate, which is potentially much higher than the rate of innovation. This fact provides an analytical explanation of the empirically observed phenomena that oversampling often provides improvement over sampling at the rate of innovation in the presence of noise.

It is interesting to examine this phenomenon from a UoS viewpoint. Denote the set of feasible signals by $\mathcal{X}$ and suppose that it can be described as a union of an infinite number of subspaces $\{\mathcal{U}_\alpha\}$ indexed by the continuous parameter α, so that $\mathcal{X} = \bigcup_\alpha \mathcal{U}_\alpha$. In this case, a finite sampling rate captures all of the information present in the signal if and only if

$$\dim\left(\sum_\alpha \mathcal{U}_\alpha\right) < \infty \qquad (15.166)$$

where $\dim(\mathcal{M})$ is the dimension of the subspace $\mathcal{M}$. By contrast, in the noise-free case, we have seen in Chapter 10 that the number of samples required to recover $x(t)$ is given by

$$\max_{\alpha_1,\alpha_2} \dim(\mathcal{U}_{\alpha_1} + \mathcal{U}_{\alpha_2}), \qquad (15.167)$$

i.e. the largest dimension among sums of two subspaces belonging to the union. In general, the dimension of (15.166) will be much higher than (15.167), illustrating the qualitative difference between the noisy and noise-free settings. For example, if the subspaces $\mathcal{U}_\alpha$ are finite-dimensional, then (15.167) is also necessarily finite, whereas (15.166) need not be.

Nevertheless, one may hope that the structure embodied in $\mathcal{X}$ will allow nearly optimal recovery using a sampling rate close to the rate of innovation. In general, the CRB for samples taken at the rate of innovation is substantially higher than the optimal, continuous-time bound. This demonstrates that the sensitivity to noise is a fundamental aspect of estimating FRI signals, rather than a limitation of existing algorithms. However, some specific FRI models, such as the semiperiodic pulse stream (15.8), exhibit considerable noise resilience, and indeed in these cases the CRB converges to the continuous-time value more rapidly. We discuss this phenomenon in the next section.

15.5.2   Periodic versus semiperiodic FRI signals

The different levels of robustness to noise can be explained when the signal models are examined in a UoS context. In this case, the vector $\boldsymbol{\theta}$ parameterizing $x(t)$ can be partitioned into parameters defining the subspace $\mathcal{U}_\alpha$ and parameters pinpointing the position within the subspace. The CRB analysis hints that estimation of the position

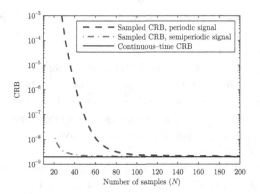

Figure 15.46 Comparison between the CRB for a periodic signal and a semiperiodic signal.

within a subspace is often easier than estimation of the subspace itself. Thus, when most parameters are used to select an intra-subspace position, estimation at the rate of innovation is successful, as occurs in the semiperiodic case (15.8). By contrast, when a large portion of the parameters define the subspace in use, a sampling rate higher than the rate of innovation is necessary; this is the situation in the nonperiodic pulse stream (15.5), wherein $\boldsymbol{\theta}$ is evenly divided among subspace-selecting parameters $\{t_\ell\}$ and intra-subspace parameters $\{a_\ell\}$.

In Fig. 15.46 we compare the CRB for reconstructing a nonperiodic pulse stream and a semiperiodic pulse stream. Both signals consist of an identical pulse $h(t)$ and have the same local rate of innovation. The sampling kernels were chosen to measure the N lowest-frequency components of the signal. For the periodic signal, we choose $L = 10$ pulses with random delays and amplitudes and a period of $\tau = 1$. This implies that the signal of interest is determined by $k = 20$ parameters (L amplitudes and L time delays). To construct a semiperiodic signal with the same number of parameters, we choose a period of $1/9$ with $L = 2$ pulses. The segment $[0, \tau]$ then contains precisely $M = 9$ periods, for a total of 20 parameters: two unknown amplitudes per period plus two unknown delays.

Since the number of parameters to be estimated is identical in both signal models, the continuous-time CRB for the two settings coincides. Consequently, for a large number of measurements, the sampled bounds also converge to the same values. However, when the number of samples is closer to the rate of innovation, the bound on the reconstruction error for the semiperiodic signal is much lower than that of the periodic signal, as shown in Fig. 15.46.

In our setting, the periodic signal contains 10 parameters for selecting the subspace and 10 additional parameters determining the position within it; whereas for the semiperiodic signal, only two parameters determine the subspace while the remaining 18 parameters set the location in the subspace. Evidently, identification of the subspace is challenging, especially in the presence of noise, but once the subspace is determined, the remaining parameters can be estimated using a simple linear operation (a projection onto the chosen subspace). Consequently, if many of the unknown parameters identify the position within a subspace, estimation may be performed more accurately.

15.5.3 Choosing the sampling kernels

An interesting (and difficult) question is how to choose an optimal set of sampling kernels in some sense for a given allotted number of samples. This problem is considered in [309] by adopting a Bayesian framework, wherein the signal $x(t)$ is a random process with the parameters distributed according to a known prior distribution. The sampling and reconstruction techniques are further assumed to be linear. While nonlinear reconstruction methods are typically used for estimating FRI signals, this assumption is required for analytical tractability, and is used only for the purpose of identifying sampling kernels. Once these kernels are chosen, they can be used in conjunction with nonlinear reconstruction algorithms (though in this case no analytic optimality conditions are provided). Given a budget of N samples, the optimal sampling kernels are given by the N eigenfunctions of the autocorrelation

$$R_X = R_X(t, T) = E\{x(t)x^*(T)\} \tag{15.168}$$

corresponding to the N largest eigenvalues.

A setting of particular interest is when the autocorrelation R_X is cyclic, in the sense that

$$R_X(t, T) = R_X((t - T) \bmod \tau) \tag{15.169}$$

for some τ. This scenario occurs, for example, in the periodic pulse stream (15.7) and the semiperiodic pulse stream (15.8) settings, assuming a reasonable prior distribution on the parameters. It is not difficult to show that the eigenfunctions of R_X are given, in this case, by the complex exponentials

$$\psi_n(t) = \frac{1}{\sqrt{\tau}} e^{j\frac{2\pi}{\tau}nt}, \quad n \in \mathbb{Z}. \tag{15.170}$$

Furthermore, the magnitudes of the eigenvalues of R_X are directly proportional to the magnitudes of the respective Fourier coefficients of the pulse shape $h(t)$. It follows that the optimal sampling kernels are the exponentials (15.170) corresponding to the largest Fourier coefficients of $h(t)$. Interestingly, this is precisely the class of sampling kernels that we have been advocating in our sub-Nyquist FRI sampling techniques.

15.6 General FRI sampling

Although we have focused primarily on sampling streams of pulses, the theory of FRI extends beyond this class of signals. In this section, we address the problem of reconstructing arbitrary FRI signals from possibly nonlinear measurements obtained at the rate of innovation, following the ideas presented in [457]. The only assumption on the sampling mechanism and signal prior is that the parameters defining the signal can be stably recovered from the samples. This assumption must be made by any practical sampling theorem that attempts to recover the signal parameters. Whereas in the previous section we focused on explicit recovery techniques, here we take an iterative approach to treat the general setting.

Since we allow for nonlinear sampling, we use algorithms similar to those introduced in the context of nonlinear sampling of subspace signals in Chapter 8. Specifically, our approach is based on an LS criterion which minimizes the error norm between the given set of samples and those of our signal estimate. Under the stability assumption, this LS criterion possesses a unique stationary point. Consequently, any optimization algorithm designed to trap a stationary point, will necessarily converge to the true parameters. In particular, we show that the steepest-descent and quasi-Newton methods, introduced in Chapter 8, can be used to recover the signal parameters in our setting as well. This approach provides a unified framework for recovering signals from samples taken at their rate of innovation without making any specific assumptions on the signal structure.

We focus on recovery of an arbitrary segment from an FRI signal of the form $x(t) = h(t, \boldsymbol{\theta})$ where $\boldsymbol{\theta} \in \mathcal{A}$ is the length-k parameter vector defining $x(t)$ and $\mathcal{A}$ is the set of possible vectors. We assume that h is Fréchet-differentiable with respect to $\boldsymbol{\theta}$. We denote the set of signals $x(t)$ by $\mathcal{X}$. In order to identify $\boldsymbol{\theta}$, we must ensure that there do not exist two vectors $\boldsymbol{\theta}_1 \neq \boldsymbol{\theta}_2$ such that $h(t, \boldsymbol{\theta}_1) = h(t, \boldsymbol{\theta}_2)$. For the recovery to be stable, we require the slightly stronger condition

$$\alpha_h \|\boldsymbol{\theta}_1 - \boldsymbol{\theta}_2\|_2 \leq \|h(t, \boldsymbol{\theta}_1) - h(t, \boldsymbol{\theta}_2)\|_2 \tag{15.171}$$

for some constant $\alpha_h > 0$ and for all $\boldsymbol{\theta}_1, \boldsymbol{\theta}_2 \in \mathcal{A}$. In the case of pulse streams, for example, (15.171) implies that $t_\ell - t_{\ell-1}$ has to be bounded above and below for every ℓ, and that $a_\ell > a > 0$.

15.6.1   Sampling method

Our goal is to recover x by observing N generalized samples $\mathbf{c} = [c_1, \ldots, c_N]^T$ obtained as $\mathbf{c} = S(x)$, where $S: L_2(\mathbb{R}) \to \mathbb{R}^N$ is some (possibly nonlinear) Fréchet-differentiable operator. For example, S can represent the samples

$$c_n = f(\langle s_n, x \rangle), \quad n = 1, \ldots, N, \tag{15.172}$$

where $f(\cdot)$ is a nonlinear sensor response and s_n are the linear sampling functions.

We say that a sampling operator S is *stable with respect to* $\mathcal{X}$ if there exist constants $0 < \alpha_s \leq \beta_s < \infty$ such that

$$\alpha_s \|x_2 - x_1\|_{L_2} \leq \|S(x_1) - S(x_2)\|_{\mathbb{R}^N} \leq \beta_s \|x_2 - x_1\|_{L_2} \tag{15.173}$$

for all $x_1, x_2 \in \mathcal{X}$. This definition is the same as that used in Chapter 10 apart from the fact that here the set $\mathcal{X}$ is not necessarily a UoS and the operator S is not required to be linear. The left-hand inequality ensures that if two signals x_1 and x_2 are sufficiently different from one another, then their samples $S(x_1)$ and $S(x_2)$ differ as well. In particular, it implies that two distinct signals $x_1, x_2 \in \mathcal{X}$ cannot produce the same set of samples, so that there is a unique recovery $x \in \mathcal{X}$ associated with every valid set of samples $\mathbf{c} = S(x) \in \mathbb{R}^N$.

Conditions (15.173) and (15.171) lie at the heart of any practical sampling theorem, whether implicitly or not. We do not discuss here in detail situations in which these conditions are satisfied, as this is rather problem-specific. The interested reader may refer to [159] and Chapter 8 for an analysis of SI signals with nonlinear samples, to [458] for linear sampling of several UoS models, and to [459] for a general theory of stability in FRI settings. In what follows, we show that conditions (15.173) and (15.171) dictate a minimal sampling rate below which perfect recovery cannot be guaranteed. More interestingly, we will also see that when these requirements are met, perfect recovery is possible at this minimal sampling rate by using a wide family of iterative algorithms.

15.6.2 Minimal sampling rate

To be able to devise a general reconstruction strategy for signals in $\mathcal{X}$ that were sampled by S, we first determine the minimal number of samples N required for perfect recovery.

Proposition 15.7. *Suppose that the function $h\colon \mathcal{A} \to L_2$ satisfies (15.171) and that the operator $S\colon L_2 \to \mathbb{R}^N$ satisfies (15.173). Then*

$$N \geq k + \max_{x_1 \in \mathcal{X}} \dim \left(\mathcal{N}\left(\left.\frac{\partial S}{\partial x}\right|^*_{x_1} \right) \right), \tag{15.174}$$

where k is the rate of innovation.

Proof: For brevity, we use the shorthand notation $h(\boldsymbol{\theta})$ instead of $h(t, \boldsymbol{\theta})$. Since $h(\boldsymbol{\theta})$ and $S(x)$ are Fréchet-differentiable, it follows that the function $\hat{\mathbf{c}}(\boldsymbol{\theta}) = S(h(\boldsymbol{\theta}))$ is Fréchet-differentiable as well. We start by showing that its derivative $\partial \hat{\mathbf{c}}/\partial \boldsymbol{\theta}$, which is an $N \times k$ matrix, has an empty null space.

By definition (15.161), the Fréchet derivative $\partial \hat{\mathbf{c}}/\partial \boldsymbol{\theta}$ at $\boldsymbol{\theta}_1$ satisfies

$$\lim_{\boldsymbol{\delta} \to 0} \frac{\left\| \hat{\mathbf{c}}(\boldsymbol{\theta}_1 + \boldsymbol{\delta}) - \hat{\mathbf{c}}(\boldsymbol{\theta}_1) - \left.\frac{\partial \hat{\mathbf{c}}}{\partial \boldsymbol{\theta}}\right|_{\boldsymbol{\theta}_1} \boldsymbol{\delta} \right\|}{\|\boldsymbol{\delta}\|} = 0. \tag{15.175}$$

In particular, for any nonzero $\mathbf{a} \in \mathbb{R}^k$,

$$\lim_{t \to 0} \frac{\left\| \hat{\mathbf{c}}(\boldsymbol{\theta}_1 + t\mathbf{a}) - \hat{\mathbf{c}}(\boldsymbol{\theta}_1) - t \left.\frac{\partial \hat{\mathbf{c}}}{\partial \boldsymbol{\theta}}\right|_{\boldsymbol{\theta}_1} \mathbf{a} \right\|}{\|t\mathbf{a}\|} = 0, \tag{15.176}$$

where t is a scalar variable. Now, assume that $\mathbf{a} \in \mathcal{N}(\partial \hat{\mathbf{c}}/\partial \boldsymbol{\theta}|_{\boldsymbol{\theta}_1})$. Then (15.176) implies that

$$\lim_{t \to 0} \frac{\|\hat{\mathbf{c}}(\boldsymbol{\theta}_1 + t\mathbf{a}) - \hat{\mathbf{c}}(\boldsymbol{\theta}_1)\|}{\|t\mathbf{a}\|} = 0. \tag{15.177}$$

However, from (15.171) and (15.173),

$$\frac{\|\hat{c}(\theta_1 + t\mathbf{a}) - \hat{c}(\theta_1)\|}{\|t\mathbf{a}\|} = \frac{\|S(h(\theta_1 + t\mathbf{a})) - S(h(\theta_1))\|}{\|t\mathbf{a}\|}$$

$$\geq \alpha_s \frac{\|h(\theta_1 + t\mathbf{a}) - h(\theta_1)\|}{\|t\mathbf{a}\|}$$

$$\geq \alpha_s \alpha_h > 0 \qquad (15.178)$$

for every $t \neq 0$. This contradicts (15.177) and therefore demonstrates that $\mathcal{N}(\partial\hat{c}/\partial\theta|_{\theta_1}) = \{0\}$, which implies that $\dim(\mathcal{R}(\partial\hat{c}/\partial\theta|_{\theta_1})) = k$.

Next, note that $\partial\hat{c}/\partial\theta|_{\theta_1} = (\partial S/\partial x|_{h(\theta_1)})(\partial h/\partial\theta|_{\theta_1})$ so that $\mathcal{R}(\partial\hat{c}/\partial\theta|_{\theta_1}) \subseteq \mathcal{R}(\partial S/\partial x|_{h(\theta_1)}) = \mathcal{N}(\partial S/\partial x|_{h(\theta_1)}^*)^{\perp}$. Therefore,

$$k \leq \dim\left(\mathcal{N}\left(\left.\frac{\partial S}{\partial x}\right|_{h(\theta_1)}^*\right)^{\perp}\right) = N - \dim\left(\mathcal{N}\left(\left.\frac{\partial S}{\partial x}\right|_{h(\theta_1)}^*\right)\right). \qquad (15.179)$$

Since (15.179) holds for every $\theta_1 \in \mathcal{A}$, it holds for the θ_1 minimizing the right-hand side, completing the proof. □

Proposition 15.7 shows that stable recovery is impossible when sampling below the rate of innovation. It also implies that if the null space of $(\partial S/\partial x)^*$ is nonempty at some $x \in \mathcal{X}$, then sampling at the rate of innovation is insufficient. We consider some examples below.

Example 15.20 Suppose that $S(x) = S^*x$ is a linear sampling operator and $\mathcal{X}$ is a subspace, spanned by vectors $\{x_\ell\}_{\ell=1}^k$. To compute the lower bound in the proposition we need to examine the dimension of $\mathcal{N}((\partial S/\partial x)^*)$. In this case, $\partial S/\partial x = S$ so that $\mathcal{N}((\partial S/\partial x)^*) = \mathcal{N}(S^*)$. To achieve the minimal rate of k we require that $\mathcal{N}(S^*)$ is 0 over $\mathcal{X}$. In other words, the $N \times k$ matrix whose $n k$th entry is $\langle s_n, x_k \rangle$ should have an empty null space, where $\{s_n\}_{n=1}^N$ are the sampling vectors.

If S is linear but $\mathcal{X}$ is not contained in any finite-dimensional subspace, then sampling at the rate of innovation necessitates that the sampling vectors $\{s_n\}_{n=1}^N$ be linearly independent. Indeed, if $\{s_n\}_{n=1}^N$ are linearly dependent, then there exists an index j such that $s_j = \sum_{n \neq j} a_n s_n$ for some coefficients $\{a_n\}_{n \neq j}$. Consequently, the sample c_j can be expressed in terms of the other samples as $c_j = \langle s_j, x \rangle = \sum_{n \neq j} \bar{a}_n \langle s_n, x \rangle = \sum_{n \neq j} \bar{a}_n c_n$ and thus can be disregarded.

Example 15.21 Suppose that one of the measurements produced by the sensing device, say c_1, is the energy $0.5\|x\|^2$ of x. In this case $(\partial c_1/\partial x)|_{x_1} = x_1$. Consequently, from Proposition 15.7, sampling at the minimal rate is impossible if the set of signals $\mathcal{X}$ contains the signal $x_1 = 0$. The intuition here follows from the observation that small perturbations in x around the signal $x_1 = 0$ do not show in c_1. Therefore, if the input to our sampling device happens to be $x = 0$ in this setting, then sampling is unavoidably unstable, as the left-hand side of condition (15.173) cannot hold.

Throughout the rest of our derivations we focus on the case in which $N = k$ samples of $x(t)$ are obtained with an operator S satisfying

$$\mathcal{N}\left(\left.\frac{\partial S}{\partial x}\right|_{x_1}^*\right) = \{0\}, \quad \forall x_1 \in \mathcal{X}. \tag{15.180}$$

This corresponds to sampling at the rate of innovation.

15.6.3   Least squares recovery

Suppose we want to recover a signal $x = h(\boldsymbol{\theta}_0) \in L_2$ from its samples $\mathbf{c} = S(x)$, where $\boldsymbol{\theta}_0 \in \mathbb{R}^k$ is an unknown parameter vector and $S \colon L_2 \to \mathbb{R}^k$ is a given sampling operator. To address this problem, it is natural to seek the minimizer of the function

$$\varepsilon(\boldsymbol{\theta}) = \frac{1}{2}\|S(h(\boldsymbol{\theta})) - \mathbf{c}\|^2 = \frac{1}{2}\|\hat{\mathbf{c}}(\boldsymbol{\theta}) - \mathbf{c}\|^2, \tag{15.181}$$

where we defined $\hat{\mathbf{c}}(\boldsymbol{\theta}) = S(h(\boldsymbol{\theta}))$. The reasoning behind this choice follows from the following observation:

Proposition 15.8.   *Suppose that the function $h \colon \mathbb{R}^k \to L_2$ satisfies (15.171) and that the operator $S \colon L_2 \to \mathbb{R}^k$ satisfies (15.173). Then $\boldsymbol{\theta}_0$ is the unique global minimizer of $\varepsilon(\boldsymbol{\theta})$.*

Proof: Clearly, $\varepsilon(\boldsymbol{\theta}) \geq 0$ for every $\boldsymbol{\theta} \in \mathbb{R}^k$ and $\varepsilon(\boldsymbol{\theta}_0) = 0$, so that $\boldsymbol{\theta}_0$ is a global minimizer of $\varepsilon(\boldsymbol{\theta})$. This minimizer is unique since, due to (15.171) and (15.173), $\varepsilon(\boldsymbol{\theta}) \geq \alpha_s \alpha_h \|\boldsymbol{\theta} - \boldsymbol{\theta}_0\|$ so that $\varepsilon(\boldsymbol{\theta}) > 0$ for every $\boldsymbol{\theta} \neq \boldsymbol{\theta}_0$.   □

The LS criterion (15.181) is also plausible when the samples $\mathbf{c}$ correspond to a perturbation of the true sample vector by white Gaussian noise. In this case, the minimizer of (15.181) is a maximum-likelihood estimate of $\boldsymbol{\theta}$ from $\mathbf{c}$.

Unfortunately, the function $\varepsilon(\boldsymbol{\theta})$ is generally nonconvex and might possess many local minima. It therefore seems that standard optimization techniques may fail in finding its global minimizer $\boldsymbol{\theta}_0$. However, as we show next, when sampling at the rate of innovation, assumptions (15.171) and (15.173) guarantee that $\boldsymbol{\theta}_0$ is the unique stationary point of $\varepsilon(\boldsymbol{\theta})$. Thus, any algorithm designed to trap a stationary point necessarily converges to the true parameter vector $\boldsymbol{\theta}_0$.

Theorem 15.3 (Stationary points).   *Suppose that the function $h \colon \mathbb{R}^k \to L_2$ satisfies (15.171), the operator $S \colon L_2 \to \mathbb{R}^k$ satisfies (15.173), and its Fréchet derivative $\partial S/\partial x$ satisfies (15.180). Then $\nabla \varepsilon(\boldsymbol{\theta}_1) = \mathbf{0}$ only if $\boldsymbol{\theta}_1 = \boldsymbol{\theta}_0$.*

Proof: The gradient $\nabla \varepsilon(\boldsymbol{\theta}_1)$ is given by

$$\nabla \varepsilon(\boldsymbol{\theta}_1) = \left.\frac{\partial \hat{\mathbf{c}}}{\partial \boldsymbol{\theta}}\right|_{\boldsymbol{\theta}_1}^* (\hat{\mathbf{c}}(\boldsymbol{\theta}_1) - \mathbf{c}). \tag{15.182}$$

We showed in the proof of Proposition 15.7 that $\mathcal{R}(\partial \hat{c}/\partial \theta|_{\theta_1}) = \mathbb{R}^k$. Since here $\partial \hat{c}/\partial \theta|_{\theta_1}$ is a $k \times k$ matrix, it follows that

$$\mathcal{N}\left(\left.\frac{\partial \hat{c}}{\partial \theta}\right|^{*}_{\theta_1}\right) = \mathcal{R}\left(\left.\frac{\partial \hat{c}}{\partial \theta}\right|_{\theta_1}\right)^{\perp} = \{\mathbf{0}\}, \tag{15.183}$$

so that $\nabla \varepsilon(\theta_1) = 0$ only if $\hat{c}(\theta_1) - c = 0$. This, by Proposition 15.8, happens only when $\theta_1 = \theta_0$, completing the proof. $\qquad\square$

The importance of Theorem 15.3 lies in the fact that it provides a unified mechanism for recovering FRI signals from samples taken at the rate of innovation. That is, rather than developing a different algorithm for every choice of signal family and sampling method, we can employ the same general-purpose optimization technique to find the stationary point of (15.181).

15.6.4　Iterative recovery

There are many optimization algorithms that can be used to find the stationary point of the objective function $\varepsilon(\theta)$ over $\mathcal{A}$. For simplicity, we focus here on unconstrained optimization methods that are designed for the case in which $\mathcal{A} = \mathbb{R}^k$. This does not limit the generality of the discussion since if $\mathcal{A} \neq \mathbb{R}^k$, then the constrained problem $\min_{\theta \in \mathcal{A}} \varepsilon(\theta)$ can be transformed into the unconstrained problem $\min_{\tilde{\theta} \in \mathbb{R}^k} \varepsilon(p(\tilde{\theta}))$, where $p \colon \mathbb{R}^k \to \mathcal{A}$ is bijective. The latter problem possesses a unique stationary point $\tilde{\theta}_0 = p^{-1}(\theta_0)$. Therefore, once $\tilde{\theta}_0$ is determined, the desired solution is $\theta_0 = p(\tilde{\theta}_0)$.

Example 15.22　To show how any constrained model can be converted into an unconstrained one, consider the stream of pulses (15.5) under the condition that $a_\ell > a > 0$ and $T_{\min} < t_\ell - t_{\ell-1} < T_{\max}$ with $0 < T_{\min} \leq T_{\max} < \infty$.
Define

$$\tilde{\theta}^{\text{Li}}_m = \ln(a_m - a_0), \quad \tilde{\theta}^{\text{N}}_m = \tan\left(\pi \frac{t_m - t_{m-1} - \bar{T}}{\Delta}\right), \tag{15.184}$$

where $\bar{T} = (T_{\max} + T_{\min})/2$ and $\Delta = T_{\max} - T_{\min}$. Here $\tilde{\theta}^{\text{Li}}_m$ denotes the linear parameters and $\tilde{\theta}^{\text{N}}_m$ the nonlinear parameters. Then,

$$a_m = e^{\tilde{\theta}^{\text{Li}}_m} + a_0, \quad t_m = t_0 + m\bar{T} + \frac{\Delta}{\pi} \sum_{i=1}^{m} \arctan\left(\tilde{\theta}^{\text{N}}_i\right). \tag{15.185}$$

With this choice, the set $\mathcal{X}$ of all feasible signals is obtained by varying $\tilde{\theta}^{\text{Li}}$ and $\tilde{\theta}^{\text{N}}$ over the entire space $\mathbb{R}^L$ and not over some subset of $\mathbb{R}^L$.

As we discussed in Section 8.4.4, many unconstrained optimization methods start with an initial guess θ^0 and perform iterations of the form

$$\theta^{\ell+1} = \theta^\ell - \gamma^\ell \mathbf{B}^\ell \nabla \varepsilon(\theta^\ell), \tag{15.186}$$

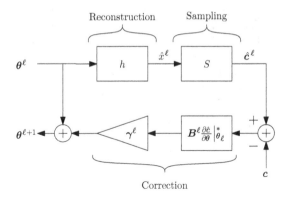

Figure 15.47 Schematic interpretation of one iteration of (15.186).

where γ^ℓ is a scalar step size obtained by means of a one-dimensional search and $\mathbf{B}^\ell$ is a positive definite matrix. Due to the structure of $\nabla\varepsilon(\boldsymbol{\theta}^\ell)$ in our case (see (15.182)), the iterations (15.186) can be given a simple interpretation, as shown in Fig. 15.47. Specifically, at the ℓth iteration, the current guess $\boldsymbol{\theta}^\ell$ of the parameters $\boldsymbol{\theta}$ is used to construct our estimate $\hat{x}^\ell$ of the signal x by applying the function h. This approximation is then sampled using the operator S to obtain an estimated sample vector $\hat{\mathbf{c}}^\ell$. Finally, the difference between $\hat{\mathbf{c}}^\ell$ and the samples $\mathbf{c}$ is multiplied by a correction matrix and added to $\boldsymbol{\theta}^\ell$ to yield the updated approximation $\boldsymbol{\theta}^{\ell+1}$.

In our setting, the objective function $\varepsilon(\boldsymbol{\theta})$ is bounded from below. The iterations (15.186) are therefore guaranteed to converge to a stationary point of $\varepsilon(\boldsymbol{\theta})$ if the following conditions hold: γ^ℓ is chosen to satisfy the Wolfe conditions (see Section 8.4.4 for more details), $\mathbf{B}^\ell$ satisfies

$$\frac{\left\langle \mathbf{B}^\ell \nabla\varepsilon(\boldsymbol{\theta}^\ell), \nabla\varepsilon(\boldsymbol{\theta}^\ell) \right\rangle}{\left\| \mathbf{B}^\ell \nabla\varepsilon(\boldsymbol{\theta}^\ell) \right\| \left\| \nabla\varepsilon(\boldsymbol{\theta}^\ell) \right\|} > \delta \tag{15.187}$$

for some constant $\delta > 0$ independent of ℓ, and the gradient $\nabla\varepsilon(\boldsymbol{\theta})$ is Lipschitz-continuous in an environment of the level set $\mathcal{N} = \{\boldsymbol{\theta} : \varepsilon(\boldsymbol{\theta}) \leq \varepsilon(\boldsymbol{\theta}^0)\}$ [180].

A step size satisfying the Wolfe conditions can be found by using the backtracking method presented in Algorithm 15.8 (this algorithm is identical to Algorithm 8.3 introduced in the context of nonlinear sampling). Equation (15.187) is trivially satisfied with $\mathbf{B}^\ell = \mathbf{I}$, which corresponds to steepest-descent iterations. It can be shown (see [457] and Section 8.4.4) that this condition also holds for $\mathbf{B}^\ell = (\partial\hat{\mathbf{c}}/\partial\boldsymbol{\theta}|_{\boldsymbol{\theta}^\ell}^* \partial\hat{\mathbf{c}}/\partial\boldsymbol{\theta}|_{\boldsymbol{\theta}^\ell})^{-1}$ if

$$\|h(\boldsymbol{\theta}_1) - h(\boldsymbol{\theta}_2)\|_{L_2} \leq \beta_h \|\boldsymbol{\theta}_1 - \boldsymbol{\theta}_2\| \tag{15.188}$$

for some $\beta_h < \infty$ and for all $\boldsymbol{\theta}_1, \boldsymbol{\theta}_2 \in \mathcal{N}$. This choice belongs to the class of quasi-Newton methods, which typically converge much faster than steepest descent. Finally, it is shown in [457] that a sufficient condition for $\nabla\varepsilon(\boldsymbol{\theta})$ to be Lipschitz-continuous over $\mathcal{N}$ is that the derivative of h is Lipschitz-continuous there.

To summarize, we have the following result.

Algorithm 15.8 Backtracking line search

Input: Function $\varepsilon(\boldsymbol{\theta})$, matrix $\mathbf{B}^\ell$, current iterate $\boldsymbol{\theta}^\ell$, constants $\rho, \eta \in (0, 1)$
Output: Step size γ^ℓ
set $\mathbf{g}^\ell = \nabla\varepsilon(\boldsymbol{\theta}^\ell)$, $\mathbf{d}^\ell = -\mathbf{B}^\ell \mathbf{g}^\ell$, $\delta = 1$
while $\varepsilon(\boldsymbol{\theta}^\ell + \delta\mathbf{d}^\ell) > \varepsilon(\boldsymbol{\theta}^\ell) + \eta\delta\langle\mathbf{d}^\ell, \mathbf{g}^\ell\rangle$ **do**
  $\delta \leftarrow \rho\delta$
end while
return $\gamma^\ell = \delta$

Theorem 15.4 (Algorithm convergence). *Suppose that the function $h\colon \mathbb{R}^k \to L_2$ satisfies (15.171), its Fréchet derivative $\partial h/\partial\boldsymbol{\theta}$ is Lipschitz-continuous over $\mathcal{N} = \{\boldsymbol{\theta}\colon \varepsilon(\boldsymbol{\theta}) \leq \varepsilon(\boldsymbol{\theta}^0)\}$, the operator $S\colon L_2 \to \mathbb{R}^k$ satisfies (15.173), and its Fréchet derivative $\partial S/\partial x$ satisfies (15.180). Consider the iterations (15.186), where the step size γ^ℓ is obtained via Algorithm 15.8 and let $\hat{\mathbf{c}}(\boldsymbol{\theta}) = S(h(\boldsymbol{\theta}))$. Then each of the following options guarantees that $\boldsymbol{\theta}^\ell \to \boldsymbol{\theta}_0$:*

1. $\mathbf{B}^\ell = \mathbf{I}$.
2. $\mathbf{B}^\ell = (\partial\hat{\mathbf{c}}/\partial\boldsymbol{\theta}|^*_{\boldsymbol{\theta}^\ell}\partial\hat{\mathbf{c}}/\partial\boldsymbol{\theta}|_{\boldsymbol{\theta}^\ell})^{-1}$ *and condition (15.188) holds.*

We now consider an application of this approach to recovery of pulse streams from nonlinear samples. Further examples can be found in [457].

Example 15.23 Consider the case in which $x(t)$ is given by (15.5). We transform the parameters using the transformation defined by (15.184) in Example 15.22. The constraints we assume on the parameters correspond to $a_0 = 0.1$, $T_{\min} = 0.3$, $T_{\max} = 0.7$, and $t_0 = -0.3$. Our goal is to recover the signal parameters from the samples (15.172), where $\{s_n(t)\}_{n=1}^N$ are sampling kernels in $L_2([0, \tau])$ and $f(\cdot)$ is a nonlinear response function. In this example, we choose $s_n(t) = s(t - T_0 - nT_s)$ with $T_0 = T_s/2, T_s = \tau/N$ and $\tau = 1$, so that the sampling functions span the entire observation segment $[0, 1]$. The pulse shape $h(t)$ and the sampling filter $s(t)$ are taken to be Gaussian functions with variances $\sigma_h^2 = 0.05$ and $\sigma_s^2 = 0.1$, respectively. The nonlinear response curve was set to be $f(c) = 100 \arctan(0.01c)$.

To apply the quasi-Newton or steepest-descent methods, we note that, with the transformation $\boldsymbol{\theta} = p(\tilde{\boldsymbol{\theta}})$ of (15.185),

$$\frac{\partial\hat{\mathbf{c}}}{\partial\tilde{\boldsymbol{\theta}}} = \frac{\partial\hat{\mathbf{c}}}{\partial\boldsymbol{\theta}}\frac{\partial p}{\partial\tilde{\boldsymbol{\theta}}}. \tag{15.189}$$

Explicit computation shows that

$$\frac{\partial\hat{\mathbf{c}}}{\partial\boldsymbol{\theta}} = \mathbf{C}\,[\mathbf{A}\ \mathbf{B}] \tag{15.190}$$

with

$$
\mathbf{A} = \begin{bmatrix} -a_1 \langle s_1, h'(t-t_1) \rangle & \cdots & -a_L \langle s_1, h'(t-t_L) \rangle \\ \vdots & & \vdots \\ -a_1 \langle s_N, h'(t-t_1) \rangle & \cdots & -a_L \langle s_N, h'(t-t_L) \rangle \end{bmatrix},
\tag{15.191}
$$

$$
\mathbf{B} = \begin{bmatrix} \langle s_1, h(t-t_1) \rangle & \cdots & \langle s_1, h(t-t_L) \rangle \\ \vdots & & \vdots \\ \langle s_N, h(t-t_1) \rangle & \cdots & \langle s_N, h(t-t_L) \rangle \end{bmatrix}
\tag{15.192}
$$

and

$$
\mathbf{C} = \mathrm{diag}\left[f'(\langle s_1, x \rangle), \ldots, f'(\langle s_N, x \rangle) \right].
\tag{15.193}
$$

Furthermore,

$$
\frac{\partial p}{\partial \tilde{\theta}} = \begin{bmatrix} \mathbf{D} & \mathbf{0} \\ \mathbf{0} & \mathbf{E} \end{bmatrix}
\tag{15.194}
$$

with

$$
\mathbf{D} = \mathrm{diag}\left[e^{\tilde{\theta}_1}, \ldots, e^{\tilde{\theta}_L} \right]
\tag{15.195}
$$

and

$$
\mathbf{E} = \frac{\Delta}{\pi} \begin{bmatrix} \frac{1}{1+\tilde{\theta}_{L+1}^2} & 0 & \cdots & 0 \\ \frac{1}{1+\tilde{\theta}_{L+1}^2} & \frac{1}{1+\tilde{\theta}_{L+2}^2} & \cdots & 0 \\ \vdots & \vdots & \ddots & \vdots \\ \frac{1}{1+\tilde{\theta}_{L+1}^2} & \frac{1}{1+\tilde{\theta}_{L+2}^2} & \cdots & \frac{1}{1+\tilde{\theta}_{2L}^2} \end{bmatrix}.
\tag{15.196}
$$

Figure 15.48 demonstrates the convergence of the Newton iterations for recovering $L = 2$ pulses over the period $[0, 1]$ from $N = 4$ samples, which is equal to the rate of innovation. The solid lines represent the true pulses while the dashed lines denote the estimated pulses. Note that all inner products in (15.191) and (15.192) can be computed analytically at every iteration. The true parameters in this experiment were $t_1 = 0.2$, $t_2 = 0.8$, $a_1 = 1$ and $a_2 = 5$. As shown in Fig. 15.48(a), the iterations were initialized at $t_1 = 1/3$, $t_2 = 2/3$, and $a_1 = a_2 = 3$. The estimated samples at this point, shown in X-marks, deviate substantially from the true samples, marked with circles. However, this gap decreases quickly in the first 15 iterations (see Fig. 15.48(b)) and almost completely vanishes after 30 iterations (Fig. 15.48(c)). Figure 15.48(d) shows the rapid decrease in the LS objective (15.181) as a function of the iterations.

Figure 15.49 demonstrates the behavior of the algorithm in the presence of noise. The setting here is the same as that of Fig. 15.48 with the distinction that white Gaussian noise is added to the samples prior to recovery. This figure depicts the

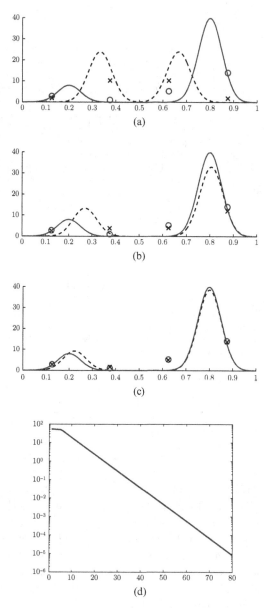

Figure 15.48 Convergence of Newton iterations for pulse stream recovery. (a) Initialization. (b) 15th iteration. (c) 30th iteration. (d) LS objective value as a function of the iterations.

MSE as a function of the SNR. The solid line corresponds to the CRB. As can be seen, the MSE of our method coincides with the CRB in high SNR scenarios and outperforms it at low SNR levels. This is a result of the fact that our technique is biased.

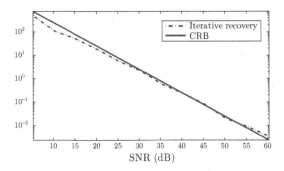

Figure 15.49 MSE as a function of SNR for pulse stream recovery in the setting of Fig. 15.48.

15.7 Applications of FRI

In this section we discuss two applications of the pulse-stream models we studied in this chapter to radar and ultrasound imaging. These applications are treated in detail in [424, 425, 426, 427, 430, 448].

There are a myriad of other possible settings in which FRI modeling can play an important role. One example is image super-resolution, which is considered in [460]. Another interesting application is to signal compression [461]. Other examples include ultra-wideband communications [462] and neuroscience [463]. Many of the ideas presented here have also been extended to multidimensional FRI signals [464, 465].

15.7.1 Sub-Nyquist radar

We begin by applying FRI theory to a pulse-Doppler based radar system following the ideas in [426]. As we show below, identifying targets' distance and closing velocity in such a radar system becomes equivalent to time-delay and (complex) amplitude estimation from pulse streams. Using the FRI framework together with appropriate processing, referred to as *Doppler focusing* below, allows sampling and processing of radar signals at rates far below Nyquist. Furthermore, Doppler focusing enjoys an optimal SNR improvement which scales linearly with the number of pulses, obtaining good detection performance even at SNR as low as -25 dB.

Radar model
Consider a radar system that detects targets by transmitting a periodic stream of pulses and processing its reflections. The transmitted signal consists of P equally spaced pulses $h(t)$:

$$x_T(t) = \sum_{p=0}^{P-1} h(t - p\tau), \quad 0 \le t \le P\tau. \tag{15.197}$$

The pulse-to-pulse delay τ is referred to as the pulse repetition interval (PRI), and the entire span of the signal in (15.197) is called the coherent processing interval (CPI).

The pulse $h(t)$ is a known time-limited baseband function with CTFT $H(\omega)$ that has negligible energy at frequencies beyond $B_h/2$.

The target scene is composed of L nonfluctuating point targets (Swerling 0 model; see [466, 467]). The pulses reflect off the L targets and propagate back to the transceiver. Each target ℓ is defined by three parameters:

1. A time delay $\tau_\ell = 2r_\ell/c$, proportional to the target's distance from the radar r_ℓ, where c is the speed of light.
2. A Doppler radial frequency $\nu_\ell = 2\dot{r}_\ell f_c/c$, proportional to the target–radar closing velocity, namely, the target's velocity radial component $\dot{r}_\ell$, and the radar's carrier frequency f_c.
3. A complex amplitude α_ℓ, proportional to the target's radar cross-section, dispersion attenuation, and all other propagation factors.

In order to simplify the received signal model, we make the following assumptions on the targets' location and motion:

1. Far targets: target–radar distance is large compared with the distance change during the CPI which allows for constant α_ℓ within the CPI:

$$\dot{r}_\ell P\tau \ll r_\ell \Rightarrow \nu_\ell \ll \frac{f_c \tau_\ell}{P\tau}. \tag{15.198}$$

2. Slow targets: low target velocity allows for constant τ_ℓ during the CPI and constant Doppler phase during pulse time. This condition holds when the baseband Doppler frequency is smaller than the frequency resolution:

$$\frac{2\dot{r}_\ell B_h}{c} \ll \frac{1}{P\tau} \Rightarrow \nu_\ell \ll \frac{f_c}{P\tau B_h}. \tag{15.199}$$

3. Small acceleration: target velocity remains approximately constant during the CPI allowing for constant ν_ℓ. This condition is satisfied when the velocity change induced by acceleration is smaller than the velocity resolution:

$$\ddot{r}_\ell P\tau \ll \frac{c}{2f_c P\tau} \Rightarrow \ddot{r}_\ell \ll \frac{c}{2f_c(P\tau)^2}. \tag{15.200}$$

Although these assumptions may seem hard to comply with, they all rely on slow enough relative motion between the radar and its targets. Radar systems tracking people, ground vehicles, and sea vessels usually comply quite easily, as we illustrate in the following example.

Example 15.24 Consider a $P = 100$ pulse radar system with PRI $\tau = 100\,\mu s$, bandwidth $B_h = 30\,MHz$, and carrier frequency $f_c = 3\,GHz$, tracking cars traveling up to $\dot{r}_\ell = 120\,km/h$. To compute the maximal distance change over the CPI we note that $\dot{r}_\ell = 33.3\,m/s$. Therefore, the maximal distance change over the observation interval is $0.33\,m$. If the target's minimal distance from the radar is a few meters, then assumption (1) is satisfied. As for (2), the maximal Doppler frequency is given by $2\dot{r}_\ell f_c/c = 667\,Hz$, which is much smaller than $f_c/P\tau B_h = 10\,kHz$.

An extreme acceleration of $10\,\mathrm{m/s^2}$ would cause a velocity change of $0.1\,\mathrm{m/sec}$ over the CPI, easily satisfying (3).

Under these three assumptions, we can write the received signal as

$$x_{\mathrm{R}}(t) = \sum_{p=0}^{P-1}\sum_{\ell=0}^{L-1} \alpha_\ell h(t - \tau_\ell - p\tau)e^{-j\nu_\ell p\tau}. \tag{15.201}$$

It will be convenient to express the signal as a sum of single frames:

$$x_{\mathrm{R}}(t) = \sum_{p=0}^{P-1} x_p(t), \tag{15.202}$$

where

$$x_p(t) = \sum_{\ell=0}^{L-1} \alpha_\ell h(t - \tau_\ell - p\tau)e^{-j\nu_\ell p\tau}. \tag{15.203}$$

In reality $x_{\mathrm{R}}(t)$ will be contaminated by additive noise and clutter. We will take this into account in the simulations below. A detailed discussion of sub-Nyquist clutter filtering is beyond our scope here and can be found in [468].

Our goal is to sample the received signal at sub-Nyquist rates and to reconstruct the $3L$ parameters $\{\tau_\ell, \nu_\ell, \alpha_\ell\}$, with $0 \leq \ell \leq L - 1$. Estimating τ_ℓ and ν_ℓ will enable approximation of each target's distance and closing velocity, as demonstrated in Fig. 15.50. The targets in the figure can be equivalently represented by a delay–Doppler map, where

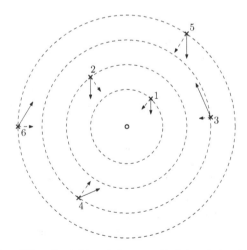

Figure 15.50 Illustration of $L = 6$ point targets in space. Each target, denoted by a cross, is located at a specific distance from the radar and moving at an unknown velocity (continuous vector). The closing velocity of each target is denoted by the dashed vectors. Our goal is to sample the received signal at the radar (in the center) at sub-Nyquist rates and to reconstruct the distance and radial velocity of each target.

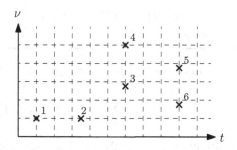

Figure 15.51 The targets of Fig. 15.50 represented by a delay–Doppler map.

each target is marked by its delay (proportional to its distance) and its Doppler frequency (proportional to the radial velocity), as shown in Fig. 15.51.

Classic sampling

Before describing our proposed sub-Nyquist processing, we review classic methods for radar processing, which typically consist of the following stages [466, 467]:

1. **ADC:** Sample each incoming frame $x_p(t)$ at its Nyquist rate B_h, equal to the bandwidth of $h(t)$, creating $x_p[n]$, $0 \leq n < N$, where $N = \tau B_h$. We assume for simplicity that N is an integer.
2. **Matched filter:** Apply a standard matched filter on each frame $x_p[n]$, as explained in Section 15.2.1. This results in the outputs $y_p[n] = x_p[n] * \overline{h[-n]}$, where $h[n]$ is the sampled version of the transmitted pulse $h(t)$, at its Nyquist rate.
3. **Doppler processing:** For each discrete time index n, perform a P-point DFT along the pulse dimension, namely, $z_n[k] = \text{DFT}_P\{y_p[n]\} = \sum_{p=0}^{P-1} y_p[n]e^{-j2\pi pk/P}$ for $0 \leq k < P$.
4. **Delay–Doppler map:** Stacking the vectors $\mathbf{z}_n$, and taking absolute value, we obtain a delay–Doppler map $\mathbf{Z} = \text{abs}[\mathbf{z}_0, \ldots, \mathbf{z}_{N-1}] \in \mathbb{R}^{P \times N}$.
5. **Peak detection:** A heuristic detection process, where knowledge of number of targets, target power, clutter location, etc. may help in discovering target positions. For example, if we know there are L targets, then we can choose the L strongest points in the map.

This process is illustrated in Fig. 15.52.

The classic processing described above requires sampling the received signal at its Nyquist rate B_h, which can be hundreds of MHz and even up to several GHz. The required computational power is P convolutions of a signal of length $N = \tau B_h$ and N FFTs of length P – both also proportional to B_h. Using the FRI framework leads to low-rate sampling and processing of radar signals, regardless of their bandwidth.

Doppler focusing

Doppler focusing is a processing technique, suggested in [426], which uses target echoes from different pulses to create a single superimposed pulse focused at a particular Doppler frequency. The techniques suggested in this chapter can then be used to

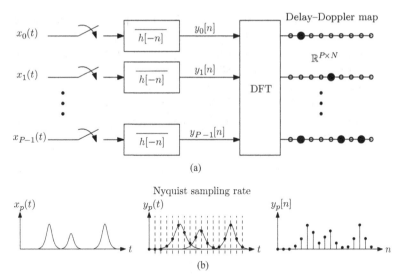

Figure 15.52 Classic processing of radar signals. (a) Each period of the signal is sampled at its Nyquist rate, followed by pulse-matched filtering. Doppler processing then consists of performing a DFT followed by peak detection in the delay–Doppler map. (b) Creating one of the frames $y_p[n]$.

recover the corresponding delays. This method results in an optimal SNR boost, and can be carried out in the frequency domain, thus enabling sub-Nyquist sampling and processing with the same SNR increase as matched filtering.

We can view the output of Doppler processing as a discrete equivalent of the following time shift and modulation operation on the received signal:

$$\Phi(t;\nu) = \sum_{p=0}^{P-1} x_p(t + p\tau)e^{j\nu p\tau} = \sum_{\ell=0}^{L-1} \alpha_\ell h(t - \tau_\ell) \sum_{p=0}^{P-1} e^{j(\nu - \nu_\ell)p\tau}, \qquad (15.204)$$

where we used (15.203). Consider the sum $g(\nu|\nu_\ell) = |\sum_{p=0}^{P-1} e^{j(\nu-\nu_\ell)p\tau}|$. For any given ν, targets with Doppler frequency ν_ℓ in a band of width $2\pi/P\tau$ around ν will achieve coherent integration and an SNR boost of approximately

$$g(\nu|\nu_\ell) = \left| \sum_{p=0}^{P-1} e^{j(\nu-\nu_\ell)p\tau} \right| \overset{|\nu-\nu_\ell|<\pi/P\tau}{\cong} P. \qquad (15.205)$$

On the other hand, since the sum of P equally spaced points covering the unit circle is generally close to zero, targets with ν_ℓ not "in focus" will approximately cancel out. We can therefore estimate the sum of exponents in (15.204) as

$$\Phi(t;\nu) \cong P \sum_{\ell\in\Lambda(\nu)} \alpha_\ell h(t - \tau_\ell) \qquad (15.206)$$

where $\Lambda(\nu) = \{\ell\colon |\nu - \nu_\ell| < \pi/P\tau\}$. In other words, the sum is only over the targets whose Doppler shifts are in the range $|\nu - \nu_\ell| < \pi/P\tau$.

For each Doppler frequency ν, $\Phi(t; \nu)$ represents a standard pulse-stream model where the problem is to estimate the unknown delays. Thus, using Doppler focusing, we have reduced our problem to delay-only estimation for a small range of Doppler frequencies, with increased SNR by a factor of P. We have seen throughout this chapter that the delays can be found from low-rate samples that are equivalent to a small set of Fourier coefficients. We therefore show next how we can perform Doppler focusing directly on these low-rate samples by viewing Doppler focusing in the frequency domain.

Suppose that we sample m values of $x_{\mathrm{R}}(t)$ over each period τ, such that we obtain m Fourier coefficients $c_p[k]$, $k \in \mathcal{K}$. Here $\mathcal{K}$ are the chosen Fourier frequencies with $|\mathcal{K}| = m$, and $c_p[k]$ is the Fourier series of the pth frame which can be obtained by replacing $t \to t + p\tau$ and $\alpha_\ell \to \alpha_\ell e^{-j\nu_\ell p\tau}$ in (15.25):

$$c_p[k] = \frac{1}{\tau} H\left(\frac{2\pi k}{\tau}\right) \sum_{\ell=0}^{L-1} \alpha_\ell e^{-j\nu_\ell p\tau} e^{-j2\pi k\tau_\ell/\tau}. \tag{15.207}$$

Performing the Doppler focusing operation results in

$$\Psi_\nu[k] = \sum_{p=0}^{P-1} c_p[k] e^{j\nu p\tau} = \frac{1}{\tau} H\left(\frac{2\pi k}{\tau}\right) \sum_{\ell=0}^{L-1} \alpha_\ell e^{-j2\pi k\tau_\ell/\tau} \sum_{p=0}^{P-1} e^{j(\nu-\nu_\ell)p\tau}. \tag{15.208}$$

Note that $\Psi_\nu[k]$ is the Fourier series of $\Phi(t; \nu)$ with respect to t. Following the same arguments as in (15.205), for any target ℓ satisfying $|\nu - \nu_\ell| < \pi/P\tau$, we have

$$\Psi_\nu[k] \cong \frac{P}{\tau} H\left(\frac{2\pi k}{\tau}\right) \sum_{\ell \in \Lambda(\nu)} \alpha_\ell e^{-j2\pi k\tau_\ell/\tau}. \tag{15.209}$$

In [426] it is shown that in the presence of white noise on $x_{\mathrm{R}}(t)$, Doppler focusing achieves an increase in SNR by a factor of P. It is also proven that this approach can operate at the minimal possible sampling rate for recovering the radar's parameters. Finally, an additional advantage of Doppler focusing is its ability to deal with certain models of clutter and target dynamic range, by adding a simple windowing operation in the sum (15.208), and by pre-whitening in frequency. See [426, 468] for details.

Equation (15.209) is identical in form to (15.25) except it is scaled by P. Thus, for each ν, we now have a standard delay estimation problem, as treated throughout the chapter. Note, however, that improved performance can be obtained by jointly processing the sequences $\{\Psi_\nu[k]\}$ for different values of ν, since we know that there are at most L targets over all possible ν. Thus, instead of searching separately for each of the delays $\tau_\ell, \ell \in \Lambda(\nu)$, we estimate the L delays by joint processing over all Doppler frequencies.

A particularly convenient method in this case is to employ a matching pursuit type approach where we begin by searching for the strongest peak over all ν assuming a single delay. For simplicity we assume that the pulse is flat, or we simply divide the values of $\Psi_\nu[k]$ by $H(2\pi k/\tau)$. We then solve

$$(\hat{t}, \hat{\nu}) = \arg\max_{t,\nu} \left| \sum_{k \in \mathcal{K}} \Psi_\nu[k] e^{j2\pi kt/\tau} \right|. \tag{15.210}$$

Once the optimal values $\hat{t}$ and $\hat{\nu}$ are found, we subtract their influence from the focused sub-Nyquist samples:

$$\Psi'_{\hat{\nu}}[k] = \Psi_{\hat{\nu}}[k] - \frac{1}{\tau} \hat{\alpha}_\ell e^{-j2\pi k\hat{t}/\tau} \sum_{p=0}^{P-1} e^{j(\nu - \hat{\nu}_\ell)p\tau}, \tag{15.211}$$

where

$$\hat{\alpha}_\ell = \frac{\tau}{P|\mathcal{K}|} \sum_{k \in \mathcal{K}} \Psi_{\hat{\nu}}[k] e^{j2\pi k\hat{t}/\tau}. \tag{15.212}$$

We then continue to iterate by finding all the desired L peaks. In practice, the search for peaks can be limited to a grid, which allows us to carry out all computations using simple FFT operations (see [426, Section VI.E]).

Example 15.25 In this example we present numerical experiments reported in [426] illustrating the recovery performance of a sparse target scene.

We corrupt the received signal $x_R(t)$ with additive white Gaussian noise $n(t)$ with power spectral density $N_0/2$, bandlimited to $x(t)$'s bandwidth B_h. We define the signal-to-noise power ratio for target ℓ as

$$\mathrm{SNR}_\ell = \frac{\frac{1}{T_p} \int_0^{T_p} |\alpha_\ell h(t)|^2 dt}{N_0 B_h}, \tag{15.213}$$

where T_p is the pulse time. The scenario parameters used were $L = 5$ targets, $P = 100$ pulses, a PRI of $\tau = 10\,\mu s$, and a signal bandwidth $B_h = 200\,\mathrm{MHz}$. Target delays and Doppler frequencies are spread uniformly at random in the appropriate unambiguous regions, and target amplitudes were chosen with constant absolute value and random phase. The classic time and frequency resolutions (Nyquist bins), defined as $1/B_h$ and $1/P\tau$, are 5 ns and 1 kHz, respectively.

The signal was subsampled at a tenth of the Nyquist rate, resulting in 200 Fourier coefficients per pulse. Doppler focusing was tested with two types of Fourier coefficient sets: a consecutive set and a random set. The performance of Doppler focusing is compared with classic processing using the hit rate, which is defined as the number of correct target detections, where a hit corresponds to a delay–Doppler estimate which is circumscribed by an ellipse around the true target position in the time–frequency plane. We used an ellipse with axes equivalent to ± 3 times the time and frequency Nyquist bins. In both Doppler focusing and classic processing the delay and Doppler frequency region was discretized with uniform steps of half a Nyquist bin. Figure 15.53 demonstrates the hit rate of the different recovery methods for various SNR values. Doppler focusing with consecutive coefficients tends to perform better under this metric for lower SNR, while choosing coefficients randomly improves performance as SNR increases. In contrast to Doppler focusing recovery

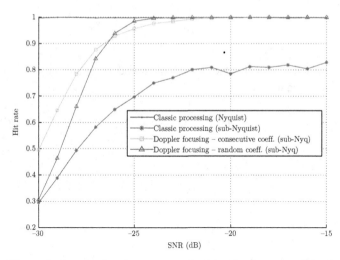

Figure 15.53 Hit rate for Doppler focusing at one-tenth of the Nyquist rate, and classic processing.

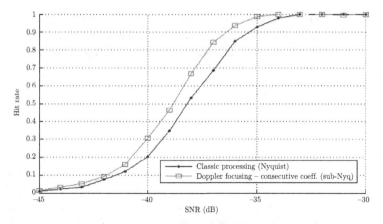

Figure 15.54 Hit rate for classic processing and Doppler focusing at one-tenth the Nyquist rate, where the waveform used for Doppler focusing has its entire energy contents concentrated in the sampled frequencies.

performance which decreases gracefully with sample rate reduction, classic processing suffers significantly when the sampling rate is reduced below Nyquist.

Figure 15.54 shows the same hit-rate graph for classic processing, but this time the waveform used for Doppler focusing had its CTFT adjusted so that energy was transferred from frequencies which were not sampled, to those that were. This was performed by passing the signal through a low-pass filter and rescaling its amplitude so that target SNR (15.213) remains constant. Since Doppler focusing imposes no restrictions on the transmitter, we can use a signal with the same total energy, but have it spread out in a manner which is more favorable to the frequency-domain sub-Nyquist sampling, so that signal energy is not lost in sampling. As performance for

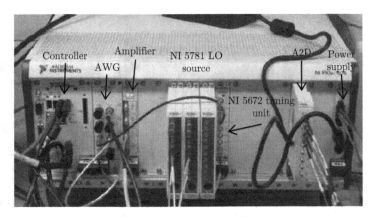

Figure 15.55 The NI chassis.

Doppler focusing improves significantly, excellent recovery results can be obtained at much lower SNR values, surpassing classic processing which uses 10 times as many samples. This shows that the performance degradation caused by sub-Nyquist sampling can be compensated for using a suitable transmitter.

Radar experiment

In [426], Doppler focusing is used in combination with the sub-Nyquist prototype of Fig. 15.40 in order to demonstrate radar reception at sub-Nyquist rates. The experimental setup is based on national instruments (NI) PXI equipment which is used to synthesize a radar environment and also ensure system synchronization. The entire component ensemble, wrapped in the NI chassis, is depicted in Fig. 15.55. Additional information regarding the system's configuration and synchronization issues can be found in [427].

The experimental process consists of the following steps. We begin by using the applied wave research (AWR) software, which enables us to examine a large variety of scenarios, consisting of different target parameters, i.e. delays, Doppler frequencies, and amplitudes. The AWR software can simulate a complete radar scenario, including pulse transmission and accurate power loss due to wave propagation in a realistic medium. We then use an arbitrary waveform generator (AWG) module to produce an analog signal which is amplified and routed to the radar receiver board of Fig. 15.40. The Nyquist rate of the signal chosen is 30 MHz. The crystal receivers in the filter have a bandwidth of 80 kHz. Each channel is sampled at a rate of 250 kHz, leading to a total sampling rate of 1 MHz. These samples are fed into the chassis controller and a Matlab function is launched that runs the Doppler focusing reconstruction algorithm. The system contains a fully detailed interface implemented in the LabView environment. Various target scenes, with different delays, Doppler frequencies, and amplitudes, are recovered successfully using this setup. Screenshots of the interface are depicted in Figs. 15.56 and 15.57.

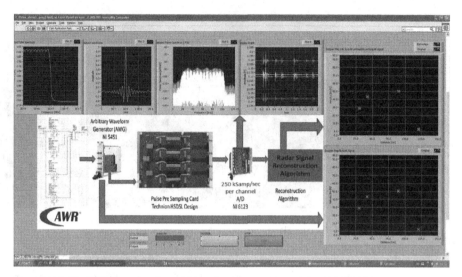

Figure 15.56 The LabView experimental interface. From left to right: $H(\omega)$, $h(t)$, the frequency response of each channel, the four signals detected in each channel; at the top, the reconstructed target scene; at the bottom, the original target scene.

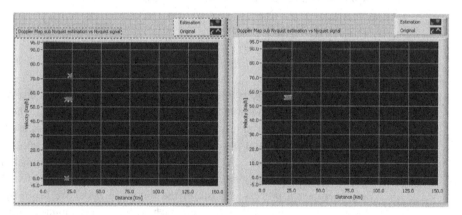

Figure 15.57 Two additional target scenes. On the left, all four targets have closely spaced delays, and two of the targets also have close Doppler frequencies. On the right, all four targets have very similar delays and Doppler frequencies. Doppler focusing based recovery is successful in both cases.

This experimental prototype demonstrates that the sub-Nyquist methodology described in this chapter is feasible in practice and can be implemented with standard RF hardware.

Before concluding this application, we note that the ideas presented here can also be extended to the case in which the pulse shape is not known. The path to sub-Nyquist sampling in this setting is by using the Gabor transform and noting that short radar pulses will have a sparse Gabor representation. These sparse coefficients can be recovered, as in the known pulse case, from a small set of Fourier coefficients. We refer the reader to [469] for details.

15.7.2   Time-varying system identification

The techniques used to treat the radar problem can also be applied to identification of underspread linear systems (ULSs), whose responses lie within a unit-area region in the delay–Doppler space. The importance of this problem stems from the fact that many physical systems can be described as linear and time-varying. The work in [430] relies on FRI theory and algorithms to show that sufficiently underspread parametric linear systems, described by a finite set of delays and Doppler shifts, are identifiable from a single observation as long as the time–bandwidth product of the input signal is proportional to the square of the total number of delay–Doppler pairs in the system.

Mathematically, identification of a given time-varying linear system $\mathcal{H}$ involves probing it with a known input signal $x(t)$ and identifying $\mathcal{H}$ by analyzing the single system output $\mathcal{H}(x(t))$. Unlike time-invariant linear systems, however, a single observation of a time-varying linear system does not lead to a unique solution unless additional constraints on the system response are imposed. This is due to the fact that such systems introduce both time shifts (delays) and frequency shifts (Doppler shifts) to the input signal. It is now a well-established fact in the literature that a time-varying linear system $\mathcal{H}$ can only be identified from a single observation if $\mathcal{H}(\delta(t))$ is known to lie within a region $\mathcal{R}$ of the delay–Doppler space such that area$(\mathcal{R}) < 1$ [470, 471, 472, 473]. Identifiable time-varying linear systems are termed *underspread*, as opposed to nonidentifiable *overspread* linear systems, which satisfy area$(\mathcal{R}) > 1$ [472, 473].[2]

The response of an ULS can be described by a finite set of delays and Doppler shifts:

$$\mathcal{H}(x(t)) = \sum_{\ell=1}^{L} \alpha_\ell x(t - \tau_\ell) e^{j \nu_\ell t} \tag{15.214}$$

where (τ_ℓ, ν_ℓ) denotes a delay–Doppler pair and $\alpha_\ell \in \mathbb{C}$ is the complex attenuation factor associated with (τ_ℓ, ν_ℓ). An important problem in this context is to characterize conditions on the bandwidth and temporal support of the input signal that ensure identification of such ULSs from a single observation. A small time–bandwidth product allows fast identification with low sampling rates.

Suppose that the delays and Dopplers are limited to the region $(\tau_i, \nu_i) \in [0, \tau_{\max}] \times [-\nu_{\max}/2, \nu_{\max}/2]$. We use $\mathcal{T}$ and $\mathcal{W}$ to denote the temporal support and the two-sided bandwidth of the known input signal $x(t)$ used to probe $\mathcal{H}$, respectively. The probing signal is chosen as a finite train of pulses:

$$x(t) = \sum_{n=0}^{N-1} x_n h(t - n\tau), \ 0 \le t \le \mathcal{T} \tag{15.215}$$

---

[2] It is still an open research question as to whether *critically spread* linear systems, which correspond to area$(\mathcal{R}) = 1$, are identifiable or nonidentifiable [473]; see [472] for a partial answer to this question for the case where $\mathcal{R}$ is a rectangular region.

where $h(t)$ is a prototype pulse of bandwidth $\mathcal{W}$ that is (essentially) temporally supported on $[0, \tau]$ and is assumed to have unit energy ($\int |h(t)|^2 dt = 1$), and $\{x_n \in \mathbb{C}\}$ is an N-length probing sequence. The parameter N is proportional to the time–bandwidth product of $x(t)$, which roughly defines the number of temporal degrees of freedom available for estimating $\mathcal{H}$ [474]: $N = \mathcal{T}/\tau \propto \mathcal{T}\mathcal{W}$. As in the radar setting, we assume that $\tau_{\max} < \tau$, and $\nu_{\max} \ll \mathcal{W}$. Under these assumptions, the Doppler focusing method developed in the context of radar can be used to prove the following result [430]:

Theorem 15.5 (Identification of parametric underspread linear systems). *Suppose that $\mathcal{H}$ is a parametric ULS that is described by a total of L triplets $(\tau_i, \nu_i, \alpha_i)$. Then, under the assumptions stated above, and as long as the probing sequence $\{x_n\}$ remains bounded away from zero in the sense that $|x_n| > 0$ for every $n = 0, \ldots, N-1$, the system can be identified if the time–bandwidth product of the known input signal $x(t)$ satisfies*

$$\mathcal{T}\mathcal{W} \geq 8\pi L_\tau L_{\nu,\max} \tag{15.216}$$

where L_τ is the number of distinct delays, and $L_{\nu,\max}$ is the maximum number of Doppler shifts associated with any of the distinct delays. In addition, the time–bandwidth product of $x(t)$ is guaranteed to satisfy (15.216) if $\mathcal{T}\mathcal{W} \geq 2\pi(L+1)^2$.

15.7.3   Ultrasound imaging

We next show how FRI modeling can be used to obtain ultrasound images at rates much lower than Nyquist, following the ideas in [425, 448]. The main novelty in this context is that not only can the sampling rate be substantially reduced, but the required digital processing, which takes on the form of *beamforming*, is performed on the low-rate samples without the need for interpolation.

Ultrasound transmission

In diagnostic ultrasound, imaging is performed by transmitting a pulse along a narrow beam from an array of transducer elements. During its propagation, echoes are scattered by acoustic impedance perturbations in the tissue and detected by the array elements. The data, collected by the transducers, is sampled and digitally integrated in a way referred to as beamforming, which results in SNR enhancement and improvement of angular localization. Such a beamformed signal forms a line in the image [475, 476].

Consider the array depicted in Fig. 15.58, comprising M transducer elements, aligned along the x axis. Denote by δ_m the distance from the mth element to the reference receiver m_0 set at the origin ($\delta_{m_0} = 0$). A pulse $h(t)$ is transmitted at $t = 0$ along a relatively narrow beam, whose central axis forms an angle θ with the axis z. Focusing the pulse along such a beam is achieved by applying appropriate time delays to modulated acoustic pulses, transmitted from the different antennas in the array. The echoes of the

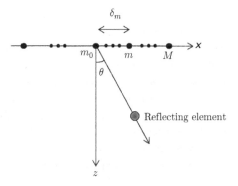

Figure 15.58 M receivers aligned along the x axis. An acoustic pulse is transmitted in a direction θ. The echoes scattered from perturbation in the radiated tissue are received by the array elements.

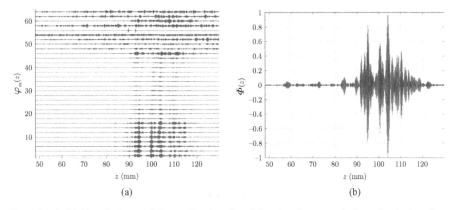

Figure 15.59 (a) Signals detected for cardiac imaging following the transmission of a single pulse. The vertical alignment of each trace matches the index of the corresponding receiver element. (b) Beamformed signal obtained by combining the detected signals with appropriate time delays. The data was acquired using a GE breadboard ultrasonic scanner [425].

transmitted pulse are received in the array elements. Ignoring noise for now, the detected signal in the mth antenna can be written as

$$\varphi_m(t) = \sum_{\ell=1}^{L} a_{\ell,m} h(t - t_{\ell,m}), \tag{15.217}$$

where L is the number of scattering elements, distributed throughout the sector radiated by the transmitted pulse, $t_{\ell,m}$ denotes the time in which the reflection from the ℓth element arrived at the mth receiver, and $a_{\ell,m}$ denotes the reflection's amplitude, as detected by the mth receiver. Figure 15.59(a) depicts signals detected by 32 out of 64 active array elements for a cardiac image of a healthy volunteer. Note that the z-axis is related to time by a scaling factor, $z = tc/2$, where c is the speed of sound in the medium.

From (15.217) it follows that the received signal in each antenna can be modeled as a stream of pulses. Therefore, in principle we may apply FRI sampling methods, as explained throughout the chapter, to each receiver element individually in order to sample the detected signals at a low rate and then determine the positions of the scattering elements along the beam. A B-mode image[3] can then be formed by transmitting over all feasible θ, and deriving a geometric model for estimating the two-dimensional position of a scattering element, based on the delays of pulses associated with it. This approach, however, faces two fundamental obstacles. The first is the low SNR at each of the detected elements, as can be seen in Fig. 15.59(a), due to the strong noise. The second is proper interpretation of the estimated signal parameters, considering the profile of the transmitted beam. Due to the difference in their geometric location, each of the array elements detects a signal with a different delay compared with the reference signal at m_0.

In standard ultrasound imaging, both these difficulties are overcome by the process of *beamforming*, described next, which increases the SNR and focuses the signal [475, 476].

Beamforming

Beamforming refers to averaging the detected signals, after their alignment with appropriate time-varying delays, as depicted in Fig. 15.59(b). This results in a single focused beam with improved angular localization of the scattering structures and improved SNR.

To obtain a mathematical description of beamforming, consider a pulse, transmitted by the array at $t = 0$, in the direction θ. The pulse propagates through the tissue at speed c, so that at time $t \geq 0$ its coordinates are $(x, z) = (ct \sin \theta, ct \cos \theta)$. A potential point reflector located at this position scatters the energy, such that the echo is detected by all array elements at a time depending on their locations. Denote by $\varphi_m(t; \theta)$ the signal detected by the mth element and by $\hat{\tau}_m(t; \theta)$ the time of detection. It is then readily seen that:

$$\hat{\tau}_m(t; \theta) = t + \frac{d_m(t; \theta)}{c}, \tag{15.218}$$

where $d_m(t; \theta) = \sqrt{(ct \cos \theta)^2 + (\delta_m - ct \sin \theta)^2}$ is the distance traveled by the reflection. Beamforming involves averaging the signals detected by multiple receivers while compensating for the differences in detection time. In that way we obtain a signal containing the intensity of the energy reflected from each point along the central transmission axis θ.

Using (15.218), the detection time at m_0 is $\hat{\tau}_{m_0}(t; \theta) = 2t$ since $\delta_{m_0} = 0$. Applying an appropriate delay to $\varphi_m(t; \theta)$, such that the resulting signal $\hat{\varphi}_m(t; \theta)$ satisfies $\hat{\varphi}_m(2t; \theta) = \varphi_m(\hat{\tau}_m(t; \theta))$, we can align the reflection detected by the mth receiver with the one detected at m_0. Denoting $\tau_m(t; \theta) = \hat{\tau}_m(t/2; \theta)$ and using (15.218), the following aligned signal is obtained:

---

[3] B-mode is a common ultrasound imaging technique, where the resulting image displays a two-dimensional cross-section of the tissue being imaged.

$$\hat{\varphi}_m(t;\theta) = \varphi_m(\tau_m(t;\theta);\theta), \tag{15.219}$$

$$\tau_m(t;\theta) = \frac{1}{2}\left(t + \sqrt{t^2 - 4(\delta_m/c)t\sin\theta + 4(\delta_m/c)^2}\right).$$

The beamformed signal results from averaging the aligned signals to form:

$$\Phi(t;\theta) = \frac{1}{M}\sum_{m=1}^{M}\hat{\varphi}_m(t;\theta). \tag{15.220}$$

Such a beam is optimally focused at each depth and hence exhibits improved angular localization and enhanced SNR.

Ultrasound imaging systems perform the beamforming process in (15.219) and (15.220) in the digital domain: the analog signals $\varphi_m(t;\theta)$ are sampled individually, so that the delays in (15.219) can be applied digitally. The Nyquist rate is generally insufficient for direct digital implementation of beamforming owing to the high delay resolution needed in implementing (15.219) on a sufficiently dense grid. Typically, the sampling interval is on the order of nanoseconds, resulting in sampling rates that are sometimes as high as several hundred MHz. To overcome the demand for such high rates, often the sampling rate is reduced to tens of MHz followed by digital interpolation. However, the processing, or beamforming rate, still remains high. As imaging systems evolve, the amount of elements participating in the imaging cycle continues to grow. Consequently, large amounts of data need to be transmitted from the system front end and digitally processed in real-time. This motivates reducing the amounts of data as close as possible to the system front end.

Sub-Nyquist beamforming
In [425, 448] it is shown that the time-domain beamforming introduced in (15.219) can be replaced by Fourier-domain beamforming. Denoting the Fourier coefficients of the beamformed signal $\Phi(t;\theta)$ and the individual signals $\varphi_m(t)$ by $c[k]$ and $\Upsilon_m[k]$ respectively, we have that

$$c[k] = \frac{1}{M}\sum_{m=1}^{M}\sum_{n}\Upsilon_m[k-n]Q_{k,m,\theta}[n], \tag{15.221}$$

where $Q_{k,m,\theta}[n]$ is a pre-computed function which depends on the array geometry and on the angle θ. As shown in [448], this function decays quite rapidly so that in practice only a small number of elements in the sum are needed to compute $c[k]$.

The relationship in (15.221) allows the beamforming process to be carried out entirely in the frequency domain. Since the beamformed signal is typically narrow band, this implies that only a small number of coefficients $\Upsilon_m[k]$ are needed to compute all nonzero values of $c[k]$, from which the beamformed signal can be determined. Throughout the chapter we showed how low-rate sampling can be used to determine a small set of Fourier coefficients. Thus, the beamformed signal may be computed by using the low-rate sampling techniques introduced in this chapter to obtain a small set of Fourier coefficients $\Upsilon_m[k]$, and then applying (15.221). This allows us to exploit the low bandwidth of the ultrasound signal and bypass the oversampling dictated by digital implementation of beamforming in time.

To further reduce the rate beyond the bandwidth of $\Phi(t; \theta)$, we exploit the fact that, as shown in [425], the beamformed signal can be well approximated by a stream of pulses of the form

$$\Phi(t; \theta) = \sum_{\ell=1}^{L} b_\ell h(t - t_\ell), \tag{15.222}$$

where t_ℓ denotes the time in which the ℓth reflection arrived at the reference receiver. Using FRI or CS techniques, this implies that the required delays t_ℓ and amplitudes b_ℓ can be determined from only a small number of Fourier coefficients $c[k]$. Thanks to (15.221), this translates into a small number of samples of each of the individual signals $\varphi_m(t)$. We therefore conclude that each signal may be sampled at a low rate using FRI sampling. After performing a DFT, the samples are combined via (15.221) to obtain the Fourier coefficients of the beamformed signal. Using FRI or CS methods, the delays t_ℓ and amplitudes b_ℓ are computed, based on which the appropriate line in the image is plotted.

Simulations and results

We now present simulation results, taken from [448], that demonstrate low-rate frequency-domain beamforming. We also show images obtained by a stand-alone ultrasound machine. These simulations prove the feasibility of sub-Nyquist processing for medical ultrasound, leading to the potential of considerable reduction in future ultrasound machines size, power consumption, and cost.

To demonstrate low-rate beamforming in frequency and evaluate the impact of rate reduction on image quality, we applied the method on in vivo cardiac data. The data acquisition setup consisted of a pulse with a carrier frequency of 16 MHz, leading to 3360 real-valued samples when sampled at the Nyquist rate. To perform beamforming in frequency we used a subset of 100 DFT coefficients. This implies 28-fold reduction in sampling compared with standard beamforming. The resulting images, corresponding to two different frames, are shown in Figs. 15.60(b) and (d). Although the images are not identical to those obtained by standard beamforming (Figs. 15.60(a) and (c)), it can easily be seen that this approach allows us to reconstruct both strong reflectors and speckle noise.

We next show the results of low-rate frequency-domain beamforming implemented on an ultrasound imaging system. The laboratory setup used for implementation and testing is shown in Fig. 15.61 and includes a GE ultrasound machine, a phantom, and an ultrasound scanning probe.

The breadboard ultrasonic scanner contains 64 acquisition channels. The radiated depth $r = 15.7$ cm and speed of sound $c = 1540$ m/s yield a signal of duration $2r/c \simeq 204$ μs. The acquired signal is characterized by a narrow bandpass bandwidth of 1.77 MHz, centered at a carrier frequency $f_0 \approx 3.4$ MHz. The signals are sampled at the rate of 50 MHz and then are digitally demodulated and downsampled to the demodulated processing rate of $f_p \approx 2.94$ MHz, resulting in 1224 real-valued samples per transducer

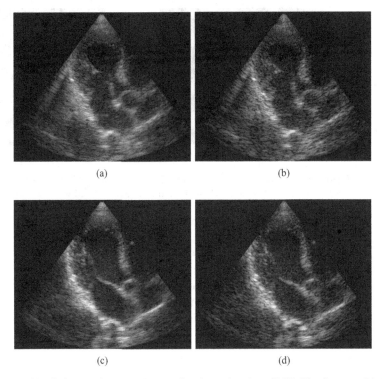

(a)   (b)

(c)   (d)

Figure 15.60 Simulation results on in vivo cardiac data taken from [448]. The first row, (a), (b), corresponds to frame 1, the second row, (c), (d), corresponds to frame 2. (a), (c) Time-domain beamforming at the Nyquist rate. (b), (d) Frequency-domain beamforming at $1/28$ of the Nyquist rate.

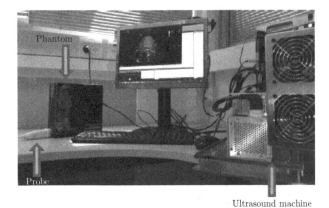

Ultrasound machine

Figure 15.61 Laboratory setup: ultrasound system, probe, and phantom.

element. Linear interpolation is then applied in order to improve beamforming resolution, leading to 2448 real-valued samples. Low-rate processing is performed on the data obtained in real-time by scanning a heart with a 64-element probe by using only 100 DFT coefficients of the beamformed signal. These are computed using 120 DFT

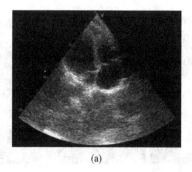
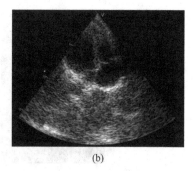

(a)   (b)

Figure 15.62 Cardiac images obtained by the demo system of [448]. (a) Time-domain beamforming. (b) Frequency-domain beamforming, obtained with 12-fold reduction in processing rate.

coefficients of each one of the detected signals. This corresponds to 240 real-valued samples used for beamforming in frequency compared with 2448 samples when sampling at the Nyquist rate. The images obtained by low-rate beamforming in frequency and standard time-domain beamforming are presented in Fig. 15.62. As can be readily seen, sufficient image quality is retained despite the significant reduction in sampling and processing rates.

15.8   Exercises

1. Determine the τ-local rate of innovation of the following signals for two choices of $h(t)$: $h(t) = e^{-t^2/2}$ and $h(t) = \beta^p(t)$ where $\beta^p(t)$ is the pth-order B-spline function.
   a. $x(t) = \sum_{n \in \mathbb{Z}} a[n]h(t - n\tau)$.
   b. $x(t) = \sum_{\ell=1}^{L} a_\ell h(t - t_\ell)$.
   c. $x(t) = \sum_{n=1}^{P} \sum_{\ell=1}^{L} a_\ell h(t - t_\ell - n\tau)$.
   d. $x(t) = \sum_{n \in \mathbb{Z}} \sum_{\ell=1}^{L} a_\ell[n]h(t - n\tau)\sin(\omega_\ell t)$.

2. 1. Consider the signal $x(t) = \sum_{\ell=1}^{L} \sum_{n \in \mathbb{Z}} a_\ell h(t - t_\ell - n\tau)$ where $h(t) = \mathrm{sinc}(t)$.
   a. Determine whether or not $x(t)$ is FRI. If so, then what is its rate of innovation?
   b. Determine the CTFT of $x(t)$.
   2. Repeat the previous part with:
   a. $x(t) = \sum_{\ell=1}^{L} a_\ell h(t - t_\ell)$.
   b. $x(t) = \sum_{\ell=1}^{L} \sum_{n \in \mathbb{Z}} a_\ell[n]h(t - t_\ell - n\tau)$.

3. Establish the equivalence between (15.13) and (15.14) by noting that $a[m] = \tilde{a}[m] - \tilde{a}[m - 1]$ and substituting this relation into (15.13).

4. Let $y(t) = \overline{h(-t)} * x(t)$ where $h(t)$ is a given pulse bandlimited to π/T, and let $1/T$ denote the Nyquist rate of $y(t)$. Show that $y[n] = \overline{h[-n]} * x[n]$ where $y[n] = y(nT)$, $h[n] = h(nT)$, and $x[n]$ are samples at rate $1/T$ of a bandlimited version of $x(t)$.

5. In Section 15.2.1 we developed the MF estimate of a single delay and amplitude by minimizing an LS error in the time domain. In this exercise we derive a frequency-domain equivalent. Suppose we are given m samples of a single exponential function: $X[k] = a_1 e^{-j2\pi k t_1/\tau}$ for $k = 0, 1, \ldots, m - 1$. We would like to estimate a_1 and t_1 from $X[k]$.

    a. Develop an expression for the values of a_1 and t_1 that minimize the LS error $\sum_{k=0}^{m-1} |X[k] - a_1 e^{-j2\pi k t_1/\tau}|^2$.

    b. Relate your answer to Equations (15.20) and (15.21).

6. Suppose that $Y[k]$ obeys (15.32) with $\omega_1 = 2\pi \cdot 0.2$, $\omega_2 = 2\pi \cdot 0.7$, and $a_1 = 1$, $a_2 = 0.2$. We are given noisy observations of $Y[k]$ where the noise is zero-mean white Gaussian noise with variance equal to 0.1. Estimate the values u_ℓ in (15.32) using Prony's method for the following choices of $\mathcal{K}$. Explain your results:

    a. $\mathcal{K} = \{0, 1, \ldots, 9\}$;

    b. $\mathcal{K} = \{5, 6, \ldots, 14\}$;

    c. $\mathcal{K} = \{0, 2, 4, \ldots, 18\}$.

7. Let $\mathbf{A}_{n \times m}$ with $n \geq m$ be a full column-rank matrix, let $\mathbf{C}_{m \times k}$ with $m \leq k$ be a full row-rank matrix, and let $\mathbf{B}$ be a matrix with rank r. Show that the rank of $\mathbf{ABC}$ is equal to r.

8. Consider a Hankel matrix $\mathbf{A}$. Show that the matrix $\mathbf{A}$ that minimizes the error $\|\mathbf{B} - \mathbf{A}\|_F^2$ is obtained by averaging the skew-diagonals of $\mathbf{B}$.

9. In this exercise we examine several methods for estimating the unknown amplitudes and frequencies in the model (15.26).

    a. Suppose that $t_1 = 0$, $t_2 = 0.1$, $t_3 = 0.41$, the amplitudes are all set to 1, and $\tau = 1$. Plot the performance of MUSIC, root-MUSIC, and matrix pencil for $m = 30$ measurements as a function of SNR between -10 to 10 dB.

    b. Repeat the previous part where now we use a compressed sensing approach with different choices of resolution step $\Delta = 0.02, 0.01$.

    c. Repeat the question where now $a_1 = 0.1$, $a_2 = 1$, $a_3 = 3$. Explain the results.

10. A special case of the model (15.26) occurs when $L = 1$ so that only one exponent is present in the sum: $Y[k] = ae^{-j2\pi t_0 k/\tau}$, $k \in \mathcal{K}$.

    a. What is the minimal number of samples $Y[k]$ needed in order to recover a and t_0?

    b. Write out Algorithm 15.1–Algorithm 15.5 explicitly for this special case.

    c. Let $a = 1$, $t_0 = 0.2$, $\tau = 1$. Suppose we are given noisy measurements $Z[k] = Y[k] + W[k]$ where $W[k]$ are zero-mean iid Gaussian variables with variance σ^2, and $\mathcal{K} = \{0, 1, 2, 3\}$. Plot the performance of the algorithms above as a function of SNR for $\sigma^2 = [0:0.01:0.5]$.

11. Suppose that $p(t)$ is a periodic waveform with period T and let $d[k]$ denote its Fourier series coefficients. We filter $p(t)$ with $g(t)$ and denote the filtered waveform by $\tilde{p}(t) = p(t) * g(t)$. Show that the Fourier series coefficients of $\tilde{p}(t)$ are given by $\tilde{d}[k] = d[k]G\left(\frac{2\pi}{T}k\right)$.

12. For each of the following choices of m and frequency set $\mathcal{K}$, indicate whether coset sampling as in Fig. 15.28 is possible. If yes, then show explicitly how the Fourier coefficients $X[k]$ are obtained from the resulting samples. If not, then explain.

Assume that the input is a periodic stream of impulse functions $h(t) = \delta(t)$, and that the filter $s(t)$ is chosen such that (15.87) is satisfied and $S(\omega) = 1$ for $\omega = 2\pi k/\tau, k \in \mathcal{K}$.

a. $\mathcal{K} = \{1, 2, 3, 4\}, m = 4$;
b. $\mathcal{K} = \{5, -2, 1, 4\}, m = 4$;
c. $\mathcal{K} = \{3, 4, -2\}, m = 3$;
d. $\mathcal{K} = \{3, 5, -2\}, m = 3$.

13. Repeat the previous exercise where now instead of using coset sampling we use a modulation-based multichannel sampler as in Fig. 15.38.

14. Suppose we obtain the Fourier coefficients $X[k]$ of a signal of the form (15.7) for $k \in \mathcal{K}$, and we wish to recover the delays and amplitudes from these values. For each of the choices of the set $\mathcal{K}$ below, indicate which of the following algorithms can be applied for this task: compressed sensing, Algorithms 15.2 and 15.4-15.6. Show in principle how each algorithm is applied where relevant:

a. $\mathcal{K} = \{0, 1, 2\}$;
b. $\mathcal{K} = \{1, 2, 7, 8\}$;
c. $\mathcal{K} = \{1, 5, 9\}$;
d. $\mathcal{K} = \{2, 3, 4, 5\}$.

15. In this exercise we illustrate the effect of noise on coset sampling of Fig. 15.28 assuming a sum-of-sincs filter. Suppose that a periodic stream of pulses $x(t)$ is contaminated by zero-mean white noise. We consider two options as to where the noise is added: In the first, the noise is added after the filter, prior to the sampler. In the second, the noise is added before the filter and the sampler. Explain which situation is preferable in terms of the output SNR of the samples.

16. Consider the MSE bound of (15.163) for estimating a single shifted pulse $x(t) = a_1 h(t - t_1)$ defined over $[0, \tau]$ from noisy measurements $y(t) = x(t) + w(t)$ as in (15.155), where $w(t)$ is continuous-time white Gaussian noise with variance σ^2.

a. Plot the bound for $h(t) = \delta(t)$ and the following sampling functions $\varphi_n(t)$ in (15.159):
   i. $\varphi_n(t) = e^{j2\pi nt/\tau}$.
   ii. $\varphi_n(t) = \beta^0(t)$, where $\beta^0(t)$ is the 0th-order B-spline function.
Suppose now that $h(t) = 1$ for $0 \le t \le \tau/2$ and $h(t) = 0$ otherwise.
b. Repeat the previous question for this choice of $h(t)$.
c. Suggest sampling functions $\varphi_n(t)$ that will yield a lower bound than those obtained by the two choices above.

17. Consider a sum of exponents of the form

$$Y[k] = \sum_{\ell=1}^{L} a_\ell e^{-j\omega_\ell k}, \quad k = 0, 1, \ldots, m - 1. \tag{15.223}$$

Suppose that $N < m$ of the values ω_ℓ are known and the remaining frequencies are unknown. The amplitudes a_ℓ are all unknown. Suggest two methods to recover the unknown amplitudes and frequencies based on the techniques discussed in this chapter, modified to take into account the known frequencies.

18. Consider a random process composed of L complex exponentials

$$Y[k] = \sum_{\ell=1}^{L} a_\ell e^{-j(\omega_\ell k + \theta_\ell)}, \quad k = 0, 1, \ldots, m - 1, \tag{15.224}$$

where θ_ℓ are iid random variables uniformly distritbuted over $[0, 2\pi)$ and the amplitudes a_ℓ are zero-mean iid random variables with variance σ_ℓ^2.
   a. Show how to estimate ω_ℓ using a covariance-based method as discussed in Sec-
      tion 15.2.7.
   b. Suppose now that the phases are known, and the amplitudes are unknown deter-
      ministic values. Can we use the annihilating filter method to recover the unknown
      frequencies?
19. In this exercise we consider Doppler focusing when the Doppler frequencies lie on
   a grid so that $\nu_\ell = 2\pi r_\ell / (P\tau)$ for some integer r_ℓ in the range $0 \le r_\ell \le P - 1$.
   a. Derive an expression for the focused coefficients $\Psi_\nu[k]$ of (15.208) where $\nu =
      2\pi\ell/(P\tau)$.
   b. Adapt the algorithm in (15.210)-(15.212) to this setting in which both the true
      frequencies and the focused frequencies are on a grid.
   c. Show how the resulting algorithm can be implemented using FFT operations.
20. Consider a pulse-Doppler radar problem in which the transmitted pulse $H(\omega)$ is
   equal to one within its bandwidth, and suppose that there is only one target in
   the scene. We would like to compare between classic processing, and sub-Nyquist
   Doppler focusing in this setting. In classic processing, we follow the approach presented in Fig. 15.52 with $N = \tau B_h$ samples. We assume that the unknown delay
   and Doppler lie on a grid so that $\tau_1 = s\tau/N$ where s is an integer in the range
   $0 \le s \le N - 1$, and $\nu_1 = 2\pi r/(P\tau)$ for some integer $0 \le r \le P - 1$.
   a. Describe the steps in classic processing to determine s and r.
   b. Describe the steps in sub-Nyquist Doppler focusing to determine s and r.
   c. How many samples are needed in order to recover s and r using Doppler focusing
      assuming there is no noise?
   d. Suppose now that we perform the Doppler focusing approach using N sam-
      ples (i.e. at the Nyquist rate). Compare between classic processing and Doppler-
      focusing in this setting.
21. Assume that the support of the signals detected at each one of M ultrasound transducer elements is contained in $[0, T)$. Prove that the support of the beamformed
   signal $\Phi(t, \theta)$ is contained in $[0, T_B(\theta))$, where

$$T_B(\theta) = \min_m \tau_m^{-1}(T; \theta), \tag{15.225}$$

   with $\tau_m(t; \theta)$ given in (15.219).
22. In this exercise we develop the relationship between the Fourier coefficients of the
   beamformed signal and the Fourier coefficients of the detected signals $\varphi_m(t)$ given
   in (15.221).

a. Prove that the Fourier coefficients of the beamformed signal $c[k]$ can be expressed as

$$c[k] = \frac{1}{M} \sum_{m=1}^{M} \frac{1}{T} \int_0^T \varphi_m(t) q_{k,m}(t; \theta) e^{-i\frac{2\pi}{T}kt} dt. \tag{15.226}$$

Provide an explicit expression for $q_{k,m}(t; \theta)$.

b. Derive (15.221) by substituting $\varphi_m(t)$ with its Fourier series expansion over the interval $[0, T)$.

Appendix A
Finite linear algebra

In this appendix we review some important concepts from matrix algebra that are used throughout the book. For a more comprehensive treatment of finite-dimensional linear algebra, the reader is referred to one of the many excellent textbooks on this topic, e.g. [24, 28, 34, 38, 223]. In our exposition, we assume that the reader is familiar with the basic notions of linear algebra. The material is meant to complement Chapter 2; therefore, we do not repeat many of the notions and definitions contained in that chapter. The proofs of results in this appendix can be found in the references above.

A.1   Matrices

Let $\mathbf{A}$ be an $m \times n$ matrix. We denote its elements by a_{ij}:

$$\mathbf{A} = \begin{bmatrix} a_{11} & a_{12} & \dots & a_{1n} \\ \vdots & \vdots & & \vdots \\ a_{m1} & a_{m2} & \dots & a_{mn} \end{bmatrix}. \tag{A.1}$$

A matrix is *square* if $m = n$. When $m < n$ we will say that the matrix is *fat*. It is referred to as *skinny* when $m > n$.

A.1.1   Matrix operations

The *transpose* $\mathbf{A}^T$ of $\mathbf{A}$ is defined as the $n \times m$ matrix with elements a_{ji}. The *Hermitian conjugate* $\mathbf{A}^*$ is the $n \times m$ matrix with elements $\overline{a_{ji}}$. Thus, the Hermitian conjugate of a column vector is a row vector, and vice versa. For a real matrix, $\mathbf{A}^T = \mathbf{A}^*$. A useful result is that $(\mathbf{AB})^* = \mathbf{B}^*\mathbf{A}^*$.

The *trace* of a square $n \times n$ matrix $\mathbf{A}$ is the sum of its diagonal elements: $\mathrm{Tr}(\mathbf{A}) = \sum_{i=1}^{n} a_{ii}$. The trace operator satisfies

$$\mathrm{Tr}(\mathbf{AB}) = \mathrm{Tr}(\mathbf{BA}), \quad \mathrm{Tr}(\mathbf{A}^*\mathbf{B}) = \sum_{k,\ell=1}^{n} \overline{a_{k\ell}} b_{k\ell}, \tag{A.2}$$

where $\mathbf{A}$ and $\mathbf{B}$ are of appropriate dimensions. For a scalar a, $a = \mathrm{Tr}(a)$. Therefore, using (A.2), for two $n \times 1$ vectors $\mathbf{x}, \mathbf{y}$ we have

$$\mathbf{y}^*\mathbf{x} = \mathrm{Tr}(\mathbf{xy}^*). \tag{A.3}$$

The *determinant* $\det(\mathbf{A})$ of a square $n \times n$ matrix $\mathbf{A}$ is given by

$$\det(\mathbf{A}) = \sum_{j=1}^{n} (-1)^{i+j} a_{ij} M_{ij}, \tag{A.4}$$

where M_{ij} is the *minor* of a_{ij} and is equal to the determinant of the submatrix of $\mathbf{A}$ obtained by deleting the ith row and jth column. For a 2×2 matrix this expression becomes

$$\left| \begin{bmatrix} a & b \\ c & d \end{bmatrix} \right| = ad - bc. \tag{A.5}$$

Let $\mathbf{A}$ be an $m \times n$ matrix and let $\mathbf{B}$ be a $q \times p$ matrix. The *Kronecker product* of $\mathbf{A}$ and $\mathbf{B}$ is the $mq \times np$ matrix defined by

$$\mathbf{A} \otimes \mathbf{B} = \begin{bmatrix} a_{11}\mathbf{B} & a_{12}\mathbf{B} & \cdots & a_{1n}\mathbf{B} \\ a_{21}\mathbf{B} & a_{22}\mathbf{B} & \cdots & a_{2n}\mathbf{B} \\ \vdots & \vdots & \vdots & \vdots \\ a_{m1}\mathbf{B} & a_{m2}\mathbf{B} & \cdots & a_{mn}\mathbf{B} \end{bmatrix}. \tag{A.6}$$

The *vec* operator creates a column vector from a matrix $\mathbf{A} = [\mathbf{a}_1 \ \mathbf{a}_2 \ \cdots \ \mathbf{a}_n]$ by stacking its columns vertically:

$$\text{vec}(\mathbf{A}) = \begin{bmatrix} \mathbf{a}_1 \\ \mathbf{a}_2 \\ \vdots \\ \mathbf{a}_n \end{bmatrix}. \tag{A.7}$$

Some important properties of the Kronecker product and vec operation are given below:

1. For conforming matrices, $(\mathbf{A} \otimes \mathbf{B})(\mathbf{C} \otimes \mathbf{D}) = \mathbf{AC} \otimes \mathbf{BD}$.
2. $(\mathbf{A} \otimes \mathbf{B})^* = \mathbf{A}^* \otimes \mathbf{B}^*$.
3. $\text{Tr}(\mathbf{A} \otimes \mathbf{B}) = \text{Tr}(\mathbf{A})\,\text{Tr}(\mathbf{B})$.
4. $\text{vec}(\mathbf{AXB}) = (\mathbf{B}^T \otimes \mathbf{A})\,\text{vec}(\mathbf{X})$.

A.1.2   Matrix properties

Rank and inverse

The *rank* of an $m \times n$ matrix $\mathbf{A}$ denoted by $r = \text{rank}(\mathbf{A})$ is defined as the largest number of linearly independent rows or columns. The matrix is said to be full row-rank if $r = m$ and full column-rank if $r = n$. Given two matrices $\mathbf{A}$ and $\mathbf{B}$, $r(\mathbf{AB}) \leq \min(r(\mathbf{A}), r(\mathbf{B}))$.

The *inverse* of a square matrix $\mathbf{A}$, when it exists, is the $n \times n$ matrix $\mathbf{A}^{-1}$ satisfying

$$\mathbf{A}^{-1}\mathbf{A} = \mathbf{A}\mathbf{A}^{-1} = \mathbf{I}, \tag{A.8}$$

where $\mathbf{I}$ is the identity matrix. We will sometimes use the notation $\mathbf{I}_n$ to denote an $n \times n$ identity matrix. The matrix $\mathbf{A}$ is *invertible* if and only if its rank is equal to n. Otherwise it is said to be *singular*. A matrix $\mathbf{A}$ is invertible if and only if $\det(\mathbf{A}) \neq 0$.

A useful formula for inverting matrices is the *matrix inversion lemma* or the *Woodbury matrix identity*:

$$(\mathbf{A} + \mathbf{BCD})^{-1} = \mathbf{A}^{-1} - \mathbf{A}^{-1}\mathbf{B}(\mathbf{DA}^{-1}\mathbf{B} + \mathbf{C}^{-1})^{-1}\mathbf{DA}^{-1}, \qquad (A.9)$$

where $\mathbf{A}$ and $\mathbf{C}$ are invertible $n \times n$ matrices, $\mathbf{B}$ is $n \times m$, $\mathbf{D}$ is $m \times n$, and the required matrices are invertible. A special case occurs when $\mathbf{B} = \mathbf{b}$ is a length-n vector, $\mathbf{D} = \mathbf{d}^*$ is a length-n row vector, and $\mathbf{C} = \mathbf{I}$:

$$(\mathbf{A} + \mathbf{bd}^*)^{-1} = \mathbf{A}^{-1} - \frac{\mathbf{A}^{-1}\mathbf{bd}^*\mathbf{A}^{-1}}{\mathbf{d}^*\mathbf{A}^{-1}\mathbf{b} + 1}. \qquad (A.10)$$

This identity is referred to as the *Sherman–Morrison formula*.

An $m \times n$ matrix $\mathbf{A}$ has full row-rank if and only if the $m \times m$ matrix $\mathbf{M} = \mathbf{AA}^*$ is invertible. Similarly, $\mathbf{A}$ has full column-rank if and only if the $n \times n$ matrix $\mathbf{G} = \mathbf{A}^*\mathbf{A}$ is invertible. A matrix formed by taking m rows (columns) of an invertible matrix will be full row- (column-) rank.

An important example of a full column-rank matrix is the *Vandermonde* matrix.

Example A.1 An $m \times n$ Vandermonde matrix $\mathbf{V}$ is defined by n roots $\lambda_\ell, \ell = 1, \ldots, n$. Its $k\ell$th element is given by $v_{k\ell} = \lambda_\ell^{k-1}$:

$$\mathbf{V} = \begin{bmatrix} 1 & 1 & \cdots & 1 \\ \lambda_1 & \lambda_2 & \cdots & \lambda_n \\ \lambda_1^2 & \lambda_2^2 & \cdots & \lambda_n^2 \\ \vdots & \vdots & & \vdots \\ \lambda_1^{m-1} & \lambda_2^{m-1} & \cdots & \lambda_n^{m-1} \end{bmatrix}. \qquad (A.11)$$

A Vandermonde matrix with $m \geq n$ has full column-rank as long as the values of λ_i are distinct. In this case, any n rows of $\mathbf{V}$ are linearly independent.

The determinant of an $n \times n$ Vandermonde matrix $\mathbf{V}$ is given by

$$\det(\mathbf{V}) = \prod_{1 \leq k < \ell \leq n} (\lambda_\ell - \lambda_k). \qquad (A.12)$$

If $\lambda_\ell \neq \lambda_k$ for all $k \neq \ell$, then $\det(\mathbf{V}) \neq 0$ and $\mathbf{V}$ is invertible.

The space spanned by the columns of a matrix $\mathbf{A}$ is referred to as the *range space* of $\mathbf{A}$. Any vector in the range space can be written as $\mathbf{y} = \mathbf{Ax}$ for an appropriate choice of $\mathbf{x}$. The dimension of the range space is equal to the rank of $\mathbf{A}$. The *null space* of $\mathbf{A}$ consists of vectors $\mathbf{x}$ such that $\mathbf{Ax} = \mathbf{0}$. The null space of an invertible $n \times n$ matrix is equal to the zero vector.

For two $n \times n$ invertible matrices $\mathbf{A}$ and $\mathbf{B}$ we have:

$$(\mathbf{A}^*)^{-1} = (\mathbf{A}^{-1})^*, \quad (\mathbf{AB})^{-1} = \mathbf{B}^{-1}\mathbf{A}^{-1}. \qquad (A.13)$$

Norms and inner products

Let $\mathbf{x}$ and $\mathbf{y}$ be two length-n vectors. The *inner product* between $\mathbf{x}$ and $\mathbf{y}$ is the scalar given by $a = \mathbf{x}^*\mathbf{y}$. The vectors are orthogonal if $a = 0$.

The inner product of $\mathbf{x}$ with itself is the *squared norm* of $\mathbf{x}$, and is denoted by

$$\|\mathbf{x}\|_2^2 = \mathbf{x}^*\mathbf{x} = \sum_{i=1}^{n} |x_i|^2. \tag{A.14}$$

We will often omit the subscript 2 when it is clear from the context that the norm is the 2-norm. Other norms are defined in Chapter 11. Clearly, $\|\mathbf{x}\| > 0$ unless $\mathbf{x} = 0$. From the Cauchy–Schwarz inequality we have that

$$|\mathbf{x}^*\mathbf{y}| \le \|\mathbf{x}\|\|\mathbf{y}\|, \tag{A.15}$$

with equality if and only if $\mathbf{x} = c\mathbf{y}$ for some scalar c.

The *outer product* between a length-m vector $\mathbf{x}$ and a length-n vector $\mathbf{y}$ is the $m \times n$ matrix $\mathbf{A} = \mathbf{xy}^*$. Clearly $\mathbf{A}$ has rank equal to 1. Conversely, any rank-1 matrix can be written as an outer product for appropriate choices of $\mathbf{x}$ and $\mathbf{y}$.

A.1.3   Special classes of matrices

Let $\mathbf{A}$ be an $n \times n$ square matrix. The matrix $\mathbf{A}$ is *Hermitian* if $\mathbf{A} = \mathbf{A}^*$ and *unitary* if $\mathbf{A}^* = \mathbf{A}^{-1}$. The columns $\mathbf{a}_i$ of a unitary matrix are orthonormal: $\mathbf{a}_i^*\mathbf{a}_j = \delta_{ij}$. Let $\mathbf{Q}$ be an $m \times n$ matrix equal to the first m rows of an $n \times n$ unitary matrix. We refer to $\mathbf{Q}$ as a *partial unitary matrix*. Clearly the matrix $\mathbf{Q}$ will have full row-rank as its rows are linearly independent.

Example A.2 An important example of an $n \times n$ unitary matrix is the *Fourier matrix* $\mathbf{F}$ defined by

$$F_{k\ell} = \frac{1}{\sqrt{n}} e^{-j\frac{2\pi k\ell}{n}}. \tag{A.16}$$

The fact that the resulting matrix $\mathbf{F}$ is unitary follows from the relation

$$\frac{1}{n} \sum_{m=0}^{n-1} e^{-j\frac{2\pi km}{n}} = \delta[k]. \tag{A.17}$$

A partial Fourier matrix is a matrix composed of (arbitrary) rows of the Fourier matrix.   •

A *projection* matrix $\mathbf{E}$ is a square $n \times n$ matrix such that $\mathbf{E}^2 = \mathbf{E}$. An *orthogonal projection* is a Hermitian projection matrix. Let $\mathbf{H}$ be an arbitrary $m \times n$ matrix with full column-rank (so that in particular, $m \ge n$). Then,

$$\mathbf{P} = \mathbf{H}(\mathbf{H}^*\mathbf{H})^{-1}\mathbf{H}^* \tag{A.18}$$

is an orthogonal projection matrix onto the range of $\mathbf{H}$.

A matrix $\mathbf{A}$ is *Toeplitz* if $a_{ij} = a_{i-j}$. Namely, the elements on all the diagonals of $\mathbf{A}$ are equal:

$$\mathbf{A} = \begin{bmatrix} a_0 & a_1 & a_2 & \cdots & a_{n-1} \\ a_{-1} & a_0 & a_1 & \cdots & a_{n-2} \\ a_{-2} & a_{-1} & a_0 & \cdots & a_{n-3} \\ \vdots & \vdots & & & \vdots \\ a_{-n+1} & a_{-n+2} & a_{-n+3} & \cdots & a_0 \end{bmatrix}. \tag{A.19}$$

A related matrix is the *Hankel matrix* which is an upside-down Toeplitz matrix – its skew-diagonals are all equal, or $a_{ij} = a_{(i-1)(j+1)}$. An example is the following matrix:

$$\mathbf{A} = \begin{bmatrix} a_0 & a_1 & a_2 & \cdots & a_{n-1} \\ a_1 & a_2 & a_3 & \cdots & a_{-1} \\ a_2 & a_3 & a_4 & \cdots & a_{-2} \\ \vdots & \vdots & & & \vdots \\ a_{n-1} & a_{-1} & a_{-2} & \cdots & a_{-n+1} \end{bmatrix}. \tag{A.20}$$

A matrix $\mathbf{A}$ is *circulant* if each row is obtained by a circular shift of the previous row. In particular, a circulant matrix is Toeplitz as can be seen in the following example:

$$\mathbf{A} = \begin{bmatrix} a_0 & a_1 & a_2 & a_3 \\ a_3 & a_0 & a_1 & a_2 \\ a_2 & a_3 & a_0 & a_1 \\ a_1 & a_2 & a_3 & a_0 \end{bmatrix}. \tag{A.21}$$

A matrix $\mathbf{A}$ is *diagonal* if all of its elements are zero besides possibly its diagonal values a_{ii}. Note that a diagonal matrix does not have to be square; for example, the following 3×2 matrix is diagonal:

$$\mathbf{A} = \begin{bmatrix} 3 & 0 & 0 \\ 0 & 1 & 0 \end{bmatrix}. \tag{A.22}$$

Its diagonal elements are given by 3 and 1. The notation $\mathbf{A} = \mathrm{diag}\,(a_1, \ldots, a_n)$ stands for an $n \times n$ diagonal matrix with diagonal elements $a_1, \ldots, a_n$.

A matrix $\mathbf{A}$ is *diagonally dominant* if

$$|a_{ii}| \geq \sum_{j=1,\, j \neq i}^{n} |a_{ij}| \tag{A.23}$$

for all i. It is strictly diagonally dominant if the inequality is strict.

A matrix $\mathbf{A}$ is *normal* if $\mathbf{A}\mathbf{A}^* = \mathbf{A}^*\mathbf{A}$. In particular, a normal matrix must be square. Clearly, both Hermitian matrices and unitary matrices are normal. Another example is a skew-Hermitian matrix, namely a matrix $\mathbf{A}$ for which $\mathbf{A}^* = -\mathbf{A}$. It can also be shown that circulant matrices are normal.

A square $n \times n$ matrix is *nonnegative definite* if $\mathbf{A}$ is Hermitian and

$$\mathbf{x}^*\mathbf{A}\mathbf{x} \geq 0 \tag{A.24}$$

for any $\mathbf{x}$. When the inequality is strict the matrix is said to be *positive definite*. Note that when $\mathbf{A}$ is Hermitian we have that $\mathbf{x}^*\mathbf{Ax}$ is real for any $\mathbf{x}$. We denote a positive (nonnegative) definite matrix by $\mathbf{A} \succ 0$ ($\mathbf{A} \succeq 0$). For two Hermitian matrices of the same dimension, the notation $\mathbf{A} \succeq \mathbf{B}$ means that $\mathbf{A} - \mathbf{B} \succeq 0$.

Suppose that $\mathbf{A}$ is positive (nonnegative) definite. Then

1. The diagonal elements of $\mathbf{A}$ are positive (nonnegative);
2. Let $\mathbf{B}$ be an arbitrary $m \times n$ matrix. Then $\mathbf{BAB}^*$ is nonnegative definite.
3. Let $\mathbf{B}$ be positive (nonnegative) definite. Then $\mathbf{A} + \mathbf{B}$ is positive (nonnegative) definite.
4. We can define a *square root* of $\mathbf{A}$ such that $\mathbf{A}^{1/2}\mathbf{A}^{1/2} = \mathbf{A}$ and $\mathbf{A}^{1/2}$ is nonnegative definite. A convenient way to define $\mathbf{A}^{1/2}$ is via the eigendecomposition of $\mathbf{A}$, which we discuss next.

A.2   Eigendecomposition of matrices

A.2.1   Eigenvalues and eigenvectors

An *eigenvector* $\mathbf{v}$ of a square $n \times n$ matrix $\mathbf{A}$ is a length-n vector satisfying

$$\mathbf{Av} = \lambda\mathbf{v}, \tag{A.25}$$

for some scalar λ which is called the *eigenvalue* of $\mathbf{A}$. It is assumed here that $\mathbf{v}$ is normalized so that $\|\mathbf{v}\|_2^2 = 1$. The eigenvalues of $\mathbf{A}$ are equal to the n roots of the *characteristic polynomial* of $\mathbf{A}$:

$$D(\lambda) = \det(\mathbf{A} - \lambda\mathbf{I}). \tag{A.26}$$

Since $D(\lambda)$ has degree n, it has n zeros. A root λ_0 has multiplicity L if $D(\lambda)$ contains a factor $(\lambda - \lambda_0)^L$.

If $\lambda = 0$ then $\mathbf{v}$ is in the null space of $\mathbf{A}$. We therefore conclude that the matrix $\mathbf{A}$ is invertible if and only if its eigenvalues are all nonzero. The rank of $\mathbf{A}$ is equal to the number of nonzero eigenvalues. It can be shown that if $\mathbf{v}_1$ and $\mathbf{v}_2$ are eigenvectors corresponding to distinct eigenvalues, then $\mathbf{v}_1$ and $\mathbf{v}_2$ are linearly independent, namely, we cannot write $\mathbf{v}_1 = a\mathbf{v}_2$ for any scalar a.

Let $\lambda_1, \ldots, \lambda_n$ be the n eigenvalues of $\mathbf{A}$ and let $\mathbf{v}_1, \ldots, \mathbf{v}_n$ be the corresponding eigenvectors: $\mathbf{Av}_i = \lambda_i\mathbf{v}_i$. Define the $n \times n$ matrix $\mathbf{V}$ with columns $\mathbf{v}_i$, and let Λ be the $n \times n$ diagonal matrix with diagonal elements λ_i. The relation (A.25) can then be written as

$$\mathbf{AV} = \mathbf{V}\Lambda. \tag{A.27}$$

Suppose that $\mathbf{A}$ is an $n \times n$ matrix with n distinct eigenvalues. It then follows that the corresponding eigenvectors are linearly independent and each eigenvector is unique (up to scaling). On the other hand, if the eigenvalues of $\mathbf{A}$ are not distinct, then in general $\mathbf{A}$ may not have n linearly independent eigenvectors. When the eigenvectors are linearly independent (which may happen even when the eigenvalues are not distinct), we can write (A.27) as

$$\mathbf{A} = \mathbf{V}\Lambda\mathbf{V}^{-1}. \tag{A.28}$$

This relation is referred to as the *eigendecomposition* of $\mathbf{A}$. The matrix $\mathbf{V}$ is said to *diagonalize* $\mathbf{A}$.

A matrix $\mathbf{A}$ is diagonalizable if it can be written as in (A.28) for some diagonal Λ and invertible $\mathbf{V}$. In this case, the values λ_i are the eigenvalues of $\mathbf{A}$ and the columns of $\mathbf{V}$ are the corresponding eigenvectors. It is important to note that not every matrix is diagonalizable. An example is given next.

Example A.3 Consider the matrix

$$\mathbf{A} = \begin{bmatrix} 0 & 1 \\ 0 & 0 \end{bmatrix}. \tag{A.29}$$

It is easy to see that this matrix has only one eigenvalue, $\lambda = 0$. Indeed, $\det(\mathbf{A} - \lambda\mathbf{I}) = \lambda^2$ so that $\lambda = 0$ has multiplicity 2. Since $\Lambda = \mathbf{0}$, for any choice of $\mathbf{V}$, $\mathbf{V}\Lambda\mathbf{V}^{-1} = \mathbf{0}$, and therefore $\mathbf{A}$ cannot be diagonalized.

Two classes of matrices that can always be diagonalized are:

1. An $n \times n$ matrix $\mathbf{A}$ with n distinct eigenvalues;
2. An $n \times n$ normal matrix.

Any normal $n \times n$ matrix $\mathbf{A}$ has n linearly independent eigenvectors (which in general are not unique). Furthermore, the eigenvectors corresponding to distinct eigenvalues are orthonormal. It follows that for any normal matrix $\mathbf{A}$ we can choose $\mathbf{V}$ in (A.27) to be a unitary matrix. We may then express $\mathbf{A}$ as

$$\mathbf{A} = \mathbf{V}\Lambda\mathbf{V}^* = \sum_{i=1}^{n} \lambda_i \mathbf{v}_i \mathbf{v}_i^*. \tag{A.30}$$

The inverse of $\mathbf{A}$ follows as

$$\mathbf{A}^{-1} = \mathbf{V}\Lambda^{-1}\mathbf{V}^* = \sum_{i=1}^{n} \frac{1}{\lambda_i} \mathbf{v}_i \mathbf{v}_i^*. \tag{A.31}$$

Evidently, the eigenvalues of $\mathbf{A}^{-1}$ are the reciprocals of those of $\mathbf{A}$. This result is true for any invertible matrix $\mathbf{A}$, not necessarily normal.

Example A.4 We have seen already that a circulant matrix is normal. Therefore, it is diagonalized by a unitary matrix. In fact, it can be shown that any circulant matrix $\mathbf{A}$ is diagonalized by the Fourier matrix defined in Example A.2. The eigenvalues of $\mathbf{A}$ are given by the DFT of the coefficients of the first row of $\mathbf{A}$.

As an example, consider the matrix

$$
\mathbf{A} = \begin{bmatrix} 0 & 1 & 2 & 3 \\ 3 & 0 & 1 & 2 \\ 2 & 3 & 0 & 1 \\ 1 & 2 & 3 & 0 \end{bmatrix}.
\tag{A.32}
$$

The DFT of the first row is given by $\{6, \ -2 + 2j, \ -2 - 2j, \ -2\}$. Computing the eigendecomposition leads to $\mathbf{A} = \mathbf{F}\,\mathrm{diag}\,(6, -2+2j, -2-2j, -2)\mathbf{F}^*$, as expected.

Properties of eigenvalues

1. The eigenvalues of a Hermitian matrix are real. If, in addition, the matrix is nonnegative (positive) definite, then the eigenvalues are nonnegative (positive). In fact, a Hermitian matrix $\mathbf{A}$ is nonnegative (positive) definite if and only if all of its eigenvalues satisfy $\lambda_i \geq 0$ ($\lambda_i > 0$).
2. The eigenvalues of a unitary matrix have magnitude equal to 1.
3. The matrix $\mathbf{A}$ has an eigenvalue equal to 0 if and only if it is singular.
4. If λ is an eigenvalue of $\mathbf{A}$, then $\overline{\lambda}$ is an eigenvalue of $\mathbf{A}^*$.
5. If $\mathbf{A}$ is invertible with eigenvalues λ_i, then the eigenvalues of $\mathbf{A}^{-1}$ are equal to $1/\lambda_i$.

Using the eigendecomposition we can define the *square root* of a nonnegative definite matrix $\mathbf{A}$. Specifically, since $\mathbf{A}$ is Hermitian it has an eigendecomposition of the form $\mathbf{A} = \mathbf{U}\Lambda\mathbf{U}^*$ for some unitary matrix $\mathbf{U}$. Furthermore, $\lambda_i \geq 0$ for all i because $\mathbf{A}$ is nonnegative definite. Therefore, we can choose $\mathbf{A}^{1/2} = \mathbf{U}\Lambda^{1/2}\mathbf{U}^*$.

An important result on the eigenvalues of a general square matrix $\mathbf{A}$ is given by the *Geršgorin disk theorem*:

Theorem A.1 (Geršgorin disk theorem).   *Let*

$$
R_i(\mathbf{A}) = \sum_{j=1, j \neq i}^{n} |a_{ij}|
\tag{A.33}
$$

be the deleted absolute row sum of $\mathbf{A}$. *Define the Geršgorin disk as*

$$
G_i(\mathbf{A}) = \{z \in \mathbb{C} : |z - a_i| \leq R_i(\mathbf{A})\}.
\tag{A.34}
$$

Then all the eigenvalues of $\mathbf{A}$ *are in the union* $G(\mathbf{A})$ *of the* n *Geršgorin disks:*

$$
G(\mathbf{A}) = \bigcup_{i=1}^{n} G_i(\mathbf{A}).
\tag{A.35}
$$

The following consequences can be deduced from the theorem: Let $\mathbf{A}$ be strictly diagonally dominant. Then

1. $\mathbf{A}$ is invertible (since $z = 0$ cannot be in $G(\mathbf{A})$);
2. If $\mathbf{A}$ is Hermitian and all the diagonal entries of $\mathbf{A}$ are positive, then $\mathbf{A}$ is positive definite (since $G(\mathbf{A})$ contains only positive elements).

Extrema properties

Eigenvalues and eigenvectors play an important role in the analysis of extrema values of the quadratic form $\mathbf{x}^*\mathbf{A}\mathbf{x}$ where $\mathbf{A}$ is a Hermitian matrix. In particular, we know that for a Hermitian matrix, $\mathbf{x}^*\mathbf{A}\mathbf{x}$ is real. The following result bounds the values of this quadratic form by the eigenvalues of $\mathbf{A}$:

Theorem A.2 (Rayleigh–Ritz theorem). *Consider the Rayleigh–Ritz quotient*

$$R(\mathbf{x}) = \frac{\mathbf{x}^*\mathbf{A}\mathbf{x}}{\mathbf{x}^*\mathbf{x}}, \tag{A.36}$$

for $\mathbf{x} \neq 0$ and a Hermitian matrix $\mathbf{A}$. Then $\lambda_{\min} \leq R(\mathbf{x}) \leq \lambda_{\max}$ where $\lambda_{\min}, \lambda_{\max}$ are the smallest and largest eigenvalues of $\mathbf{A}$. In addition, the maximum (minimum) is achieved when $\mathbf{x}$ is the eigenvector corresponding to $\lambda_{\max}$ ($\lambda_{\min}$).

A.2.2 Singular value decomposition

The eigendecomposition is very useful in applications, however, it applies only to square matrices. To extend these ideas to more general matrices we note that for an arbitrary matrix $\mathbf{A}$, both $\mathbf{M} = \mathbf{A}\mathbf{A}^*$ and $\mathbf{G} = \mathbf{A}^*\mathbf{A}$ are Hermitian and nonnegative definite. Therefore, both $\mathbf{M}$ and $\mathbf{G}$ will have a set of orthonormal eigenvectors, and the same nonzero eigenvalues. We can use their eigendecompositions to define the *singular value decomposition (SVD)* which is applicable to any $m \times n$ matrix $\mathbf{A}$.

Let $\mathbf{A}$ be an arbitrary $m \times n$ matrix with rank r. Then $\mathbf{A}$ can be written as

$$\mathbf{A} = \mathbf{U}\boldsymbol{\Sigma}\mathbf{V}^*, \tag{A.37}$$

where $\mathbf{U}$ is an $m \times m$ unitary matrix, $\mathbf{V}$ is an $n \times n$ unitary matrix, and $\boldsymbol{\Sigma} = \mathrm{diag}\,(\sigma_1, \ldots, \sigma_r, 0, \ldots, 0)$ is an $m \times n$ diagonal matrix with the first r diagonal elements equal to $\sigma_i > 0$ arranged in decreasing order (so that $\sigma_1 \geq \sigma_2 \geq \cdots \geq \sigma_r$), and the remaining elements equal to 0.

The *singular values* σ_i are given by $\sigma_i = \sqrt{\lambda_i}$ where λ_i are the nonzero eigenvalues of $\mathbf{A}\mathbf{A}^*$ or $\mathbf{A}^*\mathbf{A}$, arranged in decreasing order. The singular values are unique. The columns of $\mathbf{U}$ are the eigenvectors of $\mathbf{M}$ and are referred to as the *left singular vectors*. The columns of $\mathbf{V}$ are the eigenvectors of $\mathbf{G}$ and are referred to as the *right singular vectors*. Both columns are arranged in the same order as λ_i.

A nonnegative real number σ is a singular value of $\mathbf{A}$ if and only if there exists vectors $\mathbf{u} \in \mathbb{C}^m$ and $\mathbf{v} \in \mathbb{C}^n$ such that

$$\mathbf{A}\mathbf{v} = \sigma\mathbf{u}, \quad \mathbf{A}^*\mathbf{u} = \sigma\mathbf{v}. \tag{A.38}$$

The vectors $\mathbf{u}$ and $\mathbf{v}$ are the right and left singular vectors.

The singular values have an extrema property similar to that of the eigenvalues of Hermitian matrices:

Theorem A.3 (Extrema properties of singular values). *Let $\mathbf{A}$ be an arbitrary $m \times n$ matrix with largest singular value $\sigma_{\max}$ and smallest singular value $\sigma_{\min}$.*

The corresponding right singular vectors are denoted $\mathbf{v}_{max}$ and $\mathbf{v}_{min}$. Then

$$\max_{\mathbf{x}\neq 0} \frac{\|\mathbf{Ax}\|_2}{\|\mathbf{x}\|_2} = \sigma_{max}, \tag{A.39}$$

where the maximum is achieved with $\mathbf{x} = \mathbf{v}_{max}$. Similarly,

$$\min_{\mathbf{x}\neq 0} \frac{\|\mathbf{Ax}\|_2}{\|\mathbf{x}\|_2} = \sigma_{min}, \tag{A.40}$$

where the minimum is achieved with $\mathbf{x} = \mathbf{v}_{min}$.

Clearly for a unitary matrix it follows that $\sigma_{max} = \sigma_{min} = 1$.

The SVD can be used to define a variety of matrix norms, as we will see below in Section A.4. It is also used to obtain approximations of a given rank to arbitrary matrices. This result is incorporated in the *Eckart–Young theorem* stated below. In the theorem, $\|\mathbf{A}\|_F^2$ denotes the Frobenius norm given by $\|\mathbf{A}\|_F^2 = \mathrm{Tr}(\mathbf{A}^*\mathbf{A})$.

Theorem A.4 (Eckart–Young theorem).   *Let $\mathbf{A}$ be an arbitrary $m \times n$ matrix with SVD $\mathbf{A} = \mathbf{U\Sigma V}^*$, and consider the problem*

$$\min_{\mathrm{rank}(\mathbf{Q})=s} \|\mathbf{A} - \mathbf{Q}\|_F^2, \tag{A.41}$$

where $\mathbf{Q}$ is an arbitrary matrix of rank s, and s is given. The solution is equal to $\mathbf{Q} = \mathbf{U\Sigma}_s\mathbf{V}^*$ where Σ_s is a diagonal matrix with the first s diagonal elements equal to σ_i, and the remaining elements equal to 0.

Thus, the best rank-s approximation to a given matrix $\mathbf{A}$ is given by taking only the s largest singular values and the corresponding singular vectors of $\mathbf{A}$.

A.3   Linear equations

Consider a set of equations of the form

$$\mathbf{y} = \mathbf{Ax} \tag{A.42}$$

where $\mathbf{A}$ is an $m \times n$ matrix. Our goal is to find a solution $\mathbf{x}$ to this set of equations if it exists, and to approximate such a solution otherwise. As we will see below, a useful tool in analyzing the solutions to (A.42) is the pseudoinverse, which we studied in detail in Section 2.7.

In Section 2.7 we mentioned that one of the common ways to compute the pseudo-inverse of a matrix is via the SVD. Specifically, if $\mathbf{A}$ has an SVD of the form $\mathbf{A} = \mathbf{U\Sigma V}^*$ and rank r, then

$$\mathbf{A}^\dagger = \mathbf{V\Sigma}^\dagger\mathbf{U}^*, \tag{A.43}$$

where $\Sigma^\dagger$ is an $n \times m$ diagonal matrix with diagonal elements equal to $1/\sigma_i$ for $1 \leq i \leq r$, and 0 otherwise. It can also be shown that if $\mathbf{A}$ has full column-rank, then

$$\mathbf{A}^\dagger = (\mathbf{A}^*\mathbf{A})^{-1}\mathbf{A}^* \tag{A.44}$$

and when $\mathbf{A}$ has full row-rank,

$$\mathbf{A}^\dagger = \mathbf{A}^*(\mathbf{A}\mathbf{A}^*)^{-1}. \tag{A.45}$$

Note that in the first case, $\mathbf{A}^\dagger\mathbf{A} = \mathbf{I}$, and in the second, $\mathbf{A}\mathbf{A}^\dagger = \mathbf{I}$.

Going back to (A.42), if $\text{rank}(\mathbf{A}) < m$ then depending on $\mathbf{y}$ there may or may not be a solution. In particular, a solution exists only if $\mathbf{y}$ is in the range of $\mathbf{A}$. We then have the following:

1. If $\mathbf{A}$ has full row-rank, which implies $m \leq n$, then there exists a solution $\mathbf{x}$. When $m = n$, $\mathbf{A}$ is invertible and the solution is unique. For $m < n$ there are infinitely many solutions. The minimal-norm solution, namely the solution $\mathbf{x}$ that has minimal norm $\|\mathbf{x}\|_2$ among all possible vectors $\mathbf{v}$ satisfying $\mathbf{A}\mathbf{v} = \mathbf{y}$, is given by

$$\mathbf{x} = \mathbf{A}^\dagger\mathbf{y} = \mathbf{A}^*(\mathbf{A}\mathbf{A}^*)^{-1}\mathbf{y}. \tag{A.46}$$

2. If $\mathbf{A}$ has full column-rank, which implies $m \geq n$, and $\mathbf{y}$ is in the range of $\mathbf{A}$, then there is a unique solution $\mathbf{x}$ given by

$$\mathbf{x} = \mathbf{A}^\dagger\mathbf{y} = (\mathbf{A}^*\mathbf{A})^{-1}\mathbf{A}^*\mathbf{y}. \tag{A.47}$$

Suppose next that $\mathbf{y}$ is not in the range of $\mathbf{A}$ so that there is no solution to (A.42). We can instead seek the value of $\mathbf{x}$ for which $\mathbf{A}\mathbf{x}$ is close to $\mathbf{y}$ in some sense. A popular approach is to seek the *least squares solution* that minimizes the squared error of the differences:

$$\hat{\mathbf{x}}_{\text{LS}} = \arg\min_{\mathbf{x}} \|\mathbf{y} - \mathbf{A}\mathbf{x}\|_2^2. \tag{A.48}$$

By setting the derivative to zero it can be shown that $\hat{\mathbf{x}}_{\text{LS}}$ satisfies the *normal equations*

$$\mathbf{A}^*\mathbf{A}\hat{\mathbf{x}}_{\text{LS}} = \mathbf{A}^*\mathbf{y}. \tag{A.49}$$

If $\mathbf{A}$ has full column-rank, then the unique solution to (A.49) is

$$\hat{\mathbf{x}}_{\text{LS}} = (\mathbf{A}^*\mathbf{A})^{-1}\mathbf{A}^*\mathbf{y}. \tag{A.50}$$

Otherwise, there are infinitely many choices of $\hat{\mathbf{x}}_{\text{LS}}$ that satisfy (A.49). From all possible solutions, the one with minimal norm is given by

$$\hat{\mathbf{x}}_{\text{LS}} = \mathbf{A}^\dagger\mathbf{y}. \tag{A.51}$$

A.4   Matrix norms

Matrix norms are useful in a variety of applications. A function $\|\cdot\|: \mathbb{C}^{m \times n} \to \mathbb{R}$ is a *matrix norm* if it satisfies the following properties for any $m \times n$ matrices $\mathbf{A}, \mathbf{B}$:

1. Nonnegative: $\|\mathbf{A}\| \geq 0$ with equality if and only if $\mathbf{A} = \mathbf{0}$.
2. Homogeneous: $\|c\mathbf{A}\| = |c|\|\mathbf{A}\|$ for any $c \in \mathbb{C}$.
3. Triangle inequality: $\|\mathbf{A} + \mathbf{B}\| \leq \|\mathbf{A}\| + \|\mathbf{B}\|$.
4. Submultiplicative (this property is defined for square matrices, namely $m = n$): $\|\mathbf{A}\mathbf{B}\| \leq \|\mathbf{A}\|\|\mathbf{B}\|$. We note that in some textbooks this last property is not mandatory to define a matrix norm.

Three classes of popular matrix norms are induced norms, entrywise norms, and Schatten norms.

A.4.1 Induced norms

Many matrix norms can be expressed as *induced norms*

$$\|\mathbf{A}\| = \max_{\mathbf{x} \neq 0} \frac{\|\mathbf{A}\mathbf{x}\|}{\|\mathbf{x}\|}, \tag{A.52}$$

where different norms may be defined in the numerator and the denominator. If $m = n$ and the same norms are used then the induced norm is a submultiplicative matrix norm. Employing the p-norm in both numerator and denominator we have

$$\|\mathbf{A}\|_p = \max_{\mathbf{x} \neq 0} \frac{\|\mathbf{A}\mathbf{x}\|_p}{\|\mathbf{x}\|_p}. \tag{A.53}$$

For any induced norm it follows from the definition that $\|\mathbf{I}\| = 1$.

Different choices of p lead to different matrix norms:

1. The *spectral norm* results from choosing $p = 2$:

$$\|\mathbf{A}\|_2 = \sigma_{\max}, \tag{A.54}$$

where $\sigma_{\max}$ is the largest singular value of $\mathbf{A}$. Note that for a unitary matrix, $\|\mathbf{A}\|_2 = 1$.

2. The maximum column-sum matrix norm corresponds to $p = 1$:

$$\|\mathbf{A}\|_1 = \max_{1 \leq j \leq n} \sum_{i=1}^{m} |a_{ij}|. \tag{A.55}$$

3. The maximum row-sum matrix norm results from $p = \infty$:

$$\|\mathbf{A}\|_\infty = \max_{1 \leq i \leq m} \sum_{j=1}^{n} |a_{ij}|. \tag{A.56}$$

The spectral norm is often used to define the *condition number* $\kappa(\mathbf{A})$ of an invertible matrix $\mathbf{A}$:

$$\kappa(\mathbf{A}) = \|\mathbf{A}^{-1}\|\|\mathbf{A}\|. \tag{A.57}$$

This quantity provides an estimate on how accurately we can determine $\mathbf{x}$ in the linear equation $\mathbf{A}\mathbf{x} = \mathbf{y}$ in the presence of measurement errors. Note, that any matrix norm may be used in this definition.

For general operators A, we can define the *operator norm* in a similar fashion to (A.52):

$$\|A\| = \sup_{x \neq 0} \frac{\|Ax\|}{\|x\|}. \tag{A.58}$$

A.4.2 Entrywise norms

An alternative way to obtain matrix norms is by viewing a matrix as a vector of size mn and applying one of the familiar vector norms. For example, using the p-norm for vectors we obtain:

$$\|\mathbf{A}\| = \left(\sum_{i=1}^{m} \sum_{j=1}^{n} |a_{ij}|^p \right)^{1/p}. \tag{A.59}$$

Choosing $p = 2$ leads to the *Frobenius norm*:

$$\|\mathbf{A}\|_F = \left(\sum_{i=1}^{m} \sum_{j=1}^{n} |a_{ij}|^2 \right)^{1/2} = \text{Tr}^{1/2}(\mathbf{A}^*\mathbf{A}). \tag{A.60}$$

A.4.3 Schatten norms

The Schatten p-norms result from applying a vector p-norm to the singular values of the matrix $\mathbf{A}$. Thus,

$$\|\mathbf{A}\| = \left(\sum_{i=1}^{\min(m,n)} \sigma_i^p \right)^{1/p}. \tag{A.61}$$

Note that for $p = 2$ this norm reduces to the Frobenius norm. For $p = \infty$ we have the spectral norm. The case $p = 1$ yields the *nuclear norm* defined as

$$\|\mathbf{A}\|_* = \sum_{i=1}^{\min(m,n)} \sigma_i = \text{Tr}(\sqrt{\mathbf{A}^*\mathbf{A}}). \tag{A.62}$$

Appendix B

Stochastic signals

In this appendix we briefly summarize some of the basic tools from probability theory and random processes that we use throughout the book. Comprehensive treatment of this material can be found in many excellent textbooks, e.g. [477, 478, 479].

B.1 Random variables

We assume that the reader is familiar with the basic notions of probability theory and random variables, and therefore we do not go into details of probability definitions. Instead we focus on fundamental properties and concepts pertaining to random variables and processes.

A random variable x is a mapping from a sample space of possible outcomes to a subset of the real line. If x takes on discrete values then it is called a *discrete random variable*. If it takes on continuous values then we refer to x as a *continuous random variable*. In the book, we will primarily be needing continuous random variables. Therefore, below, when we refer to a random variable we always assume it is continuous.

B.1.1 Probability density function

The *probability density function (pdf)* $f_x(x)$ of a random variable x determines the probability that x is smaller than a given value:

$$P(x \leq a) = \int_{-\infty}^{a} f_x(x)dx. \tag{B.1}$$

Any valid pdf must satisfy that $f_x(x) \geq 0$ and $\int_{-\infty}^{\infty} f_x(x)dx = 1$. Given a function $h(x)$ of a random variable x, its *mean* or *expected value* is defined by

$$E\{h(x)\} = \int_{-\infty}^{\infty} h(x)f_x(x)dx. \tag{B.2}$$

The moments of a random variable are given by

$$E\{x^n\} = \int_{-\infty}^{\infty} x^n f_x(x)dx. \tag{B.3}$$

When $n = 1$ we have the mean of x denoted $\mu_x = E\{x\}$. The variance of a random variable can be expressed in terms of its second moment as

$$\sigma_x^2 = E\{x^2\} - \mu_x^2. \tag{B.4}$$

For a zero-mean random variable, $\sigma_x^2 = E\{x^2\}$.

Two popular pdfs are the uniform density

$$f_x(x) = \begin{cases} \frac{1}{b-a}, & a \le x \le b \\ 0, & \text{otherwise,} \end{cases} \tag{B.5}$$

and the Gaussian density

$$f_x(x) = \frac{1}{\sqrt{2\pi\sigma_x^2}} \exp\left(-\frac{(x - \mu_x)^2}{2\sigma_x^2}\right). \tag{B.6}$$

In the above pdf $\mu_x = E\{x\}$ is equal to the expected value of x, and $\sigma_x^2 = E\{(x-\mu_x)^2\}$ to the variance of x. The Gaussian pdf is also referred to as the normal distribution and is often denoted as $\mathcal{N}(\mu_x, \sigma_x^2)$.

B.1.2   Jointly random variables

Two random variables x and y on the same probability space are described by their *joint pdf* $f_{xy}(x, y)$:

$$P(x \le a, y \le b) = \int_{-\infty}^{a} \int_{-\infty}^{b} f_{xy}(x, y) dx dy. \tag{B.7}$$

The *marginal* pdf of each of the variables can be obtained by integrating over the other variable. For example,

$$f_x(x) = \int_{-\infty}^{\infty} f_{xy}(x, y) dy. \tag{B.8}$$

Clearly the joint pdf also integrates to 1: $\int_{-\infty}^{\infty} \int_{-\infty}^{\infty} f_{xy}(x, y) dx dy = 1$. These definitions extend in a straightforward way to any finite number of random variables.

The *cross-correlation* between x and y is defined as

$$r_{xy} = E\{xy\}. \tag{B.9}$$

The *cross-covariance* is given by

$$c_{xy} = E\{(x - \mu_x)(y - \mu_y)\}, \tag{B.10}$$

where μ_x, μ_y denote the expected values of x and y, respectively. It is easy to see that $r_{xy} = c_{xy} + \mu_x\mu_y$. Thus, the correlation and covariance are equal if either of the random variables has zero mean.

Two random variables are said to be *independent* if $f_{xy}(x, y) = f_x(x)f_y(y)$. They are *uncorrelated* if $r_{xy} = \mu_x\mu_y$, or equivalently, $c_{xy} = 0$. If two variables are statistically independent, then they are also uncorrelated; however, the reverse is not generally true. More broadly, if x and y are independent random variables, then $E\{g(x)h(y)\} = E\{g(x)\}E\{h(y)\}$ for any functions $g(x)$ and $h(y)$. If x and y are jointly Gaussian, then they are independent if and only if they are uncorrelated.

B.2   Random vectors

A *random vector* $\mathbf{x} \in \mathbb{R}^m$ is defined by the joint pdf of its elements $x_1, \ldots, x_m$. Its mean $\mathbf{m} = E\{\mathbf{x}\}$ is a length-m vector with components $E\{x_i\}$. The random vector $\mathbf{x}$ has zero mean if $E\{x_i\} = 0$ for all i.

The *autocorrelation* matrix of $\mathbf{x}$ is defined by $\mathbf{R} = E\{\mathbf{x}\mathbf{x}^T\}$, and its *covariance matrix* is given by $\mathbf{C} = E\{(\mathbf{x} - \mathbf{m})(\mathbf{x} - \mathbf{m})^T\}$. In particular, the ith diagonal element of $\mathbf{C}$ is the variance of x_i. Both $\mathbf{R}$ and $\mathbf{C}$ are symmetric nonnegative definite matrices. If the elements of $\mathbf{x}$ are all uncorrelated and have equal variance, then $\mathbf{C} = \sigma_x^2 \mathbf{I}$.

The pdf of a Gaussian random vector with mean $\mathbf{m}$ and covariance matrix $\mathbf{C}$ is given by

$$f(\mathbf{x}) = \frac{1}{\sqrt{(2\pi)^m |\det \mathbf{C}|}} \exp\left(-\frac{1}{2}(\mathbf{x} - \mathbf{m})^T \mathbf{C}^{-1} (\mathbf{x} - \mathbf{m})\right). \tag{B.11}$$

Here we assume that $\mathbf{C}$ is invertible. A random vector $\mathbf{x}$ is Gaussian if and only if $\sum_i a_i x_i$ is a Gaussian random variable for any choice of scalars $\{a_i\}$.

A complex random variable $z = x + jy$ is said to be complex Gaussian if x and y are jointly Gaussian real random variables. The distribution of z can then be obtained as the joint distribution of the vector $\mathbf{x} = [x \; y]^T$.

B.3   Random processes

B.3.1   Continuous-time random processes

A *random process* $x(t)$ is a mapping from a sample space to a set of real functions. It is defined by the joint pdf $P(x(t_1) \leq a_1, \ldots, x(t_m) \leq a_m)$ of its samples at times $t_1, \ldots, t_m$ for all possible times $t_1, \ldots, t_m$ and all m. The mean of a random process $x(t)$ is defined by $\mu_x(t) = E\{x(t)\}$. We further define the cross-correlation and cross-covariance functions

$$r_x(t, \tau) = E\{x(t)x(\tau)\}, \quad c_x(t, \tau) = E\{(x(t) - \mu(t))(x(\tau) - \mu(\tau))\} \tag{B.12}$$

between any two samples of the process $x(t)$ and $x(\tau)$.

A random process is Gaussian if and only if for every finite set of indices $t_1, \ldots, t_m$, the random variables $x(t_1), \ldots, x(t_m)$ are jointly Gaussian.

Stationary processes

A process $x(t)$ is *stationary* if for all T and m, and all times $t_1, \ldots, t_m$, it holds that

$$P(x(t_1) \leq a_1, \ldots, x(t_m) \leq a_m) = P(x(t_1 + T) \leq a_1, \ldots, x(t_m + T) \leq a_m). \tag{B.13}$$

Intuitively this implies that arbitrary time shifts do not change the pdf. In general, stationarity is a difficult property to verify. A relaxed and more useful notion is that of *wide-sense stationarity (WSS)* which implies stationarity of the first and second moments. More specifically, a process $x(t)$ is said to be WSS if

1. Its expected value $\mu_x(t) = \mu_x$ is independent of t;
2. Its correlation function $r_x(t, \tau) = r_x(\tau - t)$ depends only on the time difference $\tau - t$. We can then write $r_x(\tau) = E\{x(t - \tau)x(t)\}$.

For a zero-mean WSS process, $r_x(0) = E\{x^2(t)\} = \sigma_x^2$ is its variance. It is easy to see that $r_x(\tau)$ is symmetric: $r_x(\tau) = r_x(-\tau)$. In addition, it can be shown that the correlation function takes on its maximal value at 0: $|r_x(\tau)| \leq r_x(0) = \sigma_x^2$.

Stationary processes are obviously WSS. However, the reverse is not true in general. For Gaussian processes it holds that $x(t)$ is stationary if and only if it is WSS.

A random process is said to be *independent identically distributed (iid)* if $x(t)$ and $x(\tau)$ are independent for all t and τ, and in addition the pdf of $x(t)$ is independent of t. In this case $x(t)$ is WSS with $r_x(\tau) = \sigma_x^2 \delta(\tau)$, where σ_x^2 is its variance. A zero-mean WSS process $x(t)$ is said to be *white* if $r_x(\tau) = \sigma_x^2 \delta(\tau)$. It is defined as *colored* otherwise.

One of the reasons that Gaussian distributions are very popular in applications and often used to model noise in linear systems is due to the *central limit theorem*. This theorem states that the limiting distribution of a sum of a large number of iid random variables is a Gaussian distribution, under general conditions. Therefore, if we can view a particular random variable, such as noise, as a sum of a large number of iid elements, then its pdf is approximately Gaussian.

For a WSS process we can define its *power spectrum density (PSD)*, which is the CTFT of its correlation function $r_x(\tau)$:

$$\Lambda_x(\omega) = \int_{-\infty}^{\infty} r_x(\tau)e^{-j\omega\tau}d\tau. \tag{B.14}$$

Since $r_x(\tau)$ is symmetric, $\Lambda_x(\omega)$ is real-valued. In addition, $\Lambda_x(\omega) \geq 0$ for all ω. From the properties of the inverse CTFT we have that

$$r_x(\tau) = \frac{1}{2\pi} \int_{-\infty}^{\infty} \Lambda_x(\omega)e^{j\omega\tau}d\omega. \tag{B.15}$$

In particular,

$$r_x(0) = E\{x^2(t)\} = \frac{1}{2\pi} \int_{-\infty}^{\infty} \Lambda_x(\omega)d\omega, \tag{B.16}$$

so that the expected power of the random process is the integral of its PSD.

The PSD of a zero-mean white random process is equal to $\Lambda_x(\omega) = \sigma_x^2$.

Ergodic processes

When considering random processes we are often interested in their time averages. A WSS random process $x(t)$ is said to be *ergodic in the mean* if

$$\mu_x = E\{x(t)\} = \lim_{T \to \infty} \frac{1}{T} \int_{-T}^{T} x(t)dt. \tag{B.17}$$

In other words, the time average of the process converges to its (constant) expected value. It is *ergodic in the correlation* if

$$r_x(\tau) = E\{x(t-\tau)x(t)\} = \lim_{T \to \infty} \frac{1}{T} \int_{-T}^{T} x(t-\tau)x(t)dt. \tag{B.18}$$

If $x(t)$ is ergodic both in the mean and in the correlation then we refer to $x(t)$ as *wide-sense ergodic*.

Just like stationarity, ergodicity can be defined for higher moments as well. Clearly any process that is ergodic must be stationary; however, the reverse claim is not true.

B.3.2   Discrete-time random processes

A sequence $x[n]$ of random variables is defined as a discrete-time random process, and is specified by the joint m-dimensional pdf of every m of its variables $x[n_1], \ldots, x[n_m]$. The mean of the process is denoted by $\mu_x[n]$, and its cross-correlation and cross-covariance functions are given by

$$r_x[m, n] = E\{x[m]x[n]\}, \quad c_x[m, n] = E\{(x[m] - \mu_x[m])(x[n] - \mu_x[n])\}. \tag{B.19}$$

Stationary processes

In analogy to continuous-time processes, a discrete-time random process $x[n]$ is said to be WSS if:

1. Its expected value $\mu_x[n] = \mu$ is independent of n;
2. Its correlation function $r_x[m, n] = r_x[n - m]$ depends only on the difference $n - m$. We can then write $r_x[k] = E\{x[n - k]x[n]\}$. Clearly, $r_x[k] = r_x[-k]$.

The variance of a zero-mean WSS process is given by $r_x[0] = E\{x^2[n]\} = \sigma_x^2$. The process $x[n]$ is said to be iid if $x[n]$ and $x[m]$ are independent for all n and m, and in addition the pdf of $x[n]$ is independent of n. In this case $x[n]$ is WSS with $r_x[k] = \sigma_x^2 \delta[k]$, where σ_x^2 is its variance. A zero-mean WSS process $x[n]$ is white if $r_x[k] = \sigma_x^2 \delta[k]$.

The PSD of a WSS process $x[n]$ is defined as the DTFT of the correlation function $r_x[n]$:

$$\Lambda_x(e^{j\omega}) = \sum_{n \in \mathbb{Z}} r_x[n]e^{-j\omega n}. \tag{B.20}$$

As with its continuous-time counterpart, $\Lambda_x(e^{j\omega})$ is real-valued and $\Lambda_x(e^{j\omega}) \geq 0$ for all ω. From the properties of the inverse DTFT we have that

$$r_x[n] = \frac{1}{2\pi} \int_{-\pi}^{\pi} \Lambda_x(e^{j\omega})e^{j\omega n}d\omega. \tag{B.21}$$

The PSD of a zero-mean white random process is given by $\Lambda_x(e^{j\omega}) = \sigma_x^2$.

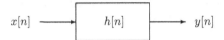

Figure B.1 LTI filtering of a random process.

Passage through LTI systems

Suppose that we pass a WSS random process $x[n]$ through a stable LTI system $h[n]$ to obtain the output $y[n] = h[n] * x[n]$ as depicted in Fig. B.1. It can then be shown that $y[n]$ is also WSS with

$$\mu_y = \mu_x \sum_{n \in \mathbb{Z}} h[n], \quad \Lambda_y(e^{j\omega}) = \Lambda_x(e^{j\omega})|H(e^{j\omega})|^2, \qquad (\text{B.22})$$

where $H(e^{j\omega})$ is the DTFT of $h[n]$. As an example, if $x[n]$ is a zero-mean white process with variance σ_x^2, then $\mu_y = 0$ and

$$\Lambda_y(e^{j\omega}) = \sigma_x^2 |H(e^{j\omega})|^2. \qquad (\text{B.23})$$

Thus, the PSD of $y[n]$ is colored by the LTI system $H(e^{j\omega})$ and the variance of $y[n]$ is amplified by the energy in the impulse response:

$$r_y[0] = \frac{\sigma_x^2}{2\pi} \int_{-\pi}^{\pi} |H(e^{j\omega})|^2 d\omega = \sigma_x^2 \sum_{n \in \mathbb{Z}} h^2[n]. \qquad (\text{B.24})$$

Similar relations hold in the case of continuous-time random processes. Namely, if a continuous-time WSS process $x(t)$ with PSD $\Lambda_x(\omega)$ passes through a stable LTI filter $h(t)$ with frequency response $H(\omega)$, then the output $y(t)$ is WSS with PSD

$$\Lambda_y(\omega) = \Lambda_x(\omega)|H(\omega)|^2. \qquad (\text{B.25})$$

Consider a random process with PSD $\Lambda_x(\omega)$ which is not constant, so that the signal is colored. In many applications it is useful to *whiten* the signal prior to further processing. Thus, we seek a filter $H(\omega)$ such that the PSD of its output $y(t)$ is constant. From (B.25) we can choose $H_W(\omega) = \Lambda_x^{-1/2}(\omega)$ assuming that the PSD is strictly positive. With this choice, $\Lambda_y(\omega) = 1$. We therefore refer to $H_W(\omega)$ as the *whitening filter* of the process $x(t)$.

B.4   Sampling of bandlimited processes

The Shannon–Nyquist theorem states that any bandlimited signal $x(t)$ can be represented by samples taken at its Nyquist rate. In this section we consider a similar theorem for random processes. To state the result, we first need to define the notion of a bandlimited random processes.

A WSS process $x(t)$ is called *bandlimited* if it has finite power $r_x(0) < \infty$ and its PSD satisfies $\Lambda_x(\omega) = 0$ for $|\omega| \geq \pi/T$.

If $x(t)$ is bandlimited, then we can apply the Shannon–Nyquist theorem to its correlation function $r(\tau)$ (which is the inverse CTFT of $\Lambda_x(\omega)$) in order to express it as

$$r(\tau) = \sum_{n \in \mathbb{Z}} r(nT) \operatorname{sinc}((\tau - nT)/T). \tag{B.26}$$

However, what we are interested in is expressing the process itself in terms of its samples. As we show below, such an expansion is possible, in the mean-squared sense [151]. We say that $x(t) = y(t)$ in a mean-squared sense if $E\{(x(t) - y(t))^2\} = 0$.

Theorem B.1 (Sampling of bandlimited WSS signals). *Suppose that $x(t)$ is a WSS process bandlimited to π/T. We can then express $x(t)$ in terms of its samples $x(nT)$ as*

$$x(t) = \sum_{n \in \mathbb{Z}} x(nT) \operatorname{sinc}((\tau - nT)/T), \tag{B.27}$$

where equality holds in the mean-squared sense.

A simple proof of this result is given in [477].

References

[1] C. E. Shannon, "Communications in the presence of noise," *Proc. IRE*, vol. 37, no. 1, pp. 10–21, Jan. 1949.

[2] H. Nyquist, "Certain topics in telegraph transmission theory," *AIEE Trans.*, vol. 47, no. 2, pp. 617–644, Jan. 1928.

[3] V. A. Kotelnikov, "On the transmission capacity of the ether and of cables in electrical communications," in *Proc. First All-Union Conference on the Technological Reconstruction of the Communications Sector and the Development of Low-current Engineering*, Moscow, 1933.

[4] E. T. Whittaker, "On the functions which are represented by the expansions of the interpolation theory," *Proc. Roy. Soc. Edinburgh*, vol. 35, pp. 181–194, Jul. 1915.

[5] J. M. Whittaker, *Interpolatory Function Theory*, Cambridge, UK: Cambridge University Press, 1935.

[6] A. Cauchy, "M'moire sur diverses formules danalyse," *C. R. Acad. Sci.*, vol. 12, pp. 283–298, 1841.

[7] J. R. Higgins, "Five short stories about the cardinal series," *Bull. Am. Math. Soc.*, vol. 12, no. 1, pp. 45–89, 1985.

[8] A. J. Jerri, "The Shannon sampling theorem – Its various extensions and applications: A tutorial review," *Proc. IEEE*, vol. 65, no. 11, pp. 1565–1596, Nov. 1977.

[9] P. L. Butzer, "A survey of the Whittaker–Shannon sampling theorem and some of its extensions," *J. Math. Res. Expo.*, vol. 1983, no. 1, pp. 185–212, Jan. 1983.

[10] M. Unser, "Sampling – 50 years after Shannon," *Proc. IEEE*, vol. 88, no. 4, pp. 569–587, Apr. 2000.

[11] Y. C. Eldar and T. Michaeli, "Beyond bandlimited sampling," *IEEE Signal Processing Mag.*, vol. 26, no. 3, pp. 48–68, May 2009.

[12] D. L. Donoho, "Compressed sensing," *IEEE Trans. Inform. Theory*, vol. 52, no. 4, pp. 1289–1306, Apr. 2006.

[13] E. Candès, J. Romberg and T. Tao, "Robust uncertainty principles: Exact signal reconstruction from highly incomplete frequency information," *IEEE Trans. Inform. Theory*, vol. 52, no. 2, pp. 489–509, Feb. 2006.

[14] Y. C. Eldar and G. Kutyniok, *Compressed Sensing: Theory and Applications*, Cambridge, UK: Cambridge University Press, 2012.

[15] R. O. Schmidt, "Multiple emitter location and signal parameter estimation," *Proc. RADC Spectral Estimation Workshop*, pp. 243–258, Oct. 1979.

[16] R. Roy and T. Kailath, "ESPRIT – estimation of signal parameters via rotational invariance techniques," *IEEE Trans. Acoust. Speech Signal Processing*, vol. 37, no. 7, pp. 984–995, Jul. 1989.

[17] P. Stoica and Y. Selen, "Model-order selection: a review of information criterion rules," *IEEE Signal Processing Mag.*, vol. 21, no. 4, pp. 36–47, Jul. 2004.

[18] S. Baker, S. K. Nayar and H. Murase, "Parametric feature detection," *Int. J. Computer Vision*, vol. 27, no. 1, pp. 27–50, Mar. 1998.

[19] J. Crols and M. S. J. Steyaert, "Low-IF topologies for high-performance analog front ends of fully integrated receivers," *IEEE Trans. Inform. Theory*, vol. 45, no. 3, pp. 269–282, Mar. 1998.

[20] N. L. Scott, R. C. Vaughan and D. R. White, "The theory of bandpass sampling," *IEEE Trans. Signal Processing*, vol. 39, no. 9, pp. 1973–1984, Sep. 1991.

[21] M. Mishali, Y. C. Eldar and A. J. Elron, "Xampling: Signal acquisition and processing in union of subspaces," *IEEE Trans. Signal Processing*, vol. 59, no. 10, pp. 4719–4734, Oct. 2011.

[22] M. Mishali and Y. C. Eldar, "Xampling: Compressed sensing of analog signals," in *Compressed Sensing: Theory and Applications*, Cambridge, UK: Cambridge University Press, pp. 88–148, 2012.

[23] S. K. Berberian, *Introduction to Hilbert Space*, New York, NY: Oxford University Press, 1961.

[24] P. R. Halmos, *Introduction to Hilbert Space*, 2nd edn. New York, NY: Chelsea Publishing Company, 1957.

[25] R. M. Young, *An Introduction to Nonharmonic Fourier Series*, New York, NY: Academic Press, 1980.

[26] O. Christensen, *An Introduction to Frames and Riesz Bases*, Boston, MA: Birkhäuser, 2003.

[27] L. Debnath and P. Mikusiński, *Hilbert Spaces with Applications*, 3rd edn. New York, NY: Academic Press, 2005.

[28] R. A. Horn and C. R. Johnson, *Matrix Analysis*, Cambridge, UK: Cambridge University Press, 1985.

[29] Y. C. Eldar, "Least-squares inner product shaping," *Linear Alg. Appl.*, vol. 348, nos. 1–3, pp. 153–174, May 2002.

[30] Y. C. Eldar, "Least-squares orthogonalization using semidefinite programming," *Linear Alg. Appl.*, vol. 412, nos. 2–3, pp. 453–470, Jan. 2006.

[31] Y. C. Eldar and G. D. Forney, Jr., "Optimal tight frames and quantum measurement," *IEEE Trans. Inform. Theory*, vol. 48, no. 3, pp. 599–610, Mar. 2002.

[32] A. Barvinok, "Measure concentration," 2005, Math 710 lecture notes, www.math.lsa. umich.edu

[33] I. Daubechies, *Ten Lectures on Wavelets*, Philadelphia, PA: SIAM, 1992.

[34] K. Hoffman and R. Kunze, *Linear Algebra*, 2nd edn. New Jersey: Prentice-Hall, Inc., 1971.

[35] S. Kayalar and H. L. Weinert, "Oblique projections: Formulas, algorithms, and error bounds," *Math. Contr. Signals Syst.*, vol. 2, no. 1, pp. 33–45, Mar. 1989.

[36] A. Aldroubi, "Oblique projections in atomic spaces," *Proc. Am. Math. Soc.*, vol. 124, no. 7, pp. 2051–2060, Jul. 1996.

[37] R. T. Behrens and L. L. Scharf, "Signal processing applications of oblique projection operators," *IEEE Trans. Signal Processing*, vol. 42, no. 6, pp. 1413–1424, Jun. 1994.

[38] G. H. Golub and C. F. Van Loan, *Matrix Computations*, 3rd edn. Baltimore, MD: Johns Hopkins University Press, 1996.

[39] A. Ben-Israel and T. N. E. Greville, *Generalized Inverses: Theory and Applications*, New York, NY: Springer Verlag, 2003.

[40] R. J. Duffin and A. C. Schaeffer, "A class of nonharmonic Fourier series," *Trans. Am. Math. Soc.*, vol. 72, no. 2, pp. 314–366, Mar. 1952.

[41] I. Daubechies, A. Grossmann and Y. Meyer, "Painless nonorthogonal expansions," *J. Math. Phys.*, vol. 27, no. 5, pp. 1271–1283, May 1986.

[42] I. Daubechies, "The wavelet transform, time-frequency localization and signal analysis," *IEEE Trans. Inform. Theory*, vol. 36, no. 5, pp. 961–1005, Sep. 1990.

[43] A. Aldroubi, "Portraits of frames," *Proc. Am. Math. Soc.*, vol. 123, no. 6, pp. 1661–1668, Jun. 1995.

[44] C. E. Heil and D. F. Walnut, "Continuous and discrete wavelet transforms," *SIAM Rev.*, vol. 31, no. 4, pp. 628–666, Dec. 1989.

[45] O. Christensen and Y. C. Eldar, "Generalized shift-invariant systems and frames for subspaces," *J. Fourier Anal. Applicat.*, vol. 11, no. 3, pp. 299–313, Jun. 2005.

[46] O. Christensen and Y. C. Eldar, "Oblique dual frames and shift-invariant spaces," *Appl. Comp. Harm. Anal.*, vol. 17, no. 1, pp. 48–68, Jul. 2004.

[47] Y. C. Eldar and O. Christensen, "Characterization of oblique dual frame pairs," *EURASIP J. Appl. Signal Processing*, pp. 1–11, Apr. 2006.

[48] A. Aldroubi and K. Gröchenig, "Non-uniform sampling and reconstruction in shift-invariant spaces," *SIAM Rev.*, vol. 43, no. 4, pp. 585–620, Dec. 2001.

[49] I. M. Gelfand and M. A. Naimark, "On the imbedding of normed rings into the ring of operators on a Hilbert space," *Math. Sbornik*, vol. 12, no. 2, pp. 197–217, 1943.

[50] K. Gröchenig, "Acceleration of the frame algorithm," *IEEE Trans. Signal Processing*, vol. 41, no. 12, pp. 3331–3340, Dec. 1993.

[51] R. N. Bracewell, *The Fourier Transform and its Applications*, 3rd edn. New York, NY: McGraw Hill, Inc., 1999.

[52] A. Papoulis, *The Fourier Integral and its Applications*, New York, NY: McGraw Hill, Inc., 1962.

[53] A. V. Oppenheim, R. W. Schafer and J. R. Buck, *Discrete-Time Signal Processing*, Englewood Cliffs, NJ: Prentice-Hall, 1999.

[54] A. V. Oppenheim, A. S. Willsky and S. Hamid, *Signals and Systems*, Englewood Cliffs, NJ: Prentice-Hall, 1997.

[55] I. W. Sandberg, "Notes on representation theorems for linear discrete-space systems," in *Proc. IEEE Int. Symp. Circuits and Systems*, vol. 5, pp. 515–518, May 1999.

[56] I. W. Sandberg, "Linear maps and impulse responses," *IEEE Trans. Circuits Systems*, vol. 35, no. 2, pp. 201–206, Feb. 1988.

[57] I. W. Sandberg, "Causality and the impulse response scandal," *IEEE Trans. Circuits Systems I: Fund. Theory Applicat.*, vol. 50, no. 6, pp. 810–813, Jun. 2003.

[58] R. Strichartz, *A Guide to Distribution Theory and Fourier Transforms*, Boca Raton, FL: CRC Press, 1994.

[59] S. G. Mallat, *A Wavelet Tour of Signal Processing*, San Diego, CA: Academic Press, Inc., 1998.

[60] H. L. Royden, *Real Analysis*, New York, NY: Macmillan, 1968.

[61] S. G. Mallat, "A theory for multiresolution signal decomposition: The wavelet representation," *IEEE Trans. Patt. Anal. Mach. Intell.*, vol. 11, no. 7, pp. 674–693, Jul. 1989.

[62] J. J. Benedetto and G. Zimmermann, "Sampling multipliers and the Poisson summation formula," *J. Fourier Anal. Applicat.*, vol. 3, pp. 505–523, Sep. 1997.

[63] C. Shannon, "A mathematical theory of communication," *Bell Labs Tech. J.*, vol. 27, pp. 379–423, Jul. 1948; 623–656, Oct. 1948.

[64] N. C. Gallagher Jr. and G. L. Wise, "A representation for band-limited functions," *Proc. IEEE*, vol. 63, no. 11, pp. 1624–1625, Nov. 1975.

[65] M. Unser, "Splines: A perfect fit for signal and image processing," *IEEE Signal Processing Mag.*, pp. 22–38, Nov. 1999.

[66] I. J. Schoenberg, "Contributions to the problem of approximation of equidistant data by analytic functions, Part A: On the problem of smoothing or graduation, a first class of analytic approximation formulas," *Quart. Appl. Math.*, pp. 45–99, 1946.

[67] I. J. Schoenberg, *Cardinal Spline Interpolation*, Philadelphia, PA: SIAM, 1973.

[68] L. L. Schumaker, *Spline Functions: Basic Theory*, New York, NY: Wiley, 1981.

[69] R. H. Bartels, J. C. Beatty and B. A. Barsky, *An Introduction to Splines for Use in Computer Graphics and Geometric Modelling*, San Francisco, CA: Morgan Kaufmann, 1998.

[70] H. Prautzsch, W. Boehm and M. Paluszny, *Bézier and B-spline Techniques*, Berlin, Germany: Springer Verlag, 2002.

[71] M. Unser, A. Aldroubi and M. Eden, "Fast B-spline transforms for continuous image representation and interpolation," *IEEE Trans. Patt. Anal. Mach. Intell.*, vol. 13, no. 3, pp. 277–285, Mar. 1991.

[72] M. Unser, A. Aldroubi and M. Eden, "B-spline signal processing: Part I – Theory," *IEEE Trans. Signal Processing*, vol. 41, no. 2, pp. 821–833, Feb. 1993.

[73] M. Unser, A. Aldroubi and M. Eden, "B-spline signal processing: Part II – Efficient design and applications," *IEEE Trans. Signal Processing*, vol. 41, no. 2, pp. 834–848, Feb. 1993.

[74] C. de Boor, R. DeVore and A. Ron, "The structure of finitely generated shift-invariant spaces in $L_2(\mathbb{R}^d)$," *J. Funct. Anal.*, vol. 119, no. 1, pp. 37–78, Jan. 1994.

[75] G. Strang and G. J. Fix, *An Analysis of the Finite Element Method*, Englewood Cliffs, NJ: Prentice-Hall, 1973.

[76] A. Papoulis, "Generalized sampling expansion," *Theor. Comput. Sci.*, vol. CAS-24, no. 11, pp. 652–654, Nov. 1977.

[77] Y. C. Eldar and A. V. Oppenheim, "Filter bank reconstruction of bandlimited signals from nonuniform and generalized samples," *IEEE Trans. Signal Processing*, vol. 48, no. 10, pp. 2864–2875, Oct. 2000.

[78] P. Nikaeen and B. Murmann, "Digital compensation of dynamic acquisition errors at the front-end of high-performance A/D converters," *IEEE J. Select. Top. Signal Processing*, vol. 3, no. 3, pp. 499–508, Jun. 2009.

[79] J. Goodman, B. Miller, M. Herman, G. Raz and J. Jackson, "Polyphase nonlinear equalization of time-interleaved analog-to-digital converters," *IEEE J. Select. Top. Signal Processing*, vol. 3, no. 3, pp. 362–373, Jun. 2009.

[80] J. S. Geronimo, D. P. Hardin and P. R. Massopust, "Fractal functions and wavelet expansions based on several scaling functions," *J. Approx. Theory*, vol. 78, no. 3, pp. 373–401, Sep. 1994.

[81] G. Kaiser, *A Friendly Guide to Wavelets*, Boston, MA: Birkhäuser, 1994.

[82] C. K. Chui, *Wavelets: A Mathematical Tool for Signal Analysis*, Philadelphia, SIAM Monographs on Mathematical Modeling and Computation, 1997.

[83] H. G. Feichtinger and T. Strohmer (Eds.), *Gabor Analysis and Algorithms: Theory and Applications*, Boston, MA: Birkhäuser, 1998.

[84] H. G. Feichtinger and T. Strohmer (Eds.), *Advances in Gabor Analysis*, Boston, MA: Birkhäuser, 2003.

[85] K. Gröchenig, *Foundations of Time-Frequency Analysis*, Boston, MA: Birkhäuser, 2001.

[86] D. Gabor, "Theory of communication," *J. IEE Radio Commun. Eng.*, vol. 93, no. 3, pp. 429–457, Nov. 1946.

[87] A. J. E. M. Janssen, "The Zak transform: A signal transform for sampled time-continuous signals," *Philips J. Res.*, vol. 43, no. 1, pp. 23–69, 1988.

[88] J. Wexler and S. Raz, "Discrete Gabor expansions," *J. Signal Processing*, vol. 21, no. 3, pp. 207–220, Nov. 1991.

[89] A. Ron and Z. Shen, "Weyl–Heisenberg frames and Riesz bases in $L_2(\mathbb{R}^d)$," *Duke Math. J.*, vol. 89, no. 2, pp. 237–282, 1997.

[90] V. Bargmann, P. Butera, L. Girardello and J. R. Klauder, "On the completeness of the coherent states," *Rep. Math. Phys.*, vol. 2, no. 4, pp. 221–228, 1971.

[91] H. Bacry, A. Grossmann and J. Zak, "Proof of completeness of lattice states in the kq representation," *Phys. Rev. B*, vol. 12, no. 4, pp. 1118–1120, Aug. 1975.

[92] J. J. Benedetto, C. Heil and D. F. Walnut, "Wavelab and reproducible research," *J. Fourier Anal. Applicat.*, vol. 1, no. 4, pp. 355–402, 1994.

[93] E. Matusiak, T. Michaeli and Y. C. Eldar, "Noninvertible Gabor transforms," *IEEE Trans. Signal Processing*, vol. 58, no. 5, pp. 2597–2612, May 2010.

[94] Y. C. Eldar, E. Matusiak and T. Werther, "A constructive inversion framework for twisted convolution," *Monatsh. Math.*, vol. 150, no. 4, pp. 297–308, Apr. 2007.

[95] T. Werther, E. Matusiak, Y. C. Eldar and N. K. Subbana, "A unified approach to dual Gabor windows," *IEEE Trans. Signal Processing*, vol. 55, no. 5, pp. 1758–1768, May 2007.

[96] J. Morlet and A. Grossman, "Decomposition of Hardy functions into square integrable wavelets of constant shape," *SIAM J. Math. Anal.*, vol. 15, no. 4, pp. 723–736, 1984.

[97] I. Daubechies, "Orthonormal bases of compactly supported wavelets," *Commun. Pure Appl. Math.*, vol. 41, no. 7, pp. 909–996, Oct. 1988.

[98] G. Strang and G. Fix, "A Fourier analysis of the finite element variational method," in *Constructive Aspects of Functional Analysis*, Rome: Edizione Cremonese, pp. 796–830, 1971.

[99] H. Landau, "Necessary density conditions for sampling and interpolation of certain entire functions," *Acta Math.*, vol. 117, no. 1, pp. 37–52, Jul. 1967.

[100] J. Mitola III, "Cognitive radio for flexible mobile multimedia communications," *Mobile Networks Applicat.*, vol. 6, no. 5, pp. 435–441, Sep. 2001.

[101] Y. M. Lu and M. N. Do, "A theory for sampling signals from a union of subspaces," *IEEE Trans. Signal Processing*, vol. 56, no. 6, pp. 2334–2345, Jan. 2008.

[102] M. Mishali, Y. C. Eldar, O. Dounaevsky and E. Shoshan, "Xampling: Analog to digital at sub-Nyquist rates," *IET Circuits Devices Syst.*, vol. 5, no. 1, pp. 8–20, Jan. 2011.

[103] M. Vetterli, P. Marziliano and T. Blu, "Sampling signals with finite rate of innovation," *IEEE Trans. Signal Processing*, vol. 50, no. 6, pp. 1417–1428, Jun. 2002.

[104] Y. C. Eldar, "Compressed sensing of analog signals in shift-invariant spaces," *IEEE Trans. Signal Processing*, vol. 57, no. 8, pp. 2986–2997, Aug. 2009.

[105] M. Unser and T. Blu, "Fractional splines and wavelets," *SIAM Rev.*, vol. 42, no. 1, pp. 43–67, Jan. 2000.

[106] A. Aldroubi and M. Unser, "Sampling procedures in function spaces and asymptotic equivalence with Shannon's sampling theory," *Num. Funct. Anal. Optim.*, vol. 15, no. 1–2, pp. 1–21, Feb. 1994.

[107] S. Ries and R. L. Stens, "Approximation by generalized sampling series," in *Constructive Theory of Functions*. B. Sendov *et al.*, Eds. Sofia, Bulgaria: Bulgarian Academy of Sciences, pp. 17–37, 1984.

[108] M. Unser and A. Aldroubi, "A general sampling theory for nonideal acquisition devices," *IEEE Trans. Signal Processing*, vol. 42, no. 11, pp. 2915–2925, Nov. 1994.

[109] P. P. Vaidyanathan, "Generalizations of the sampling theorem: Seven decades after Nyquist," *IEEE Trans. Circuit Syst. I*, vol. 48, no. 9, pp. 1094–1109, Sep. 2001.

[110] Y. C. Eldar, "Sampling and reconstruction in arbitrary spaces and oblique dual frame vectors," *J. Fourier Anal. Appl.*, vol. 1, no. 9, pp. 77–96, Jan. 2003.

[111] S. Ramani, D. Van De Ville and M. Unser, "Non-ideal sampling and adapted reconstruction using the stochastic Matern model," in *Proc. Int. Conf. Acoustics, Speech and Signal Processing (ICASSP'06)*, vol. 2, May 2006.

[112] C. A. Glasbey, "Optimal linear interpolation of images with known point spread function," *Scand. Conf. Image Anal. SCIA-2001*, Bergen, pp. 161–168, 2001.

[113] Y. C. Eldar and T. G. Dvorkind, "Minimax sampling with arbitrary spaces," *11th IEEE Int. Conf. Electronics, Circuits and Systems (ICECS-2004)*, pp. 559–562, Dec. 2004.

[114] Y. C. Eldar, "Sampling without input constraints: Consistent reconstruction in arbitrary spaces," in *Sampling, Wavelets and Tomography*, A. I. Zayed and J. J. Benedetto, Eds. Boston, MA: Birkhäuser, pp. 33–60, 2004.

[115] Y. C. Eldar and T. Werther, "General framework for consistent sampling in Hilbert spaces," *Int. J. Wavelets Multires. Inform. Proc.*, vol. 3, no. 3, pp. 347–359, Sep. 2005.

[116] R. G. Keys, "Cubic convolution interpolation for digital image processing," *IEEE Trans. Acoust. Speech Signal Processing*, vol. 29, no. 6, pp. 1153–1160, Dec. 1981.

[117] C. E. Duchon, "Lanczos filtering in one and two dimensions," *J. Appl. Meteorol.*, vol. 18, no. 8, pp. 1016–1022, Aug. 1979.

[118] C. L. Lawson and R. J. Hanson, *Solving Least Squares Problems*, Englewood Cliffs, NJ: Prentice-Hall, 1974.

[119] T. Kailath, *Lectures on Linear Least-Squares Estimation*, Wein, New York: Springer, 1976.

[120] A. Björck, *Numerical Methods for Least-Squares Problems*, Philadelphia, PA: SIAM, 1996.

[121] E. L. Lehmann and G. Casella, *Theory of Point Estimation*, 2nd edn. New York, NY: Springer, 1999.

[122] Y. C. Eldar, *Rethinking Biased Estimation: Improving Maximum Likelihood and the Cramer–Rao Bound*, Foundation and Trends in Signal Processing, Hanover, MA: Now Publishers Inc., 2008.

[123] S. Kay and Y. C. Eldar, "Rethinking biased estimation," *IEEE Signal Processing Mag.*, vol. 25, pp. 133–136, May 2008.

[124] S. Boyd and L. Vandenberghe, *Convex Optimization*, Cambridge, UK: Cambridge University Press, 2004.

[125] S. Ramani, D. Van De Ville, T. Blu and M. Unser, "Nonideal sampling and regularization theory," *IEEE Trans. Signal Processing*, vol. 56, no. 3, pp. 1055–1070, Mar. 2008.

[126] M. Unser, "Cardinal exponential splines: part II – think analog, act digital," *IEEE Trans. Signal Processing*, vol. 53, no. 4, pp. 1439–1449, Apr. 2005.

[127] Y. C. Eldar, A. Ben-Tal and A. Nemirovski, "Robust mean-squared error estimation in the presence of model uncertainties," *IEEE Trans. Signal Processing*, vol. 53, no. 1, pp. 168–181, Jan. 2005.

[128] T. G. Dvorkind, H. Kirshner, Y. C. Eldar and M. Porat, "Minimax approximation of representation coefficients from generalized samples," *IEEE Trans. Signal Processing*, vol. 55, no. 9, pp. 4430–4443, Sep. 2007.

[129] Y. C. Eldar and M. Unser, "Nonideal sampling and interpolation from noisy observations in shift-invariant spaces," *IEEE Trans. Signal Processing*, vol. 54, no. 7, pp. 2636–2651, Jul. 2006.

[130] G. H. Hardy, "Notes of special systems of orthogonal functions – IV: The orthogonal functions of Whittakers series," *Proc. Camb. Phil. Soc.*, vol. 37, pp. 331–348, Oct. 1941.

[131] W. S. Tang, "Oblique projections, biorthogonal Riesz bases and multiwavelets in Hilbert space," *Proc. Am. Math. Soc.*, vol. 128, no. 2, pp. 463–473, Feb. 2000.

[132] E. E. Tyrtyshnikov, *A Brief Introduction to Numerical Analysis*, Boston, MA: Birkhauser, 1997.

[133] S. Boyd, *Convex Optimization*, Cambridge, UK: Cambridge University Press, 2004.

[134] M. Unser and J. Zerubia, "Generalized sampling: Stability and performance analysis," *IEEE Trans. Signal Processing*, vol. 45, no. 12, pp. 2941–2950, Dec. 1997.

[135] T. Blu and M. Unser, "Quantitative Fourier analysis of approximation techniques: Part I – Interpolators and projectors," *IEEE Trans. Signal Processing*, vol. 47, no. 10, pp. 2783–2795, Oct. 1999.

[136] T. Blu and M. Unser, "Quantitative Fourier analysis of approximation techniques: Part II – Wavelets," *IEEE Trans. Signal Processing*, vol. 47, no. 10, pp. 2796–2806, Oct. 1999.

[137] C. Lee, M. Eden and M. Unser, "High-quality image resizing using oblique projection operators," *IEEE Trans. Signal Processing*, vol. 7, no. 5, pp. 679–692, May 1998.

[138] Y. C. Eldar, A. Ben-Tal and A. Nemirovski, "Linear minimax regret estimation of deterministic parameters with bounded data uncertainties," *IEEE Trans. Signal Processing*, vol. 52, no. 8, pp. 2177–2188, Aug. 2004.

[139] Y. C. Eldar and N. Merhav, "A competitive minimax approach to robust estimation of random parameters," *IEEE Trans. Signal Processing*, vol. 52, no. 7, pp. 1931–1946, Jul. 2004.

[140] Y. C. Eldar and T. G. Dvorkind, "A minimum squared-error framework for generalized sampling," *IEEE Trans. Signal Processing*, vol. 54, no. 6, pp. 2155–2167, Jun. 2006.

[141] I. Djokovic and P. P. Vaidyanathan, "Generalized sampling theorems in multiresolution subspaces," *IEEE Trans. Signal Processing*, vol. 45, no. 3, pp. 583–599, Mar. 1997.

[142] M. Unser and J. Zerubia, "A generalized sampling theory without band-limiting constraints," *IEEE Trans. Circuits Syst. II*, vol. 45, no. 8, pp. 959–969, Aug. 1998.

[143] D. Jagerman and L. Fogel, "Some general aspects of the sampling theorem," *IEEE Trans. Inform. Theory*, vol. 2, no. 4, pp. 139–146, Dec. 1956.

[144] D. A. Linden and N. M. Abramson, "A generalization of the sampling theorem," *Inform. Control*, vol. 3, no. 1, pp. 26–31, Mar. 1960.

[145] J. Yen, "On nonuniform sampling of bandwidth-limited signals," *IRE Trans. Circuit Theory*, vol. 3, no. 4, pp. 251–257, Dec. 1956.

[146] J. Brown Jr., "Multi-channel sampling of low-pass signals," *IEEE Trans. Circuits Syst.*, vol. 28, no. 2, pp. 101–106, Feb. 1981.

[147] P. P. Vaidyanathan, *Multirate Systems and Filter Banks*, Englewood Cliffs, NJ: Prentice-Hall, 1993.

[148] E. H. Lieb and M. Loss, *Analysis*, 2nd edn. American Mathematical Society, 2001.

[149] http://www.soe.ucsc.edu/~milanfar/software/sr-datasets.html

[150] T. Michaeli and Y. C. Eldar, "High-rate interpolation of random signals from nonideal samples," *IEEE Trans. Signal Processing*, vol. 57, no. 3, pp. 977–992, Mar. 2009.

[151] A. Balakrishnan, "A note on the sampling principle for continuous signals," *IEEE Trans. Inform. Theory*, vol. 3, no. 2, pp. 143–146, Jun. 1957.

[152] S. P. Lloyd, "A sampling theorem for stationary (wide sense) stochastic processes," *Trans. Am. Math. Soc.*, vol. 92, no. 1, pp. 1–12, 1959.

[153] I. W. Hunter and M. J. Korenberg, "The identifiication of nonlinear biological systems: Wiener and Hammerstein cascade models," *Biological Cybernetics*, vol. 55, nos. 2–3, pp. 135–144, 1985.

[154] M. B. Matthews, "On the linear minimum-mean-squared-error estimation of an under-sampled wide-sense stationary random process," *IEEE Trans. Signal Processing*, vol. 48, no. 1, pp. 272–275, 2000.

[155] T. Michaeli and Y. C. Eldar, "Optimization techniques in modern sampling theory," in *Convex Optimization in Signal Processing and Communications*, Y. C. Eldar and D. P. Palomar, Eds. Cambridge, UK: Cambridge University Press, 2010, pp. 266–314.

[156] Y. C. Eldar, "Robust deconvolution of deterministic and random signals," *IEEE Trans. Inform. Theory*, vol. 51, no. 8, pp. 2921–2929, Aug. 2005.

[157] S. Ramani, D. Van De Ville, T. Blu and M. Unser, "Nonideal sampling and regularization theory," *IEEE Trans. Signal Processing*, vol. 56, no. 3, pp. 1055–1070, Mar. 2008.

[158] K. Kose, K. Endoh and T. Inouye, "Nonlinear amplitude compression in magnetic resonance imaging: Quantization noise reduction and data memory saving," *IEEE AES Mag.*, vol. 5, no. 6, pp. 27–30, Jun. 1990.

[159] T. G. Dvorkind, Y. C. Eldar and E. Matusiak, "Nonlinear and nonideal sampling: theory and methods," *IEEE Trans. Signal Processing*, vol. 56, no. 12, pp. 5874–5890, Dec. 2008.

[160] V. Volterra, *Theory of Functionals and of Integral and Integro-Differential Equations*, New York, NY: Dover, 1959.

[161] E. W. Bai, "An optimal two-stage identification algorithm for Hammerstein–Wiener nonlinear systems," *Automatica*, vol. 34, no. 3, pp. 333–338, Mar. 1998.

[162] E. W. Bai, "A blind approach to the Hammerstein–Wiener model identification," *Automatica*, vol. 38, no. 6, pp. 967–979, Jun. 2002.

[163] P. A. Traverso, D. Mirri, G. Pasini and F. Filicori, "A nonlinear dynamic S/H-ADC device model based on a modified Volterra series: Identification procedure and commercial CAD tool implementation," *IEEE Trans. Instrum. Measurem.*, vol. 52, no. 4, pp. 1129–1135, Sep. 2003.

[164] F. Ding and T. Chen, "Identification of Hammerstein nonlinear ARMAX systems," *Automatica*, vol. 41, no. 9, pp. 1479–1489, Sep. 2005.

[165] Y-M. Zhu, "Generalized sampling theorem," *IEEE Trans. Circuits Systems II: Analog Digital Signal Processing*, vol. 39, no. 8, pp. 587–588, Aug. 1992.

[166] K. Yao and J. B. Thomas, "On some stability and interpolation properties of nonuniform sampling expansions," *IEEE Trans. Circuit Theory*, vol. CT-14, no. 4, pp. 404–408, Dec. 1967.

[167] F. J. Beutler, "Error-free recovery of signals from irregularly spaced samples," *SIAM Rev.*, vol. 8, no. 3, pp. 328–335, Jun. 1966.

[168] R. S. Prendergast, B. C. Levy and P. J. Hurst, "Reconstruction of band-limited periodic nonuniformly sampled signals through multirate filter banks," *IEEE Trans. Circuits Syst. I: Regular Papers*, vol. 51, no. 8, pp. 1612–1622, Aug. 2004.

[169] F. Marvasti, M. Analoui and M. Gamshadzahi, "Recovery of signals from nonuniform samples using iterative methods," *IEEE Trans. Signal Processing*, vol. 39, no. 4, pp. 872–878, Apr. 1991.

[170] H. G. Feichtinger, K. Gröchenig and T. Strohmer, "Efficient numerical methods in nonuniform sampling theory," *Num. Math.*, vol. 69, no. 4, pp. 423–440, Feb. 1995.

[171] E. Margolis and Y. C. Eldar, "Nonuniform sampling of periodic bandlimited signals," *IEEE Trans. Signal Processing*, vol. 56, no. 7, pp. 2728–2745, Jul. 2008.

[172] N. Aronszajn, "Theory of reproducing kernels," *Trans. Am. Math. Soc.*, vol. 68, no. 3, pp. 337–404, May 1950.

[173] T. Ando, *Reproducing Kernel Spaces and Quadratic Inequalities*, Japan: Sapporo, 1987.

[174] N. Aronszajn, *Theory of Reproducing Kernels*, Cambridge, MA: Harvard University, 1951.

[175] M. Z. Nashed and G. G. Walter, "General sampling theorems for functions in reproducing kernel Hilbert spaces," *Math. Control Signals Syst.*, vol. 4, pp. 373–412, Dec. 1991.

[176] H. P. Kramer, "A generalized sampling theorem," *J. Math. Phys.*, vol. 38, pp. 68–72, 1959.

[177] H. J. Landau and W. L. Miranker, "The recovery of distorted band-limited signals," *J. Math. Anal. Applic.*, vol. 2, no. 1, pp. 97–104, 1961.

[178] T. Faktor, T. Michaeli and Y. C. Eldar, "Nonlinear and nonideal sampling revisited," *IEEE Signal Processing Lett.*, vol. 17, no. 2, pp. 205–208, Feb. 2010.

[179] K. Goebel and W. A. Kirk, "A fixed point theorem for asymptotically nonexpansive mappings," *Proc. Am. Math. Soc*, vol. 35, no. 1, pp. 171–174, 1972.

[180] J. Nocedal and S. J. Wright, *Numerical Optimization*, New York, NY: Springer, 1999.

[181] V. Vapnik, *The Nature of Statistical Learning Theory*, New York, NY: Springer, 1999.

[182] M. Unser, A. Aldroubi and M. Eden, "Enlargement or reduction of digital images with minimum loss of information," *IEEE Trans. Image Processing*, vol. 4, no. 3, pp. 247–258, Mar. 1995.

[183] Y. C. Eldar and M. Mishali, "Robust recovery of signals from a structured union of subspaces," *IEEE Trans. Inform. Theory*, vol. 55, no. 11, pp. 5302–5316, Nov. 2009.

[184] Y.-P. Lin and P. P. Vaidyanathan, "Periodically nonuniform sampling of bandpass signals," *IEEE Trans. Circuits Syst. II*, vol. 45, no. 3, pp. 340–351, Mar. 1998.

[185] C. Herley and P. W. Wong, "Minimum rate sampling and reconstruction of signals with arbitrary frequency support," *IEEE Trans. Inform. Theory*, vol. 45, no. 5, pp. 1555–1564, Jul. 1999.

[186] R. Venkataramani and Y. Bresler, "Perfect reconstruction formulas and bounds on aliasing error in sub-Nyquist nonuniform sampling of multiband signals," *IEEE Trans. Inform. Theory*, vol. 46, no. 6, pp. 2173–2183, Sep. 2000.

[187] A. Paulraj, R. Roy and T. Kailath, "ESPRIT – a subspace rotation approach to signal parameter estimation," *Proc. IEEE*, vol. 74, no. 7, pp. 1044–1045, Jul. 1986.

[188] R. Walden, "Analog-to-digital converter survey and analysis," *IEEE J. Selected Areas Comm.*, vol. 17, no. 4, pp. 539–550, Apr. 1999.

[189] D. Healy, "Analog-to-information: Baa #05-35," 2005, Available online at http://www.darpa.mil/mto/solicitations/baa05-35/s/index.html.

[190] R. DeVore, "Nonlinear approximation," *Acta Num.*, vol. 7, pp. 51–150, 1998.

[191] E. Candès, J. Romberg and T. Tao, "Stable signal recovery from incomplete and inaccurate measurements," *Comm. Pure Appl. Math.*, vol. 59, no. 8, pp. 1207–1223, Aug. 2006.

[192] R. Prony, "Essai expérimental et analytique sur les lois de la Dilatabilité des fluides élastiques et sur celles de la Force expansive de la vapeur de l'eau et de la vapeur de l'alkool, à différentes températures," *J. l'École Polytechnique*, Floréal et Prairial III, vol. 1, no. 2, pp. 24–76, 1795; R. Prony is Gaspard Riche, Baron de Prony.

[193] C. Carathéodory, "Über den Variabilitätsbereich der Fourierschen Konstanten von positiven harmonischen Funktionen," *Rend. Circ. Mat. Palermo*, vol. 32, pp. 193–217, 1911.

[194] I. Gorodnitsky, B. Rao and J. George, "Source localization in magnetoencephalography using an iterative weighted minimum norm algorithm," in *Proc. Asilomar Conf. Signals, Systems, and Computers*, Pacific Grove, CA, Oct. 1992.

[195] B. Rao, "Signal processing with the sparseness constraint," in *Proc. IEEE Int. Conf. Acoustics Speech, and Signal Processing (ICASSP)*, Seattle, WA, vol. 3, pp. 1861–1864, May 1998.

[196] I. F. Gorodnitsky, J. S. George, and B. D. Rao, "Neuromagnetic source imaging with FOCUSS: A recursive weighted minimum norm algorithm," *J. Electroencephalog. Clinical Neurophysiol.*, vol. 95, no. 4, pp. 231–251, Oct. 1995.

[197] I. F. Gorodnitsky and B. D. Rao, "Sparse signal reconstruction from limited data using FOCUSS: A re-weighted minimum norm algorithm," *IEEE Trans. Signal Processing*, vol. 45, no. 3, pp. 600–616, Mar. 1997.

[198] A. Beurling, "Sur les intégrales de Fourier absolument convergentes et leur application à une transformation fonctionelle," in *Proc. Scandenatical Mathematical Congress*, Helsinki, Finland, 1938.

[199] W. B. Pennebaker and J. L. Mitchell, *JPEG: Still Image Data Compression Standard*, New York, NY: Van Nostrand Reinhold, 1993.

[200] D. Taubman and M. Marcellin, *JPEG 2000: Image Compression Fundamentals, Standards and Practice*, Dordrecht: Kluwer, 2001.

[201] D. Donoho, "Denoising by soft-thresholding," *IEEE Trans. Inform. Theory*, vol. 41, no. 3, pp. 613–627, May 1995.

[202] T. Hastie, R. Tibshirani and J. Friedman, *The Elements of Statistical Learning*, New York, NY: Springer, 2001.

[203] Y. C. Eldar, M. Davenport, M. Duarte and G. Kutyniok, "Introduction to compressed sensing," in *Compressed Sensing: Theory and Applications*, Cambridge, UK: Cambridge University Press, 2011.

[204] N. Ahmed, T. Natarajan and K. R. Rao, "Discrete cosine transform," *IEEE Trans. Comput.*, vol. 23, no. 1, pp. 90–93, Jan. 1974.

[205] L. He and L. Carin, "Exploiting structure in wavelet-based Bayesian compressive sensing," *IEEE Trans. Signal Processing*, vol. 57, no. 9, pp. 3488–3497, Sep. 2009.

[206] M. F. Duarte R. G. Baraniuk, V. Cevher and C. Hegde, "Model-based compressive sensing," *IEEE Trans. Inform. Theory*, vol. 56, no. 4, pp. 1982–2001, Apr. 2010.

[207] Y. C. Eldar, P. Kuppinger and H. Bölcskei, "Block-sparse signals: Uncertainty relations and efficient recovery," *IEEE Trans. Signal Processing*, vol. 58, no. 6, pp. 3042–3054, Jun. 2010.

[208] P. Schniter, L. C. Potter and J. Ziniel, "Fast Bayesian matching pursuit," in *Proc. Workshop on Information Theory and Applications (ITA)*, La Jolla, CA, Jan. 2008.

[209] T. Peleg, Y. C. Eldar and M. Elad, "Exploiting statistical dependencies in sparse representations for signal recovery," *IEEE Trans. Signal Processing*, vol. 60, no. 5, pp. 2286–2303, May 2012.

[210] P. J. Wolfe, S. J. Godsill and W. J. Ng, "Bayesian variable selection and regularization for time-frequency surface estimation," *J. R. Statist. Soc. B*, vol. 66, no. 3, pp. 575–589, Jun. 2004.

[211] P. J. Garrigues and B. A. Olshausen, "Learning horizontal connections in a sparse coding model of natural images," in *Advances in Neural Information Processing Systems 20*, J. C. Platt, D. Koller, Y. Singer and S. Roweis, Eds. Cambridge, MA: pp. 505–512, 2008.

[212] J. Partington, *An Introduction to Hankel Operators*, Cambridge, UK: Cambridge University Press, 1988.

[213] A. So and Y. Ye, "Theory of semidefinite programming for sensor network localization," *Math. Programm. Series A and B*, vol. 109, no. 2, pp. 367–384, Mar. 2007.

[214] D. Goldberg, D. Nichols, B. Oki and D. Terry, "Using collaborative filtering to weave an information tapestry," *Comm. ACM*, vol. 35, no. 12, pp. 61–70, Dec. 1992.

[215] E. Candès and B. Recht, "Exact matrix completion via convex optimization," *Found. Comput. Math.*, vol. 9, no. 6, pp. 717–772, Dec. 2009.

[216] B. Recht, M. Fazel and P. Parrilo, "Guaranteed minimum rank solutions of matrix equations via nuclear norm minimization," *SIAM Rev.*, vol. 52, no. 3, pp. 471–501, Aug. 2010.

[217] Z-Q. Luo, W-K. Ma, A-C. So, Y. Ye and S. Zhang, "Semidefinite relaxation of quadratic optimization problems," *IEEE Signal Processing Mag.*, vol. 27, no. 3, pp. 20–34, May 2010.

[218] Y. Shechtman, Y. C. Eldar, A. Szameit and M. Segev, "Sparsity based sub-wavelength imaging with partially incoherent light via quadratic compressed sensing," *Opt. Express*, vol. 19, no. 16, pp. 14807–14822, Aug. 2011.

[219] H. Ohlsson, A. Y. Yang, R. Dong and S. S. Sastry, "Compressive phase retrieval from squared output measurements via semidefinite programming," *16th IFAC Symp. System Identification*, vol. 16, part 1, pp. 89–94, 2012.

[220] E. J. Candes, Y. C. Eldar, T. Strohmer and V. Voroninski, "Phase retrieval via matrix completion," *SIAM J. Imaging Sci.*, vol. 6, no. 1, pp. 199–225, Feb. 2013.

[221] D. L. Donoho and M. Elad, "Optimally sparse representation in general (nonorthogonal) dictionaries via l^1 minimization," *Proc. Natl Acad. Sci.*, vol. 100, no. 5, pp. 2197–2202, Mar. 2003.

[222] X. Feng and Z. Zhang, "The rank of a random matrix," *Appl. Math. Comp.*, vol. 185, no. 1, pp. 689–694, Feb. 2007.

[223] R. A. Horn and C. R. Johnson, *Topics in Matrix Analysis*, New York, NY: Cambridge University Press, 1991.

[224] A. Cohen, W. Dahmen and R. DeVore, "Compressed sensing and best k-term approximation," *J. Am. Math. Soc.*, vol. 22, no. 1, pp. 211–231, Jan. 2009.

[225] E. J. Candès and T. Tao, "Decoding by linear programming," *IEEE Trans. Inform. Theory*, vol. 51, no. 12, pp. 4203–4215, Dec. 2005.

[226] M. Davenport, "Random observations on random observations: Sparse signal acquisition and processing," *Rice University*, Aug. 2010, http://dsp.rice.edu/publications/random-observations-random-observations-sparse-signal-acquisition-and-processing

[227] E. J. Candès, "The restricted isometry property and its implications for compressed sensing," *C. R. Acad. Sci. Paris Ser. I Math.*, vol. 346, pp. 589–592, May 2008.

[228] J. Tropp and A. Gilbert, "Signal recovery from random measurements via orthogonal matching pursuit," *IEEE Trans. Inform. Theory*, vol. 53, no. 12, pp. 4655–4666, Dec. 2007.

[229] M. Rosenfeld, "In praise of the Gram matrix," in *The Mathematics of Paul Erdős II*, R. L. Graham and J. Nešetril, Eds. Berlin: Springer, pp. 318–323, 1996.

[230] T. Strohmer and R.W. Heath, "Grassmannian frames with applications to coding and communication," *Appl. Comput. Harmonic Anal.*, vol. 14, no. 3, pp. 257–275, May 2003.

[231] J. J. Seidel, P. Delsarte and J. M. Goethals, "Bounds for systems of lines and Jacobi poynomials," *Philips Res. Rep.*, vol. 30, no. 3, pp. 91–105, 1975.

[232] J. A. Tropp, I. S. Dhillon, Jr., R. W. Heath and T. Strohmer, "Designing structured tight frames via an alternating projection method," *IEEE Trans. Inform. Theory*, vol. 51, no. 1, pp. 188–209, Jan. 2005.

[233] S. Geršgorin, "Über die Abgrenzung der Eigenwerte einer Matrix," *Izv. Akad. Nauk SSSR Ser. Fiz.-Mat.*, vol. 6, pp. 749–754, 1931.

[234] D. L. Donoho and X. Huo, "Uncertainty principles and ideal atomic decompositions," *IEEE Trans. Inform. Theory*, vol. 47, no. 7, pp. 2845–2862, Nov. 2001.

[235] M. Elad and A. M. Bruckstein, "A generalized uncertainty principle and sparse representation in pairs of bases," *IEEE Trans. Inform. Theory*, vol. 48, no. 9, pp. 2558–2567, Sep. 2002.

[236] D. L. Donoho and P. B. Stark, "Uncertainty principles and signal recovery," *SIAM J. Appl. Math.*, vol. 49, no. 3, pp. 906–931, Jun. 1989.

[237] J. Shore, "On the application of Haar functions," *IEEE Trans. Commun.*, vol. 21, no. 3, pp. 209–216, Mar. 1973.

[238] F. A. Berezin, *The Method of Second Quantization*, New York, NY: Academic Press, 1966.

[239] P. Kuppinger, G. Durisi and H. Bolcskei, "Uncertainty relations and sparse signal recovery for pairs of general signal sets," *IEEE Trans. Inform. Theory*, vol. 58, no. 1, pp. 263–277, Jan. 2012.

[240] Y. C. Eldar, "Uncertainty relations for shift-invariant analog signals," *IEEE Trans. Inform. Theory*, vol. 55, no. 12, pp. 5742–5757, Dec. 2009.

[241] R. Vershynin, "Introduction to the non-asymptotic analysis of random matrices", in *Compressed Sensing: Theory and Applications*, Cambridge, UK: Cambridge University Press, 2011.

[242] A. Garnaev and E. Gluskin, "The widths of Euclidean balls," *Dokl. An. SSSR*, vol. 277, pp. 1048–1052, 1984.

[243] W. Johnson and J. Lindenstrauss, "Extensions of Lipschitz mappings into a Hilbert space," in *Proc. Conf. Modern Anal. Prob.*, New Haven, CT, June 1982.

[244] T. Jayram and D. Woodruff, "Optimal bounds for Johnson-Lindenstrauss transforms and streaming problems with sub-constant error," in *Proc. ACM-SIAM Symp. Discrete Algorithms (SODA)*, San Francisco, CA, Jan. 2011.

[245] R. Baraniuk, M. Davenport, R. DeVore and M. Wakin, "A simple proof of the restricted isometry property for random matrices," *Construct. Approx.*, vol. 28, no. 3, pp. 253–263, Dec. 2008.

[246] F. Krahmer and R. Ward, "New and improved Johnson–Lindenstrauss embeddings via the restricted isometry property," *SIAM J. Math. Analysis*, vol. 43, no. 3, pp. 1269–1281, Jun. 2011.

[247] M. Herman and T. Strohmer, "High-resolution radar via compressed sensing," *IEEE Trans. Signal Processing*, vol. 57, no. 6, pp. 2275–2284, Jun. 2009.

[248] T. Strohmer and R. Heath, "Grassmanian frames with applications to coding and communication," *Appl. Comput. Harmon. Anal.*, vol. 14, no. 3, pp. 257–275, Nov. 2003.

[249] P. Indyk, "Explicit constructions for compressed sensing of sparse signals," in *Proc. ACM-SIAM Symp. Discrete Algorithms (SODA)*, San Francisco, CA, Jan. 2008, pp. 30–33.

[250] R. DeVore, "Deterministic constructions of compressed sensing matrices," *J. Complex.*, vol. 23, no. 4, pp. 918–925, Aug. 2007.

[251] E. Candès and Y. Plan, "Matrix completion with noise," *Proc. IEEE*, vol. 98, no. 6, pp. 925–936, Jun. 2010.

[252] T. Cai and T. Jiang, "Limiting laws of coherence of random matrices with applications to testing covariance structure and construction of compressed sensing matrices," *Ann. Statist.*, vol. 39, no. 3, pp. 1496–1525, 2011.

[253] E. Candès and Y. Plan, "Near-ideal model selection by ℓ_1 minimization," *Ann. Stat.*, vol. 37, no. 5A, pp. 2145–2177, Oct. 2009.

[254] H. Rauhut, "Compressive sensing and structured random matrices," *Theor. Found. Num. Methods Sparse Recovery*, vol. 9, pp. 1–92, 2010.

[255] H. Rauhut, G. E. Pfander and J. Tanner, "Identification of matrices having a sparse representation," *IEEE Trans. Signal Processing*, vol. 56, no. 11, pp. 5376–5388, Nov. 2008.

[256] Y. Chi, L. Scharf, A. Pezeshki and R. Calderbank, "Sensitivity to basis mismatch in compressed sensing," *IEEE Trans. Signal Processing*, vol. 59, no. 5, pp. 2182–2195, 2011.

[257] J. Tropp and S. Wright, "Computational methods for sparse solution of linear inverse problems," *Proc. IEEE*, vol. 98, no. 6, pp. 948–958, Jun. 2010.

[258] E. Candès, Y. C. Eldar, D. Needell and P. Randall, "Compressed sensing with coherent and redundant dictionaries," *Appl. Comput. Harmon. Anal.*, vol. 31, no. 1, pp. 59–73, 2011.

[259] S. Muthukrishnan, *Data Streams: Algorithms and Applications*, vol. 1 of *Found. Trends in Theoretical Comput. Science*, Boston, MA: Now Publishers, 2005.

[260] S. Chen, D. Donoho and M. Saunders, "Atomic decomposition by basis pursuit," *SIAM J. Sci. Comp.*, vol. 20, no. 1, pp. 33–61, 1998.

[261] L. I. Rudin, S. Osher and E. Fatemi, "Nonlinear total variation based noise removal algorithms," *Physica D: Nonlinear Phenom.*, vol. 60, no. 1, pp. 259–268, Nov. 1992.

[262] D. L. Donoho, I. Drori, V. Stodden, Y. Tsaig and M. Shahram, "SparseLab: Seeking sparse solutions to linear systems of equations," http://sparselab.stanford.edu/, Oct. 2007.

[263] E. Hale, W. Yin and Y. Zhang, "A fixed-point continuation method for ℓ_1-regularized minimization with applications to compressed sensing," Rice Univ., CAAM Dept., Tech. Rep. TR07-07, 2007.

[264] M. A. T. Figueiredo, R. Nowak and S. Wright, "Gradient projections for sparse reconstruction: Application to compressed sensing and other inverse problems," *IEEE J. Select. Top. Signal Processing*, vol. 1, no. 4, pp. 586–597, Dec. 2007.

[265] E. van den Berg and M. P. Friedlander, "Probing the Pareto frontier for basis pursuit solutions," *SIAM J. Sci. Comput.*, vol. 31, no. 2, pp. 890–912, Nov. 2008.

[266] A. Beck and M. Teboulle, "A fast iterative shrinkage-thresholding algorithm for linear inverse problems," *SIAM J. Imag. Sci.*, vol. 2, no. 1, pp. 183–202, Mar. 2009.

[267] J. Friedman, T. Hastie and R. Tibshirani, "Regularization paths for generalized linear models via coordinate descent," *J. Stat. Software*, vol. 33, no. 1, pp. 1–22, Jan. 2010.

[268] S. Osher, Y. Mao, B. Dong and W. Yin, "Fast linearized Bregman iterations for compressive sensing and sparse denoising," *Commun. Math. Sci.*, vol. 8, no. 1, pp. 93–111, Feb. 2010.

[269] Z. Wen, W. Yin, D. Goldfarb and Y. Zhang, "A fast algorithm for sparse reconstruction based on shrinkage, subspace optimization and continuation," *SIAM J. Sci. Comput.*, vol. 32, no. 4, pp. 1832–1857, Jun. 2010.

[270] S. Wright, R. Nowak and M. Figueiredo, "Sparse reconstruction by separable approximation," *IEEE Trans. Signal Processing*, vol. 57, no. 7, pp. 2479–2493, Jul. 2009.

[271] W. Yin, S. Osher, D. Goldfarb and J. Darbon, "Bregman iterative algorithms for ℓ_1-minimization with applications to compressed sensing," *SIAM J. Imag. Sci.*, vol. 1, no. 1, pp. 143–168, 2008.

[272] M. Grant and S. Boyd, "CVX: Matlab software for disciplined convex programming (web page and software)," March 2008, http://stanford.edu/~boyd/cvx.

[273] Y. C. Eldar, "Generalized SURE for exponential families: Applications to regularization," *IEEE Trans. Signal Processing*, vol. 57, no. 2, pp. 471–481, Feb. 2009.

[274] S. Ji, Y. Xue and L. Carin, "Bayesian compressive sensing," *IEEE Trans. Signal Processing*, vol. 56, no. 6, pp. 2346–2356, Jun. 2008.

[275] B. Logan, *Properties of High-Pass Signals*, Ph.D. thesis, Columbia University, 1965.

[276] D. Donoho and B. Logan, "Signal recovery and the large sieve," *SIAM J. Appl. Math.*, vol. 52, no. 6, pp. 577–591, Apr. 1992.

[277] H. Taylor, S. Banks and J. McCoy, "Deconvolution with the ℓ_1 norm," *Geophysics*, vol. 44, no. 1, pp. 39–52, Jan. 1979.

[278] S. Levy and P. Fullagar, "Reconstruction of a sparse spike train from a portion of its spectrum and application to high-resolution deconvolution," *Geophysics*, vol. 46, no. 9, pp. 1235–1243, Sep. 1981.

[279] C. Walker and T. Ulrych, "Autoregressive recovery of the acoustic impedance," *Geophysics*, vol. 48, no. 10, pp. 1338–1350, Oct. 1983.

[280] M. Talagrand, "New concentration inequalities in product spaces," *Invent. Math.*, vol. 126, no. 3, pp. 505–563, Nov. 1996.

[281] S. G. Mallat and Z. Zhang, "Matching pursuits with time-frequency dictionaries," *IEEE Trans. Signal Processing*, vol. 41, no. 12, pp. 3397–3415, Dec. 1993.

[282] W. Dai and O. Milenkovic, "Subspace pursuit for compressive sensing signal reconstruction," *IEEE Trans. Inform. Theory*, vol. 55, no. 5, pp. 2230–2249, May 2009.

[283] I. Daubechies, M. Defrise and C. De Mol, "An iterative thresholding algorithm for linear inverse problems with a sparsity constraint," *Comm. Pure Appl. Math.*, vol. 57, no. 11, pp. 1413–1457, Nov. 2004.

[284] D. Donoho, I. Drori, Y. Tsaig and J-L. Stark, "Sparse solution of underdetermined linear equations by stagewise orthogonal matching pursuit," *IEEE Trans. Inform. Theory*, vol. 58, no. 2, pp. 1094–1121, Feb. 2012.

[285] T. Blumensath and M. Davies, "Iterative hard thresholding for compressive sensing," *Appl. Comput. Harmon. Anal.*, vol. 27, no. 3, pp. 265–274, Nov. 2009.

[286] A. Cohen, W. Dahmen and R. DeVore, "Instance optimal decoding by thresholding in compressed sensing," in *Int. Conf. Harmonic Analysis and Partial Differential Equations*, Madrid, Spain, Jun. 2008.

[287] D. Needell and J. A. Tropp, "CoSaMP: Iterative signal recovery from incomplete and inaccurate samples," *Appl. Comput. Harmon. Anal.*, vol. 26, no. 3, pp. 301–321, May 2009.

[288] D. Needell and R. Vershynin, "Signal recovery from incomplete and inaccurate measurements via regularized orthogonal matching pursuit," *IEEE J. Select. Top. Signal Processing*, vol. 4, no. 2, pp. 310–316, Apr. 2010.

[289] J. A. Tropp, "Greed is good: Algorithmic results for sparse approximation," *IEEE Trans. Inform. Theory*, vol. 50, no. 10, pp. 2231–2242, Oct. 2004.

[290] M. Davenport and M. Wakin, "Analysis of orthogonal matching pursuit using the restricted isometry property," *IEEE Trans. Inform. Theory*, vol. 56, no. 9, pp. 4395–4401, Sep. 2010.

[291] J. A. Tropp, A. C. Gilbert and M. J. Strauss, "Algorithms for simultaneous sparse approximation. Part I: Greedy pursuit," *Signal Processing*, vol. 86, no. 3, pp. 572–588, Apr. 2006.

[292] Y. Pati, R. Rezaifar and P. Krishnaprasad, "Orthogonal matching pursuit: Recursive function approximation with applications to wavelet decomposition," in *Asilomar Conf. Signals, Systems, and Computers*, Pacific Grove, CA, Nov. 1993.

[293] D. Needell and J. A. Tropp, "CoSaMP: Iterative signal recovery from incomplete and inaccurate samples," *Appl. Comput. Harmon. Anal.*, vol. 26, no. 3, pp. 301–321, May 2008.

[294] T. Blumensath and M. Davies, "Gradient pursuits," *IEEE Trans. Signal Processing*, vol. 56, no. 6, pp. 2370–2382, Jun. 2008.

[295] A. Miller, *Subset Selection in Regression*, 2nd edn. New York, NY: Chapman & Hall, 2002.

[296] R. A. DeVore and V. N. Temlyakov, "Some remarks on greedy algorithms," *Adv. Comp. Math.*, vol. 5, pp. 173–187, Dec. 1996.

[297] J. Hogbom, "Aperture synthesis with a non-regular distribution of interferometer baselines," *Astrophys. J. Suppl. Series*, vol. 15, pp. 417–426, Jun. 1974.

[298] J. H. Friedman and J. W. Tukey, "A projection pursuit algorithm for exploratory data analysis," *IEEE Trans. Comput.*, vol. 23, no. 9, pp. 881–890, Sep. 1974.

[299] A. Beck and Y. C. Eldar, "Sparsity constrained nonlinear optimization: Optimality conditions and algorithms," *SIAM J. Optimization*, vol. 23, no. 3, pp. 1480–1509, 2012.

[300] D. L. Donoho, A. Maleki and A. Montanari, "Message-passing algorithms for compressed sensing," *Proc. Natl Acad. Sci.*, vol. 106, no. 45, pp. 18914–18919, Nov. 2009.

[301] D. Du and F. Hwang, *Combinatorial Group Testing and its Applications*, Singapore: World Scientific, 2000.

[302] A. Gilbert and P. Indyk, "Sparse recovery using sparse matrices," *Proc. IEEE*, vol. 98, no. 6, pp. 937–947, Jun. 2010.

[303] G. Cormode and S. Muthukrishnan, "Improved data stream summaries: The count-min sketch and its applications," *J. Algorithms*, vol. 55, no. 1, pp. 58–75, Apr. 2005.

[304] A. Gilbert, Y. Li, E. Porat and M. Strauss, "Approximate sparse recovery: Optimizing time and measurements," in *Proc. ACM Symp. Theory Comput.*, Cambridge, MA, Jun. 2010.

[305] A. Gilbert, M. Strauss, J. Tropp and R. Vershynin, "One sketch for all: Fast algorithms for compressed sensing," in *Proc. ACM Symp. Theory Comput.*, San Diego, CA, June 2007.

[306] S. Nam, M. E. Davies, M. Elad and R. Gribonval, "The cosparse analysis model and algorithms," *Appl. Comput. Harm. Anal.*, vol. 34, no. 1, pp. 30–56, Jan. 2013.

[307] S. Li and J. Lin, "Compressed sensing with coherent tight frames via ℓ_q-minimization for $0 < q \leq 1$," *Inverse Prob. Imaging*, vol. 8, no. 3, pp. 761–777, 2014.

[308] J. Treichler, M. Davenport and R. Baraniuk, "Application of compressive sensing to the design of wideband signal acquisition receivers," in *Proc. US/Australia Joint Workshop Defense Appl. Signal Processing (DASP)*, Lihue, Hawaii, Sep. 2009.

[309] Z. Ben-Haim, T. Michaeli and Y. C. Eldar, "Performance bounds and design criteria for estimating finite rate of innovation signals," *IEEE. Trans. Inform. Theory*, vol. 58, no. 8, pp. 4993–5015, 2012.

[310] E. Arias-Castro and Y. C. Eldar, "Noise folding in compressed sensing," *IEEE Signal Processing Lett.*, vol. 18, no. 8, pp. 478–481, Aug. 2011.

[311] D. Donoho, M. Elad and V. Temlyahov, "Stable recovery of sparse overcomplete representations in the presence of noise," *IEEE Trans. Inform. Theory*, vol. 52, no. 1, pp. 6–18, Jan. 2006.

[312] Z. Ben-Haim, Y. C. Eldar, and M. Elad, "Coherence-based performance guarantees for estimating a sparse vector under random noise," *IEEE Trans. Signal Processing*, vol. 58, no. 10, pp. 5030–5043, Oct. 2010.

[313] P. J. Bickel, Y. Ritov and A. B. Tsybakov, "Simultaneous analysis of Lasso and Dantzig selector," *Ann. Stat.*, vol. 37, no. 4, pp. 1705–1732, 2009.

[314] Z. Ben-Haim and Y. C. Eldar, "The Cramér–Rao bound for estimating a sparse parameter vector," *IEEE Trans. Signal Processing*, vol. 58, no. 6, pp. 3384–3389, Jun. 2010.

[315] C. Stein, "Inadmissibility of the usual estimator for the mean of a multivariate normal distribution," in *Proc. Third Berkeley Symp. Math. Statist. Prob.*, vol. 1, Berkeley, CA: University of California Press, 1956, pp. 197–206.

[316] W. James and C. Stein, "Estimation of quadratic loss," in *Proc. Fourth Berkeley Symp. Math. Statist. Prob.*, vol. 1, Berkeley, CA: University of California Press, pp. 361–379.

[317] Y. C. Eldar, *Rethinking Biased Estimation: Improving Maximum Likelihood and the Cramer–Rao Bound*, Foundation and Trends in Signal Processing, Hanover, MA: Now Publishers, 2008.

[318] R. Ward, "Compressive sensing with cross validation," *IEEE Trans. Inform. Theory*, vol. 55, no. 12, pp. 5773–5782, Dec. 2009.

[319] P. Wojtaszczyk, "Stability and instance optimality for Gaussian measurements in compressed sensing," *Found. Comput. Math.*, vol. 10, no. 1, pp. 1–13, Feb. 2010.

[320] D. Donoho and J. Tanner, "Counting faces of randomly-projected polytopes when the projection radically lowers dimension," *J. Am. Math. Soc.*, vol. 22, no. 1, pp. 1–53, Jan. 2009.

[321] D. Donoho and J. Tanner, "Precise undersampling theorems," *Proc. IEEE*, vol. 98, no. 6, pp. 913–924, Jun. 2010.

[322] M. E. Davies, T. Blumensath and G. Rilling, "Greedy algorithms for compressed sensing," in *Compressed Sensing: Theory and Applications*, Cambridge, UK: Cambridge University Press, 2011.

[323] E. Livshitz, "On efficiency of orthogonal matching pursuit," Preprint at arXiv: 1004.3946, Apr. 2010.

[324] T. Zhang, "Sparse recovery with orthogonal matching pursuit under RIP," *IEEE Trans. Inform. Theory*, vol. 57, no. 9, pp. 6215–6221, 2011.

[325] H. Rauhut, "On the impossibility of uniform sparse reconstruction using greedy methods," *Sampl. Theory Signal Image Processing*, vol. 7, no. 2, pp. 197–215, May 2008.

[326] S. Cotter, B. Rao, K. Engan and K. Kreutz-Delgado, "Sparse solutions to linear inverse problems with multiple measurement vectors," *IEEE Trans. Signal Processing*, vol. 53, no. 7, pp. 2477–2488, Jul. 2005.

[327] J. Chen and X. Huo, "Theoretical results on sparse representations of multiple-measurement vectors," *IEEE Trans. Signal Processing*, vol. 54, no. 12, pp. 4634–4643, Dec. 2006.

[328] J. A. Tropp, "Algorithms for simultaneous sparse approximation. Part II: Convex relaxation," *Signal Processing*, vol. 86, pp. 589–602, Apr. 2006.

[329] M. Mishali and Y. C. Eldar, "Reduce and boost: Recovering arbitrary sets of jointly sparse vectors," *IEEE Trans. Signal Processing*, vol. 56, no. 10, pp. 4692–4702, Oct. 2008.

[330] D. Malioutov, M. Cetin and A. S. Willsky, "A sparse signal reconstruction perspective for source localization with sensor arrays," *IEEE Trans. Signal Processing*, vol. 53, no. 8, pp. 3010–3022, Aug. 2005.

[331] S. F. Cotter and B. D. Rao, "Sparse channel estimation via matching pursuit with application to equalization," *IEEE Trans. Commun.*, vol. 50, no. 3, pp. 374–377, Mar. 2002.

[332] I. J. Fevrier, S. B. Gelfand and M. P. Fitz, "Reduced complexity decision feedback equalization for multipath channels with large delay spreads," *IEEE Trans. Commun.*, vol. 47, no. 6, pp. 927–937, Jun. 1999.

[333] Z. Yu, S. Hoyos and B. M. Sadler, "Mixed-signal parallel compressed sensing and reception for cognitive radio," in *IEEE Int. Conf. Acoustics, Speech, and Signal Processing (ICASSP)*, Las Vegas, NV, Apr. 2008, pp. 3861–3864.

[334] J. A. Bazerque and G. B. Giannakis, "Distributed spectrum sensing for cognitive radio networks by exploiting sparsity," *IEEE Trans. Signal Processing*, vol. 58, no. 3, pp. 1847–1862, Mar. 2010.

[335] M. Mishali and Y. C. Eldar, "Blind multiband signal reconstruction: Compressed sensing for analog signals," *IEEE Trans. Signal Processing*, vol. 57, no. 3, pp. 993–1009, Mar. 2009.

[336] M. Mishali and Y. C. Eldar, "From theory to practice: Sub-Nyquist sampling of sparse wideband analog signals," *IEEE J. Select. Top. Signal Processing*, vol. 4, no. 2, pp. 375–391, Apr. 2010.

[337] M. E. Davies and Y. C. Eldar, "Rank awareness in joint sparse recovery," *IEEE Trans. Inform. Theory*, vol. 58, no. 2, pp. 1135–1146, 2013.

[338] M. Fornasier and H. Rauhut, "Recovery algorithms for vector valued data with joint sparsity constraints," *SIAM J. Num. Anal.*, vol. 46, no. 2, pp. 577–613, Feb. 2008.

[339] S. F. Cotter, B. D. Rao, K. Engan and K. Kreutz-Delgado, "Sparse solutions to linear inverse problems with multiple measurement vectors," *IEEE Trans. Signal Processing*, vol. 53, no. 7, pp. 2477–2488, Jul. 2005.

[340] Y. C. Eldar and H. Rauhut, "Average case analysis of multichannel sparse recovery using convex relaxation," *IEEE Trans. Inform. Theory*, vol. 6, no. 1, pp. 505–519, Jan. 2010.

[341] R. Gribonval, H. Rauhut, K. Schnass and P. Vandergheynst, "Atoms of all channels, unite! Average case analysis of multi-channel sparse recovery using greedy algorithms," *J. Fourier Anal. Appl.*, vol. 14, no. 5, pp. 655–687, Dec. 2008.

[342] K. Schnass and P. Vandergheynst, "Average performance analysis for thresholding," *IEEE Signal Processing Lett.*, vol. 14, no. 11, pp. 828–831, Nov. 2007.

[343] J. Bien and R. Tibshirani, "Sparse estimation of a covariance matrix," *Biometrika*, vol. 98, no. 4, pp. 807–820, 2011.

[344] P. P. Vaidyanathan and P. Pal, "Sparse sensing with co-prime samplers and arrays," *IEEE Trans. Signal Processing*, vol. 59, no. 2, pp. 573–586, Feb. 2011.

[345] D. Ariananda, D. Dony and G. Leus, "Compressive wideband power spectrum estimation," *IEEE Trans. Signal Processing*, vol. 60, no. 9, pp. 4775–4789, Sep. 2012.

[346] C. P. Yen, Y. Tsai and X. Wang, "Wideband spectrum sensing based on sub-Nyquist sampling," *IEEE Trans. Image Processing*, vol. 61, no. 12, pp. 3028–3040, Jun. 2013.

[347] D. Cohen and Y. C. Eldar, "Sub-Nyquist sampling for power spectrum sensing in cognitive radios: A unified approach," *IEEE Trans. Signal Processing*, vol. 62, no. 15, pp. 3897–3910, Aug. 2015.

[348] H. M. Quiney, "Coherent diffractive imaging using short wavelength light sources: A tutorial review," *J. Mod. Opt.*, vol. 57, no. 13, pp. 1109–1149, Jul. 2010.

[349] N. E. Hurt, *Phase Retrieval and Zero Crossings: Mathematical Methods in Image Reconstruction*. New York, NY: Springer, 2001.

[350] R. W. Harrison, "Phase problem in crystallography," *J. Opt. Soc. Am. A*, vol. 10, no. 5, pp. 1045–1055, May 1993.

[351] A. Walther, "The question of phase retrieval in optics," *Opt. Acta.*, vol. 10, no. 1, pp. 41–49, 1963.

[352] R. W. Gerchberg and W. O. Saxton, "Phase retrieval by iterated projections," *Optik*, vol. 35, pp. 237–246, Aug. 1972.

[353] J. R. Fienup, "Phase retrieval algorithms: a comparison," *Appl. Opt.*, vol. 21, no. 15, pp. 2758–2769, Aug. 1982.

[354] M. L. Moravec, J. K. Romberg and R. G. Baraniuk, "Compressive phase retrieval," *Proc. SPIE*, vol. 6701, *Wavelets XII*, p. 670120, 2007.

[355] K. Jaganathan, S. Oymak and B. Hassibi, "Recovery of sparse 1-D signals from the magnitudes of their Fourier transform," in *IEEE Int. Symp. Inform. Theory Proc. (ISIT)*, Cambridge, MA, 2012, pp. 1473–1477.

[356] Y. Shechtman, A. Beck and Y. C. Eldar, "GESPAR: Efficient phase retrieval of sparse signals." *IEEE Trans. Signal Processing*, vol. 62, no. 4, pp. 928–938, Feb. 2014.

[357] X. Li and V. Voroninski, "Sparse signal recovery from quadratic measurements via convex programming," *SIAM J. Math. Anal.*, vol. 45, no. 5, pp. 3019–3033, 2013.

[358] Y. C. Eldar and S. Mendelson, "Phase retrieval: Stability and recovery guarantees." *Appl. Comput. Harmon. Anal.*, vol. 36, no. 3 pp. 473–494, 2014.

[359] M. Stojnic, F. Parvaresh and B. Hassibi, "On the reconstruction of block-sparse signals with an optimal number of measurements," *IEEE Trans. Signal Processing*, vol. 57, no. 8, pp. 3075–3085, Aug. 2009.

[360] Z. Ben-Haim and Y. C. Eldar, "Near-oracle performance of greedy block-sparse estimation techniques from noisy measurements," *IEEE J. Select. Top. Signal Processing*, vol. 5, no. 5, pp. 1032–1047, Sep. 2011.

[361] S. Erickson and C. Sabatti, "Empirical Bayes estimation of a sparse vector of gene expression changes," *Stati. Applic. Genet. Mol. Biol.*, vol. 4, no. 1, p. 22, Sep. 2005.

[362] F. Parvaresh, H. Vikalo, S. Misra and B. Hassibi, "Recovering sparse signals using sparse measurement matrices in compressed DNA microarrays," *IEEE J. Select. Top. Signal Processing*, vol. 2, no. 3, pp. 275–285, Jun. 2008.

[363] P. Sprechmann, I. Ramirez, G. Sapiro and Y. C. Eldar, "C-HiLasso: A collaborative hierarchical sparse modeling framework," *IEEE Trans. Signal Processing*, vol. 59, no. 9, pp. 4183–4198, Sep. 2011.

[364] F. R. Bach, "Consistency of the group Lasso and multiple kernel learning," *J. Mach. Learn. Res.*, vol. 9, pp. 1179–1225, Jun. 2008.

[365] Y. Nardi and A. Rinaldo, "On the asymptotic properties of the group lasso estimator for linear models," *Electron. J. Stat.*, vol. 2, pp. 605–633, 2008.

[366] L. Meier, S. van de Geer and P. Bühlmann, "The group lasso for logistic regression," *J. R. Stat. Soc. B*, vol. 70, no. 1, pp. 53–77, Feb. 2008.

[367] M. Yuan and Y. Lin, "Model selection and estimation in regression with grouped variables," *J. R. Stat. Soc. Ser. B Stat. Methodol.*, vol. 68, no. 1, pp. 49–67, Feb. 2006.

[368] M. S. Lobo, L. Vandenberghe, S. Boyd and H. Lebret, "Applications of second-order cone programming," *Linear Algeb. Applic.*, vol. 284, no. 1–3, pp. 193–228, Nov. 1998.

[369] R. Baraniuk, V. Cevher, M. Duarte and C. Hegde, "Model-based compressive sensing," *IEEE Trans. Inform. Theory*, vol. 56, no. 4, pp. 1982–2001, Apr. 2010.

[370] E. Elhamifar and R. Vidal "Structured sparse recovery via convex optimization," *IEEE Trans. Signal Processing*, vol. 60, no. 8, pp. 4094–4107, Aug. 2012.

[371] J. J. Hull, "A database for handwritten text recognition research," *IEEE Trans. Patt. Anal. Mach. Intell.*, vol. 16, no. 5, pp. 550–554, May 1994.

[372] R. Gribonval and P. Vandergheynst, "On the exponential convergence of matching pursuits in quasi-incoherent dictionaries," *IEEE Trans. Inform. Theory*, vol. 52, no. 1, pp. 255–261, Jan. 2006.

[373] B. A. Olshausen and D. J. Field, "Emergence of simple-cell receptive field properties by learning a sparse code for natural images," *Nature*, vol. 381, no. 6583, pp. 607–609, Jun. 1996.

[374] B. A. Olshausen and D. J. Field, "Sparse coding with an overcomplete basis set: A strategy employed by V1?" *Vision Res.*, vol. 37, no. 23, pp. 3311–3325, Dec. 1997.

[375] B. A. Olshausen and D. J. Field, "Sparse coding of sensory inputs," *Curr. Opinion Neurobiol.*, vol. 14, no. 4, pp. 481–487, Aug. 2004.

[376] K. Kreutz-Delgado, J. F. Murray, B. D. Rao *et al.*, "Dictionary learning algorithms for sparse representation," *Neural Comput.*, vol. 15, no. 2, pp. 349–396, Feb. 2003.

[377] M. S. Lewicki and T. J. Senowski, "Learning overcomplete representations," *Neural Comput.*, vol. 12, no. 2, pp. 337–365, Feb. 2000.

[378] K. Engan, S. O. Aase and J. H. Husoy, "Frame based signal compression using method of optimal directions (MOD)," *IEEE Intern. Symp. Circuits Syst.*, vol. 4, pp. 1–4, Jul. 1999.

[379] M. Aharon, M. Elad, A. Bruckstein and Y. Kats, "K-SVD: An algorithm for designing of overcomplete dictionaries for sparse representation," *IEEE Trans. Signal Processing*, vol. 54, no. 11, pp. 4311–4322, Nov. 2006.

[380] M. Aharon, M. Elad and M. Bruckstein, "On the uniqueness of overcomplete dictionaries, and a practical way to retrieve them," *Linear Algeb. Appl.*, vol. 416, no. 1, pp. 48–67, Jul. 2006.

[381] K. Rosenblum, L. Zelnik-Manor and Y. C. Eldar, "Dictionary optimization for block sparse representations," *IEEE Trans. Signal Processing*, vol. 60, no. 5, pp. 2386–2395, May 2012.

[382] R. Basri and D. W. Jacobs, "Lambertian reflectance and linear subspaces," *IEEE Trans. Patt. Anal. Mach. Intell.*, vol. 25, no. 2, pp. 218–233, Feb. 2003.

[383] J. Wright, A.Y. Yang, A. Ganesh, S. S. Sastry and Y. Ma, "Robust face recognition via sparse representation," *IEEE Trans. Patt. Anal. Mach. Intell.*, vol. 31, no. 2, pp. 210–227, Apr. 2008.

[384] R. Vidal and Y. Ma, "A unified algebraic approach to 2-D and 3-D motion segmentation and estimation," *J. Math. Imaging Vision*, vol. 25, no. 3, pp. 403–421, Oct. 2006.

[385] R. Vidal, Y. Ma and S. Sastry, "Generalized principal component analysis (GPCA)," *IEEE Trans. Patt. Anal. Mach. Intell.*, vol. 27, no. 12, pp. 1945–1959, Dec. 2005.

[386] E. Elhamifar and R. Vidal, "Sparse subspace clustering," in *IEEE Conf. Computer Vision and Pattern Recognition, 2009*. IEEE, Jun. 2009, pp. 2790–2797.

[387] J. Mairal, F. Bach, J. Ponce, G. Sapiro and A. Zisserman, "Discriminative learned dictionaries for local image analysis," in *IEEE Conf. Computer Vision and Pattern Recognition, 2008*. IEEE, Jun. 2008, pp. 1–8.

[388] S. C. Johnson, "Hierarchical clustering schemes," *Psychometrika*, vol. 32, no. 3, pp. 241–254, Sep. 1967.

[389] S. Gleichman and Y. C. Eldar, "Blind compressed sensing," *IEEE Trans. Inform. Theory*, vol. 57, no. 10, pp. 6958–6975, Oct. 2011.

[390] J. Silva, M. Chen, Y. C. Eldar, G. Sapiro and L. Carin, "Blind compressed sensing over a structured union of subspaces," Preprint at arXiv:1103.2469, 2011.

[391] R. Rubinstein, M. Zibulevsky and M. Elad, "Double sparsity: Learning sparse dictionaries for sparse signal approximation," *IEEE Trans. Signal Processing*, vol. 58, no. 3, pp. 1553–1564, Mar. 2010.

[392] K. Rosenblum, L. Zelnik-Manor and Y. C. Eldar, "Dictionary optimization for block-sparse representations," *IEEE Trans. Signal Processing*, vol. 60, no. 5, pp. 2386–2395, May 2012.

[393] T. Blu and M. Unser, "Approximation error for quasi-interpolators and (multi-)wavelet expansions," *Appl. Comput. Harm. Anal.*, vol. 6, pp. 219–251, Mar. 1999.

[394] J. G. Proakis, *Digital Communications*, 3rd edn. McGraw-Hill, Inc., 1995.

[395] Y. Xie, Y. C. Eldar and A. Goldsmith, "Reduced-dimension multiuser detection," *IEEE Trans. Inform. Theory*, vol. 59, no. 6, pp. 3858–3874, Sep. 2013.

[396] S. Verdú, *Multiuser Detection*, Cambridge, UK: Cambridge University Press, 1998.

[397] A. Duel-Hallen, "Decorrelating decision-feedback multiuser detector for synchronous code-division multiple-access channel," *IEEE Trans. Commun.*, vol. 41, no. 2, pp. 285–290, Feb. 1993.

[398] W. Hoeffding, "Probability inequalities for sums of bounded random variables," *J. Am. Stat. Assoc.*, vol. 58, no. 301, pp. 13–30, Mar. 1963.

[399] I. Budiarjo, H. Nikookar and L. P. Ligthart, "Cognitive radio modulation techniques," *IEEE Signal Processing Mag.*, vol. 25, no. 6, pp. 24–34, Nov. 2008.

[400] D. Cabric, "Addressing feasibility of cognitive radios," *IEEE Signal Processing Mag.*, vol. 25, no. 6, pp. 85–93, Nov. 2008.

[401] K. Gröchenig and H. Razafinjatovo, "On Landau's necessary density conditions for sampling and interpolation of band-limited functions," *J. London Math. Soc.*, vol. 54, no. 3, pp. 557–565, Dec. 1996.

[402] W. C. Black and D. A. Hodges, "Time interleaved converter arrays," *IEEE J. Solid-State Circuits*, vol. 15, no. 6, pp. 1022–1029, Dec. 1980.

[403] C. Vogel and H. Johansson, "Time-interleaved analog-to-digital converters: Status and future directions," in *Proc. IEEE Int. Symp. Circuits and Systems* (ISCAS), no. 4, Kos, Greece, May 2006, pp. 3386–3389.

[404] D. M. Akos, M. Stockmaster, J. B. Y. Tsui and J. Caschera, "Direct bandpass sampling of multiple distinct RF signals," *IEEE Trans. Commun.*, vol. 47, no. 7, pp. 938–988, Jul. 1999.

[405] A. Kohlenberg, "Exact interpolation of band-limited functions," *J. Appl. Phys.*, vol. 24, no. 12, pp. 1432–1435, Dec. 1953.

[406] T. I. Laakso, V. Valimaki, M. Karjalainen and U. K. Laine, "Splitting the unit delay [fir/all pass filters design]," *IEEE Signal Processing Mag.*, vol. 13, no. 1, pp. 30–60, Jan. 1996.

[407] M. E. Dominguez-Jimenez, N. Gonzalez-Prelcic, G. Vazquez-Vilar and R. Lopez-Valcarce, "Design of universal multicoset sampling patterns for compressed sensing of multiband sparse signals," in *Proc. IEEE Int. Conf. Acoustics Speech and Signal Processing* (ICASSP), Kyoto, Japan, pp. 3337–3340, Mar. 2012.

[408] T. Tao, "An uncertainty principle for cyclic groups of prime order," *Math. Res. Lett.*, vol. 12, no. 1, pp. 121–127, 2005.

[409] R. J. Evans and I. M. Isaacs, "Generalized Vandermonde determinants and roots of unity of prime order," *Proc. Am. Math. Soc.*, vol. 58, pp. 51–54, Jul. 1976.

[410] M. Mishali and Y. C. Eldar, "Sub-Nyquist sampling," *IEEE Signal Processing Mag.*, vol. 28, no. 6, pp. 98–124, Nov. 2011.

[411] R. Khoini-Poorfard, L. B. Lim and D. A. Johns, "Time-interleaved oversampling A/D converters: Theory and practice," *IEEE Trans. Circuits Syst. II*, vol. 44, no. 8, pp. 634–645, Aug. 1997.

[412] A. J. Viterbi, *CDMA Principles of Spread Spectrum Communication*, Reading, MA: Addison-Wesley Wireless Communications Series, 1995.

[413] R. Pickholtz, D. Schilling and L. Milstein, "Theory of spread-spectrum communications – A tutorial," *IEEE Trans. Commun.*, vol. 30, no. 2, pp. 855–884, May 1982.

[414] C. Kienmayer, M. Tiebout, W. Simburger, and A. L. Scholtz, "A low-power low-voltage nmos bulk-mixer with 20 GHz bandwidth in 90 nm CMOS," *Proc. IEEE Intl Symp. Circuits and Systems* (ISCAS), vol. 8, Vancouver, 2004, pp. 385–388.

[415] B. Razavi, "A 60-GHz CMOS receiver front-end," *IEEE J. Solid-State Circuits*, vol. 41, no. 1, pp. 17–22, Jan. 2006.

[416] E. Laskin and S. P. Voinigescu, "A 60 mW per lane, 4×23-Gb/s 2^7-1 PRBS generator," *IEEE J. Solid-State Circuits*, vol. 41, no. 10, 2198–2208, Oct. 2006.

[417] T. O. Dickson, E. Laskin, I. Khalid *et al.*, "An 80-Gb/s $2^{31} - 1$ pseudorandom binary sequence generator in SiGe BiCMOS technology," *IEEE J. Solid-State Circuits*, vol. 40, no. 12, pp. 2735–2745, Dec. 2005.

[418] K. Gentile, "Introduction to zero-delay clock timing techniques," Application notes AN-#0983, Analog Devices Corp. http://www.analog.com/static/imported-files/application_notes/AN-0983.pdf.

[419] Y. Chen, M. Mishali, Y. C. Eldar and A. O. Hero III, "Modulated wideband converter with non-ideal lowpass filters," in *Proc. IEEE Int. Conf. Acoustics Speech and Signal Processing (ICASSP)*, Dallas, TX, 2010, pp. 3630–3633.

[420] M. Mishali and Y. C. Eldar, "Expected-RIP: Conditioning of the modulated wideband converter," in *IEEE Information Theory Workshop ITW 2009*, Oct. 2009, pp. 343–347.

[421] L. Gan and H. Wang, "Deterministic binary sequences for modulated wideband converter," in *SAMPTA 2013*, Bremen, 2013, pp. 264–267.

[422] J. M. Jensen, H. Elbrønd Jensen and T. Hoholdt, "The merit factor of binary sequences related to difference sets," *IEEE Trans. Inform. Theory*, vol. 37, no. 3, pp. 617–626, May 1991.

[423] F. Gardner, "Properties of frequency difference detectors," *IEEE Trans. Commun.*, vol. 33, no. 2, pp. 131–138, Feb. 1985.

[424] R. Tur, Y. C. Eldar and Z. Friedman, "Low rate sampling of pulse streams with application to ultrasound imaging," *IEEE Trans. Signal Processing*, vol. 59, no. 4, pp. 1827–1842, Apr. 2011.

[425] N. Wagner, Y. C. Eldar and Z. Friedman, "Compressed beamforming in ultrasound imaging," *IEEE Trans. Signal Processing*, vol. 60, no. 9, pp. 4643–4657, Sep. 2012.

[426] O. Bar-Ilan and Y. C. Eldar, "Sub-Nyquist radar via Doppler focusing," *IEEE Trans. Signal Processing*, vol. 62, no. 7, pp. 1796–1811, Apr. 2014.

[427] E. Baransky, G. Itzhak, I. Shmuel *et al.*, "A sub-Nyquist radar prototype: Hardware and algorithms," *IEEE Trans. Aerospace Electronic Systems*, vol. 50, no. 2, pp. 809–822, Apr. 2014.

[428] A. Bruckstein, T. J. Shan and T. Kailath, "The resolution of overlapping echos," *IEEE Trans. Acoust. Speech Signal Processing*, vol. 33, no. 6, pp. 1357–1367, Dec. 1985.

[429] K. Gedalyahu and Y. C. Eldar, "Time delay estimation from low rate samples: A union of subspaces approach," *IEEE Trans. Signal Processing*, vol. 58, no. 6, pp. 3017–3031, Jun. 2010.

[430] W. U. Bajwa, K. Gedalyahu and Y. C. Eldar, "Identification of parametric underspread linear systems and super-resolution radar," *IEEE Trans. Signal Processing*, vol. 59, no. 6, pp. 2548–2561, Jun. 2011.

[431] H. Meyr, M. Moeneclaey and S. A. Fechtel, *Digital Communication Receivers: Synchronization, Channel Estimation, and Signal Processing*, New York, NY: Wiley-Interscience, 1997.

[432] A. Quazi, "An overview on the time delay estimate in active and passive systems for target localization," *IEEE Trans. Acoust. Speech Signal Processing*, vol. 29, no. 3, pp. 527–533, Jun. 1981.

[433] P. Stoica and R. Moses, *Introduction to Spectral Analysis*, Upper Saddle River, NJ: Prentice-Hall, 1997.

[434] O. Toeplitz, "Zur theorie der quadratischen und bilinearen formen von unendlichvielen veränderlichen," *Math. Ann.*, vol. 70, no. 3, pp. 351–376, 1911.

[435] G. H. Golub and C. F. Van Loan, "An analysis of the total least-squares problem," *SIAM J. Num. Anal.*, vol. 17, no. 6, pp. 883–893, Dec. 1980.

[436] M. S. Mackisack, M. R. Osborne, M. Kahn and G. K. Smyth, "On the consistency of Prony's method and related algorithms," *J. Comput. Graph. Stat.*, vol. 1, pp. 329–349, 1992.

[437] D. W. Tufts and R. Kumaresan, "Estimation of frequencies of multiple sinusoids: Making linear prediction perform like maximum likelihood," *Proc. IEEE*, vol. 70, no. 9, pp. 975–989, Sep. 1982.

[438] M. D. Rahman and K. B. Yu, "Total least squares approach for frequency estimation using linear prediction," *IEEE Trans. Acoust. Speech Signal Processing*, vol. 35, no. 10, pp. 1440–1454, Oct. 1987.

[439] J. A. Cadzow, "Signal enhancement – a composite property mapping algorithm," *IEEE Trans. Acoust. Speech Signal Processing*, vol. 36, no. 1, pp. 49–62, Jan. 1988.

[440] Y. Hua and T. K. Sarkar, "Matrix pencil method for estimating parameters of exponentially damped/undamped sinusoids in noise," *IEEE Trans. Acoust. Speech Signal Processing*, vol. 38, no. 5, pp. 814–824, May 1990.

[441] T. K. Sarkar and O. Pereira, "Using the matrix pencil method to estimate the parameters of a sum of complex exponentials," *IEEE Antennas Propag. Mag.*, vol. 37, no. 1, pp. 48–55, Feb. 1995.

[442] V. F. Pisarenko, "The retrieval of harmonics from a covariance function," *Geophys. J. Roy. Astron. Soc.*, vol. 33, no. 3, pp. 347–366, Sep. 1973.

[443] A. Barabell, "Improving the resolution performance of eigenstructure-based direction-finding algorithms," in *Proc. IEEE Int. Conf. Acoustics, Speech and Signal Processing (ICASSP'83)*, vol. 8, Apr. 1983, pp. 336–339.

[444] T-J. Shan, M. Wax and T. Kailath, "On spatial smoothing for direction-of-arrival estimation of coherent signals," *IEEE Trans. Acoust. Speech Signal Processing*, vol. 33, no. 4, pp. 806–811, Aug. 1985.

[445] G. Tang, B. Narayan Bhaskar, P. Shah and B. Recht, "Compressed sensing off the grid," *IEEE Trans. Inform. Theory*, vol. 59, no. 11, pp. 7465–7490, Nov. 2013.

[446] B. Narayan Bhaskar, G. Tang and B. Recht, "Atomic norm denoising with applications to line spectral estimation," *49th Annual Allerton Conf. Communication, Control, and Computing*, pp. 261–268, 2011.

[447] P. Stoica and P. Babu, "Sparse estimation of spectral lines: Grid selection problems and their solutions," *IEEE Trans. Signal Processing*, vol. 60, no. 2, pp. 962–967, Feb. 2012.

[448] T. Chernyakova and Y. C. Eldar, "Fourier domain beamforming: The path to compressed ultrasound imaging" *IEEE Trans. Ultrasonics, Ferroelectrics, and Frequency Control*, vol. 61, no. 8, pp. 1252–1267, Aug. 2014.

[449] P. L. Dragotti, M. Vetterli and T. Blu, "Sampling moments and reconstructing signals of finite rate of innovation: Shannon meets Strang–Fix," *IEEE Trans. Signal Processing*, vol. 55, no. 5, pp. 1741–1757, May 2007.

[450] J. A. Uriguen, T. Blu and P. L. Dragotti, "FRI sampling with arbitrary kernels," *IEEE Trans. Signal Processing*, vol. 61, no. 12, pp. 5310–5323, Nov. 2013.

[451] H. Akhondi Asl, P. L. Dragotti and L. Baboulaz, "Multichannel sampling of signals with finite rate of innovation," *IEEE Signal Processing Lett.*, vol. 17, no. 8, pp. 762–765, Aug. 2010.

[452] R. Tur, K. Gedalyahu and Y. C. Eldar, "Multichannel sampling of pulse streams at the rate of innovation," *IEEE Trans. Signal Processing*, vol. 59, no. 4, pp. 1491–1504, Apr. 2011.

[453] G. Golub and C. Van Loan, *Matrix Computations*, 2nd edn. Baltimore, MD: Johns Hopkins University Press, 1989.

[454] M. Z. Win and R. A. Scholtz, "Characterization of ultra-wide bandwidth wireless indoor channels: A communication-theoretic view," vol. 20, no. 9, pp. 1613–1627, Dec. 2002.

[455] S. Kay, *Fundamentals of Statistical Signal Processing*, Englewood Cliffs, NJ: PTR Prentice-Hall, 1993.

[456] J. A. Uriguen, Y. C. Eldar, P. L. Dragotti and Z. Ben-Haim, "Sampling at the rate of innovation: theory and applications," in *Compressed Sensing: Theory and Applications*, Cambridge, UK: Cambridge University Press, pp. 148–209, 2012.

[457] T. Michaeli and Y. C. Eldar, "Xampling at the rate of innovation," *IEEE Trans. Signal Processing*, vol. 60, no. 3, pp. 1121–1133, Mar. 2012.

[458] T. Blumensath, "Sampling and reconstructing signals from a union of linear subspaces," *IEEE Trans. Inform. Theory*, vol. 57, no. 7, pp. 4660–4671, Jul. 2011.

[459] Q. Sun, "Frames in spaces with finite rate of innovation," *Adv. Comput. Math.*, vol. 28, no. 4, pp. 301–329, May 2008.

[460] L. Baboulaz and P. L. Dragotti, "Exact feature extraction using finite rate of innovation principles with an application to image super-resolution," *IEEE Trans. Image Processing*, vol. 18, no. 2, pp. 281–298, Feb. 2009.

[461] V. Chaisinthop and P. L. Dragotti, "Centralized and distributed semi-parametric compression of piecewise smooth functions," *IEEE Trans. Signal Processing*, vol. 59, no. 7, pp. 3071–3085, Jul. 2011.

[462] K. M. Cohen, C. Attias, B. Farbman, I. Tselniker and Y. C. Eldar, "Channel estimation in UWB channels using compressed sensing," in *Proc. IEEE ICASSP-14*, Florence, Italy, May 2014.

[463] J. Onativia, S. Schultz and P. L. Dragotti, "A finite rate of innovation algorithm for fast and accurate spike detection from two-photon calcium imaging," *J. Neural Eng.*, vol. 10, no. 4, Jul. 2013.

[464] I. Maravic and M. Vetterli, "Exact sampling results for some classes of parametric non-bandlimited 2-D signals," *IEEE Trans. Signal Processing*, vol. 52, no. 1, pp. 175–189, Jan. 2004.

[465] P. Shukla and P. L. Dragotti, "Sampling schemes for multidimensional signals with finite rate of innovation," *IEEE Trans. Signal Processing*, vol. 55, no. 7, pp. 3670–3686, Jul. 2007.

[466] M. I. Skolnik, *Introduction to Radar Systems*, New York, NY: McGraw-Hill, 1980.

[467] C. E. Cook and M. Bernfeld, *Radar Signals – An Introduction to Theory and Applications*, Norwood, MA: Artech House, 1993.

[468] Y. C. Eldar, R. Levi and A. Cohen, "Clutter removal in sub-Nyquist radar," *IEEE Signal Processing Lett.*, vol. 22, no. 2, pp. 177–181, 2014.

[469] E. Matusiak and Y. C. Eldar, "Sub-Nyquist sampling of short pulses," *IEEE Trans. Signal Processing*, vol. 60, no. 3, pp. 1134–1148, Mar. 2012.

[470] T. Kailath, "Measurements on time-variant communication channels," *IRE Trans. Inform. Theory*, vol. 8, no. 5, pp. 229–236, Sep. 1962.

[471] P. Bello, "Measurement of random time-variant linear channels," *IEEE Trans. Inform. Theory*, vol. 15, no. 4, pp. 469–475, Jul. 1969.

[472] W. Kozek and G. E. Pfander, "Identification of operators with bandlimited symbols," *SIAM J. Math. Anal.*, vol. 37, no. 3, pp. 867–888, 2005.

[473] G. E. Pfander and D. F. Walnut, "Measurement of time-variant linear channels," *IEEE Trans. Inform. Theory*, vol. 52, no. 11, pp. 4808–4820, Nov. 2006.

[474] D. Slepian, "On bandwidth," *Proc. IEEE*, vol. 64, no. 3, pp. 292–300, Mar. 1976.

[475] J. A. Jensen, "Linear description of ultrasound imaging systems," *Notes for the International Summer School on Advanced Ultrasound Imaging*, Technical University of Denmark, 1999.

[476] T. L. Szabo, *Diagnostics Ultrasound Imaging: Inside Out*, J. Bronzino, Ed., Ch. 7, 10. Oxford, UK: Elsevier Academic Press, 2004.

[477] A. Papoulis, *Probability, Random Variables, and Stochastic Processes*, 3rd edn. New York, NY: McGraw Hill, Inc., 1991.

[478] W. Feller, *An Introduction to Probability Theory and its Applications*, 2nd edn., vol. 2. New York, NY: Wiley, 1971.

[479] B. Porat, *Digital Processing of Random Signals: Theory and Methods*, Englewood Cliffs, NJ: Prentice-Hall, 1994.

Index

Printed in the United States
By Bookmasters